Handbook of Landscape

Archaeology

WORLD ARCHAEOLOGICAL CONGRESS
RESEARCH HANDBOOKS IN ARCHAEOLOGY

Sponsored by the World Archaeological Congress

Series Editors:
George Nicholas *(Simon Fraser University)*
Julie Hollowell *(Indiana University)*

The World Archaeological Congress's (WAC) *Research Handbooks in Archaeology* series provides comprehensive coverage of a range of areas of contemporary interest to archaeologists. Research handbooks synthesize and benchmark an area of inquiry by providing state-of-the-art summary articles on the key theories, methods, and practical issues in the field. Guided by a vision of an ethically embedded, multivocal global archaeology, the edited volumes in this series—organized and written by scholars of high standing worldwide—provide clear, in-depth information on specific archaeological themes for advanced students, scholars, and professionals in archaeology and related disciplines. *All royalties on these volumes go to the World Archaeological Congress.*

Bruno David and Julian Thomas (eds.), *Handbook of Landscape Archaeology*

Soren Blau and Douglas Ubelaker (eds.), *Handbook of Forensic Anthropology and Archaeology*

Jane Lydon and Uzma Rizvi (eds.), *Handbook of Postcolonial Archaeology*

Handbook of Landscape Archaeology

Bruno David and Julian Thomas

Editors

Routledge
Taylor & Francis Group

LONDON AND NEW YORK

First published 2008 by Left Coast Press, Inc.
First paperback edition 2010.

Published 2016 by Routledge
2 Park Square, Milton Park, Abingdon, Oxon OX14 4RN
711 Third Avenue, New York, NY 10017, USA

Routledge is an imprint of the Taylor & Francis Group, an informa business

Library of Congress Cataloguing-in-Publication Data:

Handbook of landscape archaeology/Bruno David, Julian Thomas, editors.
p. cm. — (World Archaeological Congress research handbooks in archaeology; 1)
ISBN 978-1-59874-294-7 (hardback : alk. paper)
ISBN 978-1-59874-616-7 (paperback   )
1. Landscape archaeology—Handbooks, manuals, etc. I. David, Bruno, 1962- II. Thomas, Julian.
CC75.H35 2008
930.1—dc22
2008019849

ISBN 978-1-59874-616-7 paperback
ISBN 978-1-59874-294-7 hardcover

CONTENTS

FIGURES

TABLES

SERIES EDITORS' FOREWORD

With this inaugural volume, we are very pleased to introduce the *World Archaeological Congress (WAC) Research Handbooks in Archaeology*—a series designed to synthesize and benchmark a given field of contemporary archaeological inquiry by providing comprehensive coverage of a range of areas of interest to archaeologists. Each volume offers articles written by a widely range of scholars and specialists on key topics that outline major historical developments, current trends in research and interpretation, thorny ethical issues, and promising new directions for the future. These handbooks compile foundational concepts and theories along with practical advice and extensive bibliographies.

Guided by a vision of an ethically embedded global archaeology, this cohesive series grounds archaeological theory, method, and practice in an understanding of contemporary ethical issues surrounding each theme. As a WAC series, these handbooks emerge from the intersection of local archaeological practice and global situations and perspectives. They aspire to geographic and cultural diversity through both the contributing authors and subject matter, and include the perspectives of Indigenous peoples, source communities, and other groups or parties affected by archaeology and its practices. By grounding archaeological practice in an understanding of contemporary ethical issues, these handbooks attempt to illustrate what it means to conduct responsible, ethical archaeology that contributes to goals of social justice.

We envision that these handbooks will be useful reference books for scholars, students, and emerging professionals in heritage management fields, departments of archaeology and anthropology, museums, and research institutions throughout the world. They will have particular value as graduate-level texts for specific fields of research and inquiry. The various themes addressed will also appeal to government agencies, nongovernment organizations, historical societies, and community groups interested in archaeology and how archaeologists have interpreted the past.

All royalties generated by sales of books in this series go directly to World Archaeological Congress. Since its inception, WAC has nurtured the growth of archaeological communities and discussions, and supported participation in meetings in cases where economic and political conditions make this hard to sustain. One of the major ways WAC has accomplished this is through the donation of royalties from WAC-related publications. In addition, Left Coast Press is donating 50 copies of each volume produced in the Research Handbook series to WAC's Global Libraries Project, for distribution to libraries in need around the world. We are grateful to the indomitable Mitch Allen and his colleagues for their generosity to WAC and its activities.

We want especially to acknowledge and thank the following people: Claire Smith and Heather Burke for their initial work and foresight in conceiving the series and planting the seeds for the first volumes; Mitch Allen and Jennifer Collier for their guidance and trust; Bruno David and Julian Thomas, as the editors of this first volume, for being persistent, well-organized, and responsive to our many requests (imagine the task of gathering 65 manuscripts from 75 leaders in the field, not to mention keeping track of them over several rounds of revisions). We were also thankful to have the able assistance and careful eyes of Stacey C. Sawyer, our copyeditor and project manager, who worked under tremendous pressure, and Sarah Cavanaugh, who helped at a critical juncture. Also, we thank Rahul Rajagopalan, Supriya Sahni, and others at Sage Publications, who went out of their way to compose the pages quickly and get the book to the printer. Finally,

we are very interested in hearing from you, the reader, concerning suggestions for future volumes and any feedback regarding the series.

We dedicate this series to Peter Ucko for insisting that research in archaeology is just as much about the social as the scientific.

George Nicholas
Julie Hollowell
Series editors
May 2008

PREFACE

Implicitly or explicitly, the archaeology of landscape has been a central theme for the World Archaeological Congress (WAC) throughout its history. Volumes in the One World Archaeology series on *Sacred Sites, Sacred Places* (Carmichael et al. 1994), *The Archaeology and Anthropology of Landscape: Shaping Your Landscape* (Ucko and Layton 1999), and *The Archaeology of Drylands: Living at the Margin* (Barker and Gilbertson 2000) reflect one of the abiding preoccupations of WAC congresses and inter-congresses.

We could go even further to argue that the concept of landscape is uniquely placed to articulate the key concerns of WAC. The organization was born out of the struggle against apartheid in South Africa, and this inspired a continuing concern with the cultural construction of identity and difference, and with social exclusion (Hubert 2000; Lawrence 2003; Shennan 1989; Torrence and Clarke 2000). Any concern with the material expression of identity almost inevitably leads to the question of how difference spreads itself over *space* and thus to the formation of social landscapes. Similarly, WAC meetings have often concerned themselves with food and domestication, animal and plant exploitation, hunting, pastoralism, and agriculture (e.g., Clutton-Brock 1989; Gosden and Hather 1999; Harris and Hillman 1989). These are issues concerned with the inhabitation of the landscape, patterns of movement and settlement, and of located economic practices—and equally, problems concerned with the exploitation, management, and preservation of cultural heritage, with urban origins, with time and temporality, and with architecture and power are all nested in the investigation of landscape. Perhaps most important of all is the

question of Indigenous perspectives on the past; alongside issues of reburial and repatriation, it has been non-Western conceptions of land and landscape—in Aboriginal Australia refashioned in the notion of *country* (cf. Sutton 1995), for example—that have provided one of the most profound challenges to conventional archaeological knowledge (and the claims of archaeologists to a privileged status in the characterization of cultural heritage) over the past 30 years.

It is therefore highly appropriate that this volume on landscape archaeology should be one of the first in the series of *The World Archaeological Congress Research Handbooks in Archaeology*. The archaeology of landscape is a notably diverse field, covering as it does everything from the symbolic significance of places, to the ways people organize themselves in geographical space as *social* space, to the scientific analysis of environmental change. Furthermore, it is also an aspect of archaeology to which any of the major theoretical frameworks might potentially make a contribution: evolutionary theory, ecology, Marxism, feminism, phenomenology, structuration theory, and so on. The consequence of this is that the existing literature is vast and bewildering. Yet despite this vastness, the notion of landscape archaeology retains its usefulness as an orienting concept, one that directs the archaeologist to unpack *emplacement*, in all or any of its dimensions. In this context, the present volume is intended as a *manual* in the purest sense: it aims to present a range of different approaches to landscape in a concise and digestible package. As editors, we have attempted to make the book multivocal. That is to say, we have tried to include perspectives on landscape archaeology that are diverse and that extend well beyond our own particular preoccupations. In this respect, the objective of the book is not to be prescriptive but (we hope) to provide inspiration. As a manual, it is intended to be *useful* to the student and the professional archaeologist, as well as simply interesting.

To this end, we have recruited a group of contributors who are able to write with authority on the intellectual history of landscape archaeology and the conceptual and methodological problems involved in addressing past landscapes. We recognize that a preponderance of these authors are Australian, British, North American, or New Zealanders and that as such they focus on English-language traditions of scholarship. We do not wish to deny the existence of important schools of landscape investigation in other regions: we have simply concentrated on presenting a coherent picture of the landscape archaeologies with which we are most familiar, among an otherwise overwhelmingly large, multilanguage literature (each with its own historical traditions, a review of which would go far beyond a single volume of this size). Furthermore, with a few influential exceptions, we have decided to address landscape archaeology not through regional themes but through present and emerging approaches in historical perspective. We note in this context that were a volume such as this one to focus on regional studies for their own sake rather than on approaches to landscape, in a "world archaeology" that aims toward decolonization of the discipline and that aims also to reach beyond intellectual globalization, the question then arises as to which regions would we include or exclude. Would we target, for example, traditions of landscape archaeology in Zimbabwe by Zimbabwan archaeologists or by foreign archaeologists working in Zimbabwe? The range of examples that our contributors draw on covers North and South America, northwestern and southeastern Europe, the Near East, Australasia, Micronesia, Melanesia, Polynesia, and Africa and include both terrestrial and maritime themes. Nonetheless, our intention was not to provide global coverage through case studies of a variety of landscape types (however those might be defined) from around the world but to outline a range of modes of investigation that possess a worldwide applicability.

Structure of the *Handbook*

The *Handbook of Landscape Archaeology* includes a broad range of essays on existing approaches and methods, as well as novel and promising theoretical and practical advances, taking into account three broad themes:

1. landscapes as *fields of human engagement*, as in Heidegger's notion of dwelling. These include both explorations on conceptual *ways of approaching*, and *experiences* of, landscapes as fields of engagement, as the "in" of "being-in-the-world";

2. landscapes as *physical environmental contexts* of human behavior (such as investigations of the tree cover or topography of site environments);

3. reflections on *representations* of landscapes, such as in landscape art, or the identification of colonial tropes in landscape archaeological literature, or the analysis of textual preconceptions.

We begin the book with two introductory chapters. In Chapter 1, we address the question of what "landscape" is and how it has been approached by different generations of archaeologists. Particular attention is paid to the role of Indigenous perspectives in transforming archaeological conceptions of space, place, and topography. Following this, Chapter 2, by philosopher Edward Casey, explains how historically the West has had different understandings of what "place" is. If landscape archaeology concerns the archaeology of human emplacement, then it follows that how we see social landscapes will depend in part on what we understand place itself to be. Casey's overview, together with the Introduction, sets the scene for landscape archaeology as an archaeology of place, whether this be by focusing on human *engagement*, the *physical environment*, or *representation*.

Based on these conceptual foundations, the book is divided into six parts:

I. *Historical Perspectives*, which includes three chapters on the history of landscape archaeology in Europe, the Americas, and Africa;

II. *Encountering Humans: Mapping Place*, which discusses how primates interact with their surroundings, and the evolution of cognition of place (mapping the landscape) leading to contemporary humans;

III. *Thinking through Landscapes*, which discusses intellectual reflections and visions as to how we as archaeologists can approach landscape archaeology;

IV. *Living Landscapes: The Body and the Experience of Place,* which explores how we experience landscapes, discussing archaeological implications. The way we think of and experience landscapes will affect our archaeological methodologies; for this reason the chapters in Parts III and IV precede those in Part V. Together, Parts III and IV reflect on human *engagements* with place, one of the three major themes of landscape archaeology;

V. *Characterizing Landscapes,* which looks at human surroundings as contexts of knowing about the places of human engagement; it focuses on the *physical environment* and on methodologies for obtaining information about that physical environment, the second of the three major themes of landscape archaeology;

VI. *Nonlevel Playing Fields: Diversities, Inequalities, and Power Relations in Landscape Archaeology,* which explores how we *represent* people and place in

landscape archaeology, the third major theme of landscape archaeology.

As editors of this *Handbook,* like editors of any book, we have had to make decisions about its contents and its structure, how the parts come together as a whole that is more than the sum of its individual chapters. And in such a large topic in particular, we cannot possibly hope to cover all themes; nor can the themes covered each be addressed in the same level of detail—some repeatedly cross-cut between chapters while others sit more or less by themselves. In this context, we offer a *Handbook of Landscape Archaeology* that concerns the major themes that have already appeared in the literature, in a dual spirit of review and exploration that aims to challenge and inspire archaeological practice well into the 21st century.

Julian Thomas
Bruno David
Manchester and Melbourne
May 15, 2008

ACKNOWLEDGMENTS

Thanks to Julie Hollowell, Jane Lydon, Ian McNiven, and George Nicholas for comments on earlier drafts; Diana Arbelaez Ruiz for translating Spanish texts (that owing to limitations in word length ended up not being used in our introductory chapter); Shane Revell and Jean Newey for their considerable administrative input while on fieldwork researching aspects of this paper; and Mitch Allen for his patience and advice in the completion of this book. Thanks to Kara Rasmanis (Monash University) for producing most of the final figures in this Handbook. We also thank Nic Dolby, Jeremy Ash, Nick Araho, Joe Crouch, and Kathleen Marcoux for preparing the index. Last but not least, thank you to the many people who refereed chapters for this *Handbook*: Jeremy Ash, Wendy Ashmore, Robert Attenborough, Bryce Barker, Peter Bellwood, Soren Blau, Denis Byrne, Chris Chippindale, Chris Clarkson, James Connolly, Joe Crouch, Vicki Cummings, John Darnell, Tim Denham, Denise Donlon, David Dunkerley, Peter Dwyer, Scott Elias, Andrew Fairbairn, Chris Fowler, Clive Gamble, Colin Groves, Rodney Harrison, Lesley Head, Michael Heilen, Russell Hill, Peter Hiscock, Alan Hogg, Colin Hope, Geoff Hope, Amanda Kearney, Peter Kershaw, Julia King, Christian Kull, Marcia Langton, Malcolm Lillie, Jane Lydon, Lesley McFadyen, Ian McNiven, Ian Moffat, Meredith Orr, Fiona Petchey, Jim Peterson, Paul Rainbird, Elizabeth Reitz, Cassandra Rowe, Lynette Russell, Michael Shott, Claire Smith, Matthew Spriggs, Nicola Stern, Glenn Summerhayes, Paul Taçon, Christopher Tilley, James Weiner, Marshall Weisler, María Nieves Zedeño, Xuan Zhu.

References

Barker, G. W., and Gilbertson, D. (eds.). 2000. *The Archaeology of Drylands: Living at the Margin.* London: Routledge.

Carmichael, D. L., Hubert, J., Reeves, B., and Schanche, A. (eds.). 1994. *Sacred Sites, Sacred Places.* London: Routledge.

Clutton-Brock, J. (ed.). 1989. *The Walking Larder: Patterns of Domestication, Pastoralism and Predation.* London: Unwin Hyman.

Gosden, C., and Hather, J. (eds.). 1999. *The Prehistory of Food: Appetites for Change.* London: Routledge.

Harris, D., and Hillman, G. (eds.). 1989. *Foraging and Farming: The Origins of Plant Exploitation.* London: Unwin Hyman.

Hubert, J. (ed.). 2000. *Madness, Diasability and Social Exclusion: The Archaeology and Anthropology of "Difference."* London: Routledge.

Lawrence, S. (ed.). 2003. *Archaeologies of the British: Explorations of Identity in Great Britain and its Colonies 1600–1945.* London: Routledge.

Shennan, S. (ed.). 1989. *Archaeological Approaches to Cultural Identity.* London: Unwin Hyman.

Sutton, P. 1995. *Country: Aboriginal Boundaries and Land Ownership in Australia.* Aboriginal History Monograph 3. Canberra: Australian National University.

Torrence, R., and Clarke, A. (eds.). 2000. *The Archaeology of Difference: Negotiating Cross-Cultural Engagements in Oceania.* London: Routledge.

Ucko, P., and Layton, R. (eds.). 1999. *The Archaeology and Anthropology of Landscape: Shaping Your Landscape.* London: Routledge.

HISTORICAL PERSPECTIVES

The concept of landscape is rich and protean and has inspired a dazzling array of different archaeologies. It follows that a historical perspective is vital if we are to unravel the many ways that archaeologists have chosen to address space and place, and the reasons behind their choices. Landscape archaeology is today an outstandingly vibrant aspect of the discipline, because it brings together a series of quite distinct traditions of thought and practice, but these are by no means reconciled to a common set of objectives or approaches.

For many decades, landscape has provided archaeologists with a framework for contextualizing observations and establishing relations and parallels between sites of a particular period. Moreover, it presents the opportunity for diachronic investigations, in which the changing use and inhabitation of a particular region are the focus. In these studies, the scale of analysis and the potential for integration each provides the imperatives for a landscape perspective. However, the landscape can also be understood as an aggregation of resources, affording both opportunities and limitations for human development. In this strand of landscape archaeology, it is the spatial relationships among people, soils, raw materials, and water sources that demand attention.

More recently, a philosophical concern with landscape has become influential, with the recognition that the lived world is not simply a backdrop to everyday action but integral to all human activity. Thus, landscape becomes a source of reference and a context of meaning, central to archaeological theorizing. Consequently, "landscape archaeology" has become a terrain in which highly evolved empirical methodologies confront conceptual approaches that draw on discourses that extend beyond the discipline and sometimes achieve accommodation.

Historically, concerns with space and landscape have appeared on the archaeological agenda at times when difference, variability, and plurality have been at issue. In some cases, this has been connected with an acknowledgment of human diversity and the celebration of the particularity of both national and Indigenous communities. But equally, the mapping of difference can resonate with atavistic beliefs. It is not surprising, then, that there are regional and national differences in the ways in which archaeologists seek to put their evidence into a landscape setting—or, in some cases, decline to do so. The chapters in this first section of the volume draw out the theoretical and historical trajectories involved in the development of landscape archaeology in different parts of the world, exploring major themes that have come to influence, and at times dominate, landscape approaches to regional archaeological programs.

1

LANDSCAPE ARCHAEOLOGY: INTRODUCTION

Bruno David and Julian Thomas

[Without place] there would be neither language, nor action nor being as they have come to consciousness through time. There would be no "where" within which history could take place. "Where" is never there, a region over against us, isolated and objective. "Where" is always part of us and we part of it. It mingles with our being, so much so that place and human being are enmeshed, forming a fabric that is particular, concrete and dense. (Joseph Grange 1985: 71)

. . . a given place takes on the qualities of its occupants, reflecting these qualities in its own constitution and description and expressing them in its occurrence as event: places not only *are*, they *happen* (and it is because they happen that they lend themselves so well to narration, whether as history or as story). (Edward S. Casey 1996: 27)

The term *landscape* came into being in the final years of the 16th century, when the early Dutch landscape artists began to paint rural sceneries that incorporated reference to changing conditions of life (see Cosgrove [1988] for comparable observations on Renaissance Venetian art; see Cosgrove [1998], Daniels and Cosgrove [1988], and Schama [1995] for masterful discussions of "landscape"

in art history). It was introduced, as *The Oxford English Dictionary* (Simpson and Weiner 1989: 628) notes, "as a technical term of painters." The word "entered the English language . . . as a Dutch import. And 'landschap,' like its Germanic root, 'Landschaft,' signified a unit of human occupation, indeed a jurisdiction, as much as anything that might be a pleasing object of depiction" (Schama 1995: 10). The tension between landscape as an entity to be viewed like a painting from afar, and either analyzed or aestheticized, and landscape as a context of dwelling or inhabitation is one that has haunted landscape studies, and that was bequeathed to archaeology once it began to be concerned with the concept, much later on.

"Landscape archaeology" does not have a particularly long history. It was perhaps first used by Mick Aston and Trevor Rowley in the mid-1970s (Aston and Rowley 1974), but it was only in the mid- to late-1980s that it began to be widely cited in academic work. This is not to say that archaeologists have not long employed notions of "landscape" (see Darvill, this volume). But it is arguable that during the 1970s and 1980s "landscape" ceased to be simply a unit of analysis over and above the "site" and became instead an object of investigation in its own right. As a specialized term within the archaeological discipline, the word has witnessed a recent efflorescence, and with this a privileged if somewhat uneasy use. This is because what archaeologists have understood to be "landscape archaeology" has shifted, so that today it does not mean exactly what it used to even 20 years ago. Nor is it presently employed in quite the same way by everyone (see below). When "landscape" *has* been used by archaeologists, it cannot therefore be assumed *a priori* to refer to one particular preconceived thing or another. Indeed, even within the works of individual archaeologists, the term may shift its connotation according to context.

We return to a definition of "landscape archaeology" toward the end of this chapter. First, however, to get a better sense of its nuances and its boundaries, we begin by discussing aspects of its historical emergence, focusing on the post-1970s years (for this is when landscape archaeology began to take its present shape), and discuss also its various attributes. We direct the reader to chapters in this volume by Darvill, Patterson, and McIntosh for histories of landscape archaeology in various parts of the world.

There is little mention of the term *landscape archaeology* in any of the major archaeological journals until the mid-1980s (see Table 1.1), as far as we are aware anywhere in the world.[1] Indeed, silences are telling. In 1978, the leading international journal *World Archaeology* dedicated an entire issue of the journal to the theme "landscape archaeology," but not a single paper in that issue ever used the term. Instead, we find the papers directing their attention to site distributions in environmental settings (e.g., Hurst and Stager 1978; Marshall 1978; Stjernquist 1978), economic strategies and their interregional dynamics (Irwin 1978), economic determinants of settlement patterns (e.g., Conrad 1978), artifact distributions (e.g., Foard 1978; Hirth 1978), environmental impacts and limitations on agricultural production (Marshall 1978), and demographic processes and socio-organizational complexity (e.g., Hirth 1978) in particular regional settings.

This is not to say that there is anything inherently wrong with the approaches chosen by these authors. What these directions do highlight, however, is the way landscape archaeology was understood at that time. The focus was then firmly on human impacts on and interactions with their physical surroundings, evidencing disciplinary concerns also apparent in many contemporary books, which almost invariably used the language of "environmental" or "ecological" archaeology rather than "landscape" archaeology *per se*. (Although there are exceptions, as in Aston and Rowley's [1974] landmark text, which outlined a field of study combining archaeological fieldwork with landscape history. Here, though, the conception of landscape employed was one borrowed from another discipline, rather than representing the emergence of an "archaeology of the landscape.") We thus have during these early years, and on both sides of the Atlantic and beyond, a number of influential works all aiming to address past human historical landscapes as *environmental* archaeology. Examples abound, and include John Evans's *The Environment of Early Man in the British Isles* (1975) and *An Introduction to Environmental Archaeology* (1978), each of which approached the topic mainly by looking at the impact that people had on the land. Karl Butzer's classics, *Environment and Archaeology: An Ecological Approach to Prehistory* (1964) and his subsequent *Archaeology as Human Ecology* (1982), explore the "dynamic interactions between human groups or societies and their environments" (Butzer 1982: xi). Many of these books aimed to understand human-environment relations in terms of the *economic* and/or *adaptive* settlement-subsistence strategies adopted by people in the past, such as in Eric Higgs's (1975) *Palaeoeconomy* and Michael Jochim's (1976) *Hunter-Gatherer Subsistence and Settlement: A Predictive Model*, respectively. During the 1970s and into the 1980s, the focus was very much on "economic" (in the United Kingdom) and "adaptive" (in the United States) attitudes toward the environment.

With such a focus on relationships between people and their physical environments came ongoing calls to more accurately and systematically characterize the way people occupied and used places in the past (e.g., Clarke 1968; Foard 1978; Redman 1975). This meant refinements in field methodologies and statistical analyses, in particular as they relate to the distribution of archaeological materials and sites across the landscape. It also led to a more detailed understanding of landscape formation processes (ultimately to better assess human impacts on the environment and environmental constraints on demographic processes). These new targets of enquiry were aimed at more systematically addressing human organization and scheduling in the landscape, and, to achieve these aims, innovative analytical techniques were required. What was also necessary was a new spatial scale of approach, one that targeted relatively small and well-defined regions. The results were major developments in survey methodologies (e.g., Foard 1978), simulation and predictive modeling (e.g., Sabloff 1981), taphonomy (e.g., Wood and Johnson 1978), geoarchaeology (e.g., Hassan 1979; Neumann 1978), and bioarchaeology and palaeoecology (e.g., Shawcross 1967a, 1967b). Although

Table 1.1 Number of papers that include the phrase "landscape archaeology" or "landscape archaeologist(s)," by journal in 5-year blocks.

Years	Antiquity		Current Anthropology		Journal of Field Archaeology		World Archaeology		American Antiquity	
	Title, abstract & keywords	Text	Title, abstract & keywords	Text	Title, abstract & keywords	Text	Title, abstract & keywords	Text	Title, abstract & keywords	Text
2001–2005	7	3	0	0	1	1	1	1	1	1
1996–2000	3	1	0	1	0	4	0	3	0	0
1991–1995	0	0	0	3	0	0	1	1	0	0
1986–1990	0	0	0	1	0	1	0	0	0	0
1981–1985	0	0	0	0	0	0	0	1	0	2
1976–1980	0	0	0	0	0	0	0	0	0	0
1971–1975	0	0	0	0	0	0	0	0	0	0
1966–1970	0	0	0	0	Starts 1974		0	0	0	0
1961–1965	0	0	0	0			Starts 1969		0	0
1956–1960	0	0	0	0					0	0
1951–1955	0	0	0	0					0	0
1946–1950	0	0	Starts 1955						0	0
1941–1945	0	0							0	0
1936–1940	0	0							0	0
1931–1935	0	0							0	0
1926–1930	0	0							Starts 1935	
1921–1925	Starts 1927									

each of these specialist developments had deeper historical roots (e.g., see the papers in Brothwell and Higgs [1969]), including multidisciplinary faunal and vegetation investigations incorporating pollen, land snail, and beetle remains, the 1960s to 1970s saw an exciting explosion of ideas generally focused on how to better investigate past human-environmental relations at fine-grained geographical scales. Many of these developments were closely associated with an increasing sophistication of statistical procedures (e.g., LeBlanc 1973; Simek and Leslie 1983; Spaulding 1976; Wilcock and Laflin 1974; see reviews by Clark 1982; Clark and Stafford 1982).

In this context, both in the United Kingdom and in the United States there emerged schools of thought that systematically sought to access past spatial patterning of settlements and cultural objects across landscapes (e.g., in the United Kingdom: Clarke 1977; Hodder and Orton 1976; in the United States: Bettinger 1977; King 1978; Mueller 1975), and similar approaches to the archaeological record also emerged—not entirely independently—across the globe (e.g., Bakels 1978). One key development in the United States was a differentiation by Winters (1967, cited in Parsons 1972: 132) of the terms *settlement pattern* (the spatial distribution of sites) and *settlement system* (the way that people organized themselves in the landscape). These now-disaggregated concepts quickly took hold throughout much of the English-speaking archaeological world. However, they remained most influential in the United States, where the processual interests of the New Archaeology, as championed by Lewis Binford in particular, targeted settlement *systems* (incorporating an understanding of settlement patterns) for their ability to inform on an "archaeology of place" that was reduced largely to relationships between settlement (places where people lived and undertook economic activities) and subsistence (things that people ate). Settlement-subsistence system analyses, such as the influential and impressive investigations among the Nunamiut of Anaktuvuk Pass in Alaska by Binford (e.g., 1978, 1981), were largely strategies by which to investigate humans responding to biological needs for food and shelter in their particular environmental settings. As Binford (1982: 6) notes, in undertaking an "archaeology of place": "I am interested in sites, the fixed places in the topography when man [sic] may periodically pause and carry out actions." But these were "long-term repetitive patterns in the positioning of adaptive systems in geographic space . . . arising from the interaction between *economic zonation* . . . and tactical mobility" (Binford 1982: 6). As McNiven and colleagues (2006: 14) cogently

put it, these were landscapes "stripped . . . of their cosmological, symbolic and spiritual meaning" that failed to mention religious sites and concepts that were important to the Nunamiut themselves and "that clearly mediated ecological relationships." It was "an archaeology of place devoid of meaningful place and of meaningful emplacement, just as it is devoid of social experience and salience" (see also Insoll 2004). We return to these latter notions later in this chapter.

By focusing on settlement systems rather than settlement patterns in and for themselves, archaeological attention in the United States thus quickly turned to *process* rather than *location* of human behavior in the landscape. This is perhaps best exemplified by Binford's (1980) very influential differentiation between "collectors" who generally "map onto" resources "through residential moves and adjustments in groups size" (Binford 1980: 10) (producing three kinds of sites in the process: *field camps, stations,* and *caches*); and "logistically" organized "foragers" who bring back resources to base camps on a daily "encounter" basis (and who produce two kinds of sites along the way: residential *base camps* and resourcing *locations*). Such differentiation of what are essentially economically (and largely subsistence-) driven mobility strategies aimed to distinguish among various forms of organizational alternatives in specific environmental settings so as to better model evolutionary pathways under changing environmental conditions. Thus, "since systems of adaptation are energy-capturing systems, the strategies that they employ *must* bear some relationship to the energy or, more important, the entrophy structure of the environments in which they seek energy" (Binford 1980: 13, italics in original); changing environmental conditions will in this formulation have considerable influence on settlement-subsistence systems.

In the United Kingdom, however, archaeological interest took what initially looked like a minor turn in a different direction aimed more at characterizing the spatial *patterning* of archaeological sites and artifacts, an originally potentially insignificant turn that eventually led to what could be described as a paradigmatic shift (see below).

The 1970s into the 1980s also saw the rapid development of various multidisciplinary methodologies, principally geoarchaeological and bioarchaeological, enabling a more detailed characterization of human-environmental relations through notions of palaeoecology. Developments along these lines were apparent in many English-speaking nations, including the United States, the United Kingdom, and Australia, as is evident by the many works that appeared at that time. For example:

- in Australia, Coutts 1970; Mulvaney and Golson 1971
- in Ireland, Reeves-Smyth and Hamond 1983
- in the United Kingdom, Higgs 1975; Pryor 1980
- in the United States, Butzer 1982
- in central America, Hirth et al. 1989
- in Africa, Greenwood and Todd 1976; Stewart 1989
- in Japan, Hiroko 1986
- in New Zealand, Shawcross 1967a
- in Holland, Bakels 1978

A contemporary interest on taphonomic studies by United States and South African practitioners in particular benefited hugely from a new focus on middle range research (e.g., Behrensmeyer and Hill 1980; Binford 1981; Brain 1981). Major developments in the ecological sciences began to be systematically applied to archaeological problems, such as in New Zealand, when Wilfred Shawcross combined information on the local environmental productivity of shellfish species with archaeological deposition rates of those same species to determine the likely duration of occupation at specific campsites (Shawcross 1967a) and the carrying capacity of specific locations (Shawcross 1967b). These and other related methods were innovative, but they tended to be closely tied to an ongoing preoccupation with settlement-subsistence systems and ecological modeling, which were themselves closely associated with developments both in methods of data retrieval (for example, concerned with characterizing artifact distributions; see below) and "natural" environmental details, such as faunal and vegetation distributions (usually seen as actual exploited rather than just potential exploitable resources for local populations) across landscapes.

Settlement patterns in their environmental settings were thus an important focus at the time, along with spatial patterning of environmental variables. Together, detailed data on environmental and cultural distributions were targeted, so as to better characterize the economic nature and reasoning behind settlement-subsistence systems and patterns. This was much the underlying logic of both the adaptive thinking of the New Archaeology in the United States (and in the United Kingdom largely expressed by exponents of systems theory, in particular David Clarke [1968]) and of the British

school of economic archaeology and its focus on site catchment analyses. Such a general focus on environmental perspectives and economic parameters are well summed up in the 1978 *World Archaeology* issue dedicated to landscape archaeology, in which Stjernquist (1978: 261) concludes that a "clearly noticeable trend is the concentration on ecological archaeology studying man's [*sic*] role in and adaptation to his [*sic*] environment over time;" Irwin (1978: 306) writes of "technology, economy and environment," and Foard (1978: 372) concludes with a call for a "total archaeology" that is concerned with understanding human behavior in the environment, necessitating a multidisciplinary approach to settlement-subsistence systems. In essence, like Michael Reed's (1990: xii) in *The Landscape of Britain*, the general understanding during those early years of landscape archaeology was that "the theme of the landscape historian is the evolution of that external world in which men and women have carried on the everyday business of their lives from the remotest periods of prehistory down to the present" and the settlement-subsistence history of human societies in those environmental contexts.

It is during this period of focus on past human-environmental relations that field surveying strategies began to change from site-based to "off-site" (or "non-site" or "siteless") surveys (e.g., Dunnell and Dancey 1983; Foley 1981), with a greater emphasis on probability sampling (of both "site" and "off-site" surveys) (e.g., Cowgill 1975), because it was quickly realized that what were often effectively continuous (but varied) artifact distributions across the landscape had to be accurately characterized and cross-referenced with environmental variables. In the United States, these developments took place synergistically with the development of the "regional" approach that helped define the New Archaeology. By recording information on artifact distributions and environmental patterns at unprecedented levels of detail, new advances were made in landscape archaeology. This rethinking of surveying methodology took hold both in the United Kingdom and the United States, in many ways world leaders in landscape archaeology at that time, although such developments also took place in many other countries (in particular South Africa, Australia, and New Zealand). But in the United Kingdom, the search for functional and adaptive processes in landscape archaeology never really took hold in the same way as it did in the United States.[2] Rather, partly through developments in methods aimed at exploring settlement and artifact patterning, there came a realization that archaeologists were not simply dealing with humans adapting to environmental circumstances, but rather with people interacting among themselves as much as they were interacting with their physical environments (e.g., see Renfrew's [1983] critical first Plenary Address to the Society for American Archaeology). In this respect a key publication was Ian Hodder's 1978 edited volume, *The Spatial Organisation of Culture*, which explicitly addressed the relationship between spatial distributions of material culture and human identities. It has been argued that British archaeology (as opposed to that of the United States) has always had a deeper conviction that artifact assemblages reflect the existence of coherent and bounded social entities in the past (Binford and Sabloff 1982: 141). It was continuing unease over precisely what spatial patterning *meant* that led Hodder and others to the insight that the adoption of specific artifact types might represent a deliberate strategy of social inclusion and exclusion, rather than simply reflecting a pregiven identity. By the mid-1980s, Hodder had become one of the most important and innovative exponents of a new kind of *social* archaeology that soon came to inform landscape archaeology itself. The critique here was pervasive across the discipline, contributing significantly to the creation of a new community of culture that came to stay (although it influenced different national archaeological agendas in different ways and to various degrees): the archaeological record now signaled not so much *adaptive (biological) humans* as *interacting (social) people* who engaged with their surroundings in various ways. These included symbolic practices that required social and philosophical rather than environmental understandings to decipher. This key period heralded the beginning of contemporary notions of landscape archaeology. This broad shift toward social dimensions of landscapes expands the earlier emphasis on more behavioral modes of interaction.

Changing Directions: From Environmental to Social Landscapes

The move toward a more socially oriented landscape archaeology came from many fronts, and it came together as part of a broadly changing culture of understanding. Four major influences on archaeological practices were (1) sourcing studies; (2) the rising importance of cultural heritage management and public archaeology; (3) a developing interest in "style;" and (4) Indigenous critiques.

Sourcing Studies

Sourcing studies around the world quickly developed from the 1960s onward, although earlier moves had been made, such as in the petrology of stone axes (e.g., Keiller et al. 1941) and in trace-element analyses of faience beads from Bronze Age Britain (e.g., Hawkes and Hawkes 1947) and obsidian from the Near East in the late 1940s and 1950s (e.g., see Cann et al. 1969). In his influential paper, "Trade as action at a distance," Colin Renfrew (1975) argued that social change often occurred simultaneously over wide geographical expanses, necessitating a focus not just on individual places but on relationships between places in systems of peer polity interactions. This recognition gave new impetus for a *socially* oriented economic archaeology, in particular an emphasis on trade. In the British context, this set of concerns was given further momentum by the introduction of ideas drawn from neo-Marxist anthropology, which emphasized the importance of long-distance exchange relations in creating and reproducing patterns of alliance and positions of authority (e.g., Bradley 1982, 1984; Bradley and Edmonds 1993). Here the objects themselves were acknowledged as being the means by which social relationships were articulated, rather than necessarily being of purely pragmatic value.

Similar developments were also taking place in the Pacific on the opposite side of the globe, where Shutler and Marck (1975) and Bellwood (1978), following earlier observations (see Avias 1950; Gifford and Shutler 1956; Golson 1961), came to link the distribution of Lapita ceramics across vast seascapes into a single and unified historical sphere of interaction with people who spoke Austronesian languages. In this way, these researchers populated the archaeological record with language-speaking people rather than just material objects such as ceramic sherds (see Spriggs 1997: 67–107 for a review). This region of Austronesia subsequently became a focus for lithic and ceramic sourcing studies, in particular by archaeologists in New Zealand (e.g., Green 1987; Summerhayes 2000). Of concern here were not so much environmentally adaptive histories as social processes of colonization across vast and previously unoccupied seascapes. In Australia also, similar concerns for the sociality of interregional interaction were being voiced by Isabel McBryde (initially with Ray Binns, then with Alan Watchman), who undertook a series of sourcing studies of ground stone hatchets, at first among the stone quarries of northern New South Wales (Binns and McBryde 1972) and subsequently and more influentially among the greenstone quarries of Mt. William, Mt. Camel, and Berrambool, Baronga, Geelong, Jallukar, and Howqua in central and western Victoria (e.g., McBryde 1978; McBryde and Harrison 1981; McBryde and Watchman 1976). In these latter investigations, McBryde showed that although ground-edged hatchet heads were traded to distant lands more than 600 kilometers away, examples from the Mt. William quarry were preferentially traded toward the north and southwest, practically halting 150 kilometers away. She further identified anomalies in the distance-decay curve—especially in an absence of Mt. William axes in the Wimmera-Mallee region and in eastern Victoria—and thereby posited strong socioterritorial deterrents to past exchange relations beyond those frontiers. Her interpretations were supported by regional ethnohistoric records indicating the presence of two major and largely antagonistic social groups, the Kulin and Kurnai of central and eastern Victoria, respectively. Her study came to be significantly informed by linguistic and other ethnographic knowledge, which indicated that social groups were aligned not only in abstract geographical space but also in *territorial* space and social systems of alliance. (See Lourandos [1977] for another influential Australian example of the archaeology of socioterritorial space; see Tamisari and Wallace [2006] for discussion of the significance of this work.) This form of geographical configuration had not hitherto received a great deal of archaeological attention in Australia, although such approaches had much in common with Renfrew's (and others') contemporary concerns elsewhere, with the additional insights offered by 19th- and early 20th-century ethnohistorical texts.

Cultural Resource Management

The late 1960s into the 1970s saw major transformations to cultural heritage management, the working face of public archaeology. This was a time of increasing popular and professional awareness of the progressive dwindling of cultural sites as heritage places, including the establishment of new legal mechanisms by which sites could be protected (e.g., see Colley 2002; King 1998). Along with a major influx in the scale and rate of cultural heritage studies came an increasing need for explicit assessment of the significance of sites and landscapes as cultural resource catchments. (See Schiffer and Gummerman [1977] for an excellent contemporary assessment of the state of cultural heritage management.) The need for increased protection of archaeological sites was prefaced by new and explicit criteria for the assessment of heritage places as locations of social significance. Hence recognition of the educational, cultural, historical and aesthetic values of archaeological sites and landscapes effectively rendered them significant public places that went beyond their environmental and academic significance (e.g., King et al. 1977). These new social dimensions of significance, as expressions of public recognition, meant that the significance of archaeological sites and archaeological landscapes could no longer be reduced to environmental, ecological, or economic agendas (e.g., Lipe 1984; Moratto and Kelly 1978).

A striking example of the way that cultural resource management issues have shifted the focus toward a *social* landscape is provided by the case of Stonehenge in southern Britain. Here, the parallel debates over the upgrading of visitor facilities and the visitor "experience" in general, the rerouting of the A303 main road from its present position beside the monument, and access to the prehistoric monument (particularly at the solstices) by diverse groups, including Druids, New Age "Travelers," and Pagans, all explicitly implicate the landscape. The landscape is recognized as the aesthetic setting of Stonehenge and its attendant monuments, as the topographical context in which the visitor experience is embedded, and as the political terrain over which struggles between interests (past and present) have been played out and within which identities are negotiated (see Bender 1998; Chippindale et al. 1990; Darvill 2006; Worthington 2004). In the case of Stonehenge, it is now very difficult to imagine the monument in a landscape that is either purely "ecological" in character, or socially uncontested. These new directions are well illustrated by the United Nations Educational, Scientific and Cultural Organization's (UNESCO) criteria for the inclusion of places in the World Heritage list, which include "archaeological sites that are of outstanding universal value from the historical, aesthetic, ethnological, or anthropological points of view" (http://whc.unesco.org/opgulist.htm#para23). These criteria were adopted by UNESCO at its 17th session, held in Paris in 1972, precisely during the period when a more socially informed landscape archaeology was gaining momentum.

Another, slightly distinct aspect of cultural resource management that has also fueled the growing concern with social landscapes has been the changing character of "salvage" or "rescue" archaeology in the industrialized nations. Since the 1960s, the construction of new homes, industrial facilities, and infrastructure (particularly telecommunications) has continued apace in many areas.

This has coincided with a growing conviction on the part of national and regional governments that the archaeological heritage should be preserved, or at least recorded. Whether funded by government or by the developers themselves (as in the case of Britain, following the adoption of *Planning Policy Guidance 16: Archaeology and Planning* in 1990), the colossal scale of development has increasingly been matched by the scale of archaeological interventions in advance. To give some examples: the Aldenhovener Platte Project in the Rhineland (1965–1981) was occasioned by large-scale open-cast extraction of brown coal (Lüning 1982); the Hardinxveld sites in the Rhine/Meuse delta of Holland were excavated in advance of the expansion of the port of Rotterdam and its attendant rail link (Louwe Kooimans 2001); the urban expansion of Malmö in southern Sweden has resulted in a series of very large-scale open-area excavations, including four vast Neolithic palisaded enclosures (Brink and Hydén 2006); the new terminal for London's Heathrow Airport required similarly large investigations (Andrews et al. 2000), as did the bypass road around Dorchester in Dorset (Smith et al. 1997) and gravel extraction at Barrow Hills in Oxfordshire, England (Barclay and Halpin 1999). Each of these projects has resulted in the recovery of highly important archaeological evidence, but in each case the sheer size of the undertaking has demanded that the investigation must be conceived at the landscape (as opposed to "site") scale.

Although in some cases it has been possible to address these landscapes in purely environmental terms, for the most part the nature of these projects has required a consideration of social networks that extend beyond residential locations, and the dispersal of social practices across the landscape. Moreover, salvage projects conducted at the multisite level have inevitably often tended to be multiperiod as well, frequently prompting a consideration of landscape development over time. It may be, then, that the issue of social landscapes is one area in which the "two cultures" of the discipline, commercial field archaeology and academic field archaeology, can find a degree of common ground.

Landscapes with Style

A third approach toward the social began to take effect through concerns for the symbolic. Although the notion of "style" (especially when it is opposed to that of "function") has been the subject of searching critiques in recent years (e.g., Boast 1997), it is undeniable that it was the focus of some of the most important developments in the archaeology of the 1980s and 1990s. In particular, the concept was implicated in the rise of a socially informed landscape archaeology both in the United Kingdom and in the United States, where economic and adaptive frameworks had established a strong grip. In "style," the move was toward an understanding of the past that focused more on social relationships within and between communities of people through the way they decorated items of material culture, a move that was also happening in sourcing studies. This was not so much an environmentally as a socially informed disciplinary interest, although geographic distance and the presence of geographical barriers limiting the spread of ideas also came into play.

In stylistic studies it quickly became apparent that geographical barriers are as much social as they are environmental, as McBryde (1978) was similarly finding in her sourcing studies. Through ethnoarchaeological and ethnohistorical research across many parts of the globe—for example:

- among the Ilchamus, Tugen, and Pokot of the Lake Baringo district of western Kenya, Ian Hodder (e.g., 1982)
- among the San of the Kalahari desert in Namibia, Polly Wiessner (e.g., 1983, 1984)
- among Yugoslavian ethnic groups, Martin Wobst (1977)
- subsequently in central and northern Australia, Claire Smith (e.g., 1992)

—a new form of landscape archaeology was being fashioned, one that talked of symbolic rather than of environmental configurations (for a review see Conkey 1990).

While archaeologists with interests other than symbolic archaeology were also increasingly involved in ethnoarchaeological research—for example:

- Gould [1968, 1971] among the Ngatatjara of Western Australia
- Hayden [1979] among the Pintupi of central Australia
- Binford and O'Connell (1984) among the Alyawarra, also in central Australia
- Binford [e.g., 1978] among the Nunamiut of Alaska
- Jacobs [1979] among Fars province villagers of Iran
- White (e.g., 1967) among the Duna of New Guinea

- Yellen (1977) among Dobe !Kung San of the Kalahari desert, to name but a few; see David and Kramer (2001) for a review

—new concerns with information exchange via symbolic behavior began to refashion landscape archaeology as social archaeology. A new, largely ethnographically informed focus on the social geography of stylistic behavior among interacting communities of people effectively bridged the gap between an environmental archaeology concerned with artifact and site distributions in physical landscapes and a social and symbolic archaeology interested in the geographical spread of stylistic conventions among archaeological objects (in particular rock art) (e.g., Gamble 1982; for subsequent applications, cf. Bradley 1997).

Indigenous Critiques

Each of these new and influential directions helped to reformat landscape archaeology increasingly toward the social. But a fourth and most significant influence also made its very considerable mark: the realization that environmental notions of landscape archaeology did not by themselves accurately reflect Indigenous peoples' own notions of their landscapes or the reasons why they lived in certain ways. Such a realization came about from an increasing reading of anthropological texts by archaeologists around the globe (and the development of ethnoarchaeology as a distinct subdiscipline of its own), increasing direct engagements with the Indigenous peoples whose homelands and histories were often being studied and increasing dissatisfaction with abstract archaeological concerns that often seemed far removed from Indigenous notions of their own histories and lifeways. These changes were to some extent connected with a retreat from the more extreme positions of processual archaeology, and its demand for universal laws of human behavior, valid in all temporal and spatial contexts. Increasingly, archaeologists have come to recognize the value of multiple perspectives and perceptions. Thus, at the 1982 Australian Archaeological Association (AAA) annual conference, the Indigenous representative Ros Langford (1983) impressed on the audience how the local Aboriginal community had had enough of archaeological characterizations of Indigenous lands and Indigenous history (and through this, the Indigenous present) as an archaeological playground. "Our heritage, your playground" became a rallying point from which to make archaeologists aware of the inadequacy of then-predominant archaeological practices and

to change the discipline toward a more socially aware enterprise. This episode was a turning point in Australian archaeological practice, with the subsequent adoption by the AAA of a code of ethics that gave prominence to Indigenous rights and to the recognition of requisite ethical standards in the archaeological research of Indigenous history and Indigenous lands (see Colley [2002]; McNiven and Russell [2005] for discussions on the decolonization of archaeological practice in Australia).

Around the globe, the growing number of Indigenous archaeologists considerably influenced such developments; although initially few, Indigenous archaeological voices were increasingly heard in academic writings, at conferences and in the field (Watkins 2000). Together, these four sets of disciplinary developments—increasing concern with social landscapes as informed by sourcing studies, cultural heritage management, symbolic archaeology, and Indigenous constructions of place—signaled an increasingly *anthropological* archaeology (and not coincidently, the *Journal of Anthropological Archaeology* was founded during this period, in 1982). Colin Renfrew (1982: 6) has coined that period prior to the emergence of this more socially oriented, anthropologically informed archaeology as the "long sleep of archaeological theory."

In many Indigenous languages there is no word for landscape-as-environment. But there is a word for *country*, referring to the places of human existence in all their existential and phenomenological (experiential) dimensions (e.g., see Bradley; Teeman, both this volume). These notions of country include not just the trees and the rocks of the physical land but also the spirits of the land and the waters and the skies that others may not know. And because the ancestral spirits from whence present people came reside in place, country itself identifies history as it historicizes identity. Landscape as country concerns people's relationships with places, a landscape richly inscribed with history, agency, territorial rights, ancestral laws, and behavioral protocols. From the 1980s onward, Indigenous critiques increasingly began to seriously influence the general study of archaeological landscapes in Indigenous peoples' own terms (e.g., Langford 1983; Ross et al. 1996; Watkins 2000; see Lane, this volume; McNiven, this volume).

As a result of these critiques, the landscape increasingly began to be seen as engaged socially and culturally as much as it is engaged environmentally, and it is this engagement that defines the lie of the land, what a landscape looks like. Landscapes are topographies of the social and the cultural as much as they are physical contours. To understand a landscape one has to outline its means of engagement, the way it is understood, codified, and lived in social practice; and each of these, along with the landscape itself, have history. Engagement gives and is defined by the way we give cultural meaning to the location of our existence—so that even the trees and the rocks mean different things to different people.

The anthropologist Marcia Langton (2002) notes that in Western systems of knowledge, we look at the stars in the sky and understand that millions of light years away gigantic balls of fire emanate their light across the vast expanses of the universe. The night sky glows with innumerable lights that we understand through and that confirms to us a sense of, astronomical time. She points out that among Australian Aboriginal peoples a similar process of landscape recognition takes place: the land, as are the waters and the skies, is populated by the ancestors who are ever-present and by various Dreaming spirit beings who created the law of the land, the social codes of conduct, and who generally imbue the world with its defining features. This is a temporal landscape that combines the past and the future through the timeless truth of a codified law of conduct sanctified in an ever-present Dreaming. Here, too, as in the Western night sky, the landscape emanates a sense of time, a sense of cosmological order. In the words of Veronica Strang (1997), we can thus speak of landscape as "uncommon ground"—one land but multiple visions of that land, multiple understandings, multiple landscapes. Such an approach allows for an archaeology not only of monuments but also of so-called natural places (e.g., Bradley 2000), because they, too, are culturally inscribed in social consciousness and therefore possess archaeological signatures defined by social attitude. This move toward a more socially informed, and in this a more ethically responsible, landscape archaeology recognizes that the world has many voices. But this recognition has come at a price: some would say that archaeology's innocence has come of age, but others would say that its guilt has been found out, highlighting the discipline's inherently ethical entanglements.

Landscape Archaeology Today

In such a historical context that began largely with an *environmental* archaeology (but see also Darvill [this volume] for a longer-term history of landscape archaeology), it comes as no surprise to find little use of the term *landscape archaeology* in any of the major archaeology journals until the mid-1980s (Table 1.1), anywhere in the world. Yet a general paucity of reference to "landscape archaeology" in academic texts until the 1980s is only part of the story; both *landscape*

and *archaeology* have long been widely used by professional practitioners. The question thus remains as to why we did not find a common conjunction of the two words until the closing years of the 20th century. Indeed, this question is brought into sharp focus when one realizes that, in computer searches undertaken on *Google* and *ninemsn* between 18 and 25 July 2006, "landscape archaeology" rates sixth after "historical archaeology," "classical archaeology," "industrial archaeology," "prehistoric archaeology," and "environmental archaeology" in a long list of archaeologies (see Table 1.2). It is apparent that the four most commonly used forms each pertains to an established subdiscipline of archaeology in its own right. Based on these counts, it would appear that "landscape archaeology" is even more popular than "social archaeology," "marine archaeology," "processual archaeology," "gender archaeology," "behavioral/behavioral archaeology," and many others. The question remains: what has made "landscape archaeology" so attractive to archaeologists since the last decade of the 20th century, while despite the widespread use of the individual terms "landscape" and "archaeology," little reference was previously made to "landscape archaeology" as a unified concept.

We argue that the answer lies in three related factors: first, the recent emergence of "landscape" as something other, and more, than "environment"; second, an understanding that being-in-the-world is entangled in social process and is not entirely reducible to notions of environmental adaptation; and third, along with these changes in perception of social landscapes, the recent development of a culture of understanding that sees people and culture at the core of worldly engagements.

In this context, landscape archaeology today is much different from what it was in the 1970s and 1980s. By the first decade of the 21st century, many archaeologists around the world have turned their attention to spiritual dimensions of Indigenous landscapes:

- "ritual engines" (Gibbs and Veth 2002) in Aboriginal Australia
- "spiritscapes" (David et al. 2005; McNiven 2003) and "ritual orchestration" (McNiven and Feldman 2003) in northern Australia and Torres Strait
- "sacred geographies" in Papua New Guinea (Ballard 1994)
- "*kastom*" and the "spirit world" in Vanuatu (Wilson et al. 2000)
- "cosmovisions" in central America (Broda 1987); cosmologies in southern India (Boivin 2004)

Table 1.2 Number of hits made using two different search engines in July 2006 (all searches were made in double quotation marks).

Term	Google	ninemsn
historical archaeology	c.609,000	44,511
classical archaeology	c.400,000	46,998
industrial archaeology	c.386,000	48,813
prehistoric archaeology	c.183,000	28,829
environmental archaeology	c.133,000	18,977
landscape archaeology	**c.125,000**	**25,172**
marine archaeology	c.114,000	18,057
ethnoarchaeology/ ethno-archaeology	c.105,610	15,820
social archaeology	c.72,800	7,399
new archaeology	c.65,700	13,410
community archaeology	c.47,900	6,952
cognitive archaeology	c.27,600	3,810
processual archaeology	c.23,400	3,469
settlement archaeology	c.22,500	2,648
theoretical archaeology	c.20,900	6,669
gender archaeology	c.13,900	3,307
indigenous archaeology	c.12,100	1,805
colonial archaeology	c.10,200	1,186
postprocessual/ post-processual archaeology	c.10,944	1,057
Darwinian archaeology	c.5,260	219
behavioural/behavioral archaeology	c.1,095	400
ecological archaeology	c.752	244
symbolic archaeology	c.458	122
postcolonial/post-colonial archaeology	c.190	92
total archaeology	c.157	85
postmodern/post-modern archaeology	c.139	127

- shamanism in South Africa (e.g., Lewis-Williams and Dowson 1990) and parts of the United States (e.g., Whitley 1992)

- "sympathetic control" in southern Africa (Thackeray 2005)

- generally "ceremonial landscapes" (Ashmore this volume), "religious experience" (e.g., Dornan 2004), and liminal spaces (cf. Turner 1995) in various parts of the world

What we now have today is an archaeology of landscapes that is as much about the ontological and cosmological dimensions of places as it is about their physical characteristics. Landscape archaeology has come to refer to the places that are meaningful to people, and in so doing, to the archaeology of that meaningfulness.

What Is Landscape?

Disciplinary subdivisions are pointers to how we normalize the world; they direct our attention and enable us to approach the world through very particular frames of reference and understanding. "Landscape archaeology" does just this in ways peculiar to the post-1970s era, continuing today to inspire our archaeological endeavors and archaeological imagination in novel ways. And this is what "landscape archaeology" gives us: a conceptual framework that enables us to address human pasts in all their contexts and that goes beyond a purely environmental archaeology. In this sense, and along with other developments, it enables us to go forward from our own disciplinary pasts.

This, then, is the crux of landscape archaeology: it concerns not only the physical environment *onto* which people live out their lives but also the meaningful location *in* which lives are lived. This includes the trees and the rocks and the stars, not as abstract objects but as meaningful things that are located ontologically and experientially in people's lives and social practices (praxis). *People* lie at the core of a landscape archaeology and, befitting the general purpose of all archaeologies (in contrast to ethology, geology, botany, zoology, and the like), it is those past human dimensions that a landscape archaeology targets.

Broadly speaking, landscape archaeology is thus concerned with the things that locate human existence. A landscape archaeology is an archaeology of place, not just as defined in a set of physical nodes in space (cf. Binford 1982) but in all its lived dimensions: experiential, social, ontological, epistemological, emotional, as place and emplacement

concern social identity, as much as they concern the economic and environmental aspects of life. If, as Lefebvre (1991: 8) has it, "spatial practice consists in a projection onto a (spatial) field of all aspects, elements and moments of social practice," then landscape archaeology, in its concern with past human engagements with place, concerns the past spatiality of all aspects, elements, and moments of social practice.

Landscape archaeology is an archaeology of how people visualized the world and how they engaged with one another across space, how they chose to manipulate their surroundings or how they were subliminally affected to do things by way of their locational circumstances. It concerns the intentional and the unintentional, the physical and the spiritual, human agency and the subliminal. Landscapes concern how people scheduled their daily routines—seasons affect the rhythms of work and play, and social time is implicated in the daily rhythms of work and play, Tim Ingold's (1993) "taskscapes." Landscapes implicate social order and gender, because who lives where, who goes or works where, and the significance of places are each mediated by social structure, worldviews, and the meaningfulness of place. Landscapes are ecological, all peoples constructing frames of knowledge by which to know the world in which they live. Landscapes are institutional as space is structured and behavior normalized through codified social practice. Landscape concerns moral codes, who can go where, under which conditions, and is played out in ongoing reassessment of social rights and social wrongs. Landscapes are always territorial spaces in that they are controlled and contested in social and political practice. Landscapes are ontological in that they are always known through historically emergent worldviews. And landscapes are always engaged as the location of social and personal experience, as the place of being-in-the-world. There is, as Henri Lefebvre (1991) has pointed out, a truth of space rather than true space, and that truth is generated in social process, in the constant assessment and renegotiation of emplacement. "Social processes are also processes of interaction with the environment as a whole, which provides the medium through which values are created and expressed," writes Veronica Strang (1997: 176); "the landscape is a crucial part of this medium, and the development of an effective relationship with the natural environment depends on the location of certain values in the land."

Landscape archaeology concerns each of these dimensions of social emplacement. We concur with Torrence (2002: 766), who notes that

by definition, the term "landscape" takes in all
physical and natural components of the terrestrial environment. . . . it should be combined
with "seascape" . . . to encompass adequately
the settings where human behaviour took place.
Adding "cultural" to land- and seascapes emphasizes the role of the individuals who conceptualized these spaces and actively created and
modified them in culturally specific ways.

This, then, is what has changed from those earliest expressions of the archaeology of landscapes
that largely began with economic, environmental,
and ecological concerns (dimensions that continue
to usefully inform aspects of "landscape archaeology"; for recent volumes, see Dincauze 2000; Rapp
and Hill 1998): landscape archaeology has become
today more about the archaeology of socially and
experientially engaged place as it is an archaeology of the causes and consequences of environmental conditions on human behavior. It is less about
an absolute notion of "place" as it is about singular *senses* of place (cf. Feld and Basso 1996). And
this is the binding glue of contemporary landscape
studies: a concern for the where of all human practice, in any or all of its dimensions.

Notes

1. A digital search (followed by manual perusal)
of all papers published in the professional jour-
nals *American Antiquity, American Journal of
Archaeology, Antiquity, Current Anthropology,
Journal of Field Archaeology,* and *World
Archaeology* and a less intensive search of
the journals *Annual Review of Anthropology,
Archaeometry, Arctic Archaeology, Bulletin
of the School of Oriental and African Studies,
Cambridge Archaeological Journal, European
Journal of Archaeology, Journal of African
Archaeology, Journal of Anthropological
Archaeology, Journal of Archaeological Science,
Journal of Human Evolution, Journal of Irish
Archaeology, Journal of Near Eastern Studies,
Journal of the Royal Anthropological Institute*
(incorporating *Man*), *Latin American Antiquity,
Nyame Akuma, Post-Medieval Archaeology,
Proceedings of the Prehistoric Society,* and
the popular magazines *Archaeology, Current
Archaeology,* and *Popular Archaeology* have
failed to recover any evidence for recurrent
use of "landscape archaeology" or "landscape
archaeologist(s)" in paper titles, abstracts, key-
words, or texts until the mid-1980s. Computer
searches were made of each volume; all hits
were then individually checked for context and

to ensure that in-text citations were not limited to
reference entries, acknowledgments, and the like
(with the exception of the *American Journal of
Archaeology,* where footnotes were not searched
for the exclusion of bibliographic listings). The
latter entry types were excluded from the counts
presented in Table 1.1. Book reviews were also
excluded.

2. This is probably well illustrated by David
Clarke's (1968) *Analytical Archaeology,* a
work that is often compared to American-
style New Archaeology but that in fact
focused more on the logic and methodology of archaeological research as a means
of exploring past cultural *patterning* rather
than as a means of elucidating universal laws
of cultural *process.*

References

Andrews, G., Barrett, J. C., and Lewis, J. S. C. 2000.
Interpretation not record: The practice of archaeol-
ogy. *Antiquity* 74: 525–30.
Aston, M., and Rowley, T. 1974. *Landscape
Archaeology: An Introduction to Fieldwork
Techniques on Post-Roman Landscapes.* Newton
Abbot: David and Charles.
Avias, J. 1950. Poteries Canaque et poterie préhis-
toriques en Nouvelle-Calédonie. *Journal de la
Société des Océanistes* 6: 111–40.
Bakels, C. C. 1978. *Four Linearbandkeramik Settlements
and Their Environment: A Paleoecological Study
of Sittard, Stein, Elsloo and Hienheim. Analecta
Praehistorica Leidensia* XI. Leiden: Leiden
University Press.
Ballard, C. 1994. The centre cannot hold: Trade networks
and sacred geography in the Papua New Guinea
Highlands. *Archaeology of Oceania* 29: 130–48.
Barclay, A., and Halpin, C. 1999. *Excavations at
Barrow Hills, Radley, Oxfordshire. Volume 1: The
Neolithic and Bronze Age Monument Complex.*
Oxford: Oxford Archaeology Unit.
Behrensmeyer, A. K., and Hill, A. P. 1980. *Fossils in
the Making.* Chicago: University of Chicago Press.
Bellwood, P. S. 1978. *Man's Conquest of the Pacific.*
London: Collins.
Bender, B. 1998. *Stonehenge: Making Space.* Oxford:
Berg.
Bettinger, R. L. 1977. *The Surface Archaeology of the
Long Valley Caldera, Mono County, California.*
Riverside: University of California Archaeological
Research Unit Monographs 1.
Binford, L. R. 1978. *Nunamiut Ethnoarchaeology.*
New York: Academic Press.
———. 1980. Willow smoke and dogs' tails: Hunter-
gatherer settlement systems and archaeological site
formation. *Journal of Anthropological Archaeology*
1(1): 4–20.

Binford, L. R. 1981. *Bones: Ancient Men and Modern Myths*. New York: Academic Press.

———. 1982. The archaeology of place. *Journal of Anthropological Archaeology* 1: 5–31.

Binford, L. R., and O'Connell, J. F. 1984. An Alyawarra day: The stone quarry. *Journal of Anthropological Research* 40: 406–32.

Binford, L. R., and Sabloff, J. A. 1982. Paradigms, systematics, and archaeology. *Journal of Anthropological Research* 38: 137–53.

Binns, R. A., and McBryde, I. 1972. *A Petrological Analysis of Ground-Edge Artifacts from Northern New South Wales*. Canberra: Australian Institute of Aboriginal Studies.

Boast, R. 1997. A small company of actors: A critique of style. *Journal of Material Culture* 2: 173–98.

Boivin, N. 2004. Landscape and cosmology in the South Indian Neolithic: New perspectives on the Deccan ashmounds. *Cambridge Archaeological Journal* 14: 235–57.

Bradley, R. J. 1982. Position and possession: Assemblage variation in the British Neolithic. *Oxford Journal of Archaeology* 1: 27–38.

———. 1984. *The Social Foundations of Prehistoric Britain*. London: Longmans.

———. 1997. *Rock Art and the Prehistory of Atlantic Europe: Signing the Land*. London: Routledge.

———. 2000. *An Archaeology of Natural Places*. London: Routledge.

Bradley, R., and Edmonds, M. 1993. *Interpreting the Axe Trade*. Cambridge: Cambridge University Press.

Brain, C. K. 1981. *The Hunters or the Hunted? An Introduction to African Cave Taphonomy*. Chicago: University of Chicago Press.

Brink, K., and Hydén, S. 2006. *Hyllie Vattentorn—Delområde 4 och Palissaden—Delområde 5*. Malmö: Malmö Kulturmiljö.

Broda, J. 1987. Templo Mayor as ritual space, in J. Broda, D. Carrasco, and E. Matos Moctezuma (eds.), *The Great Temple of Tenochtitlan: Center and Periphery in the Aztec World*, pp. 61–123. Berkeley and Los Angeles: University of California Press.

Brothwell, D., and Higgs, E. (eds.). 1969. *Science in Archaeology*. New York: Praeger.

Brown, M. F. 2003. *Who Owns Native Culture?* Cambridge, MA: Harvard University Press.

Butzer, K. W. 1964. *Environment and Archaeology: An Introduction to Pleistocene Geography*. London: Methuen.

———. 1982. *Archaeology as Human Ecology: Method and Theory for a Contextual Approach*. New York: Cambridge University Press.

Cann, J. R., Dixon, J. E., and Renfrew, C. 1969. Obsidian analysis and the obsidian trade, in D. Brothwell and E. Higgs (eds.), *Science in Archaeology: A Survey of Progress and Research*, pp. 578–91. London: Thames and Hudson.

Casey, E. S. 1996. How to get from space to place in a fairly short stretch of time, in S. Feld and K. H. Basso (eds.), *Senses of Place*, pp. 13–51. Santa Fe: School of American Research Press.

Chippindale, C., Devereux, P., Fowler, P., Jones, R., and Sebastian, T. 1990. *Who Owns Stonehenge?* London: Batsford.

Conrad, G. W. 1978. Models of compromise in settlement pattern studies: An example from coastal Peru. *World Archaeology* 9: 281–98.

Clarke, D. L. 1968. *Analytical Archaeology*. London: Methuen.

Clarke, D. L. (ed.). 1977. *Spatial Archaeology*. New York: Academic Press.

Clark, G. A. 1982. Quantifying archaeological research, in M. B. Schiffer (ed.), *Advances in Archaeological Method and Theory* 5, pp. 217–73. New York: Academic Press.

Clark, G. A., and Stafford, C. 1982. Quantification in American archaeology: A historical perspective. *World Archaeology* 14: 98–118.

Colley, S. 2002. *Uncovering Australia: Archaeology, Indigenous People and the Public*. Sydney: Allen and Unwin.

Conkey, M. W. 1990. Experimenting with style in archaeology: Some historical and theoretical issues, in M. W. Conkey and C. A. Hastorf (eds.), *The Uses of Style in Archaeology*, pp. 5–17. Cambridge: Cambridge University Press.

Cosgrove, D. E. 1988. The geometry of landscape: Practical and speculative arts in sixteenth-century Venetian land territories, in D. E. Cosgrove and S. Daniels (eds.), *The Iconography of Landscape*, pp. 254–76. Cambridge: Cambridge University Press.

———. 1998. *Social Formation and Symbolic Landscape*. Madison: The University of Wisconsin Press.

Coutts, P. J. F. 1970. *The Archaeology of Wilson's Promontory*. Canberra: Australian Institute of Aboriginal Studies.

Cowgill, G. L. 1975. A selection of samplers: Comments on archaeo-statistics, in J. W. Mueller (ed.), *Sampling in Archaeology*, pp. 258–74. Tucson: University of Arizona Press.

Daniels, S., and Cosgrove, D. 1988. Introduction: Iconography and landscape, in D. Cosgrove and S. Daniels (eds.), *The Iconography of Landscape*, pp. 1–10. Cambridge: Cambridge University Press.

Darvill, T. C. 2006. *Stonehenge: The Biography of a Landscape*. Stroud: Tempus.

David, B., Crouch, J., and Zoppi, U. 2005. Historicizing the spiritual: *bu* shell arrangements on the island of Badu, Torres Strait. *Cambridge Archaeological Journal* 15: 71–91.

David, N., and Kramer, C. 2001. *Ethnoarchaeology in Action*. Cambridge: Cambridge University Press.

Dincauze, D. F. 2000. *Environmental Archaeology: Principles and Practice*. Cambridge: Cambridge University Press.

Dornan, J. L. 2004. Beyond belief: Religious experience, ritual, and cultural neuro-phenomenology in the interpretation of past religious systems. *Cambridge Archaeological Journal* 14: 25–36.

Dunnell, R. C., and Dancey, W. S. 1983. The siteless survey: A regional scale data collection strategy, in M. B. Schiffer (ed.), *Advances in Archaeological Method and Theory* 6, pp. 267–87. New York: Academic Press.

Evans, J. G. 1975. *The Environment of Early Man in the British Isles*. London: Paul Elek.

———. 1978. *An Introduction to Environmental Archaeology*. London: Granada.

Feld, S., and Basso, K. H. (eds.). 1996. *Senses of Place*. Santa Fe: School of American Research Press.

Foard, G. 1978. Systematic fieldwalking and the investigation of Saxon settlement in Northamptonshire. *World Archaeology* 9: 357–74.

Foley, R. A. 1981. Off-site archaeology: An alternative approach for the short-sited, in I. Hodder, G. Isaac, and N. Hammond (eds.), *Pattern of the Past: Essays in Honour of David Clarke*, pp. 152–84. Cambridge: Cambridge University Press.

Gamble, C. 1982. Interaction and alliance in Palaeolithic society. *Man* n.s. 17: 92–107.

Gibbs, M., and Veth, P. 2002. Ritual engines and the archaeology of territorial ascendancy, in S. Ulm, C. Westcott, J. Reid, A. Ross, I. Lilley, J. Prangnell, and L. Kirkwood (eds.), *Barriers, Borders, Boundaries: Proceedings of the 2001 Australian Archaeological Association Annual Conference*, pp. 11–19. Tempus 7. Brisbane: Anthropology Museum, The University of Queensland.

Gifford, E. W., and Shutler, R. J. 1956. *Archaeological Excavations in New Caledonia. Anthropological Records* 1(1). Berkeley and Los Angeles: University of California Press.

Golson, J. 1961. Report on New Zealand, Western Polynesia, New Caledonia and Fiji. *Asian Perspectives* 5: 166–80.

Gould, R. A. 1968. Living archaeology: The Ngatatjara of Western Australia. *Southwestern Journal of Anthropology* 24: 101–22.

———. 1971. The archaeologist as ethnographer: A case study from the Western Desert of Australia. *World Archaeology* 3: 143–77.

Grange, J. 1985. Place, body and situation, in D. Seamon and R. Mugerauer (eds.), *Dwelling, Place and Environment: Towards a Phenomenology of Person and World*, pp. 71–84. Dordrecht: Martinus Nijhoff Publishers.

Green, R. C. 1987. Obsidian results from Lapita sites of the Reef/Santa Cruz Islands, in W. R. Ambrose and J. M. J. Mummery (eds.), *Archaeometry: Further Studies in Australasia*, pp. 239–49. *Occasional Papers in Prehistory* 14. Canberra: Research School of Pacific Studies, Australian National University.

Greenwood, P. H., and Todd, E. J. 1976. Fish remains from Upper Paleolithic sites near Idfu and Isna, in F. Wendorf and R. Schild (eds.), *Prehistory of the Nile Valley*, pp. 383–89. New York: Academic Press.

Hassan, F. A. 1979. Geoarchaeology: The geologist and archaeology. *American Antiquity* 44: 267–70.

Hawkes, J., and Hawkes, C. 1947. *Prehistoric Britain*. London: Chatto and Windus.

Hayden, B. 1979. *Palaeolithic Reflections: Lithic Technology and Ethnographic Excavations Among Australian Aborigines*. Canberra: Australian Institute of Aboriginal Studies.

Higgs, E. S. 1975. *Palaeoeconomy*. Cambridge: Cambridge University Press.

Hiroko, K. 1986. Jomon shell mounds and growth-line analyses of molluscan shells, in R. J. Pearson, G. L. Barnes, and K. L. Hutterer (eds.), *Windows on the Japanese Past: Studies in Archaeology and Prehistory*, pp. 267–78. Ann Arbor: The University of Michigan.

Hirth, K. G. 1978. Teotihuacan regional population administration in eastern Morelos. *World Archaeology* 9: 320–33.

Hirth, K., Pinto, G. L., and Hasemann, G. (eds.). 1989. *Archaeological Research in the El Cajon Region: Prehistoric Cultural Ecology*. Pittsburgh: University of Pittsburgh Memoirs in Latin American Archaeology 1.

Hodder, I. 1978. *The Spatial Organisation of Culture*. London: Duckworth.

———. 1982. *Symbols in Action: Ethnoarchaeological Studies of Material Culture*. Cambridge: Cambridge University Press.

Hodder, I., and Orton, C. 1976. *Spatial Analysis in Archaeology*. Cambridge: Cambridge University Press.

Hurst, H., and Stager, L. E. 1978. A metropolitan landscape: The late Punic port of Carthage. *World Archaeology* 9: 334–46.

Ingold, T. 1993. The temporality of the landscape. *World Archaeology* 25: 152–74.

Insoll, T. 2004. *Archaeology, Ritual, Religion*. London: Routledge.

Irwin, G. J. 1978. Pots and entrepots: A study of settlement, trade and the development of economic specialization in Papuan prehistory. *World Archaeology* 9: 299–319.

Jacobs, L. 1979. Tell-I Nun: archaeological implications of a village in transition, in C. Kramer (ed.), *Ethnoarchaeology: Implications of Ethnography for Archaeology*, pp. 175–91. New York: Columbia University Press.

Jochim, M. A. 1976. *Hunter-Gatherer Subsistence and Settlement: A Predictive Model*. New York: Academic Press.

Keiller, A., Piggott, S., and Wallis, F. S. 1941. First report on the sub-committee on the petrological identification of stone axes. *Proceedings of the Prehistoric Society* 7: 50–72.

King, T. F. 1978. *The Archaeological Survey: Methods and Uses. Cultural Resources Management Studies*. Washington, DC: Office of Archaeology and Historic Preservation, Heritage, Conservation, and Recreation Service.

———. 1998. *Cultural Resource Laws and Practice: An Introductory Guide*. Walnut Creek, CA: AltaMira Press.

King, T. F., Hickman, P. P., and Berg, G. 1977. *Anthropology in Historic Preservation: Caring for Culture's Clutter*. New York: Academic Press.

Langford, R. 1983. Our heritage—your playground. *Australian Archaeology* 16: 1–6.

Langton, M. 2002. The edge of the sacred, the edge of death: sensual inscriptions, in B. David and M. Wilson (eds.), *Inscribed Landscapes: Marking and Making Place*, pp. 253–69. Honolulu: University of Hawai'i Press.

LeBlanc, S. A. 1973. Two points of logic concerning data, hypotheses, general laws, and systems, in C. L. Redman (ed.), *Research and Theory in Current Archaeology*, pp. 199–214. New York: John Wiley and Sons.

Lefebvre, H. 1991. *The Production of Space*. Oxford: Blackwell Publishing.

Lewis-Williams, J. D., and Dowson, T. A. 1990. Through the veil: San rock paintings and the rock face. *South African Archaeological Bulletin* 45: 5–16.

Lipe, W. D. 1984. Value and meaning in cultural resources, in H. Cleere (ed.), *Approaches to the Archaeological Heritage: A Comparative Study of World Cultural Resource Management Systems*, pp. 1–11. Cambridge: Cambridge University Press.

Lourandos, H. 1977. Aboriginal spatial organisation and population: South-western Victoria reconsidered. *Archaeology and Physical Anthropology in Oceania* 12: 202–25.

Louwe Kooijmans, L. P. 2001. *Hardinxveld-Giessendam Polderweg: Een Mesolithisch Jachtkamp in Het Rivierngebied (550–500 v. Chr.)*. Utrecht: NS Railinfrabeheer B.V.

Lüning, J. 1982. Research into the Bandkeramik settlement of the Aldenhovener Platte in the Rhineland. *Analecta Praehistorica Leidensia* 15: 1–29.

Marshall, A. J. 1978. Environment and agriculture during the Iron Age: Statistical analysis of changing settlement ecology. *World Archaeology* 9: 347–56.

McBryde, I. 1978. *Wil-im-ee Moor-ring*: or, where do axes come from? *Mankind* 11: 354–82.

McBryde, I., and Harrison, G. 1981. Valued good or valuable stone? Consideration of some aspects of the distribution of greenstone artifacts in southeastern Australia, in F. Leach and J. Davidson (eds.), *Archaeological Studies of Pacific Stone Resources*, pp. 183–208. Oxford: British Archaeological Reports.

McBryde, I., and Watchman, A. 1976. The distribution of greenstone axes in southeastern Australia: A preliminary report. *Mankind* 10: 163–74.

McNiven, I. J. 2003. Saltwater People: Apiritscapes, marine rituals and the archaeology of Australian indigenous seascapes. *World Archaeology* 35: 329–49.

McNiven, I. J., David, B., and Barker, B. 2006. The social archaeology of Indigenous Australia, in B. David, B. Barker, and I. McNiven (eds.), *The Social Archaeology of Australian Indigenous Societies*, pp. 2–19. Canberra: Aboriginal Studies Press.

McNiven, I. J., and Feldman, R. 2003. Ritually orchestrated seascapes: Hunting magic and dugong bone mounds in Torres Strait, NE Australia. *Cambridge Archaeological Journal* 13: 169–94.

McNiven, I. J., and Russell, L. 2005. *Appropriated Pasts: Indigenous Peoples and the Colonial Culture of Archaeology*. Walnut Creek, CA: AltaMira Press.

Moratto, M. J., and Kelly, R. E. 1978. Optimising strategies for evaluating archaeological significance, in M. B. Schiffer (ed.), *Advances in Archaeological Method and Theory* 1: 1–30. New York: Academic Press.

Mueller, J. W. 1975. *Sampling in Archaeology*. Tucson: University of Arizona Press.

Mulvaney, D. J., and Golson, J. (eds.). 1971. *Man and Environment in Australia*. Canberra: Australian National University Press.

Neumann, T. W. 1978. A model for the vertical distribution of flotation-size particles. *Plains Anthropologist* 23: 85–110.

Parsons, J. R. 1972. Archaeological settlement patterns. *Annual Review of Anthropology* 1: 127–50.

Pryor, F. 1980. *Excavation at Fengate, Peterborough, England: The Third Report*. Leicester: Northamptonshire Archaeological Society Monograph 1.

Rapp, G., Jr., and Hill, C. L. 1998. *Geoarchaeology: The Earth-Science Approach to Archaeological Interpretation*. New Haven, CT: Yale University Press.

Redman, C. L. 1975. Productive sampling strategies for archaeological sites, in J. W. Mueller (ed.), *Sampling in Archaeology*, pp. 147–54. Tucson: University of Arizona Press.

Reed, M. 1990. *The Landscape of Britain: From the Beginnings to 1914*. London: Routledge.

Reeves-Smyth, T., and Hamond, F. (eds.). 1983. *Landscape Archaeology in Ireland*. Oxford: BAR British Series 116.

Renfrew, C. 1975. Trade as action at a distance: Questions of integration and communication, in J. A. Sabloff and C. C. Lamberg-Karlovsky (eds.), *Ancient Civilisation and Trade*, pp. 3–59. Albuquerque: University of New Mexico Press.

———. 1982. Explanation revisited, in C. Renfrew, M. J. Rowlands, and B. Abbott Segraves (eds.), *Theory and Explanation in Archaeology: The*

Southampton Conference, pp. 5–23. New York: Academic Press.

———. 1983. Divided we stand: Aspects of archaeology and information. *American Antiquity* 48: 3–16.

Ross, A., and members of the Quandamooka Aboriginal Land Council. 1996. Aboriginal approaches to Cultural Heritage Management: A Quandamooka case study, in S. Ulm, I. Lilley, and A. Ross (eds.), *Australian Archaeology '95: Proceedings of the 1995 Australian Archaeological Association Annual Conference*, pp. 107–12. *Tempus* 6. St. Lucia: Anthropology Museum, University of Queensland.

Sabloff, J. A. (ed.). 1981. *Simulations in Archaeology*. Albuquerque: University of New Mexico Press.

Schama, S. 1995. *Landscape and Memory*. London: Fontana Press.

Schiffer, M. B., and Gummerman, G. J. (eds.). 1977. *Conservation Archaeology: A Guide for Cultural Resource Management Studies*. New York: Academic Press.

Shawcross, W. 1967a. An investigation of prehistoric diet and economy on a coastal site at Galatea Bay, New Zealand. *Proceedings of the Prehistoric Society* 33: 107–31.

———. 1967b. An evaluation of the theoretical capacity of a New Zealand harbour to carry a human population. *Tane* 13: 3–11.

Shutler, R. J., and Marck, J. C. 1975. On the dispersal of the Austronesian horticulturalists. *Archaeology and Physical Anthropology in Oceania* 13: 215–28.

Simek, J. F., and Leslie, P. W. 1983. Partitioning chi-square for the analysis of frequency table data: An archaeological application. *Journal of Archaeological Science* 10: 79–85.

Simpson, J. A., and Weiner, E. S. C. (eds.). 1989. *The Oxford English Dictionary* VIII (2nd ed.). Oxford: Clarendon Press.

Spaulding, A. C. 1976. Multifactor analysis of association: An application to Owasco ceramics, in C. E. Cleland (ed.), *Cultural Continuity and Change*, pp. 59–68. New York: Academic Press.

Smith, C. 1992. The articulation of style and social structure through Aboriginal Australian art. *Australian Aboriginal Studies* 1: 28–34.

Smith, R. J. C., Healy, F., Allen, M. J., Morris, E. L., Barnes, I., and Woodward, P. J. 1997. *Excavations Along the Route of the Dorchester By-Pass, Dorset, 1986–8*. Salisbury: Wessex Archaeology.

Spriggs, M. 1997. *The Island Melanesians*. Oxford: Blackwell Publishing.

Stewart, K. M. 1989. *Fishing Sites of North and East Africa in the Late Pleistocene and Holocene*. Oxford: Cambridge Monographs in African Archaeology 34, BAR International Series 521.

Stjernquist, B. 1978. Approaches to settlement archaeology in Sweden. *World Archaeology* 9: 251–64.

Strang, V. 1997. *Uncommon Ground: Cultural Landscapes and Environmental Values*. Oxford: Berg.

Summerhayes, G. 2000. *Lapita Interaction. Terra Australis* 15. Canberra: Pandanus Press.

Tamisari, F., and Wallace, J. 2006. Towards an experiential archaeology of place: From location to situation through the body, in B. David, B. Barker, and I. McNiven (eds.), *The Social Archaeology of Australian Indigenous Societies*, pp. 204–23. Canberra: Aboriginal Studies Press.

Thackeray, J. F. 2005. Eland, hunters and concepts of "sympathetic control" expressed in southern African rock art. *Cambridge Archaeological Journal* 15: 27–34.

Torrence, R. 2002. Cultural landscapes on Garua Island, Papua New Guinea. *Antiquity* 76: 776–76

Turner, V. 1995. *The Ritual Process: Structure and Anti-Structure*. Chicago: Aldine.

Watkins, J. 2000. *Indigenous Archaeology: American Indian Values and Scientific Practice*. Walnut Creek, CA: AltaMira Press.

White, J. P. 1967. Ethno-archaeology in New Guinea: two examples. *Mankind* 6: 409–14.

Whitley, D. S. 1992. Shamanism and rock art in Far Western North America. *Cambridge Archaeological Journal* 2: 89–113.

Wiessner, P. 1983. Style and social information in Kalahari San projectile points. *American Antiquity* 48: 253–76.

———. 1984. Reconsidering the behavioral basis for style: a case study among the Kalahari San. *Journal of Anthropological Archaeology* 3: 190–234.

Wilcock, J., and Laflin, S. (eds.). 1974. *Computer Applications in Archaeology*. Birmingham: University of Birmingham Computer Centre.

Wilson, M., Sanhampath, J., Senembe, P. D., David, B., Hall, N., and Abong, M. 2000. "Tufala kev blong devil": People and spirits in northwest Malakula, Vanuatu—implications for management. *Conservation and Management of Archaeological Sites* 4: 151–66.

Wobst, H. M. 1977. Stylistic behavior and information exchange, in C. Cleland (ed.), *Papers for the Director: Research Essays in Honor of James B. Griffin*, pp. 317–42. Michigan Anthropological Papers 61. Ann Arbor: Museum of Anthropology, University of Michigan.

Wood, W. R., and Johnson, D. L. 1978. A survey of disturbance processes in archaeological site formation, in M. B. Schiffer (ed.), *Advances in Archaeological Method and Theory* 1, pp. 315–83. New York: Academic Press.

Worthington, A. 2004. *Stonehenge: Celebration and Subversion*. Loughborough: Alternative Albion.

Yellen, J. E. 1977. *Archaeological Approaches to the Present: Models for Reconstructing the Past*. New York: Academic Press.

2

Place in Landscape Archaeology: A Western Philosophical Prelude

Edward S. Casey

Every body must be in a place. (Philoponus, *Aristotelis Physicorum*)

If there is no place thought about, there is no thought at all—no intelligible proposition will have been entertained. (Gareth Evans, *The Varieties of Reference*)

I

A philosophical approach to landscape archaeology must begin with a consideration of place: not as a geographical or cartographic entity but as a basic unit of lived experience. In what follows I trace out such an approach to place, which is here considered as an indispensable constituent of any landscape and its archaeology. In this way, I offer a distinctive inroad into landscape archaeology, one that is tempered with philosophical, and in particular phenomenological, analysis and description.

It is becoming increasingly clear that all events—human and nonhuman alike—*occur nowhere else than in place*. Each event has its own most appropriate, indeed unique, place—whether this is a microscopic spot where molecules collide or the mega-place of a galaxy. In between, there are the many places that suit the scale of human perception: hot tubs and houses, temples and tents, counties and countries. These constitute a veritable landscape of places that are at once situational and consolidating, challenging and orienting.

All such places—those that we can see in one sweep of the eye or traverse with our moving legs—anchor and locate even as they also resist and repel. More than this: they lend to their inhabitants (that is, people, and animals and things of many kinds) their own distinctive identities. Only ask *where* you are or have been, and I will be able to say much about *who* you are. As Carson McCullers puts it: "To know who you are, you have to have a place to come from" (McCullers 1967, cited in Basso 1996: 105).[1]

This is an extraordinary phenomenon, which needs much more acknowledgment than it has so far garnered in the controlling normative frameworks of Western science. In none of the many varieties of that science—certainly not in physics or biology, but not even in cartography or geography, ecology, or meteorology—does place receive its full due. Everywhere in Western thought, place counts for so very little, time and space for so very much. It is presumed, without further question, that time and space furnish the ultimate parameters of the animate and inanimate worlds. What is needed is a critical analysis of this dogma that allows us to grasp its limitations and to deconstruct its premises.

In the prelude I offer in this brief chapter, I trace out the philosophical roots of these premises in an effort to undo the claim that time and space are the primary dimensions of events. I consider this claim to be illusory on philosophical grounds alone. But it is also detrimental to an appreciation of the importance of place in human (and other animal) affairs. It blocks this appreciation in the name of "objectivity" and "truth." But we must ask from the beginning: which objectivity, whose truth?

I do not want to suggest that a philosophical analysis of place should simply take the place of a scientific assessment of space and time. In a fuller account elsewhere (Casey 1997), I have shown that philosophers themselves laid the foundation for the "scientific" point of view that eliminated place from serious consideration in the West. But I want here to suggest that the cautionary tale I shall tell indicates ways in which a promising discipline such as landscape archaeology can benefit from a sensitivity to place and its ramified significance, whether in the present or in the twilight of human prehistory. It is of special pertinence to the emerging field of landscape archaeology, which has begun to focus on early human settlements in their place-specificity. Under the guise of landscape, it is coming to terms with place. For this reason, it behooves us to reflect on how place came to be neglected in Western philosophical thought, taken as an exemplary instance that may bring light to other areas of research.

The move to landscape archaeology which this volume explores—and whose recent development is deftly traced out by the editors of this volume in their Introduction (David and Thomas, Chapter 1, this volume)—is a welcome and timely event. It is welcome since it affirms an indispensable dimension of prehistoric life on earth: the placed-based and place-oriented existence of early human settlements. It is timely insofar as it is confluent with the growing recognition of place in other disciplines at this historical moment: most notably, cultural anthropology, ecology and evolution, depth psychology, and philosophy (not coincidently, all of which are represented to some degree in this volume). What all these fields share is a radical and salutary paradigm shift from the presumption that space in the early modern sense of the term—homogeneous, neutral, isotropic—is no longer suitable for the understanding and modeling of concrete human activity in the world. The obsession with space that arose with the triumph of Newtonian physics and its emulation in economics, empiricist philosophy, and associationist psychology has proven to be not just barren of insightful and lasting results but inherently misguided for the comprehension of human praxis. This praxis arises in particular places—those of the family, the neighborhood, the school, the church, the workplace, the hospital, the burial place—and its construal calls for a very different model than those that imitate and internalize the abstractness and emptiness of pure space.

At the same time, the increasing acknowledgment of cultural plurality has brought with it a new awareness of place—the ways that different peoples relate to place, speaking about it differently and acting differently in regard to it. The "senses of place"—in the title of a widely cited book on the anthropology of place (Feld and Basso 1996)—are as diverse as the cultures that they infuse from below. To be in place in Aboriginal Australia is to be in a locale that bears little comparison to residing in Los Angeles, California. This does not mean, however, that there are not place variables that act to span different cultures: "lateral universals" in Merleau-Ponty's suggestive term (Merleau-Ponty 1964: 120). These are structures that, however diversely specified in detail they may be, stay virtually the same overall across cultures—that continually surface and resurface around the known world.

Landscape archaeology is taking an equally radical but parallel step in the forefront of its research: finding the primacy of place not only in different cultural and geographical locations (as happens in cultural anthropology) but more particularly in earlier times—so early that no written records or oral traditions survive, only remnants of buildings or various artifacts. This is an audacious step that is as apt to revolutionize the field of archaeology, as have comparable steps in other leading disciplines. Here, too, lateral universals are likely to emerge as basic ways of coping with other humans in one's own and other settlements, and with regional flora and fauna. All these parts of such life are beginning to be delineated with an eye to their placial properties. This is a laudable undertaking that inspires my enthusiasm as a philosopher who has concerned himself increasingly with gaining a more complete recognition of place in people's experiences, especially as they are studied in various contemporary disciplines.

My premise is that place is central in the study of different cultures, past as well as present. Here, however, I restrict myself to one episode, that of the transition from space to place in modern Western philosophy. It is my conviction that much of this same transition (and sometimes this transition itself: when philosophy has had direct influence on other fields) is shared by other research disciplines, including landscape archaeology, albeit in different formats that would call for spelling out on their own terms. In the case of landscape archaeology, I must leave such explication to others—in effect, all the authors in this volume—who write as specialists in this burgeoning and most resourceful field.

II

From the middle of the 17th century in Europe, space was regarded as infinite: an enormous unending empty totality. It is certainly true that something about space encourages this extreme view: namely, its encompassing character. We don't need philosophers to tell us this: camping in New Mexico, I am reassuringly surrounded by the spatial spread of the local landscape at all times. Yet as I stay in that landscape for a while I notice something else happening, something that does not belong simply to the order of space as sheerly extended. This is my camp itself, the place I created on the hilltop where I first pitched my tent, built a fire, talked with friends, and gazed out on the landscape itself. This place was not just an aspect or part of the total space of the situation—even if it is true that it was located in that world-space as charted by cartography and geography. The place is unique: I could pitch the same tent, talk with the same fellow campers, and even (perhaps) have the same thoughts, but if all this occurred on a neighboring and even quite similar hill, the place would be experienced as different. And it would be different even if the sense of surrounding space remained much the same. A basic divergence between space and place thus arises even in a mundane circumstance such as camping.

To mark this divergence, many languages—certainly most European languages—distinguish between "place" and "space" (for example, *lieu* or *endroit* vs. *espace*; *Platz* vs. *Raum*; *lugar* vs. *espacio*, and so on). Nevertheless, the difference between space and place is one of the best-kept secrets in the history of Western thought. The putative hegemony of space—and of time, with which it was paired as constituting God's "infinite sensoria" (Newton 1952: 370)—had everything to do with this repression of place. Especially in modern philosophy, where the very distinction came to be questioned and even discredited: one way of understanding modernity is by its very neglect of this distinction.

The ancient world knew otherwise—knew better. Indeed, the pre-premodern (that is, more than two millennia before the modern era that begins in the 17th century) and the postmodern join forces in a common recognition of the importance of place as something essentially other than space, something one cannot afford to ignore in its very difference from space. For the ancient Greeks, what I like to call the Archtyian Axiom obtained: *to be is to be in place*; to be without place is not to be.[2] Plato and Aristotle alike, their differences concerning place vs. space notwithstanding, both endorsed this axiom—as did such disparate thinkers as Gorgias and Zeno. Aristotle's endorsement is most to the point: "everyone supposes that things that are are somewhere, because what is not is nowhere—where for instance is a goat-stage or a sphinx?"[3] But beginning with a Neoplatonist such as Philoponus, who insisted on the difference between corporeal and spatial extension (the latter gesturing toward a space not congruent with, much less exhausted by, the bodies it contains), and continuing through a strong tendency in the Middle Ages to insist on the spatial infinity of God, we reach a point in the late Renaissance when a new axiom captivated philosophical (as well as scientific and theological) minds: to be is to be *in space*, where "space" meant something nonlocal and nonparticular, having little to do with exact location or close containment and everything to do with a vast homogeneous medium. Alexandre Koyré, the eminent philosopher of science, has aptly described this radical transformation of thought, this triumph of space over place, as a movement "from the closed world to the infinite universe."[4] He observed that by the 17th century we find:

> the substitution for the conception of the world as a finite and well-ordered whole, in which the spatial structure embodied a hierarchy of perfection and value, that of an indefinite or even infinite universe no longer united by natural subordination, but unified only by the identity of its ultimate and basic components and laws; and the replacement of the Aristotelian conception of space—a differentiated set of inner-wordly places—by that of Eucidean geometry—an essentially infinite and homogeneous extension—from now on considered as identical with the real space of the world. (Koyré 1957: viii)

In truth, many of the elements of the 17th century's view of space had been postulated by ancient and medieval thinkers in the West. But it took the audacity of thinkers such as Newton in science and Locke and Leibniz in philosophy to propose explicitly that space is prior to place.

Before this could happen, Descartes placed the primacy of place in question by turning a skeptical light toward it. In Descartes' decidedly ambivalent attitude toward place, it could no longer be assumed to be "the first of all things" (Archytas). Descartes' very equivocation is revealing, for it shows that "the father of modern philosophy" could not decide which term was more basic: space or place. Here, too, Descartes marks the turning point between the ancient and the modern worlds. Where place clearly figured first in the thinking of Greek philosophy—there was not yet a coherent concept of space, only a notion of the "boundless" (*to apeiron*)—its priority disappears in the rigors of Cartesian thought. He does not espouse an outright infinity of space, arguing instead that it is indefinitely extended. He retains a remarkably Aristotelian conception of place as "the surface immediately surrounding what is in the place."[5] This sense of place he calls "external," since it fits around a given physical thing as its tight surrounder. "Internal" place, in contrast, has to do with volume and thus with spatial extension in three dimensions. As such, it is a model for space. However, Descartes is unwilling to generalize internal place to any "cosmic" dimension, that is, infinite space: at most, it possess a "generic unity" that allows different bodies of the same volume to occupy it.[6] When Descartes is driven to distinguish between "place" and "space" *simpliciter*, he refuses to prioritize space and place. To split the difference, he ascribes position to place and volume to space:

> The difference between the terms "place" and "space" is that the former designates more explicitly the position, as opposed to the size or shape, while it is the size and shape that we are concentrating on when we talk of space . . . When we say that a thing is in a given place, all we mean is that it occupies such a position relative to other things; but when we go on to say that it fills up a given space or place, we mean in addition that it has precisely the size and shape of the space in question.[7]

This seemingly innocent remark—including the revealingly equivocal expression "space or place" —harbors momentous consequences. Because in singling out position as intrinsic to place, Descartes departs from Aristotle after all and opens up an issue that will preoccupy the entire early modern period. This is the issue of *location* or, more exactly, "simple location" in Whitehead's term for what is "the very foundation of the 17th-century scheme of nature" (Whitehead 1953: 58). Simple location encompasses both space *and* place—in whatever acceptation these terms assume during this foundational century—just as it bridges over the celebrated differences between absolutist and relativist

views of space and time. For it is the view that any "bit of matter"—that is, any physical body— "is where it is, in a definite region of space, and throughout a definite finite duration of time, apart from any essential reference of the relations of that bit of matter to other regions of space and to other durations of time."[8] Put in the terms just discussed by Descartes, simple location is the view that *position matters most in questions of place*. The position is a simple location in a determinate region and thus a position relative to other occupants of that region—even if, as Whitehead stresses, that region itself is considered without reference to other regions. Others in the history of philosophy, most notably Theophrastus and Aquinas, had certainly noticed the crucial role of relative position in the determination of place. But position as such began to become thematic, and not exceptional, only in the second half of the 17th century, that is, after the publication of Descartes' *Principles of Philosophy* in 1644. No longer confined to the determination of place in its uneasy equipoise with space as a matter of volume, it was soon to become an overriding conception of space itself in thinkers as diverse as Locke and Leibniz, one an arch-empiricist and the other an arch-rationalist.

John Locke, for his part, considered place and space alike in terms of measurable distance rather than any experiential quality: "each different distance is a different modification of space; and each idea of any different distance, or space, is a simple mode of this idea."[9] By concatenating particular distances, we reach the idea of "immensity," or more vividly put, "the undistinguishable inane of infinite space" (Locke 1959 [1680]: 224). In contrast with infinite space, place is "nothing else but [the] relative position of anything."[10] So powerful is the idea of relative position that it comes to dominate what Locke has to say about space and place alike: "as in simple space, we consider the relation of distance between any two bodies or points; so in our idea of place, we consider the relation of distance betwixt anything, and any two or more points, which are considered as keeping the same distance one with another, and so considered as at rest."[11] Any intuitive difference between space and place—between, say, the capacious and the situated—here begins to dissolve in the acidic solution of purely relational positions. It follows that space triumphs over place, because it alone contains the total set of such positions.

Locke, the source of associationist psychology (the forebear of behaviorism), is representative of the early modern view that space trumps place at its own game: that is, location. If to be located is merely to be positioned in relation to another location, then the nexus of all pertinent relations will be the effective reality; and space, as the collection of all such relations, will be prior to place in those respects that are held to matter most. Not surprisingly, Leibniz, who was Locke's successor in many ways, defined space as "that which *comprehends* all those places."[12] Space itself is defined by Leibniz as "an order of co-existences" (Leibniz 1976: 89). Such an order is interpreted as "situation or distance,"[13] while situation is equivalent to relative position: another instance of the reduction of all spatial phenomena to simple location.

In this progressive dissolution of place as an independent variable in human experience, the coup de grace is delivered by Kant. Responding to Leibniz, Kant agrees that space is the order of co-existences, but he argues that space (and time) belong to individual human subjects, for whom space is a form of intuition. In mental intuition, however, there is no room for place, which Kant does not deign to discuss in his *Critique of Pure Reason* (first edition, 1781). Thus, by the end of the 18th century, the high point of modernist thought, place was no longer addressed by the leading philosophers: it had become, quite literally, beneath notice.

III

For the most part (with the notable exception of Kierkegaard), 19th-century philosophers continued this pattern of neglect. The role of place, if it was noted at all, was pursued in other fields, though only as an accompaniment to supposedly more serious concerns. For example, Darwin saw clearly that for variation and selection to occur, members of a species had to become separated from one another in their respective places of habitation (for example, the Galápagos Islands) and to adapt successfully to the different environments characterizing these places. Even so, he did not thematize place as such: it was a crucial but unnamed variable in the evolutionary equation.

Twentieth-century philosophy witnessed a gradual re-appreciation of the power of place—its deferred dawning after centuries of marginalization, the return of the repressed as it were. The recognition of the value of place was intermittent, however. A curious pattern can be observed in which the acknowledgment of place comes late in the career of a given thinker. Foucault, for example, began to endorse the priority of space over time and history in his later essays and interviews, and by "space" he meant the location constituted by a given institution or set of historical practices—a location that is determined more definitively by social and political forces than by any geographical or physical parameters. Heidegger's later

essays explored place as "dwelling," taken not as a domestic residence alone but as a locus for the event of Being to emerge in the guise of language. Deleuze and Guattari take up the difference between "striated" and "smooth" space—between Cartesian gridded space and an open-ended, porous, and unbounded place—and seek to uncover the historical and social roots of this difference. Irigaray compares the place in which the sexes intermingle as undoing the model of the "interval" (*diastema*) that had been a crux for the later Greeks and that keeps men and women at an untraversable distance. My own work examines the felt features of places by means of a phenomenological description that is no longer eidetic or formal (as in Husserl) but based on concrete bodily experiences of specific places. This work is much inspired by the earlier investigations of Merleau-Ponty (see Casey 1993; Deleuze and Guattari 1987; Foucault 1986; Heidegger 1971; Husserl 1981; Irigaray 1993; and Merleau-Ponty 1962 [1945]).

Independent in provenance, these several efforts to reconsider space and place share a skepticism regarding the very idea of infinite space that had obsessed the West since the late Renaissance and early modern era. In their diverse ways, they effect a deconstruction of this idea, claiming that it is not only hugely projective and ungrounded in human experience but that it has detrimental consequences for the understanding and interpretation of this experience. In particular, the collusion between the notion of a neutral and homogeneous space and the presumptive neutrality and objectivity of the human sciences is called into question—perhaps most poignantly in the work of Foucault, who demonstrates that the institutional spaces of the neoclassical epoch of Enlightenment—the highly organized spaces of schools, hospitals, and prisons—are rife with social and political determinations, undermining their putative rationality: a critique that rejoins that of Adorno and Horkheimer in *The Dialectic of Enlightenment*.

IV

And the implications of all this for landscape archaeology? My suspicion is that the belated but ongoing recognition of the power of place in philosophy—lost sight of in the early modern period of the West but now resurfacing in new and challenging ways—will offer a fruitful underpinning of the decisive turn to landscape in archaeology research. For it sanctions in advance and at the level of concrete description the indispensability of place in studies of the prehistoric world in its landscape dimensions. These dimensions are all the more crucial to retrieve, or at least to posit, in view of the remoteness in time of the settlements under scrutiny as well as the scattered remains that constitute the only surviving evidence of earlier life. Without the imagined reconstruction of the immediate and surrounding landscapes of these settlements, we are left with scant sense of how early human beings inhabited the earth in various locales, often distant from one another.

Conclusions

A landscape, then, is inconceivable without place. It is made up of a set of discrete places and is itself a place. As such, it is an instance of a *placescape*; it is part of what I like to call a "place-world." By this I mean an historic or prehistoric world that is anchored in a given unique place—there in particular, nowhere else and certainly not in an abstract and universal space that tells us nothing about the character of a concrete locality, its layout as it bears on human habitation and in relation to the natural world in which it is situated. A focus on place, then, allows landscape studies of any kind—but especially those at stake in landscape archaeology—to tie down what would otherwise remain a matter of sheer speculation, of literally ungrounded thought. The description of place predicates helps make a lost landscape come alive again as a plausible scene of human settlement. This is a very significant step forward in the evolution of archaeology as a distinctive discipline and one that rejoins developments in philosophy and other allied fields at this historical moment.

Notes

1. Basso (1996: 146) himself adds: "selfhood and placehood are completely intertwined."

2. The statement attributed to Archytas by Simplicius is as follows: "All existing things are either in place or not without place" (as cited in Sambursky 1962: 37).

3. *Physics* 208 29–31. Plato's endorsement is at *Timaeus* 52b: "anything that is must needs be in some place and occupy some room . . . what is not somewhere in earth and heaven is nothing." On Gorgias and Zeno, see Cornford (1957: 47–8).

4. See Alexandre Koyré (1957: passim). Curiously, however, Koyré tells only the last chapters of this long tale, those that bear on the Renaissance and early modern period. For a more complete account, the reader must consult such texts as Sambursky (1962) and Sorabji (1988).

5. Descartes, 1985 [1644], I: 229. Descartes specifies that such a surface does not belong, strictly speaking, to the "surrounding body" but to "the boundary between the surrounding and surrounded bodies," being in effect the "common surface" (ibid.). Concerning the question of infinity, article 26 of the same text states that "we should never enter into arguments about the infinite. Things in which we observe no limits—such as the extension of the world, the division of the parts of matter, the number of the stars, and so on—should instead be regarded as indefinite" (ibid.: 201). As Descartes makes clear in the next article, he prefers to reserve the term "infinite" for God; but if God is co-extensive with the extended universe, then surely it, too, is infinite.

6. "In reality the extension in length, breadth, and depth that constitutes a space is exactly the same as that which constitutes a body. The difference arises as follows: in the case of a body, we regard the extension as something particular . . . but in the case of a space, we attribute to the extension only a generic unity, so that when a new body comes to occupy the space, the extension of the space is reckoned not to change but to remain one and the same" (ibid.: 227).

7. Ibid.: 229. Note also Descartes' claim that "internal place is exactly the same as place" (ibid.).

8. Whitehead, 1953: 58. Cf. also p. 49 for a more elaborate alternative formulation. Whitehead remarks that "this concept of simple location is independent of the controversy between the absolutist and the relativist views of space or of time" (ibid.: 58).

9. Locke, 1959 [1680], I: 220. Locke italicizes "simple mode." The importance of distance follows from Locke's instrumentalist conception of place: "this modification of distance we call place, being made by men for their common use . . . men consider and determine of this place by reference to those adjacent things which best served to their present purpose" (ibid.: 223).

10. Ibid.: 224. Locke says expressly that "we can have no idea of the place of the universe, though we can of all the parts of it" (ibid.).

11. The first statement occurs at ibid.: 225; the second is at p. 222.

12. Leibniz 1976 [1715]: 92, my italics. Cf. also the statement that "space is that, which results from places taken together" (ibid.).

13. Ibid.: 91. Leibniz underlines "situation." At ibid.: 97, Leibniz speaks of space as "an order of situations."

References

Basso, K. H. 1996. *Wisdom Sits in Places: Landscape and Language among the Western Apache*. Albuquerque: University of New Mexico Press.

Casey. E. 1993. *Getting Back into Place: Toward a Renewed Understanding of the Place-World*. Bloomington: Indiana University Press.

———. 1997. *The Fate of Place: A Philosophical History*. Berkeley and Los Angeles: University of California Press.

Cornford, F. M. 1957. *Plato's Cosmology*. New York: Liberal Arts Press.

Deleuze, G., and Guattari, F. 1987. *A Thousand Plateaus: Capitalism and Schizophrenia*, tr. B. Massumi. Minneapolis: University of Minnesota Press.

Descartes, R. 1985 [1644]. *Principles of Philosophy*, in J. Cottingham, R. Stoothoff, and D. Murdoch, The Philosophical Writings of Descartes. Cambridge: Cambridge University Press.

Feld, S., and Basso, K. H. (eds.). 1996. *Senses of Place*. Santa Fe: School of American Research Press.

Foucault, M. 1986. Of Other Spaces, tr. J. Miskowiec. *Diacritics* (spring issue).

Heidegger, M. 1971. Building Dwelling Thinking, in M. Heidegger, *Poetry, Language, Thought*, A. Hofstadter (tr.). New York: Harper and Row.

Husserl, E. 1981. Foundational Investigations of the Phenomenological Origin of the Spatiality of Nature, in F. Elliston and P. McCormick (eds.), *Husserl: Shorter Works*, F. Kersten (tr.). Notre Dame: University of Notre Dame Press.

Irigaray, L. 1993. Place, Interval: A Reading of Aristotle, *Physics IV*, in C. Burke and G. C. Gill (trs.), *The Ethics of Sexual Difference*. Ithaca, NY: Cornell University Press.

Koyré, A. 1957. *From the Closed World to the Infinite Universe*. Baltimore: Johns Hopkins.

Leibniz, G. W. 1976 [1715]. *The Leibniz-Clarke Correspondence*, in H. G. Alexander (ed.), (University of Manchester Press, 1956), as reprinted in J. J. C. Smart (ed.), *Problems of Space and Time*. New York: Macmillan.

Locke, J. 1959 [1680]. *An Essay Concerning Human Understanding*, A. C. Fraser (ed.). New York: Dover.

Kant, I. 1965 [1781]. *Critique of Pure Reason*, tr. N. K. Smith. New York: St. Martins.

McCullers, C. 1967. *The Heart Is a Lonely Hunter*. Boston: Houghton Mifflin.

Merleau-Ponty, M. 1962 [1945]. *Phenomenology of Perception*, tr. C. Smith. New York: Humanities Press.

———. 1964. *Signs*. tr. R. McCleary. Evanston, IL: Northwestern University Press.

Newton, I. 1952. *Opticks: Or a Treatise of the Reflections, Refractions, Inflections, and Colours of Light*. New York: Dover.

Sambursky, S. 1962. *The Physical World of Late Antiquity*. New York: Basic Books.

Sorabji, R. 1988. *Matter, Space, and Motion*. London: Duckworth.

Whitehead, A. N. 1953. *Science and the Modern World*. New York: Macmillan.

3

UNCOMMON GROUND: LANDSCAPE AS SOCIAL GEOGRAPHY

Veronica Strang

Emerging Theoretical Landscapes

This chapter is concerned with multiple notions of, and engagements in, place. It centers on "cultural landscapes," which provide a useful bridge between anthropology and archaeology, bringing together social and material worlds and acknowledging the processual nature of both.

The concept of a "cultural landscape" has been generated by some key theoretical developments in the social sciences. Its deepest etymological roots, which go back to early agricultural history in Europe, lie in the idea that a landscape is a humanized, acted-on, and defined space, as Jackson points out: "A landscape is not a natural feature of the environment but a synthetic space, a man-made system functioning and evolving not according to natural laws but to serve a community" (1986: 68).

In more recent history, reflecting the emergence of an increasingly detached and distanced vision of the environment, the term *landscape* has been used to describe a particular view or vista, most often in representational forms such as literature and landscape painting (see Cosgrove and Daniels 1988; Daniels 1993). These core meanings remain embedded in contemporary ideas about "cultural landscapes," which retain both subjective and objective ideas about landscape as something

acted on and lived within, as well as something that can be viewed and considered in more abstract terms.

In anthropology and archaeology, analyses of landscape have gone through several important shifts. Initially, physical landscapes and local ecologies appeared primarily as a backdrop to a focus on human economic endeavors. However, this led to an acknowledgment that the landscape itself was changed in the process, materially reflecting a history of human engagement (Hoskins 1955; also Aston 1997; Crouch and Ward 1988; Lowenthal 1985, 1991) and becoming what Sauer (1962) called a "palimpsest"—a layered and inscribed record of human activity in temporal and spatial terms: "We cannot form an idea of landscape except in terms of its time relations as well as its space relations. It is in continuous process of development or of dissolution and replacement (Sauer 1962: 333).

Researchers began to consider how this material transformation or "morphogenesis," as Prince called it (1971), provided insights into cultural ideas and practices, leading Meinig (1979, cited in Crang 1998: 2)[1] to observe that "if we want to understand ourselves, we would do well to take a searching look at our landscapes."

There was also a growing acknowledgment that social relations were materialized in spatial relationships (Bourdieu 1971), and that unequal power relations were similarly expressed in spatial terms (Foucault 1970; see also Bender 1993, 1998, 1999; Bender and Winer 2001; Hamnett 1996; Jackson and Penrose 1993).

Perhaps the most influential theoretical shift in landscape theory came with the advent—heavily influenced by philosophy—of a phenomenological perspective: a much more fluid and dynamic view of human-environmental interaction, which focused on human experiences of "being-in-the-world" (Bachelard 1994; Casey 1993, 1996; Hegel 1979; Malpas 1999; Merleau-Ponty 1962). In accord with Heidegger's vision of people as "existential insiders" (1971, 1977), Bourdieu's notion of *habitus* (1977) presented a vision of human actors positioned within a social and material cultural context. Researchers began to consider how this was experienced at a sensory level, inculcated, and "embodied" (see Antze and Lambek 1996; Csordas 1994; Feld and Basso 1996; Kratz 1994; Nast and Pile 1998). There was a florescence of writing about how people create "place" from abstract "space" (see Douglas 1975; Penning-Rowsell and Lowenthal 1986; Rival 1998; Strang 2004; Tuan 1974, 1977), in both imaginative and practical terms, encoding meaning in their material surroundings.[2]

There was also a new interest in the environment itself and its role in providing ecological opportunities and constraints (Reed 1988) to which people made cultural adaptations over time (Morphy 1998a [1993]). There emerged a clearer vision of human-environmental interaction as a dynamic and recursive relationship, in which societies created and interacted with particular landscapes as collective "works in progress," thus maintaining a continual process of cultural reproduction (Morphy 1995). This perspective also highlighted the importance of the material landscape as a repository of memory and history (Kuchler 1993; Read 1996; Schama 1996; Stewart and Strathern 2003) and as a theater for the expression of cultural identity (see Anderson and Faye 1992; Kearns and Philo 1993; Keith and Pile 1993; Strang 2001).

Underlying these explorations, and heavily influenced by feminist theory in anthropology and the subsequent postmodern movement, was a growing appreciation of the multiplicity of individual and cultural experiences of "being-in-place." Anthropologists, archaeologists, and human geographers began to consider the particularity of cultural landscape and to make greater efforts to match comparative abstractions with a deeper understanding of local perspectives.

It is now well understood that diverse cultural groups create and maintain quite different cultural landscapes, even in the same physical environments (Strang 1997; Trigger and Griffiths 2003). As underlined by Atkinson and associates (2005) and Crumley and colleagues (2001), particular engagements with place rely on a whole spectrum of cultural beliefs and practices. This is an important point: a "cultural landscape" is holistic, incorporating every aspect of culture and its material expression. This includes cosmological understandings of the world; religious beliefs and practices; languages and categories; social and spatial organisation; economic activities; systems of property ownership; political processes; values and their manifestation in laws; history and memory; constructs of social identity; representational forms; material culture; forms of knowledge and their intergenerational transmission; embodied experiences of the environment, and so forth.

The concept of holistic engagement has been heavily influenced by ethnographic research with Indigenous groups.[3] In contrast to the more specialized and fragmented visions of human-environmental engagement that tend to dominate in complex industrial societies, Indigenous cultural forms are more commonly seen as embedded in and mediated by the land and thus bound together into a seamless, interrelated whole. Such ethnographies have therefore contributed to landscape archaeology and anthropologies of landscape in a dual sense, both by providing comparative Indigenous perspectives and by strengthening a theoretical appreciation of the need to consider landscapes within the entirety of their particular cultural contexts. Considered systematically, each aspect of culture illuminates particular notions of and engagements with place and allows us to compare very diverse cultural landscapes. With the caveat that this is no more than the briefest of thumbnail sketches, we might, for example, consider the cultural landscapes of Indigenous communities in Far North Queensland and the non-Indigenous European Australians who live and work as pastoralists in the same geographic area.[4]

Uncommon Ground

Cosmological understandings of the world are foundational to cultural beliefs and practices. European-Australian pastoralists' visions of "how the world works," like those in other Western industrial societies, are a mix of Christian beliefs involving divine genesis, and scientific explanations of the material world as a product of evolutionary and ecological processes. In accord with this cosmological perspective, "nature" is perceived as the opposing other to "culture." The landscape is described in primarily Cartesian (material) terms, as parcels of property mapped and measured out across "natural" ecosystems, and dotted with places and features named after early explorers and settlers, or reflecting topographical attributes and the presence of economic resources.

In contrast, the Aboriginal cosmos centers on a "Dreamtime" in which ancestral beings—female and male—formed the landscape and remained immanent in it, as sentient forces that continue to generate human spiritual being and ecological resources (see Charlesworth et al. 1984; Hiatt 1978; Morphy 1984; Morton 1987; Rose 1992, 1996; Sharp 1939). The landscape and its ecosystems are therefore suffused with ancestral forces, patterned with their tracks, and known through the sacred sites where their spiritual power is concentrated. Place names refer to these totemic ancestors, and simultaneously to their clans of human descendents (see Alpher 1991; Weiner 1991). In Aboriginal terms, it is as important to maintain the spiritual and social integrity of these sites as it is to manage them as natural resources, there being no perceptual separation between "nature" and "culture."

With Durkheimian predictability,[5] the social, spatial, and political organization of each group reflects these cosmological beliefs. Aboriginal clans, and the kin networks of which they are composed, are defined and spatially located through their totemic ancestors. Each clan member has a spiritual home in the landscape from which their "spirit child" emerges and to which it must be returned upon death, and perpetual rights of ownership and use of that place and its associated tracts of land, thus creating an inalienable link between people and place. This system was (and to some extent still is) upheld by egalitarian political structures in which—reflecting the gender complementarity of the ancestral forces—male and female elders provide collective gerontocratic leadership.

Cultural landscapes are also landscapes of knowledge. Through ritual performance and the transmission of traditional knowledge, it is the Aboriginal elders' task to maintain the reproductive connections among human, spiritual, and material worlds (see Morphy 1995). In material terms, these cultural norms are revealed by many traces of long-term usage of sites: shell middens, domiculture, graveyards, traces of ritual activities, and scatterings of ancient tools, which also serve

as a record of the hunting and gathering, which was, for so many millennia, the sole economic mode for Indigenous Australians (Hamilton 1982; Myers 1986; Williams and Hunn 1986). Like other hunter-gatherers, Aboriginal communities in Cape York—though involved in introduced forms of production as well—continue to place great importance on the vast lexicon of local ecological knowledge that was vital to their precolonial livelihoods as hunter-gatherers and that remains integral to ideas about Aboriginal identity. Much of this knowledge, like the social and spatial organisation with which it articulates, is encoded in the ancestral stories, which are redolent with details about local flora and fauna, places and ecosystems, and how to make use of these (Morphy 1995; Rose 1996).

The social, spatial, and political organization of non-Aboriginal society in Australia is very different, emerging primarily from family histories of settlement and the pressures of particular economic practices. In a patriarchal social and political structure, the leasehold cattle stations belong either to the male descendents of early colonial settlers or to individuals or families with sufficient wealth and inclination to purchase them in a competitive property market. Given the low carrying capacity of the land, and the pressure for economic viability, the stations are vast in area (often more than 7,770 km^2), which means that small clusters of people—managers, stock teams, and domestic workers—inhabit isolated homesteads. In an increasingly mobile society, they do this for varying lengths of time, with only owner-managers usually remaining for more than a few years. These emergent social and spatial forms are revealed in the material traces of early settlements, in the grid of fencelines and boundaries inscribed on the landscape, and in the carefully arranged layout of contemporary homesteads, which manifest a range of ideas about social identity, status, gender, and power (Schaffer 1988; Strang 1997). These material forms serve to illustrate key differences between Aboriginal and European systems of property ownership and control (McCorquodale 1987; Reynolds 1987; Strang 2000).

The graziers' economic practices depend on highly specialized knowledge about animal husbandry, in which the landscape is categorized, assessed, and managed in accord with its feed and water provision, physical access for mustering, distance to market, and suchlike. This professional knowledge, which defines the identity of the pastoralists as a subcultural community, is applied to the immediate landscape and so dependent on some local knowledge but focuses primarily on stock management skills that can readily be applied elsewhere and on the production of non-indigenous animals for a distant market. Reflecting the fragmented nature of information exchange in complex industrial societies, it sits alongside other, specialized areas of knowledge, such as scientific understandings of ecology or technical expertise in land and water management, which are similarly unrelated to the local environment but that had—and continue to have—major implications for the cultural landscapes of Indigenous communities, replacing the authority of the ancestors with Western models of environmental management and control (see Rose and Clarke 1997; Rumsey and Weiner 2001; Stevens 1974).

Just as knowledge provides a basis for identity, so do the respective histories and memories of each group, which, as noted previously, are important components of any cultural landscape. In Aboriginal terms, history and memory divide sharply into that of the precolonial "early days," whose everyday details of human movement and action are primarily described in oral history, song, ritual, dance, and the other forms of representation, and that of postcontact history (Attwood and Markus 1999; Layton 1989; May 1994; McGrath 1987; Rowse 1998; Schweitzer, Biesele, and Hitchcock 2000), which is concerned with efforts to defend Aboriginal land and people's subsequent experiences of colonial domination and cultural change. Though recorded in a range of media, this latter history, like other Aboriginal cultural forms, is also held locally—and specifically—in the land, for example, in the particular places where previous generations fought battles or were taken as prisoners; in the grave sites marking colonial massacres; in the early homestead sites dependent on Aboriginal labor. Despite the pull of new alternate identities (see Povinelli 1999), both periods of history, and the Aboriginal identity they support, thus remain firmly located in the immediate landscape (Beckett 1988; Kapferer 1996; Myers 1986).

Leading a more mobile existence, few of the contemporary graziers know a great deal about local history,[6] though they are made aware of its traces through landscape features named after early explorers or settlers, marked graves, and site names (such as "Battle Camp") reflecting colonial conflicts. Their vision of history is more general, learned in the texts and the visual material of State educational systems. It locates them in commensurately general terms, in a wider and much more fluid form of social identity.

Much can be learned about particular cultural landscapes from peoples' representations of them.[7] There is a wealth of ethnographic literature on this topic examining how representational forms:

- transmit knowledges and values intergenerationally
- describe histories
- define and sometimes contest (see Bender 1999; Orlove 1991) rights and ownership
- express cultural identities
- depict social, spatial, and topographical relationships (e.g., Kleinert and Neale 2000; Langton 1993; Layton 1997; Morphy 1991, 1998b; Munn 1973; Taylor 1996)

A key point of difference between the representational forms of non-Aboriginal graziers and those of Aboriginal communities echoes a recurring contrast. Although the graziers make use of highly specialized representations of landscape—art, novels, maps, and such—only maps are local in their focus. Aboriginal art, in contrast, is characterized by its local specificity and—perhaps most importantly—by its holistic incorporation of all aspects of indigenous culture. Relying on multivalent images,[8] it expresses a reality in which these are wholly interdependent and located in the land.

Similar observations could be made about the material culture produced or used by the respective communities: "traditional" Aboriginal material culture is locally produced, specific, and carries a multivalency of meanings (see Griffiths 1996; Mundine et al. 2000; Strang 1999), whereas that of the graziers is largely introduced, mass produced, and more specialized, with exceptions, such as local craft production, family heirlooms, and historic monuments, that reflect the more localized and affective aspects of their relationships with place (see Csikzentmihalyi and Rochberg-Halton 1981; Milton 2002; Strang 2003; Tilley 1991, 1994). In each case, the production of material culture is a vital expression of the values of each group (see Goodey 1986; Strang 1997) and its ability to demonstrate power and agency (see Gell 1998).

Conclusions

Even in this most simple sketch, which plainly omits the complex subtleties and diversities of a proper ethnographic account, two strikingly different cultural landscapes can be glimpsed.[9] In each, particular cultural forms lead to quite different engagements with the same physical environment and to widely differing experiences of place. Both communities have incorporated massive and complex changes in the last 200 years. Indigenous groups have experienced a traumatic invasion, colonial dominance, the enforced encompassment of new economic modes, knowledges and values, attempts at assimilation, and slow progress toward greater self-determination. They have nevertheless maintained their own cultural beliefs and practices to a considerable extent. The immediate local environment is still densely encoded with cultural meanings and a highly specific history. Sensory experience continues to be informed by long-term social and ecological knowledges and by practices that rely on close "reading" of the environment. The result is a holistic, ideologically permanent and deeply affective relationship with place (Benterrak, Muecke, and Roe 1996).

The non-Aboriginal pastoralists' ties to land are more fragmented and impermanent and contain a fundamental conflict between connection and mobility. On the one hand, they are (albeit loosely), rooted in place by the exigencies of their economic, social, and spatial forms and by the layers of colonial history that provides a broad foundation to their identity (see Seddon 1972). Longer-term residents undoubtedly have close ties to particular places. On the other hand, land is alienable in European terms and readily perceived as a commodity. There are also vestiges of ideas that characterized the environmental interactions of the early settlers: an adversarial vision of hostile "Nature" and the need to control it, and a commitment to making the land productive through the imposition of non-indigenous animals and the use of material culture aimed at management, containment, and development (see Strang 2001, 2005). For the graziers, these realities, and the sensory and cognitive experiences they engender, tend to work against the development of affective connections to place.

Within each cultural landscape, people's engagements with place over time have created a material record that provides potential insights for landscape archaeologists. By examining and interpreting that record holistically, researchers can consider its relationship to contemporary cultural landscapes and, in doing so, strengthen the potential for valid statements about past human behavior.[10] As Clarke observed, archaeology is the "time dimension" of anthropology and ethnology (1968: 13),[11] and when disciplines share a model that deals systematically with each aspect of culture, acknowledging that all these

are implicated in a dynamic articulation between social and material worlds, a vision emerges of cultural landscapes transforming over time, providing a useful foundation—an analytic basis of discussion—through which human-environmental relationships can be understood both in the past and in the present.

Notes

1. See also Cresswell 2004; Hirsch and O'Hanlon 1995; Layton and Ucko 1999; Low and Lawrence-Zuniga 2003.

2. Commensurately, in a period of increasing social and geographic mobility in many parts of the world, and anxieties about modernity, people also began to consider the converse implications of "placelessness" or, as Berger and colleagues called it, "homelessness" (1973). See also Auge 1995, Relph 1976, Sack 1992.

3. I have argued elsewhere that Indigenous worldviews have had a significant influence on the development of anthropological theory (Strang 2006).

4. This example is based on my own research in Cape York (Strang 1997), as well as drawing on the work of other Australian ethnographers.

5. Durkheim (1961 [1912]) famously argued that the religious beliefs of a society were reflected in the form of its social and political organization.

6. An exception are those who have inherited stations from early colonial settlers, who often have a keen sense of family history and who refer with pride to the documentary records of their tenure.

7. As Appleton observes, art and other forms of representations of landscape have much to offer analytically: "it is in this very kind of material that we may find new insights which the rational, logical language of science cannot evoke" (cited in Penning-Rowsell and Lowenthal 1986: 28).

8. That is, those that signify many meanings simultaneously (see Morphy 1991, 1998b).

9. Note that there are major diversities even within groups, between individuals and subgroups, and also in terms of the many different contexts in which action takes place. As Moore observes, the culturally constructed environment is the expression of multiple "decision domains" (2005).

10. As Hardesty and Fowler note: "Archaeology is concerned with the things, relations, processes, and meanings of past sociocultural/environmental totalities and their temporal and spatial boundaries, organisation, operation, and changes

over time and across the world," and the central question is how to make valid interpretations of past behaviour (2001: 73).

11. There is obviously fruitful potential here for inter-disciplinary exchange with other social sciences, too: for example, cultural geographers, historians, and so forth (see Stratford 1999; Wagstaff 1987).

References

Alpher, B. 1991. *Yir-Yoront Lexicon: Sketch and Dictionary of an Australian Language.* Series: Trends in Linguistics Documentation 6. Berlin, New York: Mouton de Gruyter.

Anderson, K., and Faye, G. 1992. *Inventing Places: Studies in Cultural Geography.* London: Belhaven Press.

Antze, P., and Lambek, M. (eds.). 1996. *Tense Past: Cultural Essays in Trauma and Memory.* London: Routledge.

Appleton, J. 1986. The role of the arts in landscape research, in E. Penning-Rowsell and D. Lowenthal (eds.), *Landscape Meanings and Values,* pp. 26–47. Boston: Allen and Unwin.

Aston, M. 1997. *Interpreting the Landscape: Landscape Archaeology and Local History.* London: Routledge.

Atkinson, D., Jackson, P., Sibley, D., and Washbourne, N. (eds.). 2005. *Cultural Geography: A Critical Dictionary of Key Concepts.* London: Taurus.

Attwood, B., and Markus, A. 1999. *The Struggle for Aboriginal Rights: A Documentary History.* NSW: Allen and Unwin.

Auge, M. 1995. *Non-Places: Introduction to an Anthropology of Supermodernity.* London: Verso.

Bachelard, G. 1994. *The Poetics of Space,* M. Jolas (tr.). Boston: Beacon Press.

Beckett, J. (ed.). 1988. *Past and Present: The Construction of Aboriginality.* Canberra: Aboriginal Studies Press.

Bender, B. (ed.). 1993. *Landscape: Politics and Perspectives.* Oxford: Berg.

———. 1998. *Stonehenge: Making Space.* Oxford: Berg.

———. 1999. Subverting the Western gaze: Mapping alternative worlds, in R. Layton and P. Ucko (eds.), *The Archaeology and Anthropology of Landscape: Shaping Your Landscape,* pp. 31–45. London: Allen and Unwin.

Bender, B., and Winer, M. (eds.). 2001. *Contested Landscapes: Movement, Exile and Place.* Oxford: Berg.

Benterrak, K., Muecke, S., and Roe, P. 1996. *Reading the Country: An Introduction to Nomadology.* Liverpool: Liverpool University Press.

Berger, P., Berger, B., and Kellner, H. 1973. *The Homeless Mind.* New York: Vintage.

Bourdieu, P. 1971. The Berber house or the world reversed, in J. Pouillon and P. Maranda (eds.), *Exchanges and Communications.* The Hague: Mouton.

———. 1977. *Outline of a Theory of Practice,* R. Nice (tr.). Cambridge: Cambridge University Press.

Casey, E. 1993. *Getting Back Into Place: Toward a Renewed Understanding of the Place World.* Bloomington: Indiana University Press.

———. 1996. How to get from space to place in a fairly short stretch of time: Phenomenological prolegomena, in S. Feld and K. Basso (eds.), *Senses of Place,* pp. 14–52. Santa Fe: School of American Research Press.

Charlesworth, M., Morphy, H., Bell, D., and Maddock, K. (eds.). 1984. *Religion in Aboriginal Australia: An Anthology.* St. Lucia: University of Queensland Press.

Clarke, D. 1968. *Analytical Archaeology.* London: Methuen.

Cosgrove, D., and Daniels, S. 1988. *The Iconography of Landscape: Essays on the Symbolic Representation, Design and Use of Past Environments.* Cambridge: Cambridge University Press.

Crang, M. 1998. *Cultural Geography.* London, New York: Routledge.

Cresswell, T. 2004. *Place: A Short Introduction.* Oxford: Blackwell Publishing.

Crouch, D., and Ward, C. 1988. *The Allotment: Its Landscape and Culture.* London: Faber and Faber.

Crumley, C., Deventer, A., and Fletcher, J. 2001. *New Directions in Anthropology and Environment: Intersections.* New York, Oxford: AltaMira Press.

Csikzentmihalyi, M., and Rochberg-Halton, E. 1981. *The Meaning of Things: Domestic Symbols and the Self.* Cambridge: Cambridge University Press.

Csordas, T. (ed.). 1994. *Embodiment and Experience: The Existential Ground of Culture and Self.* Cambridge: Cambridge University Press.

Daniels, S. 1993. *Fields of Vision: Landscape Imagery and National Identity in England and the U.S.* Cambridge: Polity Press.

Douglas, M. 1975. *Implicit Meanings: Essays in Anthropology.* London: Routledge and Kegan Paul.

Durkheim, E. 1961 [1912]. *The Elementary Forms of the Religious Life.* New York: Collier Books.

Feld, S., and Basso, K. (eds.). 1996. *Senses of Place.* Santa Fe: School of American Research Press.

Foucault, M. 1970. *The Order of Things.* London: Tavistock.

Gell, A. 1998. *Art and Agency: An Anthropology Theory.* Oxford: Clarendon Press.

Goodey, B. 1986. Spotting, squatting, sitting or setting: Some public images of landscape, in E. Penning-Rowsell and D. Lowenthal (eds.), *Landscape Meanings and Values,* pp. 82–101. London: Allen and Unwin.

Griffiths, T. 1996. *Hunters and Collectors: The Antiquarian Imagination in Australia.* Cambridge: Cambridge University Press.

Hamilton, A. 1982. The unity of hunting-gathering societies: Reflections on economic forms and resource management, in N. Williams and E. Hunn (eds.), *Resource Managers: North American and Australian Hunter Gatherers.* Canberra: Australian Institute of Aboriginal Studies.

Hamnett, C. (ed.). 1996. *Social Geography: A Reader.* London: Arnold.

Hardesty, D., and Fowler, D. 2001. Archaeology and environmental changes, in C. Crumley, A. Deventer, A. and J. Fletcher (eds.), *New Directions in Anthropology and Environment: Intersections,* pp. 72–89. Walnut Creek, CA: AltaMira Press.

Hegel, G. 1979. *The Phenomenology of Spirit.* Oxford: Oxford University Press.

Heidegger, M. 1971. *Poetry, Language, Thought.* New York: Harper and Row.

———. 1977. Building, Dwelling, Thinking, in D. Krell (ed.), *Martin Heidegger: Basic Writings,* pp. 319–39. New York: Harper and Row.

Hiatt, L. R. (ed.). 1978. *Australian Aboriginal Concepts.* Canberra: Australian Institute of Aboriginal Studies.

Hirsch, E., and O'Hanlon, M. (eds.). 1995. *The Anthropology of Landscape: Perspectives on Place and Space.* Oxford: Clarendon Press.

Hoskins, W. 1955. *The Making of the English Landscape.* London: Penguin.

Jackson, J. 1986. The vernacular landscape, in E. Penning-Rowsell and D. Lowenthal (eds.), *Landscape Meanings and Values,* pp. 65–81. London: Allen and Unwin.

Jackson, P., and Penrose, J. (eds.). 1993. *Constructions of Race, Place and Nation.* London: UCL Press.

Kapferer, J. 1996. *Being All Equal: Identity, Difference and Australian Cultural Practice.* Oxford: Berg.

Kearns, G., and Philo, C. (eds.). 1993. *Selling Places: The City as Cultural Capital, Past and Present.* Oxford: Pergamon.

Keith, M., and Pile, S. (eds.). 1993. *Place and the Politics of Identity.* London: Routledge.

Kleinert, S., and Neale, M. (eds.). 2000. *The Oxford Companion to Aboriginal Art and Culture.* Oxford: Oxford University Press.

Kratz, C. 1994. *Affecting Performance: Meaning, Movement and Experience in Okiek Women's Initiation.* Washington, DC: Smithsonian Institute Press.

Kuchler, S. 1993. Landscape as memory: The mapping of process and its representation in a Melanesian society, in B. Bender (ed.), *Landscape, Politics and Perspectives.* Oxford: Berg.

Langton, M. 1993. *"Well I heard it on the radio and I saw it on the television": An essay for the Australian Film Commission on the politics and aesthetics of filmmaking by and about Aboriginal people and things.* NSW: Australian Film Commission.

Layton, R. 1989. *Uluru: An Aboriginal history of Ayers Rock.* Canberra: Aboriginal Studies Press.

———. 1997. Representing and translating people's place in the landscape of northern Australia, in A. James, J. Hockey, and A. Dawson (eds.), *After Writing Culture: Epistemology and Praxis in Contemporary Anthropology,* pp. 122–43. ASA Monograph 34. London: Routledge.

Layton, R., and Ucko, P. (eds.). 1999. *The Archaeology and Anthropology of Landscape: Shaping Your Landscape.* London: Routledge.

Low, S., and Lawrence-Zuniga, D. (eds.). 2003. *The Anthropology of Space and Place: Locating Culture.* Oxford: Blackwell Publishing.

Lowenthal, D. 1985. *The Past Is a Foreign Country.* Cambridge: Cambridge University Press.

———. 1991. British national identity and the English landscape. *Rural History* 2: 205–30.

Malpas, J. 1999. *Place and Experience: A Philosophical Topography.* Cambridge: Cambridge University Press.

May, D. 1994. *Aboriginal Labour and the Cattle Industry: Queensland from White Settlement to the Present.* Cambridge: Cambridge University Press.

McCorquodale, J. 1987. *Aborigines and the Law: A Digest.* Canberra: Aboriginal Studies Press.

McGrath, A. 1987. *Born in the Cattle: Aborigines in Cattle Country.* Sydney: Allen and Unwin.

Meinig, D. 1979. *The Interpretation of Ordinary Landscapes.* New Haven, CT: Yale University Press.

Merleau-Ponty, M. 1962. *Phenomenology of Perception.* London: Routledge and Kegan Paul.

Milton, K. 2002. *Loving Nature: Towards an Ecology of Emotion.* London: Routledge.

Moore, J. 2005. *Cultural Landscapes in the Ancient Andes: Archaeologies of Place.* Florida: University Press of Florida.

Morphy, H. 1984. *Journey to the Crocodile's Nest: An Accompanying Monograph to the Film* Madarrpa Funeral at Gurka'wuy. Canberra: Australian Institute of Aboriginal Studies/Humanities Press.

———. 1991. *Ancestral Connections: Art and an Aboriginal System of Knowledge.* Chicago: Chicago University Press.

———. 1995. Landscape and the reproduction of the ancestral past, in E. Hirsch and M. O'Hanlon (eds.), *The Anthropology of Landscape,* pp. 184–209. Oxford: Clarendon Press.

———. 1998a [1993]. Cultural adaptation, in G. Harrison and H. Morphy (eds.), *Human Adaptation,* pp. 99–150. Oxford: Berg.

———. 1998b. *Aboriginal Art.* London: Phaidon.

Morton, J. 1987. The effectiveness of totemism: Increase rituals and resource control in Central Australia. *Man* 22: 453–74.

Mundine, J., Murphy, B., and Rudder, J. 2000. *The Native Born: Objects and Representations from Ramanging, Arnhem Land*. Sydney: Museum of Contemporary Art in association with Bula'bula Arts, Ramingining.

Munn, N. 1973. *Walbiri Iconography: Graphic Representation and Cultural Symbolism in a Central Australian Society*. Ithaca, NY: Cornell University Press.

Myers, F. 1986. *Pintupi Country, Pintupi Self: Sentiment, Place and Politics among Western Desert Aborigines*. Canberra: Australian Institute of Aboriginal Studies.

Nast, H., and Pile, S. (eds.). 1998. *Places through the Body*. London: Routledge.

Orlove, B. 1991. Mapping reeds and reading maps: The politics of representation in Lake Titicaca. *American Ethnologist* 1991: 3–40.

Penning-Rowsell, E., and Lowenthal, D. (eds.). 1986. *Landscape Meanings and Values*. London: Allen and Unwin.

Prince, H. 1971. Real, imagined and abstract worlds of the past. *Progress in Geography* 3: 1–18.

Povinelli, E. 1999. Settler modernity and the quest for an Indigenous tradition, in D. Gaonkar (ed.), *Alter/Native Modernities*. Public Culture Series Volume 1. Millennial Quartet. Durham, NC: Duke University Press, pp.19–48.

Read, P. 1996. *Returning to Nothing: The Meaning of Lost Places*. Cambridge: Cambridge University Press.

Reed, E. 1988. The affordances of the animate environment: Social science from the ecological point of view, in T. Ingold (ed.), *What Is an Animal?* London: Allen and Unwin.

Relph, R. 1976. *Place and Placelessness*. London: Pion.

Reynolds, H. 1987. *The Law of the Land*. London, New York: Penguin.

Rival, L. (ed.). 1998. *The Social Life of Trees: Anthropological Perspectives on Tree Symbolism*. Oxford: Berg.

Rose, D. 1992. *Dingo Makes Us Human: Life and Land in an Aboriginal Australian Culture*. Cambridge: Cambridge University Press.

———. 1996. *Nourishing Terrains: Australian Aboriginal Views of Landscape and Wilderness*. Australia: Australian Heritage Commission.

Rose, D., and Clarke, A. (eds.). 1997. *Tracking Knowledge in North Australian Landscapes: Studies in Indigenous and Settler Ecological Knowledge Systems*. Casuarina: North Australia Research Unit.

Rowse, T. 1998. *White Flour, White Power: From Rations to Citizenship in Outback Australia*. Cambridge: Cambridge University Press.

Rumsey, A., and Weiner, J. (eds.). 2001. *Mining and Indigenous Lifeworlds in Australia and Papua New Guinea*. Hindmarsh: Crawford House Publishing.

Sack, R. 1992. *Place, Consumption and Modernity*. Baltimore, MD: Johns Hopkins University Press.

Sauer, C. 1962. *Land and Life: A Selection from the Writings of Carl Sauer*, J. Leighley (ed.), Berkeley and Los Angeles: University of California Press.

Schaffer, K. 1988. *Women and the Bush: Forces of Desire in the Australian Cultural Tradition*. Cambridge: Cambridge University Press.

Schama, S. 1996. *Landscape and Memory*. London: Fontana Press.

Scheffler, H. 1978. *Australian Kin Classification*. Cambridge: Cambridge University Press.

Schweitzer, P., Biesele, M., and Hitchcock, R. (eds.). 2000. *Hunters and Gatherers in the Modern World: Conflict, Resistance and Self-Determination*. New York: Berghahn Books.

Seddon, G. 1972. *Sense of Place: A Response to an Environment, the Swan Coastal Plain, Western Australia*. Nedlands: University of Western Australia Press.

Sharp, L. 1939. Tribes and totemism in north-east Australia. *Oceania* 9: 254–75, 439: 61.

Stevens, F. 1974. *Aborigines in the Northern Cattle Industry*. Canberra: Australian National University Press.

Stewart, P., and Strathern, A. (eds.). 2003. *Landscape, Memory and History*. Cambridge: Cambridge University Press.

Strang, V. 1997. *Uncommon Ground: Cultural Landscapes and Environmental Values*. Oxford: Berg.

———. 1999. Familiar forms: Homologues, culture and gender in northern Australia. *Journal of the Royal Anthropological Society* 5: 75–95.

———. 2000. Not so black and white: The effects of Aboriginal law on Australian legislation, in A. Abramson and D. Theodossopoulos (eds.), *Mythical Lands, Legal Boundaries: Rites and Rights in Historical and Cultural Context*, pp. 93–115. London: Pluto Press.

———. 2001. Of human bondage: The breaking in of stockmen in Northern Australia. *Oceania* 72: 53–78.

———. 2003. Moon shadows: Aboriginal and European heroes in an Australian landscape, in P. Stewart and A. Strathern (eds.), *Landscape, Memory and History*, pp. 108–35. Cambridge: Cambridge University Press.

———. 2004. *The Meaning of Water*. Oxford: Berg.

———. 2005. Knowing me, knowing you: Aboriginal and Euro-Australian concepts of nature as self and other. *Worldviews* 9: 25–56.

Strang, V. 2006. A happy coincidence? Symbiosis and synthesis in anthropological and Indigenous knowledges. *Current Anthropology* 47(6): 981–1008.

Stratford, E. (ed.). 1999. *Australian Cultural Geographies*. Oxford: Oxford University Press.

Taylor, L. 1996. *Seeing the Inside: Bark Painting in Western Arnhem Land*. Oxford: Clarendon Press.

Tilley, C. 1991. *Material Culture and Text: The Art of Ambiguity*. New York: Routledge.

———. 1994. *A Phenomenology of Landscape: Places, Paths and Monuments*. Oxford: Berg.

Trigger, D., and Griffiths, G. (eds.). 2003. *Disputed Territories: Land, Culture and Identity in Settler Societies*. Hong Kong: Hong Kong University Press.

Tuan, Y. 1974. *Topophilia: A Study of Environmental Perception, Attitudes and Values*. Englewood Cliffs, NJ: Prentice Hall.

———. 1977. *Space and Place: The Perspective of Experience*. Minneapolis: University of Minnesota Press.

Wagstaff, J. (ed.). 1987. *Landscape and Culture: Geographical and Archaeological Perspectives*. Oxford: Blackwell Publishing.

Weiner, J. 1991. *The Empty Place: Poetry, Space, and Being among the Foi of Papua New Guinea*. Bloomington: Indiana University Press.

Williams, N., and Hunn, E. (eds.). 1986. *Resource Managers: North American and Australian Hunter-Gatherers*. Canberra: Australian Institute of Aboriginal Studies.

4

Pathways to a Panoramic Past: A Brief History of Landscape Archaeology in Europe

Timothy Darvill

What we nowadays label *landscape archaeology* emerged as a distinct subdiscipline during the early 1970s. Widespread use of the term can be traced back to Mick Aston and Trevor Rowley's book *Landscape Archaeology,* in which they attempted to expand the field of landscape history through the promotion of fieldwork techniques for the investigation of post-Roman landscapes (Aston and Rowley 1974). In a very real sense, the term caught the spirit of the moment, consolidating various strands of thinking about how archaeology could better connect broad notions of time and space within the prevailing processualist paradigm that had emerged out of the New Archaeology on both sides of the Atlantic in the late 1960s. Very quickly, landscape archaeology advanced beyond a methodology and became much more than simply about doing archaeology over a wide area or with an emphasis on time-depth. Embedded within the very idea of landscape archaeology right from the start was an interest in the eponymous defining concept of "landscape" itself, a matter that during the last quarter of the 20th century kept it alive and now provides the motor that powers the subject forward. Uniquely, landscape archaeology has kept pace with changing theoretical and philosophical positions over the last 30 years or so and now displays considerable heterogeneity in its practice and articulation (Darvill 1999; Lemaire 1997; Sherratt 1996).

In the following sections, three main phases to the development and expansion of landscape archaeology in Europe are briefly considered. First, attention is directed to the roots of the subject in Britain especially, mainly in terms of the influences, traditions, and pressures that contributed to the flood of interest evident from the mid-1970s. Second, emphasis is given to the changing theoretical paradigms within archaeology over the past 30 years as the tenets of processual archaeology were challenged and new tracks developed for the pluralistic approaches of postprocessual archaeology. And third, something of the multiplicity of traditions old and new that now subsist under the banner of landscape archaeology is sketched out as the subject achieves what for practitioners and theorists alike has become a Golden Age, with implications for archaeology across our planet.

Before the Flood

A school of landscape archaeology did not come about suddenly, nor for a single reason; it emerged in parallel with work taking place in other disciplines and ultimately depends for sustenance on several deep intellectual roots. One prerequisite was the idea that human activity, societies, and culture have a spatial dimension. The application of geographical models and principles to an understanding of the past can be traced back to the early 20th century in the works of H. J. MacKinder (1907), Cyril Fox (1932), and O. G. S. Crawford (1953), among others. Gordon Childe's clear expression of culture as an entity situated in space and time played a major part in articulating what archaeologists find (artifacts, monuments, and so on) with recognizable communities and the occupation of specific areas and territories (Childe 1929; Renfrew 1977: 92). Carl Sauer was among the first geographers to express the view that under the influence of a given culture the landscape became the repository of that culture's striving against its environment and the tangible record of human adaptation to their physical milieu. Culture was the agent, the natural environment was the medium, and the cultural landscape the result (Gold 1980: 34; Sauer 1925). Time and space became fundamental components of many social models (e.g., Gurevich 1969).

Another formative influence was an appreciation of the aesthetic and perceptual dimensions of landscape through language, placenames, folklore, literature, and poetry. Occasionally, artists such as Heywood Sumner (1913, 1917) and Edward J. Burrow (1919, 1924) used archaeologically rich landscapes as themes for illustration, continuing a tradition that Peter Howard has traced well back into the 18th century (Howard 1991).

A third influence was the recognition that so much archaeology was visible in the contemporary landscape of fields, heathland, pasture, upland, and woods and that its future was bound up with the very evolving fabric of the countryside. As far back as the 18th century, William Stukeley was lamenting the impact of agriculture on the ancient monuments around Stonehenge (1740: 1), and in the 19th century much attention was given to the reconstruction of monuments as a means of preserving them while also enhancing the visual aesthetic of the landscape. Luckily, much remained untouched, and in the early 20th century Adrian Allcroft noted in the introduction to his rather neglected volume *Earthwork of England* "that not a year passes without the discovery of new earthworks which have waited for hundreds, nay, thousands of years for mere recognition" (1908: 22). It was a view that in many ways cleared the ground for the development of an archaeology of landscape.

Through the late 1950s, 1960s, and 1970s these interests in "reading the land" (Fowler 2001) combined with three far-reaching developments in archaeology to provide the first focus for landscape archaeology. The developments are expanding horizons, cross-disciplinary exchanges, and access to source materials.

Expanding Horizons

By the mid-20th century, attention began to be paid to relationships between sites and their environmental setting, a concern that also spilled over into the wider interpretation of settlement patterns and processes of long-term change. Investigations at Star Carr, Yorkshire, in 1949–1951, for example, clearly show the emergence of these approaches wherein ecological setting, vegetation history, and the local lake stratigraphy were given as much prominence as the more traditional aspects of the site (Clark 1954).

The fast developing field of Rescue Archaeology (or Salvage Archaeology) during the early postwar period also had an influence. Although work initially focused on specific sites under immediate threat of destruction, as developments grew bigger, and the full extent of the threatened archaeology became better understood, things began to change. Work in Britain on the M4 and M5 motorways especially highlighted the density of sites even in unprepossessing areas (Fowler 1979), while gravel extraction at Mucking, Essex, for example, unpicked a zone of cropmarks 350 m wide and over 1 km long to reveal an almost continuous human presence from Neolithic times up to the present day with all the activity sites of a single community represented in many periods (Jones 1973; Jones, Evison, and Myres 1968; Jones and Jones 1974).

Also relevant was the gradual switch from reactive responses to proactive planning and the developing tradition of strategic resource surveys such as those for the river gravels (Benson and Miles 1974; Gates 1975; RCHM 1960) and new town development (e.g., RCHM 1969). The preparation of these documents served to broaden perspectives and provided the context for introducing the concept of landscape to provide an academic structure to the collection and presentation of information that was applied to very practical ends.

Cross-Disciplinary Exchanges

Stimuli from other disciplines continued, especially geography, anthropology, local history, and placename studies. In Britain, the work of Maurice Beresford (1954, 1957) was a landmark in the study of medieval settlement and landscape, but the single most widely acknowledged stimulant was undoubtedly W. G. Hoskins's seminal book *The Making of the English Landscape* (Hoskins 1955). The way in which Hoskins charted the development of the countryside, drawing mainly on local history and geography, provided a context for archaeological sites and monuments and struck a chord with a number of young archaeologists (Aston 2000: 49). However, Hoskins was not the only commentator on landscape in the 1950s. Contributions from those with interests in the artistic and aesthetic elements of landscape continued and in one particular case found expression within an archaeological context through the pen of Jacquetta Hawkes in her book *A Land* (1951).

Access to Source Materials

A third contribution was the increasing availability of relevant source material and the innovative application of appropriate techniques to their analysis. Landscape studies relied on an ability to see large tracts of countryside or townscape at one time and to recognize and document patterns and relationships in the data. Aerial photographs and early cartographic sources provide just such views. Aerial photographs began to become widely available and easily accessible from the early 1950s onward. O. G. S. Crawford, J. K. St. Joseph, G. W. G. Allen, D. N. Riley, and others pioneered the use of aerial photography, but it was John Bradford in his book *Ancient Landscapes* (1957) who provided one of the first systematic accounts and eloquent demonstrations of how the techniques could be applied to the same kinds of problem that Hoskins was addressing. In Bradford's own words, his book arose from "an ultimate desire to explore thoroughly *complete* social units, advancing from single sites to regions" (1957: 3, original emphasis). At about the same time, public record offices adopted a higher profile by making archives of early maps and documents easily accessible.

Blood on the Tracks

Landscape archaeology as it emerged in Britain during the mid-1970s drew on the rapidly accumulating wealth of archaeological and historical material by emphasizing the time-depth inherent to every patch of countryside. Mick Aston and Trevor Rowley (1974: 14) noted that:

> The landscape is a palimpsest on to which each generation inscribed its own impressions and removes some of the marks of earlier generations. Constructions of one age are often overlain, modified or erased by the work of another. The present patchwork nature of settlement and patterns of agriculture has evolved as a result of thousands of years of human endeavor, producing a landscape which possesses not only a beauty associated with

long and slow development, but an inexhaustible store of information about many kinds of human activities in the past.

Such approaches provided academic underpinning for many extensive survey projects yielding high-resolution understandings of defined territories and regions with reference to prehistoric and historic times, and could also be tied to questions about subsistence strategies, economic organization, changing belief systems, and the development of social complexity. Dartmoor (Fleming 1978, 1983, 1988), the Somerset Levels (Coles and Coles 1986), and the Stonehenge Environs (RCHM 1969) were among many such areas studied in this way. In the Americas, similar objectives were being pursued at this time, albeit with rather different kinds of baseline data drawn from formal sampling procedures and predictive modeling to document surface scatters and monuments alike (Flannery 1976: 31–224), and these made their influence felt (e.g., Shennan 1985). Throughout much of mainland Europe, archaeological interest remained firmly focused on individual sites and monuments, although as in Britain and America the number and scale of investigations increased dramatically through the 1970s and the 1980s.

Even while many of these projects were unfolding, the hard empiricism of the New Archaeology was giving way to the softer processualist approaches of Social Archaeology (Renfrew 1973, 1977), with its inherent focus on people and places and its interest in relationships between "man [*sic*] and the landscape." Contributions from geography and various branches of the social sciences, especially anthropology and ethnography, added further dimensions to the conceptualization and interpretation of ancient landscapes. Spatial archaeology in particular drew heavily on the work of the New Geography and provided a wealth of tools and approaches that were widely explored and tested by Ann Ellison and J. C. Harriss (1972), David Clarke (1977a, 1977b, 1978), Ian Hodder (1978a, 1978b, 1981, 1982), Claudio Vita-Finzi (1978), and others. Site catchment analysis (SCA), central place theory (CPT), Thessian Polygons (TPs), and trend surface analysis (TSA) and more found applications that provided thought-provoking understandings (Hodder and Orton 1976) but shared an essential detachment from the landscape itself.

Rather different were the emergent humanistic approaches. In a perceptive paper at a conference in 1974, Frances Lynch examined the way that people had real and sensitive appreciations of the landscape, noting that people's relationship with their environment "was not simply one of economic exploitation and struggle but was also one in which the beauty and grandeur of the rocks and mountains and the broad views over valleys and plains had an importance and value in their own right" (1975: 124). It was a perspective that chimed with emerging postmodernist views elsewhere in academe (Cosgrove and Daniels 1988; Nuttgens 1972; Tuan 1977; Wheatley 1971) and would eventually find a place in archaeological discourse (see below), but not before other avenues had been explored. Some understanding was needed of the way that landscapes might have been perceived, understood, and utilized by extinct communities, and how spatial patterning of material culture might relate to such matters. Ethnoarchaeology at the landscape scale provided some of the answers—Lewis Binford's work on the archaeology of place (1982) and on hunter-gatherer communities (1983: 109–143) being fundamental.

Among the most important lessons of ethnoarchaeology and spatial archaeology was the recognition that, in social terms, space is essentially continuous and that what varies is the way that people differentially value, categorize, subdivide, and use the spaces available to them. The implication, of course, is the need to move away from looking at monuments in isolation and instead consider much larger tracts of land. As Peter Fowler once suggested, Britain should perhaps be seen as one enormous archaeological site (1977: 48)—a sentiment applicable to many parts of the world. In this sense, the gaps between what in conventional thinking might be called *sites* are just as important as the hotspots themselves. Rob Foley put his finger right on the problem when he spoke of "off-site" archaeology (1981a), pointedly subtitling one of his papers "an alternative approach for the short-sited" (1981b). His analysis examined both the incremental pattern of artifact accumulation within space, and the systematic and nonsystematic taphonomy of postdepositional transformations.

Less than a decade after Aston and Rowley introduced their methodologically grounded view of landscape archaeology, the field had not only expanded rapidly but also diversified markedly. Such was the multiplicity of interest that Coones (1985) raised the question of just how many landscapes there were. In archaeology it became widely accepted that although there was one land available for study, there were many landscapes that could be developed and explored (Darvill and Gojda 2001). Through the later 1980s, 1990s, and the first few years of the 21st century, traditions of landscape archaeology have multiplied still further, and the overall approach spread to studies in all corners of the globe.

Planet Waves

There is no single consolidated tradition of landscape archaeology in Europe, rather a series of related approaches under a single banner (Gojda 2001). Some are grounded firmly in conventional empirical traditions, whereas others articulate with changing visions of the landscape across a wide range of intellectual frontiers. Three interconnected factors appear to be driving an interest in landscape archaeology forward, each deriving insights from the other.

First is the continuing imperative to understand past societies and the recognition that people exist within worlds far larger than the confines of a particular "site" or "monument," even though aspiring to a "totalizing" perspective is fraught with paradoxes and tensions (Johnston 1998a, 1998b). Reflexive archaeologies provide one solution, well exemplified at Çatalhöyük in Turkey, where the examination of a Neolithic tell provides the focus for more extensive surveys and investigations (Hodder 2000). The balance between theoretical and ideological positions is also highly relevant to the realization of shared objectives in archaeology, especially in the case of landscape archaeology, where real differences are visible. Smyntyna (2006) has usefully compared and contrasted the physiographic paradigms that dominate thinking in eastern Europe (especially post-Soviet Europe) with the social paradigms employed widely in western Europe. Terminology remains problematic, and debates about the analytical categories identified with such studies continue (Meier 2006). However, the biggest single development in recent decades surrounds the recognition that natural places in the landscape have specific meaning and significance for prehistoric and later communities and that they are often connected with rock art, votive offerings, special sources of raw material, and monuments (Bradley 2000).

Second is the increasing scale of the opportunities available for the detailed investigation of large slices of landscape. Research projects have certainly adapted to this possibility, but the largest single contribution arises from the pan-European success in integrating archaeological work with the spatial planning system so that predetermination studies can be undertaken, often under the rubric of the European Union's environmental impact assessment regulations (EC 1985, 1997; Jones and Slinn 2006) and mitigation strategies agreed and implemented as part of a broader management cycle (Darvill 2004a: 415–20; Darvill and Gerrard 1994; Waugh 2006; Willems 1998: 195–96). Major infrastructure and energy projects, sometimes involving more than one country, stand at the top of the list in terms of their scale, as with the Betuweroute rail-freight line between Rotterdam and Zevenaar in the Netherlands (Rijksdienst voor Archeologie, Cultuurlandschap en Monumenten [RAM] 2000–2002), the motorway building program in Ireland (O'Sullivan 2003; O'Sullivan and Stanley 2005, 2006), and the open-cast lignite mining around Cottbus in eastern Germany (Bönisch 2001), to mention just three. But the value of smaller schemes must not be underestimated. Many states have signed and ratified the European Convention on the Protection of the Archaeological Heritage, opened for signature in Valletta (Malta) in January 1992, whose concern is with all kinds of archaeological materials, sites, and landscapes "as a source of the European collective memory and as an instrument for historical and scientific study" (CoE 1992: article 1.1).

Third is a deepening concern for the management of the European landscape through the structured deployment of conservation, preservation, protection, and controlled exploitation. A Council of Europe recommendation on the integrated conservation of cultural landscape areas was approved in 1995 (CoE 1995; Darvill 1996), followed by the more powerful and wide-ranging *European Landscape Convention* opened for signature at Firenze in October 2000 (CoE 2000; Fairclough 2006; Högberg 2006). All are powerful instruments not just in advancing the interests of archaeology but also in fully integrating archaeological interests with broader environmental concerns and the creation of social policy and a European identity (Fairclough 2006; Lozny 2005; Machat 1993; Tzanidaki 2000; Willems 2000). Arising from these developments is an increased concern for the establishment of research networks and dedicated periodicals. The Man and Nature Centre at Odense University, Denmark, was one organization that provided an innovative interdisciplinary approach to landscape between 1993 and 1997 (Juel 1997). Its place has partly been taken by Landscape Europe, based in the Netherlands. This comprises a network of about 20 national research institutes in more than 15 countries with expertise in landscape assessment, planning, and management at the interface of policy implementation, education, and state-of-the-art science in support of sustainable landscapes (LE 2006). Mention may also be made of the interdisciplinary journals *Landscape Research*, published by Taylor and Frances for the Landscape Research Group at Oxford Brookes University, and *Landscapes*, published since 1999 by Windgather Press.

To classify the myriad strands of investigation now taking place with a landscape archaeology

perspective would require enumeration of all that is happening, but seven very broad and partly overlapping strands can be recognized and deserve attention in terms of their recent history and development. Many of the themes and approaches covered by these strands are explored in more detail in later chapters of this Handbook.

Total Archaeology

This traditional approach, the name for which was coined by Christopher Taylor (1974a), developed the fundamental axiom of historical geography that the landscape we see today is the product of prolonged evolution involving both human and natural agencies and that to understand and decipher it requires, at the very least, "reading" all available archaeological and related evidence including fieldwork, aerial photography, cartographic sources, historical documents, placenames, folklore, and so on (Aston 1985; Aston and Rowley 1974; Rippon 2004; Steane and Dix 1978; Taylor 1974b). Throughout, it is assumed that human impact increases from the almost insignificant in early prehistoric times through to very significant in recent centuries, while the effect of the natural environment on human activity decreases inversely.

The result of such landscape archaeology is typically a series of maps summarizing the distribution of sites and land-use for each defined phase. As a general rule, many overlays tend to be rather bare of sites until the relatively modern period when detailed cartographic sources become available. A commentary outlining the historical evolution of the block of landscape in question is usually provided to accompany the maps and plans. An early example is Taylor's study of Whiteparish, a forest edge parish in Wiltshire (1967). More recent is a study of medieval and later agriculture and tin working in St. Neot Parish, Cornwall (Austin, Gerrard, and Greeves 1989). This analysis was confined to a particular chronological period (A.D. 1100–1700) and just one community (an upland parish), with a view to understanding human interactions with natural resources.

Palaeo-Environments and Palaeo-Land-Use

A variation on total archaeology is the development of models that focus on the changing environment, land-use patterns, and settlement systems within a given territory. In a sense, such work emphasizes the relationships between the physical environment and human communities, although recognizing that the relationship is two-way (Evans, Limbrey, and Cleere 1975; Limbrey and Evans 1978). Factors such as the nature and extent of tree cover, and the composition and abundance of the natural fauna, soil fertility, microtopography, geomorphology, and climate are fundamental dimensions of the landscape addressed through such work, although not necessarily all together.

Excavation and survey-based projects that integrate archaeological and palaeo-environmental evidence within a landscape framework are increasingly common and have moved from being multidisciplinary in their execution to being rather more interdisciplinary. Examples in Britain include Fisherwick, Staffordshire (Smith 1979), the Essex coast (Wilkinson and Murphy 1995), the Stonehenge area of Wiltshire (Allen 1997), and the Severn Estuary (Bell, Caseldine, and Neumann 2000). This integrative approach is also an increasingly recognized in the study of landscapes in central and eastern Europe (Suhr 2006)—for example, studies of the Sobiejuchy area of Poland (Ostoja-Zagórski 1993), the Ljubljansko Barje region of Slovenia (Budja 1997), the Upper Tisza valley of northeastern Hungary (Gillings 1997), the Gyomaendrőd Project in southeast Hungary (Bökönyi 1992), and the hinterland of Novgorod in northwest Russia (Brisbane and Gaimster 2001).

Cultural Landscapes and Community Areas

Almost the converse of palaeo-environmental studies of landscape are those that foreground human communities and the way that they inhabit a world of their own creation. Often based on the results of total archaeology, regional surveys, or palaeo-environmental archaeology, the cultural landscape is built as a "snapshot" image at some defined moment in time.

Relatively recent landscapes lend themselves to such study, especially historic parks and gardens (Pattison 1998). A rather good, and very detailed, piece of work on an extensive cultural landscape is the study of Blenheim Park at Woodstock in Oxfordshire by James Bond and Kate Tiller (1987). Very large-scale excavations such as those in the upper Thames Valley around Claydon Pike, Gloucestershire (Miles 1983) and Yarnton, Oxfordshire (Hey 2004; Hey, Bayliss, and Boyle 1999), and in the Welland Valley at Fengate, Cambridgeshire (Pryor and French 1985), provide exceptionally detailed insights and high-resolution views of cultural landscapes. Broader surveys coupled with selective excavation give a still bigger picture, as with the study of 450 km² of chalkland around Danebury, Hampshire (Cunliffe 1995, 2000; Palmer 1984), and the Fenland survey, which looked at a 60%

sample of a study area covering 4200 km² in eastern England (Hall 1987; Hall and Chippindale 1988: 305–80; Hall and Coles 1994). In Ireland, the Discovery Programme's work around the prehistoric and early medieval site of Tara provides a benchmark for the execution and presentation of extensive landscape surveys (Bhreathnach 1995, 2005; Newman 1997).

Many studies of cultural landscapes are essentially regional surveys closely tied to broader research agendas. Examples are legion, and in many parts of Europe this approach is the mainstay of landscape archaeology (Barker 1996: Dommelen and Prent 1996). Although this approach is rooted in processual archaeology, there are of course opportunities to apply other perspectives:

- In England: the central Welsh Marches (Whimster 1989), the Yorkshire Wolds (Stoertz 1997), the Berkshire Downs (Ford 1987), East Hampshire (Shennan 1985), Bodmin Moor, Cornwall (Johnson and Rose 1994), and Cranborne Chase, Dorset (Barrett, Bradley, and Green 1991)

- In Scotland: Perthshire (RCAHMS 1990, 1994)

- In Ireland: the Neolithic of country Sligo (Bergh 1995)

- In Denmark: the Bronze and later maritime landscapes of the island of Fyn (Crumlin-Pedersen, Porsmose, and Thrane 1996)

- In the Netherlands: the Maaskant Project in the Meuse Valley (Fokkens 1996)

- In Italy: the Biferno Valley (Barker 1995); later prehistory of the Pontine region (Attema 1993); the middle Tiber Valley (Patterson 2004); and Etruria and Umbria (Christie 2004; Terrenato 1995)

- In Spain: the Tarragona hinterland survey (Carreté, Keay, and Millett 1995)

- In Turkey: the Amuq Valley Regional Project in the Plain of Antioch and the Orontes Delta (Yener 2005)

- In Greece: the Aegean Islands of Keos (Cherry, Davis, and Mantzourani 1991); Melos (Renfrew and Wagstaff 1982); and Laconia (Cavanagh et al. 1996)

Aerial photography established in some parts of Europe during the 1920s plays an increasingly important role in the location and the mapping of relict cultural landscapes. In England, the National Mapping Programme will eventually provide full plotting of earthwork and cropmark features based on the analysis of more than a million images (Bewley 2001). In Belgium, several thousand pictures have been analyzed to provide a database for the study of prehistoric landscapes, notably the Bronze Age and the early Iron Age (Bourgeois and Verlaeckt 2001). And in eastern Europe, the easing of restrictions imposed during communist times is now beginning to pay off with extensive work in Bohemia (Gojda 2006) and Romania (Hanson and Oltean 2005), among others, and innovative approaches in Armenia (Faustmann and Palmer 2002).

Geophysical and geochemical survey increasingly plays a role in understanding landscape patterning for particular periods where it can be applied extensively and where the signatures relevant to particular arrangements are distinctive (Spoerry 1992). Investigations in the Walton Basin of the Welsh Borderland (Gibson 1999) and Billown in the Isle of Man (Darvill 2004b with earlier references) illustrate the potential for Neolithic landscapes, while work around Wroxeter, Shropshire, has begun to unravel a very detailed picture of the Roman landscape (BUFAU 1996; van Leusen 1999). Overall, the range of approaches that can be brought to bear on the mapping and the investigation of cultural landscapes is now very considerable. Indeed, the EU-funded POPULUS project had as one of its aims the development of coherent research goals, methods, and standards for work on landscapes around the Mediterranean (Barker and Mattingly 1999–2000).

In central Europe, studies of cultural landscapes have found a theoretical background in the idea of "community areas" as spatial units occupied by identifiable social groups in prehistoric and later times and reconstructable through studying distributions of archaeological sites and materials (Neustupný 1991). Such units do not necessarily coincide with modern administrative areas, and indeed the development of this thinking was in part an attempt to move away from the common practice of imposing modern patterns on the archaeological evidence to support the idea of continuity in a fast-changing landscape. Martin Kuna's (1991) investigation of prehistoric habitation areas in Bohemia provides an example of community area analysis applied from the household scale through to the settlement pattern found in sampled river valleys.

Social Use of Space

In the course of developing the spatial theme, we have given some attention to modeling the ways that human communities subdivide, utilize, and conceptualize space at the landscape level in a way that shifts the focus of analyses away from the purely functional aspects of landscape and its development through time toward the realm of cognition and meaning. Developments in the field of geography (Wagstaff 1987) and anthropology (Hirsch and O'Hanlon 1995) have been influential here, as well as case studies such as that by Christine Hugh-Jones of the Tukanoan Indians in northwest Amazonia (1979). Building on some of this, Tim Ingold (1993) has usefully introduced the idea of the "taskscape" as the entire ensemble of tasks or actions that a society, community, or individual performs—a seamless spread of events and experiences.

In thinking about differences in the form and layout of occupation sites of the later first millennium B.C. in the upper Thames Valley, for example, Richard Hingley (1984) showed that differences in settlement type and the use of material culture could be related to contrasting modes of production within the landscape. Elsewhere, attention has focused on the way that people and places were bound together through belief systems, cosmologies, views of the world, and many other determinants of social action (Barrett 1994; Bradley 1993, 1998). Experience, structuration, memory, and the creation of place have become themes running though much postprocessual discussion of landscape—for example, the way that people perceive space as they move through it and have different views of significant structures, objects, and places (Thomas 1993). The question of cosmological referencing in the landscape and the resultant patterning of activity has been explored with reference to Stonehenge (Darvill 1997) and the arrangement of Neolithic monuments on Orkney (Richards 1996). More generally, Cooney (2000), in his book on Neolithic Ireland, has utilized the concept of landscape as a constructed relationship between people and the places they inhabit, how they perceived the physical world of soil, water, rocks, and air and made it a lived-in place to explore. And, more recently, he has extended the vision of landscapes to embrace the idea of "seascapes," too (Cooney 2003).

Landscape Theory

In the United States, the postmodernist critique of earlier views of human-environment relationships opened up the possibility of new approaches to the landscape. The history of landscape theory has been discussed in some detail (Jackson 1984; Norton 1989) and draws on four key sources (see Whittlesey 1997: 18–19 for summary). First, cultural geography and environmental design; second, bio-logical approaches to landscape ecology and dialectical biology; third, the Marxist notion that human history is part of natural history; and fourth, the use of encultured approaches to human-environment relationships. Cognition and the symbolic meanings attached to landscapes through art and language are important ingredients, as is the elimination of artificial barriers between nature and culture and between culture and environment.

Following Zedeño and colleagues (1997), landscapes are seen as having three basic dimensions: formal (physical characteristics and properties); historical (sequential network links that result from transformational processes); and relational (interactive elements such as behavioral, social, and symbolic links that connect people and the land). Thus, in seeking to explore explicit research themes in relation to particular landscapes, suitable methodologies can be developed to explore each. This has successfully been achieved, for example, in the case of work by the Lower Verde Archaeological Project between 1991 and 1996 in central Arizona, where the story of a vanishing river and its place in the lives of Hohokan, Sinagua, Yavapai, Western Apache, and Euroamerican peoples over a period of more than 1,500 years is unfolded (Whittlesey, Ciolek-Torrello, and Altschul 1997).

In Europe, landscape theory as such is less visible than in north America, but a variant based on the cultural biography of a particular landscape has emerged, including, for example, studies of the Avebury (Pollard and Reynolds 2002) and Stonehenge (Darvill 2006) areas in southern Britain and the urnfield landscapes of the Meuse-Demer-Scheldt of southern Netherlands and northern Belgium (Roymans 1995). In these works attention focuses on the unfolding life of the area, relationships among people and the world they have created for themselves, an acceptance that cultural and natural worlds are indivisible, and an interest in tradition, memory, and way in which past events give meaning to later actions.

Phenomenology

The publication by Christopher Tilley in 1994 of *A Phenomenology of Landscape* opened a whole new range of possibilities for landscape archaeology that many have since followed. At its core is an essentially humanist view of landscape, not so very different from that espoused by Frances Lynch, Denis Cosgrove, Stephen Daniels, and others a decade or more earlier, in which simple binary divisions between culture and nature are broken down (Descola and Pálsson 1996) and

Cartesian rationalism put to one side. As Tilley notes (1994: 26):

> A fundamental part of the daily experience in non-industrial societies is the physical and biological experience of landscape—earth, water, wood, stone, high places and low places, wind, rain, sun, stars and sky. The rhythms of the land and the seasons correspond to and are worked into the rhythms of life. A landscape has ontological import because it is lived in and through, mediated, worked on and altered, replete with cultural meanings and symbolism—and not just something looked at or thought about, an object merely for contemplation, description, representation and aestheticization.

What differs is the means by which such matters are explored with reference to past societies. Unlike landscape theory, the methodology adopted by phenomenologists involves using a contemporary experience of the landscape to consider how earlier people might have thought about the world as a socially constructed reality involving the embodiment and the communication of cosmologies, economic relations, power structures, and social order. Tilley himself reviewed three contrasting landscapes in Britain—southwest Wales, the Black Mountains, and the chalklands of northwest Dorset—to look at changing themes of ancestral power and meanings, and their appropriation by individuals and groups through the construction of monuments (1994). Subsequent studies include the use by Neolithic communities of distinctive rocks on Bodmin Moor, Cornwall (Tilley 1996a), and the study of stone in the landscape (Tilley 2004). Elsewhere, cursus monuments of the 3rd millennium B.C. have been used widely as case studies in phenomenology, physically (Johnson 1999) and through GIS-based modeling (Chapman 2005). Beyond Britain, phenomenology has variously been applied to the problem of megalithic tombs in Galicia, Spain (Criado Boado and Villoch Vázquez 2000) and the Neolithic of southern Scandinavia (Tilley 1996b), and from a central European perspective the potential for using aerial photography as a way of seeing lost worlds in this way has been highlighted by Włodzimierz Rączkowski (2001).

Walking through monuments to record views (Tilley 1994: 172), mapping skylines (Cummings, Jones, and Watson 2002; Cummings and Whittle 2004), or replicating doorways to assess intervisibility (Bender, Hamilton, and Tilley 1997) offers new ways of investigating old problems despite obvious practical difficulties relating to viewpoints, the effect of vegetation cover (Chapman and Gearey 2000), and the physical changes that have taken place over intervening millennia. Inhabitation and bodily engagement are central to all applications of the phenomenological approach and provide exciting interpretative narratives (e.g., Chadwick 2004). To some, however, hyper-interpretative approaches to landscape archaeology lack rigor and are increasingly being criticized for their selective and misleading use of observations (Brück 2005; Fleming 1999, 2005, 2006). There is no doubt, however, that the phenomenological approach has much potential and that perhaps what is needed to make it more useful in archaeological situations is further middle-range research on perception and experience under a variety of social conditions.

Historic Landscapes

In conceptualizing past landscapes as sets of meanings and differentially valued spaces, structures, and things, archaeology has equipped itself, both intellectually and practically, with the means to contribute to larger ongoing debates about landscapes now and in the future. Barbara Bender has explored the issue of contested landscapes past and present, focusing on political and social issues (Bender 1993; Bender and Winer 2001). Criado and Parcero (1997) look at how landscapes and archaeology relate to heritage in Galicia, northwest Spain, while Zedeño and colleagues (1997) consider similar issues of landscape in relation to Native Americans.

In Britain and Ireland, the focus of much cultural resource management work in the 1980s and 1990s moved away from what might be called large sites or archaeologically rich countryside and townscape toward a more holistic approach with broadly defined "historic landscapes" (Aalen 1996; Countryside Commission 1996; Darvill 1987; Darvill, Gerrard, and Startin 1993; Goodchild 1990; Haynes 1983; Swanwick 1982; Wager 1981). Instead of subdividing the landscape into particular segments or blocks on the basis of some special interest, there is increasing emphasis on the integration of archaeology with other environmental and conservation disciplines. This is closely bound up with two trends. First is the integration of archaeology into the green movement (Mcinnes and Wickham-Jones 1992; Swain 1993). Second is the way in which multidisciplinary consultancies and government departments at all levels have put archaeologists alongside their counterparts from disciplines such as ecology, nature conservation, and countryside access, in order to provide advice

to planners, land managers, and the public. Similar developments are happening elsewhere in Europe as interest in the broader issue of managing change replaces earlier simple but stultifying ideas of landscape protection (Tzanidaki 2000; Willems 2000).

Much emphasis is now being placed on the recognition of local "landscape character" (Clark, Darlington, and Fairclough 2004; Countryside Commission 1991, 1996; Fairclough 1999; Herring 1998), while regionally and nationally the role of parks and extensive designations has come to the fore in Wales (Cadw 1998, 2001) and Spain (Méndez 1997, 2000), among other places. In more than a dozen countries across Europe attempts are being made to characterize and to model historic landscapes as an aid to future management (Clark, Darlington, and Fairclough 2003; Fairclough and Rippon 2002). Much of this depends on direct observation of the modern landscape in relation to inventories and records of archaeological finds, but in the Netherlands attention is also being given to predictive modeling in order to develop a more robust archaeological resource management (Leusen and Kammermans 2005).

At a global scale, the introduction of cultural and natural landscapes to the designations possible within UNESCO's World Heritage Convention (UNESCO 1972) brought new and wide-ranging discussions about the conceptualization, definition, and valuation of potential examples of international importance. Three categories of cultural landscape were eventually defined: intentionally designed and created landscapes; organically evolved landscapes; and associative cultural landscapes (Cleere 1995). In 2003, the Convention for the Safeguarding of the Intangible Heritage extended this interest from the physical remains to the practices, representations, expressions, knowledge, skills, and cultural spaces that communities, groups and individuals recognize as part of their cultural heritage (UNESCO 2003), which together represent another key strand in the understanding and appreciation of landscape.

Conclusions

The development of landscape archaeology during the later 20th century has been one of the most exciting and dynamic developments within the discipline as a whole. As the papers in collective works such as those edited by Ashmore and Knapp (1999) and Ucko and Layton (1999) show, landscape archaeology is a central component of most large-scale archaeological projects, a widely taught subject in university courses, and a perspective underpinning much research. Landscape as the unifying concept has been widely, and differently, defined according to the orientation and the purpose of the work being carried out. Such diversity in the way that the idea of "the landscape" is conceptualized within archaeological research is healthy to the debate and serves to sharpen thinking and expand ways of understanding. Uniting them in their application to archaeological problems is dependent on our ability as archaeologists to recognize pattern, order, context, association, and repetition in the various constituent components (whether physical, ideational, or experiential) defined through any preferred theoretical perspective.

Three issues connected with questions of scale and temporality emerge from discussion of the scheme set out above that in future perhaps deserve more attention:

1. Landscapes for archaeologists are as much about spaces and gaps in the archaeological record defined in the traditional way as about defined sites and monuments.

2. Order, structure, and pattern may be perceived from many different directions according to the position of the observer.

3. Landscapes do not have defined physical limits either in time or space, except where imposed by analytical procedures and intellectual traditions.

References

Aalen, F. H. A. (ed.). 1996. *Landscape Study and Management*. Dublin: The Office of Public Works.

Allcroft, A. H. 1908. *Earthwork of England*. London: Macmillan and Co.

Allen, M. J. 1997. Environment and land-use: The economic development of the communities who built Stonehenge, in B. Cunliffe and C. Renfrew (eds.), *Science and Stonehenge*, pp. 115–44. Proceedings of the British Academy 92. London: The British Academy.

Ashmore, W., and Knapp, A. B. (eds.). 1999. *Archaeologies of Landscape: Contemporary Perspectives*. Oxford: Blackwell Publishing.

Aston, M. 1985. *Interpreting the Landscape*. London: Batsford.

———. 2000. *Mick's Archaeology*. Stroud: Tempus.

Aston, M., and Rowley, T. 1974. *Landscape Archaeology: An Introduction to Fieldwork Techniques on Post-Roman Landscapes*. Newton Abbot: David and Charles.

Attema, P. A. J. 1993. *An Archaeological Survey in the Pontine Region: A Contribution to the Early Settlement History of South Lazio, 900–100 B.C.* Groningen: Rijksuniversiteit Archeologisch Centrum.

Austin, D., Gerrard, G. A. M., and Greeves, T. A. P. 1989. Tin and agriculture in the middle ages and beyond: Landscape archaeology in St. Neot Parish, Cornwall. *Cornish Archaeology* 28: 5–25.

Barker, G. 1995. *A Mediterranean Valley. Landscape Archaeology and* Annales History *in the Biferno Valley*. Leicester: Leicester University Press.

———. 1996. Regional archaeological projects: Trends and traditions in Mediterranean Europe. *Archaeological Dialogues* 3: 160–75.

Barker, G., and Mattingly, D. (eds.). 1999–2000. *The Archaeology of Mediterranean Landscapes*. Oxford: Oxbow Books (5 volumes).

Barrett, J., Bradley, R., and Green, M. 1991. *Landscape, Monuments and Society: The Prehistory of Cranborne Chase*. Cambridge: Cambridge University Press.

Barrett, J. C. 1994. *Fragments from Antiquity*. Oxford: Blackwell Publishing.

Bell, M., Caseldine, A., and Neumann, H. 2000. *Prehistoric Intertidal Archaeology in the Welsh Severn Estuary*. CBA Research Report 120. York: Council for British Archaeology.

Bender, B. 1993. Introduction: Landscape—meaning and action, in B. Bender (ed.), *Landscape: Politics and Perspectives*, pp. 1–18. Oxford: Berg.

Bender, B., Hamilton, S., and Tilley, C. 1997. Leskernick: Stone worlds; alternative narratives; nested landscapes. *Proceedings of the Prehistoric Society* 63: 147–78.

Bender, B., and Winer, M. (eds.). 2001. *Contested Landscapes: Movement, Exile and Place*. London: Berg.

Benson, D., and Miles, D. 1974. *The Upper Thames Valley: An Archaeological Survey of the River Gravels*. Oxford: Oxford Archaeological Unit.

Beresford, M. W. 1954. *The Lost Villages of England*. London: Lutterworth.

———. 1957. *History on the Ground*. London: Lutterworth.

Bergh, S. 1995. *Landscape of the Monuments: A Study of the Passage Tombs in the Cúil Irra Region, Co. Sligo, Ireland*. Arckeologiska undersökningar Skrifter nr 6. Stockholm: Riksantikvarieämbetet.

Bewley, B. 2001. Understanding England's historic landscapes: An aerial perspective. *Landscapes* 2: 74–84.

Bhreathnach, E. 1995. *Tara: A Select Bibliography*. Discovery Programme Reports 3. Dublin: Royal Irish Academy.

Bhreathnach, E. (ed.). 2005. *The Kingship and Landscape of Tara*. Dublin: Four Courts Press.

Binford, L. R. 1982. The archaeology of place. *Journal of Anthropological Archaeology* 1: 5–31.

———. 1983. *In Pursuit of the Past*. London: Thames and Hudson.

Bökönyi, S. 1992. *Cultural and Landscape Changes in South-East Hungary I*. Budapest: Archaeolingua.

Bond, J., and Tiller, K. 1987. *Blenheim: Landscape for a Palace*. Gloucester: Alan Sutton and Oxford University Department of External Studies.

Bönisch, E. 2001. Von der Schwerpunktgrabung zum Schwerpunkt komplexe Forschung—Zur Konzeption der "Braunkohlenarchäologie" im Niederlausitzer Revier. *Archäologisches Nachrichtenblatt* 6: 138–51.

Bourgeois, J., and Verlaeckt, K. 2001. The Bronze Age and early Iron Ages in Western Flanders (Belgium): Shifting occupation patterns, in M. Lodewijckx (ed.), *Belgian Archaeology in a European Setting II*, pp. 13–25. Act Archaeologica Lovaniensia Monographiae 13. Leuven: Leuven University Press.

Bradford, J. 1957. *Ancient Landscapes*. London: Bell and Sons.

Bradley, R. 1993. *Altering the Earth*. Society of Antiquaries of Scotland Monograph Series 8. Edinburgh: Society of Antiquaries of Scotland.

———. 1998. *The Significance of Monuments*. London: Routledge.

———. 2000. *An Archaeology of Natural Places*. London: Routledge.

Brisbane, M., and Gaimster, D. (eds.). 2001. *Novgorod: The Archaeology of a Russian Medieval City and Its Hinterland*. British Museum Occasional Paper 141. London: The British Museum.

Brück, J. 2005. Experiencing the past? The development of a phenomenological archaeology in British prehistory. *Archaeological Dialogues* 12: 45–72.

Budja, M. 1997. Landscape changes in the Neolithic and Copper Ages in Slovenia. Case study: The Ljubljansko Barje Region, in J. Chapman and P. Dolukhanov (eds.), *Landscapes in Flux: Central and Eastern Europe in Antiquity*, pp. 77–87. Oxford: Oxbow Books.

BUFAU. 1996. *The Wroxeter Hinterland Project Internet Site*. www.bufau.bham.ac.uk/projects/WH/base.html.

Burrow, E. J. 1919. *The Ancient Entrenchments and Camps of Gloucestershire*. Cheltenham: Burrow and Co.

———. 1924. *The Ancient Earthworks and Camps of Somerset*. Cheltenham: Burrow and Co.

Cadw. 1998. *Landscapes of Historic Interest in Wales. Part 2.1*. Cardiff: Cadw.

———. 2001. *Landscapes of Historic Interest in Wales. Part 2.2*. Cardiff: Cadw.

Carreté, J.-M., Keay, S. J., and Millett, M. 1995. *A Roman Provincial Capital and Its Hinterland: The Survey of the Territory of Tarragona, Spain, 1985–90*. Journal of Roman Archaeology Supplementary Series 15. Michigan: Ann Arbor.

Cavanagh, W., Crouwel, R. M. V., Catling, H., and Shipley, G. (eds.). 1996. *Continuity and Change in a Greek Rural Landscape: The Laconia Survey Volume II*. Annual of the British School at Athens Supplementary Volume 27. London and Athens: British School at Athens.

Chadwick, A. M. (ed.). 2004. *Stories from the Landscape: Archaeologies of Inhabitation*. BAR International Series 1238. Oxford: Archaeopress.

Chapman, H. P. 2005. Rethinking the "Cursus Problem": Investigating the Neolithic landscape archaeology of Rudston, East Yorkshire, UK, using GIS. *Proceedings of the Prehistoric Society* 71: 159–70.

Chapman, H. P., and Gearey, B. R. 2000. Palaeoecology and the perception of the prehistoric landscapes: Some comments on visual approaches to phenomenology. *Antiquity* 74: 316–19.

Cherry, J. F., Davis, J. L., and Mantzourani, E. 1991. Landscape archaeology as long-term history: Northern Keos in the Cycladic Islands from earliest settlement until modern times. *Monumenta Archaeologica* 16. Los Angeles: UCLA Institute of Archaeology.

Childe, G. V. 1929. *The Danube in Prehistory*. Oxford: Clarendon Press.

Christie, N. (ed.). 2004. *Landscapes of Change: Rural Evolutions in Late Antiquity and the Early Middle Ages*. Aldershot: Ashgate.

Clark, J., Darlington, J., and Fairclough, G. 2004. *Using Historic Landscape Characterization*. London and Lancaster: English Heritage and Lancashire County Council.

———. (eds.). 2003. *Pathways to Europe's Landscapes*. London and Lancaster: English Heritage and Lancashire County Council.

Clark, J. G. D. 1954. *Star Carr: An Early Mesolithic Site at Seamer Near Scarborough, Yorkshire*. Cambridge: Cambridge University Press (reprinted 1971).

Clarke, D. L. (ed.). 1977a. *Spatial Archaeology*. London and New York: Academic Press.

Clarke, D. L. 1977b. Spatial information in archaeology, in D. L. Clarke (ed.), *Spatial Archaeology*, pp. 1–32. London and New York: Academic Press.

———. 1978. *Analytical Archaeology* (2nd ed., revised by R. Chapman). London: Methuen.

Cleere, H. 1995. Cultural landscapes as World Heritage. *Conservation and Management of Archaeological Sites* 1: 63–8.

Coles, B., and Coles, J. 1986. *Sweet Track to Glastonbury: The Somerset Levels in Prehistory*. London: Thames and Hudson.

Coones, P. 1985. One landscape or many? A geographical perspective. *Landscape History* 7: 5–12.

Cooney, G. 2000. *Landscapes of Neolithic Ireland*. London: Routledge.

———. 2003. Introduction: Seeing Land from the Sea. *World Archaeology* 35: 323–28.

Cosgrove, D., and Daniels, S. 1988. *The Iconography of Landscape*. Cambridge: Cambridge University Press.

Council of Europe (CoE). 1992. *European Convention on the Protection of the Archaeological Heritage* (revised). *Valletta.* European Treaty Series 143. Strasbourg: Council of Europe.

———. 1995. *Recommendation of the Committee of Ministers to Member States on the Integrated Conservation of Cultural Landscape Areas as Part of Landscape Policies* (Recommendation R[95]9). Strasbourg: Council of Europe.

———. 2000. *European Landscape Convention. Florence* (European Treaty Series 176). Strasbourg: Council of Europe.

Countryside Commission. 1991. *Assessment and Conservation of Landscape Character: The Warwickshire Landscape Project Approach.* Cheltenham: Countryside Commission CCP332.

———. 1996. *Views from the Past: Historic Landscape Character in the English Countryside.* Cheltenham: Countryside Commission CCW4.

Countryside Commission, English Heritage, and English Nature. 1996. *The Character of England: Landscape, Wildlife and Natural Features.* Cheltenham: Countryside Commission CCX41.

Crawford, O. G. S. 1953. *Archaeology in the Field.* London: Phoenix House.

Criado Boado, F., and Parcero, C. (eds.). 1997. *Landscape, Archaeology, Heritage.* Trabojos en Arqueología del Paisaje 2. Santiago de Compostela: Grupo de Investigación en Arqueología del Paisaje.

Criado Boado, F., and Villoch Vázquez, V. 2000. Monumentalizing landscape: From present perception to the past meaning of Galician megalithism (north-west Iberian Peninsula). *European Journal of Archaeology* 3: 188–216.

Crumlin-Pedersen, Porsmose, E., and Thrane, H. 1996. *Atlas over Fyns kyst I jernalder, vikingetid og middelalder.* Odense: Odense Universitetsforlag.

Cummings, V., Jones, A., and Watson, A. 2002. Divided places: Phenomenology and asymmetry in the monuments of the Black Mountains, southeast Wales. *Cambridge Journal of Archaeology* 12: 59–70.

Cummings, V., and Whittle, A. 2004. *Places of special virtue: Megaliths in the Neolithic landscapes of Wales.* Oxford: Oxbow Books.

Cunliffe, B. 1995. *Danebury: An Iron Age Hillfort in Hampshire. Volume 6: A Hillfort Community in Perspective.* CBA Research Report 102. London: Council for British Archaeology.

———. 2000. *The Danebury Environs Programme. Volume 1: Introduction.* Oxford University Committee for Archaeology Monograph 48. Oxford: Oxford University Committee for Archaeology.

Darvill, T. 1987. *Ancient Monuments in the Countryside: An Archaeological Management Review.* Historic Buildings and Monuments Commission for England Archaeological Report 5. London: English Heritage.

———. 1996. Council of Europe: Heritage landscapes and sites, in F. H. A. Aalen (ed.), *Landscape Study and Management,* pp. 173–82. Dublin: The Office of Public Works.

———. 1997. Ever-increasing circles: The sacred geographies of Stonehenge and its landscape, in B. Cunliffe and C. Renfrew (eds.), *Science and Stonehenge,* pp. 167–202. Proceedings of the British Academy 92. London: The British Academy.

———. 1999. The historic environment, historic landscapes, and space-time-action models in landscape archaeology, in P. J. Ucko and R. Layton (eds.), *The archaeology and Anthropology of Landscape,* pp. 104–18. London: Routledge.

———. 2004a. Public archaeology: A European perspective, in J. Bintliffe (ed.), *A Companion to Archaeology,* pp. 409–34. Oxford: Blackwell Publishing.

———. 2004b. *Billown Neolithic Landscape Project, Isle of Man. Eighth Report: 2003.* Bournemouth University School of Conservation Sciences Research Report 12. Bournemouth: Bournemouth University.

———. 2006. *Stonehenge: The Biography of a Landscape.* Stroud: Tempus Publishing.

Darvill, T., and Gerrard, C. 1994. *Cirencester: Town and Landscape.* Cirencester: Cotswold Archaeological Trust.

Darvill, T., Gerrard, C., and Startin, B. 1993. Identifying and protecting historic landscapes. *Antiquity* 67: 563–74.

Darvill, T., and Gojda, M. (eds.). 2001. *One Land, Many Landscapes.* BAR International Series 987. Oxford: Archaeopress.

Descola, P., and Pálsson, G. (eds.). 1996. *Nature and Society: Anthropological Perspectives.* London: Routledge.

Dommelen, P., and Prent, M. 1996. The history, theory and methodology of regional archaeological projects: An introduction. *Archaeological Dialogues* 3: 13739.

European Commission. 1985. Council Directive 85/337/EEC of the 27th June 1985 on the assessment of the effects of certain public and private projects on the environment. *Official Journal of the European Communities* 5.7.85, L175: 40–48.

———. 1997. Council Directive 91/11/EC of 3rd March 1997 amending Directive 85/337/EEC on the assessment of the effects of certain public and private projects on the environment. *Official Journal of the European Communities* 14.3.97, L73: 5–15.

Ellison, A., and Harriss, J. 1972. Settlement and land use in the prehistory and early history of southern England: A study based on locational models, in D. L. Clarke (ed.), *Models in Archaeology,* pp. 911–62. London: Methuen.

Evans, J., Limbrey, S., and Cleere, H. (eds.). 1975. *The Effect of Man on the Landscape: The Highland Zone.* CBA Research Report 11. London: Council for British Archaeology.

Fairclough, G. (ed.). 1999. *Historic Landscape Characterisation: Papers Presented at an English Heritage Seminar, 11th December 1998*. London: English Heritage.

Fairclough, G. 2006. Our place in the landscape? An archaeologist's ideology of landscape perception and management, in T. Meier (ed.), *Landscape Ideologies*. Archaeolingua Series Minor 22, pp. 177–97. Budapest: Archaeolingua.

Fairclough, G., and Rippon, S. (eds.). 2002. *Europe's Cultural Landscape: Archaeologists and the Management of Change*. Europae Archaeologiae Consilium Occasional Paper 2. Brussels: Europae Archaeologiae Consilium.

Faustmann, A., and Palmer, R. 2002. Use of a paramotor for archaeological aerial survey in Armenia. *Antiquity* 79: 402–10.

Flannery, K. V. (ed.). 1976. *The Early Mesoamerican Village*. Academic Press: New York.

Fleming, A. 1978. The prehistoric landscape of Dartmoor. Part 1: South Dartmoor. *Proceedings of the Prehistoric Society* 44: 97–124.

———. 1983. The prehistoric landscape of Dartmoor. Part 2: North and East Dartmoor. *Proceedings of the Prehistoric Society* 49: 195–242.

———. 1988. *The Dartmoor Reaves*. London: Batsford.

———. 1999. Phenomenology and the megaliths of Wales: A dreaming too far? *Oxford Journal of Archaeology* 18: 119–26.

———. 2005. Megaliths and post-modernism: The case of Wales. *Antiquity* 79: 921–32.

———. 2006. Post-processual landscape archaeology: A critique. *Cambridge Archaeological Journal* 16: 267–80.

Fokkens, H. 1996. The Maaskant Project: Continuity and change of a regional research project. *Archaeological Dialogues* 3: 196–215.

Foley, R. 1981a. A model of regional archaeological structure. *Proceedings of the Prehistoric Society* 47: 1–18.

———. 1981b. Off-site archaeology: An alternative approach for the short-sited, in I. Hodder, G. Isaac, and N. Hammond (eds.), *Pattern of the Past: Studies in Honour of David Clarke*, pp. 157–84. Cambridge: Cambridge University Press.

Ford, S. 1987. *East Berkshire Archaeological Survey*. Reading: Berkshire County Council.

Fowler, P. J. 1977. *Approaches to Archaeology*. London: A. and C. Black.

———. 1979. Archaeology and the M4 and M5 motorways, 1965–78. *Archaeological Journal* 136: 12–26.

———. 2001. Reading the land. *British Archaeology* 62: 15–19.

Fox, C. 1932. *The Personality of Britain: Its Influence on Inhabitant and Invader in Prehistoric and Early Historic Times*. Cardiff: National Museum of Wales.

Gates, T. 1975. *The Middle Thames Valley: An Archaeological Survey of the River Gravels*. Reading: Berkshire Archaeological Committee.

Gibson, A. 1999. *The Walton Basin Project: Excavation and Survey in a Prehistoric Landscape 1993–97*. CBA Research Report 118. York: Council for British Archaeology.

Gillings, M. 1997. Spatial organization in the Tisza Flood Plain: Dynamic landscapes and GIS, in J. Chapman and P. Dolukhanov (eds.), *Landscapes in Flux: Central and Eastern Europe in Antiquity*, pp. 163–79. Oxford. Oxbow Books.

Gold, J. R. 1980. *An Introduction to Behavioural Geography*. Oxford: Oxford University Press.

Gojda, M. 2001. Archaeology and landscape studies in Europe: Approaches and concepts, in T. Darvill and M. Gojda (eds.), *One Land, Many Landscapes*, pp. 9–18. BAR International Series 987. Oxford: Archaeopress.

———. 2004. Prehistoric Bohemia: Landscapes and settlement in the heart of Europe. *Landscapes* 5: 35–54.

———. 2006. The archaeology of lowlands: A few remarks on the methodology of aerial survey, in T. Meier (ed.), *Landscape Ideologies*, pp. 117–23. Archaeolingua Series Minor 22. Budapest: Archaeolingua.

Goodchild, P. H. 1990. *Some Principles for the Conservation of Historic Landscapes: Draft Discussion Paper*. York: Centre for the conservation of Historic Parks and Gardens.

Gurevich, A. Y. 1969. Space and time in the *Weltmodell* of the old Scandinavian peoples. *Medieval Scandinavia* 2: 42–53.

Hall, D. 1987. The Fenland Project, Number 2: Cambridgeshire survey, Peterborough to March. *East Anglian Archaeology* 35: 1–77.

Hall, D., and Chippindale, C. (eds.). 1988. Special section: Survey, environment and excavation in the English Fenland. *Antiquity* 62: 303–80.

Hall, D., and Coles, J. 1994. *Fenland survey: An essay in landscape and persistence*. English Heritage Archaeological Report 1. London: English Heritage.

Hanson, B., and Oltean, I. 2005. Go east again. *ARRG news Supplement* 1: 48–49.

Hawkes, J. 1951. *A Land*. London: Cresset Press (reprinted 1978 by David and Charles).

Haynes, J. S. 1983. *Historic Landscape Conservation*. Gloucestershire Papers in Local and Rural Planning 20. Gloucester: Gloucestershire College of Arts and Technology.

Herring, P. 1998. *Cornwall's Historic Landscape: Presenting a Method of Historic Landscape Character Assessment*. Truro: Cornwall Archaeological Unit.

Hey, G. 2004. *Yarnton: Saxon and Medieval Settlement and Landscape Results of Excavations 1990–96*. Oxford: Oxford Archaeology.

Hey, G., Bayliss, A., and Boyle, A. 1999. Iron Age inhumation burials at Yarnton, Oxfordshire. *Antiquity* 73: 551–62.

Hingley, R. 1984. Towards social analysis in archaeology: Celtic society in the Iron Age of the upper Thames Valley, in B. Cunliffe and D. Miles (eds.), *Aspects of the Iron Age in Central Southern Britain*, pp. 52–71. Oxford University Committee for Archaeology Monograph 2. Oxford: Oxford University Committee for Archaeology.

Hirsch, E., and O'Hanlon, M. (eds.). 1995. *The Anthropology of Landscape: Perspectives on Place and Space*. Oxford: Clarendon Press.

Hodder, I. (ed.). 1978a. *The Spatial Organization of Culture*. London: Duckworth.

———. 1978b. Simple correlations between material culture and society: A Review, in I. Hodder (ed.), *The Spatial Organization of Culture*, pp. 1–24. London: Duckworth.

———. 1981. Society, economy and culture: An ethnographic case study among the Lozi, in I. Hodder, G. Isaac, and N. Hammond (eds.), *Pattern of the past: Studies in honour of David Clarke*, pp. 67–96. Cambridge. Cambridge University Press.

———. 1982. *Symbols in Action: Ethnoarchaeological Studies of Material Culture*. Cambridge: Cambridge University Press.

Hodder, I. (ed.). 2000. *Towards Reflexive Method in Archaeology: The Example at Çatalhöyük*. British Institute of Archaeology at Ankara Monograph 28. Oxford: Oxbow Books/McDonald Institute Monographs.

Hodder, I., and Orton, C. 1976. *Spatial Analysis in Archaeology*. Cambridge: Cambridge University Press.

Högberg, A. 2006. The EU: In need of a supranational view of cultural heritage, in T. Meier (ed.), *Landscape Ideologies*, pp. 199–208. Archaeolingua Series Minor 22. Budapest: Archaeolingua.

Hoskins, W. G. 1955. *The Making of the English Landscape*. London: Hodder and Stoughton.

Howard, P. 1991. *Landscapes: The Artists' Vision*. London: Routledge.

Hugh-Jones, C. 1979. *From the Milk River: spatial and temporal processes in northwest Amazonia*. Cambridge: Cambridge University Press.

Ingold, T. 1993. The temporality of the landscape. *World Archaeology* 25: 152–74.

Jackson, J. B. 1984. *Discovering the vernacular landscape*. New Haven: Yale University Press.

Johnson, N. 1985. Archaeological field survey: A Cornish perspective, in S. Macready and F. H. Thompson (eds.), *Archaeology and Field Survey in Britain and Abroad*, pp. 51–66. Society of Antiquaries of London Occasional Paper (ns) VI. London: Society of Antiquaries.

Johnson, N., and Rose, P. 1994. *Bodmin Moor: An Archaeological Survey. Volume 1: The Human Landscape to c. 1800*. Truro and London: English Heritage, RCHME and Cornwall Archaeological Unit.

Johnston, R. 1998a. The paradox of landscape. *European Journal of Archaeology* 1: 313–25.

———. 1998b. Approaches to the perception of landscape. *Archaeological Dialogues* 5: 54–68.

———. 1999. An empty path? Processions, memories and the Dorset Cursus, in A. Barclay and J. Harding (eds.), *Pathways and Ceremonies: The Cursus Monuments of Britain and Ireland*, pp. 39–48. Neolithic Studies Group Seminar Papers 4. Oxford: Oxbow Books.

Jones, C., and Slinn, P. 2006. *Cultural Heritage and Environmental Impact Assessment in the Planarch Area of Northwest Europe*. Maidstone: Kent County Council and the Planarch Partnership.

Jones, M. U. 1973. An ancient landscape palimpsest at Mucking. *Essex Archaeology and History* 5 (3rd Series): 6–12.

Jones, M. U., Evison, V. I., and Myres, J. N. L. 1968. Crop-mark sites at Mucking, Essex. *Antiquaries Journal* 48: 210–30.

Jones, M. U., and Jones, W. T. 1974. An early Saxon landscape at Mucking, Essex, in T. Rowley (ed.), *Anglo-Saxon Settlement and Landscape*, pp. 20–35. BAR British Series 6. Oxford: British Archaeological Reports.

Juel, H. 1997. Calendar. *Man and Nature Newsletter* 6: 2–4.

Kuna, M. 1991. The structuring of prehistoric landscape. *Antiquity* 65: 332–47.

Landscape Europe (LE). 2006. Landscape Europe Home Page, www.landscape-europe.net (accessed September 18, 2006).

Lemaire, T. 1997. Archaeology between the invention and the destruction of the landscape. *Archaeological Dialogues* 4: 5–21.

Limbrey, S., and Evans, J. (eds.). 1978. *The Effect of Man on the Landscape: The Lowland Zone*. CBA Research Report 21. London: Council for British Archaeology.

Lozny, L. R. (ed.). 2005. *Landscape under Pressure: Theory and Practice of Cultural Heritage Research and Preservation*. New York: Springer.

Lynch, F. M. 1975. The impact of landscape on prehistoric man, in J. Evans, S. Limbrey, and H. Cleere (eds.), *The Effect of Man on the Landscape: The Highland Zone*, pp. 124–26. CBA Research Report 11. London: Council for British Archaeology.

Machat, C. (ed.). 1993. *Historische Kulturlandschaften*. München: ICOMOS Deutschland.

MacKinder, H. J. 1907. *Britain and the British seas* (2nd ed.). Oxford: Clarendon Press.

Macinnes, L., and Wickham-Jones, C. R. (eds.). 1992. *All Natural Things. Archaeology and the Green Debate*. Oxbow Monograph 21. Oxford: Oxbow Books.

Meier, T. 2006. On landscape ideologies: An introduction, in T. Meier (ed.), *Landscape Ideologies*, pp. 11–50. Archaeolingua Series Minor 22. Budapest: Archaeolingua.

Méndez, M. G. 1997. Landscape archaeology as a narrative for designating archaeological parks, in F. Criado and C. Parcero (eds.), *Landscape, Archaeology, Heritage*, pp. 47–51. Trabojos en Arqueología del Paisaje 2. Santiago de Compostela: Grupo de Investigación en Arqueología del Paisaje.

———. 2000. *La revalorización del patrimonio arqueológico* (Arqueoloxia Investigación 8). Santiago de Compostela: Xunta de Galicia.

Miles, D. 1983. An integrated approach to the study of ancient landscapes: The Claydon Pike Project, in G. S. Maxwell (ed.), *The Impact of Aerial Reconnaissance on Archaeology*, pp. 74–84. CBA Research Report 49. London: Council for British Archaeology.

Murray, M. L. 2006. Place names and folk landscapes in southern Germany as archaeological resources, in T. Meier (ed.), *Landscape Ideologies*, pp. 155–73. Archaeolingua Series Minor 22. Budapest: Archaeolingua.

Neustupný, E. 1991. Community areas of prehistoric farmers in Bohemia. *Antiquity* 65: 326–31.

Newman, C. 1997. *Tara: An archaeological survey* (Discovery Programme Monographs 2). Dublin: Royal Irish Academy.

Norton, W. 1989. *Explorations in the Understanding of Landscape*. New York: Greenwood Press.

Nuttgens, P. 1972. *The Landscape of Ideas*. London: Faber and Faber.

O'Sullivan, J. (ed.). 2003. *Archaeology and the National Roads Authority*. Dublin: National Roads Authority Monograph Series 1.

O'Sullivan, J., and Stanley, M. (eds.). 2005. *Recent Archaeological Discoveries on National Road Schemes 2004*. Dublin: National Roads Authority Monograph Series 2.

———. 2006. *Settlement, Industry and Ritual*. Dublin: National Roads Authority Monograph Series 3.

Ostoja-Zagórski, J. 1993. Changing paradigms in the study of the prehistoric economy: An example from east-Central Europe, in P. Bugucki (ed.), *Case Studies in European Prehistory*, pp. 207–28. Boca Raton: CRC Press.

Palmer, R. 1984. *Danebury: An Iron Age Hillfort in Hampshire: An Aerial Photographic Interpretation of Its Environs*. Royal Commission on Historical Monuments (England) Supplementary Series 6. London: RCHM(E).

Patterson, H. (ed.). 2004. *Bridging the Tiber: Approaches to Regional Archaeology in the Middle Tiber Valley*. Archaeological Monograph of the British School at Rome 13. London: British School at Rome.

Pattison, P. (ed.). 1998. *There by Design: Field Archaeology in Parks and Gardens*. BAR British Series 267. Oxford: Archaeopress.

Pollard, J., and Reynolds, A. 2002. *Avebury: The Biography of a Landscape*. Stroud: Tempus.

Pryor, F., and French, C. 1985. The Fenland Project No. 1: Archaeology and environment in the Lower Welland Valley. *East Anglian Archaeology* 27: 1–339 (2 volumes).

Rączlowski, W. 2001. Post-processual landscape: The lost world of aerial photography, in T. Darvill and M. Gojda (eds.), *One Land, Many Landscapes*, pp. 3–8. BAR International Series 987. Oxford: Archaeopress.

Rapportage Archeologische Monumentenzorg (RAM). 2000–2002. *Archaeologie in de Betuweroute. Volumes 1–11*. Rapportage Archeologische Monumentenzorg 80–90. Amersfoort: Rijksdient voor het Oudheidkundig Bodemonderzoek.

———. 1979. *Stonehenge and Its Environs*. Edinburgh: Edinburgh University Press.

Renfrew, C. 1973. Monuments, mobilisation and social organization in Neolithic Wessex, in C. Renfrew (ed.), *The Explanation of Culture Change: Models in Prehistory*, pp. 539–58. London: Duckworth.

———. 1977. Space, time and polity, in J. Friedman and M. J. Rowlands (eds.), *The Evolution of Social Systems*, pp. 89–114. London: Duckworth.

Renfrew, C., and Wagstaff, M. (eds.). 1982. *An Island Polity: The Archaeology of Exploitation in Melos*. Cambridge: Cambridge University Press.

Richards, C. 1996. Monuments as landscape: Creating the centre of the world in late Neolithic Orkney. *World Archaeology* 28: 190–208.

Rippon, S. 2004. *Historic Landscape Analysis: Deciphering the Countryside*. York: Council for British Archaeology.

Royal Commission on the Ancient and Historical Monuments of Scotland (RCAHMS). 1990. *North-East Perth: An Archaeological Landscape*. Edinburgh: HMSO.

———. 1994. *South-East Perth: An Archaeological Landscape*. Edinburgh: HMSO.

Royal Commission on Historical Monuments (England) (RCHM). 1960. *A Matter of Time: An Archaeological Survey of the River Gravels of England Prepared by the Royal Commission on Historical Monuments (England)*. London: HMSO.

———. 1969. *Peterborough New Town: A Survey of the Antiquities in the Areas of Development*. London: HMSO.

Roymans, N. 1995. The cultural biography of urnfields and the long-term history of a mythical landscape. *Archaeological Dialogues* 2: 2–24.

Sauer, C. O. 1925. The morphology of landscape, in J. Leighley (ed. 1969), *Land and Life: A Selection from the Writings of Carl Ortwin Sauer*, pp. 315–50.

Berkeley and Los Angeles: University of California Press.

Sherratt, A. 1996. Settlement patterns or landscape studies? Reconciling reason and romance. *Archaeological Dialogues* 3: 140–59.

Shennan, S. 1985. *Experiments in the Collection and Analysis of Archaeological Survey Data: The East Hampshire Survey*. Sheffield: Department of Archaeology and Prehistory.

Smith, C. 1979. *Fisherwick: The Reconstruction of an Iron Age Landscape*. BAR British Series 61. Oxford: British Archaeological Reports.

Smyntyna, O. V. 2006. Landscape in prehistoric archaeology: Comparing Western and eastern Paradigms, in T. Meier (ed.), *Landscape Ideologies*, pp. 81–96. Archaeolingua Series Minor 22. Budapest: Archaeolingua.

Spoerry, P. (ed.). 1992. *Geoprospection in the Archaeological Landscape*. Oxbow Monograph 18. Oxford: Oxbow Books.

Steane, J. M., and Dix, B. F. 1978. *Peopling Past Landscapes: A Handbook Introducing Archaeological Fieldwork Techniques in Rural Areas*. London: Council for British Archaeology.

Stoertz, C. 1997. *Ancient Landscapes of the Yorkshire Wolds*. London: Royal Commission on the Historical Monuments of England.

Stukeley, W. 1740. *Stonehenge: A Temple Restored to the British Druids*. London: Innys and Manby.

Suhr, G. 2006. Settlement-, environmental-, and landscape archaeology in eastern central Europe between Anglo-American influence and Communist ideology, in T. Meier (ed.), *Landscape Ideologies*, pp. 97–114. Archaeolingua Series Minor 22. Budapest: Archaeolingua.

Sumner, H. 1913. *The Ancient Earthworks of Cranborne Chase*. London: Chiswick Press.

———. 1917. *The Ancient Earthworks of the New Forest*. London: Chiswick Press.

Swain, H. (ed.). 1993. *Rescuing the Historic Environment: Archaeology, The Green Movement and Conservation Strategies for the British Landscape*. Hertford: RESCUE.

Swanwick, C. (ed.). 1982. *Conserving Historic Landscapes*. Castleton: Peak National Park Study Centre.

Taylor, C. C. 1967. Whiteparish: A study of the development of a forest-edge parish. *Wiltshire Archaeological and Natural History Magazine* 62: 79–102.

———. 1974a. Total archaeology, in A. Rogers and T. Rowley (eds.), *Landscapes and Documents*, pp. 15–26. Bury St. Edmunds: Standing Conference on Local History.

———. 1974b. *Fieldwork in Medieval Archaeology*. London: Batsford.

Terrenato, N. 1996. Field survey methods in central Italy (Etruria and Umbria). *Archaeological Dialogues* 3: 216–30.

Thomas, J. 1993. The politics of vision and the archaeologies of landscape, in B. Bender (ed.), *Landscape: Politics and Perspectives*, pp. 19–48. Oxford: Berg.

Tilley, C. 1994. *A Phenomenology of Landscape*. Oxford: Berg.

———. 1996a. The power of rocks: topography and monument construction on Bodmin Moor. *World Archaeology* 28: 161–76.

———. 1996b. *An Ethnography of the Neolithic: Early Prehistoric Societies in Southern Scandinavia*. Cambridge: Cambridge University Press.

———. 2004. *The Materiality of Stone: Explorations in Landscape Phenomenology*. London: Berg.

Tuan, Y.-F. 1977. *Space and Place: The Perspectives of Experience*. Minneapolis: University of Minnesota Press.

Tzanidaki, J. D. 2000. Rome, Maastricht and Amsterdam: The common European heritage. *Archaeological Dialogues* 7: 20–33.

Ucko, P. J., and Layton, R. (eds.). 1999. *The Archaeology and Anthropology of Landscape*. London: Routledge.

United Nations Educational, Scientific and Cultural Organization. 1972. *Convention Concerning the Protection of the World Cultural and Natural Heritage, 16th November 1972*. Paris: UNESCO.

———. 2003. *Convention for the Safeguarding of the Intangible Cultural Heritage, 17th October 2003*. Paris: UNESCO.

van Leusen, M. 1999. The Viroconium Cornoviorum Atlas: High resolution, high precision non-invasive mapping of a Roman civitas capital in Britain. *European Journal of Archaeology* 2: 393–405.

van Leusen, M., and Kamermans, H. 2005. *Predictive Modelling for Archaeological Heritage Management: A Research Agenda*. Nederlandse Archeologische Rapporten 29. Amersfoort: Rijksdienst voor het Oudheidkundig Bodemonderzoek.

Vita-Finzi, C. 1978. *Archaeological Sites in Their Setting*. London: Thames and Hudson.

Wager, J. F. 1981. *Conservation of Historic Landscapes in the Peak District National Park*. Bakewell: Peak Park Joint Planning Board.

Wagstaff, J. M. (ed.). 1987. *Landscape and Culture: Geographical and Archaeological Perspectives*. Oxford: Blackwell Publishing.

Waugh, K. 2006. *Archaeological Management Strategies in the Planarch Area of Northwest Europe*. Maidstone: Kent County Council and the Planarch Partnership.

Wheatley, P. 1971. *The Pivot of the Four Quarters: A Preliminary Enquiry into the Origins and Character of the Ancient Chinese City*. Chicago: Aldine.

Whimster, R. 1989. *The Emerging Past: Air Photography and the Buried Landscape*. London: Royal Commission on the Historical Monuments of England.

Wilkinson, T. J., and Murphy, P. L. 1995. The archaeology of the Essex coast, volume I: The Hullbridge Survey. *East Anglian Archaeology* 71: 1–23.

Willems, W. 1998. Archaeological heritage management in Europe. *European Journal of Archaeology* 1: 293–312.

Willems, W. (ed.). 2000. *Challenges for European Archaeology*. Zoetermeer/The Hague: European Archaeologiae Consilium.

Whittlesey, S. M. 1997. Archaeological landscapes: a methodological and theoretical discussion, in S. M. Whittlesey, R. Ciolek-Torrello, and J. H. Altschul (eds.), *Vanishing River—Landscapes and Lives of the Lower Verde Valley: The Lower Verde Archaeological Project Overview, Synthesis and Conclusions*, pp. 17–28. Tucson: Statistical Research Inc. Press.

Whittlesey, S. M., Ciolek-Torrello, R., and Altschul, J. H. (eds.). 1997. *Vanishing River–Landscapes and Lives of the Lower Verde Valley: The Lower Verde Archaeological Project Overview, Synthesis and Conclusions*. Tucson, AZ: Statistical Research Inc. Press.

Yener, K. A. 2005. *The Amuq Valley Regional Projects, volume 1: Surveys in the Plain of Antioch and Orontes Delta, Turkey, 1995–2002*. Chicago: Oriental Institute of the University of Chicago.

Zedeño, M. N., Austin, D., and Stoffle, R. 1997. *Landmark and Landscape: A Contextual Approach to the Management of American Indian Resources*. Tucson: Bureau of Applied Research in Anthropology, University of Arizona.

5

THE HISTORY OF LANDSCAPE ARCHAEOLOGY IN THE AMERICAS

Thomas C. Patterson

Archaeologists trained or working in the Americas have long been interested in the relations of ancient societies to the landscapes they created and inhabited. These spaces—their natural features, settlements, ruins, and resources—are infused with meaning. Archaeologists have conceptualized and sought to understand the significance of landscapes in various ways. As Knapp and Ashmore (1999: 1) have noted, what has changed significantly in the last decade is how they think about the place of landscapes in the theory and practice of archaeology.

Most Americanist archaeologists are familiar with the writings of geographers on landscapes, broadly defined. There are four reasons for this. The first results from the interconnections of archaeology, anthropology, and geography in 19th-century Germany, echoes of which still reverberate today (e.g., Marchand 1996; Ryding 1975; Smith 1991). The second is the historic linkage of archaeology and anthropology in university curricula throughout the Western Hemisphere. The third is the important role that anthropology, archaeology, and geography played in the development of area studies programs from the late 1930s onward (Steward 1950; Wallerstein et al. 1996: 36–48). The fourth reflects the sequential publication of three widely cited compendia, each with significant articles on the cultural, physical, and historical geography of the hemisphere:

the *Handbook of South American Indians* (Steward 1944–1959); the *Handbook of Middle American Indians* (Wauchope 1964–1975); and the *Handbook of North American Indians* (Sturtevant 1978-onward). Furthermore, many Americanists were also acquainted to varying degrees and in different ways with the writings of particular geographers such as Carl Sauer (1925/1963), William Denevan (1966, 1992), and Karl Butzer (1982), to name only three.

Americanist Conceptions of Archaeological Landscapes

In the last 70 years, Americanist archaeologists have conceptualized landscape in at least seven different ways that are not necessarily mutually exclusive. These perspectives see them as (1) ecological habitats; (2) settlement patterns; (3) subsistence-settlement systems; (4) encompassing both the terrestrial and celestial spheres; (5) materializations of worldview; (6) built or marked environments; and (7) stages for performance.

Ecological Habitats

This perspective views landscapes as the ecological theater in which the drama of everyday life has been performed and involves a cultural or human ecology that focuses on the interaction of communities and their physical environment (Sanders 1962, 1963; Wedel 1953). In some studies, archaeologists saw environmental landscapes as playing roles that shaped, limited, or determined the ways in which the drama of everyday life was or even could be performed (Meggers 1954); here, the actors either adapted to the environmental setting in which they lived or transformed it in some manner, usually through the development or adoption of agriculture (Flannery 1968; MacNeish 1971; Moseley 1975; Murra 1972, 1985; Napton 1969; Steward 1930).

Other studies sought to determine the structure and conditions of the environmental setting in which the everyday life of a particular ancient community was enacted (Johnson 1942). Some viewed these settings as dynamic ones that changed through time; others assumed, often implicitly, that distributional features of the modern landscape were more or less representative of past distributions and that landforms themselves had not been affected in any significant way by erosion or postdepositional burial; more than a few studies combined these potentially incommensurate viewpoints, while others have critiqued that stance as oversight (Joyce and Mueller 1997; Lanning 1963; MacNeish et al. 1983). In recent years, archaeologists are beginning to point out how dramatically ancient peoples transformed their environmental settings and built the landscapes in which they lived (Erickson 2000).

Settlement Patterns

Here, landscapes are viewed in terms of settlement patterns, which Gordon Willey (1953: 1), in the foundational work of this viewpoint, defined as

> the way in which man disposed himself over the landscape in which he lived. It refers to the dwelling, to their arrangement, and to the nature and disposition of other buildings pertaining to community life. These settlements reflect the natural environment, the level of technology on which the builders operated, and various institutions of social interaction and control which the culture maintained.

Regional settlement pattern studies in the Americas multiplied steadily from the late 1950s onward after Willey's initial formulation of the perspective (Ashmore 1981; Lekson 1991; Parsons 1972; Sanders, Parsons, and Santley 1979; Smith 1978; Willey 1956). Willey's definition of settlement patterns was sufficiently broad that it gave

archaeologists the opportunity to pursue studies concerned with both the relationships of people to their ecological settings and their social relationships with one another. It was also not wed to a particular social-theoretical standpoint (although Willey had one), but rather to the collection and analysis of empirical evidence. New questions, methodological innovations, and clarifications of theoretical standpoint and of the interrelationships of theory and data followed in its wake.

These post-1950s approaches to regional settlement strategies reflect archaeology's appropriation of logical positivism and neoclassical economic models, on the one hand, and reactions to them, on the other. Notable areas of debate were concerned with the interconnections of settlement systems and subsistence activities (Flannery 1976; Struever 1971), settlement patterns as expressions of social inequality (Crumley 1976, 1979; Paynter 1982), and the ongoing interrelations of groups residing in different regions (Gledhill 1978; Mathien and McGuire 1986; MacNeish, Patterson, and Browman 1975; Schortman and Urban 1992).

Subsistence-Settlement Systems

The third perspective, which is concerned with the relationships between subsistence activities of a past community and their environment, does not see landscape exclusively in terms of settlement (Thomas 1975). This view was elaborated from the mid-1970s onward and took traditional notions of an archaeological site as problematic. The problem of archaeological practice that arose when traditional definitions were implemented was that

> most sites in a traditional sense represent domestic or activity loci from which the exploitation of the surrounding environment took place. Using *site* to structure recovery limits data collection to a small fraction of the total area occupied by any past cultural system and systematically excludes nearly all direct evidence of the actual articulation between people and their environment. As a result, we are forced to puzzle out the connections from the grossly incomplete, complex, multifunctional deposits called *sites*. (Dunnell and Dancey 1983: 271–72)

Important consequences of this perspective were a shift of attention away from the settlements themselves to the larger environmental settings in which they were situated and a refocusing of attention on "the distribution of archaeological artifacts and features relative to elements of the landscape (and not merely the spatial relation-ships among artifacts and features)" (Rossignol and Wandsnider 1992: viii). Many advocates of this approach adopted "base-superstructure" models of society and tended to focus their attention on explanations of the interactions between subsis-tence economies and resources provided by land-scapes, all of which were changing through time. However, there seems to be no good reason for limiting investigations of landscapes viewed from this perspective to activities associated with the economic base of the society involved. It should be possible to use this perspective on landscape to elucidate, for example, gender relations, divisions of labor, performance, disciplinary practices, his-torical ecology, regional dynamics, or the impor-tance of history (Crumley and Marquardt 1987; Epperson 1990; Handsman and Lamb Richmond 1995; Patterson 1985, 1991).

Integrating the Terrestrial and Celestial Spheres

Archaeoastronomers extended the notion of land-scape beyond terrestrial, aquatic, and underground environments to include the skies, as well as the cyclical and long-term changes that occur in the heavens. Although there are clear connections with scattered, earlier writings on the astrono-mies of pre-Columbian societies (Bowditch, 1910; Nuttall 1906; Posnansky 1942), the perspective gained new levels of credibility in the late 1970s and early 1980s as the significance of the cosmos was understood in new ways. This was largely a result of astronomer Anthony Aveni's collaborative projects with archaeologists and anthropologists (Aveni 1975, 1977, 2001; Aveni and Urton 1982). Since that time, it has fueled a steadily grow-ing number of important studies throughout the Americas (Aveni 2003).

Extending landscape to the celestial sphere and to periodic astronomical events—such as the rising or the setting of the sun or the moon, as well as various planets and stars on the horizon, or the visibility of particular constellations (which are cultural constructions)—afford new insights into the timing and significance of practices (of daily, annual, or longer cyclical duration) and their meaning in terms of particular belief systems. In some instances, ancient societies oriented struc-tures—like the medicine wheels sites on the Great Plains, the "woodhenges" associated with Cahokia near St. Louis, or the diversity of astroarchaeologi-cal features in Chaco Canyon—to mark astronomi-cal events (Eddy 1977; Kehoe and Kehoe 1977; Liebmann 2002; Sofaer 1997; Wittry 1977). Such studies are closely related to other investigations that were concerned primarily with the calendric organization of work and with cosmology (Coggins 1980; Freidel et al. 1993; Sherbondy 1977, 1986; Urton 1982; Zuidema 1982).

Materializations of Worldview

This perspective, which crystallized in the late 1980s, is concerned with the ways in which world-view, cosmology, and history are materialized and expressed in the plans of buildings, civic centers, and settlements as well as in archaeological land-scapes themselves (Ashmore 1986, 1989, 1991; Ashmore and Sabloff 2002; Brady and Ashmore 1999; Coggins 1967; Lekson 1999; Snead and Preucel 1999; Sugiyama 1993). In this perspective, civic plans and landscapes are seen as complex spatial manifestations of culturally and historically contingent views about the cosmological order. Here, features of the natural environment, as well as buildings, gain significance as the peoples who inhabited those places continually incorporated them into their ideational landscapes, assigned meaning to them, and transformed them in the process. Thus, buildings, civic centers, and even the landscapes themselves are seen as "works in progress" that were continually under construction and, sometimes never completed.

The range of studies making use of this con-ceptualization of landscapes is quite broad. This breadth encompasses studies of the cultural meanings that people assigned to features of their landscapes—such as fields, forests, moun-tains, caves, springs, or the abundance of spiders during particular times of the year (Salazar-Burger and Burger 1983; Schele and Freidel 1990; Taube 2003); investigations of the continuities and chang-es in the spatial order of civic plans as revealed by their superposition (Ashmore and Sabloff 2002); examinations of the roles ancestors and cemeteries played in the construction of community (Buikstra and Charles 1999); and considerations of resis-tance as identities were constructed, reproduced, renewed, and changed over periods of time of greater or lesser duration (Preucel 2002).

Built or Marked Environments

The recognition of landscapes as marked or built environments that developed in the 1980s had roots in earlier research. For example, studies of rock art in the Americas originate in 19th-century commentaries (Bostwick 2001; Bray 2002; Greer 2001; Schobinger and Strecker 2001; Turpin 2001; Whitley 2001), but development of cultural resource management archaeology in the 1970s and the passage of the Native American Graves Protection and Repatriation Act of 1990 (NAGPRA) fueled growing interest in rock art in the United States and also helped to underwrite new relationships between archaeologists and Indian peoples. The interpretation of rock art has involved diverse theoretical standpoints ranging from empiricism and structuralism, on the one hand, to symbolic and neuropsychological approaches, on the other. American rock art involves an array of images and scenes, many of which commemorated significant events for the artists and peoples who made them—for example, representations in the American Southwest of the supernova that occurred in A.D. 1054 (Brandt and Williamson 1977).

If rock art is viewed as dotting a landscape, then geoglyphs, such as the famous Nazca Lines (ca. 200 B.C.–A.D. 500) of coastal Peru, mark the landscape even more dramatically. These geoglyphs, some of which are nearly 300 meters in length, structure the interfluvial environment between the Nazca and Ingenio Valleys (Aveni 2000; Silverman and Proulx 2002: 163–92); they have been viewed in terms of a series of potentially complementary interpretations—most notably ceremonial roads, arenas for performance, and physical and cognitive maps of subsurface water (Johnson 1999; Johnson et al. 2002).

If rock art marks one end of an imaginary spectrum and the Nazca Lines some middle point, then the heavily modified environments of Amazonian lowlands or the Titicaca Basin in southern Peru and northern Bolivia (ca. 500 B.C. onward) represent the other extreme. For more than 50 years, scholars have noted the existence of almost continuous deposits of "man-made" soils along many rivers in the Amazon basin (Lathrap 1968; Sauer 1950, 1952). Clark Erickson (2000) has suggested that the amount of labor involved in the construction of agricultural terraces, field walls, canals, reservoirs, roads, raised fields, irrigated pastures, and sunken gardens in the Lake Titicaca Basin far exceeds the time and effort that were required to build all the ceremonial centers in the region.

Landscapes as Stages for Performance

The final perspective comes full circle and returns to the "theater-play" metaphor, although the allusions and meanings of the metaphor have changed dramatically. Here landscape is seen as a space of public performance, the transcendence of the ordinary, communication, and the cultural reproduction of social relations, replete with spectacle, theatricality,

ritual, impersonation, movement, and meaning, on the one hand, and sights, sounds, smells, textures, light intensities, temperatures, humidities, and so forth, on the other. This view began to emerge in the late 1990s, although it also has roots in early works (Burger and Salazar-Burger 1998; Carrasco 1991; Inomata and Coben 2006; Moore 2004; Stone 1992).

These spaces of performance are diverse. Two examples provide a glimpse into the range of landscapes that have been conceptualized in this manner. For instance, each year the peoples of Huarochirí in Peru participated in a series of ceremonies that took place at shrines located from the humid, overcast coastal plain and sunny, dry mid-valleys to the windswept alpine grasslands of the Andes Mountains in central Peru; these were highly charged performances that reaffirmed identity, political relations, and tensions with both kin and neighbors (Spalding 1984). A second example consists of the theaters of power that were replicated at provincial Inca capitals scattered throughout the imperial state that stretched from northern Ecuador to central Chile and northwestern Argentina (Coben 2006). The writers who are currently developing and refining this perspective draw inspiration from various theoretical standpoints: performance theory, Peircean semiotics, phenomenology, and Marxism, to name only a few.

Conclusions

This survey of landscape archaeology in the Americas indicates that it has developed in diverse directions over the past 70 years. It also suggests that the proliferation of perspectives has been especially dramatic since the mid to late 1970s. This is undoubtedly correlated with the elaboration of both internal and external critiques of processual archaeology such as behavioral archaeology, on the one hand, and various postpositivist standpoints, on the other (Preucel 1995). These critiques provided the theoretical, epistemological, and conceptual space that was required to think about landscapes in new ways, to move away from perspectives that saw landscapes largely or exclusively in terms of ecology or settlements. This shift was most dramatic in the United States where processual archaeology has been an important standpoint, if not the hegemonic one, since the late 1950s, and less dramatic in Latin American countries, where processualism was at best only one of a series of competing viewpoints and was never a dominant one (Politis 2003).

Acknowledgments

I want to thank Bruno David, Julian Thomas, and Mitch Allen for their invitation to contribute to this volume. I also want to thank Wendy Ashmore for her thoughtful, constructive commentary and critique while I was writing the chapter; any errors of fact or lack of clarity, of course, are mine and result from not heeding her sound advice.

References

Ashmore, W. (ed.). 1981. *Lowland Maya Settlement Patterns*. Albuquerque: University of New Mexico Press.

———. 1986. Peten cosmology in the Maya southeast: An analysis of architecture and settlement patterns at Classic Quiriguá, in P. A. Urbaan and E. M. Schortman (eds.), *The Southeast Maya Periphery*, pp. 35—49. Austin: University of Texas Press.

———. 1989. Construction and cosmology: Politics and ideology in lowland Maya settlement tatters, in W. F. Hanks and D. S. Rice (eds.), *Word and Image in Maya Culture: Explorations in Language, Writing, and Representation*, pp. 272–86. Salt Lake City: University of Utah Press.

———. 1991. Site-planning principles and concepts of directionality among the ancient Maya. *Latin American Antiquity*, 2: 199–226.

Ashmore, W., and Sabloff, J. A. 2002. Spatial orders in Maya civic plans. *Latin American Antiquity* 13: 201–16.

Aveni, A. F. (ed.). 1975. *Archaeoastronomy in Pre-Columbian America*. Austin: University of Texas Press.

———. 1977. *Native American Astronomy*. Austin: University of Texas Press.

———. 2000. *Between the Lines: The Mystery of the Giant Ground Drawings of Ancient Nasca, Peru*. Austin: University of Texas Press.

———. 2001. *Skywatchers*. Austin: University of Texas Press.

———. 2003. Archaeoastronomy in the Ancient Americas. *Journal of Archaeological Research* 11: 149–91.

Aveni, A. F., and Urton, G. (eds.). 1982. *Ethnoastronomy and Archaeoastronomy in the American Tropics*. Annals of the New York Academy of Sciences, vol. 385. New York.

Bostwick, T. W. 2001. North American Indian agriculturalists, in D. S. Whitley (ed.), *Handbook of Rock Art Research*, pp. 414–58. Walnut Creek, CA: AltaMira Press.

Bowditch, C. P. 1910. *The Numeration, Calendar Systems and Astronomical Knowledge of the Mayas*. Cambridge: Cambridge University Press.

Brady, J. E., and Ashmore, W. 1999. Landscapes of the ancient Maya, in W. Ashmore and A. B. Knapp (eds.), *Archaeologies of Landscape: Contemporary Perspectives*, pp. 124–45. Oxford: Blackwell Publishing.

Brandt, J. C., and Williamson, R. A. 1977. Rock art representations of the A.D. 1054 supernova: A progress report, in A. F. Aveni, *Native American Astronomy*, pp. 171–78. Austin: University of Texas Press.

Bray, T. L. 2002. Rock art, historical memory, and ethnic boundaries: A study from the northern Andean highlands, in H. Silverman and W. H. Isbell (eds.), *Andean Archaeology II: Art, Landscape, and Society*, pp. 333–54. New York: Kluwer Academic/ Plenum Publishers.

Buikstra, J. E., and Charles, D. K. 1999. Centering the ancestors: Cemeteries, mounds, and sacred landscapes of the ancient North American midcontinent, in W. Ashmore and A. B. Knapp (eds.), *Archaeologies of Landscape: Contemporary Perspectives*, pp. 201–28. Oxford: Blackwell Publishing.

Burger, R. L., and Salazar-Burger, L. 1998. A sacred effigy from Mina Perdida and the unseen ceremonies of the Peruvian Formative. *Res* 33: 28–53.

Butzer, K. W. 1982. *Archaeology as Human Ecology*. Cambridge: Cambridge University Press.

Carrasco, D. (ed.). 1991. *To Change Place: Aztec Ceremonial landscapes*. Boulder: University of Colorado Press.

Coben, L. S. 2006. Other Cuzcos: Replicated theaters of Inka power, in T. Inomata and L. S. Coben (eds.), *Archaeology of Performance: Theaters of Power, Community, and Politics*, pp. 223–59. Walnut Creek, CA: AltaMira Press.

Coggins, C. C. 1967. Palaces and the planning of ceremonial centers in the Maya lowlands. Unpublished manuscript. Cambridge: Tozzer Library, Peabody Museum, Harvard University.

———. 1980. The shape of time: Some political implications of a four-part figure. *American Antiquity* 45: 727–39.

Crumley, C. L. 1976. Toward a locational definition of State systems of settlement. *American Anthropologist* 78: 59–73.

———. 1979. Three locational models: An epistemological assessment for Anthropology and archaeology, in M. B. Schiffer, *Advances in Archaeological Method and Theory* 2, pp. 143–73. New York: Academic Press.

Crumley, C. L., and Marquardt, W. H. (eds.). 1987. *Regional Dynamics: Burgundian Landscapes in Historical Perspective*. New York: Academic Press.

Denevan, W. 1966. *The Aboriginal Cultural Geography of the Llanos de Mojos of Bolivia*. Berkeley: Ibero-Americana 48.

Denevan, W. 1992. The pristine myth: The landscape of the Americas in 1492. *Annuals of the Association of American Geographers* 82: 369–85.

Dunnell, R. C., and Dancey, W. S. 1983. The siteless survey: A regional scale data collection strategy, in M. B. Schiffer (ed.), *Advances in Archaeological Method and Theory* 6, pp. 267–87. New York: Academic Press.

Eddy, J. A. 1977. Medicine wheels and Plains Indian astronomy, in A. Aveni (ed.), *Native American Astronomy*, pp. 147–69. Austin: University of Texas Press.

Erickson, C. L. 2000. The Lake Titicaca Basin: A Precolumbian built landscape, in D. Lentz (ed.), *Imperfect Balance: Landscape transformations in the Precolumbian Americas*, pp. 311–56. New York: Columbia University Press.

Epperson, T. W. 1990. Race and the disciplines of the plantation. *Historical Archaeology* 24(4): 29–36.

Flannery, K. V. 1968. Archaeological systems theory and early Mesoamerica, in B. J. Meggers (ed.), *Anthropological Archaeology in the Americas*, pp. 67–87. Washington, DC: Anthropological Society of Washington.

———. (ed.). 1976. *The Early Mesoamerican Village.* New York: Academic Press.

Freidel, D. A., Schele, L., and Parker, J. 1993. *The Maya Cosmos: Three Thousand Years on the Shaman's Path.* New York: William Morrow.

Gledhill, J. 1978. Formative development in the North American South West, in D. Green, C. Haselgrove, and M. Spriggs (eds.), *Social Organisation and Settlement*, pp. 247–90. Oxford: BAR International Series 47 (Part 2).

Greer, J. 2001. Lowland South America, in D. S. Whitley (ed.), *Handbook of Rock Art Research*, pp. 665–706. Walnut Creek, CA: AltaMira Press.

Handsman, R. G., and Lamb Richmond, T. 1995. Confronting colonialism: The Mahican and Schaghticoke peoples and us, in P. R. Schmidt and T. C. Patterson (eds.), *Making Alternative Histories: The Practice of Archaeology and History in Non-Western Settings*, pp. 87–118. Santa Fe: School of American Research Press.

Inomata, T., and Coben, L. S. 2006. Overture: An invitation to the archaeological theater, in T. Inomata and S. Coben (eds.), *Archaeology of Performance: Theaters of Power, Community, and Politics*, pp. 11–44. Latham: AltaMira Press.

Johnson, D. W. 1999. Die Nasca-Linien als Markierungen für unterirdische Wasservorkommen, in J. Rickenbach (ed.), *Nasca: Geheimnisvolle Zeichen im Alten Peru*, pp. 157–64. Zürich: Museum Reitberg.

Johnson, D. W., Proulx, D. A., and Mabee, S. B. 2002. The correlation between geoglyphs and subterranean water resources in the Río Grande de Nazca drainage, in H. Silverman and W. H. Isbell (eds.), *Andean Archaeology II: Art, Landscape, and Society*, pp. 307–32. New York: Kluwer Academic/Plenum Publishers.

Johnson, F. 1942. *The Boylston Street Fishweir: A Study of the Archaeology, Biology, and Geology of a Site on Boylston Street in the Back Bay District of Boston, Massachusetts.* Andover, MA: Papers of the Robert S. Peabody Foundation for Archaeology, vol. II.

Joyce, A. A., and Mueller, R. G. 1997. Prehispanic human ecology of the Río Verde drainage basin, Mexico. *World Archaeology* 29: 75–94.

Kehoe, T. F., and Kehoe, A. B. 1977. Stones, solstices and Sun Dance structures. *Plains Anthropologist* 22: 85–97.

Knapp, A. B., and Ashmore, W. 1999. Archaeological landscapes: Constructed, conceptualized, ideational, in W. Ashmore and A. B. Knapp, *Archaeologies of Landscape: Contemporary Perspectives*, pp. 1–30. Oxford: Blackwell Publishing.

Lanning, E. P. 1963. A pre-agricultural occupation on the central coast of Peru. *American Antiquity* 28: 360–71.

Lathrap, D. 1968. The "hunting" economies of the tropical forest zone of South America: An attempt at historical perspective, in R. B. Lee and I. DeVore (eds.), *Man the Hunter*, pp. 23–29. Chicago: Aldine Publishing Company.

Liebmann, M. J. 2002. Demystifying the Big Horn medicine wheel: A contextual analysis of symbolism, meaning, and function. *Plains Anthropologist* 47: 46–56.

Lekson, S. H. 1991. Settlement patterns in the Chaco region, in P. L. Crown and W. J. Judge (eds.), *Chaco and Hohokam: Prehistoric Regional Systems in the American Southwest*, pp. 31–55. Santa Fe: School of American Research Press.

———. 1999. *The Chaco Meridian: Centers of Political Power in the Ancient Southwest.* Walnut Creek, CA: AltaMira Press.

MacNeish, R. S. 1971. Speculations about how and why food production and village life developed in the Tehuacan valley, Mexico. *Archaeology* 24: 307–15.

MacNeish, R. S., Patterson, T. C., and Browman, D. L. 1975. *The Central Peruvian Prehistoric Interaction Sphere.* Andover, MA: Papers of the Robert S. Peabody Foundation for Archaeology, vol. VII.

MacNeish, R. S., Vierra, R. K., Nelkin-Turner, A., Luric, R., and García-Cook, A. 1983. *Prehistory of the Ayacucho Basin, Peru IV: The Preceramic Way of Life.* Ann Arbor: University of Michigan Press.

Marchand, S. L. 1996. *Down from Olympus: Archaeology and Philhellenism in Germany, 1750–1970.* Princeton, NJ: Princeton University Press.

Mathien, F. J., and McGuire, R. H. (eds.). 1986. *Ripples in the Chichimeca Sea: New Considerations of Southwestern-Mesoamerican Interactions.* Carbondale: Southern Illinois University Press.

Meggers, B. J. 1954. Environmental limitations on the development of culture. *American Anthropologist* 56: 801–24.

Moore, J. D. 2004. The social basis of sacred spaces in the Prehispanic Andes: Ritual landscapes of the dead in Chimú and Inka societies. *Journal of Archaeological Method and Theory* 11: 83–124.

Moseley, M. E. 1975. *The Maritime Foundations of Andean Civilization.* Menlo Park, CA: Cummings Publishing Company.

Murra, J. V. 1972. El control vertical de un máximo de pisos ecológicos en la economía de las sociedades andinas, in J. V. Murra (ed.), *Iñigo Ortiz de Zúñiga, Visita de la Provincia de Huánuco* 2, pp. 429–76. Huánuco: Universidad Nacional Hermilio Valdizán.

———. 1985. "El Archipiélago Vertical" revisited, in S. Masuda, I. Shimada, and C. Morris (eds.), *Andean Ecology and Civilization: An Interdisciplinary Perspective on Andean Ecological Complementarity,* pp. 15–20. Tokyo: University of Tokyo Press.

Napton, L. 1969. The lacustrine subsistence pattern in the desert west. Berkeley: *Kroeber Anthropological Society Papers, Special Publication* 2, pp. 28–98.

Nuttall, Z. 1906. The astronomical methods of the ancient Mexicans, in B. Laufer (ed.), *Boas Anniversary Volume: Anthropological Papers Written in Honor of Franz Boas,* pp. 290–98. New York: G. E. Stechert and Company.

Parsons, J. R. 1972. Archaeological settlement patterns. *Annual Review of Anthropology* 1: 127–50.

Patterson, T. C. 1985. Pachacamac—An Andean oracle under Inca rule, in M. Thompson, M. T. Garcia, and F. J. Kense (eds.), *Status, Structure, and Stratification,* pp. 13–18. Calgary: Proceedings of the Sixteenth Annual Conference of the Archaeological Association of the University of Calgary.

———. 1991. El desarrollo de la agricultura y el surgimiento de la civilización en los Andes centrales. *Boletín de Arqueología Americana* 4: 7–23.

Paynter, R. 1982. *Models of Spatial Inequality: Settlement Patterns in Historical Archaeology.* New York: Academic Press.

Politis, G. G. 2003. The theoretical landscape and methodological development of archaeology in Latin America. *American Antiquity* 68: 245–72.

Posnansky, A. 1942. Los conocimientos astronómicos de los constructors de Tihuanacu y su applicación en el templo del sol para la determinación exacta de fechas agricolas. *Boletín de la Sociedad Geográfica de La Paz,* año 53, no. 64: 40–49.

Preucel, R. W. 1995. The Postprocessual Condition. *Journal of Archaeological Research* 3: 147–75.

———. (ed.). 2002. *Archaeologies of the Pueblo Revolt: Identity, Meaning, and Renewal in the Pueblo world.* Albuquerque: University of New Mexico Press.

Rossignol, J., and Wandsnider, L. A. 1992. Preface, in J. Rossignol and L. A. Wandsnider (eds.), *Space, Time, and Archaeological Landscapes,* pp. vii–ix. New York: Plenum Press.

Ryding, J. N. 1975. Alternatives in nineteenth-century German ethnology: A case study in the sociology of science. *Sociologus* 25(1): 1–28.

Salazar-Burger, L., and Burger, R. L. 1983. La arána en la iconografía del Horizonte Temprano en la costa norte del Perú. *Beitrage zur Allgemeinen und Vergleichenden Archaeologie,* band 4: 213–53.

Sanders, W. T. 1962. Cultural ecology of the Maya lowlands, Part 1. *Estudios de Cultura Maya* 2: 79–121.

———. 1963. Cultural ecology of the Maya lowlands, Part 2. *Estudios de Cultura Maya* 3: 203–41.

Sanders, W. T., Parsons, J. R., and Santley, R. S. 1979. *The Basin of Mexico: Ecological Processes in the Evolution of a Civilization.* New York: Academic Press.

Sauer, C. O. 1925/1963. The morphology of landscape, in *Land and Life: A Selection from the Writings of Carl Ortwin Sauer,* pp. 315–50. Berkeley and Los Angeles: University of California Press.

———. 1950. Geography of South America, in J. Steward (ed.), *Handbook of South American Indians,* pp. 319–44. Bureau of American Ethnology Bulletin 143(6). Washington, DC: United States Government Printing Office.

———. 1952. *Agricultural Origins and Dispersals.* New York: American Geographical Society.

Schele, L., and Freidel, D. A. 1990. *A Forest of Kings.* New York: William Morrow.

Schobinger, J., and Strecker, M. 2001. Andean South America, in D. S. Whitley (ed.), *Handbook of Rock Art Research,* pp. 707–59. Walnut Creek, CA: AltaMira Press.

Schortman, E. M., and Urban, P. A. (eds.). 1992. *Resources, Power, and Interregional Interaction.* New York: Plenum Publishers.

Sherbondy, J. 1977. Les réseaux d'irrigation dans la géographie politique de Cuzco. *Journal de la Société des Américanistes* LXVI: 45–66.

———. 1986. Los ceques: Código de canales en el Cusco Incáico. *Allpanchis* 18(27): 39–74.

Silverman, H., and Proulx, D. A. 2002. *The Nasca.* Oxford: Blackwell Publishing.

Smith, B. D. (ed.). 1978. *Mississippian Settlement Patterns.* New York: Academic Press.

Smith, W. D. 1991. *Politics and the Sciences of Culture in Germany, 1840–1920.* Oxford: Oxford University Press.

Snead, J. E., and Preucel, R. W. 1999. The ideology of settlement: Ancient Keres landscapes in the northern Rio Grande, in W. Ashmore and A. B. Knapp (eds.), *Archaeologies of Landscape:*

Contemporary Perspectives, pp. 169–99. Oxford: Blackwell Publishing.

Sofaer, A. 1997. The primary architecture of the Chacoan culture: A cosmological expression, in B. H. Morrow and V. B. Price (eds.), *Anaxazi Architecture and American Design*, pp. 88–132. Albuquerque: University of New Mexico Press.

Spalding, K. 1984. *Huarochirí: An Andean society under Inca and Spanish rule*. Stanford: Stanford University Press.

Steward, J. H. 1930. Irrigation without agriculture. Ann Arbor: *Papers for the Michigan Society of Science, Arts, and Letters* 12: 149–56.

———. 1950. *Area Research: Theory and Practice*. New York: Social Science Research Council Bulletin 63.

———. (ed.). 1944–1959. *Handbook of South American Indians*. Bureau of American Ethnology Bulletin 143, 7 vols. Washington, DC: United States Government Printing Office.

Stone, A. 1992. From ritual in the landscape to capture in the urban center: The recreation of ritual environments in Mesoamerica. *Journal of Ritual Studies* 6(1): 109–32.

Struever, S. (ed.). 1971. *Prehistoric Agriculture*. Garden City, NY: Natural History Press.

Sturtevant, W. C. (ed.). 1978-onward. *Handbook of North American Indians*, 15 vols. Washington, DC: Smithsonian Institution Press.

Sugiyama, S. 1993. Worldview materialized in Teotihuacan. *Latin American Antiquity* 4: 103–29.

Taube, K. A. 2003. Ancient and contemporary Maya conceptions about field and forest, in A. Gómez-Pompa, M. F. Allen, S. L. Fedick, and J. J. Jiménez-Osorníó (eds.), The Lowland Maya Area: Three Millennia at the Human-Wildland Interface, pp. 461–94. Binghamton, NY: Haworth Press.

Thomas, D. H. 1975. Nonsite sampling in archaeology: Up the creek without a site? in J. W. Mueller (ed.), *Sampling in Archaeology*, pp. 61–81. Tucson: University of Arizona Press.

Turpin, S. 2001. Archaic North America, in D. S. Whitley (ed.), *Handbook of Rock Art Research*, pp. 361–413. Walnut Creek, CA: AltaMira Press.

Urton, G. 1982. Astronomy and calendrics on the coast of Peru, in A. F. Aveni and G. Urton (eds.), *Ethnoastronomy and Archaeoastronomy in the American Tropics*, pp. 231–48. New York: Annals of the New York Academy of Sciences 385.

Wallerstein, I., Juma, C., Keller, E. F., Kocka, J., Lecourt, D., Mudimbe, V. Y., Mushakoji, K., Prigogine, I., Taylor, P. J., and Trouillot, M.-R. 1996. *Open the Social Sciences: Report of the Gulbenkian Commission on the Restructuring of the Social Sciences*. Stanford: Stanford University Press.

Wauchope, R. W. (ed.). 1964–1975. *Handbook of Middle American Indians*, 16 vols. Austin: University of Texas Press.

Wedel, W. R. 1953. Some aspects of human ecology in the Central Plains. *American Anthropologist* 55: 499–514.

Whitley, D. S. 2001. Rock art and rock art research in worldwide perspective: An introduction, in D. S. Whitley, *Handbook of Rock Art Research*, pp. 7–51. Walnut Creek, CA: AltaMira Press.

Willey, G. R. 1953. *Prehistoric Settlement Patterns in the Virú Valley, Perú*. Bureau of American Ethnology Bulletin 155. Washington, DC: United States Government Printing Office.

———. (ed.). 1956. *Prehistoric Settlement Patterns in the New World*. New York: Viking Fund Publications in Anthropology 23.

Wittry, W. 1977. The American Woodhenge, in M. L. Fowler (ed.), *Explorations in Cahokia Archaeology*, pp. 43–48. Urbana: Illinois Archaeological Survey Bulletin 7.

Zuidema, T. 1982. Catachillay: The role of the Pleiades and of the Southern Cross and α and β Centauri in the calendar of the Incas, in A. F. Aveni and G. Urton (eds.), *Ethnoastronomy and Archaeoastronomy in the American Tropics*, pp. 203–30. New York: Annals of the New York Academy of Sciences 385.

6

THINKING OF LANDSCAPE ARCHAEOLOGY IN AFRICA'S LATER PREHISTORY: ALWAYS SOMETHING NEW

Rod McIntosh

There is always something new out of Africa.
(Pliny the Elder 1949, *Natural History*)

In the spirit of Pliny the Elder's hoary chestnut, African researchers and their expatriate colleagues have tended to dismiss as irrelevant much of the sequential research typologies (in particular the notion that research and understanding progress sequentially from survey and site discovery to environmental interpretation to symbolic concerns) and adaptationist or functional explanations so prevalent in much English language archaeology. Although such approaches have their own logic, one located in local and national philosophical and social agendas, Africa has always taken a different path: it has always been its own master. Many landscape archaeologists here have thus entered into data and interpretative fields that remain largely unexplored elsewhere, not so much because of differences in archaeological records but differences in the way things are understood to operate in the world.

It is all the more unfortunate, then, that the astonishing African archaeological record remains, for the most part, unknown by most researchers beyond the continent. After all, the identification of archaeological sites for purposes of what are often thinly disguised functional or environmental explanations, triumphantly evolving into the symbolic (cf. Ashmore and Knapp 1999: 2), more often speak to abstract theorizing that has little to do with historical circumstances, trends, or social understandings and more to do with contemporary power struggles within the discipline. However, it is worth reiterating that most Africanists believe that the literature coming from Africa is largely neglected internationally, because the African literature deeply challenges various established positions of the systemic schools (including environmental adaptationist models).

Africanists tend to deem such debates as irrelevant to their more insistent need to document what's on the ground before it is effaced by rapid development. But what really is new from Africa is a willingness to conceive of landscapes as layered social and symbolic—and physical—transformations, as holistic, deep-time fields of multiple perceptions, as well as reciprocal exploitations, arenas where peoples co-adapt and co-evolve. It is these more socially inspired forms of landscape archaeology that Africanists usually favor. Unlike most "middle-range-research" thinking, whereby method is tailored for the retrieval of general principles or laws usually concerned with environmental adaptation, in Africa symbolic interests are not seen as handmaiden to evolutionary and adaptationist perspectives but stand as key dimensions of social behavior. Granted, this stance is in common with other recent innovations in archaeological thinking elsewhere, such as Indigenous critiques in Australia (e.g., see David and Thomas, this volume) and a concern for the spiritual dimensions of life (e.g., see also McNiven, this volume), some of these innovations being at least partly inspired by African research.

However, I suggest that two attitudes toward archaeological research have influenced African archaeology to a significant degree. I refer to these as the "Tyranny of Hot Spots" and the "Handmaiden's Tragedy," the former indicating the way that statements about Africa signaled colonialist attitudes about the commentators as much as they reflected on Africa's past, and the latter referring in particular to the attribution of impressive African cultural developments to external origins by early writers. I examine each in detail below.

"Tyranny of Hot Spots"

The idea of an *African* history, as opposed to one based on external sources, barely predated Independence from Western colonial powers, essentially coming into its own as a discipline in the mid-1960s. Well into the 1970s, many archaeologists focused on "hot spots" mentioned in Arab or European sources. I use this term to mean individual points in the landscape so important to these external authors that they merited mention (when all others were ignored), often telling us more about the outsider's interests in, or prejudices about, Africa, rather than about events and processes happening locally (Knapp and Ashmore 1999: 2).

Often archaeologists or other commentators were content simply to locate the archaeological sites referenced by these early writers. For example, after all the research at Koumbi Saleh in Mauritania, we still have only those north-african-like ruins presumed to have been the capital of the Ghana Kingdom. Driving research at Koumbi Saleh was a conviction that these ruins must be the remains of the capital of the Ghana Kingdom (a state so early that it was believed *had* to be derived from elsewhere [Mauny 1961, 1970; Robert and Robert 1972]). Arguably, the most famous site south of the Sahara, Great Zimbabwe, is still better known both in the public imagination and to professionals in isolation, abstracted, as it were, from its immediate hinterland (but see Pikirayi 2001). In addition, the Egyptian Predynastic was until recently conceptualized as a series of discontinuous points, be they necropoli (such as Naqada and Abydos, sources of seriated ceramic cultures) or remains of deeply disturbed settlements (Hierakonpolis or Buto). Our view of the pre-Pharaonic landscape has been revolutionized recently, as we shall see.

"The Handmaiden's Tragedy"

If not focused on such discontinuous archaeological "hot spots" as "Great Zimbabwe" or Koumbi Saleh, archaeologists historically often investigated a tableaux of sites (sometimes with their surrounding resources) deemed important to an historical developmental or evolutionary logic external to the continent. Classically, projects designed to locate named sites and the trade networks linking them were handmaidens to historical explanations that (at least implicitly) reaffirmed the presumed cultural passivity (and "unchanging" ethos) of the continent.

Since West Africa was *obviously* shaken from its Iron Age slumber by mediaeval Arabs who established a line of trade entrepots immediately south of the Sahara, it followed that archaeologists had to look for those caravan colonies. Archaeologists read the names and some details of this trade in the mediaeval Arabic geographies of, for example, al-Bakri or Ibn Battuta, and then went into the Sahara to find the places so mentioned. Were the Mauritanian ruins of Tegdaoust indeed the Awdaghost of the chroniclers al-Bakri and Ibn Battuta's (Couq 1975; Levtzion and Hopkins 1981)? Details of contacts with Indigenous peoples were irrelevant to archaeologists of this ilk, who could scarcely fathom that cultural interactions could go both ways. Similarly, the East African Swahili zone was *obviously* a seaward-looking outlier of the vibrant Indian Ocean interaction sphere for many early Africanist archaeologists. Kilwa and other Swahili cities were seen not as bubbling cauldrons of cultural hybrid vigor (as they are now), but as backwaters (e.g., Chittick 1965, 1975).

This (mis)conception of African regions as developmental backwaters extends even to the Classical African Mediterranean littoral. There is still virtually no interest in, for example, Carthaginian relations with the free or servile Berbers of the hinterland, nor much investigation of the client-peoples' craft production (e.g., iron) or Saharan exchange relations that underpinned Carthage's might.

Always Something Novel

Reaction against the racialism lurking just beneath the "Tyranny of Hot Spots" and the "Handmaiden's Tragedy" may, in fact, be the reason for Africanists' insistence on leapfrogging the second stage in the standard investigative and explanatory sequence beginning with (1) site identification, followed by (2) functional or environmental explanations, and finishing with (3) symbolic aspects of behavior. By directly addressing symbolic aspects within their cultural, historical, and environmental settings, local issues, rather than more abstract and universal notions of environmental adaptation, are explored. So little archaeology has been done in sub-Saharan Africa that *aliquid novi*-[always something new] is relevant here also—Murdock's (1959: 73) observation that one ounce of earth having been lifted on the Niger (or Congo, or Limpopo) for each ton along the Nile rings as true today as it did nearly 50 years ago. Because of this, there has been a general tendency for Africanist archaeologists to explore more than an exclusively "environmental" or even isolated "symbolic" archaeology, as often apparent on other continents, and embrace a more holistic notion of landscape. Generally, African archaeologists today tend not to make the artificial distinction between "environmental" versus "symbolic" behavior, instead investigating each in broader context. Although some research programs may indeed have a "biophysical" or a "political economy" thrust, rarely do even these eschew the ideological or "cognitive" directions now so prevalent in African archaeology (see examples below). Most exciting, many of these radically redefine the ancient world in terms of external relations that came to birth premodern globalizations.

One example is the ongoing Asmara Plateau research, jointly conducted by the Universities of Asmara and Florida (Schmidt and Curtis n.d. [2007]). Intensive archaeological coverage of the highlands west of Eritrea's capital has begun to correct a general neglect of the Horn of Africa. Taking the perspective that this region was not just a passive provider of goods and peoples to more brilliant classical (especially Hellenistic) and "late antique" civilizations across the Red Sea, this survey supports the idea of a social basin of interacting, competing polities of a very innovative stripe across the northern Horn, Arabian Peninsula and northeastern Africa—think multiple polities birthed by, not separated by, the Red Sea. Here, a "political economy" approach (although that term does the novelty of the interpretation a disservice) is adopted by one of the pioneers of a gendered and symbolic view of landscape, Peter Schmidt, who decades ago brought what was at the time a controversial ideological perspective to the Tanzanian iron-producing landscape (Schmidt 1983, 1997). Part of that landscape is now understood to be not an isolated hinterland but a productive interior critical to the emergence of the Swahili world. Once basic survey (and then detailed excavations, as at Shanga and on Pemba Island) addressed the relationship of the aforementioned Swahili coastal cities and the interior, it became clear that Swahili

"civilization" was as much a product of local and regional integration and exchanges as of "radiance" from across the Indian Ocean (Chami 1998; Horton 1996; LaViolette 2004). Now East African archaeologists have joined with Indian colleagues to look at the equally partnered relationships linking landscapes on both sides of the Indian Ocean—pioneering study of premodern globalization stripped of its heretofore (Western) presumptions of asymmetries.

Even African landscape studies taking predominantly an "environmental" focus generally are hardly deterministic or adaptationist in approach. Three classic cases illustrate this more integrated ecological thrust (a fourth, the Middle Niger is discussed below). On the one hand, the UNESCO Libyan Valley Surveys (Barker et al. 1996a, 1996b; Barker and Gilbertson 2000) document in fine ecological detail the failure of Romano-Libyan (and later) farming in the Tripolitanian predesert. However, this was a long and complex story that involved not only technological (successes and) failures but also long-term changes in attitudes toward the land, which contributed to the once lush landscapes within the Roman *limes* becoming today's inhospitable, degraded desert. Similarly, intensive prehistoric and biophysical surveys along the southern margins of the Chad Basin exquisitely detail the interplay of climate change and human occupation along the southern margin of that vast lake (Gronenborn 1998; Krings and Platte 2004). Here, linked ethno-graphic and ethnohistorical studies suggest the long-term local perspectives on change (including abrupt climate change) that underpin the region's complex ethnic relations. Finally, the vast southern African grassland belt, from Namibia through Botswana and the Limpopo Valley of South Africa and Zimbabwe (and into Mozambique), continues to be the focus of a coordinated fine-resolution climatic and settlement investigation (Huffman 1996; Leslie and Maggs 2000; Manyanga, Pikirayi, and Ndoro, 2000; Pikirayi 2001). Particularly along the Limpopo Corridor, an integrated approach to the climate and human exploitation of the local environment and interest in symbolic landscapes has come to contribute significantly to debates about (a) the production anthropogenic or "natural" of the vast southern African grasslands, (b) the appearance of a really extensive polities (antecedent to and including Great Zimbabwe and its southern contemporaries; Huffman 2000), and (c) the origins of ethnohistoric perceptions of climate change (such as, but not limited to, persistent associations of rainmaking rituals and political authority).

The last three examples—the Tripolitanian predesert, the southern Lake Chad Basin, and the Limpopo Corridor—all contribute to the broader "anthropocene" debate (namely, that human land use has had a far longer, far more dramatic effect on climate change than previously appreciated; McIntosh and Tainter 2006; Ruddiman 2003). At the same time, these three examples demonstrate why any profound approach to long-term landscape change must necessarily factor in changing "perceptions" of reasons for changes to the physical environment (an approach generally known as historical ecology; for example, more fully developed in McIntosh, Tainter, and McIntosh 2000).

Conjoining the Symbolic and Biophysical Conceptions of the Landscape

I end with three cases, very different in scale but parallel in their insistence that there should be no separation of the symbolic or the ideological from the functional (political or economic or subsistence), or from the biophysical conception of the landscape, as African researchers have been propounding over the last few years. One of the early and most cohesive studies of multi-ethnic, pluri-specialist (and often refugia) landscapes is that of the Mandara Mountains in northeast Nigeria and Cameroon (David and Sterner 1999; MacEachern 2002; Sterner 1998). Although at one level an analysis of exchange, interaction and production specialization in a "marginal" zone, the researchers have experimented with an explicitly "symbolic reservoir" perspective on land use and land conception. Multiple interacting peoples share differentially in a fluid reservoir of beliefs and symbols (about themselves, one another, and the land) that allows rules of peaceful interaction as well as separation (leading to ethnogenesis: historical genesis of different peoples, with their own self-definition and critical defining of ethnic characteristics) to be invented, maintained, and perpetuated in a state of creative flux.

On a vastly larger scale is the recent reassessment of impulses to complexity, leading eventually to unification (by This/Abydos) as early as 3250 B.C., during the later Predynastic of Upper Egypt. No longer seen as a backwater, Darnell (n.d.[a] and n.d.[b]) and colleagues (see Friedman 2002; Hendrickx and Friedman 2004; Wilkinson 1999) look to trade and cultural connections between the Upper Nile Valley, the oasis landscape of the Western Desert, and lower Nubia (and probably further south; see Grzymski 2004) as an arena of

unitary ethos—expansive, opportunistic, sporting a desert worldview of solar cycles, totemic animals, and deep-time symbols of power. This reservoir of symbols and concepts, etched in the desert landscape, historically was subsequently appropriated and manipulated by the emerging elites of Hierakonopolis, This/Abydos, and Naqada in a drama of political consolidation, alliances, exploitation of mineral and agricultural wealth, and invention of traditions of (eventual) kingship (Wilkinson 1999). What a very different dynamic of antecedents to Pharaonic civilization than that current among Egyptologists even a decade ago.

Finally, since the early 1980s, R. McIntosh (1998, 2005) and S. McIntosh (1999a, 1999b) have taken a similarly broad approach to the landscape archaeology of the vast (170,000 km^2) Middle Niger of Mali. Abrupt climate change, a mosaic of landforms, the complementary productive potential of the Middle Niger's several basins all were certainly important factors encouraging of a *tell*-littered urban landscape rivaling that of Mesopotamia. But, as years of excavation and survey progressed, it became very clear that a materialist, functional explanation for the pluri-ethnic settlement and land-use practices and beliefs (of these specialists with an elaborate template of reciprocity) could be understood only by reference to the rich ethnohistorical literature concerning Mande (local Indigenous) beliefs about the landscape as a network of sociopolitical and moral obligations. For the peoples of the Middle Niger, an occult fabric of causation is spread out over a highly stressful, unpredictable biophysical landscape (McIntosh 2005). To try to understand the ancient landscape while ignoring this conceptual framework of deep-time beliefs about causation of change in the world would be as futile as would be a modern Mande person's attempts to navigate her or his world without the guideposts of always-changing but deeply rooted understandings of *nyama*, the moral power of, among other things, landscape.

Landscapes of Causation

What I hope has become evident in this too-short survey is that, on the one hand, Africist understandably resent any inference that basic, "exploratory" surveys need to go through a stage of environmental-adaptationist explanation to render the archaeological record meaningful. On the other hand, Africanist researchers also resent the notion that landscape archaeologies that focus more on the symbolic are somehow lesser, somehow more shallow than are other forms of archaeology.

What makes African archaeology most significant to Africanists is how the archaeological record can inform us about the *African* past, in all its diversity, and without automatically assuming that all impressive innovation comes from the outside. Without discovering the continent's innovations, Africa's past will forever be considered a shadow of, or even worse, derivative of, the brilliant pasts of other lands. Lacking space, I have not even mentioned the imperfectly known but truly astonishing fields of earthen tumuli and megaliths in Senegal and Gambia (McIntosh and McIntosh 1993) or the world's largest ancient engineering landscape—the urban wall-and-ditch complexes of southwestern Nigeria (Darling 1998); or the exquisitely detailed tracing of the seasonal exploitation of Late Stone Age peoples in the western Cape of South Africa, anchored on changing coastal exploitations chronicled in John Parkington's (1999) excavations at Eland's Bay Cave. However, the African

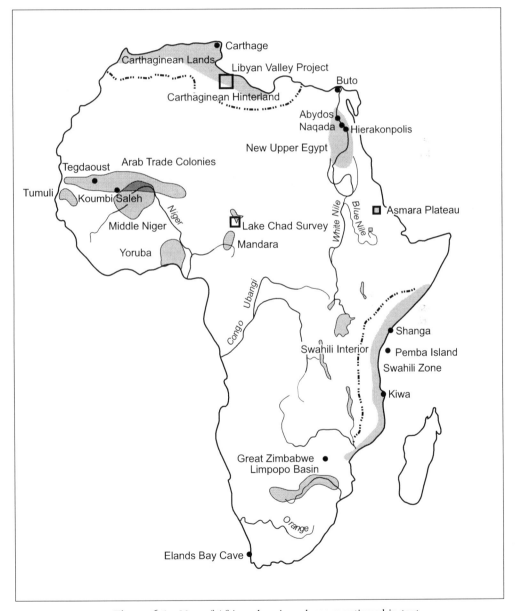

Figure 6.1 Map of Africa, showing places mentioned in text.

contribution to a rethinking of landscape archaeology is perhaps best showcased by those studies
such as the last three presented here that integrate
the functional with a symbolic treatment. Not the
static "ideational" or "symbolic" treatment of landscapes as adopted by adaptationist archaeologies;
or even the "moral" ethnographic landscapes of,
for example, Basso (1984; Feld and Basso 1996),
but rather all of these wrapped together in a deep-
time treatment of those ever-changing social practices that give long-term trajectories to cultural
systems.

As evident in the Mandara, Upper Nile
Predynastic, and Middle Niger cases, such a holistic
approach to landscape archaeology allows people
to "make sense" of their environment, certainly,
but especially to "make sense" of all interacting
agencies of causation—biophysical, human, and
supernatural—today and in the times of the ancestors. To speak of African "landscapes of causation"
is perhaps not going too far—the term summarizes the synergistic logic of settlements and the
"reasons" for changing resources and perceptions
of landform opportunities, and it provides hints
to archaeologists of local and socially constructed
logics as to why things happen.

References

Ashmore, W., and Knapp, A. B. (eds.). 1999. *Archaeologies of Landscape: Contemporary Perspectives*. Oxford: Blackwell Publishing.

Barker, G., and Gilbertson, D. D. (eds.). 2000. *The Archaeology of the Drylands: Living at the Margin*. One World Archaeology, Volume 39. London: Routledge.

Barker, G., Gilbertson, D. D., Jones, G. D. B., and Mattingly, D. J. 1996a. *Farming the Desert: The UNESCO Libyan Valleys Archaeological Survey*, Volume 1. Synthesis. Paris: UNESCO Publishing.

———. 1996b. *Farming the Desert: The UNESCO Libyan Valleys Archaeological Survey*, Volume 2. Gazetteer and Pottery. Paris: UNESCO Publishing.

Basso, K. H. 1984. "Stalking with stories": Names, plac-es and moral narratives among the Western Apache, in S. Plattner and E. M. Bruner (eds.), *Text, Play, and Story: The Construction and Reconstruction of Self and Society*, pp. 19–55. Washington, DC: Proceedings of the American Ethnological Society.

Chami, F. A. 1998. A review of Swahili archaeology. *African Archaeological Review* 15(3): 199–218.

Chittick, N. 1965. The "Shirazi" colonization of East Africa. *Journal of African History* 6(3): 275–94.

———. 1975. The peopling of the East Africa coast, in N. Chittick and R. Rotberg (eds.), *East Africa and the Orient: Cultural Syntheses in Pre-Colonial Times*, pp. 16–43. New York: Africana.

Cuoq, J. M. 1975. *Recueil des Sources Arabes Concernant l'Afrique Occidentale du VIIIe Siècle*. Paris: Editions du Centre National de la Recherche Scientifique.

Darling, P. 1998. Legacy in earth: Ancient Benin and Ishan, southern Nigeria, in K. W. Wesler (ed.), *Historical Archaeology in Nigeria*, pp. 143–97. Trenton, NJ: Africa World Press.

Darnell, J. C. n.d. (a). The deserts, in T. A. H. Wilkinson (ed.), *The Egyptian World*. London: Routledge.

———. n.d. (b). Graffiti and rock Inscriptions, in J. Allen and I. Shaw (eds.), *Oxford Handbook of Egyptology*. Oxford: Oxford University Press.

David, N., and Sterner, J. 1999. Wonderful society: The Burgess Shale Creatures: Mandara polities and the nature of prehistory, in S. K. McIntosh (ed.), *Beyond Chiefdoms: Pathways to Complexity in Africa*, pp. 96–109. Cambridge: Cambridge University Press.

Feld, S., and Basso, K. H. 1996. *Senses of Place*. Santa Fe, NM: School of American Research Press.

Friedman, R. F. (ed.). 2002. *Egypt and Nubia: Gifts of the Desert*. London: British Museum.

Hendrickx, S., and Friedman, R. F. (eds.). 2004. *Egypt at Its Origins*. Dudley, MA: Peeters.

Gronenborn, D. 1998. Archaeological and ethnohis-torical investigations along the southern fringes of Lake Chad, 1993–1996. *African Archaeological Review*. 1x(4): 225–59.

Grzymski, K. 2004. Landscape archaeology of Nubia and central Sudan. *African Archaeological Review* 21(1): 7–30.

Horton, M. 1996. *Shanga: The Archaeology of a Muslim Trading Community on the Coast of East Africa*. London: British Institute in Eastern Africa.

Huffman, T. N. 1996. Archaeological evidence for cli-mate change during the last 2,000 years in south-ern Africa. *Quaternary International*. 33: 55–60.

———. 2000. Mapungubwe and the origins of the Zimbabwe culture, in M. Leslie and T. Maggs (eds.), *African Naissance: The Limpopo Valley 1,000 Years Ago*, pp. 14—29. The South African Archaeological Society Goodwin Series. Volume 8.

Krings, M., and Platte, E. 2004. *Living with the Lake: Perspectives on History, Culture and Economy of Lake Chad*. Colgne: Ruediger Koeppe Verlag.

LaViolette, A. 2004. Swahili archaeology and history on Pemba, Tanzania: A critique and case study of the use of written and oral sources in archaeol-ogy, in A. M. Reed and P. J. Lane (eds.), *African Historical Archaeologies*, pp. 125—62, Dordrecht: Kluwer Academic.

Leslie, M., and Maggs, T. 2000. *African Naissance: The Limpopo Valley 1,000 Years Ago*. The South African Archaeological Society Goodwin Series. Volume 8.

Levtzion, N., and Hopkins, J. F. P. 1981. *Corpus of Early Arabic Sources for West African History*. Cambridge: Cambridge University Press.

MacEachern, S. 2002. Beyond the belly of the house: Space and power in the Mandara Mountains. *Journal of Social Archaeology* 2(2): 179–219.

Manyanga, M., Pikirayi, I., and Ndoro, W. 2000. Coping with dryland environments: Preliminary results from Mapungubwe and Zimbabwe phase sites in the Mateke Hills, south-eastern Zimbabwe, in M. Leslie and T. Maggs (eds.), *African Naissance: The Limpopo Valley 1,000 Years Ago.* The South African Archaeological Society Goodwin Series. Volume 8, pp. 69–77.

Mauny, R. 1961. *Tableau Géographique de l'Ouest Africain au Moyen Age d'après les Sources Ecrites, la Tradition, et l'Archéologie.* Mémoire de l'Institut Fondamental d'Afrique Noire 61. Dakar: Institut Fondamental d'Afrique Noire.

———. 1970. *Les Siécles Obscurs de l'Afrique Noire.* Paris: Fayard.

McIntosh, R. J. 1998. *The Peoples of the Middle Niger.* Oxford: Blackwell Publishing.

———. 2005. *Ancient Middle Niger: Urbanism and the Self-Organizing Landscape.* Cambridge: Cambridge University Press.

McIntosh, R. J., and McIntosh, S. K. 1993. Field survey in the tumulus zone of Senegal. *African Archaeological Review* 11: 73—107.

McIntosh, R. J., and Tainter, J. A. 2006. Palaeoclimates and the Mande, in R. J. McIntosh and J. A. Tainter (eds.), *Climates of the Mande, Mande Studies,* Volume 6, pp. 1–10.

McIntosh, R. J., Tainter, J. A., and McIntosh, S. K. (eds.). 2000. *The Way the Wind Blows: Climate, History and Human Action.* New York: Columbia University Press.

McIntosh, S. K. 1999a. *Beyond Chiefdoms: Pathways to Complexity in Africa.* Cambridge: Cambridge University Press.

———. 1999b. Floodplains and the development of complex society: Comparative perspectives from the West African semi-arid tropics, in E. A. Bacus and L. J. Lucero (eds.), *Complex Polities of the Ancient Tropical World,* Archaeological Papers of the American Anthropological Association, No. 9, pp. 151–65. Washington, DC: American Anthropological Association.

Murdock, G. P. 1959. *Africa: Its People and the Culture History.* New York: McGraw-Hill.

Parkington, J. E. 1999. Western Cape landscapes. *Proceedings of the British Academy* 99: 25–35.

Pliny the Elder. 1949. *Natural History.* Cambridge, MA: Harvard University Press.

Robert, S., and Robert, D. 1972. Douze années de recherches archéologiques en République Islamique de Mauritanie. *Annales de la Faculté des Lettres (Dakar)* 2: 195–233.

Ruddiman, W. F. 2003. The anthropogenic greenhouse era began thousands of years ago. *Climatic Change* 61: 261–93.

Pikirayi, I. 2001. *The Zimbabwe Culture: Origins and Decline of Southern Zambezian States.* Walnut Creek, CA: AltaMira Press.

Schmidt, P. R. 1983. An alternative to a strictly materialist perspective: A review of historical. Archaeology, ethnoarchaeology, and symbolic approaches in African archaeology. *American Antiquity* 48: 62–79.

———. 1997. *Iron Technology in East Africa: Symbolism, Science, and Archaeology.* Bloomington: Indiana University Press and James Curry.

Schmidt, P. R., and Curtis, M. C. n.d. [2007]. *The Archaeology of Ancient Eritrea.* Asmara: Red Sea Press.

Sterner, J. 1998. The Ways of the Mandara Mountains: A Comparative Regional Approach. Unpublished Ph.D. dissertation, School of Oriental and African Studies, University of London.

Wilkinson, T. A. H. 1999. *Early Dynastic Egypt.* London: Routledge.

ENCOUNTERING HUMANS: MAPPING PLACE

Because animals possess physical form, they occupy space. However, understanding the culture of animals—humans included—concerns more than a preoccupation with space as world somewhere "out there." Rather, what is at stake are the specific ways that organisms strategize their existence in the world, how they come to map themselves in a landscape that, in addition to "resources," includes interspecies and intraspecies relations.

Humans have evolved specific forms of mapping that are a step removed from their immediate physical domains: we have evolved *symbolic landscapes* that at once enable us to effectively communicate information about distant and changing conditions, and new conceptualizations and categorizations of the world in which we live. We may think of these cognitive mapping processes both as forms of *displacement* (in the sense that landscape engagements concern mediations through symbolic representation) and *placement* (in the sense that through representation we *find* our place in the world).

If archaeology concerns the history of the human species—in all of its specifics, as well as in its varied and local expressions—then one aspect of the archaeological project is to track and understand how Hominini (humans and their ancestors) have come to take a particular road to placiality (cognitively meaningful reflections of spatiality). How do individuals and groups organize themselves in place? How do we map ourselves in a world with physical, cultural, and social topographies? How do our particular cognitive abilities allow us to organize ourselves in, and in doing so inhabit, the world? The chapters in this section explore such questions by reviewing nonprimate and pre-*Homo sapiens* place-worlds and evolutionary concepts that enable us to better understand how we have come to be where we are.

7

NONHUMAN PRIMATE APPROACHES TO LANDSCAPES

Russell A. Hill

Landscape archaeology focuses on the relationship between archaeological data and its natural setting or environment, and one facet of this approach is aimed at understanding the impact of the environment on human thought, action, and interaction. Studies of our closest living relatives, primates, are important in this respect to help us contextualize the relationships between humans and their landscape.

The Primate Order consists of some 233 species (Goodman et al. 1998), the vast majority of which are found in tropical and subtropical regions across the continental landmasses of Africa, Asia, and Central and South America. Although equatorial rainforest is home to the greatest number of taxa (Hacker, Cowlishaw, and Williams 1998), primates inhabit a variety of other ecosystems, including woodlands, savannahs, and deserts, as well as colder temperate forests and montane regions. Most species are diurnal, arboreal, and frugivorous (fruit-eating) (Cowlishaw and Dunbar 2000), although primates occupy a range of ecological guilds defined by combinations of activity (arboreal vs. terrestrial; nocturnal vs. diurnal) and diet (frugivores, folivores, insectivores) (Bourlière 1985). As a consequence, primates inhabit a diverse array of ecological niches and communities and are exposed to a wide variety of selective pressures.

This chapter reviews three elements of how primates interact with their environments. Initially it examines the factors that shape primate distribution and community structure and the position of primates in ecosystems. The second part then goes on to explore how ecological factors shape how primates use their environment at the home range scale. Finally, the chapter considers how primates perceive the landscape and whether primates possess a cognitive map of their environment.

Primate Community Ecology

A primate community can be broadly defined as the primate species that live together and utilize resources at one geographic location (Reed and Bidner 2004). Understanding the fundamental causes of the structure of primate communities is essential to studies of primate evolutionary history and behavioral ecology as well as the development of conservation strategies. Previous research has highlighted a range of abiotic and biotic factors shaping community structure such as historical circumstances (for example, land area, isolation, and colonization), climate, habitat heterogeneity, productivity and food sources, and species interactions. A number of these factors are briefly reviewed here.

In general, the number of species in a region increases with the area of that region (Reed and Fleagle 1995), in a logarithmic relationship that is referred to as the species-area curve. Across Africa, the richness of primate species increases with closed forest area (Cowlishaw 1999). This relationship is probably explained by larger areas holding bigger populations that, in turn, have lower extinction risks (Cowlishaw and Dunbar 2000). Isolation is also important, however, with regions distant from a colonizing source having fewer species than those closer to a source, because local sources result in higher immigration rates and greater buffering against extinction. For example, diurnal primate species richness in East African forests is strongly related to the distance of the forests from a Pleistocene refuge (Struhsaker 1981). Isolation effects are particularly important in accounting for lower species richness in island communities (Cowlishaw and Dunbar 2000), although islands colonized millions of years ago, such as Madagascar, have undergone multiple speciation, extinction, and radiation events and thus have more primate species than might be expected (Reed and Bidner 2004).

Environmental factors also influence primate communities and species richness. Cold, seasonal environments tend to have fewer species than warmer, less seasonal regions. As a consequence, species richness tends to decline at higher altitudes and with increasing distance from the equator (Cowlishaw and Hacker 1997). The most important determinant of fine-grained patterns of species richness, however, appears to be the diversity of the habitat. Across a range of locations, there is a positive association between primate species richness and habitat heterogeneity (Madagascar: Ganzhorn 1994; South America: Schwarzkopf and Rylands 1989; Africa: Skorupa 1986). Diverse habitats have a greater number of ecological niches permitting more species to coexist in a region. Regional diversity at a broader scale also appears driven by habitat heterogeneity, given that vegetationally complex forests harbor more primate species than does woodland or savannah (Cowlishaw and Dunbar 2000). Forest habitats may contain as many as 10 to 15 primate species, whereas savannahs typically contain just 1 to 5 (Bourlière 1985). Since the degree of forest cover is dependent on climate, particularly the degree of rainfall (Reed 1998), primate species richness at individual sites is highly correlated with mean annual rainfall for South America, Africa, and Madagascar (Reed and Fleagle 1995). Rainfall is also important, since it correlates with resource productivity. In fact, primate community structure is more directly affected by productivity (and thus food availability) than by rainfall (Kay et al. 1997).

Rather than overall productivity, however, it is the seasonality in resource availability that is critical to community dynamics, since changes in seasonality can profoundly affect the vegetation production cycle (Reed and Bidner 2004). Latitude is important in this respect, since it influences patterns of temperature, rainfall, and day length with seasonal differences affecting habitat type. Low-latitude forest habitats usually experience long wet and short dry seasons, whereas woodlands at the same latitudes have shorter wet and longer dry seasons. These variations shape the production of leaves and fruits in tropical forests and thus affect the resources available to the species within primate communities.

Latitude also has indirect importance, since the distribution of tropical forests is centered on the equator and gives way to woodland and savannah at higher latitudes (Cowlishaw and Dunbar 2000). Forest alone, however, is insufficient to promote high primate species richness, since dominance by particular tree species in forests may limit the number of species. For example, certain tree species may limit primate biomass by not supporting primates (Thomas 1991), through limiting the number of frugivorous species (Caldecott 1986) or through contributing to low density or biomass through productivity patterns (Terborgh and van Schaik 1997). Coupled with this, the stratification of the canopy and other vegetative structures within forests is also important for determining primate niche space (Reed and Bidner 2004).

Ecological competition among members of the same community is a critical factor in determining community structure. In many primate communities, different species compete for the same food sources (Terborgh 1983), and thus diet is the most important determinant of niche partitioning (Struhsaker 1978). Both intraspecific and interspecific competition with nonprimates may also influence community structure (Ganzhorn 1999), and birds and other mammals can be important competitors of primates (Terborgh 1986). Predation is also likely to have an important role in structuring primate communities (Terborgh et al. 2001) and may increase species interactions through encouraging the formation of polyspecific associations (Gartlan and Struhsaker 1972). Red colobus (*Piocolobus badius*) actively seek associations with Diana monkeys (*Cercopithecus diana*) in response to hunting pressure from chimpanzees (*Pan troglodytes*), because chimpanzees refrain from hunting associated groups (Bshary and Noë 1997). Although polyspecific associations probably intensify competitive interactions among primate communities through increased competition for resources, predation could potentially serve to reduce competition through influencing the density of competitors (Reed and Bidner 2004). The interplay between resources and predation is thus important in shaping the structure of primate communities. The trade-off between acquiring food and avoiding predation is also central to understanding the behavior of primates within the landscape.

Primate Behavioral Ecology

Some of the earliest papers on primate behavior and ecology (e.g., Alexander 1974) take predation pressure to be the primary force favoring group living in primates. Cohesive groups serve to reduce predation risk through the benefits of increased vigilance, dilution effects,[1] and the potential for group defense. Such benefits, however, come at the expense of increased feeding competition. The ecological constraints model predicts that as group size increases, total food requirements also increase. As a consequence, the distance traveled each day to forage, as well as the total size of the home range exploited by the group, should expand accordingly (Chapman and Chapman 2000; Clutton-Brock and Harvey 1977). Where daily travel paths or home range areas do not (or cannot) increase, individuals may face reduced reproductive success owing to the effects of within-group competition (Dunbar 1988; Janson and Goldsmith 1995).

The relationship between group size and daily travel distances depends on the distribution and the abundance of food resources in the environment. A number of studies have found larger groups to travel further each day than smaller groups for frugivorous primates. In long-tailed macaques (*Macaca fascicularis*), the daily travel distance, as well as the time spent traveling and searching for dispersed food items, shows a linear increase with group size (van Schaik et al. 1983). Conversely, far fewer relationships are reported in folivorous species (Ganas and Robbins 2005). These differences are explained by the fact that frugivorous primates face higher levels of feeding competition than folivorous species to the costs of feeding on discrete, monopolizable food patches (fruit trees) that occur at a lower density than leaves or herbaceous vegetation (Dunbar 1988; Janson and Goldsmith 1995). Diet plays less of a role in determining the total ranging area of a group, however, and across primates there is a consistent relationship for home range sizes to increase with group size for both frugivores (for example, red colobus: Gillespie and Chapman 2001) and folivores (for instance, mountain gorilla [*Gorilla beringei*

beringei], Ganas and Robbins 2005). Such relationships are not always present though, and differences in food availability and the density of intra- and interspecific competitors between ranges may obscure any direct relationship between group size and home range size in primates. In chimpanzees, the number of adult males is the best predictor of home range size, since males play an important role in defending the home range against neighbors (Lehmann and Boesch 2003).

Within home ranges, numerous studies have reported a relationship between habitat preferences and the spatial and temporal variability in food distribution (Barton et al. 1992; Marsh 1981; Watts 1991). In general, primates forage in areas of greatest food availability, although a number of factors serve to modify this general relationship. Both yellow *(Papio hamadryas cynocephalus)* and olive *(P. h. anubis)* baboons show increased intensities of home range use around water sources and sleeping sites (Altmann and Altmann 1970; Barton et al. 1992), and parasite avoidance has also been suggested as important (Freeland 1980). Predation risk is also a significant determinant of ranging behavior and habitat use, although it has received considerably less attention in the literature. In arboreal species, predation is thought to play an important role in canopy height selection with animals avoiding feeding in exposed locations where they are exposed to the risk of attack from raptors. Similarly, small groups of wedge-capped capuchin monkeys (*Cebus olivaceus*), which are likely to perceive a greater risk of predation than larger groups do, spend more time scanning and occupy greater heights in the canopy where they can avoid terrestrial predators such as felids (de Ruiter 1986). In terrestrial species, chacma baboons (*P.h. ursinus*) preferentially select low predation-risk (high visibility) habitats for foraging, since leopards (*Panthera pardus*), their principal predator, prefer to attack from ambush. This leads to a trade-off between predation and food availability in habitat choice, since high visibility habitats are often of low food availability, and this trade-off is again most evident in smaller groups (Cowlishaw 1997a; Hill and Weingrill 2006). Small baboon groups also spend more time on or close to refuges (trees and cliff faces) that they can use to escape from terrestrial predators (Cowlishaw 1997b; Stacey 1986). Such strategic use of habitats and topographic features of the landscape to balance the conflicting demands of food acquisition and predator avoidance is an essential element of the survival strategy of most animals.

Cognitive Maps and Spatial Information

Obtaining sufficient access to food resources is a critical component of a primate's foraging strategy, and thus the ability for an animal to move efficiently between resources such as feeding sites or water sources is directly linked to individual fitness. Many primatologists have argued that monkeys and apes maintain a detailed knowledge of the spatial location and phenology[2] of resources in their home range such that from any point they can assess the distance and the direction to potential feeding, resting and drinking sites (Altmann and Altmann 1970; Menzel 1991; Milton 1981; Sigg and Stolba 1981). In most instances, however, the evidence for spatial learning is weak or anecdotal and based principally on observations of relatively direct travel to distant food patches (Garber 2000). In fact, based on the errors humans make when estimating angles and distances in familiar urban environments, Byrne (1978, 2000) suggested that human cognitive mapping capacities of large-scale space may also be overestimated.

The need for animals to solve spatial problems has been viewed as a potential trigger of cognitive evolution (Menzel 1997; Milton 1981, 1988, 2000), and the relationship between brain size and intelligence in primates is a topic of considerable debate (Barton 2000; Dunbar 1992; van Schaik and Deaner 2003). Nevertheless, there is evidence that enlargement of the neocortex in primates may be related to spatial competence (Byrne 1995). Strangely, however, despite the diverse array of studies from other taxa (arthropods: Collet and Zeil 1998; birds: Shettleworth and Hampton 1998; rodents: Save, Poucet, and Thinus-Blanc 1998), studies of how primates find their resources are comparatively scarce.

Since the earliest studies introducing the concept of cognitive maps (O'Keefe and Nadel 1978; Tolman 1948), the idea that animals represent space in a maplike way has developed considerable appeal. A cognitive map is a set of mental representations that encode the metric properties of large-scale space (that is, distances and directions) so that novel routes to unseen goals can be planned. Byrne (1978, 2000) has refined this to distinguish *vector* maps and *network* maps. Vector maps are synonymous with cognitive maps, whereby animals place objects onto a map that is isomorphic to the three-dimensional world and can plot novel routes between points. Network maps in contrast, consist of a series of commands that contain only minimal direction (left, half right) and distance information, and, although they preserve topological relations among locations, they lack Euclidian or vector information. In contrast to vector maps, animals navigating by network maps are unlikely to plot optimal routes between known locations or take novel shortcuts. In essence the world is encoded as a series of nodes; locations where a change of movement is required in a specific direction in order to move between two locations.

Although suggestions of primates using complex cognitive maps are relatively common in the literature (Altmann and Altmann 1970; Menzel 1991; Milton 1981; Sigg and Stolba 1981), there is in fact little evidence to differentiate these reports from the animals using computationally simpler network maps (Byrne 2000). There is, however, evidence that certain primates may view their world as a network of paths.

Hamadryas baboons (*P.h. hamadryas*) are large, terrestrial primates that live in multilayered societies where aggregations on sleeping sites split up into smaller foraging parties (often one-male units) that re-aggregate at intervals throughout the day. Not only do these baboons appear to know about the layout of their range, but they can rely on the spatial knowledge of other individuals when "negotiating" a future aggregation site such as water holes in their Ethiopian desert environment (Sigg and Stolba 1981). In moving between re-aggregation points, the hamadryas tend to use familiar paths rather than novel shortcuts. However, the interconnecting nature of their pathways allows parties to wander independently of one another en route to the aggregation sites. Similar observations on the use of familiar foraging pathways are evident in the arboreal New World monkeys of South America. Milton (1981, 1988) describes spider monkeys (*Ateles geoffroyi*) moving along the same arboreal pathways through the forest while howler monkeys (*Alouatta palliata*) travel using traditional arboreal pathways that connect important clusters of food trees. In all these examples, the animals are using regular pathways between landmarks, behavior that supports the idea that primates navigate predominantly by means of network maps.

The observation that many primates appear to navigate using network maps does not preclude the ability to use cognitive or vector maps. The difficulty is that the two models make similar predictions for animals foraging under most conditions in their natural environments. Since nonhuman primates are generally long-lived and live in cohesive groups that range over large proportions of their home range each day, immature individuals have ample time to learn the network of paths that their group traditionally uses. By adulthood, therefore, they would possess a richly interconnected network map of their home range, and novel routes or shortcuts would rarely be beneficial or required. Detecting vector map knowledge of large-scale space in free-ranging primates is thus likely to be virtually impossible (Byrne 2000). For example, although a number of studies have shown primates to use the most efficient routes in navigating their environment (Boesch and Boesch 1984; Garber 1989; MacKinnon 1974; Milton 1981; Sigg and Stolba 1981), one cannot distinguish routes of greatest efficiency from a well-rehearsed network map. Similarly, the experimentally induced travel of Japanese macaques (*Macaca fuscata*) between out-of-season fruit trees (Menzel 1997) does not demonstrate vector map ability. In fact, it is only in humans that vector map knowledge is certainly present, and even here its use is probably a lot less common or valuable than is traditionally supposed (Byrne 2000).

Conclusions

Nonhuman primates are widely distributed throughout the tropical and subtropical regions of the world, inhabiting ecosystems ranging from equatorial rainforest and woodland to deserts and cold montane grassland. Their distribution is constrained by a range of abiotic and biotic factors that serve to shape community structure. These same ecological pressures also influence how groups of primates exploit their home range. Group size, reproductive rates, and patterns of habitat use are all driven by the ubiquitous trade-off between food acquisition and predator avoidance. Within their environment, there is no evidence that nonhuman primates possess a cognitive map that is isomorphic to the world as viewed from above. Rather, patterns of travel appear to be most parsimonious with primates using a network map and navigating via a set of interconnected and familiar pathways. Nevertheless, vector maps cannot be completely refuted, and a challenge for future studies of the ecology of group movement in primates is to distinguish the ways in which spatial information is represented. Only then will we start to understand how primates perceive the landscapes in which they live.

Notes

1. The dilution effect simply states that the more individuals there are in a group the lower the probability that any one animal will be the target during a particular predator attack.

2. *Phenology* refers to the relationship between climate and periodic (seasonal) biological phenomena, such as the date of emergence of leaves, flowers, and fruits.

References

Alexander, R. D. 1974. The evolution of social behaviour. *Annual Review of Ecology and Systematics* 5: 325–83.

Altmann, S. A., and Altmann, J. 1970. *Baboon Ecology*. London: University of Chicago Press.

Barton, R. A. 2000. Primate brain evolution: Cognitive demands of foraging or of social life, in S. Boinski and P. A. Garber (eds.), *On the Move: How and Why Animals Travel in Groups*, pp. 204–37. London: University of Chicago Press.

Barton, R. A., Whiten, A., Strum, S. C., Byrne, R. W., and Simpson, A. J. 1992. Habitat use and resource availability in baboons. *Animal Behaviour* 43: 831–44.

Boesch, C., and Boesch, H. 1984. Mental map in wild chimpanzees: An analysis of hammer transports for nut cracking. *Primates* 25: 160–70.

Bourlière, F. 1985. Primate communities: Their structure and role in tropical ecosystems. *International Journal of Primatology* 6: 1–26.

Bshary, R., and Noe, R. 1997. Anti-predation behaviour of red colobus monkeys in the presence of chimpanzees. *Behavioral Ecology and Sociobiology* 41: 321–33.

Byrne, R. W. 1978. Plans and errors in human memory for urban geography, in M. Grunberg, P. Morris, and R. Sykes (eds.), *Practical Aspects of Memory*, pp. 93–100. London: Academic Press.

———. 1995. *The Thinking Ape: Evolutionary Origins of Intelligence*. Oxford: Oxford University Press.

———. 2000. How monkeys find their way: Leadership, coordination and cognitive maps of African baboons, in S. Boinski and P. A. Garber (eds.), *On the Move: How and Why Animals Travel in Groups*, pp. 491–518. London: University of Chicago Press.

Caldecott, J. O. 1986. An ecological and behavioral study of the pig-tailed macaque. *Contributions to Primatology* 21: 1–259

Chapman, C. A., and Chapman, L. J. 2000. Determinants of group size in primates: The importance of travel costs, in S. Boinski and P. A. Garber (eds.), *On the Move: How and Why Animals Travel in Groups*, pp, 24–42. London: University of Chicago Press.

Clutton-Brock, T. H., and Harvey, P. H. 1977. Species differences in feeding and ranging behaviour in primates, in T. H. Clutton-Brock (ed.), *Primate Ecology: Studies of Feeding and Ranging Behaviour in Lemurs, Monkeys and Apes*, pp. 557–84. London: Academic Press.

Collett, T., and Zeil, J. 1998. Places and landmarks: An arthropod perspective, in S. Healy (ed.), *Spatial Representation in Animals*, pp. 18–53. Oxford: Oxford University Press.

Cowlishaw, G. C. 1997a. Trade-offs between foraging and predation risk determine habitat use in a desert baboon population. *Animal Behaviour* 53: 667–86.

———. 1997b. Refuge use and predation risk in a desert baboon population. *Animal Behaviour* 54: 241–53.

———. 1999. Predicting the decline of African primate diversity: An extinction debt from historical deforestation. *Conservation Biology* 13: 1183–93.

Cowlishaw, G., and Dunbar, R. 2000. *Primate Conservation Biology*. London: University of Chicago Press.

Cowlishaw, G., and Hacker, J. E. 1997. Distribution, diversity and latitude in African primates. *American Naturalist* 150: 505–12.

de Ruiter, J. R. 1986. The influence of group size on predator scanning and foraging behaviour of wedge-capped capuchins (*Cebus olivaceus*). *Behaviour* 98: 240–58.

Dunbar, R. I. M. 1988. *Primate Social Systems*. London: Chapman and Hall.

———. 1992. Neocortex size as a constraint on group size in primates. *Journal of Human Evolution* 28: 287–96.

Freeland, W. J. 1980. Mangabey (*Cercocebus albigena*) movement patterns in relation to food availability and fecal contamination. *Ecology* 61: 1297–303.

Ganas, J., and Robbins, M. M. 2005. Ranging behaviour of the mountain gorillas (*Gorilla beringei beringei*) in Bwindi Impenetrable National Park, Uganda: A test of the ecological constraints model. *Behavioural Ecology and Sociobiology* 58: 277–88.

Ganzhorn, J. U. 1994. Lemurs as indicators for habitat change, in B. Thierry, J. R. Anderson, J. J. Roeder, and N. Herrenschmidt (eds.), *Current Primatology*, Volume 1: *Ecology and Evolution*, pp. 51–56. Strasbourg: Université Louis Pasteur.

———. 1999. Body mass, competition and the structure of primate communities, in J. G. Fleagle, C. H. Janson, and K. E. Reed (eds.), *Primate Communities*, pp. 141–57. Cambridge: Cambridge University Press.

Garber, P. A. 1989. Role of spatial memory in primate foraging patterns: *Saguinus mystax* and *Saguinus fuscicollis*. *American Journal of Primatology* 19: 203–16.

———. 2000. Evidence for the use of spatial, temporal, and social information by primate foragers, in S. Boinski and P. A. Garber (eds.), *On the move: How and Why Animals Travel in Groups*, pp. 261–98. London: University of Chicago Press.

Gartlan, J. S., and Struhsaker, T. T. 1972. Polyspecific associations and niche separation of rain-forest anthropoids in Cameroon, West Africa. *Journal of Zoology, London* 168: 221–66.

Gillespie, T. R., and Chapman, C. A. 2001. Determinants of group size in the red colobus monkey (*Procolobus badius*): An evaluation of the ecological constraints model. *Behavioural Ecology and Sociobiology* 50: 329–38.

Goodman, M., Porter, C. A., Czelusniak, J., Page, S. L., Schneider, H., Shoshani, J., Gunnell, G., and Groves, C. P. 1998. Toward a phylogenetic classification of primates based on DNA evidence complemented by fossil evidence. *Molecular Phylogenies and Evolution* 9: 585–98.

Hacker, J. E., Cowlishaw, G., and Williams, P. H. 1998. Patterns of African primate diversity and their evaluation for the selection of conservation areas. *Biological Conservation* 84: 251–62.

Hill, R. A., and Weingrill, T. 2006. Predation risk and habitat use in chacma baboons (*Papio hamadryas ursinus*), in S. Gursky and K. A. I. Nekaris (eds.), *Primates and their predators*, pp. 339–54. New York: Kluwer Academic Publishers (Developments in Primatology Series).

Janson, C. H., and Goldsmith, M. L. 1995. Predicting group size in primates: Foraging costs and predation risks. *Behavioral Ecology* 6: 326–36.

Kay, R. F., Madden, R. H., van Schaik, C. P., and Higdon, D. 1997. Primate species richness is determined by plant productivity: Implications for conservation. *Proceedings of the National Academy of Science USA* 94: 13023–27.

Lehmann, J., and Boesch, C. 2003. Social influences on ranging patterns among chimpanzees (*Pan troglodytes verus*) in the Tai National Park, Cote d'Ivoire. *Behavioural Ecology* 14: 642–49.

MacKinnon, J. 1974. The behaviour and ecology of wild orangutans (*Pongo pygmaeus*). *Animal Behaviour* 22: 3–74.

Marsh, C. W. 1981. Ranging behaviour and its relation to diet selection in Tana River red colobus (*Colobus badius rufomitratus*). *Journal of Zoology, London* 195: 473–92.

Menzel, C. R. 1991. Cognitive aspects of foraging in Japanese monkeys. *Animal Behaviour* 41: 397–402.

———. 1997. Primates' knowledge of their natural habitat as indicated in foraging, in A. Whiten and R. W. Byrne (eds.), *Machiavellian Intelligence II: Extensions and Evaluations*, pp. 207–39. Cambridge: Cambridge University Press.

Milton, K. 1981. Diversity of plant foods in tropical forests as a stimulus to mental development in primates. *American Anthropologist* 83: 534–48.

———. 1988. Foraging behaviour and the evolution of primate behaviour, in R. W. Byrne and A. Whiten (eds.), *Machiavellian Intelligence: Social Evolution of Intellect in Monkeys, Apes and Humans*, pp. 285–305. Oxford: Clarendon Press.

———. 2000. Quo vadis? Tactics of food search and group movement in primates and other animals, in S. Boinski and P. A. Garber (eds.), *On the Move: How and Why Animals Travel in Groups*, pp. 375–417. London: University of Chicago Press.

O'Keefe, J., and Nadel, L. 1978. *The Hippocampus as a Cognitive Map*. Oxford: Oxford University Press.

Reed, K. E. 1998. Using large mammal communities to examine ecological and taxonomic organization and predict vegetation in extant and extinct assemblages. *Paleobiology* 24: 384–408.

Reed, K. E., and Bidner, L. R. 2004. Primate communities: Past, present and future. *Yearbook of Physical Anthropology* 47: 2–39.

Reed, K. E., and Fleagle, J. G. 1995. Geographic and climatic control of primate diversity. *Proceedings of the National Academy of Science USA* 96: 7874–76.

Save, E., Poucet, B., and Thinus-Blanc, C. 1998. Landmark use and the cognitive map in the rat, in S. Healy (ed.), *Spatial Representation in Animals*, pp. 119–32. Oxford: Oxford University Press.

Schwarzkopf, L., and Rylands, A. B. 1989. Primate species richness in relation to habitat structure in Amazonian rainforest fragments. *Biological Conservation* 48: 1–12.

Shettleworth, S., and Hampton, R. 1998. Adaptive specializations of spatial cognition in food-storing birds? Approaches to testing a comparative hypothesis, in R. Balda, I. Pepperberg, and A. Kamil, (eds.), *Animal Cognition in Nature: The Convergence of Psychology and Biology in Laboratory and Field*, pp. 65–98. San Diego: Academic Press.

Sigg, J., and Stolba, A. 1981. Home range and daily march in a hamadryas baboon troop. *Folia Primatologica* 36: 40–75.

Skorupa, J. P. 1986. Responses of rainforest primates to selective logging in Kibale Forest, Uganda: A summary report, in K. Benirshcke (ed.), *Primates: The Road to Self-Sustaining Populations*, pp. 57–70. Berlin: Springer.

Stacey, P. B. 1986. Group size and foraging efficiency in yellow baboons. *Behavioural Ecology and Sociobiology* 18: 175–87.

Struhsaker, T. T. 1978. Food habits of five monkey species in the Kibale Forest, Uganda, in D. J. Chivers, and J. Herbert (eds.), *Recent Advances in Primatology, Vol. 1. Behaviour*, pp. 225–47. London: Academic Press.

———. 1981. Forest and primate communities in East Africa. *African Journal of Ecology* 19: 99–114.

Terborgh, J. 1983. *Five New World Primates*. Princeton: Princeton University Press.

———. 1986. Community aspects of frugivory in tropical forests, in A. Estrada and T. H. Flemming (eds.), *Frugivores and seed dispersal*, pp. 371–84. Dordrecht: W. Junk Publishers.

Terborgh, J., Lopez, L., Nunez, V. P., Rao, M., Shahbuddin, G., Orihuela, G., Riveros, M., Ascanio, R., Adler, G. H., Lambert, T. D., and Balbas, L. 2001. Ecological meltdown in predator-free forest fragments. *Science* 294: 1923–26.

Terborgh, J., and van Schaik, C. P. 1997. Convergence vs. non-convergence in primate communities, in J. H. R. Gee and P. S. Gillar (eds.), *Organisation of communities*, pp. 205–66. Oxford: Blackwell Science.

Thomas, S. C. 1991. Population densities and patterns of habitat use among anthropoid primates of the Ituri Forest, Zaire. *Biotropica* 23: 68–83.

Tolman, E. 1948. Cognitive maps in rats and men. *Psychology Review* 55: 189–208.

van Schaik, C. P., and Deaner, R. 2003. Life history and cognitive evolution in primates, in F. de Waal and P. Tyack (eds.). *Animal Social Complexity: Intelligence, Culture and Individualised Societies*, pp. 5–25. Cambridge MA: Harvard University Press.

van Schaik, C. P., van Noordwijk, M. A., de Boer, R. J., and den Tonkelaar, I. 1983. The effect of group size on time budgets and social behaviour in wild long-tailed macaques (*Macaca fasicularis*). *Behavioural Ecology and Sociobiology* 13: 173–81.

Watts, D. P. 1991. Strategies of habitat use by mountain gorillas. *Folia Primatolgica* 56: 1–16.

8

PRE-*HOMO SAPIENS* PLACE-WORLDS

Andrew Chamberlain

The fossil record of human evolution spans over 6 million years, and at least a dozen extinct species of hominid have been identified from palaeontological sites in Africa and Eurasia. The natural world inhabited by these premodern hominids was largely the same as the world that we inhabit today—if we ignore the effects that *Homo sapiens* has had on the earth's biota, topography, and climate during 100,000 years of stewardship. With the emergence of modern humans came the capacity first to drive other species to extinction through the effects of over-exploitation, followed by the ability to further modify the nature of the earth's flora, fauna, and habitats through domestication, plant and animal husbandry, and deliberate habitat modification. More recently, the irreversible impacts of human industrial activities on the natural world at the global scale have become apparent; nonetheless, in some parts of the modern world there still exist a few examples of natural landscapes and habitats that resemble those occupied by our distant Plio-Pleistocene hominid ancestors. Furthermore, according to some views, *Homo sapiens* has retained some elements of its original biological and cognitive adaptation to a small-scale, mobile hunting and gathering lifestyle (Tooby and Cosmides 1990), an evolutionary legacy that is amenable to investigation

as a potential guide to possible environmental preferences in earlier hominid species.

The techniques of palaeoenvironmental reconstruction allow the recreation, at least on a broad scale, of the physical and biotic conditions pertaining to particular localities in earlier times, and the palaeoanthropological record, including hominid fossils, artifacts, and humanly modified faunal and palaeobotanical assemblages, provides the framework for populating those past environments with hominid species. However, our understanding of how early hominids occupied and exploited their natural world suffers from a dearth of "traditional" archaeological evidence such as recognizable occupation sites with indications of spatially organized activities, functional and symbolic artifacts, deliberate human modifications to landscapes, and instead depends heavily on modeling and use of analogues from natural history and human behavioral ecology (Fedigan 1986). Inevitably, this lack of material cultural evidence and dependence on theoretical modeling gives a severely functional flavor to reconstructions of early hominid life and reinforces preconceived notions that early hominids were aesthetically unaware, were lacking in symbolic capacity, and were unable to plan beyond life's immediate necessities. Yet the notion that premodern hominids had limited cognitive abilities is hard to reconcile with the palaeoecological evidence for successful long-term adaptation to life in hostile environments and with the demographically driven expansion of hominid populations to many parts of Africa, Europe, and Asia by about 1 million years ago (mya). What is clear is that pre-modern hominids must have been able to perceive and to understand those qualities and attributes of natural landscapes that were of critical importance to survival, either for their positive (resource) or for their negative (hazard) value (Kaplan 1992), and that their abilities to do so greatly exceeded those of contemporary primates and carnivores occupying the same environments.

In this chapter, I review a variety of recent work that may shed light on early hominid landscape preferences, ranging from theoretical modeling of landscape occupation to studies of the spatial distribution of archaeological and palaeontological discoveries. The latter studies, however, provide only broad indicators of habitat preferences in early hominids, since it is not until the emergence of our own species, anatomically modern *Homo sapiens*, that we can detect regular, detailed, and informative patterns of activities at specific occupation sites. As is the case in many areas of palaeoanthropological research, the extent of debate and speculation about hominid utilization of space and resources is in inverse proportion to the amount of reliable evidence; nevertheless, some useful pointers to avenues of fruitful investigation can be discerned.

The Cognitive Psychology of Landscape Preference

Present-day *Homo sapiens* is a ubiquitous species with a predominantly sedentary pattern of landscape occupation, but even urban-dwelling humans are reported to express a preference for visual landscapes that have properties resembling those of savanna habitats (that is, low-relief, sparsely wooded tropical grasslands). This preference, which exists cross-culturally and is thought to be innate, has been attributed by evolutionary psychologists to selection pressures operating during early human evolution (Balling and Falk 1982; Orians and Heerwagen 1992). Human individuals are able to express a rapid and often unconscious affective response to those general properties of a landscape that are perceived on initial visual encounter. Preferred landscapes are those containing features indicative of environmental conditions favorable for survival, such as an abundance of subsistence resources or a reduced threat from predators or other natural hazards. The preference for savanna-style visual landscapes is most strongly expressed in children (Balling and Falk 1982) and is also manifest in the deliberate design of artificial landscapes as exemplified by modern (that is, post-Renaissance) ornamental parks and gardens (Kaplan 1992).

A more specific hypothesis of human landscape preference stems from prospect-refuge theory, which predicts that within a given landscape, preferred locations are found at interfaces between prospect-dominant and refuge-dominant areas (Appleton 1996; Hudson 1992). These vantage points combine unimpeded visual prospects with a ready opportunity for concealment and/or withdrawal to a safe refuge. Thus a treeless landscape is less visually attractive than a habitat containing isolated trees that can provide opportunities to hide or to escape from potential predators.

Exponents of the human preference for savanna-like habitats have reasoned that the human environment of evolutionary adaptiveness (EEA) was located in the Plio-Pleistocene savannas of sub-Saharan Africa. Evolutionary psychologists have argued that habitual occupation of the savanna biome by *Australopithecus* and early species of *Homo* provided an extended period of selection for the reinforcement of intuitive preferences for certain topographic, botanical, and faunal features of the savanna landscape (cf. Orians and Heerwagen 1992: 556). However, this scenario of human evolution is oversimplified, and there is increasing consensus among palaeoanthropologists that there is no single unitary environment to which earlier human species were optimally adapted (Foley 1996). Palaeoenvironmental reconstructions for early hominid sites in Africa indicate that closed and open woodland, rather than open savanna, was the predominant biome occupied by the earliest species such as *Ardipithecus ramidus* and *Australopithecus anamensis* (see below), and it is therefore more likely that the visual preference for savanna landscapes is either of more recent evolutionary origin or reflects a narrower and perhaps context-specific attribute of human spatial perception and environmental aesthetics.

Modeling Past Landscape Usage

Primate Models

Living species of primates (monkeys and apes) differ substantially from humans in their diet, social structure, and ranging behavior; nonetheless, these animals provide some relevant and useful analogies for early hominid landscape usage. Great apes such as chimpanzees are often characterized as forest animals, but, although they depend on daily access to arboreal refuges, these animals in fact occupy a wide range of natural habitats from closed-canopy forests through to more open savanna habitats (Hunt and McGrew 2002). Like humans, great apes need to sleep in a recumbent position, but most apes are also dependent on resting in trees to avoid the risk of predation by large carnivores. They solve the problem of arboreal dormancy by weaving branches into sleeping nests, and by analogy this practice may have been retained by early hominids (Fruth and Hohmann 1996; Sabater Pi, Veà, and Serrallonga 1997; Sept 1998). Chimpanzees select tree nesting sites that are close to their preferred feeding areas (Fruth and Hohmann 1996), but they also appear to show a preference for nesting in elevated topographical locations that provide good prospects over the surrounding terrain (Furuichi and Hashimoto 2004).

Estimates of the home range size for groups of chimpanzees living in open habitats vary from about 50 km² in the Semliki Wildlife Reserve in Uganda to around 300 km² at Mont Assirik in Senegal (Hunt and McGrew 2002: 41–42). These estimates are substantially higher than values for forest-dwelling chimpanzees and probably reflect the more dispersed distribution of savanna food resources, since other savanna-dwelling primates such as baboons and the patas monkey also have relatively large home range areas. Early hominids may have utilized even larger home ranges, based on modeling of hominid group sizes and on parallels with the ranging behavior of large-bodied carnivores. Australopithecines and early species of *Homo* are predicted to have had social group sizes at the upper end or above the ranges observed in living primates (Dunbar 2003), and the increasing contribution of meat to hominid diets may have necessitated even larger home ranges, since territories are substantially larger in carnivores than in omnivores or herbivores of similar body size (Jetz et al. 2004).

Great apes have often been viewed as the "default" primate model for early hominids, because the apes share a close genetic relationship as well as some advanced cognitive and behavioral attributes with modern humans (Elton 2006). However, from an ecological perspective the cercopithecid monkeys, especially the larger-bodied baboons and geladas, exploit a broader spectrum of generally more open habitats than do the present-day great apes. The baboons in particular, with their larger body size, adaptation to terrestrial locomotion, behavioral flexibility, and responsiveness to environmental change and maintenance of large, complex yet stable social groups have provided a fertile source of models for early hominid subsistence strategies and social behavior (Elton 2006; Jolly 2001; Strum and Mitchell 1987; Washburn and Devore 1961). Looking more specifically at foraging and territorial ranging strategies, researchers have proposed another cercopithecid species, the Patas monkey *Erythrocebus patas*, as a suitable analogue for the evolution of long leg length and increased mobility in the Lower Pleistocene hominid *Homo erectus* (Isbell et al. 1998). Patas monkeys, like baboons, are among the most terrestrial of primate species, and they spend a substantial proportion of their activity in ranging and mobile foraging.

Ethnographic Models

Ethnographic studies of hunter-gatherer societies have provided a useful source of referential models for early hominid subsistence behavior in the sense that both hunter-gatherers and early hominids obtained their foods from undomesticated sources, but with two important caveats. First, few documented hunter-gatherer groups can be regarded as pursuing a lifestyle that has been unaffected by the activities of neighboring pastoralist and agriculturalist communities, and, second, in the historical period, hunter-gatherers have been increasingly confined to marginal or less favorable land that is of low value to food producers (but see Terrell and Hart, this volume). In the tropical regions of Africa, hunter-gatherer communities are characterized by small groups or bands of 20–50 individuals living in temporary camps, with camp locations chosen to optimize the availability of food and/or water resources and often relocated within the group's territory several times per year (Lee and Daly 1999). Territories of individual hunter-gatherer groups typically extend for several hundred km², somewhat larger than is the case for nonhuman primate groups. Foraging behaviors of mobile hunter-gatherers are also conditioned by seasonal availability of key resources: at higher latitudes, where more pronounced climatic seasonality creates narrower windows of resource availability, logistical foraging from a permanent base camp is a more common strategy than is residential relocation (Kelly 1983).

Effects of Food-Sharing

As is the case in other mammals, primate mothers actively provision their dependent offspring, but the extent to which primates willingly share food among other nondependent individuals in their group is variable. Chimpanzees are perhaps the most willing food-sharers among nonhuman primates, exhibiting this behavior both in the wild and in captivity, and they appear to share food spontaneously rather than solely in response to demands from other members of their group (Feistner and McGrew 1989). Chimpanzees are nonetheless selective in allocating shared food, favoring individuals with whom they have had close interactions through grooming (de Waal 1997), and the incidence of food-sharing increases in direct response to levels of harassment, showing that sharing among these animals is not entirely voluntary (Stevens 2004).

As voluntary food-sharing is an important feature of human societies, the evolution of this distinctive behavior has been incorporated into models of hominid landscape utilization. Accumulations of worked stones and modified animal bones at East African sites associated with early *Homo* have been interpreted as central places or "home bases" where hominids gathered to share and consume resources obtained through individual or small-group foraging (e.g., Isaac 1978; Rose and Marshall 1996). The validity of the "home base" model has been challenged on the basis that alternative site formation processes could have led to the observed accumulations, and on the theoretical grounds that competition from large carnivores would have necessitated active defense and rapid consumption of any acquired meat resources. Nonetheless, meat was likely to have been an important component in Pleistocene hominid diets, and the acquisition and consumption of meat from large mammals, whether by hunting or scavenging, probably involved a level of collective activity and a degree of planning and cooperation among individuals that exceeded that typically observed among living primates.

Fossil Hominids and Palaeolithic Archaeology

Pliocene and Lower Pleistocene

The hominids are a diverse group of species of bipedal apes that are phylogenetically closer to the human lineage than to other living primate species. The earliest hominid fossils, which date to the Late Miocene and Pliocene, between 6 and 4 million years ago, are assigned to the genera *Ardipithecus, Orrorin, Sahelanthropus,* and *Australopithecus* (White et al. 2006; see Table 8.1 at the end of the chapter). All hominid species with adequately preserved limb bones show adaptations to upright posture and bipedal locomotion,

but the earlier species also exhibit morphological features that are compatible with arboreal life. *Ardipithecus ramidus* and *Australopithecus anamensis*, examples of the earliest known hominids, show characteristically apelike morphological adaptations to tree-climbing in the form of robust distal humeri, relatively long forearms, and curved hand phalanges with strong flexor attachments (Leakey et al. 1998; Ward, Leakey, and Walker 2001; White, Suwa, and Asfaw 1994; White et al. 2006). The fossils of *Ardipithecus ramidus* have been recovered from depositional contexts characteristic of closed canopy woodland rather than open savanna (Wolde Gabriel et al. 1994), and palaeoenvironmental reconstructions of *Australopithecus anamensis* sites at Kanapoi and Allia Bay in Kenya and Asa Issie in Ethiopia indicate that this species favored terrain with open woodland (Ward, Leakey, and Walker 2001; White et al. 2006). Later species of *Australopithecus*, including *A. afarensis, A. africanus,* and *A. robustus/boisei*, also retain some of the hypothesized arboreal adaptations in their forelimb skeletons, and it is only with the emergence of the genus *Homo* that the hominid skeleton appears to be fully adapted to terrestrial life.

There has been sustained interest among palaeoanthropologists in investigating ecological differences among hominid species, because this may explain some of the morphological diversity observed among hominid species (Behrensmeyer 1978; White 1988; Wood and Strait 2004). Evidence of variation in hominid landscape and habitat preferences has been sought in studies of the geological and ecological contexts within which hominid fossils have been discovered. There is a substantial and difficult taphonomic problem here: are the habitats that are conducive to the deposition and preservation of hominid fossils representative of the environments to which the hominids were originally and optimally adapted? Notwithstanding issues of taphonomy and site formation processes, it has been established that many East and South African Pleistocene hominid sites have palaeoenvironmental profiles that are not characteristic of savanna, indicating instead that early hominids occupied sites with a "mosaic" of different habitat types, with a mixture of flora and fauna that are individually adapted to open grassland, woodland, and proximity to water (Potts 1998; Reed 1997; Wood and Strait 2004). Even though some of this palaeoenvironmental evidence may be "time-averaged" (that is, reflecting short-term environmental fluctuation that is temporally unresolved), the prevalence of mosaic habitats associated with the early hominid fossil record provides indirect support for the prospect-refuge theory of human landscape preference: if hominid occupation sites were preferentially located at habitat interfaces, such as the margins between woodland and grassland, then the palaeoenvironmental evidence would be expected to evince a mixture of habitat types.

Alternative theories of early hominid landscape preference that compete with the savanna hypothesis are the variability selection hypothesis (Potts 1998) and the riparian woodland scavenging model (Blumenschine 1987). Potts (1998) proposed that early hominids were evolutionarily adapted to the occupation of a diverse range of environments, because they had evolved during a time when environmental conditions in Africa were varying rapidly and unpredictably in response to global climate change. Potts views the development in early hominids of bipedality, the emergence of tool use, and the enhancement of cognitive and social capacities as co-related adaptations that helped hominids to cope with environmental novelty and allowed them to survive in varied and fluctuating environments. Blumenschine's riparian woodland scavenging model relates primarily to a later phase in human evolution, after about 2.5 million years ago, when hominids first developed the capacity (through stone tool use) to compete successfully with carnivores to gain access to large mammal carcasses. Blumenschine (1987) argued that these scavenging opportunities arose first in gallery woodlands alongside the margins of lakes and permanent watercourses, where visibility of carcasses to competing scavengers was reduced and the proximity of arboreal refuges gave hominids some security from nonclimbing carnivores such as hyenas. The location of these woodland habitats adjacent both to more open grassland and to aquatic habitats might explain the diversity of palaeoenvironmental evidence recovered from fossil hominid sites.

Middle and Upper Pleistocene

During the Lower Pleistocene (approximately 1.7 to 0.8 mya), the australopithecine hominid genera were replaced by the single genus *Homo*, which initially contained a diversity of species but was increasingly dominated by the African species *Homo ergaster* and subsequently (in both Africa and Asia) by *Homo erectus*. The boundary between the Lower and Middle Pleistocene corresponds more or less to the decline of *Homo erectus* and the emergence of *Homo heidelbergensis*, the presumed ancestor of *Homo neanderthalensis* and *Homo sapiens*. *Homo heidelbergensis* shows

range expansion to latitudes higher than the regions colonized by *Homo erectus*, including the first occupation of sites in northern Europe (Parfitt et al. 2005), and in general the Middle Pleistocene hominids were able to occupy a wider range of environments than their Lower Pleistocene forbears (Roebroeks, Conard, and van Kolfschoten 1992).

It is at Middle Pleistocene hominid sites that the first reasonably convincing evidence emerges for small-scale spatial organization of activities. In addition to numerous examples of artifact concentrations at flint-knapping and animal butchery sites (Gamble 1999), there are instances of possible shelters and spatially demarcated working areas suggestive of more formally structured and longer duration activities at sites such as Bilzingsleben in Germany and the Grotte du Lazaret Cave in France. The proposal that controlled (as opposed to opportunistic) use of fire occurred in this time period is controversial (James 1989), but distributions of burned organic materials and flint have been used to identify hearth localities at the site of Gesher Benot Ya'aqov in Israel, dating to about 800,000 years ago (Goren-Inbar et al. 2004). Similarly, the first instances of the deliberate deposition of human remains at specific locations in the landscape emerges during the Middle Pleistocene (Carbonell et al. 2003), although clear evidence for ritualized burial of the dead is not apparent until the emergence of modern humans in the Upper Pleistocene.

Conclusions

Table 8.1 presents in summary form a synthesis of the key attributes of technology, subsistence, ranging behavior, and hypothesized use of space by premodern hominids. The evidence suggests a gradual and progressive increase in the capacity of hominids to range through, to occupy, and to exploit a diversity of landscape types, as well as an increasing ability to migrate and to colonize over long distances. Traditionally, this pattern has been explained in terms of increasing independence from environmental constraints delivered by technological innovations. The existence of landscape preferences among modern humans that mirror those of our early Pleistocene ancestors, despite hundreds of thousands of years of hominid adaptation to the temperate climates of Eurasia, is explicable if the origin of modern humans as a species took place in tropical habitats similar to those existing today in sub-Saharan Africa.

Table 8.1 Technology, subsistence activities, ranging behavior, and use of space by hominids.

Period	Hominid Taxa	Technology/Subsistence	Ranging and Use of Space
Pre-Paleolithic 8–2.5 mya	*Ardipithecus* *Australopithecus* *Paranthropus* *Orrorin* *Sahelanthropus*	Use of unmodified tools, localized foraging, secondary access to animal carcasses, food-sharing at point of acquisition	Dependence on arboreal refuges, localized ranging behavior, restricted use of open landscapes, short-distance migrations
Lower Paleolithic 2.5–0.2 mya	*Australopithecus robustus* *Homo habilis* *Homo rudolfensis* *Homo ergaster* *Homo erectus*	Hand-held stone tools, opportunistic (?) use of fire, longer-distance foraging, primary access to animal carcasses, food distribution at point of consumption	Transport of raw materials over short distances to central or favored places, greater use of open landscapes, longer-distance (transcontinental) migrations and colonizations
Middle Paleolithic 0.2–0.05 mya	*Homo heidelbergensis* *Homo neanderthalensis*	Hafted stone tools, controlled use of fire, unrestricted foraging, full primary access to animal carcasses, food storage and delayed consumption	Access to inhospitable and hazardous landscapes, long-term occupation of sites with spatially demarcated living and working areas, ritualized burials, long-distance transport

References

Appleton, J. 1996. *The Experience of Landscape* (rev. ed.). New York: Wiley.

Balling, J. D., and Falk, J. H. 1982. Development of visual preference for natural environments. *Environment and Behavior* 14: 5– 28.

Behrensmeyer, A. K. 1978. The habitat of Plio-Pleistocene hominids in East Africa, in C. Jolly, (ed.), *Early Hominids of Africa*, pp. 165–89. London: Duckworth.

Blumenschine, R. J. 1987. Characteristics of an early hominid scavenging niche. *Current Anthropology* 28: 383–407.

Carbonell, E., Mosquera, M., Ollé, A., Rodríguez, X. P., Sala, R., Vergès, J. M., Arsuaga, J. L., and Bermúdez de Castro, J. M. 2003. Did the earliest mortuary practices take place more than 350,000 years ago at Atapuerca? *Anthropologie* 107: 1–14.

De Waal, F. B. M. 1997. The chimpanzee's service economy: Food for grooming. *Evolution and Human Behaviour* 18: 375–86.

Elton, S. 2006. Forty years on and still going strong: The use of hominin-cercopithecid comparisons in paleoanthropology. *Journal of the Royal Anthropological Institute* 12: 19–38.

Fedigan, L. M. 1986. The changing role of women in models of human evolution. *Annual Review of Anthropology* 15: 25–66.

Feistner, A. T. C., and McGrew, W. C. 1989. Food-sharing in primates: A critical review, in P. K. Seth and S. Seth (eds.), *Perspectives in Primate Biology*, pp. 21–36. New Delhi: Today and Tomorrow's Printers and Publishers.

Foley, R. A. 1996. An evolutionary and chronological framework for human social behaviour. *Proceedings of the British Academy* 88: 95–117.

Fruth, B., and Hohmann, G. 1996. Nest building behavior in the great apes: The great leap forward? in W. C. McGrew, L. F. Marchant, and T. Nishida (eds.), *Great Ape Societies*, pp. 225–40. Cambridge: Cambridge University Press.

Furuichi, T., and Hashimoto, C. 2004. Botanical and topographical factors influencing nesting-site selection by chimpanzees in Kalinzu Forest, Uganda. *International Journal of Primatology* 25: 755–65.

Gamble, C. 1999. *The Palaeolithic Societies of Europe*. Cambridge: Cambridge University Press.

Goren-Inbar, N., Alperson, N., Kislev, M. E., Simchoni, O., Melamed, Y., Ben-Nun, A., and Werker, W. 2004. Evidence for hominin control of fire at Gesher Benot Ya'aqov, Israel. *Science* 304: 725–27.

Hudson, B. J. 1992. Hunting or a sheltered life: Prospects and refuges reviewed. *Landscape and Uban Planning* 22: 53–57.

Hunt, K. D., and McGrew, W. C. 2002. Chimpanzees in the dry habitats of Assirik, Senegal and Semliki Wildlife Reserve, Uganda, in C. Boesch, G. Hohmann, and L. Marchant (eds.), *Behavioural Diversity in Chimpanzees and Bonobos*, pp. 35–51. Cambridge: Cambridge University Press.

Isaac, G. L. 1978. The food-sharing behaviour of protohuman hominids. *Scientific American* 238: 90–108.

Isbell, L. A., Pruetz, J. D., Lewis, M., and Young, T. P. 1998. Locomotor activity differences betwen sympatric Patas monkeys (*Erythrocebus patas*) and vervet monkeys (*Cercopithecus aethiops*): Implications for the evolution of long hindlimb length in *Homo*. *American Journal of Physical Anthropology* 105: 199–207.

James, S. R. 1989. Hominid use of fire in the Lower and Middle Pleistocene. *Current Anthropology* 30: 1–26.

Jetz, W., Carbone, C., Fulford, J., and Brown, J. H. 2004. The scaling of animal space use. *Science* 306: 266–68.

Jolly, C. J. 2001. A proper study for mankind: Analogies from the papionin monkeys and their implications for human evolution. *Yearbook of Physical Anthropology* 44: 177–204.

Kaplan, S. 1992. Environmental preference in a knowledge-seeking, knowledge-using organism, in J. H. Barkow, L. Cosmides, and J. Tooby (eds.), *The Adapted Mind*, pp. 581–98. Oxford: Oxford University Press.

Kelly, R. L. 1983. Hunter-gatherer mobility strategies. *Journal of Anthropological Research* 39: 277–306.

Leakey, M. G., Feibel, C. S., McDougall, I., Ward, C., and Walker, A. 1998. New specimens and confirmation of an early age for *Australopithecus anamensis*. *Nature* 393: 62–66.

Lee, R. B., and Daly, R. 1999. *The Cambridge Encyclopedia of Hunters and Gatherers*. Cambridge: Cambridge University Press.

Orians, G. H., and Heerwagen, J. H. 1992. Evolved responses to landscapes, in J. H. Barkow, L. Cosmides, and J. Tooby (eds.), *The Adapted Mind*, pp. 555–79. Oxford: Oxford University Press.

Parfitt, S. A., Barendregt, R. W., Breda, M., and 16 others. 2005. The earliest record of human activity in northern Europe. *Nature* 438: 1008–12.

Potts, R. 1998. Environmental hypotheses of hominin evolution. *Yearbook of Physical Anthropology* 41: 93–136.

Reed, K. E. 1997. Early hominid evolution and ecological change through the African Plio-Pleistocene. *Journal of Human Evolution* 32: 289–322.

Roebroeks, W., Conard, N. J., and van Kolfschoten, T. 1992. Dense forests, cold steppes and the Palaeolithic settlement of northern Europe. *Current Anthropology* 33: 551–86.

Rose, L., and Marshall, F. 1996. Meat eating, hominid sociality, and home bases revisited. *Current Anthropology* 37: 307–38.

Sabater Pi, J., Veà, J. J., and Serrallonga, J. 1997. Did the first hominids build nests? *Current Anthropology* 38: 914–16.

Sept, J. 1998. Shadows on a changing landscape: Comparing nesting patterns of hominids and chimpanzees since their last common ancestor. *American Journal of Primatology* 46: 85–101.

Stevens, J. R. 2004. The selfish nature of generosity: Harassment and food sharing in primates. *Proceedings of the Royal Society of London B* 271: 451–56.

Strum, S. C., and Mitchell, W. 1987. Baboon models and muddles, in W. G. Kinzey (ed.), *The Evolution of Human Behavior: Primate Models*, pp. 87–104. New York: State University of New York Press.

Tooby, J., and Cosmides, L. 1990. The past explains the present: Emotional adaptations and the structure of ancient environments. *Ethology and Sociobiology* 11: 375–424.

Ward, C. V., Leakey, M. G., and Walker, A. 2001. Morphology of *Australopithecus anamensis* from Kanapoi and Allia Bay, Kenya. *Journal of Human Evolution* 41: 255–368.

Washburn, S. L., and Devore, I. 1961. Social behavior of baboons and early man. *Yearbook of Physical Anthropology* 9: 91–105.

White, T. D. 1988. The comparative biology of "robust" *Australopithecus*: Clues from context, in F. E. Grine (ed.), *Evolutionary History of the "Robust" Australopithecines*, pp. 449–79. New York: Aldine de Gruyter.

White, T. D., Suwa, G., and Asfaw, B. 1994. *Australopithecus ramidus*, a new species of early hominid from Aramis, Ethiopia. *Nature* 371: 306–12.

White, T. D., Wolde Gabriel, G., Asfaw, B., and 19 others. 2006. Asa Issie, Aramis and the origin of *Australopithecus*. *Nature* 440: 883–89.

Wolde Gabriel, G., White, T. D., Suwa, G., Renne, P., de Heinzelin, J., Hart, W. K., and Heiken, G. 1994. Ecological and temporal placement of early Pliocene hominids at Aramis, Ethiopia. *Nature* 371: 330–33.

Wood, B., and Strait, D. S. 2004. Patterns of resource use in early *Homo* and *Paranthropus*. *Journal of Human Evolution* 46: 119–62.

EVOLUTIONARY PSYCHOLOGY AND ARCHAEOLOGICAL LANDSCAPES

Herbert D. G. Maschner and Ben C. Marler

The relationship between humans and their environments (internal, external, and social) is central to many, if not all, of the human sciences. Humans create their environments and, in turn, are created by them (Crumley 1999; Odling-Smee, Laland, and Feldman 2003). The study of this relationship has taken many different and interesting forms and is fundamental for both archaeology and evolutionary theory. We would expect that, in the process of becoming human, adaptive mechanisms would have evolved that represent a long history of human-landscape interactions and that these mechanisms would reflect the most dominant landscapes in human evolutionary history. Central to understanding the interplay between landscape and our evolutionary history is the burgeoning field of evolutionary psychology. Some of the most basic and fundamental aspects of evolutionary psychology suggest there are common cognitive algorithms or mental domains or modules that are ubiquitous throughout human history (Mithen 1996; Tooby and Cosmides 2005). These mental modules are thus adaptations that evolved over millennia as natural selection acted on our hominid ancestors as they interacted with the environments of the east African landscape. Thus, we would expect our brains to be adapted to landscapes quite different from those most of us live in today; yet, we would also expect that

these adaptations influence how we perceive landscapes everywhere. The archaeological implications for this approach are profound. If the basic cognitive foundations for landscape perception and responses are a product of our evolutionary history and are, in fact, shared by most humans, then this may have a powerful effect on the means by which we symbolize and interpret landscapes in a diversity of cultural and phenomenological contexts. The focus of this chapter is, therefore, on what evolutionary psychology can tell us about the cognitive mechanisms responsible for landscape perception, the role of an evolutionary history of the sexual division of labor in human-landscape interactions, and what these, in turn, can do to help us evaluate the evolutionary history of *homo sapiens* and the archaeological record.

Landscape Archaeology

Landscape archaeology is as old as archaeology itself. As Bernard Knapp and Wendy Ashmore (1999) tell us: "As long as archaeologists have studied human past, they have been interested in space, and consequently in landscapes" (Knapp and Ashmore 1999: 1). This is because, at the very least, landscapes are the framework that contains the archaeological record. The way we understand landscapes is very much influenced by our theoretical perspective (for example, see Darvill, this volume). Landscapes are investigated as economic, social, political, and symbolic entities that often are seen as mutually exclusive areas of inquiry. But most can recognize that landscapes (as most things) are both constructed and objective, internal and external—in John Searle's (1995) terminology, both institutional facts and brute facts, and in Bruno Latour's (2002) words, they are "factishes." As George Children and George Nash have said: "One might ask; what is landscape? We know its there; the hills, mountains, rivers, streams, trees and so on. These features, although external, are socially constructed within our minds, that is, given meaning" (Children and Nash 1997: 1).[1]

Archaeological approaches to landscapes have changed from viewing landscapes as passive objects that humans leave their mark on to dynamic human-constructed environments. Some view landscapes as a part of the feedback loop between humans and their environments or, in some cases, even between genes, their phenotypes, and their environment (Dawkins 1982). It has also been suggested that landscapes and other constructed environments constitute a fully independent transmission system and that these aspects of environments have their own evolutionary trajectory and impact on the larger system of human evolution (Odling-Smee, Laland, and Feldman 2003; Sterelny 2003, 2005, 2007).

The fact that landscapes are both socially constructed and objective realities creates dialogue that moves between their social constructedness and their ontological objectivity (for example see Gamble, this volume). Although traditional approaches to the archaeology of landscape have emphasized functional relationships between people and their environments (e.g., Binford 1982), Knapp and Ashmore again tell us that "today, the most prominent notions of landscape exist by virtue of its being perceived, experienced, and contextualized by people" (1999: 1). The current interest in landscape archaeology is in phenomenological and/or semiotic approaches to landscapes (Children and Nash 1997; Nash 1997) or cognitive approaches (Zube, Pitt, and Evans 1995). Although we believe that no single approach to landscape fully represents the complex interaction between people and their worlds, we find Darwinian theory a potent and unifying framework for investigating human-landscape interactions and quite possibly a means of integrating cognitive and phenomenological approaches.

Darwinian theory is capable of emphasizing and redefining many of the above-mentioned issues in landscape archaeology into a more coherent, continuous, and predictive system. Evolutionary psychology in particular will help with elaborating, contextualizing, and stabilizing phenomenological and cognitive approaches to archaeology by virtue of its emphasis on uniformity of human cognition. The application of this framework will allow archaeologists to interpret how earlier societies perceived landscapes and for these interpretations to become testable hypotheses. From these data, a model of human cognition can be created and sent back in time, much like Steven Mithen's (1996) working hypothesis in which *Homo sapiens*'s cognition is contextualized relative to other hominids by the level of cognitive fluidity, which is the ability to share information between cognitive domains. Here we take a much more specific approach looking at the role of cognitive algorithms in how humans interact and perceive landscapes.

A Darwinian Foundation for Landscape Archaeology

We believe that a Darwinian foundation for a holistic landscape archaeology requires three fundamental components. The first is a broad understanding of the evolution of our cognitive adaptations to landscapes, which will set boundaries on all symbolic interpretations of landscapes. This is done through evolutionary psychology and recognizing the conditions under which the human brain evolved. The second requirement is an approach that integrates multiple levels of selection, from the genetic to the phenomenological, and that creates hierarchies of integrated and embodied systems that evolve simultaneously, each acting on the evolution of the other systems. This means that both evolved responses to landscapes, and symbolic representations of landscapes, are codependent, because both are a product of a hierarchy of selective processes involving human-landscape interactions. Last, we argue that a hierarchically evolved cognitive mechanism is what generates human phenomenology. A linking concept at this level is the schema (defined and elaborated below), which allows phenomenology to be grounded to the human phenotype.

Evolutionary Psychology

Two of the earliest, most vocal, and well-known evolutionary psychologists, John Tooby and Leda Cosmides, tell us that the fundamental assumption of evolutionary psychology is that the "programs comprising [*sic*] the human mind were designed by natural selection to solve the adaptive problems regularly faced by our hunter-gatherer ancestors";

some of these problems were "finding a mate, cooperating with others, hunting, gathering, protecting children, navigating, avoiding predators, [and] avoiding exploitation" (Tooby and Cosmides 2005: 16). In and of itself, this argument is not controversial. Few in archaeology doubt that humanity is a product of evolutionary processes. But from this basic premise follow several other fundamental points. The presence of these mental modules implies, much like any other aspect of the human phenotype, that these programs evolved, and are a function of, human interactions with their environment. It is also argued that these mental modules are so complex as to not have been modified over short evolutionary time scales and are, rather, a product of hundreds of thousands or millions of years of evolution, much, for example, like the human hand. Owing to the fact that these programs were developed over long evolutionary time scales, their evolution has difficulty keeping up with or tracking the evolution of culture. Therefore, we expect that these mental modules are a byproduct of adapting to a Plio-Pleistocene or older world yet are being used to interact with the landscapes of villages, states, and empires. Thus, there is no *a priori* reason to assume modern humans are behaving in an optimal or adaptive sense in the modern world because we are so far removed from the environments where these adaptations arose (Maschner 1996a; Tooby and Cosmides 1987, 2005). This argument does not imply that humans never behave adaptively; evolutionary ecologists have repeatedly demonstrated that foraging societies do create behavioral and often occasionally optimal adaptations to local landscapes but that these behavioral adaptations are built with the complex cognitive adaptations of our evolutionary past.

Much research has gone into supporting the thesis that large portions of human cognition and behavior not only *happen* to be, but *must* be, innate.[2] There are "strong" versions of this argument that suggest that the brain is relatively inflexible (e.g., Pinker 1997) and "weak" versions that allow for this structure to be much more flexible (Dennett 1991; Plotkin 1995; Tomasello 1999). These views are becoming very difficult to overlook in any analysis of human behavior. It is increasingly obvious that human cognition and behavior are to some extent "hardwired" and innate and that the form this "hardwiring" takes is the product of millions of years of human evolution and, on a larger scale, the history of evolution planet-wide.

The *environment of evolutionary adaptedness* (EEA) is the hypothetical environment (social and physical landscape) where humans evolved their cognitive capacities (see also Chamberlain, this volume). Thus, we have a Pleistocene brain and a Neolithic social world, and it is the interaction of these two domains, the evolved and the symbolic, that creates our modern landscape perceptions as well as how we investigate the archaeology of landscapes (Barkow, Cosmides, and Toob 1993; Bentley et al. 2007). Evolutionary psychology has been criticized as having devalued or entirely removed culture from the analysis of human behavior, and in some of the stronger versions of this argument this may be the case. It has also been argued that these foundational points are supported by ethnocentric studies. As is discussed below, these criticisms do not take into account that culture is itself an evolved mechanism. R. B. Hull and G. R. B Revell (1995) give a fair treatment of both the criticisms and the defenses of this issue as it applies to landscape perception, specifically scenic beauty evaluations (discussed below) and identify several key problems with some of these studies. A criticism they discuss is that many of the studies conducted use North American or European participants. They argue that these samples cannot be generalized to other cultures, because it is likely that most of these will have similar experiences of non-urban landscapes, which are generally either romanticized or fictionalized via any number of different media (Hull and Revell 1995). These criticisms (if methodologically sound) may, however, be misdirected, and the problems could be better blamed on lack of cross-cultural studies rather than on a theoretical problem with evolutionary psychology. An approach that reemphasizes culture and integrates cultural and individual evolution into a larger picture of human evolution already containing genetic evolution may help to illustrate this.

The G-T-R Heuristic

For a comprehensive analysis of the archaeology of landscapes to be successful, theory must integrate all units of variation from the phenotypic to the phenomenological. Henry Plotkin's (1995) version of universal Darwinism does just this and is particularly interesting for two related and linked reasons. Plotkin's model is an adaptable version of evolutionary logic that emphasizes and makes room for multiple interacting systems, each of which is evolving in its own way and simultaneously as part of a larger co-evolutionary system. Plotkin's theoretical approach works on two general levels. The first is evolution as a knowledge-acquiring-and-integrating system. The second is the treatment of knowledge-acquiring-and-integration as a system of evolution. The most important concept in Plotkin's conceptualization of evolution is what he calls the "g-t-r heuristic." This is a simplified way of writing the notion that evolution has phases. There is the "*g*eneration" phase, a "*t*esting" phase, and a "*r*egeneration" phase:

> The g-t-r refers to three consecutive and continuous phases in the overall evolutionary process: the g phase is the generation of variants (the nature of which need not be specified, but which in the paradigmatic case would be genotypes, phenotypes or parts of phenotypes); the t phase is the test or selection phase (natural selection in the standard case); and the r phase refers to the regeneration of variants, combining previously selected variants and newly arising ones. (Plotkin 1993: 84)

Any system capable of doing these three things is considered an evolving system. This g-t-r is recursive. Any output from the system has the potential to be brought in as part of the system; therefore, the system affects its own future evolution by acquiring and integrating knowledge (that is, adapting).

Important to this analysis is that the human brain/mind is seen as a product of evolution and as a system of evolution. The evolving brain gives rise to Plotkin's third heuristic (culture) and social reality and is, in part, one of the mediums in which culture is acted on and, in that sense, is one place where we can expect to find selection pressures on cultural behavior. Each of Plotkin's heuristics is capable of tracking change in the environment at different rates. Genetic evolution is the slowest; individual evolution is faster than genetic but slower than cultural evolution, which is the fastest. Most important, each of these heuristics is fundamental and foundational to human behavior and human cognition, and each one relies on the others for its information and its generation of difference. As can be seen with Plotkin's model, what an evolved psychology does represent is the macro features of human cognition and behavior. Culture, either as a biological or nonbiological entity, would fine-tune human cognition and behavior and generate some of the diversity with which evolution would act on the macro-scale features. We further suggest, therefore, that the similarities found between people should be an important starting point for analysis and a powerful tool for prediction and hypothesis generation when one is discussing the archaeology of landscape. For an example of how cultural systems are knowledge-acquiring systems, see Torrence (this volume), and for an example of how humans construct their environments, see McNiven (this volume).

Schemas and Neural Architectures

As was mentioned above, evolutionary and cognitive psychology can be used to unify many different approaches to human behavior. In this section, we link the phenomenological and Darwinian analyses using the concept of a schema. The term *schema* has been defined as a "bounded, distinct and unitary representation" (quoted in D'Andrade 1995: 122). This representation is a mental one that is supposed by Roy D'Andrade to be one of the primary centripetal forces that cause distinct if evolving cultures. A schema is a complex mental map that people develop about objects, people, and situations that allows them to more easily navigate their world. The schema is "activated" by context—that is, the internal experience (schema) is a response to the external experience, and the schema informs one's understanding of and behavior toward that external experience. Thus, to some extent, a schema is what makes complex meaning arise out of experience. One note, however; schemas are not one-to-one maps of experience. That is, they are not direct copies of the object, people, or situations that they refer to and are activated by. A decent analogy could be a large computer file that has been compressed or "Zipped" so that internal information (a schema or the compression program's key) is necessary to fill in the gaps and to make sense out of the larger file (experience or reality).

Schemas also tend to be hierarchically structured (D'Andrade 1995; Lakoff and Johnson 1999). That is, large schemas contain smaller subschemas that contain still smaller schemas and so forth. One interesting byproduct of this situation is that schemas then become an analytical concept that allows us to understand the process of abstraction as hierarchical and "embodied" (Lakoff and Johnson 1999) and to understand how relationships between nonsimilar items can be created if they are contained within a larger schema (D'Andrade 1995). The same neural structures that generate our experience of the world generate our categories of things, as well as our metaphors (Lakoff and Johnson 1999).

Schemas are created by joint attentional scenes (Tomasello 1999). Joint attentional scenes are the ability for individuals to share an experience and to share the fact that the experience is being shared—that is, not only to realize that you and another are experiencing the same thing but also to realize that the other has an experience of that thing that is similar to yours. This realization is thought to be a necessary component of the use of symbols and icons. Symbols and icons allow for similar schemas to activate in more than one individual allowing for shared group meaning and social reality (Searle 1995)—hence, D'Andrade's assertion that schemas help create distinct cultures.

The idea of a schema is a useful generalized concept that has several benefits. It is situated at a nexus of numerous fields that allow it the ability of bridging some gaps that can be quite tricky to deal with. The schema framework is grounded; that is, it fits with the dominant model of how our brain processes information. This model is known as the *computational model of mind*. One aspect of this model is that our brain is both a serial processor *and* a massively parallel processor. In other words, not only does the brain process symbols in a syntactical fashion, but also it does so very rapidly, and it processes many symbols at once (D'Andrade 1995; Dennett 1991; Lakoff and Johnson 1999; Minsky 1988; Pinker 1997). The particularities of this are not worked out yet, as has been humorously pointed out by the notorious antirelativist, Jerry Fodor (2001), and a bit more soberly by Noam Chomsky (2000). However, counterintuitively, this apparent failing is actually another strength of the schema framework, since it is not necessary for this massively parallel processor to behave in any more specific a fashion than that which has already been described for the schema framework to be valid. Whatever more specific information we do get from neurologists will help to further develop the schema concept.

Theoretical Discussion and Importance to Landscape Archaeology

The archaeology of landscape must be seen as the interaction between our evolved mental algorithms and human-created symbolic meaning for that landscape. While symbolic and poststructuralist archaeologists have made it potently clear that there is significant variation in the role of landscape and how it is perceived, evolutionary psychologists have demonstrated that this variation is not limitless, and, thus, there are bounds on the potential for landscape perception among humans (for a different approach to understanding landscape perceptions, see Rainbird, this volume).

Any mental modules evolved for dealing with landscape among humans are a product of a long evolutionary history, and these adaptations are of sufficient complexity that it is unlikely selection

will alter them in cultural time frames. Therefore, we would expect that any individual or cultural variation in the role of landscape in society would be embodied in our evolved responses to landscape. This expectation requires that we create evolutionary constructs that incorporate and appreciate both evolved adaptations and the phenomenological characteristics of human-landscape interactions.

The G-T-R approach does just this by creating a model whereby mental modules and symbolic representations are seen as interrelating evolving systems. It also illustrates how each new system is embodied in the previous and generates some of the difference required to make the more primary system evolve. Schemas are units of analysis that allow the mental machinery to be linked to social/symbolic levels of analysis by virtue of the fact that schemas are created by syntactical neurological systems that have evolved for particular purposes. Thus, schemas are also responsible for linking symbols together and for the generation of shared meaning by virtue of this. (For more on the implications of shared meaning see both Gamble and Cummings, this volume.)

Having addressed some of the current issues in both landscape archaeology and evolutionary psychology and illustrated how the two disciplines may be used to help each other at a conceptual level, we now turn to the more specific scope of this chapter, that of spatial cognition and its relationship to the sexual division of labor in our evolutionary history. Keeping Plotkin in mind, we hope that we can attempt this in the least ethnocentric way possible and develop hypotheses that can be generalized to other populations than just modern Western university students and more specifically to nonmodern people, thus allowing us to interpret and to predict the archaeological record.

Three Case Studies

Evolved Responses to Landscapes

Aesthetics has been debated by philosophers throughout recorded history and likely even further back to whatever point humans began to argue with other humans. One of the ubiquitous questions is the origin and residence of beauty. Recent studies in evolutionary and cognitive psychology that address this question find interesting similarities in human aesthetics. However, this fact should not be interpreted to suggest that human aesthetics are reducible to evolutionary psychology *as it now stands*. Developmental and social psychology will be needed to understand human preference. This need is illustrated by the fact that younger test subjects tend to prefer what are commonly seen as landscapes that would have been important in our evolutionary history, but adult subjects show a similar preference for those landscapes and also for the landscapes they were in during their youth (Dutton 2003; Kaplan 1992; Orians and Heerwagen 1992).

Most archaeologists are well aware that there seems to be a cross-cultural pattern in the choice of elevated ground with a view over lowlands without views for human settlement location and as a measure of status and prestige. What archaeologists are generally not aware of is the immense number of tests that have been performed by cognitive psychologists to investigate innate psychological mechanisms of landscape choice. In a paper entitled "Evolved Responses to Landscapes," Gordon Orians and Judith Heerwagen (1992) have demonstrated interesting and perhaps revolutionary human responses to certain landscape features that may have a profound effect on how we build and interpret the archaeological record.

First, in a number of cross-cultural and cross-environmental tests, photographs of landscapes were shown to a statistically valid sample of individuals of different ages. These photographs were of different kinds of landscapes, landscape features, and environments. At first, some of the photographs showed fresh water or animals, and it became immediately clear that photographs with animals or fresh water were chosen over landscape shots without them, regardless of the environment depicted.

In a second series of tests, only landscape shots were used that had no fresh water or wildlife. They found in their less than 10-year-old group that cross-culturally, regardless of the environment in which the test subjects were living, open woodland and savanna-like landscapes were consistently selected over other landscapes. In their adult samples, these same landscapes were selected, along with the local landscape of the test subjects as favorite locations. Further, they found that there were particular types of savanna-like landscapes being selected. On the African savanna there are two dominant zones, one with tall trees with straight trunks and no low branches, and another with many small trees with trunks branching low to the ground. They found consistent cross-cultural preference for low-branching trees, trees that during our evolutionary history would have provided fruits and forage within reach and that would have been accessible for defense. This

preference for selecting savanna-like trees and trees for defensive purposes has been supported by further research conducted by Summit and Sommer (1999).

This basic model has been used to assess the choices that European colonizers made 500 years ago during their 350-year-spread from Europe, which has been linked to the population doubling that happened at that time (Fox, Hoobs, and Loneragan 2000). The authors illustrate how the writings from these colonizers show a particular preference for open areas and scattered trees. They choose as the primary topic for this discussion the Swan River area, which was Western Australia's first European colony, a fact that had perplexed ecologists. The authors admit that colonizers could be poor choices for this type of research, since it is equally possible that "naturally, explorers concentrated on open-grassy areas: they were looking for grazing country and found this type of land easy to move through" (Fox, Hoobs, and Loneragan 2000: 207). However, it is just as likely that the overrepresentation of reports on these savanna-like areas is a direct response to subconscious evolved preferences to these landscapes, especially when the somewhat nomadic life of European settlers is compared to the nomadic life humans had during the environment of evolutionary adaptedness and before the development of cities, more than 10,000 years ago.

It must be stressed that evolutionary analyses are necessarily statistical, meaning that most humans demonstrate a particular preference most of the time. The finding that the most common favorite color worldwide is blue, with green being second, does not mean that any particular individual's favorite color will be blue (Dutton 2003), just that the person's favorite color is more likely to be blue than any other color. When these preferences are shown to be statistically significant cross-culturally, we must look for a pan-human causative mechanism to explain these commonalities, and it is becoming more likely that this mechanism takes a biological evolutionary form.

This example of aesthetics, art, and other virtual environments helps to illustrate the concept of a schema as discussed above by clearly showing how evolved preferences are activated by particular contexts, and how these preferences are hierarchical. For example, in the same research that showed the above-mentioned color preference, conducted on people living in 10 countries (including countries from Africa, Asia, Europe, and the Americas), it was found that respondents expressed a liking for realistic representative paintings. Preferred elements included water, trees and other plants, human beings (with a preference for women and children and also for historical figures such as Jomo Kenyatta or Sun Yat-sen), and animals, especially large mammals, both wild and domestic. (Dutton 2003: 696)

Even when removed from immediate life choices, in a massively cross-cultural survey, most people would prefer images and colors of the objects or settings that would be found in our EEA most of the time. These representations can be decontextualized and recombined to form abstract art, and the preference still remains. We must remember that a schema is an *analytical* and not an *empirical* unit. When a schema is activated for something or in relation to something, the neural architecture, with all its various influences, is being activated in a particular way.

Wayfinding

Much research has focused on wayfinding and route learning (Choi and Silverman 2002, 2003; Chown, Caplin, and Kortenkamp 1995; Silverman and Choi 2005; Silverman et al. 2000). The geographer Reginald Golledge gives a useful definition of wayfinding:

> the ability to determine a route, learn it, and retrace or reverse it from memory. Wayfinding is universal to all cultures. It is involved in a myriad of daily and longer-term episodic activities ranging from a search of local areas for food sources to the large-scale and long-term international migrations that first populated the world. (Golledge 2002: 25).

The cognition of wayfinding has been usefully synthesized with the acronym PLAN (Prototypes, Location, and Associative Networks) by Chown and associates (1995). Their model is based on cognitive, developmental, and evolutionary psychology. Thus, it is supported, defined, and constrained by these fields. They make the argument that evolution is constrained to work in particular ways and to build things under these constraints. They identify three important evolutionary constraints operating on cognitive maps. The first is "simplicity of structure" or parsimony, meaning that evolution will not create an organ in a way that is overcomplicated when a simpler one could suffice, since a simple structure is less likely to

break down than is a more complicated one. The next is "consistency," which "mandates" that the brain is unlikely to develop unique representational structures for every task but rather would use existing structures. These would be, at least to some extent, general-purpose and could be used in several domains. The third constraint they discuss is "economy," which suggests that evolution is more likely to create a brain that stores and processes information in less costly ways (Chown, Caplin, and Kortenkamp 1995). These three evolutionary constraints are closely related to one another and are used by the authors to develop a model of human wayfinding cognition that utilizes landmarks as being particularly important to wayfinding. Landmarks are conceived not as the prototypes—since a useful landmark would not be something that is represented as the generalization of the sum of all trees—but rather as the deviation from the prototype—a unique tree.

Gender differences in spatial cognition are not only accessible by behavioral analyses; MRI (magnetic resonance imagery) and fMRI (functional magnetic resonance imagery) can be used to map the physical responses of the brain to stimulus. In a series of tests on humans and other animals, Grön and colleagues (2000) demonstrated that there is a gender difference between males and females in response to maze navigation. The researchers used a three-dimensional virtual-reality maze and an fMRI to map the differences. Navigation activated several areas of the brain in both males and females, including the medial occipital gyri, lateral and medial parietal regions, posteriorcingulate and the parahippocampal gyri, and the right hippocampus proper. However, in males there was distinct activation of the left hippocampus and in females of the right parietal and right prefrontal cortex. Computer simulations and virtual environments (VEs) have been used to investigate gender differences in wayfinding techniques and abilities (Cutmore et al. 2000; Grön et al. 2000). Although this method does contain its own difficulties, and VEs are not the same as other environments, when coupled with fMRI these techniques are a useful tool for exploring modern populations and their abilities in this realm. Current research using these techniques are showing that males tend to have a higher ability at acquiring wayfinding techniques using landmarks and that the right hemisphere is more active than the left during wayfinding experiments in VEs (Cutmore et al. 2000).

The example of wayfinding helps to illustrate how a schema is a complex mental map and how this map is dependent on experience as well as the evolved mental machinery in a manner similar to a

Zipped file on a computer, as well as the influence of joint attentional scenes on schemas. The creation of a schema for a route and a schema for a landmark is particularly interesting. Neither one is useful without external stimulus from the landscape, yet both contain information about this landscape that is vital for navigating it. Humans utilize landmarks, viewed as a deviation from a schema, while instructing another person on a particular route. Thus, the necessity is there for both or all involved to share the general schema of the landscape as well as the same schema for whatever it is that the landmark is a deviation from. The utilization of landmarks itself is very amenable to the schema concept. For a landmark to work at all, it is required to be hierarchically linked to other schemas via a mental map of a particular landscape.

[B*Spatial Mapping and Evolutionary Histories*
Recent developments in cognitive and evolutionary psychology have found that there are specific and patterned human responses to landscapes and landscape features. It has further been found that there are gender-based differences in how individuals map their local landscapes. Here we argue that the implications of these recent studies will have long-term consequences for evaluating human responses to landscapes. Spatial cognition is important to both evolutionary psychology and to landscape archaeology, and thus a discussion of it may help to illustrate our argument up to this point. Hypotheses have been generated by archaeologists and tested by psychologists, and these have been refined and retested and can, in turn, be reapplied to the archaeological record. Let us begin with a discussion of spatial cognition and a review of some of the literature and studies that have been conducted. This discussion will, in turn, help us to explain how humans utilized their landscapes in the past and how their perception of these landscapes was formed by natural selection.

Irwin Silverman and Marion Eals begin their (1992) account of spatial differences with the statement that "the near universality of sex differences in spatial abilities across human cultures and their occurrence in other species indicates the feasibility of an evolutionary approach" (Silverman and Eals 1992: 533). It was relatively common for investigators to approach this topic from the point of view that males would need greater spatial abilities for use in navigation to maintain their territories and for purposes of hunting so they could compete for, attract, and defend females. Silverman and Eals, however, modified this and looked for dimorphism in spatial cognition rather than male superiority:

We hold that the critical factor in selection for spatial dimorphism in humans was sexual division of labor between hunting and gathering during hominid evolution. Although there has, undoubtedly, been overlap in sexes in these functions, archaeological and paleontological data show that across evolutionary time, males predominantly hunted and females predominantly foraged. (Silverman and Eals 1992: 534)

This hypothesis is logically consistent with Darwinian theory. We reemphasize that there is indeed variation and overlap in the sexual division of labor, but data from both anthropology and psychology can be seen to support this thesis (e.g., Bird 1999; Brightman 1996; Choi and Silverman 2003; Chown, Caplin, and Kortenkamp 1995; Elston and Zeanah 2002; James and Kimura 1997; Lee 1979; McBurney et al. 1997; Panter-Brick 2002; Phillips and Silverman 1997; Silverman and Choi 2005; Silverman and Eals 1992; Silverman et al. 1999; Silverman et al. 2000). The tests conducted to separate out male specializations and female specializations are seen as proxy measures for tasks that would have been performed by our hunter-gatherer ancestors. These tests have been performed cross-culturally, including in North America, Japan, England, Scotland, India, South Africa, Australia, Ghana, and Sierra Leone, with several of these studies not only showing that the direction of the sex differences is similar but also showing the magnitude (Choi and Silverman 1996).

One test in which males consistently perform higher is the three-dimensional object mental rotation task. A series of four or five three-dimensional objects are shown to the participants, who are tested for the speed in which they can mentally rotate the first object and then say which of the other objects are the same but in a different position. Several rows of shapes are given and a time limit set. Time is subtracted for each incorrect answer when calculating the total time. Males also performed better on a similar test in which objects were shown and the participant had to identify which of the objects could be made from the initial object. These studies were seen as testing for mental abilities necessary for hunting (Choi and Silverman 1992).

Other tests were developed to test for mental operations necessary in gathering, such as object memory and location memory. These tasks included showing participants two separate pictures with numerous objects on them. In the study on object memory, the first picture contains fewer items than the second, and the second contains all the objects from the first with several added. The task is to identify which items are in both pictures. The study designed to test location memory again had two pictures but both with the same items; but in one picture, some of the objects had been moved. The task was to identify which items had been moved. A third test was constructed that put these participants into a more familiar environment, an office cubicle. All external stimuli were removed before the subjects entered, and the subjects were left for a defined amount of time to sit in this room and to try to memorize item locations and types. Upon exiting the room, they were asked to identify as many items and their locations as they could. These tasks all showed a higher ability in females (Choi and Silverman 1992).

Further studies replicating these tests have criticized Choi and Silverman for having too small a sample size and for not representing the differences between location memory and object memory accurately, finding instead that men show a similar aptitude at location memory (James and Kimura 1997; McBurney et al. 1997; Stumpf and Eliot 1995). However, this finding does not challenge the hypothesis that these abilities in memory tasks are consistent with the hunter-gatherer sexual division of labor (James and Kimura 1997; McBurney et al. 1997). Some researchers, in fact, considered this outcome to be more supportive of the gatherer hypothesis (McBurney et al. 1997). Thus, there is a clear male bias toward mental rotations, map reading, and maze learning and a clear female bias toward object location and object memory, including the spatial relationships among objects.

While untested, these findings imply that males and females used the landscape quite differently for millennia and that these differences resulted in different cognitive adaptations to the EEA. If males and females generally used landscapes differently, and if the scale of landscape-use varied between the sexes, then we would certainly expect that these differences would be imprinted on our evolutionary history.

The implications of these studies are profound. If females do indeed have detailed maps and mapping abilities for local spaces as a product of gathering in restricted, near-camp landscapes and men have generalized, long-distance maps, then we might expect that in environments where the bulk of the resources are collected by women, such as in sub-Saharan Africa, site location is more likely to be based on women's perceptions of landscape. In the arctic, in contrast, where resources are widely distributed and largely harvested by men, site location should be based substantially on male perceptions of the landscape (Maschner 1996b). This hypothesis is fully testable using the ethnographic and archaeological records.

Conclusions

For several million years, humans lived in small, mobile, mostly egalitarian social formations that, for much of that time, existed on the open woodlands and grasslands of Africa, the Near East, and adjacent regions. These hominids lived during a period of cognitive expansion and development that has led evolutionary psychologists to argue that many aspects of the modern mind represent innate abilities that developed in the context of living on this environment of evolutionary adaptedness, or EEA (Barrett, Dunbar, and Lycett 2002). We also now recognize that this is not simply binary, whereby either all humans have these adaptations or all do not. Rather, there is a growing body of evidence that differing landscape-uses by males and females may have resulted in different, but complementary, cognitive adaptations to landscapes—adaptations that we have manipulated into our modern interactions with the 21st-century world and that are probably not adaptive, or are accidentally adaptive, in these modern contexts.

These findings may seem in conflict with social, structuralist, and phenomenological approaches to landscape, or even in conflict with modern evolutionary ecology, but they are not. They are not, because we see symbolic representations and interpretations of landscape as nested within these Darwinian programs resulting in a hierarchy of selective processes that transcend and incorporate all alternative interpretations. Building on Plotkin's g-t-r heuristic, we find it critical to recognize these hierarchies as nested evolutionary domains from the most conservative (mental algorithms) to the most flexible (phenomenology). The concept of schema provides the bridging argument that allows the symbolic/phenomenological level of analysis to be connected to and supported by the evolutionary/cognitive level of analysis. It does this by creating a phenomenological category that follows all the syntactical rules that have been neurologically hardwired by our evolution.

Thus, human perceptions of landscape cannot simply be seen as adaptations to a current natural environment, as functionalists would have us believe, or as the navigation of a social world, as structuralists would have us believe, or as someone standing in a field contemplating the universe, as some phenomenologists and palaeophilosophers would like us to believe. Rather, all these different approaches provide data on nested levels of landscape perceptions, critical details as to our innate landscape predispositions, and, ultimately, evidence of our evolutionary history.

Notes

1. For a more in-depth discussion of these different approaches to landscape in archaeology and how they interact with one another, see David and Thomas (this volume) and Strang (this volume).

2. For examples, of these diverse attempts, see Chomsky 2000; Dennett 1991; Fodor 1983; Gabora 2001; Lakoff and Johnson 1999; Pinker 1997; Plotkin 1995; Tomasello 1997, 1999.

References

Barkow, J., Cosmides, L., and Tooby, J. 1993. *The Adapted Mind: Evolutionary Psychology and the Generation of Culture*. New York: Oxford University Press.

Barrett, L., Dunbar, R., and Lycett, J. 2002. *Human Evolutionary Psychology*. Princeton, NJ: Princeton University Press.

Bentley, R. A., Lipo, C., Maschner, H. D. G., and Marler, B. 2007. Darwinian archaeologies, in R. A. Bentley, H. D. G. Maschner, and C. Chippinale (eds.), *The Handbook of Archaeological Theory*. Walnut Creek, CA: AltaMira Press.

Binford, L. R. 1982. The archaeology of place. *Journal of Anthropological Archaeology* 1: 1–31.

Bird, R. 1999. Cooperation and conflict: The behavioral ecology of the sexual division of labor. *Evolutionary Anthropology: Issues, News and Reviews* 8: 65–75.

Brightman, R. 1996. The sexual division of foraging labor: Biology, taboo, and gender politics. *Comparative Studies in Society and History* 38: 687–729.

Children, G., and Nash, G. 1997. Establishing a discourse: The language of landscapes, in G. Nash (ed.), *Semiotics of Landscapes: The Archaeology of Mind*. Oxford: Archaeopress.

Choi, J., and Silverman, I. 1996. Sexual dimorphism in spatial behaviours: Applications to route learning. *Evolution and Cognition* 2: 165–71.

———. 2002. The relationship between testosterone and route-learning strategies in humans. *Brain and Cognition* 50: 116–20.

———. 2003. Processes underlying sex differences in route-learning strategies in children and adolescents. *Personality and Individual Differences* 34: 1153–66.

Chomsky, N. 2000. *New Horizons in the Study of Mind and Language*. Cambridge, MA: MIT Press.

Chown, E., Caplin, S., and Kortenkamp, D. 1995. Prototype, location and associative networks (PLAN): Toward a unified theory of cognitive mapping. *Cognitive Science* 19: 1–51.

Crumley, C. L. 1999. Sacred landscapes: Constructed and conceptualized, in A. B. Knapp and W. Ashmore (eds.), *Archaeologies of Landscape*. Oxford: Blackwell Publishing.

Cutmore, T. R. H., Hine, T. J., Maberly, K. J., Langford, N. M., and Hawgood, G. 2000. Cognitive and gender factors influencing navigation in a virtual environment. *International Journal of Human-Computer Studies* 53: 223–49.

D'Andrade, R. 1995. *The Development of Cognitive Anthropology*. Cambridge: Cambridge University Press.

Dawkins, R. 1982. *The Extended Phenotype*. New York: Oxford University Press.

Dennett, D. 1991. *Consciousness Explained*. Boston: Back Bay Books.

Dutton, D. 2003. Aesthetics and evolutionary psychology, in J. Levinson (ed.), *The Oxford Handbook for Aesthetics*. New York: Oxford University.

Elston, R. G., and Zeanah, D. W. 2002. Thinking outside the box: A new perspective on diet breadth and sexual division of labor in the Prearchaic Great Basin. *World Archaeology* 34: 103–30.

Fodor, J. 1983. *The Modularity of Mind: An Essay on Faculty Psychology*. Cambridge, MA: MIT Press.

———. 2001. *The Mind Doesn't Work That Way*. Cambridge, MA: MIT Press.

Fox, P., Hoobs, R. J., and Loneragan, W. A. 2000. The lie of the land: The role of perception in our relationship to the environment. *Ecological Management and Restoration* 1: 105–10.

Gabora, L. 2001. Cognitive mechanisms underlying the origin and evolution of culture. Unpublished Ph.D. thesis. Brussels: Free University of Brussels.

Golledge, R. G. 2002. Human wayfinding and cognitive maps, in M. Rockman and J. Steele (eds.), *Colonization of Unfamiliar Landscapes: The Archaeology of Adaptation*. London: Routledge.

Grön, G., Wunderlich, A. P., Spitzer, M., Tomczak, R., and Riepe, M. V. 2000. Brain activation during human navigation: Gender-different neural networks as substrate of performance. *Nature Neuroscience* 3: 404–08.

Hull, R. B. IV, and Revell, G. R. B. 1995. Cross-cultural comparison of landscape scenic beauty evaluations: A case study in Bali, in A. Sinha (ed.), *Readings in Environmental Psychology: Landscape Perception*. San Diego: Academic Press.

James, T. W., and Kimura, D. 1997. Sex differences in remembering the locations of objects in an array: Location-shifts versus location-exchanges. *Evolution and Human Behavior* 18: 155–73.

Kaplan, S. 1992. Environmental preference in a knowledge-seeking, knowledge-using organism, in J. Barkow, L. Cosmides, and J. Tooby (eds.), *The Adapted Mind: Evolutionary Psychology and the Generation of Culture*. New York: Oxford University Press.

Knapp, A. B., and Ashmore, W. 1999. Archaeological landscapes: Constructed, conceptualized, ideational, in A. B. Knapp and W. Ashmore (eds.), *Archaeologies of Landscape*. Oxford: Blackwell Publishing.

Lakoff, G., and Johnson, M. 1999. *Philosophy in the Flesh: The Embodied Mind and Its Challenge to Western Thought*. New York: Basic Books.

Latour, B. 2002. *We Have Never Been Modern*, C. Porter (trans.). Cambridge, MA: Harvard University Press.

Lee, R. B. 1979. *The !Kung San: Men, Women, and Work in a Foraging Society*. New York: Cambridge University Press.

Maschner, H. D. G. 1996a. The politics of settlement choice on the Northwest Coast: Cognition, GIS, and coastal landscapes, in M. Aldenderfer and H. D. G. Maschner (eds.), *Anthropology, Space, and Geographic Information Systems*, pp. 175–89. Oxford: Oxford University Press.

———. 1996b. Theory, Technology, and the Future of Geographic Information Systems in Archaeology, in H. D. G. Maschner (ed.), *New Methods, Old Problems: Geographic Information Systems in Modern Archaeological Research*, pp. 301–08. Carbondale, IL: Center for Archaeological Investigations Press.

McBurney, D. H., Gaulin, S. J. C., Devineni, T., and Adams, C. 1997. Superior spatial memory of women: Stronger evidence for the gathering hypothesis. *Evolution and Human Behavior* 18: 165–74.

Minsky, M. 1988. *The Society of Mind*. New York: Touchstone Books.

Mithen, S. 1996. *Prehistory of the Mind: The Cognitive Origins of Art and Science*. New York: Thames and Hudson.

Nash, G. 1997. Experiencing space and symmetry: The use, destruction and abandonment of La Hougue Bie Neolithic passage grave, Jersey, in G. Nash (ed.), *Semiotics of Landscapes: The Archaeology of Mind*. Oxford: Archaeopress.

Odling-Smee, F. J., Laland, K. N., and Feldman, M. W. 2003. *Niche Construction*. Princeton, NJ: Princeton University Press.

Orians, G. H. 2001. An evolutionary perspective on aesthetics, in G. J. Feist (ed.), *Bulletin of Psychology and the Arts: Evolution, Creativity, and Aesthetics* 2(1): 25–29.

Orians, G. H., and Heerwagon, J. H. 1992. Evolved responses to landscapes, in J. Barkow, L. Cosmides, and J. Tooby (eds.), *The Adapted Mind: Evolutionary Psychology and the generation of Culture*. New York: Oxford University Press.

Panter-Brick, C. 2002. Sexual division of labor: Energetic and evolutionary scenarios. *American Journal of Human Biology* 14: 627–40.

Phillips, K., and Silverman, I. 1997. Differences in the relationship of menstrual cycle phases to spatial performance on two- and three- dimensional tasks. *Hormones and Behavior* 32: 167–75.

Pinker, S. 1997. *How the Mind Works*. New York: Norton.

Plotkin, H. 1993. *Darwin Machines and the Nature of Knowledge*. Cambridge, MA: Harvard University Press.

Searle, J. R. 1995. *The Construction of Social Reality*. New York: The Free Press.

Silverman, I., and Choi, J. 2005. Locating places, in D. M. Buss (ed.), *The Handbook of Evolutionary Psychology*. Hoboken, NJ: John Wiley and Sons, Inc.

Silverman, I., Choi, J., Mackewn, A., Fisher, M., Moro, J., and Olshansky, E. 2000. Evolved mechanisms underlying wayfinding: Further studies on the hunter-gatherer theory of spatial sex differences. *Evolution and Human Behavior* 21(3): 201–13.

Silverman, I., and Eals, M. 1992. Sex differences in spatial abilities: Evolutionary theory and data, in J. Barkow, L. Cosmides, and J. Tooby (eds.), *The Adapted Mind: Evolutionary Psychology and the generation of Culture*. New York: Oxford University Press.

Silverman, I., Kastuk, D., Choi, J., and Phillips, K. 1999. Testosterone levels and spatial ability in men. *Psychoneuroendocrinology* 24: 813–22.

Silverman, I., Phillip, K., and Silverman, L. K. 1996. Homogeneity of effect sizes for sex across spatial tests and cultures: Implications for hormonal theories. *Brain and Cognition* 31: 90–94.

Sterelny, K. 2003. Thought in a hostile world: The evolution of human cognition. Malden, MA: Blackwell.

———. 2005. Made by each other: Organisms and their environment. *Biology and Philosophy* 20: 21–36.

———. 2007. An alternative evolutionary psychology? in S. A. Gangestad and J. A. Simpson (eds.), *The Evolution of Mind: Fundamental Questions and Controversies*. New York: Guilford Press.

Stumpf, H., and Eliot, J. 1995. Gender-related differences in spatial ability and the k factor of general spatial ability in a population of academically talented students. *Personality and Individual Differences* 19: 33–45.

Summit, J., and Sommer, R. 1999. Further studies of preferred tree shapes. *Environment and Behavior* 31: 550–76.

Tomasello, M. 1997. *Primate Cognition*. Oxford: Oxford University Press.

———. 1999. *The Cultural Origins of Human Cognition*. Cambridge, MA: Harvard University Press.

Tooby, J., and Cosmides, L. 2005. Conceptual foundations of evolutionary psychology, in D. M. Buss (ed.), *The Handbook of Evolutionary Psychology*. Hoboken, NJ: John Wiley and Sons, Inc.

Zube, E. H., Pitt, D. G., and Evans, G. W. 1995. A lifespan developmental study of landscape assessment, in A. Sinha (ed.), *Readings in Environmental Psychology: Landscape Perception*. San Diego: Academic Press.

THINKING THROUGH LANDSCAPES

In recent years, many archaeologists have been drawn to perspectives that break down the divisions between mind and body, thought and action, whether these be phenomenological or inspired by the evolutionary concept of the "extended phenotype," whereby all aspects of an organism's behavior and environmental engagements are relevant to its reproductive fitness and survival. These approaches often seek to argue that thought is itself a kind of action, or at least that it is always undertaken by an embodied being in a contingent material context (Ingold 1998; O'Brien and Holland 1995; Thomas 2006). It follows that thinking "about" the landscape is also thinking "in" the landscape and that our physical situation can be said to contribute to our conceptual argumentation. And thinking about landscape is also finding our way through the landscape.

In this section, the authors are concerned both with the ways that we as modern-day archaeologists conceptualize past landscapes and with the understanding of landscape on the part of Indigenous dwellers, past and present. One hopes that there is a potential for the former to be informed by the latter, so that our archaeology does not remain railroaded to a single, more or less narrow Western frame of reference, and rather can become more aware of the many ways that

people find themselves in the midst of their worlds and make sense of them.

Thinking about landscape explicitly or implicitly embraces cosmology and its concerns with materials, substances, spirits, and mortals. The landscape itself is the totality that links all these elements. When people engage in some action in the world, they often consider themselves to be operating on some part of a greater whole, just as archaeologists try to make their observations meaningful by putting them into the framework of landscape. The landscape is also the physical context for the circulation of objects in exchange from person to person, the circulation of livestock between pastures, and the circulation of persons between dwellings and taskplaces. Thinking about landscape requires that we recognize the complementarity of physical movement and inhabitation to cosmology. This recognition means that there is no distinction between the physical world "as it is" and the world-as-cognized: people inhabit and move through the landscape of understanding and experience.

Moreover, thinking about landscape involves the ways in which people divide the world into territories and zones, culturally regulated spaces appropriate for particular kinds of conduct, associated with particular beings, or reserved for particular communities. Historically, they have done this in a variety of ways, one being understanding landscape in the abstract as property that can be negotiated and contested, bought and sold.

References

Ingold, T. 1998. From complementarity to obviation: On dissolving the boundaries between social and biological anthropology, archaeology and psychology. *Zeitschrift für Ethnologie* 123: 21–52.

O'Brien, M. J., and Holland, T. D. 1995. Behavioural archaeology and the extended phenotype, in J. M. Skibo, W. M. Walker, and A. E. Nielsen (eds.), *Expanding Archaeology*, pp. 143–61. Salt Lake City: University of Utah Press.

Thomas, J. S. 2006. Phenomenology and material culture, in C. Tilley, W. Keane, S. Küchler, M. Rowlands, and P. Spyer (eds.), *Handbook of Material Culture*, pp. 43–59. London: Sage.

10

THE SOCIAL CONSTRUCTION OF WATER

Veronica Strang

In recent years, water has become an important topic in landscape archaeology and anthropology. As an integral part of every landscape, however arid,[1] water is literally essential to human existence and to all human-environmental engagements. As water issues have become increasingly urgent in many parts of the world, concerns have risen about the sustainability of many forms of land-use, and conflicts over water resources have increased. There is a pressing need for a fuller understanding of the social and environmental consequences of these problems.

Because water is the "lifeblood" of all organic organisms and essential to every form of material production, the ownership and control of water resources is often presented as the quintessential symbol of enfranchisement, fundamental to democracy and to the ethical distribution of resources. In the social sciences, there has been a long-term interest in the politics of water, reflecting its potential to confer power and influence on particular groups (Holmberg 1952; Orlove 2002; Strang In press; Wittfogel 1957; Worster 1992) and to play a central role in political and moral economies (Bennett 1995; Blatter and Ingram 2001; Lowi 1993; Ward 2003).[2]

An understanding of the relationship between social and environmental change is provided by ethnohistories focused on the use and management of major rivers such as the Columbia (Golay 2003; Meinig 1968; Worster 2001), the Mississippi (Harvey 2004), and the Colorado (Nye 1999). There is some overlap with research concerned with development and environmental issues that, in relation to water, have focused on the "taming" of "wild" rivers (Harden 1996; Pearson 2002; Ulrich 1999) and the social and ecological effects of vast river engineering projects, such as the building of major dams, in the United States (Evenden 2004; Khagram 2004; Nye 1999; Scarpino 1999) and more recently in China, where the damming of the Yangtze River has had an extreme impact on its dependent ecosystems and on the riparian communities dispossessed of their homes and lands (see Barber and Ryder 1993; Priyanka 2004). Clearly, as Leif Ohlsson, points out (1995), conflicts over water are a major development constraint. There is now growing concern about the effects of widespread over-irrigation; for example, in the United States, massive efforts to "green" the western deserts have had little success and have produced major environmental degradation (see Reisner 2001 [1986]).[3] In Australia, a cautionary example is provided by the Murray-Darling basin, where irrigation has led to critical water shortages, the loss of numerous aquatic species, and the salination of large areas of land.[4] This situation has placed considerable social and

123

economic pressure on its farming community, leading to intense competition for water allocations and
leaving many small farmers unable to maintain their
way of life. It has also heightened tensions between
rural and urban communities, since the latter are
seen to sympathize with the environmental movement's increasingly vocal critique of farmers' land
and water management.

As conflicts over water escalate and environmental problems become more extreme, social
scientists must get "under the surface" of the
issues. Clearly all engagements with water entail
the material enactment of particular ideas: about
human relationships with the environment; about
who (or what) should have water; and about how
water should be used and managed. It is visions
of development, growth, and empowerment that
drive major infrastructural endeavors and lead
to contests over control; and it is the beliefs and
values of water users, and the meanings that they
encode in water, that direct everyday practices.
There is excellent potential, in social archaeology,
for researchers to consider the materialization of
these ideas and processes, as they are expressed
both in artifacts and in cultural landscapes (see
Allen, Hey, and Miles 1997).

Fluid Metaphors

Water is also of interest at a theoretical level. Many
of the issues surrounding it recur in diverse cultural
contexts, enabling fruitful ethnographic comparisons. The cross-cultural commonality in the themes
of meaning encoded in water raises key questions
about universalities in human experience. This is
particularly relevant, for example, in considering
sensory and cognitive processes and their influence on cultural ideas and practices (see Bloch
1998; Strang 2005a; Strang and Garner 2006).

Theoretical models are inevitably reliant on
metaphors, and a focus on water has gained relevance as theory in the social sciences has become
more appreciative of the fluid nature of human
and nonhuman processes of change and adaptation. In both archaeology and anthropology there
has been a move away from rooted, land-based
analyses of economic and political practices to
a more phenomenological and processual view,
encompassing shifting "fluidscapes" of identities
and experiences.

Water provides vast imaginative potential to
carry cultural meanings (see Douglas 1973).[5] Illich
observes that "water has a nearly unlimited ability to carry metaphors" (1986: 24; also Bachelard
1942; Lakoff and Johnson 1980), and as I have
noted elsewhere:

[Water's] characteristics of transmutability and
fluidity make it the perfect analogue for describing complex ideas about change, transformation,
mood and movement. Because it can transform
from one extreme to another it can readily
convey all of the binary oppositions through
which people construct meanings and values.
Of all the elements in the environment, it is the
most suited to convey meaning in every aspect
of human life. (2004a: 61)

The "flow" of water is regularly employed, for
example, to articulate ideas about time and the
spiritual progression of human lives. Concepts of
time are culturally specific, ranging from circular visions of local movement (for example, the
human spirit emerging and returning to a spiritual
home in Aboriginal landscapes [see Allen 1997;
Barber 2005; Langton 2006; Magowan 2001; Strang
2002]) to metaphorical ideas about journeys down-
river/through life to "the great sink" of the sea,
and to larger and abstract concepts of temporal
movement, linked with vast hydrological cycles
(see Tuan 1968).

The use of water to imagine spiritual being and
time leads to its widespread use in religious rituals
(see Eliade 1958; Rothenburg and Ulvaeus 2001;
Strang 2004a; Wild 1981) to signify social and spiritual "congregation" (see Daniélou 1961; Davies
1994; Pocknee 1967; Schmemann 1976), processes
of transformation (Somé 1994), and the generation
of life itself (e.g., Furst 1989; Merlan 1998; Tacon,
Wilson, and Chippindale 1996). Concurrent with
concepts of spiritual, social, and moral well-being
come many ideas about physical health and balance. Visions of "nature" in harmony, or "healthy"
flows in ecosystems, match more immediate concepts of human physical health as a matter of
"proper" circulation and flow (Giblett 1996; Strang
2004a). Associations between water and health
are multiple, ranging from the practical and social
(e.g., Astrup 1993; Goubert 1986; Shove 2003)
to more spiritually orientated concepts of health,
cleanliness, and morality (e.g., Anderson and Tabb
2002; Forty 1986).

Awareness that water is the most vital "substance" of the self, as well as encouraging associations between spiritual, social, and physical
well-being, provides powerful metaphors about
human identity. An important aspect of human
"being in the world" is the sense of the self as
a physical being, composed of particular sub
stances that have a definable identity (e.g., Caplan
1997; Fischler 1988; Lupton 1996; Magowan
2001; Strang 2002, 2005a). Thus, discourses
about race, ethnicity, and cultural identity are

heavily reliant on fluid metaphors about blood and the potential for purity to be adulterated by otherness: a cross-cultural concept of pollution that has long been of interest in anthropology (Douglas 1966; Strang 2004b [2001]). This notion of pollution also links with religious notions of social and spiritual identity and the potential for its (social and moral) purity to be compromised or adulterated.

Alongside ideas about "substance" or "essence" lies a reality that human engagements with water involve intense sensory experiences: thirst, its relief and the taste of water; the pleasures of bathing, and the excitement or restfulness of immersion; the seductive sounds of water; and the mesmeric effects of gazing on its glittering surfaces (e.g., Damasio 1999; Feld 1982, 1996; Strang and Garner 2006; Gell 1992; Howes 1991, 2003; Sprawson 1992; Stoller 1989; Strang 2005b). All these experiences offer stimuli that have real physiological effects and, more importantly, generate ideas about water that are highly imaginative and influential. "Being-in-place" in and around water engenders affective responses that are commonly spiritual or aesthetic in their nature. In concert with the particular qualities of water—its constant circulation through the environment and the body; its visual fluidity and numinous shimmer—these responses encourage the location of particular meanings in water that are specifically cultural but also readily comparable cross-culturally. Water appears as a life-containing element: for example, as the pool of ancestral force that generates Australian Aboriginal spirit children; as the substance of "supernatural potency" that raises the dead in Kwakiutl mythology; and as the "living water" that carries the Holy Spirit in the Christian cosmos.

Water is thus the most vital of substances. It is the most essential element for survival, health, and wealth; the inspiration for metaphors of life, time, movement, and transformation; the source of powerful sensory and aesthetic experiences; and the fluid of social and spiritual identity. Imbued with such powerful meanings, water bodies—of any kind—unsurprisingly tend to be among the most important elements in every cultural landscape/fluidscape.

Locating Water

Water is never entirely still: even captured and contained, it shimmers and transforms itself from one form to another. Locating it in a cultural fluidscape is equally elusive, forcing us to approach the task from a phenomenological perspective that is fully appreciative of the shifting social, economic,

political, and religious dynamics of a particular cultural frame and the relationship between these and a local ecological context. From the perspective of landscape archaeology, it is especially relevant to consider the more materially "concrete" aspects of that environment through which water is controlled, distributed, managed, and used.

There are some recent ethnographic forays into this kind of approach. Lansing's influential analysis of the relationship between religious beliefs and practices and the management of irrigation in Bali (1991) was useful in highlighting the multiplicity of engagements that people have with water and how these act on one another. Other work has stressed the importance of a holistic analysis of the way that water permeates all aspects of culture, both ideationally and in material terms (see Lansing, Lansing, and Erazo 1998; Mosse 2003; Rigg 1992; Strang 2004a; Toussaint, Sullivan, and Yu 2005). Some recent ethnographic research in the south of England maps a particular waterscape (Strang 2004) that bears mention here.

The Stour Valley

The Stour River, in Dorset, runs for about 110 kilometers, through rich farmland and chalk downs, ancient mills and pretty thatched villages, and finally into the growing conurbations on the south coast. Upstream, its population is composed mainly of retirees from London, or relatively wealthy landowners, whereas the younger generations (and the less well-off retirees) congregate in the south, where there are jobs in tourism and light industry, less expensive housing, and the joys of the seaside. The river joins and connects these communities, allowing them to "locate" themselves in relation to it and to each other, while the containing valley further defines its inhabitants as a local community, albeit one that has—as elsewhere in the United Kingdom—become increasingly mobile and transient.

The social and spatial organization of the inhabitants of the valley reflects long-term changes and adaptations that began with Celtic hunter-gatherers and that remained intensely rural through centuries of invasion and settlement by Romans, Saxons, Danes, and Normans. For many centuries, Dorset—and its water supplies—fell under the control of powerful abbeys and manors who disbursed land and water to dependent communities. The River Stour was central to local economic activities, and the Domesday Book records 166 mills along the river. Population growth was steady: during the industrial revolution, the water power that had enabled milling was converted to other forms of production, and (as machines replaced farm labor) many people moved into the towns and cities. In addition to powering a variety of small industries, the river continued to support a shift toward intensification in farming, supplying the "water meadows"[6] of the 16th century and, in subsequent centuries, providing resources for the mechanized irrigation of fodder and arable crops.

With these processes of rural and urban industrialization, the ownership of land and water underwent equally radical changes. Collective forms of property rights, such as the use of "common land," were supplanted by waves of enclosure, which excluded and disenfranchised the majority of the population. However, the situation was more complicated in relation to water, which had, over the centuries, generally been regarded as a "common good." There was considerable technological change, as communal wells and village pumps were replaced by a wider infrastructure of bores, pumps, reservoirs, and pipes, particularly in urban areas,[7] but though much of this early supply technology belonged to private water companies, these were small, local, and largely set up by Victorian philanthropists whose aim was to improve the (often poor) health and sanitation of the lower classes. By the early 20th century, most water companies were publicly owned by municipal bodies, and, in a postwar desire for democratization, this collective ownership was broadened and formalized with the nationalization of the water industry in the 1940s. This situation was then reversed, in 1989, with the Thatcher government's controversial privatization of the water industry, which handed the water treatment and supply infrastructure to large private corporations, over 40% of which are now owned by foreign corporations.

There is a detailed material record of these changes and developments in the local fluidscape, which also reflects the shifting cosmological perspectives that accompanied each military and ideological invasion, as well as the accelerating drive toward modernity that Dorset shared with the rest of the nation. The holy/healing wells and springs that had provided a focus for rituals and votive offerings in a sentient and polytheistic Celtic landscape were taken over first by the Romans, who imported their own water gods and goddesses and celebratory rituals (such as Fontanalia[8]), and then by Christians, who built churches and fonts over them, celebrated their "miraculous" healing abilities, and renamed them after their saints. Worship of the Mother Goddess and an array of nature Gods were thus replaced by Christian visions of spiritual cleansing and rebirth.

With the Enlightenment came more secular cosmological explanations that initiated the transformation of water from holy essence to H_2O, creating the potential for its commodification as a mere material "resource." It is this vision that underlies a contemporary idea of water as the "product" of a private water industry, to be abstracted, treated, distributed, and paid for in measured quantities. Although the water industry and successive governments committed to privatization have made strenuous efforts to foreground this vision, it remains, for most people, only a thin overlay of ideas over a vast historical well of beliefs about water as the essence of social and spiritual identity, as a substance that can be compromised both literally and metaphorically, and as the flow of life and life-time. This vision is further challenged by contemporary engagements with water, which provide intense sensory and cognitive experiences and thus continue to generate a stream of meanings and metaphors employing images of water and fluidity.

All these tributaries of thought are readily evident in contemporary Dorset, running alongside one another and mingling as people try to balance spiritual concepts with the pragmatics of everyday life. Neolithic henges, originally positioned in relation to sacred water sources,[9] have become tourist destinations.[10] Although many of the ancient wells have succumbed to modern drainage and are memorialized only in place names and documents, some continue to provide a focus for neo-Pagan well-dressing rituals and historical interest. Many villages retain—even if they no longer rely on—their communal pumps as a visible reminder of highly localized identities. Christian groups use ancient fonts, or in some cases the river itself, to baptize people into their congregations. Ornate Victorian pumphouses provide an abiding testimony to the pride of their philanthropic sponsors, and modern water company employees exhibit with equal pride their technologically sophisticated water and sewage treatment plants.

In Dorset today, as elsewhere in the United Kingdom, the control of water lies in the hands of a small elite, composed of private corporations, regulatory bodies,[11] and wealthy landowners. Although the control of the regulators is less tangible, there is now a sophisticated infrastructure—pipelines, pumps, treatment plants, and so forth—that (though largely controlled now by computers in centralized offices) embeds the ownership of the water corporations in the material landscape. Some quite large tranches of riparian land along the Stour are also still owned by the local aristocracy, whose inheritance of manorial houses and their gracious parklands, gardens, and fountains, along with income from their dependent farms, helps to maintain their status and influence in the pecking order of village life, parish and county councils, and other local organizations. Although, as in other parts of the United Kingdom, there has been a demographic shift off the land into urban centers, villages have not lost their importance as focal points for social events, and this situation has been boosted by the influx of wealthy retirees. Some of these are hopeful newcomers to the water-owning elite, having been able to buy riverside homes and land or ancient watermills where, almost without exception, they direct time and resources toward the creative construction of lakes, ponds, fountains, and other water features.

Despite the decline of farming in the United Kingdom, the Stour Valley, with its rich soils and kind southerly climate, remains a relatively wealthy farming area in which even tenant farmers can prosper. Farmers have retained their access to water for irrigation, and there has been some reintroduction of the water meadows that enriched the area in the 1700s. However, farmers are not immune from contemporary difficulties in the industry: many small-holdings have been amalgamated, and there are growing complaints about economic pressures—even as farmers are trying to intensify production with more irrigation, their position as water owners has become less secure, with increasing control over abstraction licenses and more competition for resources from water companies focused on more profitable domestic supplies. Similar issues face the small number of riparian industries, mostly dairies and breweries, that rely on reliable access to clean water.

The private ownership of water is challenged at a local level by environmental organizations and community groups, who strive for access to and some measure of control over the river and its tributaries. It is also countered, to some degree, by the wider population of water users, who, even if they no longer own or manage water supplies collectively, and can only visit the prestigious lakes and fountains of the rich, make what water companies describe as "profligate" use of supplies in their own homes. Here, in the small amount of space that individuals and families can still control, water is used to create an aspirational "lifestyle" echoing that of the water owning elites. Thus, householders invest in luxurious bathing facilities and spas, ponds, water features, and green lawns to manifest, in material culture, the flow of their own agency and creative power (see Strang 2004a, 2005c; Symmes 1998).

Conclusions

What emerges from this account is that human engagements with water permeate every aspect of culture. In each of these, water is socially constructed, perceived not just through the senses but also through ideas and imagery in which water's particular qualities enable culturally mediated visions of social, economic, religious, ecological, and cosmological systems flowing through time and space.

For landscape archaeologists, the major challenge is to use the material record of water use and management to make a comprehensive analysis of past relationships with water. As the fluidscape described above makes plain, just as the Stour has cut its particular course through the downs, this material record reveals the contours of people's wider engagements with one another and their material surroundings. Such an analytic approach can therefore provide insights into a whole human-environmental interaction over time, connecting the past with the present and enabling archaeologists to make major contributions to contemporary analyses of water issues.

Notes

1. One might say that in arid landscapes the rarity of water sources gives them a particularly powerful influence.

2. Other relevant texts include Cruz-Torres 2004, Cummings 1990, Donahue and Johnson 1998, Elhance 1999, Kinnersley 1994, Oakley 1990, Strang 2001, 2005a, Ward 1997.

3. As Reisner notes, the diversion of much of the Colorado River for irrigation has left water users on the Mexican side of the border with little more than saline "liquid death" (2001 [1986]).

4. Salination renders soils unable to support agricultural crops or even a normal range of native plants.

5. Water imagery has also played a central symbolic role in psychological analyses (e.g., Freud 1961; Wittels 1982).

6. A watermeadow is created by the managed flooding of grazing areas near the river, to prevent them from freezing, thus extending the period when the land will produce fodder and enabling farms to support larger herds of cattle.

7. Many rural dwellings in Dorset didn't get piped water supplies until the 1940s and 1950s, and even after that many remained dependent on individual springs and pumps, and some villages continued to use (and defend) their own local sources of supply.

8. This is a well-dressing ritual involving the decoration of the well-head and the throwing of flowers and other votive offerings into the water. In recent decades, it has been revived in many villages in Britain as a celebration of community life.

9. Richards (1996) observes that water is a crucial element in the architectural layout of henges and argues that "the relationship between henges and rivers provides a metaphorical conjunction between the natural flowing of water and human movement into the monuments" (1996: 313).

10. As Bender's work on Stonehenge illustrates (1998), there are diverse perspectives on such sites, and their use and management are much contested.

11. The water industry is regulated primarily by the Office of Water Services (OFWAT), whose function is to regulate pricing; the Drinking Water Inspectorate (DWI), which tries to maintain standards in the quality of supplies; and the Environment Agency (EA), whose task is to protect the environment.

References

Allen, H. 1997. Conceptions of time in the interpretation of the Kakadu landscape, in D. Rose and A. Clarke (eds.), *Tracking Knowledge in North Australian Landscapes: Studies in Indigenous and Settler Ecological Knowledge Systems*, pp. 141—52. Casuarina, NT: North Australia Research Unit.

Allen, T., Hey, G., and Miles, D. 1997. A line of time: Approaches to archaeology in the upper and middle Thames Valley, England. *World Archaeology* 29: 114–29.

Anderson, S., and Tabb, B. (eds.). 2002. *Water, Leisure and Culture: European Historical Perspectives*. Oxford: Berg.

Astrup, P. 1993. *Salt and Water in Culture and Medicine*. Copenhagen: Munksgaard International Publishers.

Bachelard, G. 1942. *L'Eau et les Reves—Essai sur l'Imagination et la Matiere*. Paris: Jose Corti.

Barber, M. 2005. Where the clouds stand: Australian Aboriginal relationships to water, place and the marine environment in Blue Mud Bay, Northern Territory. Unpublished Ph.D. thesis, Australian National University, Canberra.

Barber, M., and Ryder, G. (eds.). 1993. *Damming the Three Gorges: What the Dam Builders Didn't Want You to Know*. London: Earthscan Publications.

Bender, B. 1998. *Stonehenge: Making Space*. Oxford: Berg.

Bennett, V. 1995. *The Politics of Water: Urban Protest, Gender and Power in Monterrey, Mexico*. Pittsburgh, PA: University of Pittsburgh Press.

Blatter, J., and Ingram , H. (eds.). 2001. *Reflections on Water: New Approaches to Transboundary Conflicts and Cooperation*. Cambridge, MA: MIT Press.

Bloch, M. 1998. Why trees, too, are good to think with: Towards an anthropology of the meaning of life, in L. Rival (ed.), *The Social Life of Trees: Anthropological Perspectives on Tree Symbolism*, pp. 39–55. Oxford: Berg.

Caplan, P. (ed.). 1997. *Food, Health and Identity*, London: Routledge.

Cruz-Torres, M. 2004. *Lives of Dust and Water: An Anthropology of Change and Resistance in Northwestern Mexico*. Tucson: University of Arizona Press.

Cummings, B. J. 1990. *Dam the Rivers, Damn the People: Development and Resistance in Amazonian Brazil*. London: Earthscan Publications.

Damasio, A. 1999. *The Feeling of What Happens: Body and Emotion in the Making of Consciousness*. London: Heinemann.

Daniélou, J. 1961. *Primitive Christian Symbols*. London: Burns and Oates.

Davies, D. 1994. Christianity, in J. Holm and J. Bowker (eds.), *Attitudes to Nature*, pp. 28–52. London: Pinter Publishers.

Donahue, J., and Johnston, B. (eds.). 1998. *Water, Culture and Power: Local Struggles in a Global Context*. Washington, DC: Island Press.

Douglas, M. 1966. *Purity and Danger: An Analysis of Concepts of Pollution and Taboo*. London: Routledge.

———. 1973. *Natural Symbols*. London: Random House.

Elhance, A. 1999. *Hydropolitics in the Third World: Conflict and Cooperation in International River Basins*. Washington, DC: U.S. Institute of Peace Press.

Eliade, M. 1958. *Patterns in Comparative Religion*. London: Sheed and Ward.

Feld, S. 1982. *Sound and Sentiment: Birds, Weeping, Poetics and Song in Kaluli Expression*. Philadelphia: University of Philadelphia Press.

———. 1996. Waterfalls of song: An acoustemology of place resounding in Bosavi, Papua New Guinea, in S. Feld and K. Basso (eds.), *Senses of Place*, pp. 91–113. Santa Fe: School of American Research Press.

Fischler, C. 1988. Food, self and identity. *Social Science Information* 27: 275–92.

Forty, A. 1986. *Objects of Desire*. London: Thames and Hudson.

Freud, S. 1961. *The Interpretation of Dreams*. London: George Allen and Unwin.

Furst, P. 1989. The water of life: Symbolism and natural history on the Northwest Coast. *Dialectical Anthropology* 14: 95–115.

Gell, A. 1992. The technology of enchantment and the enchantment of technology, in J. Coote and

A. Shelton (eds.), *Anthropology, Art and Aesthetics*, pp. 40–63. Oxford: Clarendon Press.

Giblett, R. 1996. *Postmodern Wetlands: Culture, History, Ecology*. Edinburgh: Edinburgh University Press.

Goubert, J.-P. 1986. *The Conquest of Water: The Advent of Health in the Industrial Age*. Princeton, NJ: Princeton University Press.

Harvey, M. 2004. The river we have wrought: A history of the upper Mississippi. *Technology and Culture* 45: 432–33.

Hays, S. 1987. *Beauty, Health and Permanence: Environmental Politics in the United States, 1955–1985*. Cambridge: Cambridge University Press.

Holmberg, A. 1952. The wells that failed: An attempt to establish a stable water supply in the Viru Valley, Peru, in E. Spicer (ed.), *Human Problems in Technological Change: A Casebook*, pp. 113–26. New York: Russell Sage Foundation.

Howes, D. (ed.). 1991. *The Varieties of Sensory Experience: A Sourcebook in the Anthropology of the Senses*. Toronto: University of Toronto Press.

———. 2003. *Sensual Relations: Engaging the Senses in Culture and Social Theory*. Ann Arbor: University of Michigan Press.

Illich, I. 1986. *H$_2$O and the Waters of Forgetfulness*. London: Marion Boyars.

Khagram, S. 2004. *Dams and Development: Transnational Struggles for Water and Power*. Ithaca, NY: Cornell University Press.

Kinnersley, D. 1994. *Coming Clean: The Politics of Water and the Environment*. Harmondsworth, Middlesex: Penguin.

Lakoff, G., and Johnson, M. 1980. *Metaphors We Live By*. Chicago: University of Chicago Press.

Langton, M. 2006. Earth, wind, fire and water: The social construction of water in Aboriginal societies, in B. David, B. Barker, and I. J. McNiven (eds.), *The Social Archaeology of Australian Indigenous Societies*, pp. 139–60. Canberra: Aboriginal Studies Press.

Lansing, S. 1991. *Priests and Programmers: Technologies of Power in the Engineered Landscape of Bali*. Princeton, NJ: Princeton University Press.

Lansing, S., Lansing, P., and Erazo, J. 1998. The Value of a River. *Journal of Political Ecology* 5: 1–22.

Lowi, M. 1993. *Water and Power: The Politics of a Scarce Resource in the Jordan River Basin*. Cambridge: Cambridge University Press.

Lupton, D. 1996. *Food, the Body and the Self*. London: Sage Publications.

Magowan, F. 2001. Waves of knowing: Polymorphism and co-substantive essences in Yolngu sea cosmology. *The Australian Journal of Indigenous Education* 29(1): 22–35.

Merlan, F. 1998. *Caging the Rainbow*. Honolulu: University of Hawai'i Press.

Mosse, D. 2003. *The Rule of Water: Statecraft, Ecology, and Collective Action in South India*. Oxford: Oxford University Press.

Oakley, R. 1990. *The Waters of the Nile: Hydropolitics and the Jonglei Canal 1900–1988*. Oxford: Clarendon Press.

Ohlsson, L. (ed.). 1995. *Hydropolitics: Conflicts over Water as a Development Constraint*. Dhaka: University Press Ltd; London: Zed Books.

Orlove, B. 2002. *Lines in the Water: Nature and Culture at Lake Titicaca*. Berkeley and Los Angeles: University of California Press.

Pocknee, C. 1967. *Water and the Spirit: A Study in the Relation of Baptism and Confirmation*. London: Darton, Longman and Todd.

Priyanka, A. 2004. *Three Gorges Dam: An Economic Success or an Environmental and Social Failure?* Manchester: University of Manchester Press.

Reisner, M. 2001 [1986]. *Cadillac Desert: The American West and Its Disappearing Water*. London: Pimlico.

Richards, C. 1996. Henges and water: Towards an elemental understanding of monumentality and landscape in late Neolithic Britain. *Journal of Material Culture* 1: 313–35.

Rigg, J. (ed.). 1992. *The Gift of Water: Water Management, Cosmology and the State in South East Asia*. London: School of Oriental and African Studies, University of London.

Rothenberg, D., and Ulvaeus, M. (eds.). 2001. *Writing on Water*. Cambridge, MA: MIT Press.

Scarpino, P. 1999. The organic machine: The remaking of the Columbia River. *Technology and Culture* 40: 419–20.

Schmemann, A. 1976. *Of Water and the Spirit: A Liturgical Study of Baptism*. London: SPCK.

Shove, E. 2003. *Comfort, Cleanliness and Convenience: The Social Organization of Normality*. Oxford: Berg.

Somé, M. 1994. *Of Water and the Spirit: Ritual, Magic, and Initiation in the Life of an African Shaman*. New York: Putnam.

Sprawson, C. 1992. *Haunts of the Black Masseur: The Swimmer as Hero*. London: Jonathan Cape.

Stoller, P. 1989. *The Taste of Ethnographic Things: The Senses in Anthropology*. Philadelphia: University of Pennsylvania press.

Strang, V. 2001. Negotiating the river: Cultural tributaries in far north Queensland, in B. Bender and M. Winer (eds.), *Contested Landscapes: Movement, Exile and Place*, pp. 69–86. Oxford: Berg.

———. 2002. Life down under: Water and identity in an Aboriginal cultural landscape. *Goldsmiths College Anthropology Research Papers* 7. London: Goldsmiths College.

———. 2004a. *The Meaning of Water*. Oxford: Berg.

———. 2004b [2001]. Poisoning the rainbow: Cosmology and pollution in Cape York, in A. Rumsey and J. Weiner (eds.), *Mining and Indigenous Lifeworlds in Australia and Papua New Guinea*, pp. 208–25. Wantage: Sean Kingston Publishing.

Strang, V. 2005a. Common senses: Water, sensory experience and the generation of meaning. *Journal of Material Culture* 10: 93–121.

———. 2005b. Taking the waters: Cosmology, gender and material culture in the appropriation of water resources, in A. Coles and T. Wallace (eds.), *Water, Gender and Development*, pp. 21–38. Oxford: Berg.

———. 2005c. Water works: Agency and creativity in the Mitchell River catchment. *The Australian Journal of Anthropology* 16: 366–81.

———. In press. *Gardening the World: Agency, Identity, and the Ownership of Water.* Oxford: Berghahn.

Strang, V., and Garner, A. (eds.). 2006. *Fluidscapes: Water, Identity and the Senses.* Special issue of *Worldviews* 10(2). Leiden: Brill Academic Publishers.

Symmes, M. 1998. *Fountains, Splash and Spectacle: Water and Design from the Renaissance to the Present.* London: Thames and Hudson, Smithsonian Institution.

Tacon, P., Wilson, M., and Chippindale, C. 1996. Birth of the Rainbow Serpent in Arnhem Land rock art and oral history. *Archaeology in Oceania* 31: 103–24.

Toussaint, S., Sullivan, P., and Yu, S. 2005. Water ways in Aboriginal Australia: An interconnected analysis. *Anthropological Forum* 15: 61–74.

Tuan, Y.-F. 1968. *The Hydrologic Cycle and the Wisdom of God: A Theme in Geoteleology.* Toronto: University of Toronto Press.

Ward, C. 1997. *Reflected in Water: A Crisis in Social Responsibility.* London: Cassell.

Wild, R. 1981. *Water in the Cultic Worship of Isis and Sarapsis.* Leiden: Brill Academic Publishers.

Wittels, B. 1982. Interpretation of the 'Body of Water' metaphor in patient artwork as part of the diagnostic process. *The Arts in Psychotherapy* 9: 177–82.

Wittfogel, K. 1957. *Oriental Despotism: A Comparative Study of Total Power.* New Haven, CT: Yale University Press.

Worster, D. 1992. *Rivers of Empire: Water, Aridity and the Growth of the American West.* Oxford: Oxford University Press.

11

READING BETWEEN THE LANDS: TOWARD AN AMPHIBIOUS ARCHAEOLOGICAL SETTLEMENT MODEL FOR MARITIME MIGRATIONS

Joe Crouch

Everything in the sea move on tide—turtle, dugong, fish, cray—everything on tide.
(Torres Strait Islander, Mabuyag, 1976, cited in Nietschmann 1989: 73)

In this chapter I suggest archaeological sites also move on the tides, with the argument that canoes are mobile sites. I explore the implications this has for modeling seascape settlement systems and maritime migrations associated with specialized marine-oriented island societies, such as Torres Strait Islanders in an Indigenous Australian "salt-water people" context (see Sharp 2002) and Lapita colonists in the western Pacific. A preliminary review of how islands are conventionally conceptualized by archaeologists—as variously marginal terrestrial habitats—highlights a paradox in island settlement modeling, where a focus on terrestrial dynamics limits our potential understanding of the holistic nature of social seascape construction (see McNiven, this volume) and island settlement that is a part of it. Expanding our spatial context beyond terrestrial habitation, and reorienting our approach to pay more attention to the vital role of nonresidential islands, coral reefs, and canoes, provides a more appropriate and insightful context to interpret the complexities of seascape settlement systems and better comprehend the relationship between available site patterning on islands and the missing sites between them.

Background: Landscapes, Islands, Seascapes

Islands are ambiguous terrestrial landscapes; they are someplace between continents and the sea. On the one hand, there are continents that are also islands (that is, Australia) and subcontinental islands (for example, New Guinea). On the other hand—and more the focus of this chapter—there are islands, such as coral atolls, that cannot support fully terrestrial economies. These are alien terrestrial landscapes that can be more sensibly conceived as being part of the seascape. Islands can be defined simply as "a piece of land surrounded by water" (Rowland 2002: 62), yet ultimately every piece of land is surrounded by water, and, as Nunn (1987: 228) suggests, the "definition of 'small' when applied to an island is as subjective as the definition of what is an island and what is not." Islands can be classified according to various natural processes associated with geography and geology, but there are also such things as "artificial islands" (Ivens 1927: 23).

One point of clarity, however, is that referenced to terrestrial continental landscapes, islands are inevitably perceived as marginal: spatially (remote, inaccessible); temporally ("among the last parts of the globe to be settled" Broodbank 2000: 7); socially (insular, isolates); ecologically and economically (tenuous, depleted). Small islands, especially, are perceived as some of the most illogical, risky, and precarious places to live on the planet, and there is a general expectation that relatively small, low, remote islands will be settled *after* larger, nearer, higher, less risky islands more suited to cultivation and permanent habitation. However, radiocarbon date distributions do not always support this logic (Keegan and Diamond 1987: 52). For example:

- The lower atolls and raised limestone islands of Eastern Micronesia appear to have been settled around the same time as the higher rockier islands of Western Micronesia (Rainbird 2004: 86–97);

- Polynesia's "mystery islands" likewise appear to have been systematically settled around the same time as larger, less remote islands in the region (Irwin 1992: 177);

- In a Torres Strait context, Barham's (2000) tentative model for settlement of the larger islands closer to Australia and Papua New Guinea prior to the settlement of more remote islands is also now in doubt: small remote islets currently provide the earliest evidence for specialized maritime settlement in the region (see Crouch et al. 2007; McNiven et al. 2006a).

Conventional island settlement models, however, largely focus on patterning associated with the process of islands becoming *permanent habitats* (e.g., Graves and Addison [1995]: discovery → colonization → establishment; or Anderson [1995]: visitation → colonization → habitation). These can be understood as "habitation space approaches" (see Kirch 1984: 96–122) that tend to exclude rather than integrate the role of nonresidential islands. However, identifying permanently inhabited sites (Kintigh 1994) or islands (Terrell 1986: 14) is far from straightforward, and islands are not always socially designed specifically as a locus for permanent habitation. These sorts of islands are variously described throughout the archaeological literature in terms of their role in subsistence settlement systems, as "economic satellites" (Irwin 1992); "utilized" vs. "occupied" (Cherry 1990); and "seasonal" vs. "residential" (Jones 1977). Their identification lies in *not* being the locus and focus of permanent settlement—they are spatially and economically marginal. Yet McNiven (2000) has shown that this is precisely why some islands are used as

social retreats. And as Weisler (2001: 3) explains: "Despite the marginal position Utrōk Atoll occupies as a terrestrial landscape, the marine environment is clearly at the opposite extreme". Furthermore, *land* shouldn't be assumed the "locus of settlement"—or "home"—for highly mobile sea-oriented communities (Gosden and Pavlides 1994: 162) or the most suitable frame of reference for modeling centrality and marginality in island networks (cf. Hage and Harary 1996). Broodbank (2000: 238) rejects focusing on "unitary islands" and instead suggests modeling centrality in terms of "patch-works of land and sea." And along with land and sea we need to acknowledge the role of coral, because although island landscapes are generally regarded as potential habitats and coral reefs are not, reef settlement is a reality. Stilt villages situated on island fringing reefs are key Lapita settlement sites (see Gosden and Webb 1994; Gosden and Pavlides 1994; Kirch 2000: 107; Rainbird 2004: 92–95). For example, during the Lapita period the Arawe Island group in the Bismarck Archipelago (Papua New Guinea) provides "evidence of a clustered settlement pattern in the form of stilt villages built in shallow water on the lee sides of islands" (Gosden and Webb 1994: 29). Here the stilt villages actually caused and nucleated beach formation—the seabed itself is a cultural deposit (see also Felgate [2001] regarding intertidal settlement sites across Near Oceania). In addition to the leeward patterning of Lapita stilt village settlements, they are also commonly located proximal to reef passes facilitating canoe access to the sea (Kirch 2000: 107). Thus intra-island settlement patterning (leeward-windward, inside-outside) is highly influenced by the *shape* of the seascape—the interrelationship between land, sea, coral, winds, tides, channels, and so on. Ultimately, marginality can be seen as a "socially perceived construct" (Lourandos 1997: 23), and in a socially constructed seascape, the most marginal place you can be is in the middle of a large island.

As seascapes, islands are no less ambiguous, only here they are just the tip of the iceberg. A growing understanding of systems of Indigenous customary marine tenure (a key book is Cordell's [1989] *A Sea of Small Boats*) warn archaeologists that "the water is not empty" (Jackson 1995). So, whereas Kirch (2000: 304) suggests that "while the human settlement of Oceania is often thought of in maritime terms, as constituting a succession of great voyaging feats, these voyages were primarily means to an end: the discovery of new landscapes that could be claimed, named, divided, planted, and inherited," we must recognize that for many Indigenous maritime societies, the seascape is likewise claimed, named, divided, and inherited (see McNiven's definition of "seascapes," this volume; see also Magowan 2001; Nietschmann 1989; Sharp 2002). So, there is a paradox: why do we model terrestrial habitation and migration triggers for specialized maritime island societies? If the goal of modeling island settlement *patterning* is to identify "rules" or structures of the settlement *system* (Flannery 1976), then we need to look beyond terrestrial "rules" and patterning for maritime migrations and examine the holistic settlement of patches of sea. We need to better integrate the quintessential role of coral reefs and non-residential islands into the settlement system, and pay more attention to the coordinated use of entire island clusters and patches of sea. For example, Di Piazza and Pearthree (2004) adopt a palaeo-economic approach (following Anderson 1997) to modeling settlement of the Phoenix Islands where "the *domestication of the sea* by these ancient navigators fostered the domestication of the entire Phoenix archipelago and even beyond" (Di Piazza and Pearthree 2004: 105, my emphasis). I posit that a platform to reorient our approach is in conceptualizing canoes themselves as mobile and manifold archaeological sites: they are the glue that binds the seascape settlement system together, and although all but invisible to archaeological survey, they are a key to contextualizing the available patterning of nonmobile sites they complement.

Centrality in Patches of Sea

Tenacious debate in island archaeology was sparked by the notion of "islands as laboratories" (developed initially for island biogeography by MacArthur and Wilson 1967) over whether insularity is a key concept for understanding cultural processes on islands (e.g., Broodbank 2000; Evans 1973; Gosden and Pavlides 1994; Rainbird 1999; Terrell 1986; Terrell et al. 1997; Walker 1972). This debate is largely beyond the scope of this chapter, but within it a key series of interrelated dichotomies have crystallized: bridge/barrier; connected/insular; seascape/landscape. Ultimately, there is general agreement that the concept of *relative* insularity (see Terrell, this volume) and principles of island biogeography can generate significant insights into settlement phases for island societies, but primarily in terms of colonization as dispersal (one-way mobility, cf. migration as two-way mobility, Anderson 2003) and thus via phylogenetic models (e.g., Kirch 1984). However, the focus here is on maritime migrations by highly mobile island societies and how they build bridges, connections and social seascapes. Binford (2001: 166–67) neatly expresses what is perhaps the

antithesis of my view: "Although human actors are capable of direct participation, ecologically speaking, in a terrestrial setting, they may be thought of as outsiders in aquatic biomes; they intrude at times, but always at very restricted locations and under rather specific conditions." Torres Strait Islanders are "one of the most marine-oriented and sea-life dependant indigenous societies on the planet" (Cordell 1993: 159), documented as exploiting over 450 species of marine organisms (McNiven and Hitchcock 2004). They are not "outsiders" or "intruders" in their seas, ecologically, or in any sense. Instead we can ask: what conditions specialized maritime "intrusions" into terrestrial island biomes?

Tudu (Figure 11.1a) is a small, low sandy cay measuring 1.6 by 0.75 kilometers, and with a maximum elevation above sea level of 4.6 meters (Fuary 1991: 443–44). Tudu is "merely a sandbank" (Haddon 1935: 27), yet it was calculated by Hage and Harary (1996: 165–203) as the most "central" island in the Torres Strait network. Here "centrality" has three definitions: number of trade routes connected; nearness to every other island in the network; and degree of "betweenness." The argument is presented alongside another theory, where the relative potential of islands to act as agricultural versus foraging locations is seen as the architect for resulting trade relations (for example, trade/exchange of marine resources for plant resources), following Harris's (1979) trade-horticulture hypothesis. However, the ca.1892 resettling of people from Tudu (the Tudulgal) to Iama (or Yam)—owing to water shortages (Shnukal 2004: 330) and sociopolitical factors associated with the colonial regime (McNiven, Crouch, and David 2004)—creates a situation wherein the most central and strategic island in the Strait was abandoned as a residential base. Now although from an island-centric perspective this is puzzling, if the highly proximal location of Tudu to Warrior Reefs (the largest coral reef system in the Strait, adjacent to the Great North East Channel voyaging route) is taken into account, the residential emigration to Iama is not a case of abandoning the most central settlement node in the seascape network, because in reality (Figure 11.1b) Warrior Reefs surely deserves that title. There is the theory that the Kulkulgal (the Central Islanders of Torres Strait, including the Tudulgal) operated as specialist traders or "middlemen" (Hage and Harary 1996; Harris 1979; Vanderwal 1973) in the elaborate regional customary exchange network documented ethnographically (see Lawrence 1994). While I do not dispute this view, this is only part of *how*—and not *why*—Tudu was settled.

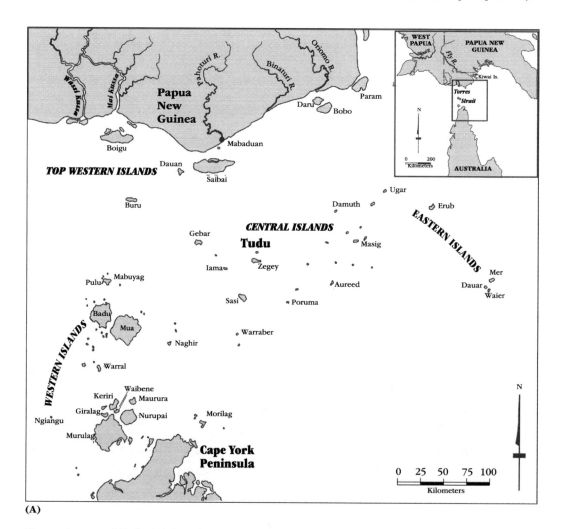

(A)

The settlement of Tudu needs to be understood and contextualized as part of the settlement of Warrior Reefs within the social construction of the Kulkulgal seascape.

So the small *size* of Tudu, and associated demographic situation, was not what triggered the ca.1892 emigration, and evidently size did not deter initial colonization (migration). Perhaps island *shape* is more pertinent than size here: an elongated flat sandbank provides ideal *canoe anchorage*, which is a vital commodity for maritime migrations and not something you can import (cf. freshwater, imported by canoe to Tudu from Iama, in long thick lengths of bamboo, Shnukal 2004: 320). With specialized maritime island migrations, large canoes are already operating as mobile and manifold sites in the seascape. So a prime question is this: how do mobile canoe sites structure the subsequent patterning of terrestrial nonmobile sites?

Canoes as Mobile Sites

Canoes are usually discussed in terms of maritime technology. For Torres Strait, canoes have been identified as the single most significant item of material culture, allowing vital trade/exchange systems to operate, and ultimately making island settlement viable (Barham 2000). In the context of Pacific island colonization, their virtual archaeological invisibility yet fundamental role has led to "contentious conjecture" over seafaring hypotheses (Anderson 2003: 78–81) and decades of experimental voyaging (e.g., Finney 1977, 1996; Irwin 1992) and computer simulations (pioneered by Levison, Ward, and Webb 1973, but see also Irwin 1992). If Irwin's (1992) model for continuous and systematic exploration and colonization of the Pacific islands through "return voyaging" is accepted (but see Anderson et al. [2006] regarding "ENSO Forcing" and cf. Sharp's

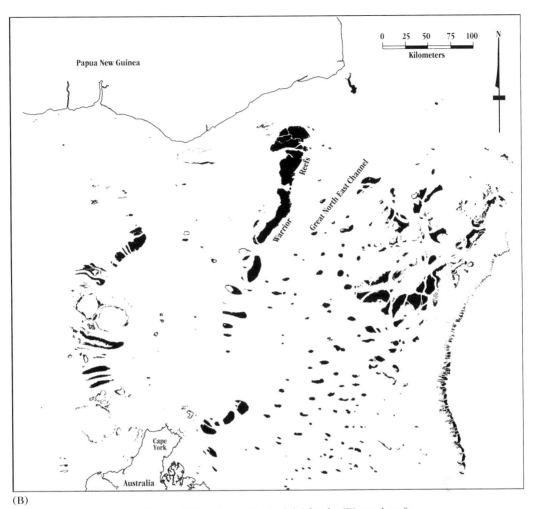

(B)

Figure 11.1 Torres Strait. (A) islands; (B) coral reefs.

[1956] argument for "accidental colonization"), a key point is that colonists were spending vast amounts of time in canoes within "a frontier of exploration." So, there is much more to canoes than just getting from A to B, and the focus here, following Ames (2002), is how their *capacity*—combined with their mobility (and motility)—has profound implications for structuring island settlement patterning.

An extreme example of canoe capacity is the long-distance trading expeditions—the *hiri*—of Motu groups in southern Papua New Guinea, where it is estimated between 26,000 and 32,500 pots loaded on 20 *lagatoi* canoes with 600 men would leave Western Motu villages in single expeditions, returning around five months later with 600 tons of sago carbohydrate (Allen 1984: 426–27). Furthermore, Ames (2002) shows that canoe capacity concerns more than just freight; it facilitates the movement of large and varied task groups that can include dogs, children, men, and women. We need to be thinking about canoes as manifold sites: moving hearths and water reservoirs; sites of learning and "enskilment at sea" (Palsson 1994); hunting sites and fishing sites; sites of residence and trade/exchange. And in a seascape context of "amity and enmity" such as Torres Strait (McNiven 1998), where trade/exchange alliances are entangled with headhunting prospects, it seems logical to expect canoes were often loaded with a variety of cargos and prepared for various kinds of negotiations—including open water ritual negotiation of the seascape itself as a spiritscape (see McNiven 2003, this volume).

In Binford's (1982) renowned paper "The Archaeology of Place," a key conceptual outcome for settlement models is the argument that although places are fixed in the landscape, sites (for example, "field camps") *move*. Importantly, archaeology can model this—through the nature and the degree of superimposition of site-specific material assemblages across fixed places. But for seascape settlement systems, we have a situation in which both sites *and* places move. Although a full discussion of how canoes are moving places is largely beyond the present scope, I would suggest—in an archaeological context at least—that canoes are mobile places in a similar sense to the way Casey (2001) discusses the body as a dimension of place: the outgoing body "goes out to meet the place-world," and the "place-world is energized and transformed by the bodies that inhabit it, while these bodies are in turn guided and influenced by this world's inherent structures" while the incoming body "is shaped by the places it has come to know and that have come to it—come to take up residence in it, by a special kind of placial incorporation" (Casey 2001: 413–14). Thus archaeology has access to canoes as outgoing bodies—we can investigate how canoes have "energized and transformed" the place-world—but only through disciplines such as ethnography and anthropology can we see them as incoming bodies. Notions of embodiment and perception also encourage comparisons between driving and voyaging: Dant's (2004) notion of the "driver-car" assemblage "as a form of social being that produces a range of social actions that are associated with the car; driving, transporting, parking, consuming, polluting, killing, communicating, and so on" (Dant 2004: 61–62), highlights how people and canoes *together* (that is, "crew-canoes") structure—and are structured by—the material world in a very different way to pedestrian bodies; to the "world perceived through the feet" (Ingold 2004).

Discussion

So if canoes are mobile sites *and* places, the question is (how) can archaeology model this (cf. "the archaeology of place")? I suggest that it is possible if we examine amphibious processes, where the best resolution must come from non-residential islands, owing to a more direct (less conflated) relationship between mobile and non-mobile sites. First, it is necessary to differentiate between *types* of mobility (for example, "residential," "logistical," and "tactical," [see Kelly 1992]) and mobility itself—(that is, "moveableness" [Close 2000]). Close (2000) is concerned with fine-grained modeling of actual patterns of movement in the past and how this can be achieved through refitting artifacts. When we identify imported material culture on islands, and their source locations, we are also inherently refitting. However, movement between the source site and the import site also

involves intermediate mobile canoe sites. In a Torres Strait context, McNiven (1998) has shown that stone-headed clubs (*gabagaba*) were actually moving back and forth across the Strait; sometimes in canoes operating as mobile trade/exchange sites but also in canoes operating as mobile raiding (headhunting) sites.

Canoes are certainly "moveable," and their mobile capacity makes them also a key factor in terms of conceptual types of mobility for archaeology. Kelly (1992: 60) suggests that a continuum between sedentary and mobile is inadequate, because mobility is "not merely variable but multidimensional" and that the focus should be on disentangling various dimensions of mobility, and modeling how they interrelate. Ames (2002) likewise questions the relevance of the forager-collector continuum for modeling subsistence settlement patterns associated with "aquatic hunter-gatherers," where canoe (boat) capacity can generate widespread multitask (resembling residential) signatures throughout "foraging"[1] areas—where "it becomes hard to distinguish a logistical foray from a residential shift involving an entire settlement" (Ames 2002: 43). However, there is more to settlement than simply subsistence, and, moreover, aspects of ritual and subsistence practices can be deeply entangled in social seascapes and island archaeological records (in a Torres Strait context, see Barham, Rowland, and Hitchcock 2004; McNiven 2006; McNiven and Feldman 2003).

Binford's (1980) forager-collector continuum polarizes residential mobility (that is, "foragers" who are "mapping on" to resources) against logistical mobility (that is, collectors who employ complex logistics), argued to be a product of environmental adaptation. Thus, the ends of his continuum are exemplified by San ("foragers" in an arid setting) and Nunamuit ("collectors" in an arctic setting). The model suggests collectors will generate a greater range of archaeologically discernable (homogeneous) site types outside the "residential base" (heterogeneous) site range. Thus "foragers" produce only "residential bases" and "locations," while collectors produce at least three additional site types: "stations," "caches," and "field camps." But canoes facilitate simultaneous "mapping on" and "complex logistics"—they are akin to Nunamuit sleds in terms of logistic potential and can facilitate various types of residential migrations (colonizations, seasonal shifts, evacuations, and so on). Mobile canoe sites potentially encompass all of Binford's site types, and many more.

Besides Binford's framework of environmental adaptation, and his secular focus (Insoll 2004: 48; McNiven, David, and Barker 2006: 14), there are some other problematic aspects to this model for both landscape and seascape settings. In an Aboriginal Australian context, Lourandos (1997: 20) suggests the continuum is far from environmentally determined; it can clinally shift at any time in any environment, owing to sociodemographic circumstances and strategies. In terms of Lapita colonization in the western Pacific, where there is an established inventory of over 180 Lapita sites (for example, containing dentate-stamped pottery [see Anderson et al. 2001]), "the density of sites on the coasts of large islands is lower than on small islands throughout the Lapita range" (Anderson 2001: 16), leading Anderson (2001)—in conjunction with chronological patterning—to question whether there might in fact be binary phasing in migration mobility: "released and space-transgressive"— evident in Remote Oceania (cf. Groube's (1971) "strandlooping" model) versus "tethered and time-transgressive"—evident in Near Oceania (cf. population growth model).

Synthesizing Anderson's idea of binary migration mobility with the concept of canoes as mobile sites, and with attention to trends associated with shifting sociodemographic circumstances (for example, increased degrees of social closure, territoriality, boundary maintenance, and alliance formation [Lourandos 1993, 1997]), we can begin to formulate an amphibious seascape settlement model. When specialized maritime societies migrate to new patches of sea, constructing and propagating their social seascape frontier, mobile canoe sites (and their nonmobile terrestrial counterparts) will have a spatially extensive distribution. At this stage, canoes operating as mobile sites provide a *risk-minimization strategy*; even if remote islands are found to be highly marginal terrestrial habitats, this is offset by the already established canoe sites: advanced mobile canoe sites are the secret to remote island colonization. As the patch of sea develops into a "sea territory" (see Nietschmann 1989), migration mobility shifts to a more time-transgressive phase, where the distribution of mobile canoe sites becomes funneled and "tethered" into a more formalized system of "canoe traffic" (Lawrence 1990) associated with the development of more complex systems of regional alliance, as well as increased ritual and territorial marking (for example, rock art sites, ritual and ceremonial sites). Last, there is the prospect of more intensive use of ecological niches previously considered as more marginal environments (Lourandos 1997: 23), which in the case of specialized maritime island societies may relate to island interiors, and larger islands generally.

Conclusions

I have attempted to demonstrate that mobile canoe sites occupy an integral, fundamental, and central place in seascape settlement systems. They are not *outside* the settlement system, and for specialized maritime peoples such as Micronesian navigators— for whom the sea is a moving frame of reference (see Gladwin 1970; Turnbull 1991)—canoe sites are in fact perceived as "the centre of the world" (Ingold 2000: 240), moving in temporal and locational harmony with the sea. Islands, in contrast, are perpetually morphing their shapes and sizes, oscillating and reforming in tempo with the tides, moons, seasons, and so on. Amphibious settlement of such a dynamic world, launched from a platform of canoe sites, makes the intertidal zone a highly complex boundary—spatially, temporally, socially, and physically—and a crucial locus for ongoing archaeological attention. Nonresidential islands, with their more direct relationship with canoes, crews, and cargos, offer exciting prospects and special insights into the complex nonresidential world of voyaging, customary marine tenure, management, and trade/exchange.

Acknowledgments

Ideas in this chapter were aided by many discussions with Ian McNiven, who also provided valuable comments on earlier drafts. Thanks also to Lynette Russell and Bruno David for their comments; and to Gary Swinton and Phil Scamp (School of Geography and Environmental Science, Monash University) for drafting Figure 11.1.

Note

1. I use the words "foraging" and "forager" to facilitate engagement with literature about mobility, where those terms are established and widespread. However, as Ingold (2000: 58) suggests, "foraging" derives "from the field of ecology, to denote the feeding behaviour of animals" and is a problematic and potentially highly offensive abbreviation for understanding how *people* feed themselves.

References

Allen, J. 1977. Pots and poor princes: A multidimensional approach to the role of pottery trading in coastal Papua, in S. E. Van der Leeuw and A. C. Pritchard (eds.), *The Many Dimensions of Pottery: Ceramics in Archaeology and Anthropology*, pp. 407–63. Amsterdam: University of Amsterdam.

Ames, M. K. 2002. Going by boat: The forager-collector continuum at sea, in B. Fitzhugh and J. Habu (eds.), *Beyond Foraging and Collecting: Evolutionary Change in Hunter-Gatherer Settlement Systems*, pp. 19–52. New York: Kluwer Academic/Plenum Publishers.

Anderson, A. J. 1995. Current approaches in East Polynesian colonization research. *Journal of the Polynesian Society* 104: 110–32.

———. 1997. Prehistoric Polynesian impact on the New Zealand environment: Te Whenua Hou, in P. V. Kirch and T. L. Hunt (eds.), *Historical Ecology in the Pacific Islands: Prehistoric Environmental and Landscape Change*. New Haven, CT: Yale University Press.

———. 2001. Mobility models in Lapita migration, in G. R. Clarke, A. J. Anderson, and T. Vunidilo (eds.), *The Archaeology of Lapita Dispersal in Oceania: Papers from the Fourth Lapita Conference, June 2000, Canberra, Australia*, pp. 15–23. Canberra: Pandanus Books.

———. 2003. Initial human dispersal in remote Oceania: Pattern and explanation, in C. Sand (ed.), *Pacific Archaeology: Assessments and Prospects. Proceedings of the International Conference for the 50th anniversary of the first Lapita excavation*, pp. 71–84. Nouméa: Le Cahiers de l'Archéologie en Nouvelle-Calédonie.

Anderson, A. J., Bedford, S., Clarke, G., Lilley, I., Sand, C., Summerhayes, G., and Torrence, R. 2001. An inventory of Lapita sites containing dentate-stamped pottery, in G. R. Clarke, A. J. Anderson, and T. Vunidilo (eds.), *The Archaeology of Lapita Dispersal in Oceania: Papers from the Fourth Lapita Conference, June 2000, Canberra, Australia*, pp. 1–13. Canberra: Pandanus Books.

Anderson, A. J., Chappell, J., Gagan, M., and Grove, R. 2006. Prehistoric maritime migration in the Pacific Islands: An hypothesis of ENSO forcing. *The Holocene* 16: 1–6.

Barham, A. J. 2000. Late Holocene maritime societies in the Torres Strait Islands, northern Australia: Cultural arrival or cultural emergence? in S. O'Connor and P. Veth (eds.), *East of the Wallace Line: Studies of Past and Present Maritime Cultures of the Indo-Pacific Region*, pp. 223–314. *Modern Quaternary Research in Southeast Asia* 16. Rotterdam: Balkema Press.

Barham, A. J., Rowland, M., and Hitchcock, G. 2004. Torres Strait *Bepotaim*: An overview of archaeological research and ethnoarchaeological investigations and research, in I. J. McNiven and M. Quinnell (eds.), *Torres Strait Archaeology and Material Culture*, 3(1): 1–72. Brisbane: Memoirs of the Queensland Museum Cultural Heritage Series.

Binford, L. R. 1980. Willow smoke and dogs' tails: Hunter-gatherer settlement systems and archaeological site formation. *American Antiquity* 45: 4–20.

———. 1982. The archaeology of place. *Journal of Anthropological Archaeology* 1: 5–31.

———. 2001. *Constructing Frames of Reference*. Berkeley and Los Angeles: University of California Press.

Broodbank, C. 2000. *An Island Archaeology of the Early Cyclades*. Cambridge: Cambridge University Press.

Casey, E. S. 2001. Body, self, and landscape: A geophilosophical inquiry into the place-world, in P. C. Adams, S. Hoelscher, and K. E. Till (eds.), *Textures of Place: Exploring Humanist Geographies*, pp. 403–25. Minneapolis: University of Minnesota Press.

Close, A. E. 2000. Reconstructing Movement in Prehistory. *Journal of Archaeological Method and Theory* 7: 49–77.

Cherry, J. F. 1990. The first colonization of the Mediterranean islands: A review of recent research. *Journal of Mediterranean Archaeology* 3: 145–221.

Cordell, J. (ed.). 1989. *A Sea of Small Boats*. Cambridge, MA: Cultural Survival, Inc.

———. 1993. Indigenous people's coastal-marine domains: Some matters of cultural documentation. *Turning the Tide: Conference on Indigenous Peoples and Sea Rights*, pp. 159–74. Darwin: Northern Territory University.

Crouch, J., McNiven, I. J., David, B., Rowe, C., and Weisler, M. 2007. Berberass: Marine resource specialisation and environmental change in Torres Strait during the past 4,000 years. *Archaeology in Oceania* 42(2): 49–64.

Dant, T. 2004. The driver-car. *Theory, Culture and Society* 21(4/5): 61–79.

Di Piazza, A., and Pearthree, E. 2004. Sailing routes of old Polynesia: The prehistoric discovery, settlement and abandonment of the Phoenix Islands. *Bishop Museum Bulletin in Anthropology II*. Honolulu: Bishop Museum Press.

Evans, J. D. 1973. Islands as laboratories for the study of culture processes, in A. C. Renfrew (ed.), *The Explanation of Culture Change: Models in Prehistory*, pp. 517–20. London: Duckworth.

Felgate, M. 2001. A Roviana ceramic sequence and the prehistory of Near Oceania: Work in progress, in G. R. Clarke, A. J. Anderson, and T. Vunidilo (eds.), *The Archaeology of Lapita Dispersal in Oceania: Papers from the Fourth Lapita Conference, June 2000, Canberra, Australia*, pp. 39–60. Canberra: Pandanus Books.

Finney, B. R. 1977. Voyaging canoes and the settlement of Polynesia. *Science* 196(4296): 1277–85.

———. 1996. Colonizing an island world. *Transactions of the American Philosophical Society* 86(5): 71–116.

Flannery, K. V. (ed.). 1976. *The Early Mesoamerican Village*. Orlando: Academic Press.

Fuary, M. 1991. In So Many Words: An Ethnography of Life and Identity on Yam Island, Torres Strait. Un published Ph.D. thesis, Department of Anthropology and Archaeology, James Cook University, Townsville.

Gladwin, T. 1970. *East Is a Big Bird*. Cambridge, MA: Harvard University Press.

Gosden, C., and Pavlides, C. 1994. Are islands insular? Landscape vs. seascape in the case of the Arawe Islands, Papua New Guinea. *Archaeology in Oceania* 29: 162–71.

Gosden, C., and Webb, J. 1994. The creation of a Papua New Guinean landscape: Archaeological and geomorphological evidence. *Journal of Field Archaeology* 21: 29–51.

Graves, M. W., and Addison, D. J. 1995. The Polynesian settlement of the Hawaiian archipelago: Integrating models and methods in archaeological interpretation. *World Archaeology* 26: 380–99.

Groube, L. M. 1971. Tonga, Lapita pottery, and Polynesian origins. *Journal of the Polynesian Society* 80: 278–316.

Hage, P., and Harary, F. 1996. Centrality, in *Island Networks: Communication, Kinship and Classification Structures in Oceania: Structural Analysis in the Social Sciences*, pp. 165–203. Cambridge: Cambridge University Press.

Harris, D. R. 1979. Foragers and farmers in the Western Torres Strait Islands: An historical analysis of economic, demographic and spatial differentiation, in P. C. Burnham and R. F. Ellen (eds.), *Social and Ecological Systems*. London: Academic Press.

Ingold, T. 2000. *The Perception of the Environment: Essays in Livelihood, Dwelling and Skill*. London: Routledge.

———. 2004. Culture on the ground: The world perceived through the feet. *Journal of Material Culture* 9: 315–40.

Insoll, T. 2004. *Archaeology, Ritual, Religion*. London: Routledge.

Irwin, G. 1992. The *Prehistoric Exploration and Colonisation of the Pacific*. Cambridge: Cambridge University Press.

Ivens, W. G. 1927. *Melanesians of the South-east Solomon Islands*. London: Kegan Paul, Trench, Trubner and Co., Ltd.

Jackson, S. E. 1995. The water is not empty: Cross-cultural issues in conceptualising sea space. *Australian Geographer* 26(1): 87–96.

Jones, R. 1977. Man as an element of a continental fauna: The case of the sundering of the Bassian bridge, in J. Allen, J. Golson, and R. Jones (eds.), *Sunda and Sahul: Prehistoric studies in Southeast Asia, Melanesia and Australia*, pp. 317–86, New York: Academic Press.

Keegan, W. F., and Diamond, J. M. 1987. Colonization of Islands by humans: A Biogeographical Perspective, in M. B. Schiffer (ed.), *Advances in Archaeological Method and Theory* 10, pp. 49–92, New York: Academic Press.

Kelly, R. L. 1992. Mobility/Sedentism: Concepts, Archaeological Measures, and Effects. *Annual Review of Anthropology* 21: 43–66.

Kintigh, K. W. 1994. Contending with contemporaneity in settlement-pattern studies. *American Antiquity* 59: 143–48.

Kirch, P. V. 1984. *The Evolution of the Polynesian Chiefdoms*. Cambridge: Cambridge University Press.

———. 2000. *On the Road of the Winds: An Archaeological History of the Pacific Islands before European Contact*. Berkeley and Los Angeles: University of California Press.

Lawrence, D. 1990. "Canoe Traffic" of the Torres Strait and Fly Estuary, in J. Siikala (ed.), *Culture and History in the Pacific* 27: 184–201. Helsinki: The Finnish Anthropological Society Transactions.

———. 1994. Customary exchange across Torres Strait. *Memoirs of the Queensland Museum* 34(2): 241–446. Brisbane: Queensland Museum.

Levison, M., Ward, R. G., and Webb, J. W. 1973. *The Settlement of Polynesia: A Computer Simulation*. Minneapolis: University of Minnesota Press.

Lourandos, H. 1993. Hunter-gatherer Cultural Dynamics: Long- and Short-term trends in Australian prehistory. *Journal of Archaeological Research* 1: 67–88.

———. 1997. *Continent of Hunter-Gatherers: New Perspectives in Australian Prehistory*. Cambridge: Cambridge University Press.

MacArthur, R. H., and Wilson, E. O. 1967. *The Theory of Island Biogeography. Princeton Landmarks in Biology*. Princeton, NJ: Princeton University Press.

Magowan, F. 2001. Waves of knowing: Polymorphism and co-substantive essences in Yolngu cosmology. *Australian Journal of Indigenous Education* 29(1): 22–35.

McNiven, I. J. 1998. Enmity and amity: Reconsidering stone-headed club (*gabagaba*) procurement and trade in Torres Strait. *Oceania* 69: 94–115.

———. 2000. Treats or Retreats? Aboriginal island usage along the Gippsland coast. *The Artefact* 23: 22–34.

———. 2003. Saltwater people: Spiritscapes, maritime rituals and the archaeology of Australian indigenous seascapes. *World Archaeology* 35(3): 329–49.

———. 2006. Dauan 4 and the emergence of ethnographically known social arrangements across Torres Strait during the last 600–800 years. *Australian Archaeology* 62: 1–11.

McNiven, I. J., Crouch, J., and David, M. 2004. Tudu (Warrior Island), Torres Strait: Historical and archaeological survey. Report to the Tudu/Yama community, Torres Strait. *Cultural Heritage Report Series* 4, Programme for Australian Indigenous Archaeo-logy, School of Geography and Environmental Science, Monash University, Melbourne.

McNiven, I. J., David, B., and Barker, B. 2006. The social archaeology of Indigenous Australia, in B. David, B. Barker, and I. J. McNiven (eds.), *The Social Archaeology of Australian Indigenous Societies*, pp. 2–19. Canberra: Aboriginal Studies Press.

McNiven, I. J., Dickinson, W. R., David, B. Weisler, von Gnielinski, M. F., Carter, M., and Zoppi, U. 2006. Mask Cave: Red-slipped pottery and the Australian-Papuan settlement of Zenadh Kes (Torres Strait). *Archaeology in Oceania* 41: 49–81.

McNiven, I. J., and Hitchcock, G. 2004. Torres Strait marine subsistence specialisation and terrestrial animal translocation, in I. J. McNiven and M. Quinnell (eds.), *Torres Strait Archaeology and Material Culture, Memoirs of the Queensland Museum, Cultural Heritage Series* 3: 105–62.

Nietschmann, B. 1989. Traditional sea territories, resources and rights in Torres Strait, in J. Cordell (ed.), *A Sea of Small Boats*, pp. 60–93. Cambridge: Cultural Survival, Inc.

Nunn, P. D. 1987. Small islands and geomorphology: Review and prospect in the context of historical geomorphology. *Transactions of the Institute of British Geographers* 12: 227–39.

Palsson, G. 1994. Enskilment at Sea. *Man* 29: 901–27.

Rainbird, P. 1999. Islands out of time: Towards a critique of island archaeology. *Journal of Mediterranean Archaeology* 12: 216–34.

———. 2004. *The Archaeology of Micronesia*. Cambridge: Cambridge University Press.

Rowland, M. 2002. "Crows," swimming logs and auditory extoses: Isolation on the Keppel Islands and broader implications, in S. Ulm, C. Westcott, J. Reid, A. Ross, I. Lilley, J. Prangnell, and L. Kirkwood (eds.), *Barriers, Borders, Boundaries. Proceedings of the 2001 Australian Archaeological Association*

Annual Conference, 2001, Tempus 7, pp. 61–73. Brisbane: Anthropology Museum, University of Queensland.

Sharp, A. 1956. *Ancient Voyagers in the Pacific*. Baltimore: Penguin Books.

Sharp, N. 2002. *Saltwater People: The Waves of Memory*. Crows Nest: Allen and Unwin.

Shnukal, A. 2004. The post-contact created environment in the Torres Strait Central Islands, in I. J. McNiven and M. Quinnell (eds.), *Torres Strait Archaeology and Material Culture*, 3: 317–46. Brisbane: Memoirs of the Queensland Museum Cultural Heritage Series.

Terrell, J. E. 1986. *Prehistory in the Pacific Islands*. Cambridge: Cambridge University Press.

Terrell, J. E., Hunt, T.L., Gosden, C., Bellwood, P., Finney, B., Ward, H., Goodenough, H., Grace, G. W., Harms, V., Heathcote, G. M., Diego, V. P., Camacho, F. A., Taisipic, T. F., Kirch, P. V., and Ross, M. 1997. The dimensions of social life in the Pacific: Human diversity and the myth of the primitive isolate [and comments and reply]. *Current Anthropology* 38: 155–95.

Turnbull, D. 1991. *Mapping the World in the Mind: An Investigation of the Unwritten Knowledge of the Micronesian Navigators*. Geelong: Deakin University Press.

Vanderwal, R. L. 1973. The Torres Strait: Protohistory and beyond. *University of Queensland Anthropology Museum Occasional Papers* 2: 173–94.

Walker, D. (ed.). 1972. *Bridge and Barrier: The Natural and Cultural History of Torres Strait*. Canberra: Australian National University.

Weisler, M. I. 2001. On the margins of sustainability: Prehistoric settlement of Utrōk Atoll Northern Marshall Islands. *BAR International Series* 967. Oxford: Archaeopress.

12

ISLAND BIOGEOGRAPHY: IMPLICATIONS AND APPLICATIONS FOR ARCHAEOLOGY

John Edward Terrell

In his ornithological notes prepared in 1835 while he was homeward bound, young Charles Darwin made what is now seen as a prescient remark about the mockingbird specimens he had gathered in the Galápagos Islands: "When I see these Islands in sight of each other, & possessed of but a scanty stock of animals, tenanted by these birds, but slightly differing in structure & filling the same place in Nature, I must suspect they are only varieties. . . . If there is the slightest foundation for these remarks the zoology of Archipelagoes will be well worth examination; for such facts would undermine the stability of Species" (quoted in MacArthur and Wilson 1967: 3, note 1). Such was Darwin's later fame and the value of his evolutionary theory that the Galápagos Islands are now iconic symbols of the power of islands to open our minds to the mysteries of nature and the wisdom of evolution. However, islands as archaeological landscapes have long been problematic. Has the circumscription and possible remoteness of islands made island societies exceptional or deviant in some respects from the normal run of things in continental regions such as Asia, Europe, or the Americas? Or can we use islands as naturally defined "units of study" in landscape archaeology to explore questions about the ecology and evolution of human societies that might be overwhelming or utterly confusing if tackled on the scale of continents?

Rhetoric is a goodly part of scholarship (Gross 1990). It is currently fashionable in archaeology to complain about island biogeography and the idea of islands as "laboratories" as anachronisms that should probably be shoveled into the grave of failed ideas and discredited worldviews as too inhuman and disrespectful of human agency to be taken seriously.[1] In this chapter I want to challenge current wisdom to suggest not only that islands are fine places to do archaeology but also that island biogeography is alive and well—although it could be argued that to survive it has had to morph into something called *metapopulation biology.*

Anthropology and Islands

What islands have meant to human beings has differed with time and place as well as with the texture of historical experience. As John Donne's poetry and Sir Thomas More's *Utopia* alike inform us, islands can be imagined as "other worlds" cut off from the corruption of foreign influences, invasions, and social responsibilities. Yet, life tells us otherwise. It is their geographic *definitiveness* and the challenges of *access* that define them, not their isolation—certainly not if isolation is taken as "other-worldliness," or life lived in Nature's equivalent to a laboratory petri dish (see Broodbank 1999).

Islands may be places you want to escape to, especially if the islands have palm trees and it is the dead of winter in higher latitudes, but this tourist's fantasy won't work if, having reached their shores, a sailor cannot come ashore. The claim made by some that a previous generation of archaeologists thought islands could be used as natural laboratories, because they are isolated cases of human adaptation to presumed peculiarities or simplicities of island ecosystems is either exaggerated or false (Keegan 1999). Remember this: when it comes to archaeology, landscapes, and island biogeography, the word "isolation" needs to be read as relative, not absolute, and certainly not as "out of the ordinary" (Terrell 1986: Chapter 6).

If we must have a villain in our tale, then anthropologists rather than archaeologists will do (Kuklick 1996). It is not farfetched to say that after World War II, especially in the Pacific, the islands-as-isolates idea did have apparent standing among anthropologists for a while (e.g., see Goodenough 1957; Mead 1957; Sahlins 1957, 1958; Sharp 1956; Suggs 1961: 194; Terrell, Hunt, and Gosden 1997; Vayda and Rappaport 1963; Watson 1963). This was the heyday of positivism (Popper 1959), the quantitative revolution in the social and biological sciences (Haggett 1966; Kingsland 1985), and of functionalism, structuralism, and comparative studies in anthropology (Lévi–Strauss 1963; Murdock 1949). In sociocultural anthropology, fieldwork was often reported to the world in narrative ways that made it sound as if anthropologists were privileged persons: each one had "my village" to talk about, and these little microcosms could be treated for comparative purposes as if they were indeed human isolates— cut off not only from the world but evidently from their neighbors, as well.

However appealing these imagined unworldly places may have been to anthropologists and other social scientists for a few decades after World War II, most would nowadays acknowledge that human societies, like human languages, are not discrete entities sealed off from contact with others (e.g., Leach and Leach 1983). Minimally, it is accepted that cultures and societies are "mixed" in the sense that their characteristics and distinctive features stem both from their own history and from their participation in broader ecological, social, economic, and political relationships (Lape 2004; Terrell, Hunt, and Gosden 1997). This is not to say, of course, that everyone in the academe has yet heard the word. In one of the more transcendental corners of biology, the field of phylogenetics, there are those who still unblushingly announce that for their purposes "languages and cultures are treated as being analogous to species,

although there has been a vigorous debate about how far we can treat cultures as discrete, bounded units, similar to species" (Mace and Holden 2005: 116). This shows that there is nothing new under the sun.

Looking over what was written about island archaeology 20 or 30 years ago (e.g., Clark and Terrell 1978; Evans 1977; Glover 1977: 59; Hunt 1986, 1987; Irwin 1973; Kaplan 1976; Kirch 1986, 1987; Rouse 1977: 7; Shutler and Marck 1975; Terrell 1976, 1977a, 1977b) lends little to the claim (e.g., Rainbird 1999a: 223–26) that archaeologists thought very differently then about such places than they do nowadays (see Keegan 1999). As I commented in the mid 1970s:

> The value of islands as natural laboratories does not lie in their supposed uniqueness, isolation or divergence from the main paths of history and evolution. On the contrary, islands are good places to study the world because they are numerous, because they occur in many sizes, shapes and degrees of ecological complexity, and because they can be found both in secluded locations where evolution may proceed virtually free from outside intervention and also near continents which are sources of new immigrants and ideas. (Terrell 1977b: 79)

In other words, there is scant evidence supporting the idea that archaeologists once upon a time were naïvely seeing islands as laboratories merely on the shaky grounds that such places are cut off from the main currents of history. Even those whose theoretical orientation toward data then as now led them to privilege isolation as an analytical simplification (e.g., Kirch 1984) were normally careful not to overdo the thrust of the oversimplification (e.g., Kirch 1986; Kirch and Green 1987, 2001).

Island Biogeography

Island biogeography (Whittaker 1998) is a scientific calling nested within the broader field of biogeography (Lomolino and Heaney 2004; Lomolino et al. 2005). Both have long been a mixed bag of concerns, field methods, laboratory procedures, and analytical techniques (Lomolino 2000; Whittaker 2000). They are alike readily defined as the study of the geographic distributions of organisms, past and present (Brown and Gibson 1983: 557). Both now add up to a more coherent body of thought than was certainly the case a generation ago (cf. Lomolino and Heaney

2004; Sauer 1977). I would add that neither island biogeography nor biogeography writ large is a simple or singular template (Brown and Lomolino 2000) for defining what an archaeological island biogeography is or should strive to become as a discipline among the other sciences (or among the humanities, if one prefers).

Critiques of island biogeography and of seeing islands as natural laboratories for studying the patterning and processes of biological evolution have long been a concurrent part of the business of evolutionary biology (Gilbert 1980; Hanski and Simberloff 1997; Sauer 1977; Walter 2004; Williamson 1989). These critiques are worth reading. It is my guess, however, that this criticism is not well understood outside the biological sciences. The thought, for example, that gene pools, groups of organisms, or places can be usefully modeled as island-like (for example, as in metapopulation biology today; see Hanski 1999; Hanski and Gilpin 1997) exploits rather than reifies the idea of islands. In any case, it is relationships *among* such seemingly definitive things or places, not their presumed character as closed systems, high-security laboratories, or inaccessible plots of ground, that lend this kind of model building its biological cachet. In other words, the landscape of choice in island biogeography—as Darwin commented—is normally *not an isolated island but an archipelago*, real or imagined (Lomolino 2000; Terrell 1999).

Furthermore, it makes a difference whose biogeography you are talking about. Ernst Mayr (1963) pushed the paradigm of allopatric speciation as far as it would go, but not all agreed with him a generation ago (Sauer 1977: 322); nor is it now any longer advisable to dismiss sympatric speciation as improbable and therefore basically irrelevant to the evolution of life's diversity (Jiggins and Bridle 2004; Via 2001). Yes, Mayr's kind of biogeography has informed anthropology (e.g., Sahlins 1957; Watson 1963) and archaeology both then and now (e.g., Kirch and Green 1987, 2001). Yet I suspect that for many social science practitioners, old and new, the concept of allopatric speciation and the citation of processes such as adaptive radiation have probably always sounded far too biological to be employed productively in what they do.

However, it is true that archaeologists working in the Mediterranean (e.g., Broodbank 2000; Cherry 1981; Evans 1977), the Caribbean (Keegan and Diamond 1987), and the Pacific (Hunt 1987, 1988; Irwin 1992; Kaplan 1976; Terrell 1974, 1976, 1977a, 1977b) were captivated at least for a while, as many biogeographers then were, too, by

Robert MacArthur's and Edward Wilson's kind of biogeography as heralded in their famous equilibrium theory of island biogeography (1967).

MacArthur's and Wilson's basic proposition was simple (as it turned out, too simple). They suggested that island species diversity could be predicted mathematically as a dynamic equilibrium between the immigration of species from elsewhere and local species extinction. Their theory, of course, had much more to offer than just this (see Whittaker 1998: Chapter 7; also Quammen 1996). It became famous for its apparent elegance, its mathematical models, and its emphasis on such seemingly direct and easily acquired field data as island size (area) and inventories of species found on islands under scrutiny.

What attracted me to this brand of biogeography was not so much its mathematical side or its seemingly predictive powers, but rather the emphasis this approach placed on being honest and direct about the ideas being advanced and used. As it turned out, this was a prescient position to adopt, for in a devastating article published in 1980 in the *Journal of Biogeography*, F. S. Gilbert (1980: 230–31) gave the *coup de grâce* to seeing this theory as an accurate explanatory device with these now classic closing words: "The qualitative use of the equilibrium concept has stimulated a great deal of valuable research, and is clearly of use as a way of approaching an appropriate problem . . . Quantitatively, however, it would seem that the model has little evidence to support its application to any situation."

This may sound like a weak endorsement of a bad idea, but when even important science journals such as *Nature* are willing to publish poorly explained models purporting to detail how ancient "express trains" or "slow boats" once upon a time carried prehistoric colonists out to virgin islands in the Pacific (e.g., Austin 1999; Diamond 1988; Gibbons 2001; Gray and Jordan 2000), I continue to admire how open MacArthur and Wilson were in going public with their willingness to analyze obviously complex issues by first describing them in clear, relevant, and simple terms so that actual work could be done to start teasing them apart.

However tarnished the MacArthur–Wilson equilibrium theory of island biogeography may now be (Brown and Lomolino 2000), the lesson for me remains unchanged: archaeologists ought to be as direct and open in showing us their cards and playing their hands. The fact that some archaeologists now criticize the efforts of their predecessors, because they feel they had not paid sufficient attention to an elusive causal variable called "human agency" tells me such modern commentators miss the point I had thought was obvious. No, of course, the simple models favored by MacArthur and Wilson, mathematical or graphical, are not entirely realistic models. They weren't really intended to be (see Levins 1966, 1993; Orzack and Sober 1993). They were mostly what C. H. Waddington (1977) in the 1970s called "tools for thought."

Implications for Archaeology

Here is a basic question: *what is an island?* As biogeographers see them, islands are what they are because they are *living spaces (habitats) surrounded by radical shifts in habitat*—so radical that (1) few species of plants and animals are able to live for long in more than one of these radically different habitats; and (2) consequently we must pay close attention not only to what these habitat islands are like at any one moment but also to how and how often what is living there (plants, animals, and humans) comes and goes. This, at least as MacArthur and Wilson saw things, is basic island biogeography. "Islandness" may thus be seen as a common property of life rather than as a peculiar property of unusual places.

If islands are habitats surrounded by radical shifts in habitat, then islands have much to offer us. We may use the property we are calling "a radical shift in habitat" to determine the boundaries of "our places of study" rather than, say, using more artificial parameters such as latitude, longitude, or the dimensions of an arbitrary "sampling grid" or geopolitical unit. Seen this way—contrary to what the famous poet John Donne told us—all men *are* islands. So, too, are berry bushes in a cow pasture; cow pastures beside an interstate; cornfields great and small; and so on. All are "islands," since the world, as ecologists like to say, is "patchy." Once more, this is basic island biogeography. From a biogeographer's point of view, paying scientific attention to this patchiness is one way to get our arms around our earth's diversity. In short, islands are not special because of what they are; they are just great places to think about and study.

By the 1980s, MacArthur's and Wilson's equilibrium theory of island biogeography had fallen from favor in population biology. Depending on your point of view, it had by then either been replaced by, or had begun evolving into, a related way of thinking about and mathematically modeling Nature called *metapopulation biology* (Hanski 1999; Hanski and Gilpin 1997). Loosely described, this alternative approach relies on the idea that organisms are unlikely to be uniformly distributed in space. If so, then they can often be usefully modeled as assemblages of separated local populations that are spatially related to one another by migration (or in other ways, too) into larger regional populations. When this is the case, efforts to analyze and understand what is happening locally (or archaeologically speaking, happened) probably need to take into account these broader interactions.

Two of the leaders of this relatively new approach wrote a decade ago: "An important reason for the appeal of the metapopulation concept comes from our subjective conviction that natural landscapes truly are, for many species, patchworks of one or several habitat types" (Hanski and Gilpin 1997: 3). Given this statement, it may be obvious that this way of thinking about biogeography has much in common both with older MacArthur–Wilson island biogeography (Hanski 1999: 11) and also with modern landscape ecology (e.g., Lomolino and Perault 2001). Like the former and unlike the latter, metapopulation models often focus on idealized habitat patches in more or less featureless worlds (Urban and Keitt 2001); like the latter and possibly less like the former, students of metapopulation biology favor real-world applications, especially in the arena of conservation biology (e.g., Patterson and Atmar 2000)—an arena within which classical island biogeography tried to help but largely failed (Gilbert 1980: 230; Hanski and Simberloff 1997: 17–19).

It might be possible for landscape archaeologists to morph along with island biogeography and try to emulate metapopulation biology, but instead of proposing that what archaeologists now need is metapopulation archaeology, I would like to return to a thought I had when I first came across *The Theory of Island Biogeography* in Staver's Bookshop on the south side of Chicago in 1971. Recently Hartmut Walter (2004: 189), who is highly critical of MacArthur's and Wilson's ideas as expressed in this classic text, observed nevertheless that both were right to stress the importance of islands. I thought so in 1971, and still think this. I also continue to believe—as I did when I bought a copy of Haggett's *Locational Analysis in Human Geography* (1966) in Cambridge, Massachusetts, in 1968—that the formation processes at all levels that have led to what is now the "archaeological record" obviously had their real-world spatial dimensions.

Combine these two thoughts—the analytical importance of islands (Renfrew 2004) and the spatial character of life's processes—and you have a basis for island biogeography.

A few years ago Robert O'Neill (2001) at Oak Ridge National Laboratory took a critical look at a related "tool for thought," the ecosystem idea that, like equilibrium island biogeography, was once all the rage but that is now enjoying a much more muted coexistence with rival ways of thinking about Nature. Like MacArthur–Wilson modeling, this concept once looked like a practical approach to the complexity of natural systems. In both cases, however, it was too easy; some would say necessary, to leave history out of our equations, since both approaches made much of equilibrium, feedback, and homeostasis.

O'Neill observes that using a tool such as the ecosystem concept is not cost free: such tools are grounded on assumptions that can limit our thinking and predetermine the research questions we think to ask. In the case of ecosystems, he writes, the spatial distribution of the component populations may be much larger than the presumed boundaries of the system being examined, and this may lead to anomalies. Said differently, the stability properties of an ecosystem may not be explicable by a concept that considers only the dynamics occurring within those boundaries. Ecosystems, like islands, may be more "open" than is scientifically convenient.

At first, it might seem that MacArthur–Wilson equilibrium models do not suffer from this limitation, because they make so much of species coming and going from one place to another. This difference, however, is more semantic than real. It is just a question of how you draw your boundaries, or if you like, it's just a question of scale. In this regard, O'Neill has offered a number of observations about ecosystem modeling that can be applied more broadly. Here are several of them restated in somewhat more generic terms (2001: 3280–82):

1. An ecosystem works on an array of spatial scales going from the local up to the full potential dispersal range of the species within the local system being considered;

2. The potential scale of an ecosystem is set by the environmental constraints on each species, the character of barriers to their dispersal, and the particular ways that the various species involved get about from place to place;

3. The potential size or scale of an ecosystem is not constant or uniform over time and can change, for example, with changes in climate, geological events, and the like—these changes may be slow and thus fairly easy to adapt to, or they may be rapid and cataclysmic;

4. Furthermore, conditions all across the potential range of the constituent species are unlikely to be uniform, either locally or on a larger scale;

5. Over the long run, what maintains the sustainability of ecosystems is heterogeneity in time and space. "Stability to smaller scale impacts depends on the system's ability to resist change and recover with resilience. But long-term stability or sustainability depends on a flexibility of response that can only be maintained in an environment that varies with time and space" (2001: 3281).

It is unnecessary, I think, to spell out how directly these observations can be applied also to human systems of interaction, economic engagement, and political control.

These are not the full measure of what O'Neill has to say about the character of ecosystems. I have selected these observations because by making these observations O'Neill is trying to make this concept, this tool for thought, work better. As he concludes: "Is it time to bury the ecosystem concept? Probably not. But there is certainly need for improvement before ecology loses any more credibility" (2001: 3282).

By way of summary, he identifies three key elements that he suggests cannot be ignored in model building if we want to understand how the world works, because the real-time processes and constraints thus hallmarked are unlikely to be encompassed within the boundaries of the local ecological system: *spatial pattern*, *extent*, and *heterogeneity*. Each of these dimensions, he maintains, is critical to ecosystem stability (2001: 3282). He is probably right. The biogeographer Lomolino (2000) would add that individuals, groups of individuals, and species differ in their abilities to respond to variation in the spatial patterning, extent, and heterogeneity of the world around them, and, furthermore, their abilities to do so evolve over time. Said colloquially, individuals, groups, and species do not stand still—and the importance of including a dimension that many nowadays call "agency" in our model building holds true for species other than just *Homo sapiens* (Terrell 1986: Chapter 7; Terrell and Hart 2002; Terrell et al. 2003)

One major implication for landscape archaeology, island or otherwise, that may be drawn from these observations is fairly certain. In thinking about the stability of ecosystems, it won't do to see *Homo sapiens* as an *external* disturbing force. Instead, we are a keystone species *within* ecosystems. Looking to our own endurance as a species, he observes: "In the long term, it may not be the magnitude of extracted goods and services that will determine sustainability. It may well be our disruption of ecological recovery and stability mechanisms that determines system collapse" (O'Neill 2001: 3282).

I think this observation may also be applied retrospectively. We are taught as young archaeologists that to be successful at explaining prehistoric cultural stability and change we cannot neglect that we are part of—not external to—the world we live in (Terrell et al. 2003). Therefore the challenge we face cannot be ignored: how are we to grapple with this fact in a successful way? If we are part of the problem, so to speak, how can we expect to see our way to a solution?

This is why I think it is a mistake to throw away tools for modeling the issues we want to study as archaeologists on the grounds that they have become no longer fashionable or overlook something we now emphasize more strongly than we once did (Terrell 1997). Speaking personally, I do not consider the education of young archaeologists complete without advising them to read about and evaluate for themselves such venerable (or new) tools for thought as the ecosystem concept, the equilibrium theory of island biogeography, locational geography, landscape ecology, and metapopulation biology. A fly fisherman knows that different fish and different fishing conditions call for different kinds of flies. Instead of throwing away the investigatory tools we have acquired over the years, we need to focus instead on deciding which of these tools may be able to help us do whatever it is we want to do right.

Conclusions

The great thing about islands for archaeology is that they invite, even beg, for comparative study—which in the eyes of many remains a primary concern of anthropology broadly defined (e.g., Curet 2005; Kirch and Green 2001; Rainbird 2004). In the right hands (e.g., Curet 2005; Irwin 1992, 1999), islands lead us to confront the essential diversity that is so characteristic of our species. And as Broodbank has said: "Although island biogeography must not (and has no pretensions to) set the main agenda for island archaeology, we merely impoverish ourselves by entirely denying its relevance out of a faddish distaste for Darwinism in the social sciences" (1999: 237; also 1999: 28).

Acknowledgments

I thank Ethan Cochrane and John Hart for comments on the manuscript for this chapter.

Note

1. For a sampling of opinions, see Broodbank 1999, 2000; Curet 2005: 31–32, 222; Fitzpatrick 2004; Irwin 1999: 253; Lape 2004; Rainbird 1999a, 1999b, 2004.

References

Austin, C. 1999. Lizards took express train to Polynesia. *Nature* 397: 113–14.

Broodbank, C. 1999. The insularity of island archaeologists: Comments on Rainbird's "Islands out of time." *Journal of Mediterranean Archaeology* 12: 235–39.

———. 2000. *An Island Archaeology of the Early Cyclades*. Cambridge: Cambridge University Press.

Brown, J., and Gibson, A. 1983. *Biogeography*. St. Louis, MO: Mosby.

Brown, J., and Lomolino, M. 2000. Concluding remarks: Historical perspective and the future of island biogeography theory. *Global Ecology and Biogeography* 9: 87–92.

Cherry, J. 1981. Pattern and process in the earliest colonization of the Mediterranean islands. *Proceedings of the Prehistoric Society* 47: 41–68.

Clark, J., and Terrell, J. 1978. Archaeology in Oceania. *Annual Review of Anthropology* 7: 293–319.

Curet, A. 2005. *Caribbean Paleodemography: Population, Culture History, and Sociopolitical Processes in Ancient Puerto Rico*. Tuscaloosa: University of Alabama Press.

Diamond, J. 1988. Express train to Polynesia. *Nature* 336: 307–08.

Evans, J. 1977. Island archaeology in the Mediterranean: Problems and opportunities. *World Archaeology* 9: 1–11.

Fitzpatrick, S. 2004. Synthesizing island archaeology, in S. Fitzpatrick (ed.), *Voyages of Discovery: The Archaeology of Islands*, pp. 1–18. Westport, CT: Praeger.

Gibbons, A. 2001. The peopling of the Pacific. *Science* 291: 1735–37.

Gilbert, F. 1980. The equilibrium theory of island biogeography: Fact or theory? *Journal of Biogeography* 7: 209–35.

Glover, I. 1977. The Late Stone Age in eastern Indonesia. *World Archaeology* 9: 42–61.

Goodenough, W. 1957. Oceania and the problem of controls in the study of cultural and human evolution. *Journal of the Polynesian Society* 66: 146–55.

Gray, R., and Jordan, F. 2000. Language trees support the express-train sequence of Austronesian expansion. *Nature* 405: 1052–55.

Gross, A. 1990. *The Rhetoric of Science*. Cambridge, MA: Harvard University Press.

Haggett, P. 1966. *Locational Analysis in Human Geography*. New York: St. Martin's Press.

Hanski, I. 1999. *Metapopulation Ecology*. Oxford: Oxford University Press.

Hanski, I., and Gilpin, M. (eds.). 1997. *Metapopulation Biology: Ecology, Genetics, and Evolution*. San Diego: Academic Press.

Hanski, I., and Simberloff, D. 1997. The metapopulation approach, its history, conceptual domain, and application to conservation, in I. Hanski and M. Gilpin (eds.), *Metapopulation Biology: Ecology, Genetics, and Evolution*, pp. 5–26. San Diego: Academic Press.

Hunt, T. 1986. Conceptual and substantive issues in Fijian prehistory, in P. Kirch (ed.), *Island Societies: Archaeological Approaches to Evolution and Transformation*, pp. 20–32. Cambridge: Cambridge University Press.

———. 1987. Patterns of human interaction and evolutionary divergence in the Fiji Islands. *Journal of the Polynesian Society* 96: 299–334.

———. 1988. Graph theoretic network models for Lapita exchange: A trial application, in P. Kirch and T. Hunt (eds.), *Archaeology of the Lapita Cultural Complex: A Critical Review*, pp. 135–55. Research Report 5. Seattle: Thomas Burke Memorial Washington State Museum.

Irwin, G. 1973. Man-land relationships in Melanesia: An investigation of prehistoric settlement in the islands of the Bougainville Strait. *Archaeology and Physical Anthropology in Oceania* 8: 226–52.

———. 1992. *The Prehistoric Exploration and Colonisation of the Pacific*. Cambridge: Cambridge University Press.

———. 1999. Commentary on Paul Rainbird, "Islands out of time: Towards a critique of island archaeology." *Journal of Mediterranean Archaeology* 12: 252–54.

Jiggins, C., and Bridle, J. 2004. Speciation in the apple maggot fly: A blend of vintages? *Trends in Ecology and Evolution* 19: 111–14.

Kaplan, S. 1976. *Ethnological and Biogeographical Significance of Pottery Sherds from Nissan Island, Papua New Guinea. Fieldiana: Anthropology* 66(3). Chicago: Field Museum of Natural History.

Keegan, W. 1999. Comment on Paul Rainbird, "Islands out of time: Towards a critique of island archaeology." *Journal of Mediterranean Archaeology* 12: 255–58.

Keegan, W., and Diamond, J. 1987. Colonization of islands by humans: A biogeographical perspective. *Advances in Archaeological Method and Theory* 10: 49–92.

Kingsland, S. 1985. *Modeling Nature: Episodes in the history of population ecology*. Chicago: University of Chicago Press.

Kirch, P. 1984. *The Evolution of the Polynesian Chiefdoms*. Cambridge: Cambridge University Press.

———. 1986. Introduction: The archaeology of island societies, in P. Kirch (ed.), *Island Societies: Archaeological Approaches to Evolution and Transformation*, pp. 1–5. Cambridge: Cambridge University Press.

———. 1987. *The Lapita Peoples: Ancestors of the Oceanic world*. Oxford: Blackwell Publishing.

Kirch, P., and Green, R. 1987. History, phylogeny, and evolution in Polynesia. *Current Anthropology* 28: 431–56.

———. 2001. *Hawaiki, Ancestral Polynesia: An essay in historical anthropology*. Cambridge: Cambridge University Press.

Kuklick, H. 1996. Islands in the Pacific: Darwinian biogeography and British anthropology. *American Ethnologist* 23: 611–38.

Lape, P. 2004. The isolation metaphor in island archaeology, in S. Fitzpatrick (ed.), *Voyages of Discovery: The Archaeology of Islands*, pp. 223–32. Westport, CT: Praeger.

Leach, E., and Leach, J. (eds.). 1983. *The Kula: New Perspectives in Massim Exchange*. Cambridge: Cambridge University Press.

Lévi-Strauss, C. 1963. *Structural Anthropology*. New York: Basic Books.

Levins, R. 1966. The strategy of model building in population biology. *American Scientist* 54: 421–31.

———. 1993. A response to Orzack and Sober: Formal analysis and the fluidity of science. *Quarterly Review of Biology* 68: 547–55.

Lomolino, M. 2000. A call for a new paradigm of island biogeography. *Global Ecology and Biogeography* 9: 1–6.

Lomolino, M., and Heaney, L. (eds.). 2004. *Frontiers of Biogeography: New Directions in the Geography of Nature*. Sunderland, MA: Sinauer Associates.

Lomolino, M., and Perault, D. 2001. Island biogeography and landscape ecology of mammals inhabiting fragmented, temperate rain forests. *Global Ecology and Biogeography* 10: 113–32.

Lomolino, M., Riddle, B., and Brown, J. 2005. *Biogeography* (3rd ed.). Sunderland, MA: Sinauer Associates.

MacArthur, R., and Wilson, E. 1967. *The Theory of Island Biogeography*. Princeton, NJ: Princeton University Press.

Mace, R., and Holden, C. 2005. A phylogenetic approach to cultural evolution. *Trends in Ecology and Evolution* 20: 116–21.

Mayr, E. 1963. *Animal Species and Evolution*. Cambridge, MA: Harvard University Press.

Mead, M. 1957. Introduction to Polynesia as a laboratory for the development of models in the study of cultural evolution. *Journal of the Polynesian Society* 66: 145.

Murdock, G. 1949. *Social Structure*. New York: Macmillan.

O'Neill, R. 2001. Is it time to bury the ecosystem concept? (with full military honors, of course!). *Ecology* 82: 3275–84.

Orzack, S., and Sober, E. 1993. A critical assessment of Levins's the strategy of model building in population biology. *Quarterly Review of Biology* 68: 533–46.

Patterson, B., and Atmar, W. 2000. Analyzing species composition in fragments, in G. Rheinwald (ed.), *Isolated Vertebrate Communities in the Tropics*, pp. 9–24. Proceedings of the 4th International Symposium, Bonn. *Bonner Zoologische Monographen* 46, Bonn, Germany.

Popper, K. 1959. *The Logic of Scientific Discovery*. London: Hutchinson.

Quammen, D. 1996. *The Song of the Dodo: Island Biogeography in the Age of Extinctions*. New York: Scribner.

Rainbird, P. 1999a. Islands out of time: Towards a critique of island archaeology. *Journal of Mediterranean Archaeology* 12: 216–34.

———. 1999b. Nesophiles miss the boat? A response. *Journal of Mediterranean Archaeology* 12: 259–60.

———. 2004. *The Archaeology of Micronesia*. Cambridge: Cambridge University Press.

Renfrew, C. 2004. Islands out of time? Toward an analytical framework, in S. Fitzpatrick (ed.), *Voyages of Discovery: The Archaeology of Islands*, pp. 276–94. Westport, CT: Praeger.

Rouse, I. 1977. Pattern and process in West Indian archaeology. *World Archaeology* 9: 12–26.

Sahlins, M. 1957. Differentiation by adaptation in Polynesian societies. *Journal of the Polynesian Society* 66: 291–300.

———. 1958. *Social Stratification in Polynesia*. Seattle: University of Washington Press.

Sauer, J. 1977. Biogeographical theory and cultural analogies. *World Archaeology* 8: 32031.

Sharp, A. 1956. *Ancient voyagers in the Pacific*. Polynesian Society Memoir 32. Wellington: The Polynesian Society.

Shutler, R., and Marck, J. 1975. On the dispersal of the Austronesian horticulturalists. *Archaeology and Physical Anthropology in Oceania* 10: 81–113.

Suggs, R. 1961. *The Archaeology of Nuku Hiva, Marquesas Islands, French Polynesia*. Anthropological Papers, 49(1). New York: American Museum of Natural History.

Terrell, J. 1974. *Comparative Study of Human and Lower Animal Biogeography in the Solomon Islands*. Solomon Island Studies in Human Biogeography 3. Chicago: Department of Anthropology, Field Museum of Natural History.

———. 1976. Island biogeography and man in Melanesia. *Archaeology and Physical Anthropology in Oceania* 11: 1–17.

———. 1977a. Geographic systems and human diversity in the North Solomons. *World Archaeology* 9: 62-81.

———. 1977b. Human biogeography in the Solomon Islands. *Fieldiana: Anthropology* 68(1): 1–47.

———. 1986. *Prehistory in the Pacific Islands*. Cambridge: Cambridge University Press.

———. 1997. The postponed agenda: Archaeology and human biogeography in the twenty-first century. *Human Ecology* 25.

———. 1999. Comment on Paul Rainbird, "Islands out of time: Towards a critique of island archaeology." *Journal of Mediterranean Archaeology* 12: 240–45.

Terrell, J., and Hart, J. 2002. Introduction, in J. Hart and J. Terrell (eds.), *Darwin and Archaeology: A Handbook of Key Concepts*, pp. 1–13. Westport: Bergin and Garvey.

Terrell, J., Hart, J., Barut, S., Cellinese, N., Curet, A., Denham, T., Kusimba, C., Latinis, K., Oka, R., Palka, J., Pohl, M., Pope, K., Williams, P., Haines, H., and Staller, J. 2003. Domesticated landscapes: The subsistence ecology of plant and animal domestication. *Journal of Archaeological Method and Theory* 10: 323–68.

Terrell, J., Hunt, T., and Gosden, C. 1997. The dimensions of social life in the Pacific: Human diversity and the myth of the primitive isolate. *Current Anthropology* 38: 155–95.

Urban, D., and Keitt, T. 2001. Landscape connectivity: A graph-theoretic perspective. *Ecology* 82: 1205–18.

Vayda, A., and Rappaport, R. 1963. Island cultures, in F. Fosberg (ed.), *Man's Place in the Island Ecosystem*, pp. 133–44. Honolulu: Bishop Museum Press.

Via, S. 2001. Sympatric speciation in animals: The ugly duckling grows up. *Trends in Ecology and Evolution* 16: 381–90.

Waddington, C. 1977. *Tools for Thought*. St. Albans: Paladin.

Walter, H. 2004. The mismeasure of islands: Implications for biogeographical theory and the conservation of nature. *Journal of Biogeography* 31: 177–97.

Watson, J. 1963. A micro-evolution study in New Guinea. *Journal of the Polynesian Society* 72: 188–92.

Whittaker, R. 1998. *Island Biogeography: Ecology, evolution, and conservation*. Oxford: Oxford University Press.

———. 2000. Scale, succession and complexity in island biogeography: Are we asking the right questions? *Global Ecology and Biogeography* 9: 75–85.

Williamson, M. 1989. Guest editorial: The MacArthur and Wilson theory today: True but trivial. *Journal of Biogeography* 16: 3–4.

13

Sentient Sea: Seascapes as Spiritscapes

Ian J. McNiven

the ocean is not merely a space used *by* society;

it is one component of the space *of* society. (Steinberg 2001: 20)

Seas cover 70% of Earth's surface, yet for all of recorded history most people have spent their lives on land. Landscapes are normative—hills, forests, rivers, buildings, roads, and myriad other marked places structure, order, frame, constrain, and give meaning to our lives as social beings sharing a lived space. Whether your landscape is the tropical rainforest of lowland Papua New Guinea or the concrete jungle of New York, every morning you can wake up and traverse your landscape and experience through visual, olfactory, acoustic, and tactile senses familiar and fixed surroundings resonating with memories, emotions, and deep symbolic meaning. The main phenomena you experience as ever changing are the weather and social engagements with people. But what if your landscape was morphologically dynamic and fluid? It is such a dynamic realm that maritime peoples engage with day to day. If landscapes are so important to social behavior and identity, how is it that maritime peoples whose social realm is mostly the sea construct their identity, through seascapes, resonating

with memories, deep knowledge, and symbolic meaning? Maritime people's intimacy with the sea is illustrated well by Torres Strait Islanders of northeast Australia who have "more than 80 terms for different tides and tidal conditions" (Nietschmann 1989: 69) and whose history includes use and ecological knowledge of 450 species of marine animals (McNiven and Hitchcock 2004).

This chapter explores the cosmological construction of seascapes and the special ways maritime peoples engage with and control seascapes. The fluidity of seascapes presents unique challenges for cultural inscription and place-marking not generally encountered for landscapes. Focusing on Australian Indigenous seascapes, I argue that a key cosmological element of seascape construction and engagement is spiritscapes and the imbuement of seas with anthropomorphic spiritual entities. For archaeologists, such spiritscapes—as a relational nexus between people, spirits, and the sea—are the key to defining, understanding, and appreciating ancient seascapes (cf. Thomas 2001). I posit that ancient seascapes as spiritscapes can be accessed and historicized, in part, through archaeological analysis of places of ritual orchestration located strategically across tidal flats and the adjacent coastal zone. An archaeology of seascapes, therefore, necessitates an archaeology of spiritscapes. This is the major theme of this chapter.

Seascapes: What Are They?

My spiritscapes approach responds to and aims to redress limitations of Western definitions and constructions of seascapes in encompassing the broader cosmological and spiritual dimensions of past Australian Indigenous seascapes. It emerged out of attempts to develop a conceptual framework that could accommodate the cultural meaning and significance of two seemingly enigmatic archaeological site types from northeastern Australia: dugong bone mounds on Torres Strait islands (cf. McNiven and Feldman 2003) and stone arrangements across tidal mudflats along the central Queensland coast (cf. McNiven 2003). So what are Western conceptions of seascapes, and how have they limited understanding of Indigenous seascapes?

The *Oxford English Dictionary* (OED) defines seascape as "a picture of the sea" and "a picturesque view or prospect of the sea" (Trumble and Stevenson 2002: 2727). The English term *seascape* is most commonly used in reference to visual representations of the sea (mostly paintings and photographs), usually depictions of natural coastlines or ships. Beyond this authoritative and generally accepted view, it is difficult to find more nuanced definitions. In recent years, seascapes have been incorporated into environmental and heritage management, reflecting both the rise of environmentalism and concerns for heritage in general (cf. *UNESCO Convention for the Protection of the World Cultural and Natural Heritage* 1972) and increased interest in seas following the *United Nations Convention on the Law of the Sea* 1982 and more recently the *UNESCO Convention for the Protection of the Underwater Cultural Heritage* 2001. Although none of these conventions mentions "seascapes," various government agencies around the world have incorporated the term into heritage management discourse. The *Guide to Best Practice in Seascape Assessment* (Hill et al. 2001), a joint Irish/Welsh initiative, provides a "definition of seascape" based on the OED definition but "broadened the concept and assumed the definition to include: views from land to sea, views from sea to land, views along coastline, [and] the effect on landscape of the conjunction of sea and land" (Hill et al. 2001: 1). However, this definition of seascapes remains superficial for our purposes.

Why do Westerners tend not to conceptualize seascapes beyond sea vistas? Key to understanding this limitation is appreciating the history of European capitalism and the associated production of sea-space. The orthodox European conception of

seascapes as little more than areas of sea, devoid of deeper cosmological, cultural, and political meaning, may be linked with European legal notions of the "freedom of the seas" and seas as public space. In *Saltwater People: The Waves of Memory*, anthropologist Nonie Sharp (2002) links the development of the "freedom of the seas" notion with the European desire for unmitigated access to the world's waters for trade and commerce during the 17th and 18th centuries. Critical was the c. 1604 treatise *Mare Liberum* (*Freedom of the Seas*) by Dutch jurist Hugo Grotius. According to Philip Steinberg (2001), *mare liberum* was central to capitalist constructions of the ocean-space as an empty, nonterritorial domain defined as an antithetical counterpoint to land-space. Furthermore, the ocean was seen "as an empty surface between the terrestrial places that 'mattered'. . . . the sea was constructed as an asocial space between societies" (Steinberg 2001: 208). This emptiness and placelessness established a cartographic tradition of representing sea-space two dimensionally as homogenized blue space and necessitated a new (arbitrary) georeferencing grid system of latitude and longitude for European ocean navigation. Steinberg (2001: 122) points out that *mare liberum* is encapsulated in Jules Verne's 1869 famous novel *20,000 Leagues under The Sea*, wherein Captain Nemo describes the sea as "an immense desert. . . . Only there do I have no master! There I am free" (Verne 2000: 68–69).

Sharp (2002) argues that a key consequence of *mare liberum* was that it blinded most Europeans to those maritime societies, usually small-scale Indigenous communities, with complex seascapes. However, such a view also ignored complex traditional seascapes of various small-scale maritime societies in Europe, especially in Scandinavia (e.g., Westerdahl 2005). But what are these complex seascapes? Remarkably, scholarly insights into the nature and the construction of complex seascapes are restricted to the past 40 years.

A major conceptual leap was the publication, in 1989, of *A Sea of Small Boats,* edited by anthropologist John Cordell. In his introduction to this seminal volume, Cordell (1989: 1) provides a clear and useful definition of complex seascapes: "Seascapes are blanketed with history and imbued with names, myths, and legends, and elaborate territories that sometimes become exclusive provinces partitioned with traditional rights and owners much like property on land." The impetus for the volume was the plight of many small-scale maritime fishing communities around the world whose fishing rights and broader systems of "customary maritime tenure" were being impinged on, and largely ignored, by government agencies and bureaucracies imposing open-access, common property doctrines regarding in-shore areas and resources. In the settler colonies of Australia and the Americas, such impingement reflected the doctrine of *mare liberum* and ignorance that rendered Indigenous seascapes invisible to be mapped as vacant "blue" space.

A key ethical consequence of these actions is that it not only effaces legal recognition of seascapes and strips Indigenous communities of their ancestral rights to customary marine tenure but also delegitimizes those rights and represents them as legally and morally counter to the common law right of "freedom of the seas." It is a moral philosophy to legitimize colonial expansion. In other contexts, for example, Japan and Oceania, customary marine tenure remains strong and supported by government agencies (Ruddle and Akimichi 1984, 1989). However, fisherfolk of European heritage in North America have developed informal marine tenure systems that often clash with government formal policies that juggle open access and regulation (Bowles and Bowles 1989; McCay 1989) and in some cases Native American treaty rights (Knutson 1989; Langdon 1989). *A Sea of Small Boats* was the first major step toward filling the knowledge void on complex seascapes in terms of customary marine tenure. The past decade has seen a focus of research on Australian Indigenous systems of sea tenure and cosmological construction of seascapes. Key volumes are *Li-anthawirriyarra, People of the Sea* by John Bradley (1997); *Customary Marine Tenure in Australia,* edited by anthropologists Nicolas Peterson and Bruce Rigsby (1998); *Saltwater People,* compiled by the Buku-Larrngay Mulka Centre (1999); and *Saltwater People* by Nonie Sharp (2002). Key journal papers include Jackson (1995), Mulrennan and Scott (2000), Magowan (2001) and Morphy and Morphy (2006).

Much of our awareness of the above-mentioned issues and problems has come about from Indigenous calls for recognition of their Indigenous rights not simply to *land*, as defined by Western legal custom and following Western agendas of recognition, but from broader conceptions of place, as defined by Indigenous peoples' own customary legal frameworks. Demands for legal recognition of Indigenous seascape rights has resulted in major advances in understanding customary marine tenure, the territorial dimensions of seascapes, and the cosmological construction of seascapes by Western academics (mainly anthropologists). That these advances are beginning to have broader impact is revealed by a recent issue of the English-based journal *World Archaeology* devoted to "seascapes." In his introductory essay, editor Gabriel Cooney (2003: 323)

observed that "seeing and thinking of the sea as seascape—contoured, alive, rich in ecological diversity and in cosmological and religious significance and ambiguity—provides a new perspective on how people in coastal areas actively create their identities, sense of place and histories."

This new perspective similarly informs my conceptual approach to seascape archaeology (McNiven 2003; McNiven and Feldman 2003). It also provides the basis of my definition of seascapes as *the lived sea-spaces central to the identity of maritime peoples. They are owned by right of inheritance, demarcated territorially, mapped with named places, historicized with social actions, engaged technologically for resources, imbued with spiritual potency and agency, orchestrated ritually, and legitimated cosmologically.* This definition concerns communities who are best described as "maritime peoples" or "sea peoples" or in the Indigenous Australian context—"Saltwater peoples" (Sharp 2002). For such peoples the sea is central to identity. For example, the Yanyuwa people of the Gulf of Carpentaria (NE Australia) "use as a metaphor for their existence and their identity the term *Li-anthawirriyarra,* which means 'those people whose spiritual and cultural heritage comes from the sea'" (Bradley 1998: 131). For the remainder of this chapter, I explore the spiritual and ritual dimensions of seascapes of maritime peoples and what I consider to be a key aspect to the cosmological construction of seascapes as spiritscapes—anthropomorphism.

Anthropomorphic Construction of Seascapes as Spiritscapes

Cosmological construction of seascapes as spiritscapes socializes seas and marine environments so they are

1. explicable and comprehensible;

2. domesticated and familiar;

3. historical and transformative;

4. sociable and engagable.

Separation of these four dimensions is difficult, because they exhibit considerable overlap. The key aspect of overlap and the essential basis of the seascapes as spiritscapes construction is animism through anthropomorphism whereby certain marine *features, forces,* and *fauna* are imbued with sentience expressed largely as human cognitive, emotional, and social qualities. Thus the sea, as a sentient realm, is capable of reacting consciously to human presence and action in much the same way that people are capable of reacting consciously to the presence and the action of the sea (see Anderson 2000 for an extended discussion of "sentient ecology"). Anthropomorphism imbues the sea with agency and intentionality such that it can be engaged sensuously and socially and controlled ritually in certain circumstances.

Explicable and Comprehensible

A key aspect of human cosmological construction of seascapes is anthropomorphic imbuement of "inanimate" (for example, waves) and "animate" (for instance, animals) phenomena with various degrees of human physical, cognitive, emotional, and/or social qualities. In *Faces in the Clouds: A New Theory of Religion*, Guthrie (1993) argues that animism and anthropomorphism are central to religious belief and to human perception and cognitive capacities more generally. The intuitive capacity and tendency for anthropomorphism by humans is revealed by its constant employment in animal behavior studies by scientists (Kennedy 1992). "We habitually anthropomorphize about animal behaviour, using our own mental processes as models to 'explain' the behaviour in terms of intentions" (Kennedy 1992: 89). Kennedy (1992: 167) goes so far as to suggest that anthropomorphism "is probably programmed into us genetically" (cf. Boyer 1996). The attribution of spiritual entities to phenomena, largely through a process of anthropomorphism, provides an ontological framework to structure, to understand, and to explain the "natural" world. This framework is in direct contrast to the ontological dualism between society and nature—the cornerstone of Western science.

Nietschmann (1989: 60) cogently observes that "people conceptually produce the environment they use, delimit, and defend." Environment encapsulates the engaged spaces and phenomena of society while cosmologies account for what is relevant and important to know and understand in the environment. "Anthropomorphism may best be explained as a result of an attempt to see not what we want to see or what is easy to see, but what is *important* to see: what may affect us for better or worse" (Guthrie 1993: 82–83). The desire to make phenomena and processes comprehensible and explicable is needs-based and engagement-dependent. I suggest that marine cosmologies tend to focus on dynamic forces (for example, tides, winds, currents) and associated topographical features (for instance, channels, reefs, sandbanks), and mobile fauna (for example, fish, mammals),

because understanding processes that influence and determine spatial and temporal variability of these phenomena is critical for maritime people's livelihoods. To this I would add that what is considered dynamic is culturally contingent and based on culturally specific degrees of engagement and intimacy with dynamic phenomena.

Anthropomorphic ascription concerns both "static" and "dynamic" marine phenomena. For example, the Goemulgal of Mabuyag island (Torres Strait, NE Australia) know that granite boulders emerging from the sea to the west of their residential island are Kamutnab and her children who "sat down in the sea" after a family argument (see Lawrie 1970: 85–86 for details). In terms of dynamic phenomena, Burarra and Yan-nhangu peoples of northeast Arnhem Land (Australia) use human anatomical and behavioral "referents for various saltwater features," such as "knees" for waves, "teeth" or "mouth" for "the shoreline edge of the sea," "abdomen" for "moderately distant waters," "habitually speaking" for "the constant crashing of beach surf and of inshore waves," and "rumble" or "growl" for "the more distant (and therefore less clearly audible) waters of open sea" (Bagshaw 1998: 159).

Domesticated and Familiar

A key aspect of anthropomorphic imbuement of seascapes is a cosmological cartography of creation narratives and placenames. In conjunction with dwelling and engagement (see Thomas this volume), such narratives and placenames socialize spaces by making them part of one's familiar, historicized and ancestral environment. As Nietschmann (1989) notes, "places used are places named." For example, the Yolngu of northeast Arnhem Land (northern Australia) understand that numerous places such as "rocks, sandbanks, mud banks, channels, tidal eddies or reefs" were created by ancestral beings and may even be places where such ancestral beings still reside (Davis 1989: 51; see also Buku-Larrngay Mulka Centre 1999; Morphy and Morphy 2006). An important part of seascape narratives, as for landscape narratives, is a vocabulary of placenames that help to structure and to punctuate narrative events and engagements. In the context of the Aegean Sea during Hellenic times, Loukatos (1976: 468) suggested (albeit in a somewhat simplistic and functionalist sense) that anthropomorphism of coastal features was due to "man's [sic] need for a milieu of 'human-like' beings, and to his [sic] fear of solitude." Furthermore, the "capes, promontories, rocks, and islets, because of their strange forms or of the caprice of the seas, because either friends or enemies of the sailors, and sometimes, veritable monsters to be avoided. But, in merciless solitude, any animated presence is acceptable, even if it is an agent of the devil" (Loukatos 1976: 469). Yet Indigenous Australians reveal that placenames in the sea can take on greater referential significance and spiritual potency. For example, the Yanyuwa of the Gulf of Carpentaria recount the mythic journey of the Spirit Ancestor Dugong Hunters for a distance of 110 kilometers through their seas; a journey marked by more than 50 named places on islands and in the sea (Figure 13.1). Yanyuwa hunters have a strong sense that past and present become one as their own dugong hunting journeys are a recitation and reenactment of the "mythic" journey of the Spirit Ancestor Dugong Hunters (Bradley 1998: 137). Thus, dugong hunting at once recreates the past in the present and reaffirms a cosmology that cocreates a seascape and a maritime identity. In short, cosmological cartographies help legitimate one's identity and place in the world.

Historical and Transformative

All ethnographically known seascapes dealing with the relatively near-shore environment developed within the past 6,000 years given flooding of continental shelves with the postglacial marine transgression. As such, cosmological construction of seascapes, along with associated creation narratives and placenames, all date to some time within the past 6,000 years. While this eustatic reality provides a maximum date for ethnographically known seascapes, it is clear that seascapes, like landscapes, are historically dynamic and ever-evolving.

Part of this dynamism reflects the inherent dynamism of marine environments and concomitant socially mediated changes in marine resource use strategies. When historical events and changing social circumstances are included, it is clear that seascapes represent palimpsests of old and new engagements and perceptions of the sea. Yet such palimpsests are not necessarily cumulative as places can become disengaged and forgotten. Maritime peoples maintain seascapes as dynamic cosmological entities so that they can be kept alive with relevance and resonance. History construction, as a teleological process, constantly reinterprets those sites and places that are relevant and thus remembered (as heritage) and those sites and places that become irrelevant and eventually forgotten. Thus, relevance and resonance extend

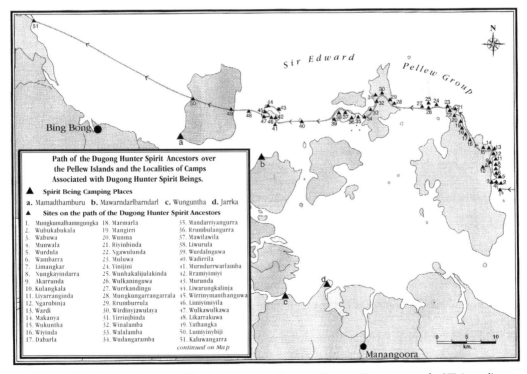

Figure 13.1   Path and camps of the Spirit Ancestor Dugong Hunters, Yanyuwa people, NE Australia (after Bradley 1998: fig. 8.3).

beyond superficial engagement for raw material and food resources. As discussed above, seascapes are cosmologically constructed and are central to the identity of maritime peoples and their ancestral connections to place. In this connection, Cordell (1989: 34) notes that Torres Strait Islanders "imbue their marine environment with a density of names far in excess of anything that would be required purely by the logistics of fishing. Territories, sub-surface features, rocks, and reef clefts are named after events and mythical characters, providing local people with a constant, visible historical anchor."

Sociable and Engagable

If seascapes are understood cosmologically as anthropomorphized spiritscapes, they can be engaged socially by people. In this conception, the sea is a participant in, not a contextual backdrop to, social engagements (cf. Ingold 2000: 40–60). Thus, maritime people technologically travel *across* the sea but socially negotiate their way *through* a seascape. A "Saltwater person can never be alone while out on the sea" (McNiven 2003: 334). In some circumstances, sensing the presence of a malevo-lent spiritual entity brings on a psychological state of fear and stress.

In other circumstances, spiritual engage-ments—phenomenal and sensual, predictable and unpredictable—can have impacts that are benign or neutral (Langton 2002). However, a key quality of spiritscapes is that anthropomorphized phenom-ena are engaged socially and may be controlled (or contained within the rules of societal frameworks for human behavior in a world shared with spirits) by people through formal, codified ritual perfor-mances—sometimes at specially designated places (that is, ritual sites) and sometimes using portable objects (that is, ritual paraphernalia). In marine contexts, I refer to such engagements as the "ritual orchestration of seascapes" (McNiven and Feldman 2003), and it is such ritual orchestrations that can leave rich material signatures amenable to archaeo-logical enquiry.

Ritual Orchestration of Seascapes

Maritime peoples undertake a range of rituals to engage with marine spiritscapes. These ritual engagements relate to the following six domains:

1. Continue the availability of key subsistence species (that is, "increase" or "maintenance" rites);

2. Conveying spirits of the dead to the spiritual realm of the sea (for example, mortuary rites);

3. Celebrating the creative acts of ancestral beings (for instance, through songs and dances);

4. Introduce people to the spirits who inhabit the sea, for their own security or protection (as with certain forms of initiation);

5. Capture of key subsistence species (that is, hunting magic);

6. Controlling elements important to use of the sea (for example, wind, waves, and tides).

Domains 1 to 3 relate to maintenance and renewal of seascape spiritscapes to reaffirm mari-time cosmology and identity. Domains 5 and 6 relate more to strategic manipulation of seascape spiritscapes to aid food procurement and sea travel and are often associated with behavioral restric-tions (taboos). Domain 4 concerns and links both.

Rituals of orchestration usually take place in the following marine contexts—on land behind the edge of the sea (for example, inland of the high tide mark); across the intertidal zone; and/or across the sea (isolated tidal banks/reefs or on the water surface). Yet, at the outset it is acknow-ledged that this geographical tripartite structuring is from a Western perspective; the high-tide mark is imbued with ontological significance because it represents the meeting place of two major realms—the land and the sea. Yet for the Burarra and Yan-nhangu peoples of northeast Arnhem Land, the intertidal zone down to the Lowest Astronomical Tide is designated "land" (Bagshaw 1998: 158). Alternatively, for the Gunggalida people of the Gulf of Carpentaria (NE Australia), the "land" or "main-land" starts 3–10 km "inland" from the beach with sand ridges across saltpans considered "islands" in the same sense as off-shore "islands" (Trigger 1987; see also Bradley 1998: 131). Thus marine ritual sites located many kilometers "inland" of the high tide mark may well be considered shoreline sites under certain cosmological and ontological frameworks. As such, my tripartite division is used simply as a heuristic and not as a universalizing ontological statement. Indeed, identifying how cul-tures of the past ontologically constructed the land-sea interface is necessarily an empirical question that needs to be demonstrated and not assumed (e.g., Helskog 1999). Geographical issues of ontol-ogy aside, Australian Indigenous ethnographic examples of each of these three contexts of marine rituals are provided by McNiven (2003: 336–38). To summarize, examples of open water rituals include returning parts of marine animals (food items) back to the sea for spiritual renewal of these animals, manipulation of sea animal parts (e.g., incorporation of body parts within harpoons when hunting) to attract a species to the hunter or fisher, sea burial of people, and propitiatory offerings to placate sea spirits for safe sea travel. Examples of tidal flat rituals include stone-walled tidal fishtraps with ritual offerings to attract fish and stone arrangements to control the elements. Examples of shoreline rituals include burials and arrangements of stones, bones, and/or shells to attract marine animals or control the elements.

I argue that a key direction for any attempts to access past seascapes through archaeological investigation is focusing on the cosmological construction of seascape spiritscapes as expressed through rituals of seascape orchestration. To date, little archaeological attention has been directed toward seascape rituals. The potential for archaeological insights into open water rituals seems remote given their ephemeral nature and aquatic context (cf. Lindenlauf 2003). Much greater promise exists for ritual sites located across tidal flats and adjacent coastal areas. For example, various locations along the tropical coast of central Queensland, northeast Australia, exhibit Aboriginal stone arrangements of unknown antiquity across intertidal flats. While no ethnographic information has been forthcoming about these sites, I hypothesize that at least some of these sites may be associated with ritual control of extreme tidal regimes critical for scheduling of local marine mobility and subsistence (see McNiven 2003 for details). Case studies of terrestrial sites associated with marine rituals are more common and usually concern hunting magic (mostly associated with marine mammals) and mortuary activities. For example, elaborately constructed mounds of dugong bones are located on various islands in Torres Strait, northeast Australia (see McNiven and Feldman 2003 for details). The number of dugongs in these mounds ranges from a few hundred to many thousands. These mounds are linked to hunting magic rituals with specific items (for example, dugong ear bones) used to establish sensory contact with dugongs to attract these animals to hunters. These mounds date mostly to the past 300–400 years with a maximum known antiquity of 600 years (David and Mura Badulgal Committee 2006; McNiven 2006). Perhaps the best example of mortuary activities that inform marine cosmologies are the boat burials and associated rock art of Scandinavia, Southeast Asia, and Melanesia (e.g., Ballard et al. 2003).

Conclusions

People dwell in a world where negotiating spiritscapes is part of the normal rhythm of daily life (Thomas 2001: 175–76). Spiritscapes are pervasive and all encompassing. While particular places across land- and seascapes can be the focus of spiritual energies and spiritual beings, spiritual lifefulness can be an essential feature of most phenomena. Often spiritual forces are engaged unpredictably and informally. At other times deliberate attempts are made to engage spiritual forces through formal ritual performances. Seascapes, as the space of maritime peoples, are cosmologically constructed as sentient realms of spiritual energies and entities with anthropomorphic characteristics where mutually recognizable, formal and informal engagements are possible both sensuously and socially. Yet, as archaeologists our conceptual frameworks and methodologies restrict our ability to access marine cosmologies, because we tend to read material culture terrestrially. By conceptually framing seascapes as anthropomorphized spiritscapes, I have attempted to provide an analytical process whereby engaging with formal ritual sites of the coastal and adjacent intertidal zones puts us in a better position to extend our gaze seaward to appreciate in some small way the seascapes that formed and informed the everyday lives of ancient maritime peoples.

Acknowledgments

Helpful comments on previous drafts of this chapter were kindly provided by John Bradley, Joe Crouch, Bruno David, and Lynette Russell.

References

Anderson, D. G. 2000. *Identity and Ecology in Arctic Siberia*. Oxford: Oxford University Press.

Bagshaw, G. 1998. *Gapu Dhulway, Gapu Maramba*: Conceptualisation and ownership of saltwater amongst the Burarra and Yan-nhangu peoples of northeast Arnhem Land, in N. Peterson and B. Rigsby (eds.), *Customary Marine Tenure in Australia*, pp. 154–77. Sydney: University of Sydney Oceania Publications.

Ballard, C., Bradley, R., Nordenborg Myhre, L., and Wilson, M. 2003. The ship as symbol in the prehistory of Scandinavia and Southeast Asia. *World Archaeology* 35(3): 385–403.

Boyer, P. 1996. What makes anthropomorphism natural: intuitive ontology and cultural representations. *The Journal of the Royal Anthropological Institute* 2(1): 83–97.

Bowles, F. P., and Bowles, M. C. 1989. Holding the line: Property rights in the lobster and herring fisheries of Matinicus Island, Maine, in J. Cordell (ed.), *A Sea of Small Boats*, pp. 228–57. Cambridge, MA: Cultural Survival, Inc.

Bradley, J. J. 1997. *Li-anthawirriyarra*, People of the Sea: Yanyuwa Relations with their Maritime Environment. Unpublished Ph.D. thesis, Northern Territory University, Darwin.

———. 1998. "We always look north": Yanyuwa identity and the marine environment, in N. Peterson and B. Rigsby (eds.), *Customary Marine Tenure in Australia*, pp. 125–41. Sydney: University of Sydney Oceania Publications.

Buku-Larrngay Mulka Centre. 1999. *Saltwater People: Yirrkala Bark Paintings of Sea Country. Recognising Indigenous Sea Rights*. Buku-Larrngay Mulka Centre in association with Jennifer Isaacs Publishing.

Cooney, G. 2003. Introduction: Seeing land from the sea. *World Archaeology* 35: 323–28.

Cordell, J. 1989. Introduction: Sea tenure, in J. Cordell (ed.), *A Sea of Small Boats*, pp. 1–32. Cambridge, MA: Cultural Survival, Inc.

David, B., and Mura Badulgal Committee. 2006. What happened in Torres Strait 400 years ago? Ritual transformations in an island seascape. *Journal of Island and Coastal Archaeology* 1: 123–43

Davis, S. 1989. Aboriginal tenure of the sea in Arnhem Land, northern Australia, in J. Cordell (ed.), *A Sea of Small Boats*, pp. 37–59. Cambridge, MA: Cultural Survival, Inc.

Guthrie, S. E. 1993. *Faces in the Clouds: A New Theory of Religion*. Oxford: Oxford University Press.

Helskog, K. 1999. The shore connection: Cognitive landscape and communication with rock carvings in northernmost Europe. *Norwegian Archaeological Review* 32(2): 73–94.

Hill, M., Briggs, J., Minto, P., Bagnall, D., Foley, K., and Williams, A. 2001. *Guide to Best Practice in Seascape Assessment*. Maritime Ireland/Wales INTERREG Report No. 5.

Ingold, T. 2000. *The Perception of the Environment: Essays on Livelihood, Dwelling and Skill*. London: Routledge.

Jackson, S. E. 1995. The water is not empty: Cross-cultural issues in conceptualising sea space. *Australian Geographer* 26(1): 87–96.

Kennedy, J. S. 1992. *The New Anthropomorphism*. Cambridge: Cambridge University Press.

Knutson, P. R. 1989. The unintended consequences of the Bold Decision, in J. Cordell (ed.), *A Sea of Small Boats*, pp. 263–303. Cambridge, MA: Cultural Survival, Inc.

Langdon, S. 1989. From communal property to common property to limited entry: Historical ironies in the management of southeast Alaska salmon, in

J. Cordell (ed.), *A Sea of Small Boats*, pp. 304–32. Cambridge, MA: Cultural Survival, Inc.

Langton, M. 2002. The edge of the sacred, the edge of death: Sensual inscriptions, in B. David and M. Wilson (eds.), *Inscribed Landscape: Marking and Making Place*, pp. 253–69. Honolulu: University of Hawai'i Press.

Lawrie, M. 1970. *Myths and Legends of Torres Strait*. St. Lucia: University of Queensland Press.

Lindenlauf, A. 2003. The sea as a place of no return in ancient Greece. *World Archaeology* 35: 416–33.

Loukatos, D. 1976. Personifications of capes and rocks in the Hellenic seas, in A. Bharati (ed.), *The Realm of the Extra-Human: Agents and Audiences*, pp. 467–74. The Hague: Mouton.

Magowan, F. 2001. Waves of knowing: Polymorphism and co-substantive essences in Yolngu sea cosmology. *The Australian Journal of Indigenous Education* 29(1): 22–35.

McCay, B. J. 1989. Sea tenure and the culture of the Commoners, in J. Cordell (ed.), *A Sea of Small Boats*, pp. 203–27. Cambridge, MA: Cultural Survival, Inc.

McNiven, I. J. 2003. Saltwater People: Spiritscapes, maritime rituals and the archaeology of Australian indigenous seascapes. *World Archaeology* 35: 329–49.

———. 2006. Dauan 4 and the emergence of ethnographically-known social arrangements across Torres Strait during the last 600–800 years. *Australian Archaeology* 62: 1–12.

McNiven, I. J., and Feldman, R. 2003. Ritual orchestration of seascapes: Hunting magic and dugong bone mounds in Torres Strait, NE Australia. *Cambridge Archaeological Journal* 13(2): 169–94.

McNiven, I. J., and Hitchcock, G. 2004. Torres Strait Islander marine subsistence specialisation and terrestrial animal translocation, in I. J. McNiven and M. Quinnell (eds.), *Torres Strait Archaeology and Material Culture*. Memoirs of the Queensland Museum, Cultural Heritage Series 3: 105–62.

Morphy, H., and Morphy, F. 2006. Tasting the waters: Discriminating identities in the waters of Blue Mud Bay. *Journal of Material Culture* 11(1/2): 67–85.

Mulrennan, M. E., and Scott, C. H. 2000. *Mare Nullius*: Indigenous rights in saltwater environments. *Development and Change* 31: 681–708.

Nietschmann, B. 1989. Traditional sea territories, resources and rights in Torres Strait, in J. Cordell (ed.), *A Sea of Small Boats*, pp. 60–93. Cambridge, MA: Cultural Survival, Inc.

Peterson, N., and Rigsby, R. (eds.). 1998. *Customary Marine Tenure in Australia*. Sydney: University of Sydney Oceania Publications.

Ruddle, K., and Akimichi, T. (eds.). 1984. *Maritime Institutions of the Western Pacific*. Senri Ethnological Studies No. 17. Osaka: National Museum of Ethnology.

Ruddle, K., and Akimichi, T. 1989. Sea tenure in Japan and the southwestern Ryukyus, in J. Cordell (ed.), *A Sea of Small Boats*, pp. 337–70. Cambridge, MA: Cultural Survival, Inc.

Sharp, N. 2002. *Saltwater People: The Waves of Memory*. Crows Nest: Allen and Unwin.

Steinberg, P. E. 2001. *The Social Construction of the Ocean*. Cambridge: Cambridge University Press.

Thomas, J. 2001. Archaeologies of place and landscape, in I. Hodder (ed.), *Archaeological Theory Today*, pp. 165–86. Cambridge: Polity.

Trigger, D. S. 1987. Inland, coast and islands: Traditional Aboriginal society and material culture in a region of the southern Gulf of Carpentaria. *Records of the South Australian Museum* 21(2): 69–84.

Trumble, W. R., and Stevenson, A. (eds.). 2002. *Shorter Oxford English Dictionary* (5th ed.), Vols. 1 & 2. Oxford: Oxford University Press.

Verne, J. 2000 [1869]. *20,000 Leagues under the Sea*. New York: HarperCollins.

Wessex Archaeology. 2005. *England's Historic Seascapes: Historic Environment Characterisation in England's Intertidal and Marine Zones. Project Design*. Report to English Heritage. Wessex: The Trust for Wessex Archaeology Limited.

Westerdahl, C. 2005. Seal on land, elk at sea: Notes on and applications of the ritual landscape at the seaboard. *The International Journal of Nautical Archaeology* 34(1): 2–23.

14

LIVING LANDSCAPES OF THE DEAD: ARCHAEOLOGY OF THE AFTERWORLD AMONG THE RUMU OF PAPUA NEW GUINEA

Bruno David, Max Pivoru, William Pivoru, Michael Green, Bryce Barker, James F. Weiner, Douglas Simala, Thomas Kokents, Lisa Araho, John Dop

All over the world, cultural groups understand the world to operate in different ways. Central to these various philosophies of life is how people make sense of the mysteries of their lived environments: how do we connect what we know (and can control), with what we do not know (and cannot truly control). An answer that all cultures have—our own included—is the spirit world, that powerful, nebulous realm that enables us to make sense of things and that touches us at the deepest existential levels while always remaining somehow distant and mysterious.

Given that all cultural landscapes contain spiritual presences, we would be remiss to try to explain past cultural practices without considering such existential and operational dimensions. Yet how do we attempt landscape archaeology of the intangibles of life? (see McNiven; Ashmore; Bradley, this volume; also David 2002 for detailed discussion). The first step, we suggest, is to recognize the central importance of spiritscapes in everyday social and environmental engagement. This chapter is dedicated toward this aim, by focusing on spiritscapes in one ethnographic community, the Rumu of Papua New Guinea. We outline various kinds of spiritual connections to place to show how everyday life is embedded in sacred geographies. We then focus on a single dimension, the *kepe* and the land of the dead, to highlight how an archaeology of the dead itself opens doorways to an archaeology of spiritual landscapes.

The subject of what governs mortuary practices, and how these relate to social life, is a vast anthropological and archaeological subject in its own right, with a long history. Archaeologists have long been aware that mortuary practices do not simply represent cultural ways of disposing of recently deceased bodies. Mortuary practices and understanding the particular contexts of archaeological human remains may be affected by specific circumstances of death and invariably reflect a variety of social processes, including:

1. varied processes of social power (e.g., King [2004], who argues that early Saxon grave goods acted to establish ongoing relations of social expectation and due between survivors and donors);

2. particular circumstances of death (e.g., Spindler [1994], who suggested that 5,000 years ago during the Copper Age, Ötzi, "the ice-man," was a traveling distant villager caught during a severe snow storm on the Schnalstal glacier of the Italian-Austrian Alps and died, soon to be covered by snowfall; or Fleckinger's [2003] view that Ötzi was an old man who had been fatally injured during a fight, only to escape and die in the Schnalstal glacier);

3. political negotiation and contestation (e.g., Pardoe [1988], who argues that cemeteries are an expression of territory-building by member corporate groups calling on the historical hegemony of place);

4. attitudes of social order (e.g., Glob [1969], who argues that [in the main] Iron Age, bog bodies in northern Europe may have been executed criminals ritually sacrificed to the goddess of fertility);

5. cosmological understandings (e.g., Greber and Ruhl [1989: 273] who suggest that the archaeological remains at the Hopewell burial site signal an understanding of how the universe operates and how to keep the cosmos in good working order);

6. the social manipulation of psychological states of remembering (e.g., Williams [2005], who in part argues that European Medieval funerary goods, such as weapons, knives, and iron buckles, were biographic aids to remembering past social circumstances, whereas cremation acted to selective forgetting; see also Hallam and Hockey [2001]); and

7. religious beliefs that concern the human soul (e.g., Solecki [1971], who argued that the deliberate burial of Neanderthals at Shanidar Cave in Iraq indicates religious beliefs in a life after death).

Burial practices are, therefore, about preparing the once-living and meaningful person for a life in the afterworld (however defined) by living members of social, political, and spiritual communities. In short, burial practices are mediated by the ontological constitution of landscapes of the dead by the living.

It is, of course, the material body and associated material remains with which the archaeologist dealing with mortuary practices usually commences her or his studies—although this is only partly true, because when we aim to address religious pasts through mortuary practices we also assume that the world is animated by spiritual concerns, since these are variously expressed in the earth's many cultures and beliefs, past and present. Perhaps the most celebrated archaeological case for a belief in life after death concerns the much-debated Neanderthal burial practices (for examples of differing views, see Solecki 1971; Stringer and Gamble 1993; Turner and Hannon 1988), widely thought to indicate that Neanderthals conceived of an afterworld. Other examples abound, such

as Predynastic and Dynastic Egyptian attempts to preserve the body undertaken in the expectation of enabling the person to go through the process of rebirth into the afterlife (e.g., see Museum of Science 2003; also Ikram 2003 for details). The use of amulets and sacred texts among burial goods enabled communication between (or on behalf of) the dead and the spirits in many cultures of the world, including ancient Egyptian, Abyssinian Christian, and Muslim (e.g., Wallis Budge 1973: 282–87). In the early years of Christianity, burials were oriented west-east, with the head facing upward and the feet to the east "in order to rise facing the Son of Man on the day of resurrection" (Davies 1999: 199). Among the Kerewo of Papua New Guinea, decapitated enemy heads were placed on sacred *gobi* boards during the 19th and early 20th centuries A.D. to enhance the spiritual power of the holder (e.g., Haddon 1918). Among the Yolngu of northern Australia, coffins are painted for the well-being of the *birrimbirr* [soul] during its return journey to the clan's well (from whence clan "souls" enter the physical world of the living) (Morphy 1984, especially pp. 69–75). With more general application, Dimitrov (2002, cited in Bailey and Hofmann 2005: 220–22) makes the oft-made observation that the inclusion of grave goods with human burials, in this case at the Neolithic cemetery of Durankulak in Bulgaria, signals the preparation of worldly goods for the benefit of the recently departed in the afterworld (see also, for example, Ikram [2003] and Ikram and Dodson [1998] for similar notions of ancient Egyptian funerary goods).

In this chapter, we present one example of Rumu burial practices to illustrate the notion that what is at stake in the archaeology of systematic "human disposal" concerns how the living perceive the dead and life after death. Through bodily preparation, maintenance, and "disposal" in the land of the living, archaeology allows us to historicize living landscapes of the dead.

Rumu Landscapes

The Rumu people of Utiti Creek and the Kikori River in the Gulf Province of Papua New Guinea live in a lush green landscape. Gardens are cleared on creek edges, villages are built on higher ground by waterways, hunting camps are made against limestone cliffs, large logs are dug out for canoe hulls in the midst of the rainforest, and sago palms are processed for their rich starch among the sago groves at the edges of swamps. Here the dense lowland forest temporally opens up as Rumu clan members—such as the Parua'uki, Himaiyu, and

Yoto Uti clans—work the land. But soon enough, the 6 meters of rain that yearly falls quickly help the forest grow back until all signs of human activity disappear, at least as far as their obvious traces are concerned.

It would be easy to think of these Rumu lands in the language of forest landscapes. Here can be found the massive fig trees with their high, spreading roots; the dense stands of sago palms, the major source of food that during the 19th century helped determine irregular but seasonal residential locations for family groups; the *arave* (rosewood), *tao bokore* (red cedar), *niniho* (pencil cedar), *aniki* (wildnut), *yeni,* and *apiko* (*irimo*) (in the Motu language) trunks that are sought after for the making of indispensable canoe hulls by which the river systems afford the major means of travel; the black palms commonly used for floorboards in house construction; the bamboos whose tubes are used as carrying and cooking containers and whose sharp edges are relied on as cutting knives; the many food and medicinal plants, such as the fruiting *emehi* (breadfruit), *makahe* and *yahoo* trees (these two have no English names); and all around the rainforest closes in dense and extensive blankets of undergrowth, tall trees, and vines (Jack Kaiwari, Yoto Uki clan, personal communication, 2006). Today these trees are located within one of the largest rainforests in the world, second only to the Amazon. But for the members of the Himaiyu clan, life and death revolves not simply around a forested landscape but around the way people read their surroundings as a mysterious yet tangible world inhabited by spirits that mediate the known and the unknown; the domestically benign, numinous, and often dangerous realms; and the land of the living and the land of the dead.

Yakari

As is the case throughout the Papuan lowlands, these Rumu lands are inhabited by spirits. *Yakari* spirits are all around, informing clan members of the state of the world. In the treetops are the *yakari pai* with which people communicate. When young men are being initiated, sometimes the initiation helps pull *yakari* spirits to their body. Initiates attract *yakari* emotionally and thus become friends with specific *yakari*. Among the Rumu, every man is expected to have a *yakari* "guardian spirit" to help him look after his land or to help his clanspeople in the daily run of life. Some *yakari* are healers who help cure people. Some *yakari* are expert gardeners, or fishers, or hunters or warriors. Some *yakari* are dangerous, some are helpful. The power of the *yakari* will depend on its character and on the kind of place in which it resides. But the spirits are everywhere, and Rumu people call them *yakari* (John Soba, Para [Wauri Moro] clan, personal communication 2005; see David 2005).

Kowoi

Among the Rumu, *kowoi* sorcerers' spirits are dangerous spirits that people fear, because it is they who bring death to the unsuspecting. *Kowoi* is the name for the sorcerer. "The *yakari* is the thing that a *kowoi* man gets to help him," says John Soba (personal communication 2005). "The *yakari* is initiated into the *kowoi* man's body to help him. A *kowoi* is his own being made into spirit-form, to go out in a spiritual way. So a *kowoi* always uses the *yakari* spirit to incarnate into its spirit." But *yakari* itself does not have the strength to act on its own, nor is it visible to ordinary eyes. The *yakari* is the spirit that *kowoi* uses, but the *yakari* are not themselves ruthless, acting only on the orders of the *kowoi* man; he uses them. *Yakari* is mostly a helpful spirit that does help even if *kowoi* has this *yakari*. A *yakari* may be a gardening spirit who helps a person make more gardens; or it may be a *yakari* who knows how to make canoes, and in this capacity helps a person make canoes better than any other person. Or it may be a *yakari* who can protect a person from enemies. When a *kowoi* wants to attack someone, the *yakari* around a person are there to protect that person— *yakari* make signs to protect people against danger, including against the danger of the *kowoi* who use the *yakari* spirits in their own misdemeanors.

Nania

There are also *nania* animal spirits that people regard as personal guardians. *Nania* spirits are like *yakari*, but they are powerful spirits in themselves. *Nania* tend to live in rivers, in stones, and sometimes they can be seen as they turn into snakes, crocodiles, or cassowaries. These spirits help people, or they can punish people by making them sick. *Nania* spirits are powerful and therefore can make a person sick if the rules of the sacred are not obeyed in an area where *nania* reside. "Sometimes when we go out," says John Soba, "we are not allowed to cook a small fish called *yuhini* in any of those places, or when going with children. Because these are spirits. *Nania* will turn around and give pain to the children or person when this kind of rule is broken. But not *yakari*. *Yakari* are the ones who sometimes try to make friends with human beings. But they can be dangerous if so ordered by the *kowoi*. The *kowoi* man can use his spiritual soul to go out and use the spiritual snake or pig or dog to help him as his transport, to go out and do [mischievous acts]." So *nania* are powerful by themselves, but *yakari* become powerful only under the control of *kowoi*. *Yakari* do not hurt people of their own volition, only under the orders of the *kowoi* men. But *nania* have their own order and can hurt people when their rule is broken.

Kupi

Kupi are another spirit form. *Kupi* live in the caves of the limestone karst pinnacles, and each clan has its special name for their cave men or *kupi*. These are thought of as real physical beings possessing supernatural powers. *Kupi* are generally dangerous, and sometimes *kowoi* can make friends with particular *kupi*. It is only *kowoi* who make friends with *kupi*; ordinary people do not. *Kupi* are more powerful than *kowoi* because of their imposing physical presence; when *kupi* and *kowoi* unite, it is the *kupi* who will lead and be the most dangerous partner. *Kupi* are generally of human form; sometimes they even enter local villages. They have the characteristics of rascals or petty criminals, and sometimes come to steal things from villagers. Few people have seen them, but they are generally said to be real physical beings, often described today as "apes" or "gorillas."

Kepe

Kepe, the spirits of the dead, are yet another spirit form in the Rumu landscape. They are the spirits of the loved ones who have passed away. "Sometimes," says John Soba, "even when a person goes hunting, they might call out their dead father's name, or their mother's or brother's. And sometimes calling the spirits of the dead is an omen, and so when someone goes hunting they will never call the name of the family. But sometimes, a few months after the dead has been buried, there will be signs of communication, such as when food has been left out for them, and you will hear some kind of whistling, you will know that the spirit of that dead person is talking to the clan members or families." These are the *kepe*, the spirits of dead family members.

Throughout Papua, from Mt. Karimui to Mt. Bosavi, elaborate ritual procedures existed before the era of Christianization to formally detach the spirit of a dead person from the precincts of community life (see, for example, Schieffelin 1976; Wagner 1972; Weiner 1988). Among the Rumu, on the death of a relative a special song called *yey* is sung by the grieving relatives, a song that sends the spirit of the dead person to the afterlife. When it is sung, the song will start from the deceased person's house. If that person's house has a clan name, they will call that person's name, and in the song they will tell him "we are taking you out from the house now." They will say "we are taking your axe;" they will call the name of the axe. Or they will call the name of the bow, the arrows, the *bilum* (this is a pidgin word) bag, and other objects that belonged to the deceased person. And they will sing, "we are packing your things, and it is time for you to go down the steps of your house." They will call the step, what kind of step the dead spirit will walk down. When the dead person's spirit walks down the step, after going a short distance, in the song they will call the dead person's name and say, "you turn around and have a last look at your house before you go." Each clan has its own names for the objects in question, and each clan also has its own spiritual afterworld, named according to its respective location accessible only by members of that particular clan.

The route followed by the deceased's spirit will be that particular clan's spirit road, the one that leads to that clan's afterworld. Among the Parua'uki clan of the Kopi area, that afterworld is in the sacred mountain Ru, a large limestone pinnacle that lies but a kilometer or so from the present-day Kopi village. (Here spirits travel along the underground waters that link subterranean caverns in limestone karst outcrops.) Among the Himaiyu clan, the afterworld lies in the sacred mountain Hoimu, on Himaiyu lands more than 10 kilometers from Kopi village. (Kopi was established as a centralized colonial village during the 20th century A.D.; Himaiyu clan members previously lived in villages on their own clan lands.) Each of these clans, and others, sings the *kepe* spirit to its clan afterworld during ritual performances held within the village itself, and principally directed by the closest members of the deceased person's family.

The passing of the dead involves ritual performances for the safe journey of *kepe* to the afterworld. In the process, particular acts are required for the proper treatment of the physical remains of the dead, and it is here that Rumu spiritscapes attain their greatest archaeological potential. Prior to Christianity—that is, until the early years of the 20th century—the bodies of the dead were not buried but rather underwent other forms of complex ritual and then display. The displayed bones of the person were repeatedly accessed, from the moment of initial display soon after death (when living people knew the recently deceased person) into times well past personal memories of that person.

The following description of traditional mortuary practices, although described as for an adult male, applied to all people regardless of age, sex, or social position. Following the death of a Rumu person, the deceased's body is left in its own house for the first three or four days, during which the widow removes all the skin and hair from the body. During this ritual, clan members ask[1] the dead person's *kepe* spirit: who killed him, who made the magic that caused his or her death? The body is then removed and laid on a wooden platform or shelter called *tete*, with the skin and hair placed beneath the platform. A fire is lit on the ground beneath the shelter so that the smoke rises and covers the body. The fire is kept burning for one to two months, until all the flesh has rotted away, leaving only the skeleton.

Once the body has completely decomposed, the bones are gathered and left in the sun to dry. While the bones are drying, the male relatives weave a special *tumbuna* (a pidgin word) net from bamboo, which is given to the women. The widow, or another close female relative, places the dried bones into the net, and carries them in procession to another specially constructed house, this time much smaller. The path to this house is strewn with the leaves of a special tree, known locally as *kamu*. Inside this house the *tumbuna* net is hung over a small fire, to further smoke the dry bones. This smoking process continues for about two weeks, until the bones are thoroughly dried. Once dry, the bones are taken, still wrapped in the *tumbuna* net, to the widow's house; then preparations for a feast begin. The men collect delicacies such as sago grubs, while the women weave a special *bilum* bag into which the bones of the deceased will later be placed.

Once this feast has been prepared, the women collect oil from the *ate* tree and gather flowers known as *kiwau*. A bowl-shaped cut is made in the tree for the oily resin to drip and settle into. After a week, white spots in the liquid begin to clear, and the white creamy liquid turns into a shiny, oily liquid. The "oil" is poured into bamboo tubes, and the human bones from the recently deceased are rubbed with a mixture of oil and flowers to give them a clean, fragrant smell. When this washing occurs, the female relatives gather and mourn the deceased by crying and wailing. After washing, the bones are placed in the specially prepared *bilum*, which has been decorated with small branches from the *kamu* tree. The top of the *bilum* is covered with a *tapa* cloth and hung up in the village longhouse for public display.

All the villagers gather and sing mourning songs into the night, while close relatives continue to cry and mourn until daybreak. The following day, the widow retrieves the *bilum* and hangs it in her own house. If she must leave her house for more than one week, she must carry the bones of her husband with her.

After a period of one to two years, preparations are made for removal of the bones to a special cave such as Rupo among the Himaiyu clan (see below). Female relatives weave a number of *bilum*s while the men make bows and arrows. Some of these newly made artifacts are then burnt in a fire, others are thrown into the river, and the remainder are given to those people who help in preparing the feast. The feast begins early in the morning, when cooked sago is placed in split black palm tubes and laid over bamboo racks, while a male relative of the deceased calls out to his spirit, asking him to come and eat. After this, half of the prepared food is eaten by the friends and relatives of the deceased. Three days later, the bones of the deceased and the remainder of the food are taken to the cave, where the people sit and eat the remainder of the food, along with gifts that include *bilum*s, *tapa* (a pidgin word of Polynesian origin) cloth, and *kina* (a pidgin word) shells, bows, arrows, and spears. The bones, now placed in new *bilum*s, are then laid on the *kipa* (black palm) beside the gifts.

At the time the skull is placed in the cave, it is given a special identifying mark, either painted with ochre or by a special cloth tied to it, which indicates the clan affiliation of the deceased. Some skulls can be further decorated by filling the orbits and the nasal aperture with a mixture of ochre, plant fiber, and soil. Sometimes small cowrie shells were impressed into this orbital filling.

Each clan has its own cave; among the Himaiyu clan, the bones were until recently taken to a small cave called Rupo, high up a limestone karst tower at the headwaters of Utiti Creek. Upon depositing the bones of a recently deceased person at Rupo, a group of clan members would camp at Rupo and/or on the nearby banks of Utiti Creek, near the cave. Rupo Cave is very small and cannot accommodate many people; therefore, a camping site is established by the creek near the cave, (David Kupere Snr, Amou-uki clan, personal communication 2006). That cave was discovered long ago by the ancestors of the present-day Himaiyu clan members—the ossuary was located thus because it was some distance from the village, and it was rather uncomfortable, meaning that villagers would not be tempted to use it as a residential cave. Upon depositing the bones at Rupo, Himaiyu and other clan members would visit the site out of respect for the deceased, leaving there also items of material culture that belonged to the deceased person, such as bows and arrows and *bilum* bags. A feast would then ensue at the deceased person's village. The owner of the feast would be the clan to whom the deceased belonged; if other clans wished to help contribute food to the feast, then they would do so, or assist in hunting or the like for the feast.

In the years following the deposition of a person's bones in the ossuary, Rumu individuals repeatedly visited the bones of their deceased family members. "In the caves [ossuaries]," says William Pivoru, a member of the Himaiyu clan, "you can see the bone, so you remember. Today with burials you don't see the bones, so you forget the person. That's why bones were put in the cave, to not forget the person." In one way, the caves are as effectively outside the vicinity of everyday life as being 6 feet under; but in another, crucial way, the bones—as the continuing presence of once-living beings—can be accessed and thus can mediate interactions between the living and the dead, between the material and the spiritual.

The Archaeology of the Afterlife among the Rumu

It is customary among archaeologists to see ossuaries as places within which human skeletal remains are placed. But in effect these are not only places of disposal; rather, they are also places that mediate the spirit flow from the land of the living to the land of the dead. At the same time, they are also places that enable memorialization of the ancestors, remembrances of how clan members are unified into communities of landed peoples, each with ancestral claims to territory and with ancestral claims to mutual obligations and responsibilities as social beings who are required by ancestral privilege to ensure the safety of the clan, in the present, among the living, and by seeing the spirit of the dead through to the afterlife. An archaeology of ossuaries in Rumu lands is thus more than an archaeology of skeletal remains; it represents a historicizing of beliefs about the structure of the world, a cultural inscription of biography onto the landscape, including the way that the living organize themselves socially, territorially, and spiritually. It is also a historicizing of cosmology, of the way that people see the world as a fully operating system, and one that identifies how people are located ontologically in the world. An archaeology of Rumu ossuaries is thus an archaeology of the cognate world as much as it is an archaeology of social relationships among the living and between the living and the ancestors.

Rupo Cave

In March and April 2006, three ossuaries within Rumu lands near Utiti Creek and Kikori River were archaeologically excavated and radiocarbon dated as part of a community-based cultural heritage consultancy to do with planned developments in the region. Two of these are small ossuaries located in Parua'uki clan lands to the immediate south of present-day Kopi village, south also of Utiti Creek. The third is the cave of Rupo, the Himaiyu clan ossuary. Each of these caves is small; contains a damp, clayey floor; is located some way up a limestone karst tower; and is generally uncomfortable for camping purposes. In short, these are specialized ossuaries, although this is not to say that occasional rest or camps were not sometimes held within these sites, particularly at times of refuge during intertribal wars.

Rupo is a small cave, 1.8 meters wide and 1.8 meters high at the cave entrance, with floor area of 7.5 by 6.5 meters. To access the site requires a steep vertical climb of 14 meters across a horizontal distance of 22 meters from Utiti Creek to the cave; the entrance faces east and overlooks this steep and heavily vegetated slope down to the creek.

Today the cave surface is dominated by open pits from previous archaeological excavations undertaken by James Rhoads (1980) during the mid-1970s. These pits show sediment deposits from surface to bedrock to be about 120 centimeters at their deepest. In the southwest corner of the cave are the arranged skulls of some 17 individuals, with many hundreds and possibly thousands of postcranial human bones placed behind them and in a deep natural niche in the limestone rock just to the north. One of the skulls contains a small metal ring around the zygomatic arch. This marked skull also had an unmarked red plastic tag attached to the ring. Kuiari, a senior Kopi resident in 1984–1985, told one of us (MG) that the ring had been attached to some skulls in ossuaries (such as at Rupo and another ossuary called Urapo) by his grandfather's people, and that they marked the skulls of important chiefs. Scattered around the cave at Rupo are fragments of human bone and shell as well as European contact artifacts such as bottle glass, nails, and even a rusted old bush knife (machete) (Figure 14.1). This site ceased to be used as an ossuary soon after the commencement of government administration in the area (after which the dead were buried).

James Rhoads's (1980) original excavations pointed to a first use of Rupo as an ossuary around the end of the 19th century A.D., and this was

Figure 14.1 Rupo, with Rhoads excavation pit. Among the Himaiyu, the display of human bones in clan ossuaries is a customary*practice that highlights affiliation to land and testifies to ancestral connections. Such displays are a source of historical pride and proof of social, territorial, and spiritual ties among descendents, ancestors, and clan places. (Clan elder Max Pivoru, also a co-author of this chapter, verified that permission was granted for this photo to be used in the context of this volume.)

confirmed by our 2006 excavations, when finer-grained investigations revealed the presence of human bones only following the introduction of glass beads. The cave was first used some 2,000 years ago (as indicated by radiocarbon dating) and witnessed pulses of use and abandonment.

The glass beads at Rupo indicate major cultural changes with the coming of European colonial powers about 100 years ago. At that time, Rupo began to be used as *the* major Himaiyu clan ossuary, from whence the *kepe* spirits of recently deceased family members proceeded to the afterworld at the sacred mountain of Hoimu. These changes, detected archaeologically, implicate not so much cosmological changes in the way that the spirits of the dead were sent to the afterlife among the Rumu clanspeople of upper Utiti Creek and this part of the Kikori River, but a new focus on a single, large ossuary marking Himaiyu clan lands. Rupo is the largest ossuary we have encountered in the region, whereas very small ossuaries, often containing the remains of one or a few individuals on other clan lands (such as sites KG141 and KG142 of the Parua'uki clan), appear to predate the use of Rupo as an ossuary. The emerging archaeological evidence points to the presence of numerous small ossuaries predating the coming of Europeans, followed by the onset of large, centralized ossuaries around 100 years ago, signaling broader changes in regional cultural systems associated with post-European shifts toward river-bordering villages and increased levels of village nucleation.

An archaeology of social and cultural landscapes by definition implicates, in the first instance, an archaeology of cosmological systems, because all human groups live and breath in and through beliefs about how the world operates. These operational dimensions of life are, by definition, key components of social landscapes and, as such, core dimensions also of landscape archaeology. The results from Rupo on Himaiyu lands, and ossuary sites KG141 and KG142 on Parua'uki land (among other small ossuaries), announce significant changes in the way various clans expressed views of the afterlife in the mid-Kikori River lowlands. What is at stake is more than settlement-subsistence patterns, for such ossuaries are important clan locations and symbols that allow the recently deceased to access the afterworld. We suggest that late in the 19th century, with the commencement of Rupo as a key clan ossuary, increasingly centralized mortuary practices emerged and, with this, new Rumu understandings of how to send the recently deceased to the afterworld (however defined). The timing of commencement of use of Rupo and other ossuaries on Rumu clan lands indicates not simply a centralization of mortuary practices but also local community responses to the coming of Europeans, the ensuing shifts of villages along the banks of major rivers, increasing village nucleation, and possibly also shifting opportunities and threats (from both European/missionary and neighboring Indigenous groups) to access of clan lands, via new cosmological configurations concerned with the land of the dead accessed through the land of the living.

Historical Contexts

The first sustained contacts between Rumu peoples and Europeans took place in 1887 "when the explorer Theodore Bevan visited the region twice, traveling as far up as the Kikori at its junction with the Sirebi, just above where Kopi village stands today" (Busse, Turner, and Araho 1993: 17). In 1912, the government station was built at Kikori, and a London Missionary Society mission was established in the Aird Hills to the immediate east in 1913. Soon after, the Ogamobu coconut and rubber plantation was begun to the immediate northwest of Kikori in 1914, and in 1922 a sawmill at Kikori and a distillery along the Paibuna River were established. Regular government patrols, commercial employment, and government administration had become established across Rumu lands by the second decade of the 20th century (Busse, Turner, and Araho 1993: 18). These contacts caused local Rumu settlements to relocate, to take advantage of the new trade and resource opportunities created by the newly established government stations, missions, and plantations—at times by order or pressure of government authorities who aimed to administer the region in colonial terms and as a result of the decreased threats of external headhunting raids that government intervention (and protection) entailed. Additionally, government policies enforced changes in rubbish disposal from inside the village (as was customary practice) into the rivers, further encouraging a shift to closer riverside residence locations following early European contact.

The onset of Rupo as a centralized clan ossuary coincides in timing with the first use or reestablishment of a number of Rumu hunting camps and villages around the time of European arrival or shortly beforehand when population levels were rapidly increasing—sites such as Puriau, Barauni, Ihi Kaeke, Kikiniu, Epe, Wokoi Amoho, Leipo cave, KG124, KG125, and KG143, among others. We are uncertain at this stage whether the major regional changes are direct responses to the onset of Europeans or whether they relate to ritual responses of a spiritual kind, to responses to actual or perceived threats and opportunities, or to indirect responses to demographic changes caused farther away to the west, only eventually to be felt in Rumu lands, although at Rupo the association of the first human remains with glass beads is clear (confirming James Rhoads's earlier observations).

One implication of these results is that at times of social change, uncertainty and/or new opportunities, people consolidate territory. The mortuary practices evidenced on Rumu lands suggest that one way of doing this is to create focal locations legitimated by cosmological necessity— places of activity where particular types of practice are repeatedly undertaken and in the process strengthen claims to place. People establish an ongoing history in places by reference to cosmological emplacement—in the case of Rupo and

the Himaiyu clan, the entryway to the clan land of the dead being in the clan land of the living. Establishing a place as the gateway to a clan's afterworld, incorporating the physical (skeletal) remains of recently deceased clan members in the process, is a strong marking of earthly place with sanctity, furthering that same process of territorial consolidation.

This, we suggest, is what happened at Rupo: at a time of population increase (possibly due in part at least to immigration) and of spreading populations, each establishing new homelands across parts (or much of) the Gulf Province lowlands, and at a time also when Europeans were establishing new colonial administrative bases and mission stations in the region, the locations of Rumu villages were shifting. It is at this time that the Himaiyu clan established its major gateway to the afterworld in one cave at the core of its territory, farther away from what became smaller numbers of larger and more nucleated villages than was previously the case. Toward the end of the 19th century, at a time of initial European contact, village locations and residential strategies were shifting as a result of increasing administrative influence and a decreased need for strategically locating settlements out of sight of headhunting raiding parties who came by canoe along the major waterways. Recently deceased members of the Himaiyu clan were brought to Rupo to allow them access to the afterworld. Around this time or soon after, Rumu villages were relocated to Kopi and elsewhere (mainly along the banks of the Kikori River), and in the space of a few years following missionization, Christian burial practices were introduced, with major influences on local cosmologies (including the imminent cessation of use of Rupo for the placement of human bones). But in the land of the living, it is Rupo that was, and continues to this day to be, the symbolic heartland of life after death for the Himaiyu clan, a clan with land and an ongoing local history and sacred geography that continues to be recognized and entrenched throughout Rumu territory.

Conclusions

Rumu landscapes can be properly understood not only by reference to spirits but more critically *as* spiritscapes. Historicizing Rumu culture must therefore aim to historicize Rumu sacred landscapes: how people have in the past engaged with a world of meaning. In places where ethnographic information is available, this may be attempted by reference to the Dual Historical Method—systematically tracking back in time material expressions of particular ethnographic practices and beliefs (the backward approach to history), while ping-ponging with an exploration of how things have unfolded through the course of history (the forward-movement of history) (see David 2002). Although the archaeology of the intangibles of life may be a challenge, all the more reason to recognize their central importance to cultural landscapes of the present and the past.

Acknowledgments

This chapter combines information recorded by Bruno David in 2005–2006 on Rumu spiritscapes and mortuary practices (including information given by, among others, Max Pivoru and William Pivoru, clan representatives and co-authors) and by Michael Green in 1984–1985 on mortuary practices. We thank Soren Blau, Michael Boiru, Tim Denham, John Soba, and Ian McNiven for comments on various aspects of this paper, and David Kupere for helping with the Rupo excavations. Michael Green recorded details of Rumu mortuary practice during interviews with Kuiari and other senior men in the company of many residents of Kopi village in 1985; we thank all clan representatives and Kopi villagers who participated in these discussions.

Note

1. The interrogator can be a man or a woman clan member.

References

Bailey, D. W., and Hofmann, D. 2005. Review of H. Todorova (ed.), *Durankulak 2: Die Prähistorischen Gräberfelder von Durankulak*. Sofia: Deutsches Archäologisches Institut. *Antiquity* 79(303): 220–22.

Busse, M., Turner, S., and Araho, N. 1993. *The People of Lake Kutubu and Kikori: Changing Meanings of Daily Life*. Port Moresby: Papua New Guinea National Museum and Art Gallery.

David, B. 2002. *Landscapes, Rock-Art and the Dreaming: An Archaeology of Preunderstanding*. London: Leicester University Press.

———. 2005. Four interviews at Kopi, Gulf Province, Papua New Guinea. Unpublished Cultural Heritage Report Series 15, Programme for Australian Indigenous Archaeology, School of Geography and Environmental Science, Monash University, Clayton.

Davies J. 1999. *Death, Burial and Rebirth in the Religions of Antiquity*. London: Routledge.

Fleckinger, A. 2003. *Ötzi, the Ice Man: The Full Facts at a Glance*. Bolzano: South Tyrol Museum of Archaeology.

Glob, P. V. 1969. *The Bog People: Iron-Age Man Preserved*. Ithaca, NY: Cornell University Press.

Greber, N. B., and Ruhl, K. C. 1989. *The Hopewell Site: A Contemporary Analysis Based on the Work of Charles C. Willoughby*. Boulder, CO: Westview Press.

Haddon, A. C. 1918. The Agiba cult of the Kerewo culture. *Man* (OS) 18: 177–83.

Hallam, E., and Hockey, J. 2001. *Death, Memory and Material Culture*. Oxford: Berg.

Ikram, S. 2003. *Death and Burial in Ancient Egypt*. London: Longman.

Ikram, S., and Dodson, A. 1998. The mummy in Ancient Egypt: Equipping the dead for eternity. London: Thames and Hudson

King, J. M. 2004. Grave goods as gifts in early Saxon burials (ca. A.D. 450–600). *Journal of Social Archaeology* 4: 214–38.

Morphy, H. 1984. *Journey to the Crocodile's Nest*. Canberra: Australian Institute of Aboriginal Studies.

Museum of Science. 2003. Mummification in ancient Egypt. Electronic document: www.mos.org/quest/mummyegypt.php. Accessed Oct. 23, 2007.

Pardoe, C. 1988. The cemetery as symbol: The distribution of prehistoric Aboriginal burial grounds in southeastern Australia. *Archaeology in Oceania* 23: 1–16.

Polach, H. 1980. Appendix 10: ANU Radiocarbon Laboratory report on C-14 dates, in J. Rhoads, Through a Glass Darkly, p. A68. Unpublished Ph.D. thesis, Australian National University, Canberra.

Schieffelin, E. 1976. *The Sorrow of the Lonely and the Burning of the Dancers*. New York: St. Maartens Press.

Solecki, R. S. 1971. *Shanidar: The First Flower People*. New York: Knopf.

Spindler, K. 1994. *The Man in the Ice*. London: Weidenfeld and Nicolson.

Stringer, C., and Gamble, C. 1993. *In Search of the Neanderthals: Solving the Puzzle of Human Origins*. London: Thames and Hudson.

Turner, C., and Hannon, G. E. 1988. Vegetational evidence for late Quaternary climatic changes in southwest Europe in relation to the influence of the North Atlantic Ocean. *Philosophical Transactions of the Royal Society of London* B 318: 451–85.

Wagner, R. 1972. *Habu*. Chicago: University of Chicago Press.

Wallis Budge, E. A. 1973 [1911]. *Osiris and the Egyptian Resurrection*, volume 1. New York: Dover Publications.

Weiner, J. 1988. *The Heart of the Pearl Shell: The Mythological Dimension of Foi Sociality*. Berkeley and Los Angeles: University of California Press.

Williams, H. 2005. Keeping the dead at arm's length: Memory, weaponry and early Medieval mortuary technologies. *Journal of Social Archaeology* 5: 253–75.

15

Visions of the Cosmos: Ceremonial Landscapes and Civic Plans

Wendy Ashmore

Peoples around the world regularly materialize conceptions of the cosmos. The issue for archaeologists is not whether this is so but determining what evidence is needed to recognize and interpret such expressions, and with what reliability we can do so. Certainly, a culture-specific worldview may be perceived in an existing landscape by its inhabitants, acknowledged in narratives of ancestral mapping, and in customary reverence for or avoidance of specific places (e.g., Colwell-Chanthaphonh and Ferguson 2006; Morphy 1995; Richards 1999; Taçon 1999; see also Bradley, this volume). Cosmological ideas might also be inscribed materially in varied media, prominently (but not exhaustively) including rock art, isolated monuments, and such other material categories as architecture—from individual buildings through bounded cityscapes and beyond. This chapter focuses on archaeological study of cosmology and worldview in ceremonial landscapes, and especially in civic plans (that is, public works) as a frequent category within the range of such landscapes. Although analysts sometimes consider the two highlighted domains separately, they are considered jointly here, principally but not solely with reference to considering societies for which the archaeological record is a fundamental interpretive resource.

Nature of Inquiry

Ceremonial landscapes are defined here as settings in which arrangements of specific features situate the cosmos on earth, and where ritualized movements to and among these features are means to evoke and reinforce understandings of cosmic order. The features in question may or may not be easily recognizable to outsiders and can be quite subtle in material manifestation (e.g., van de Guchte 1999). Such landscapes are often expressions of sacred geographies or spiritscapes (e.g., Crumley 1999; also David and Thomas; McNiven, this volume), but their components need not always be designated as "sacred" (e.g., Ashmore and Blackmore 2008; Knapp and Ashmore 1999; Smith 1987). Nor are they always readily distinct from "mundane" landscapes of everyday activities (e.g., Bender 2002; Ingold 1993; Mack 2004; Moore 2005; Tuan 1977). Although most would agree that all these landscapes exist through people's interaction with them, the term *ceremonial landscape* shifts emphases squarely onto the practices of interaction. *Civic plans* are arrangements of public buildings, open spaces, and monuments in villages, towns, or cities—landscapes that are arenas for public interaction, ceremonial and otherwise, in sedentary societies.

Inquiry in these domains finds greatest, though not exclusive, receptivity among postpositivist analysts. To clarify (if not resolve) issues of interpretive disagreement, archaeologists with diverse epistemic perspectives come into conflict about the legitimate and appropriate ways of knowing the past, especially when the aim is to infer ancient meaning. Those grounded in positivist theoretical traditions generally stipulate hypothesis testing, with rigorously defined standards for proof, expressed in material measures of archaeological data (e.g., Moore 2005; M. Smith 2003). From broadly postpositivist vantages, however, acceptance of multiple ways of knowing make possible alternative understandings of past phenomena, within limits of evidential constraint (e.g., Colwell-Chanthaphonh and Ferguson 2006; McGuire 2004; Preucel 1991; Wylie 2002). Among such approaches prominent within studies examined here, structuralism and phenomenology have been especially subject to exploration and critique, and with productive response (e.g., Ashmore 2004a; Barrett 1988; Bender 2002, 2006; Bender, Hamilton, and Tilley 1997; Blake 2004; Brück 2006; Hamilakis, Pluciennik, and Tarlow 2002; Thomas 1991; Tilley 1994, 2004). As Anschuetz and his colleagues note: "Some of the most highly productive landscape research draws from complementary theoretical perspectives" (2001: 159), and the paragraphs that follow here select critically among contributions from sometimes seemingly disparate approaches in the literature.

Ceremonial Landscapes

As foundation, concepts about the human body often inform spatial understandings, at multiple scales (e.g., Brown 2004; Whitridge 2004; see also Gamble, this volume). Mapping of worldview may be grounded explicitly in the form of the human body (e.g., Gillespie 1991, 2000; Low 2003; Whitridge 2004). The body is also, of course, the starting point for phenomenological understandings of the world (e.g., Brück 2005; Tilley 1994; see Tilley, this volume). Cosmic referents may be inscribed physically on the body, as tattoos or other corporeal tangible modification (e.g., Meskell and Joyce 2004; Rainbird 2002). Spatial order at larger (and sometimes smaller) scales reiterates cosmological conceptions and inscriptions, commonly—and significantly—merging notions of time with beliefs about the structure of space (e.g., Hirsch 2006; Mathews and Garber 2004; Parker Pearson et al. 2006; Rice 2004; Thomas 1991, 1995; Zaro and Lohse 2005). Such mergers can usefully be referenced by the term *cosmovision*, which Broda describes concisely for Mesoamerica: "Borrowed from the common Spanish and German usage, [cosmovision] denotes the *structured view* in which the ancient Mesoamericans combined their notions of cosmology *relating time and space into a systematic whole*. This term is thus somewhat more specific than the English terms 'cosmology' and 'worldview'" (Broda 1987: 108, emphasis added).

Ceremonial landscapes acknowledge concepts of cosmovision through repeated human practices in space and time. The literature on such landscapes is immense, its scope global, and its time range encompassing deep antiquity through the present day (e.g., Ashmore and Knapp 1999; Aveni 2001; Bender 1998; Bradley 1993, 1997; Carrasco 1991; Thomas 1991, 1995; Ucko and Layton 1999). Whether with reference to past times or present, inquiries about ceremonial landscapes focus on the places documented or inferred to be landmarks of cosmic mapping and ritual practices, and on the practices by which inhabitants enact cosmovision. Taçon (1999: 36–37) notes the following as features most likely to "invoke . . . feelings of awe, power, majestic beauty, respect, [or] enrichment": (1) places "where the results of great acts of natural transformation can best be seen, such as mountain ranges, volcanoes, steep valleys or gorges;" (2) points of relatively abrupt transition in geology, hydrology, vegetation, or some combination of these; (3) unusual elements that "one comes upon suddenly;" and (4) vantage points with dramatic views. These same kinds of points can suggest junctions between mundane and supernatural realms, often identified further as *axis mundi*, or the center of the world (Eliade 1959; Taçon 1999; Tuan 1977). Caves, mountaintops, and bodies of water appear with particular frequency in such a role (e.g., Brady and Ashmore 1999; Hall 1976; Parker Pearson 2004). Some materialize the bodies of mythic or legendary beings, or the results of their actions (e.g., Layton 1999). Sometimes natural places are modified to mimic or reestablish appropriate features, arguably to assert the *axis mundi* more emphatically (e.g., Brady and Ashmore 1999; Richards 1996b, 2003; Stein and Lekson 1992). Sounds, odors, and other senses accentuate the visual experience of these features in ways whose potentials archaeologists are still just beginning to explore (e.g., Houston, Stuart, and Taube 2006; Moore 2005; Rainbird 2002; Stein, Friedman, and Blackhorse 2007; Stoddart 2002).

Once established, these same places readily become destinations for visitation or, alternatively, places of danger, taboo, and avoidance. Movement among the stations can take the form of pilgrimage, procession, or individual observance (e.g., Bauer and Stanish 2001; Bradley 1993; Cole 2004; Moore 2005; Snead and Preucel 1999; Sofaer, Marshall, and Sinclair 1989; Zedeño 2000). Songlines of Aboriginal Australia are well known in this regard. In a very different context, Central Mexico of imperial Aztec times, ritual movements among designated mountaintops, springs, and other stations across the ceremonial landscape were scheduled in accordance with annual feasts and sacrifices of important deities, or significant events of mytho-historical time (e.g., Carrasco 1991). The practice of ritual, through movement and other sensory experiences, activates the places visited, both reiterating and reinforcing the cosmovision that structures the whole. In Neolithic Britain, Stonehenge and the ceremonial landscape of which it is a part have attracted extraordinary scrutiny (e.g., Bender 1993, 1998; Bradley 1993, 1998a; Parker Pearson et al. 2006; Parker Pearson and Ramilisonina 1998). For example, from the cumulative record and their own new research program, Parker Pearson and his colleagues contend that Stonehenge and the great earthen henge at Durrington Walls are end points for mutually complementary circuits of the encompassing ceremonial landscape, perpetually embodying balance of transitions between life and death in an orderly cosmos.

Mortuary practices and their material traces yield what are often among the most significant landmarks in ceremonial landscape. Barrows in the greater Stonehenge landscape, for example, cluster at a major visual transition along the formal avenue, a topographic threshold at which Stonehenge comes into view.[1] In other contexts, such as the North American mid-continent, burial tumuli and their locations become signposts for mapping social, economic, and political history and its transformations, as well as enduring cosmovision (e.g., Buikstra and Charles 1999; Charles and Buikstra 2002; Charles, Van Nest, and Buikstra 2004; see also Hall 1976). The placement of Egyptian royal tombs at Abydos both reflected and affirmed the Old Kingdom cosmovision mapped in the valley around (Richards 1999).

Celestial phenomena commonly define rhythms for experiencing sacred landscapes, just as they often structure more mundane aspects of daily life (e.g., Urton 1981). The annual rounds of ritual in Aztec Central Mexico were mentioned earlier; timing of specific events was determined by the calendar, as confirmed by predictable astronomical phenomena (e.g., Aveni 1991, 2001, 2003; Broda 1991). Deeper in the past, solstice sunsets inferentially choreographed the processions along prescribed routes anchored at Durrington Walls in mid-summer and Stonehenge in mid-winter. Predictability of solar and lunar transitions relative to lived space likewise confirmed cosmovision in ceremonial landscapes as disparate as pharaonic Egypt and ancestral Puebloan southwest (e.g., Richards 1999; Sofaer 1997; Stein, Suiter, and Ford, 1997).

Archaeologists increasingly recognize the meaning-laden roles of physical elements—what analysts know prosaically as stone, soil, timber, fire, and water—in experience of ceremonial landscapes (e.g., Boivin and Owoc 2004; Tilley 2004). For example, water is commonly seen as portal to what we as analysts usually consider a supernatural otherworld, often linked more generally to (among other notions) an array of mutually overlapping concepts such as boundary, barrier, conduit, transition, and transformation (e.g., Brady and Ashmore 1999; Hall 1976; Parker Pearson et al. 2006; Richards 1996a; Urton 1981). Properties of stone, including color, texture, or origin, potentially punctuate the cosmovision perceived and enacted in ceremonial landscapes (e.g., Jones and Bradley 1999; Scarre 2002, 2004; Spence 1999; Taçon 1991, 2004; Tilley 2004; Whitley et al. 1999). Similarly, soil takes on elemental meaning in construction contexts, as when careful selection and deposition of discrete bits of turf are used in precise complement to other sediment types to recreate an ordered cosmos within individual mortuary mounds, in settings as culturally divergent as Hopewellian sites and barrows at Stonehenge (e.g., Charles, Van Nest, and Buikstra 2004; Parker Pearson et al. 2006). In all the works cited parenthetically, inference of specific meaning is contingent on marshaling converging lines of evidence consistent with that interpretation, although it is not necessarily reliant on close ethnographic analogy.

Cosmovision in Civic Plans

Turning now to the specific case of civic plans, many have inferred cosmovision as structuring the layout of town and city spaces and architecture, in culturally diverse instances of antiquity as well as in more recent times (e.g., Urton 1981). Such expressions are broadly congruent with those of ceremonial landscapes, with the added impacts of politics and local or regional history. Wheatley's (1971) masterful analysis of ancient Chinese city planning remains widely influential, as does Rykwert's (1988) discussion of Roman town planning; both authors make stimulating cross-cultural comparisons. Not all civic plans have cosmovision underpinnings, however. An assemblage of articles in the *Cambridge Archaeological Journal* illustrates the diversity of possibilities (e.g., Carl 2000; Cowgill 2000; Kemp 2000). Some of the contention involves analytic scale. In a treatise on the Mesoamerican city, Marcus (1983) draws on Constantinos Doxiadis (1913–1975) to remind us that cities and towns everywhere have multiple components and that these components can be organized spatially by quite different principles. Although the civic center may be tightly ordered by deliberate planning, adjacent, largely residential precincts may reflect more organic growth, something Carl (2000: 329) refers to as the "messiness" of undirected practices in daily life. Still, even if one focuses on the civic center alone, variation in civic plan—in arrangements of buildings, outdoor spaces, and their orientations—makes some doubt that something as putatively internally unified as a culture-specific cosmovision can be responsible (e.g., Kemp 2000; Nalda 1998; M. Smith 2003).

Some individual cases of cosmovision in civic plans are widely accepted, as for the central Mexican metropolis of Teotihuacan (e.g., Cowgill 2000; Heyden 2000; Sugiyama 1993). But even where cosmovision might be cited as a factor, variation *among civic plans within a single culture* can ensue from differing political histories. The structuralist regularity supported by the analyses of Wheatley and others is modified by history and practice. Often civic plans combine expressions of cosmovision with state propaganda, extolling a particular dynasty or ruling house materially, using grand public works to situate temporal leaders in primordially sanctioned positions of authority. Such shaping is evident in civic plans for capital cities of ancient China, as it is for those of the Classic Maya and other state civilizations of Mesoamerica, as well as for Caesar's Rome (e.g., Ashmore 1991, 2005; Ashmore and Sabloff 2002, 2003; Aveni and Hartung 1986; Carl 2000: 330; Joyce 2004; Keller 2006; Rice 2004; Steinhardt 1986). Sounds, colors, smells, dramatic performance, and other sensory perceptions would have augmented lived experience of the place, in the moment, and in social memory (e.g., Inomata and Coben 2006; Jones and Bradley 1999; Moore 2005; A. Smith 2003; see also Rainbird, this volume; Van Dyke, this volume).

Civic plans may also incorporate natural features strategically, so as to "demonstrate" the sovereign's control over natural forces in the cosmos. The Classic Maya city of Dos Pilas is an apt example, where the principal palace was placed atop an important cave; each year, after the onset of the rainy season, a thunderous rush of water out of the cave proclaimed the royal resident's ability to call forth the rains and with them, agricultural prosperity and social well-being (Brady 1997). His place in the cosmos was thus secured, renewed each year through the civic plan and the ceremonial landscape. Natural features could also be modified to link worldly authority with cosmically sanctioned control of water and other resources (e.g., Fash 2005; Scarborough 1998). Returning to propaganda in overt construction, multiple architectural assemblages common in Maya civic centers have been proposed specifically as stages for regular royal performance in ritual time. These assemblages, labeled technically by analysts as ballcourts, "E-Groups," and twin-pyramid complexes, were relatively standardized arenas for marking solar, agricultural, or longer cycles, often punctuating observances with human sacrifice or other potent offerings and always implicating the sovereign as he who maintains cosmovision within and among polities (e.g., Aimers and Rice 2006; Aveni 2001; Gillespie 1991; Rice 2004).

Like the rushing water at Dos Pilas, other hierophanies, or manifestations of the sacred (Eliade 1959) used architecture to manipulate light and shadow, as well as sound, as a seemingly natural affirmation of cosmically sanctioned authority (e.g., Brady and Ashmore 1999; Stein, Friedman, and Blackhorse 2007; Stein, Suiter, and Ford 1997). Or materials acquired at great distance could ratify royal or imperial authority, as in the sands imported to surface the main plaza in Inca Cuzco (Protzen and Rowe 1994). Arenas beyond the civic centers could be linked in political and cosmic celebration, as well, as in the ritual circuits and processions at Vijayanagara, joining the worldly ruler of that South Asian empire with the ruling beings and forces of the cosmic realm (Fritz 1986; cf. Keller 2006; Mack 2004; Rice 2004). Ceremonial landscapes incorporate civic centers in expressions of cosmovision, if in these cases with a powerful political edge (cf. Bender 2002).

In long-enduring civic centers, however, the effects of history can blur the structure behind the practices, as well as the distinctions between civic and wider frames. Sometimes this involves revisionist moves, materially rewriting history by replacing public works of earlier regimes. Sometimes it involves what Barrett (1999) calls *inhabitation*, social practices that acknowledge extant monuments (or buildings) and their meanings, and sometimes it involves as well the process of ruin formation and the ambiguous place of ruins in social memory (e.g., Ashmore 2004b; Bradley 1993; Bradley and Williams 1998; Scarre 2002; Stanton and Magnoni 2008; Van Dyke and Alcock 2003).

Conclusions

As the preceding discussion indicates, consideration of cosmovision in civic plans merges with that of ceremonial landscapes more broadly. That is, both ceremonial landscapes and civic plans constitute arenas in which cosmovision can be expressed, in ways that archaeologists are coming to detect and at a scale we treat as landscape. In both sets of cases, meaning is materialized in the land, by perception and by intervention of practice. Many have noted that landscapes are "quasi-artifacts" (e.g. Tilley 2006), and the boundary between built and "natural" features in even an urban landscape can become blurred over the long term (e.g., Barrett 1999; Bradley 1987, 1993, 1998b; Tilley et al. 2000; but see Stone 1992 and Taube 2003).[2]

In all cases, positivist calls for falsifiable hypotheses stand in counterpoint to postpositivist interpretations. From either epistemic perspective, the most compelling works tend to be data-rich, with much empirical information brought to bear on inferences made. As noted earlier, however, the disparities need not become (or remain) polarized oppositions; when allowed to complement each other, the diversity can be mutually stimulating and quite productive (Anschuetz, Wilshusen, and Scheick, 2001; Ashmore 2005; Bender 1999; Fisher and Thurston 1999).

Studying ceremonial landscapes and civic plans is integral to landscape archaeology as practiced in the 21st century. Although not all practitioners choose such topics as primary focus of inquiry, comprehensive archaeological research into people's relations with the land around them requires understanding how cosmology and political history shaped lived experience in that land.

Acknowledgments

For stimulus and critical discussion, I thank Bruno David, Julian Thomas, George Nicholas, Julie Hollowell, Chelsea Blackmore, Jane Buikstra, Angie Keller, Jim Brady, Bernard Knapp, Tom Patterson, and a very helpful anonymous reviewer.

Note

1. See, for example, Parker Pearson et al. (2006), as well as compare inferences of visual choreography elsewhere: Bender, Hamilton, and Tilley 1997; Bradley 1993, 1998a; Richards 1996a; Scarre 2002; Sofaer, Marshall, and Sinclair, 1989; Thomas 1993a, 1993b.

2. See, for example, Barrett 1999.

References

Aimers, J. J., and Rice, P. M. 2006. Astronomy, ritual, and the interpretation of Maya "E-Group" architectural assemblages. *Ancient Mesoamerica* 17: 79–96.

Anschuetz, K. F., Wilshusen, R. H., and Scheick, C. L. 2001. An archaeology of landscapes: Perspectives and directions. *Journal of Archaeological Research* 9: 157–211.

Ashmore, W. 1991. Site planning principles and concepts of directionality among the ancient Maya. *Latin American Antiquity* 2: 199–226.

———. 2004a. Social archaeologies of landscape, in L. Meskell and R. W. Preucel (eds.), *A Companion to Social Archaeology*, pp. 255–71. Oxford: Blackwell Publishing.

———. 2004b. Ancient Maya landscapes, in C. W. Golden and G. Borgstede (eds.), *Maya Archaeology at the Millennium*, pp. 95–109. London: Routledge.

———. 2005. Sacred landscapes, political spaces. Paper presented at Coloquio Bosch Gimpera VI, Instituto de Investigaciones Antropológicas, Universidad Autónoma de México, Mexico.

Ashmore, W., and Blackmore, C. 2008. Landscape archaeology, in D. Pearsall (ed.), *Encyclopedia of Archaeology*, pp. 1569–78. Oxford: Elsevier.

Ashmore, W., and Knapp, A. B. (eds.). 1999. *Archaeologies of Landscape: Contemporary Perspectives*. Oxford: Blackwell Publishing.

Ashmore, W., and Sabloff, J. A. 2002. Spatial order in Maya civic plans. *Latin American Antiquity* 13: 201–15.

———. 2003. Interpreting ancient Maya civic plans: Reply to Smith. *Latin American Antiquity* 14: 229–36.

Aveni, A. F. 1991. Mapping the ritual landscape: Debt payment to Tlaloc during the month of Atlcahualo, in D. Carrasco (ed.), *To Change Place: Aztec Ceremonial Landscapes*, pp. 58–73. Niwot: University Press of Colorado.

———. 2001. *Skywatchers*. Austin: University of Texas Press.

———. 2003. Archaeoastronomy in the ancient Americas. *Journal of Archaeological Research* 11: 149–91.

Aveni, A. F., and Hartung, H. 1986. Maya city planning and the calendar. *Transactions of the American Philosophical Society* 76(1). (Whole issue).

Barrett, J. C. 1988. Fields of discourse: Reconstituting a social archaeology. *Critique of Anthropology* 7: 5–16.

———. 1999. The mythical landscapes of the British Iron Age, in W. Ashmore and A. B. Knapp (eds.), *Archaeologies of Landscape: Contemporary Perspectives*, pp. 253–65. Oxford: Blackwell Publishing.

Bauer, B. S., and Stanish, C. 2001. *Ritual and Pilgrimage in the Ancient Andes: The Islands of the Sun and the Moon*. Austin: University of Texas Press.

Bender, B. 1993. Stonehenge: Contested landscapes (medieval to present-day), in B. Bender (ed.), *Landscape: Politics and Perspectives*, pp. 245–79. Oxford: Berg.

———. 1998. *Stonehenge: Making Space*. Oxford: Berg.

———. 1999. Introductory comments. *Antiquity* 73: 632–34.

———. 2002. Time and landscape. *Current Anthropology* 43 (Supplement): S103–S112.

———. 2006. Place and landscape, in C. Tilley, W. Keane, S. Küchler, M. Rowlands, and P. Spyer (eds.), *Handbook of Material Culture*, pp. 303–14. London: Sage.

Bender, B., Hamilton, S., and Tilley, C. 1997. Leskernick: Stone worlds, alternative narratives, nested landscapes. *Proceedings of the Prehistoric Society* 63: 147–78.

Blake, E. C. 2004. Space, spatiality, and archaeology, in L. Meskell and R. W. Preucel (ed.), *A Companion to Social Archaeology*, pp. 230–54. Oxford: Blackwell Publishing.

Boivin, N., and M. A. Owoc (eds.). 2004. *Soils, Stones, and Symbols: Cultural Perceptions of the Mineral World*. London: UCL Press.

Bradley, R. 1987. Time regained: The creation of continuity. *Journal of the British Archaeological Association* 140: 1–17.

———. 1993. *Altering the Earth: The Origins of Monuments in Britain and Continental Europe*. The Rhind Lectures 1991–1992. Monograph Series, no. 8. Edinburgh: Society of Antiquaries of Scotland.

———. 1997. *Rock Art and the Prehistory of Atlantic Europe: Signing the Land*. London: Routledge.

———. 1998a. *The Significance of Monuments: On the Shaping of Human Experience in Neolithic and Bronze Age Europe*. London: Routledge.

———. 1998b. Ruined buildings ruined stones: Enclosures, tombs and natural places in the Neolithic of southwest England. *World Archaeology* 30: 13–22.

Bradley, R., and Williams, H. (eds.). 1998. The past in the past: The reuse of ancient monuments. *World Archaeology* 30(1) (Whole issue).

Brady, J. E. 1997. Settlement configuration and cosmology: The role of caves at Dos Pilas, *American Anthropologist* 99: 602–18.

Brady, J. E., and Ashmore, W. 1999. Mountains, caves, water: Ideational landscapes of the ancient Maya, in W. Ashmore and A. B. Knapp (eds.), *Archaeologies of Landscape: Contemporary Perspectives*, pp. 124–45. Oxford: Blackwell Publishing.

Broda, J. 1987. Templo Mayor as ritual space, in J. Broda, D. Carrasco, and E. Matos Moctezuma (eds.), *The Great Temple of Tenochtitlan: Center and Periphery in the Aztec World*, pp. 61–123. Berkeley and Los Angeles: University of California Press.

———. 1991. The sacred landscape of Aztec calendar festivals: Myth, nature, and society, in D. Carrasco (ed.), *To Change Place: Aztec Ceremonial Landscapes*, pp. 74–120. Niwot: University Press of Colorado.

Brown, L. A. 2004. Dangerous places and wild spaces: Creating meaning with materials and space at contemporary Maya shrines on El Duende Mountain. *Journal of Archaeological Method and Theory* 11: 31–58.

Brück, J. 2005. Experiencing the past? The development of a phenomenological archaeology in British prehistory. *Archaeological Dialogues* 12: 45–72.

Buikstra, J. E., and Charles, D. K. 1999. Centering the ancestors: Cemeteries, mounds and sacred landscapes of the North American Midcontinent, in W. Ashmore and A. B. Knapp (eds.), *Archaeologies of Landscape: Contemporary Perspectives*, pp. 201–28. Oxford: Blackwell Publishing.

Carl, P. 2000. City-image versus topography of *praxis*. *Cambridge Archaeological Journal* 10: 328–35.

Carrasco, D. (ed.). 1991. *To Change Place: Aztec Ceremonial Landscapes*. Niwot: University Press of Colorado.

Charles, D. K., and Buikstra, J. E. 2002. Siting, sighting, and citing the dead, in H. Silverman and D. B. Small (eds.), *The Space and Place of Death*, pp. 13–26. Archeological Papers of the AAA 11. Arlington: American Anthropological Association.

Charles, D. K., Van Nest, J., and Buikstra, J. E. 2004. From the earth: Minerals and meaning in the Hopewellian world, in N. Boivin and M. A. Owoc (eds.), *Soils, Stones and Symbols: Cultural Perceptions of the Mineral World*, pp. 43–70. London: UCL Press.

Cole, S. G. 2004. *Landscapes, Gender, and Ritual Space: The Ancient Greek Experience*. Berkeley and Los Angeles: University of California Press.

Colwell-Chanthaphonh, C., and Ferguson, T. J. 2006. Memory pieces and footprints: Multivocality and the meanings of ancient times and ancestral places among the Zuni and Hopi. *American Anthropologist* 108: 148–62.

Cowgill, G. L. 2000. Intentionality and meaning in the layout of Teotihuacan, Mexico. *Cambridge Archaeological Journal* 10: 358–61.

Crumley, C. L., 1999. Sacred landscapes: Constructed and conceptualized, in W. Ashmore and A. B. Knapp (eds.), *Archaeologies of Landscape: Contemporary Perspectives*, pp. 269–76. Oxford: Blackwell Publishing.

Eliade, M. 1959. *The Sacred and the Profane*. New York: Harcourt Brace.

Fash, B. W. 2005. Iconographic evidence for water management and social organization at Copán, in E. W. Andrews and W. L. Fash (eds.), *Copán: The History of an Ancient Maya Kingdom*, pp. 103–38. Santa Fe: SAR Press.

Fisher, C. T., and Thurston, T. L. (eds.). 1999. Special Section: Dynamic landscapes and socio-political process: The topography of anthropogenic environments in global perspective. *Antiquity* 73: 630–88.

Fritz, J. M. 1986. Vijayanagara: Authority and meaning of a south Indian imperial capital. *American Anthropologist* 88: 44–55.

Gillespie, S. D. 1991. Ballgames and boundaries, in V. L. Scarborough and D. R. Wilcox (eds.), *The Mesoamerican Ballgame*, pp. 317–45. Tucson: University of Arizona Press.

———. 2000. Maya "nested houses": The ritual construction of place, in R. A. Joyce and S. D. Gillespie (eds.), *Beyond Kinship: Social and Material Reproduction in House Societies*, pp. 135–60. Philadelphia: University of Pennsylvania Press.

Hall, R. L. 1976: Ghosts, water barriers, corn, and sacred enclosures in the Eastern Woodlands. *American Antiquity* 41: 360–64.

Hamilakis, Y., Pluciennik, M., and Tarlow, S. 2002. Introduction: Thinking through the body, in Y. Hamilakis, M. Pluciennik, and S. Tarlow (eds.), *Thinking through the Body: Archaeologies of Corporeality*, pp. 1–21. New York: Kluwer Academic/Plenum Publishers.

Heyden, D. 2000. From Teotihuacan to Tenochtitlan: City planning, caves, and streams of red and blue waters, in D. Carrasco, L. Jones, and S. Sessions (eds.), *Mesoamerica's Classic Heritage: From Teotihuacan to the Aztecs*, pp. 165–84. Boulder: University Press of Colorado.

Hirsch, E. 2006. Landscape, myth and time. *Journal of Material Culture* 11: 151–65.

Houston, S. D., Stuart, D. and Taube, K. 2006. *The Memory of Bones: Body, Being, and Experience among the Classic Maya*. Austin: University of Texas Press.

Ingold, T. 1993. The temporality of the landscape. *World Archaeology* 25: 152–74.

Inomata, T., and Coben, L. S. (eds.). 2006. *Archaeology of Performance: Theaters of Power, Community, and Politics*. Walnut Creek, CA: AltaMira Press.

Jones, A., and Bradley, R. 1999. The significance of colour in European archaeology. *Cambridge Archaeological Journal* 9: 112–14.

Joyce, A. A. 2004. Sacred space and social relations in the Valley of Oaxaca, in J. A. Hendon and R. A. Joyce (eds.), *Mesoamerican Archaeology: Theory and Practice*, pp. 192–216. Oxford: Blackwell Publishing.

Keller, A. H. 2006. Roads to the center: A study of the design, use, and abandonment of the roads of Xunantunich, Belize. Unpublished Ph.D. thesis, University of Pennsylvania.

Kemp, B. 2000. Bricks and metaphor. *Cambridge Archaeological Journal* 10: 335–46.

Knapp, A. B., and Ashmore, W. 1999. Archaeological landscapes: Constructed, conceptualized, ideational, in W. Ashmore and A. B. Knapp (eds.), *Archaeologies of Landscape: Contemporary Perspectives*, pp. 1–30. Oxford: Blackwell Publishing.

Layton, R. 1999. The Alawa totemic landscape: Ecology, religion and politics, in P. J. Ucko and R. Layton (eds.), *The Archaeology of Landscape: Shaping Your Landscape*, pp. 219–39. London: Routledge.

Low, S. M. 2003. Anthropological theories of body, space, and culture. *Space and Culture* 6: 9–18.

Mack, A. 2004. One landscape, many experiences: Differing perspectives of the Temple Districts of Vijayanagar. *Journal of Archaeological Method and Theory* 11: 59–81.

Marcus, J. 1983. On the Nature of the Mesoamerican City, in E. Z. Vogt and R. M. Leventhal (eds.), *Prehistoric Settlement Patterns: Essays in Honor of Gordon R. Willey*, pp. 195–242. Albuquerque: University of New Mexico Press, and Cambridge: Peabody Museum, Harvard University.

Mathews, J., and Garber, J. F. 2004. Models of cosmic order: Physical expression of sacred space among the ancient Maya. *Ancient Mesoamerica* 15: 49–59.

McGuire, R. H. 2004. Contested pasts: Archaeology and Native Americans, in L. Meskell and R. W. Preucel (eds.), *A Companion to Social Archaeology*, pp. 374–95. Oxford: Blackwell Publishing.

Meskell, L., and Joyce, R. A. 2004. *Embodied Lives: Figuring Ancient Maya and Egyptian Experience*. London: Routledge.

Moore, J. D. 2005. *Cultural Landscapes in the Ancient Andes: Archaeologies of Place*. Gainesville: University Press of Florida.

Morphy, H. 1995. Landscape and the reproduction of the ancestral past, in E. Hirsch and M. O'Hanlon (eds.), *The Anthropology of Landscape: Perspectives on Place and Space*, pp. 184–209. Oxford: Clarendon Press.

Nalda, E. 1998. The Maya city, in P. Schmidt, M. de la Garza, and E. Nalda (eds.), *Maya*, pp. 102–29. New York: Rizzoli International Publications.

Parker Pearson, M. 2004. Earth, wood, and fire: Materiality and Stonehenge, in N. Boivin and M. A. Owoc (eds.), *Soils, Stones and Symbols: Cultural Perceptions of the Mineral World*, pp. 71–89. London: UCL Press.

Parker Pearson, M., Pollard, J., Richards, C., Thomas, J., Tilley, C., Welham, K., and Albarella, U. 2006. Materializing Stonehenge: The Stonehenge Riverside Project and new discoveries. *Journal of Material Culture* 11: 227–61.

Parker Pearson, M., and Ramilisonina. 1998. Stonehenge for the ancestors: The stones pass on the message. *Antiquity* 72: 308–26.

Preucel, R. W. 1991. The philosophy of archaeology, in R. W. Preucel (ed.), *Processual and Postprocessual Archaeologies: Multiple Ways of Knowing the Past*, pp. 17–29. Center for Archaeological Investigations, Occasional Paper, 10. Carbondale: Southern Illinois University.

Protzen, J.-P., and Rowe, J. H. 1994. Hawkaypata: The terrace of leisure, in Z. Çelik, D. Favro, and R. Ingersoll (eds.), *Streets: Critical Perspectives on Public Space*, pp. 235–46. Berkeley and Los Angeles: University of California Press.

Rainbird, P. 2002. Marking the body, marking the land: Body as history, land as history: Tattooing and engraving in Oceania, in Y. Hamilakis, M. Pluciennik, and S. Tarlow (eds.), *Thinking through the Body: Archaeologies of Corporeality*, pp. 233–47. New York: Kluwer Academic/Plenum Publishers.

Rice, P. M. 2004. *Maya Political Science: Time, Astronomy, and the Cosmos*. Austin: University of Texas Press.

Richards, C. 1996a. Henges and water: Towards an elemental understanding of monumentality and landscape in Late Neolithic Britain. *Journal of Material Culture* 1: 313–36.

———. 1996b. Monuments as landscape: Creating the centre of the world in Late Neolithic Orkney. *World Archaeology* 28: 190–208.

Richards, C. (ed.). 2003. *Dwelling among the Monuments: The Neolithic Village of Barnhouse, Maeshowe Passage Grave and Surrounding Monuments at Stenness, Orkney*. Cambridge: McDonald Institute for Archaeological Research.

Richards, J. E. 1999. Conceptual landscapes in the Egyptian Nile Valley, in W. Ashmore and A. B. Knapp (eds.), *Archaeologies of Landscape: Contemporary Perspectives*, pp. 83–100. Oxford: Blackwell Publishing.

Rykwert, J. 1988. *The Idea of a Town: The Anthropology of Urban Form in Rome, Italy and the Ancient World*. Cambridge, MA: MIT Press.

Scarborough, V. L. 1998. Ecology and ritual: Water management and the Maya. *Latin American Antiquity* 9: 135–59.

Scarre, C. 2002. A place of special meaning: Interpreting pre-historic monuments in the landscape, in B. David and M. Wilson (eds.), *Inscribed Landscapes: Marking and Making Place*, pp. 154–75. Honolulu: University of Hawai'i Press.

———. 2004. Choosing stones, remembering places: Geology and intention in the megalithic monuments of western Europe, in N. Boivin and M. A. Owoc (eds.), *Soils, Stones and Symbols: Cultural Perceptions of the Mineral World*, pp. 187–202. London: UCL Press.

Smith, A. T. 2003. *The Political Landscape: Constellations of Authority in Early Complex Polities*. Berkeley and Los Angeles: University of California Press.

Smith, J. Z. 1987. *To Take Place: Toward Theory in Ritual*. Chicago: University of Chicago Press.

Smith, M. E. 2003. Can we read cosmology in ancient Maya city plans? Comment on Ashmore and Sabloff. *Latin American Antiquity* 14: 221–28.

Snead, J. E., and Preucel, R. W. 1999. The ideology of settlement: Ancestral Keres landscapes in the northern Rio Grand, in W. Ashmore and A. B. Knapp (eds.), *Archaeologies of Landscape: Contemporary Perspectives*, pp. 169–97. Oxford: Blackwell Publishing.

Sofaer, A. 1997. The primary architecture of the Chacoan culture, in B. H. Morrow and V. B. Price (eds.), *Anasazi Architecture and American Design*, pp. 88–132. Albuquerque: University of New Mexico Press.

Sofaer, A., Marshall, M. P., and Sinclair, R. M. 1989. The Great North Road: A cosmographic expression of the Chaco culture of New Mexico, in A. F. Aveni (ed.), *World Archaeoastronomy*, pp. 365–76. Oxford: Oxford University Press.

Spence, K. 1999. Red, white and black: Colour in building stone in ancient Egypt. *Cambridge Archaeological Journal* 9: 114–17.

Stanton, T. W., and Magnoni, A. (eds.). 2008. *Ruins of the Past: The Use and Perception of Abandoned Structures in the Maya Lowlands*. Boulder CO: University Press of Colorado.

Stein, J. R., Friedman, R., and Blackhorse, T. 2007. Revisiting downtown Chaco, in S. H. Lekson (ed.), *Architecture of Chaco Canyon*, pp. 199–223. Salt Lake City: University of Utah Press.

Stein, J. R., and Lekson, S. H. 1992. Anasazi ritual landscapes, in D. E. Doyel (ed.), *Anasazi Regional Organization and the Chaco System*, pp. 87–100. Anthropological Papers No. 5. Albuquerque: Maxwell Museum of Anthropology.

Stein, J. R., Suiter, J. E., and Ford, D. 1997. High noon in Old Bonito: Sun, shadow, and the geometry of the Chaco complex, in B. H. Morrow and V. B. Price (eds.), *Anasazi Architecture and American Design*, pp. 133–48. Albuquerque: University of New Mexico Press.

Steinhardt, N. S. 1986. Why were Chang'an and Beijing so different? *Journal of the Society of Architectural Historians* 45: 339–57.

Stoddart, S. 2002. Monuments in the pre-historic landscape of the Maltese islands, in B. David and M. Wilson (eds.), *Inscribed Landscapes: Marking and Making Place*, pp. 176–86. Honolulu: University of Hawai'i Press.

Stone, A. 1992. From ritual in the landscape to capture in the urban center: The recreation of ritual environments in Mesoamerica. *Journal of Ritual Studies* 6: 109–32.

Sugiyama, S. 1993. Worldview materialized in Teotihuacan, Mexico. *Latin American Antiquity* 4: 103–29.

Taçon, P. S. C. 1991. The power of stone: Symbolic aspects of stone use and development in Western Arnhem Land, Australia. *Antiquity* 65: 192–207.

———. 1999. Identifying ancient sacred landscapes in Australia: From physical to social, in W. Ashmore and A. B. Knapp (eds.), *Archaeologies of*

Landscape: Contemporary Perspectives, pp. 33–57. Oxford: Blackwell Publishing.

———. 2004. Ochre, clay, stone and art: The symbolic importance of minerals as life-force among Aboriginal peoples of northern and central Australia, in N. Boivin and M. A. Owoc (eds.), *Soils, Stones and Symbols: Cultural Perceptions of the Mineral World*, pp. 31–42. London: UCL Press.

Taube, K. 2003. Ancient and contemporary Maya conceptions about field and forest, in A. Gómez-Pompa, M. F. Allen, S. L. Fedick, and J. J. Jiménez-Osornio (eds.), *The Lowland Maya Area: Three Millennia at the Human-Wildland Interface*, pp. 461–92. New York: Food Products Press.

Thomas, J. 1991. *Rethinking the Neolithic*. Cambridge: Cambridge University Press.

———. 1993a. The politics of vision and the archaeologies of landscape, in B. Bender (ed.), *Landscape: Politics and Perspectives*, pp. 19–48. Oxford: Berg.

———. 1993b. The hermeneutics of megalithic space, in C. Tilley (ed.), *Interpretative Archaeology*, pp. 73–97. Oxford: Berg.

———. 1995. Reconciling symbolic significance with Being-in-the-world, in I. Hodder, M. Shanks, A. Alexandri, V. Buchli, J. Carman, J. Last, and G. Lucas (eds.), *Interpreting Archaeology: Finding Meaning in the Past*, pp. 210–11. London: Routledge.

Tilley, C. 1994. *A Phenomenology of Landscape*. Oxford: Berg.

———. 2004. *The Materiality of Stone: Explorations in Landscape Phenomenology*. Oxford: Berg.

———. 2006. Introduction: Identity, place, landscape and heritage. *Journal of Material Culture* 11: 7–32.

Tilley, C., Hamilton, S., Harrison, S., and Anderson, E. 2000. Nature, culture, clitter: Distinguishing between cultural and geomorphological landscapes–the case of Hilltop Tors in south-west England. *Journal of Material Culture* 5: 197–224.

Tuan, Y.-F. 1977. *Space and Place: The Perspective of Experience*. Minneapolis: University of Minnesota Press.

Ucko, P. J., and Layton, R. (eds.). 1999. *The Archaeology and Anthropology of Landscape: Shaping Your Landscape*. London: Routledge.

Urton, G. F. 1981. *At the Crossroads of the Earth and the Sky*. Austin: University of Texas Press.

van de Guchte, M. 1999. The Inca cognition of landscape: Archaeology, ethnohistory, and the aesthetic of alterity, in W. Ashmore and A. B. Knapp (eds.), *Archaeologies of Landscape: Contemporary Perspectives*, pp. 149–68. Oxford: Blackwell Publishing.

Van Dyke, R. M., and Alcock, S. E. (eds.). 2003. *Archaeologies of Memory*. Oxford: Blackwell Publishing.

Wheatley, P. 1971. *The Pivot of the Four Quarters: A Preliminary Inquiry into the Origins and Character of the Ancient Chinese City*. Chicago: Aldine.

Whitley, D. S., Dorn, R. I., Simon, J. M., Rechtman, R., and Whitley, T. K. 1999. Sally's Rockshelter and the archaeology of the vision quest. *Cambridge Archaeological Journal* 9: 221–47.

Whitridge, P. 2004. Landscapes, houses, bodies, things: "Place" and the archaeology of Inuit imaginaries. *Journal of Archaeological Method and Theory* 11: 213–50.

Wylie, A. 2002. *Thinking from Things: Essays in the Philosophy of Archaeology*. Berkeley and Los Angeles: University of California Press.

Zaro, G., and Lohse, J. C. 2005. Agricultural rhythms and rituals: Ancient Maya solar observations in hinterland Blue Creek, northwestern Belize. *Latin American Antiquity* 16: 81–98.

Zedeño, M. N. 2000. On what people make of places: A behavioral cartography, in M. B. Schiffer (ed.), *Social Theory in Archaeology*, pp. 97–111. Salt Lake City: University of Utah Press.

16

QUARRIED AWAY: THINKING ABOUT LANDSCAPES OF MEGALITHIC CONSTRUCTION ON RAPA NUI (EASTER ISLAND)

Sue Hamilton, Susana Nahoe Arellano, Colin Richards, and Francisco Torres H.

The Megalithic Monument and Landscapes of Construction

Monuments have featured strongly within "interpretative" approaches to landscape archaeology (e.g., Richards 1996; Thomas 1992; Tilley 1994, 2004). Within these studies, the "monument" is assumed to be architecturally unproblematic and is experienced as a completed entity. In this chapter, we approach landscapes as contexts of construction, whereby the monument is constantly enmeshed within an ongoing process of alteration and transformation.

The megalithic monument is an attractive entity. Its sheer magnitude creates an allure that reaches out to both archaeologist and non-archaeologist alike. The physical encounter with the scale and presence of the monumental creates a sense of overpowering awe. Of course, this is precisely the intended outcome of such constructions, which in the main act as mnemonics through spectacle. The experience also fascinates, because lurking behind monumental architecture is the imagined toil and ingenuity of its creation. Yet such a conception of labor is inhibited, because through the archaeological lens the view of the monument is inevitably that of a completed entity. It is this entity that is frequently fossilized and presented to the public by national heritage agencies, and consequently assumes iconic form. However, as Ingold reminds us, "the final form is but a fleeting moment in the life of any feature" (2000: 188).

Archaeology as a discipline was founded on typological reasoning and stratigraphic principles. In other words, all material evidence is classified and sequentially ordered. As a result of this legacy, archaeologists find it difficult, even within a field context, to transform analytical strategies that privilege "phases" (slices of time) over things. In the context of monumental architecture, which frequently appears to incorporate a series of alterations, additions and reconstructions, the idea of phases has been an appealing form of representation. Such an interpretative framework has been employed, for instance, to understand the excavated features and deposits at monuments such as Stonehenge (Atkinson 1956: 58–77; Cleal, Walker, and Montague 1995; Lawson 1997). Because each archaeological "phase" is recognized as a self-contained unit, it is deemed to have both integrity and coherence. In effect, each "phase" is assumed to represent a discrete architectural entity produced by a constructional episode.

There is clearly an attraction in viewing monuments in this manner. At one level it simply reflects our day-to-day understanding and experience of architecture. We are accustomed to see the building process as simply one of facilitation where architecture is built to be used. Almost overnight, what was a building site suddenly becomes a school, prison, house, or other such entity. Consequently, we tend to think about monuments, as we do other buildings, by privileging the completed form. There is an additional factor here, that of the building assuming a social role only upon completion. At another level, it allows archaeologists to impose a degree of coherence and order to what are frequently very incoherent bodies of evidence.

If, however, the creation of monumental architecture is recognized as having strategic social value and potentially representing an arena of social reproduction, as opposed to being merely a process of facilitation, then the whole nature of monumental construction requires rethinking. Lurking behind final forms or "phases" is the idea of construction as a *process* (see also McFadyen, this volume). Indeed, it could be argued that if the social strategies manifest in the construction process are a primary mechanism of social reproduction, then the creation of monumental architecture is simply an ongoing project of unlimited duration. Other discussions of the creation of monumental architecture have taken similar routes in emphasizing the building process as a continually unfolding project (e.g., Barrett 1994; Evans 1988; McFadyen 2006; Richards 2004) or a "messy" form of experimental practice (Turnbull 1993: 315–17). In each case, the potency of the ultimate monumental form is dissipated within the practical acts of building. Importantly, the building process is not restricted to any given place but is manifest as a series of tasks spreading outwards, weblike from the monument. Here, Ingold's (1993) idea of "taskscape" is of relevance as it not only emphasizes the temporality of engagements by people with the inhabited world but also stresses the specificity of the spatial and temporal rhythms of different tasks (Edmonds 1999).

Rather than minimizing the importance of monumental architecture, we wish to follow the implications of its construction and constitution as a process embodying weblike social strategies radiating from the monument. Here, the social relationships bound up in disparate practices ranging from weaving rope to quarrying stone, inevitably far removed in space and time, achieve a physicality in the monumental. Such practices also give rise to landscapes of construction.

So the challenge when rethinking "the monument" is to link processes of construction with the materiality and fluidity of "taskscapes" (Ingold 2000: 194–200) and the social relationships that are constituted and shaped through human labor. This is an interesting avenue of research in many

ways, not least because it ultimately takes us away from the supposed object of study; the monument, to the wider material world of everyday practice and experience, into a particular domain or view of the world conceived as landscapes of construction, and how in this context the idea of "taskscape" meshes with that of a richer understanding of landscape as a social medium (cf. Tilley 2004). Here we explore these issues with reference to Rapa Nui (Easter Island), which is world-famed for its spectacular megalithic architecture. In particular, we focus attention on a major element of the construction process: the megalithic quarry.

Composite Monuments: The *ahu* of Rapa Nui

Rapa Nui is one of the most isolated inhabited islands on earth. Lying as the easternmost island of Polynesia, approximately 3,200 kilometers west of the South American coast, Rapa Nui is actually quite small, having an area of 166 kilometers[2]. Taking triangular form, the island uplift is generated by three great volcanoes, the cones of which constitute each apex (Figure 16.1). From the high cliffs bounding the seaward edges of the three volcanoes, the land drops to form low coastal plains. The coastline is composed of black volcanic lava flows that rise up from the powerful Pacific surf. Inland, the steep cones of smaller volcanoes project upward across the island.

It is along the low barren coastline that the great monuments of Rapa Nui are situated. They take the form of elongated rectangular stone platforms (*ahu*), situated parallel to the shore. Over 300 *ahu* have been identified (Love 1993: 103) along the three low lying coastal strips to the southeast, northwest, and northeast of the island. Of these, 164 have been recognized as statue *ahu* (Martinsson-Wallin 1994: 52). These statues are the famed stone figures (*moai*), which make the monumentality of Rapa Nui one of the best known and recognized throughout the world. The heads of the *moai* were frequently adorned with cylinder-shaped stone topknots (*pukao*). The vast majority of *moai* are sculpted from a distinctive volcanic tuff quarried from the inner and outer surfaces of the Rano Raraku crater, situated to the southeast of the island (Figure 16.1). Conversely, the *pukao* are sculpted from red scoria quarried from the crater of Puna Pau, situated in the southwest of the island (Figure 16.1).

Chronologies for *moai* and *pukao* sculpting are inconclusive, because the quarries have produced little dating evidence. A lengthy period of quarrying activities at Rano Raraku is suggested by the

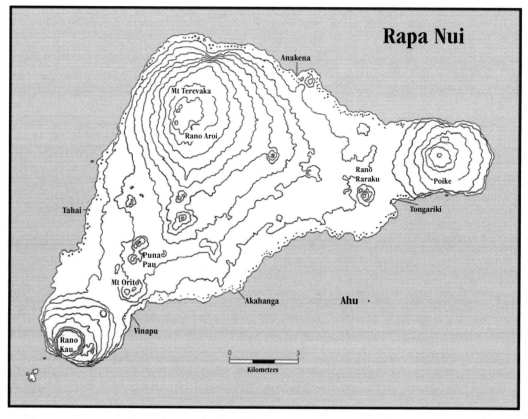

Figure 16.1 Map of Rapa Nui.

early date of A.D. 894–1035, obtained from contexts associated with the kneeling statue *Tukuturi* (Skjølsvold and Figueroa 1989: 32). No investigation has yet been undertaken of the *pukao* quarry at Puna Pau. If the "short chronology" for Rapa Nui, recently forwarded by Hunt and Lipo (2006), is accepted, then *moai* and *pukao* sculpting could have occurred only after initial colonization ca. A.D. 1200. This late date for colonization is all the more intriguing if quarrying at Rano Raraku ceased toward the end of the 14th century A.D. (Dumont et al. 1998).

Against a background of typological (e.g., Ferdon 1961: 527–33; Heyerdahl 1961: 497–502; Mulloy 1961: 159–60; Smith 1961: 218–19) and evolutionary ordering (Kirch 1984: 266–64; Mulloy and Figueroa 1978: 137–38; Stevenson 1986) of *ahu* construction, there is clear evidence for more localized and disjointed sequences of monument building (Love 1993: 110). This occurrence led to a high degree of structural complexity at particular *ahu*. Such complexity was clearly encountered by Mulloy (1961) during excavations at *ahu* Vinapu, and again at *ahu* Tahai, causing problems

of deciding what "phase" to actually conserve for public presentation (Mulloy 1997: 14). The continual sequence of building is nowhere better attested than at *ahu* Akahanga, on the south coast (Figure 16.2). Here, a lengthy sequence of building has been recognized from surface survey and mapping (Love 1993: 106–10; Van Tilburg 1994: 79–81). There are several features of this *ahu* that are of particular interest.

Ahu Akahanga actually consists of at least four different *ahu* (and probably more), the construction of which was well under way by ca. A.D. 1300 (Van Tilburg 1994: 79). Two further *ahu* platforms are present within 50 meters. Love (1993: 106) notes that "at least 4 easily visible Image *ahu* construction events are exposed, with architectural hints that these were originally built over 2 or 3 more." Just as with the attempted recognition of "phases," the attribution of "construction events" is equally problematic when applied to the building of *ahu*. The inadequacy of this form of analysis is latent in Love's statement: "From . . . the example of ahu Akahanga . . . it is obvious that currently there are so many undated construction events for

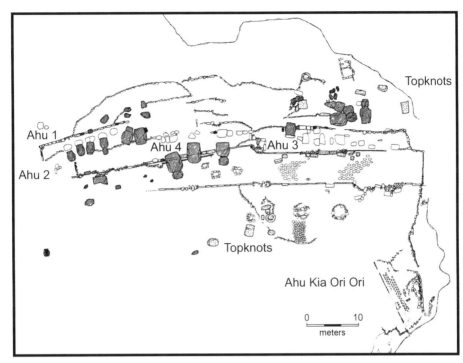

Figure 16.2 Plan of *Ahu* Akahanga (after Van Tilburg 1994 and Love 1993).

any given *ahu*, as well as a continuation and diversity of architectural elements, that even a sample of 14 excavated *ahu* cannot yet yield a significant evolution of Image *ahu* form" (1993: 110). Here the idea of process, the construction of particular *ahu* as ongoing projects, seems to better capture their architectural complexity.

Red scoria topknots (*pukao*), for the *moai*, and fascia for *ahu* embellishment, appear later in the building sequence (ibid.; Van Tilburg 1994: 81). Hence, while the *ahu* is constituted of local basaltic rock, the *moai* of volcanic tuff are being transported from Rano Raraku, and, later, red scoria *pukao* is added from the quarry at Puna Pau. In this respect, the monumental *ahu* are composite structures with the quarrying and transportation of materials derived from different places in the island. Consequently, the *process* of construction was central to the lives of prehistoric Rapa Nui islanders.

Rano Raraku: The Quarry as Monument

There is nothing on Rapa Nui that rivals the spectacular imagery of Rano Raraku, a broad volcanic cone rising from an open plain in the southeast of the island (Figure 16.3). This is the place where the vast majority of *moai* that adorned many of the *ahu* were created out of a distinctive brown-green volcanic tuff. The volcano has been sculpted in an exquisite manner by the shaping and extraction of many *moai*, but approximately 400 statues still remain in various forms at the quarry. Almost 80 stand erect on the lower slopes of the inner and outer crater in its southern sector. These *moai* face outward, gazing away from the quarry. Above, a belt of discrete quarries traverse the outer and inner slopes of the southern portion of the volcano. Approximately 160 apparently unfinished *moai*, lie in the quarries in various states of shaping.

Invariably, discussions of the quarry tend toward that of *moai* production as an almost "industrial" procedure (e.g., Flenley and Bahn 2002: 116–19; Skjølsvold 1961). After conducting excavations at the quarry in 1955–1956, Skjølsvold (1961) outlined several stages of production. First, the *moai* were shaped *in situ*, mostly in a supine position. Second, they were taken down to the lower slopes and erected vertically, where, third, their backs were dressed and the carving completed. Finally, they were dragged away from the quarry to adorn an awaiting *ahu*. In commenting on the imagery of Rano Raraku, Thor Heyerdahl concludes: "We are left with nothing but a series of production stages" (1961b: 504).

Figure 16.3 View of Rano Raraku from the south.

Significantly, the volcano slopes of Rano Raraku were not exploited as a single massive quarry face (as seen in many modern-day stone quarries) but were carefully subdivided into discrete units or compartments, which can be termed *quarry "bays"* (Figure 16.4). The bays, or niches, were described as having "been worked differently, and each has a character of its own" (Routledge 1919: 178). Equally, after having excavated in the quarry over a period of several months during 1955–1956, Skjølsvold came to similar conclusions: "each individual quarry . . . has its own particular character. In some places only the surface of the rock has been touched, while in others the work has gone relatively deep" (1961: 365).

The differences between quarry bays observed by Routledge and Skjølsvold clearly relate to variable working practices. However, such discrepancies are more than a product of differential "working" practices, and the quarry bays are better appreciated as architecture with very particular forms of spatial representation.

Overall, each quarry bay conforms to one of three discernable spatial arrangements. First, architecture of separation and exclusion where passage into substantial bays is highly restricted. This frequently takes the form of a narrow passage leading into a large open chamber, as seen in Bays

4 and 6 (Figure 16.5). It is this arrangement that led Routledge to comment that it "recalls the side-chapel of some old cathedral, save that nature's blue sky forms the only roof" (1919: 178). Second, a more open, deeply quarried space is presented where *moai* have been removed over a broader area. The imagery here is of an outcrop completely covered by *moai* in various states of carving, such as Bay 3 (Figure 16.6). The stunning spectacle of this architectural representation is clearly more than simply unfinished statues abandoned. Indeed, such imagery drew Metraux (1957: 156) to question the intention of their removal, likening them to huge petroglyphs. Routledge (1919: 181–82) also doubted that all statues shaped in the quarry were intended for removal, and despite the "production-line" perspective of Skjølsvold, he, too, considered that "it is possible to agree with Routledge that these are rock carvings" (1961: 365).

Third is an almost "shrinelike" architectural arrangement whereby the rock is left unworked to frame partially carved *moai*. This architecture is most apparent in Bay 1 (Figure 16.7), where two adjacent (but reversed) *moai* lay in a horizontal position. Metraux succinctly describes this arrangement as "a crypt patiently hollowed out with picks," where "a 50-foot colossus sleeps on a bed of stone" (1957: 156).

Figure 16.4   Quarry "bays" on outer crater of Rano Raraku.

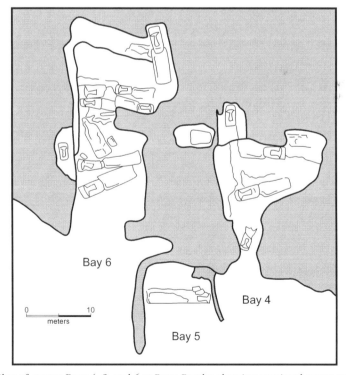

Figure 16.5   Plan of quarry Bays 4, 5, and 6 at Rano Raraku showing restricted access and internal *moai* (modified from Cristino, Vargas, and Izaurieta 1981).

Figure 16.6   Quarry Bay 3 at Rano Raraku.

Figure 16.7   Approach to Bay 1 at Rano Raraku (note the flanking spoil heaps).

These observations of the actual contexts of quarrying *moai* are interesting in that we cannot assume homogeneity of either imagery or practice, nor can we assume equivalence between bays. The encounter with each form of architecture clearly represented a different experience. Some bays have restricted access, others highly visible monstrous petroglyphs, both framed and unframed. Each bay is therefore a discrete entity being effectively severed from its surroundings. Exclusivity and secrecy were strong features. On approach, surrounding workings are gradually lost from sight, and on entering the bay nothing can be seen apart from the imagery of the immediate quarry—only the constant noise of chipping echoing around the slopes would betray the existence of other workers toiling in their bays.

Of the forms of shrinelike architecture, such as Bay 1, where the recumbent statue is "stretched out like a sleeping giant beneath a kind of dais carved in the tufa" (Metraux 1957: 156), additional features serve to enhance the approach. Through the excavations of Skjølsvold in 1955–1956, the entire topography at the base of the volcano is known to be composed of mounds of quarry debris. However, not all of these spoil heaps are randomly sited but are placed strategically. In the case of Bay 1, several spoil heaps flank the approach, providing an avenue effect. This not only channels movement up to Bay 1 but also directs vision toward the framed, darkened cavelike "shrine" (Figure 16.7). Approaching this bay would involve passage up the "avenue" and moving between several erect statues.

For some reason, the erect statues lining the lower slopes of Rano Raraku are generally considered to be unfinished examples, awaiting further sculpting and transportation (Heyerdahl 1961: 504; Routledge 1919: 188; Skjølsvold 1961: 369). However, Routledge (1919: 189) also considered, "a large number of the statues were intended to remain *in-situ*." This view is also favored by Van Tilburg (1994: 146) who observed: "The numbers and sizes of most of these statues could not have been accommodated on the existing *ahu*, none of which are prepared to receive them." A consequence of this suggestion is that the statues at Rano Raraku were erected permanently and strategically at the quarry. That these *moai* were integral to the architecture of Rano Raraku is demonstrated by examples standing within the inner crater. Here statues stand, partially covered by stone debris directly outside visible lower quarry bays.

The presence of upright *moai* at the base of the quarry was not a product of incomplete or unfinished working practices (*contra* Heyerdahl

1958: 88). Instead, they presented a monumental facade, structuring and grading the pathways to the quarry bays. Passage to work the stone clearly involved a change in state, effected by a series of architectural devices incorporating statues, spoil heaps, and bay architecture. But when and where did this transformatory process begin and end?

Dwelling within Landscapes of Construction

Today, the vegetation cover is mainly restricted to low grasses providing magnificent island-wide vistas and making such topographic features clearly visible. Yet, this was not always the case. Through root moulds first discovered in excavations by Mulloy and Figueroa (1978: 22), and botanical work (Flenley 1993; Flenley and Bahn 2002: 78–88; Flenley et al. 1991), a very different island is disclosed when the great monuments were being constructed. This island was covered by swaying groves of palm trees, if not forests, with palms up to a meter in trunk diameter and reaching up to 20 meters in height (Flenley and Bahn 2002: 84; Van Tilburg 1994: 47). Indeed, when Rapa Nui was first inhabited, some time between A.D. 800 and 1200 (Hunt and Lipo 2006; Martinsson-Wallin and Crockford 2002), "it must be concluded that rainforest existed on Easter Island" (Flenley and Bahn 2002: 87).

Potentially through this partial vegetation, roads were built radiating from Rano Raraku to different areas of the island (see Lipo and Hunt 2005). Much attention has been placed on their role in the transportation of *moai* from the quarry to *ahu* (e.g., Lipo and Hunt 2005: 164; Love 2000: 118). However, a reverse logic may be applied where the roads actually all lead toward the great quarry, thereby fixing Rano Raraku at the center of the Rapa Nui world: an *axis mundi*. Passing different landmarks and changing vistas, the route toward the sacred center is a highly structured journey that becomes monumentalized as the goal is approached. Frequently, described as "in transit" (e.g., Van Tilburg 1987: 33), many *moai* lie apparently abandoned across the land and along the network of roads (Shepardson 2005: 171; Figure 16.8). On one road alone, up to 27 statues were observed by Routledge (1919: 194). Through excavation, Routledge (1919: 195) eventually concluded that many of the *moai* had been set upright adjacent to the roads, facing away from the quarry. This situation was confirmed by Heyerdahl's (Heyerdahl, Skjølsvold, and Pavel 1989: 45–56) later excavations, where at least one *moai* was found to have stood on a stone platform. As Routledge (1919: 196) evocatively observes:

"Rano Raraku was, therefore, approached by at least three magnificent avenues, on each of which the pilgrim was greeted at intervals by a stone giant guarding the way to the sacred mountain."

However, those making this journey were not all "pilgrims," they were ordinary people engaging in the everyday tasks and activities that constituted their lives. However, the world through which they passed was ordered by social and spatial networks grounded in the process of monument construction. The roadways radiating to and from Rano Raraku may have been constructed for statue transportation but would also operate as routes for ceremonial journeys and more mundane activities. In potentially isolating points of changing and heightened visual and other sensory experience, to travel these formal routes was of far greater consequence than to merely follow functional transport roads. Each route had a unique character, and in many ways within this network was "sedimented the activity of an entire community, over many generations. It is the taskscape made visible" (Ingold 2000: 204).

For those who lived and died on Rapa Nui during the statue-erecting period, the experienced world was organized within a cosmological system that was constituted and revealed by the never-ending projects of quarrying stone, moving huge statues, and building great monuments. To dwell in Rapa Nui was essentially to labor with monumentality.

Conclusions

In this chapter, we have portrayed the monuments of Rapa Nui as being constantly in a state of flux. Indeed, all the evidence points toward *ahu* platforms being constantly enlarged and embellished. Even *moai* appear to be involved in an ongoing process of displacement and replacement, with previous *moai* being incorporated in the body of the *ahu*. This process of construction is not restricted to the "monument" but on the small island of Rapa Nui effectively constituted an understanding of landscape as a web of practices associated with monument building. Not only did such practices radiate across the land from the monument to the source quarries within a formalized network of roads, but also the organization of monuments, roadways, and quarries provided a material manifestation of more abstract social and cosmological interrelationships. Construction is a fluid process and consequently landscapes of construction are also

Figure 16.8 Collapsed *moai* adjacent to a roadway.

diverse and fluid; moreover, within their composition is a physicality of the more abstract and intangible aspects of social life. In this respect, landscapes of construction are not merely taskscapes but also provide revelatory experiences to those who inhabit them.

References

Atkinson, R. 1956. *Stonehenge*. London: Hamish Hamilton.

Barrett, J. 1994. *Fragments from Antiquity: An Archaeology of Social Life in Prehistoric Britain*. Oxford: Blackwell Publishing.

Cleal, R., Walker, K., and Montague, R. 1995. *Stonehenge in Its Landscape*. London: English Heritage Monographs.

Cristino, C., Vargas, P., and Izaurieta, R. 1981. *Atlas arqueológica de Isla de Pascua*. Santiago: Facultad de Arquitectura y Urbanismo, Instituto de Estudios, Universidad de Chile.

Dumont, H. J., Cocquyt, C., Fontugne, M., Arnold, M., Reyss, J.-L., Bloemendal, J., Oldfield, F., Steenbergen, C. L. M., Korthals, H. J., and Zeeb, B. A. 1998. The end of *moai* quarrying and its effect on Lake Rano Raraku, Easter Island. *Journal of Paleolimnology* 20: 409–22.

Edmonds, M. 1999. *Ancestral Geographies of the Neolithic*. London: Routledge.

Evans, C. 1988. Acts of enclosure: A consideration of concentrically organized causewayed enclosures, in J. C. Barrett and I. Kinnes (eds.), *The Archaeology of Context in the Neolithic and Bronze Age*, pp. 85–96. Sheffield: J.R. Collis Publications.

Ferdon, E. N., Jr. 1961. A summary of the excavated record of Easter Island prehistory, in T. Heyerdahl and E. Ferdon, Jr. (eds.), *Reports of the Norwegian Archaeological Expedition to Easter Island and the East Pacific*, Vol. 1: *The Archaeology of Easter Island*, pp. 527–36. London: Allen and Unwin.

Flenley, J. R. 1993. The present flora of Easter Island and its origins, in S. R. Fisher (ed.), *Easter Island Studies*, pp. 7–15. Oxford: Oxbow Monograph 32.

Flenley, J., and Bahn, P. 2002. *The Enigma of Easter Island*. Oxford: Oxford University Press.

Flenley, J. R., Teller, J. T., Prentice, M. E., Jackson, J., and Chew, C. 1991. The late Quaternary vegetational and climatic history of Easter Island. *Journal of Quaternary Science* 6: 85–115.

Heyerdahl, T. 1958. *Aku-Aku: The Secret of Easter Island*. London: George Allen and Unwin Ltd.

———. 1961a. Surface artefacts, in T. Heyerdahl and E. Ferdon, Jr. (eds.), *Reports of the Norwegian Archaeological Expedition to Easter Island and the East Pacific*, Vol. 1: *The Archaeology of Easter Island*, pp. 397–489. London: Allen and Unwin.

———. 1961b. General discussion, in T. Heyerdahl and E. Ferdon, Jr. (eds.), *Reports of the Norwegian Archaeological Expedition to Easter Island and the East Pacific*, Vol. 1: *The Archaeology of Easter Island*, pp. 493–526. London: Allen and Unwin.

Heyerdahl, T., Skjølsvold, A., and Pavel, P. 1989. The "walking" *moai* of Easter Island, in A. Skjølsvold (ed.), *Occasional Papers of the Kon Tiki Museum* 1: 36–64.

Hunt, T. L., and Lipo, C. P. 2006. Late colonization of Easter Island. *Science* 311: 1603–06.

Ingold, T. 1993. The temporality of the landscape. *World Archaeology* 25: 152–74.

———. 2000. *The Perception of the Environment: Essays in Livelihood, Dwelling and Skill*. London: Routledge.

Kirch, P. V. 1984. *The Evolution of Polynesian Chiefdoms*. Cambridge: Cambridge University Press.

Lawson, A. J. 1997. The structural history of Stonehenge, in B. Cunliffe and C. Renfrew (eds.), *Science and Stonehenge*, pp. 15–38. Oxford: Oxford University Press for the British Academy.

Lipo, C. P., and Hunt, T. L. 2005. Mapping prehistoric statue roads on Easter Island. *Antiquity* 79: 158–68.

Love, C. M. 1993. Easter Island *ahu* revisited, in S. R. Fisher (ed.), *Easter Island Studies*, pp. 103–11. Oxford: Oxbow Monograph 32.

———. 2000. More on moving Easter Island statues, with comments on the Nova program. *Rapa Nui Journal* 14(4): 115–18.

Martinsson-Wallin, H. 1994. *Ahu—The ceremonial stone structures of Easter Island*. Uppsala: Aun 19.

Martinsson-Wallin, H., and Crockford, S. 2002. Early settlement of Rapa Nui (Easter Island). *Asian Perspectives* 40: 244–78.

McFadyen, L. 2006. Material culture as architecture. *Journal of Iberian Archaeology* 8: 91–102.

Metraux, A. 1957. *Easter Island*. London: André Deutsch.

Mulloy, W. T. 1961. The ceremonial centre of Vinapu, in T. Heyerdahl and E. Ferdon, Jr. (eds.), *Reports of the Norwegian Archaeological Expedition to Easter Island and the East Pacific*, Vol. 1: *The Archaeology of Easter Island*, pp. 93–180. London: Allen and Unwin.

———. 1997. *The Easter Island Bulletins of William Mulloy*. Houston, TX: Easter Island Foundation.

Mulloy, W. T., and Figueroa, G. 1978. *The A Kivi—Vai Teka Complex and Its Relationship to Easter Island Architectural Prehistory*. Asian and Pacific Archaeology Series 8. Honolulu: University of Hawai'i Press.

Richards, C. 1996. Monuments as landscape: Creating the center of the world in late Neolithic Orkney. *World Archaeology* 28: 190–208.

———. 2004. A choreography of construction: Monuments, mobilization and social organisation in late Neolithic Orkney, in J. Cherry, C. Scarre, and S. Shennan (eds.), *Explaining Social Change: Studies in Honour of Colin Renfrew*, pp. 103–14. Cambridge: McDonald Institute Research Monograph.

Routledge, K. 2005 [1919]. *The Mystery of Easter Island*, Rapa Nui: Museum Press.

Shepardson, B. L. 2005. The role of Rapa Nui (Easter Island) statuary as territorial boundary markers. *Antiquity* 79: 169–78.

Skjølsvold, A. 1961. The stone statues and quarries of Rano Raraku, in T. Heyerdahl and E. Ferdon, Jr. (eds.), *Reports of the Norwegian Archaeological Expedition to Easter Island and the East Pacific,* Vol. 1: *The Archaeology of Easter Island*, pp. 339–79. London: Allen and Unwin.

Skjølsvold, A., and Figuerroa, G. 1989. An attempt to date a unique, kneeling statue in Rano Raraku, Easter Island. *Occasional Papers of the Kon Tiki Museum* 1: 7–35.

Smith, C. S. 1961. A temporal sequence derived from certain ahu, in T. Heyerdahl and E. Ferdon, Jr. (eds.), *Reports of the Norwegian Archaeological Expedition to Easter Island and the East Pacific,* Vol. 1: *The Archaeology of Easter Island*, pp. 181–218. London: Allen and Unwin.

Stevenson, C. M. 1986. The socio-political structure of the southern coastal area of Easter Island: A.D. 1300–1864, in P. V. Kirch (ed.), *Island Societies: Archaeological Approaches to Evolution and Transformation*, pp. 69–77. Cambridge: Cambridge University Press.

Thomas, J. 1992. The politics of vision and the archaeologies of landscape, in *Landscape, Politics and Perspectives*, B. Bender (ed.), pp. 19–48. Oxford: Berg.

Tilley, C. 1994. *The Materiality of Stone*. Oxford: Berg.

———. 2004. *A Phenomenology of Landscape*. Oxford: Berg.

Turnbull, D. 1993. The ad hoc collective work of building gothic cathedrals with templates, string and geometry. *Science, Technology and Human Values* 18(3): 315–40.

Van Tilburg, J. A. 1987. Symbolic archaeology on Easter Island. *Archaeology* 40(2): 26–33.

———. 1994. *Easter Island: Archaeology, Ecology and Culture*. London: British Museum Press.

17

OBJECT FRAGMENTATION AND PAST LANDSCAPES

John Chapman

The core idea of this chapter is that people, places, and things are self-constituting in a material network symbolizing collective memory and reinforcing social relations. I address this idea by introducing the notion of fragmentation studies to the domain of landscape archaeology. I begin by defining two key, articulating concepts, *fragmentation* and *enchainment*, and then turn to the evidence for deliberate acts of fragmentation and enchainment in prehistory.

To Fragment or Not to Fragment?

One of the principal characteristics of material culture that archaeologists discover is that it is usually broken. We have become so accustomed to this state of affairs—either as excavators or when studying museum collections—that broken things do not appear to be abnormal, interesting, or curious (Chapman and Gaydarska 2006). Any concept that disturbs an idea that is so deeply rooted in our *habitus*—our unspoken set of assumptions about how our world operates—will inevitably provoke resistance, scorn, or worse. One such deeply rooted idea is the notion that broken things are nothing but the result of accidental breakage or taphonomic processes—in other words, processes unrelated to human intentionality. This idea persists in many archaeologists' minds, despite the increasing

acceptance of the active use of material culture, a notion that has been one of the main breakthroughs of postprocessualism (Hodder 1982). When it comes to broken things, agency and social practices are rapidly forgotten in favor of the old chestnut that "archaeology is concerned with the rubbish of past generations" (quoted in Thomas 1991: 56). Instead of using this outdated foundation myth, I have maintained that archaeology is, rather, the "science of deposition" (Chapman 2000) and that a vital part of the reorientation that follows from this reconceptualization of our discipline is that things that are broken are not necessarily just discarded, tossed away, or dumped. That "things are merely rubbish" may be treated as a null hypothesis, but it is one that is increasingly capable of falsification.

Fragmentation

In my first book on fragmentation, I listed and discussed five possible causes of breakage (Chapman 2000b: 23–27):

1. Accidental breakage;
2. Objects buried because they are broken (e.g., Garfinkel 1994);
3. Ritual "killing" of objects (e.g., Grinsell 1960; Hamilakis 1998);

4. Dispersion to ensure fertility (e.g., Bausch 1994);

5. Deliberate breakage for reuse in enchainment.

I would accept that all these causes operated in the past, as well as the obvious taphonomic processes that can and do break things. However, the key point that arises at a certain scale of spatial closure—a grave, a hoard, a burnt house assemblage, a pit, a bounded midden, and so forth—is that none of the first four processes or practices can explain the absence of parts of the broken thing. Hamilakis's (1998) study of ritual killings of swords and spears in Minoan Crete describes complete, if broken, objects in graves and pits; the same is true for Garfinkel's (1994) study of the breakage of objects to remove their diminishing ritual power. I would continue to maintain that, for closed contexts, the phenomenon of the missing part is a good indication of deliberate object breakage. For more open settlement contexts, the complications are far greater (cf. Hayden and Cannon 1983), and this is an important area of ongoing research.

Enchainment

The notion and significance of "enchainment" can be summarized in three main points:

1. Enchainment mobilizes the identity triad of persons, places, and things through presencing;

2. Enchainment is the best, and sometimes the only, explanation for deliberate fragmentation;

3. Enchained relations subsume concepts such as curation, tokens, ancestral veneration, heirlooms, and relics.

In a recent discussion of the uses I have made of the concept of enchainment, Fowler (2004: 68) correctly observes one fundamental difference between Melanesian enchainment and fragment enchainment. In Melanesia, the object enchained through gift exchange is not held by different persons at the same time, but its materiality creates a sequence of giving and counter-giving, while fragments are held coevally by different persons. This does not mean that whole-object enchainment does not exist in societies that practice fragment enchainment—merely that there are two different but related social practices embodied in the material remains. In each case, enchainment remains at the general level of social practice—the challenge is to refine the links between persons and things for each specific cultural context. In the course of this chapter, I shall seek to exemplify each of these points. But, in general, it is the contention of much of my recent research that deliberate fragmentation is a fundamental feature of not only later Balkan prehistory but also of communities living in many other times/places (Chapman 2000b, 2000c, 2000d, 2000e). The evidence for deliberate fragmentation is increasing each year, both at the level of intersite data and intrasite data, such that the social practice can no longer be ignored by anyone seriously interested in material culture (Chapman and Gaydarska 2006).

Fragmentation and the Landscape

How do the concept and the social practice of deliberate object (and body) fragmentation map onto landscape archaeology—onto landscapes? Landscapes consist of a network of places—some natural, some culturally constituted, some created by human manipulation of the landscape. It is this network of places that gives human lives their meaning, through an identification of past activities and present embodiment. The key element of landscape archaeology is thus the relationships between different places. Whenever fragment dispersion is mapped onto places, the practice of fragmentation can be linked to landscape archaeology. This practice is but one of a series of practices constituting "inhabitation" (see Thomas, this volume).

According to John Barrett, inhabitation is not merely "occupying" a place but understanding the relevance of actions executed at that place by reference to other frames of reference, other time, other places (Barrett 1999: 258–60). Enchained social relations provide one such key frame of reference, because, following Mauss (1954), each gift carries within it the history of all previous gift exchanges. If enchainment presences absent people, fragments of things, and places, it is fundamental to the process of inhabitation as described by Barrett. The symbolism of *pars pro toto* sustains a form of fractal personhood that is dispersed across the landscape and that itself acts as a means of bringing that landscape into being. But how can the notion of enchainment dispersed across the landscape be demonstrated?

Direct Evidence: Intersite Refitting

At the methodological level, the key linking concept between fragmentation and landscape studies is that of fragment dispersion. Although the previous claim that we should at least consider the possibility of trade and exchange based on fragmentary objects (for example, exotic sherds rather than complete exotic vessels) (Chapman 2000b: 63–65) is still valid, if difficult to prove, there are two important methods that implicate movement of fragments across the landscape—at spatial scales that are becoming increasingly possible to define. The first method is the intersite refitting of fragments from the same object (see below). Here, two places in the landscape are linked by the parts of a broken object in an enchained, dispersed relationship. It has also been found that, even after refitting, the object is still missing some parts, so the enchained dispersion can be assumed to go still further, linking a third or more places. A second method is to focus on completely excavated sites or phases with good to excellent recovery and to examine the Completeness Index of the objects. If there are no other practices that would destroy ceramics (for example, the use of grog temper) or removal from the site (for instance, manuring scatters), one can argue that the missing fragments were taken off the site and dispersed across the landscape. In practice, each of these methods depends on the completeness and the recovery standards of the excavations in question; therefore selection of potential study collections requires enormous care. But both methods have recently been shown to demonstrate fragment dispersion across the landscape, giving added precision to the generally accepted notions of exchange networks and/or mobility. In this way, we can begin to link fragmentation studies to the landscape concerns that form the focus of this research handbook.

The most convincing example of intersite fragmentation concerns objects larger than the usual artifacts, such as geometrically decorated stone slabs forming the most elaborately decorated parts of megalithic monuments in Brittany; these represent some of the earliest monumental remains in western Europe (Scarre 2002). Le Roux (1992) has demonstrated that some of the motifs on the decorated stones at Gavrinis were carved before the incorporation of the stones into the passage. Scarre (2005) suggests that these stones once stood elsewhere and were "re-erected" in the gallery of motifs that formed Gavrinis. Moreover, there are several examples of large stone slabs whose engraved patterns were broken across the image, with one part built into one monument and the other half used to construct a second tomb. The best-studied is the menhir decorated with a bovid, fragmented into three pieces, one of which was built into each of the megaliths of Gavrinis, Er-Grah, and La Table des Marchand, respectively, 3 kilometers apart (L'Helgouach and Le Roux 1986; Figure 17.1), but other pairs of megaliths

linked through fragments of engraved stones have also been found (Calado 2002: 26, 30). A recent discovery next to La Table des Marchand is a menhir-shaping site, where débitage from the rechipping of menhirs brought onto the site from elsewhere has been found in postholes in the posthole row near the passage grave (Loïs Langouet, personal communication 2002). Thus, the four great fragments of Le Grand Menhir Brisé may have been broken deliberately, ready for onward transport to four different megaliths—except for a change of plan.

What are the implications of these material links, the most monumental examples of enchainment yet discovered in Europe? We can integrate the three elements of people, places, and things by emphasizing the embodied nature of these practices. The first implication concerns the design of the paired megaliths. Since it is clear that not every stone block would have fitted into a place in the passage of these megaliths, an agreed-on design was required in advance for the part of the tombs incorporating the broken blocks. This meant several meetings and several trips between the pairs of megaliths for several builders to ensure the design would produce the desired effect and that the stones were broken to approximately the correct dimensions. Second, the transport of the stone blocks—perhaps from a third site, certainly from one megalith to the other and presumably to one more hitherto undiscovered megalithic monument (the missing third piece of the menhir). In the case of Gavrinis, Er-Grah, and La Table des Marchand, the blocks of stone each weighed several tons and required land and river transport over 3 kilometers (at the time, Gavrinis was not an island). This would have brought together numerous individuals—perhaps 20 to 30 people, likely mostly males—from several dispersed communities, with the task of bringing all team members together in a coordinated display of embodied skills. The enchained relations developed through these tasks were surely not a one-off practice but led to longer-term social relations cemented by the paired stones. The places with enchained links included not only the settlements of the team members but also the source of the rock, the places visited *en route,* and the final burial places of the decorated stones. The processions across the landscape, embodying the formal movement of the stones, linked other megaliths with their own ancestral place-values, as well as integrating stretches of other paths perhaps not related before in a single route. The people whose bones were later stored in the paired megaliths were also enchained to those who made the link between the megaliths material in the first place. What is

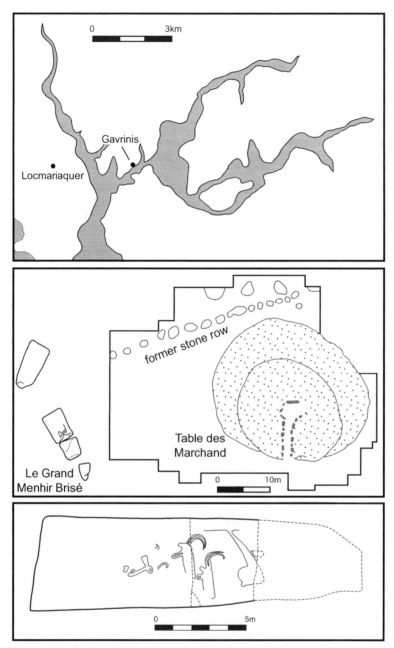

Figure 17.1 The refitted menhir from Gavrinis, Er-Grah, and Le Table des Marchand.

implied, therefore, by megalithic-scale enchain-ment is a complex network of social relations, practices, places, and things that had temporal and spatial scales and limits, while at the same time emphasizing specific ancestral and lineage con-nections (see also McFadyen, this volume). These monuments are the largest example yet known of fragment dispersion across the landscape.

How widespread is this practice? It is not yet possible to give an adequate answer to this question, because no systematic investigations into fragmentation of stones have been made in places such as Val Camonica, Monte Bego, and other major rock art sites. However, new research by Emmanuel Mens (2008) has demonstrated the potential of linking standing stones in the Carnac alignments to the sequence of extraction from neighboring outcrops; surely, the next step is to make links between stones in a mega-refitting operation! Moreover, the incorporation of decorated, cup-and/or-ring-marked stones into Late Neolithic pits in henges and into Early Bronze Age cist graves in Britain has revealed a number of fragmented stones broken across the motif(s) (Bradley 2002; Waddington 1998: 43–45), indicating that the practice of megalithic fragmentation is by no means limited to Neolithic Brittany.

Turning to smaller objects, I next discuss five cases of artifact dispersion across landscapes: four prehistoric and one from the early historic period. The earliest case of intersite refitting is reported from the Gravettian of the Achtal, in Germany, where fragments of the same lithics refitted from four different cave sites, two of which—the Geissenklösterle and the Hohler Fels—are 3 kilometers apart (Gamble 1999: 326–27, fig. 6.18; Scheer 1990; see Figure 17.2). The excavators interpreted this practice as a way of characterizing the spatial dispersion of the *chaîne opératoire* of Gravettian core reduction by mobile groups moving from cave to cave. What it also shows is that the enchained identity of these groups was materialized by discard in several linked places across the landscape.

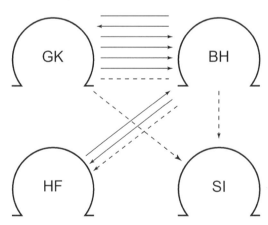

Figure 17.2 Refitting Gravettian flints in the Achtal (source: Scheer 1990: Abb. 9).

A later lithic refitting study showed similar refits between sites occupied for short periods at the Mesolithic-Neolithic transition in the South Norwegian Highlands (Schaller-Åhrberg 1990). Total or almost complete excavation of six small sites located around the shore of Lake Gyrinos (lake dimensions: 5 x 2 kilometers) produced radiocarbon dates spanning the 7th–5th millennia cal B.C. Overall, 35% of the flint material could

be sorted into 12 different groups, each deriving from a single core. Apart from the off-white flint for axe making, each of the other 11 groups had at least two and up to 48 conjoining flakes (Schaller-Åhrberg 1990: 617–20, figs. 4–7). Lithics from ten of the groups were found at two or more sites, with stronger links between several pairs of sites (Schaller-Åhrberg 1990, fig. 8; see Figure 17.3). The excavator interprets these refits as indicating coeval occupations of sites rather than the use of earlier sites as raw material sources (Schaller-Åhrberg 1990: 620–21). No social interpretations were proposed, although the materialization of enchained relations between groups not necessarily in daily contact would be a valuable social practice.

The refitting of two anthropomorphic figurine fragments from two different settlement foci (whether these foci, 230 meters apart, are two separate "sites" is a matter for local definition) has already been reported for the Shakado complex in Japan, dating to the Middle Jomon phase (Bausch 1994: 92, 108; Chapman 2000b: 26–27). Fifteen other cases of refitting figurine fragments have been found at Shakado, but the spatial dispersion is less than 230 meters. But, at 3500–2500 B.C., this remains one of the earliest examples in which the human form is deliberately broken and the fragments used in different places. It seems improbable that we are dealing with accidental refuse or the result of children's playing, since the conjoining fragments were carefully placed, together with other unusual things, in different special disposal areas, termed *dokisuteba* (Bausch 1994).

A startling example of intersite refitting at the landscape scale concerns the Trent Valley Ewart Park bronze sword fragments from Hanford and Trentham (Bradley and Ford 2004, fig. 20.1; see Figure 17.4). The lower part of a bronze sword was found at Trentham in the early 1990s by a metal-detector user who passed the find to the Burton-on-Trent Museum. Thirteen years later, another person with a metal detector, working on a hilltop on the other side of the Trent Valley at Hanford, found part of a bronze sword and also brought it into the museum. The two fragments fitted to make an almost complete bronze sword (the hilt is still missing). The two hilltops were 5 kilometers apart and intervisible across the valley. The deposition of bronze swords is characteristic of the British Late Bronze Age, but this is the first time that anyone has tried refitting sword fragments together from different "sites." The intervisibility of the places makes it possible to conceive of a simultaneous deposition of the sword fragments, constituting a landscape link between two ritual

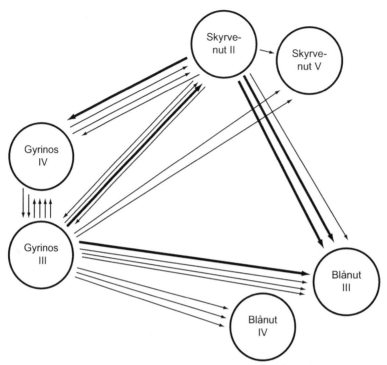

Figure 17.3 Chipped stone refitted between Mesolithic-Neolithic sites around Lake Gyrinos, Norway
(source: Schaller-Åhrberg 1990: fig. 8).

foci in the Trent Valley. But such coeval practices are not necessary to support the enchained links among people, places, and objects. There were significant differences in the life histories of the two fragments—the swordpoint was worn and the fracture was more rounded in the lower part than in the upper. This suggests that breakage had preceded deposition by a significant period of time—perhaps years—suggesting that sword fragments had an independent life of their own in the routine practices of the valley. An intriguing possibility is that Late Bronze Age swords may have acted as metaphors for people, indicating the parallels between deliberate fragmentation of things and of the human body (Williams 2001).

In a final example, the elaboration of pottery decoration offers the potential for different persons to recognize that "their" fragment, if broken across the motif, is linked to another piece of the same vessel. The refitting process has been used extensively in intrasite ceramic studies, most frequently for purposes of stratigraphic linkage (see Hoffman and Enloe 1992; Macfarlane 1985: Appendix B for a good example). But intersite refitting of decorated sherds is a rare discovery, requiring much time and effort. The only instance known to this author is the example of refitted Roman-period Samian bowls in Holland (Brandt 1983; Vons and

Bosman 1988). The two Early Roman forts established at Velsen were occupied for only a short time (Velsen I between A.D. 15 and 28; Velsen II between A.D. 40 and 50), before changes to the defense of the *limes* elsewhere (Brandt 1983: 132). Refits have been made at 80% probability between (1) decorated fragments of a Samian bowl (Dragendorff Type 27) found, respectively, at Velsen I and Hoogovens site 21 (Vons and Bosman 1988: figs. 2–4); (2) decorated fragments of a Samian bowl (Dragendorff Type 29) found at *two* sites—Velsen I, 't Hain site 39 and Roman Nijmegen (Vons and Bosman 1988: figs. 5–8); and (3) decorated body sherds of another example of the Dragendorff 29 type found at Velsen I and on the Roman site beneath the Medieval castle of Brederode, on a sand ridge 4.5 kilometers southwest of Velsen (Bosman 1994). The refits between Velsen I and the three local sites indicate movement across some 3 to 8 kilometers.

The initial interpretation has been that the local population raided the abandoned fort for valuable materials and, in the process, also removed small pottery fragments to take back to their settlements as trophies—the "pick-up" explanation (Bogaers 1968). This notion would indicate the value placed on incomplete objects, perhaps for reuse. Research on ethnographic discard among the Maya has shown that large sherds were often reused as

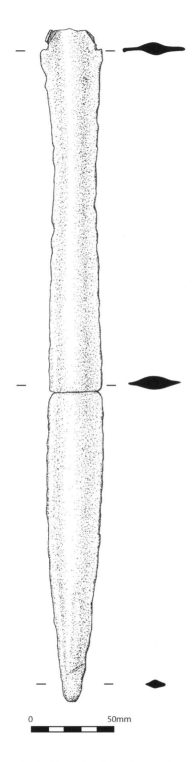

Figure 17.4 The refitted Late Bronze Age sword from the Trent Valley, English Midlands (source: Bradley and Ford 2004: fig. 20.1).

scoops or spoons (cf. Hayden and Cannon 1983). But these highly decorated, distinctively colored Samian sherds were so different from the local potting tradition that a symbolic element in the appropriation of the material culture of the "Other" is surely present. They were also too small for reuse as scoops.

A second explanation is that soldiers at the Velsen fort entered into enchained relations with local populations through exchange of things complete *and* incomplete (Brandt 1983). Brandt (1983: 137) lists the classes of Roman objects found on "local" sites: glass, fibulae, "scrap" metalwork, and pottery, including 350 decorated Samian sherds—all small and many eroded. Interestingly, each small Samian sherd came from a different vessel, as did the glass fragments. These observations prompted Brandt to discuss the possibility of deliberate fragmentation of Samian vessels, perhaps for use as primitive money (their wear suggests frequent use), as ritual objects (their red color and sheen) or for redistribution as fragments by local elites (1983: 140–42). The refitting demonstrates that at least some Samian vessels were broken at Velsen II but this does not rule out further redistribution of fragments in local communities. Bosman (1997) and Vons and Bosman (1988) return to the issue, criticizing Brandt's explanations of primitive money and exchange by arguing that the dating evidence does not place the sherds on local sites at the same date as the occupation of the Velsen forts. The latter prefer a "pick-up" explanation for all the Roman finds found on what they claim to be *later* (that is, post-Velsen I and II) local sites, things that now also include amphora, mortaria, and grinding stone fragments. The chronological issue can be settled only by multiple AMS radiocarbon dating of the local sites.

In one sense, it does not matter whether the refitted Samian sherds were traded or picked up. Here is a well-documented case of intersite refitting across an "ethnic" division, by which the material identity of an invading population is valued sufficiently by the local Indigenous groups to promote the further use of *fragments* as trinkets or heirlooms after breakage. These refittings document links across this low-lying landscape that enable the closer identification of the mechanism of dispersal.

The most extensive interstice refit of all is the famous lithic refit from the Chuckwalla Valley, Southern California (Singer 1984). Detailed technological analysis of lithics from a series of surface sites characterizing the local prehistoric groups, without clear dating, revealed a refit between an almost complete quartzite pebble at a quarry site and a decortifiction flake at a discard site 63 kilometers down the valley. To my knowledge, this is the longest intersite refit in the world. The implications include widespread movement of lithics by groups with enchained relations to both sites.

There are several important implications of intersite refitting. The first is that, because of the distance between sites, we can exclude accidental discard, the movement of objects through cleaning of structures, dumping, and children's play. This means that we have a number of documented cases of deliberate fragmentation with subsequent discard in different places. Second, although the rationale behind the movement of fragments between sites is different (contrast Velsen I and the Achtal caves), there remains an important commonality of links among people, places, and broken things, by which the identity of each of the three elements is mutually constituted in relation to the other two elements. Third, in several cases (the Achtal caves, Shakado, Trent Valley), the notion of enchainment applies to the fragmentation practices, suggesting a certain kind of relationship between people and things that involves the objectification and fractality of personhood. This has major consequences for the way we view past material and social worlds. Fourth, and of particular relevance to this volume, intersite refitting means that fragmentation has a landscape dimension—a result leading to the theorization of materiality and movement *across* the landscape, not only at specific places *in* the landscape. Issues include the extent to which broken objects are enchained to places along the routes between starting-point and destination (for example, of a decorated megalithic panel). Finally, the broad time/place span of these practices suggests that, even if the examples are as yet few in number, this could be a significant practice over a much wider range of Eurasian prehistory than we currently have evidence for. Naturally, this cannot be documented at the present time but the questions raised set an archaeological agenda for future investigations.

It would be disingenuous to deny the problems any researcher faces in seeking intersite refitting: (a) a large, if not huge, labor input—with the potential for no results; (b) few examples of previous good practice; and (c) the potential obstacles placed in the way of the research by museum or other institutional directors not permitting collections to be moved to nearby locations for refitting tests. Nonetheless, none of these problems is insuperable; in any case, they are surely not much greater than those working in Eurasian archaeology face on an everyday basis. Of particular value for refitting studies are national museums in smaller nation-states (for instance, Slovenia, Eire, and Latvia), whose collections include the majority of archaeological finds from most of the key sites.

Indirect (Intrasite) Evidence from Completely Excavated Sites

The evidence for intersite refitting currently derives from a restricted number of cases—primarily because few archaeologists have looked for the material evidence indicating the practice of deliberate fragment dispersion. Since the complexities of research are somewhat reduced, there are more examples of intrasite investigations into object fragmentation than of intersite studies. In this chapter, only those examples detailing the results of complete or nearly complete excavations of sites or phases are discussed. If high levels of object recovery can be demonstrated, and this is not always possible, a case can be made for several fascinating social practices centered on the differential use of fragments and, indeed, fragment dispersion across the landscape.

There are three social arenas in which intrasite refitting has been attempted: the separate domestic domain, the separate mortuary arena, and conjoint studies of both domains. Study of the domestic domain is made particularly challenging by the combination of the diversity of contexts, few of which are "closed," and the variety of taphonomic processes in operation. The assumption of intentional deposition in contexts such as megalithic tombs, barrows, cairns, and graves, where there can be a greater degree of closure, would appear to make refitting in the mortuary arena more straightforward. I present an example from each domain, beginning with a domestic settlement.

Dolnoslav. The Dolnoslav is a tell located in the Maritsa Valley, in the southern part of the Thracian plain. There are two main prehistoric horizons—the earlier, dating to the Early Neolithic, has not been investigated at all, whereas the Final Eneolithic horizon has been almost totally excavated by A. Raduntcheva and B. Koleva in the 1980s (Koleva 2001, 2002; Raduntcheva 1996, 2002). Hand excavation was used, but no sieving or flotation was carried out; the size of the recovered figurine fragments leads us to believe that only small fragments (< 2 cm in size) may have been missed. The vast majority of anthropomorphic fired clay figurines (over 480 out of 500) were deposited in the final process of burning the 28 structures and the creation of four large middens. Most of the finds represent deliberate collection of objects for deposition in abandonment contexts (see Chapman and Gaydarska 2006).

The vast majority of figurines at Dolnoslav had been deposited in a broken condition. A refitting experiment on all of the anthropomorphic figurines produced an increase in conjoined figurine fragments of 25 cases, most refits leading to complete or nearly complete bodies but with four figurines still incomplete, even after refitting (Figure 17.5). These included 25 complete figurines (5% of the total), including 3 refitted from 8 fragments on the dig; 422 "orphan" fragments (84%); 53 fragments refitted in 2004 (11%) to make 25 more refitted cases; 21 refits with a Completeness Index of 90–100%; 1 refit with a Completeness Index of 80%; and 3 refits with a Completeness Index of 50–60%. Many of the figurine fragments, including some refitted examples, show signs of life "after the break," indicating that figurine fragmentation was not the end of a figurine's life but one stage

in the total figurine biography (Gaydarska et al. 2007).

The depositional contexts of the refitted figurines indicate a significant degree of fragment dispersion after breakage. It could be maintained that the figurine fragments were moved around the site by various processes of attrition, or through children's play, but the relatively "closed" and special character of the depositional contexts of the figurine fragments—burnt structures and middens—suggests that the fragments were too significant to be considered simply as "rubbish." It is more probable that figurine fragments at Dolnoslav embodied enchained relations between people and objects, acting as tokens in the same way as fragments refitted between sites.

The significance of the Dolnoslav refitting study is that it demonstrates that over 10% of all figurine

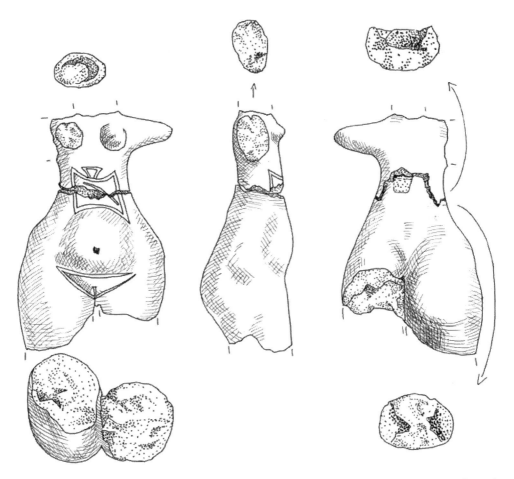

Figure 17.5   Refitted fired clay anthropomorphic figurine (refit 6), Dolnoslav Late Copper Age tell, South Bulgaria: the upper torso was deposited in Building 24, the head in Midden 1, 28 m apart (drawing by Elena Georgieva).

fragments have been refitted on-site but that a much higher proportion of fragments are, *pace* Schiffer (1976), "orphan" fragments. Where are the missing fragments?

Given the good, if not complete recovery rate of figurines at Dolnoslav, and the total excavation of the final phase of structured deposits, one could conclude that the missing fragments were used and deposited somewhere off the tell. Although deposits of figurine parts have been found in arable land near a settlement (Chapman 2000b: 64), the spatial scale of the fragment dispersion remains one of the hardest points to establish. In particular, it is difficult to distinguish between two possible scenarios to the issue of missing parts at Dolnoslav: (1) figurines were made and used whole at another settlement, fragmented there, and one of the fragments was brought to Dolnoslav for deposition; or (2) the figurine was made and used whole at Dolnoslav, and, following fragmentation, one or more parts were removed for deposition on another settlement. This issue can, indeed, be generalized to all cases of parts missing from deliberately fragmented objects. Whatever the direction of movement, the Dolnoslav example provides further support for fragment dispersion across the landscape, potentially connecting several settlements over a distance of several kilometers. In other words, this case resembles the movement of decorated megalithic panels in Neolithic Brittany.

The lower frequency of sites that combine both the domestic and the mortuary domains (in contrast to one or other arena: cf. Chapman 1992) means that relatively few studies have focused on possible refits between these two domains. I present one indisputable example from the Balkan Chalcolithic and refer to another example from the British Bronze Age.

Durankulak. The Durankulak complex, on the Northern part of the Bulgarian Black Sea coast, comprises a long-lived cemetery with both Neolithic (Early Hamangia) and Chalcolithic (Middle-Late Hamangia and Varna groups) burials in the largest-known cemetery in the Balkans, lying on the shores of a lagoon (Todorova 2002). The Neolithic settlement lies near the cemetery on the lagoon shore (Todorova and Dimov 1989); the Chalcolithic settlement was moved onto an island in the lagoon in the early 5th millennium B.C. (Todorova 1997; for new AMS radiocarbon dates for Durankulak, see Honch et al. In press).

A single example of conjoined pottery has been published, concerning half of a decorated vessel from the Varna group Grave 584, that refits to a large decorated sherd deposited in a house in horizon VII of the tell on the island (Todorova et al. 2002: 59–60, tab. 99/11; see Figure 17.6). The use of an elaborately decorated necked carinated bowl underlines the

visual importance of fragmentation practices—that those insiders who know the story will recognize the whole from which the part has been separated and reconstruct it in their mind's eye as the part symbolizing the whole. The refitting of vessel fragments from both the mortuary and the domestic domains underlines the importance of maintaining enchained links between the dead and the living, even though the spatial scale is no more than 200 meters. Once again, it should be emphasized that the Durankulak example is an unequivocal case of deliberate fragmentation followed by fragment dispersion. As with some Dolnoslav anthropomorphic figurines, the conjoined vessel is still missing a substantial fragment possibly deposited on the Tell in an as yet unexcavated area or, yet again, on another settlement or cemetery. A comparable case of conjoined vessel parts—one in a grave, the other in a house—has been recognized for the Middle Bronze Age of Southern Britain at the Itford Hill complex (Burstow and Holleyman 1957; Holden et al. 1972; for details, see Chapman and Gaydarska 2006).

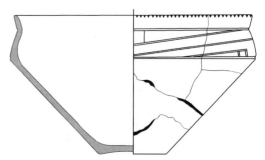

Figure 17.6 Refitted Late Copper Age vessel, Durankulak complex, northeast Bulgaria (source: Todorova et al. 2002, tab. 99/11): the upper part was deposited in Grave 584, the lower part in a house in Level VII on the "Big Island" tell.

The frequently closed nature of individual burial contexts provides fragmentation studies with an opportunity to identify object fragments whose missing parts have not been placed in the grave and that can therefore establish enchained relations between different domains. Nevertheless, care must be taken to document the closed nature of the context, especially since the work of Buko has demonstrated the ease with which more recent deposits can be trapped in middle and upper grave fill (Buko 1998).

Tiszapolgár-Basatanya. A good example of a collection of published graves that shows a clear pattern of deliberate fragmentation is the Copper Age cemetery of Tiszapolgár-Basatanya, in Eastern Hungary (Bognár-Kutzian 1963, 1972). Individual inhumation burials of men, women, and children were dated to the Early and Middle phases of the Copper Age, ca. 4500–3600 B.C. (Chapman 2000f). In both phases, ceramic grave goods were common, varying with the life-stage of both sexes (Sofaer Derevenski 1997). The vessels could be classified in three groups according to their Completeness Index: (1) fully complete vessels (termed "C"); (2) vessels that had been restored to a complete profile but with minor or substantial parts missing (termed "R"); and (3) orphan sherds (termed "S").

In the Early Copper Age graves, only 6% of the graves with ceramic grave goods contained complete vessels without any Restored pots or orphan Sherds (Figure 17.7a). The remaining graves showed a complex pattern of vessels, dominated by graves with complete and restored vessels. Most of the pots had missing fragments that must have been deposited elsewhere—most probably in the domain of the

living. In the Middle Copper Age, the percentage of graves with no Restored vessels or orphan Sherds had dropped to 4% (Figure 17.7b), with the graves dominated by assemblages with all ceramic classes— Complete, Restored and orphan Sherds. In both phases, there was a complex relationship between the completeness of the vessels deposited as grave goods and the age/sex category of the person buried (Chapman 2000b: 51–53, tab. 3.2).

Although incomplete, the fragmentation study of the Tiszapolgár-Basatanya cemetery provides an example of the intensity of enchained relations connecting those buried in a single cemetery and settlements dispersed across the landscape. Only four graves in the entire cemetery did not rely on enchained relations to those outside the graves. In the Copper Age in Eastern Hungary, the dominant settlement form is the dispersed farm or homestead

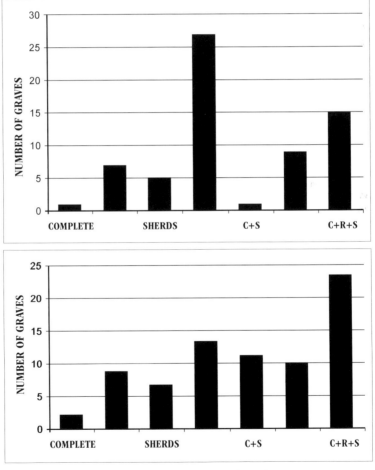

Figure 17.7 (*Top*) Proportion of graves with incomplete vessels, Early Copper Age cemetery, Tiszapolgár-Basatanya, northeast Hungary; (*bottom*) proportion of graves with incomplete vessels, Middle Copper Age cemetery, Tiszapolgár-Basatanya, northeast Hungary (source: the author).

(Bognár-Kutzian 1972; Chapman 1994, 1995). In intensive systematic fieldwalking of the Basatanya area in 1991, the Upper Tisza Project defined a number of surface scatters that probably constituted the material remains discarded from such homesteads (Chapman, Gillings, and Shiel 2003). The homesteads radiated out from the central cemetery—at that time the dominant landscape feature in that part of the Hungarian Plain—to a radius of 5 kilometers.

If the enchained relations implied at Basatanya linked the cemetery to those homesteads, the landscape implications are clear. The family of the deceased would have carried the newly dead from the homestead across the Plain to Basatanya, placing complete vessels in the grave to symbolize the integration of the lineage group, the completeness of the deceased's life journey or her/his integrated identities finally achieved in death, or a combination of these ideas. More distant kin members could have broken a vessel by the graveside, leaving the greater part of the vessel in the grave and carrying off the remaining fragment as a memento of the dead. Those with still less intimate links to the deceased may have broken a vessel at home and carried only an "orphan sherd" to Basatanya, to be thrown into the grave as a token of respect. In this way, the Completeness Index of the vessels in the Basatanya graves may have stood metaphorically for the closeness of the social relations of the mourner to the deceased. If found, conjoint fragments between two or more different graves could indicate the enchained relations between mourners and two or more newly dead and between the dead themselves. It was not only by the people buried there but also by the material left in memory of the dead that the Basatanya cemetery related a group of homesteads dispersed across the landscape. The local spatial scale of this enchained network— perhaps in the order of 5 kilometers—resembles that of the Breton megalithic network and perhaps is not dissimilar to the network of figurine fragments dispersed around the Dolnoslav landscape.

There are several important implications of intrasite refitting. Each of them depends on a good understanding of the taphonomic processes affecting the site in question (that is, children's play, off-site disposal in manuring scatters, weathering and other forms of attrition, and reuse of fragments), and a good to excellent recovery rate. First, it has been possible to isolate practices of deliberate fragmentation through the demonstration of missing parts of objects. The missing parts indicate enchained relations between persons on the site and persons off the site, in other parts of the landscape. Second, enchainment is practiced with many types of material culture—pottery, lithics, shell, amber,

and jet ornaments, and so forth. Initial results indicate the likelihood of different types of enchainment practices with each kind of raw material. Third, one of the hardest things to tell is whether a fragment has been broken away from the site and then brought onto the site *or* a whole object on the site was broken and parts of that object removed off the site. This question is just as problematic in the domestic as in the mortuary context. Fourth, the spatial scale of enchained fragment dispersion is variable: there does seem to be some support for networks of enchainment at the local scale (up to 5 kilometers), while, at supralocal centers such as Varna I and Mycenae, the spatial scale of enchainment was much greater. Hence the exchange of sherds over long distances should not be ruled out at this stage of research and understanding. Fifth, the concepts of heirlooms and commemorative pieces—both with ancestral significance—could with advantage be integrated into the interpretative framework of fragmentation studies as subsets of enchained relations.

Discussion

While all parts of the landscape can, and often do, take on cultural significance (Bradley 2000), the places of inhabitation known as settlements and locales of commemoration known as cemeteries can often express a sociocultural identity through what Richard Bradley has called "special attention markers"—elaborately decorated structures or naturally distinctive features—or through what I term "timemarks"—the association of a place with a significant event that took place there at a specific time. Often in a more concentrated way than natural places, inhabited places are sites of accumulation, with the accumulated things bringing with them associations, experiences, and histories, creating memory and place-value.

Some of the things that are particularly effective in the creation and the maintenance of cultural memory and place-value are those objects embodying enchained relations between kinsfolk or non-kin across the landscape. Of the many aspects of enchainment, the one most relevant here is its ability to enable people to presence absent people, objects, and places. What this chapter has sought to demonstrate is that the objects embodying enchained relations across the landscape were often broken objects that could have been refitted to fragments from the same once-complete object that were deposited in another place. This claim is supported by two kinds of evidence: (1) the deposition in two different places of refitting fragments from the same object; and

(2) the dispersion of fragments of objects from one place to (an)other(s), documented on completely excavated sites with good to excellent recovery rates and containing many orphan fragments. In both cases, it is hard to resist the conclusion that fragment dispersion across the landscape was one of the important social practices through which enchained relations were maintained at the local and sometimes wider level. There is currently an increasing acceptance, based on isotopic evidence for the sourcing of people, of greater mobility among prehistoric individuals. If for *persons*, then why not for *things* and *fragments of things*?

We are now in a position to identify a variety of forms of enchained relations, some of which are more relevant to specific cultural contexts than others. Here, it is worth emphasizing the important point that fragment dispersion implicates temporal as well as landscape distance (see also McFadyen, this volume, for implications of temporal spacing in construction). The fragmentation of objects for use as tokens implies a temporal distance until validation occurs through the re-presentation of the token. On perhaps a longer time scale, local curation strategies can ensure the availability of fragments for exchange or deposition at significant events, such as keeping sherds broken from vessel fragments buried with a household leader for a later burial of a cherished relative. The circulation of items of ancestral veneration is closely related to such a practice. Moreover, long-lasting curation of vessel or necklace fragments can convert enchained items into heirlooms, whereas relics would have a more distant social relationship to the person and a closer relationship to a generalized past. A final case concerns the collection of items from the abandoned site of another ethnic group—a case with implications for temporal as well as ethnic distance bridged by enchained object relations. It is important to develop ways of distinguishing between these forms of enchainment in future case studies supporting deliberate fragmentation.

The documentation of fragment dispersion raises certain interesting issues and problems. First, intersite refitting is beginning to give us an idea of the spatial scale of this practice: several studies indicate refitting within up to a 5-kilometer radius of a site, though much longer enchained networks have been documented. Second, even after inter- or intrasite refitting, the conjoint object is still often incomplete—suggesting an even more complex object biography than we can currently document. Third, there are tantalizing hints that, after the break, fragments follow separate biographical pathways before they are reunited, often in a burial. Fourth, with the exception of lithic *chaînes opératoires*, the direction of movement of the fragment dispersion is still resolutely resistant to analysis, both for the domestic and the mortuary domains. Each of these issues merits greater attention in future research.

Conclusions

The increasing emphasis on dwelling practices in landscape archaeology (see chapters by Fowler, McFadyen, and Thomas) foregrounds the contribution of fragmentation studies to landscape archaeology. To the extent that the dispersion of fragments across past landscapes concurrently disperses memories, persons, and places, landscape-oriented fragmentation is a vital tool for the recognition of past identities. The creation of person- or place-identities from accepted elements, rejected elements, and absent but presenced elements cannot but rely on enchained social relations, especially for that elusive third category. There is a growing body of theory relating to enchained relations of various kinds and an emerging suite of methods for the identification of fragment refits at both the intersite and the intrasite level. Although site-based taphonomic issues are the starting point for any landscape-oriented analysis of fragmentation, they should no longer be considered as the end point. The evidence for deliberate object (and body) fragmentation across past landscapes can no longer be overlooked or dismissed as an irrelevant and time-consuming curiosity. The future horizons for fragmentation and refitting studies are as broad and as open as the landscapes presented in this book.

Acknowledgments

I am most grateful to Bruno David for the opportunity to publish a shortened version of a Keynote Address presented to the 2nd Chicago Eurasian Archaeology Conference in this research handbook. I am very grateful to Sanja Raduntcheva and Bistra Koleva for all their help with the Dolnoslav study. Grateful thanks to Stephen Aldhouse-Green, Chris Scarre, and Willem Willems for introducing me to important research findings; to Linda Bosveld for the translation of an article from Dutch; to Richard Bradley for continuing inspiration; and to Bisserka for being the ideal partner in fragmentation research.

References

Barrett, J. 1999. The mythical landscapes of the British Iron Age, in W. Ashmore and A. B. Knapp (eds.), *Archaeologies of Landscape: Contemporary Perspectives*, pp. 253–65. Oxford: Blackwell Publishing.

Bausch, I. 1994. Clay figurines and ritual in the Middle Jomon period: A case study of the Shakado site in the Kofu Basin. Unpublished M.A. thesis, University of Leiden.

Bogaers, J. E. 1968. Waarnemingen in Westerheem, I: In het spoor van Verritus en Malorix? *Westerheem* 17: 173–79.

Bognár-Kutzian, I. 1963. *The Copper Age Cemetery of Tiszapolgár-Basatanya*. Archaeologia Hungarica 42. Budapest: Akadémiai Kiadó.

———. 1972. *The Early Copper Age Tiszapolgár Culture in the Carpathian Basin*. Budapest: Akadémiai Kiadó.

Bosman, A. V. A. J. 1994. Een onverwachte ontdekking bij de Ruïne de Brederode. *Velisena, Velsen in historisch perspectief* 3: 4–6.

———. 1997. Het culturele vondstmateriaal van de vroeg-Romeinse versterking Velsen 1. Unpublished Ph.D. thesis, University of Amsterdam.

Bradley, R. 2000. *An Archaeology of Natural Places*. London: Routledge.

———. 2002. *The Past in Prehistoric Societies*. London: Routledge.

Bradley, R., and Ford, D. 2004. A long distance connection in the Bronze Age: Joining fragments of a Ewart Park sword from two sites in England, in H. Roche, E. Grogen, J. Bradley, J. Coles, and B. Raftery (eds.), *From Megaliths to Metal: Essays in Honour of George Eogan*, pp. 174–77. Oxford: Oxbow Books.

Brandt, R. 1983. A brief encounter along the Northern frontier, in R. Brandt and J. Slofstra (eds.), *Roman and Native in the Low Countries. Spheres of interaction*, pp. 129–43. BAR International Series 184. Oxford: BAR.

Buko, A. 1998. Pottery, potsherds and the archaeologist: An approach to the pottery analyses, in S. Tabaczyński (ed.), *Theory and Practice of Archaeological Research* Vol. III, pp. 381–408. Warszawa: PAN.

Burstow, G. P., and Holleyman, G. A. 1957. Late Bronze Age settlement on Itford Hill, Sussex. *Proceedings of the Prehistoric Society* 23: 167–212.

Calado, M. 2002. Standing stones and natural outcrops: The role of ritual monuments in the Neolithic transition of the Central Alentejo, in C. Scarre (ed.), *Monuments and Landscape in Atlantic Europe*, pp. 17–35. London: Routledge.

Chapman, J. 1992. The creation of arenas of social power in Serbian prehistory. *Zbornik Narodnog Muzeja (Beograd)* 14: 305–17.

———. 1994. The living, the dead, and the ancestors: Time, life cycles and the mortuary domain in later European prehistory, in J. Davies (ed.), *Ritual and Remembrance: Responses to Death in Human Societies*, pp. 40–85. Sheffield: Sheffield Academic Press.

———. 1995. Social power in the early farming communities of Eastern Hungary: Perspectives from the Upper Tisza region. *A Josa András Múzeum Évkönyve* 36: 79–99.

———. 2000a. "Rubbish-dumps" or "places of deposition?": Neolithic and Copper Age settlements in Central and Eastern Europe, in A. Ritchie (ed.), *Neolithic Orkney in its European context*, pp. 347–62. Cambridge: McDonald Institute.

———. 2000b. *Fragmentation in archaeology: People, places and broken objects in the prehistory of South Eastern Europe*. London: Routledge

———. 2000c. The fractality of personal relations in the Mesolithic and Early Neolithic of South East Europe, in R. Kertész and J. Makkay (eds.), *From Mesolithic to Neolithic*, pp. 145–66. Proceedings of the Szolnok Conference, September 1996. Budapest: Archaeolingua.

———. 2000d. Fragmentation analysis and social relations in later Balkan prehistory, in B. Ginter et al. (eds.), *Problems of the Stone Age in the Old World*, pp. 369–88. Kraków: Jagiellonian University Institute of Archaeology.

———. 2000e. Object fragmentation in the Neolithic and Copper Age of Southeast Europe, in P. F. Biehl, F. Bertemès, and H. Meller (eds.), *The Archaeology of Cult and Religion*, pp. 89–106. Budapest: Archaeolingua.

———. 2000f. *Tensions at Funerals: Mortuary Archaeology in Later Hungarian Prehistory*. Budapest: Archaeolingua.

Chapman, J., and Gaydarska, B. 2006. *Parts and Wholes: Fragmentation in Later Prehistoric Context*. Oxford: Oxbow Books.

Chapman, J., Gillings, M., and Shiel, R. 2003. *The Upper Tisza Project: Studies in Hungarian Landscape Archaeology. E-book 1*. http://ads.ahds.ac.uk/catalogue/projArch/uppertisza_ba_2003/index.cfm

Fowler, C. 2004. *The Archaeology of Personhood: An Anthropological Approach*. London: Routledge.

Gamble, C. 1999. *The Palaeolithic Settlement of Europe*. Cambridge: Cambridge University Press.

Garfinkel, Y. 1994. Ritual burial of cultic objects: The earliest evidence. *Cambridge Archaeological Journal* 4: 159–88.

Gaydarska, B., Chapman, J., Raduntcheva, A., and Koleva, B. 2007. The châine opératoire approach to prehistoric figurines: An example from Dolnoslav, Bulgaria, in C. Renfrew and I. Morley (eds.), *Image and Imagination: A Global Prehistory of Figurative Representation*, pp. 171–84. Cambridge: McDonald Institute.

Grinsell, L. 1960. The breaking of objects as a funeral rite. *Folklore* 71: 475–91.

Hamilakis, Y. 1998. Eating the dead: Mortuary feasting and the politics of memory in the Aegean Bronze Age societies, in K. Branigan (ed.), *Cemetery and Society in the Aegean Bronze Age*, pp. 115–31. Sheffield: Sheffield Academic Press.

Hayden, B., and Cannon, A. 1983. Where the garbage goes: Refuse disposal in the Maya Highlands. *Journal of Anthropological Archaeology* 2: 117–63.

Higham, T., Chapman, J., Slavchev, V., Gaydarska, B., Honch, N., Yordanov, Y., and Dimitrova, B. 2007. New perspectives on the Varna cemetery (Bulgaria): AMS dates and social implications. *Antiquity* 81: 640–54.

Hodder, I. 1982. *Symbols in Action*. Cambridge: Cambridge University Press.

Hoffman, J. L., and Enloe, J. G. (eds.). 1992. *Piecing Together the Past: Applications of Re-Fitting Studies in Archaeology*. BAR I-578. Oxford: BAR.

Holden, E. W., Bradley, R., Ellsion, A., and Ratcliffe-Densham, H. B. A. 1972. A Bronze Age cemetery-barrow on Itford Hill, Beddingham, Sussex. *Sussex Archaeological Collections* 110: 70–117.

Honch, N., Higham, T., Chapman, J., Gaydarska, B., Todorova, H., Slavchev, V., Yordanov, Y., and Dimitrova, B. (In press). Pontic chronologies and diets: A scientific framework for understanding the Durankulak and Varna I cemeteries, Bulgaria, in L. Nikolova (ed.), *Circumpontica*. Oxford: Archaeopress.

Koleva, B. 2001. Certaines observations sur le complexe de culte énéolithique près du village de Dolnoslav (région de Plovdiv). *Godishnik na Arheologicheski Muzei Plovdiv* X: 5–19.

———. 2002. Spatial model of the objects of first construction horizon of the Late Eneolithic complex near the village of Dolnoslav. *Godishnik na Arheologicheski Muzei Plovdiv* IX/1: 120–30.

Le Roux, C.-T. 1992. The art of Gavrinis presented in its Armorican context and in comparison with Ireland. *Journal of the Royal Society of Antiquaries of Ireland* 122: 79–108.

L'Helgouach, J., and Le Roux, C.-T. 1986. Morphologie et chronologie des grandes architectures de l'Ouest de la France, in J.-P. Demoule and J. Guilaine (eds.), *Le Néolithique de la France*, pp. 181–91. Paris: Picard.

Macfarlane, C. 1985. Analysis of join linkages, in C. Renfrew (ed.), *The Archaeology of Cult: The Sanctuary at Phylakopi*, pp. 453–68. London: Thames and Hudson.

Mauss, M. 1954. *The Gift: Forms and Functions of Exchange in Archaic Societies*. London: Cohen and West.

Mens, E. 2008. Re-fitting megaliths in western France. *Antiquity* 82(315): 25–36.

Raduntcheva, A. 1996. Dolnoslav: A temple centre from the Eneolithic. *Godishnik Nov Bulgarski Universitet* II-III: 168–81.

———. 2002. Eneolithic temple complex near the village of Dolnoslav, District of Plovdiv, and the system of rock sanctuaries with prehistoric cultural strata in Rodopi Mountains and outside its territory. *Godishnik na Arheologicheski Muzei Plovdiv* IX/1: 96–119.

Scarre, C. 2002. Coast and cosmos: The Neolithic monuments of Northern Brittany, in C. Scarre (ed.), *Monuments and Landscape in Atlantic Europe*, pp. 84–105. London: Routledge.

———. 2005. Displaying the stones: The materiality of "megalithic" monuments, in E. DeMarrais, C. Gosden, and C. Renfrew (eds.), *Rethinking Materiality: The Engagement of Mind with the Material World*, pp. 141–52. Cambridge: McDonald Institute.

Schaller-Åhrberg, E. 1990. Refitting as a method to separate mixed sites: A test with unexpected results, in E. Cziesla, S. Eickhoff, N. Arts, and D. Winter (eds.), *The Big Puzzle: International Symposium on Refitting Stone Artefacts*, pp. 611–22. Bonn: Holos.

Scheer, A. 1990. Von der Schichtinterpretation bis zum Besiedlungsmuster: Zusammensetzungen als absoluter Nachweis, in E. Cziesla, S. Eickhoff, N. Arts, and D. Winter (eds.), *The Big Puzzle: International Symposium on Refitting Stone Artefacts*, pp. 623–51. Bonn: Holos.

Schiffer, M. B. 1976. *Behavioral Archaeology*. New York: Academic Press.

Singer, C. 1984. The 63-kilometer fit, in J. E. Ericson and B. A. Purdy (eds.), *Prehistoric Quarries and Lithic Production*, pp. 35–48. Cambridge: Cambridge University Press.

Sofaer Derevenski, J. 1997. Age and gender at the site of Tiszapolgár-Basatanya, Hungary. *Antiquity* 71: 875–89.

Thomas, J. 1991. *Rethinking the Neolithic*. Cambridge: Cambridge University Press.

Todorova, H. 1997. Tellsiedlung von Durankulak. *Fritz Thiessen Stiftung Jahresbericht* 1995/96: 81–84.

———. (ed.). 2002. *Durankulak Band II. Die prähistorischen Gräberfelder*. DAI, Berlin. Sofia: Anubis.

Todorova, H., and Dimov, T. 1989. Ausgrabungen in Durankulak 1974–1987. *Varia Archaeologica Hungarica* II: 291–306.

Todorova, H., Dimov, T., Bojadžiev, J., Dimitrov, K., and Avramova, M. 2002. Katalog der prähistorischen Gräber von Durankulak, in H. Todorova (ed.), *Durankulak Band II. Die prähistorischen Gräberfelder*, pp. 31–87. DAI, Berlin. Sofia: Anubis.

Vons, P., and Bosman, A. V. A. J. 1988. Inheemse boeren bezochten de verlaten Romeinse versterkingen Velsen I and II. *Westerheem* 37: 1–17.

Waddington, C. 1998. Cup and ring marks in context. *Cambridge Archaeological Journal* 8: 29–54.

Williams, M. 2001. Shamanic interpretations: reconstructing a cosmology for the later prehistoric-period of north-western Europe. Unpublished Ph.D. thesis, University of Reading.

18

BOUNDARIES AND THE ARCHAEOLOGY OF FRONTIER ZONES

Mike McCarthy

The study of boundaries and frontiers has been an important focus in geopolitical and historical geography and historical studies of social and cultural landscapes for almost a century (Bartlett and MacKay 1989; Jones 1964; Minghi 1970; Parker 2002; Pohl, Wood, and Reimitz 2001; Power 2004). Early geographical and geopolitical studies of borders had to contend with issues such as the relative efficacy of natural versus anthropogenic boundaries, while in recent years geographers have paid particular attention to functional aspects of boundaries and considered their impact on existing societies in modern times. For some historians and geographers, frontiers can also be "processes," by which is generally meant the changes that take place during a period of colonization transforming a zone from border to heartland (Burns 1989). For politicians and civil servants, the growth of empire building and nationalism, proceeding hand in hand with the rise of map-making agencies and colonial administration, was an increasingly important focus of frontier attention in the 19th and 20th centuries (Given 2004: 70; Lamb 1968; Mellor 1989).

Fascinated by the processes of colonization on the American frontiers (Canny 1998; Dyson 1985; Mancall 1988), and mesmerized by the Roman Empire (Birley 1961, 1974; Whittaker 1994), as well as China and South-East Asia (Allard 2006; Lattimore 1962; Stark 2006), archaeologists have necessarily

taken a longer view of frontiers and boundaries focusing on subsistence strategies, social identities, colonialism, the growth of nation- states, and defense (Baker 1993; Covey 2003; Elton 1996; Green and Perlman 1985; Hunter 2001; Rollason 2003; Wells 1999; Wigg 1999). Some of the most important work has taken place at regional and local levels and include city walls, the enclosing of private properties, areas of urban jurisdiction, religious units such as temple complexes, dioceses and parishes, private parks and estates, townships, burial mounds, deer parks, and royal forests (Beresford 1971: 23–62; Creighton and Higham 2005; Griffith, Reynolds, and Semple 2003; Moorhouse 1981; Reynolds 2002; Winchester 2000).

The archaeological literature also reflects a concern with boundaries in a wider sense of the term, although it is not always explicitly stated. For example, there is the transfigurative sense of boundaries whereby the process of dying, death, and burial crosses a line between this world and another. The location of burial mounds, execution cemeteries and, in Ireland at least, the deposition of bog bodies and metalwork on boundaries are cases in point (Kelly 2006; Reynolds 2002). Boundaries are also alluded to in myths and annals as being places of particular significance, sometimes supernatural, but sometimes as the venues of important political or diplomatic events. The labors of Cúchulainn in the *Táin Bó Cuailgne* (Tarzia 1987), the meeting of Aethelstan and an array of kings at Eamont Bridge in Cumbria in A.D. 926, and the famed ancient oak of Gisors on the Epte in Normandy are examples.

There are many other ways of thinking about boundaries. Not only are achievements in sport or in science lauded, but pushing at the boundaries of physical attainment or knowledge is positively encouraged. Equally, the idea of same-sex marriages or transvestism may compel a reassessment of inherited social boundaries. In complex societies, behavioral limits and punishments for infringements may be set out in law codes, sometimes inscribed on stelae as if to say, you are now in the territory of X and these are our laws. Such boundaries are so deeply ingrained in the collective psyche that they have become linguistic clichés—"the boundaries of acceptable behavior," "the line in the sand," "the frontiers of knowledge." What is important here is that these examples challenge the notion that rules are always to be obeyed, that boundaries should not be crossed.

Types of Frontier

Within academia there are increasing trends toward challenging the rules and fostering cross-disciplinary research, as Parker (2006) has recently observed, and definitions of frontiers and boundaries, borders, and borderlands are numerous. Parker's useful model, "Continuum of Boundary Dynamics," attempts to "characterize specific boundary situations" as an aid to scholars wishing to make comparisons (2002, 2006).

To me, frontiers and boundaries are territorial edges. By definition, they are at the margins of a heartland or a core territory that could be as small as a single valley containing a group of farms or imperial in scale. They also imply ownership or the presence of a leader, such as a king and government, intent on defense in the face of an aggressor, and the control of cross-border traffic. Frontiers, however, were not necessarily viewed as permanent arrangements, and although permanence requires international agreements (treaties) between countries together with the administrative infrastructure to police them, temporary frontiers could be defined by truces, as in Spain during the Middle Ages (Jiménez 1989). Temporary arrangements also require policing, usually by the military or by lordships centered on castles.

The word beloved of Roman archaeologists to represent a frontier capable of being defended is *limes,* but, as Isaac (1987) has shown, it was rarely used by ancient authors before the 4th century A.D. Indeed, the contexts of the words *limes* or *limites*

in Roman sources in the 1st to 3rd centuries A.D. show that they refer either to roads or a land border, but not a defensive frontier, and *limes* is not a word used in literary or epigraphic sources in connection with Hadrian's Wall or the Antonine Wall. By the 4th century and afterward, the words are more frequently used, although by this time in the sense of a border district under the command of a *dux.* In fact the idea of a broad zone is not only a normal feature of the edges of the Roman Empire (Isaac 1990: 396–97; Wells 1999: 126), but, under the descriptions of client kingdoms, vassal states, buffer states, margraveships, marks, or marcher lordships, often politically subservient to a larger adjacent polity, they became a regular feature in the growth of empires and nation-states. Thus, at the northern extremity of the Carolingian Empire, Saxony formed a buffer between the heartland of Charlemagne's lands against the Slavs to the east, and the Inka empire in Peru expanded from a core around Cuzco to include additional territories (Covey 2003).

In sociopolitical terms, the classification of frontiers and boundaries is fraught with problems. Kristof (1970: 134–35) cited the British Association Geographical Glossary Committee, according to whom *frontiers* are either (a) border regions, zones, or tracts separating two political units or (b) a demarcated boundary between states. *Boundaries,* the Committee held, are either synonymous with (a), or lines delimiting administrative units, or geographical regions. Kristof went further, claiming that *frontiers* look outward and constitute zones from which settlers looked beyond with future colonization in mind. American frontiersmen in the 18th and 19th centuries epitomized the concept, while nowadays we speak of space as "the final frontier." *Boundaries,* however, are "inner oriented" and mark the edges of sovereign units. The difference can also be expressed as that between two opposed forces—the frontier being centrifugal in nature, whereas boundaries are centripetal.

Working from an anthropological perspective, Green and Perlman (1985: 3–4) defined frontier studies as being concerned with the peripheries of societies, while boundaries are more concerned with the interaction between societies at political, economic, and other levels. This is not very far from Lord Curzon's (1907) definition in a famous lecture entitled "Frontiers" delivered in 1907. In this he argued that linear divisions between polities should be called "boundaries," whereas the word "frontier" should be reserved for less precisely defined edges, buffer zones, and marcher lordships.

Many of these issues are as germane to post-medieval and modern societies as to the ancient world where frontier zones were frequently, in Parker's definitions, porous, if not fluid. He instances the Assyrian Empire setting up buffer states, esta-blishing strongly fortified towns, ruling their vassals by way of networks of officials obtaining tribute, monitoring activity, and gathering information on potentially hostile neighbors. A similar system was employed in Mesoamerica, where the Spanish appointed local native leaders to keep the peace, collect taxes, and provide a variety of services (Jones 2000: 366), but the natives could escape across the frontier into jungle or the mountains and desert.

Herodotus and Xenophon rarely referred to boundaries in their writings on long-distance travels around the eastern Mediterranean and Eurasia in the 5th and 4th centuries B.C. Exceptionally, in Book 1 Herodotus mentions the River Halys as the boundary between Cappadocia and Paphlagonia, but his description of the road to Susa in Book 5 makes no mention of boundaries, even though the road passes through many territories. The mention of rivers as boundaries is not confined to Herodotus. The River Tigris in northern Iraq acted as both a boundary and a major means of communication for the Assyrians (Parker 2002, 2006), and the Romans made use of the Rhine and Danube in a similar way, as did the Inka in the Vilcanota valley in Peru (Covey 2003). Xenophon wrote of "the country of the Carduchi" (IV, 3) and the "country of the Taochi" (IV, 7). Lands were identified by the people who lived there, and in some cases stelae or other monuments erected in public places, and incorporated the names of kings, telling travelers and others whose land they were in.

By the same token, Roman writers, concerned with emphasizing the differences between civilization and barbarism, or what was acceptable and what was beyond the pale, hardly ever refer to imperial frontiers. The empire comprised the peoples of Rome—*populi Romani*. To Caesar, Tacitus, and other Roman writers, polities were thought of in terms of the *gens* rather than the territories they inhabited. Rome negotiated with people, not states or land, and the people would, it was hoped, enrich it. Indeed, to Orosius in the 5th century the Empire had no need of a frontier because it was the world—*orbis terrarum* (Goetz 2001).

In the post-Roman world, Europe melted into an extraordinarily complex palimpsest of states, kingdoms, and tribal areas with varying degrees of political and ethnic identity, which did not necessarily correspond with one another. Although we may suppose that many kings from the 6th century on were well aware as to their territorial limits, the idea of marking them out on the ground evolved

slowly. Offa, king of Mercia (A.D. 757–796), adopted the linear approach against the men of Wales, but his contemporary, Charlemagne, seems to have preferred "buffer territories." Three centuries later, when William Duke of Normandy was crowned King of England in 1066, the northern limit of his kingdom was only vaguely defined.

Imperial and State Boundaries

The classic examples of imperial boundaries include the Great Wall of China and the limits of the Roman Empire, but in terms of scale they were the exception rather than the rule. In both cases, linear works, natural features, and buffer territories were used. Again, in both cases the Chinese and the Roman emperors were concerned with defense along the borders and the periodic threat of incursions threatening the stability of the core areas.

The Great Wall of China is over 7,000 kilometers in length, making it far and away the most impressive of boundaries in the ancient world. It comprises several walls, not all of which are linked but that served as frontiers separating the "barbarians," "peoples without history" (Linduff 1995: 133) in Mongolia and Manchuria, from the Central Kingdom to the south. Although originating during the Zhou dynasty, one of the earliest Walls was built during the short-lived Qin dynasty (221–207 B.C.) and later modified during Han times (202 B.C.–A.D. 220), and again during the Ming dynasty, to which period many of the remains visible today belong. It was built over extremely varied terrain, including the Gobi Desert, river valleys, plateaus and mountains, as well as the coastal plains north and east of Beijing, and was constructed from materials obtained locally, including limestone, granite, brick, and timber. Essentially, it is a series of curtain walls with towers and battleforts at frequent intervals.

It is hardly surprising that the purpose of the walls changed over time given the enormous scale of the works. The walls certainly had a defensive function that was to delay, deflect, and hinder attacking armies and nomadic hordes until such time as Chinese forces could be mustered to counter the threat. However, the Chinese came not to rely on this so much as to establish a number of buffer states, many of which accepted an overriding Chinese authority but that were otherwise nominally independent (Lamb 1968: 21–38; Lattimore 1962: 97–118). Much archaeological attention has inevitably been directed toward "core" areas of China, but the peripheral kingdoms have also been examined. Cultural assemblages from these areas demonstrate that although sinicization may have taken place in the higher echelons of society, it was far from uniform at lower levels in border regions, and this may have contributed to periodic unrest (Allard 2006: 250–51; Stark 2006).

The Roman Empire also utilized a wide range of natural features, mountains, rivers, desert, and the sea as well as walls, palisades, and ditches. In Lower Germany, the River Rhine became the boundary after the defeat of Varus in A.D. 9, legions backed up by auxiliary regiments being deployed in fortresses and forts linked by roads along its length (Carroll 2001; Millar 1981). In Upper Germany the frontier deviated at times from the Rhine in the Wetterau and Taunus regions where, during the reign of Trajan, the boundary comprised fortlets and towers to which was added a palisade. As the Empire expanded eastward, the Danube effectively became a frontier during the Dacian wars of Domitian beginning in A.D. 85, but Trajan also took in lands to the north. After his accession Hadrian redrew the administrative boundary at the Danube itself, designating the northern lands as *extraprovinciam*. Here, in the province of Moesia, it has been suggested that the frontier system was complex, with up to three lines of defense including one to the north, the Danube with the bulk of military personnel, and troops guarding the Balkan passes (Zahariade 1977). In Africa, climate and terrain helped determine the location and nature of the frontier works. The main period of construction of the Numidian frontier, the *fossatum Africae*, belongs to the reign of Hadrian, who used a combination of ditch, forts, watchtowers, roads, and the mountains to mark administrative limits (Fentress 1979: 111). In the Middle East, the changing political geography was determined by external threats from the Parthians, nomads, and others who were contained by fortified cities such as Dura Europos, and Resafa in Iraq, as well as networks of forts (Isaac 1990; Kennedy and Riley 1990; Woolf 1998: 180).

A key feature of Roman frontier policy was the cultivation of tribal loyalties along the borders. In what are sometimes known as vassal kingdoms or buffer states, the Assyrians changed the nature of settlement and archaeological assemblages by importing new agricultural settlers and ceramics as part of their territorial expansion (Parker 2002, 2006: 85, 93). The Romans also adopted the idea with client kingdoms such as Judaea, Palmyra, Armenia, and many others existing throughout the Empire from Arabia to Britain, the idea first being promoted by Augustus (Ball 2000; Isaac 1990: 396–401; Wells 1999: 125ff). Whereas the activities of Assyrian client kings were monitored, their Roman counterparts were generally allowed

a substantial degree of autonomy provided they retained a subservient relationship to Rome. Part of their treaty relationships was based on the need to maintain a peaceful frontier with minimal military interference by Rome. However, part of the relationships undoubtedly concerned profit and the ways in which the kingdom could be exploited for the good of Rome and its citizens. This is reflected in the archaeology, especially finds and assemblages containing Roman artifacts located well beyond the frontier, as can be seen in Scotland, or in princely graves in Thuringia, Poland, and Denmark (Carroll 2001: 97–101; Hedeager 1992: 156–7; Hunter 2001).

While archaeologists, understandably, attempt to construct cultural groupings from the data they have available—that is, the artifacts—a difficulty in attempting to distinguish buffer zones on the basis of artifactual evidence is that archaeological distributions do not necessarily reflect other facets of local cultures. We certainly cannot assume that ancient societies lacked dynamism and models need to build in factors that take this into account, as Hedeager (1992) has attempted. In other words, since any frontier zone is likely to contain a variety of overlapping but not necessarily congruent frontiers, there is a clear need to understand the social contexts within which different categories of artifacts were moved about (Elton 1996; Parker 2002). Parts of that argument will, in turn, require us to understand the need for and purpose of borders in complex societies.

Fragmentation of the Roman Empire resulted in the emergence of numerous polities and regional groupings presided over by officials or kings, many of whom did not invest in expensive construction works on the fringes of their lands, partly because they did not need them. In Merovingian Gaul, especially after the division of the kingdom following the death of Clovis in A.D. 511, differences from kingdom to kingdom could be subtle or even non-existent. Across Gaul and neighboring lands, as in the ancient Middle East, distinctions between groups may have been apparent in linguistic terms, styles of dress, or greeting, or by the nature of public posts and administrative machinery, as Wood and others have discussed in relation to the Franks, Jutes, Alamanni, and others (Halsall 1998: 141–65; Wood 1994, 1998). Doubtless such distinctions worked for contemporaries, but they present problems for archaeologists, especially for those dealing with peoples who borrowed cultural traits from neighboring groups. Although much earlier in date, Herodotus's disarming comment that "no race is so ready to adopt foreign ways as the Persian" (Book 1, 70) is as applicable for the 6th and 7th centuries A.D. as when it was written.

Nevertheless, physical boundaries that may have had a political function become a feature of the landscape in some areas. A number of substantial linear earthworks in the United Kingdom fit this category, although few are datable (Arnold 1997: 224). One such is the Wansdyke in Wiltshire and Somerset, some 45 kilometers in length; part of it overlies a Roman road, showing that some of it at least is Roman or later in date. The Aberford Dykes in West Yorkshire may have defined the British kingdom of Elmet (Faull 1981: 172–74). One of the most well-known linear works is Offa's Dyke between England and Wales (Hill 1974, 1985). At 103 kilometers (64 miles) in length, and with a bank 10 m wide at its base and a 2-m-deep ditch on the Welsh side, texts associate it with Offa, king of Mercia in the 8th century. Asser, and the later *Brut y Twysogyon*, record that Offa ordered a great *vallum* to be made from sea to sea and that, if true, must include Wat's Dyke, some 62 km in length, taking the line to the estuary of the River Dee.

There are numerous dyke systems in Ireland, many of which are also difficult to date accurately. The Black Pig's Dyke extends in a discontinuous fashion from County Down in the east to County Sligo in the west (Lynn 1989a, 1989b; Raftery 1994: 83–97; Walsh 1987; Williams 1987). The Doon of Drumsna, which extends over 1.6 kilometers and consists of a rampart up to 30 meters wide at the base and some 6 meters high in places, is another example (Condit and Buckley 1989: 12–14). The precise function of these earthworks remains uncertain, although some are thought to be provincial boundaries, as between Ulster and Connacht. The construction of The Dorsey in County Armagh, dated by radiocarbon determinations to the early 1st century B.C., may have been linked in some way with the building of the royal site at Navan (Lynn 1989: 9, 18; Raftery 1994: 97).

One of the greatest continental examples is the Danevirke in Schleswig. It is 30 kilometers long and was first constructed in A.D. 640–650 (cited in Hamerow 2002: 112) but was modified by Godfred in A.D. 804, according to the *Annales regni Francorum*. It linked western Jutland with Schleswig next to the major Baltic trading port of Hedeby. The huge scale of the Danevirke is one of the most impressive post-Roman monuments and is testimony to the emergence of the state of Denmark and the consolidation of a single leadership in the 8th century (Hedeager 1992: 2, 250–55).

It was not the only artificial frontier in Europe. There is growing archaeological evidence, supported by radiocarbon and dendrochronological dates, that Charlemagne, alarmed at movements of Slavs beyond the Elbe, had refortified a series of La Tène hillforts along the Elbe and the Saale rivers. One site, Höhbeck near Magdeburg, even resembles Roman fortifications and raises the question of whether Charlemagne attempted to emulate the Roman model (Hardt 2001: 223). Further east, Tsar Symeon defined his Slavic Bulgarian state from that of Byzantium with inscribed boundary stones (Stephenson 2000: 18).

In northern England, William the Conqueror and his successors consolidated their respective claims with a series of strong castles as well as secular and ecclesiastical lordships. From A.D. 1080, what was essentially a buffer state, the Palatinate of Durham, ruled with regalian authority by the Prince Bishops and earls of Northumberland, was formed in the east, and in 1133 the See of Carlisle was carved out of the former diocese of Glasgow in the west. Eventually, the Anglo-Scottish border was formally defined in the Treaty of York in 1237 as a line between the Tweed at Berwick to the Solway. On either side of that line, from 1249, the frontier zone comprised buffer territories in the form of lordships of the Western, Middle, and Eastern Marches in which the Wardens ruled as *de facto* kings (Jack 2004). The union of the crowns in 1603 finally resolved the precise position of the frontier.

Discussion

It is probably true to say that societies have been criss-crossed with frontiers and boundaries in all walks of life for the entire span of human existence. Nonetheless, we can identify monuments defining territories through the use of walls, cairns, or other markers, clearly beginning at the domestic level in circumstances, such as the need to exploit and to manage animals and plants. Good examples are Bronze Age garden plots and early field systems. Over time, the need to mark out estates or other units on the ground can be linked with chiefs establishing their own territories, as well as the need to control the movement of livestock and protect crops. In some cases, these required more substantial earthworks. Unlike many portable artifacts, the distributions of which are very difficult to interpret, such territorial markers are contextually rooted in that they are tied to a landscape with specific topographic, soil, drainage, and climatic elements.

It is arguable that at the level of the state, the most sophisticated means of territorial and administrative definition was the solution adopted by the Chinese and Roman Empires. Yet, as impressive as monuments such as Hadrian's Wall may be, the key frontier element was not so much a wall or a ditch but the maintenance of good relations with client states beyond and the stationing of military units along its length.

With the collapse of Empire and the removal of the military, notwithstanding the problems of dating, we see the periodic use of dykes, ramparts, and hillforts in the second half of the 1st millennium A.D. apparently acting as political boundaries, but, in some cases, they were relatively short lived. In the case of linear boundaries the questions are these: Why were they constructed? How did they work? What message was being conveyed by the builders?

Territorial definition in medieval Britain was not dissimilar to that of the Roman period. The castle supplanted the fort, lordships the *civitates*, and in some places new towns, as in north Wales or *bastides* in Gascony, the *civital* capitals. Buffer territories, the equivalent of Roman client kingdoms, were also created, but neither the great Palatinate of Durham nor the marcher lordships lasted indefinitely. From the 11th century on, populations grew; new states were formed; aristocratic, crusading, colonizing adventurers established wide networks of family influence; towns proliferated; trade routes moved toward a worldwide network; and money and the growth of banking facilities assumed an ever more prominent role. Peoples were also moved, not least in order to populate the new towns, but the emphasis gradually shifted toward the protection of territorial resources and the control of goods. Regulation and the payment of tolls probably controlled this more effectively than walls, palisades, and ditches. From the 16th century, changing views as to the nature of monarchy and its relationship with the Pope and increasing cartographic expertise gradually gave rise to concepts of international law and the rights of governments to annex territory that it had not previously held (Jack 2004). This is the point from which a phase of greater precision in the delineation of national boundaries is ushered in.

Conclusions

Despite the considerable archaeological and historical research achievements on borders and frontiers in the ancient world, there remain many problems for archaeologists aiming to explore the history of social landscapes, as a number of writers have pointed out. Among these definitions the adoption of a common language is a frequently raised issue, but perhaps more important is the need for writers concerned with state formation and socioeconomic dynamics to shift their emphasis from the sometimes spectacular cores to the hinterlands and regions peripheral to the heartlands. Here, archaeology certainly faces challenges. Themes adopted by many scholars of recent frontier and border issues include the question of Indigenous compliance to imposed structures with their border structures, as well as identity, and the problems they create (Hartshorne 1970). However, as an archaeologist, it is fair to say that we need many more large, well-dated, stratified datasets capable of serving as "benchmarks" against which other material can be matched. Scholars of the 19th- and 20th-century frontiers and boundaries have this, but the archaeological database is slighter, different in character, and more difficult to penetrate.

References

Allard, F. 2006. Frontiers and boundaries: The Han empire from its southern periphery, in M. T. Stark (ed.), *An Archaeology of Asia*, pp. 233–54. Oxford: Blackwell Publishing.

Arnold, C. J. 1997. *An Archaeology of the Early Anglo-Saxon Kingdoms*. London: Routledge (2nd ed.).

Baker, F. 1993. The Berlin Wall: Production, preservation and consumption of a 20th century monument. *Antiquity* 67: 709–33.

Ball, W. 2000. *Rome in the East: The Transformation of an Empire*. London: Routledge.

Bartlett, R., and MacKay, A. (eds.). 1989. *Medieval Frontier Societies*. Oxford: Oxford University Press.

Beresford, M. 1971. *History on the Ground: Six Studies in Maps and Landscapes*. Methuen: London.

Birley, E. 1961. *Research on Hadrian's Wall*. Kendal: Titus Wilson.

———. 1974. Twenty Years of *Limesforschung*, in E. Birley, B. Dobson, and M. Jarrett (eds.), *Roman Frontier Studies 1969: Eighth International Congress of Limesforschung*, pp. 1–4. Cardiff: University of Wales Press.

Burns, R. I. 1989. The significance of the frontier, in R. Bartlett and A. MacKay (eds.), *Medieval Frontier Societies*, pp. 307–30. Oxford: Oxford University Press.

Canny, N. 1998. England's New World and the Old 1480s to 1630s, in N. Canny (ed.), *The Origins of Empire: The Oxford history of the British Empire, 1*. Oxford: Oxford University Press.

Carroll, M. 2001. *Romans, Celts and Germans: The German Provinces of Rome*. Stroud: Tempus.

Condit, T., and Buckley, V. M. 1989. The "Doon" of Drumsna—Gateways to Connacht. *Emania* 6: 12–14.

Covey, R. A. 2003. A processual study of Inka state formation. *Journal Anthropological Archaeology* 22: 333–57.

Creighton, O., and Higham, R. 2005. *Medieval Town Walls: An Archaeology and Social History of Urban Defence*. Stroud: Tempus.

Curzon, Lord. 1907. *Frontiers: Romanes Lecture.* Oxford: Clarendon Press.

Dyson, S. L. (ed.). 1985. *Comparative Studies in the Archaeology of Colonialism.* Oxford: BAR International Series 233.

Elton, H. 1996. *Frontiers of the Roman Empire.* Batsford: London.

Faull, M. L. 1981. The post-Roman period, in M. L. Faull and S. A. Moorhouse (eds.), *West Yorkshire: An Archaeological survey to A.D. 1500*, pp. 171–78. West Yorkshire Metropolitan County Council.

Fentress, E. W. B. 1979. *Numidia and the Roman Army.* Oxford: BAR International Series 53.

Given, M. 2004. *The Archaeology of the Colonized.* London: Routledge.

Goetz, H.-W. 2001. Concepts of realm and frontier from late antiquity to the early Middle Ages, in W. Pohl, I. Wood, and H. Reimitz (eds.), *The Transformation of Frontiers from Late Antiquity to the Carolingians*, pp. 73–82. Brill: Leiden.

Green, S. W., and Perlman, S. M. (eds.). 1985. *The Archaeology of Frontiers and Boundaries.* Orlando: Academic Press.

Griffith, D., Reynolds, A., and Semple, S. (eds.). 2003. *Boundaries in Early Medieval Britain.* Anglo-Saxon Studies in Archaeology and History 12. Oxford: Oxford University School of Archaeology.

Halsall, G. 1998. Social identities and social relationships in early Merovingian Gaul, in I. Wood (ed.), *Franks and Alamanni in the Merovingian Period: An Ethnographic Perspective*, pp. 141–65. Woodbridge: Boydell Press.

Hamerow, H. 2002. *Early Medieval Settlements: The Archaeology of Rural Communities in North-West Europe 400–900.* Oxford: Oxford University Press.

Hardt, M. 2001. Hesse, Elbe, and Saale and the frontiers of the Carolingian empire, in W. Pohl, I. Wood, and H. Reimitz (eds.), *The Transformation of Frontiers from Late Antiquity to the Carolingians*, pp. 219–32. Brill: Leiden.

Hartshorne, R. 1970. The functional approach in political geography, in R. E. Kasperson and J. V. Minghi (eds.), *The Structure of Political Geography*, pp. 34–49. London: University of London Press.

Hedeager, L. 1992. *Iron Age Societies: From Tribe to State in Northern Europe 500 B.C. to A.D. 700.* Oxford: Blackwell Publishing.

Herodotus. 1966. *The Histories.* Harmondsworth: Penguin.

Hill, D. 1974. Offas's and Wat's Dykes: some exploratory work on the frontier between Celt and Saxon, in T. Rowley (ed.), *Anglo-Saxon Settlement and Landscape*, pp. 102–07. Oxford: BAR British Series 6.

———. 1985. The construction of Offa's Dyke. *Antiquaries Journal* 65: 140–42.

Hunter, F. 2001. Roman and native in Scotland: New approaches. *Journal Roman Archaeology* 14: 289–309.

Isaac, B. 1987. The meaning of the terms *limes* and *limitanei*. *JRS* LXXVII: 125–47.

———. 1990. *The Limits of the Empire: The Roman Army in the East.* Oxford: Clarendon Press.

Jack, S. M. 2004. The "debateable lands," terra nullius, and natural law in the sixteenth century. *Northern History* XLI (2): 289–300.

Jiménez, M. G. 1989. Frontier and settlement in Castile, in R. Bartlett and A. MacKay (eds.), *Medieval Frontier Societies*, pp. 49–74. Oxford: Oxford University Press.

Jones, G. D. 2000. The lowland Maya, from the conquest to the present, in R. E. W. Adams and M. J. MacLeod (eds.), *Mesoamerica, Part II: The Cambridge History of the Native Peoples of America*, pp. 346–91. Cambridge: Cambridge University Press.

Jones, S. B. 1964. Boundary concepts in the setting of space and time, in W. A. D. Jackson (ed.), *Politics and Geographic Relationships*, pp. 119–34. New Jersey: Prentice Hall.

Kelly, E. P. 2006. Secrets of the bog bodies: The enigma of the Iron Age explained. *Archaeology Ireland* 20: 26–30.

Kennedy, D., and Riley, D. 1990. *Rome's Desert Frontier From the Air.* London: Batsford.

Kristof, L. K. D. 1970. The nature of frontiers and boundaries, in R. E. Kasperson and J. V. Minghi. (eds.), *The Structure of Political Geography*, pp. 126–31. London: University of London Press.

Lamb, A. 1968. *Asian Frontiers: Studies in a continuing problem.* London: Pall Mall Press.

Lattimore, O. 1962. *Studies in Frontier History: Collected papers 1928–1958.* London: Oxford University Press.

Linduff, K. M. 1995. Zhukaigou, steppe culture and the rise of Chinese civilization. *Antiquity* 69: 133–45.

Lynn, C. J. 1989a. An Interpretation of "The Dorsey." *Emania* 6: 5–10.

———. 1989b. A bibliography of Northern Linear Earthworks. *Emania* 6: 18–21.

Mancall, P. C. 1988. Native Americans and Europeans in English America, 1500–1700, in N. Canny (ed.), *The Origins of Empire: The Oxford history of the British Empire, 1*, pp. 328–50. Oxford: Oxford University Press.

Mellor, R. E. H. 1989. *Nation, State and Territory: A Political Geography.* London: Routledge.

Millar, F. 1981. *The Roman Empire and Its Neighbour* (2nd ed.). London: Duckworth.

Minghi, J. 1970. Boundary studies in political geography, in R. E. Kasperson and J. V. Minghi (eds.), *The Structure of Political Geography*, pp. 140–60. London: University of London Press.

Moorhouse, S. A. 1981. Boundaries, in M. L. Faull and S. A. Moorhouse (eds.), *West Yorkshire: An Archaeological Survey to A.D. 1500*, pp. 265–89. West Yorkshire Metropolitan County Council.

Parker, B. J. 2002. At the edge of empire: Conceptualizing Assyria's Anatolian Frontier ca. 700 B.C. *Journal of Anthropological Archaeology* 21: 371–95.

———. 2006. Towards an understanding of borderland processes. *American Antiquity* 71(1): 77–100.

Pohl, W., Wood, I., and Reimitz, H. (eds.). 2001. *The Transformation of Frontiers from Late Antiquity to the Carolingians*. Brill: Leiden.

Power, D. 2004. *The Norman Frontier in the Twelfth and Thirteenth Centuries*. Cambridge: Cambridge University Press.

Raftery, B. 1994. *Pagan Celtic Ireland: The Enigma of the Irish Iron Age*. London: Thames and Hudson.

Reynolds, A. 2002. Burials, boundaries and charters in Anglo-Saxon England: A reassessment, in S. Lucy and A. Reynolds, *Burial in Early Medieval England and Wales*, pp. 171–94. Society for Medieval Archaeology Monograph 17.

Rollason, D. 2003. *Northumbria A.D. 500–1100: Creation and the Destruction of a Kingdom*. Cambridge: Cambridge University Press.

Stark, M. T. 2006. Introduction, in M. T. Stark (ed.), *An Archaeology of Asia*, pp. 3–13. Oxford: Blackwell Publishing.

Stephenson, P. 2000. *Byzantium's Balkan Frontier: A Political Study of the Northern Balkans 900–1204*. Cambridge: Cambridge University Press.

Tarzia, W. 1987. No trespassing: Border defence in the Tain Bo Cuailnge. *Emania* 3: 28–33.

Walsh, A. 1987. Excavating the Black Pig's Dyke. *Emania* 3: 5–11.

Wells, P. S. 1999. *The Barbarians Speak: How the conquered peoples shaped Roman Europe*. Princeton: Princeton University Press.

Wigg, A. 1999. Confrontation and interaction: Celts, Germans and Romans in the central German highlands, in J. D. Creighton and R. J. A. Wilson (eds.), *Roman Germany: Studies in Interaction*, pp. 35–53. JRA Supplementary Series 32. Portsmouth: Rhode Island.

Winchester, A. J. L. 2000. Dividing lines in a moorland landscape: Territorial boundaries in upland Britain. *Landscapes* 2: 16–32

Whittaker, C. R. 1994. *Frontiers of the Roman Empire*. Baltimore: John Hopkins University Press.

Williams, F. 1987. The Black Pig and Linear Earthworks. *Emania* 3: 12–19.

Wood, I. 1994. *The Merovingian Kingdoms 450–751*. London: Longman.

———. (ed.). 1998. *Franks and Alamanni in the Merovingian Period: An Ethnographic Perspective*. Woodbridge: Boydell Press.

Woolf, G. 1998. *Becoming Roman: The Origins of Provincial Civilization in Gaul*. Cambridge: Cambridge University Press.

Xenophon. 1972. *The Persian Expedition*. Harmondsworth: Penguin.

Zahariade, M. 1977. The structure and functioning of the Lower Danube *limes* in the 1st to 3rd centuries A.D., in J. Fitz (ed.), *Limes: Akten des XI Internationalen Limeskongresses*. Budapest.

19

THE ARCHAEOLOGY OF TERRITORY AND TERRITORIALITY

María Nieves Zedeño

The concept of *territory* has been variously used by scholars of many fields to denote a specific space or spaces to which individuals or groups of animals and humans are attached on a relatively exclusive and permanent basis. Here, discussion of territory is confined to modern humans, even though many useful things may be gleaned from the spatial frameworks of other species. The conceptual treatment of territory and territoriality, as well as relevant examples, focuses on non-industrial or non-nation-state societies, because these provide the most parsimonious analogs for interpreting the majority of Indigenous archaeological contexts.

Fields of Inquiry

Germane to any discussion of human territories is a consideration of the field of inquiry. Which discipline can best provide the theories and methods needed to unpack territory? Geography, sociology, psychology, ecology, anthropology, and biology are among the disciplines wherein territories have been defined and territorial behaviors systematically studied (e.g., Bakker and Bakker-Rabdau 1973; Casimir and Rao 1992; Graham 1998; Harvey 2000; Kelso and Most 1990; Malmberg 1980, 1983; Sack 1983, 1986, 1997; Saltman 2002; Soja 1971). In his treatise on human territoriality, geographer

Torsten Malmberg (1980: 16) went as far as to propose that the study of territory or "territorology" could become a legitimate field if it were approached with a solid comparative and multidisciplinary framework for the study of human behavior with respect to spaces of various shapes, sizes, and qualities. A science of territory, he thought, should encompass a broad scientific base, in which ethology, animal psychology, and ecology could be successfully combined with ethnology, physical anthropology, and sociology.

That such a field of inquiry would require the intellectual contribution of so many disciplines attests to the complexity of territory-making behaviors and partially explains why no discipline has, on its own, fully addressed all aspects of human territoriality and territory formation; nor has there been a strong push toward achieving this goal. One notable exception, and perhaps also a promise for the future of studies of territory, is archaeology, which not only uses material traces to identify human territories but also adapts models from "living disciplines" to reconstruct actions, events, and processes associated with the emergence, maintenance, and transformation of a territory (Zedeño 1997). The ability to draw concepts and models from a broad and sometimes disparate range of scientific and humanistic fields and build them into a single interpretive framework about human societies is a unique characteristic of archaeological practice that manifests itself in contemporary studies of land use and territoriality.

Although territories encompass a vast range of human actions, archaeologists generally seek solutions to the problem of identifying territories from the material record by adapting the scale, content, and historical relevance of frameworks lent by other disciplines in order to fit them into particular theoretical perspectives and research topics. For example, territorial models that employ principles of evolutionary biology and evolutionary and behavioral ecology (e.g., Allen and Hoekstra 1992; Dyson-Hudson and Smith 1978; Winterhalder and Smith 1981) are popular among archaeologists interested in territory formation vis-à-vis hunter-gatherer adaptations (e.g., Bettinger 1991; Binford 1982; Eerkens 1999; Kelly 1995; Lee and DeVore 1968) and adoption of agricultural economies (e.g., Rosenberg 1990, 1998). Inquiries into long-term change in land-use strategies incorporate geological and ecological models (e.g., Rossignol and Wandsnider 1992), whereas spatial analyses of land and resource use draw heavily from geography (e.g., Holl and Levy 1993; Morehouse 1996). Geographic Information Systems (GIS), in particular, have opened new avenues for comprehending and interpreting land and resource use at unanticipated scales (e.g., Aldenderfer and Maschner 1996; Graham 1998; Heilen and Reid 2006). Recent advances also include agent-based modeling of land use efficiency (Kohler et al. 2000).

Those interested in sociopolitical organization, in contrast, approach the study of territories and territorial behaviors from neoevolutionary perspectives to address the effect of spatial circumscription, social and environmental stresses, conflict, and warfare (e.g., Bender 2001; Chrisholm and Smith 1990; Keeley 1996; Kim 2003; Saltman 2002; Walsh 1998). For the most part, archaeologists have combined anthropological models with elements from geography, ecology, and biology to interpret differential spatial distributions of material items as indicators of social, political, or ethnic boundaries (De Atley and Findlow 1984; Graves 1994; Provansal 2000; Stark 1998; Sampson 1988; Wobst 1974). Postmodern social theories of structuration, power, and identity (e.g., Calhoun 1994; Forsberg 2003; Giddens 1984; Saltman 2002) also contribute to developing an understanding of the social and political construction of territories as well as the development of territorial boundaries and identities.

An innovative trend in contemporary archaeology is the integration of cultural landscape research into the study of territory (Garraty and Ohnersorgen In press; Heilen and Reid 2006; Whittlesey 1998). Research on power struggles, inequality, and contested landscapes (Bender 2001) also tackles the development of territorial strategies in the face of class and ethnic differences. Symbols and memory, too, are strongly linked to the prevalence of attachments to land and resources and to the maintenance of status quo in power relations (e.g., Knapp and Ashmore 1999; Lane 2003; Meskell 2003 Van Dyke and Alcock 2003). In this chapter, connections between territory and landscape are explored, as are other perspectives relevant to understanding human territoriality.

Concepts and Frameworks

Chief among key concepts used to discuss human-nature interactions is *territory as object aggregate* (land + natural resources + human modifications) (Zedeño 1997: 69) and *territoriality* as the sum of actions and emotions toward a specific space, with an emphasis toward influence, control, and differential access (Malmberg 1980: 10; Sack 1983: 55; Soja 1971: 19). Commonsense usage of territory presupposes the existence of more or less homogeneous spaces with recognizable boundaries or at least some type of distinctive marking intended to prevent access by those who do not own or possess them. This view derives from modern Western geopolitical thought; however, the existence of diverse forms of human territoriality observed by anthropologists, geographers, and ecologists show that this usage does not directly apply to non-nation-state societies (Zedeño 1997, 2000).

Also key in conceptualizing archaeological territories is the distinction between territory as *space* and territory as *land*. Scholars of various disciplines who focus on territorial behaviors address territory in terms of space, which allows them to discuss a broad range of behavioral contexts, from personal space to a state's territorial base (Cieraad 1999; Malmberg 1980; Sack 1983; Valentine 2001). Land is, therefore, a type of space. Inconveniently enough, land has many an ambiguous meaning. Here, land is used as synonym of terrain, upon which lie natural resources and objects of human manufacture. Perhaps the most useful result of decoupling land from resources is finding that human actions vary in nature, extent, and intensity according to the properties and significance of specific resources and singular landforms (Zedeño 1997, 2000; Zedeño et al. 1997). At least in principle, ties to land or resources subsequently lead to the emergence of different forms of territoriality and corresponding object aggregates. Furthermore, and as demonstrated by Bradley (this volume), land is rich in meaning, and meanings differ among peoples and cultures. Land is thus always more than abstract space (see Casey, this volume) and always more than a universally recognized source of resources. More is thus at stake than "space" and "land" when it comes to defining territory (see below).

It is important to keep in mind, particularly in archaeological research, that territorial actions and emotions do not necessarily result in the formation of a territory as a material manifestation or object aggregate. By the same token, a territory inferred archaeologically by the presence of differential distributions of human modifications on the terrain does not necessarily imply exclusive behaviors or effective control over that specific space. Although human modifications are generally interpreted as a form of ordering and claiming land, research by Whitelaw (1994), Belanger (2001), Stanner (1965), and Myers (1988) among many others, demonstrates that there are enduring principles of spatial order that neither require the construction of permanent facilities, nor necessarily result in the patterned distribution of portable artifacts. As Thomson (1939) has also shown for parts of northeastern Australia, neighboring peoples of different territorial groups (such as those divided by a river that marks a territorial boundary) have more in common culturally than geographically distant individuals of the same territorial group.

Contemporary studies of territoriality in past societies have moved beyond spatial distributions of portable artifacts to evaluate previously ignored features such as shrines (Mather 2003), megaliths and statuary (Kalb 1996; Shepardson 2005), and rock art (Dematte 2004; Taçon, this volume) as material signals of ancient territories. In reality, it is the combination of a host of human modifications as well as natural features that allow archaeologists to identify not only territories in general but also specific forms and stages of territoriality (Zedeño 1997)— hence the importance of carefully and thoroughly incorporating salient characteristics of the natural setting, as Bradley (2000) and Williamson (1982) suggest, into archaeological inferences of land use, order, and territory formation.

Unpacking Object Aggregates

A weak understanding of territories as the material manifestations of human territoriality may be blamed for the lack of a useful framework for reconstructing territories in archaeology. Additionally, overemphasis on spatial structures has overshadowed

the dynamic nature of human-land interactions that form territories. Conceiving a territory as an object aggregate allows one to integrate formal, spatial, and temporal dimensions in a single "empirical" life history framework (adapted from Schiffer 1987: 13, as the sequence of formation, use, and transformation of objects and aggregates), where each object in the aggregate has its own life history. It may be said that territories follow specific trajectories that result from the combined natural history of land and resources and social history of land and resource users (Zedeño 1997: 73). In this non-anthropocentric approach, any object in the aggregate can potentially affect behavior; land and resources (for example, volcanoes, seasonal floods, or migratory herds), in fact, have played active roles in shaping and changing human territories. Even objects of human manufacture, such as ruins left by previous occupants, can inform and determine territory formation among many other activities (Schiffer 1999).

Dimensionality of Territory

Human territoriality, in all its economic, social, political, and ritual realms, is enacted in three dimensions: (1) the formal or material dimension, which refers to the physical characteristics of land, resources and human modifications; (2) the spatial or relational dimension, which encompasses the loci of human action, as well as the inter-active links that, through the movement of actors, connect loci to one another; and (3) the temporal of historical dimension, which is characterized by sequential links resulting from successive use of land and resources by individuals and groups (Zedeño 2000: 107). Specific properties of territories, including structure (for example, continuous or discontinuous), organizing principles (kinds of activity loci; classificatory systems; layering or nesting of activity loci; boundaries), and transformative processes (for instance, expansion, contraction, consolidation, abandonment, reclamation) may, in turn, be identified across one or more dimensions.

Territorial strategies of mobile hunters and foragers, which have generated so much scholarly debate over the years, illustrate the importance of examining territories as object aggregates dimensionally and inclusively. A basic and sound observation is that foragers' territories relate primarily with resource distributions rather than fixed land tracts, so foragers may control noncontiguous resource patches or water sources (Dyson Hudson and Smith 1978; Kelly 1995; Malmberg 1983). This is the case for territories of historic buffalo hunters of the North American Plains (Milloy 1991) and trappers of the Subartic interlakes (Belanger 2001), which were defined by the habitats of target species. The wide-ranging buffalo herds of the 19th century, for example, frequently forced hunting groups to anticipate the seasonal movement of herds; hunting often required penetration beyond their familiar and exclusive hunting grounds, leading to violence along territorial boundaries of competing groups (Bowers 2004; Ewers 1958). Facilities associated with resource uses—including, for instance, buffalo cairns and offering locales, drive lines, impoundments, jumps, and camping circles—were scattered across the landscape where they would be most useful logistically to approach, hunt, and process game. Key landscape features, such as buffalo jumps, may have been used over long periods of time by successive hunting groups.

Cross-cutting buffalo herd ranges of 19th-century northern Plains hunters were territories of other animal species that required alternative territorial strategies. For example, bear dens and trails were considered exclusive to bears and thus were avoided by those groups with religious taboos against bear consumption (Ewers 1958: 85). However, eagle trapping rights belonged to individuals with ceremonial rights to them and their trapping territories were strictly respected by the community at large (Wilson 1928). Many plant habitats (for example, berry patches), as well as mineral sources (for instance, flint and pigments, salt licks), were also approached from the perspective of individual and group use rights (Bowers 2004). Among historic interlake hunters and trappers, ownership of wild rice beds was a family affair; however, the actual harvesting process was orchestrated by supra-family leaders (Veenum 1988: 176) just as buffalo hunts were in the plains. Territories formed around specific resources and, therefore, criss-crossed land tracts and often overlapped, each representing a particular realm of life—economic, social, political, and spiritual. Spirit beings, too, had exclusive rights to spaces (ravines, certain springs, certain landforms, river pools), to which people did not have granted or unconditional access. Nevertheless, for these and many other hunter and forager societies (e.g., Williams 1982: 151), rules of knowledge acquisition and boundary-crossing regulations (which are rooted in social and in religious principles), facilitated individual or group access to important places and resources owned or controlled by another entity. When combined, all these relations among people, land, and resources paint a complex, fluid, and highly dynamic picture of territorial strategies and corresponding object aggregates.

Territory Life Histories

Perhaps because many fields of inquiry lack the benefit of long-term views that archaeology furnishes, territories are rarely conceived as dynamic units. Yet, territoriality is a process whereby individuals or groups develop and even inscribe attachments to a particular place over a given period of time (see Taçon, this volume). Although it is true that not all actions are territorial in the conventional sense that presupposes control, exclusion, and defense (Sack 1983: 55), for humans to be able to interact in a three-dimensional space that may eventually become a territory, they must at least possess access, opportunity, and freedom of disposal of that space.

Territory aggregates, then, are the material expression of the territorial process, or what Ingold (1986) and Mather (2003) call appropriation or domestication of nature. Territory formation may also involve the process of appropriation and domestication of another's territory. A generalized territory life history, as sketched in Figure 19.1, shows relationships among three main formation stages—establishment, maintenance, and transformation—as well as processes within each stage that may or may not result in the successful generation of a territorial unit (Zedeño 1997: 86). A further consideration for future studies of territory is the application of life history approaches to the modeling of territory life spans relative to specific forms of territoriality that do not follow the old social evolutionary model but that consider agency, practice, and historical contingency alongside systemic processes of territory formation.

Landscape and Territory

Human-nature relations that result in the social construction of landscape, including individual and group identity as well as memory, imply that individuals and groups were able to engage in direct interactions with their surroundings. These, in turn, may have introduced temporary or permanent modifications into the natural setting (Gil García 2003; Zedeño and Stoffle 2003). Heritage landscapes, for example, assume that the ancestors of the group who inherited a landscape had access and opportunity for effective interaction (e.g., Freire 2003; Larsen 2003). There is little doubt in the mind of contemporary Palestinian, Siouan, Athapaskan speakers or of any conquered, dispossessed, or relocated people for that matter, that the landscape, as expressed in oral tradition, historic documents, sacred texts, maps, monuments or memories, was

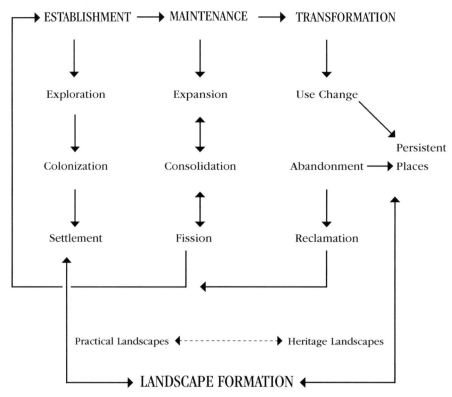

Figure 19.1 Territory life history and landscape formation.

once theirs, that it *belonged* (or still belongs) to them, and that they could traverse it at will and use it on their own terms.

Although archaeological and historic landscapes contain the history of past interactions among humans and the natural and supernatural worlds, landscapes belong in the present, as they are reified in memories about past interactions and present practices aimed at preserving ancestral connections and land-based identities (see also Bradley, this volume). Hence, a landscape begins at the time people come into contact with land and resources; it extends as people develop territorial attachments and strive to possess land and resources; and, through memory and action (e.g., pilgrimage, storytelling), it continues to change long after people have rescinded possession of a territory (Zedeño and Stoffle 2003: 75). Landscapes tend to be cumulative, incorporating past and present territories. Thus, landscapes and territories have parallel life histories with common beginnings rooted

in actual experiences, with overlapping spatial and formal dimensions, but with distinctive temporal scales and *territory* histories being generally shorter or narrower than *landscape* histories.

Examples of territory-to-landscape transition abound in origin and migration traditions among North American tribes. The Great Lakes Ojibwa, for example, have clearly mapped on the land the journeys they took, people they fought, spirits they encountered, and many other actions and events that took place along the journey to their present reservations. In fact, Ojibwa maps, such as Redsky's bark scroll of the migration myth (Dewdney 1975), depict actual loci of interaction among humans, nature, and the supernatural in a not-so-distant past (1700s–1800s), when the historic Ojibwa bands were advancing west and contesting territories then possessed by the Dakota Sioux and other groups (Zedeño and Stoffle 2003). Redsky's bark scroll illustrates how places were added to the migration story as bands arrived at certain destinations, took possession of the land peacefully or through war, and began to form their own territorial units. At the same time, the old territories became part of that large, storied, and mythical space that is the fabric of a human landscape.

Conclusions

Essential to any archaeological study of territories is the understanding that these spaces encompass the historical record of human interactions with land and resources and are multidimensional and even non-anthropocentric. Through time, humans may create different kinds of object aggregates that require specific forms of access and that represent multifarious individual and social relationships with land and resources. Although territories undergo transformations often leading to abandonment, attachments to territorial units of different ages and geographies generally remain and evolve in the history, memory, and practice of individuals and groups—these are socially constructed landscapes.

Thus, *landscape* formation cannot be fully understood without explicit reference to *territory*. While it is tempting to explain territory simply as a special kind of human landscape—that which represents effective use, influence, and control of land and resources over a specific period of time, it is more useful to argue, instead, that for landscapes to exist they had to have been effectively and even exclusively used or experienced by individuals and groups. Given the ambiguity and multifacetedness of the landscape concept as it is used in archaeology, this argument is all the more compelling because it proposes that the study of territory can furnish insights into the ways in which humans socially construct landscapes as places rich in meaning and experience.

References

Aldenderfer, M., and Maschner, H. D. G. 1996. *Anthropology, Space, and Geographic Information Systems*. New York: Oxford University Press.

Allen, T. F., and Hoekstra, T. W. 1992. *Toward a Unified Ecology*. New York: Columbia University Press.

Bakker, C. B., and Bakker-Rabdau, M. K. 1973. *No Trespassing! Explorations in Human Territoriality*. San Francisco: Chandler and Sharp Publishers.

Belanger, Y. 2001. The region "teemed with abundance": Interlake Saulteaux concept of territory and sovereignty. *Papers of the Algonkian Conference* 32: 17–34.

Bender, B. 2001. Introduction, in B. Bender (ed.), *Contested Landscapes: Movement, Exile, and Place*, pp. 1–18. Oxford: Berg.

Bettinger, R. L. 1991. *Hunter-Gatherers: Archaeological and Evolutionary Theory*. New York: Plenum Press.

Binford, L. 1982. The archaeology of place. *Journal of Anthropological Archaeology* 1: 5–31.

Bowers, A. W. 2004. *Mandan Social and Ceremonial Organization*. Lincoln: University of Nebraska Press.

Bradley, R. 2000. *An Archaeology of Natural Places*. London: Routledge.

Calhoun, C. 1994. *Social Theory and the Politics of Identity*. Malden: Blackwell Publishing.

Casimir, M. J., and Casimir, A. R. 1992. *Mobility and Territoriality: Social and Spatial Boundaries among Foragers, Fishers, Pastoralists, and Peripatetics*. New York: Berg.

Chrisholm, M., and Smith, D. M. 1990. *Shared Space, Divided Space: Essays on Conflict and Territorial Organization*. Boston: Unwin Hyman.

Cieraad, I. 1999. *At Home: An Anthropology of Domestic Space*. Syracuse: Syracuse University Press.

De Atley, S. P., and Findlow, F. J. 1984. *Exploring the Limits: Frontiers and Boundaries in Prehistory*. BAR International Series 223. Oxford: BAR.

Dematte, P. 2004. Beyond shamanism: Landscape and self-expression in the petroglyphs of Inner Mongolia and Ningxia (China). Cambridge Archaeological Journal 14: 5–23.

Dewdney, S. 1975. *Sacred Scrolls of the Southern Ojibway*. Toronto: University of Toronto Press.

Dyson-Hudson, I., and Smith, E. A. 1978. Human territoriality: An ecological reassessment. *American Anthropologist* 80: 21–41.

Eerkens, J. W. 1999. Common pool resources, buffer zones, and jointly owned territories: Hunter-gatherer land and resource tenure in Fort Irwin, southeastern California. *Human Ecology* 27: 297–318.

Ewers, J. C. 1958. *The Blackfeet: Raiders of the Northwestern Plains*. Norman: University of Oklahoma Press.

Forsberg, T. 2003. The ground without foundation: Territory as a social construct. *Geopolitics* 8(2): 7–24.

Freire, G. 2003. Tradition, change, and land rights: Land use and territorial strategies among the Piaroa. *Critique in Anthropology* 23: 349–72.

Garraty, C. P., and Ohnersorgen, M. A. In press. Negotiating the imperial landscape: The geopolitics of Aztec control in the Outer Provinces of the empire, in B. Bowser and M. N. Zedeño (eds.), *Archaeologies of Meaningful Places*. Salt Lake City: University of Utah Press.

Giddens, A. 1984. *The Constitution of Society: Outline of the theory of Structuration*. Berkeley and Los Angeles: University of California Press.

Gil García, F. M. 2003. Manejos espaciales, construcción de paisajes, y legitimación territorial: En torno al concepto de monumento. *Complutum* 14: 19–38.

Graham, S. 1998. The end of geography or the explosion of place? Conceptualizing space, place, and information technology. *Progress in Human Geography* 22: 16585.

Graves, M. W. 1994. *Kalinga Social and Material Culture Boundaries: A case of spatial convergence*, W. Longacre and J. Skibo (eds.), *Kalinga Ethnoarchaeology*. Washington, DC: Smithsonian Institution Press.

Harvey, D. C. 2000. Landscape organization, identity and change: Territoriality and hagiography in medieval west Cornwall. *Landscape Research* 25: 201–12.

Heilen, M., and Reid, J. J. (In press). A Landscape of gambles and guts: Commodification of land in the Arizona frontier, in B. Bowser and M. N. Zedeño (eds.), *Archaeologies of Meaningful Places*. Salt Lake City: University of Utah Press.

Holl, A., and Levy, T. E. 1993. *Spatial Boundaries and Social Dynamics: Case studies from Food-Producing Societies*. Ann Harbor, MI: International Monographs in Prehistory.

Ingold, T. 1986. *The Appropriation of Nature*. Manchester: Manchester University Press.

Kalb, P. 1996. Megalith-building, stone transport and territorial markers: Evidence from Vale de Rodrigo, Evora, south Portugal. *Antiquity* 70.

Keeley, L. 1996. *War before Civilization*. New York: Oxford University Press.

Kelly, R. 1995. *The Foraging Spectrum: Diversity in hunter-gatherer lifeways*. Washington, DC: Smithsonian Institution Press.

Kelso, W. M., and Most, R. 1990. *Earth Patterns: Essays in Landscape Archaeology*. Charlottesville: University Press of Virginia.

Kim, J. 2003. Land-use conflict and the rate of the transition to agricultural economy: A comparative study of southern Scandinavia and central-western Korea. *Journal of Archaeological Method and Theory* 10: 277–321.

Knapp, A. B., and Ashmore, W. 1999. Archaeological landscapes: Constructed, conceptualized, and ideational, in W. Ashmore and A. B. Knapp (eds.), *Archaeologies of Landscape: Contemporary Perspectives*, pp. 1–30. Malden: Blackwell Publishing.

Kohler, T. A., Kresl, J., and Van West, C. 2000. *Be There Then: A Modeling Approach to Settlement Determinants and Spatial Efficiency among Late Ancestral Pueblo Populations of the Mesa Verde Region, U.S. Southwest*, T. Kohler and G. Gumerman (eds.), New York: Oxford University Press.

Lane, R. 2003. History, mobility, and land use interests of Aborigines and farmers in north-west Australia, in P. Stewart and A. Strathern (eds.), *Landscape, Memory and History*, pp. 136–65. London: Pluto Press.

Larsen, S. 2003. Promoting aboriginal territoriality through interethnic alliances: The case of the Cheslatta T'en in northern British Columbia. *Human Organization* 62: 74–84.

Lee, R. B., and De Vore, I. 1968. *Man the Hunter*. Chicago: Aldine.

Malmberg, T. 1980. *Human Territoriality*. New York: Mouton.

———. 1983. Water, rhythm and territoriality. *Geografiska Annaler* 66B(2): 73–89.

Meskell, L. 2003. Memory's materiality: Ancestral presence, commemorative practice, and disjunctive locales, in R. Van Dyke and S. Alcock (eds.), *Archaeologies of Memory*. Oxford: Blackwell Publishing.

Milloy, J. 1991. Our country: The significance of the Buffalo resource for a Plains Cree sense of territory, in *Aboriginal Resource Use in Canada*. Winnipeg: University of Manitoba Press.

Morehouse, B. 1996. A Functional approach to boundaries in the context of environmental issues. *Journal of Borderland Studies* 10(2).

Myers, F. 1988. Burning the truck and holding the country: Property, time, and the negotiation of identity among Pintupi Aborigines, in D. R. T. Ingold and J. Woodburn (eds.), *Hunters and Gatherers*, vol. 2, pp. 52–74. Oxford: Berg.

Parker Pearson, M., and Richards, C. 1994. *Architecture and Order*. London: Routledge.

Provansal, D. 2000. *Espacio y Territorio: Miradas antropológicas*. Barcelona: Department d'Antropología, Universidad de Barcelona.

Rosenberg, M. 1990. Mother of invention: Evolutionary theory, territoriality, and the origins of agriculture. *American Anthropologist* 92: 399–415.

———. 1998. Cheating at musical chairs: Territoriality and sedentism in an evolutionary context. *Current Anthropology* 39: 653–81.

Rossignol, J., and Wandsnider, L. 1992. *Space, Time, and Archaeological Landscapes*. New York: Plenum Press.

Sack, R. D. 1983. Human territoriality: A theory. *Annals of the Association of American Geographers* 73(1): 55–74.

———. 1986. *Human Territoriality: Its Theory and History*. Cambridge: Cambridge University Press.

———. 1997. *Homo Geographicus: A Framework for Action, Awareness, and Moral Concern*. Baltimore: Johns Hopkins University Press.

Saltman, M. 2002. *Land and Territoriality*. Oxford: Berg.

Sampson, G. 1988. *Stylistic Boundaries among Mobile Hunter-Gatherers*. Washington, DC: Smithsonian Institution Press.

Schiffer, M. B. 1987. *Formation Processes of the Archaeological Record*. Albuquerque: University of New Mexico Press.

Schiffer, M. B. 1999. *The Material Life of Human Beings: Artifacts, Behavior, and Communication*. London: Routledge.

Shepardson, B. 2005. The role of Rapa Nui (Easter Island) statuary as territorial boundary markers. *Antiquity* 79: 169–79.

Soja, E. W. 1971. *The Political Organization of Space*. Commission on College Geography Resource Paper 8. Washington, DC: Association of American Geographers.

Stark, M. T. 1998. *The Archaeology of Social Boundaries*. Washington, DC: Smithsonian Institution Press.

Thomson, D. F. 1939. The seasonal factor in human culture, illustrated from the life of a contemporary nomadic group. *Proceedings of the Prehistoric Society* 5: 209–21.

Valentine, G. 2001. *Social Geographies: Space and society*. Englewood Cliffs, NJ: Prentice Hall.

Van Dyke, R., and Alcock, S. 2003. *Archaeologies of Memory*. Oxford: Blackwell Publishing.

Veenum, T., Jr. 1988. *Wild Rice and the Ojibway People*. Minneapolis: Minnesota Historical Society Press.

Walsh, M. R. 1998. Lines in the sand: Competition and stone selection on the Pajarito Plateau, New Mexico. *American Antiquity* 63: 573–95.

Whitelaw, T. M. 1994. Order without architecture: Functional, social and symbolic dimensions in hunter-gatherer settlement organization, in M. Parker Pearson and C. Richards (eds.), *Architecture and Order*. London: Routledge.

Whittlesey, S. 1998. Archaeological landscapes: A theoretical and methodological discussion, in S. Whittlesey, R. Ciolek-Torrello, and J. Alschul (eds.), *Vanishing River*. Tucson: SRI Press.

Williams, N. M. 1982. A boundary is to cross: Observations on Yolngu boundaries and permission, in N. M. Williams and E. S. Hunn (eds.), *Resource Managers: North American and Australian Hunter-Gatherers*. Boulder, CO: Westview Press.

Wilson, G. L. 1928. Hidatsa eagle trapping. *Anthropological Papers of the American Museum of Natural History* 30: 99–245.

Winterhalder, B., and Smith, E. A. 1981. *Hunter-Gatherer Foraging Strategies*. Chicago: University of Chicago Press.

Wobst, H. M. 1974. Boundary conditions for Paleolithic social systems: A simulation approach. *American Antiquity* 39: 147–78.

Zedeño, M. N. 1997. Landscape, land use, and the history of territory formation: An example from the Puebloan Southwest. *Journal of Archaeological Method and Theory* 4: 67–103.

———. 2000. On what people make of places: A behavioral cartography, in M. B. Schiffer (ed.), *Social Theory in Archaeology*. Salt Lake City: University of Utah Press.

Zedeño, M. N., Austin, D., and Stoffle, R. 1997. Landmark and landscape: A contextual approach to the management of American Indian resources. *Culture and Agriculture* 19(3): 123–29.

Zedeño, M. N., and Stoffle, R. W. 2003. Tracking the role of pathways in the evolution of a human landscape: The St. Croix riverway in ethnohistorical perspective, in M. Rockman and J. Steele (eds.), *Colonization of Unfamiliar Landscapes: The Archaeology of Adaptation*. London: Routledge.

20

MARKS OF POSSESSION: THE ARCHAEOLOGY OF TERRITORY AND CROSS-CULTURAL ENCOUNTER IN AUSTRALIA AND SOUTH AFRICA

Paul Taçon

There are many different ways in which humans have marked, mapped, and managed territory, and a concern with relationships to land appears to be an ancient as well as contemporary activity (see Taçon 2002 for detailed list). When we look around the world today we see recent evidence of humans defining, dividing, and describing parcels of land almost everywhere, from fences to garden beds, from patrolled borders to lines on maps, from elaborate signposts to subtle changes in architecture. In the archaeological record multiple overlapping layers of boundary marking have accumulated to form historical patterns of group and individual relationships to place, space, and landscape. Some areas are convoluted and complex, challenging to decipher or to tease apart. Others are subtle or contain forms of territorial association that quickly dissolve into the dust, forests, jungles, or deserts that take over when humans move off.

The active defense of territory, including through both prominent boundary marking and warfare, is commonly associated with agriculturalists, cities, states, and organized religion. However, both recent and ancient hunter-gatherers also engaged in these activities, with evidence emerging in various parts of the world. In this chapter, some of the archaeological and ethnographic signatures of territorial marking, defense of territory, and relationships to territory are briefly explored for parts of Australia and southern Africa. Territorial association during times of change is a major focus.

Northern and Southeastern Australia

One of the ways Aboriginal people have and sometimes continue to connect to both land and to other people right across Australia is through Dreaming tracks, original paths and travel routes of powerful Ancestral Beings (e.g., see Taçon 2005a). There also were hundreds, possibly thousands, of shared ceremony sites, sacred sites, and special meeting places in the recent and more distant past. As well, we know of at least 125,000 rock art sites that contain imagery and depictions reflecting Aboriginal identity, experience, and relationships to land, to other creatures, to other people, to the past, and to Ancestral Beings who are said to have created all these things. There is debate as to how old aspects of the Dreaming cosmology might be, with some arguing it is of relatively recent origin (e.g., David 2002), whereas others maintain at least some aspects have great antiquity (e.g., Layton 1992, 2005; Taçon 2005b; Taçon and Chippindale 2001; Taçon, Wilson, and Chippindale 1996). Across much of Australia, links to land are also commonly expressed through kinship. Kinship and other land relationships are encoded into recent and contemporary painting across northern and central Australia, as well as parts of the southeast, and knowledgeable elders suggest that many sites also feature people-place connections in aspects of their rock art imagery.

The late Big Bill Neidjie, for instance, once told me that in more traditional times most fights and battles between large groups of Aboriginal people were over land and women. In reference to a large mural-like painted rock art panel, with dozens of opposed human figures arranged in two groups and armed with spears, Bill made special note of a prominent crack in the rock between the two troops. This natural feature, that divides one of the largest and most spectacular recent Kakadu warrior panels in half, represented the nearby East Alligator River, according to Bill, and the scene depicted a major battle of 200 years ago. This led to a major study of western Arnhem Land rock art with depictions of fighting, battles, and, arguably, warfare (Taçon and Chippindale 1994). Several previously unknown and quite early panels were discovered in the process, at least 4,000–6,000 years of age, with rows of armed stick figures arranged in opposing formations and volleys of spears shown in flight overhead. Some figures on the front lines are riddled with spears while a few farther back are also injured. These are the oldest

battle scenes from anywhere in the world and are quite different from instances of formalized ritualized combat recorded ethnographically.

Rock art can be used to elucidate changing relationships to land in many parts of the country, for instance, shifts from more "shamanistic" to "totemic" orientations beginning about 4,000 years ago (Taçon and Chippindale In press) and changes in views about caves between the Pleistocene and Holocene (Taçon et al. In press). One of the main ways we can examine some aspects of the Aboriginal response to early contact with Europeans is also through rock art, perhaps the most enduring indigenous archival record from the early colonial period. However, rarely have contact period rock art images been studied comprehensively in order to gain insight into how Aboriginal people depicted their responses to the arrival of Europeans. Layton (1992) was the first to provide useful summaries of contact rock art of various areas, and Frederick is one of the few in Australia who has focused research on early contact imagery, but her work has largely been confined to a small area of central Australia, Watarrka (Kings Canyon) National Park (1997, 1999) and to depictions of ships (2003). As Frederick notes: "Contact and cross-cultural studies remain a relatively unexplored theme in Australian

Figure 20.1 Recent battle scene at Ngarradj-Warde-Djobkeng, Kakadu National Park, N.T. Australia, showing two groups on the sides of a crack in the rock confronting each other.

Figure 20.2   Part of one of two known complex Simple Figure battle scenes, Arnhem Land plateau, N.T., Australia, between 4,000 and 6,000 years of age.

rock art research" (1999: 133), yet "the rock art of contact provides generous scope for a convergence of archaeological, anthropological, and historical research designs" (1999: 132). Few studies have built on Frederick's pioneering work, although McNiven and Russell (2002) have noted a focus by most rock art researchers on secular interpretations of contact rock art and contact material culture (2002: 32–33), when it has been interpreted by previous researchers at all. They conclude that "by extending a counter-reading of sketchy historical sources to include archaeological evidence such as contact rock art, we have revealed the existence of a post-contact Indigenous landscape that was regulated by ceremonial strategies and systems of place marking designed to combat European colonization" (2002: 37; see also David and Wilson 2002: 57–58).

It did not take long for Europeans to dramatically change the landscapes of Australia after their arrival in the late 1700s. Fences, boundary markers, signposts, forts, buildings, and other domineering expressions of possession, conquered territory, and defense soon popped up like a plague of invading mushrooms that eventually swept across the continent. Historic records, historical archaeology, and oral history provide dramatic accounts of the affects on Indigenous populations but also show just how different traditional European and

traditional Aboriginal Australian relationships to land were. There is an extensive and fast-growing literature on colonialism, especially Aboriginal responses and impacts on Australian landscapes. This is not the place to review these in detail, but a few particular aspects should be highlighted.

First of all, although the effects of the early contact period on rock art production have not been well studied in southeast Australia, and only slightly and selectively elsewhere, it is becoming evident that there were a range of Aboriginal responses to colonization. These include not only the incorporation of introduced subject matter into rock art bodies—such as Europeans, ships, horses, sheep, cattle, rifles, and hand guns—but also an increase in depictions of Ancestral Beings and other spiritual subject matter.

In much of southeast Australia, people were removed from their traditional lands and banned from ceremonial sites. Restricted to reserves, missions, and jails, rock and ground-based art motifs were transferred to wooden objects (e.g., Kleinert 1997; Taçon, South, and Hooper 2003) or sheets of paper (e.g., Sayers 1994). Both scenes of traditional and contemporary life were depicted, and many decorated wooden objects, such as boomerangs and shields, were made both to express experience and to earn money to supplement meager rations. A strong attachment to land can be found throughout this iconography, and it continues in much contemporary urban Aboriginal art. Today, it also is more strongly political, with protest art, land rights themes, and statements about living conditions mixed with other land and identity expressions.

Today rock art sites, ceremonial grounds, middens, camping places, meeting places, missions, and massacre sites are all important places for Aboriginal people of southeast Australia, with each telling a different story about territory, attachment to land, combat, dispossession, and change. Massacre sites and battlefields are probably the most contentious of these places, but they are especially significant for Aboriginal Australians right across the country (e.g., see Elder 1988; Grassby and Hill 1988; as well as extensive recent literature by dueling historians such as Keith Windschuttle [2002] and Henry Reynolds [e.g., 1981, 1987, 1995]). Among other things, these sites highlight the profound differences between Aboriginal and invading European senses of territoriality, land attachment, land use, and each other. As Elder (1988: 200) concludes:

> In the case of Aborigines, their "reason to exist," both as individuals and as an entire race, has been systematically leached away by 200 years of dispossession. The intimate love of the land, the subtle ecological balance that recognized that there was a time to pick bush fruits and kill animals and a time to refrain from picking and killing, the careful response to the seasons, the powerful acknowledgment of the land's spirituality, the careful cycle of ritual and initiation that was at the centre of every life, the clear definitions of tribal land, these were all part of an elabor-ate and beautiful part of every Aborigine's "reason to exist."
>
> We, the invaders, took all that away. We destroyed it. We took the land as if it was our own. We destroyed the native fruit-bearing trees to create pastures for cattle and sheep. We killed native wildlife so that it would not compete for the pastures. We replaced ecology with aggressive 19th-century exploitation capitalism. We built roads over sacred sites. We denuded the land its spirituality. We killed off Aborigines with guns and poison and disease. We refused, through ignorance and arrogance, to see any tribal differentiation in those Aborigines who survived our insidious, long-term holocaust. Those Aborigines who survived were herded into reserves or "allowed" to live in humpies on the fringes of towns. We took away their reason to exist, and

when, in their despair, they took to the bottle or simply threw up their hands in hopelessness and gave up life, we had the arrogance to accuse them of drunkenness and laziness.

Southern Africa

As in Australia, the arrival of Europeans in southern Africa very quickly and radically transformed the land, but there had already been a long history of arrivals with farmers, raiders, and herders from the north settling "Bushman" or "San," here referred to as "Bushfolk" following recent precedent (e.g., Walker 1996; see also Taçon and Ouzman 2004: 42), lands beginning over 2,000 years ago. When early invaders from the north arrived, Indigenous Bushfolk already had a long history of land marking, and this continued in new ways in response to resulting changes that occurred to their lands. At least 100,000 painted rock shelters and open-air engraving sites can be found across southern Africa. As with the rock art of any area, there were many motivations and forms of meaning associated with imagery (including gender, identity, landscape, and politics to name a few), although there are convincing arguments that much of it is also associated with aspects of spirituality, belief, and ritual (e.g., Lewis-Williams 1981). And although the marking of landscapes with rock art always involves some form of connection to place and expression of individual and group identity, the particular boundaries that indigenous Bushfolk were most interested in marking were those between their everyday world and the spirit world.

Lewis-Williams and Dowson (1990), along with many others since, have noted this particularly for the painted sites. However, many engraved sites are also concerned with this form of boundary marking (Taçon and Ouzman 2004). In both cases, depictions of symbolically important animals, such as eland, mythical composite beings, and sometimes human-like figures, are arranged so that they appear to be moving into or out of cracks in the rock. In some cases they are aligned next to cracks with some body parts missing, giving the appearance that they are frozen in a state of transition—in the process of passing from one world to the next. Given that there are so many painted and engraved sites illustrating this phenomenon, spread right across southern Africa, it obviously was particularly important for the Bushfolk to mark these boundaries and to use these focal points to interact with other dimensions. Indeed, much of the early rock art of southern Africa seems to be oriented to this particular theme of boundary marking, with other

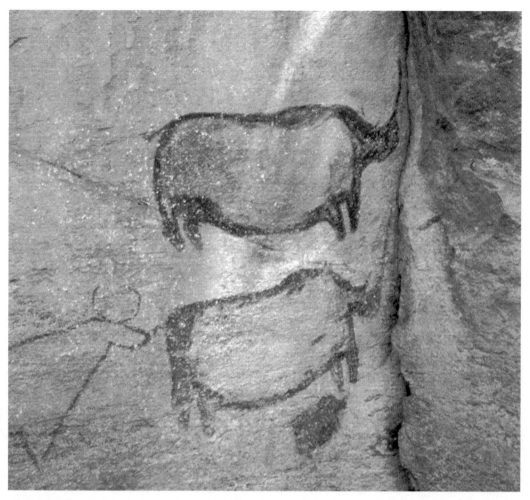

Figure 20.3 Two painted rhinos from Tsodilo Hills, northern Botswana, were deliberately placed to suggest that they are moving toward a crack in the rock face. If the observer moves from left to right, they appear to disappear inside the rock.

sorts of locations, such as high peaks, large rock holes, and a range of other prominent landscape features, marked with imagery that emphasizes boundaries with another world.

When new groups of people arrived in southern Africa they brought very different marking traditions with them, and there are now four other traditions recognized alongside that of the Bushfolk. Each group marked the landscape in different ways and in different contexts, and with differing concerns about boundaries. For instance, recent research has identified a widespread primarily geometric form of rock art. Much of it is engraved, whereas some is finger painted. Smith and Ouzman (2004) have presented compelling evidence to link this form of art to the Khoekhoen herders, rather

than to Bushfolk, Bantu-speaking farmers, Korana raiders, or European colonists. This geometric rock art is "banded along watercourses" (Smith and Ouzman 2004: 509) that herders are thought to have followed south, mostly in the central interior. At many locations, both herder and Bushfolk rock art is found, superimposed over each other in varying combinations. Herder rock art techniques and motifs, such as finger painting or fat-tailed sheep, are sometimes incorporated into Bushfolk art and vice versa. This shows, among other things, that both groups were using some of the same sites during certain periods and that cultural interaction and change were resulting in various ways. Whether the superimpositioning of one culture's art over another's at sites reflects actual thoughts of land possession and dispossession is debatable, but certainly this is an area of research worth exploring now that the Khoekhoen tradition has been isolated and defined.

Bantu-speaking farmers also produced rock art, but oral history has revealed it was mainly made in the context of initiation, protest, and Late Iron Age settlement (Smith and Ouzman 2004: 502). Finger-painted rock art was made by the Korana, usually in hidden cavelike locations. Over 450 paintings at 31 of these locations show influence from a range of ethnic groups; it primarily appears to be related to magic and military conquest (Ouzman 2005; Wadley 2001) and in this sense reflects concerns over possessing land. Bushfolk responses to invading black farmers and herders varied, with rock art indicating that some became enemies while others were "trading partners, friends and relatives" (Jolly 1998: 247), especially within southeastern Nguni and Sotho communities (Jolly 1995, 1996, 1998).

The rock art of European colonists is more letter- and number-based, with "the names and dates of early travelers seeking to inscribe themselves on the land, inscriptions made during the Anglo-Boer War (Ouzman 1999), quotidian images made by workers during the Great Depression, and prison inscriptions" (Smith and Ouzman 2004: 502). Some of it, such as that of early settlers and travelers, is very much concerned with possession of and connection to land, whereas other forms, such as the Anglo-Boer War inscriptions, reflect the acquisition, division, possession, and fighting over land more directly. As Ouzman (1999: 4) notes for the latter: "The art is found over a wide area—virtually everywhere the War was fought." However, it does not glorify war and is not deeply symbolic. "Rather, it shows how ordinary people experienced a brutal war that claimed at least 70,000 lives and cost in excess of £250 million" (Ouzman 1999: 5).

A war over land of a different kind is reflected in the final phase of Bushfolk rock art, which is a pale, ghostly white that depicts the end of a way of life. Violence, death, and destruction are key themes, along with distorted and grotesque imagery. "Images of the Apocalyptic Phase are found not only in large, spectacular rock shelters but also in small, often scrappy sites hidden behind boulder tumbles or tucked away deep in river valleys . . . Human and animal figures are combined in impossible

Figure 20.4 This purposely incomplete engraving of an eland appears to emerge from the edge of a boulder. It is one of four incomplete animal engravings around a central peak at Wildebeestkuil, South Africa (see Taçon and Ouzman 2004: 54–56).

ways. Frightening hallucinatory visions and explicit sexual scenes are recorded. Bizarre monsters with vicious teeth and protrusions from their heads lurk menacingly on shelter walls" (Ouzman and Loubser 2000: 41–42). About 800 paintings are associated with this phase, and Ouzman and Loubser argue that they represent a "millenarianism movement" that followed a failure of armed resistance against Black and White colonists, spear and rifle, cattle and horses: "Like the paintings, they faded into hidden rock shelters. They became an insubstantial presence in their homeland. Their previously stable spirit world became moribund even as the ordinary world became strange and violent" (Ouzman and Loubser 2000: 44). Some of this violence is reflected both historically and symbolically in rock art battle scenes, as Jolly (1998: 247) notes:

> Unambiguous depictions of black farmers in the rock art of the Drakensberg and adjacent areas can be found in the scenes of battle between San and these people. . . . The differences in weaponry and physical sizes of the groups have made it relatively easy to distinguish black farmers from San here, although we cannot be certain that all figures with bows and arrows in the art are San.

Many parallels with Australia can be drawn, especially in terms of both symbolic/spiritual and actual physical resistance to the invasion of traditional hunter-gatherer lands, showing that concepts of "territory" and "defense of territory" were common to hunter-gatherers of both regions. This pattern is also reflected in other aspects of the material culture/archaeological records (e.g., see Akerman and Brockwell 2007; Harrison and Williamson 2004; Mitchell 2002; Murray 2004; and various papers in Stahl 2004). Important parallels can also be made with North America. One of the best examples is "Writing-On-Stone," in the province of Alberta, Canada, a rock art complex with more than 280 panels at 93 sites in the proximity. There are numerous scenes of combat: horses, armed warriors, arrows in flight, and figures with shields, guns and spears. Many sites show figures engaged in fighting, whereas other complex panels contain narrative battle scenes. Michael Klassen (1995, 1998; Keyser and Klassen 2003) has convincingly demonstrated that the area was long considered a sacred place by local peoples and that "cosmic and historical conceptions of the world structured and controlled pictorial expression within Plains cultures, leading to the presence of both iconic and narrative imagery in rock art . . . While a shift towards increased historicality

and narrativity in the rock art of Writing-On-Stone occurred during the equestrian period, underlying continuities allied even the most narrative scenes to an overarching cosmic conception of the world" (Klassen 1998: 68). In other words, in Australia, southern Africa, and parts of North America, land was actively defended by Indigenous peoples physically, spiritually, and symbolically, all of which is evident in their rock art.

Conclusions

These case studies highlight many aspects of an archaeology of landscape that need to be better researched and understood. First, is the importance of rock art as sign, symbol, expression of identity, reflection of experience, and boundary marker all in one. Each of these signals the physical marking of the landscape as a social phenomena, one that leaves behind a material trace. In particular times and places, one of these might be emphasized more than others, but in essence all rock art functions at these and other levels at the same time. This makes specific interpretation difficult without informed knowledge from artists and/or their direct descendants, but more general interpretation, mainly accessible through formal analysis, including "landscape archaeology," can be undertaken (Taçon and Chippindale 1998).

Second, there are many different ways in which people have used rock art in Australia and southern Africa to express relationships to, and sometimes possession of, land. These need to be researched further, especially in relation to periods of culture contact and change. For instance, Akerman and Brockwell (2007) have recently linked the results of excavation, an analysis of change in point technology and rock art depictions of hunting in the Top End of the Northern Territory, showing that other forms of archaeological investigation support aspects of rock art interpretation. The implications of such research can then be applied to other areas, such as the Americas, northern Africa, Asia, and even Europe, in order to develop fresh hypotheses about ways in which people depicted culture change and their relationships to land. Indeed, the examples from Australia, southern Africa, and Alberta, Canada, suggest that people of very different backgrounds react in a similar range of ways to the invasion of their land and resulting culture change.

Third, the warfare rock art panels of western Arnhem Land, Australia, are significant for many reasons—especially because of their age, their territorial theme, and the fact the two opposing groups are undoubtedly both hunter-gatherers.

Until their discovery, it was thought that ancient hunter-gatherers rarely, if at all, engaged in organized territorial disputes, such activity being more characteristic of agriculturalists, kingdoms, city-states, and modern nations. The evidence from Arnhem Land suggests that arguments over land, resources, and perhaps ethnic difference may be very old human problems. This leads us to wonder if the earliest modern humans who ventured out of Africa over 100,000 years ago were also territorial from the start (see Keeley 1996; LeBlanc and Register 2003; van der Dennen 1995) and, if so, whether this helped them overwhelm and replace Neanderthals and *Homo erectus* across Europe and Asia. Evidence of territorial behavior prior to the arrival of people in Australia from any part of the world would be extremely important in confirming or refuting whether organized aggression over land is a recent phenomenon or something that began with the rise of modern humans, perhaps helping to make them so successful at occupying diverse and extreme environments.

Finally, in these times of ever-increasing culture contact, climate change, and communication, what can we learn from the ways people have coped with rapidly transforming worlds in the past? Will we build bigger barricades? Will we unite to fight problems of poverty, ignorance, climate instability, and political domination, or will we fight to defend concepts of territory and freedom? Will we descend into madness and millenarianism, with monsters, violence, and distortion increasingly dominating our environments? Some people might say that this is already happening, with a range of media delivering us the grotesque in many ways on a daily basis. Perhaps we need to develop new ways of looking at and relating to land. Instead of possessing, hoarding, and sectioning off land, we should focus on big pictures and landscapes, like Australian Dreaming tracks that remind us of where our sustenance comes from (Taçon 2005a). Indeed, we need to stop the slaughter, rape, and pillage of not only the land and other creatures but also ourselves. Ultimately, long-term human survival may well depend on it.

Acknowledgments

Bruno David and Julian Thomas are thanked for the invitation to contribute to this remarkable handbook. The Indigenous peoples of northern Australia, southeast Australia and southern Africa are especially thanked for sharing knowledge, time, friendship, and experience. David Canari, Shaun Hooper, and Oscar Motsumi deserve particular credit, although many Indigenous Australians and Africans, past and present, contributed to my growing knowledge of Indigenous time, place, relationship, and expression. I am also indebted to Christopher Chippindale, Ken Mulvaney, Sven Ouzman, and Wayne Brennan for both fieldwork and campfire/conference discussions in many parts of the world. The Australian Museum (Sydney) and Griffith University (Gold Coast) supported research and writing time that led to this chapter.

References

Akerman, K., and Brockwell, S. 2007. Bone points from the Adelaide River. *Australian Aboriginal Studies* 1: 83–97.

David, B. 2002. *Landscapes, Rock Art and the Dreaming: An Archaeology of Preunderstanding*. London: Leicester University Press.

David, B., and Wilson, M. 2002. Spaces of resistance: Graffiti and indigenous place markings in the early European contact period of northern Australia, in B. David and M. Wilson (eds.), *Inscribed Landscapes: Marking and Making Place*, pp. 42–60. Honolulu: University of Hawai'i Press.

Elder, B. 1988. *Blood on the Wattle: Massacres and Maltreatment of Australian Aborigines since 1788*. Frenchs Forest (Sydney): National Book Distributors and Publishers.

Frederick, U. 1997. Drawing in differences: Changing social contexts of rock art production in Watarrka (Kings Canyon) National Park, Central Australia. Unpublished Ph.D. thesis, Australian National University, Canberra.

———. 1999. At the centre of it all: Constructing contact through the rock art of Watarrka National Park, central Australia. *Archaeology in Oceania* 34: 132–44.

———. 2003. Considering contact rock art. Paper presented at the Australian Archaeological Association Conference, Jindabyne, NSW on 6 December, 2003.

Grassby, A., and Hill, M. 1988. *Six Australian Battlefields: The Black Resistance to Invasion and the White Struggle against Colonial Oppression*. North Ryde: Angus and Robertson.

Harrison, R., and Williamson, C. 2004. *After Captain Cook: The Archaeology of the Recent Indigenous Past in Australia*. Walnut Creek, CA: AltaMira Press.

Jolly, P. 1995. Melikane and Upper Mangolong revisited: The possible effects on San art of symbiotic contact between south-eastern San and southern Sotho and Nguni communities. *South African Archaeological Bulletin* 50: 68–80.

———. 1996. Symbiotic interaction between black farming communities and the south-eastern San:

Some implications for southern African rock art studies, the use of ethnographic analogy, and the cultural identity of hunter-gatherers. *Current Anthropology* 37: 277–305.

———. 1998. Modelling change in the contact art of the south-eastern San, southern Africa, in C. Chippindale and P. Taçon (eds.), *The Archaeology of Rock Art*, pp. 247–67. Cambridge: Cambridge University Press.

Keeley, L. H. 1996. *War before Civilization: The myth of the Peaceful Savage*. Oxford: Oxford University Press.

Keyser, J. D., Klassen, M. 2003. Every detail counts: More additions to the Plains biographic rock art lexicon. *Plains Anthropologist* 48: 7–20.

Klassen, M. 1995. Icons of power, narratives of glory: Ethnic continuity and cultural change in the contact period rock art of Writing-On-Stone. Unpublished M.A. thesis, Trent University, Peterborough, Ontario.

———. 1998. Icon and narrative in transition: Contact-period rock art at Writing-On-Stone, southern Alberta, Canada, in C. Chippindale and P. S. C. Taçon (eds.), *The Archaeology of Rock Art*, pp. 42–72. Cambridge: Cambridge University Press.

Kleinert, S. 1994. "Jacky Jacky was a smart fella": A study of art and Aboriginality in south east Australia 1900–1980. Unpublished Ph.D. thesis, Australian National University, Canberra.

Layton, R. 1992. *Australian Rock Art: A New Synthesis*. Cambridge: Cambridge University Press.

———. 2005. Review of Bruno David's "Landscapes, rock art, and the Dreaming: An archaeology of preunderstanding." *Antiquity* 79: 215–16.

Leblanc, S. A., with Register, K. 2003. *Constant Battles: The Myth of the Peaceful Savage*. New York: St. Martins Press.

Lewis-Williams, J. D. 1981. *Believing and Seeing: Symbolic Meanings in Southern San Rock Paintings*. London: Academic Press.

Lewis-Williams, J. D., and Dowson, T. A. 1990. Through the veil: San rock paintings and the rock face. *South African Archaeological Bulletin* 45: 5–16.

McNiven, I., and Russell, L. 2002. Ritual response: Place marking and the colonial frontier in Australia, in B. David and M. Wilson (eds.), *Inscribed Landscapes: Marking and Making Place*, pp. 27–41. Honolulu: University of Hawai'i Press.

Mitchell, P. 2002. *The Archaeology of Southern Africa*. Cambridge: Cambridge University Press.

Murray, T. (ed.). 2004. *The Archaeology of Contact in Settler Societies*. Cambridge: Cambridge University Press.

Ouzman, S. 1999. "Koeka ka kie, hents op bokkor of ik schiet!" Introducing the rock art of the South African Anglo-Boer War, 1899–1902. *The Digging Stick* 16(3): 1–5.

———. 2005. The magical arts of a raider nation: Central South Africa's Korana rock art. *South African Archaeological Bulletin Goodwin Series*. 9: 101–13.

Ouzman, S. and J. Loubser 2000. Art of the apocalypse: Southern Africa's Bushmen left the agony of their end time on rock walls. *Discovering Archaeology* 2(5): 38–45.

Reynolds, H. 1981. *The Other Side of the Frontier*. Harmondsworth: Penguin Books.

———. 1987. *Frontier: Aborigines, Settlers and Land*. Sydney: Allen and Unwin.

———. 1995. *Fate of a Free People: A Radical Re-Examination of the Tasmanian Wars*. Harmondsworth: Penguin Books.

Sayers, A. 1994. *Aboriginal Artists of the Nineteenth Century*. Melbourne: Oxford University Press.

Smith, B., and Ouzman, S. 2004. Taking stock: Identifying Khoekhoen herder rock art in southern Africa. *Current Anthropology* 45: 499–526.

Stahl, A. B. (ed.). 2004. *African Archaeology: A Critical Introduction*. Oxford: Blackwell Publishing.

Taçon, P. S. C. 2002. Rock art and landscapes, in B. David and M. Wilson (eds.), *Inscribed Landscapes: Marking and Making Place*. Honolulu: University of Hawai'i Press.

———. 2005a. Chains of connection. *Griffith Re-view Edition 9—Up North: Myths, Threats, and Enchantment*. Electronic document, URL: www.griffith.edu.au/griffithreview, accessed November 5, 2007.

———. 2005b. Marks on and of land: The relationship of rock and bark painting to people, places and the ancestral past, in C. Kaufmann and B. Luthi (eds.), *John Mawurndjul: Image and Land in Northern Australia*. Basel: Museum Tinguely.

Taçon, P. S. C., Brennan, W., Hooper, S., Kelleher, M., and Pross, D. In press. Differential Australian cave and rock-shelter use during the Pleistocene and Holocene, in H. Moyes (ed.), *Journeys into the Dark Zone: Throwing Light on Cross-Cultural Ritual Cave Use*. Boulder: University of Colorado Press.

Taçon, P. S. C., and Chippindale, C. 1994. Australia's ancient warriors: Changing depictions of fighting in the rock art of Arnhem Land, N.T. *Cambridge Archaeological Journal* 4: 211–48.

———. 1998. An archaeology of rock art through informed methods and formal methods, in C. Chippindale and P. S. C. Taçon (eds.), *The Archaeology of Rock Art*, pp. 1–10. Cambridge: Cambridge University Press.

———. 2001. Transformation and depictions of the First People: Animal-headed beings of Arnhem Land, N.T., Australia, in K. Helskog (ed.), *Theoretical Perspectives in Rock Art Research*, pp. 175–210. Oslo: Instituttet for sammenlignende kulturforskning.

———. In press. Changing places: North Australian rock art transformations 4000–6000 B.P., in D.

Papagianni and R. Layton (eds.), *Time and Change: Archaeological and Anthropological Perspectives on the Long Term in Hunter-Gatherer Societies.* Oxford: Oxbow Books.

Taçon, P. S. C., and Ouzman, S. 2004. Inner and outer landscapes: Rock art views of north Australian and southern African worlds within stone, in C. Chippindale and G. Nash (eds.), *The Figured Landscapes of Rock Art.* Cambridge: Cambridge University Press.

Taçon, P. S. C., South, B., and Hooper, S. 2003. Depicting cross-cultural interaction: Figurative designs in wood, earth and stone from south-east Australia. *Archaeology in Oceania* 38: 89–101.

Taçon, P. S. C., Wilson, M., and Chippindale, C. 1996. Birth of the Rainbow Serpent in Arnhem Land rock art and oral history. *Archaeology in Oceania* 31: 103–24.

van der Dennen, J. M. G. 1995. *The Origin of War.* Groningen: Origin Press.

Wadley, L. 2001. Who lived in the Mauermanshoek shelter, Korannaberg, South Africa. *African Archaeological Review* 18: 153–79.

Walker, N. 1996. *The Painted Hills: Rock Art of the Matopos.* Gweru: Mambo Press.

Windschuttle, K. 2002. The Fabrication of Aboriginal History, Vol. I: Van Diemen's Land 1803–1847. Paddington: Macleay Press.

21

FROM PHYSICAL TO SOCIAL LANDSCAPES: MULTIDIMENSIONAL APPROACHES TO THE ARCHAEOLOGY OF SOCIAL PLACE IN THE EUROPEAN UPPER PALAEOLITHIC

Jean-Michel Geneste, Jean-Christophe Castel, and Jean-Pierre Chadelle

In southwestern France, Paleolithic studies have long relied on the study of archaeological sites largely devoid of proper consideration of their environmental settings. Over a period lasting approximately ten years, initially researchers focused much of their attention to attributing cultural finds into chronological phases and sequences. It was only in the 1960s that serious attention began to be given to environmental contexts and raw material distributions so as to allow stone artifact sourcing studies to proceed (Valensi 1960). Subsequently, such studies on the provenance of archaeological raw materials proliferated (Borde 2002; Bressy et al. 2006; Geneste 1985; Tiffagom 2006), as did those concerned with the origins of faunal resources (Castel 1995; Costamagno 2006; Fontana 2001) and those comparing technological and subsistence provisioning strategies (Castel et al. 1998).

In contrast to these earlier and largely techno-economic programs, here we present a multidisciplinary study of past landscape engagements by systemically investigating the regional environment and its relationships with the material culture of Solutrean deposits in southwestern France. We combine these various sources of information to explore strategies of past landscape engagement, in particular as these relate to subsistence practices, technology, and social organization.

The Solutrean, an archaeological "culture" or phase (20,000–17,000 B.P.) of the European Upper Paleolithic (40,000–10,000 B.P.), is amenable to a range of functional and spatial research approaches, since its restricted temporal and spatial extension limits possible confusions between the synchronic and dynamic perspectives emphasized by Binford (Binford 1982a).

It is thus possible to approach different archaeological perceptions of Solutrean territories through patterning in the exploitation of animal and mineral resources (Castel et al. 1998, 2005, 2006). This analysis, which is synchronic owing to the short duration of the Solutrean, interrogates diverse data concerned with subsistence economy and symbolism to inform us on ecological contexts of territorial behavior through the spatial distribution of sites. In contrast to other recent studies that favor a limited disciplinary field or that concern limited data collected a long time ago, our approach concerns an interdisciplinary perspective of the human ecology of the Upper Pleistocene (Binford 1981; Ellen 1979; Guille-Escuret 1989).

In this chapter, we explore how what is by now a detailed archaeological understanding of the temporal and spatial distribution of Solutrean archaeological materials can inform us on Solutrean *territorial* networks—that is, how material distributions across the land can be used to inform us as to past social arrangements as sociopolitical landscapes.

Territories of Alimentary Resource Acquisition

To understand ancient territories, at least in part from animal resource acquisition activities, one must have an ensemble of well-preserved, contemporaneous sites, knowledge of the seasonality and duration of occupation at these sites, as well as an understanding of the zones of geographical distribution and patterns of movement of the fauna in question.

The Sites

In the Aquitaine region of France, the Solutrean sites that have well-preserved faunal remains contain palimpsests of different occupations; sites with specialized activities attributable to individual *events* are rare. All faunal remains were collected at four recently excavated sites: Combe Saunière, Cuzoul de Vers, Les Peyrugues, and Cave XVI. At five other sites, all fauna were collected from a limited area of the site: Le Placard, Jeanblancs, Casserole, Petit-Cloup-Barrat, and Sainte-Eulalie. The faunal assemblages of sites that were excavated a long time ago (Badegoule, Laugerie Haute, Pech de la Boissière, Fourneau du Diable, among others) have been altered by successive selection of the excavated material and by loss of parts of those collections.

Seasonality

The period of occupation at a site can be determined through analysis of seasonality. One method is based on tooth eruption and tooth wear among the fauna. Another consists of the analysis of dental cementum annuli for seasonality of death, a method applied mostly to ungulates (Castanet, Meunier, and Francillon-Vieillot 1992; Sergeant and Pimlott 1959). Other factors, such as the presence of fetuses, can also provide information concerning the seasonality of use at a site. These methods can be used on fauna where births are restricted to particular seasons.

We currently have seasonality information for five sites: Combe Saunière; Les Peyrugues; Cuzoul de Vers (cementum annuli, dental eruption, fetus); Badegoule; and Fourneau-du-Diable (dental wear and eruption, fetus). Seasonal hunting is characteristic of sites that were *occasionally* occupied and indicates geographic movements among human groups. Year-round hunting indicates either

a site of permanent residence or a palimpsest of shorter duration occupations at different times of the year (for example, seasonal occupation). It can thus be difficult to infer the movement of populations across space based only on attempts to interpret the seasonal presence of people at a given site.

On the southwestern border of the Massif Central in France, the southern sites of Cuzoul de Vers and Les Peyrugues were occupied during summer (Allard, Chalard, and Martin 2005; Martin In press). In the north, Combe-Saunière (Castel 1999; Pike-Tay in Castel 1999) and Fourneau-du-Diable (Fontana 2001) were occupied during late winter, spring, and perhaps summer. Badegoule was occupied year-round (Bouchud 1966).

A synthesis based only on these preliminary analyses would lead us to conclude that the sites in the south were occupied in summer, whereas those to the north were occupied year-round, notably including winter. The data are still too limited, however, and it is more important to observe that different site occupational strategies were probably followed in different parts of the Massif Central during the Solutrean. To further explore this issue, one must integrate archaeozoological results with other forms of archaeological and multidisciplinary information.

Distribution of Hunted Animals

We consider different animal species to be associated with different resource zones, which can for the purposes of our work be divided into three major environmental forms: (1) an open arctic zone; (2) a non-arctic open zone; (3) a forested zone (Delpech et al. 1983; Griggo 1995). Each of these environmental zones tends to be found at some distance from the others, rather than in short-spaced patchwork. The representation in the same site of species belonging to different environmental zones would thus attest to the circulation of animal bones across relatively large distances.

For the arctic environment, we use as a reference the model defined by Binford (1978, 1982b)—namely, exploitation of a territory from a residence base camp surrounded by specialized activity sites. Following this model, the species hunted within given locations will differ according to seasonal and geographical availability.

The different species identified at a site reveal choices made within an integrated, annual cycle of activities. In the case of human populations circulating over vast territories, the presence of a species represents its rank in an optimization of hunting activities, which is primarily related to the relative abundance of different species, their spatial distributions, and the economic profitability of the hunt.

From the perspective of the optimization of acquisition and exploitation (cf. Binford 1978; Keene 1981; Lee and De Vore 1968; Winterhalder and Smith 1981), one can distinguish three types of species representation within individual sites (Castel et al. 1998):

1. *Species brought back to the site whole then completely exploited.* These species are hunted near the sites. They correspond to specific ecological circumstances to which the human subsistence strategies are well adapted.

2. *Species of which only certain parts of the skeleton are introduced into the site.* This reduction in transport costs is sought in diverse circumstances, such as when hunting locations are far from occupation sites, when the animal carcass is particularly large, or in the absence of sufficient human means to exploit and transport the animal (e.g., Bartram 1993; Binford 1978; Bunn 1982). This is the case for bovids and mammoth, whose presence in the local environment is often difficult for archaeologists to determine. Such a representation of only partial skeletal presence could also be found among poorly represented small-sized species. We note that poorly represented species do not necessarily represent a lack of exploitation of a given environmental zone, because absence of evidence could relate to delayed consumption, among other possible explanations.

3. *Species represented exclusively by body parts of symbolic or technological value.* Such skeletal evidence cannot be taken as evidence for food consumption at those sites, because portable "art" made on animal bones could (and indeed in some circumstances at least to) were likely have been carried around over potentially large distances and over long periods of time.

For these reasons, our present analyses of the distribution of animal populations during the Solutrean concern only locally exploited ungulates.

The geographical distribution of animal populations 20,000 years ago was informed by reference to modern ethological data in order to identify the environments favorable to different species. Being unable to reconstruct the climate of the last glacial maximum with sufficiently fine-grained precision, researchers estimated faunal distribution zones from landscape relief and soil type. This argument can be applied only to species dependent on particular physical environments, such as horse and ibex. Reindeer, however, are present in large quantities at all Solutrean sites. This species was hunted everywhere within a local range and exploited for both alimentary and technological purposes. Bovids also appear to have been ubiquitous, and for them there is also the question of transport, which was forcibly selective between the kill site and the cave or rockshelter sites studied here. These species do not contribute to the question of the circulation of bone remains over long distances.

Despite the small sample size, some significant tendencies can be observed. Reindeer are largely dominant, both in terms of the number of individuals and the quantity of meat they provided. For the other species, there is a zonal variation from north to south: ibex are present only in the south, where horse and saiga antelope appear to be absent; chamois is present further north than ibex; red deer are present only in the central zone. At the sites in the northern half, the Solutreans hunted ten reindeer for every horse. Because horse bone remains are associated with greater quantities of meat than are reindeer bones, the alimentary contribution of this animal seems to be more important than apparent from ratios of the numbers of represented fauna alone. These examples show that the relative representation of fauna within archaeological sites does not necessarily indicate the relative importance of environmental zones as food resource zones.

The Geographical Distribution and Circulation of Technological Objects

The exploitation of non-alimentary animal and mineral resources (for technological and/or symbolic reasons) does not follow the same model of land use and geographical organization as that of alimentary resources.

Animal Resource Exploitation

The ibex remains at Combe Saunière (Castel 1999), Badegoule (Cheynier 1949), Le Placard (Griggo In press), and probably Fourneau du Diable (Stéphane Madeleine, personal communication), exist only in the form of pierced or unpierced incisors. Their presence attests to territorial relations with the southeast, as this is where ibexes came from. At Cuzoul de Vers, horse remains are represented almost exclusively in the form of retouchers (fragments of long bone diaphyses used to retouch flint tools), which could have been introduced into the site in the same way as lithic tools made from exogenous materials (Castel 1999). The situation at Les Peyrugues seems to be analogous (Michel Allard, personal communication). These remains could have originally come from some considerable distance away, if horse was absent in the Quercy region, or from earlier times, if horse was not exploited during the period of site occupation, or a combination of both.

Do the unique saiga antelope remains present in the Solutrean assemblage of Jeanblancs (Drucker et al. 2000) indicate a displacement from the north, or mixing with the Magdalenian levels in which this species has been identified, or perhaps an introduction by carnivores? This example illustrates the difficultly of interpreting the status of a species represented by very few remains. In any case, it does not appear to result from confusion with chamois because this species is absent from the deposits.

Mammoth is represented in several sites in the form of ivory fragments that are sometimes worked. In this case, it seems that we can attribute the presence of this material to an exploitation of a raw material source rather than to the exploitation of an animal for its meat. At Laugerie Haute, the presence of dozens of mammoth molars, numerous ivory fragments, and a few diaphysis fragments of standardized dimensions also indicate the transport of materials for purposes other than alimentary needs.

These different elements attest to the presence of technological and symbolic products made from animal materials in regions far from those where archaeozoological analyses show the presence of corresponding species, and which were intensively exploited following transport over short distances. These displacements of animal materials in the form of technological and symbolic products attest either to the movement of human groups or exchanges between partners within interacting groups.

Mineral Resource Exploitation

The study of the origins of lithic materials allows us to perceive a locality of social and environmental engagement, exploited daily (the "local zone"); a "neighboring zone," exploited episodically; and a "distant zone," which sometimes simply represents the preceding residence site(s). However, the raw materials exploited do not always represent the geographical patterning and physical availability of natural resources in the environment. For example, Bergeracois flint, an abundant and high quality material, appears to have been minimally used during the Solutrean.

Technological studies of lithic industries allow us to characterize the duration of occupation of sites. The presence of remains from all phases of the operational sequence indicates the exploitation of nearby flint sources, diverse activities, and prolonged occupation. On the other hand, the presence of only the initial phases of a reduction sequence reveals a knapping workshop and a brief, sometimes unique occupation at the flint source. The presence of remains from only the last phases of tool production indicates the preliminary exploitation of distant sources, specific activities, brief and possibly repeated occupations.

At Combe Saunière, the intensive and repeated exploitation of sources near the site represent 90% of the retouched objects, whereas distant sources (Aubry 1991; Servelle and Vaquer 2000) are attested only by the most efficient hunting equipment (Chadelle, Geneste, and Plisson 1991; Geneste and Plisson 1986; Plisson and Geneste 1989). The choice, availability, and optimized modes of transport of these diverse products seem to reveal a tradition common to all Solutrean groups within a range extending over more than 300 kilometers, from the Tarn region in the south to the Indre-et-Loire region in the north.

The economic strategies adopted within a local zone, regularly and intensively exploited within a 5-kilometer range around the residence sites, differ from those developed within the broader range that includes the distant zone. In the latter case where the distant zone is exploited, only selected sources are exploited and the products are diffused in the optimized form of raw blanks or roughouts within a vast interregional zone according to the rules, now classic for the Palaeolithic, of long-distance diffusion (Renfrew 1977). The distances of between 50 and 200 kilometers in the region considered at such broader spatial scales are far greater than those that can be covered in one or two days of walking. At the intermediary scale that incorporates a consideration of the neighboring zone, which varies according to the relative position of resource sources and habitats, we encounter intermediate economic solutions. It is also possible that within zones of social interaction (contacts and exchanges) located at medium distances from residence sites (one or two days by foot), unworked raw materials and roughouts may have circulated through exchanges between groups. Detailed studies of these questions are in progress.

The multiple territorial entities related to the economy of lithic materials are distributed over a vast geographic zone constituted by the middle and high valleys of the rivers circulating over the highlands of the Massif Central toward the Atlantic. This multiplicity has already been observed (Peyrony 1932) and remains difficult to interpret. Though a systematic confrontation of data related to the exploitation of the physical and faunal environments in this ensemble of Solutrean sites is still lacking, it already appears that geographical zones of engagement determined on exclusively economic bases were integrated within a single environmental zone. Human movements across long and medium distances are oriented according to a large north-south axis, whereas the east-west movements, following the river courses and natural paths of circulation, concern only short and medium distance movements, which archaeologically appear less frequently as these are more difficult to detect because of the natural downstream transport of flint blocks by rivers.

The Circulation of Objects of Sociosymbolic Value

To compare and contrast the geographical circulation of alimentary and technological objects with that of sociosymbolic objects (artistic, aesthetic, or ritual items), we now consider body ornaments (worked teeth, pendants on horse hyoid bones, beads, buttons, bracelets, shells, and fossils). Such an approach will enable us to consider the circulation of people across space in terms of relations of symbolic value.

Briefly, the method used here to understand the circulation of "non-economic"[1] symbolic objects, for which we do not have localized sources identifiable through the natural sciences, consists of an analysis of the spatial distribution of objects with highly determined characteristics (nature, technique, morphology, decoration) and which are found only in Solutrean contexts. Bilobed ivory buttons, ivory "bracelets" and "rings" with notched decoration, quadrangular beads with geometric motifs (Geneste and Rigaud In preparation), and notched pendants made on horse hyoid bones materialize contacts and exchanges between groups.

The geographical distributions of these objects indicate human movements across a maximum distance of 80 kilometers. These distributions are well integrated within the range of transported lithic raw materials; they may thus also correspond geographically to the alimentary subsistence zones superposed on the distribution zones of the exploited animals.

The origins of marine shells found in the majority of Solutrean sites indicate two constant but diametrically opposed sources: the Atlantic coast (both along the coastline and at fossil deposits in the Bordeaux region) and the Mediterranean coast.

These contact zones, geographically the most distant identified during the Solutrean, forcibly imply intergroup exchange zones that overlap at their margins with those of other contemporary sociocultural entities.

Conclusions

Certain aspects of the economic organization of prehistoric human groups across space and through time will always be out of reach; data relevant to the recurrence and seasonality of occupation are perceptible within archaeological sequences only in a relative and generalized time frame. Consequently, our conclusions must necessarily be restricted in scope. Although in this chapter our results are preliminary, this study, which is above all methodological in its attempt to integrate the results of archaeozoological analyses with those of lithic and osseous technological productions, has aimed to explore the spatial and territorial organization of people across the landscape for a very distant past, one that is entirely without local ethnography. Similar approaches have been undertaken for other periods of the Upper Palaeolithic, but in the majority of cases these have relied on single lines of evidence rather than multidisciplinary datasets, albeit aiming at the eventual multidisciplinary integration of results (Bressy et al. 2006; Chalard, Guillermin, and Jarry 2006; Costamagno 2006).

The sites studied here correspond essentially to sheltered sites that could have been regularly and frequently occupied, perhaps seasonally, during the Solutrean. This would explain the thickness of some archaeological deposits.

The archaeological assemblages of these rock-shelter sites are divided into two functionally significant categories. Those containing the longest sequences, with the richest and most diversified assemblages in terms of raw materials and especially domestic tools, correspond to prolonged periods of residence, or perhaps aggregation sites following the criteria defined by Conkey (1980). The other sites, with assemblages dominated by hunting equipment—Combe Saunière in particular—appear to be related to more temporary, probably repeated, hunting activities (acquisition, butchery, preservation, tool maintenance, and so on). The artifact assemblages at these sites attest to repeated use, sometimes to distant contacts, and occupations of varied duration and more or less complex organization.

The sites studied within the regional scales considered in this chapter would thus be linked by a network of contacts and exchanges and participate in a common system of regional mobility in the framework of an ecological exploitation differentiated in space and time, and by varied spatial and temporal scales. Technological and symbolic relations between places allow us to trace past contacts between sites during the Solutrean. In this way, the conventions employed in the manufacture of body ornaments indicate close social relationships between Combe Saunière, Fourneau du Diable, and Le Placard in the north (ivory beads, rings and bilobed buttons). Combe Saunière could

thus be considered as a satellite to the larger confirmed occupation sites, such as Fourneau du Diable (1 day by foot) and Le Placard (2 or 3 days by foot). The contents of these latter sites is much more diverse, and projectile points are proportionally less numerous. These larger occupation sites could also be defined as *aggregation sites* (Conkey 1980). This difference in status of site use seems to be confirmed by an integration of data related to parietal and portable art. Among these three sites,

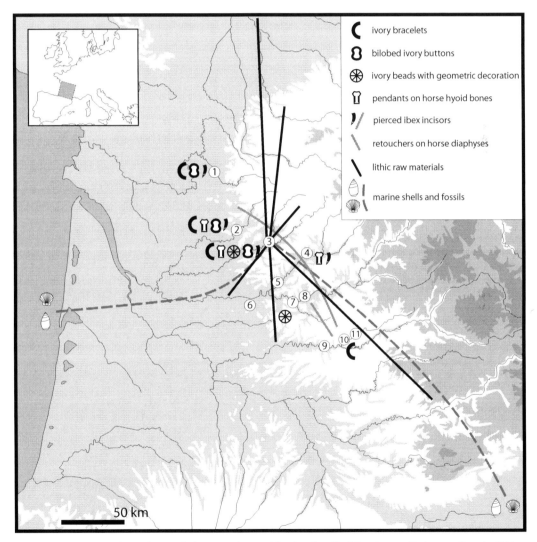

Figure 21.1 Spatial circulation of objects with a social value for all of the sites studied, as well as the lithic raw materials and shells of the Solutrean of level IV at Combe Saunière. The sites: 1 = Le Placard, 2 = Le Fourneau du Diable, 3 = Combe Saunière, 4 = Badegoule, 5 = Laugerie Haute, 6 = Les Jeanblancs, 7 = La Grotte XVI, 8 = Pech de la Boissière, 9 = Le Cuzoul de Vers, 10 = Les Peyrugues, 11 = Sainte-Eulalie. The shoreline shown is today's, not that of the Solutrean.

Combe Saunière is the only one without evidence of portable and parietal art.

The strategies for the exploitation of alimentary animal resources, lithic resources, as well as non-utilitarian objects with a primarily social and symbolic function, show that different ways of relating to place prevail at each domain.

The spatial inscription of these different subsistence activities through the deposition of material items within sites and regions allows us to identify social landscapes of human interaction and land use. The environmental homogeneity (of geology, climate, biomass, physical environment, habitat in natural rockshelter) in the geographical distribution of Solutrean sites along the Atlantic zone, at the limit of the plains and the western border of the Massif Central, is in fact remarkably consistent from the Loire River to the Pyrenean Piedmont, to the Basque Country, and beyond until the Asturian coastal zone. It is probable that an analogous situation exists also along the French and Spanish Mediterranean coast.

Despite the reservations formulated above concerning the criteria used to define faunal resource zones, the Solutrean sites discussed here appear to represent more or less prolonged residence sites where a range of activities related to the acquisition, exploitation and consumption of hunted species took place. Based on the skeletal remains of these diverse species, we can estimate human movements within this landscape. Through a determination of the ages of the animals hunted or fished, we can obtain indications of the seasonality of occupation within these sites, and of regional land use.

Finally, the results of this study show that lithic raw material exploitation is organized in relation to the intended end products (domestic needs, commonly used tools, projectile points, hearth stones, and so on). Similarly, before any archaeozoological, economic, or alimentary understandings can be claimed, the animal species must be understood in relation to their accessibility and subsistence, technological and economic importance, as well as, if possible, their symbolic roles for particular peoples at particular times.

Acknowledgments

The authors wish to thank Michel Allard and Francis Juillard, who contributed their unpublished data on the Solutrean of Les Peyrugues; Hélène Martin, Jean-Philippe Rigaud, Jean-Jacques Cleyet-Merle, Jean Cattaliotti, and Christophe Griggo for authorizing us to use their unpublished information and results; and Stéphane Madelaine for sharing his data concerning the collections of Solutrean sites conserved at the National Museum of Prehistory in Les Eyzies. Finally, we thank Magen O'Farrell for the English translation and Bruno David for his stimulating suggestions in contribution to the final version of this chapter.

Note

1. We use the term *non-economic* for objects whose primary reason for existence concerned their symbolic and/or aesthetic values, rather than alimentary or technological values. This is not to deny that such symbolic and aesthetic values may also be components of broader economic strategies or networks.

References

Allard, M., Chalard, P., and Martin, H. 2005. Témoins de mobilité humaine aux Peyrugues (Orniac, Lot) durant le Paléolithique supérieur. Signification spatio-temporelle, in M. Barbaza and J. Jaubert (eds.), *Territoire, déplacements, mobilité, échanges durant la préhistoire. 126e congrès des sociétés historiques et scientifiques, Toulouse, 9–14 avril 2001*, pp. 219–31. CTHS.

Aubry, T. 1991. L'exploitation des ressources en matières premières lithiques dans les gisements solutréens et badegouliens du bassin versant de la Creuse (France). Unpublished Ph.D. thesis, Université Bordeaux I.

Aubry, T., Walter, B., Robin, E., Plisson, H., Benhabdelhadi, M. 1998. Le site solutréen de plein-air des Maitreaux (Bossay-sur-Claise, Indre-et-Loire): Un faciès original de production lithique. *Paléo* 10: 163–84.

Bartram, L. E. 1993. Perspectives on Squeletal part profiles and utility curves from Eastern Kalahari ethnoarchaeology, in J. Hudson (ed.), *From Bones to Behaviour: Ethnoarchaeeological and Experimental Contributions to the Interpretation of Faunal Remains*, pp. 115–37. Carbondale, IL: Center for Archaeological Investigations, Southern Illinois University (Occasional Paper, 21).

Binford, L. R. 1978. *Nunamiut Ethnoarchaeology.* New York: Academic Press.

———. 1981. *Bones. Ancient Men and Modern Myths.* New York: Academic Press.

———. 1982a. Some thoughts on the Middle to Upper Paleolithic transition, in L. Binford, *Working at Archaeology*, pp. 423–33. New York: Academic Press.

———. 1982b. The archaeology of place, in L. Binford, *Working at Archaeology*, pp. 357–78. New York: Academic Press.

Bouchud, J. 1966. *Essai sur le Renne et la climatologie du Paléolithique moyen et supérieur.* Périgueux: Imprimerie Magne.

Bressy, C., Burke, A., Chalard, P., Martin, H., (eds.). 2006. *Notions de territoire et de mobilité. Exemples de l'Europe et des premières nations en Amérique du Nord avant le contact européen*, Actes des sessions présentées au Xe congrès annuel de l'Association Européenne des Archéologues (EAA), Lyon, 8–11 septembre 2004 116, Liège, ERAUL, 169 pp.

Bunn, H. T. 1982. Meat-eating and human evolution: Studies on the diet and subsistence activities at Olduvai Gorge, Tanzania, and Koobi Fora, Kenya. Unpublished Ph.D. thesis, Berkeley, University of California.

Castanet, J., Meunier, F., and Francillon-Vieillot, H. 1992. Squelettochronologie à partir des os et des dents chez les vertébrés, in H. Baglinière, J. Castanet, J. Conand, and F. Meunier (eds.), *Tissus dur et âge individuel des vertébrés. Actes du colloque de Bondy (1991)*. ORSTOM/INRA.

Castel, J.-C. 1999. Comportements de subsistance au Solutréen et au Badegoulien d'après les faunes de Combe Saunière (Dordogne) et du Cuzoul de Vers (Lot). Unpublished thesis, Université de Bordeaux I.

Castel, J.-C., Chadelle, J.-P., and Geneste, J.-M. 2005. Nouvelle approche des territoires solutréens du Sud-ouest de la France, in M. Barbaza and J. Jaubert (eds.), *Territoire, déplacements, mobilité, échanges durant la préhistoire. 126e congrès des sociétés historiques et scientifiques, Toulouse, 9–14 avril 2001*, pp. 279–94. CTHS.

Castel J.-C., Chauvière, F.-X., L'Homme, X., and Camus, H. 2006. Un nouveau site du Paléolithique supérieur récent: Le Petit Cloup Barrat (Cabrerets, Lot, France). *Bulletin de la SPF* 103: 263–73.

Castel, J.-C., Liolios, D., Chadelle, J.-P., and Geneste, J.-M., 1998. De l'alimentaire et du technique: La consommation du renne dans le Solutréen de la grotte de Combe Saunière, in J.-P. Brugal, L. Meignen, and M. Patou-Mathis (eds.), *Economie préhistorique: Les comportements de subsistance au Paléolithique*, pp. 433–50. XVIIIe rencontres internationales d'Archéologie et d'Histoire d'Antibes. Actes des rencontres, 23-24-25 octobre 1997. Sophia-Antipolis, APDCA.

Castel J.-C., Liolios, D., Laroulandie, V., Chauvière, F.-X., Chadelle, J.-P., Pike-Tay, A., and Geneste, J.-M. 2006. Solutrean Animal resource exploitation at Combe Saunière (Dordogne, France), in M. Maltby (ed.), *Integrating Zooarchaeology*, pp. 138–52. 9th ICAZ Conference, Durham, 2002. Oxford: Oxbow Books.

Chadelle, J.-P., Geneste, J.-M., and Plisson, H. 1991. Processus de formation des assemblages technologiques dans les sites du Paléolithique supérieur. Les pointes de projectiles lithiques du Solutréen de la grotte de Combe-Saunière (Dordogne, France), in *25 ans d'études technologiques en Préhistoire. Bilan et perspectives*, pp. 275–87. XIes rencontres internationales d'archéologie et d'histoire d'Antibes. Actes des rencontres, 18-19-20 octobre 1990, Juan-les-Pins, APDCA.

Chalard, P., Guillermin, P., and Jarry, M. 2006. Acquisition et exploitation des silex allochtones au Gravettien. L'exemple de la couche E du gisement des Fieux (Lot, France), in C. Bressy, A. Burke, P. Chalard, and H. Martin (eds.), *Notions de territoire et de mobilité. Exemples de l'Europe et des premières nations en Amérique du Nord avant le contact européen*, Actes de sessions présentées au Xe congrès annuel de l'Association des Archéologues (EAA), Lyon, 8–11 septembre 2004, 116, Liège, ERAUL, pp. 29–40, 10 fig., 2 tabl.

Cheynier, A. 1949. *Badegoule, station solutréenne et proto-magdalénienne*. Paris : Masson et Cie. Archives de l'Institut de Paléontologie Humaine. Vol. 23.

Conkey, M. 1980. The identification of prehistoric hunter-gatherer aggregation sites: The case of Altamira. *Current Anthropology* 21: 609–30.

Costamagno, S. 2006. Territoires de chasse paléolithiques : Des méthodes d'études à l'application archéologique, in C. Bressy, A. Burke, P. Chalard, and H. Martin (eds.), *Notions de territoire et de mobilité. Exemples de l'Europe et des premières nations en Amérique du Nord avant le contact européen*. Actes des sessions présentées au Xe congrès annuel de l'Association Européenne des Archéologues (EAA), Lyon, 8–11 septembre 2004, 116, Liège, ERAUL, pp. 63–70, 7 fig., 3 tabl.

Delpech, F., Donard, E., Gilbert, A., et al. 1983. Contribution à la lecture des paléoclimats quaternaires d'après les données de la paléontologie en milieu continental, in *Paléoclimats*, Bulletin de l'Inst. Géol. Bassin d'Aquitaine, 34 et CNRS, Cahiers du Quaternaire, n° spécial, pp. 165–77.

Drucker, D., Bocherens, H., Cleyet-Merle, J.-J., Madelaine, S., and Mariotti, A. 2000. Implications paléoenvironnementales de l'étude isotopique (^{13}C, ^{15}N) de la faune des grands mammifères des Jamblancs (Dordogne, France). *Paléo* 12: 127–40.

Ellen, R. F. 1979. Introduction: anthropology, the environment and ecological systems, in P. C. Burnham and R.F. Ellen (eds.), *Social and Ecological systems*, pp. 1–17. London: Academic Press (A.S.A. Monographies 18).

Fontana, L. 2001. Étude archéozoologique des collections du Fourneau du Diable (Bourdeilles, Dordogne): un exemple de potentiel des faunes paléolithiques issues des fouilles anciennes. *Paléo* 13: 159–82.

Geneste, J.-M., and Rigaud, J.-P. (In preparation). Les boutons cruciformes du Solutréen du Périgord.

Geneste, J.-M., and Plisson, H. 1986. Le Solutréen de la grotte de Combe Saunière 1 (Dordogne). *Gallia Préhistoire* 29(1): 9–27.

Griggo, C. 1995. Significations paléoenvironnementales des communautés animales pléistocènes reconnues dans l'abri Suard (Charente) et la grotte de Bois-Ragot (Vienne). Unpublished thesis, Université de Bordeaux I.

Griggo, C. In press. La faune de la grotte du Placard: Etudes paléontologique, paléoenvironnementale et archéozoologique, in J. Clottes, L. Duport and V. Féruglio (eds.), *Le gisement solutréen et badegoulien du Placard (Charente)*.

Guille-Escuret, G. 1989. *Les sociétés et leur nature*. Paris, Armand Colin, Anthropologie du présent.

Keene, A. S. 1981. Optimal Foraging in a Nonmarginal Environment: A Model of Prehistoric Subsistence Strategies in Michigan, in B. Winterhalder and E. A. Smith (eds.), *Hunter-Gatherer Foraging Strategies. Ethnographic and Archaeological Analyses*, pp. 171–93. Chicago: The University of Chicago Press:

Lee, R. B., and De Vore, I. (eds.). 1968. *Man the Hunter*. Chicago: Aldine Publishing.

Martin, H. In press. Le Cuzoul de Vers: Analyse cémentochronologique des restes dentaires issus des niveaux solutréens et badegouliens, in J. Clottes et al. (eds.), *L'abri Solutréen et Badegoulien du Cuzoul de Vers*. Documents d'Archéologie Française.

Peyrony, D. 1932. *Les gisements préhistoriques de Bourdeilles (Dordogne)*. Paris: Masson et Cie. Archives de l'Institut de Paléontologie Humaine. Vol. 10.

Pike-Tay, A., and Castel, J.-C. In preparation. Saisonnalité des occupations solutréennes de Combe Saunière (Dordogne).

Plisson, H., and Geneste, J.-M. 1989. Analyse technologique des pointes à cran solutréennes du Placard (Charente), du Fourneau du Diable, du Pech de la Boissière et de Combe-Saunière (Dordogne), *Paléo* 1: 65–105.

Renfrew, C. 1977. Alternative models for exchange and spatial distribution, in T. K. Earle and J. E. Ericson (eds.), *Exchange Systems in Prehistory*, pp. 71–89. New York: Academic Press.

Sergeant, D. E., and Pimlott, D. H. 1959. Age determination in moose from sectioned incisor teeth. *Journal of Wildlife Management* 23: 315–21.

Servelle, C., and Vaquer, J. 2000. Les haches polies en cinérite du Rouergue, des producteurs aux consommateurs. *Rencontres méridionales de préhistoire récente, IIIème session. Toulouse, 1998*, pp. 81–100. Ed. Archives d'Ecologie Préhistorique.

Valensi, L. 1960. De l'origine des silex protomagdaléniens de l'abri Pataud, les Eyzies, in *Bulletin de la Société Préhistorique Française, 1960* 57: 80–84.

22

THE USE OF ETHNOGRAPHY IN LANDSCAPE ARCHAEOLOGY

Paul J. Lane

As with other elements of the broader discipline, ethnographic information has been used to aid the interpretation of various aspects of archaeological landscapes since the very beginnings of the subfield. The way in which such information has been used and incorporated within the interpretive process has also followed broader temporal trends, starting with the use of simple ethnographic parallels, often picked seemingly at random from the available literature, to the formulation of more explicit formal and, most recently, relational analogies (see Wylie [1985] for a discussion of these terms). However, the use of ethnographic information has not featured as widely in landscape archaeology as it has in many other aspects of the archaeological process, such as the analysis and interpretation of stylistic behavior, intrasite spatial patterning, spatial and architectural symbolism, butchery practices, and artifact technologies. There have also been remarkably few ethnoarchaeological studies explicitly concerned with notions of "landscape" as opposed to attempts to document the material traces of different activities and mobility patterns at a landscape scale through either direct ethnoarchaeological observation (see below) or synthesis of ethnographic studies (e.g., Allen 1996). This is also well illustrated by a recent and relatively comprehensive review of the ethnoarchaeological literature of the last four to five decades (David and Kramer 2001),

which only considers "landscape" in terms of the analysis of settlement patterning and distributions. Here, I aim to summarize these various trends with examples and go on to make some suggestions for possible future research.

Beginnings

O. G. S. Crawford, widely regarded as a pioneer of landscape approaches in British archaeology, was one of the first to make explicit use of ethnographic information in the interpretation of archaeological landscapes. Although this was due undoubtedly to his abilities as a field archaeologist and his "eye" for detail, it can also be attributed in part to his own career history, which included a visit to the Sudan in 1950. This is especially evident in his book, *Archaeology in the Field* (1953), which was published shortly after he had left the Sudan and contains several ethnographic parallels from Africa (see especially pp. 226–31). In common with the strongly empirical focus of British landscape archaeology at the time, Crawford's primary concern was with demonstrating how recourse to ethnography could aid in the correct identification of specific landscape features, rather than with how landscapes might have been perceived or understood by people in the past, or with the cultural meanings associated with specific elements of a particular landscape. Archaeologists during this era also often used brief ethnographic vignettes to illustrate how archaeological landscapes may have operated when they were in use, often likening the modern conditions in parts of the non-Western world with those that possibly prevailed during medieval, Roman, or later prehistoric times in Western Europe. This is illustrated by Crawford's use of an aerial view of Luo homesteads and farms around Kisumu, Kenya in c. 1950 to envision the form and appearance of rural landscapes in parts of medieval Europe (1953: plate 15, following p. 184).

With the rise of more critical approaches to the use of analogical reasoning in the 1960s and 1970s (e.g., Freeman 1968; Gould 1978; Wobst 1978), one might have expected the issue of how to improve the use of ethnographic information about landscapes to have become a central component of ethnoarchaeological research. That this did not happen can perhaps be attributed to the more general lack of consideration given to understanding archaeological landscapes during the heyday of the New Archaeology and also to the continuing dominance of the conception of "landscape" as being broadly equivalent to "the environment" (see David and Thomas, this volume). This said, some ethnoarchaeological studies have been conducted at a landscape scale, especially those concerned with understanding settlement systems and dynamics among hunter-gatherers (e.g., Binford 1982; Yellen 1977), pastoralists (e.g., Cribb 1991; Hole 1978), and agriculturalists (e.g., Loubser 1991; Stone 1991). These provide some valuable insights into the manner in which societies utilize different areas of land and distribute their activities across space.

This is especially well illustrated by Binford's (1982) descriptions of the way in which different Nunamiut hunter-gatherer task groups and residential clusters moved around their landscape in the recent past, at times reusing previously occupied places for alternative purposes, at others using certain spaces for the first time in living memory albeit perhaps just for a few weeks or even days, as part of annual and seasonal settlement mobility and the changing social composition of work groups. The archaeological consequences arising from the operating of such systems, as Binford amply illustrates, is a series of palimpsests of discarded materials, specially deposited items, abandoned site furniture, and, at least by implication since this aspect is not discussed by Binford, modified ecosystems. Binford's ultimate purpose, however, was to illustrate how these composite archaeological traces created by hunter-gatherer communities with delayed-return subsistence strategies (termed *collectors* by Binford) differ in a fundamental way from the kinds of material traces groups of immediate-return hunter-gatherers (termed *foragers* by Binford, and including groups such as the !Kung San communities of northern Botswana with whom Yellen [1977] worked) might leave across the landscape.

This model (for a tabulated summary of the key differences, see David and Kramer 2001: 235) has been particularly influential in hunter-gatherer/Paleolithic archaeology (e.g., Bamforth 1991; Dale, Marshall, and Pilgrim 2004; Kornfeld 1996; Marean 1997; Savelle 1987) and has resulted in some positive improvements in archaeological understanding of the diversity of land-use practices and settlement systems among both past and ethnographic hunter-gatherers. However important it may be for archaeologists to recognize that immediate-return foragers and delayed-return collectors might utilize the land in different ways, the rich corpus of anthropological literature on both low- and high-latitude hunter-gatherers indicates that they also imbue their world with complex symbolic meanings, spiritual values, and mythological associations (e.g., Bodenhorn 1993; Layton 1995; Morphy 1995; Silberbauer 1981). Of course, this is equally true of other types of society, and in the last few decades archaeologists have increasingly turned to a different range of ethnographic data concerning landscape in their attempts to access these more intangible elements for a diverse range of groups including herders, farmers, maritime communities, and urban dwellers as well as hunter-gatherers (e.g., Boivin 2004; Breen and Lane 2003; Dunning et al. 1999; Thomas 1993; Tilley 1994; Ucko and Layton 1999; Young 2000). Below, I sketch out some of the more common approaches to the use of ethnographic data in terms of the construction of relational analogies, the influence of natural places, socializing the land, and notions of dwelling and "being-in-the-land."

Relational Analogies

Unlike the use of simple ethnographic parallels, in the construction of relational analogies much greater emphasis is placed on establishing the *relevance* of the observed similarities between the archaeological phenomena under analysis and the ethno-graphic case material used to infer other, non-observable similarities. Thus, it is not just the existence of observable similarities between the subject (that is, archaeological) and source (that is, ethnographic) sides of the comparison that matter, but the fact that these similarities directly relate to the inferences being drawn (see Wylie 1985). There are many ways of demonstrating relevance. One common way, sometimes known as the "direct historical approach," is to draw on the ethnographic and ethnohistoric information associated with the presumed (or actual) descendants of the archaeological community under investigation. Another strategy, which is by no means incompatible with the direct historical approach, is to enumerate the degree of similarity between the subject and source sides of the equation in such a way that other possible interpretations can be shown to be invalid.

Gartner's (1999) study of Late Woodland Period "effigy mounds" found in various parts of Wisconsin is a good recent example of the "direct historical approach" within a landscape archaeology context. Dated to between A.D. c. 700 and 1050, effigy mounds are earthen and gravel constructions with plan forms in the shape of stylized birds and terrestrial animals (especially bears and panthers) that occasionally, but infrequently, contain burials and/or artifacts. They occur in groups, sometimes with the terrestrial and celestial types carefully segregated, and in some areas also set in distinct alignments with local topographic features. Other clusters exhibit no such obvious structure, with mounds of both terrestrial and celestial form intermingled and seemingly placed at random within the broader landscape. Drawing on Winnebago ethnography, and especially information about Winnebago clans, territoriality, and rituals, Gartner argues that the effigy mounds were probably a kind of territorial marker associated with different clans and used to "convey notions of social and geographic boundary within and across generations" (1999: 680). The typical "empty" nature of effigy mounds, he suggests, implies that the act of their construction was more important than their contents. This is also supported by the use of special kinds of "exotic" earths in the construction of some mounds, and also the extensive geoarchaeological evidence for the ritual use of fire and the admixture of various natural substances such as crushed bone, clam shell, and different minerals to the ashes, in a manner consistent with the ethnohistoric information for the region. Taken together with their spatial positioning within the landscape, typically overlooking streams or lakes, or situated close to cross-country trails, the effigy mounds seem to have served as material manifestations of Late Woodland mythological and cosmological principles that simultaneously served to classify areas, communicate social boundaries and enforce control (Gartner 1999: 681).

Ian McNiven's recent paper on the archaeological dimensions of indigenous Australian marine specialists (including Torres Strait Islanders), whom he terms "Saltwater Peoples" (2003), is a fine example of the second strategy mentioned above for establishing the relevance of an analogy. As he notes, archaeological surveys of coastal zones, including foreshore and inter-tidal areas, around mainland Australia and on the off-shore islands have located a wide range of different archaeological traces that can be associated with different types of maritime activity. These include the more obvious, and more extensively investigated, traces associated with the exploitation of marine resources, such as shell middens, faunal assemblages dominated by marine fish and mammal bones, artifacts associated with fish- and shellfish-processing and tidal fish-traps. There is also more indirect evidence of seafaring activities and abilities, such as the presence of archaeological traces on the Percy Islands, some 58 kilometers off the shores of Queensland, which can be reached even with "island hopping" only via an open sea voyage of at least 27 kilometers (Rowland 1984). However, surveys of intertidal zones along parts of the Queensland coast have also revealed numerous arrangements of small stones, typically set in curvilinear lines, that can cover anything from less than 100 m² to several thousand square meters. All these sites are located on later Holocene marine muds and sands, typically in backwater areas behind the fringing mangrove forest and close to the current high-water mark. They are neither terrestrial sites that have been inundated by sea-level rise; nor are they likely to have been used as "subsistence" facilities, even though some bear formal similarities to ethnographic and archaeological fish traps.

As McNiven explains (2003: 238–89), there are four reasons for this: (1) the V- and U-shaped arrangements that most closely resemble tidal fish-traps are small in area and have their openings facing the sea, not away from it as required by fish traps; (2) many other forms, including circles and stone piles, also occur that would have provided no technical advantage for fishing; (3) the sites tend to be located close to the upper limits of high water, and so would have been inundated only rarely, and hence their location would be inappropriate for use as fish traps; (4) the arrangements often lack interlinking walls, which would have been necessary for trapping fish on a falling tide.

Having discounted more functional explanations, McNiven turns to the extensive ethnographic information on Australia's indigenous Saltwater Peoples concerning their spiritual engagements with the sea. He notes, in particular, that although some rituals were performed on water, many were also practiced on tidal flats in precisely the same kind of location as the archaeological features he describes. Moreover, the form and structure of the marine stone arrangements are also comparable to terrestrial stone arrangements found across Australia and known to have ritual and sacred connotations. Taken together with the broader ethnography concerning the rights and responsibilities towards the sea and marine resources, and the rules governing customary marine tenure, as encapsulated in Dreaming cosmologies, the stone arrangements found in the intertidal zone along the Queensland coast are all consistent with their role as places of ritual significance and spiritual engagement with the sea and especially the ontological significance of the fluidity of the tides and water (McNiven 2003: 344).

Research by Klara Kelley and Harris Francis on Navajo landscapes (1994) can also be mentioned here. Although primarily motivated by concerns with cultural rights protection and historic preservation concerns, their study provides detailed accounts of how members of thirteen Navajo local communities in New Mexico and Arizona consider landscape to be culturally significant, why this is the case and what constitutes landscape for these Navajo. Kelley and Francis also noted some significant differences between Navajo conceptions of landscape and those typically employed by cultural resources managers. Specifically, instead of thinking about their land as being made up of a number of isolated, albeit culturally significant, places that deserved protection, for Navajo a "place is usually important because it is part of a larger landscape constituted by a story, customary activities, or both" (Kelley and Francis 1993: 157). Each place draws its "particular distinct significance qualities" (p. 157) from its interrelations with other places that are themselves created partly by the repeat performance of customary (typically ceremonial) activities but also partly through the act of telling.

Landscape for these Navajo is thus not only the whole land and its functioning elements but also comprises multiple and overlapping entities that are in constant need of reconstitution through daily practice and telling stories about that practice, with each repeat performance having the capacity to equally reconfigure or reconfirm the relationships of significance. Significantly, perhaps, the notion that landscapes are revealed only through a combination of dwelling and storytelling is not unique to the Navajo but has also been documented among other Native American peoples of North America, including the Algonkian (Bruchac 2005) and Iroquois (Carson 2002).

Ethnography of Natural Landscapes

As the previous examples illustrate, ethnographic literature on landscapes commonly indicates that people throughout the world not only often have an intimate knowledge of their physical surroundings but also frequently associate certain "natural places" such as prominent rock formations, various watery contexts, caves, mountain peaks, and trees (among others) with the supernatural or mythological world. In the last few decades, landscape archaeologists have turned increasingly toward these aspects of the literature (e.g., Boivin 2004; Cummings and Whittle 2003; Richards 1996; Tilley 1996).

Bradley, for instance, in a recent book on the archaeology of natural places (2000), draws initially on an extended review of Saami sacred geography. Of the more than 500 known examples of sacred natural places across Finland and other parts of northern Scandinavia, most stand out as distinctive features in the wider terrain, and the majority also contain small idols made from either unworked pieces of stone or carved sections of living trees. These natural places were also the focus of various sacrifices associated with the exploitation of local resources including its fish stocks and wild and domesticated animals. The sacrifices were also dedicated to natural forces such as the wind, thunder, the sun and water, and a wide range of votive objects were often deposited nearby. These sacrifices were commonly made to secure divine protection from the elements, to secure food supplies, and as offerings to the dead (Bradley 2000: 5–13). At one level, therefore, Saami could be said to have adopted a fairly practical approach to the supernatural, aimed at securing livelihoods and protection from the natural and human spirit world. However, as Bradley notes, for the people who used and designated such sacred elements of the topography, these sites "would have been only the outward embodiment of a wider system of belief that had profound consequences for the ways in which the landscape was perceived" (2000: 11).

Drawing inspiration from this ethnography, and also examples from other parts of the world, Bradley proceeds to offer a range of worked examples of the possible importance of natural places and spaces at different times and in different landscapes in Neolithic and Bronze Age Europe. These include the changing patterning of votive deposits over the course of the Bronze Age in Denmark, the location and meaning of rock engravings in Scandinavia, and the specific positioning of Neolithic stone axe production sites in Britain and Ireland. In some cases, Bradley uses the insights drawn from ethnography simply as a demonstration of general principles, but elsewhere, he suggests that some prehistoric practices, such as those that structured votive deposits in watery bogs in Denmark during the Neolithic, may be directly ancestral to the later, ethnographically recorded systems found among the Saami (Bradley 2000: 60–63).

Socializing Landscapes

One class of archaeological evidence that is most obviously associated with natural places, and hence open to dual symbolic loading in terms of its stylistic and/or formal content and its place within the landscape, as Bradley's work also illustrates, is rock art. Indeed, as Whitley has observed, "the defining characteristic of 'rock art' is its placement on geological substrates" (1998: 11) and not the manner in which it was created or the motifs employed. Taçon makes a similar point where he argues that "in the process of marking and mythologizing landscapes humans socialized them" (1994: 117). As he goes on to observe, one effect of marking landscapes through rock art is to make them part of social activity, which in turn brings with it implications of access, control, ownership, and the significance of such marks to different individuals and groups. Whitley's landscape study of the rock art of far western North America (California, the Great Basin and the Columbia Plateau), in which ethnographic, ethnohistorical, environmental, and archaeological data are integrated provides a useful illustration of some of these points.

The rock art of this area is dominated by two regional traditions (although others have been identified), known as the Californian Tradition which occurs mostly in the wetter, western portions of California, and the Great Basin Tradition that is typically found in the dryer desert regions of eastern California and the Great Basin in Nevada. The former tradition is dominated by geometric paintings, mostly monochrome and in small panels in caves and rock shelters. Human and animal figures are comparatively rare. Great Basin Tradition rock art, however, is dominated by rock engravings, which often occur in large panels with hundreds or even thousands of individual motifs. Although geometric forms occur, there is a greater proportion of figurative art, the most common being bighorn sheep, snakes, lizards, and felines. The art is typically marked on open cliff faces and boulders. Despite such differences, the combined ethnohistoric and ethnographic evidence strongly supports the view that, as in some other parts of North America and

elsewhere, rock art was associated with shamanistic activity (although some was also produced in connection with female puberty rites) and relates especially to altered states of consciousness (see e.g., Whitley 1992; also Lewis-Williams and Dowson 1988, 1989). As such, rock art sites were widely regarded as symbolic portals into the supernatural world. This is substantiated by the regular placing of motifs close to cracks and openings on the rock face, and also in the naming of the landscape features on which rock art occurs.

Another feature of this art is its emphasis on gender symbolism. Again, this is reflected in the generic terms used to refer to rock art sites, which typically have strong feminine referents, and in the range of motifs and their symbolic referents. However, it is also evident from the placing of rock art sites in the landscape. Most obviously, caves and rockshelters in the local ethnography are regarded as symbolic wombs, and mountain peaks as penises. Other landscape features, such as the U-shaped sandstone outcrop on the Carrizo Plain in south-central California, on which a very rich rock art site known as Painted Rock occurs, are explicitly likened to a natural vulva. At the apex of this site is a panel depicting human figures standing in a plank canoe, which can also be read as a sexual metaphor, since in Chumash ethnography "getting into the canoe" is a metaphor for intercourse (Whitley 1998: 20).

A particular feature of this rock art is that it is dominated by female sexual symbolism and referents. This is also evident from the predominant selection of lower topographic localities for the placement of rock art than on the highest elevations, since in the local ethnography height was associated with males and masculinity and low spots with females and femininity (Whitley 1998: 22–23). Yet, as Whitley remarks, the art was almost certainly exclusively produced by male shamans. To explain this apparent paradox, Whitley suggests that because female sexuality was regarded as dangerous and threatening, at one level rock art sites can be regarded as centers of power controlled by shamans for both beneficial and malign purposes that the majority of the population would have avoided. However, the ethnographic literature also indicated that in the local mindscape, the supernatural world was regarded as the direct opposite to the human world. Thus, through the principle of symbolic inversion the rock art sites, as portals to the supernatural world, and despite the overt female symbolism, could have been read also as potent centers of masculinity and heightened male sexuality (Whitley 1998: 24–25).

Being-in-the-Land

An additional feature of more recent approaches that has emerged across the disciplines, which has obvious resonance with several of the examples given above, has been the recognition that landscapes are not simply viewed from afar or acted on but are also *lived in*. As David Cohen and E. S. Atieno Odihambo observe in their historical anthropology of Siaya, western Kenya, for its Luo inhabitants, landscape "means existence" (1989: 9). Here, the concept refers simultaneously to the physical terrain, its occupants, and the habits and customs that allow its exploitation in a particular way (cf. Crawford's perspective on what constituted the Luo landscape, cited above). As such, the term evokes both possibilities and constraints. In a more general essay, Ingold makes much the same point: "Landscape is the world as it is known to those who dwell therein" (1993: 156). Being "dwelt in" necessarily means that landscapes are not the static, inscribed forms conventionally documented by cartographers, archaeologists, and historical geographers but are instead temporal phenomena with multiple and often overlapping rhythms that come into being through the process of human occupancy—in other words, by a "*being in*" the land. This perspective has been developed further with reference to archaeological landscapes by Tilley (1996), who suggests that there are four defining characteristics to the concept. First, landscape is holistic, in the sense that it comprises a series of locales, each with its own associated meanings, and a series of relationships that link these locales into a composite whole. Second, landscape comes to be known and is experienced through the human body and its variable spatiotemporal placement within that landscape. Third, learning about a landscape is a form of socialization, the practical mastery of which provides an important component of a person's sense of ontological security. Finally, control over the mechanisms of learning the "lie of the land" can be an important source of power and a basis for social domination (Tilley 1996: 161–62).

In marked contrast to the manner in which ethnography has often been used in other landscape studies, as illustrated above, many of the recent phenomenological approaches draw less on the ethnography of others, except perhaps as a way to illustrate general principles, and more on the ethnographic experiences of the researcher through her/his own physical engagement with an archaeological landscape (e.g., Thomas 1993; Tilley 2004). Although this experiential engagement has obvious advantages over a purely map- and text-based analysis of landscape, since it provides the archaeologist with direct bodily and visual encounters with the land, it is not without its problems. Most notably, the landscapes of today are often vegetated differently than they were when they were occupied in the past (Cummings and Whittle 2003), and hence many points that may be intervisible today may have been obscured from view in the past (or *vice versa*). Equally, as Barrett has noted (2004), the landscapes of today are structured by entirely different rights of access, familial, and political obligations, and even bodily habitus. Modern landscapes also have their own historicity, which overlies and subsumes that of the older landscapes that archaeologists seek to investigate—by which is meant that round barrows and dykes in Wiltshire today, for instance, exist alongside other features such as modern farm buildings, enclosure-era hedgerows, Roman roads, and medieval droveways that were not there during the Bronze and Iron Ages. As palimpsests, modern landscapes are also sites of memory open to contestation, debate, and resistance (Bender 1993; Given 2004: 138–61; Küchler 1993), of which the archaeological process is also a part.

Conclusions

A wide variety of ethnographic sources and types of information has been used to aid the interpretation of archaeological landscapes and landscape features from the very beginnings of the subfield. As in the broader discipline, these applications have ranged from the use of rather piecemeal ethnographic parallels to more carefully argued formal and relational analogies. The insights gained from more general anthropological principles concerning non-Western perceptions of landscape and the concept of "dwelling in the landscape" have also been informative. In recent years, there has been a trend toward more phenomenological approaches, wherein the experience of the fieldworker's engagement with a landscape is treated as the primary source of ethnographic imagination. The aspects of ethnographic landscapes that archaeologists have found particularly inspirational in recent years include the significance of natural places, the links between landscape and memory, and the politics of place. Although not discussed here, ethnographic data regularly feature in more historical ecology oriented studies of landscape change and continuity. Although there have been some landscape-scale ethnoarchaeological studies, these have tended to focus on the dynamics of settlement and site distributions and their material signatures, with little or no discussion of the symbolism and meanings of these landscapes and their archaeological traces to the ethnographic actors concerned. There is certainly considerable scope for more integrated ethnoarchaeological studies that consider these latter elements in conjunction with the more material aspects.

References

Allen, H. 1996. Ethnography and prehistoric archaeology in Australia. *Journal of Anthropological Archaeology* 15: 137–59.

Bamforth, D. B. 1991. Technological organization and hunter-gatherer land use: A California example. *American Antiquity* 56: 216–34.

Barrett, J. 2004. Comment on C. Tilley "Round barrows and dykes as landscape metaphors." *Cambridge Archaeological Journal* 14: 199.

Bender, B. 1993. Stonehenge: Contested landscapes (medieval to present-day), in B. Bender (ed.), *Landscape: Politics and Perspectives*, pp. 245–79. Oxford: Berg.

Binford, L. R. 1982. The archaeology of place. *Journal of Anthropological Archaeology* 1: 5–31.

Bodenhorn, B. 1993. Gendered spaces, public places: Public and private revisited on the North Slope of Alaska, in B. Bender (ed.), *Landscape: Politics and Perspectives*, pp. 169–203. Oxford: Berg.

Boivin, N. 2004. Landscape and cosmology in the South Indian Neolithic: New perspectives on the Deccan ashmounds. *Cambridge Archaeological Journal* 14: 235–57.

Bradley, R. 2000. *An Archaeology of Natural Places.* London: Routledge.

Breen, C., and Lane, P. J. 2003. Archaeological approaches to East Africa's changing seascapes. *World Archaeology* 35: 469–89.

Bruchac, M. M. 2005. Earthshapers and placemakers: Algonkian Indian stories and the landscape, in C. Smith and H. M. Wobst (eds.), *Indigenous Archaeologies: Decolonizing Theory and Practice,* pp. 56–80. London: Routledge, One World Archaeology 47.

Carson, J. T. 2002. Ethnogeography and the Native American past. *Ethnohistory* 49: 769–88

Crawford, O. G. S. 1953. *Archaeology in the Field.* London: Phoenix House.

Cribb, R. L. D. 1991. *Nomads in Archaeology.* Cambridge: Cambridge University Press.

Cohen, D. W., and Atieno Odhiambo, E. S. 1989. *Siaya: The Historical Anthropology of an African Landscape.* Oxford: James Currey.

Cummings, V., and Whittle, A. 2003. Tombs with a view: Landscape, monuments and trees. *Antiquity* 77: 255–66.

Dale, D., Marshall, F., and Pilgrim, T. 2004. Delayed-return hunter-gatherers in Africa? Historic perspectives from the Okiek and archaeological perspectives from the Kansyore, in G. M. Crothers (ed.), *Hunters and Gatherers in Theory and Archaeology*, pp. 340–75. Carbondale: Southern Illinois University, Center for Archaeological Investigations, Occasional Paper No. 31.

David, N., and Kramer, C. 2001. *Ethnoarchaeology in Action.* Cambridge: Cambridge University Press.

Dunning, N., Scarborough, V., Valdez, Jr., F., Luzzadder-Beach, S., Beach, T., and Jones, J. G. 1999. Temple mountains, sacred lakes, and fertile fields: Ancient Maya landscapes in northwestern Belize. *Antiquity* 73: 650–60.

Freeman, L. 1968. A theoretical framework for interpreting archaeological materials, in R. B. Lee and I. DeVore (eds.), *Man the Hunter*, pp. 262–67. Chicago: Aldine.

Gartner, W. G. 1999. Late Woodland landscapes of Wisconsin: Ridged fields, effigy mounds and territoriality. *Antiquity* 73: 671–83.

Given, M. 2004. *The Archaeology of the Colonized.* London: Routledge.

Gould, R. A. 1978. Beyond analogy in ethnoarchaeology, in R. A. Gould (ed.), *Explorations in Ethnoarchaeology*, pp. 249–93. Albuquerque: University of New Mexico Press.

Hole, F. 1978. Pastoral nomadism in western Iran, in R. A. Gould (ed.), *Explorations in Ethnoarchaeology*, pp. 127–67. Albuquerque: University of New Mexico Press.

Kelley, K. B., and Francis, H. 1993. Places important to Navajo people. *American Indian Quarterly* 17: 151–69.

———. 1994. *Navajo Sacred Places.* Bloomington: University of Indiana Press.

Kornfeld, M. 1996. The big-game focus: Reinterpreting the archaeological record of Cantabrian Upper Paleolithic economy. *Current Anthropology* 37: 629–57.

Küchler, S. 1993. Landscape as memory: The mapping process and its representation in a Melanesian society, in B. Bender (ed.), *Landscape: Politics and Perspectives*, pp. 85–106. Oxford: Berg.

Layton, R. 1995. Relating to the country in the Western Desert, in E. Hirsch and M. O'Hanlon (eds.), *The Anthropology of Landscape: Perspectives of Place and Space*, pp. 210–31. Oxford: Clarendon.

Lewis-Williams, J. D., and Dowson, T. A. 1988. The signs of all times: Entoptic phenomena in Upper Paleolithic art. *Current Anthropology* 29: 201–45.

———. 1989. *Images of Power: Understanding Bushman Rock Art.* Johannesburg: Southern Books.

Loubser, J. 1991. The ethnoarchaeology of Venda-speakers in southern Africa. *Navorsinge van die Nasionale Museum, Bloemfontein* 7: 146–464.

Marean, C. W. 1997. Hunter-gatherer foraging strategies in tropical grasslands: Model building and testing in the East African Middle and Later Stone Age. *Journal of Anthropological Archaeology* 16: 189–225.

McNiven, I. J. 2003. Saltwater People: Spiritscapes, marine rituals and the archaeology of Australian indigenous seascapes. *World Archaeology* 35: 329–49.

Morphy, H. 1995. Landscape and the reproduction of the ancestral past, in E. Hirsch and M. O'Hanlon (eds.), *The Anthropology of Landscape: Perspectives of Place and Space*, pp. 184–209. Oxford: Clarendon.

Richards, C. 1996. Henges and water: Towards an elemental understanding of monumentality and landscape in Late Neolithic Britain. *Journal of Material Culture* 1: 313–36.

Rowland, M. J. 1984. Long way in a bark canoe: Aboriginal occupation of the Percy Islands. *Australian Archaeology* 18: 17–31.

Savelle, J. M. 1987. *Collectors and Foragers: Subsistence-settlement system change in the Central Canadian Arctic, A.D. 1000–1960*. Oxford: BAR International Series.

Silberbauer, G. B. 1981. *Hunter and Habitat in the Central Kalahari Desert*. Cambridge: Cambridge University Press.

Stone, G. D. 1991. Settlement ethnoarchaeology: Changing patterns among the Kofyar of Nigeria. *Expedition* 33: 16–23.

Taçon, P. S. C. 1994. Socialising landscape: The long-term implications of signs, symbols and marks on the land. *Archaeology in Oceania* 29: 117–29.

Thomas, J. 1993. The politics of vision and archaeologies of landscape, in B. Bender (ed.), *Landscape: Politics and Perspectives*, pp. 19–48. Oxford: Berg.

Tilley, C. 1994. *A Phenomenology of Landscape*. Oxford: Berg.

———. 1996. The powers of rocks: Topography and monument construction on Bodmin Moor. *World Archaeology* 28: 161–76.

———. 2004. Round barrows and dykes as landscape metaphors. *Cambridge Journal of Archaeology* 14: 185–203.

Ucko, P. J., and Layton, R. (eds.). 1999. *The Archaeology and Anthropology of Landscape: Shaping Your Landscape*. London: Routledge.

Whitley, D. S. 1992. Shamanism and rock art in far western North America. *Cambridge Archaeological Journal* 2: 89–113.

———. 1998. Finding rain in the desert: Landscape, gender and far western North American rock art, in C. Chippindale and P. S. C. Taçon (eds.), *The Archaeology of Rock Art*, pp. 11–29. Cambridge: Cambridge University Press.

Wobst, M. 1978. The archaeo-ethnography of hunter-gatherers or the tyranny of the ethnographic record in archaeology. *American Antiquity* 43: 303–09.

Wylie, A. 1985. The reaction against analogy. *Advances in Archaeological Method and Theory* 8: 63–111.

Yellen, J. 1977. *Archaeological Approaches to the Present: Models for Reconstructing the Past*. London: Academic Press.

Young, A. L. (ed.). 2000. *Archaeology of Southern Urban Landscapes*. Tuscaloosa: University of Alabama Press.

IV

LIVING LANDSCAPES: THE BODY AND THE EXPERIENCE OF PLACE

The contributions in this section together build the conviction that inhabiting, discovering, and remembering a landscape are fundamental to the business of being human. Any awareness of the landscape that human beings can develop depends entirely on our incarnation as embodied creatures. The most "distanced" perspective that we can experience is generally considered as being a visual one, but even this is reliant on having eyes to see with. Of course, it is a mistake to separate vision from the other senses, since people generally experience their world through forms of *synaesthesia,* in which sight, sound, smell, and touch blur into one another, forming complementary aspects or understandings that cannot be reduced to atomized sensory data.

Moreover, the world in general, and the landscape in particular, reveal themselves to us in specific ways according to conditions: light, humidity, and the mood that we happen to find ourselves in. The embodied experience of place is always unique to the circumstances and the particular embodiment that we live through. It is not just any-body that encounters the world: the body is gendered, aged, and enabled in such a way that biological and cultural components are often not easy to separate.

Landscapes may be familiar or unfamiliar, but they are continually disclosed to us, through the body and through material things. Tools provide us with insights into particular places (the hard soil that resists the spade; the machete cutting through the undergrowth), while architecture at once transforms the landscape and renders it memorable. Because this revelation and learning of the land do not take place in a single location, our relationship with the landscape is dispersed. Inhabitation or dwelling is spread through space, and our personal identity (as a combination of past experience, present residence and future projects) is not fixed or bounded in one location. We *come* to know the world, and ourselves in the world, by moving through it. In these embodied landscapes, time itself sets and unsets the scene, for we do not merely build on what has come before. Rather, we also silence the past, and with this our previous understanding and experience of our location, continuously performing cultural amnesias through shifting experiences and social performances in very particular places. Thus, landscapes are sensual performances that incorporate not only sight, hearing, smell, taste, and touch but also memory, sentiment, morality, and imagination.

23

GENDER IN AUSTRALIAN LANDSCAPE ARCHAEOLOGY

Amanda Kearney

The importance of gender, both past and present, is now widely recognized in archaeology. There is, in fact, a substantial literature—example, Balme and Beck 1995; Classen 1992; Conkey 1991; Di Leonardo 1991; Du Cros and Smith 1993; Gero and Conkey 1991; Gilchrist 1994, 1999; Hartman and Messer-Davidow 1991; Hays-Gilpin and Whitley 1998; Meskell 1995, 1999, 2000; Moore 1994, 1999; Nelson 1997; Sorenson 2000; Warren and Hackney 2000.

Gender is distinguished from sex on the grounds that it goes beyond the physical distinction between men and women, and is constructed as a set of roles, activities, and behaviors (Nelson 1997: 15). While sharing a "performative" character with sexuality, it extends beyond those aspects of the life journey that pertain to erotic significance. Gender is socially and variably assigned in today's world, and gender archaeology builds on the premise that this was also the case in the past. Gender is not pregiven. It is defined differentially across cultural groups, with culturally determined "frameworks of intelligibility" often delimiting the possibilities of sex, gender, and sexuality within socially permitted and naturalized categories (e.g., Butler 1990). It has been given considerable attention in the fields of social anthropology, history, cultural studies, and women's studies, and each of these has certainly influenced the incorporation of gender within landscape archaeology. In line with poststructuralist and critical, postmodern concerns, each discipline has adopted gender analysis and explicit excursions into the gendered world in ongoing commitments to often-marginalized peoples, bodies, and emotions.

This chapter offers a study of gender in lived and living cultural landscapes, set within the context of Indigenous Australian archaeology. Although it is specifically concerned with gender and gendered expressions in landscape archaeology, there are a range of research areas in gender in archaeology. These include not only the study of men, women, and children in the past but also contemporary treatments of the issue, by turning to study the androcentric assumptions and preconceptions in aspects of archaeology and the political aspects of workplace structures and interactions in archaeology (Conkey 1993).

Experiential aspects of the human past have also come to dominate landscape archaeologies and signal the shift away from "high-level systemic explanations" of human groups and the "reinsertion of agency into archaeological social theory" (Hamilakis, Pluciennik, and Tarlow 2002: 3). These dimensions of landscape archaeology have been inspired by the pivotal theoretical works of Bourdieu (1977, 1990), Derrida (1978), Giddens (1979, 1984), and Merleau-Ponty (1962,

1964) and from key contributors in interpretative archaeology, in particular, Bender (1992, 1993), Hodder (1982a, 1982b, 1982c, 1989, 1991), Shanks and Tilley (1987a, 1987b), Tilley (1990, 1993) and Thomas (1996, 2001). In landscape archaeologies, we now engage some of the ambiguities of human histories, such as gender in past and present worlds, childbirth and childhood (Baxter 2005; Beausang 2000), emotion (Tarlow 1999, 2000), commemoration, memory (Hamilakis 1998; Jones 1997; Van Dyke and Alcock 2003), and identity and bereavement (Tarlow 1999), all of which represent a commitment to humanity and the esoteric.

Gender Is All Around and Yet Nowhere to Be Seen

Landscape archaeology is a form of archaeology that attempts to engage the totality of "country," "place," "homeland," or "territory" by reading the temporality of these domains.[1] In this chapter I am concerned with how people engage with and experience the time, place, and people of the landscapes in which they dwell. Gender is but one part of this. I illustrate these concerns by reference to Yanyuwa gendered worlds, an Aboriginal group from northern Australia. Here gender expresses itself in daily and ceremonial lives, and the fluidity it is granted in the daily workings and engagements of life, Law, land, and sea (see Bradley, this volume for discussion of the interrelatedness of these concepts). Gender is omnipresent in the Yanyuwa world of meaning and experience. It is manifest in country and objects, among natural phenomena, in people and ancestors, within language and knowledge forms. At once, gender is all around and yet nowhere to be seen. How do we begin to grasp the meaning of gender and its historicity in such dynamic landscapes as Yanyuwa country? How does the gendered world relate to social notions of time? Is there more to gender in archaeology than may at first be apparent?

Here I explore threads of gender as they run through past and present landscapes. I consider some of the tangible and intangible expressions gender can have, as they concern the landscape archaeologist. I reflect on gender as a point of human engagement and experience, marking it as a fundamental yet socially defined part of the human condition. I look to Yanyuwa people and places and my experience with both to help me realize the historicity and character of gender in social landscapes and in landscape archaeology. Gender in landscape archaeology requests of us an interpretive archaeology that embraces the senses, the subliminal, the engaged and often the intangible. In essence we are concerned with social patterning to access social meaning and vitality.

Across Yanyuwa country, gender is not presented in binary terms. Yanyuwa people rarely speak of gender in terms of fixed, dichotomized constructions and often express the view that all is negotiable. This view reflects, in part, the dynamics of everyday life. It speaks of contemporary shifts in knowledge, increased pressures to "open up" the landscape, to "free up" sacred and gendered locations in light of colonial experiences and the subsequent relocation of Aboriginal people, contemporary non-Indigenous land-use practices, and tourism.

Yanyuwa country appears to have been gendered more explicitly in the past than it is today. Although meanings in the landscape may have shifted, among Yanyuwa elders there persists a memory of the sacred places and ceremonies of men and of women. These sacred places are spoken of with emotion and great care. People recall and understand country to be gendered, especially through means of ceremony and ancestral activity, but at the same time they recognize that land and seascapes are now much more "free," in that restrictions of access and interaction with key places are no longer enforced in the manner with which they were in the pre- and postcontact past. To enforce Law relating to country (for a discussion of the concept of Yanyuwa Law, see Bradley, this volume), whether in terms of gender restrictions or permissions to land, sea, and resource access, would be impossible for Yanyuwa people today. Their daily world is now frequented by tourists, and non-Indigenous landowners are not sensitive of the *ngalki*[2] and Law of the land.

Overall there is an understanding among many Yanyuwa that some parts of country must become more open. Thus today such places are granted the status of *warruki*—partially sacred, and released from secret and sacred restrictions. An understanding of this fluidity of landscape and its gendered meanings that *lie between* pre-established categories is framed by a colonized and traumatized landscape, one that is best described as a *wounded space* (Rose 2004: 34). And it is precisely this fluidity that enables an archaeology of gendered Yanyuwa landscapes to take place, because not only do we find spatially gendered domains in the landscape, but such gendered domains can be expected to have emerged in time and to have unfixed, traceable histories amenable to archaeological inquiry.

Throughout Yanyuwa country, places and Dreaming ancestors are all granted names and gender. They are often attributed with characteristics that are considered expressions of a particular gender identity. This fact is still evident today; it is part of the Law that forms the basis of Yanyuwa existence, a Law that, according to Yanyuwa elder Dinah Norman Marrngawi (personal communication 2003), "has never changed from the beginning." Dreaming spirits such as the ancestral Rock Wallaby—*a-buluwardi*—are ascribed female gender and often described in overtly feminine terms. *A-buluwardi* encounters the Tiger Shark Dreaming, which is male, that passes along the coastline of Vanderlin Island. This encounter, translated from the Yanyuwa language, is described as follows:

> He [Tiger Shark ancestor] looked onto Vanderlin Island and saw his sister there, the Rock Wallaby (*a-buluwardi*); she was standing on top of the hill *Wubuwarrarnngu*. The Rock Wallaby spoke to the Tiger Shark, saying, "Hey! That bundle of food on your head, what is it?" The Tiger Shark answered saying, "Sister! I have carried this cycad food a long way from the east, from Dumbarra. Can I place this food here as I am tired and my shoulders are cramped as I have carried this bundle a long way?" The Rock Wallaby grew angry and shouted back "No! Never! Go away from here to the south to the mainland, I will stay here by myself, I belong here, I am bitter with feelings, I am dangerous, I am heated. I will stay here and eat shellfish that I break from the rocks. Here I will stay, I belong here by myself, I have few possessions and I have no relations!" Her words were really heated and she threatened the Tiger Shark Dreaming with her fighting stick. She moved all of the west coast of Vanderlin in her anger, striking the ground with her fighting stick, that she created the twisted coastline that is there today. (Yanyuwa Families, Bradley, and Cameron 2003: 67)

The expression given to the Rock Wallaby is of a heated, bitter, and dangerous woman. In visual depictions the Rock Wallaby ancestral spirit is wearing a headpiece of string and feathers, the same style of headdress worn by women in ceremony.[3] When people retell this Dreaming narrative and adopt the voice of the Rock Wallaby, it is in a high-pitched tone, not unlike the voice of a little girl. Further, *a-buluwardi* is described by Yanyuwa people as enacting a ritual haranguing technique called *jijijirla*. As John Bradley notes, anyone who has seen *jijijirla* or participated in it,

will have been awed and staggered by the emotional physicality of the movements (Bradley, personal communication 2006). The term *jijijirla* is also used to denote a defensive movement that is a threatening display used by women before dueling with fighting or digging sticks. The Rock Wallaby ancestor expresses herself in many ways that align her with the socially feminine, and therefore with a certain range of gender expressions.

Potentials of gender expression and gender identity are also found in the Dreaming spirits of the Two Young Initiates and the Rainbow Serpent. The initiates are young boys traveling through country on and in the body of the Rainbow Serpent. Their encounters and experiences are used to measure and encode types of behavior that are appropriate and adhering to Law. The initiates are mischievous young boys who cause great trouble for the Rainbow Serpent and are punished accordingly. They dance ceremony and instill danger and potency in parts of the landscape, shaping the spiritual essence, or *ngalki* of several places. Such is the danger associated with the wayward initiates that only very senior men who have been through their second initiation are allowed to travel the places visited by the young initiates (Yanyuwa Families, Bradley, and Cameron 2003: 72). These ancestral spirits marked country and set it apart as the country of initiated and senior men.

This is but a glimpse at the Dreaming spirit-scape of Yanyuwa country. It is a vital snapshot of the ancestral heritage that defines life and Law for Yanyuwa people. These ancestors have expressions and characters that accord with gender identity in Yanyuwa culture. There are many more instances in which spirits and places are feminized and masculinized, but now I turn to a more localized discussion of gender expression in the lives of Yanyuwa people today.

In the last decade, ceremonial activity among the Aboriginal women of Borroloola has effloresced and come to represent a clear point of power and prestige for Yanyuwa women and their neighbors, Garrwa and Mara women. During this time many male elders within the Yanyuwa and Garrwa communities have died, leaving in their wake a small group of younger men who can speak of men's ceremony but who are regarded as having only passive knowledge of ceremonial activity. In contrast, when asking Dinah Norman Marrngawi (personal communication 2001) how she feels as a woman today, carrying out ceremony, she says unequivocally that "women really strong now, we've been here since little girl and now really big woman." In the same discussion, Annie Karrakayn (personal communication 2001), a senior Yanyuwa woman, made it clear that in doing ceremony, "old people look up at you from a long way, they just listen what you doing up there. Old people—they help you, Old people from a long way." She (personal communication 2001) followed up by saying: "All the women go high people now, women always strong, old lady used to teach us." By asserting a gendered form of cultural knowledge through ceremony, women actively ascribe existing and new meanings on the physical places and objects with which they engage in ceremony. In recent times this has brought about the reinscription of places of importance, including a women's ceremonial ground, practice ground, and other places associated with song origins, song lines, ancestral spirits, and individual performers. [4]

The desire on behalf of women to enliven and relive women's Business and Law has come about as part of a commitment to remembering and managing the landscapes of times past and present. It, at times, moves beyond the maintenance of women's Business and Law and has seen the gentle incorporation of men's Business and Law into the arms of women's knowledge. This expresses a fundamental point that often governs gendered and exclusionary knowledge systems—namely, that *knowledge* and *practice* or *Being* are actions of a very different nature. It is occasionally the case that Yanyuwa women know of men's Business and Law, song, and ceremony; however, they do not practice this Law, or if they do, they do so in a cautionary manner, recognizing its place in the lives of men. In turn, men have some knowledge of women's ceremony and exhibit the same healthy respect and distance from an intimacy that is seen to be the domain of women. Men and women also share elements of ceremonial life, with each occupying a different but complementary role. It is necessary to remember that many forms of ceremony were shared, and both men and women occupied essential roles in the performance and maintenance of ceremony. An example of this is in the initiation of young boys—*rdaru* (initiates) in the *a-Marndiwa* ceremony. The *a-Marndiwa* ceremony is a ritual during which boys aged 9–13 are circumcised and brought to age. The ritual lasts up to two weeks and includes the whole community, both men and women. In some ceremonial settings, women are responsible for part of the food getting and preparation, offering this and other support to performers and their families. In other instances men alone are responsible for negotiating across families and kin groups for access to key objects—for example, stones—that are needed for the production of key ceremonial objects and other items of spiritual potency (see Kearney and Bradley 2006).

In addition to the ceremonial aspects of Yanyuwa life, women and men share knowledge of gendered language forms, birthing songs, birthing places, objects, and places. Sharing knowledge is one part of the continuum of complementary duality that characterizes the gendered landscapes of Yanyuwa country. The careful inclusion of one gender in the knowledge system of the other is that end of the continuum in which men and women are the same, but different, in which case people know it is possible to mediate the danger of this knowledge by simply holding it and not putting it into practice. To do so would be in breach of Law, something that is taken very seriously by Yanyuwa people.

Finally Seeing Gender

In October, 2003, while resident in Borroloola[5] for the purpose of doctoral fieldwork, I had the opportunity to participate in a part of Yanyuwa life that allowed me to see gender expressions and gender identity at work in Yanyuwa country. I was able to participate in a week of women's ceremonial activity outside the Borroloola township. This would prove to be my introduction to women's ceremony, and in June 2004 I was invited back to participate in a much larger and more powerful ceremony at a more isolated ceremonial ground.

The place where the 2003 ceremony occurred was located outside the Borroloola township, on Yanyuwa country. This location had recently been given to the women of Borroloola, as a sacred site—courtesy of the Northern Territory Sacred Sites Authority. From initial discussions several women asserted that the area was too open and too close to the road. The turn-off to the ground was marked by a star picket, which was located at the edge of the road, signaling the turning point to access the site.[6] Women spoke often of the location's proximity to the road and the danger this posed for men who traveled within range of the site, and for any men who might prove to be too curious for their own good. For this reason, it was later agreed that the star picket marking the turn-off should be removed, and in due course it was.

As a participant and worker in the ceremony, I had cause to be present at the ceremonial ground for the setting up and for performance. I was also allocated the almost daily task of going into town to collect supplies. On one occasion, I had to head back into town to collect food, medication, and tobacco for several women. As I arrived at the local shop, I saw my *kangku*, paternal grandfather, old Pyro. I approached him to say hello, at which point he quickly stepped back against the wall and spoke strongly, saying: "Don't you touch me!" As a man with whom I have always had a very relaxed and nurturing relationship, I was astonished by the manner he adopted on this particular day. His engagement with me on this occasion was of a completely different nature to any we had had in the past. He avoided eye contact with me and briefly commented that I was, at that point, considered dangerous, because of my involvement in the women's ceremony. Sensing his great discomfort, I moved away and went about my shopping with haste. I was clearly wearing my gender loudly on that day, and as a result of the encounter I was rather unnerved by the massive shift that had apparently occurred in my status, quite unbeknown to me. Not only had the ceremonial ground gained in potency from women's activities, but also the women and their bodies (including mine) took on potency and a marking that stood them in contrast to and exclusive of men's bodies and men's knowledge. Bodies and place have gender, know gender, and can signal gender.

After several days of setting up at the ceremonial ground, and the singing of "fun songs,"[7] several elder women decided that there was to be an end to people's coming and going from the ceremonial ground. The place was gradually "closing up"—in light of shifts in ceremonial activity and a shift into more secret and powerful song and dance.[8] From this point, women and children's movements were to be restricted, and visits into the township of Borroloola were avoided. Over the week, women did continue to travel away from the ceremonial ground, but not into the township of Borroloola itself. They would go hunting and fishing across other parts of Yanyuwa country. These visits to country were part of an introduction for visiting women performers and their families to Yanyuwa country and involved efforts to get turtle and other foods to be cooked at the ceremonial ground. The shift to an altogether different level of performance and knowledge enacting not only was signaled by a recreation of place meaning and engagements but also was marked by the introduction of new physical elements to place, at once geographical, spatial, and material.[9] The place was changed and marked in ways that engaged and signed the lives, bodies, and histories of women.

All this formed part of an entire dialogue that was orchestrated among women—both Yanyuwa and visitor—country, ancestral spirits, deceased kin, and objects. This dialogue speaks a women's dialect of remembering and reliving the landscape, holding it up, strengthening it, testing the limits of ceremony in today's world, maintaining and sharing knowledge and recreating the emotional links between women and place.

The structure that governs the Law of ceremony creates many of the potentials for how women engage with place, objects, food, and people—all of which step gently across the gulf of what is intangible and tangible. Hence, I speak of gender being everywhere and yet nowhere to be seen (even though I have discussed these issues largely in relation to women, similar processes also occur in relation to men). Ceremony creates the opportunity for *knowing* and *being* to come together and be expressed through action and form, both of which, as concepts, preoccupy the imaginations of archaeologists and often govern the manner in which the archaeologist engages a given landscape. As I witnessed over the days of ceremony, several tangible signatures became evident, including the following:

- People set up camp within the ceremonial ground according to familial and kin relations;
- People visited other parts of the camp according to dancing partnerships—partnerships that reflected kinship;
- Physical barriers and structures were erected within the grounds;
- Ground ovens and surface fires were used for the purpose of cooking;
- Tree bark was collected from many trees in the immediate area and burnt down to ash to accompany chewing tobacco;
- Debris and rubbish was spread across the site in various discard patterns, including cooked items, bones, tin cans, and other food scraps;
- People repeatedly swept the site in an effort to keep the immediate living area clean;
- Certain objects were introduced and placed at certain locations within the site;
- These introductions initiated a shift in people's settlement and living patterns on the site.

At the completion of the week, most structures were left in place, and relatively minor effort was made to clean up the area. All in all the ceremonial ground looked lived in. There were signatures of the week that was, a week that left gendered debris all around. As such, this place has great potential as an "archaeological site," in an ethnoarchaeological sense. Kent (1984) has explored, through ethnoarchaeology, the basic assumptions that underpin archaeological engagements with spatial and sex-specific behavior such as found at the women's ceremonial location. In the tradition of this earlier work, the study of gender in landscape archaeology raises a number of questions and challenges to how we view the living spaces of cultural groups. The ceremonial ground was marked in such a manner as to reveal something of gendered site structure and occupational episodes, abandonment, gendered material culture and engagements, with sacred space. There is, however, as highlighted by Kent (1984) also a cautionary note to be made. When concerned with sex or gender specific spatial behavior and the multifunctional nature of places across human landscapes we must recognize the potential for complex cultural determinants (see Kent 1984, 1998).

It is with great subtlety that *knowing* (*knowing gender*) and *being* (*being gender*) come to express themselves on country, and the degree to which they are encoded by Law and sociality often makes it difficult to disentangle them with broad sweeps of materialist logic. When the physicality—namely, the tangible—can be traced back to the intangible—namely, in expression, memory, engagement, or experience—then gender is everywhere to be seen. Having the eyes to see gender in the landscape requires, first, knowledge of the manner in which gender can express itself (as a precursor to gender identity itself) and, second, recognition of potential gender identities. Even in those instances when the intangibility of gender cannot be accessed through direct engagements with landscape descendants, such as the Yanyuwa, then one must interrogate the fathomable range of gender expressions and gender identities among human groups. It does not suffice to conclude that men do one thing, women do another, for such things are not pregiven. The range of potentials for gender expression and gender identity may be far and wide, and this is what should be the point of interest for the landscape archaeologist.

Conclusions: Knowing Gender and Being Gender

Unlike many discussions of gender in archaeology, this chapter has not so much been a child of feminist archaeology, nor has it presented a prevailing discussion of gender as power and prestige. Rather, it is an expression of gender as complementary forms (usually dualities or more) and of individual and group relationships. Gender archaeology is not simply about looking for men or women in the past, but rather concerns understanding *terms of engagement* that inform interactions with places, objects, and people. This knowledge can manifest itself archaeologically—for example, as physical traces, such as in women's and men's ceremonial locations, material culture items associated with gendered practices and life rituals, or discard patterns of food goods that were consumed on the basis of gender exclusions and inclusions. In reality, these traces are often difficult to discern, and the ability to identify such patterns in the tangible record often comes only from engagements with the intangible record—namely, oral histories that reveal something of the terms of engagement of a particular human group. In other words, gender archaeology requires us to explore the experience of something that relates to the lived cultural domain (see Kearney 2005).

Today, as in the past, Yanyuwa society retains distinctions between genders and their associated bodies of knowledge, ceremony, and economy. This is most commonly expressed in terms of "women's Business" and "women's Law" and "men's Business" and "men's Law." In this case, Business and Law are terms that capture a wealth of meaning that is expressed and embodied in moral, jural, social, and ceremonial rules and practices (see Bradley, this volume). The enacting of Yanyuwa Law requires that people engage these meanings and activate and maintain relationships between people, ancestors, and place. All this cannot be done in isolation: men alone cannot do it, and women alone cannot hold up country without the support of their male counterparts. Without kin and recognized gender you are nothing; you cannot engage or be engaged with. Kinship and gender provide vital linkages between people and groups. At no point are people and their engagements and experiences of place objectified and autonomous of these fundamental points of relation. This is best understood as an "interdependence-dependence" relationship, one in which difference is seen as complimentary.

And so it is that gender is all around and yet nowhere to be seen. The best way for us to grasp the meaning of gender and its historicity in dynamic landscapes is to propose a range of potentials for gender expression and gender identity. From here we can seek to ethnographically document and ethnoarchaeologically interrogate relived and living landscapes as gendered phenomena.

Acknowledgments

I wish to acknowledge and thank the women of Borroloola—Yanyuwa, Garrwa, and Mara—for setting into action the chain of events that have informed this chapter. Thank you to Dinah Norman Marrngawi, Annie Karrakayn, and Pyro Didiyalma for guiding me toward the gentle nuances of gender difference. I also acknowledge and thank John Bradley for his ongoing contributions to my work over the years and Bruno David for his assistance with this manuscript.

Notes

1. For an extended discussion of "country," see Bradley (this volume).

2. *Ngalki* is a Yanyuwa word that refers to the "essence or quality that identifies and gives distinction to its owner or owners. The *ngalki* of a flower is its perfume; of food its taste and a person's *ngalki* can equally be the individual's social or landowning (semi-moiety) group or the sweat from under their armpits" (Bradley 1997: 150).

3. Only in recent years has the Rock Wallaby been depicted visually. This coincides with the compilation and the publication of the Yanyuwa Visual Atlas (Yanyuwa Families, Bradley, and Cameron 2003). In this document, all Dreaming ancestors are given a physical form and drawn onto maps that track their ancestral journey. The manner in which each ancestor was visually represented was the subject of lengthy negotiation and discussion among Yanyuwa elders. The act of committing an image of the ancestor to page was taken very seriously and people were committed to representing the Rock Wallaby in the correct manner.

4. Mackinlay's work provides an insight into acts of cultural maintenance that have occurred in the last ten years across Yanyuwa country (see 1995, 1996, 1998a, 1998b, 2000a, 2000b).

5. The township of Borroloola was established in the late 1880s and became the central point of settlement for the region's Aboriginal people in the early stages of European invasion. Today it is home to many Aboriginal people, including the Yanyuwa and their neighbors.

6. A star picket is a metal fencing post, approximately 5 cm wide and over 1 meter long.

7. "Fun songs" include the *a-Ngadiji*. This name is given to a specific fun dance performed by women. It relates to the Mermaid Dreaming spirits (Bradley, Kirton, and the Yanyuwa Community 1992: 116).

8. By the expression "closing up" I mean that country and this particular place were coming to be governed by stronger Law—Law that restricted people's coming and going from the actual location, restriction that forbids speaking about the ceremonial activity both to men and outsiders and that demands profound respect for the performers, their knowledge, and their actions during the course of ceremonial activity.

9. I cannot elaborate on this point, because these parts of the ceremony are governed by Law that pertains to secret and sacred ceremony.

References

Balme, J., and Beck, W. (eds.). 1995. *Gendered Archaeology: The Second Australian Women in Archaeology Conference*. Canberra: Division of Archaeology and Natural History, Australian National University.

Baxter, J. 2005. *The Archaeology of Childhood: Children, Gender and Material culture*. Walnut Creek, CA: AltaMira Press.

Beausang, E. 2000. Childbirth in prehistory: An introduction. *European Journal of Archaeology* 3: 69–87.

Bender, B. 1992. Theorising landscapes, and the prehistoric landscapes of Stonehenge. *Man* 27: 735–56.

———. 1993. *Landscape: Politics and Perspectives*. Oxford: Berg Publishers.

Bourdieu, P. 1977. *Outline of a Theory of Practice*. Cambridge: Cambridge University Press.

———. 1990. *In Other Words: Essays towards a reflexive sociology*. Stanford, CA: Stanford University Press.

Bradley, J., with J. Kirton and the Yanyuwa Community. 1992. Yanyuwa Wuka: Language from Yanyuwa country, a Yanyuwa Dictionary and Cultural Resource. Unpublished document.

Butler, J. 1990. *Gender Trouble: Feminism and the Subversion of Identity*. London: Routledge.

Classen, C. 1992. Questioning gender: An introduction, in C. Classen (ed.), *Exploring Gender through Archaeology*, pp. 1–10. Madison, WI: Prehistory Press.

Conkey, M. 1991. Original narratives: The political economy of gender in archaeology, in M. Di Leonardo (ed.), *Gender at the Crossroads of Knowledge: Feminist Anthropology in the Postmodern Era*, pp. 102–39. Berkeley and Los Angeles: University of California Press.

———. 1993. Making the connections: Feminist theory and archaeologies of gender, in H. DuCros and L. J. Smith (eds.), *Women in Archaeology: A Feminist Critique*, pp. 3–15. Canberra: Department of Prehistory, Research School of Pacific Studies, Australian National University.

Derrida, J. 1978. *Writing and Difference*. London: Routledge.

Di Leonardo, M. (ed.). 1991. *Gender at the Crossroads of Knowledge: Feminist Anthropology in the Postmodern Era*. Berkeley and Los Angeles: University of California Press.

Gero, J., and Conkey, M. (eds.). 1991. *Engendering Archaeology: Women and Prehistory*. Oxford: Blackwell Publishing.

Giddens, A. 1979. *Central Problems in Social Theory: Action, Structure and Contradiction in Social Analysis*. Berkeley and Los Angeles: University of California Press.

———. 1984. *The Constitution of Society: Outline of a Theory of Structuration*. Berkeley and Los Angeles: University of California Press.

Gilchrist, R. 1994. *Gender and Material Culture: The Archaeology of Religious Women*. London: Routledge.

———. 1999. *Gender and Archaeology: Contesting the Past*. London: Routledge.

Hamilakis, Y. 1998. Eating the dead: Mortuary feasting and the politics of memory Aegean Bronze Age societies, in K. Branigan (ed.), *Cemetery and Society Aegean Bronze Age*, pp. 115–32. Sheffield: Sheffield Academic Press.

Hamilakis, Y., Pluciennik, M. and Tarlow, S. 2002. *Thinking through the Body: Archaeologies of Corporeality*. New York: Kluwer Academic/Plenum Press.

Hartman, J., and Messer-Davidow, E. 1991. *Gendering Knowledge*. Knoxville: University of Tennessee Press.

Hays-Gilpin, K., and Whitley, D. 1998. *Reader in Gender Archaeology*. London: Routledge.

Hodder, I. 1982a. *The Present Past*. London: B. T. Batsford.

———. (ed.). 1982b. *Symbolic and Structural Archaeology*. Cambridge: Cambridge University Press.

———. (ed.). 1982c. *Symbols in Action: Ethnoarchaeological Studies of Material Culture*. Cambridge: Cambridge University Press.

———. (ed.). 1989. *The Meanings of Things: Material Culture and Symbolic Expression*. London: Unwin Hyman.

———. 1991. *Reading the Past: Current Approaches to Interpretation in Archaeology*. Cambridge: Cambridge University Press.

Jones, S. 1997. *The Archaeology of Ethnicity: Constructing Identities in the Past and Present*. London: Routledge.

Kearney, A. 2005. An Ethnoarchaeology of Engagement: Yanyuwa Country and the Lived Cultural Domain in Archaeology. Unpublished Ph.D. thesis, University of Melbourne, Australia.

Kearney, A., and Bradley, J. 2006. Landscapes with shadows of once living people: *Kundawira* and the challenge for archaeology to understand, in B. David, I. J. McNiven, and B. Barker (eds.), *The Social Archaeology of Australian Indigenous Societies*. Canberra: Aboriginal Studies Press.

Kent, S. 1984. *Analyzing Activity Areas: An Ethnoarchaeological Study of the Use of Space*. Albuquerque: University of New Mexico Press.

Kent, S. 1998. *Gender in African Prehistory*. Walnut Creek, CA: AltaMira Press.

Mackinlay, E. 1995. "Im been dance la me!": Yanyuwa women's song creation of the Northern Territory. *Australian Women's Composing Festival*. Sydney: Australian Music Centre.

———. 1996. "Men Don't Talk Much Anymore": The changing status of women in society and possible implications for Yanyuwa women as keepers, composers and performers of *a-Nguyulnguyul. Context* 12: 45–50.

———. 1998a. For Our Mother's Song We Sing: Yanyuwa Women Performers and Composers of *A-nguyulnguyul*. Unpublished Ph.D. thesis, University of Adelaide, Adelaide.

———. 1998b. Traditional Australian music—Gulf of Carpentaria, in A. Kaeppler and J. Love (eds.), *Garland Encyclopaedia of World Music: Australia and the Pacific Islands,* Vol. 9, pp. 427–28. New York: Garland Publishing.

———. 2000a. Music for dreaming: Aboriginal lullabies in the Yanyuwa community at Borroloola, Northern Territory. *British Journal of Ethnomusicology* 8: 97–111.

———. 2000b. Blurring boundaries between restricted and unrestricted performance: A case study of the Mermaid song of Yanyuwa women of Borroloola. *The Pacific Journal of Research into Contemporary Music and Popular Culture* 4(4): 73–84.

Merleau-Ponty, M. 1962. *Phenomenology of Perception*. London: Routledge.

———. 1964. *The Visible and the Invisible*. Evanston IL: Northwestern University Press.

Meskell, L. 1995. Goddesses, Gimbutas and "New Age" archaeology. *Antiquity* 69: 74–87.

———. 1999. *Archaeologies of Social Life: Age, Sex, Class, etcetera in Ancient Egypt*. Oxford: Blackwell Publishing.

Meskell, L. 2000. Writing the body in archaeology, in E. Alison (ed.), *Reading the Body: Representation and Remains in the Archaeological Record*, pp. 13–21. Philadelphia: University of Pennsylvania Press.

Moore, H. 1994. *A Passion for Difference: Essays in Anthropology and Gender.* Cambridge: Polity Press.

———. (ed.). 1996. *The Future of Anthropological Knowledge.* London: Routledge.

Nelson, S. M. 1997. *Gender in Archaeology: Analyzing Power and Prestige.* Walnut Creek, CA: AltaMira Press.

Rose, D. B. 2004. *Reports from a Wild Country: Ethics for Decolonisation.* Sydney: University of New South Wales Press.

Shanks, M., and Tilley, C. 1987a. *Reconstructing Archaeology: Theory and Practice.* Cambridge: Cambridge University Press.

———. 1987b. *Social Theory and Archaeology.* Cambridge: Polity Press.

Sorensen, M. 1998. Is there a feminist contribution to archaeology? *Archaeological Review from Cambridge* 7(1): 9–20.

———. 2000. *Gender Archaeology.* Cambridge: Polity Press.

Tarlow, S. 1999. *Bereavement and Commemoration: An Archaeology of Mortality.* Oxford: Blackwell Publishing.

———. 2000. Emotion in archaeology. *Current Anthropology* 41: 713–46.

Thomas, J. 1996. *Time, Culture and Identity: An Interpretive Archaeology.* London: Routledge.

———. 2001. Archaeologies of place and landscape, in I. Hodder (ed.), *Archaeological Theory Today*, pp. 165–86. Cambridge: Polity Press.

Tilley, C. (ed.). 1990. *Reading Material Culture: Structuralism, Hermeneutics and Post-Structuralism.* Oxford: Basil Blackwell.

———. (ed.). 1993. *Interpretive Archaeology.* Oxford: Berg Publishers.

VanDyke, R., and Alcock, S. 2003. *Archaeologies of Memory.* Melbourne: Blackwell Publishers.

Warren, C., and Hackney, J. 2000. *Gender Issues in Ethnography.* Sage University Papers Series on Qualitative Research Methods, Vol. 9. Thousand Oaks, CA: Sage Publications.

Yanyuwa Families, Bradley, J., and Cameron, N. 2003. *Forget about Flinders: A Yanyuwa Atlas of the Southwest Gulf of Carpentaria.* Brisbane: Private Publication.

24

HIDDEN LANDSCAPES OF THE BODY

Clive Gamble

> Yet there is no use pretending that all we know about time and space, or rather history and geography, is more than anything else imaginative. (Said 1978: 55)

Is there an archaeology of the soul? Of course the question is preposterous and most archaeologists would dismiss it as irrelevant: how can the immortal and immaterial be studied by a discipline wedded so strongly to the corporeal and material? And yet archaeologists outlay considerable effort on understanding the workings of other concepts that are also hidden and internal and that work in mysterious ways; for example, socio-economic systems, cultures, and, at another scale, individual identity comprising personhood and the self. These concepts are hidden because they are inferred from material remains, rather than directly observed, and they are internal because such understanding would not exist without our knowledge of life as layers of experience contained in memories and objects (Knappett 2006).

In this chapter I concentrate on the hidden landscape of human identity rather than the more familiar landscapes of organizational systems and cultural behavior. I am not interested in souls but instead in the rhetorical devices that infuse material

with bodily experience thereby giving legs to the immanent in the form of metaphors. As Martin Hollis (1985: 227) put it from a philosopher's perspective, we need metaphors to express what goes on inside us in order to analyze the hidden, inner sense that makes the whole system of human relations run. What Hollis identified was an imaginary geography located inside the body, which I explore here as a suite of hidden landscapes, or, more accurately, *scapes*. My task as an archaeologist interested in such inner landscapes is to recognize their metaphorical basis in the form of ancient material culture.

The essence of metaphor lies in understanding one thing in terms of another. Metaphors, moreover, are experiential (Lakoff and Johnson 1980). Investigating these inner landscapes does not therefore depend upon language or the cognitive capacity for rational, discursive thought. Rather, they depend upon the metaphorical relationship that exists between material culture and the experience of the body (Gamble 2004, 2007). Consequently, I argue for a relational perspective on these infusions of bodily experience with material metaphors. I will show that the value for archaeologists lies in the exploration of almost three million years of changing human identity, through inner states of self and personhood.

Landscape, Scapes, and Material Culture

Landscape is a contested, imaginative space, defined by Daniels and Cosgrove (1988: 1) as "a cultural image, a pictorial way of representing or symbolizing surroundings." However, Ingold (2000: 191) disagrees, rejecting their division between inner and outer worlds implicit in the notion of representation. Instead, landscape is "the familiar domain of our dwelling, it is *with* us, not *against* us, but it is no less real for that. And through living in it, the landscape becomes a part of us, just as we are a part of it."

These contrasted views of inner and outer landscape are played out in many disciplines and in more extreme forms. Central to the distinction is the difference between *rational* and *relational* approaches to knowledge. These polar views are common in archaeology with the former still commanding majority support in many periods and especially when the economy is considered. For example, from a rational perspective, human adaptation is often presented as an attritional "game *against* the environment" (Jochim 1976) and where the economic analogy of the corner-store (Earle 1980) sets the tone of the enquiry for such encounters with landscape. An alternative, relational view sees landscape as an aspect of *habitus*, what Bourdieu regarded as a "feel for the game" (Gosden 1999: 125–26). An example of

this perspective is provided by Bird-David's (1999) study of contemporary hunters and gatherers via the primary metaphor (Gudeman 1986; Ortner 1973) of the *giving* environment.

For the archaeologist, landscapes begin and end with material culture. The affect and testimony of the people who dwelt in those landscapes are unavailable in familiar ethnographic form. But although the rhetorical nuances of language may be missing, we should not be hoodwinked by the apparent silence of the actors to the metaphorically deafening volume of material remains. As Tilley (1999) and others (Gamble 2007; Jones 2002; Parker-Pearson and Ramilisonina 1998) have shown, a productive approach to such landscapes is to view their material culture as solid metaphors. The ancient objects, monuments, villages, and fields that inhabit the landscapes archaeologists investigate are therefore as potent for a relational perspective as they are for a rational.

But these are not simple landscapes and a road map, or rather mapscape (Figure 24.1) is needed to organize such complex imaginary geographies. Many scapes exist as a result of the relationship of the body to material culture (Hamilakis, Pluciennik, and Tarlow, 2002; Thomas 1996); among them bodyscapes, sensescapes and taskscapes (Ingold 1993), as well as the broader terms of landscapes of habit and social landscapes (Gamble 1999) and the environment of development, or *childscape* (Gamble 2007). The intersection of some of these

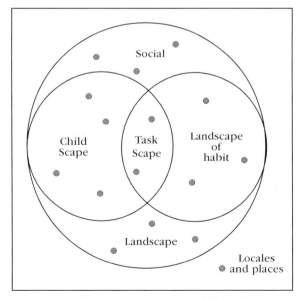

Figure 24.1 An imaginary geography, or *mapscape*, of several hidden landscapes that are accessible to archaeologists through the use of material metaphors. In this mapscape, they form layers that are sampled at locales and places. The layers are explained in the text (after Gamble 2007: fig. 8.5).

is indicated diagrammatically in Figure 24.1. The important point is not the list or the scales of such scapes but rather their layered character, because this sedimentation qualifies them as hidden. I shall return to the childscape later since it is this, the most hidden of all the scapes, that is important for understanding a fundamental change such as the transition to agriculture.

Missing Persons and the "Darkness of the Body"

An apt analogy to the archaeologist's pursuit of the hidden is provided by one of Andrew Wyeth's much reproduced paintings, *Sea Boots* (www.museumsyndicate.com/item.php?item=15796). The body of the boots' owner, Walter Anderson, is hidden and yet very strongly presented in this realistic, strutting image. The boots clearly have a history. They are biographical objects at ease in their familiar landscape. A person extends from those battered boots and by juxtaposing shoreline with roofline, settles into a landscape of habit as into an old shoe (Adams 2006).

Wyeth dwells on the mundane—boots, beach and roof—and composes them into a common artistic trope, the still life. In doing so, he tugs the missing person into the picture. We want to see their flesh and blood owner to confirm the image we have of his identity from his boots inhabiting their landscape. That might be the rational desire when confronted by such a realistic image. However, the boots also embody a relational perspective. They are not simply a proxy for the absent person but possess their own agency through the structure of material relationships (Gell 1998: 123). They are one layer of identity as thin as the paint on the canvass or the skin on the body.

The sculptor Antony Gormley (2004: 134) has made a similar point but in solid material form when he talks of "the darkness of the body" being revealed through "the other side of appearances:" "I am very aware as I speak to you now that where I am is behind my face; my face and my body in some way belong more to you than they do to me, and vice versa" (Gormley 2004: 134). These appearances consist of layers. They continue the journey into the dark, hidden landscapes of the body. A journey that began metaphorically by taking off a pair of old boots.

Layerings

My point here is that our understanding of these inner landscapes, as conceptual devices for identity based on self and personhood, depends on material metaphors such as a pair of sea boots. Of course, we talk about how we feel inside and so construct elaborate geographies of our inner workings. These can then be shared in word and print. But these identities and their interior location are not dependent on language alone. Material metaphors such as Wyeth's boots or Van Gogh's chair (www.nationalgallery.org.uk/cgi-bin/WebObjects.dll/CollectionPublisher.woa/wa/work?workNumber=NG3862) are referenced to bodily experience, as are the more familiar linguistic forms. And as with their linguistic counterparts, material metaphors experience one thing in terms of another. Moreover, as Chapman (2000; Tilley 1999) has shown, material metaphors can be just as sophisticated in terms of the targets they reach through analogy, metonymy, and synecdoche.

If this is the case, then the way to understand material culture is not to search for semiotics and meaning but rather to see how material metaphors are referenced by the body. Elsewhere (Gamble 2004), I have done this through a basic division into containers and instruments, those material proxies for the routine actions of the body. Hence in Wyeth's picture, the boots and the house are containers. They change the shape of the body when worn and they condition the structure of action when inhabited. When wearing the boots and sitting inside the house the individual concerned is layered, his agency structured and informed by landscapes that are assimilated by dwelling in them (Ingold 2000). The bodily experience associated with living in boots and houses becomes a material metaphor of those inner states enacted in everyday routines.

But let me be clear on one point. Those sea boots are not akin to the "wrong trousers" of Wallace and Gromit fame that control the actions of their unfortunate wearer. With material culture, the agency of the one (a pair of sea boots) depends on the agency of the other (the missing person). This is because such objects stand in networks of relationships to the landscape of the beach and the house as much as they stand in relation to the boots and the missing body. As Carl Knappett (2005, 2006) argues, these networks are layerings that entangle us with objects and them with us.

Wyeth's painting suggests a further avenue for exploration. Containers form biographical traps, described by Hoskins (1998: 5) as memory boxes, not only of the mundane, such as fishing and keeping the feet dry, but also of the relationships between varied materials that a concept of landscape brings into relation. In this respect, the image suggests an approach to understanding that Lévi-Strauss (1966) championed as the distinction between the bricoleur and the engineer. The latter works by design, solving problems such as how to keep dry by rational means, projecting thoughts formed in an inner landscape onto an external world. By contrast, the bricoleur is more concerned with constructing meaning by bringing things together, most importantly for Lévi-Strauss in the form of myths. The distinction is between rational and relational attitudes to material culture, and, while never mutually exclusive, these have changed in the course of human evolution.

2.5 Million Years of Identity

An analysis of hominin technology, from the earliest stone tools to the appearance of writing in the Near East, reveals a slow gradient of change (Figure 24.2). During this time, innovations in technology changed the balance among material proxies of the body from instruments to containers (Fagan 2004; Gamble 2007; Troeng 1993). But this was not a question of replacement of one proxy by the other. Containers always existed, but for long periods of time they were rare. They can be found in the technologies of the great apes, such as sleeping pallets and leaf sponges (McGrew 1992). However, instruments dominate primate technologies and now include observations of New World monkeys (de Amoura and Lee 2004), indicating a truly deep evolutionary ancestry for hominid/hominoid relationships with material culture.

Two further observations need to be added to this overview of hominin involvement with technology. First, instruments and containers existed long before language. Exactly when words were first uttered is still contested, but an exponential rise in the size of hominin neocortex 500,000 years ago can be parsimoniously explained by the social brain hypothesis (Dunbar 2003) that links selection for new forms of communication to the demands of integrating individuals into larger group sizes. At this time, language offered advantages to increasingly busy hominins as they went about their social lives. But this early appearance of language did not equate with the origins of imaginary geographies constructed by metaphor. Hominins had been using such concepts for at least two million years previously, as shown by the archaeology of instruments and containers. Objects preceded words, and for a long time instruments dominated as proxies for the body.

In an interesting rerun of ontogeny recapitulating phylogeny (Gould 1977), child psychologists (Bloom 2004; Hespos and Spelke 2004) have discovered a similar object-word sequence.

Prelanguage children learn to think by experiencing the material world. They come to recognize concepts through the shapes, textures, and fit of objects rather than first learning them through language and only later transferring this knowledge to the material world. Our first appreciation of metaphor, understanding one thing in terms of another, is material.

The second observation concerns the invisibility of children in the standard histories of technology (Fagan 2004). Indeed, children are a hidden category in most archaeologists' descriptions of the past (Sofaer Deverenski 2000). They can be identified from their tiny footprints and to a limited extent through the study of apprenticeship (Pigeot 1987). They occur as burials, but here the tendency is to regard them solely as biological rather than cultural categories. Those landscapes of infancy and childhood will not, however, be revealed by correctly identifying toys or even a children's technology such as cradle or sling. The situation is comparable to engendering the past by trying to find more objects that can be directly associated with women instead of considering how identities of self and personhood are constructed out of local conditions (Spector 1993).

The Childscape

The history of technology therefore needs to consider what is currently hidden. To this end, the *mapscape* (Figure 24.1) recognizes an imaginary geography, a space where these hidden layers can be revealed through their metaphorical associations. One solution is to consider a childscape intersecting with the adult spaces of the taskscape and the landscape of habit. The childscape is the environment of growth. It is a particular landscape of experience that is important in the construction of identity, the material associations of self and personhood. The childscape acts as any other archaeological landscape in that, as discussed above, it begins and ends with material culture. It therefore follows that the material arrays within the childscape, those instruments and containers acting as material proxies for the body, will have significance in structuring our metaphorical understanding. The difference between a childscape dominated by instruments to one where containers are the commonest artifact proxy (Figure 24.2) now assumes significance for the construction of those hidden landscapes of the body. Other technologies, such as language and literacy, have shaped our understanding of these inner landscapes. But in both cases, they were preceded by material metaphors and, in the case of writing, by the culmination of the long

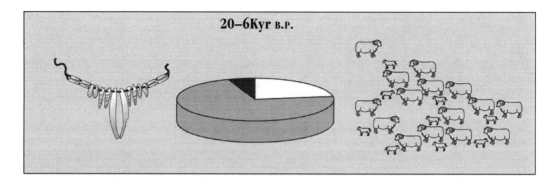

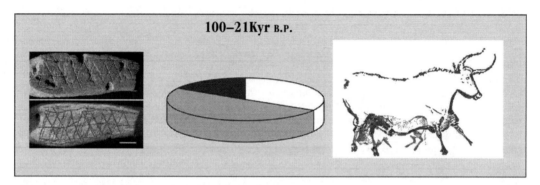

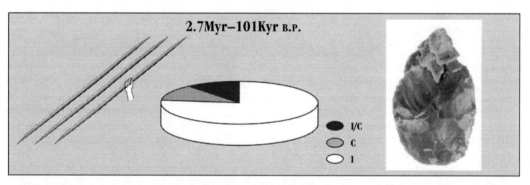

Figure 24.2   Changing material proxies during hominin evolution. Key: I = instrument, C = container and I/C hybrid proxies. (after Gamble 2007: fig, 7.1). The temporal divisions were chosen to illustrate the slow gradient in such changes to the understanding of hidden landscapes through material metaphors. Some examples of containers and instruments are shown. (engraved Blombos ochre container, courtesy C. Henshilwood; Boxgrove handaxe instrument, courtesy G. Marshall).

gradient that saw a change in the dominant material proxy for the body. Such changes were based on a different metaphorical understanding of our interior states stated not in words but in the relationship of things to the experiences of the body. As a result, the world of hominins became increasingly layered and elaborated through accumulation and enchainment (Chapman 2000).

But such a general trend should not obscure the immense local variation that occurred. Archaeological cultures did not vary as a result of their role in supporting economic adaptations (for example, hunter, fisher, herder and farmer) but because of the knowledge that comes from inhabiting a world of objects that resisted some interpretations and encouraged others (Parker-Pearson and Ramilisonina 1998: 310). An economic change as fundamental as that between hunting and farming therefore required a restructuring of that knowledge. This change was largely hidden because it occurred in the childscape (Figure 24.1). It is in this layer of the mapscape that concepts are learned through material metaphors and structured by root metaphors (Ortner 1973), and so open to all sorts of interpretation. For example, Ingold (2000: 86) has commented that "growing plants and raising animals are not so different, in principle, from bringing up children." The move to containers (Figure 24.2) identifies the local shift from one root metaphor, the giving environment to a new one, growing the body. These root metaphors provide the metaphorical associations for that layer of identity I call the childscape. The seeds of agriculture were first established in this environment of development, the childscape. And that identity changed the material world in ways that are now all too familiar to us. A process as fundamentally transforming as agriculture therefore derives from the layerings in the mapscape (Figure 24.1), of which the childscape is the most hidden but of great importance.

But who made these changes to the array of objects in the childscape? Not the child, an external change in temperature, or the return of deciduous forests to formerly glaciated lands. Instead, this is where the bricoleur rather than the engineer plays an evolutionary role. Materials were brought together in locally constructed patterns of meaning. Long before there were containers in the form of houses there were houselike constructions where bones, stones, earth, and bodies were placed, in association, by bricoleurs. This may explain why the evidence for architecture before the Neolithic is scant and problematic (Kolen 1999; Verpoorte 2001), because we approach it as engineers looking in the acts of accumulation for a rational purpose. When we come looking for relationships made from accumulation and enchainment, we find a different picture altogether and within such relationships there is space for the hidden landscape of the child.

Conclusions

The craft of the archaeologist has provided many disciplines with a metaphor for discovery, and none more so than psychology where digging into the unconscious is an exercise in peeling back the layers of accumulated experience. When the hidden landscapes of the body are considered on an evolutionary timescale, then it is these disciplines that, in turn, provide archaeologists with useful rhetorical devices. Sedimentation, layering, accumulation, and the deposition of memories are all dependent on notions of landscape. From personal experience, our earliest childhoods remain hidden. For us, they are an imaginary geography that depends on what others tell us or by leafing through the family album long after the events they record. And yet the childscape is known to us

through our developing skills of forming concepts based on the experience of our bodies as they grew and inhabited the material world. These hidden landscapes of memory and inspiration were not static during human evolution. They varied from place to place as indicated by the proxies (Figure 24.2). Although we have always entertained the bricoleur and the engineer in our identities, we are having to rediscover more of the former in order to understand those changing landscapes in the past. Through the material metaphors they contain, we have always engaged with a coherent world and made it ours.

Acknowledgments

This paper was supported by the British Academy Centenary Project "From Lucy to Language: The Archaeology of the Social Brain." I would like to thank Fiona Coward and Matt Grove for their very helpful comments and criticisms.

References

Adams, H. 2006. Wyeth's world. *Smithsonian* 37: 85–92.

Bird-David, N. 1999. "Animism" revisited: Personhood, environment, and relational epistemology. *Current Anthropology* 40: 67–91.

Bloom, P. 2004. Children think before they speak. *Nature* 430: 410–11.

Chapman, J. 2000. *Fragmentation in Archaeology: People, Places and Broken Objects in the Prehistory of South-Eastern Europe*. London: Routledge.

Daniels, S., and Cosgrove, D. 1988. Introduction: Iconography and landscape, in D. Cosgrove and S. Daniels (eds.), *The Iconography of Landscape*, pp. 1–10. Cambridge: Cambridge University Press.

de Amoura, A. C., and Lee, P. C. 2004. Caphuchin stone tool use in Caatinga dry forest. *Science* 5703: 1909.

Dunbar, R. I. M. 2003. The social brain: Mind, language, and society in evolutionary perspective. *Annual Review of Anthropology* 32: 163–81.

Earle, T. 1980. A model of subsistence change, in T. Earle and A. L. Christenson (eds.), *Modeling Change in Prehistoric Subsistence Economies*, pp. 1–29. New York: Academic Press.

Fagan, B. (ed.). 2004. *The Seventy Great Inventions of the Ancient World*. London: Thames and Hudson.

Gamble, C. S. 1999. *The Palaeolithic Societies of Europe*. Cambridge: Cambridge University Press.

———. 2004. Materiality and symbolic force: A Palaeolithic view of sedentism, in E. DeMarrais, C. Gosden, and C. Renfrew (eds.), *Rethinking Materiality: The Engagement of Mind with the Material World*, pp. 85–95. Cambridge: McDonald Institute of Archaeological Research.

Gamble, C. S. 2007. *Origins and Revolutions: Human Identity in Earliest Prehistory*. New York: Cambridge University Press.

Gell, A. 1998. *Art and Agency: Towards a New Anthropological Theory*. Oxford: Clarendon Press.

Gormley, A. 2004. Art as process, in C. Renfrew, C. Gosden, and E. DeMarrais (eds.), *Substance, Memory, Display: Archaeology and Art*, pp. 131–51. Cambridge: McDonald Institute Monographs.

Gosden, C. 1999. *Anthropology and Archaeology: A Changing Relationship*. London: Routledge.

Gould, S. J. 1977. *Ontogeny and Phylogeny*. Cambridge: Harvard University Press.

Gudeman, S. 1986. *Economics as Culture: Models and Metaphors of Livelihood*. London: RKP.

Hamilakis, Y., Pluciennik, M., and Tarlow, S. (eds.). 2002. *Thinking through the Body: Archaeologies of Corporeality*. New York: Kluwer Academic/ Plenum.

Hespos, S. J., and Spelke, E. S. 2004. Conceptual precursors to language. *Nature* 430: 453–56.

Hollis, M. 1985. Of masks and men, in M. Carrithers, S. Collins, and S. Lukes (eds.), *The Category of the Person: Anthropology, Philosophy, History*, pp. 217–33. Cambridge: Cambridge University Press.

Hoskins, J. 1998. *Biographical Objects: How Things Tell the Story of People's Lives*. New York: Routledge.

Ingold, T. 1993. The temporality of the landscape. *World Archaeology* 25: 152–73.

———. 2000. *The Perception of the Environment: Essays in Livelihood, Dwelling and Skill*. London: Routledge.

Jochim, M. A. 1976. *Hunter-Gatherer Settlement and Subsistence*. New York: Academic Press.

Jones, A. 2002. *Archaeological Theory and Scientific Practice*. Cambridge: Cambridge University Press.

Knappett, C. 2005. *Thinking through Material Culture: An Interdisciplinary Perspective*. Pittsburgh: University of Pennsylvania Press.

———. 2006. Beyond skin: Layering and networking in art and archaeology. *Cambridge Archaeological Journal* 16: 239–51.

Kolen, J. 1999. Hominids without homes: On the nature of Middle Palaeolithic settlement in Europe, in W. Roebroeks and C. Gamble (eds.), *The Middle Palaeolithic Occupation of Europe*, pp. 139–75. Leiden: University of Leiden and European Science Foundation.

Lakoff, G., and Johnson, M. 1980. *Metaphors We Live By*. Chicago: University of Chicago Press.

Lévi-Strauss, C. 1966. *The Savage Mind*. Chicago: University of Chicago Press.

McGrew, W. C. 1992. *Chimpanzee Material Culture: Implications for Human Evolution*. Cambridge: Cambridge University Press.

Ortner, S. B. 1973. On key symbols. *American Anthropologist* 75: 1338–46.

Parker-Pearson, M., and Ramilisonina 1998. Stonehenge for the ancestors: The stones pass on the message. *Antiquity* 72: 308–26.

Pigeot, N. 1987. *Magdaléniens d'Étiolles: Économie de débitage et organisation sociale (l'unité d'habitation U5)*, Vol. XXV supplement à *Gallia Préhistoire*. Paris: CNRS.

Said, E. 1978. *Orientalism*. New York: Pantheon.

Sofaer Deverenski, J. 2000. Material culture shock: Confronting expectations in the material culture of children, in J. Sofaer Deverenski (ed.), *Children and Material Culture*, pp. 3–16. London: Routledge.

Spector, J. 1993. *What This Awl Means: Feminist Archaeology at a Wahpeton Dakota Village*. St. Paul: Minnesota Historical Society Press.

Thomas, J. 1996. *Time, Culture and Identity: An Interpretive Archaeology*. London: Routledge.

Tilley, C. 1999. *Metaphor and Material Culture*. Oxford: Blackwell.

Troeng, J. 1993. *Worldwide Chronology of Fifty-Three Prehistoric Innovations*. Stockholm: Acta Archaeologica Lundensia 21.

Verpoorte, A. 2001. *Places of Art, Traces of Fire: A Contextual Approach to Anthropomorphic Figurines in the Pavlovian (Central Europe, 29–24 kyr B.P.)*. Leiden: Faculty of Archaeology, University of Leiden.

25

The Body and the Senses: Implications for Landscape Archaeology

Paul Rainbird

Landscape archaeology has had at its foundation the requirement of observation. These observations have usually been derived from patterns identified on aerial photographs, maps and plans, revealing palimpsests of relative chronology in landscape use. These two dimensional perspectives have been added to by a concern with how the landscapes are actually experienced on the ground in the present and how this may allow for a new understanding of landscapes at various points in the past. Such approaches have been influenced by phenomenology, but, like their predecessors, the observations have usually privileged vision over other sensory experiences. In this chapter, I consider the possibilities of going beyond vision to incorporate other sensory perceptions of the landscape. I consider the body in landscape archaeology and landscape in sociocultural anthropology and move on to present examples of the roles of sound in the landscapes of the ancient past.

The Body's Career in Landscape Archaeology

Christopher Tilley (1994) introduced archaeology to the exciting possibilities of a phenomenology of landscape. The starting point of his analyses was that the human body was a universal phenomenon that in physical attributes and abilities differed little from our prehistoric ancestors. Thus, Tilley argued, we could adopt the anthropological method of "participant observation" by bodily engagement with archaeological landscapes and the monuments within it. Walking along monuments such as the 10-kilometer-long Dorset cursus,[1] and assessing the views allowed by the location and alignments of monuments such as long barrows, would provide a three-dimensional experience of space similar to that of the original constructors and users of the monuments and landscape. Lines of sight and intervisibility of monuments thus became a key concern for scholars researching the landscape context of the British Neolithic and Early Bronze Age, the same focus as Tilley himself experienced. The ocular-centric nature of such studies—that is, the focus on vision—caused concern on a number of levels: for example, was the past environment really the same as in the present day (Fleming 1999; cf. Cummings and Whittle 2004); was the meaning of the location of a monument really linked only to what could be seen from there (Brück 2005; Fleming 1999, 2005); and why give primacy to vision when the other senses could be of equal or greater importance (e.g., Hamilakis 2001, 2002, In preparation)?

The Neolithic monuments themselves became subject to a suite of studies that aimed to give greater consideration to the full haptic (that is, synaesthetic) potential of bodily experience. For example, beyond lines of sight, color was considered in relation to stones chosen for monument building (Lynch 1998), and texture of the stones was observed (Cummings 2002), and Alisdair Whittle (1997) observed that the renowned trilithon stones at Stonehenge were paired rough and smooth. Although most of these observations appear to be based on visual assessment, Gavin MacGregor's (2002) consideration of texture of Recumbent Stone Circles in northeast Scotland does indicate the likely use of touch in using adjectives such as "waxy" and "smooth" in describing particular stones. The sound propensities of monuments also became a focus of study (Lawson et al. 1998), with findings that even apparently very open monuments, such as stone circles, affected the free flow of sound within them (Watson and Keating 1999).

These studies, although usefully drawing attention beyond vision, did not go much beyond individual monuments and therefore were of reduced significance in relation to *landscape* archaeology. Recognizing that landscapes indeed had the potential to go beyond vision, Tilley (1999) urged a consideration of other "scapes" for touch, sound, and smell. Curiously, he omitted tastescapes, although I have suggested elsewhere that such may exist. For example, when approaching the sea from the onshore breeze, the salt from the sea may be tasted before the sea is actually seen (Rainbird 2007). Yannis Hamilakis (2002) has in particular looked at the role of the sense of taste in archaeology, discussing it as an "oral history," whereby the tastes of different foods invoke memories of events or act as mnemonics for particular social behavior.

The body's role in landscape archaeology has so far been a limited one; its place in the phenomenology of landscape has been questioned (Brück 2005; Hodder 1999). We know, for example, that anthropologists have long recognized that the body is socially constituted (see Csordas 1999), and its cultural and historical mutability means that its status as an individual and independent observer is not likely to have been possible in the past either. Julian Thomas (2002) has used the depositional evidence from Neolithic monuments to propose that what we would consider the live person was only one stage in the biography of the body as constituted in the Neolithic. Placing modern conceptions of experiencing the landscape through modern sensibilities may not tell us any more than something about the psyche of the individual archaeologist conducting the exercise (cf. Brück 2005), especially as we know that the five senses are a construct of modern Western

categorization (see Ingold 2000). However, the evidence from anthropology also teaches us that a decentering of vision and an attentiveness to other sensual possibilities, as a necessary consideration of agency, rather than "the individual" (cf. Barrett 2000; *contra* Hodder 2000), opens up interpretative possibilities that can serve only to enhance our imaginings about real experiences of inhabiting landscapes in the past.

Landscape's Career in Social Anthropology

It may be no coincidence that the phenomenological aspects of landscape archaeology were being acknowledged at the same time as social anthropology as a discipline recognized landscape as a useful field to conceptualize (e.g., papers in Bender 1993a and Hirsch and O'Hanlon 1995). The sacred landscape had been recognized by cultural geographers, as well as ethnographers, but the different perceptions of landscape that ordered and made sense of the world in everyday experience had not often been expressed; this is now being addressed (Feld and Basso 1996; Ucko and Layton 1999).

Anthropologists are aware that people perceive, and thus experience, their existence in many different ways. Barbara Bender (1993b) has highlighted the fact that perception of the landscape depends on the status of the person in relation to such variables as gender, social standing, and previous experience. Anthropologists have recognized that all the senses are key to understanding the embodiment of experience. For example, David Sutton (2001) has worked on taste in relation to the social role of food and memory; the Canadian school has explored how smell is culturally constituted (Classen, Howes, and Synott 1994); and all agree that each of the identifiable senses ought not to be considered in isolation but regarded as a total body experience with all senses working in combination and one or some heightened in relation to the others depending on previous experience. As we have seen for the phenomenology of the Neolithic above, scholars have found it difficult to describe this total experience and in preliminary studies have gone beyond vision in landscape studies by assessing sound, the supposed second of the "higher" human senses.

Steven Feld (1996) has researched sound in the landscape of the Kaluli people of Boasavi, Papua New Guinea. He found through a study of Kaluli singing that the poetics of place merged with a sensuousness of the locale. Places in the Kaluli landscape were heard and felt through bodily performances, onomatopoeic utterances in song brought out recognizable landscape features, such as the sound of water running over a waterfall. The close connection between the people and their landscape is revealed through the sounds and the body as an organ for both producing those sounds and feeling the landscape resonating within. Feld (1996) found that the landscape and the body were further linked by the Kaluli when the contours of the hills were compared with curves of the body, and the passage of water with the flows of both bodily fluids and a physical sense of the voice:

> Voice flows by resounding through the human body, feelingfully connecting its spatially contiguous physical segments, resonating so as to sensually link and stress the whole. Likewise, when water flows through land, it is always multiply connected, always multiply present across and along a variety of relatively distinct, contiguous landforms, linking them and revealing their wholeness. (Feld 1996: 104)

Here, then, sound connects in a direct way the inhabitants and their landscape. The analogies that are drawn indicate a oneness between lived bodies and a vibrant landscape. Alfred Gell (1995), also working in Papua New Guinea among the Umeda, recalls his initial frustration in conducting ethnography in a densely forested environment. In searching for a vista from which he would be able to gaze over the village, he began to realize that, in contrast with the Umeda, his own perception was ocular-centric. In recognizing this, he was then able to identify that the Umeda live in a soundscape rather than a *land*scape or visionscape. In the forest, vision is of less use than sound in regard to spatial orientation and knowledge of activities. Groups of people on the move mark their presence by various shouts, the chopping of wood, or the pounding of sago—each having distinctive sounds. A heightened sense of sound is the normal sensory experience in the forest.

Among the Yanyuwa community of the Gulf of Carpentaria country in northern Australia, John Bradley and Elizabeth Mackinlay (2000) find that the music of these Indigenous Australians is intimately linked to the traversing and making of the landscape by their Spirit Ancestors. This is a landscape where the histories of features are known through song, and concentrations of power derived from the Spirit Ancestors can be found:

> The sharpest concentrations of the Spirit Ancestors' powers are found in such marks: places where they created a landform, left an object behind, raised a tree or entered the ground. These are powerful places, or in contemporary English usage, the sacred sites, the places where important knowledge is said to reside. This knowledge, much of it associated with music and performance, provides a rich soundscape which can still be used by the Yanyuwa to assist in the maintenance of the life-order which is derived from the events of what they call the *Yijan*, a word that generally and confusingly is translated into English as "Dreaming." (2000: 4)

The soundscape that Bradley and Mackinlay describe here is one that, although rich in meaning, does not necessarily communicate that meaning by resounding across the landscape in the way that Gell described for the Umeda noises. But in the performative and musical aspects that communicate the knowledge necessary to maintain social life, the sounds of instruments can have particularly important roles. For example, in describing percussion instruments used by Yanyuwa people in postfuneral ceremonies, Bradley and Mackinlay (2000) were told that the heavy thudding of the instruments had the purpose of directing the spirit of deceased people back to their country.

Of course, the sacred landscape of Indigenous Australians is well known in the anthropological literature and provides examples of how different people were able to perceive such places (e.g. Morphy 1993; Munn 1996). The longevity of such perceptions has also been the subject of archaeological analysis and related rock art sites (David 2002), and it also provided much inspiration for Tilley's (1994) phenomenology of landscape discussed above. I look in more detail at rock art below but first conclude this section with a brief consideration of the sense of touch and landscape in anthropology.

Tim Ingold (2000), drawing on a wide range of disciplinary traditions, has criticized the anthropology of the senses as perpetuation of the potentially limited Western conception, limited to the five senses but also as detached from environmental conditions that most people in modern Western societies have little experience with (see also Finney 1995). Following from this critique, Ingold (2004) has explored the "groundlessness" of Western sensory perceptions of living in the world by the development and now constant and expected use of footwear, machines for walking, and, to a certain extent, chairs, all of which detach sensory experience from the ground. He notes records of non-Western societies in which the foot has prehensile qualities and in which squatting rather than sitting is the norm, as it is in much of the world. In particular, for our purposes here, Ingold points out that the formation of the landscape through walking is quite different for people normally unshod and people wearing sandals, shoes, or boots.

Picturing and Perceiving the Social Landscape

Rock art studies have had various and at times strained relations with the discipline of archaeology, as they have often in the past been pursued as a type or branch of art history that neglected the archaeological context of its production and use. In recent years there has been a significant rapprochement, and rock art often features as a facet of archaeology and archaeological interpretation more generally (e.g., Bradley 1997; Chippindale and Taçon 1998). In Australia, the early recognition through discussion with traditional custodians that the rock art often acted as a mnemonic for the creation stories of the Spirit Ancestors led to an immediate and obvious relationship with the landscape as a social landscape (Taçon 1994). Although most work on rock art has remained overtly concerned with the visual symbols, there is a growing body of literature that has considered other sensory manifestations of rock art production and associated practices (e.g., Boivin 2004; Fagg 1997; Waller 2002).

Elsewhere I have explored sensory experiences in relation to the production and meaning of petroglyphs at two sites widely separated by space and time (Rainbird 2002a, 2002b). The first is a site named Pohnpaid and located on the Pacific island of Pohnpei in the eastern Caroline Islands of the Federated States of Micronesia. This site consists of more than 750 petroglyphs in two types of location: the first is a large outcrop overlooking wooded lowland areas, and the second is a nearby series of boulders scattered over a grassy plateau (Rainbird and Wilson 1997; Rainbird In preparation). The boulders are known locally to represent human body parts, whereas the motifs on the large outcrop have been related in oral history accounts of foundation myths to the patterns found on a blanket taken from people on the other side of the island (other stories have also been collected). The marks consist of anthropomorphs, footprints, handprints, fish, bows, nets, and various abstract and unidentified shapes. They are currently undated, but a specific motif type known as an "enveloped cross," which has a wide distribution in the corpus of western Pacific rock art, may suggest an antiquity for at least some of

the marks of more then 2,000 years (Rainbird and Wilson 2002).

This site has been considered in relation to the sensory perception of feeling and the sense of sound. In some Oceanic societies, tattooing was traditionally a painful process that brought the unsullied body in to a state of hardness and readiness for the period of existence as a corporeal being, only part of the body's biography (cf. Thomas 2002). The tattooing also inscribed a biography, a history, of the community that could be read and understood by members of the group. The markings on rock, and potentially other media such as ceramics (see discussion in Rainbird 1999), may be related to tattooing and share some of its meaning. For example, penetrating the surface of the stone may be preparing the landscape for social action in the same way that the body is prepared for social life by tattooing, and inscribing motifs on the rock may also be the inscription of history. Observing, or perhaps more importantly touching the engravings may act as a mnemonic for feeling the pain, a pain that is both personal and historical and inscribed in the landscape and on the body. A sense of pain in the landscape may be too much to bear, and we should always keep in mind that memories invoked by such landscapes may also be actively forgotten (see Rainbird 2000 for a 20th-century example).

The second sensory experience explored at Pohnpaid relates to the recognition that the main outcrop had sonic qualities, meaning that when hit (as local children regularly did with sticks of wood during recent times) the sound resonated across the valley floor. The valley floor where settlements are located is shrouded in dense forest when observed from the outcrop, and this forest also inhibits views of any distance, evoking the type of response described by Gell (see above) in his fieldwork in New Guinea. Sound in these environments takes on a much more significant role than vision, and on Pohnpei a rich ethnographic literature provides indications to the significant use and meanings of various patterns of sound created by stone against stone. Through these noises, the location in the landscape of senior members of the community and the stage of certain activities, such as the preparation of kava, a (mildly narcotic drink consumed in proscribed group situations), could be identified. This soundscape may have a greater chronological depth, which has gone unrecorded, such as indicated at the rock art site at Pohnpaid where the motifs might be soundmarks, and the production of the petroglyphs would be sound events perhaps equal to those recorded ethnographically.

Figure 25.1 Fagg (1997) noted that "rock gongs," as she called them, typically have cupules at their edges. Such an arrangement can be seen on this boulder at Pohnpaid, Pohnpei Island, Federated States of Micronesia.

In a British context, I related the soundings from the Pacific to the corpus of rock marks, usually dated to the late Neolithic or early Bronze Age, found in the area of Ilkley Moor in West Yorkshire (Rainbird 2002b). Here the engravings of abstract cup and ring marks had been interpreted in a strongly visual sense in relation to views over lowland valleys and intervisibility between markings, perhaps linking significant pathways across the Pennine uplands (Bradley 1997). In considering the recent history of quarrying that has significantly altered the landscape, the location of a number of panels on the valley sides rather than the tops, and palaeoenvironmental evidence indicating a wooded environment when these marks were made some 5,000 to 4,000 years ago, one surmises that vistas were not significant in their placement. I concluded that these rock markings could also be interpreted as sound-marks representing the soundscape; although the gritstone geology was not amenable to high resonance sounds, communal pounding and associated singing and other group activities would provide a less vision-based interpretation. Similar communal activities in this period could include quarrying, especially for stone prized in axe making, and it is clear that communal events were significant in this period for maintaining social relations (Cooney 2000; Edmonds 1999).

Sven Ouzman (2001) has also pursued issues of sound as basic to the production of San rock engravings in southern Africa. In a paper rich with the detail of direct ethnographic analogy, he argues that the production of the engravings was in the ritual context of the medicine dance (itself often depicted on the engravings), whereby participants attempted to reach altered states of consciousness. As Ouzman (2001: 244) opines:

[T]he only constant in an inverse altered state of reality is the insistent, percussive sound and rhythm of the medicine dance that reminds the shaman of the Ordinary World and guides her/him back to it. . . . Consider also the role of the "audience" at such a medicine dance. . . . They too became caught up in the peristaltic rhythms and sounds of the dance and may have wished to add to the dance's rhythm or accentuate particularly important passages or phases of the dance. San societies were thus deeply concerned with producing sound—by singing, clapping, dancing and by hammering certain rocks and engraved images.

Beyond sound, Ouzman (2001) also notes that the actual hammering produced a physical sensation in the person's boys; a "thrumming vibration," as he describes it. Such a sensation adds to the possibility of reaching an altered state of reality and also into the haptic sense, which Ouzman briefly explores in regard to the presence of "touchstones" at San rock art sites. He finds evidence of human rubbing at a number of San rock art sites, which is typically associated with engravings of large eland-like animals and rhinoceroses, both being spiritually important animals in San society. More than 5% of the 762 engraving sites assessed by Ouzman had been used to flake stone from the rocks on which the rock art was present. Ouzman argues that the engraved panels for stone removal were selected carefully as preeminent rocks within an engraving site. The removal of stone flakes from these sites, according to Ouzman, was in order to collect some of the spiritual potency of these locales. He says (2001: 250, references removed): " These flakes or 'righteous rocks' are not fetishes but metonyms, with the flake a fragment capable of indirectly but powerfully evoking a compound totality comprising image, site, personal relations, the Spirit World and so forth."

Ouzman describes the collection of the flakes as "harvesting potency," and in a rather more speculative turn he suggests that the debris produced by hammering and flaking may have been ingested to further connect the person and the locale. Such practices are not unknown elsewhere, and in this context Thomas, Sheppard, and Walter (2001) provide an exploration of the interplay between bodily experience and landscape in relation to an architecturally embellished hill top site in the Solomon Islands. Here, the hilltop has many stone-built shrines often associated with human skulls, also varieties of marine shells and coral brought from the shore. One such shrine, named Liqutu, is constructed from imported basalt, and oral tradition indicates that "[h]ere they scratched stone and fed it to babies so that they would become strong warriors." (2001: 557).

Conclusions: Thinking through the Body

David Howes (2003) has identified a need to explore the interplay of the senses, and it is this complete bodily understanding of being to which Chris Gosden (2001) refers in introducing a volume dedicated to the study of archaeology and aesthetics. He says that "our sensory apprehension of the world is not purely a physiological matter of impulses reaching the brain from the body, but rather it is something we have to engage in actively, albeit unconsciously. . . . The locus of sensory activity is as much cultural as bodily, so that various cultures apprehend the world in different ways" (2001: 163). To fully appreciate bodily presencing in the landscape, there is a need not to exclude vision over the other senses; rather we need to recognize that vision is not the only mode by which being in the world is achieved. Our stories of the past will be greatly enhanced through the simple acceptance of a multisensory present and a multisensory past.

Such an acceptance will allow us to pursue questions in regard to such things as how did the presence of mortuary enclosures or platforms where the dead were excarnated smell and alter perception of the landscape? How did it sound in relation to presence of carrion? And did the smell carry with it a certain potency? In earlier times, did gatherers recognize a tastescape related to subtle differences in fruits and nuts owing to underlying geology? Did coast-dwellers recognize differential salt content in the waters in the vicinity of estuaries, and could perhaps such things allow us to consider a phenomenology of the sea alongside that of the land? (Rainbird 2007). What role might have smell played in a landscape of hunting? In the outback of Australia, on a number of occasions in the sparsely vegetated but undulating landscape I have smelled the presence of a kangaroo, known its direction, and, if it was moving, could tell which way it was going.

Some archaeologists will resist such a landscape archaeology claiming the lack of quantification and substantiation. Computing and virtual reality may in the future allay some of these fears, although current technology is highly visual in orientation with sound an improving addition. The full embodied sensory experience is a long way off, and what the landscapes ought to include, and which senses ought to be heightened in relation to one another, will still be as problematic as ever, even if the presence of computers and their output act as reassuringly empirical.

There is no simple answer to how we get to or how we represent multisensory landscapes of the past, but if we accept that human life is experienced in this manner in the present, then ignoring it in the past reduces our ability to interpret fully past lives and will always leave us falling short of the mark. Thinking about such past landscapes is hard enough, applying it is harder still, but I believe it is an effort that is not only worthwhile but absolutely necessary.

Note

1. *Cursus* is Latin for "circus"—the long parallel earthen banks with ditches often used prehistorically as procession ways, the longest known being the Dorset cursus.

References

Barrett, J. C. 2000. A thesis on agency, in M.-A. Dobres and J. E. Robb (eds.), *Agency in Archaeology*, pp. 61–68. London: Routledge.

Bender, B. (ed.). 1993a. *Landscape: Politics and Perspectives*. Oxford: Berg.

———. 1993b. Introduction: Landscape—meaning and action, in B. Bender (ed.), *Landscape: Politics and Perspectives*, pp. 1–17. Oxford: Berg.

Boivin, N. 2004. Rock art and rock music: Petroglyphs of the south Indian Neolithic. *Antiquity* 78: 38–53.

Bradley, J., and Mackinlay, E. 2000. Songs from a plastic water rat: An introduction to the musical traditions of the Yanyuwa Community of the Southwest Gulf of Carpentaria. *Ngulaig* 17.

Bradley, R. 1997. *Rock Art and the Prehistory of Atlantic Europe*. London: Routledge.

Brück, J. 2005. Experiencing the past? The development of a phenomenological archaeology in British prehistory. *Archaeological Dialogues* 12: 45–72.

Chippindale, C., and Taçon, P. S. C. (eds.). 1998. *The Archaeology of Rock Art*. Cambridge: Cambridge University Press.

Classen, C., Howes, D., and Synott, A. 1994. *Aroma: The Cultural History of Smell*. London: Routledge.

Cooney, G. 2000. *Landscapes of Neolithic Ireland*. London: Routledge.

Csordas, T. J. 1999. The body's career in anthropology, in H. L. Moore (ed.), *Anthropological Theory Today*, pp. 172–205. Cambridge: Polity.

Cummings, V. 2002. Experiencing texture and transformation in the British Neolithic. *Oxford Journal of Archaeology* 21: 249–61.

Cummings, V., and Whittle, A. 2004. *Places of Special Virtue: Megaliths in the Neolithic Landscape of Wales*. Oxford: Oxbow.

David, B. 2002. *Landscapes, Rock Art and Dreaming: An Archaeology of Preunderstanding*. Leicester: Leicester University Press.

Edmonds, M. 1999. *Ancestral Geographies of the Neolithic: Landscapes, Monuments and Memory*. London: Routledge.

Fagg, M. C. 1997. *Rock Music*. Pitt Rivers Museum, University of Oxford, Occasional Paper on Technology 14.

Feld, S. 1996. Waterfalls of song: An acoustemology of place resounding in Bosavi, Papua New Guinea, in S. Feld and K. Basso (eds.), *Senses of Place*, pp. 91–135. Santa Fe, NM: School of American Research Press.

Feld, S., and Basso, K. (eds.). 1996. *Senses of Place*. Santa Fe, NM: School of American Research Press.

Finney, B. 1995. A role for magnetoperception in human navigation? *Current Anthropology* 36: 500–06.

Fleming, A. 1999. Phenomenology and the megaliths of Wales: A dreaming too far? *Oxford Journal of Archaeology* 18: 119–25.

———. 2005. Megaliths and post-modernism: The case of Wales. *Antiquity* 79: 921–32.

Gell, A. 1995. The language of the forest: Landscape and phonological iconism in Umeda, in E. Hirsch and M. O'Hanlon (eds.), *The Anthropology of Landscape: Perspectives on Place and Space*, pp. 232–54. Oxford: Oxford University Press.

Gosden, C. 2001. Making sense: Archaeology and aesthetics. *World Archaeology* 33: 163–67.

Hamilakis, Y. 2001. Art and the representation of the past: Commentary. *Journal of the Royal Anthropological Institute* 7: 153–54.

———. 2002. The past as oral history: Towards an archaeology of the senses, in Y. Hamilakis, M. Pluciennik, and S. Tarlow (eds.), *Thinking through the Body: Archaeologies of Corporeality*, pp. 121–36. New York: Plenum/Kluwer Academic.

———. In preparation. *The Archaeology of the Senses*. Cambridge: Cambridge University Press.

Hodder, I. 1999. *The Archaeological Process*. Oxford: Blackwell.

———. 2000. Agency and individuals in long-term processes, in M.-A. Dobres and J. E. Robb (eds.), *Agency in Archaeology*, pp. 21–33. London: Routledge.

Howes, D. 2003. *Sensual Relations: Engaging the Senses in Culture and Social Theory*. Chicago: University of Michigan Press.

Ingold, T. 2000. *The Perception of the Environment: Essays in Livelihood, Dwelling and Skill*. London: Routledge.

———. 2004. Culture on the ground: The world perceived through the feet. *Journal of Material Culture* 9: 315–40.

Lawson, G., Scarre, C., Cross, I., and Hills, C. 1998. Mounds, megaliths, music and mind: Some thoughts on the acoustic properties and purposes of archaeological spaces. *Archaeological Review from Cambridge* 15: 111–34.

Lynch, F. 1998. Colour in prehistoric architecture, in A. Gibson and D. Simpson (eds.), *Prehistoric Ritual and Religion*, pp. 62–67. Stroud: Sutton.

MacGregor, G. 2002. Making monuments out of mountains: The role of colour and texture in the constitution of meaning and identity at Recumbent Stone Circles, in A. Jones and G. MacGregor (eds.), *Colouring the Past: The Significance of Colour in Archaeological Research*, pp. 141–58. Oxford: Berg.

Morphy, H. 1993. Colonialism, history and the construction of place: The politics of landscape in northern Australia, in B. Bender (ed.), *Landscape: Politics and Perspectives*, pp. 205–43. Oxford: Berg.

Munn, N. 1996. Excluded spaces: The figure in the Australian aboriginal landscape. *Critical Inquiry* 22: 446–65.

Ouzman, S. 2001. Seeing is deceiving: Rock art and the non-visual. *World Archaeology* 33: 237–56.

Rainbird, P. 1999. Entangled biographies: Western Pacific ceramics and the tombs of Pohnpei. *World Archaeology* 31: 214–24.

———. 2000. "Round, black and lustrous": A view of encounters with difference in Chuuk Lagoon, Federated States of Micronesia, in R. Torrence and A. Clarke (eds.), *The Archaeology of Difference: Negotiating Cross-Cultural Engagements in Oceania*, pp. 32–50. London: Routledge.

———. 2002a. Marking the body, marking the land—body as history, land as history: Tattooing and engraving in Oceania, in Y. Hamilakis, M. Pluciennik, and S. Tarlow (eds.), *Thinking through the Body: Archaeologies of Corporeality*, pp. 233–47. New York: Plenum/Kluwer Academic.

———. 2002b. Making sense of petroglyphs: The sound of rock art, in B. David and M. Wilson (eds.), *Inscribed Landscapes: Marking and Making Place*, pp. 93–103. Honolulu: University of Hawai'i Press.

———. 2007. *The Archaeology of Islands*. Cambridge: Cambridge University Press.

———. In preparation. *Interpreting Rock Art: A Social Archaeology of the Petroglyphs of Pohnpei, Federated States of Micronesia*. Oxford: Oxford University Press.

Rainbird, P., and Wilson, M. 1999. Pohnpaid Petroglyphs, Pohnpei. Unpublished report prepared for the Pohnpei State Historic Preservation Office.

———. 2002. Crossing the line: The enveloped cross in Pohnpei, Federated States of Micronesia. *Antiquity* 76: 635–36.

Sherratt, A. 1991. Sacred and profane substances: The ritual use of narcotics in Later Neolithic Europe, in P. Garwood, D. Jennings, R. Skeates, and J. Toms (eds.), *Sacred and Profane: Proceedings of a Conference on Archaeology, Ritual and Religion*, pp. 50–64. Oxford: Oxford University Committee for Archaeology Monographs 32.

Sutton, D. E. 2001. *Remembrance of Repasts: An Anthropology of Food and Memory*. Oxford: Berg.

Taçon, P. S. C. 1994. Socialising landscapes: The long-term implications of signs, symbols and marks on the land. *Archaeology in Oceania* 29: 117–29.

Thomas, J. 2002. Archaeology's humanism and the materiality of the body, in Y. Hamilakis, M. Pluciennik, and S. Tarlow (eds.), *Thinking through the Body: Archaeologies of Corporeality*, pp. 29–45. New York: Plenum/Kluwer Academic.

Thomas, T., Sheppard, P., and Walter, R. 2001. Landscape, violence and social bodies: Ritualized architecture in a Solomon Islands society. *Journal of the Royal Anthropological Institute* (N.S.) 7: 545–72.

Tilley, C. 1994. *A Phenomenology of Landscape*. Oxford: Berg.

———. 1999. *Metaphor and Material Culture*. London: Routledge.

Ucko, P. J., and Layton, R. (eds.). 1999. *The Archaeology and Anthropology of Landscape*. London: Routledge.

Waller, S. J. 2002. Rock art acoustics in the past, present and future. *The Proceedings of the 1999 International Rock Art Congress, Vol. 2*, pp. 11–20.

Watson, A., and Keating, D. 1999. Architecture and sound: An acoustic analysis of megalithic monuments in prehistoric Britain. *Antiquity* 73: 325–36.

Whittle, A. 1997. Remembered and imagined belongings: Stonehenge and its traditions and structures of meaning. *Proceedings of the British Academy* 92: 145–66.

26

PHENOMENOLOGICAL APPROACHES TO LANDSCAPE ARCHAEOLOGY

Christopher Tilley

The purpose of this chapter is to outline some of the key elements of a phenomenological approach to landscape archaeology. From a phenomenological perspective, knowledge of landscapes, either past or present, is gained through perceptual experience of them from the point of view of the subject (for some general theoretical and philosophical discussions, see Thomas 2006; Tilley 1994, 2004a, 2005). A phenomenologist attempts to describe these experiences as fully as possible. The objective is to provide a rich or "thick" description allowing others to comprehend these landscapes in their nuanced diversity and complexity, and to enter into these experiences through their metaphorical textual mediation.

Embodiment

Embodiment is a central term. A phenomenologist's experience of landscape is one that takes place through the medium of his or her sensing and sensed carnal body. This involves participant observation which means being a part of what one is attempting to describe and understand. A phenomenologist works and studies landscapes from the "inside." This may be contrasted with mediated or abstracted "outside" experiences of landscapes, such as those that might be gained from texts, maps, photographs, paintings, or any computer-based technologies, simulations, or statistical analyses. The claim is that studying landscapes through such *representations* can provide only a relatively superficial and abstracted knowledge. There is no substitute for personal experience.

It follows that for the phenomenologist his or her body is the primary research tool. He or she experiences and observes the landscape through the body. As far as is possible, landscapes are studied without "prejudice." In other words, the phenomenologist does not start out with a list of hypotheses to be "tested" or a set of prior assumptions about what may, or may not, be significant or important. Rather, he or she enters into the landscape and allows it to have its own impact on his or her perceptive understandings. This is to accept that there is a dialogic relationship between person and landscape. Experiencing the landscape allows insights to be gained through the subject observer's immersion in that landscape. This is to claim that landscapes have *agency* in relation to persons. They have a profound effect on our thoughts and interpretations because of the manner in which they are perceived and sensed through our carnal bodies. We cannot, therefore, either represent or understand them in any way we might like. This is an approach that stresses the *materiality* of landscapes: landscapes as real and physical rather than simply cognized or imagined or represented. The physicality of landscapes acts as a foundation for all thought and social interaction. It profoundly affects the way we think, feel, move, and act. The phenomenologist is a figure immersed within the ground of landscape. Landscape is fundamental for human existence because it provides both a medium for, and an outcome of, individual and social practices. The physicality of landscapes grounds and orientates people and places within them, a physical and sensory resource for living and the social and symbolic construction of life-worlds.

Temporality

A phenomenological study takes time. In principle, the longer one experiences a landscape the more it will be understood. First of all, this is because only familiarity can produce a structure of feeling for the landscape, which a phenomenological account attempts to evoke. Second, such a study is based on the understanding that landscapes, unlike their representations, are constituted in space-time; they are always changing, in the process of being and becoming, never exactly the same twice over. Places alter according to natural rhythms, such as the progression of the seasons, the time of day, or qualities of light and shade. The weather, for which an entire archaeology might be developed, is a fundamental medium surrounding and affecting both people and their landscapes (see the discussion in Ingold 2006). Temporality is thus at the heart of a phenomenological study in which we must learn how to see and how to experience and try to learn about the experiences of others in this way (Thomas 1996).

Places and Paths

At their simplest and most abstract conceptualization, human (and humanized) landscapes consist of two elements: (1) places and their properties and (2) paths or routes of movement between these places and their properties (Tilley 1994). There can be no noncontextual definition of either landscape or place. All depends on the scale of analysis (see also Head, this volume). A place might be a rock outcrop, a hill, the point at which two streams converge, a field, a dwelling, or a settlement. A phenonmenologist attempts to both describe the individual experiences of different kinds of places and the paths or routes between them. The concern is with both stasis and movement. He or she recognizes that there are multiple understandings of both. Places alter with regard to how they are experienced, as do the paths or routes of movement within or between them. So according to the manner in which one senses and experiences landscapes, one ends up with a different descriptive understanding of them.

You and I encounter places and paths from a *point of view*, in both the literal and metaphorical sense of this term, through the medium of our bodies, and the character of this experience changes in relation to both the directionality of our movement and the postures of our bodies. The manner in which we understand places differs inevitably according to how we encounter it from within and the routes we take to reach a place and the sequences of other places we experience along the way. These structure our perceptive experience. Our experience is "colored" by the manner in which we encounter landscapes. Memory is thus fundamental to the nature of our experience. This is simply to accept our own embodied humanity. There can be no "objective" (in the sense of impersonal) experience of landscape. We are infallible humans and can never aspire to be of the status of gods who might comprehend and understand everything from every possible point of view. In our common humanity, we share biologically similar perceptive bodies with others in both the past and the present. We also differ significantly in relation to the cross-cutting divisions of gender, age, class, ethnicity, culture, and knowledges. These together with the physicality of our bodies provide both essential resources and limitations for our understanding of landscapes.

Sensory Experience

To understand landscapes phenomenologically requires the art of walking in and through them, to touch and be touched by them. An experience of landscape mediated by trains or cars or airplanes is always partial or distanciated. The view from the airplane is, of course, inhuman. We do not normally see or experience landscapes in this manner. The view from the car or train window is sensorily deprived: experience is reduced to vision. The phenomenologist acknowledges the multisensorial qualities of our human experiences of landscape, that a landscape is simultaneously a visionscape, a touchscape, a soundscape, a smellscape, and a tastescape. These different perceptive experiences occur all at once. Thus, our experience is always synaesthetic (a mingling or blending of the senses), whether we realize or acknowledge this or not. Landscapes reside as much in the tastes of their wines, or the odors of their flowers, as in their visual experiences. Such a multisensory approach in archaeology, in which discussions of the visual in relation to landscape have always dominated, is only just beginning to be developed (e.g., Cummings 2002; Fowler and Cummings 2003; Goldhahn 2002; Jones 2006; Jones and MacGregor 2002; Tilley 2004; Watson and Keating 1999).

The phenomenologist undertakes a task that is simultaneously very simple and incredibly difficult. He or she "resides" in places and walks between them. This is a humble, potentially subversive, and democratic project open to student or teacher alike, requiring no fancy technical equipment or expertise in using it, or money beyond that required for subsistence. Archaeological excavations, by contrast, are fertile breeding grounds for institutionalized power and the egos of their directors (Bender, Hamilton, and Tilley 2007). For the phenomenologist, technical equipment, as often as not, gets in the way, because it always mediates and limits experience. Beyond a notebook and pencil, a camera may be useful in capturing some aspects of visual experience, but little else is usually required.

Poetics and the Body

A phenomenological study is always limited, and the limits are essentially the limits of one's own body. Landscape studies conducted in this manner are inevitably small scale. It would not be possible to conduct such a study of the world or even of a nation-state such as France. Clearly, this is beyond human possibility, but we could build up a comparative global phenomenological study through comparing and contrasting the accounts of different social scientists. Phenomenological landscape studies are inevitably particularistic rather than generalizing. They attempt to capture the poetics and politics of paths and places (Bender 1998; Cummings 2003; Edmonds 2006; Edmonds and Seabourne 2002; Scarre 2002; Tilley 1996, 1999).

The human perceptive experience of landscape is inevitably structured in relation to basic bodily dyads: things that are to the front or the back of an observer, those that are above or below, to the left or right of the body, near or far away. These dualisms are directly related to basic body symmetries. It is necessary, therefore, that experiential qualities of landscape be described and discussed in these terms. In relation to the body, vision is the most distanciated of the senses: we can often see much farther than we can hear or smell landscapes. For us to touch things they must be in reach; taste (apart from sticking out the tongue) requires taking things into our bodies and is thus the most intimate of the senses.

It has been claimed that different hierarchies of the senses exist in different cultures, vision most important in Western modernity, smell or sound in other cultures. However, the very attempt to single out any particular sensory dimension and suggest it has all-pervasive significance in one culture rather than in another is an unhelpful simplification. Which of the senses is most significant depends both on context and the practices being undertaken; smells may be relatively more important in one context, sounds or sight or touch in another, and analysis needs to be sensitive to these variations rather than the scenario of one culture and one dominant sense (Tilley 2006). For example, I have recently argued that in many areas of prehistoric lowland Europe, the advent of the Neolithic ushered in a sensory revolution in relation to the *perception* of landscape. The removal of forest cover allowed vision to become a distant and dominant sense in relation to the landscape for the first time. Without the trees, the contours and shapes of the land could be seen in a completely different manner, as could people, monuments, and places within it. By contrast, in a densely forested Mesolithic landscape, smell and sound might be far more important in relation to orientation and resource exploitation, with sight being a far more intimate bodily sense (Tilley 2007).

"Nature" and "Culture"

Landscapes themselves influence forms of perception and activity, but they do not determine thought and action and not anything can be made of them. They offer a series of affordances for living and acting in the world, and a series of constraints. We cannot determine in advance what may be of particular significance in any specific case. In one landscape, rock outcrops may be the most significant reference points; in another, river valleys and so on (Tilley 2004b). One of our most common prejudices in landscape archaeology is to assume that the most important places in the landscape are those that have been humanly created, such as settlement sites and monuments. One of the most obvious phenomenological questions we try to answer is this: why was this place chosen rather than another? However, such a question cannot be answered in isolation. We need to consider the monument or settlement in relation to others, (searching for locational patterns) and with respect to its landscape context, which requires analyzing its sensory affordances or constraints and the ways in which it might be experienced differently if approached from one direction rather than another. We cannot assume that the places for which we have no evidence of human presence were not important (Bradley 2000). The peculiar hill or ridge without a monument may be of equal significance. A "natural" stone may be as, if not more significant than those deliberately erected and there may exist both mimetic and contrastive relationships between humanly created and unaltered places (Rowlands and Tilley 2006; Tilley 1996; Tilley, Hamilton, and Bender 2000). A phenomenological study of landscape thus requires a holistic approach in which we pay as much attention to the "natural" as the "cultural," to places with and without evidence of human alteration or activity (Tilley and Bennett 2002; Tilley et al. 2000).

Our experience of any unfamiliar landscape is that of a child. Gradually we need to explore, to learn how to look, to hear, to smell, to touch, and to taste. We need to open out our bodies to all these sensory dimensions as much as is possible, to try and experience landscapes from the inside. In relation to past as opposed to contemporary landscapes, the task is inevitably difficult, since so much has irrevocably altered. But much also remains in the form of the geological and topographic "bones" of the land: the character of the rocks, the mountains and hills, valleys and the river courses, sometimes the coastline. The deafening sound of the waterfall (Goldhahn 2002) or the smell of rotting seaweed or meadowsweet, the sight of the conical hill, the way in which a stone feels to touch and its color, experiences of light and darkness within monuments, or the taste of honey may remain almost the same now as then; we do, in this limited sense, still have a direct bodily connection with the past.

Methodological Implications

There can be no rulebook method to undertaking "good" phenomenological research. What I offer is a sketch of the basic stages involved in my own style of phenomenological research:

1. Familiarizing oneself with the landscape through walking within and around it, developing a feeling for it and opening up oneself to it;
2. Visiting known places of prehistoric significance and recording the sensory affordances and constraints they provide. This requires *writing* and then visually recording, through still or video photography, these experiences in the place, creating a written and visual text (rather than a series of abbreviated notes), because the very process of writing is a primary aid and stimulus to perception;
3. Revisiting the same places during different seasons or times of the day as far as is possible, experiencing them in and through the weather;
4. Approaching these places from different directions and recording the manner in which their character alters as a result;
5. Following paths of movement through the landscape and recording the manner in which this may change the manner in which places within it are perceived in relation to each other. These paths of movement will usually be suggested by features of the landscape itself, such as, for example, following the lines of ridges or the courses of valleys, or prehistoric monuments within it, such as, for instance, walking along the line of a stone row, a cursus monument, a cross-ridge dyke, a Roman road, or between nearby groups of barrows or settlements (Barclay and Harding 1999; Bradley 2002; Parker-Pearson et al. 2006; Tilley 1994, 1999, 2004b; Witcher 1998);
6. Visiting and exploring and recording "natural" places within the landscape for which there is little or no archaeological evidence of human activity (Bradley 2000; Tilley and Bennett 2002; Tilley et al. 2000);
7. Drawing together all these observations and experiences in the form of a synthetic text and imaginatively interpreting them in terms of possible prehistoric life-worlds: how people in the past made sense of, lived in, and understood their landscapes (e.g., Bender, Hamilton, and Tilley 2007; Tilley 2004).

Conclusions

All landscapes have profound significance and meaning for persons and groups. These are, as often as not, variable and contested: related to different interests and practices (Bender and Winer 2001). Although landscapes have meanings whose significance we can attempt to interpret phenomenologically, they also *do* things and have experiential effects in relation to persons—the two are intimately linked. For example, prehistoric rock carvings or monuments undoubtedly had specific sets of meanings that we can try to semiotically decode. They also have specific somatic effects that we can describe, such as having to move in one direction or another, within and between them, and in terms of light and sound and touch (Goldhahn 2002; Jones 2006; Tilley 2004). The significance these places had, and the emotional and kinaesthetic effects they produced in relation to the body, are likely to be intimately related, because meaning and doing work exist both through the body and through the mind. Because our minds and thoughts are embodied, the manner in which we think is profoundly structured by the kinds of bodies and the sensory apparatus we possess.

Phenomenological approaches to landscape archaeology remain in their infancy. There is an enormous amount of comparative field research to be undertaken. Phenomenological studies attempt to explore landscapes on the basis of the full depth of their human sensory experience. To be a good phenomenologist is to try to develop an intimacy of contact with the landscape akin to that between lovers.

References

Barclay, A., and Harding, J. (eds.). 1999. *Pathways and Ceremonies: The Cursus Monuments of Britain and Ireland*. Oxford: Oxbow Books.

Bender, B. 1998. *Stonehenge: Making Space*. Oxford: Berg.

Bender, B., Hamilton, S., and Tilley, C. 2007. *Stone Worlds: Narrative and Reflexive Approaches to Landscape Archaeology*. Walnut Creek, CA: Left Coast Press.

Bender, B., and Winer, M. (eds.). 2001. *Contested Landscapes*. Oxford: Berg.

Bradley, R. 2000. *An Archaeology of Natural Places*. London: Routledge.

———. 2002. *The Past in Prehistoric Societies*. London: Routledge.

Cummings, V. 2002. Experiencing texture and transformation in the British Neolithic. *Oxford Journal of Archaeology* 21: 249–61.

———. 2003. *Places of Special Virtue: Megaliths in the Neolithic Landscapes of Wales*. Oxford: Oxbow Books.

Edmonds, M. 2006. Who said romance was dead? *Journal of Material Culture* 11: 167–88.

Edmonds, M., and Seabourne, T. 2002. *Prehistory in the Peak*. Stroud: Tempus Publishing.

Fowler, C., and Cummings, V. 2003. Places of transformation: Building monuments from water and stone in the Neolithic of the Irish sea. *Journal of the Royal Anthropological Institute* 9: 1–20.

Goldhahn, J. 2002. Roaring rocks: An audiovisual perspective on hunter-gatherer engravings in northern Sweden and Scandinavia. *Norwegian Archaeological Review* 35: 29–61.

Ingold, T. 2006. Comments on Christopher Tilley: The Materiality of Stone: Explorations in Landscape Phenomenology. *Norwegian Archaeological Review* 38: 122–29.

Jones, A. 2006. Animated images: Images, agency and landscape in Kilmartin, Argyll, Scotland. *Journal of Material Culture* 11: 211–25.

Jones, A., and MacGregor, G. (eds.). 2002. *Colouring the Past: The Significance of Colour in Archaeological Research*. Oxford: Berg.

Parker-Pearson, M., Pollard, J., Richards, C., Thomas, J., Tilley, C., Welham, K., and Albarella, U. 2006. Materializing Stonehenge: The Stonehenge riverside project and new discoveries. *Journal of Material Culture* 11: 227–61.

Pollard, J., and Reynolds, A. 2002. *Avebury: The biography of a landscape*. Stroud: Tempus Publishing.

Rowlands, M., and Tilley, C. 2006. Monuments and memorials, in C. Tilley, W. Keane, S. Kuechler, M. Rowlands, and P. Spyer (eds.), *Handbook of Material Culture*, pp. 500–15. London: Sage.

Scarre, C. (ed.). 2002. *Monuments and Landscape in Atlantic Europe*. London: Routledge.

Thomas, J. 1996. *Time, Culture and Identity*. London: Routledge.

———. 2006. Phenomenology and material culture, in C. Tilley, W. Keane, S. Kuechler, M. Rowlands, and P. Spyer (eds.), *Handbook of Material Culture*, pp. 43–59. London: Sage.

Tilley, C. 1994. *A Phenomenology of Landscape: Places, Paths and Monuments*. Oxford: Berg.

———. 1996. The powers of rocks: Topography and monument construction on Bodmin Moor. *World Archaeology* 28: 161–76.

———. 1999. *Metaphor and Material Culture*. Oxford: Blackwell.

———. 2004a. *The Materiality of Stone: Explorations in Landscape Phenomenology 1*. Oxford: Berg.

———. 2004b. Round barrows and dykes as landscape metaphors. *Cambridge Archaeological Journal* 14: 185–203.

———. 2005. Phenomenological archaeology, in C. Renfrew and P. Bahn (eds.), *Archaeology: The Key Concepts*, pp. 201–07. London: Routledge.

———. 2006. The sensory dimensions of gardening. *The Senses and Society* 1(3): 311–30.

———. 2007. The Neolithic sensory revolution: Monumentality and the experience of landscape, in V. Cummings and A. Whittle (eds.), *Going Over: The Mesolithic-Neolithic Transition in North-West Europe*. London: British Academy Monographs 144: 327–43.

Tilley, C. In press. Iconographic and kinaesthetic approaches to rock art, in L. Forsberg and E. Walderhaug (eds.), *Cognition and Signification in Northern Landscapes*. Bergen: University of Bergen Press.

Tilley, C., and Bennett, W. 2002. An archaeology of supernatural places: The case of West Penwith. *Journal of the Royal Anthropological Institute* 7: 335–62.

Tilley, C., Hamilton, S., and Bender, B. 2000. Art and the re-presentation of the past. *Journal of the Royal Anthropological Institute* 6: 35–62.

Tilley, C., Hamilton, S., Harrison, S., and Andersen, E. 2000. Nature, culture, clitter: Distinguishing between cultural and geomorphological landscapes—The case of hilltop tors in southwest England. *Journal of Material Culture* 5: 197–224.

Watson, A., and Keating, D. 1999. Architecture and sound: An acoustic analysis of megalithic monuments in prehistoric Britain. *Antiquity* 73: 325–36.

Witcher, R. 1998. Roman roads: Phenomenological perspectives on roads in the landscape, in C. Forcey, J. Hawthorne, and R. Witcher (eds.), *TRAC 97: Proceedings of the Seventh Annual Theoretical Roman Archaeology Conference, Notttingham 1997*, pp. 60–70. Oxford: Oxbow Books.

27

MEMORY, PLACE, AND THE MEMORIALIZATION OF LANDSCAPE

Ruth M. Van Dyke

As I write these words in May 2006, construction is soon to begin in lower Manhattan for the World Trade Center Memorial, a monument that will commemorate the destruction of the trade center's twin towers and the deaths of nearly 3,000 people at the hands of al-Qaeda terrorists on September 11, 2001. Everyone agrees there should be a monument, but there has been little consensus as to the specific form the monument should take. What should the monument look like? Should the monument be under or above ground? How much of the original World Trade Center Foundations should be preserved? Should the names of those who perished on 9/11 be incorporated, and, if so, in what order? Should there be a museum, and, if so, what texts, objects, and other representations should it contain? These issues and others have been at the forefront of a great deal of emotional controversy among factions that include survivors, relatives of the deceased, historians, architects, businessmen, politicians, and developers. At stake is the collective memory of the events of 9/11. Behind the struggle over the appearance and content of the monument lie deeper conflicts—which aspects of that event will be remembered, and which forgotten? To what ends (political, commercial, emotional, aesthetic?) shall these memories be used?

Memory involves the selective preservation, construction, and obliteration of ideas about the way things were in the past. A contemporary fascination with memory is rooted in larger cultural phenomena, including postmodern dislocation, nostalgia for an imagined past, millennial angst, and struggles over interpretations of global events (see, for example, Lowenthal 1985; Nora 1989; Schama 1996). The discipline of archaeology is itself one way in which our own society constructs memory for contemporary social and political ends. Archaeology is frequently to be found in the service of nationalist or other political agendas (Bender 1998; Dietler 1998; Kohl and Fawcett 1995; Meskell 1998; Trigger 1984). Recently, archaeologists have turned attention to the ways in which past peoples viewed, interpreted, memorialized, utilized, and obliterated their own, more distant pasts.

Memory is closely integrated with place and landscape (Bachelard 1964; Casey 1987; de Certeau 1984; Nora 1989). Landscapes are meaningfully constituted physical and social environments, and meaning is inscribed on landscapes through experience. As humans create, modify, and move through a spatial milieu, the mediation between spatial experience and perception reflexively creates, legitimates, and reinforces social relationships and ideas (see Bourdieu 1977;

Lefebvre 1991; Soja 1996). Places, meanings, and memories are intertwined to create a "sense of place" that rests on, and reconstructs, a history of social engagement with the landscape and is thus inextricably bound up with remembrance, and with time (Basso 1996; Feld and Basso 1996). Place might be defined as the intersection of memory and landscape.

The construction of memory frequently leaves material traces, so memory can be accessed archaeologically through the study of monuments and shrines, or burial practices, or successive remodeling events, or life histories of specific artifacts, for example. Memory studies have helped archaeologists to think about "how and why things changed as they did as active process-es of adoption and/or rejection of ideas, and not as inevitable outcomes of social or environmen-tal circumstances" (Yoffee 2007: 4). These studies have emerged out of several related strands of theory, all of which share the recognition that the negotiation of human reality is an experiential, relational, and altogether messy business. In this chapter, I review the concept of social memory and summarize current archaeological studies that employ landscapes, monuments, and other aspects of material culture to study the "past in the past."

Some Ways to Think about Memory: Theory and Method

Most memory studies in archaeology are ground-ed in the work of Durkheim's student Maurice Halbwachs (1925, 1992). Halbwachs moved the discussion of memory beyond the Freudian bounds of the individual, arguing that memory is a social phenomenon. Collective memory is historically situated, as people remember or forget the past according to the needs of the present. Halbwachs also recognized the spatial dimension of memory, using the term *cadre matériel* to refer to the topog-raphy, architecture, and ruins that carry meanings about the past.

There are multiple ways to think about the passage of time. Braudel (1969) distinguishes between the *longue durée*, or the passage of deep social time, and the immediate history of events. Gell (1992) contrasts the linear construction of past, present, and future with a relational perspec-tive that locates time with reference to a particular event. Ingold (1993) explores the ways in which traces of human activity in landscapes link past and present. Following these authors, archaeolo-gists have recognized that memory can involve direct connections to ancestors in a remembered

past, or it can involve more general links to a vague mythological antiquity, often based on the reinterpretation of monuments (Gosden 1994; Gosden and Lock 1998; Thomas 1996).

The poststructuralist concept of citationality can help archaeologists think about how memory is transformed over time. Citationality is derived from Derrida's (1977) notion of iterability. Derrida's classic example is a signature—because it has a recognizable form, a signature can be replicated, but it can also be counterfeited. A signature denotes authenticity, but at the same time it sets up the possibility for an inauthentic copy. Judith Butler (1990) applied the concept of citationality to gender theory, developing the idea of performativity, arguing that gender is a performance that cites all previous performances of gender (see also Austin 1962; Butler 1993; Hall 2000). Jones (2001) and Pauketat (2008) have employed these concepts to understand how past meanings were shared and transformed. Jones uses citationality to describe how similarities in design create connections among diverse cat-egories of Bronze Age artifacts. Pauketat (2008) translates citationality as the construction of knowledgeable references to the past, and he contrasts citationality with more general, aes-thetic references to antiquity that do not neces-sarily assume past meanings. Through a process of enchainment (Chapman 2000), citations lead to more citations, until the original meanings are lost or transformed.

Memory is an ongoing process that dis-guises ruptures, creating the appearance of a seam-less social whole. Collective or social memory is often employed to naturalize or legitimate author-ity (Hobsbawm and Ranger 1983), or to support a sense of community identity (Halbwachs 1925, 1992). It is often easiest to see the top-down machinations of elite groups using memory to these ends, but memory is also employed in the service of resistance. Multiple and conflicting ver-sions of events can coexist and can be wielded by marginalized genders, classes, ethnicities, or other social groups (Alonso 1988; Burke 1989; Laqueur 2000). Memory coexists with its alter ego, forgetting, as pasts are selectively reconstructed, obliterated, consumed, conquered, and dismantled (Forty and Küchler 1999; Küchler 1993; Mills 2008; Papalexandrou 2003; Pauketat 2008).

Monuments—Nora's (1989) *lieux de mem-oire* (places of memory)—are conscious state-ments about what to remember. As such, they are logical foci for archaeological investigations into memory (Ashmore 2002; Thomas 2001). But monuments are only one of the many ways that memory leaves material traces for archaeologists to interpret. Memory is made through repeated, engaged social practices. Connerton (1989) distinguishes between inscribed memory, involving monuments, texts and representations, and embodied or incorporated memory, encompassing bodily rituals and behavior. Similar distinctions among prescriptive, formulaic, repetitive, and materially visible acts on the one hand and performative, mutable, transitory behavior on the other have been made by Bloch (1985), Sahlins (1985), and Rowlands (1993). It is perhaps most obvious for archaeologists to think about memory in terms of monuments, but incorporated memory leaves material traces as well.

Van Dyke and Alcock (2003: 4–5) subdivide memory's materiality into several overlapping categories: narratives, representations, objects, ritual behaviors, and places. Where *narratives* are transmitted through texts or other verbal media, archaeologists have histories against which to contrast the material record. *Representations* such as rock art may depict ancient mythic events while locating them on the landscape (Bradley 1997; Taçon 1999). Following Kopytoff (1986), *objects*, including masks, plaques, figures, and other votive items, have life-histories that illuminate both memory and obliteration (R. Joyce 2003; Küchler 1993; Lillios 1999, 2003; Walker 1999). *Ritual behaviors* are materially visible in mortuary treatments (Chesson 2001; Mizoguchi 1993; Williams 2003), feasting (Hamilakis 1999; Prent 2003), votive deposition (Bradley 1990; Mills 2008), abandonments (Walker 1995), and procession routes (Barrett 1994; Tilley 1994: 173–200). *Places* can encompass not only stelae, shrines, and buildings but also landscapes, tombs, trees, mountain peaks, and caves (Alcock 2002; Ashmore and Knapp 1999; Bender 1993; Blake 1998; Bradley 1998, 2000; Holtorf 1998; Thomas 2001; Williams 1998).

Places—whether monuments, domestic structures, tombs, or natural features—may be repeatedly inhabited, modified, and imbued with changing meanings. Archaeologists working from diverse perspectives have recognized the potency of place for investigating social transformations across time. Schlanger's (1992) "persistent places," Barrett's (1999) "inhabitation," Bradley's (1998) "afterlife of monuments," and Ashmore's (2002) "life histories of place" all provide ways in which to think about the intersections of social memory and monuments or other structures. A. Joyce (n.d.) has explicitly adopted a "life histories of place" approach for Monte Albán, the Oaxacan city occupied for 2,500 years.

Stonehenge is perhaps the best-known example of a monument with changing yet related functions and meanings reaching across nearly five millennia (Bender 1998).

Place is a sensual experience, with the body, social identity, and shifting perceptions of society intersecting through daily, lived spatial experiences (Bourdieu 1977). The experiential nature of place provides one starting point to retrieve social memory. A *phenomenological* approach in archaeology (Gosden 1994; Thomas 1996; Tilley 1994) allows us to think about the ways in which landscapes and built forms were experienced, perceived, and represented by ancient subjects, working from the starting point of a contemporary body in the same space. In my own work (Van Dyke 2003, 2004, 2007), visibility and embodied spatial experiences are key to interpretations of place and landscape among Ancestral Pueblo Chacoan peoples of the North American Southwest.

Bourdieu's (1977, 1990) recognition that meaning is embodied and transmitted through lived, daily practices is behind the concept of materiality—another jumping-off point from which some archaeologists have addressed memory. Materiality (R. Joyce 2000; Meskell 2004, 2005; Miller 2005) refers to the reciprocal interactions among humans and the material, embedded within a set of cultural relationships. Following Gell (1998) and LaTour (1993, 2005), agency may be extended beyond human actors to animals, buildings, and objects—these "secondary agents" may be perceived as animate and may cause actions with consequences (see, for example, Walker 2008).

Les Lieux de Mémoire (Places of Memory): Recent Archaeological Studies

Over the past decade or so, archaeologists have begun to discuss social memory as one dimension of larger investigations into landscape (e.g., Alcock 1993; Ashmore 2002; Barrett 1999; Zedeño 2000), rock art (Bradley 1997; Taçon 1999), mortuary ritual (Chesson 2001), heirlooms (Joyce 2000; Lillios 1999), and identity (Joyce and Gillespie 2000). A growing body of literature specifically directed toward social memory is currently emerging, much of it centered on monuments and places (Alcock 2002; Bradley 1998, 2002). Recent and forthcoming compilations have been edited by Bradley and Williams (1998), Williams (2003), Van Dyke and Alcock (2003), Yoffee (2007), Mills and Walker (2008), and Barber and Larkin (n.d.). These works represent diverse methods and theoretical perspectives, but they are united in recognition of the potent role played by material remains in assigning meanings to the past.

In 1998, Richard Bradley and Howard Williams edited an issue of *World Archaeology* entitled "The Past in the Past," the first collection of studies specifically directed toward the ways archaeological societies viewed time and appropriated more distant pasts. Some of these Old World authors examined the rather benign ways that landscapes are incorporated into memorials of real or imagined past events, some looked at the roles of monuments in identity formation over time, and still others focused on the role of memory in moments of rupture as new orders challenged old.

Bradley followed this edited journal issue with a book on memory in Neolithic and Bronze Age Europe entitled *The Past in Prehistoric Societies* (2002). Drawing on Gell (1992) and Connerton (1989), Bradley postulated that past societies likely used both linear and relational time, and both inscribed and incorporated memory practices. He then employed these ideas in a series of case studies. Prehistoric Europeans constructed memory on the landscape through the creation of burial barrows, transformed the past through the reuse of passage graves, and obliterated memory through the destruction of votive objects. Archaeological landscapes are palimpsests—during the Roman and later periods, Bronze Age and earlier sites were repeatedly reinterpreted as societies engaged with the past for their own purposes.

Tombs and burial grounds are particularly potent loci for memory studies. Death and memory are the focus of *Archaeologies of Remembrance*, a recent volume edited by Howard Williams (2003). Mortuary rituals are often complex occasions involving memorialization as well as selective forgetting. As the dead are mourned, memories and identities are created. Tombs tie the living to ancestors and imbue specific places on the landscape with all that this connection entails (McAnany 1998). Bodies and grave goods may be buried—hidden from view and forgotten—while long-term visible markers such as stelae may be erected. Tombs and grave markers can be reinterpreted over time. Authors in Williams's volume investigate the ways in which past peoples constructed memory through mortuary practices in European prehistoric and historic archaeological contexts, using such material remains as monuments, tombstones, grave goods, and bog bodies.

Susan Alcock (2001, 2002) has employed memory as one lens through which to understand the shifting, conflicting ideologies of factions of Greek (and Roman) society over time. Romans in Greece, for example, appropriated a Bronze Age tholos tomb called the "Treasury of Minyas," converting it into a shrine to the Roman emperor. This not only helped legitimate the emperor through a direct link to Minyas, a legendary ruler in the local, mythic past, but it also generated community pride in local past grandeur. At the same time, however, the connection between Roman imperial rule and elite families of the past emphasized class distinctions within the local community, making the Romans targets for resistance. This careful consideration of the multiple, conflicting operations of social memory is one of the great strengths of Alcock's work. In *Archaeologies of the Greek Past* (2002), Alcock expanded her examination of the Roman use of Greek antiquity; she also investigated the appropriation of the Minoan past by Hellenic Cretes and the ways in which Messenian slaves forged a common identity through creation of an imagined, collective past.

Alcock and I brought classicists and anthropological archaeologists together in a series of conference sessions on memory that culminated in *Archaeologies of Memory* (Van Dyke and Alcock 2003). We organized case studies in the volume not along disciplinary lines but into literate and prehistoric contexts. This division reflected the more nuanced kinds of analyses undertaken by archaeologists with access to texts, versus those undertaken by prehistorians without such access. Our case studies spanned the globe, from India (Sinopoli 2003), Egypt (Meskell 2003), and Iberia (Lillios 2003), across the Classical world (Papalexandrou 2003; Prent 2003), to the New World (R. Joyce 2003; Pauketat and Alt 2003; Van Dyke 2003). Many of the case studies dealt in some fashion with the reuse or reinterpretation of monumental architecture to promote the interests of elites, as leaders seeking legitimacy created both real and mythic connections with pasts both remembered and imagined.

Other archaeologists similarly began to bring together groups of scholars working on memory issues in diverse archaeological contexts. Classical and anthropological archaeologists in Yoffee's (2007) volume participated in a graduate seminar at the University of Michigan. These authors examine shifting identities, memories, and landscapes primarily in the ancient Mediterranean world. They use memory as an entry point for understanding social change and power relationships. The authors investigate the ways in which pasts were constructed, contested, and at times erased to imbue landscapes with meaning and to further specific social or political ends.

Constructed, social memory is not only employed to bolster continuity—it may also be used to subvert the social order. Barber and Larkin's (n.d.) forthcoming edited volume emerged out of a 2006 Society for American Archaeology session on this topic. Authors of a series of case studies focus on the roles of memory and tradition during periods of social transformation in ancient New World societies, from the Andes to the U.S. Southwest.

Mills and Walker (2008) have edited a volume on memory, based on an advanced seminar at the School of Advanced Research (SAR) and a subsequent American Anthropological Association conference session. Most of the case studies in this volume deal with New World societies, although Neolithic Britain and sub-Saharan Africa are also represented. Many of the authors in Mills and Walker's volume explicitly use a practice-based approach, invoking the concepts of materiality and agency to exhume prehistoric memory from the material traces of ritual behaviors and depositional practices, as well as in the construction and obliteration of monuments.

Conclusions: Whither Memory?

Monuments and landscapes are potent material venues through which archaeologists can access social memory in the past. Memory studies can provide archaeologists with an innovative window through which to think about the creation, maintenance, and transformation of power and identity. Because multiple, mutable, competing pasts can coexist, memory encourages us to think about the relationships and negotiations among different social factions. However, "if memory is as pervasive as it is starting to seem, and if, more importantly, it can be used to explain just about anything, is there a danger that it can end up explaining nothing at all?" (Low 2004: 934).

Not all reuse of older sites is intended to create meaningful links with antiquity. For example, Blake (2003) demonstrated that Byzantine reoccupation of Bronze Age rock-cut tombs in Sicily had little to do with referencing the past but, rather, was part of the creation of a forward-looking, pan-Mediterranean identity. As archaeologists enthusiastically search for social memory in the past, we must take care to avoid circular reasoning. We need independent evidence—beyond just the simple reuse of a building, or the discontinuation of an architectural style—that indicates the

memorialization or obliteration was intentional and meaningful to the participants. Otherwise, memory risks becoming merely a construct of the archaeologist's gaze, devoid of any real explanatory power.

As archaeological memory studies begin to mature, scholars need to move beyond the simplistic recognition that the past was referenced in antiquity. Who was referencing the past, and what did they seek to accomplish? How was memory employed by different factions, to different ends? Who commemorated, who obliterated, and why? The contemporary struggles over the World Trade Center Memorial remind us that memories can serve different ends and advance different agendas for multiple social groups. Archaeologists studying social memory should first and foremost keep this question in mind: what does a focus on memory tell us about larger issues—identities, ideologies, power, class—that we would not otherwise know?

References

Alcock, S. E. 1993. *Graecia Capta: The Landscapes of Roman Greece*. Cambridge: Cambridge University Press.

———. 2001. Reconfiguration of Memory in the Eastern Roman Empire, in S. E. Alcock, T. N. D'Altroy, K. D. Morrison, and C. M. Sinopoli (eds.), *Empires: Perspectives from Archaeology and History*, pp. 323–50. Cambridge: Cambridge University Press.

———. 2002. *Archaeologies of the Greek Past: Landscape, Monuments and Memories*. Cambridge: Cambridge University Press.

Alonso, A. M. 1988. The effects of truth: Re-presentations of the past and the imagining of community. *Journal of Historical Sociology* 1: 33–57.

Ashmore, W. 2002. "Decisions and Dispositions:" Socializing spatial archaeology. *American Anthropologist* 104: 1172–83.

Ashmore, W., and Knapp, A. B. (eds.). 1999. *Archaeologies of Landscape: Contemporary Perspectives*. Oxford: Blackwell.

Austin, J. L. 1962. *How to Do Things with Words*. Cambridge: Harvard University Press.

Bachelard, G. 1964. *The Poetics of Space*. Boston: Beacon Press.

Barber, S., and Larkin, K. (eds.). In preparation. *Uses of the Past: Negotiating Social Change through Memory and Tradition*. Boulder: University of Colorado Press.

Barrett, J. C. 1994. Moving beyond the monuments: Paths and peoples in the Neolithic landscapes of the "Peak District." *Northern Archaeology* 13.

Barrett, J. C. 1999. The mythical landscapes of the British Iron Age, in W. Ashmore and A. B. Knapp (eds.), *Archaeologies of Landscape: Contemporary Perspectives*, pp. 253–65. Oxford: Blackwell.

Basso, K. H. 1996. *Wisdom Sits in Places: Landscape and Language among the Western Apache*. Albuquerque: University of New Mexico Press.

Bender, B. 1993. Stonehenge: Contested landscapes (medieval to present day), in B. Bender (ed.), *Landscape: Politics and Perspectives*, pp. 245–79. Oxford: Berg.

———. 1998. *Stonehenge: Making Space*. Oxford: Berg.

Blake, E. 1998. Sardinia's Nuraghi: Four millennia of becoming. *World Archaeology* 30: 59–71.

———. 2003. The familiar honeycomb: Byzantine era reuse of Sicily's prehistoric rock-cut tombs, in R. M. Van Dyke and S. E. Alcock (eds.), *Archaeologies of Memory*, pp. 203–20. Oxford: Blackwell.

Bloch, M. 1985. From cognition to ideology, in R. Fardon (ed.), *Power and Knowledge*, pp. 21–48. Edinburgh: Scottish Academic Press.

Bourdieu, P. 1977. *Outline of a Theory of Practice*. Cambridge: Cambridge University Press.

———. 1990. *The Logic of Practice*. Palo Alto, CA: Stanford University Press.

Bradley, R. 1990. *The Passage of Arms: An Archaeological Analysis of Prehistoric Hoards and Votive Deposits*. Cambridge: Cambridge University Press.

———. 1997. *Rock Art and the Prehistory of Atlantic Europe: Signing the Land*. London: Routledge.

———. 1998. *The Significance of Monuments: On the Shaping of Human Experience in Neolithic and Bronze Age Europe*. London: Routledge.

———. 2000. *An Archaeology of Natural Places*. London: Routledge.

———. 2002. *The Past in Prehistoric Societies*. London: Routledge.

Bradley, R., and Williams, H. (eds.). 1998. The past in the past. *World Archaeology* 30.

Braudel, F. 1969. *Écrits sur l'Histoire*. Paris: Flammarion.

Burke, P. 1989. History as social memory, in T. Butler (ed.), *Memory: History, Culture, and the Mind*, pp. 97–113. Oxford: Blackwell.

Butler, J. 1990. *Gender Trouble: Feminism and the Subversion of Identity*. New York: Routledge.

———. 1993. *Bodies That Matter: On the Discursive Limits of "Sex."* New York: Routledge.

Casey, E. S. 1987. *Remembering: A Phenomenological Study*. Bloomington: Indiana University Press.

Chapman, J. 2000. *Fragmentation in Archaeology: People, Places, and Broken Objects in the Prehistory of South Eastern Europe*. London: Routledge.

Chesson, M. (ed.). 2001. *Social Memory, Identity, and Death: Anthropological Perspectives on Mortuary Rituals*. Arlington, VA: Archeological Papers of the American Anthropological Association 10. American Anthropological Association.

Connerton, P. 1989. *How Societies Remember*. Cambridge: Cambridge University Press.

De Certeau, M. 1984. *The Practice of Everyday Life*. Berkeley and Los Angeles: University of California Press.

Derrida, J. 1977. Signature, event, context. *Glyph* I: 172–97.

Dietler, M. 1998. A tale of three sites: The monumentalization of Celtic oppida and the politics of collective memory and identity. *World Archaeology* 30: 72–89.

Feld, S., and Basso, K. H. (eds.). 1996. *Senses of Place*. Santa Fe, NM: School of Advanced Research Press.

Forty, A., and Küchler, S. (eds.). 1999. *The Art of Forgetting*. Oxford: Berg.

Gell, A. 1992. *The Anthropology of Time*. Oxford: Berg.

———. 1998. *Art and Agency: An Anthropological Theory*. Oxford: Clarendon Press.

Gosden, C. 1994. *Social Being and Time*. Oxford: Blackwell.

Gosden, C., and Lock, G. 1998. Prehistoric histories. *World Archaeology* 30: 2–12.

Halbwachs, M. 1925. *Les Cadres Sociaux de la Mémoire*. New York: Arno.

———. 1992. *On Collective Memory*. Chicago: University of Chicago Press.

Hall, K. 2000. Performativity. *Journal of Linguistic Anthropology* 9: 184–87.

Hamilakis, Y. 1999. Food technologies/technologies of the body: The social context of wine and oil production and consumption in Bronze Age Crete. *World Archaeology* 31: 38–54.

Hobsbawm, E. J., and Ranger, T. (eds.). 1983. *The Invention of Tradition*. Cambridge: Cambridge University Press.

Holtorf, C. J. 1998. The lifehistories of megaliths in MecklenburgVorpommern (Germany). *World Archaeology* 30: 23–38.

Ingold, T. 1993. The temporality of the landscape. *World Archaeology* 25: 152–74.

Jones, A. 2001. Drawn from memory: The archaeology of aesthetics and the aesthetics of archaeology in earlier Bronze Age Britain and the present. *World Archaeology* 33: 334–56.

Joyce, A. In press. The main plaza of Monte Alban: A life history of place, in B. J. Bowser and M. N. Zedeño (eds.), *The Archaeology of Meaningful Places*. Salt Lake City: University of Utah Press.

Joyce, R. 2000. Heirlooms and houses: Materiality and social memory, in R. A. Joyce and S. D. Gillespie (eds.), *Beyond Kinship: Social and Material Reproduction in House Societies*, pp. 189–212. Philadelphia: University of Pennsylvania Press.

Joyce, R. 2003. Concrete memories: Fragments of the past in the Classic Maya present (500–1000 A.D.), in R. M. Van Dyke and S. E. Alcock (eds.), *Archaeologies of Memory*, pp. 104–25. Oxford: Blackwell Publishing.

Joyce, R. A., and Gillespie, S. D. (eds.). 2000. *Beyond Kinship: Social and Material Reproduction in House Societies*. Philadelphia: University of Pennsylvania Press.

Kohl, P., and Fawcett, C. (eds.). 1995. *Nationalism, Politics, and the Practice of Archaeology*. Cambridge: Cambridge University Press.

Kopytoff, I. 1986. The cultural biography of things: Commoditization as process, in A. Appadurai (ed.), *The Social Life of Things: Commodities in Cultural Perspective*, pp. 65–91. Cambridge: Cambridge University Press.

Küchler, S. 1993. Landscape as memory: The mapping of process and its representation in a Melanesian society, in B. Bender (ed.), *Landscape: Politics and Perspectives*, pp. 85–106. Oxford: Berg.

Laqueur, T. W. 2000. Introduction. *Representations* 69: 1–8.

LaTour, B. 1993. *We Have Never Been Modern*. Cambridge: Harvard University Press.

———. 2005. *Reassembling the Social: An Introduction to Actor-Network Theory*. Oxford: Oxford University Press.

Lefebvre, H. 1991. *The Production of Space*. Oxford: Blackwell Publishing.

Lillios, K. 1999. Objects of memory: The ethnography and archaeology of heirlooms. *Journal of Archaeological Method and Theory* 6: 235–62.

———. 2003. Creating memory in prehistory: The engraved slate plaques of northwest Iberia, in R. M. Van Dyke and S. E. Alcock (eds.), *Archaeologies of Memory*, pp. 129–50. Oxford: Blackwell Publishing.

Low, P. 2004. Ancient uses of the past. *Antiquity* 78: 930–34.

Lowenthal, D. 1985. *The Past Is a Foreign Country*. Cambridge: Cambridge University Press.

McAnany, P. 1998. Ancestors and the Classic Maya built environment, in S. D. Houston (ed.), *Function and Meaning in Classic Maya Architecture*, pp. 271–98. Washington DC: Dumbarton Oaks.

Meskell, L. (ed.). 1998. *Archaeology under fire: Nationalism, politics, and heritage in the eastern Mediterranean and the Middle East*. London: Routledge.

———. 2003. Memory's materiality: Ancestral presence, commemorative practice and disjunctive locales, in R. M. Van Dyke and S. E. Alcock (eds.), *Archaeologies of Memory*, pp. 34–55. Oxford: Blackwell Publishing.

———. 2004. *Object Worlds: Material Biographies Past and Present*. Oxford: Berg.

———. 2005. *Archaeologies of Materiality*. Oxford: Blackwell Publishing.

Miller, D. (ed.). 2005. *Materiality*. Durham, SC: Duke University Press.

Mills, B. J. 2008. Remembering while forgetting: Depositional practices and social memory at Chaco, in B. J. Mills and W. H. Walker (eds.), *Memory Work: Archaeologies of Material Practices*. Santa Fe, NM: School of Advanced Research Press.

Mills, B. J., and Walker, W. H. (eds.). 2008. *Memory Work: Archaeologies of Material Practices*. Santa Fe, NM: School of Advanced Research Press.

Mizoguchi, K. 1993. Time in the reproduction of mortuary practices. *World Archaeology* 25: 223–35.

Nora, P. 1989. Between memory and history: Les lieux de mémoire. *Representations* 26: 7–25.

Papalexandrou, A. 2003. Memory tattered and torn: Spolia in the heartland of Byzantine Hellenism, in R. M. Van Dyke and S. E. Alcock (eds.), *Archaeologies of Memory*, pp. 56–80. Oxford: Blackwell.

Pauketat, T. R. 2008. Founders' cults and the archaeology of Wa-ka[n]-da, in B. J. Mills and W. H. Walker (eds.), *Memory Work: Archaeologies of Material Practices*. Santa Fe, NM: School of Advanced Research Press.

Pauketat, T. R., and Alt, S. 2003. Mounds, memory, and contested Mississippian history, in R. M. Van Dyke and S. E. Alcock (eds.), *Archaeologies of Memory*, pp. 151–79. Oxford: Blackwell Publishing.

Prent, M. 2003. Glories of the past in the past: Ritual activities at palatial ruins in early Iron Age Crete, in R. M. Van Dyke and S. E. Alcock (eds.), *Archaeologies of Memory*, pp. 81–103. Oxford: Blackwell Publishing.

Rowlands, M. 1993. The role of memory in the transmission of culture. *World Archaeology* 25: 141–51.

Sahlins, M. 1985. *Islands of History*. Chicago: University of Chicago Press.

Schama, S. 1996. *Landscape and Memory*. New York: Vintage Books.

Schlanger, S. 1992. Recognizing persistent places in Anasazi settlement systems, in J. Rossignol and L. Wandsnider (eds.), *Space, Time, and Archaeological Landscapes*, pp. 91–112. New York: Plenum Press.

Sinopoli, C. 2003. Echoes of empire: Vijayanagara and Historical Memory, Vijayanagara as Historical Memory, in R. M. Van Dyke and S. E. Alcock (eds.), *Archaeologies of Memory*, pp. 17–22. Oxford: Blackwell Publishing.

Soja, E. W. 1996. *Thirdspace*. Oxford: Blackwell.

Taçon, P. S. 1999. Identifying ancient sacred landscapes in Australia: From physical to social, in W. Ashmore and A. B. Knapp (eds.), *Archaeologies of Landscape: Contemporary Perspectives*, pp. 33–57. Oxford: Blackwell Publishing.

Thomas, J. 1996. *Time, Culture, and Identity*. London: Routledge.

Thomas, J. 2001. Archaeologies of place and landscape, in I. Hodder (ed.), *Archaeological Theory Today*, pp. 165–86. Cambridge: Polity Press.

Tilley, C. 1994. *A Phenomenology of Landscape*. Oxford: Berg.

Trigger, B. G. 1984. Alternative archaeologies: Nationalist, colonialist, imperialist. *Man* 19: 355–70.

Van Dyke, R. M. 2003. Memory and the construction of Chacoan Society, in R. M. Van Dyke and S. E. Alcock (eds.), *Archaeologies of Memory*, pp. 180–200. Oxford: Blackwell Publishing.

———. 2004. Memory, meaning, and masonry: The Late Bonito Chacoan landscape. *American Antiquity* 69: 413–31.

———. 2007. *The Chaco Experience: Landscape and Ideology at the Center Place*. Santa Fe, NM: School of Advanced Research Press.

Van Dyke, R. M., and Alcock, S. E. 2003. Archaeologies of memory: An introduction, in R. M. Van Dyke and S. E. Alcock (eds.), *Archaeologies of Memory*, pp. 1–13. Oxford: Blackwell Publishing.

Walker, W. H. 1995. Ceremonial trash? in J. M. Skibo, W. H. Walker and A. E. Nielson (eds.), *Expanding Archaeology*, pp. 67–79. Salt Lake City: University of Utah Press.

———. 1999. Ritual life histories and the afterlives of people and things. *Journal of the Southwest* 41: 383–405.

———. 2008. Practice and the afterlife histories of witches and dogs. in B. J. Mills and W. H. Walker (eds.), *Memory Work: Archaeologies of Material Practices*. Santa Fe, NM: School of Advanced Research Press.

Williams, H. 1998. Monuments and the past in early AngloSaxon England. *World Archaeology* 30: 90–108.

Williams, H. (ed.) 2003. *Archaeologies of Remembrance: Death and Memory in Past Societies*. New York: Kluwer Academic/Plenum.

Yoffee, N. 2007. Peering into the palimpsest: An introduction to the volume, in N. Yoffee (ed.), *Identity, Memory, and Landscape in Archaeological Research: Negotiating the Past in the Past*. Tucson: University of Arizona Press.

28

VIRTUAL REALITY, VISUAL ENVELOPES, AND CHARACTERIZING LANDSCAPE

Vicki Cummings

In this chapter, I consider how archaeologists have used visual representations to accompany discussions of landscape. In particular, I focus on recent "postprocessual" considerations of landscape and the forms of representation that have been used to illustrate both theoretical points and landscape case studies. I go on to consider how new virtual methods of representation offer a way forward for landscape studies.

Theoretically, landscape is now understood to be a key element of the experienced and engaged world. All human activity, past or present, takes place in the landscape; it is not simply a backdrop to life but is directly involved in how people undertake their everyday activities. Landscape plays an active role, structuring and structured by human agency. All landscapes are meaningful and dwelt in, and people assign meanings to the landscape. Landscape is therefore active, historical, and directly associated with a person and with a community's identity (Ashmore and Knapp 1999; Bender 1993; Hirsch and O'Hanlon 1995).

These key theoretical points have been the background against which a number of landscape case studies have taken place. Scholars have been attempting to understand the potential significance of landscape in specific periods or involving particular types of evidence. One area that has seen considerable debate has been the setting of built architecture, in particular Neolithic monuments in Britain and Ireland. This includes Tilley's original groundbreaking studies in *A Phenomenology of Landscape* (Tilley 1994), Bradley's *The Significance of Monuments* (1998) and, more recently, a detailed consideration of the setting of chambered tombs throughout the Irish Sea region (Cummings Forthcoming; Cummings and Whittle 2004). The conclusions of this work are that Neolithic monuments are very carefully located in very specific parts of the landscape. They seem to be purposefully positioned in order to afford views of mountains, in particular visually distinctive mountains (Figure 28.1). Monuments are also positioned so that there are views of water, in particular the sea, lochs, and rivers, as well as other natural features such as rocky outcrops. It has been argued that these places were of key significance to the people who built and used these sites, probably associated with belief systems and ancestor or creation myths, as well as creating connections with a wider Neolithic community (Cummings Forthcoming; Cummings and Whittle 2004).

The Problems of Characterizing Landscape

One of the key problems facing landscape archaeologists, however, is how we represent these

Figure 28.1   Photograph of the landscape from the chambered tomb of Pentre Ifan, southwest Wales.

landscapes to the reader. Undertaking a landscape study involves visiting and experiencing archaeological sites and their associated landscapes firsthand. However, this firsthand experience cannot be easily replicated for others. There are other problems, too. First, all landscapes are experienced differently by different people. At the most simple level, attempting to characterize landscape is problematic: what one person considers to be an upland and remote landscape may not be the same for other people. At its most extreme, the example of the Umeda of Papua New Guinea may be cited (Gell 1995). These people live in the forest and struggle to perceive depth of field, especially distant landscapes. This raises the possibility that how we characterize landscape is culturally and contextually specific.

Second, there is a tendency to treat landscapes as visual phenomenon. This is almost certainly because landscape is for the most part a visual experience in today's society. Although archaeologists have now begun to write about the significance of the other senses when engaging with landscape (e.g., Cummings 2002; MacGregor 2002; Mills 2005; Watson and Keating 1999), it is almost always the visual aspect of these landscapes that has been illustrated. One must remember that landscapes are multisensual and engage all the senses. How we characterize or represent a landscape and its engagements, however, is problematic.

The third problem facing landscape archaeologists is that landscapes change: many have changed radically since the periods we are studying, but landscapes also change over short periods of time—seasonally, for example. The presence of vegetation

does not necessarily mean the landscape is not visible from a site (see Cummings and Whittle 2003), but it does mean that there would have been noticeable differences in the view and experience of place over time. With all these problems, how can we as archaeologists represent the lived-in, changing, temporal, and experienced landscape to a wider audience?

Archaeology, Landscape, and Representation

I consider the range of ways in which we can characterize or represent the landscape, discussing the full range of techniques currently available to us. The traditional way of representing landscape is through a map. Archaeological distribution maps did not commonly appear before the 20th century; prior to the pioneering culture-historical study of cultures across a wide area, there was neither the data nor the desire to show the distribution of material culture or structures. The growth of aerial photograph and the high resolution mapping by the Ordnance Survey of the British Isles in the early 20th century, however, enabled archaeologists to utilize this form of representation. Fox's (1932) *The Personality of Britain* can be seen as an early example of the prominent use of the distribution map, and it remains one of the key ways of showing the landscape location of specific types of object.

It is fair to say that the experiential qualities of landscapes were not one of the primary interest areas of processualists, although certain landscape analysis techniques were employed, such as site catchment analysis (Higgs and

Vita-Finzi 1972), thiessen polygon assessment (e.g., Renfrew 1973), and other forms of settlement pattern analysis (Hodder and Orton 1976). Postprocessual archaeology did not really begin to consider the significance of landscape until the early 1990s, and the favored method of representation in these early studies was the distribution map, in combination with photos and plans (e.g., Bender 1993; Tilley 1994). Although the extensive use of landscape photography was pioneering at the time, these "traditional methods" were rather inadequate at illustrating the range of experiences such landscapes can generate and have been critiqued by a number of authors (Fleming 1999; Thomas 1993). Furthermore, distribution maps are characterized by Cartesian notions of space, which is abstracted, timeless, and passive (Thomas 1996: 88), and the observer is removed from a lived-in world and ends up seeing literally everything from nowhere, what Haraway (1991: 189) describes as the "God-trick." Therefore, with the sudden growth in the study of archaeological landscapes, particularly as part of a broader experiential archaeology (e.g., Rodaway 1994), and the critique of traditional forms of representation, new methods were sought. However, it is significant that the vast majority of landscape studies still use distribution maps, plans, and photos as their main way of representing landscape, almost certainly a reflection of the dominance of the published article/book.

One obvious alternative to the static photograph is video footage. This medium offers a dynamic method of showing the whole landscape and has the advantage of including sound, which can be the ambient background or a voiced narrative. There are problems with video, however: the image is as fixed in time as a photograph, and the viewer has no control over what is shown. With regard to accessibility, video can now be streamed over the internet, meaning that it is becoming an increasingly feasible means of viewing data, but storage can be a problem, because video is memory intensive. A few years ago, I considered some alternative methods (Cummings 2000b). One simple alternative, which has been in use for a number of years, is the 360° photographic montage. Quite simply, one takes a series of photos from a fixed point in the landscape and then uses a computer package to stitch the images together seamlessly. They can then be viewed on the internet (see Cummings 2000b for a series from the chambered tombs of South Uist). These panoramas are quicker to view than video, smaller to store, and enable users to look around from a fixed point. Another

advantage is that the photos can be labeled, so that particular landscape features can be highlighted, for example, www.english-heritage.org.uk/stonehengeinteractivemap/sites/stonehenge/swf.html). This technology (QuickTime Virtual Reality: QTVR) is now regularly used on commercial websites to provide a quasi-immersive visual experience. However, the vast majority of landscape archaeologists have not employed this technology, possibly because publishers still wish books or papers to "stand alone" and not rely on a website for some of the information.

QTVR may well have been derived from an earlier form of landscape representation that may have its origins in Alfred Wainwright's landscape drawings of the Lake District (e.g. Wainwright 1960). Landscape panoramas appear in a number of publications (e.g., Bradley 1998; Cleal, Walker, and Montague 1995: 38–39). However, these strips are often unwieldy. It is also often difficult to make out the finer details of the landscapes they depict. A variation of these landscape strips has been published that compresses and stretches the landscape strip into a single image (Cummings 2000a: fig. 3). This method is also problematic, since it compresses the entire 360° panorama into a single diagram, and so one tends to think of this as a single view. One alternative method of representing the landscape has interesting parallels with the earliest antiquarian drawings of monumental landscape settings (Peterson 2003). These landscape representations are schematics that show the entire 360° in a single diagram; it has the key advantage of also showing directionality, and large numbers of panoramas can be easily compared for broader patterns (see Figure 28.2). Another variation on traditional method is the photographic "montage." Inspired by the work of David Hockney (1984) among others, these are an alternative way of illustrating a sense of place and also embed temporality into an image (see for example Shanks 1992: 123). Intimate details of particular places are also included, and this photographic style has recently been used successfully by others (e.g., Chadwick 2004; Edmonds and Seaborne 2001).

Following on from the artistic use of photographs in landscape archaeology is the use of artwork as a way of illustrating the experience of landscape. This is particularly interesting, because the term *landscape* was introduced into the English language as a technical term used by painters to refer to the genre of landscape painting (Hirsch 1995). Early paintings of archaeological sites frequently included their wider landscape setting, from Constable's Stonehenge to Hodge's (Captain

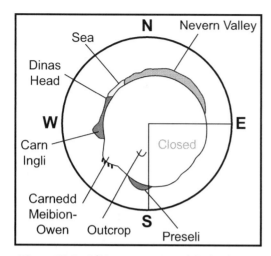

Figure 28.2   360° representation of the landscape from the chambered tomb of Pentre Ifan A.

Cook's artist) painting of Easter Island (cf. Bahn 1996). More recently, art has again been used as part of the repertoire of landscape representation. In particular, the works of such artists as Richard Long and Antony Gormley have been cited as ways into representing experience, engagement, and landscape (see papers in Renfrew, Gosden, and DeMarrais 2004).

Geographic Information Systems (GIS) are increasingly being used as a key tool in landscape archaeology (Gillings, Mattingly, and van Dalen 1999; Johnson and North 1997; Wheatley, Earle and Poppy 2002; and see Conolly, this volume). One key tool in a GIS is "viewshed" analysis (Gillings 1998: 118), which shows which parts of the landscape are visible from a particular point of view (Figure 28.3). To support the observations made on the ground, I generated viewsheds for all the megaliths in Wales (Cummings and Whittle 2004: chs. 4–6; and see www.cf.ac.uk/hisar/people/vc/megaliths/megaliths.html). A more dynamic use of viewshed analysis was conducted at Stonehenge, which demonstrated which parts of the landscape were visible as one moves around the landscape (Exon et al. 2000). Although a useful tool in analyzing visibility, these viewsheds suffer all the problems of abstracted two-dimensional Cartesian views of the landscape and give no real sense of what particular views look like in reality (and see Cummings and Whittle 2004, ch. 3, for a further critique). GIS does enable archaeologists to assemble data from across a landscape, useful for the management or manipulation of existing data (Wheatley and Gillings 2002). In this sense, GIS is a useful tool for the organization

of data but at present remains limited for the representation and characterization of landscape, particularly from an experiential point of view.

Another way of representing landscape has begun to be explored with virtual reality, with some interesting results. A virtual Avebury was created as part of a broader project aimed at how we look at and see this World Heritage site (Pollard and Gillings 1998). This enabled the viewer to negotiate around the Avebury landscape without the problem of having all the modern features in the way. There are a number of good examples of virtual worlds now available on the internet (and see Barceló, Forte, and Sanders 2000). However, the vast majority do not address the wider landscape setting, although there is clearly scope for this in the future. Furthermore, virtual reality is, of course, just that: a representation of a world that does not, and did not, exist. However, it enables a whole new set of experiences of place and landscape that in themselves may assist in the broader aim of representing archaeological landscapes.

Conclusions

How we characterize landscape, represent it, then present our findings to a wider audience can be considered one of the most difficult problems facing landscape archaeologists. In this chapter, I have considered a suite of techniques that can be employed by archaeologists wishing both to illustrate their points and to enable the reader to get some sense of the landscapes in question. There are quite clearly advantages and disadvantages to all the methods that have been outlined here. There is no denying the longevity of the distribution map, and photographs are effective because of their effectiveness in communicating data in a relatively easy and inexpensive way. However, nontraditional ways of representing the landscape can also offer us new ways of experiencing and dwelling in the landscape. Technology offers us alternatives to traditional text-based narratives. At this particular time, it seems that web-based resources offer a way forward for landscape characterization and representation. They offer the potential to use all of the visual methods of representation that have been discussed in this chapter, which, when embedded within a GIS or similar interface, can also incorporate video or sound clips. It seems clear that the most effective way of illustrating our case studies is to use as many different representations of landscape as are available. This should enable the reader to gain a range of experiences that may lead to additional alternative interpretations.

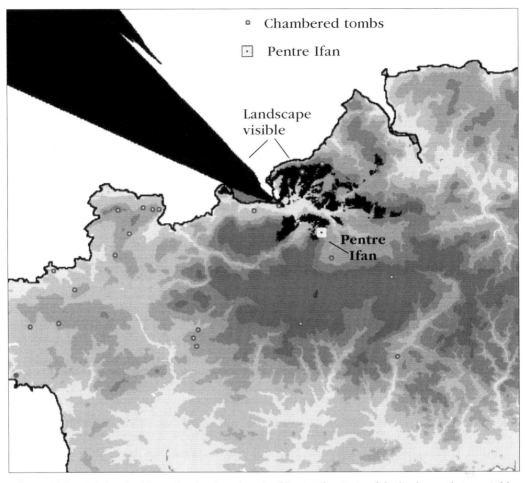

Figure 28.3   GIS viewshed from the chambered tomb of Pentre Ifan. Parts of the landscape that are visible from the site are in black.

References

Ashmore, W., and Knapp, A. B. (eds.). 1999. *Archaeologies of Landscape: Contemporary Perspectives*. Oxford: Blackwell Publishing.

Bahn, P. 1996. *Cambridge Illustrated History of Archaeology*. Cambridge: Cambridge University Press.

Barceló, J., Forte, M., and Sanders, D. (eds.). 2000. *Virtual Reality in Archaeology*. Oxford: British Archaeological Reports.

Bender, B. (ed.). 1993. *Landscape: Politics and perspectives*. Oxford: Berg.

Bradley, R. 1998. *The Significance of Monuments: On the Shaping of Human Experience in Neolithic and Bronze Age Europe*. London: Routledge.

Chadwick, A. (ed.). 2004. *Stories from the Landscape*. Oxford: British Archaeological Reports.

Cleal, R., Walker, K., and Montague, R. 1995. *Stonehenge in its Landscape: Twentieth-Century Excavations*. London: English Heritage.

Cummings, V. 2000a. Landscapes in motion: Interactive computer imagery and Neolithic landscapes of the Outer Hebrides, in C. Buck, V. Cummings, C. Henley, S. Mills, and S. Trick (eds.), *UK Chapter of Computer Applications and Quantitative Methods in Archaeology*, pp. 11–20. Oxford: British Archaeological Reports.

———. 2000b. The world in a spin: Recreating the Neolithic landscapes of South Uist. *Internet Archaeology* 8.

———. 2002. Experiencing texture and touch in the British Neolithic. *Oxford Journal of Archaeology* 21(3): 249–61.

———. Forthcoming. *A View from the West: The Neolithic of the Irish Sea Zone*. Oxford: Oxbow Books.

Cummings, V., and Whittle, A. 2003. Tombs with a view: Landscape, monuments and trees. *Antiquity* 77: 255–66.

———. 2004. *Places of Special Virtue: Megaliths in the Neolithic Landscapes of Wales*. Oxford: Oxbow Books.

Edmonds, M., and Seaborne, T. 2001. *Prehistory in the Peak*. Stroud: Tempus Publishing.

Exon, S., Gaffney, V., Woodward, A., and Yorston, R. (eds.). 2000. *Stonehenge Landscapes: Journeys through Real and Imagined Worlds*.

Fleming, A. 1999. Phenomenology and the megaliths of Wales: A dreaming too far? *Oxford Journal of Archaeology* 18: 119–25.

Fox, C. 1932. *The Personality of Britain*. Cardiff: National Museum of Wales.

Gell, A. 1995. The language of the forest: Landscape and phonological iconism in Umeda, in E. Hirsch and M. O'Hanlon (eds.), *The Anthropology of Landscape: Perspectives on Place and Space*, pp. 232–54. Oxford: Oxford University Press.

Gillings, M. 1998. Embracing uncertainty and challenging dualism in the GIS-based study of a palaeo-flood plain. *European Journal of Archaeology* 1: 117–44.

Gillings, M., Mattingly, D., and van Dalen, J. 1999. *Geographical Information Systems and Landscape Archaeology*. Oxford: Oxbow Books.

Haraway, D. 1991. *Simians, Cyborgs and Women: The Reinvention of Nature*. London: Free Association.

Higgs, E., and Vita-Finzi, C. 1972. Prehistoric economies: A territorial approach, in E. Higgs (ed.), *Papers in Economic Prehistory*, pp. 27–36. Cambridge: Cambridge University Press.

Hirsch, E. 1995. Landscape: Between place and space, in E. Hirsch and M. O'Hanlon (eds.), *The Anthropology of Landscape: Perspectives on Place and Space*, pp. 1–29. Oxford: Oxford University Press.

Hirsch, E., and O'Hanlon, M. (eds.). 1995. *The Anthropology of Landscape: Perspectives on Place and Space*. Oxford: Oxford University Press.

Hockney, D. 1984. *Cameraworks*. London: Thames and Hudson.

Johnson, I., and North, M. 1997. *Archaeological Applications of GIS*. Sydney: University of Sydney.

Mills, S. 2005. Sensing the place: Sounds and landscape archaeology, in D. Bailey, A. Whittle, and V. Cummings (eds.), *(Un)settling the Neolithic*, pp. 79–89. Oxford: Oxbow Books.

Peterson, R. 2003. William Stukeley: An eighteenth-century phenomenologist? *Antiquity* 77: 394–400.

Pollard, J., and Gillings, M. 1998. Romancing the stones: Towards an elemental and virtual Avebury. *Archaeological Dialogues* 5: 143–64.

Renfrew, C. 1973. Monuments, mobilisation and social organisation in Neolithic Wessex, in C. Renfrew (ed.), *The Explanation of Culture Change*, pp. 539–58.

Renfrew, C., Gosden, C., and DeMarrais, E. (eds.). 2004. *Substance, Memory, Display: Archaeology and Art*. Cambridge: McDonald Institute for Archaeological Research.

Rodaway, P. 1994. *Sensuous Geographies*. London: Routledge.

Shanks, M. 1992. *Experiencing the Past: On the Character of Archaeology*. London: Routledge.

Thomas, J. 1993. The politics of vision and the archaeologies of landscape, in B. Bender (ed.), *Landscape: Politics and Perspectives*, pp. 19–48. Oxford: Berg.

———. 1996. *Time, Culture and Identity: An Interpretive Archaeology*. London: Routledge.

Tilley, C. 1994. *A Phenomenology of Landscape*. Oxford: Berg.

Wainwright, A. 1960. *The Southern Fells: A Pictorial Guide to the Lakeland Fells*. Kendal: Westmoreland Gazette.

Watson, A., and Keating, D. 1999. Architecture and sound: An acoustic analysis of megalithic monuments in prehistoric Britain. *Antiquity* 73: 325–36.

Wheatley, D., Earle, G., and Poppy, S. (eds.). 2002. *Contemporary Themes in Archaeological Computing*. Oxford: Oxbow Books.

Wheatley, D., and Gillings, M. 2002. *Spatial Technology and Archaeology: The Archaeological Applications of GIS*. London: Routledge.

29

LANDSCAPE AND PERSONHOOD

Chris Fowler

Personhood is the condition of being a person, and this condition is defined culturally. Attempts to grasp personhood in the past have often focused on interpreting the body and have frequently been directed toward reconstructing individual lives. Personhood has commonly been conceptualized as the state of being a singular, individual person bounded within the body (cf. Fowler 2004a; Thomas 2004), but persons can also be recognized as multifaceted beings formed through relations with others, consisting of different aspects and extending throughout the material world. In many contexts, the latter may be stressed over the former (e.g., Strathern 1988; cf. LiPuma 1998), and archaeologists have recently turned toward examining the relational characteristics of personhood (e.g., Brück 2004; Fowler 2004a). At the same time, a phenomenological trend investigating ways of "being in the world" explores the embeddedness of body and world. (See Brück 1998, 2006 and Joyce 2005 for recent reviews of work in this area.) Combining these perspectives supports investigation of landscape and personhood together and provides new opportunities to reflect on the relationships between them. Such studies may include a consideration of how an individual's experience of his or her own self is generated by personal engagement with and within specific places, but it also requires us to

consider how personhood is invested in broader cultural engagements with the landscape. In this chapter, I explore some different instances of how communities have inhabited landscapes in ways that materialize distinct senses of personhood and specific ways of conceptualizing the material world and valuing its components. Although this relationship can be approached by considering how landscape becomes embedded within personal identity, here I look at how personhood is distributed throughout, and understood with reference to, the landscape. I focus most heavily on how landscape is inhabited in ways that reflect on the temporal characteristics of personhood.

The Distribution of Personhood throughout the Landscape

Personhood is a composite state, with persons consisting of a variety of aspects—examples might include mind, soul, image, breath, spirit, biography, individuality, memory, experience, and name—these aspects may reside in features of the body (see, for example, Meskell and Joyce 2003: ch. 2), and other things and places (see below). The configuration of the person out of such features varies from context to context and is often a matter of debate and controversy. Some personal qualities may be temporary or changeable while others are eternal, ancestral, or otherwise pre-exist the individual person. These might also extend after the person's death. When and after a person dies, aspects of their personhood become transformed, and may be relocated: souls may go to heaven, ghosts may emerge.

The idea of spirits or souls being separated from bodies after death and relocating to another part of the cosmos is familiar to Western thought—other traditions may consider aspects of the person to become associated with places in this plane of existence. Among the Papuan Orokaiva, death involves a process of diminishing a person's biography or social person (*hamo*) and dealing with their emergent image (*ahihi*), as well as their transformation into an anonymous aspect (*onderi*) that roams the forest in the form of a wild pig (Iteanu 1988, 1995). Features of the deceased's personal biography are remembered through the curation of their intimate possessions by survivors, and their image will be encountered through prestigious gifts displayed at mortuary practices or circulated between individuals and communities. The qualities of personhood may therefore be extended for the Orokaiva to things, animals, and zones of the landscape, as well as to the human body. The boundaries of the person fluctuate throughout life through interactions via objects, animals, and places. Eventually the person becomes unbounded through mortuary practices (cf. Fowler 2004a: 87–92).

For the Madagascan Merina, people are "'hairs growing out of the head,' which is the ancestral land" (Bloch 1982: 211). After death, the body is interred on a hill near to the village where its soft parts, which were contributed through the female line, leach into the soil. The dry, desiccated remains are exhumed years later and taken to a distant tomb in a communal *necropolis* associated with the male lineage that contributed to that person's body. These remains become "earth," associated with fertility and ongoing social generation (Bloch 1982: 215). The Merina, Bloch reports, consider their kinship endogamous, and so the ancestral energies are maintained within the ongoing community by this practice, while flesh, redolent of distinct individual bodies and carrying the pollution of death, is removed in the early stages of mortuary rituals. Different aspects of the body are therefore distributed to appropriate landscape locations with distinct associations (including temporal and gendered ones), and ancestral substance from people who may have lived far apart from one another in life is combined in the tombs. Merina personhood therefore requires the direction of essences not only through human bodies but also through the landscape.

In both these examples, persons are constituted from features or elements of the material world. In such cases, these elements may be conveyed through the human body and through other personal media; they are often drawn together in conceiving and growing a person (and initiating him or her into adulthood) and dispersed during deconception. A person encapsulates the relationships constituting the world and through mortuary practices becomes decomposed into aspects resituated in the landscape.

Person, Place, and Landscape

The human body may be thought to combine a range of substances, which may themselves be identified with elements in the landscape. For instance, bodies have often been thought to consist of "humours" that exist elsewhere in the broader world, and changes in the ratios of humours are believed to affect character, mood, and even gender (Busby 1997; Rublack 2002). Shrines, tombs, or houses may be considered to have bodies like human beings and may be composed of materials found in different landscape features. Madagascan Zafimaniry wooden houses built following a marriage begin life flimsy and, in our terms, "green," hardening and drying out over time as they are turned into more permanent structures; human bodies are initially "wet" and flimsy in the same way (Bloch 1995; cf. the progressive drying of Merina bodies above).

Creating a parallel between house and body illustrates the immersion of the body in its world, stresses the relationships forming both, and can serve to personify houses and households or illustrate their physical embodiment. It can demonstrate the nesting of person, place, and landscape. For example, the floors of trapezoid Mesolithic/Neolithic houses at Lepenski Vir in the Danube Gorges were made from soft red limestone river sediment that hardened once laid down, the walls bedded with white rock cut from the gorge edge behind the village. We could suggest these houses were laid out according to a plan of the human body with skulls and sculpted busts at the narrow "head" end, the hearth at the heart of the house, some lined with stone "ribs" that intriguingly resemble jawbones—one actually being a human jawbone (Radovanović 1996: 134, 2000: 337)—and the long front walls as shins from two legs meeting as feet pressed together in the doorway. One of the earliest burials at Lepenski Vir (burial 69) was laid out in this way, shins parallel to the river and head propped up to look at the river and face the rising sun (Radovanović 1996: 178, 2000: 337; Srejović 1972: 153).[1]

In this place, directions of natural movement were loaded with meaning: the river teeming with life in front of the village, ancient burial grounds and the rock behind; upstream and downstream (where bodies were laid out in houses, they ran parallel to the course of the river); sunrise over the hills opposite and the river (sun shining into each house), and sunset behind the village; and the remains of the dead placed at times behind the hearth (Borić 2002a; Radovanović 2000). These directions referred to cycles, as well as linear paths: the daily cycle; the life passage from childhood to death, burial, and ancestry (children's remains were laid down through house floors that were then resealed, and human remains from the ancient burial grounds, as well as recently deceased bodies, were deposited in the structures [Radovanović 2000: 340–41]); household abandonment and renewal possibly related to periodic flooding; annual cycles including beluga and sturgeon migration patterns (adult fish swim upstream in Spring, returning downstream with young in Autumn), and other animal lifecycles. Borić (2002a) associates the use of red and white coloring at the site not only with human bodily substance and the red ochre used to cover the remains of the dead but also the colors of autumn foliage against white cliffs in the gorge. Life in such places reflected on relationships between human bodies and the landscape, framing personhood through a knowledgeable experience of place and landscape at differing points in time (Borić 2002a, 2002b; Chapman 2000: 194–203). The human body was situated in this world through patterns of personal movement. The personal life cycle was situated alongside animal lifecycles, the seasons, and the time span of the house, household and village.

Through continual interreferentiality between place, landscape and bodies, each can come to exhibit a complimentary "fractal" pattern in the arrangement of features or elements where the same relationships are manifest at differing scales (cf. Chapman 2000; Fowler 2004a: 48–50, 108–111, In press; Wagner 1991). Such fractal patterns may be foundational to understanding both personhood and the landscape; therefore, patterns in the production and use of places within landscapes may provide clues about contemporary understandings of personhood. I develop this point mainly through examples from Neolithic and Bronze Age Britain.

Personal Transformation and the Landscapes of Neolithic and Early Bronze Age Britain

The landscapes of Britain are rich in Neolithic and Bronze Age monuments—here I consider their roles in how past people made sense of personhood. Mortuary structures within the earlier Neolithic often channeled the movement of human remains along linear chambers (Lucas 1996). Later in the earlier Neolithic, many were covered with earthen mounds that were also thrown up over midden and other places. Causewayed enclosures—large gathering places visited seasonally—were often located at the boundary between two landscape zones. Many earlier to middle Neolithic cursus monuments—linear paths defined by banks, ditches, pits or posts—were located to cut streams or run parallel to rivers, which they resembled (Brophy 2000). They directed movement through landscapes, and some of these route ways became monumentalized by cutting ditches and raising mounds of earth as banks (Johnston 1999; Tilley 1994: 170–201). Later Neolithic henges created circular spaces located at the center of broad landscapes, some hemmed by a circle of hills on the horizon (Watson 2001). Passage graves and stone circles frequently directed attention toward regular celestial phenomena. Personal movement through the landscape while, for instance, herding animals, carrying relic remains of the dead, or transporting axe heads intended to be hafted and given away at specific places, connected cycles of daily time, annual time, the time of exchange relations, and generational time in the journeys of personal life. Exactly how these interactions with and within the landscape took place changed throughout these millennia. I will trace the relationship between landscape and personhood more closely in just a few cases.

Vicki Cummings, Andy Jones, and Aaron Watson (2002) have demonstrated that the location of a group of Neolithic long cairns in southern Wales places them in an "asymmetrical" landscape with views of the Black Mountain escarpment in one direction and low open land in another. These asymmetrical places divide the landscape along the long axis of the monument (for example, Penywyrlod) (Figure 29.1). Their interpretation of this phenomenon is that the spatial liminality of these places between upland and lowland was emblematic of their role in the transformation of bodies between life and ancestry. Like the landscape experienced at this place, corpses secluded here became asymmetrical as they decayed in this liminal state, before bodily symmetry was returned when the bones of the skeleton were fully exposed. The arrangement of place, landscape perception from that place, and personal transformation effected there were intelligible through a consistent logic.

Liminal phases of ritualized personal transformations are located in liminal places and times. These transformations include not only mortuary

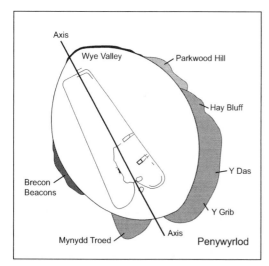

Figure 29.1 The landscape setting of Penywyrlod, the Black Mountains, southwest Wales (detail of fig. 2, Cummings, Jones, and Watson, 2002: 60, reproduced with kind permission of the authors and *Cambridge Archaeological Journal*).

practices but other rites of passage such as initiation into adulthood and acts intended to heal, purify, fulfill a debt or relationship, or bring on personal revelation. Other phases of such personal transformation may involve movement through the wider landscape. Landscape is experienced as narrative sequence, and such narratives may be equated with journeys of personal transformation. For instance, water-rounded quartz pebbles have been found deposited at the Cashtal yn Ard chambered cairn and at the Billown enclosure on the Isle of Man. These were probably brought from the shore or streams. Cashtal yn Ard is located on a round hill with views of the high hills in one direction and sea in the other—the hill is bounded on three sides by streams that flow to a quartz-rich beach (Cummings and Fowler 2004; Fowler 2004b) (Figure 29.2). Viewed from Cashtal yn Ard, the sun rises from the sea and sets behind the high hills. From the horseshoe-shaped forecourt, the passage of the monument provided a "path" through the stone chambers toward the sea, and looking inland away from the entrance to the site a valley provided a path into the high hills that face the open end of this forecourt. Being there located a person at the center of the world, on a conceptual island within the Isle of Man, punctuating a journey from coast to highland with an encounter with the relic remains of the past. That journey linked land, shore, and sea, while bringing living and dead and water and stone (particularly white water and white stone, such as a

quartz nodule above the entrance to the chambers) into physical juxtaposition with one another (see Fowler and Cummings 2003: 12–16). Such journeys could be placed in the context of personal pilgrimage or initiation rites that spurred reflections on the fluid, ephemeral nature of individual life, the potential for personal transformation, and the enduring nature of the world and community, including the dead. I have argued that in the Manx earlier Neolithic bodies were seen as vessels permeable to vital essences including milk and water. After death, when the vessel was broken, some of the essences contained could be released (for example, as bone) and transferred to other vessels, such as chambered tombs (Fowler 2004b). Ceramic bowls were tempered with quartz, exposed when they were broken. The monument at Cashtal yn Ard could be seen as an enduring vessel at a larger scale—a conduit to potent essences composed by and housing the remains of past generations.

The location of Cashtal yn Ard could be seen as liminal (between uplands and sea; cf. Cummings and Fowler 2004; Fowler and Cummings 2003) but also as central (people probably carried out many routine activities in the low hills) and as a microcosm of the world. There is no contradiction between a place being at once liminal and central—the two positions fluctuate depending on the place of the agent in sequences of activity (cf. Bourdieu 1970). Providing a place from which to stand, and a perspective on landscape and personhood together, such places are at some times liminal and at others encapsulate the entire cosmos. The location of sites to include views of outcrops, sea and hills, or the construction of monuments out of a wide range of materials (for example, stone from inland and brought from the seashore) and the architectural fusion of stones, sky, fire, and water assist in marking such a place as an *axis mundi* (center of the world [e.g., Richards 1996a]) and/or *imago mundi* (world model [e.g., Owoc 2005]). Combining the aspects of the world (which may convey, or be parallel to, aspects of the person) makes a monument special—but it is intelligible through previous experience of daily routine. Indeed, the power of cosmological principles are most persistently felt through parallel patterns in daily life. Colin Richards's studies of tombs, houses and henges in later Neolithic Orkney trace the same conceptual logic shaping experience in each forum (Richards 1993, 1996b). The hearth of each house in the village and fire at the center of the Stones of Stenness were presented as *axis mundi* in daily life and during large gatherings. Such light would enter the tombs such as Maeshowe—houses of the dead—only at Midwinter, suggesting a time

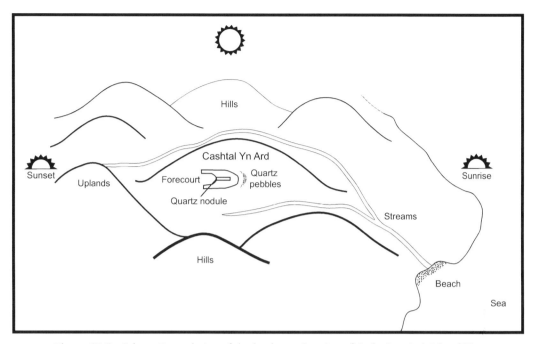

Figure 29.2   Schematic rendering of the landscape location of Cashtal yn Ard, Isle of Man
(produced by the author).

when the center of the world shifted from the living to the dead.

Joanna Brück (2004: 321) has discussed how one of the early Bronze Age barrows from the Yorkshire Wolds included clay brought to the vicinity from elsewhere—something Mortimer (1905) suggests for a number of barrows in the area—and Mary-Ann Owoc (2002: 135) has noted how distinctive yellow clay was brought to Caerloggas Downs III in Cornwall from at least 3 km away. Communities composed each of these monuments from multiple points of origin, just as any body deposited may have been multiply constituted. Human remains deposited at such barrows were therefore involved in reflecting on cosmological patterns (and perhaps in cosmogenic acts) alongside soil, wood, and other materials (Owoc 2005). These actions drew the world together in one place and referenced ties of belonging to landscape locations, and potentially ties of kinship as well (Brück 2004: 322). Owoc (2005) reminds us that the treatment of the dead should here be contextualized within a broader cosmological order taken up and transformed at each site. She examines successive phases of activity at another site, Trelen 2, and notes how the changing architecture referenced the movement of the sun, first through monument siting and alignment, and later through materials, color, and patterns of renewal. It is within the context of these

cosmological principles and the nested solar cycles (for example, day and year) that statements about personhood were framed. For instance, cremation, increasingly common in this period, may have distributed flesh into the sky and bone under the earth at different times of the daily cycle or even longer cycles. The idea that barrows commemorated dead individuals may put the cart before the horse if cremated bones, for instance, were a symbolic resource (Brück 1995) belonging to the community and deposited only at appropriate times in a cosmogenic sequence. Barrows were one device through which persons were situated within temporal landscapes.

Clusters of barrows also provided media for ongoing relations between person and landscape. Some Early Bronze Age barrow groups were arranged in lines along ridgeways visible from near to henges or located along route ways passing near to henges (Watson 2001). As people passed these lines of barrows they may have recounted genealogies or lineages. Then, as people left these rows of barrows and headed toward henges such as Avebury, they would have seen the barrow cemeteries grouped on the horizons. Finally, as people entered the henge, any distinctions traced between these barrow groups were partially secluded as the henge banks hid this community of the dead from view and enclosed the living community (cf. Watson 2001: 208–09). We can also imagine a reversal in sequence through the journey away from the gathering. Where the relational nature of personhood was brought to the fore, this monumentalized landscape provided a strong referent for considering different relations between people past and present.

Inheriting and Negotiating Ways of Being in the World

Neither regional diversity nor change through time should be overlooked—in the previous section I have not produced a seamless continuous chronological narrative but rather discussed a few examples of relations between landscape and personhood from different times and places in British prehistory. Personhood was negotiated differently with regard to the landscape and its features in each period. Some trends did endure over the long term, although they were exhibited through differing transformations of the landscape (see Bradley 1998). Many earlier Neolithic monuments seem to lend themselves to an experience of *local* place memorable relative to time of year (see below), but they became a series of similar places in the process (Cummings 2003). The architecture of later monuments more frequently extended beyond the local to draw on the paths of celestial bodies during events such as solstices. Henges in particular operated on a grand scale. Early Bronze Age monuments scaled the cosmos down to small ring ditches, timber circles, stone circles, and barrows, with even individual bodies arguably standing for community relations at times (Fowler 2005; Jones 2002). Paths, directions, and axes of personal journeys changed gradually. So, too, did the appropriate destination of personal attributes such as bone, flesh, and associated artifacts after death. A range of metaphors concerning the process of becoming, being, and ceasing to be a person (for example, bodies as vessels; relations between solid and fluid; or decay and regrowth) were manifested through differing engagements with places and landscapes, each of which served to situate the person in a community and cosmos.

Changing patterns of mobility provided different experiences of place—some communities were familiar with place during only one season, others familiar with fewer places for more seasons (Jones 1996; Pollard 1999). At certain times, places may carry different affective qualities tied to cultural beliefs and practices. Elsewhere, I have argued that cycles of decay, regrowth, and harvest at causewayed enclosures and mortuary structures in southern Britain were deployed in naturalizing

certain personal interactions and transformations. In a similar vein, Cummings and Whittle (2003: 262–63, 2004: 71–72) have also argued for seasonal differences in experience at chambered tombs in woodlands, suggesting an association between fleshed corpses and leafy trees and between bare winter trees and defleshed bone. These patterns demonstrate a link between cycles of personal transformation and seasonal changes with gatherings at monuments potentially conjoining key moments in both (cf. Borić's 2002 interpretation of red autumn foliage against the white cliffs around Lepenski Vir, above). Harrison's discussion on the "mood" of a contemporary Papuan village throughout the seasons is pertinent here, illustrating that seasonal differences in how landscapes look and feel bring to the fore distinctive reflections on personhood, the dead, and the passing of time (Harrison 2001). Memories and moods, like other features of personhood, are inextricable from social interaction with others (living and dead) through places and objects as much as they are felt in a body or perceived and recalled by a mind. Recent archaeological studies have focused on how personhood is connected to landscape through such patterns in experience and memory—Andy Jones's focus on different "technologies of remembrance" being particularly evocative here (e.g., Jones 2001, 2003).

The examples I have explored here differ from one another, but the generation of personhood, materiality, and landscape in other contexts may exhibit a far wider degree of difference than these (see also, for example, Jones 2005a). The nature of landscapes differs widely, as do ways of living in them and knowing about them. For the English, the landscape is known as *natural*, and the cultural actions of people are swallowed by nature over time—archaeologists recover the traces of such actions and so help "remember" a forgotten past (Harrison 2004: 144, 149). It is anticipated that people and events will be remembered through cultural inscription on the landscape. Among the Avatip, their riverine forest landscape is continually changing through a blend of the actions of human and nonhuman agents. Memory of the past is not involved in maintaining places by monumentalizing them—encountering the remains of the dead, including past places, artifacts, and architecture is inauspicious. Instead, the dead are remembered through oral traditions. Harrison (2004) highlights that in two cases such as these, landscapes are known, inhabited, and experienced in different ways.

Although I have focused on the cultural intersections between landscape and personhood, the means by which they come to be brought to bear on each other are ultimately dependent on the social and political negotiation of such cultural factors. No matter how basic a feature of existence personhood might seem to be, personhood and schemes of materiality are socially and politically negotiated. A vital and challenging area for ongoing research lies in tracing strategies through which different interest groups pursued personhood (Chapman 2000; Fowler 2005). This includes struggles over how personhood was negotiated through different ways of valuing and manipulating bodies, objects, and landscapes. The role of things, substances, and human and animal bodies in constituting personhood and tracing relations between people are crucial areas of study that need to be situated alongside the kind of landscape studies discussed here (e.g., Brück 2004; Chapman 2000; Fowler 2001, 2002; Jones 1998, 2002, 2005b; Thomas 1996: ch. 6, 1999). Equally, as Brück (2001) argues, consideration needs to be given to the use of monumentalized places in different contexts—from ceremonial gatherings to herding livestock to acts of deposition involving only one or two individuals. Each afforded different contexts for knowing a landscape and negotiating a sense of personhood.

Conclusions

Being a person is, in a Western world, generally differentiated from being an animal, an object, or a place; being a person is a matter of being human. However, personhood need not be conceptualized as solely the province of human beings, and the qualities that compose or characterize persons might be shared with animals, objects, and landscape features (Fowler 2004). The landscape may be rich in the presence of nonhuman persons and other kinds of beings; landscapes may be seen as the residue of their passing, made sacred by the presence of their bodies. Personhood and landscape are linked through specific schemes of materiality, acted on through particular practices and routines. A phenomenological approach that did not reflect on past ways of valuing (and negotiating the value of) bodies and material things but proceeded directly to recording experience of the landscape through contemporary Western understandings of materiality and embodiment would be insufficient to the task of interpreting past relationships between personhood and landscape. Being aware of this, many contemporary studies recognize that being a person is not an isolated state; it involves being-with-others and being-in-the-world. There are many different such ways of being, and the study of inhabited landscapes plays a crucial role in interpreting both past people's lives and past cultural worlds.

Acknowledgments

I would like to thank Kevin Greene and Sam Turner for discussions relating to this topic, and Vicki Cummings, Bruno David, and Elizabeth Kramer for comments on a draft of the chapter.

Note

1. One of the sculptures from house XLIV found lying on its back could be interpreted as having feet trussed together in the same way (a feature usually interpreted as genitalia), commemorating such a mortuary practice.

References

Bloch, M. 1982. Death, women and power, in M. Bloch and J. Parry (eds.), *Death and the Regeneration of Life*, pp. 211–30. Cambridge: Cambridge University Press.

———. 1995. Questions not to ask of Malagasy carvings, in I. Hodder, M. Shanks, A. Alexandri, V. Buchli, J. Carman, J. Last, and G. Lucas (eds.), *Interpreting Archaeology*, pp. 212–15. London: Routledge.

Borić, D. 2002a. Apotropaism and the temporality of colours: Colourful Mesolithic-Neolithic seasons in the Danube Gorges, in A. Jones and G. MacGregor (eds.), *Colouring the Past: The Significance of Colour in Archaeological Research*, pp. 23–43. Oxford: Berg.

———. 2002b. The Lepenski Vir conundrum: Reinterpretation of the Mesolithic and Neolithic sequences in the Danube Gorges. *Antiquity* 76: 1026–39.

Bourdieu, P. 1970. The Berber house or the world reversed. *Social Science Information* 9: 151–70.

Bradley, R. 1998. *The Significance of Monuments: The Shaping of Human Experience in Neolithic and Bronze Age Europe*. London: Routledge.

Brophy, K. 2000. Water coincidence? Cursus monuments and rivers, in A. Ritchie (ed.), *Neolithic Orkney in its European Complex*, pp. 59–70. Cambridge: McDonald Institute for Archaeological Research.

Brück, J. 1995. A place for the dead: The role of human remains in the Late Bronze Age. *Proceedings of the Prehistoric Society* 61: 245–77.

———. 1998. In the footsteps of the ancestors: A review of Christopher Tilley's "A Phenomenology of Landscape: Places, Paths and Monuments." *Archaeological Review from Cambridge* 15: 23–36.

Brück, J. 2001. Monuments, power and personhood in the British Neolithic. *Journal of the Royal Anthropological Institute* 7: 649–67.

———. 2004. Material metaphors: The relational construction of identity in Early Bronze Age burials in Ireland and Britain. *Journal of Social Archaeology* 4: 307–33.

———. 2006. Experiencing the past? The development of a phenomenological archaeology in British prehistory. *Archaeological Dialogues* 12: 45–72.

Busby, C. 1997. Permeable and partible persons: A comparative analysis of gender and the body in South India and Melanesia. *Journal of the Royal Anthropological Institute* 3: 261–78.

Chapman, J. 2000. *Fragmentation in Archaeology: People, places and broken objects in the prehistory of south-eastern Europe.* London: Routledge.

Cummings, V., and Fowler, C. 2004. The form and setting of Manx chambered cairns: Cultural comparisons and social interpretations, in V. Cummings and C. Fowler (eds.), *The Neolithic of the Irish Sea: Materiality and Traditions of Practice*, pp. 113–22. Oxford: Oxbow books.

Cummings, V., Jones, A., and Watson, A. 2002. Divided places: Phenomenology and assymetry in the monuments of the Black Mountains, southeast Wales. *Cambridge Archaeological Journal* 12: 57–70.

Cummings, V., and Whittle, A. 2003. Tombs with a view: Landscape, monuments and trees. *Antiquity* 77: 255–66.

———. 2004. *Places of Special Virtue: Megaliths in the Neolithic Landscapes of Wales.* Oxford: Oxbow Books.

Fowler, C. 2001. Personhood and social relations in the British Neolithic, with a study from the Isle of Man. *Journal of Material Culture* 6: 137–63.

———. 2002. Body parts: Personhood and materiality in the Manx Neolithic, in Y. Hamilakis, M. Pluciennik, and S. Tarlow (eds.), *Thinking through the Body: Archaeologies of Corporeality*, pp. 47–69. London: Kluwer/Academic Press.

———. 2003. Rates of (ex)change: Decay and growth, memory and the transformation of the dead in early Neolithic southern Britain, in H. Williams (ed.), *Archaeologies of Remembrance—Death and memory in past societies*, pp. 45–63. London: Kluwer Academic/Plenum Press.

———. 2004a. *The Archaeology of Personhood: An Anthropological Perspective.* London: Routledge.

———. 2004b. In touch with the past? Bodies, monuments and the sacred in the Manx Neolithic, in V. Cummings and C. Fowler (eds.), *The Neolithic of the Irish Sea: Materiality and Traditions of Practice*, pp. 91–102. Oxford: Oxbow.

———. 2005. Identity politics: Personhood, kinship, gender and power in Neolithic and early Bronze Age Britain, in E. Casella and C. Fowler (eds.), *The Archaeology of Plural and Changing Identities:*

Beyond Identification, pp. 109–134. London: Kluwer Academic/Plenum Press.

———. In press. Fractal bodies in the past and present, in D. Borić and J. Robb (eds.), *Past Bodies*. Oxford: Berghahn Press.

Fowler, C., and Cummings, V. 2003. Places of transformation: Building monuments from water and stone in the Neolithic of the Irish Sea. *Journal of the Royal Anthropological Institute* 9: 1–20.

Harrison, S. 2001. Smoke rising from the villages of the dead: Seasonal patterns of mood in a Papua New Guinea society. *Journal of the Royal Anthropological Institute* 7: 257–74.

———. 2004. Forgetful and memorious landscapes. *Social Anthropology* 12: 135–51.

Iteanu, A. 1988. The concept of the person and the ritual system: An Orokaiva view. *Man* 25: 35–53.

———. 1995. Rituals and ancestors, in D. de Coppet and A. Iteanu (eds.), *Cosmos and Society in Oceania*, pp. 135–63. Oxford: Berg.

Johnston, R. 1999. An empty path? Processions, memories and the Dorset Cursus, in A. Barclay and J. Harding (eds.), *Pathways and Ceremonies: The Cursus Monuments of Britain and Ireland*, pp. 39–48. Oxford: Oxbow Books.

Jones, A. 1996. Food for thought: Material culture and the transformation in food use from the Mesolithic to Neolithic, in T. Pollard and A. Morrison (eds.), *The Early Prehistory of Scotland*, pp. 291–300. Edinburgh: Edinburgh University Press.

———. 1998. Where eagles dare: Landscape, animals and the Neolithic of Orkney. *Journal of Material Culture* 3: 301–24.

———. 2001. Enduring images? Image production and memory in Earlier Bronze Age Scotland, in J. Brück (ed.), *Bronze Age Landscapes: Tradition and Transformation*, pp. 217–28. Oxford: Oxbow.

———. 2002. A biography of colour: Colour, material histories and personhood in the early Bronze Age of Britain and Ireland, in A. Jones and G. MacGregor (eds.), *Colouring the Past*, pp. 159–74. Oxford: Berg.

———. 2003. Technologies of remembrance: Memory, materiality and identity in Early Bronze Age Scotland, in H. Williams (ed.), *Archaeologies of Remembrance: Death and Memory in Past Societies*, pp. 65–88. London: Kluwer Academic/Plenum.

———. 2004. By way of illustration: Art, memory and materiality in the Irish Sea and beyond, in V. Cummings and C. Fowler (eds.), *The Neolithic of the Irish Sea: Materiality and Traditions of Practice*, pp. 202–13. Oxford: Oxbow Books.

———. 2005a. Lives in fragments?: Personhood and the European Neolithic. *Journal of Social Archaeology* 5: 193–224.

———. 2005b. Natural histories and social identities in Neolithic Orkney, in E. Casella and C. Fowler (eds.), *The Archaeology of Plural and Changing*

Identities: Beyond Identification, pp. 233–59. London: Kluwer Academic/Plenum Press.

Joyce, R. 2005. Archaeology of the body. *Annual Review of Anthropology* 34: 139–58.

Lucas, G. 1996. Of death and debt: A history of the body in Neolithic and Early Bronze Age Yorkshire. *Journal of European Archaeology* 4: 99–118.

LiPuma, E. 1998. Modernity and forms of personhood in Melanesia, in M. Lambek and A. Strathern (eds.), *Bodies and Persons: Comparative views from Africa and Melanesia*, pp. 53–79. Cambridge: Cambridge University Press.

Meskell, L., and Joyce, R. 2003. *Embodied Lives: Figuring Ancient Maya and Egyptian Experience.* London: Routledge.

Mortimer, J. 1905. *Forty Years Researches in British and Saxon Burial Mounds of East Yorkshire.* London: A. Brown and Sons.

Owoc, M. A. 2002. Munselling the mound: The use of soil colour as metaphor in British Bronze Age funerary ritual, in A. Jones and G. MacGregor (eds.), *Colouring the Past*, pp. 127–40. Oxford: Berg.

———. 2005. From the ground up: Agency, practice, and community in the southwestern British Bronze Age. *Journal of Archaeological Method and Theory* 12: 257–81.

Radovanović, I. 1996. *The Iron Gates Mesolithic.* International monographs in prehistory, Archaeological Series 11, Ann Arbor, Michigan.

———. 2000. Houses and burials at Lepenski Vir. *European Journal of Archaeology* 3: 330–49.

Richards, C. 1993. Monumental choreography: Architecture and spatial representation in Late Neolithic Orkney, in C. Tilley (ed.), *Interpretative Archaeology*, pp. 143–78. Oxford: Berg.

———. 1996a. Henges and water: Towards an elemental understanding of monumentality and landscape in late Neolithic Britain. *Journal of Material Culture* 1: 313–36.

———. 1996b. Monuments as landscape: Creating the centre of the world in late Neolithic Orkney. *World Archaeology* 28: 190–208.

Rublack, U. 2002. Fluxes: The Early Modern body and the emotions. *History Workshop Journal* 53: 1–16.

Srejović, D. 1972. *Europe's First Monumental Sculpture: New Discoveries at Lepenski Vir.* Aylesbury: Thames and Hudson.

Strathern, M. 1988. *The Gender of the Gift: Problems with Women and Problems with Society in Melanesia.* Berkeley and Los Angeles: University of California Press.

Thomas, J. 1996. *Time, Culture and Identity.* London: Routledge.

———. 1999. An economy of substances in earlier Neolithic Britain, in J. Robb (ed.), *Material Symbols: Culture and Economy in Prehistory*, pp. 70–89. Carbondale: Southern Illinois University Press.

———. 2004. *Archaeology and Modernity.* London: Routledge.

Tilley, C. 1994. *A Phenomenology of Landscape.* Oxford: Berg.

Wagner, R. 1991. The fractal person, in M. Strathern and M. Godelier (eds.), *Big Men and Great Men: Personifications of Power in Melanesia*, pp. 159–73. Cambridge: Cambridge University Press.

Watson, A. 2001. Round barrows in a circular world: Monumentalizing landscapes in Early Bronze Age Wessex, in J. Brück (ed.), *Bronze Age Landscapes: Tradition and Transformation*, pp. 207–16. Oxford: Oxbow.

30

ARCHAEOLOGY, LANDSCAPE, AND DWELLING

Julian Thomas

In this chapter, I argue that phenomenology cannot simply be drawn on for a methodology for landscape archaeology, commensurate with other techniques and methods. Rather, it requires that we should think about landscape in a wholly unfamiliar way: only then can the insights of an experiential approach enlighten or challenge more conventional perspectives. Over the past 15 years or so, phenomenological thought has come to exercise a considerable influence over the way that archaeologists address past landscapes. The phenomenological tradition in philosophy concerns itself with the conditions that make possible the human experience of the world, and it maintains that experience and interpretation are fundamental to human existence (Thomas 2006). It is not simply that people experience their world and make sense of it, as one kind of activity among others: we are distinguished by being interpreting beings. This perspective has informed approaches to space, place, and architecture that focus on the ways that topographies and structures might have been physically encountered and negotiated by people in the past, whether as a complement, or as an alternative, to more conventional landscape archaeologies (see Tilley, this volume). More recently, Andrew Fleming (2006) has presented a series of criticisms of this form of "postprocessual landscape archaeology" that demand serious consideration. Fleming contends that experiential archaeologies are unwise to neglect (or reject) the field skills that have been gradually perfected within orthodox landscape studies; that "phenomenological" fieldwork is often subjective, personal, and consequentially difficult to test or replicate; and that the written products of this kind of work gravitate toward a "hyper-interpretive" style that exceeds the capacity of the evidence to substantiate it (Fleming 2006: 276).

Fleming's observation that landscape archaeologists have always cultivated a fine-grained documentation of physical traces, which relies on immersion in the field over long periods of time, is entirely correct. Yet the problem that was originally identified with this kind of investigation is that the sense of the archaeologist's having "inhabited" and experienced a landscape in the course of fieldwork has often been missing from the written accounts that were offered, while the detailed empirical observations that were made tended to be subsumed within narratives that took the landscape itself as their object (Thomas 1993: 26). Thus, where the landscape was treated as a palimpsest of traces of changing economic regimes or systems of land tenure, and where the focus was on "things that have been done to the land" (Bender, Hamilton, and Tilley 1997: 148), the intimate scale of analysis that has traditionally provided archaeology's advantage over written history was in peril of being lost. Of course, a concern with landscape has always provided a potential antidote to a myopic concentration on arbitrarily defined (and reified) "sites." It is salutary to remember that people do not spend their entire lives on a single settlement site and that their routine activities may be dispersed over a wide area, linking a variety of different kinds of locales. In British archaeology, the practice of nesting excavations in projects conceived at the landscape scale goes back at least as far as Pitt Rivers's work at the Bokerley Junction (Bowden 1991: 118–19). Yet the danger is that the ability to focus on the rich texture of everyday life in the past will be surrendered to a concern with the long-term behavior of large-scale (and metaphysical) entities: social formations, estates, marketing systems, subsistence regimes, and so on. Furthermore, the perceived virtue of a landscape approach in archaeology has often been that it establishes an integrating framework that can bring together a variety of different classes of evidence: built structures, faunal remains, artifact distributions, pollen analysis, soils, written documents, and so forth. The problem is that this integration necessarily implies a quite particular understanding of what landscape is: a set of things or entities that can be objectively described.

If a phenomenological archaeology amounted to the substitution of a subjective for an objective investigation of this same *kind* of a landscape, Fleming's criticisms would be entirely justified. If a "postprocessual landscape archaeology" were no more than a consideration of how an independent and self-contained world of objects can be experienced by a human subject, it would have very little to recommend it. And indeed, it has to be admitted that a certain proportion of the flood of "phenomenological archaeologies" that followed in the wake of Tilley's *A Phenomenology of Landscape* (1994) amounted to little more than this: the reduction of phenomenology to a "technique" that could be applied to a given terrain alongside geographical information systems, fieldwalking survey, pedological analysis, and geomorphology. Once the archaeologist's question becomes "how might this landscape, which we have described by analytical means, have been perceived by past people," then the act of going out into the field to "have experiences" is simply license to an unbridled subjectivism, which is ultimately narcissistic. This kind of project sees as its objective the recovery or replication of the thoughts inside the heads of past people, which might be reconstructed from contemporary encounters with things and places.

In this chapter, a rather more radical view is proposed. An experiential analysis of landscape is unworkable unless it is placed in the context of an entirely different conception of landscape from that conventionally employed in archaeology. This is not to say that it cannot be utilized alongside established methods of analysis, or even that it cannot fruitfully make use of information gathered through the robust field methodologies that Fleming approves. Nonetheless, I suggest that traditional landscape archaeology and "postprocessual landscape archaeology" are not complementary, alternative ways of investigating the same phenomenon, and that the latter necessarily connects with what Tim Ingold (1995: 75) characterizes as "the dwelling perspective."

From Building to Dwelling

The force of the argument for rejecting the definition of landscape as an aggregation of cultural and natural features built up over time is that it is a distinctively modern Western notion that is anachronistic when applied to the distant past. That such a "landscape" is primarily apprehended visually, by a distanced observer, is perhaps only a symptom of a more fundamental problem. In the period since the Renaissance, space has come to be understood as an internally homogeneous medium within which objects are contained, such that their relationships with one another can be expressed in geometrical terms (Jay 1993: 114; see also Casey, this volume). Yet such a view has displaced an earlier, Aristotelian cosmology in which meaning and moral value were considered to be intrinsic to the world. Consequentially, the world required interpretation rather than description (Thomas 2004: 8–9).

The Scientific Revolution and the Cartesian philosophy that it drew on prioritized vision both because they valued the disengagement of the impartial observer and because they considered meaning to be an exclusive property of the inner realm of the mind. The external world was characterized by relations of extension rather than of intelligibility, and the senses provided the means by which the mind acquired raw information, which it might then render meaningful. Among these, sight had the privilege of gathering data from a position of detachment, appreciating spatial relationships and identifying classificatory order. So the emphasis on the visual in the modern conception of landscape (e.g., Cosgrove 1984; Olwig 1993) relates at once to the "disenchantment" of the world, the valorization of objectivity, and the separation of the mind from physical reality. And yet in everyday life, it is not just sight that identi-fies what we make of the world and that identifies our landscapes. Rather, our place in the world is made manifest, apparent, and capable of reflec-tion through varied and ongoing cues: the sounds of the birds may announce the waking morning; rhythmic poundings may announce the prepara-tion of kava and sacred men's business in various Melanesian societies (e.g., Rainbird 2002); the fall-en, dry crispy leaves underfoot signal the changing seasons. Each of these cues, and more, alerts us to the nature of the world in which we live. These are not distanced cues, but they rather identify each of us with-others-in-the-world. There is, therefore, a tension between the distanced visual landscapes theorized in Western systems of knowledge and the landscapes that we experience.

These developments stemming from Enlighten-ment thinking are characteristic of the emergence of the "building perspective," a way of understand-ing the world in which empirical reality is under-stood to be entirely independent of and prior to any degree of human involvement (Ingold 1995: 66). This implies that the landscape is first of all given as a set of material resources that people subsequently begin to exploit or inhabit. Equally, this view holds that a distinction can be drawn between the landscape "as it really is" and the landscape as it is perceived, which is presumably built up in the minds of people on the basis of their observations (Ingold 2000: 168). Where the landscape is a purely material given, it is likely to be identified with "nature," while its representa-tion in the mind (and subsequent material repre-sentation or transformation through art or craft) is identified as "culture." The building perspective therefore combines the culture/nature dichotomy with a form of cognitivism in which human beings construct internal symbolic worlds that are quite distinct from reality. An important corollary of this view is that the perception of the world is always considered to be indirect or mediated, with sense-impressions being filtered through a cognitive apparatus or a symbolic order of some kind.

These views are pervasive within archae-ology. For instance, it is commonplace to distinguish between "space" and "place," whereby the former refers to the undifferentiated condition of the world before human beings encounter and utilize it, and the latter refers to the outcome of human "social-ization" or "enculturation" (e.g., Chapman 1988). In other words, human beings bring meaning to unformed space, and in the process transform it into significant and distinctive place—by naming, build-ing structures, or making clearings, for example (for a critique of this perspective, see Casey 1996). What

this would mean is that when human beings first come across a particular location, their experience of it takes the form of the acquisition of sense-data, which have an information content but no mean-ing. Only latterly does a location gather its signifi-cance. Similarly, environmental archaeologists often talk of the "human impact" on the landscape. This gives the impression that human beings are extrin-sic to the natural world and that they discontinu-ously inflict "impacts" on it. Although not causing such impacts, they presumably occupy some other purely cultural sphere.

If this "building perspective" is characteristic of Western modernity, it can be contrasted with the notion of "dwelling" found in the work of the philosopher Martin Heidegger. Heidegger con-tends that it is unhelpful to imagine that we must first build structures before we can begin to dwell; rather, dwelling is the condition that humans experience when they are at home in the world (Heidegger 1993: 350). Ideally, architecture (and construction in general) should be an outgrowth and embodiment of this state of dwelling, rather than representing an imposition onto the mater-ial world of designs that have been produced in the abstract. For Heidegger, dwelling is a relation-ship with the world characterized by equanimity, in which one cares for and preserves one's sur-roundings while also allowing them to *be them-selves*, without bending them to one's design (Young 2002: 99). It follows that the willful for-cing of materials into some kind of template, as in the mass production of commodities, amounts to a form of violence. Dwelling is at once *caring for* and *being cared for*, a reciprocal relationship that allows the physical world to reveal its sacred char-acter. Of course, "sacredness" is precisely the kind of meaningful content that a post-Cartesian ontol-ogy would reject outright in the case of the object world. Perhaps more important, this point hints at the distinction between a landscape that is under-stood as a collection of isolated entities, and one that is fundamentally relational, an issue to which we will return. For Heidegger, the predicament of humanity in the late modern age is that we have ceased to be dwellers and to be at home in our sur-roundings (Young 2002: 74).

This brings us to an apparent contradiction in Heidegger's arguments, for he suggests that dwelling is characteristic of human existence, that humans are fundamentally *dwellers*, but that also under modern conditions we have lost sight of dwelling. For Heidegger the "loss of dwelling" is the key symptom of the destitution of modernity (Young 2002: 33), and this would appear to intro-duce the possibility of a state of "nondwelling."

However, the argument is actually that modern metaphysics restricts the ways in which the world can reveal itself to us, by reducing people and things to subjects and objects. The landscape, for instance, appears to be composed exclusively of resources that are at our disposal for consumption and gratification (as lumber, fuel, hardcore, real estate, building stone, and so on), so that we assume a position of exteriority toward it and can no longer truly inhabit it. It may be helpful to compare this argument to Marx's conception of alienation (see Ollman 1976), in which capitalist economic processes obscure the relationship between people and the products of their labor, so that the latter become reified as asocial, freestanding entities. Heidegger is effectively describing a more thoroughgoing form of alienation in which dwelling is occluded, and a caring and nonviolent relationship with worldly things becomes more difficult to achieve. We might say that this is not so much a state of nondwelling as a deprivative, etiolated, or inauthentic form of dwelling in which people experience rootlessness and continual anxiety.

Relational Landscapes

If we accept that the view of the landscape as an aggregation of self-contained entities is a modern imposition, which may not even adequately convey the way that most people today experience their surroundings, we should explore the implications of a landscape that is at once an internally interconnected whole, and inherently meaningful. The first point is that "places" are not created through a human bestowal of meaning, but emerge from the *background* of a landscape that people always *already* understand to some degree. People do not perceive worldly things through the misty gauze of a cultural filter, but nor are they capable of an entirely innocent reading of the land, severed from history and tradition. As Tilley (1994: 13, 2004: 10) has argued, the "discovery" of places is achieved through the human body. We encounter places when we are already in the midst of them, corporeally and spatially located and living through our own senses. However, it should be apparent that this is not the same thing as saying that a human subject encounters and appropriates a worldly object. On the contrary, it is the inherited and sedimented familiarity of the landscape as the context within which everyday activities are performed that enables specific places to reveal themselves in ways that are readily intelligible. Even those places that we have not visited before are generally experienced in the context of everyday engagement

(whether in strolling, searching for firewood or mushrooms, tending livestock, or seeking a path through the woods). Their meaning is not generated out of an abstract and distanced observation of an array of discrete entities, or pieced together from an accumulation of nuggets of information. Rather, our knowledge of the landscape develops in an implicit way, a general understanding of the whole preceding and contextualizing any specific observation.

We might say that people's understanding of their landscape is "pre-intentional" (Wrathall 2000: 94), in that it is not held in the head but realized only in a specific concrete setting. Following a forest path is something that is achieved in the practical event of placing one's feet on the path itself, and under these circumstances it does not need to be "thought through" in abstract terms at all. Yet, it is this kind of nonrepresentable understanding that renders intentional acts within the landscape possible. Actions such as shooting a deer, building a shelter, or knapping a flint blade are able to be conducted because at any given time people have at their disposal a range of traditions, skills, cues, and understandings that would be impossible to verbalize in their entirety. This unarticulated background inheres in the physical presence of the landscape as much as in the human body or the mind.

In the philosophical literature, there has been considerable discussion of the way that implicit bodily skills and habitual practices provide the unconsidered background for human action (Taylor 1993, 2000; Wrathall 2000). These skills and practices are hard to represent as algorithms or explicit instructions, and this undermines the credibility of cognitivist accounts of human functioning. But it is arguable that too little attention has been directed toward the status of the material world, and specifically the lived landscape, as part of this background. If we are to argue that the understandings that people routinely employ in coping with their world do not amount to a set of representations contained in the head, then it is clear that places and objects are implicated in the way that our explicit projects are formulated. This is not the same thing as saying that the objects that surround us at any given point in time constitute a kind of store, from which ideas or representations can be withdrawn (as in the notion of "external symbolic storage:" [Donald 1991: 316]). On the contrary, our existence permeates the places we presently occupy, have occupied, and plan to occupy in future. It is the physical world as it is known to us, the landscape of recalled past happenings and planned future doings, that provides an integral element in all human action. Modern Western thought often implicitly assumes that human beings are entirely self-contained entities, whose faculties are logically independent of their surroundings. But any person who did not inhabit a material world (imagine a body floating in a dark void—or even a creature with no corporeal existence—surrounded by no material things at all) would find it impossible to act and would be unable to formulate any projects for the future. There would be no motion *toward* anything, temporal or spatial, as there would be no emplacement in any meaningful location nor any emergence in any meaningful temporality. Relationality (spatial and temporal) would disappear.

When Ingold refers to the "temporality of the landscape," and stresses that it is unrelated to clock-time (1993: 158), he has in mind this embedding of protention and retention in the lived environment. As he says, any activity that is conducted in the present is situated and rendered comprehensible in relation to past happenings, which are manifested or called to mind by the landscape. In this sense, people inhabit the past (although I would add that they also inhabit the future, in the sense of dwelling in the locations of their projected actions, and sensing a momentum of life beyond the now). Barrett (1999: 256) illustrates the archaeological importance of such a perspective by pointing out that while we conventionally assign monuments of particular types to separate chronological horizons, and present them as representative of specific stages of social evolution or socioeconomic structures, their persistence over time forms the preexisting context in which subsequent construction is undertaken. Thus for Barrett, the monumental cemeteries of the British Bronze Age formed the background against which the constructional activities of the Iron Age were played out. Similarly, I have argued elsewhere that the beginning of the Neolithic period in Britain (ca. 4000 B.C.) involved the introduction of a series of unfamiliar material and symbolic resources (ceramic vessels, polished stone tools, a variety of forms of public architecture including large timber halls and megalithic chambers, and domesticated plants and animals) into a variety of landscapes that had previously been formed and understood through the rhythms and know-how of the Mesolithic (Thomas 2007: 431). In this setting, these innovations were recontextualized, and yet their presence gradually brought about the transformation of the British landscape to a pattern dominated by stock-rearing and (eventually) the cultivation of cereals.

The landscape is historical in that erosion, flooding, vegetational change, deforestation, grazing, cultivation, and building are continually transforming it. However, rather than representing a succession of static phases that are to be unpeeled from the earth's surface by the archaeologist, the changes wrought by these processes are the means by which the landscape hands itself down to the present. At any given point, the landscape is in temporal motion and presents a shifting horizon within which actions are carried out, and places "show up" as distinct and comprehensible locations. Any place is disclosed within the totality of a landscape, and this means that the significance of each place is subtly altered as its constitutive background is transformed. For instance, a stone burial cairn constructed in a woodland clearance becomes an entirely different kind of place when the trees surrounding it are felled, and its position on a hilltop can be appreciated from a distance. Nonetheless, the continuing presence of the cairn provides an enduring reminder of events that took place before the clearance occurred.

A Hermeneutics of Landscape

The subject/object relationship exercises a powerful influence in the contemporary world. Phenomenological thought has inspired modes of archaeological investigation that privilege the immediacy of experience over abstract description, yet it is disturbingly easy for analyses that concern themselves with the ways that places and landscapes can be occupied and moved through to be reduced to exercises in replicating the perceptions of past people (a tension that is evident in the exchange between Ingold and Tilley [2005]). I have argued that this problem cannot be resolved if we imagine that this experiential archaeology is an alternative means of investigating landscapes that are conceived as arrangements of inert matter. There is nothing wrong in mapping, photographing, digitizing, and surveying an area of the earth's surface, or incorporating it into a geographical information system (but see Byrne). But it is a mistake to suppose that these techniques provide information about the same kind of entity as that which a phenomenological archaeology might address. If we ask how such objective, meaning-free terrains were perceived by prehistoric communities, the answer must necessarily be expressed in the form of a cognized model—the content of a prehistoric mind. It follows that if we wish to eschew the mind-body dichotomy, we need to recognize that the "objective" topography is quite different from the landscape that makes up the context of human dwelling. It is presumably the latter that an archaeology that concerns itself with experience, occupation, and bodily practice seeks to investigate. Furthermore, we should not imagine that the physical terrain that constitutes the analytical object for conventional archaeologies of landscape represents an established, preinterpretive bedrock, upon which the lived and interpreted landscape is built. On the contrary, it is more like an analytical construct, extracted from the lived and experienced landscape of the present day through a process of conceptual reduction.

It follows that questions such as "what was the symbolic meaning of this structure?" may be inappropriate for an "archaeology of dwelling." Instead, the central concern would be with how a landscape was occupied and understood, and how it provided the context for the formulation and enactment of human projects. "How did people relate to it?" in different contexts of engagement may be a more appropriate question. This is nonetheless an interpretive enterprise, because it seeks to develop an holistic understanding of the past landscape in the present, analogous to (if distinct from) the understanding that past people themselves may have developed. Our own experiences in walking landscapes and encountering the physical traces that past communities have left behind undoubtedly have a place in this kind of investigation. However, we should be mindful that the "background" against which ancient objects and structures reveal themselves to us is a largely modern one, composed of contemporary skills, understandings, and practices of which we may be only partially aware. Our experience of a place or artifact in its landscape context is of value because the thing itself is more than the product or outcome of an extinct pattern of social life. On the contrary, it represents an integral and still-extant element of that pattern. But appreciating its significance requires that the part should placed into as complete as possible a whole. An "experiential archaeology" should therefore be conceived as only one aspect of a kind of "hermeneutics of landscape," in which *how a phenomenon presents itself to us in the present* is only one step in attempting to understand *how it might have presented itself in a past context*. This means that our field observations need to be set against a picture of the subsistence practices, technological capabilities, patterns of mobility, gender roles, ritual observances, and conceptions of personhood that characterized past societies. If the landscape ("natural" and built) forms a significant part of the framework of meaning that contextualizes action and renders entities comprehensible, it does so only as human practices and projects are threaded through it. The task of the archaeologist is to reintroduce these to the past landscape through the work of interpretation. *Dwelling* is what happens when traditions of practice find themselves at home in a landscape, producing a climate of expectation and assumption within which future projects can be devised and carried forward. Archaeology imagines the past by placing contemporary observations and experiences into as complete as possible a reconstruction of the factors that informed their ancient counterparts.

References

Barrett, J. C. 1999. The mythical landscapes of the British Iron Age, in W. Ashmore and B. Knapp (eds.), *Archaeologies of Landscape: Contemporary Perspectives*, pp. 253–68. Oxford: Blackwell.

Bender, B., Hamilton, S., and Tilley, C. Y. 1997. Leskernick: Stone worlds; alternative narratives; nested landscapes. *Proceedings of the Prehistoric Society* 63: 147–78.

Bowden, M. 1991. *Pitt Rivers: The Life and Archaeological Work of Augustus Henry Lane Fox Pitt Rivers*. Cambridge: Cambridge University Press.

Casey, E. S. 1996. How to get from space to place in a fairly short stretch of time, in S. Feld and K. H. Basso (eds.), *Senses of Place*, pp. 13–51. Santa Fe, NM: School of American Research Press.

Chapman, J. C. 1988. From "space" to "place": A model of dispersed settlement and Neolithic society, in C. Burgess, P. Topping, and D. Mordant (eds.), *Enclosures and Defences in the Neolithic of Western Europe*, pp. 21–46. Oxford: British Archaeological Reports 403.

Cosgrove, D. 1984. *Social Formation and Symbolic Landscape*. London: Croom Helm.

Donald, M. 1991. *Origins of the Modern Mind*. Cambridge, MA: Harvard University Press.

Fleming, A. 2006. Post-processual landscape archaeology: A critique. *Cambridge Archaeological Journal* 16: 267–80.

Heidegger, M. 1993. Building, dwelling, thinking, in D. F. Krell (ed.), *Martin Heidegger: Basic Writings*, pp. 423–40. London: Routledge.

Ingold, T. 1993. The temporality of the landscape. *World Archaeology* 25: 152–74.

———. 1995. Building, dwelling, living: How animals and people make themselves at home in the world, in M. Strathern (ed.), *Shifting Contexts: Transformations in Anthropological Knowledge*, pp. 57–80. London: Routledge.

———. 2000. *The Perception of the Environment*. London: Routledge.

Ingold, T. (and response by C. Tilley). 2005. Comments on C. Tilley: The Materiality of Stone. *Norwegian Archaeological Review* 38L: 122–29.

Jay, M. 1993. *Force Fields: Between Intellectual History and Cultural Critique.* London: Routledge.

Ollman, B. 1976. *Alienation: Marx's Conception of Man in Capitalist Society* (2nd ed.). Cambridge: Cambridge University Press.

Olwig, K. 1993. Sexual cosmology: Nation and landscape at the conceptual interstices of nature and culture; or what does landscape really mean?, in B. Bender (ed.), *Landscape: Politics and Perspectives,* pp. 307–43. London: Berg.

Rainbird, P. 2002. Making sense of petroglyphs: The sound of rock art, in B. David and M. Wilson (eds.), *Inscribed Landscapes: Marking and Making Place,* pp. 93–103. Honolulu: University of Hawai'i Press.

Taylor, C. 1993. Engaged agency and background in Heidegger, in C. Guignon (ed.), *The Cambridge Companion to Heidegger,* pp. 317–36. Cambridge: Cambridge University Press.

———. 2000. What's wrong with foundationalism? Knowledge, agency and world, in M. Wrathall and J. Malpas (eds.), *Heidegger, Coping, and Cognitive Science,* pp. 115–34. Cambridge, MA: MIT Press.

Thomas, J. S. 1993. The politics of vision and the archaeologies of landscape, in B. Bender (ed.), *Landscape: Perspectives and Politics,* pp. 19–48. London: Berg.

———. 2004. *Archaeology and Modernity.* London: Routledge.

———. 2006. Phenomenology and material culture, in C. Tilley, W. Keane, S. Küchler, M. Rowlands, and P. Spyer (eds.), *Handbook of Material Culture,* pp. 43–59. London: Sage.

———. 2007. Mesolithic-Neolithic transitions in Britain: From essence to inhabitation, in A. Whittle and V. Cummings (eds.), *Going Over: The Mesolithic-Neolithic Transition in North-West Europe.* London: British Academy.

Tilley, C. Y. 1994. *A Phenomenology of Landscape: Places, Paths and Monuments.* London: Berg.

———. 2004. *The Materiality of Stone: Explorations in Landscape Phenomenology.* Oxford: Berg.

Wrathall, M. 2000. Background practices, capacities, and Heideggerian disclosure, in M. Wrathall and J. Malpas (eds.), *Heidegger, Coping, and Cognitive Science,* pp. 93–114. Cambridge, MA: MIT Press.

Young, J. 2002. *Heidegger's Later Philosophy.* Cambridge: Cambridge University Press.

31

BUILDING AND ARCHITECTURE AS LANDSCAPE PRACTICE

Lesley McFadyen

Over the last two decades, British prehistorians have created a distinctive social archaeology, and archaeological enquiry has increasingly focused on how people in the past perceived their world. Borrowing heavily from both anthropology (e.g., Hirsch and O'Hanlon 1995; Ingold 1993) and social geography (e.g., Hägerstrand 1970; Pred 1981), there has been a move within prehistoric research toward more people-centered, experiential, and phenomenological approaches in which landscape exists as a medium created through narrative, inhabitation, taskscape, social memory, and sensory experience (e.g., see Thomas, this volume). Landscapes have been considered as a network of places, as a medium through which places were socially constructed and made meaningful.

Within early Neolithic studies there exists a strong interest in developing a landscape approach into which is incorporated studies of monuments and material culture (e.g., Barrett, Bradley, and Green 1991; Cleal and Pollard 2004). However, architecture has traditionally dominated studies of the period, and landscape is too frequently regarded as little more than a setting for monuments. Although the biography of a long barrow or causewayed enclosure site can be discussed in great detail, as an object this architecture has become situated very firmly *in* the landscape. When architecture is viewed as an object in this way, attention is focused on an overall

orientation, shape, and final form. Such a perspective presumes a unilinear trajectory from design to end product. In due course, these architectures then provide another kind of setting for material culture, so by the time artifacts are incorporated into the study they can be understood only in terms of how they were deposited inside and/or around the built architectural object.

The latter approach has an evident nested quality to its working, with each object located inside the next like a series of Russian dolls: landscape, architecture, material culture (see Figure 31.1). Rather than confronting this question of spatial scale as integral to understanding and enquiry (see also Head, this volume), this approach instead fixes scale as archaeological accounts coalesce around the resultant objects (e.g., Barrett 1994a; Cummings and Whittle 2003; Thomas 1999; Tilley 1994, 2004). The problem with establishing and enforcing these scales at an analytical level is that the resultant rigid divisions cut across an understanding of the diverse scales at which people themselves operated. This is incommensurate with the lives that social archaeology is trying to investigate in the first place.

In asking whether division into three discrete scales of inquiry—landscape (large), architecture (medium), and material culture (small)—is the best or only relationship through which to approach articulating scales of working and how

Figure 31.1 "Russian dolls" (photo by Marisa Clements).

things and people connected together in the past, this chapter proposes a new approach to thinking about this problem. Rather than treating each scalar level as separate, I look at how they intersect. I examine the notion of *architectures as landscape* in order to consider more directly how past people themselves concretized different levels of action.

Questioning Located Objects

Suggesting a more fluid approach to the question of changing scales is one thing; putting it into practice is another. For example, what do you make your point of departure if landscape is not to exist as a more or less regional container? One approach would be to consider archaeological sites as points along paths and not as central places in the landscape. For example, John Barrett has written that:

> The non-megalithic and megalithic long mounds and the causewayed enclosures did not lie at the centers of areas of land

surface, but were instead places at the ends and at the beginnings of paths. These were the intersections of paths; places of meeting and departure which may have been part of a wider and seasonal cycle of movement. (Barrett 1994b: 93)

To expand this would mean that they are constituted not through their point of location in a landscape container but through a relational context of engagement with their surroundings. In this chapter, I take this idea a step farther, with the aim of destabilizing our existing (and rigid) scalar approaches by arguing that architectures as landscapes were made from fragments of distributed practice.

Central to this is an account of what I term *dispersed space*, a space made through a series of tasks that range from the making and using of flint tools, the setting of fires, the working of wood into posts, the cutting of quarry pits, the butchery of animals, and the ways in which these things were also entwined with "recognized" building materials such as turf and chalk. The evidence presented here suggests that there should be no endpoint to what we perceive architecture or indeed landscape to be, because architecture extended outward and so was caught up in other parts of the landscape. Areas of flint working were needed for tools; areas of tree-fall and woodland clearance were needed for stakes; pathways and pasture were required for people to manage their herds; and so these features were also part of these distributed sites.

Architecture as Distributed Practice

What if architecture is not defined exclusively in terms of an object-like building that is then used but is truly considered as a medium for action? What of an architecture that exists through construction, through practices of making? Archaeologists have long been confused about how to explain the earliest recognized activities at long barrow sites, so

these practices have often been termed "prebarrow" activities and then simply sidelined or ignored. I argue that one aspect to our study of these sites should focus on precisely these early practices and how they generated time gaps between events of construction, gaps that can shed crucial light on factors such as mobility and how people went about living their lives. This may be illustrated by the long-barrow site of Beckhampton Road in Wiltshire (Ashbee, Smith, and Evans 1979).

On a small chalky ridge of glacial drift deposits located in a gentle valley, long-standing grassland had been cut into and stakes erected. After the stakes had rotted, a large fire setting was created, over 4 meters in length and nearly 2 meters wide, which was then smothered by dumps of soil and turf (see Figure 31.2). Time was marked in evident ways at this site. There was, for example, grassland that became established over a long period

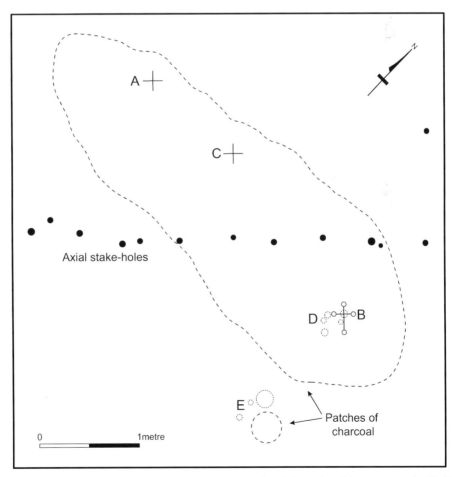

Figure 31.2   Plan showing "prebarrow" features and axial stakeline of Beckhampton Road, Wiltshire (Ashbee, Smith, and Evans 1979).

of time, and then came the cutting down of wood into stakes, which was followed by a period when things were allowed to decay. The setting of a fire and the burning of wood, then the addition of soil and turf and the quick smothering of the flames followed this period of inactivity. Then there was another delay or gap in the record of the past at this place.

These time gaps were not simply fortuitous but were a central part of the architectural process. The generation of a time gap was also a part of "building" as an active practice. It would be wrong to understand the context in which these actions took place as simply a series of progressive activities that produced events in construction with gaps in between. As the architect Bernard Tschumi has put it, *both* event and disjunction are crucial parts of architecture (Tschumi 1996).

> In architecture such disjunction implies that at no moment can any part become a synthesis or self-sufficient totality; each part leads to another, and every construction is off-balance, constituted by the traces of another construction. (Tschumi 1996: 212)

If gaps and disjunction are also a part of architecture, then building activities always have a disconnection and an elsewhere. To put this another way, the force of action goes out and on in the world, and in this way the space that emerges from the act of construction comes to be distributed (after Deleuze and Guattari 1992).

Architectures as Landscape

From the buried soil at the site of Beckhampton Road, land snail fauna suggest an initial cover of woodland that was cleared, giving rise to a grassland environment (Ashbee, Smith, and Evans 1979: 281). From the buried soil at the nearby site of South Street long barrow, there was evidence for woodland from tree-root casts and their land snail fauna. The pollen suggested a dominance of hazel, with oak, elm, birch, alder, and pine (ibid: 295). At Beckhmapton Road, the stakes were driven through long-standing grassland, so the wood for the stakes was chopped down from trees elsewhere. There was then a temporal disconnection while these timbers decayed. Oak trees were chopped down, or tree-fall managed, and this wood was brought to the area and burned in a large fire setting. In tandem with these activities, but from elsewhere, soil or turf was quarried or cut and dumped on the fire. The precise location of where these "off-site" tasks took place cannot be known; it is enough to know that the many tasks taking place here incorporated materials that were distanced spatially from the area of grassland or from one and other temporally.

This situation can be further complicated by thinking about the flint tools, such as axes, that would have been needed during any kind of subtle woodland management. Artifacts recovered from the buried soil of Beckhampton Road comprised only eight flakes of flint, so there was no evidence for axe manufacture or the management of such tools in the form of axe-sharpening flakes directly at the site. So woodland had been cleared at Beckhampton Road, and specific flint tools would have been required to do this, but although evidence for the working of flint can be found at nearby sites such as South Street (ibid: 270-72), it is not taking place at Beckhampton Road, and the only site where any part of axes was recovered was Windmill Hill , over 3 km away (Whittle, Pollard, and Grigson 1999: 331).

It is important to look again at the way time was marked at this site: this was not as clear succession of events but was instead interrupted by significant gaps and delays, which were then followed by practices of making. Early practices generated time gaps between events in construction, and so we need to consider what these conditions might be saying about mobility and how people went about living their lives. If people were living mobile lives at this time, as has been argued by Alasdair Whittle (1990) and Joshua Pollard (2005), then should not these various and noncontinuous activities be considered as a part of architecture?

Although the time gaps were not drawn out over as long a period at earlier Neolithic pit clusters, recent work on these sites provides a useful way of thinking about how to interpret the kinds of contexts in which these actions took place. From the pit cluster site of Kilverstone in Norfolk, parallel tempos to past lives have been noted, and a disjunction has been identified between them. For example, there was the relatively slow tempo and complex process of accumulation of different materials in the prepit context, with a period of delay, before the fast tempo of pits being cut and then immediately backfilled (Garrow, Beadsmoore, and Knight 2005: 13). Of the time gap between these parallel tempos, Garrow and colleagues write: "The picture we have of the site involves extended occupation overall. It has been argued that this occupation was repeated and persistent but not continuous" (ibid: 18). The gap marks the time when people were absent from the site.

Earlier Neolithic sites in Britain include a high proportion of domesticates, and so we must be careful, as Richard Bradley has pointed out, not to conflate hunter-gatherer and pastoral ways of living just because they practice a mobile economy involving animals (Bradley 2003: 220). But we do have to consider the nature of the mobility of pastoralists and the kinds of architectures or spaces that they would have produced. What is clear from all of this is that if you consider architecture as practice rather than object, with disjunction as a vital part of that practice, then you are forced to take a different landscape approach, one created through points of departure and dispersal rather than emerging from a consideration of nested qualities of a location and staying put.

Let us return to Beckhampton Road, at the point where further stakes were erected obliquely across the smothered fire setting (see Figure 31.2). After the stakes had been set, turf was stacked on the other side of the wooden partition. A fleshed cattle skull was pinned within a matrix of chalk rubble, and then turves were stacked around it. Twenty meters northeast from here another fleshed cattle skull (with the cervical vertebrae still attached) was also incorporated into chalk rubble (see Figure 31.3). This time, connections were made quickly because the cattle skulls, which were hanging from stakes, were still articulated when they were incorporated into the rubble. In turn, the rubble and the stake line were pinned in place by stacks of turf.

On first consideration, what I have described would seem to be an intensely entwined assemblage of things that would have marked out this space as a specific place or location. Yet, I would argue that disjunction is still a part of practice, albeit of a different nature to the previous discontinuities outlined above. For example, between the cattle and the grassland there was now an accumulation of things that included a smothered fire setting, dumps of soil, and then chalk rubble. From spaces where their herds moved over established areas of grassland to dead cattle with hides and fleshed skulls hanging from stakes over scorched earth concealed by dump after dump of quarried rubble, people's perceptions were changed. Space had shifted because the quality of engagement people had with things had altered; put very simply, people were treating their cattle in a very different way and incorporating them into the building process—someone was holding the animal's carcass on a pole while another person put additional material on top of this. The effects that this shift in space would have had on people must not be ignored, because it made concrete the disconnection between the living animal and its body as architectural material. Chris Gosden has written of the effects of objects:

> An object with new or subversive qualities will send social relations off down a new path, not through any intention on the part of the object, but through its effects on the sets of social relations attached to various forms of sensory activity. (Gosden 2001: 165)

I now extend this way of thinking into further research that operates beyond the extent of one long barrow. By what means could it have been

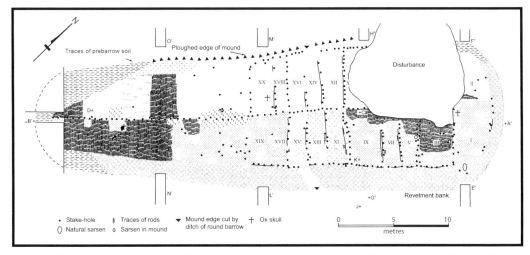

Figure 31.3 Beckhampton Road, Wiltshire: Plan showing "prebarrow" and barrow features (Ashbee, Smith, and Evans 1979).

possible for fragments of distributed practice to matter at a larger landscape scale? There are several other long barrow sites in the region (Ashbee, Smith, and Evans 1979; Whittle 1994; Whittle, Rouse, and Evans 1993), as well as causewayed enclosure sites (e.g., Whittle et al. 1999) (see Figure 31.4). All these sites have evidence of earlier practices of making that have been termed "prebarrow" or "preenclosure" and to date largely ignored (although see Pollard 2005: 106–07 for a more integrated approach to earlier activities and "recognizable" architectural practices). Similarly, there are large numbers of flint scatters that have been recorded at the Wiltshire Sites and Monuments Record, and sites with evidence for occupation (Smith 1965a, 1965b; Smith and Simpson 1966; Whittle et al. 2000). The next step is to break down differences that exist between these data sets that have been created through general phasing and site typology, and then to start to connect various tasks across sites while simultaneously noting discontinuities of practice at the site level.

For example, pits were cut at the Millbarrow and Horslip long barrow sites, and at the site of Windmill Hill causewayed enclosure. At Millbarrow, pits 401 and 548 had fragments of human bone incorporated into their matrix (Whittle 1994: 18), and at Windmill Hill there was the exposure of a least one body in a pit (Whittle, Pollard, and Grigson 1999: 350–52). One of the most complicated and interconnected assemblages comes from the earlier practices of pit-cutting at the site of Horslip long barrow, where a group of nine pits formed an arc at the proximal end of the site. It is particularly interesting to note that material culture was involved in the constructional process only in the most connective areas, that is, where pits were cut or recut to physically connect other pits together. Material culture constituted a major component of the matrix of pits 2 and 7, where flint was knapped *in situ* when two pits were joined together through the cutting of a third pit (McFadyen 2006). Material culture was also involved in the act of connecting features together at Windmill Hill, where large quantities of material, including worked flint, chalk, and sarsen, came from pits 36, 37, 41, and 42 (Whittle, Pollard, and Grigson 1999: 352).

It is therefore possible to connect various tasks across sites, but I also want to draw attention to the

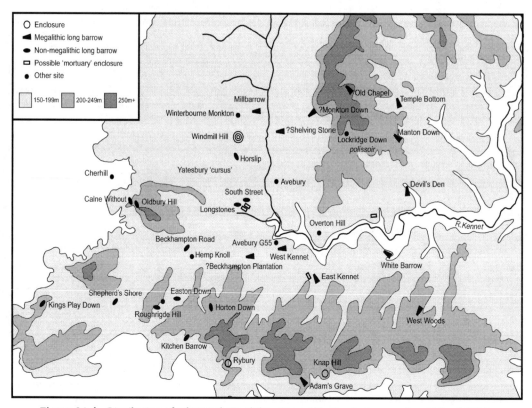

Figure 31.4   Distribution of other Early Neolithic sites within the Avebury region (Pollard 2005).

way space is created and how it is given dimension alongside these events. Evidence of site level "space making" takes the form of fragments of distributed practice, and spatial and temporal discontinuities. Rather than sites being seen solely as central places acting as foci for activity, they must instead be seen as constituted through discontinuities and connections, the latter made through practices extending elsewhere into other times. Therefore, only the disjunctions are impressed on the site (Tschumi 1996: 209). As archaeologists, we can and should attempt to trace the ways in which different kinds of practice intersect with other people and extend into additional activities across sites. The challenge is to develop an archaeological program that can trace exactly where and at what time specific connections were made.

Conclusions

In treating building and architecture not *in* landscapes but *as* elements of landscape engagement, my aim has been to demonstrate how attending to architecture as practice might be a useful way to engage with concepts of architecture beyond the particular and unified architectural object that has been the focus of previous archaeological accounts. I hope I have made clear that if you consider architecture as practice rather than object, with disjunction as a vital part of that practice, then you are forced to take a different landscape approach: an approach created through points of departure and dispersal rather than being about the nested quality of a location and staying put. There is evidence at the sites I have described for several different kinds of disjunction—those created through spatial shifts, temporal gaps, and changes in aesthetic effects.

Past people did not live their lives *with* their material culture *in* architecture *in* the landscape. Material culture, architecture, and landscape all intersected in the early Neolithic of Britain, and only through careful and attentive analysis of this will we start to understand the diverse scales at which past people operated.

Acknowledgments

I thank Bruno David and Julian Thomas for encouraging me to write this chapter, and Duncan Garrow, Mark Gillings, Andy Merrills, Josh Pollard, and Jeremy Taylor for their important comments and suggested changes.

I gratefully acknowledge the following for permission to reproduce illustrative material in this paper: The Prehistoric Society for Figures 31.2 and 31.3, and Josh Pollard for Figure 31.4.

References

Ashbee, P., Smith, I. F., and Evans, J. G. 1979. Excavation of three long barrows near Avebury, Wiltshire. *Proceedings of the Prehistoric Society* 45: 207–300.

Barrett, J. C. 1994a. *Fragments from Antiquity: Archaeology of Social life in Britain, 2900–1200 B.C.* Oxford: Blackwell Publishing.

———. 1994b. Defining domestic space in the Bronze Age of southern Britain, in M. Parker Pearson and C. Richards (eds.), *Architecture and Order: Approaches to Social Space*, pp. 87–97. London: Routledge.

Barrett, J., Bradley, R., and Green, M. 1991. *Landscape, Monuments and Society: The Prehistory of Cranborne Chase*. Cambridge: Cambridge University Press.

Bradley, R. 2003. Neolithic expectations, in I. Armit, E. Murphy, E. Nelis, and D. Simpson (eds.), *Neolithic Settlement in Ireland and Western Britain*, pp. 218–22. Oxford: Oxbow Books.

Cleal, R., and Pollard, J. (ed.). 2004. *Monuments and Material Culture: Essays on Neolithic and Bronze Age Britain*. East Salisbury: Hobnob Press.

Cummings, V., and Whittle, A. 2003. *Places of Special Virtue: Megaliths in the Neolithic landscapes of Wales*. Oxford: Oxbow Books.

Deleuze, G., and Guattari, F. 1992. *A Thousand Plateaus: Capitalism and Schizophrenia*. London: Continuum.

Garrow, D., Beadsmoore, E., and Knight, M. 2005. Pit clusters and the temporality of occupation: An earlier Neolithic site at Kilverstone, Thetford, Norfolk. *Proceedings of the Prehistoric Society* 71: 139–57.

Gosden, C. 2001. Making sense: Archaeology and aesthetics. *World Archaeology* 32: 163–67.

Hägerstrand, T. 1970. What about people in regional science? *Papers of the Regional Science Association* 24: 7–21.

Hirsch, E., and O'Hanlon, M. (eds.). 1995. *The Anthropology of Landscape: Perspectives on Place and Space*. Oxford: Clarendon Press.

Ingold, T. 1993. The temporality of the landscape. *World Archaeology* 25: 152–74.

McFadyen, L. 2006. Material culture as architecture. *Journal of Iberian Archaeology* 8: 91–102.

Pollard, J. 2005. Memory, monuments and middens in the Neolithic landscape, in G. Brown, D. Field, and D. McOmish (eds.), *The Avebury Landscape: Aspects of the Field Archaeology of the Marlborough Downs*, pp. 103–14. Oxford: Oxbow Books.

Pred, A. 1981. Power, everyday practice and the discipline of human geography, in A. Pred (ed.), *Space and Time in Geography: Essays Dedicated to*

Torsten Hägerstrand, pp. 30–55. Lund: The Royal University of Lund Department of Geography Press.

Smith, I. F. 1965a. *Windmill Hill and Avebury: Excavations by Keiller 1925–1933*. Oxford: Clarendon Press.

———. 1965b. Excavation of a bell barrow, Avebury G55. *Wiltshire Archaeological and Natural History Society Magazine* 60: 24–46.

Smith, I. F., and Simpson, D. D. A. 1966. Excavation of a round barrow on Overton Hill, North Wiltshire. *Proceedings of the Prehistoric Society* 32: 122–55.

Thomas, J. 1999. *Understanding the Neolithic*. London: Routledge.

Tilley, C. 1994. *A Phenomenology of Landscape*. Oxford: Berg.

———. 2004. *The Materiality of Stone: Explorations in Landscape Phenomenology*. Oxford: Berg.

Tschumi, B. 1996. *Architecture and Disjunction*. Cambridge, MA: MIT Press.

Whittle, A. 1990. A model for the Mesolithic-Neolithic transition in the upper Kennet Valley, North Wiltshire. *Proceedings of the Prehistoric Society* 56: 101–10.

———. 1994. Excavations at Millbarrow Neolithic chambered tomb, Winterbourne Monkton, North Wiltshire. *Wiltshire Archaeological Magazine* 87: 1–53.

Whittle, A., Rouse, A. J., and Evans, J. G. 1993. A Neolithic Downland monument in its environment: Excavations at the Easton Down long barrow, Bishops Cannings, North Wiltshire. *Proceedings of the Prehistoric Society* 59: 197–239.

Whittle, A., Davies, J. J., Dennis, I., Fairbairn, A. S., Hamilton, A., and Pollard, J. 2000. Neolithic activity and occupation outside Windmill hill causewayed enclosure, Wiltshire: Survey and excavation 1992–1993. *Wiltshire Archaeological and Natural History Society Magazine* 93: 131–80.

Whittle, A., Pollard, J., and Grigson, C. 1999. *The Harmony of Symbols: The Windmill Hill Causewayed Enclosure, Wiltshire*. Oxford: Oxbow Books.

32

FARMING, HERDING, AND THE TRANSFORMATION OF HUMAN LANDSCAPES IN SOUTHWESTERN ASIA

Ofer Bar-Yosef

The Region and the Palaeoclimatic Sources

This paper explores the changes in settlement patterns caused by the Neolithic Revolution, as well as the perception of landscape of various communities as much as it is expressed in the archaeological records. For this purpose, we need to examine differences between cultural markers of terminal Pleistocene foragers and those that stem from the evolution of agropastoral societies. The case examined here concerns a particular region within southwestern Asia known as the Levant. It is dealt with in its widest geographic definition that includes portions of northern Mesopotamia.

The geographic variability of this region should be taken into consideration, but not in today's image. Although past topographic features were essentially the same as today, the distribution of vegetational belts was different during the Upper Pleistocene and early Holocene. What we see today are the results of long-term impacts on the landscape of some ten millennia of agricultural and pastoral activities. We also know that the Holocene climate was not as stable as was once believed (Mayewski et al. 2004). Somewhat lower level of temperatures when compared to today's characterized the terminal Pleistocene, and in particular the Younger Dryas. Both temperatures and precipitation started to increase during the early Holocene. The climatic evidence has been gathered from east Mediterranean deep sea cores, marine and terrestrial pollen cores, the study of cave speleothems, geomorphological observations, fluctuations of lake levels, and, to a lesser degree, the fluctuating distributions of prehistoric hunter-gatherer sites who moved into and out the arid areas. In addition, correlations of the regional palaeoclimatic sequence with ice core records from the northern hemisphere provide a more global yardstick that facilitates tentative comparisons with socioeconomic processes in neighboring regions. However, as a cautionary note I should stress that chronological correlations between palaeoclimatic fluctuations and cultural changes are rather approximate and cannot be considered as simple cause and effect interactions. Direct climatic evidence from archaeological sequences is urgently needed in order to justify hypotheses based on environmental changes as triggers for new cultural adaptations. Yet the lack of this kind of evidence does not necessarily mean that intrinsic social changes were the sole cause for socioeconomic changes.

Timing the Neolithic Revolution

The emergence of cultivation of plants and the ensuing tending of animals with the eventual domestication of both took place as a process within changing environmental conditions since at least 11,700–11,500 cal B.P. until about 8,200 cal B.P.

This time span is also known as the Pre-Pottery Neolithic A and B periods (PPNA, PPNB), names coined by Kenyon (1957) when she excavated at Jericho, and both were later adopted by most archaeologists across the Levant. This generalized chronological subdivision is now under revision owing to the fast increasing number of radiocarbon dates. The dates from the northern Levant, Anatolia, and Greece are obtainable through the web page of CANeW (www.CANeW.org); we are still missing a similar web page for most of the Levant. Dating layers of Neolithic sites and particular assemblages, especially when based on short-lived samples (seeds and bones), are crucial for chronological correlations.

Testing competing hypotheses as regards plant domestication—that is, one Levantine "core area" versus "multiple centers" in the region—depends on accurate radiocarbon dates with small standard deviations. Unfortunately, published overviews do not discriminate between dates of short-lived samples such as seeds and those obtained from wood charcoal. The potential difference in age between charcoal from local oak trees, available everywhere in this region, and seeds can easily amount to several hundred years. Indeed, with currently available dates we may still fall into the pitfall of what I have called elsewhere the McDonald's chrono-model (Bar-Yosef 2004). The model states that radiocarbon dates for the earliest and the latest known as McDonald's restaurants from Ohio (1948) to China (2000), respectively, would obtain a date of 1975 ± 35 for both of them. This apparent age could be easily interpreted as indicating the contemporaneous emergence of McDonald's restaurants across the globe. However, within a small region such as the Levant, contemporary dates of layers in Neolithic mounds or single layer sites serve us in recognizing a larger population that survived, exploited plant resources, and was in contact within a particular landscape. Good examples for this situation are sites in the middle Euphrates River valley, along the Jordan River valley or across the landmass of the trans-Jordanian plateau and the Syro-Arabian desert. The material culture from sites located in such smaller areas reflects the degrees of socioeconomic interaction and, by reference, the landscape perception shared by the inhabitants of the different hamlets and villages.

The Interaction Sphere

An additional issue that needs to be taken into account is the evidence for long-range connections among Near Eastern human communities and their willingness to adopt imported plants, animals, and technologies or improve long-held local subsistence technologies, and try to domesticate regional species of plants and animals. This kind of evidence, when collected and compared within a large region as the Levant, is often referred to as "the interaction sphere." Among the best documented routes of exchange were those that facilitated the dispersal of obsidian products brought from Anatolia into the Levant, the exchange of chlorite bowls connecting the Zagros foothills with the northern Levant, and the Red Sea shells that reached far away Neolithic groups (e.g., Aurenche and Kozlowski 1999). Examples for transport of obsidian artifacts, Red Sea shells, and chlorite bowls between far-apart Neolithic sites were recorded by archaeologists (e.g., Cauvin 2002, and see below). The best-known case of repeated movements of entire groups is the process of colonization of Cyprus (e.g., Vigne and Cucchi 2005). Farmers, employing an unknown type of sea craft, brought across the sea the animals that were later considered as clearly among the domesticated species together with the fallow deer and the cat. This was not the first time that the island was visited by humans, as there is evidence of part-time inhabitants or short-term visitors during the Epi-Palaeolithic who came from the coast of Anatolia (Simmons 1999). Hence, we should consider coastal navigation around the Mediterranean Sea as another efficient means of communication and transportation (see also Crouch, this volume). Additional evidence for maritime routes comes from the late PPNB colonization of Thessaly (e.g., Perlès 2001, 2005) and Crete (Efstratiou 2005), whereas terrestrial movements took place from western Anatolia into the Balkans (Figure 32.1).

Thus, the practical knowledge of the region, whether shared among all humans or controlled by certain groups or by individuals, led to the expansion of the Interaction Sphere, as evidenced in rare imported or exchanged items. It may also indicate that the perception of landscape had a double meaning, including both "homeland" and "regional" landscapes.

Early Cultivation

As stated earlier, a major issue of the Neolithic Revolution is the understanding of the process of domestication of both plants and animals (Zeder 2006). There is no doubt today that this was an ongoing process and that different human communities played different roles in the formation of the "agricultural package" that was then transported from the Levant to Europe and other parts of Western Asia at a variable pace (e.g., Bellwood 2005; Cavalli-Sforza 2002; Colledge, Conolly, and Shennan 2004; Pinhasi,

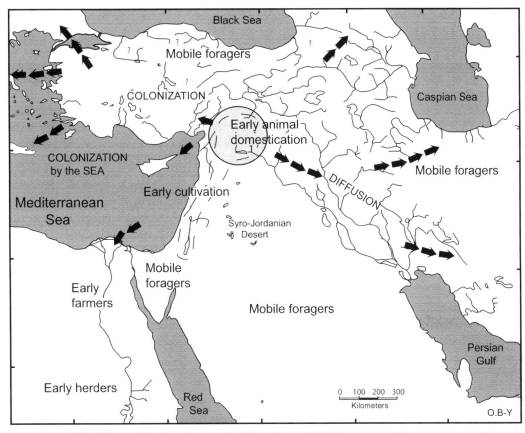

Figure 32.1 The dispersal of the agricultural package from the Northern Levant. The map represents a summary of the first ca. 4,000 years (PPNA and PPNB), as well as the uncertainties involved in identifying where the domestication of the plants took place (see text).

Fort, and Ammerman 2005). A major contribution to this discussion is the latest studies indicating that the cultivation of certain crops was tried several times and failed (such as rye; see Weiss, Kislev, and Hartman 2006). Other scholars point to the domestication of different species in different localities across the region (Willcox 2005). However, whereas barley was cultivated in most known Levantine sites (Willcox 2005; Table 32.1), Emmer wheat that appears in the southern Levant (from Damascus to the Dead Sea area) is genetically considered to originate in the foothills of the Taurus region (Ozkan et al. 2002). The discrepancy between the archaeobotanical evidence and the genetic observation requires further research. Einkorn wheat is known from Neolithic sites in the northern Levant and with geographically corresponding genetic evidence of the wild varieties (Heun et al. 1997). Thus, there is generally a correlation between what could have been the natural distribution of the cereals and their ensuing intentional cultivation and domestication.

The Climatic Impact

In sum, it seems that the current debates concerning the first steps of plant domestication are far from reaching an accepted consensus. This adds to the disagreements about the climatic impacts of the Younger Dryas. Global changes in CO_2 indicate that there was a decrease in temperatures and precipitation during the Younger Dryas. Such a reduction often plays a role in the success of C3 plants that flourish owing to large amounts of precipitation (while C4 plants characterize desert environments), an issue yet not dealt with by the archaeobotanists working in the Near East. The absence of barley, one of the most resistant cereal species from Hallan Çemi, a village site of foragers that was occupied during the Younger Dryas (Savard, Nesbitt, and Jones 2006), supports the notion that the cold and dry period had an impact on the distribution of the cereals.

Palaeoclimatic records indicate a rise in the amount of precipitation during the early Holocene

(Mayewski et al. 2004). The trend for improved environmental conditions is expressed by the genetic study of the Einkorn in southeast Turkey (Heun et al. 1997) that demonstrated the eastward spread of the younger subspecies.

Last but not least is the issue of the changing social context within which the shifts in resource exploitation techniques, accompanied by technological innovations, and realignments of cosmological configurations, took place. The most basic aspect is the nature of human occupation and the social organization of groups ranging from mobile to fully sedentary communities (e.g., Bar-Yosef 2001; Bellwood 2005; Richerson and Boyd 2001). As stressed many times before, perhaps for no avail, the evidence for the length of annual stay in one location by a given group should be based on biological markers such as the presence of commensals (for example, mice and sparrows) and not on the basis of archaeological remains such as building activities. Proposed interpretations are sometimes not necessarily based on principles derived from ethnographic examples but are simply stated as such. For example, clustered compounds such as in Bouqras, Basta, Ba'aja (Gebel, Hermansen, and Jensen 2002) could be interpreted as resulting from social pressures for living close in order to remove anxiety. Do these well-attached buildings reflect the fear of attack? The evidence for physical conflict is still missing but a proliferation in the production of arrowheads during the PPNB may indicate preparations and/or participation in limited warfare or raiding campaigns and not necessarily intensified hunting.

Early Neolithic Villages

The First Steps: Pre-Pottery Neolithic A

The archaeological sequence of the Neolithic Revolution begins with the cultivation of a selection of "founder crops" (e.g., Bar-Yosef 1998; Harris 1998; Hillman 1996; Weiss, Kislev, and Hartman 2006; Willcox 2005; Zohary and Hopf 2002). The most prominent and successful in cultivation/ domestication were Einkorn and Emmer wheat, barley, and legumes. The process of domestication— namely, when entire fields comprised already the mutation varieties of nonshattering rachis, and seeds that germinated annually—took some 800–1,000 years to complete (Kislev 1997; Tanno and Willcox 2006). Fields were probably located around the villages, and the villages themselves were of different sizes. Villages of 1.5 to 2.5 hectares in size could have accommodated a fully viable biological social unit (Bar-Yosef and Belfer-Cohen 1989) and thus

determine the size of their controlled territory, separated from other villages' territories.

Villages were spaced in the Jordan Valley some 15 to 25 km apart. Within each territory that was smaller than the average Natufian one, the perception of landscape began to change. Although we have no idea whether or not the village stood at the center of a land-owning group's territory, the village location was probably on a low hill, in close proximity to a water source. PPNA domestic buildings were rounded or oval with stone foundations; walls were built of unbaked, plano-convex bricks, with flat roofs; households comprised courtyards with grinding equipment; and above-ground rounded silos were constructed with mud bricks or clay mixed with marls—these were among the physical elements in the common look of these villages (e.g., Bar-Yosef and Gopher 1997; Kuijt 2000 and papers therein; Stordeur and Abbés 2002). The first communities of cultivators continued to hunt, trap, fish, and gather a large array of wild plant and fruit species. They also tended fruit trees such as figs, as suggested recently (Kislev et al. 2006). Maintaining and propagating wild fruit trees is a known technique among recent foragers in both Africa and South America (Laden 1992).

It is important to stress that PPNA farmers continued to operate also as hunters, probably organized by members of the same kin system. Farmers being also hunters are known historically and ethnographically (Kent 1989). It is a strategy embedded in the lifeways of semi- and fully sedentary cultivators and even pastoralists. PPNA hunters used bows and arrows and hunted the same species as their Late Natufian predecessors (e.g., Helmer 1992; Tchernov 1994). In addition to mammals such as gazelles, onagers, aurochs, wild goat and sheep, wild boar, and foxes, they collected reptiles, waterfowl, and some fish. An increase in catching waterfowl, as demonstrated by Tchernov (1994), could be explained, in part, as an increase in demand for meat and especially for feathers. Trapping birds was facilitated as fresh water bodies existed in closed basins in the Levant (e.g., in the Jordan Valley, and locations across its eastern margins). These were formed during the rapid climatic improvement immediately following the termination of the Younger Dryas. In addition, it should be noted that the Jordan Valley and the Levantine coastal plain are among the main routes of migratory birds between Europe and Africa. Thus, for example, the inhabitants of Netiv Hagdud, Gilgal, Hatoula, and probably other sites were able to exploit these seasonal resources (e.g., Pichon 1991).

Given the size of the newly formed communities, a different pattern of territoriality emerged. There is no equivocal evidence for continued residential mobility as suggested for the Natufian population. These new villages were ten times larger than the Natufian base camps. The increase in the number of people was the unconscious result of becoming cultivators/farmers with permanent supplies of staple food. Cereals were undoubtedly among the most suitable weaning foods, thus increasing the chances of survival of newborns by making porridge from crashed cereals. The process caused settlements to become villages with 300–400 people living contemporaneously in the same location. One could imagine the amount and tempo of garbage accumulation. In the known Neolithic mounds, reworked ashes, fire-cracked rocks, and discarded mud bricks would have been responsible for the nature of the rapid accumulation. However, villages survived for no more than 200–400 years at most (based on the calibrated radiocarbon dates for various sites) and then were abandoned. Discussion of the reasons for abandonment is beyond the scope of this paper, but soil erosion owing to felling trees and clearing new fields for planting; salinization; environmental deterioration such as that caused by successive droughts owing to climatic fluctuations; and group conflict come in mind as more common causes than were the potentials for local epidemics.

The social organization of the PPNA groups is not well known. However, even if it was somewhat loose with shifting responsibilities among a suite of local temporary chiefs or a few "Big Man"-type personalities (variously defined decision makers), public operations such as the building of the tower and wall in Jericho testify to a certain degree of centralized social organization and cooperation, as well as the need to guard the identity of the social unit, whether for practical (for example, defense; food production) or symbolic purposes (Bar-Yosef 1986; Naveh 2003; Ronen and Adler 2001). Other architectural evidence indicates that communal needs were expressed in buildings of the "kiva-type" semisubterranean structures. Several clear examples of this type of building were exposed in PPNA and early PPNB Mureybet III and Jerf el-Ahmar (Cauvin 2000; Goring-Morris and Belfer-Cohen 2002; Stordeur and Abbés 2002; Stordeur et al. 2001). These are thought to be places for ceremonial activities and meeting of the elders of the settlement.

Public rituals or ceremonies could have been organized in open spaces, better known from the so-called plaza in PPNB Çayönü (Özdogan 1999). Domestic rituals are reflected in female figurines that mark a major change from the preceding foragers' Natufian society, where animal figures and rare schematic presentations of humans were available (Bar-Yosef 1997; Belfer-Cohen 1991; Cauvin

2000; Özdogan 2003; Weinstein-Evron and Belfer-Cohen 1993).

It is hypothesized that, similarly to changes known from historical technological revolutions, the onset of cultivation by the inhabitants of villages, whether small or large, resulted in numerous social changes (e.g., Bar-Yosef and Belfer-Cohen 1992; Cauvin 2000; Gebel, Hermansen, and Jensen 2002; Kuijt 2000 and papers therein; Verhoeven 2002). Not least of these changes was the shifting of male and female roles within the farming communities. Presumably women as gatherers knew well the plants, and we may therefore assume that they were probably the initiators of plant cultivation. On a domestic level they were the food processors and users of grinding tools. Males were probably responsible for felling trees, building (while females could have been the makers of the bricks), tilling small plots with hand picks, and building fences, all the while continuing to hunt. Women and children gathered small game, such as tortoises and lizards, and trapped birds. The contention that females had increasing work loads, including grinding and pounding that require continuous energy expenditure, is based on ethnographic examples, but with archaeological evidence from PPNA-PPNB skeletal relics (Molleson 1994; Peterson 1997).

Analysis of mortuary practices demonstrates that Early Neolithic communities tried hard to keep their society as egalitarian as possible (Kuijt 1996). In addition, the appearance of female figurines mentioned above (e.g., Cauvin 2000; Kuijt and Chesson 2005; Voigt 2000) heralds a departure from the Natufian tradition that was based on depiction of animals. Although attributed to the domestic level of rituals, the small human figurines reflect a new ideological trend within the cosmological configurations with the human image occupying a major position (Cauvin 2000; Marshack 1997). The change epitomizes a departure from being equal partners within natural surroundings to humans as major players on the scene. Sedentism of small or large communities of cultivators definitely resulted in changing attitudes toward the immediate environments and by inference, to nature as a whole.

Within Farming Societies: From Corralling to the Domestication of Mammals

Cultivation continued to be the main source of staple food during what we call the PPNB (starting ca. 10,500 cal b.p.) with additional plants in the fields, which surrounded the village, such as broad bean, flax, and chickpeas. Palaeoclimatic data indicate a generally higher annual precipitation than exists today (Bar-Mathews et al. 1999). Interestingly,

Van Zeist (1986), when examining the carbonized cereal seeds from the northern Levant already suspected that winter precipitation was higher, or irrigation was practiced in the marginal eastern belt. Wheat seeds from PPNB Tell Halula on the banks of a tributary of the Euphrates River were tested for the ratios of stable carbon isotopes, the results supporting this contention (Araus et al. 2001). Prehistoric crops enjoyed better water supplies than those available today through winter rain or irrigation. A detailed field research around several PPNB mounds through test pits might uncover the oldest irrigation canals that could not yet be seen in aerial photos, thus increasing our knowledge concerning the changes in the landscape on the outskirts of the settlements.

Villages in the Levant accommodated rectangular buildings built on stone foundations with mud brick or adobe walls. In south Jordan, two-storey houses are common, such as in Basta, Ghwair, and Ba'aja, as well as "Corridor type" houses, such as in Beidha (Kuijt and Goring-Morris 2002), and had an elevated floor above what seems to have been a basement. The unanswered question is—why did people at that time need two-storey houses? Was there a limitation on village space and the wish not to build over agricultural land? Does it reflect defensive needs, since many of such buildings were arranged in compounds?

Along the Euphrates River (for example, Bouqras) and mainly on the Anatolian plateaux in PPNB sites such as Asikli höyük or Çatalhöyük are the walls of the houses with one or two storeys attached to each other or leaving a very narrow space in between that was used for dumping "garbage," creating a dense settlement plan that could be explained as a reflection of defensive requirements (Hodder 2006). Hence, the inhabitants' perception of their village and the territory around would resemble what we would see as the dichotomy of "inside" versus "outside."

The domestic toolkits of the early farmers included axe/adzes, either bifacially shaped with transverse blows producing a sharp cutting edge common in the southern Levant, and sometimes polished celts that were the standard tool in the northern Levant and Anatolia. All these tools were employed in wood-working activities, including tree-felling and clearing, shaping wooden objects, and building of sea craft that was used to cruise the Mediterranean Sea. Harvesting equipment included simple sickles, V-shaped bone tools for stripping the seed heads from straw, and later the threshing board or *tribulum* (Anderson 1998).

Storage facilities were special built-in installations and minirooms in houses or courtyards. Changes in the sizes and locations of storage facilities mark the shift in several PPNB sites from nuclear family consumption to larger social units and, perhaps, to an institutionalized control of public granaries.

In these early villages, penning of wild animals could have been an additional strategy for securing meat and hides. What was not a practical strategy for mobile foragers was a viable option for semisedentary or sedentary hunter-gatherers. Perhaps it is in this context that pigs were penned in Hallan Çemi around 11,000–10,500 cal B.P. (Rosenberg et al. 1998), although the proposal that these pigs were en route to be domesticated is unclear. Previously, a somewhat similar proposition was made concerning human intervention with gazelles, but their domestication was not a viable option, owing to a very short imprinting time since the birth of the kids. Therefore, most authorities agree that the paucity of goat, cattle, and pigs in Levantine PPNA lacks any morphometric signs for incipient domestication. During the same period, the proliferation of fox bones in these early villages speaks for their attraction to the garbage of human settlements. Fox hunting for their furs and canine teeth (extensively used already by the Natufians), near the villages, would be a possible explanation.

The recurrent questions of "when" and "where" the process of animal domestication began should direct us to "where" these species were frequently hunted by Epi-Palaeolithic foragers and whether modern or ancient DNA will point to the same locus (e.g., Albarella et al. 2006; Bradley and Magee 2006; Zeder 2006). The best evidence comes from cave sites such as Ökuzini at the foothills of the Taurus (Albrecht et al. 1992), where wild goat and sheep were hunted by local foragers. The "when" issue is today more complicated, because the faunal assemblages from early PPNB (ca.10, 200 cal B.P.) in Cyprus indicate that wild goat, sheep, cattle, and pigs (as well as the fallow deer and dogs) were transported over the Mediterranean Sea to the site of Shillourocambos (Vigne and Cucchi 2005; Vigne et al. 1999) with a later introduction of fox and cat. Hence, the corralling and tending of these species was the first step in a long process. Having goat, sheep, pig, and cattle as part of the daily or ritual meat supplies of early farmers (who continued to hunt) tells us that morphologically the shift from "wild" to "domesticated" took a long time and may not represent a simple dichotomy.

This discovery has more than one implication. First, characterizing Neolithic villages (based on their material culture and dates) as settlements of hunter-gatherers because the major portion of the animal bone spectra are composed of morphologically wild caprines, pigs, and cattle is an inaccurate socioeconomic definition (see also Terrell and Hart, this volume for a different slant on this issue). Obviously, the overall picture is more complex in Anatolia, where those species were corralled or hunted in their natural habitats. The situation was different in the Levant. Wild goats were common in mountain areas such as the Lebanese mountains. The aurochs was present but was not an important source of meat, as in Anatolia. Wild boar was common, and sheep were absent altogether. Second, the Cypriote findings support an earlier contention that goat, sheep, and cattle were herded south from the foothills of the Taurus into the Levant (Bar-Yosef 2000). The process of adopting animal husbandry in the northern Levant began during the Early PPNB (Peters et al. 1999; Zeder 1999) and proceeded later, mainly in the course of the Middle and especially the Late PPNB into the central and southern Levant (Horwitz et al. 1999; Martin 1999). The herding of these animals could have been part and parcel of the long-distance exchange of obsidian. Needless to stress, the visible signs of animal domestication become available during mid or late PPNB with the introduction of goat and sheep.

Cattle domestication, whether for religious reasons (the "bull-cult" [Cauvin 2000]) or for economic ones occurred mainly during the PPNB period. Similar to other animals, cattle (mostly bulls) were slaughtered, possibly during ceremonial feasts, as clearly indicated in the assemblage at Göbekli Tepe (Peters et al. 1999).

In sum, the fully developed Neolithic economy, with its domesticated species of plants and animals, seems to have emerged earlier in the northern Levant (southeast Turkey and northern Syria) than elsewhere. Owing to geographic proximity, innovations did not escape notice by the inhabitants of the central and southern Levant, and the resulting regional network formed the PPNB interaction sphere (Bar-Yosef and Belfer-Cohen 1989).

Interactions between farmer-herders and contemporary mobile foragers (Figure 32.2) possibly played a role, although this is not yet fully researched. Special type of rocks, such as limestone from the Syro-Arabian desert, the Negev, and Sinai, and marine shells from the Red Sea could have been exchanged for other products such as grain (Bar-Yosef Mayer 1997). Mutualistic interactions between these two societies could have been in constant flux, either amicable, which may have led to intermarriage, perhaps with forager women marrying into farming communities as suggested by ethnographic records, or hostile, leading to acts of violence and fighting.

Other archaeological markers of farmer-forager interactions are the game drives known as "desert kites." These were probably laid out during the PPNB period by foragers to hunt gazelles or onagers. Employing this technique, which is also known from other locations in the world, must mean that there was a demand for meat, hides, and horns. In one Jordanian site (Garrard et al. 1994), a rectangular house in a foragers' camp of rounded dwellings might be interpreted as a "merchant's temporary home." Indeed, the boundaries between farmer-herders and foragers in semiarid or mountain regions were probably rather fluid. Thus it can be suggested that the perception of the landscape, between what is "ours" and that of "others," became more entrenched than the boundaries between neighboring tribes of farmers.

The economic dichotomy between farmer-herders and pastoral nomads had emerged and became more established in the following millennia after the collapse of the PPNB. Members of both economic regimes continued to hunt and trap, as well as to gather wild plants, seeds, and fruits for various purposes, including medicine.

In sum, from the villages of the PPNA period through the complex social system during the PPNB period when clearly regular villages of farmer-herders coexisted with central ceremonial settlements such as Göbekli Tepe or Ba'aja (Kuijt and Goring-Morris 2002), a considerable number of socioeconomic changes can be recognized archaeologically. Population increase along with economic improvements led to the appearance of numerous PPNB sites of different sizes and a territorial distribution that is seen as several "amphyctionies," each being a loose territorial organization of villages and towns sharing religious centers (Belfer-Cohen and Goring-Morris 2002). Individual settlements kept ownership of land—that is, the individual village (along with its surrounding lands) was itself a territorial unit. Thus boundaries were created, and although being permeable those of the "sown land" differed from those of the "arid lands" where hunter-gatherers continued to survive, possibly through the 7th millennium cal B.P.

Pre-Pottery Neolithic B Socioeconomic Territories

Employing the accumulated evidence concerning palaeoclimatic fluctuations during the Holocene, calibrated radiocarbon dates, archaeobotanical and genetic studies of successful or failed "founder crops" when available (Weiss, Kislev, and Hartmen 2006), as well as studies of ancient DNA of domesticated species, the overall picture of the Levantine and

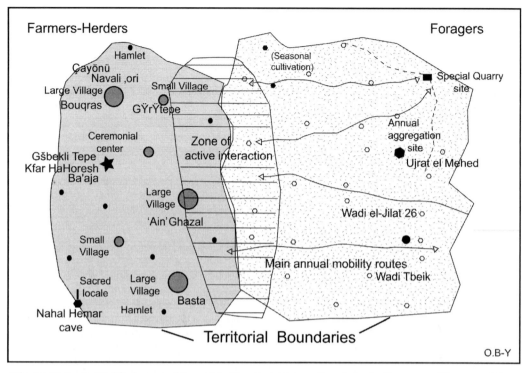

Figure 32.2 A simplified geographic model of territorial interactions between farmers and foragers. A few examples of archaeological sites are added.

eastern Anatolian landscapes are more complex. As done elsewhere (Bar-Yosef 2001), mapping the proposed tribal territories within the PPNB interaction sphere becomes a feasible option and a base for further elaborations and corrections (Figure 32.3).

The prominent markers of each territory are the ceremonial centers. In addition, subregional domestic house types, technological variations in the flaking or polishing of heavy duty tools such as axes and adzes, variations in the frequencies of projectile point types, occurrences of modeled skulls, stone masks, "white ware" vessels, stone bangles, and the like are elements in defining the territories of ethno-linguistic groups. Within the PPNB interaction sphere, one can identify similar beliefs and shared cosmologies. The modeled skulls found in various sites may hint at the presence of elite members or chiefly families. However, no uniquely rich tombs have been discovered to date, and thus we cannot classify PPNB societies as chiefdoms. Still, with ongoing fieldwork, this situation may change in a few years.

Evidence for what should have been organized efforts is available. The quarrying, shaping, and carving in low relief of mammals, birds, and reptiles on the numerous T-shaped pillars at Göbekli Tepe (Schmidt 2000), and the more

limited objects in the rural village of Nevali Çori (Hauptman 1999), were not just family affairs but the concerted efforts of many people. Later, the filling of house buildings and ceremonial centers with village garbage, such as in Çayönü or Göbekli Tepe, testify not only to their increasing isolation from "life" but also to the energy expenditure that although great was religiously required (Özdogan and Özdogan 1998).

An additional example is the complex operation of colonizing Cyprus that speaks for the presence of leaders. The building of seafaring craft, transporting of animals, and crossing of waters by several groups is attested at several early PPNB sites on the island (Vigne and Cucchi 2005).

Marking of personal property, whether of individuals or extended families, is probably indicated by the rare engraved flat pebbles, and stamp seals, in the PPNB. The engravings on these objects, as noted by Cauvin (2000), resemble pictographs used in early writing. Similarly, tokens in PPNB contexts are interpreted as elements of a counting system (Schmandt-Besserat 1990). Traded or exchanged items indicate a much wider interaction sphere in which sources and producers were located beyond the permeable boundaries of the PPNB Levantine-Anatolian civilization (e.g., Aurench and Kozlowski

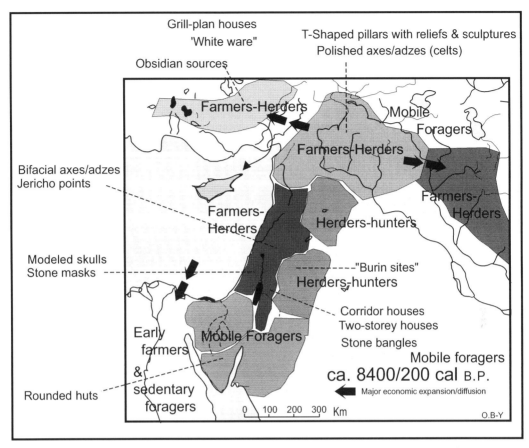

Figure 32.3   The spatial distribution of Neolithic tribes at the end of the PPNB period(ca. 8,600–8,200 cal B.P.) in the Levant and southeast Anatolia. Each entity is identified on the basis of material culture markers.

1999). Among the better-known exchanged materials are obsidian, chlorite bowls, asphalt, cinnabar, and marine shells.

The Collapse of the Levantine PPNB and the Emergence of New Territorial Concepts

During the two millennia of the PPNB, only a few settlements survived through many centuries. The issue of why villages had a limited life span has hardly been discussed. Options vary from the role of diseases or even epidemics of zoonotic diseases caused by the advanced and spread of domesticated cattle, over-exploitation of the fields, climatic fluctuations of decadal scale, conflicts among neighboring villages, and the like. A well-recorded stratigraphic gap between the PPNB and the Pottery Neolithic is established for certain parts of the Levant and the Anatolian plateau (Aurenche and Kozlowski 1999; CANeW Project 2007; Özdogan and Basgelen 1999), indicating widespread village abandonment in those areas. However, river valleys continued to be occupied. Good examples are Mezra'a Tlelat on the bank of the Euphrates River (Karul, Ayhan, and Özdogan 2002; Özdogan 2003) and Sha'ar HaGolan in the Jordan Valley (Garfinkel and Miller 2002).

Given this observation, the proposal to explain the collapse of the major village of 'Ain Ghazal in Jordan as due to local environmental over-exploitation by felling trees and herding goats (e.g., Rollefson, Simmons, and Kafafi 1992) seems inadequate. The most likely explanation for the general abandonment across the Levant is the abrupt and adverse climatic change around 8,400–8,200 cal B.P. (Bar-Yosef 2001) that is well-recorded in ice cores (Alley et al. 1997) and in the stalagmites of Soreq cave (Bar-Mathews et al. 1999), as well as in eastern Mediterranean deep sea cores (Rohling et al. 2002). Tribal societies that subsisted on farming and herding, in which the demands of better-off individuals (or families) drove the flow of prestige goods, could not continue to accumulate the needed surpluses when hit by a series of droughts. Shifts in the patterns of seasonal precipitation necessitated a search for pastures further away and resulted in lower yields of summer harvests. Digging wells, a technology known from Shillourocambous, Miloutkhia, Atlit-Yam, and Sha'ar HaGolan, could solve the problem of drinking water only when the water table was not too deep for the available techniques. Shifts in the size of territories were expected. The advent and expansion of certain cultural groups is evident from the beginning of the 8th millennium cal B.P. (Cruells and Nieuwenhuyse 2004; Flannery 1999). A good example is the Halafian culture that occupied a larger area across the northern Levant and northern Mesopotamia. From that time the presence of Mesopotamian cultures in the northern Levant becomes a permanent phenomenon that reflects the rapid development of complex social structures in this vast region, and a decrease in the economic and social importance of the Levant as a region vis-à-vis Mesopotamia, or the Nile Valley.

The new conditions in the Levant enhanced reliance on a more flexible subsistence strategy of farming and herding and increased the presence of pastoral societies after 8,000 cal B.P. The collapse of the PPNB could have been the trigger for several major phenomena. Among these were the colonization of new areas such as the Nile Delta by sea (Bar-Yosef 2002), long after Thessaly was settled by Neolithic farmers from Anatolia (Perlès 2001, 2005). Terrestrial routes through the Balkans led Neolithic farmers into some other new lands as they probably did while entering the Caucasus inter-montane valleys or moving to the east of the Caspian Sea.

Conclusions

Owing to the processes of frequent socioeconomic and territorial changes described above, we may expect shifts in the perception of immediate and distant landscapes among the late foragers, farmers-hunters, farmers-herders, and pastoral societies that emerged in a rather fast evolutionary trajectory over the first millennia of the Holocene.

The emergence of new territorial concepts allowed Neolithic farmers to incorporate the sea into their unbounded homelands, or extended interaction sphere. The inclusion of different landscapes such as the high mountains (Zagros and Taurus) into their territorial perceptions possibly led to the creation of common worlds' beliefs that later became the basis for Near Eastern cosmologies. For example, certain images of ancestors in

their natural environment included "boat people," mountain hunters, and the forest animals, while the villages of those farmers were built along riverbanks in Egypt or Mesopotamia.

In sum, the outcome of the Neolithic Revolution within the social realm was a major change in the concept and perception of territories and their actual landscapes. From the more resilient perceptions of late Palaeolithic mobile foragers in their variable landscapes, a world of bounded entities of farmers and herders emerged over six or five millennia.

References

Albrecht, G., Albrecht, B., Berke, H., Burger, D., Moser, J., Rähle, W., Schoch, W., Storch, G., Uerpmann, H. P., and Urban, B. 1992. Late Pleistocene and early Holocene finds from Öküzini: A contribution to settlement history of the bay of Antalya, Turkey. *Paléorient* 18: 123–41.

Alley, R. B., Mayewski, P. A., Sowers, T., Stuiver, M., Taylor, K. C., and Clark, P. U. 1997. Holocene climatic instability: A prominent, widespread event 8,200 years ago. *Geology* 25: 483–86.

Albarella, U., Dobney, K., and Rowley-Conwy, P. 2006. The domestication of the pig (*Sus scrofa*): New challenges and approaches, in M. A. Zeder, D. G. Bradley, E. Emshwiller, and B. Smith (eds.), *Documenting Domestication: New Genetic and Archaeological Paradigms*, pp. 209–27. Berkeley and Los Angeles: University of California Press.

Anderson, P. C. 1998. History of harvesting and threshing techniques for cereals in the prehistoric Near East, in A. B. Damania, J. Valkoun, G. Willcox, and C. O. Qualset (eds.), *The Origins of Agriculture and Crop Domestication*, pp. 145–59. Aleppo: ICARDA.

Araus, J. L., Slafer, G. A., Romagosa, I., and Molist, M. 2001. Estimated wheat yields during the emergence of agriculture based on the carbon isotope discrimination of grains: Evidence from a 10th millennium B.P. site on the Euphrates. *Journal of Archaeological Science* 28: 341–50.

Aurenche, O., and Kozlowski, S. 1999. *La Naissance du Néolithique au Proche Orient ou Le Paradis Perdu*. Paris: Editions Errance.

Aurenche, O., Kozlowski, S., and Le Mière, M. 2004. La notion de frontière dans le Protonéolithique et le Néolithique du Proche-Orient, in O. Aurenche, M. L. Mière, and P. Sanlaville (eds.), *From the River to the Sea: The Palaeolithic and the Neolithic on the Euphrates and in the northern Levant*, pp. 355–66. Oxford: Archaeopress.

Bar-Mathews, M., Ayalon, A., Kaufman, A., and Wasserburg, G. J. 1999. The eastern Mediterranean palaeoclimate as a reflection of regional events: Soreq cave, Israel. *Earth and Planetary Science Letters* 166: 85–95.

Bar-Yosef, O. 1986. The walls of Jericho: An alternative interpretation. *Current Anthropology* 27: 157–62.

———. 1997. Symbolic expressions in later prehistory of the Levant: Why are they so few? in M. W. Conkey, O. Soffer, D. Stratmann, and N. G. Jablonski (eds.), *Beyond Art: Pleistocene Image and Symbol*, pp. 161–87. San Francisco: Memoirs of the California Academy of Science.

———. 2001. From sedentary foragers to village hierarchies: The emergence of social institutions, in G. Runciman (ed.), *The Origin of Human Social Institutions*, pp. 1–38. Proceedings of the British Academy, Vol. 110. Oxford: Oxford University Press.

———. 2004. East to west—Agricultural origins and dispersal into Europe: Guest editorial. *Current Anthropology* 45 supplements1-s3.

Bar-Yosef, O., and Belfer-Cohen, A. 1989. The Levantine "PPNB" interaction sphere, in I. Hershkovitz (ed.), *People and Culture in Change*, pp. 59–72. BAR International Series 508(i), Oxford.

———. 1992. From foraging to farming in the Mediterranean Levant, in A. B. Gebauer and T. D. Price (eds.), *Transitions to Agriculture in Prehistory*, pp. 21–48. Madison, WI: Prehistory Press.

Bar-Yosef, O., and Gopher, A. 1997. The Excavations of Netiv Hagdud: Stratigraphy and architectural remains, in O. Bar-Yosef and A. Gopher (eds.), *An Early Neolithic Village in the Jordan Valley, Part I: The Archaeology of Netiv Hagdud*, pp. 41–69. American School of Prehistoric Research, Bulletin 43. Cambridge, MA: Peabody Museum of Archaeology and Ethnology, Harvard University.

Bar-Yosef Mayer, D. E. 1997. Neolithic shell bead production in Sinai. *Journal of Archaeological Science* 24: 97–112.

Belfer-Cohen, A. 1991. The Natufian in the Levant. *Annual Review of Anthropology* 20: 167–86.

Belfer-Cohen, A., and Goring-Morris, N. 2002. Recent developments in Near Eastern Neolithic research. *Paléorient* 28: 143–56.

Bellwood, P. 2005. *First Farmers: The origins of agricultural societies*. Oxford: Blackwell Publishing.

Bradley, D. G., and Magee, D. A. 2006. Genetics and origins of domestic cattle, in M. A. Zeder, D. G. Bradley, E. Emshwiller, and B. Smith (eds.), *Documenting Domestication: New Genetic and Archaeological Paradigms*, pp. 317–28. Berkeley and Los Angeles: University of California Press.

CANeW Project. 2007. The Project CANeW (www.canew.org).

Cavalli-Sforza, L. 2002. Demic diffusion as the basic process of human expansions, in P. Bellwood and C. Renfrew (eds.), *Examining the Farming/Language Dispersal Hypothesis*, pp. 79–88. Cambridge: McDonald Institute for Archaeological Research.

Cauvin, J. 2000a. *The Birth of the Gods and the Origins of Agriculture*. Cambridge: Cambridge University Press.

———. 2000b. The symbolic foundations of the Neolithic revolution in the Near East, in I. Kuijt (ed.), *Life in Neolithic Farming Communities: Social Organization, Identity, and Differentiation*, pp. 235–51. New York: Plenum Press.

Colledge, S., Conolly, J., and Shennan, S. 2004. Archaeobotanical evidence for the spread of farming in the eastern Mediterranean [and comments]. *Current Anthropology* 45: 35–58.

Cruells, W., and Nieuwenhuse, O. 2004. The proto-Halaf period in Syria: New sites, new data. *Paléorient* 30: 47–68.

Davis, S. J. M. 2005. Why domesticate food animals? Some zoo-archaeological evidence from the Levant. *Journal of Archaeological Science* 32: 1408–16.

Efstratiou, N. 2005. Tracing the story of the first farmers in Greece: A long and winding road. *BYZAS* 2: 143–53.

Flannery, K. V. 1999. Chiefdoms in the early Near East: Why it's so hard to identify them, in A. Alizadeh, Y. Majidzadeh, and S. M. Shahmirzadi (eds.), *The Iranian World: Essays on Iranian Art and Archaeology*, pp. 44–61. Tehran: Iran University Press.

Garrard, A. N., Baird, D., Colledge, S., Martin, L., and Wright, K. 1994. Prehistoric environment and settlement in the Azraq Basin: An interim report on the 1987 and 1988 excavation seasons. *Levant* 26: 73–109.

Gebel, H. G., Hermansen, B. D., and Jensen, C. H. (eds.). 2002. *Magic Practices and Ritual in the Near Eastern Neolithic*. Studies in Early Near Eastern Production, Subsistence, and Environment 8, ex oriente, Berlin.

Gebel, H. G., Hermansen, B. D., and Kinzel, M. 2006. Ba'ja 2005: A two-storied building and collective burials. Results of the 6th season of excavations. *Neo-Lithics* 6(1): 12–19.

Goring-Morris, A. N., and Belfer-Cohen, A. 2002. Symbolic behaviour from the Epipalaeolithic and early Neolithic of the Near East: Preliminary observations on continuity and change, in H. G. K. Gebel, B. D. Hermansen, and C. H. Jensen (eds.), *Magic Practices and Ritual in the Near Eastern Neolithic*, pp. 67–79. Studies in Early Near Eastern Production, Subsistence, and Environment 8, ex oriente, Berlin.

Harris, D. R. 1998. The origins of agriculture in southwest Asia. *The Review of Archaeology* 19(2): 5–12.

Hauptmann, H. 1999. The Urfa Region, in M. Özdogan and N. Basgelen (eds.), *Neolithic in Turkey: Cradle of Civilization—New Discoveries*, pp. 65–86. Istanbul: Arkeoloji ve Sanat Yayinlari.

Helmer, D. 1992. *La Domestication des Animaux par les Hommes préhistoriques*. Paris: Masson.

Heun, M., Schäfer-Pregl, R., Klawan, D., Castagna, R., Accerbi, M., Borghi, B., and Salamini, F. 1997. Site of einkorn wheat domestication identified by DNA fingerprinting. *Science* 278: 1312–14.

Hillman, G. 1996. Late Pleistocene changes in wild plant-foods available to hunter-gatherers of the Northern Fertile Crescent: Possible preludes to cereal cultivation, in D. Harris (ed.), *The Origins and Spread of Agriculture and Pastoralism in Eurasia*, pp. 159–203. London: UCL Press.

Hillman, G. C., Hedges, R., Moore, A., Colledge, S., and Pettitt, P. 2001. New evidence of Lateglacial cereal cultivation at Abu Hureyra on the Euphrates. *Holocene* 11: 383–93.

Hodder, I. 2006. *Çatalhöyük; The Leopard Tale*. London: Thames and Hudson.

Horwitz, L., Tchernov, E., Ducos, P., Becker, C., Von Den Driesch, A., Martin, L., and Garrad, A. 1999. Animal domestication in the southern Levant. *Paléorient* 25: 63–80.

Karul, N., Ayhan, A., and Özdogan, M. 2002. Mezraa Tleilat 2002, in N. Tuna and J. Velibeyyoglu (eds.), *Salvage Project of the Archaeological Heritage of the Illisu and Carchemish Dam Reservoirs: Activities in 2000*. Ankara: TAÇADAM.

Kent, S. (ed.). 1989. *Farmers and Hunters*. Cambridge: Cambridge University Press.

Kenyon, K. 1957. *Digging Up Jericho*. London: Benn.

Kislev, M. E. 1997. Early agriculture and paleoecology of Netiv Hagdud, in O. Bar-Yosef and A. Gopher (eds.), *An Early Neolithic Village in the Jordan Valley Part I: The Archaeology of Netiv Hagdud*, pp. 209–36. Cambridge: Peabody Museum of Archaeology and Ethnology, Harvard University.

Kislev, M. E., Hartman, A., and Bar-Yosef, O. 2006. Early domesticated fig in the Jordan Valley. *Science* 312: 1372–74.

Kozlowski, S. K., and Aurenche, O. (eds.). 2005. *Territories, Boundaries and Cultures in the Neolithic Near East*. Oxford: Archaeopress.

Kuijt, I. 1996. Negotiating equality through ritual: A consideration of Late Natufian and Prepottery Neolithic A period mortuary practices. *Journal of Anthropological Archaeology* 15: 313–36.

———. 2000. Life in Neolithic farming communities: An introduction, in I. Kuijt (ed.), *Life in Neolithic Farming Communities: Social Organization, Identity, and Differentiation*, pp. 3–13. New York: Plenum Press.

Kuijt, I., and Chesson, M. S. 2005. Lumps of clay and pieces of stone: Ambiguity, bodies, and identity as portrayed in Neolithic figurines, in S. Pollock and R. Bernbeck (eds.), *Archaeologies of the Middle East: Critical Perspectives*, pp. 152–83. Malden, MA: Blackwell Publishing.

Kuijt, I., and Goring-Morris, N. 2002. Foraging, farming, and social complexity in the Pre-Pottery Neolithic of the southern Levant: A review and synthesis. *Journal of World Prehistory* 16: 361–440.

Laden, G. 1992. Ethnoarchaeology and Land Use Ecology of the Efe (Pygmies) of the Ituri Rain Forest, Zaire: A Behavioral Ecological Study of Land-Use Patterns and Foraging Behavior. Unpublished Ph.D. thesis, Harvard University.

Marshack, A. 1997. Paleolithic image making and symboling in Europe and the Middle East: A comparative review, in M. Conkey, O. Soffer, D. Stratmann, and N. G. Jablonski (eds.), *Beyond Art: Pleistocene Image and Symbol*, pp. 53–91. San Francisco: Memoirs of California Academy of Sciences.

Martin, L. 1999. Mammal remains from the eastern Jordanian Neolithic, and the nature of caprine herding in the steppe. *Paléorient* 25: 87–104.

Mayewski, P. A., Rohlingb, E. E., Karlen, S. J. C., Maascha, K. A., Meekere, L. D., Meyersona, E. A., Gasse, F., van Kreveld, S., Holmgrend, K., Lee-Thorp, J., Rosqvist, G., Rack, F., Staubwasser, M., Schneider, R. R., and Steig, E. J. 2004. Holocene climate variability. *Quaternary Research* 62: 243–55.

Molleson, T. 1994. The eloquent bones of Abu Hureyra. *Scientific American* August: 70–75.

Naveh, D. 2003. PPNA Jericho: A socio-political perspective. *Cambridge Archaeological Journal* 13: 83–96.

Özdogan, A. 1999. Çayönü, in M. Özdogan and N. Basgelen (eds.), *Neolithic in Turkey: The Cradle of Civilization—New Discoveries*, pp. 35–64. Istanbul: Arkeoloji ve Sanat Yayinlari.

Özdogan, M. 2003. A group of Neolithic stone figurines from Mezraa-Teleilat, in M. Özdogan, H. Hauptmann, and N. Basgelen (eds.), *Köyden Kente: Yakindogu'da ilk Yerlesimler—From Village to Cities: Early Villages in the Near East*, pp. 511–23. Ufuk Esin'e Armagan—Studies Presented to Ufuk Esin, Vol. 2. Istanbul: Arkeoloji ve Sanat Yayinlari.

Özdogan, M., and Basgelen, N. (eds.). 1999. *Neolithic in Turkey: The Cradle of Civilization—New Discoveries*. Istanbul: Arkeoloji ve Sanat Yayinlari.

Özdogan, M., and Özdogan, A. 1998. Buildings of cult and the cult of buildings, in G. Arsebük, M. Mellink, and W. Schirmer (eds.), *Light on Top of the Black Hill: Studies Presented to Halet Çambel*, pp. 581–601. Istanbul: Ege Yayinlari.

Ozkan, H., Brandolini, A., Schäfer-Pregl, R., and Salamini, F. 2002. AFLP analysis of a collection of tetraploid wheats indicates the origin of *T. dicoccoides* and hard wheat domestication in southeast Turkey. *Molecular Biology and Evolution* 19: 1797–1801.

Perlès, C. 2001. *The Early Neolithic in Greece*. Cambridge: Cambridge University Press.

———. 2003. Le rôle du Proche-Orient dans la néolithisation de la Grèce, in B. Vandermeersch (ed.), *Èchanges et Diffusion dans la Préhistoire Mediterranéenne*, pp. 91–104. Paris: Comité des travaux historiques et scientifiques.

Peters, J., Helmer, D., von den Driesch, A., and Segui, M. S. 1999. Early animal husbandry in the Northern Levant. *Paléorient* 25(2): 27–47.

Peterson, J. 1997. Tracking activity patterns through skeletal remains: A case study from Jordan and Palestine, in H. G. Gebel, Z. Kafafi, and G. O. Rollefson (eds.), *The Prehistory of Jordan, II: Perspectives from 1997*, pp. 475–92. Berlin: ex oriente.

Pichon, J. 1991. Les oiseaux au Natoufien, avifaune et sédentarité, in O. Bar-Yosef and F. R. Valla (eds.), *The Natufian Culture in the Levant*, pp. 371–80. Ann Arbor, MI: International Monographs in Prehistory.

Pinhasi, R., Fort, J., and Ammerman, A. J. 2005. Tracing the origin and spread of agriculture in Europe. *PLoS Biology* 3: 2220–28.

Richerson, P. J., and Boyd, R. 2001. Institutional evolution in the Holocene: The rise of complex societies, in G. Runciman (ed.), *The Origin of Human Social Institutions*, pp. 197–234. Oxford: Oxford University Press.

Rohling, E. J., Casford, J., Abu-Zied, R., Cooke, S., Mercone, D., Thomson, J., Croudace, I., Jorissen, F. J., Brinkhuis, H., Kallmeyer, J., and Wefer, G. 2002. Rapid Holocene climate changes in the eastern Mediterranean, in F. A. Hassan, *Droughts, Food and Culture: Ecological Change and Food Security in Africa's Later Prehistory*, pp. 35–46. New York: Kluwer Academic/Plenum Publishers.

Rollefson, G. O., Simmons, A. H., and Kafafi, Z. 1992. Neolithic cultures at 'Ain Ghazal, Jordan. *Journal of Field Archaeology* 19: 443–70.

Ronen, A., and Adler, D. 2001. The walls of Jericho were magical. *Archaeology, Ethnology, and Anthropology of Eurasia* 2(6): 97–103.

Rosenberg, M., Nesbitt, R., Redding, R. W., and Peasnall, B. L. 1998. Hallan Çemi, pig husbandry, and post Pleistocene adaptations among the Taurus-Zagros Arc (Turkey). *Paléorient* 24(1): 25–41.

Savard, M., Nesbitt, M., and Jones, M. K. 2006. The role of wild grasses in subsistence and sedentism: New evidence from the northern Fertile Crescent. *World Archaeology* 38: 179–96.

Schmidt, K. 2000. Göbekli Tepe and the rock art of the Near East. *Turkish Academy of Sciences Journal of Archaeology (TÜBA-AR)* 3: 1–14.

Simmons, A. H. 1999. *Faunal Extinction in an Island Society*. New York: Kluwer Academic/Plenum Publishers.

Stordeur, D., and Abbés, F. 2002. Du PPNA au PPNB: Mise en lumière d'une phase de transition à Jerf el Ahmar (Syrie). *Bulletin de la Société Préhisorique Française* 99: 563–95.

Stordeur, D., Brenet, M., der Aprahamian, G., and Roux, J.-C. 2001. Les bâitements communautaires de Jerf el Ahmar et Mureybet horizon PPNA (Syrie). *Paléorient* 26(1): –29–44.

Tanno, K. I., and Willcox, G. 2006. The origin of cultivation of *Cicer arietinum* L. and *Vicia faba* L. early finds from Tel el-Kerkh, northwest Syria, late 10th millennium B.P. *Vegetation History and Archaeology* 15: 197–204.

Tchernov, E. 1994. *An Early Neolithic Village in the Jordan Valley II: The Fauna of Netiv Hagdud*. American School of Prehistoric Research Bulletin 44. Cambridge: Peabody Museum of Archaeology and Ethnography, Harvard University.

Verhoeven, M. 2002. Transformations of society: The changing role of ritual and symbolism in the PPNB and the PN in the Levant, Syria and south-east Anatolia. *Paléorient* 28(1): 5–14.

Vigne, J.-D., Buitenhuis, H., and Davis, S. 1999. Les premiers pas de la domestication animale à l'Ouest de l'Euphrate: Chypre et l'Anatolie centrale. *Paléorient* 25(2): 49–62.

Vigne, J.-D., and Cucchi, T. 2005. Prèmieres navigations au Proche-Orient: Les informations indirects de Chypre. *Paléorient* 31(1): 186–94.

Voigt, M. M. 2000. Çatal Höyük in context: Ritual at early Neolithic sites in central and eastern Turkey, in I. Kuijt (ed.), *Life in Neolithic Farming Communities: Social Organization, Identity, and Differentiation*, pp. 253–93. New York: Plenum Press.

Weinstein-Evron, M., and Belfer-Cohen, A. 1993. Natufian figurines from the new excavations of the el-Wad cave, Mt. Carmel, Israel. *Rock Art Research* 10: 102–06.

Weiss, E., Kislev, M. E., and Hartman, A. 2006. Autonomous cultivation before domestication. *Science* 312: 1608–10.

Willcox, G. 2005. The distribution, natural habitats and availability of wild cereals in relation to their domestication in the Near East: Multiple events, multiple centres. *Vegetational History and Archaeology* 14: 53454.

Zeder, M. A. 2006. Central questions in the domestication of plants and animals. *Evolutionary Anthropology* 15: 105–17.

Zohary, D., and Hopf M. 2002. *Domestication of Plants in the Old World* (2nd ed.). Oxford: Clarendon Press.

33

DOMESTICATED LANDSCAPES

John Edward Terrell and John P. Hart

One of the most persistent and possibly pernicious ideas in Western thought is the distinction commonly made between things seen as natural, wild, uncultivated, and undomesticated, and things seen instead as unnatural, tamed, cultivated, and domesticated. This distinction is easy to visualize as the difference, for example, between a large field of wheat in the American Midwest and an Amazonian rain forest. The former is a landscape with all the hallmarks of domestication, including the crop itself, mechanized plowing, and the use of petroleum-derived fertilizers. In stark contrast, the latter is perceived as a landscape in a pristine state of being that has been little, if at all, changed or corrupted by human intervention to make such a seemingly wild place suitable for human settlement and land use.

This conceptual divide between the wild and the domesticated, the natural and the unnatural, has been a part of Western intellectual heritage for longer than anyone can say. Elliot Sober (1994) has noted, for instance, that a key idea behind Aristotle's worldview is one that he refers to as Aristotle's *natural state model*. This is the thought that it is possible to distinguish between the natural and unnatural states of any given kind or type. According to Aristotle, variation arises when something is subjected to *interfering forces* that keep it from realizing its normal or *natural state of being*.

Variation can thus be seen as deviation from what is natural (Sober 1994: 210). Newton's first law of motion is an example of such thinking in the physical sciences; so, too, are certain statements about the behavior of objects in the geometry of space-time made under the general theory of relativity in post-Newtonian physics.

Given how entrenched in our Western cultural heritage is the distinction between the natural and the unnatural, it is hardly surprising that domestication as a condition or a phenomenon is still commonly being defined as "a process of increasing mutual dependence between human societies and the plant and animal populations they target" (Zeder et al. 2006: 1). This process may be true, but must human beings always be part of the equation before a species—or a place—can be labeled as "domesticated?"

From a scientific perspective, rather than just from a humanistic point of view, domestication is a form of biological mutualism. What, if anything, sets human domestication apart from other kinds of mutualistic relationships among species except that human beings are somehow involved, intentionally or unintentionally? There is no denying that for human beings, adding us to the picture is key. Yet why limit the study of domestication even in the social sciences so drastically? Why make the circular assertion that what sets domestication "apart from other successful mutualistic relationships is the role of sustained human agency in the propagation and care of plants and animals within the anthropogenic context of domestication" (Zeder et al. 2006: 1)?

It is true that how plants and animals are domesticated by human beings must be unique in at least some respects. Yet every known instance of mutualism in the biological world is undoubtedly unique, or at any rate distinctive. All species are by definition to some degree distinguishable or unique. Therefore, how any two or more species mutually interact with one another is more than likely to be similarly distinguishable. But is "sustained agency" solely characteristic of instances of *human* mutualism with other species? Or is this just another example of the dubious claim, much beloved by our kind, that what we do as human beings has to be exceptionally unique, because human beings are involved?

In the social sciences, separating out what is judged to be natural out from what is seen as unnatural (or "artificial" [e.g., Simon 1996]) continues to make perfect sense to many—for instance, in framing research on the "origins" of agriculture (see Hart 1999, 2001). However, such a distinctly humanistic understanding of plant and animal domestication weakens the value of archaeological research in the evolutionary and ecological sciences. After all, Aristotle's natural state model was long ago discredited in biology precisely because it has long been obvious to many scientists that variation within sexually reproducing species is phenomenal. Hence "there is no biologically plausible way to single out some genetic characteristics as natural while viewing others as the upshot of interfering forces" (Sober 1994: 225). For comparable reasons, we think there is much to be gained by reconfiguring archaeological approaches to the study of domestication to make what archaeologists do and say more compatible with the biological, ecological, and evolutionary sciences. This reconfiguration will in turn inform our understandings of how humans interact with their landscapes and how archaeologists approach human-plant and human-animal interactions on those landscapes.

The Conundrum

For many social scientists—and many conservationists would agree—*Homo sapiens* is the quintessential disturbing force. From such a perspective, what we as a species have been doing to Mother Nature at least since the end of the Pleistocene has been a powerfully disruptive force in what

would otherwise be the Earth's normal pace and course of historical development. For example, as the archaeologist Bruce Smith (2001) writes, many people continue to see the beginnings of agriculture after the Pleistocene as a revolution in history, a turning point marked by two alternative states or end points, one natural (hunting and foraging), the other unnatural (domestication and farming).

Unfortunately there is little agreement today on what are the best constituent definitions of foraging and farming as distinct states or stages of human subsistence life. Without agreement on what these terms mean (Bailey and Headland 1991: 266), there is no dependable way to sort people or societies into one or the other of these two categories—foragers *versus* farmers—or place any given society, modern or ancient, in a sensible way somewhere along what many now concede is the logical continuum between these two ostensibly polarized end-states (Smith 2001: 27).

It is not hard to see why labeling people as foragers or farmers is hard to do. Most people, except perhaps for modern urban dwellers who forage almost exclusively in the supermarket, do both. Even more fundamental, when looked at closely, foraging is not as different from farming as popularly believed; farming, too, is a hazy category that covers a truly diverse range of human behaviors and relationships with other species (for additional discussion and references, see Terrell et al. 2003).

In spite of this ambiguity, when it comes to talk about such long-established issues in world prehistory as the origins of domestication and the beginnings of agriculture, the research agenda in archaeology has changed little over the years. It is still widely taken for granted that archaeologists should be able to pin down when and where some of the Earth's prehistoric inhabitants finally stopped behaving like foragers long enough and successfully enough to be called farmers. And the lucky souls who recover vestiges of those ancient activities are the fortunate few who will be able to announce to the rest of us that they have successfully tracked down the culinary innovators who can be designated posthumously as "the world's first farmers." At the very least, these happy scholars will be able to proclaim that they have discovered a "new cradle of agriculture" (Neumann 2003).

As glamorous as such archaeological discoveries may be, looking for the beginnings of domestication (and we would add, agriculture) is a research pursuit doomed from the start. Why? Because (a) species do not have to be discernibly altered, morphologically or genetically, before they can be domesticated; (b) morphological and genetic changes that sometimes may be taken as "signs of domestication" take time to develop, and consequently they show up, if they are going to show up at all, *after the fact of domestication* by human beings; and (c) concluding that *only* plants and animals exhibiting plainly detectable signs of human use and cultivation can be called "domesticated' risks underestimating the generality and force of human domestication in the world we live in.

Here, then, is an archaeological conundrum. If identifying the origins of domestication and the beginnings of agriculture is as pivotal an issue in archaeology as many still maintain (e.g., Price 2000; Smith 2001), then how are archaeologists to get beyond the concern that they are looking for something they cannot find?

One Solution: Seeing Domestication for What It Is

It is a credit to archaeologists that they have been so persistent in looking for the beginnings of agriculture and domestication, but we think it would help if they now opted to take roads less traveled by. One way to do so would be to begin with the wisdom at the heart of these four basic observations (Terrell et al. 2003):

1. How human beings domesticated other species varies, and has always varied, depending on the species in question and on how extensively people want, or wanted, to exploit them;

2. It follows that domestication can be *gauged more consistently by its performance*—by the manipulative skills characterizing it for each species in question—than by its (only sometimes discernible) *consequences*, that is, the morphological and genetic changes that in due course may or may not become apparent as an upshot of human exploitation;

3. As counterintuitive as it may at first seem, when gauged by its performance, it also follows that *any species may be called domesticated when another species knows how to exploit it*; in spite of what the Book of Genesis tells us, domestication is a *generic fact of life* not a peculiar human endowment; and

4. Finally, and perhaps most important from a scientific point of view, it follows that since people usually exploit *not just a few but in fact many different species* of plants and animals, human beings domesticate (that is, know how to exploit) not only many

particular species but also in effect entire *landscapes*—a word that in this instance should be taken to mean not only certain *places*, or *types of places* such as estuaries, coastal plains, and tropical forests, but also the *species pool*, or *range of species* inhabiting such places that a particular species (in this case *Homo sapiens*) exploits.

Variation Is Real

Two obvious points arise. First, there is no doubt that people over time, intentionally or unintentionally, have altered some species genetically and behaviorally to such a marked degree that nowadays these hapless organisms are no longer viable on their own if they do not receive human care and protection (Gepts 2004). Classic examples of such dependent, or symbiotic, species are maize (*Zea mays* ssp. Mays) and bananas (*Musa* spp.). Second, it is absolutely true that for some species of plants and animals, archaeologically visible signs that they have been the focus of a great deal of human attention are, for plants, increasing seed size over time, and for animals, decreasing bone size (Gepts 2004).

Both of these observations, however, point directly to the hidden defect of traditional ways of thinking about domestication and the origins of agriculture. What is being overlooked or underrated is the overarching truth that domestication in such instances is *transformative*—that is, *evolutionary*—and there are "all degrees of plant and animal association with man" (Harlan 1992: 64). The oft-cited continuum between foraging and farming has more than one axis or dimension. One axis is *behavioral*. Different people use different mixes of what might be labeled as farming and foraging behaviors to make their living. A second axis or dimension is *genetic*. The impact of human exploitation changes different species in different ways and to different degrees ranging from nothing obvious at all (as in the case of domesticated elephants, for instance) to the opposite extreme (for example, sunflowers and the many breeds of dogs).

We would add that if you are one of those who absolutely insist that only species unable to survive without human intervention may be properly called "domesticated," then you should keep in mind that there are even pathogens meeting this restrictive definition of domestication—for example, the virus that causes human acquired immune deficiency syndrome (AIDS). Being stubborn about what is and what is not "true" domestication is a tricky highroad to take.

It is ironic that those conventionally labeled "hunter-gatherers" are perhaps the people who best show us how to see domestication for what it is. As the renowned cultural evolutionist Leslie White (1959) once observed, hunter-gatherers have and have always had abundant and accurate knowledge of the flora and fauna of the places they inhabit. This being so, the "the origin of agriculture was not, therefore, the result of an idea or discovery; the cultivation of plants required no new facts or knowledge" (White 1959: 284). We would add that knowing how to hunt and gather is not all that different from knowing how to plant and cultivate. In both situations, what counts most of all is *knowing what works effectively to put food on the table and a roof over your head.*

The Subsistence Spreadsheet

When it is agreed that variation is real and people have varying ways of making a living, then it follows that archaeological research protocols emphasizing only the retrieval and study of the remains of "truly domesticated species" are misleading ways of exploring the evolution of human subsistence. What is needed instead are research protocols directing us to document not just the co-occurrence and morphological (and genetic) state of a handful of species now considered to be the focus of human domestication, but the full range of varying and variable subsistence strategies that have supported our survival in different places on earth and at different times.

We call one such protocol the "subsistence spreadsheet." Sketched succinctly, this sort of spreadsheet builds on these basic directives (Terrell et al. 2003):

1. **Goal**: *provisioning* of food, shelter, and raw materials;
2. **Observations to be made**: the *occurrence* (presence/absence), *number of individuals*, or *amount* of each species harvested for food or shelter;
3. **Primary variables**: *yield, accessibility*, and *reliability* or *yield stability* (Cleveland 2001: 252) of each available species including:
   a. The specific *yield* provided by each resource being harvested, perhaps measured either in terms of *calories* and *profitability* (energy gain/time) when what is at stake is survival, and when not, then perhaps in locally specific terms of *social value* (measured possibly in locally defined "portions" or *units per person*);

   b. The specific *accessibility* of each resource harvested, both *temporal* (for example, its availability from season to season) and *spatial* or *geographic* (possibly assessed as the time and effort needed to find and harvest a specific resource);
   c. The *reliability*, or *yield stability*, of each resource harvested—how likely it is that each will live up to expectations over time. In evolutionary ecology, this variable is often described as "risk," and models exploring alternative risk management strategies are predicated on the assumption that the suite of resources harvested ought to be a mix of more or less reliable foods that optimizes the likelihood of survival during times of scarcity.
4. **Secondary variables**: *skills* used to achieve the specified goal (behaviors to change or adapt to the yield, accessibility, and reliability of available species populations).

Conclusions

Once it is accepted that people throughout history have been exploiting not only a few but, in fact, many kinds of plants and animals in varying ways and to varying degrees—only some of which might now be described as "true domesticates," then both in effect as well as in practice *Homo sapiens* has been domesticating not just a few species for untold years but entire landscapes for the provisioning of food, useful materials, and shelter. What is challenging then is not finding precisely when and where a few reference species evidently became morphologically or genetically altered enough (according to some formal scale) to tag them as domesticates and allow us to label those associated with such visibly altered species as "farmers" rather than as "foragers"; instead, the real challenge is developing ways of improving how successfully archaeologists can use what they discover to learn about what people in the past were actually doing on the landscapes they inhabited to put food on the table and a roof over their head.

References

Bailey, R., and Headland, T. 1991. The tropical rain forest: Is it a productive environment for human foragers? *Human Ecology* 19: 261–85.

Cleveland, D. 2001. Is plant breeding science objective truth or social construction? The case of yield stability. *Agriculture and Human Values* 18: 251–70.

Gepts, P. 2004. Crop domestication as a long-term selection experiment, in J. Janick (ed.), *Plant Breeding Reviews*, Vol. 24, Pt. 2, pp. 1–44. New York: John Wiley and Sons.

Harlan, J. 1992. *Crops & Man* (2nd ed.). Madison, WI: American Society of Agronomy, Crop Science Society of America.

Hart, J. 1999. Maize agriculture evolution in the Eastern Woodlands of North America: A Darwinian perspective. *Journal of Archaeological Method and Theory* 6: 137–80.

———. 2001. Maize, matrilocality, migration, and Northern Iroquoian evolution. *Journal of Anthropological Method and Theory* 8: 151–82.

Neumann, K. 2003. New Guinea: A cradle of agriculture. *Science* 301: 180–81.

Price, T. D. 2000. Europe's first farmer's: An introduction, in T. Price (ed.), *Europe's First Farmers*, pp. 1–18. Cambridge: Cambridge University Press.

Simon, H. 1996. *Sciences of the Artificial* (3rd ed.). Cambridge. MA: MIT Press.

Smith, B. 2001. Low-level food production. *Journal of Archaeological Research* 9: 1–43.

Sober, E. 1994. *From a Biological Point of View: Essays in Evolutionary Philosophy*. Cambridge: Cambridge University Press.

Terrell, J., Hart, J., Barut, S., Cellinese, N., Curet, A., Denham, T., Kusimba, C., Latinis, K., Oka, R., Palka, J., Pohl, M., Pope, K., Williams, R., Haines, H., and Staller, J. 2003. Domesticated landscapes: The subsistence ecology of plant and animal domestication. *Journal of Archaeological Method and Theory* 10: 323–68.

White, L. 1959. *The Evolution of Culture: The Development of Civilization to the Fall of Rome*. New York: McGraw-Hill.

Zeder, M., Emshwiller, E., Smith, B., and Bradley, D. 2006. Documenting domestication: The intersection of genetics and archaeology. *Trends in Genetics* 22: 139–55.

34

PUNCTUATED LANDSCAPES: CREATING CULTURAL PLACES IN VOLCANICALLY ACTIVE ENVIRONMENTS

Robin Torrence

Understanding how ancient societies coped with volcanic disasters has contemporary relevance since modern population growth and displacement is pushing many groups into high-risk areas (Sheets 2007; Tobin and Montz 1997: 2). Current disaster managers note that effective planning depends on historical knowledge about the behavior of the volcano and also the strategies that human societies have adopted to cope with volcanic activity, both on the long and short term (e.g., Berger 2006). Volcanically active regions provide intriguing opportunities for using the broad lens of landscape archaeology to study persistence, adaptation, and culture change in the face of natural disasters. This chapter highlights the advantages of studying ancient landscapes in environments where humans have experienced volcanic activity. The archaeological potential offered by buried landscapes, tephrochronology, and the (re)creation of spaces is discussed and then illustrated through a case study that has revealed 40,000 years of human response to relatively frequent volcanic activity.

Potentials of Volcanic Landscapes

Many archaeological studies of cultural landscapes focus on the distribution of artifacts, features, art, and monuments across space, but these generally rely on modern surfaces, most of which have had long and complex taphonomic histories. The resulting palimpsests can be very difficult to disentangle in terms of the time scales appropriate for reconstructing change within human societies (e.g., Bradley 2002; Holdaway, Fanning, and Shiner 2005; Wandsnider 1992). In contrast, volcanic activity can freeze the archaeological record at one "moment" in time. The early hominid footprints at Laetoli (Agnew, Demans, and Leakey 1996) and the settlements at Pompeii, Italy (e.g., Allison 2002) or Ceren, El Salvador (Sheets 1992, 2002) are famous examples of the special conditions that have preserved cultural remains intact under volcanic flows and/or airborne ash. Although most studies of the effects of volcanic activity on ancient societies focus on single events such as these, which are usually viewed from the perspective of a single or very few discrete locations (e.g., Allison 2002; see chapters by Driessen and Macdonald, Bicknell, and Plunket and Uruñuela in McGuire, Griffiths, and Stewart 2000; Manning and Sewell 2002; Shimoyama 2002), broadening the spatial and temporal scales of the studies significantly enhances our knowledge about human responses to natural disasters.

The volcanic flows that have buried sites such as Ceren or Pompeii are spatially quite restricted. In contrast, highly explosive volcanic eruptions (called *plinian*) can result in the spread of thick layers of airborne ash (more correctly termed *tephra*) that extend over very large areas and preserve an extensive, continuous surface measured on the appropriate scale of a landscape. A combination of field and laboratory-based methods is used to identify and trace the diagnostic volcanic covering layer across large regions (Cronin and Neall 2000; Elson et al. 2007; Lowe, Newnham, and McCraw 2000; Newnham et al. 1998; Sheets 1983; Sheets and McKee 1994; Torrence et al. 2000; Zeidler and Isaacson 2003). The burial event is usually dated by radiocarbon determinations of material trapped in the volcanic tephra or through a comparison of multiple dates from above and below the volcanic layer (e.g., Buck, Higham, and Lowe 2003), although luminescence dating can also be used (e.g., Torrence et al. 2004). The use of such well-dated, tephra marker layers has enabled scholars to track contemporary patterns of human settlement across large regions (e.g., Cordova, Martin del Pozzo, and Camacho 1994; Cronin and Neall 2000; Elson et al. 2002; Lowe et al. 2002; Machida and Sugiyama 2002; Santley et al. 2000; Sheets 1979, 1983; Sheets and McKee 1994; Torrence et al. 2000; Zeidler and Isaacson 2003; Zeidler and Pearsall 1994).

Reconstruction of cultural landscapes before and after volcanic events enables comparison of how the same geographical "space" was recreated as a social "place," especially if people were forced to flee the region and decolonize later. Since the creation of cultural landscapes from virgin territory is a rare occurrence, comparative studies of old places imbued with memory versus spaces colonized anew following a volcanic event could provide important insights about how cultural landscapes are shaped. Within highly active volcanic areas, eruptions occur relatively frequently. The resulting stratified series of buried ground surfaces dating to different periods provides an intriguing opportunity to study the history of cultural constructions of landscape within a single geographic setting.

The way memory is used within unstable volcanic environments may be different from that in other areas because places and features are frequently altered, obliterated, or buried and are therefore relatively short-lived. Rather than build on and/or reuse older elements of the natural and built environments, people who colonize or return to areas altered by volcanic activity must rework their transported concepts of place within an unfamiliar

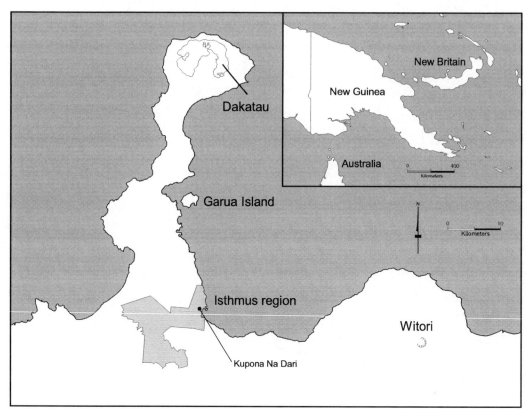

Figure 34.1 Archaeological study areas at Garua Island and the Isthmus region in the Willaumez Peninsula, Papua New Guinea, and the location of Holocene volcanoes.

setting. In contrast, the memory of the disaster can be a powerful element in the construction of cultural landscapes. Volcanoes themselves together with their eruptive products, such as bombs, pumice, and obsidian, may provide potent symbols or become elements of cosmology or religious beliefs, and these are often reflected in the location and the content of rock art and shrines, as well as meanings assigned to volcanic materials (e.g., Allison 2002; Blong 1982; Chester and Duncan 2007; Cronin and Cashman 2007; Dillian 2004, 2007; Elson et al. 2002; Hoffman 1999: 306; Holmberg 2007; Plunket and Uruñuela 1998a, 1998b).

A series of buried landscapes facilitates a comparative perspective that can address significant questions about the effects of natural disasters on human societies (cf. Hoffman 1999). Archaeologists have begun investigating the vulnerability or resilience of societies organized in different ways, the potential effects of differing levels of severity of volcanic event on the nature of culture change, and how cultural landscapes have been (re)colonized, (re)conceived, and (re)constructed after differing degrees of environmental destruction and change. Excellent studies of buried cultural landscapes have been conducted in Central America (e.g., Sheets 1994, 1999, 2007; Sheets et al. 1991; Siebe et al. 1996), South America (Zeidler and Isaacson 2003), Japan (Machida 1984; Shimoyama 2002), and Papua New Guinea (Pavlides 1993, 2004, 2006; Torrence 2002a, 2002b; Torrence and Doelman 2007; Torrence et al. 2000).

A Case Study

Current archaeological research in the Willaumez Peninsula region of New Britain, Papua New Guinea, illustrates how a series of well-dated, short-lived, buried surfaces can facilitate the study of cultural landscapes (Figure 34.1). Throughout the entire period of human history here (ca. 45,000 years), the region has experienced a series of volcanic eruptions, each of which has repeatedly preserved areas at the scale of tens of thousands of square kilometers underneath a thick blanket of airfall tephra. These marker layers are visually and chemically distinctive and temporally secure, and so individual ancient ground surfaces can be traced across the Willaumez Peninsula and beyond (Machida et al. 1996; Pavlides 1993, 2004; Torrence et al. 2000) (Figure 34.2). Since many of these volcanic disasters forced abandonment for periods ranging from several to tens of generations, the resulting punctuated history of occupation can be viewed from the perspective of a variety of cultural landscapes constructed

at different times within the same geographical region. This case study also illustrates the complex interplay between persistence and change in the way people repeatedly (re)created their cultural landscapes following devastating volcanic events.

At the Pleistocene site of Kupona na Dari (Figure 34.1; Torrence et al. 2004), stone artifacts and fire-cracked stones are found on surfaces buried under many of the 14 volcanic layers. The reappearance of cultural material after each disaster indicates that people either survived these devastating events or, more likely, reoccupied the hilltop after a period of abandonment (Figure 34.3). Despite these catastrophic interruptions, the flaked stone tool industry persisted unchanged. It is therefore possible to hypothesize that the various late Pleistocene colonizers may have used the region, and perhaps even conceived of it, in similar ways.

During the most recent 6,000 years, humans experienced four very severe volcanic eruptions from the Witori volcano (named W-K1, W-K2, W-K3, and W-K4) and one major event from Dakataua (named Dk), as well as a host of smaller, localized events (Figures 34.1 and 34.2). The timing of the eruptions has been pinpointed as closely as possible using a Bayesian analysis of radiocarbon dates from within tephras and cultural deposits that bracket them (Petrie and Torrence 2008) (Table 34.1). Given the explosive nature of these five eruptions, the enormous volume of material they produced, and the spatial scale of their impacts (Boyd, Lentfer, and Luker 1999; Machida et al. 1996), there can be no doubt that each generated a disaster resulting in widescale disruption of human populations and potentially large fatalities. For example, the depth of tephras in the study region indicates that immediately following the event the dusty air would have been dangerous to breathe, and virtually all the vegetation would have been destroyed. Geomorphological studies have also shown that the coastline was modified, particularly after W-K2 (Boyd, Lentfer, and Parr 2005) (cf. A and B in Figure 34.4).

Table 34.1 Calibrated dates (B.P.) at two standard deviations for volcanic eruptions in the Willaumez Peninsula. Based on a Bayesian analysis of radiocarbon dates by Petrie and Torrence (2008).

W-K4	1310–1170
Dk	1350–1270
W-K3	1740–1540
W-K2	3480–3160
W-K1	6160–5740

Figure 34.2   The stratigraphy within this test pit in the Willaumez Peninsula, Papau New Guinea, comprises a series of landscapes (dark layers), each of which has been buried by a volcanic tephra (light layers). Since each of the tephras has a very distinctive color and texture, they all act as diagnostic chronometric units over a large region.

Figure 34.3   Fire-cracked cooking stones and obsidian artifacts are preserved between layers of volcanic tephra at Kupona na Dari.

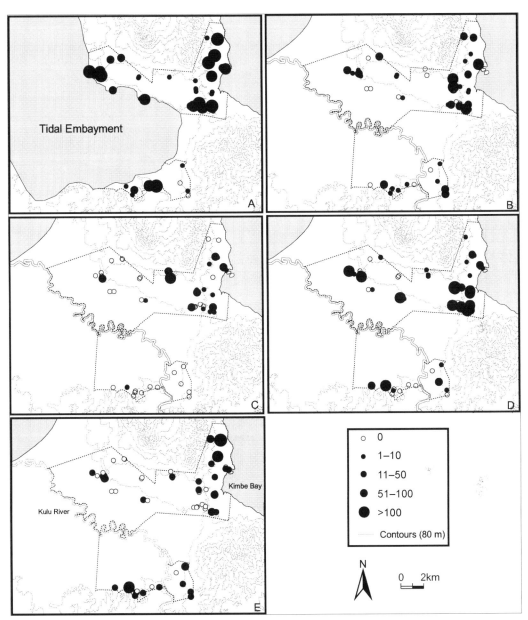

Figure 34.4 Changes in distribution patterns of obsidian artifacts recovered from 1-meter-square test pits in the Isthmus region of the Willaumez Peninsula. The size of the dots reflects the number of obsidian artifacts recovered from a test pit; (A) post-W-K1, (B) post-W-K2, (C) post-W-K3, (D) post-W-K4, (E) most recent 500 years.

Based on an analysis of radiocarbon dates taken from 126 test pits spread across the two study areas on Garua Island (Torrence 2002a; Torrence and Stevenson 2000) and the Isthmus region of the Willaumez Peninsula (Torrence 2002b, 2004a; Torrence and Doelman 2007), the immediate human reaction to all the volcanic disasters appears to have been abandonment. Whether large numbers of people perished in place or found refuge elsewhere is not known, since there has not been adequate research to look for rapid changes in the surrounding regions, although Lilley (2004a, 2004b) has possibly identified an influx of people in the Siassi Islands to the west following the W-K3 eruption. These volcanic events provide opportunities to compare and to contrast how the societies (re)created cultural landscapes within devastated environments.

Dating

Radiocarbon dating cannot pinpoint the exact date when people returned after each disaster, but some interesting patterns can be identified from the Bayesian analysis (Petrie and Torrence 2008). Given that tropical forests decimated by similar volcanic events with comparable tephra thicknesses have been witnessed to regenerate in less than 100 years (Lentfer and Boyd 2001; Thornton 1997), the estimated periods for abandonment (Table 34.2), are surprisingly long, particularly after W-K1. The character and the rate of (re)colonization after each of the volcanic disasters differ. Following the W-K1 event, there is a long gap in time with no evidence of occupation. Through time, the number of places with evidence for human presence increases slowly and erratically. This pattern suggests that small groups moved in sporadically at first and that the population increased gradually. In contrast, after W-K2, which was a larger eruption, a relatively large number of places were first colonized almost simultaneously. This pattern would be consistent with the sudden arrival of a relatively large population. The even greater population that returned to the Isthmus region immediately following the Dk event, together with a large gap in time on Garua Island, suggests the arrivals of refugees who had fled the more seriously affected region near the volcano. With the exception of the very disastrous Dk eruption on Garua, the speed of recolonization increased through time. Part of the explanation for differences in the character of recolonization is due to the particular character of the different events, including variations in tephra depths, subsequent modifications through coastal change, and patterns

Table 34.2 Length of abandonment (calendar years) in the two study areas following volcanic eruptions; based on a Bayesian analysis of radiocarbon dates at two standard deviations by Petrie and Torrence (2008).

PERIOD	LOCATION	
	Garua	**Isthmus**
Post-W-K4		0–170
Post-Dk	0–260	
Post-W-K3	0–270	0–160
Post-W-K2	0–300	0–300
Post-W-K1	0–260	1350–2000

of tephra erosion, but some of the variation is also likely to result from cultural adaptation to disasters, population growth, and cultural conceptions regarding previous "homelands" or "unexplored" areas.

Artifact Assemblages

Changes in the character and the composition of artifact assemblages could indicate that a different cultural group moved into the area following a gap in settlement after the volcanic disaster (Sheets 2007; cf. Sheets et al. 1991). Following the W-K1 event, the reappearance of large retouched obsidian artifacts, which are identical to those made before the eruption (Araho et al. 2002), suggests the same cultural group had survived somewhere else, continued the same cultural tradition of making these artifacts, and then eventually recolonized the area. After W-K2, these artifacts disappeared, and highly decorated Lapita pottery arrived as part of a much wider, regional cultural change and reorganization of landscape. Whether these pottery makers represent the arrival of a new cultural group to this area or the introduced pottery simply indicates change within the previously established group is the subject of intense debate (cf. Lilley 2004b). Interestingly, the subsequent disappearance of pottery in this region is probably not related to volcanic activity. Artifact assemblages do not alter after later events (cf. Torrence 2002a, 2002b).

Patterns of land-use can be inferred from the spatial distribution of obsidian artifacts recovered from the test pits (Figures 34.4 and 34.5) together with studies of plant microfossils (Boyd, Lentfer, and Parr 2005; Lentfer and Torrence 2007). The spatial distribution of artifacts from both study regions demonstrates that the inland areas, which have normally been under-researched by Pacific archaeologists, were consistently used throughout prehistory. In a preliminary analysis of the distributional data, it was shown that there was a steady change through time in the distribution of obsidian artifacts on Garua Island. Following a homogeneous spread of material across the island, there was a shift to a more clustered pattern in which most of the material was discarded in fewer places. A similar process can be identified for the Isthmus region, although it is not so pronounced. This trend has been interpreted as the consequence of a gradual shift toward reduced mobility that was associated with an increase in the intensification of land use (Torrence 2002a; Torrence et al. 2000).

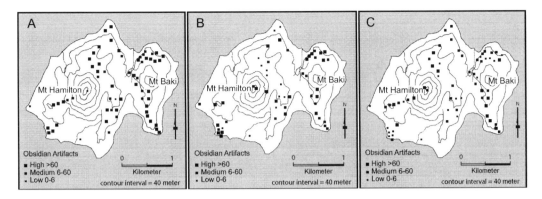

Figure 34.5 Changes in distribution patterns of obsidian artifacts recovered from 1-meter-square test pits on Garua Island. The size of the squares reflects the number of obsidian artifacts recovered from a test pit; (A) post-W-K1, (B) post-W-K2, (C) post-Dk.

It is also important to track the relationship between major environmental changes due to volcanic activity and land use. In the Isthmus area (Figure 34.4), a large tidal embayment was overwhelmed by massive flooding of volcanic debris following the W-K2 eruption. Despite the conversion of the surrounding area from coastal to inland floodplain/swamp, the same specific locations were reoccupied, possibly reflecting memory. In contrast, on Garua Island there are shifts in the distribution of artifacts between the coast and inland between different periods. In what ways the variations between the two study areas are due to local environmental differences, to land-use patterns, or relate to cultural concepts represent an important topic for continuing research.

Social Networks and Exchange

Exchange networks appear to have been an integral part of social life in the Willaumez Peninsula (e.g., Torrence et al. 1996; Torrence and Summerhayes 1997). Social links can be inferred from in the widescale movement of obsidian and the presence at many localities of material derived from multiple sources. Changes in obsidian distribution are partially tied to the effects of volcanic activity, but the role of social processes can also be inferred (Torrence 2004a; Torrence et al. 2004). Volcanic activity restricted the physical access to the sources. The Mopir obsidian source, situated close to Witori, was put out of action by the W-K2 eruption, which cut off its access by sea and buried the surrounding region under tens of meters of pyroclastic flows and airfall tephras. In contrast, following the Dk event when the region surrounding the other sources was abandoned, Mopir became popular again.

Turning to social and cultural factors in the use of obsidian, we see the Kutau source was obviously highly valued, because it was continually used on Garua, although the island has its own local obsidian. The consumption of obsidian imported to the island demonstrates that proximity to a resource was a less important factor in raw material choice than were the social relations forged by exchange. In fact, after the W-K2 eruption, there is a change from the widespread use of multiple sources to a situation in which Kutau dominated the obsidian assemblages in both Garua and the Isthmus region. This pattern suggests that special meaning may have been assigned to the Kutau obsidian outcrops themselves or possibly to the people who owned and exchanged this material. After W-K4, obsidian source use broadened out again to include all the sources, signaling yet another shift in the meaning and the role of the sources, their owners, and/or exchange relationships.

Conceptions of Landscapes

Although environmental and demographic variables must have played an important role in shaping the punctuated history of settlement of the region (Torrence and Doelman 2007), for colonization to take place, people require appropriate concepts of the new lands: for example, as extensions of their homelands, as empty, or as a new opportunity. The devastated area may not have been conceived as an empty space but as a culturally meaningful place that was physically and/or spiritually "dangerous" because of the volcanic activity, as is the case for volcanically active areas in many Pacific Islands today (Cronin and Cashman 2007; Lowe et al. 2002), as "not our territory" or as unknown and therefore "risky." Since the volcanic eruptions in the Willaumez Peninsula created empty lands, the nature of recolonization and the associated patterns of land use and social networks reintroduced or created anew may provide evidence for colonizers' conceptions of landscape.

Given archaeological evidence for the timing of settlement, Pleistocene colonizers must have had conceptions of place (as well as appropriate patterns of land use, mobility, among other responses) that enabled groups to colonize inland areas of New Britain very rapidly (Pavlides 2004) and to maintain stability in the face of ongoing volcanic activity (Torrence et al. 2004). By the time of the W-K1 eruption, however, there appears to have been a shift in focus from flexibility or fluidity to continuity. This suggestion is supported by the maintenance of the repertoire of the large stemmed obsidian artifacts even after the W-K1 disaster and the possibility that these could have been used as primitive valuables to support a self-replicating system of inheritance or prestige (Araho, Torrence, and White 2002; Torrence 2004b). At this time, it seems likely that population levels were low and people may not have envisaged themselves as deliberately moving into a different place; rather, they slowly drifted back. Following the W-K2 event, however, the rapid colonization indicates that these groups had very different idealizations of landscape. As noted previously, at this time special significance appears to have been assigned to the Kutau obsidian sources that were used preferentially. Finally, in the most recent period, large numbers fled Garua Island to the Isthmus region to escape the Dk eruption, and the long abandonment suggests they were reluctant to return.

This history of five colonizations of the same geographical space yields tantalizing variability that provides an opportunity to understand the different ways people could have envisioned and therefore created their cultural landscapes. It is difficult to discern whether similar conceptions motivated the rapid Pleistocene and post-W-K2 colonizations, but both might have incorporated ideas about place that generated population expansion. In contrast, the rapid movement to the Isthmus after the Dk disaster may have simply been the consequence of the loss of livelihood and the desire to persevere, whereas the slow return to Garua hints at a reluctance to return to an area rendered "dangerous" by volcanic activity. Finally, the very long gap following W-K1 suggests that the people who returned had a very different conception of landscape than the later more rapid colonizers.

Conclusions

As illustrated in this brief summary of ongoing research in the Willaumez Peninsula, well-preserved, buried archaeological contexts characteristic of volcanically active regions provide unique windows into past. Opening the "window" as far as possible to view human interactions with their environment and the creation of cultural landscapes across large areas and through long punctuated sequences may yield a rich understanding of conceptions and constructions of place. The advantages of volcanic settings, however, come with a cost. Reconstructing the severity and the character of the volcanic disasters people faced, as well as ancient environments before and after volcanic activity, and tracking their physical recovery demand rigorous interdisciplinary research (e.g., Boyd, Lentfer, and Parr 2005; Sheets 1983; Sheets and McKee 1994; Zeidler and Pearsall 1994). Intensive, systematic fieldwork is also required to locate, study, and date an adequate sample of the often deeply buried archaeological material. Despite such demands, the growing number of studies in volcanic environments show that when these challenges are met, archaeology among the volcanoes can make a significant contribution to our understanding of cultural landscapes.

Acknowledgments

The case study was funded by the Australian Research Council, Australian Museum, Australia and Pacific Foundation, Pacific Biological Foundation, Earthwatch Institute, ANSTO, and New Britain Palm Oil, Ltd., and supported in Papua New Guinea by the National Museum and Art Gallery, National Research Institute, University of Papua New Guinea, West New Britain Provincial Cultural Centre, Kimbe Bay Shipping Agencies, Walindi Resort, and Mahonia Na Dari Research Station. I am grateful for assistance from the many hard-working volunteers and my long-term colleagues Bill Boyd, Hugh Davies, Trudy Doelman, John Grattan, Carol Lentfer, Chris McKee, Ken Mulvaney, Vince Neall, Jeff Parr, Cameron Petrie, Ed Rhodes, Jim Specht, Glenn Summerhayes, and Peter White.

References

Agnew, N., Demans, M., and Leakey, M. 1996. The Laetoli footprints. *Science* 271: 1651–52.

Allison, P. 2002. Recurring tremors: The continuing impact of the A.D. 79 eruption of Mt. Vesuvius, in R. Torrence and J. Grattan (eds.), *Natural Disasters and Cultural Change*, pp. 101–25. London: Routledge.

Araho, N., Torrence, R., and White, J. P. 2002. Valuable and useful: Mid-Holocene stemmed obsidian artifacts from West New Britain, Papua New Guinea. *Proceedings of the Prehistoric Society* 68: 61–81.

Berger, A. 2006. Abrupt geological changes: Causes, effects, and public issues. *Quaternary International* 151: 3–9.

Blong, R. 1982. *The Time of Darkness: Local Legends and Volcanic Reality in Papua New Guinea.* Canberra: Australian National University.

Boyd, W., Lentfer, C., and Luker, G. 1999. Environmental impacts of major catastrophic Holocene volcanic eruptions in New Britain, PNG, in J. Kesby, J. Stanley, R. McLean, and R. Olive (eds.), *Geodiversity: Readings in Australian geography at the close of the 20th century,* pp. 361–72. Canberra: School of Geography and Oceanography, Australian Defence Force Academy.

Boyd, W., Lentfer, C., and Parr, J. 2005. Interactions between human activity, volcanic eruptions and vegetation during the Holocene at Garua and Numundo, West New Britain, PNG. *Quaternary Research* 64: 384–98.

Bradley, R. 2002. *The Past in Prehistoric Societies.* London: Routledge.

Buck, C., Higham, T., and Lowe, D. 2003. Bayesian tools for tephrochronology. *The Holocene* 13: 639–47.

Chester, D., and Duncan, A. 2007. Geomythology, theodicy and the continuing relevance of religious worldviews on responses to volcanic eruptions, in J. Grattan and R. Torrence (eds.), *Living Under the Shadow: Cultural Impacts of Volcanic Eruptions,* pp. 203–23. Walnut Creek, CA: Left Coast Press.

Cordova, C., Martin del Pozzo, A., and Camacho, J. 1994. Palaeolandforms and volcanic impact on the environment of prehistoric Cuicuilco, southern Mexico City. *Journal of Archaeological Science* 21: 585–96.

Cronin, C., and Cashman, K. 2007. Volcanic oral traditions in hazard assessment and mitigation, in J. Grattan and R. Torrence (eds.), *Living Under the Shadow: Cultural Impacts of Volcanic Eruptions,* pp. 175–202. Walnut Creek, CA: Left Coast Press.

Cronin, S., and Neall, V. 2000. Impacts of volcanism on pre-European inhabitants of Taveuni, Fiji. *Bulletin of Volcanology* 62: 199–213.

Dillian, C. 2004. Sourcing belief: Using obsidian sourcing to understand prehistoric ideology in northeastern California, U.S.A. *Mediterranean Archaeology and Archaeometry* 4: 33–52.

———. 2007. Archaeology of fire and glass: The formation of Glass Mountain obsidian, in J. Grattan and R. Torrence (eds.), *Living Under the Shadow: Cultural Impacts of Volcanic Eruptions,* pp. 253–73. Walnut Creek, CA: Left Coast Press.

Elson, M., Ort, M., Hesse, J., and Duffield, W. 2002. Lava, corn, and ritual in the northern Southwest, *American Antiquity* 67: 119–35.

Elson, M., Ort, M., Anderson, K., and Heidke, J. 2007. Living with the volcano: The 11th century A.D. eruption of Sunset Crater, in J. Grattan and R. Torrence (eds.), *Living under the Shadow: Cultural Impacts of Volcanic Eruptions,* pp. 133–52. Walnut Creek, CA: Left Coast Press.

Hoffman, S. 1999. After Atlas shrugs: Cultural change or persistence after a disaster, in A. Oliver-Smith and S. Hoffman (eds.), *The Angry Earth: Disaster in Anthropological Perspective,* pp. 302–26. London: Routledge.

Holdaway, S., Fanning, P., and Shiner, J. 2005. Absence of evidence or evidence of absence? *Archaeology in Oceania* 40: 33–49.

Holmberg, K. 2007. Beyond the catastrophe: The volcanic landscape of Baru, western Panama, in J. Grattan and R. Torrence (eds.), *Living Under the Shadow: Cultural Impacts of Volcanic Eruptions,* pp. 274–98. Walnut Creek, CA: Left Coast Press.

Lentfer, C., and Boyd, B. 2001. *Maunten Paia: Volcanoes, People and Environment.* Lismore: Southern Cross University Press.

Lentfer, D., and Torrence, R. 2007. Holocene volcanic activity, vegetation succession, and ancient human land use: Unraveling the interactions on Garua Island, Papua New Guinea. *Review of Palaeobotany and Palynology* 143: 83–105.

Lilley, I. 2004a. Diaspora and identity in archaeology: Moving beyond the Black Atlantic, in L. Meskell and R. Preucel (eds.), *A Companion to Social Archaeology,* pp. 287–312. Oxford: Blackwell Publishing.

———. 2004b. Trade and culture history across the Vitiaz Strait, Papua New Guinea, in V. Attenbrow and R. Fullagar (eds.), *A Pacific Odyssey: Archaeology and Anthropology in the Western Pacific—Papers in Honour of Jim Specht.* Records of the Australian Museum, Supplement 29: 89–96. Sydney: Australian Museum.

Lowe, D., Newnham, R., and McCraw, J. 2002. Volcanism and early Maori society in New Zealand, in R. Torrence and J. Grattan (eds.), *Natural Disasters and Culture Change,* pp. 126–61. London: Routledge.

Lowe, D., Newnham, R., McFadgen, B., and Higham, T. 2000. Tephras and New Zealand archaeology. *Journal of Archaeological Science* 27: 859–70.

McGuire, B., Griffiths, D., and Stewart, I. (eds.). 2000. *The Archaeology of Geological Catastrophes.* Geological Society Special Publications 171. London: Geological Society of London.

Machida, H. 1984. The significance of explosive volcanism in the prehistory of Japan. *Geological Survey of Japan Report* 263: 301–13.

Machida, H., Blong, R., Moriwaki, H., Hayakawa, Y., Talai, B., Lolock, D., Specht, J., Torrence, R., and Pain, C. 1996. Holocene explosive eruptions of Witori and Dakataua volcanoes in West New Britain, Papua New Guinea, and their possible impact on human environment. *Quaternary International* 35–36: 65–78.

Machida, H., and Sugiyama, S. 2002. The impact of the Kikai-Akahoya explosive eruptions on human societies, in R. Torrence and J. Grattan (eds.), *Natural Disasters and Culture Change*, pp. 313–25. London: Routledge.

Manning, S., and Sewell, D. 2002. Volcanoes and history: a significant relationship? The case of Santorini, in R. Torrence and J. Grattan (eds.), *Natural Disasters and Culture Change*, pp. 264–91. London: Routledge.

Newnham, R., Lowe, D., McGlone, M., Wilmshurst, J., and Higham, T. 1998. The Kaharoa Tephra as a critical datum for earliest human impact in northern New Zealand. *Journal of Archaeological Science* 25: 533–44.

Pavlides, C. 1993. New archaeological research at Yombon, west New Britain, Papua New Guinea. *Archaeology in Oceania* 28: 55–59.

———. 2004. From Misisil Cave to Eliva hamlet: Rediscovering the Pleistocene in interior West New Britain, in V. Attenbrow and R. Fullagar (eds.), *A Pacific Odyssey: Archaeology and Anthropology in the Western Pacific. Papers in Honour of Jim Specht*. Records of the Australian Museum, Supplement 29: 97–108. Sydney: Australian Museum.

———. 2006. Life before Lapita: New developments in Melanesia's long-term history, in I. Lilley (ed.), *Archaeology of Oceania: Australia and the Pacific Islands*, pp. 205–27. Oxford: Blackwell Publishing.

Petrie, C., and Torrence, R. 2006. The radiocarbon chronology of volcanic eruption, human occupation, abandonment and reoccupation in West New Britain. *The Holocene* 18(5), In press.

Plunket, P., and Uruñuela, G. 1998a. The impact of Popocatepetl Volcano on Preclassic settlement in Central Mexico. *Quaternaire* 9: 53–59.

———. 1998b. Preclassic household patterns preserved under volcanic ash at Tetimpa, Puebla, Mexico. *Latin American Antiquity* 9: 287–309.

Santley, R., Nelson, S., Reinhardt, B., Pool, C., and Arnold, P. 2000. When day turned to night, in G. Bawden and R. Reycraft (eds.), *Environmental Disaster and the Archaeology of Human Response*, pp. 143–62. Anthropological Papers No. 7. Albuquerque, NM: Maxwell Museum of Anthropology.

Sheets, P. (ed.). 1983. *Archaeology and Volcanism in Central America: The Zapotitan Valley of El Salvador*. Austin: University of Texas Press.

———. 1992. *The Ceren Site: A Prehistoric Village Buried by Volcanic Ash in Central America*. Fort Worth, Harcourt Brace.

———. 1994. Summary and conclusions, in P. Sheets and B. McKee (eds.), *Archaeology, Volcanism, and Remote Sensing in the Arenal Region, Costa Rica*, pp. 312–25. Austin: University of Texas.

———. 1999. The effects of explosive volcanism on ancient egalitarian, ranked, and stratified societies in Middle America, in A. Oliver-Smith and S. Hoffman (eds.), *The Angry Earth: Disaster in Anthropological Perspective*, pp. 36–58. New York: Routledge.

———. (ed.). 2002. *Before the Volcano Erupted: The Ancient Ceren village in Central America*. Austin: University of Texas Press.

———. 2007. People and volcanoes in the Zapotitan Valley, El Salvador. In J. Grattan and R. Torrence (eds.), *Living Under the Shadow: Cultural Impacts of Volcanic Eruptions*, pp. 67–89. Walnut Creek, CA: Left Coast Press.

Sheets, P., Hoopes, J., Melson, W., McKee, B., Sever, T., Mueller, M., Cheanult, M., and Bradley, J. 1991. Prehistory and volcanism in the Arenal area, Costa Rica. *Journal of Field Archaeology* 18: 445–65.

Sheets, P., and McKee, B., (eds.). 1994. *Archaeology, Volcanism, and Remote Sensing in the Arenal Region, Costa Rica*. Austin: University of Texas.

Shimoyama, S. 2002. Volcanic disasters and archaeological sites in Southern Kyushu, Japan, in R. Torrence and J. Grattan (eds.), *Natural Disasters and Cultural Change*, pp. 326–42.

Siebe, C., Abrama, M., Macias, J., and Obenholzner, J. 1996. Repeated volcanic disasters in prehispanic time at Popocatepetl, Central Mexico: Past key to the future? *Geology* 24: 399–402.

Thornton, I. 1997. *Krakatau: The Destruction and Reassembly of an Island Ecosystem*. London: Harvard University Press.

Tobin, G., and Montz, B. 1997. *Natural Hazards: Explanation and Integration*. London: The Guilford Press.

Torrence, R. 2002a. What makes a disaster? A long-term view of volcanic eruptions and human responses in Papua New Guinea, in R. Torrence and J. Grattan (eds.), *Natural Disasters and Cultural Change*, pp. 292–310. Walnut Creek, CA: Left Coast Press.

———. 2002b. Cultural landscapes on Garua Island, PNG. *Antiquity* 76: 776–76.

———. 2004a. Now you see it, now you don't: Changing obsidian source use in the Willaumez Peninsula, Papua New Guinea, in J. Cherry, C. Scarre, and S. Shennan (eds.), *Explaining Social Change: Studies in Honour of Colin Renfrew*, pp. 115–25. Cambridge: McDonald Institute for Archaeological Research.

———. 2004b. Pre-Lapita valuables in Island Melanesia, in V. Attenbrow and R. Fullagar (eds.), *A Pacific Odyssey: Archaeology and Anthropology in the Western Pacific. Papers in Honour of Jim Specht*, pp. 163–72. Records of the Australian Museum, Supplement 29. Sydney: Australian Museum.

Torrence, R., Bonetti, R., Guglielmetti, A., Manzoni, A., Oddone, M. 2004. Importance of source availability and accessibility: A case study from Papua New Guinea. *Mediterranean Archaeology and Archaeometry* 4: 53–65.

Torrence, R., and Doleman, T. 2007. Chaos and selection in catastrophic environments: Willaumez Peninsula, Papua New Guinea, in J. Grattan and R. Torrence (eds.), *Living Under the Shadow: Cultural Impacts of Volcanic Eruptions*, pp. 42–66. Walnut Creek, CA: Left Coast Press.

Torrence, R., Neall, V., Doelman, T., Rhodes, E., McKee, C., Davies, H., Bonetti, R., Guglielmetti, A., Manzoni, A., Oddone, M., Parr, J., and Wallace, C. 2004. Pleistocene colonisation of the Bismarck Archipelago: New evidence from West New Britain. *Archaeology in Oceania* 39:101–30.

Torrence, R., Pavlides, C., Jackson, P., and Webb, J. 2000. Volcanic disasters and cultural discontinuities in the Holocene of West New Britain, Papua New Guinea. in B. McGuire, D. Griffiths, and I. Stewart (eds.), *The Archaeology of Geological Catastrophes*, pp. 225–44. Geological Society Special Publications 171. London: Geological Society of London.

Torrence, R., Specht, J., Fullagar, R., and Summerhayes, G. 1996. Which obsidian is worth it? A view from the West New Britain sources, in G. Irwin, J. Davidson, A. Pawley, and D. Brown (eds.), *Oceanic Cultural History: Essays in Honour of Roger Green*, pp. 211–24. Wellington: New Zealand Journal of Archaeology Special Publication.

Torrence, R., and Stevenson, C. 2000. Beyond the beach: Changing Lapita landscapes on Garua Island, PNG, in A. Anderson and T. Murray (eds.), *Australian Archaeologist: Collected Papers in Honour of Jim Allen*, pp. 324–45. Canberra: Coombs Press.

Torrence, R., and Summerhayes, G. 1997. Sociality and the short distance trader: Intra-regional obsidian exchange in the Willaumez region, Papua New Guinea. *Archaeology in Oceania* 32: 74–84.

Wandsnider, L. 1992. The spatial dimension of time, in J. Rossignol and L. Wandsnider (eds.), *Space, Time, and Archaeological Landscapes*, pp. 257–82. New York: Plenum.

Zeidler, J., and Isaacson, J. 2003. Settlement process and historical contingency in the western Ecuadorian Formative, in J. Raymond and R. Burger (eds.), *Archaeology of Formative Ecuador*, pp. 69–123. Washington, DC: Dumbarton Oaks Research Library and Collection.

Zeidler, J., and Pearsall, D. (eds.). 1994. *Regional Archaeology in Northern Namabi, Ecuador: Environment, Cultural Chronology, and Prehistoric Subsistence in the Jama River valley*, Vol. I. Pittsburgh, PA: University of Pittsburgh.

CHARACTERIZING LANDSCAPES

Archaeologists concerned with how people in the past were positioned and engaged in *place* need to somehow locate social actions in their spatial and temporal contexts. In their ontological and epistemological positionings, such characterizations comprise the nuts and bolts of archaeological research, the here and the now of past activity and their behavioral milieux. Our ability to understand spatial history rests on our ability to accurately characterize *what* happened in the past, which in turn requires us to understand the *when* and *where* of behavior. But by themselves, "when" and "where" will not enable us to fully understand the *hows* and *whys* of those sociospatial engagements. To do this, we also need to know about the broader contexts of social action—environmentally, communally, nearby, and further afield. We need to also understand what else was going on at and around the time of those human engagements we are ultimately interested in. The reliability of such investigations relies on the chronometric hygiene and spatial integrity of the data we use.

The chapters in this section explore methodologies and techniques by which researchers describe and explain both the archaeological record and the environmental contexts of social action. They outline methods that can be used to archaeologically investigate past relationships between people and their environments—how sites, stone artifacts, plant and faunal remains, charcoal, sediments, human biological traits, and genetic markers found within archaeological sites can be mapped spatially, temporally, and conceptually so as to give us an archaeology of spatial engagement. The technical dimensions of such archaeological analyses are important, but they cannot on their own produce a "landscape archaeology." We need to keep in mind that the use of analytical techniques is from the onset positioned in ontological (how we understand the world to operate) and epistemological (how we understand the world's operation to be capable of investigation) frameworks.

There is an active intellectual and experiential dynamism involved here, as these ontological and epistemological understandings of the world themselves continue to change through ongoing interactions with unfolding research results. For this reason, it is not enough just to know how to measure things in the field or the laboratory—say, sites, stone artifacts, or faunal remains—but, rather, we need to recognize that individual techniques are part of a broader process of understanding and investigation about the past that involves present-day world views and ethical standards (the hermeneutic process). Each of the techniques employed must thus be understood to function ontologically and epistemologically, and, reflexively, the techniques emerge from and affect how we see the world—concepts explored throughout this handbook.

35

DATING IN LANDSCAPE ARCHAEOLOGY

Richard G. Roberts and Zenobia Jacobs

A common need in archaeology is for objects and events of interest to be dated, so that they can be arranged in the correct temporal sequence. Here we briefly review several methods available to the archaeologist to assist in age determination of the "target" object or event—comprehensive treatments of the subject are given by Aitken (1990, 1999), Wagner (1998), and Walker (2005).

We have divided the dating methods into two broad categories: "relative" and "numerical." The latter category includes methods that produce quantitative age estimates that can be placed on a standard timescale, commonly expressed as years before present. Such methods are entirely (or mostly) self-contained, so that an archaeologist can date an artifact, for example, without needing to know where it fits into a typological sequence. By contrast, relative dating methods produce ages that can be positioned on the standard timescale only by reference to a numerical age estimate. So, for example, artifacts in a typological sequence can be chronologically ordered relative to one another, but their age in years before present requires that the sequence is anchored somewhere using a numerical dating method.

We recommend that the dating terminology of Colman, Pierce, and Birkeland (1987) be followed to avoid confusion. Hence, no dating method should be referred to as "absolute," because each age estimate has an associated uncertainty,

and "dates" should be reserved for specific points in time (for example, A.D. 1950). Archaeologists mostly obtain "age estimates" or "ages," which are intervals of time measured back from the present. It is also recommended practice to date a variety of materials using multiple methods to avoid pitfalls in any single method or material and to facilitate a comparison between independent chronologies. The approximate age ranges of the various methods discussed here are summarized in Figure 35.1.

Some of the methods that we describe here are applicable to entire landscapes, as well as to individual archaeological sites. For example, optically stimulated luminescence dating is commonly employed to establish the time of formation of landforms composed of wind-blown and water-lain sediments, whereas argon-argon and fission track methods are used routinely to date regionally extensive ash layers deposited by volcanic eruptions. Many erosional and depositional features of the landscape are also amenable to dating using terrestrial cosmogenic nuclides. Such methods are capable of meeting the needs of landscape archaeologists at all spatial scales. Most dating methods, however, are more restricted in their spatial scope, being applicable to items of archaeological interest recovered from particular sites. For example, electron spin resonance can be used to date individual teeth, radiocarbon methods can be employed to date certain organic materials, and obsidian hydration may be suitable for artifacts made

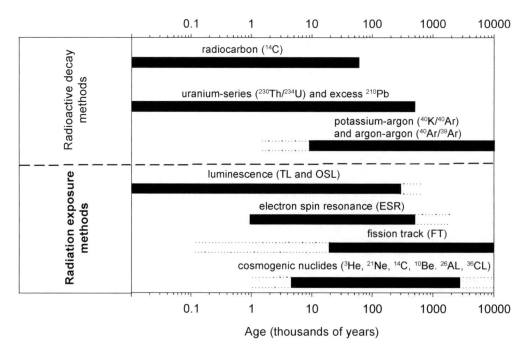

Figure 35.1 Age ranges of the different numerical dating methods applicable to landscape archaeology. The age limits are approximations; they are dictated not only by time-dependent processes but also by the nature of the materials dated. In the case of luminescence dating, for example, the amount of U, Th, and K in the material surrounding each sample will, to a large degree, govern the upper and lower age limits of the method. The dotted lines indicate the maximum and/or minimum limits reached using the methods in ideal situations.

of this type of volcanic glass. Only by comparing the sequence of ages from each archaeological site, along with those obtained from deposits in the surrounding area, can the pattern of human activity be discerned across the landscape. In practical terms, therefore, all dating methods should be viewed as germane to studies of human-landscape interactions.

Relative Dating Methods

Relative dating methods are employed routinely by archaeologists to reconstruct the order in which events occurred. For example, time is commonly implied by placement of artifacts in a typological sequence, and consideration of relative age is also implicit in the study of superposed sedimentary layers (stratigraphy) exposed during archaeological excavation (see also Stern, this volume). In this section, we discuss a range of relative dating methods under four headings: geomorphic, biological, chemical, and physical.

Geomorphic

Stratigraphic sequences containing recognizable boundaries between different strata are testimony

to the episodic accumulation of sediments. In general, younger deposits overlie older deposits, but there are exceptions to this "Law of Superposition" that may have archaeological consequences. For example, sediment mixing after primary deposition may be accompanied by the displacement of artifacts, so additional evidence of stratigraphic integrity must be sought from preserved sedimentary structures. Relative age may also be established from geographically widespread "marker horizons," such as layers of volcanic ash (tephra) produced from the fallout of volcanic eruptions; consolidated tephra is known as tuff. Tephra horizons have distinctive geochemical "fingerprints," so sites in the path of the fallout can be correlated. Numerical dating of tephra (tephrochronology) has been applied in numerous archaeological contexts, including the timing of initial human settlement of New Zealand (Lowe et al. 2000) and the earliest evidence for anatomically modern humans in Africa (McDougall, Brown, and Fleagle 2005). Varved (annually layered) sediments can also be used as relative-age indicators. Varves can form in lake and marine basins with such regularity that the incorporated biological materials can be dated and used to calibrate the radiocarbon timescale

(Hughen et al. 2006; Kitagawa and van der Plicht 1998). Detailed records of human activities and vegetation changes in the surrounding landscape can also be gleaned from changes in the physical, chemical, and biological characteristics of varved lake sediments, as illustrated by Zolitschka, Behre, and Schneider (2003) for a late Holocene site in Germany.

Biological

Tree rings are the biological equivalent of varves, providing a record of annual growth of tree trunks that currently extends back to about 12,400 years before present (Reimer et al. 2004). Dendrochronology is the most precise dating method available to archaeologists. It can be applied to wood from live and dead trees to determine the pattern of ring-width, which varies in response to the prevailing climatic conditions. By comparing the ring-width patterns for trees that overlap in age, a "floating" or relative chronology can be developed and then anchored to a numerical timescale by means of radiocarbon dating.

Dendrochronology has been deployed at Holocene archaeological sites around the world, the first applications being made to ancient timbers from pueblo dwellings in the American Southwest (Towner 2002). The growth rate of other plants, notably lichens, has also been exploited as a relative dating method. Lichens rapidly colonize many hard, bare surfaces, and some species exhibit a regular pattern of growth over several hundreds or thousands of years. Lichenometry has been used mostly to estimate exposure ages for geomorphic features related to late Holocene deglaciation, such as moraine ridges and raised shorelines (e.g., Loso and Doak 2006; O'Neal 2006). Lichenometry has been used occasionally to date archaeological sites found in association with such features, and the method could also be applied to recent rock art—and not only in arctic and alpine regions (Bednarik 2002).

Chemical

Chemical changes can also be used to track time. Perhaps the best known of these in an archaeological context is amino acid racemization (AAR), which is based on the gradual conversion of amino acids from the exclusively left-handed (L) form in living proteins to an equilibrium ratio of left- and right-handed (D) forms in organisms after death. The D/L ratio is, thus, an indication of relative age, but the correlation is not simple, because other factors (particularly temperature) also influence the rate of racemization, which must be calibrated against a numerical dating method (usually radiocarbon) (Clarke and Murray-Wallace 2006). The most reliable materials are those that remain chemically "closed" postmortem, such as mollusk shell and eggshell, whereas "open" systems (for example, bone) are much less reliable (Grün 2006). The amino acids trapped within the crystal structure of mollusk shell (Penkman et al. 2008) were used to constrain the age of the oldest artifacts yet discovered in northern Europe (Parfitt et al. 2005). AAR dating has also been applied to ostrich eggshell found at early modern human sites in southern Africa (Miller et al. 1999a) and to cassowary and brush-turkey eggshells from occupation sites in Australia and New Guinea (Clarke et al. 2007). Other examples include the measurement of D/L ratios in eggshells of the extinct giant birds *Genyornis* and *Aepyornis* to provide a timescale for megafaunal extinction events—which are commonly attributed to human impact—in Australia and Madagascar, respectively (Clarke et al. 2006; Miller et al. 1999b).

The oxygen-isotope composition of mollusk shells has also been used as a relative dating method for archaeological sites (Shackleton 1982). In this instance, changes in the ratios of ^{18}O to ^{16}O (denoted as $\delta^{18}O$) can be matched to $\delta^{18}O$ variations in deep-sea sediments and polar ice caps for which numerical timescales have been established, sometimes at annual resolution (e.g., North Greenland Ice Core Project members 2004).

Other chemical changes, that form the basis of relative dating methods, can be broadly categorized as products of weathering processes. For example, the degree of soil development (pedogenesis) can be used as a proxy for time, provided account is taken of the many other factors that also affect their formation (Birkeland 1999). Likewise, many rock surfaces become gradually altered over time, acquiring weathering rinds (which extend inwards from the surface) or mineral coatings (which accrete on to the rock surface). Relative ages for archaeological rock-structures in Canada were obtained from the relative hardness of their surfaces, which reflects the extent of weathering (Betts and Latta 2000).

In desert regions, rock surfaces are commonly encrusted in iron and manganese oxides, which form a "varnish." These sometimes occur on top of rock engravings, for which minimum ages could be obtained by dating the time of varnish formation. In the 1980s, cation-ratio (CR) dating was proposed as a means of estimating the age of rock varnish by comparing the relative amounts of potassium and calcium to titanium (Dorn 1983; Dorn, Nobbs, and Cahill 1988). The first two cations are leached more readily than titanium by weathering processes, so the (K+Ca)/Ti ratio will generally decrease over time. But varnish chemistry is highly variable at a range of spatial scales (Bierman et al. 1991), making it necessary to calibrate the CR timescale (using a numerical dating method) for the varnish directly overlying the engraving of interest. This is not always practicable, and the method has since fallen out of favor, not least because of difficulties in obtaining reliable radiocarbon ages for CR calibration (Beck et al. 1998; Bednarik 2002). Another approach, called varnish microlamination (VML) dating, exploits the optical and chemical properties of microscopic varnish layers, which form in response to climate variations. This technique has been applied to stone tools exposed on dryland surfaces in the western USA, but it requires some form of calibration to obtain numerical ages (Liu and Broecker 2007).

An alternative method of estimating the relative age of rock engravings is "microerosion dating" (Bednarik 1992, 2002). This refers to the progressive increase in roundness of individual mineral crystals, as a function of exposure time to chemical weathering processes, following crystal breakage during the act of engraving. Measurements of crystal roundness, preferably for two or more minerals, are made in the field using an optical microscope. The time elapsed since an engraving was made is then estimated by reference to calibrations performed on known-age engravings on similar rock type exposed to similar environmental conditions. Because of these calibration constraints, and the restricted application of the method to comparatively resistant rock types, microerosion dating has received scant attention, although it does target the archaeological event of interest.

Another form of chemical alteration is the slow absorption of moisture by freshly flaked surfaces of obsidian, a type of volcanic glass used for tool manufacture in many regions of the world (e.g., North America, Papua New Guinea, and New Zealand). Obsidian hydration dating is based on the increases in thickness of the hydration layer over time, but other variables, particularly temperature, also control the rate of hydration (Hull 2001; Rogers 2007). This temperature dependence provides a means of reconstructing past climates (Anovitz et al. 2006), as done previously using D/L ratios in emu eggshells (Miller et al. 1997).

Physical

Many rocks and sediments contain iron-bearing minerals, which have magnetic properties. Archaeologists can exploit mineral magnetism as a relative dating method from measurements of magnetic susceptibility and, more commonly, from records of temporal variations in the Earth's magnetic field. The latter application is known generally as palaeomagnetic dating, and as archaeomagnetism when applied to archaeological sites and artifacts no more than a few thousand years old. The Earth's magnetic field has varied in both intensity and direction over time, and these changes can be reconstructed from the orientation of magnetic minerals, which align themselves with the geomagnetic field prevailing at the time of rock formation or sediment deposition. The Quaternary has been punctuated by several major magnetic reversals, including the most recent inversion to "normal" polarity about 780,000 years ago. Short-lived magnetic "excursions" (for example, Laschamp and Mono Lake) have occurred since then, and these can be used as marker horizons to correlate between sedimentary sequences that are geographically distant. Smaller, secular geomagnetic variations occur over timescales of a few hundred years and are of limited spatial extent, so an archaeomagnetic "master curve" must be established for each region, as Lengyel and Eighmy (2002) have done for the American Southwest. At the opposite end of the Quaternary, the magnetostratigraphy of sediments containing stone tools has been used to infer a human presence in northeast Asia by 1.66 million years ago (Zhu et al. 2004), and the age of the earliest-known hominin remains in western Asia (Calvo-Rathert et al. 2008) and Europe (Carbonell et al. 2008). In contrast to palaeomagnetism, magnetic susceptibility is based on the abundance, size, and shape of magnetic minerals in sediments, rather than their orientation. Variations in magnetic susceptibility reflect environmental changes and are widely used for regional palaeoclimatic reconstructions. Environmental magnetism can, therefore, be used to correlate between archaeological sites affected by synchronous environmental changes (Dalan and Banerjee 1998).

Numerical Dating Methods

Archaeologists ultimately want to know the number of years that have elapsed since artifacts were made or other events of interest occurred. To achieve this degree of temporal resolution requires the application of numerical dating methods. None of these methods, however, can match the potential for annual precision offered by dendrochronology, varved sediments, and ice cores, owing to the uncertainty associated with the various measurements made to determine the numerical age. Uncertainties on numerical ages are most commonly expressed as standard errors at 1σ or 2σ, which correspond to the 68% and 95% confidence limits in the case of normal (Gaussian) probability distributions. Hence, there is a one-in-3 chance that the true age will lie outside ± 1σ, and a one-in-20 chance that it will fall beyond ± 2σ. These statistical considerations should always be borne in mind by the archaeologist, because "the uncertainty of a date is no less significant than the date itself" (Ludwig 2003: 632). It is also important to recognize that the quoted uncertainty denotes the statistical *precision* of an age estimate—that is, a measure of experimental reproducibility—but not its *accuracy*. The latter indicates how closely the measured age approximates the true age of the sample. Ideally, ages should be both accurate and precise, but accuracy is paramount. Aitken (1990) and Wagner (1998) provide introductions to accuracy and precision in geochronology, and explain when and how age estimates can be combined to calculate a pooled (weighted) mean.

Each of the following methods requires that samples be collected and stored in an appropriate manner to minimize the risk of contamination by older or younger materials and then prepared ("pretreated") for dating using suitable procedures to isolate the target materials from potential laboratory contaminants. Methods of sample measurement and data analysis likewise follow strict conventions, which differ between methods, and interlaboratory comparisons provide additional assurance that reliable procedures have been implemented (e.g., Scott et al. 2004). Guidelines on sample collection procedures and details of laboratory protocols for each method can be found in the references cited below. We recommend that a qualified dating practitioner be consulted before sampling to ensure that samples are not contaminated during collection, transport, or storage and that all the information needed for accurate and precise age determination is gathered in the field.

Radioactive Decay

Three of the most widely used numerical dating methods in archaeology are based on the radioactive decay of naturally-occurring elements: ^{14}C (radiocarbon dating), ^{238}U (uranium-series dating), and ^{40}K (potassium-argon and argon-argon dating).

Radiocarbon Dating. Radiocarbon dating was the first to be developed, in the 1940s, by a team led by Nobel laureate Willard Libby, once it had been realized that interaction between ^{14}N and cosmic rays in the upper atmosphere results in the production of ^{14}C. This then oxidizes to a particular form of carbon dioxide, which becomes incorporated into the living tissues of plants and animals,

and into carbonates formed in the oceans and on land. When the plant or animal dies, ^{14}C uptake ceases, and the amount of radiocarbon decreases by 50% every 5,730 ± 40 years (the so-called "Cambridge" half-life of ^{14}C), owing to radioactive decay. Each decay is accompanied by the emission of a beta particle, which can be counted to determine the concentration of ^{14}C (in what is variously referred to as conventional, radiometric, or classical ^{14}C dating). Alternatively, the number of ^{14}C atoms can be counted directly using an accelerator mass spectrometer (AMS); hence, the term "AMS ^{14}C dating." In either case, ^{14}C dating is restricted to the last 60,000 years or so, beyond which too little ^{14}C remains to be measured and ages are reported as "infinite."

Radiocarbon dating can be applied to a great variety of organic materials encountered at archaeological sites, including the remains of plants (for example, wood, charcoal, and seeds) and animals (for instance, bone and shell). But other sources of carbon amenable to ^{14}C dating may also occur in archaeological contexts, such as speleothems in caves, residues on pottery, artifacts made of iron, and mineral accretions on rock art. Walker (2005) describes a number of routine and unusual applications of ^{14}C dating, illustrated by archaeological examples from around the world, and Bronk Ramsey (2008) offers much useful advice about ^{14}C dating in archaeology. For each type of sample, the ^{14}C related to the target event must be separated from any older or younger ^{14}C. Contamination by older, water-borne carbon needs to be considered when dating speleothems and shell, for example, and there is also the potential problem of artifacts being made from preexisting old shells (Rick, Vellanoweth, and Erlandson 2005). The "old wood" effect (Wagner 1998) may be of concern in archaeological contexts where preexisting old wood or charcoal is dated instead of the remains of plants that died during the period of human occupation (e.g., Armitage et al. 2001; Kennett et al. 2002). In such cases, the ^{14}C age may be substantially older than the target archaeological event. The use of heartwood from long-lived tree species will increase the age discrepancy—especially for Holocene samples—so preference should be given to wood from directly below the bark or to short-lived plant remains (for example, twigs and seeds).

The most common concern, however, is the possible postdepositional contamination by younger carbon of charcoal and bone, which form the basis for many archaeological ^{14}C chronologies. Techniques of charcoal and bone pretreatment have improved greatly in the last decade, with the development of acid-base wet oxidation and stepped-combustion (ABOX-SC) procedures for charcoal (Bird et al. 1999) and molecular ultrafiltration procedures for bone collagen (Higham, Jacobi, and Bronk Ramsey 2006). The ^{14}C ages of older samples are particularly affected by younger contaminants. So, by removing more of the latter, these techniques have increased the accuracy of the resulting ^{14}C ages—compared to less rigorous sample pretreatments, which can yield ^{14}C ages that are too young by several millennia—and extended the limit of the method from about 40,000 to 60,000 years (e.g., Bird et al. 2003; Jacobi et al. 2006; Turney et al. 2001). It is also feasible, but not straightforward, to isolate and to date specific biomolecules containing carbon, such as amino acids and lipids (Stafford et al. 1991; Stott et al. 2003), to further increase confidence that a ^{14}C age relates directly to the target event.

Efforts have also been made to improve the reliability of ^{14}C dating of rock art, using a variety of sample types. In some cases, carbon can be extracted from the material used to make the picture, as with beeswax figures (Watchman and Jones 2002), charcoal drawings (Armitage et al. 2001; Valladas et al. 2001), and plant-fiber binders in paints (Mazel and Watchman 1997). A range of carbon-bearing mineral coatings—variously referred to as "crusts," "skins," and "varnishes"—have also been investigated. Where these cover paintings and engravings, minimum ages for the underlying art can, in principle, be obtained by dating the organic component of the accretions. Silica skins and, in particular, calcium oxalate crusts have yielded seemingly reliable ^{14}C ages (Watchman et al. 2000, 2005), whereas those obtained from desert varnish and calcite have proven more equivocal (Beck et al. 1998; Plagnes et al. 2003). But no sample type is without potential complications (Bednarik 2002; Pettitt and Pike 2007), so alternative approaches to numerical dating of rock art have also been explored (Plagnes et al. 2003; Roberts et al. 1997; Watanabe et al. 2003).

Radiocarbon ages are conventionally reported in "radiocarbon years B.P.," where B.P. means "before present." For historical reasons, "present" refers to the year A.D. 1950 and the incorrect "Libby" half-life for ^{14}C (5,568 ± 30 years, which is 3% too short) is knowingly used to calculate the ages! Archaeologists should check if the ^{14}C ages received from the dating laboratory have been calculated using the Cambridge or Libby half-life. But correction for the latter is insignificant compared to the adjustment needed to convert, or "calibrate," conventional ages into calendar-year ages, owing to long-term variations in the production rate of

^{14}C. IntCal04 is the latest international effort to calibrate the radiocarbon timescale (Reimer et al. 2004). Calibration can be achieved to annual resolution over the last 12,400 years, where dendrochronology can be applied, and with much lower resolution to 26,000 years ago using varved sediments and fossil corals, the latter dated by both ^{14}C and uranium-series. Before 26,000 years, there is disagreement about the extent of correction needed (van der Plicht et al. 2004), although a consensus data set is beginning to emerge (Hughen et al. 2006). To distinguish conventional ^{14}C ages (in radiocarbon years) from calibrated ^{14}C ages (in calendar years), the latter are expressed as "cal years B.P." For the period 0–1,000 cal years B.P., contemporaneous samples in the Northern and Southern Hemispheres differ by between 8 and 80 radiocarbon years. Older ages are obtained from the Southern Hemisphere, for which a separate calibration (SHCal04) is available back to 11,000 cal years B.P. (McCormac et al. 2004).

Some archaeological samples, usually shells of terrestrial and marine mollusks, may require age calibration for the "reservoir effect," which is the uptake of noncontemporaneous (usually older) carbon by the sample while alive. Terrestrial and freshwater snails, for example, may incorporate older carbon in their shells from sources such as dissolved limestone, while modern marine mollusks can give apparent ages of several hundred years (or more) because the carbon reservoir in seawater is a mixture of modern and ancient ^{14}C. Because the mixing ratio varies with location, it is necessary to estimate the difference (defined as ΔR) in ^{14}C years between a known-age marine sample from the region of interest and the average global ocean reservoir age to calibrate ages for marine samples using programs such as Marine04 (Hughen et al. 2004). Site-specific ΔR values are commonly estimated from modern samples, but these may not compensate for fluctuations in marine reservoir ages in the past (Fontugne et al. 2004).

A different effect on the carbon reservoir, but one with potential uses for historical and forensic archaeologists, is the "bomb" pulse in ^{14}C concentration produced by above-ground testing of nuclear weapons between 1955 and 1963. These tests injected large amounts of ^{14}C into the atmosphere, with a peak in the mid-1960s followed by an exponential decline after implementation of the international test ban treaty (Hua and Barbetti 2004). The bomb pulse can be used to estimate time since death with a precision of 1–2 years for tooth enamel, hair, and bone lipids (Geyh 2001; Spalding et al. 2005; Wild et al. 2000); other archaeological applications include dating of rock art in Vanuatu (Wilson et al. 2001). Except at the bomb peak, the measured ^{14}C concentration will yield two ages for a sample—corresponding to the points of equal activity on the rising and falling limbs of the pulse—resulting in an "either/or" choice of age.

Uranium-Series Dating. Uranium-series dating refers to a number of related techniques, all of which are based on the radioactive decay of natural uranium. This consists of two forms of uranium, that differ in their atomic mass: ^{238}U (which accounts for almost 96% of the radioactivity) and ^{235}U, each of which is the "parent" nuclide at the head of a decay chain of "daughter" products resulting from successive radioactive decays. The final product in each chain is a stable form of lead.

The parent nuclides have half-lives of about 4.5 and 0.7 billion years for ^{238}U and ^{235}U, respectively, which are too long for most archaeological purposes. But several of their decay products have much shorter half-lives, and that of ^{230}Th (75,700 years), a daughter nuclide in the ^{238}U chain, has proven especially useful for archaeological dating. Uranium is soluble in water but thorium is not, so the activity ratio of ^{230}Th to ^{234}U, its immediate parent, can be used to estimate when certain materials were formed. The most reliable ^{230}Th/^{234}U ages are obtained from "closed systems" in which uranium, but not thorium, was present at the time of formation and all of the ^{230}Th has resulted from radioactive decay of ^{234}U. This is generally the case for speleothems and travertines formed from clean crystals of calcite, and for unaltered corals. Some mollusk shells and eggshells may also be suitable, but these archaeological materials, as well as bones, teeth, and any "dirty" samples (that is, carbonates contaminated by thorium-enriched detritus), require special measures to obtain accurate ages.

As with ^{14}C dating, techniques of U-series dating have evolved over time; see Ivanovich and Harmon (1992) and Bourdon et al. (2003) for comprehensive reviews of the method and its applications. The original measurement technique involved counting the alpha particles emitted by radioactive decay, but this has been largely superseded by the atom-counting methods of thermal ionization mass spectrometry (TIMS) and inductively coupled plasma mass spectrometry (ICP-MS), which can generate high-precision ages for the last 500,000 years or so. Dating applications in archaeology span the full age range of the technique, from coral offerings at temples in protohistoric Hawaii (Kirch and Sharp 2005) to speleothems interbedded with *Homo erectus* fossils in China (e.g., Shen et al.

2004). Efforts at dating calcite accretions on cave paintings in Southeast Asia have met with mixed success (Aubert et al. 2007; Plagnes et al. 2003).

Bone has generally been avoided in U-series dating because of its "open system" behavior, but recent developments have renewed the possibility of obtaining reliable ^{230}Th/^{234}U ages from bone and, to a lesser extent, teeth (Eggins et al. 2005; Grün 2006; Pike, Hedges, and van Calsteren 2002). The approach involves measuring the uranium concentrations across sectioned pieces of bone and comparing the U-profiles with models of uranium diffusion and adsorption. Reliable ages can be obtained from bones with certain U-profiles, but these may occur in only a minority of cases. Laser-ablation multicollector ICP-MS permits the rapid analysis of a large number of samples and is almost nondestructive, so precious samples (for example, hominin fossils) can be considered for dating. Nondestructive uranium-series dating by high-resolution gamma-ray spectrometry has also been tested on certain hominin fossils, but measurement of ^{234}U (in particular) is neither simple nor fast (Grün 2006). Also, doubts have been raised about the accuracy of some gamma spectrometric ages, as illustrated by the dispute over the age of the human burials at Lake Mungo in Australia (Bowler et al. 2003; Thorne et al. 1999).

Support for "closed system" behavior can be gained from protactinium (^{231}Pa), the only long-lived daughter product of ^{235}U decay. The half-life of ^{231}Pa is 34,300 years, so comparisons between ^{230}Th/^{234}U and ^{231}Pa/^{235}U ages can, in principle, be made for samples up to 200,000 years old (Cheng et al. 1998). In practice, however, the low natural abundance of ^{235}U limits the precision of protactinium dating and restricts its application to samples with high uranium concentrations.

Much older archaeological events may be amenable to uranium-lead dating (Richards and Dorale 2003). The basis of the technique is the same as above, with the age being obtained from the activity ratios of ^{206}Pb to ^{238}U and of ^{207}Pb to ^{235}U—the parent nuclide and final (stable) daughter product in each decay chain—measured by TIMS, multicollector ICP-MS (Woodhead et al. 2006) or with a sensitive high-resolution ion microprobe (SHRIMP). In theory, the age range of U-Pb dating extends significantly beyond that of ^{230}Th/^{234}U dating, but the limits are governed in practice by the extent of sample contamination by lead-enriched detritus. Walker, Cliff, and Latham (2006) used the U-Pb method to obtain ages of about 2.2 million years for speleothems above and below hominin fossils at Sterkfontein in South Africa.

It is also possible to use ^{210}Pb, one of the decay products late in the ^{238}U chain, to date recent lake, estuarine, and marine sediments. The method is based on the continuous, natural fallout of "excess" ^{210}Pb from the atmosphere on to exposed surfaces, and its subsequent incorporation into sedimentary deposits (Ivanovich and Harmon 1992). The term "excess" refers to the fact that the ^{210}Pb of interest is not supported by radioactive decay of *in situ* parent nuclides, and so will decay away over time. With a half-life of 22 years, ^{210}Pb can be used to date sediments deposited in the last 120 years by making some assumptions about how much unsupported ^{210}Pb was deposited at the site over time and if there has been any postdepositional mixing. Ages are inferred from vertical profiles of excess ^{210}Pb activity versus depth, by assuming either a constant rate of supply (CRS) or a constant initial concentration (CIC) of excess ^{210}Pb. But the accuracy of the chosen model can be validated only by comparing the ^{210}Pb chronology with some independent measure of age (Smith 2001). The latter is commonly obtained from the A.D. 1963 peak in concentration of the short-lived radionuclide ^{137}Cs (half-life of 30 years), which was deposited worldwide as atmospheric fallout from nuclear weapons testing. Recent archaeological applications of ^{210}Pb dating include studies of human impacts on vegetation and landscape development using lake sediments in northeast England (Oldfield et al. 2003) and shallow marine deposits off western Java (van der Kaars and van den Bergh 2004).

Potassium-Argon and Argon-Argon Dating. Potassium-argon and argon-argon dating are closely related techniques, both based on the decay of the radioactive form of potassium (^{40}K) to argon (^{40}Ar), a stable, gaseous daughter product. Potassium is present in many natural minerals at the time of formation, being especially abundant in volcanic minerals, and ^{40}K accounts for a very small proportion (0.012%) of this. The principle of the method is simple. When molten, volcanic minerals contain ^{40}K but not ^{40}Ar, because all of the argon gas escapes into the atmosphere. After cooling below a certain "closure temperature" (which depends on the particular mineral), ^{40}Ar will slowly accumulate inside the crystal owing to the decay of ^{40}K, and the ^{40}Ar/^{40}K ratio can be used to estimate the time elapsed since crystallization. The potassium-argon method is applicable to most of Earth history, because of the long half-life of ^{40}K (1250 million years), but dating of archaeological events during the last 100,000 years is limited by the low precision on the ages.

Archaeological applications at the younger end of timescale require the use of the argon-argon technique, whereby the age is obtained from the ratio of ^{40}Ar to another form of argon (^{39}Ar) in the crystal (McDougall and Harrison 1999). The basic premise of the method is the same as for potassium-argon dating, with the difference being that some of the ^{40}K is converted into nonradioactive ^{39}Ar by bombarding the sample with fast neutrons in a nuclear reactor. The ^{40}Ar/^{39}Ar ratio can then be measured using the same instrument (a mass spectrometer), with a resulting improvement in the precision of the ages. Their accuracy can also be improved by measuring ^{40}Ar/^{39}Ar ratios for individual crystals to identify samples composed of crystals of differing age.

There are three possible complications with potassium-argon and argon-argon dating: sample or instrument contamination with atmospheric ^{40}Ar (as opposed to the tiny amount of radiogenic ^{40}Ar produced in the crystal by decay of ^{40}K), the presence of "excess" ^{40}Ar that never escaped from the crystal at the time of its formation, and the loss of radiogenic ^{40}Ar after crystallization due to weathering, for example. The first of these concerns is addressed by also measuring the amount of atmospheric ^{36}Ar and making the necessary corrections, while the other two concerns are dealt with by heating the sample to successively higher temperatures to release the ^{40}Ar and ^{39}Ar in a series of steps. The age is determined for each increment, with the "plateau" region taken as the most reliable indicator of crystallization age. This approach can be applied to individual crystals, using a laser to release the argon, so that any contaminant grains in a sample can be identified and discarded.

Argon-argon dating has been used in many archaeological contexts, especially in east Africa where fossils of early hominins (e.g., Leakey et al. 2001) and the genus *Homo*, including the oldest-known remains of our species (McDougall, Brown, and Fleagle 2005), have been found in stratigraphic association with tuffs. Early *Homo* fossil sites on the island of Java in Indonesia (Swisher et al. 1994) and at Dmanisi in the Republic of Georgia (Gabunia et al. 2000) also have ^{40}Ar/^{39}Ar age constraints, while recent volcanic events, such as the eruption that destroyed Pompeii around 2,000 years ago (Renne et al. 1997), can be dated if the deposit contains large crystals of potassium-rich sanidine. However, when artifacts and human remains are not situated between *in situ* volcanic deposits, and cannot be provenanced with certainty, archaeologists should bear in mind that the age obtained from ^{40}Ar/^{39}Ar dating is that of mineral crystallization and not that of any subsequent episode(s) of reworking.

Radiation Exposure

Five techniques make up this group of numerical dating methods: thermoluminescence (TL), optically stimulated luminescence (OSL), electron spin resonance (ESR), fission track (FT), and terrestrial cosmogenic nuclide (TCN) dating. Ages are obtained by measuring the cumulative effect of ionizing radiation on the crystal structure of certain minerals, some of which (for example, hydroxyapatite) occur in teeth and bones. The greater the amount of energy absorbed (TL, OSL, ESR, and TCN) or crystal damage (FT), the longer the duration since first exposure to radiation and, consequently, the greater the age of the material being dated.

Luminescence Dating. Luminescence dating is a term that embraces a range of related methods that can be applied over different time periods to different minerals and in different archaeological settings. When one method is not appropriate for a particular situation, then another may be. The technique is based on the fact that natural minerals (such as quartz and feldspar) are not perfectly formed but contain defects in their crystal lattices that are able to trap negatively charged electrons and positively charged vacancies ("holes"). Some of these charges can remain trapped for millions of years and provide a measure of the energy absorbed by the crystal owing to exposure to ionizing radiation. The latter consists mainly of gamma rays, beta particles, and alpha particles, which are emitted during the nuclear decay of uranium, thorium (and their daughter products), and potassium contained inside the mineral grains and dispersed throughout the surrounding deposit; there is also a lesser contribution from cosmic rays. The energy absorbed increases with the amount of radiation exposure, and estimation of the corresponding radiation dose is the "clock" that forms the basis of all luminescence dating methods. These have been shown to work for samples as young as a few years, to sediments deposited several hundreds of millennia ago. Luminescence ages are obtained in calendar years and typically have precisions of 5–10%.

Luminescence dating relies on the fact that the charge-traps of interest were emptied ("zeroed") at, or close to, the time of the target event, such as that of pottery or flint tool manufacture or the time of burial of artifacts and human remains. For archaeological applications, there are two main zeroing processes: heating of the crystal to at least 300°C by, for example, the intentional firing of pottery or the accidental burning of stones in a hearth; or exposure of the crystal to sufficient sunlight ("bleaching"), as occurs during transport of many wind-blown sediments, for example. Once the crystal has been emptied of its preexisting trapped charge and then buried, the traps become refilled at a predictable and measurable rate. They can be emptied in the laboratory, with an accompanying emission of light, using an appropriate means of stimulation. This can involve heating the sample to a red-hot temperature to generate the TL signal (Aitken 1985), or exposing the sample to a light source of specified wavelength to generate an OSL signal (Huntley, Godfrey-Smith, and Thewalt 1985). Blue or green light is typically used for quartz, and infrared excitation for feldspars; the latter induces infrared stimulated luminescence (IRSL). Age estimation using OSL and IRSL signals is commonly referred to as optical, or OSL, dating.

To determine the energy absorbed by quartz or feldspar grains since they were last exposed to heat or to sunlight, the TL or OSL signal from a "natural" sample is compared to the signals induced in the laboratory after administering known doses, using a calibrated radiation source, to portions ("aliquots") of the same sample. The latter signals are used to construct a TL or OSL dose-response curve, from which an estimate can be made of the equivalent dose (D_e), or "palaeodose," defined as the amount of radiation needed to generate a TL or OSL signal equal to that produced by the natural sample (Lian and Roberts 2006).

To calculate a luminescence age, one further parameter has to be determined: the environmental dose rate, which is the rate of supply of ionizing radiation to the sample in the natural environment, integrated over the entire period of sample burial. This involves assessing the radioactivity of the sample and the surrounding material, to a distance of about 30 cm, using chemical or physical methods, such as gamma and alpha spectrometry, alpha and beta counting, neutron activation and ICP-MS. Some forms of disequilibrium between parent and daughter nuclides are commonplace in the U-series— such as the loss of radon gas to the atmosphere—and should be taken into consideration (e.g., Olley, Roberts, and Murray 1997). An age estimate is obtained by dividing the equivalent dose (D_e) by the environmental dose rate.

TL dating was developed in the late 1960s for dating of fired pottery (Aitken 1985), but its main application in archaeological contexts today is dating of burnt stones, especially flint (e.g., Mercier et al. 1995, 2007). The method can cover the time span of the controlled use of fire by humans. Flint is an ideal material because it is homogeneous in composition and preserves evidence of past heating in the form of characteristic cracks and potlids, thus assisting the identification of burnt artifacts in the field or museum. When flints are not present at a site, other materials, such as burnt quartz, quartzite, silcrete, and hornfels, may be used (e.g., Henshilwood et al. 2002; Valladas et al. 2005). However, these materials do not necessarily exhibit burning characteristics, so only a small fraction of such samples may be suitable for TL dating. Also, radionuclides are not always homogeneously distributed in stones made of these materials, with radioactive "hot spots" that increase scatter in the ages (e.g., Tribolo et al. 2006). TL dating of unheated sediments has been largely superseded by OSL dating because the OSL charge-traps are emptied by sunlight far more rapidly and completely than those that give rise to the TL signal. But there are some circumstances in which OSL dating cannot be applied and the use of light-sensitive red TL emissions is a possible option (e.g., Morwood et al. 2004).

OSL dating can be applied to naturally-deposited sediments in which artifacts and faunal remains are buried, giving an age estimate for the last time the sediments were exposed to sunlight (Aitken 1998; Lian and Roberts 2006). By association, the age of the buried archaeological materials can be inferred. Quartz is generally preferred to feldspar, which suffers from a physical phenomenon known as "anomalous fading" that can result in age shortfalls unless corrections are made. Methodological and technological developments in OSL dosimetry (Bøtter-Jensen, McKeever, and Wintle 2003) now enable the routine measurement of small, individual aliquots and sand-sized grains of quartz, so a statistically significant number of independent estimates of D_e can be obtained. This has allowed issues such as insufficient sunlight exposure and postdepositional mixing to be addressed (Jacobs and Roberts 2007). For example, Jacobs and colleagues (2006) investigated the possibility of sediment mixing at Blombos Cave in South Africa, and single-grain methods were used at Jinmium rock shelter in northern Australia to determine the extent of contamination of the archaeological deposit with unbleached grains liberated by *in situ* disintegration of roof spall (Roberts et al. 1998). Although deposits in rock shelters and cave entrances may be exposed to scattered, rather than direct, sunlight, this is usually sufficient to zero the OSL "clock." By contrast, deposits in deep cave systems are poorly suited to OSL dating, because the sediment may have been reworked in the darkness of the cave, long after initial deposition.

OSL dating is now applied routinely in archaeology (Feathers 2003; Roberts 1997; Wintle 2008) and has proven especially useful for deposits that lack suitable materials for ^{14}C dating and for events that lie beyond the reliable age range of the ^{14}C method. Some examples of the latter include OSL dating of the earliest-known evidence for modern human behavior (Bouzouggar et al. 2007; Henshilwood et al. 2002; Marean et al. 2007), the oldest artifacts and human remains discovered in Australia (Bowler et al. 2003; Roberts et al. 1994), and the timing of megafaunal extinction in Australia (Roberts et al. 2001).

Electron Spin Resonance. Electron spin resonance dating is founded on the same basic principles as luminescence dating—namely the measurement of the time-dependent accumulation of trapped charge in particular minerals, some of which are found in fossils (Grün 2006; Rink 1997). In ESR dating, however, the number of trapped electrons is estimated from their paramagnetic properties, rather than from their release by heat or light. This is accomplished using an ESR spectrometer, whereby the sample is placed in a microwave cavity and exposed to a strong, external magnetic field. A paramagnetic center (that is, the trapped electron or hole) has a magnetic moment of the same orientation as the magnetic field in which it is placed. By slowly changing the direction of the magnetic field, electrons will resonate ("spin") at a certain frequency. This process involves the magnetic moment reversing direction due to absorption of microwave energy administered to the sample by the microwave generator. The amount of absorbed microwave energy is directly proportional to both the strength of the ESR signal and the number of paramagnetic centers and, hence, the age of the sample; for this reason, the technique is sometimes known as electron paramagnetic resonance (EPR) dating. As in luminescence dating, a dose-response curve is constructed and the signal intensities are compared to that of the "natural" sample to estimate the equivalent dose. The dose rate, internal and external to the sample, must also be determined. In archaeological contexts, ESR is mostly used for dating of hydroxyapatite in the enamel of teeth formed between a few thousand years and several hundreds of millennia ago; in exceptional circumstances, the age range can be extended to a few million years (Schwarcz, Grün, and Tobias 1994).

There are some potential complications with ESR dating, of which resetting of the ESR signal and uncertainties in dose rate determination are the most noteworthy. The former is not a problem when dating teeth: in this case, the zeroing event is assumed to be the time of death and burial, as very few (or no) electrons are trapped in the tooth enamel of living animals. In many situations, however, the dose rate cannot be determined easily

(Grün 2006; Rink 1997). Ideally, the dated sample and surrounding deposits should form a geochemically "closed system" with regard to the relevant radionuclides. But teeth are notorious (as are bone and shell) for the uptake of uranium after burial, and the history of U-uptake (and sometimes later U-loss) commonly cannot be reconstructed in any simple manner. To address this problem, tooth enamel is used in preference to dentine, which typically accumulates between 10 and 100 times as much uranium (Grün and Taylor 1996), and attempts are routinely made to model the process of U-uptake. Two main models have been suggested for teeth (Ikeya 1982): the early-uptake (EU) model assumes that uranium accumulated soon after burial of the tooth, whereas the linear-uptake (LU) model is based on the premise that uranium accumulated steadily throughout the period of tooth burial. Typically, the EU and LU model ages are both reported in publications. It is commonly assumed that the ages of most samples lie somewhere between these estimates, although many samples do not satisfy this assumption (Grün 2006). The EU model will yield a minimum age if there has been no recent U-loss, but a combination of ESR and $^{230}Th/^{234}U$ dating—both of which are dependent on U-uptake, but to different extents—can help narrow the likely age range. Two such approaches have been developed for teeth: the "coupled" ESR/U-series model (Grün et al. 1988) and the closed-system U-series/ESR (CSUS-ESR) model (Grün 2000). These yield the youngest and oldest possible ages for a tooth, respectively, provided no U-loss has occurred.

ESR dating has been used in many archaeological contexts but is perhaps best known for its application to Middle Palaeolithic sites in Israel (Grün and Stringer 1991; Schwarcz et al. 1988) and Middle Stone Age sites in southern Africa (Grün et al. 1996, 2003). The ESR chronologies constructed for both regions showed that the evolutionary history of early modern humans was far more complex than had hitherto been appreciated.

Fission Track Dating. Fission track dating is based on the presence of micrometer-sized, linear damage tracks that occur in minerals due to the spontaneous fission of ^{238}U. Instead of emitting an alpha particle during radioactive decay, the nucleus of a ^{238}U atom may split into two smaller nuclei, which recoil and cause damage to the crystal lattice. Spontaneous fission occurs at a known rate, so the age of a sample can be estimated by measuring the uranium content and the density of spontaneous fission tracks. The latter is determined by chemically etching the sample with sodium hydroxide or hydrofluoric acid to make the tracks visible under an optical microscope, and then counting the number of tracks. The uranium content is measured by irradiating the sample with thermal neutrons in a nuclear reactor to induce fission of the less abundant ^{235}U. The induced fission tracks are etched and counted in the same manner as for the spontaneous ^{238}U tracks, to provide a measure of ^{235}U abundance. The concentration of ^{238}U can then be obtained, because natural uranium has a known, constant $^{235}U/^{238}U$ ratio of 0.073. The most commonly used materials in FT dating are zircon, apatite, and obsidian. The age range of the technique depends on the uranium concentration in the sample, and extends from well beyond the archaeological timescale to some lower limit, which is dictated by track density. Ages as young as 100 years have been achieved using zircons with high uranium concentrations.

The main complication with FT dating is track fading or "annealing" (Wagner 1998; Westgate et al. 2007). Fission tracks are thermally unstable and disappear partially or totally over time at ambient temperature, owing to restoration of the crystal lattice. This leads to fewer fission tracks and, therefore, an underestimation of age. Fading is especially problematic with volcanic glass (for example, obsidian). In archaeological applications, consideration must also be given to the relationship between the zeroing event and the archaeological event of interest. Because heat anneals the tracks and sets the FT "clock" to zero, the heating of obsidian artifacts in a hearth, for example, will date the time of this event. The age of the last significant heating event is also obtained for minerals extracted from *in situ* volcanic tuffs. But in such cases—where artifacts are not dated directly—the association between the dated mineral and the target archaeological event must be secure, because fission tracks are not erased when minerals are reworked from the original deposits unless they are also reheated.

Archaeological applications of FT dating have been limited. The method was used to confirm the K-Ar ages of Bed I at Olduvai Gorge in Tanzania (Fleischer 1965) and the KBS tuff in Kenya (Gleadow 1980), both deposits being associated with Early Pleistocene hominin remains. In Indonesia, FT ages of up to 840,000 years were obtained for zircons buried alongside stone tools in Flores (Morwood et al. 1998; O'Sullivan et al. 2001). This island lies east of Wallace's Line, suggesting that some early hominin species, possibly *Homo floresiensis*, was capable of repeated water crossings long before *Homo sapiens*. At the younger end of the timescale, FT dating has been used to constrain the age of artifacts associated with the tephra deposited across India following the Toba super-eruption 74,000 years ago (Westgate et al. 1998), while in South America FT dating has been applied to obsidian artifacts and to trace trade routes in the Holocene (Miller and Wagner 1981).

Terrestrial Cosmogenic Nuclide Dating. Cosmogenic nuclide dating is based on the principle that high-energy cosmic rays entering the atmosphere collide with atomic nuclei and generate a surge of fast neutrons, as well as some protons and muons, which reach the Earth's surface. When these energetic particles then collide with the nuclei of atoms in certain minerals, such as quartz and calcite, the impacted nuclei are broken apart and the lighter, residual nuclei are known as "terrestrial *in situ* cosmogenic nuclides" (to distinguish them from the vastly more abundant cosmogenic nuclides created in the atmosphere, such as ^{14}C, which is incorporated into living tissues and forms the basis for ^{14}C dating). Terrestrial cosmogenic nuclides include two noble gases—^{3}He and ^{21}Ne (both of which are stable)—and four radioactive elements: ^{10}Be, ^{14}C, ^{26}Al, and ^{36}Cl. The production of these nuclides is largely restricted to rocks within the upper meter or so of the ground surface, so their concentrations (and, for the radioactive nuclides, those of their daughter products) are related to the length of time that the rocks have been exposed to cosmic rays. The greater the elapsed time since initial exposure, the higher the concentration of cosmogenic nuclides within the minerals. Muons penetrate much more deeply than neutrons and are important for TCN dating of transported sediments. The age range of TCN dating extends from a few thousand years to a few million years, depending on the nuclide used.

TCN dating is a rapidly developing field; recent reviews of the technique, and its limitations, are given by Gosse and Phillips (2001), Granger and Muzikar (2001), and Blard and associates (2006). The method has two main potential uses of relevance to archaeologists: to estimate either the exposure age of a rock surface, or the burial age of sediments eroded from the ground surface and deposited deep underground (for instance, in a cave). Determining an age is not straightforward in either case, because it is generally assumed that the "exposure clock" started at zero, such that the dated minerals did not inherit any cosmogenic nuclides produced by previous exposure events. If this assumption does not hold then the age of the most recent exposure event will be overestimated. For exposure dating of a rock surface, it is also commonly assumed that no erosion or weathering has occurred since its initial exposure to cosmic rays, and that it has been exposed for a sufficient period of time to accumulate measurable amounts of cosmogenic nuclides. The validity of these assumptions rests largely on careful field interpretations, although the use of two or more nuclides with different half-lives (for example, ^{26}Al and ^{10}Be in quartz) can provide some constraints on the first two assumptions.

The dual-nuclide approach also enables the estimation of burial ages for sediments that have entered deep caves with cosmogenic ^{26}Al and ^{10}Be. The two nuclides will decay at different rates, so the burial age can be obtained from the ratio of their concentrations, although the initial ratio at the time of sediment entry into the cave must be assumed (Granger et al. 2001). Accurate estimation of TCN production rates is a further complication, because they vary with altitude, latitude, depth below the ground surface, and the angle of inclination of the rock surface (as the cosmic-ray flux is greatest overhead and decreases to the horizon). For sediments deposited less than 30 m underground, continued TCN production due to muons must be taken into account. Also, in TCN dating—as with ^{14}C dating—consideration should be given to past variations in the intensity of the Earth's magnetic field, which affects the fraction of cosmic rays that reach the atmosphere. Given these uncertainties, TCN exposure and burial ages typically have precisions of 5–15%.

TCN dating has been used extensively to study the evolution of landscapes but has had limited application in archaeology. Partridge and associates (2003) used ^{26}Al and ^{10}Be to determine burial ages for cave breccia containing early hominin remains at Sterkfontein in South Africa, and the same approach was taken to date the oldest hominin remains in Europe (Carbonell et al. 2008). Phillips and associates (1997) used ^{36}Cl to obtain maximum ages for rock engravings in the Côa valley in Portugal, and maximum ages for two chert artifacts exposed on the ground surface at Luxor, in Egypt, were estimated using ^{10}Be (Ivy-Ochs et al. 2001). Be-10 has also been used to determine burial ages for early hominids at open-air sites in Chad—including the locality at which the world's earliest known hominid, *Sahelanthropus tchadensis*, was discovered (Lebatard et al. 2008)—but in this instance the ^{10}Be represents atmospheric fallout and not terrestrial *in situ* production.

Conclusions

Landscape archaeologists have a wide choice of dating methods at their disposal, including several numerical-age methods that complement one another and add much needed rigor to archaeological chronologies. Some of these techniques (for example, $^{40}Ar/^{39}Ar$, OSL, FT, and TCN) are applicable to major landforms, entire catchments and to geographically widespread marker horizons (for instance, tephras), as well as to archaeological deposits at individual sites. Such methods are, therefore, of immense value for understanding human-environment interactions at a range of temporal and spatial scales. We have also described a number of dating options for individual artifacts and for associated organic and inorganic remains at specific sites, but it is the task of the archaeologist to collate these disparate data sets to reveal patterns of human activity across the landscape. Given the potential pitfalls of each method, it is recommended practice that a *sequence* of ages be obtained at any given site to check that the ages are in correct stratigraphic order, taking into account their quoted uncertainties. In addition, *multiple* methods and materials should be investigated (e.g., Morwood et al. 2004; Turney et al. 2001) to avoid shortcomings in any single technique or sample type. If these precautions are taken, then landscape archaeologists can construct reliable chronological frameworks with a high degree of confidence.

References

Aitken, M. J. 1985. *Thermoluminescence Dating*. London: Academic Press.
———. 1990. *Science-Based Dating in Archaeology*. London: Longman.
———. 1998. *An Introduction to Optical Dating: The Dating of Quaternary Sediments by the Use of Photon-stimulated Luminescence*. Oxford: Oxford University Press.
———. 1999. Archaeological dating using physical phenomena. *Reports on Progress in Physics* 62: 1333–76.
Anovitz, L. M., Riciputi, L. R., Cole, D. R., Fayek, M., and Elam, J. M. 2006. Obsidian hydration: A new palaeothermometer. *Geology* 34: 517–20.
Armitage, R. A., Brady, J. E., Cobb, A., Southon, J. R., and Rowe, J. M. 2001. Mass spectrometric radiocarbon dates from three rock paintings of known age. *American Antiquity* 66: 471–80.
Aubert, M., O'Connor, S., McCulloch, M., Mortimer, G., Watchman, A., and Richer-LaFlèche, M. 2007. Uranium-series dating rock art in East Timor. *Journal of Archaeological Science* 34: 991–96.
Beck, W., Donahue, D., Jull, A. J. T., Burr, G., Broecker, W. S., Bonani, G., Hajdas, I., and Malotki, E. 1998. Ambiguities in direct dating of rock surfaces using radiocarbon measurements. *Science* 280: 2132–35.
Bednarik, R. G. 1992. A new method to date petroglyphs. *Archaeometry* 34: 279–91.
———. 2002. The dating of rock art: a critique. *Journal of Archaeological Science* 29: 1213–33.

Betts, M. W., and Latta, L. 2000. Rock surface hardness as an indication of exposure age: An archaeological application of the Schmidt Hammer. *Archaeometry* 42: 209–23.
Bierman, P. R., Gillespie, A. R., and Kuehner, S. 1991. Precision of rock-varnish chemical analyses and cation-ratio ages. *Geology* 19: 135–38.
Bird, M. I., Ayliffe, L. K., Fifield, L. K., Turney, C. S. M., Cresswell, R. G., Barrows, T. T., and David, B. 1999. Radiocarbon dating of "old" charcoal using a wet oxidation, stepped-combustion procedure. *Radiocarbon* 41: 127–40.
Bird, M. I., Fifield, L. K., Santos, G. M., Beaumont, P. B., Zhou, Y., di Tada, M. L., and Hausladen, P. A. 2003. Radiocarbon dating from 40 to 60 ka B.P. at Border Cave, South Africa. *Quaternary Science Reviews* 22: 943–47.
Birkeland, P. W. 1999. *Soils and Geomorphology*. Oxford: Oxford University Press.
Blard, P.-H., Bourlès, D., Lavé, J., and Pik, R. 2006. Applications of ancient cosmic-ray exposures: Theory, techniques and limitations. *Quaternary Geochronology* 1: 59–73.
Bøtter-Jensen, L., McKeever, S. W. S., and Wintle, A. G. 2003. *Optically Stimulated Luminescence Dosimetry*. Amsterdam: Elsevier Science.
Bourdon, B., Turner, S., Henderson, G. M., and Lundstrom, C. C. 2003. Introduction to U-series geochemistry, *Uranium-series Geochemistry*, in B. Bourbon, G. M. Henderson, C. C. Lundstrom, and S. P. Turner (eds.), *Reviews in Mineralogy and Geochemistry*, Vol. 52, pp. 1–22. Washington, DC: Mineralogical Society of America.
Bouzouggar, A., Barton, N., Vanhaeren, M., d'Errico, F., Collcutt, S., Higham, T., Hodge, E., Parfitt, S., Rhodes, E., Schwenninger, J.-L., Stringer, C., Turner, E., Ward, S., Moutmir, A., and Stambouli, A. 2007. 82,000-year-old shell beads from North Africa and implications for the origins of modern human behavior. *Proceedings of the National Academy of Sciences of the USA* 104: 9964–69.
Bowler, J. M., Johnston, H., Olley, J. M., Prescott, J. R., Roberts, R. G., Shawcross, W., and Spooner, N. A. 2003. New ages for human occupation and climatic change at Lake Mungo, Australia. *Nature* 421: 837–40.
Bronk Ramsey, C. 2008. Radiocarbon dating: Revolutions in understanding. *Archaeometry* 50: 249–75.
Calvo-Rathert, M., Goguitchaichvili, A., Sologashvili, D., Villalaín, J. J., Bógalo, M. F., Carrancho, A., and Maissuradze, G. 2008. New palaeomagnetic data from the hominin bearing Dmanisi palaeo-anthropologic site (southern Georgia, Caucasus). *Quaternary Research* 69: 91–96.
Carbonell, E., Bermúdez de Castro, J. M., Parés, J. M., Pérez-González, A., Cuenca-Bescós, G., Ollé, A., Mosquera, M., Huguet, R., van der Made, J., Rosas, A., Sala, R., Vallverdú, J., García, N., Granger,

D. E., Martinón-Torres, M., Rodríguez, X. P., Stock, G. M., Vergès, J. M., Allué, E., Burjachs, F., Cáceres, I., Canals, A., Benito, A., Díez, C., Lozano, M., Mateos, A., Navazo, M., Rodríguez, J., Rosell, J., and Arsuaga, J. L. 2008. The first hominin of Europe. *Nature* 452: 465–69.

Cheng, H., Edwards, R. L., Murrell, M. T., and Benjamin, T. M. 1998. Uranium-thorium-protactinium dating systematics. *Geochimica et Cosmochimica Acta* 62: 3437–52.

Clarke, S. J., and Murray-Wallace, C. V. 2006. Mathematical expressions used in amino acid racemisation geochronology—A review. *Quaternary Geochronology* 1: 261–78.

Clarke, S. J., Miller, G. H., Fogel, M. L., Chivas, A. R., and Murray-Wallace, C. V. 2006. The amino acid and stable isotope biogeochemistry of elephant bird (*Aepyornis*) eggshells from southern Madagascar. *Quaternary Science Reviews* 25: 2343–56.

Clarke, S. J., Miller, G. H., Murray-Wallace, C. V., David, B., and Pasveer, J. M. 2007. The geochronological potential of isoleucine epimerisation in cassowary and megapode eggshells from archaeological sites. *Journal of Archaeological Science* 34: 1051–63.

Colman, S. M., Pierce, K. L., and Birkeland, P. W. 1987. Suggested terminology for Quaternary dating methods. *Quaternary Research* 28: 314–19.

Dalan, R. A., and Banerjee, S. K. 1998. Solving archaeological problems using techniques of soil magnetism. *Geoarchaeology* 13: 3–36.

Dorn, R. I., Nobbs, M., and Cahill, T. A. 1988. Cation-ratio dating of rock-engravings from the Olary Province of arid South Australia. *Antiquity* 62: 681–89.

Eggins, S. M., Grün, R., McCulloch, M. T., Pike, A. W. G., Chappell, J., Kinsley, L., Mortimer, G., Shelley, M., Murray-Wallace, C. V., Spötl, C., and Taylor, L. 2005. *In situ* U-series dating by laser-ablation multi-collector ICPMS: new prospects for Quaternary geochronology. *Quaternary Science Reviews* 24: 2523–38.

Feathers, J. K. 2003. Use of luminescence dating in archaeology. *Measurement Science and Technology* 14: 1493–1509.

Fleischer, R. L. 1965. Fission-track dating of Bed I, Olduvai Gorge. *Science* 148: 72–74.

Fontugne, M., Carré, M., Bentaleb, I., Julien, M., and Lavallée, D. 2004. Radiocarbon reservoir age variations in the south Peruvian upwelling during the Holocene. *Radiocarbon* 46: 531–37.

Gabunia, L., Vekua, A., Lordkipanidze, D., Swisher III, C. C., Ferring, R., Justus, A., Nioradze, M., Tvalchrelidze, M., Antòn, S. C., Bosinski, G., Joris, O., de Lumley, M., Majsuradze, G., and Mouskhelishvili, A. 2000. Earliest Pleistocene hominid cranial remains from Dmanisi, Republic of Georgia: taxonomy, geological setting, and age. *Science* 288: 1019–25.

Geyh, M. A. 2001. Bomb radiocarbon dating of animal tissues and hair. *Radiocarbon* 43: 723–30.

Gleadow, A. J. W. 1980. Fission track age of the KBS tuff and associated hominid remains in northern Kenya. *Nature* 284: 225–30.

Gosse, J. C., and Phillips, F. M. 2001. Terrestrial *in situ* cosmogenic nuclides: theory and applications. *Quaternary Science Reviews* 20: 1475–560.

Granger, D. E., and Muzikar, P. F. 2001. Dating sediment burial with in situ-produced cosmogenic nuclides: theory, techniques, and limitations. *Earth and Planetary Science Letters* 188: 269–81.

Granger, D. E., Fabel, D., and Palmer, A. N. 2001. Pliocene-Pleistocene incision of the Green River, Kentucky, determined from radioactive decay of cosmogenic ^{26}Al and ^{10}Be in Mammoth Cave sediments. *Geological Society of America Bulletin* 113: 825–36.

Grün, R. 2000. An alternative for model for open system U-series/ESR age calculations: (Closed system U-series)-ESR, CSUS-ESR. *Ancient TL* 18: 1–4.

———. 2006. Direct dating of human fossils. *Yearbook of Physical Anthropology* 49: 2–48.

Grün, R., Beaumont, P., Tobias, P. V., and Eggins, S. 2003. On the age of Border Cave 5 human mandible. *Journal of Human Evolution* 45: 155–67.

Grün, R., Brink, J. S., Spooner, N. A., Taylor, L., Stringer, C. B., Franciscus, R. G., and Murray, A. S. 1996. Direct dating of Florisbad hominid. *Nature* 382: 500–11.

Grün, R., Schwarcz, H. P., and Chadam, I. 1988. ESR dating of tooth enamel: coupled corrections for U-uptake and U-series disequilibrium. *Nuclear Tracks and Radiation Measurements* 14: 237–42.

Grün, R., and Stringer, C. B. 1991. ESR dating and the evolution of modern humans. *Archaeometry* 33: 153–99.

Grün, R., and Taylor, L. 1996. Uranium and thorium in the constituents of teeth. *Ancient TL* 14: 21–25.

Henshilwood, C. S., d'Errico, F., Yates, R., Jacobs, Z., Tribolo, C., Duller, G. A. T., Mercier, N., Sealy, J. C., Valladas, H., Watts, I., and Wintle, A. G. 2002. Emergence of modern human behavior: Middle Stone Age engravings from South Africa. *Science* 295: 1278–80.

Higham, T. F. G., Jacobi, R. M., and Bronk Ramsey, C. 2006. AMS radiocarbon dating of ancient bone using ultrafiltration. *Radiocarbon* 48: 179–95.

Hua, Q., and Barbetti, M. 2004. Review of tropospheric bomb ^{14}C data for carbon cycle modeling and age calibration purposes. *Radiocarbon* 46: 1273–98.

Hughen, K. A., Baillie, M. G. L., Bard, E., Beck, J. W., Bertrand, C. J. H., Blackwell, P. G., Buck, C. E., Burr, G. S., Cutler, K. B., Damon, P. E., Edwards, R. L., Fairbanks, R. G., Friedrich, M., Guilderson, T. P., Kromer, B., McCormac, F. G., Manning, S., Bronk Ramsey, C., Reimer, P. J., Reimer, R. W., Remmele, S., Southon, J. R., Stuiver, M., Talamo, S., Taylor, F.

W., van der Plicht, J., and Weyhenmeyer, C. E. 2004. Marine04 radiocarbon age calibration, 0-26 cal kyr B.P. *Radiocarbon* 46: 1059–86.

Hughen, K., Southon, J., Lehman, S., Bertrand, C., and Turnbull, J. 2006. Marine-derived ^{14}C calibration and activity record for the past 50,000 years updated from the Cariaco Basin. *Quaternary Science Reviews* 25: 3216–27.

Hull, K. L. 2001. Reasserting the utility of obsidian hydration dating: a temperature-dependent empirical approach to practical temporal resolution with archaeological obsidians. *Journal of Archaeological Science* 28: 1025–40.

Huntley, D. J., Godfrey-Smith, D. I., and Thewalt, M. L. W. 1985. Optical dating of sediments. *Nature* 313: 105–07.

Ikeya, M. 1982. A model of linear uranium accumulation for ESR age of Heidelberg (Mauer) and Tautavel bones. *Japanese Journal of Applied Physics* 21: L690–92.

Ivanovich, M., and Harmon, R. S. 1992. *Uranium-series Disequilibrium: Applications to Earth, Marine, and Environmental Sciences*. Oxford: Clarendon Press.

Ivy-Ochs, S., Wust, R., Kubik, P. W., Müller-Beck, H., and Schlüchter, C. 2001. Can we use cosmogenic isotopes to date stone artifacts? *Radiocarbon* 43: 759–64.

Jacobi, R. M., Higham, T. F. G., and Bronk Ramsey, C. 2006. AMS radiocarbon dating of Middle and Upper Palaeolithic bone in the British Isles: improved reliability using ultrafiltration. *Journal of Quaternary Science* 21: 557–73.

Jacobs, Z., Duller, G. A. T., Wintle, A. G., and Henshilwood, C. S. 2006. Extending the chronology of deposits at Blombos Cave, South Africa, back to 140 ka using optical dating of single and multiple grains of quartz. *Journal of Human Evolution* 51: 255–73.

Jacobs, Z., and Roberts, R. G. 2007. Advances in optically stimulated luminescence (OSL) dating of individual grains of quartz from archaeological deposits. *Evolutionary Anthropology* 16: 210–23.

Kennett, D. J., Ingram, B. L., Southon, J. R., and Wise, K. 2002. Differences in ^{14}C age between stratigraphically associated charcoal and marine shell from the Archaic Period site of Kilometer 4, southern Peru: Old wood or old water? *Radiocarbon* 44: 53–58.

Kirch, P. V., and Sharp, W. D. 2005. Coral [230]Th dating of the imposition of a ritual control hierarchy in precontact Hawaii. *Science* 307: 102–04.

Kitagawa, H., and van der Plicht, J. 1998. Atmospheric radiocarbon calibration to 45,000 yr B.P. Late Glacial fluctuations and cosmogenic isotope production. *Science* 279: 1187–90.

Leakey, M. G., Spoor, F., Brown, F. H., Gathogo, P. N., Kiarie, C., Leakey, L. N., and McDougall, I. 2001. New hominin genus from eastern Africa shows diverse middle Pliocene lineages. *Nature* 410: 433–40.

Lebatard, A.-E., Bourlès, D. L., Duringer, P., Jolivet, M., Braucher, R., Carcaillet, J., Schuster, M., Arnaud, N., Monié, P., Lihoreau, F., Likius, A., Mackaye, H. T., Vignaud, P., and Brunet, M. 2008. Cosmogenic nuclide dating of *Sahelanthropus tchadensis* and *Australopithecus bahrelghazali*: Mio-Pliocene hominids from Chad. *Proceedings of the National Academy of Sciences of the USA* 105: 3226–31.

Lengyel, S. N., and Eighmy, J. L. 2002. A revision to the U.S. Southwest archaeomagnetic master curve. *Journal of Archaeological Science* 29: 1423–33.

Lian, O. B., and Roberts, R. G. 2006. Dating the Quaternary: progress in luminescence dating of sediments. *Quaternary Science Reviews* 25: 2449–68.

Liu, T., and Broecker, W. S. 2007. Holocene rock varnish microstratigraphy and its chronometric application in the drylands of western USA. *Geomorphology* 84: 1–21.

Loso, M. G., and Doak, D. F. 2006. The biology behind lichenometric dating curves. *Oecologia* 147: 223–29.

Lowe, D. J., Newnham, R. M., McFadgen, B. G., and Higham, T. F. G. 2000. Tephras and New Zealand archaeology. *Journal of Archaeological Science* 27: 859–70.

Ludwig, K. R. 2003. Mathematical-statistical treatment of data and errors for [230]Th/U geochronology, in *Uranium-series Geochemistry*, B. Bourbon, G. M. Henderson, C. C. Lundstrom, and S. P. Turner (eds.), *Reviews in Mineralogy and Geochemistry*, Vol. 52, pp. 631–56. Washington, DC: Mineralogical Society of America.

Marean, C. W., Bar-Matthews, M., Bernatchez, J., Fisher, E., Goldberg, P., Herries, A. I. R., Jacobs, Z., Jerardino, A., Karkanas, P., Minichillo, T., Nilssen, P. J., Thompson, E., Watts, I., and Williams, H. M. 2007. Early human use of marine resources and pigment in South Africa during the Middle Pleistocene. *Nature* 449: 905–08.

McCormac, F. G., Hogg, A. G., Blackwell, P. G., Buck, C. E., Higham, T. F. G., and Reimer, P. J. 2004. SHCal04 Southern Hemisphere calibration, 0–11.0 cal kyr B.P. *Radiocarbon* 46: 1087–92.

McDougall, I., Brown, F. H., and Fleagle, J. G. 2005. Stratigraphic placement and age of modern humans from Kibish, Ethiopia. *Nature* 433: 733–36.

McDougall, I., and Harrison, T. M. 1999. *Geochronology and Thermochronology by the [40]Ar/[39]Ar Method*. Oxford: Oxford University Press.

Mazel, A. D., and Watchman, A. 1997. Accelerator radiocarbon dating of Natal Drakensberg paintings: Results and implications. *Antiquity* 71: 445–49.

Mercier, N., Valladas, H., Froget, L., Joron, J.-L., Reyss, J.-L., Weiner, S., Goldberg, P., Meignen, L., Bar-Yosef,

O., Belfer-Cohen, A., Chech, M., Kuhn, S. L., Stiner, M. C., Tillier, A.-M., Arensburg, B., and Vandermeersch, B. 2007. Hayonim Cave: A TL-based chronology for this Levantine Mousterian sequence. *Journal of Archaeological Science* 34: 1064–77.

Mercier, N., Valladas, H., and Valladas, G. 1995. Flint thermoluminescence dates from the CFR laboratory at Gif: Contributions to the study of the chronology of the Middle Palaeolithic. *Quaternary Science Reviews* 14: 351–64.

Miller, D. S., and Wagner, G. A. 1981. Fission-track ages applied to obsidian artefacts from South America using the plateau-annealing and the track-size age-correction techniques. *Nuclear Tracks* 5: 147–55.

Miller, G. H., Beaumont, P. B., Deacon, H. J., Brooks, A. S., Hare, P. E., and Jull, A. J. T. 1999a. Earliest modern humans in southern Africa dated by isoleucine epimerization in ostrich eggshell. *Quaternary Science Reviews* 18: 1537–48.

Miller, G. H., Magee, J. W., Johnson, B. J., Fogel, M. L., Spooner, N. A., McCulloch, M. T., and Ayliffe, L. K. 1999b. Pleistocene extinction of *Genyornis newtoni*: human impact on Australian megafauna. *Science* 283: 205–08.

Miller, G. H., Magee, J. W., and Jull, A. J. T. 1997. Low-latitude glacial cooling in the Southern Hemisphere from amino-acid racemization in emu eggshells. *Nature* 385: 241–44.

Morwood, M. J., O'Sullivan, P. B., Aziz, F., and Raza, A. 1998. Fission-track ages of stone tools and fossils on the east Indonesian island of Flores. *Nature* 392: 173–76.

Morwood, M. J., Soejono, R. P., Roberts, R. G., Sutikna, T., Turney, C. S. M., Westaway, K. E., Rink, W. J., Zhao, J.-X., van den Bergh, G. D., Due, R. A., Hobbs, D. R., Moore, M. W., Bird, M. I., and Fifield, L. K. 2004. Archaeology and age of a new hominin from Flores in eastern Indonesia. *Nature* 431: 1087–91.

North Greenland Ice Core Project members 2004. High-resolution record of Northern Hemisphere climate extending into the last interglacial period. *Nature* 431: 147–51.

Oldfield, F., Wake, R., Boyle, J., Jones, R., Nolan, S., Gibbs, Z., Appleby, P., Fisher, E., and Wolff, G. 2003. The late-Holocene history of Gormire Lake (NE England) and its catchment: A multiproxy reconstruction of past human impact. *The Holocene* 13: 677–90.

Olley, J. M., Roberts, R. G., and Murray, A. S. 1997. Disequilibria in the uranium decay series in sedimentary deposits at Allen's Cave, Nullarbor Plain, Australia: Implications for dose rate determinations. *Radiation Measurements* 27: 433–43.

O'Neal, M. A. 2006. The effects of slope degradation on lichenometric dating of Little Ice Age moraines. *Quaternary Geochronology* 1: 121–28.

O'Sullivan, P. B., Morwood, M. J., Hobbs, D., Aziz, F., Suminto, Situmorang, M., Raza, A., and Maas, R.

2001. Archaeological implications of the geology and chronology of the Soa basin, Flores, Indonesia. *Geology* 29: 607–10.

Parfitt, S. A., Barendregt, R. W., Breda, M., Candy, I., Collins, M. J., Coope, G. R., Durbidge, P., Field, M. H., Lee, J. R., Lister, A. M., Mutch, R., Penkman, K. E. H., Preece, R. C., Rose, J., Stringer, C. B., Symmons, R., Whittaker, J. E., Wymer, J. J., and Stuart, A. J. 2005. The earliest record of human activity in northern Europe. *Nature* 438: 1008–12.

Partridge, T. C., Granger, D. E., Caffee, M. W., and Clarke, R. J. 2003. Lower Pliocene hominid remains from Sterkfontein. *Science* 300: 607–12.

Penkman, K. E. H., Kaufman, D. S., Maddy, D., and Collins, M. J. 2008. Closed-system behaviour of the intra-crystalline fraction of amino acids in mollusc shells. *Quaternary Geochronology* 3: 2–25.

Pettitt, P., and Pike, A. 2007. Dating European Palaeolithic cave art: Progress, prospects, problems. *Journal of Archaeological Method and Theory* 14: 27–47.

Phillips, F. M., Flinsch, M., Elmore, D., and Sharma, P. 1997. Maximum ages of the Côa valley (Portugal) engravings measured with chlorine-36. *Antiquity* 71: 100–04.

Pike, A. W. G., Hedges, R., and van Calsteren, P. 2002. U-series dating of bone using the diffusion-adsorption model. *Geochimica et Cosmochimica Acta* 66: 4273–86.

Plagnes, V., Causse, C., Fontugne, M., Valladas, H., Chazine, J., and Fage, L. 2003. Cross dating (Th/U-^{14}C) of calcite covering prehistoric paintings in Borneo. *Quaternary Research* 60: 172–79.

Reimer, P. J., Baillie, M. G. L., Bard, E., Bayliss, A., Beck, J. W., Bertrand, C. J. H., Blackwell, P. G., Buck, C. E., Burr, G. S., Cutler, K. B., Damon, P. E., Edwards, R. L., Fairbanks, R. G., Friedrich, M., Guilderson, T. P., Hogg, A. G., Hughen, K. A., Kromer, B., McCormac, F. G., Manning, S. W., Bronk Ramsey, C., Reimer, R. W., Remmele, S., Southon, J. R., Stuiver, M., Talamo, S., Taylor, F. W., van der Plicht, J., and Weyhenmeyer, C. E. 2004. IntCal04 terrestrial radiocarbon age calibration, 26-0 ka B.P. *Radiocarbon* 46: 1029–58.

Renne, P. R., Sharp, W. D., Deino, A. L., Orsi, G., and Civetta, L. 1997. ^{40}Ar/^{39}Ar dating into the historical realm: calibration against Pliny the Younger. *Science* 277: 1279–80.

Richards, D. A., and Dorale, J. A. 2003. Uranium-series chronology and environmental applications of speleothems, in *Uranium-series Geochemistry*, B. Bourbon, G. M. Henderson, C. C. Lundstrom, and S. P. Turner (eds.), *Reviews in Mineralogy and Geochemistry*, Vol. 52, pp. 407–60. Washington, DC: Mineralogical Society of America.

Rick, T. C., Vellanoweth, R. L., and Erlandson, J. M. 2005. Radiocarbon dating and the "old shell" problem: Direct dating of artifacts and cultural chronologies in coastal and other aquatic regions. *Journal of Archaeological Science* 32: 1641–48.

Rink, W. J. 1997. Electron spin resonance (ESR) dating and ESR applications in Quaternary science and archaeometry. *Radiation Measurements* 27: 975–1025.

Roberts, R. G. 1997. Luminescence dating in archaeology: From origins to optical. *Radiation Measurements* 27: 819–92.

Roberts, R., Bird, M., Olley, J., Galbraith, R., Lawson, E., Laslett, G., Yoshida, H., Jones, R., Fullagar, R., Jacobsen, G., and Hua, Q. 1998. Optical and radiocarbon dating at Jinmium rock shelter in northern Australia. *Nature* 393: 358–62.

Roberts, R. G., Flannery, T. F., Ayliffe, L. K., Yoshida, H., Olley, J. M., Prideaux, G. J., Laslett, G., Baynes, A., Smith, M. A., Jones, R., and Smith, B. L. 2001. New ages for the last Australian megafauna: Continent-wide extinction about 46,000 years ago. *Science* 292: 1888–92.

Roberts, R. G., Jones, R., Spooner, N. A., Head, M. J., Murray, A. S., and Smith, M. A. 1994. The human colonisation of Australia: Optical dates of 53,000 and 60,000 years bracket human arrival at Deaf Adder Gorge, Northern Territory. *Quaternary Science Reviews* 13: 575–83.

Roberts, R., Walsh, G., Murray, A., Olley, J., Jones, R., Morwood, M., Tuniz, C., Lawson, E., Macphail, M., Bowdery, D., and Naumann, I. 1997. Luminescence dating of rock art and past environments using mud-wasp nests in northern Australia. *Nature* 387: 696–99.

Rogers, A. K. 2007. Effective hydration temperature of obsidian: A diffusion theory analysis of time-dependent hydration rates. *Journal of Archaeological Science* 34: 656–65.

Schwarcz, H. P., Grün, R., and Tobias, P. V. 1994. ESR dating studies of the australopithecine site of Sterkfontein, South Africa. *Journal of Human Evolution* 26: 175–81.

Schwarcz, H. P., Grün, R., Vandermeerch, B., Bar-Yosef, O., Valladas, H., and Tchernov, E. 1988. ESR dates from the hominid burial site of Qafzeh in Israel. *Journal of Human Evolution* 17: 733–37.

Scott, E. M., Bryant, C., Carmi, I., Cook, G., Gulliksen, S., Harkness, D., Heinemeier, J., McGee, E., Naysmith, P., Possnert, G., van der Plicht, H., and van Strydonck, M. 2004. Precision and accuracy in applied ^{14}C dating: Some findings from the Fourth International Radiocarbon Inter-comparison. *Journal of Archaeological Science* 31: 1209–13.

Shackleton, N. J. 1982. Stratigraphy and chronology of the Klasies River Mouth deposits: oxygen isotope evidence, in *The Middle Stone Age at Klasies River Mouth in South Africa*, R. Singer and J. Wymer (eds.), pp. 194–99. Chicago: Chicago University Press.

Shen, G., Gao, X., Zhao, J., and Collerson, K. D. 2004. U-series dating of Locality 15 at Zhoukoudian,

China, and implications for hominid evolution. *Quaternary Research* 62: 208–13.

Smith, J. N. 2001. Why should we believe ^{210}Pb sediment geochronologies? *Journal of Environmental Radioactivity* 55: 121–23.

Spalding, K. L., Buchholz, B. A., Bergman, L.-E., Druid, H., and Frisén, J. 2005. Age written in teeth by nuclear tests. *Nature* 437: 333.

Stafford, T. W., Hare, E., Currie, L., Jull, A. J. T., and Donahue, D. 1991. Accelerator radiocarbon dating at the molecular level. *Journal of Archaeological Science* 18: 35–72.

Stott, A. W., Berstan, R., Evershed, R. P., Bronk Ramsey, C., Hedges, R. E. M., and Humm, M. J. 2003. Direct dating of archaeological pottery by compound-specific ^{14}C analysis of preserved lipids. *Analytical Chemistry* 75: 5037–45.

Swisher III, C. C., Curtis, G. H., Jacob, T., Getty, A. G., and Widiasmoro, A. S. 1994. Age of the earliest known hominids in Java, Indonesia. *Science* 263: 1118–21.

Thorne, A., Grün, R., Mortimer, G., Spooner, N. A., Simpson, J. J., McCulloch, M. T., Taylor, L., and Curnoe, D. 1999. Australia's oldest human remains: Age of the Lake Mungo 3 skeleton. *Journal of Human Evolution* 36: 591–612.

Towner, R. H. 2002. Archaeological dendrochronology in the southwestern United States. *Evolutionary Anthropology* 11: 68–84.

Tribolo, C., Mercier, N., Selo, M., Valladas, H., Joron, J.-L., Reyss, J.-L., Henshilwood, C., Sealy, J., and Yates, R. 2006. TL dating of burnt lithics from Blombos Cave (South Africa): Further evidence for the antiquity of modern human behaviour. *Archaeometry* 48: 341–57.

Turney, C. S. M., Bird, M. I., Fifield, L. K., Roberts, R. G., Smith, M., Dortch, C. E., Grün, R., Lawson, E., Ayliffe, L. K., Miller, G. H., Dortch, J., and Cresswell, R. G. 2001. Early human occupation at Devil's Lair, southwestern Australia 50,000 years ago. *Quaternary Research* 55: 3–13.

Valladas, H., Tisnérat-Laborde, N., Cachier, H., Arnold, M., Bernaldo de Quirós, F., Cabera-Valdés, V., Clottes, J., Courtin, J., Fortea-Pérez, J. J., Gonzáles-Sainz, C., and Moure-Romanillo, A. 2001. Radiocarbon AMS dates for Paleolithic cave paintings. *Radiocarbon* 43: 977–86.

Valladas, H., Wadley, L., Mercier, N., Froget, L., Tribolo, C., Reyss, J. L., and Joron, J. L. 2005. Thermoluminescence dating on burnt lithics from Middle Stone Age layers at Rose Cottage Cave. *South African Journal of Science* 101: 169–74.

van der Kaars, S., and van den Bergh, G. D. 2004. Anthropogenic changes in the landscape of west Java (Indonesia) during historic times, inferred from a sediment and pollen record from Teluk Banten. *Journal of Quaternary Science* 19: 229–39.

van der Plicht, J., Beck, J. W., Bard, E., Baillie, M. G. L., Blackwell, P. G., Buck, C. E., Friedrich, M.,

Guilderson, T. P., Hughen, K., Kromer, B., McCormac, F. G., Bronk Ramsey, C., Reimer, P. J., Reimer, R. W., Remmele, S., Richards, D. A., Southon, J. R., Stuiver, M., and Weyhenmeyer, C. E. 2004. NOTCAL04-comparison/calibration ^{14}C records 26–50 cal kyr B.P. *Radiocarbon* 46: 1225–38.

Wagner, G. A. 1998. *Age Determination of Young Rocks and Artifacts: Physical and Chemical Clocks in Quaternary Geology and Archaeology.* Heidelberg: Springer-Verlag.

Walker, M. 2005. *Quaternary Dating Methods.* Chichester: John Wiley and Sons.

Walker, J., Cliff, R. A., and Latham, A. G. 2006. U-Pb isotopic age of the StW 573 hominid from Sterkfontein, South Africa. *Science* 314: 1592–94.

Watanabe, S., Ayta, W. E. F., Hamaguchi, H., Guidon, N., La Salvia, E. S., Maranca, S., and Filho, O. B. 2003. Some evidence of a date of first humans to arrive in Brazil. *Journal of Archaeological Science* 30: 351–54.

Watchman, A., and Jones, R. 2002. An independent confirmation of the 4 ka antiquity of a beeswax figure in western Arnhem Land, northern Australia. *Archaeometry* 44: 145–53.

Watchman, A., David, B., McNiven, I. J., and Flood, J. M. 2000. Micro-archaeology of engraved and painted rock surface crusts at Yiwarlarlay (the Lightning Brothers site), Northern Territory, Australia. *Journal of Archaeological Science* 27: 315–25.

Watchman, A., O'Connor, S., and Jones, R. 2005. Dating oxalate minerals 20-45 ka. *Journal of Archaeological Science* 32: 369–74.

Westgate, J. A., Naeser, N. D., and Alloway, B. 2007. Fission-track dating, in *Encyclopedia of Quaternary Science*, S. A. Elias (ed.), pp. 651–72. Oxford: Elsevier.

Westgate, J. A., Shane, P. A. R., Pearce, N. J. G., Perkins, W. T., Korisettar, R., Chesner, C. A., Williams, M. A. J., and Acharyya, S. K. 1998. All Toba tephra occurrences across peninsular India belong to the 75,000 yr B.P. eruption. *Quaternary Research* 50: 107–12.

Wild, E. M., Arlamovsky, K. A., Golser, R., Kutschera, W., Priller, A., Puchegger, S., Rom, W., Steier, P., and Vycudilik, W. 2000. ^{14}C dating with the bomb peak: an application to forensic medicine. *Nuclear Instruments and Methods in Physics Research B* 172: 944–50.

Wilson, M., Spriggs, M., and Lawson, E. 2001. Dating the rock art of Vanuatu: AMS radiocarbon determinations from abandoned mud-wasp nests and charcoal pigment found in superimposition. *Rock Art Research* 18: 24–32.

Wintle, A. G. 2008. Fifty years of luminescence dating. *Archaeometry* 50: 276–312.

Woodhead, J., Hellstrom, J., Maas, R., Drysdale, R., Zanchetta, G., Devine, P., and Taylor, E. 2006. U-Pb geochronology of speleothems by MC-ICPMS. *Quaternary Geochronology* 1: 208–21.

Zhu, R. X., Potts, R., Xie, F., Hoffman, K. A., Deng, C. L., Shi, C. D., Pan, Y. X., Wang, H. Q., Shi, R. P., Wang, Y. C., Shi, G. H., and Wu, N. Q. 2004. New evidence on the earliest human presence at high northern latitudes in northeast Asia. *Nature* 431: 559–62.

Zolitschka, B., Behre, K., and Schneider, J. 2003. Human and climatic impact on the environment as derived from colluvial, fluvial and lacustrine archives: Examples from the Bronze Age to the Migration period, Germany. *Quaternary Science Reviews* 22: 81–100.

36

STRATIGRAPHY, DEPOSITIONAL ENVIRONMENTS, AND PALAEOLANDSCAPE RECONSTRUCTION IN LANDSCAPE ARCHAEOLOGY

Nicola Stern

This chapter explores how the stratigraphic and sedimentary context of archaeological debris affects the practice of landscape archaeology. Only a fraction of the archaeological record consists of material remains that accumulated on (or close to) the surface of the modern landscape. Material remains from the remote past are preserved only because they were covered over by sediments, and those sediments usually represent fragments of ancient landscapes that bear little or no relationship to the topographic features or habitats that can be observed in an area today. For this reason, documenting and interpreting the geological context of material remains are fundamental to the practice of landscape archaeology, whatever interpretive approaches are employed to explicate the distribution, density, and characteristics of those remains.

Stratigraphy, or the order and arrangement of strata, provides the basic framework for identifying sets of material remains and sediments that were deposited during the same time intervals. Geochronology provides the age determinations for individual beds within a stratigraphic sequence (see Roberts and Jacobs, this volume); thus, it is the springboard for estimating not only the age of the remains being studied but also how much time it took for those material remains and their encasing sediments to accumulate. Facies analysis,

which involves the identification and interpretation of distinctive bodies of sediment within a particular stratigraphic interval, is the basis on which the ancient landscape itself is described. Some idea of the habitats supported by that palaeolandscape may be gleaned from the geochemistry of the sediments themselves, as well as from the array of micro- and macro-fossils they contain (see Denham, this volume; Dolby, this volume; Porch, this volume; Rowe and Kershaw, this volume).

Over the past three decades, archaeologists have been coming to grips with the interpretative consequences of recognizing that the palaeolandscapes they study are a complex interplay between the physical features of a landscape and people's perceptions of them. Not so widely acknowledged are the interpretative consequences of recognizing that the distribution of activity traces results from a complex interplay between the processes of net sediment accumulation and the activities that generated durable material remains (see also Heilen et al., this volume). Material remains and their encasing sediments accumulated through the operation of different processes, and they did not necessarily accumulate at the same rate or over the same time intervals. The interplay between the two exerts a strong influence on the empirical structure of the data being studied and the categories of behavioral information that may be generated from a regional archaeological record (see Head, this volume for a discussion of such issues in relation to investigative scale). For this reason, understanding the way in which archaeological debris and sediments accumulated and the way in which palaeolandscape features are reconstructed is fundamental to the practice of landscape archaeology.

This chapter illustrates the role stratigraphic relationships and sedimentary context play in reconstructing palaeolandscapes at two contrasting scales, drawing on examples from the Shungura, Nachukui, and Koobi Fora Formations, Plio-Pleistocene geological formations that outcrop along the northern, western, and eastern shores of modern Lake Turkana in southern Ethiopia and northern Kenya (Figure 36.1). These formations are well known to archaeologists because of the extraordinary abundance of early hominin remains and activity traces they preserve. Together, they represent more than 700 meters of mostly fluvial and lacustrine sediments laid down between > 4.3 and 0.74 million years ago by different components of the same depositional regime, in a single, large depression formed through uplift of the Kenyan and Ethiopian domes. Archaeological remains appear in the record from about 2.4–2.3 million years ago and occur in varying abundance in different parts of the basin at different times (Howell, Haesaerts, and de Heinzelin 1987; Isaac 1997; Kibunjia et al. 1992). The two case studies presented here illustrate some of the critical interpretative issues involved in palaeolandscape reconstruction. More detailed information about how to describe and interpret stratigraphic sequences and the landforms they represent can be gleaned from both geology and geoarchaeology texts (e.g., Goldberg and McPhail 2006; Miall 2000; Reading 1996).

Which Strata? What Scale the Palaeolandscape?

How the boundaries of a palaeolandscape study are delineated depends on the questions driving the research and the characteristics of the sedimentary sequence under investigation. Obviously, palaeolandscapes can be reconstructed at different scales of spatial and temporal resolution, but there is an inverse relationship between the two. Broader stratigraphic intervals sample larger portions of an ancient landscape, but they also represent longer time spans. Vertically discrete stratigraphic horizons may represent relatively fine time lines, but only rarely can they be traced any distance and consequently they sample only a part of the broader landscape (Figure 36.2). The greater

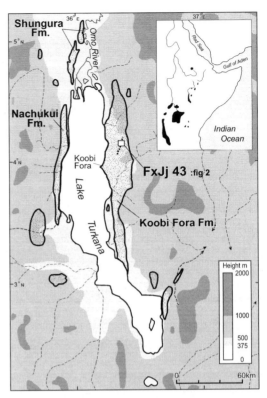

Figure 36.1 Map showing the location of the Shungura, Nachukui, and Koobi Fora Formations, which outcrop respectively along the northern, western, and eastern shores of the modern Lake Turkana, in northern Kenya, and southern Ethiopia (modified from the International Atlas published by Rand McNally in 1979). The location of FxJj43, one of the two case studies discussed here, is also marked.

the spatial and temporal discontinuities in net sediment accumulation, the more marked the trade-off between the spatial and temporal dimensions of the palaeolandscape.

There is also an inverse relationship between the quantities of archaeological debris that accumulated on a landscape and the amount of time represented by the sediments encasing the archaeological debris. Beds that accumulated rapidly (such as air-fall or water-lain ashes) rarely formed the surface for long enough to accumulate any quantity of archaeological debris (e.g., Isaac 1997: 33–41). Soil horizons representing prolonged periods of landscape stability may accrue more debris but often represent a single landform. Stratigraphic intervals representing gradual and/or intermittent sediment accumulation across a larger area will preserve greater quantities of material remains from a broader array of depositional

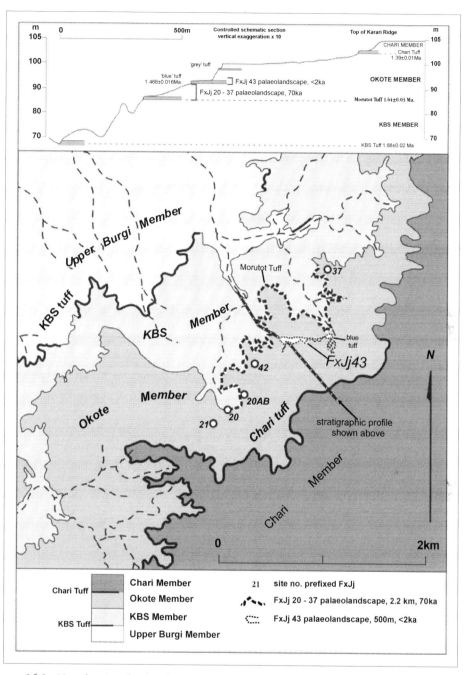

Figure 36.2 Map showing the distribution of four members of the Koobi Fora Formation, where two different palaeolandscape studies have been undertaken (the distribution of geological units is modified from Isaac 1997: fig. 15). The FxJj20-37 palaeolandscape is based on a 2.2-km section of outcrops representing a variety of channel systems and abutting, but unrelated, bank and floodplain settings. The minimum archaeological stratigraphic interval that could be traced out across this distance is up to 9 m thick and represents a time span of approximately 70,000 years (Stern 1993). The outcrops making up FxJj43 can be traced out for only 500 meters, but they represent a related set of palaeotopographic features (a channel, its southern bank, levee, and adjacent floodplain) and archaeology-bearing strata that accumulated in less than 2,000 years (see text).

settings. However, those archaeological traces are derived from a three-dimensional wedge of sediments that may have built up over thousands or tens of thousands of years (e.g., Stern 1993).

Although it is possible to identify discrete archaeological occurrences that arguably represent single events—such as the knapping of water-worn cobble(s)—the contemporaneity of any two such occurrences can be determined only by the amount of time it took to accumulate the sediments within which they are sandwiched (Stern 1993, 1994). Thus, to sample the debris that accumulated on a palaeolandscape, the material traces of many different activities and events must be aggregated, and in most Pleistocene sequences there is no way of establishing the precise temporal relationships of any two, geographically separated occurrences.

In response to the initial characterization of palaeolandscape, samples as agglomerations of debris accumulated over thousands or tens of thousands of years (e.g., Stern 1993), some researchers sought out those rare archaeology-bearing strata that represent fine time lines. It was believed that a shorter time span of accumulation would result in less overprinting of debris from different activities and thus would be more amenable to interpretation using ethnographic or ecological models (e.g., Conard 1994). However, further consideration of the way in which archaeological debris accumulates on palaeolandscapes suggests that fine time lines actually provide quite inscrutable records, because the probability that they will have captured debris resulting from less frequently occurring activities is low. Although it is obvious that some time had to pass in order to accumulate sufficient quantities of debris on a landscape surface to create an observable archaeological record, it is not clear how much time is required to generate patterns that archaeologists can identify, document, and interpret, and undoubtedly this time span varies from one context to another. However, it is evident that in many circumstances the patterned distributions that become the subject of archaeological investigation would not have been evident to the individuals whose activity traces are being studied.

Ideally, the choice of stratigraphic interval for investigation would involve considered assessment of the likelihood of generating from those sediments the information needed to answer specific research questions. However, because archaeologists are still grappling to identify the categories of behavioral information that can be generated from patterned distributions of debris generated over the time spans involved in the formation of palaeolandscape samples, in practice these assessments are part of the analytical and interpretative process. As a result, the boundaries of palaeolandscape studies are delineated most often by identifying a stratigraphic interval bounded by distinctive marker horizons that can be traced out across the study area and that contains sufficient quantities of archaeological debris to allow the use of statistical techniques to identify patterned associations. If the marker horizons are datable isochronous units, then it is possible to establish the total time span represented by the stratigraphic interval under investigation. The time span represented by the palaeolandscape provides the basis for developing an interpretative framework commensurable with the resolution of the archaeological data under investigation.

Basinwide Reconstructions of Regional Palaeogeography

Basinwide reconstructions of the palaeolandscape involve the application of familiar stratigraphic principles at scales ranging from kilometers to tens or hundreds of kilometers. In the Turkana depression, geographically scattered outcrops accumulated varying thicknesses of sediment over varying time intervals. These scattered sequences have been correlated using widespread volcanic ashes that were ejected during single eruptive events and therefore have distinct geochemical signatures. These ashes have also been used to divide each formation into a series of members and to provide a time frame for each sequence (Brown and Feibel 1986; Harris et al. 1988; de Heinzelin 1983). Only 11 of the 130 chemically distinct ashes that have been identified are found in all three formations (Figure 36.3), although each formation also contains 30–35 ashes that are found in one of the other formations (Brown and McDougall 1993). These correlations identify where sediment was accumulating during specific time intervals, so once the depositional environments represented by those sediments have been established (e.g., Feibel 1988), it is possible to reconstruct the landscape features that existed during that time span (Brown and Feibel 1988; Feibel and Brown 1993). There is, of course, an inherent bias to this landscape sample, which is restricted to low-lying areas that experienced stable, net sediment accumulation during the time period under investigation. This bias is common to most early Pleistocene landscape studies, since only in rare circumstances (such as the blanketing of an entire landscape by air-fall ash) were the more elevated, eroding portions of a landscape draped with sediment (Isaac 1972).

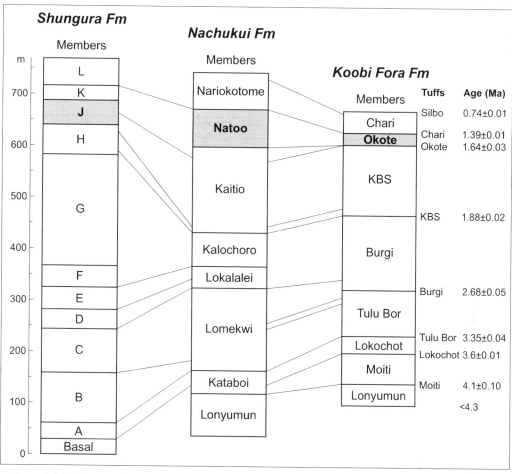

Figure 36.3 Stratigraphic summary of the Shungura, Nachukui and Koobi Fora Formations showing the volcanic ashes that are the basis for dividing each sequence up into members and for correlating the three formations (modified from Feibel et al. 1989: fig. 3).

Once the palaeogeographic features that made up the preserved portions of the landscape during specific time intervals have been established, it is then possible to investigate the distribution of archaeological sites in relation to those features. However, two issues need to be considered in such analyses. The first is the scale at which the palaeogeographic features were changing versus the time span of the stratigraphic interval under investigation (e.g., Feibel and Brown 1993: 37). It can be difficult to establish the landscape setting of sites that formed on dynamic landscapes that were changing on a faster time scale than the time span of the stratigraphic interval under investigation. This is because later erosion removes the deposits laid down by preexisting depositional systems, resulting in preservation of unconnected palaeogeographic features; for example, preserved

channel bodies may not relate to the bank and floodplain deposits they abut. The second consideration is how the palaeotopographic settings of individual sites can be related to regional palaeogeographic features. Analysis of facies at a scale intermediate between the basinwide studies that are usually the purview of geologists, and the detailed palaeotopographic settings of sites that are usually the purview of archaeologists, may be required.

Rogers, Harris, and Feibel (1994) examined the distribution of activity traces across the Turkana depression for three successive time intervals (2.3 million years ago; 1.9–1.8 mya; 1.6–1.5 mya; Figure 36.4) and observed that debris was discarded more widely, in a greater variety of topographic settings and in greater quantities after 1.6–1.5 million years ago. This change is explained by the appearance of

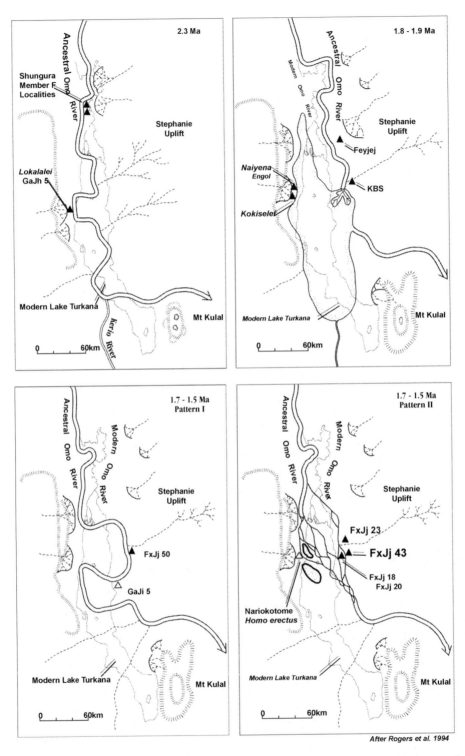

Figure 36.4   The distribution of archaeological traces in relation to palaeogeographic features for 3 successive time intervals, based on Brown and Feibel's palaeogeographic reconstructions for the Turkana basin (modified from Rogers et al. 1994: figs. 4–7).

Homo ergaster (ca. 1.9 million years ago), the first truly committed terrestrial biped that also exhibits a number of adaptations for engaging in high activity levels in hot, arid environments (Walker and Leakey 1993) and for endurance running (Bramble and Lieberman 2004). However, there is a much tighter correlation between the more widespread distribution of activity traces and the appearance of the earliest Acheulian artifact assemblages (ca. 1.65 million years ago), suggesting that it would be worth investigating whether the technological change is related to concomitant shifts in foraging

strategies and mobility patterns that were part of an adaptive shift.

Palaeolandscape Reconstruction of Smaller Stratigraphic Intervals. To acquire the information needed to investigate such problems archaeologists focus on the distribution, density, and characteristics of individual artifacts and fossils found in vertically discrete, laterally continuous sedimentary horizons found in one part of a longer stratigraphic sequence and in one part of a larger sedimentary basin (e.g., Isaac 1981). Because net sediment accumulation in terrestrial depositional

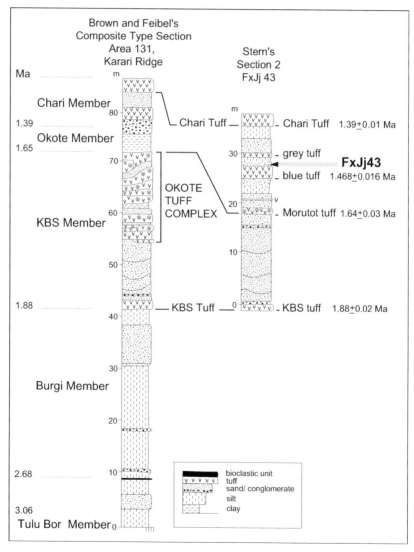

Figure 36.5 The stratigraphic position of FxJj43 within the Koobi Fora Formation, based on a measured section from FxJj43 and its correlation to Brown and Feibel's type section for Area 131 (modified from Brown and Feibel 1985: fig. 1).

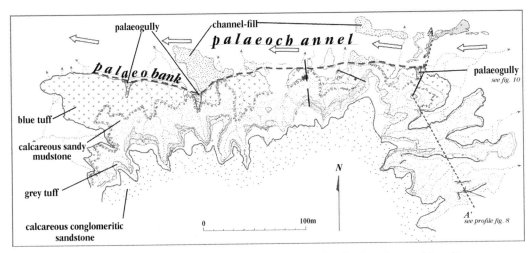

Figure 36.6   Geological map of FxJj43, showing the distribution of mappable units.

systems tends to be intermittent and localized, and because facies are time-transgressive phenomena, palaeolandscapes are represented most often in the geological record by a three-dimensional wedge of sediments, rather than a single surface or a single sedimentary layer. The "minimum-archaeological-stratigraphic unit" that can be used to reconstruct the palaeolandscape is defined by the two most closely spaced marker horizons that can be identified in the stratigraphic sections scattered across the study area (Stern 1993). Reconstruction of landscape features is facilitated if one or both of those marker horizons is an isochronous depositional unit that cuts across facies boundaries and identifies the location of specific topographic features at a particular moment in time. This is the situation at FxJj43, an Early Stone Age locality in the Okote Member of the Koobi Fora Formation (Figure 36.5) that preserves a well-defined set of palaeotopographic features that accumulated archaeological debris over a relatively short span of time (Stern 2004; Stern et al. 2002).

The Stratigraphic Sequence. FxJj43 is a collection of outcrops bounded by two distinctive and widespread marker horizons that can be traced around the edge of the modern erosion shelf for almost half a kilometer (Figure 36.6). Both are relatively more resistant to erosion than the underlying or overlying sediments, so the wedge of sediments sandwiched between them forms a prominent topographic feature on the modern landscape and provides an obvious boundary for a palaeolandscape study. Up to 8 meters of sands, silts, mud, and ashes representing a number of fluvial depositional settings are sandwiched between these marker horizons.

The lower marker horizon is a thick bed of volcanic ash whose chemical composition identifies it as one of a series of ashes of similar (though not identical) chemical composition erupted over a 50,000-year time interval from a single source in the Ethiopian highlands. It was deposited originally as an air-fall ash in the Ethiopian highlands and was washed into the Turkana depression by the river systems draining those highlands, where it was redeposited by overbank floods. The upper marker horizon is a thin but laterally extensive bed of calcareous sandstone representing a widespread sand sheet deposited by shallow, shifting braided channel systems. It is not an isochronous unit, but it represents the establishment of a different, depositional regime, and it is readily traced throughout the study area.

The Landscape Features Preserved. Landscape reconstruction began with a series of stratigraphic sections measured at intervals along the outcrops to establish the sequence of beds sandwiched between the marker horizons and to identify a series of distinctive beds that can be used to correlate those sections (Figure 36.7). It also helped to identify sets of beds representing different depositional environments that make up mappable units (Figure 36.6). The three-dimensional geometries of these mappable units, together with their textural characteristics and bedding features, provided the information needed to identify the palaeotopographic settings they represent. Geological trenches were dug to establish those three-dimensional relationships where they were not visible in the exposed outcrops or in existing gully sections. The sequence and three-dimensional relationships of those sedimentary units are, of

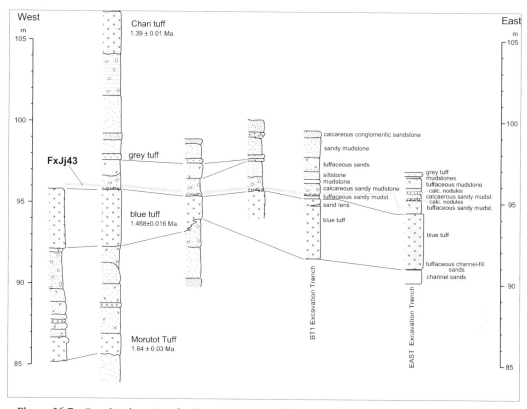

Figure 36.7   Correlated sections for the outcrops at FxJj43, showing the stratigraphic relationships of the main mappable units and the location of the archaeology-bearing horizon, immediately above the blue tuff.

course, a crucial source of information about the depositional history of the outcrops. But additional information about how the deposits were built up can be gleaned from detailed descriptions of the sections exposed in both the geological trenches and the excavations.

The oldest sediments at FxJj43 can be seen in outcrop at either end of the site but in between are visible only in the geological trenches (Figure 36.7). At the eastern edge of the site these are unconsolidated channel sands whose bedding features indicate that they were laid down by a westerly flowing river, whereas at the western end of the site they are fine-grained floodplain deposits. Both are overlain by a thick bed of ash (up to 2.5 meters thick), known as the "blue" tuff. Detailed sections through the blue tuff show that beds of pure ash preserving distinctive bedding features are intercalated with numerous lenses of sand and silt dumped by successive flood peaks. Like other widespread ashes in the Turkana depression the "blue" tuff is believed to represent a rare, high-magnitude flood event. It would have destroyed the local vegetation cover, and there are numer-

ous pieces of evidence (discussed below) to indicate that an episode of bank erosion followed its deposition.

The blue tuff has a gradational contact with the overlying tuffaceous sandy mudstone, which grades up-section to a calcareous sandy mudstone. Together, these beds record the gradual resumption of normal terrigenous sedimentation, as the channel that had carried the blue tuff gradually cleared itself of the viscous slurry of ash and water and a stable, vegetated floodplain was reestablished. Intermittent overbank floods laid down another 1.7 meter of floodplain sediments before a second episode of volcaniclastic deposition took place, represented by the "grey" tuff (Figure 36.7). It is also overlain by a tuffaceous sandy mudstone that fines up-section as its terrigenous component also increases; again, these beds represent distal floodplain sedimentation. These are overlain by a thin and laterally extensive bed of calcareous sandstone that caps the sequence. It contains lenses of poorly sorted, gravely, or conglomeratic sands and represents a widespread sand sheet laid down by shallow,

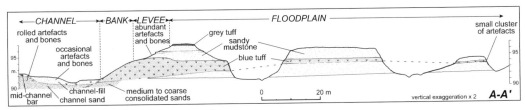

Figure 36.8   Cross-section showing the relationships between distinctive sedimentary bodies and the landforms they represent and summarizing the stratigraphic and palaeotopographic contexts of the archaeological debris.

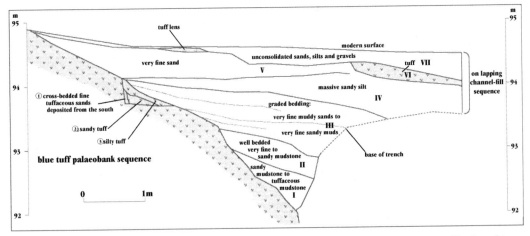

Figure 36.9   Detailed sections from one of the geological trenches, showing the sequence of beds making up the blue tuff palaeobank and the channel-fill.

braided channels that shifted frequently across a low-relief landscape.

This stratigraphic sequence has the advantage of containing an isochronous unit that can be used to establish the relationships that existed between facies at the time of the massive flood event that deposited it (Figure 36.8). Geological trenches dug across the northern edge of the outcrops show that the surface of the blue tuff dips down steeply to the north, indicating that the tuff has draped over a palaeobank. Tuffaceous channel-fill sands are banked up against the steeply sloping surface of the blue tuff palaeobank. Both overlie interfingering, unconsolidated channel sands and consolidated, medium to coarse sands representing the underlying channel-bank. Detailed sections recorded in the geological trenches that cut through the palaeobank and channel-fill deposits reveal numerous episodes of deposition and erosion (Figure 36.9).

The blue tuff palaeobank can be traced along the northern edge of the FxJj43 outcrops and in places actually has a topographic expression on the modern landscape (Figure 36.10). Outcrops, gully sections, and geological trenches all indicate

that the blue tuff is thickest at the palaeobank and that it lenses out to the south, where it is sandwiched between calcareous mudstones that are characteristic of vegetated, distal floodplain settings. In these locations, the pumices found in the upper levels of the tuff are significantly smaller than they are in outcrops adjacent to the palaeobank, reflecting lower-energy flow on the distal floodplain (Figure 36.8). The most southerly exposures of the blue tuff approximate the far reach of the floodwaters that deposited it.

In between the northerly and southerly limits of the blue tuff there are few exposures of it, so, where necessary, its position was determined by auger holes. These revealed that approximately 8–12 meters from the edge of the channel the blue tuff forms a levee that is about 7–8 meters wide.

A number of palaeogullies cut across the blue tuff palaeobank (Figures 36.6 and 36.10). They are marked by a sharp boundary between the pure ash layers making up the palaeobank and the brown tuffaceous sands filling the ancient gullies. The palaeogullies vary in width and depth, and some contain isolated artifacts and/or bones.

Figure 36.10 A view across the eastern end of the outcrops, where the blue tuff palaeobank has a topographic expression on the modern landscape. A palaeogully, filled with brown tuffaceous sand, cuts across the blue tuff palaeobank (photo: Rudy Frank, Archaeology Program, La Trobe University).

The northern bank of the sandy channel that was draped by the blue tuff has long since been eroded away; however, some estimate of the width of that channel can be gauged from an isolated outcrop of blue tuff (at the northeastern edge of the site) that has been interpreted as a mid-channel bar. It is elliptical in plan view and is overlain and abutted by a thin bed of calcareous sand, which, in turn, overlies a low bank of well-sorted, unconsolidated medium sands. It lies about 10–15 meters from the southern bank of the palaeo-channel (Figure 36.8).

The Palaeolandscape Context of the Archaeological Debris. The blue tuff provides a wonderful insight into the topographic features that existed at FxJj43 when it was deposited, but its relationship to the archaeological material that accumulated there also has to be established. Excavations and *in situ* finds in the gully sections and measured geological sections all indicate that the archaeological material is derived from three distinct beds representing different palaeotopographic settings in the same stratigraphic interval. There is a narrow horizon of archaeological debris immediately overlying the blue tuff; it straddles the tuffaceous sandy mudstone and the lower levels of the overlying calcareous sandy mudstones. The tuffaceous sandy mudstone was deposited as the river gradually cleared itself of ash, while the overlying calcareous sandy mudstone records the resumption of normal terrigenous sedimentation and the eventual establishment of a vegetated floodplain and the mobilization of carbonates through soil-forming processes. The stratigraphic distribution of the archaeological debris indicates that it accumulated only while normal terrigenous sedimentation was being reestablished. Impregnation of the archaeological horizon by carbonates (which helped to preserve the bones) took place after hominins had abandoned the area.

Immediately after the blue tuff draped the landscape, an episode of bank erosion ensued. Evidence for this includes the nick-points on the blue tuff palaeobank (Figure 36.9), the scoured surface of the blue tuff (visible in the excavation floors), the formation of the palaeogullies, and the presence of eroded clods of blue tuff at the interface between the underlying channel sands and in the channel-fill sequence. Comminuted

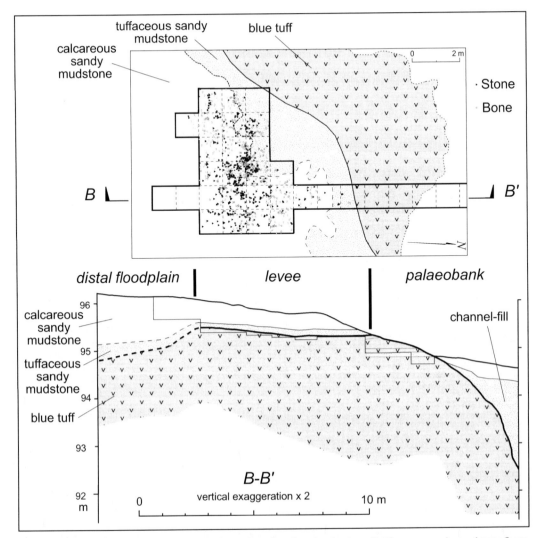

Figure 36.11 Plan of an excavation into the archaeology-bearing horizon (tuffaceous sandy mudstone/base of calcareous sandy mudstone) showing the distribution of artifacts and bones with 3D coordinates and a cross-section showing the relationship between the excavation squares and the topography of the underlying blue tuff palaeolandcape.

clods of blue tuff are found throughout the channel-fill sequence, indicating that the erosion of the blue tuff palaeolandscape and the infilling of the channel were coeval. Traces of hominin activities are scattered through both the palaeogully- and channel-fill deposits but are not found in any of the overlying floodplain deposits. So, hominins were in the area only while water continued to flow down this channel segment, albeit intermittently.

This suggests that the features of the blue tuff palaeolandscape probably still existed during the period in which the archaeological material accumulated. However, spectacular confirmation of

this comes from an excavation that sampled the abundant archaeological debris that accumulated in the sediments overlying the blue tuff palaeobank and levee. The northern edge of the artifact and bone scatter has been truncated through erosion, but its southern edge was buried and remains intact. This scatter of archaeological debris has an abrupt boundary that coincides with the edge of the underlying blue tuff levee (Figure 36.11).

The Age and Time Span of the Archaeological Debris. The age of the blue tuff palaeolandscape has been established from Argon-Argon age determinations (see Roberts and Jacobs, this volume) on feldspar crystals extracted from pumices recovered from the top of the ash, 1.468 ± 0.016 mya (Stern et al. 2002). There is no way of actually measuring the time it took for the archaeology-bearing sediments to accumulate, but contemporary observations of the time it takes for river systems to resume normal terrigenous sedimentation after a major ash-fall suggest that it may have been decades, whereas stratigraphic scaling sets an upper limit of < 2,000 years. FxJj43 provides archaeologists with a window into a 1.5 million-year-old palaeolandscape and an opportunity to document the activity traces that accumulated on a small portion of the landscape over a limited period of time.

The Implications of Palaeolandscape Context for Site Formation Processes. Not surprisingly, archaeological debris accumulated in different ways and at different rates in different parts of this palaeolandscape (Stern 2004). Small, discrete clusters of debris representing the knapping of a single block of material have been found on the distal floodplain. Isolated artifacts and/or bones are found in both the gully-fill and channel-fill deposits, but although some of those from the channel-fill exhibit surface abrasion and/or have rounded edges, there are also artifacts that show no signs of abrasion and that retain sharp edges. Clearly, some artifacts and bones were transported downstream during periodic channel flow, whereas others were lost or discarded more or less where they were found.

Artifacts and bones from the gully fills are neither abraded nor rounded suggesting that they were not transported any great distance. These isolated finds are most likely to have originated on the adjacent levee, which preserves the most abundant and varied set of activity traces at FxJj43. These include low-density scatters of debris, small, discrete clusters of artifacts and/or bones, and larger, denser agglomerations of both artifacts and bones. The bones recovered so far from one small discrete cluster, and from one larger, denser agglomeration, exhibit a range of weathering stages, suggesting that neither occurrence represents a single event. However, the larger, denser agglomeration does contain both a wider range of artifacts and a wider range of taxa and body parts than the small discrete cluster, which may reflect a longer time span of accumulation.

Thus, an understanding of the depositional context of the archaeological debris at FxJj43 provides scope for investigating not only differences in the debris-generating activities undertaken in different parts of the blue tuff palaeolandscape but also differential rates of accumulation and the impact of those on the composition and characteristics of

the resulting archaeological assemblages. Together these provide a basis for considered assessment of the behavioral information that can be generated from the different types of assemblages preserved in different settings.

Conclusions

The stratigraphic and depositional context of a related set of archaeological occurrences establishes the landscape settings in which traces of past activities accumulated. However, the geological context of archaeological materials is not simply a neutral backdrop for investigating the context and characteristics of those activity traces. The way in which, and the time span over which, strata and activity traces accumulated structures the scale and resolution of the palaeolandscape sample. Establishing the time span represented by the sediments making up the palaeolandscape is fundamental to any attempt to study the activity traces accumulated on that landscape. In particular, it provides the basis for identifying the categories of behavioral information that can be generated from the archaeological debris preserved in the palaeolandscape under investigation. This, in turn, helps to identify the analytical and interpretative frameworks considered most appropriate to the empirical structure of the record under investigation.

Acknowledgments

Fieldwork at Koobi Fora was undertaken with permission from the Government of Kenya, under the auspices of the National Museums of Kenya. It benefited from the support of successive Directors/Chief Executives of the National Museums of Kenya, Dr. Mohammed Isahkia and Dr. George Abungu, and by successive Heads of the Archaeology division, Dr. Hélène Roche and Dr. Karega-Munene. The fieldwork at FxJj43 was funded by the Australian Research Council and the Leakey Foundation.

References

Bramble, D. M., and Lieberman, D. E. 2004. Endurance running and the evolution of *Homo*. *Nature* 432: 345–52.

Brown, F. H., and Feibel, C. S. 1986. Revision of lithostratigraphic nomenclature in the Koobi Fora region, Kenya. *Journal of the Geological Society* 143: 297–310.

———. 1988. "Robust" hominids and Plio-Pleistocene palaeogeography of the Turkana basin, Kenya and Ethiopia, in F. Grine (ed.), *Evolutionary History of* the "Robust" Australopithecines, pp. 325–41. New York: Aldine de Gruyter.

Brown, F. H., and McDougall, I. 1993. Geologic setting and age, in A. Walker and R. E. F Leakey (eds.), The Nariokotome Homo erectus Skeleton, pp. 9–20. Cambridge, MA: Harvard University Press.

Conard, N. 1994. Discussion and criticism: On the prospects for an ethnography of extinct hominids. Current Anthropology 35: 281–82.

de Heinzelin, J. 1983. The Omo Group. Musée Royale de l'Afrique Centrale, Tervuren, Belguique. Annales, Série in 8°, Science Géologiques, No. 85.

Feibel, C. S. 1988. Palaeoenvironments of the Koobi Fora Formation, Turkana Basin, Northern Kenya. Unpublished doctoral dissertation, University of Utah, Salt Lake City.

Feibel, C. S., and Brown, F. H. 1993. Microstratigraphy and paleoenvironments, in A. Walker and R. E. F Leakey (eds.), The Nariokotome Homo erectus Skeleton, pp. 21–39. Cambridge, MA: Harvard University Press.

Goldberg, P., and McPhail, R. I. 2006. Practical and Theoretical Geoarchaeology. Oxford: Blackwell Publishing.

Howell, F. C., Haesaerts, P., and de Heinzelin, J. 1987. Depositional environments, archaeological occurrences and hominins from Members E and F of the Shungura Formation (Omo Basin, Ethiopia). Journal of Human Evolution 16: 665–700.

Harris, J. M., Brown, F. H., and Leakey, M. G. 1988. Stratigraphy and Paleontology of Pliocene and Pleistocene Localities West of Lake Turkana, Kenya. Contributions in Science No. 399. Natural History Museum of Los Angeles County.

Isaac, G. Ll. 1972. Comparative studies of Pleistocene site locations in East Africa, in P. J. Ucko, R. Tringham, and G. W. Dimbleby (eds.), Man Settlement and Urbanism, pp. 165–176. London: Duckworth.

———. 1981. Stone Age visiting cards: Approaches to the study of early land-use patterns, in I. Hodder, G. Ll. Isaac, and N. Hammond (eds.), pp. 131–55. Pattern of the Past. Cambridge: Cambridge University Press.

———. (ed.). 1997. Koobi Fora Research Project, Vol. 5. Plio-Pleistocene Archaeology. Oxford: Clarendon Press.

Kibunjia, M., Roche, H., Brown, F. H., and Leakey, R. E. 1992. Pliocene and Pleistocene archaeological sites of Lake Turkana, Kenya. Journal of Human Evolution 23: 432–38.

Miall, A. D. 2000. Principles of Sedimentary Basin Analysis (3rd ed.). New York: Springer-Verlag.

Reading, H. G. 1996. Sedimentary Environments: Facies, Processes, and Stratigraphy (3rd ed.). Oxford: Blackwell Science.

Rogers, M. J., Harris, J. W. K., and Feibel, C. S. 1994. Changing patterns of land use by Plio-Pleistocene hominids in the Lake Turkana Basin. Journal of Human Evolution 27: 139–58.

Stern, N. 1993. The structure of the Lower Pleistocene archaeological record: A case study from the Koobi Fora Formation in northwest Kenya. Current Anthropology 34: 201–25.

———. 1994. The implications of time-averaging for reconstructing the land-use patterns of early tool-using hominids. Journal of Human Evolution 27: 89–105.

———. 2004. Early hominin activity traces at FxJj43, a one and a half million year old locality in the Koobi Fora Formation, in northern Kenya: A field report. Proceedings of the Prehistoric Society 70: 233–58.

Stern, N., Porch, N., and McDougall, I. 2002. FxJj43: A window into a 1.5 million year old palaeo-landscape in the Okote Member of the Koobi Fora Formation. Geoarchaeology 17: 349–92.

Walker, A., and Leakey, R. E. F. (eds.). 1993. The Nariokotome Homo erectus Skeleton. Cambridge, MA: Harvard University Press.

37

Geographical Scale in Understanding Human Landscapes

Lesley Head

Recent reconceptualizations of scale in geography are relevant to the theory and practice of landscape archaeology. If we want to understand past landscapes as fields of human engagement, then the way we think about scale in both space and time matters. As Raper and Livingstone (1995: 364) argued, "the way that spatio-temporal processes are studied is strongly influenced by the model of space and time that is adopted." A large literature in geography, particularly human geography, in the last few decades has focused on the theorizing of scale. The particular context of these debates has been the economics of globalization, and the ways in which it is expressed and worked through at local, regional, and national levels. This is not to say that archaeologists have not thought about scale, but rather that they have not theorized it quite so explicitly. Further, juxtaposing discussions from very different empirical contexts to archaeological ones is a useful way to identify commonalities and differences.

Scale is defined in *The Dictionary of Human Geography* as "one or more levels of representation, experience, and organization of geographical events and processes" (Johnston et al. 2000: 724). Geographers commonly distinguish among *cartographic* scale ("the level of abstraction at which a map is constructed"), *methodological* scale ("the

choice of scale made by a researcher" to answer a research problem), and *geographical* scale (724–25). The last term "refers to the dimensions of specific landscapes: geographers might talk of the regional scale, the scale of a watershed, or the global scale, for example. These scales are also of course conceptualized, but the conceptualization of geographical scale here follows specific processes in the physical and human landscape rather than conceptual abstractions lain over it" (Johnston et al. 2000: 725).

This chapter is primarily concerned with the latter two uses of scale, and the interactions between them; that is, what scale choices do and should landscape archaeologists make in trying to understand landscapes of prehistoric human interaction? To answer this, I discuss three important themes in the literature on scale; first, its social construction; second, its relational qualities; and third, the relationship between space and time. These themes should be thought of as applying to the scales of both prehistoric behavior and contemporary research. I illustrate the challenges and the potential of operationalizing a more dynamic understanding of scale with examples from a collaborative landscape archaeology project in northwestern Australia. In this case, the landscapes of prehistoric human interaction are hunter-gatherer ones in northwestern Australia.

The Social Construction of Scale

Within both the social and natural sciences there has been a shift over recent decades away from essentialist understandings of categories and concepts toward contingency. The critique of scale should be understood as part of this trend.

> In these recent social theoretical studies, the fundamental point being made is that scale is not necessarily a preordained hierarchical framework for ordering the world—local, regional, national and global. It is instead a contingent outcome of the tensions that exist between structural forces and the practices of human agents. (Marston 2000: 220)

Archaeologists may not have theorized much explicitly about scale, but the idea of social construction is not a foreign concept and is present in many empirical studies.

For example, there is the issue of how the landscape of the prehistoric inhabitants meshes or otherwise with the analytical landscape of the contemporary archaeologist. There is both implicit and explicit discussion of scale in the literature on sites. What is the appropriate scale of analysis? What is a site? How does it sit in the landscape? Where will I put the boundaries for my study? Within landscape archaeology in particular, there is an awareness of processes that can occur at the scale of the individual body, right up to the broadest of physical landscapes.

Scale is constituted in the process of social relations:

> All social relations (capital accumulation, politics, social and cultural development) have a geographic expression, and it is the complex expression of these relations in space that produces scale. The geographical expression of varied social relations establishes a range of differentiated geographic scales that become temporarily fixed and are reproduced through their expression in the built environment and through social networks . . . Therefore, socially produced scale both establishes and is established through the geographical structure of social relations. (McGuirk 1997: 483)

As geographers, then, our goal with respect to scale should be to understand how particular scales become constituted and transformed in response to social-spatial dynamics. (Marston 2000: 221)

Archaeologists can take these points and then need to extend them to consider how we read off the scale of past social experience from the material evidence. It is also important to consider how the disciplinary conventions of archaeology pin certain understandings of scale in place. As Wishart (2004) points out, these ideas apply also to temporal scale; the idea of historical period needs as much critical scrutiny as region.

Scale as Relational

If scale is contingent and worked out through human processes, it follows that it is relational—scales are constituted in relationship with one another rather than as discrete entities. Doreen Massey (2001) has referred to this as the "Russian Doll" problem. The idea that we might be able to take scales apart right down to the micro level misses the point that they are constituted in relation to one another.

To see scale as relational has implications for understandings of causation, since "the notion of nesting assumes or implies that the sum of all the small-scale parts produces the large-scale total" (Howitt 1993: 36). Relationality challenges the idea that we can "identify discrete scales from which causes originate and at which effects are felt. In such an approach processes, outcomes, and responses are categorized into distinct 'boxes' that are seen as discrete entities originating at a particular level in an indisputable hierarchy of scales" (McGuirk 1997: 482).

Thus the relationships between scale and order, or scale and causation, should not be assumed but be the subject of empirical enquiry. In the context of economic debates, this challenge is usually applied to the assumption that the global is the primary scale. In the context of hunter-gatherer archaeology, the assumption needing critique might be the primacy of the local. Gille and O'Riain (2002: 286) make the further point that level of analysis should not be confused with the level of abstraction—the global is not necessarily universal, and the local is not necessarily particular. Extending these arguments to the relationship between space and time, we see that the long term is not necessarily global, and the short term not necessarily local.

To say that scale is both socially produced and relational does not deny that particular scales can become fixed, reproduced, and influential—they can come to be seen as natural. Indeed, this is what had happened to the local, the regional, and the global to require the critique within economic geography. We might ask then, what scales have become fixed in any particular field of landscape archaeology? What scales and assumed causal relationships have become fixed at the analytical level? What scales might have become fixed in any particular past context that we are studying?

Space, Time, and Space-time

It follows that these issues apply to the relationships between space and time, a point argued from a different direction by Albert Einstein. Doreen Massey (1999) reviewed critiques of separation of space and time, and the privileging of time as the engine of history. In this view, space was simply the container in which history took place. Using an image with obvious relevance to archaeology, Massey argued that "the closed-system/slice-through-time imagination of space denies the possibility of a real temporality—for there is no mechanism for moving from one slice to the next" (264):

> time needs space to get itself going; time and space are born together, along with the relations that produce them both. Time and space must be thought together, therefore, for they are inextricably intermixed. A first implication, then, of this impetus to envisage temporality/history as genuinely open is that spatiality must be integrated as an essential part of that process of "the continuous creation of novelty." (Massey 1999: 274)

She used the term *space/time* to denote the inextricability of the two. I follow that usage here, while noting that it has been used in various configurations (for example, time/space) by a number of writers, including in anthropology.

An issue I return to below is the extent to which our methods, particularly our methods of visual representation, inadvertently reinforce the false separation of space and time. With temporal cross-sections that hold space static, and vice versa, can we really imagine the more dynamic understanding that Massey advocates:

> for time genuinely to be held open, space could be imagined as the sphere of the existence of multiplicity, of the possibility of the existence of difference. Such a space is the sphere in which distinct stories coexist, meet up, affect each other, come into conflict or cooperate. This space is not static, not a cross-section through time; it is disrupted, active and generative. It is not a closed system; it is constantly, as space-time, being made. (1999: 274)

This understanding would not be foreign to any archaeologist who has tried to understand past landscapes as fields of human engagement; the problem is the intractability of the archaeological data. Massey did acknowledge that "it can make your head hurt to think in this way" (1999: 262).

A powerful expression of this concept of space/time is the recent discovery and analysis of the Willandra Lakes Mungo footprints (Webb et al. 2006). Found across 700 m² of exposed hardpan, near the stranded shoreline of a relict lake basin, are footprints left by at least eight individuals, including adults, adolescents, and children. OSL dating puts this space/time between 23,000 and 19,000 B.P. This is a very embodied landscape. The mud squeezes between their toes. Two tall individuals are inferred to have been fast runners. Some of the tracks disappear beneath overlying sediments; there could well be more to be found. It is a captured moment, albeit the long sort of moment provided by archaeological dating. It is a moment from which we can spin out in both space and time to understand the broader context. We can understand, to reprise both William Blake and Howitt (1993), how "to see a world in a grain of sand."

But could Webb and colleagues have pinned this space/time together without first understanding the cross-sections that are such familiar archaeological signifiers of Lake Mungo, for example, in the work of Bowler (1998) or in their own figure 3? It seems unlikely. And how do the rest of us tease out embodied landscapes when we have the cross-sections but not the footprints? These are perennial issues with the intractable evidence of hunter-gatherer archaeology. In the following section, I illustrate how we have grappled with these dilemmas in a collaborative landscape archaeology project.

Case Study: Landscape Archaeology in Northwestern Australia

The Keep River region of northwestern Australia is a landscape of considerable and continuing Aboriginal significance. As in other sandstone landscapes of northern Australia, there are many difficulties in terms of archaeological preservation and visibility, although the rock art provides a spectacular above-ground archaeology. The methodological challenge has been to take landscapes designated separately for analytical purposes and to try to pin them together to get an embodied landscape prehistory. Rather than assume a total cultural "package" that all operates together, we aim to explicitly identify convergences and divergences in the temporal and spatial rhythms of elements with very different archaeological signatures.

I compare what I have called "the geoarchaeological landscape" and the "landscape of fruit seed processing." In each case, as researchers we chose a *methodological* scale that matched the available evidence and that we hoped would get us closer to *geographical* scale, that is, to "specific processes in the prehistoric physical and human landscape" (Johnston et al. 2000: 725). In the geoarchaeological landscape, we were attempting a landscape overview, and the specific processes of interest were the context of archaeological site deposition and preservation. The fruit seed landscape focused on a set of plant remains preserved very well in some times and places, but not at all in others. The remains connected to detailed present-day indigenous ecological knowledge.

These two examples are emblematic of wider issues in the project, but I do not go into any detail here about what we can think of as the stone artifact and the rock art landscapes (Akerman, Fullagar, and van Gijn 2002; Head and Fullagar 1997; Taçon et al. 1997). Nor do I present new data or describe old data in any detail; readers are referred to recent publications for further details on the data (Atchison, Head, and Fullagar 2005; Head, Atchison, and Fullagar 2002; Ward et al. 2005, 2006).

The Geoarchaeological Landscape — The False Promise of Holism

The geoarchaeological landscape was the subject of Ph.D. research by Ingrid Ward, who examined sedimentation patterns in the context of long-term landscape evolution and climate change (Ward et al. 2005). Particularly important to understanding occupation patterns across the landscape was the ability to compare the more visible rockshelter occupation with that on the adjacent sand plains. Rockshelter deposits at the base of sandstone outcrops were visible, easily identifiable, and often associated with rock art. They were also places that Aboriginal people knew about and took us to. This was a significant scale issue for the project as a whole: what was "on-site" and "off-site" occupation, and how might we understand prehistoric settlement differently if we could compare the two (Ward et al. 2006)? In the event, this scale question became one of specifying conditions of invisibility as well as visibility. (It gave us a particular take on what Donna Haraway [1991] has called "situated knowledge.")

The labor of geoarchaeology produces as its results cross-sections such as that seen in Figure 37.1, which pins together the archaeological sites

of Jinmium and Goorurarmum by slicing through the landscape in between. This widespread geoarchaeological practice cuts through space and presents various time horizons—in this case, on the basis of luminescence dates. It enables a further interpretive step to be made such as that presented in Figure 37.2. This figure holds space, albeit a stylized space, constant, and cutting through time. It summarizes important findings that enhance the archaeological interpretations. There is no evidence to date of human occupation in the oldest sand sheet layers, which date beyond 100,000 years ago. The rockshelters and adjacent sand sheet occupation display very different patterns, most particularly that rockshelter occupation generally does not date beyond 4,000 years ago, whereas on the sand sheets nearby there is occupation evidence dating to and beyond 19,000 years ago. This is interpreted to indicate that the rockshelters are less perfect archaeological traps—that is, they do not "hold" sediments until surrounds have built to that level and/or local scree forms a trap. The sand plains have imperfections of their own: notably, organic materials are not preserved there as well as they are in the rockshelters.

The most important practical finding of this work is that there is a whole suite of sand plain occupation, dating back to at least 19,000 years ago, that would have been missed if we had focused exclusively on those parts of the landscape that were more obviously "sites." Most of it has very little expression at the surface, even though it may be as little as 20 meters from the adjacent rockshelter. This has considerable implications for settlement models of northern Australia, such as those that emphasize the increase in late Holocene site numbers but that are all dominated by rockshelter excavations (Ulm 2004).

This is a significant finding in landscape archaeology considered at the broad scale, and it reiterates the importance of taking a landscape perspective, in the sense of understanding the broad depositional and erosional contexts. More specifically, it is a situation in which the scale of the "site" was not assumed (that is, the rockshelter as a site type was not given total primacy, but was investigated empirically).

That said, it is difficult to go beyond this to a more dynamic understanding of human landscapes. At the risk of being negative about our own data, how do we go beyond presenting the geoarchaeology as background, as the stage on which social life is acted out? This is an ongoing challenge, but three points present themselves.

First, it is important not to confuse a broad-scale landscape approach with holism or synthesis.

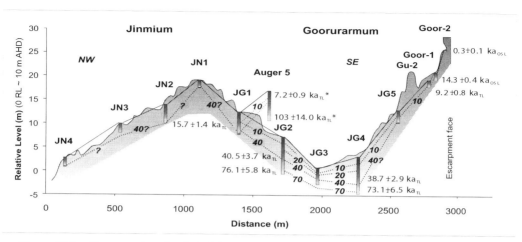

Figure 37.1   Chronostratigraphy (ka) of the Jinmium-Goorurarmum transect (Ward et al. 2005: fig. 3).

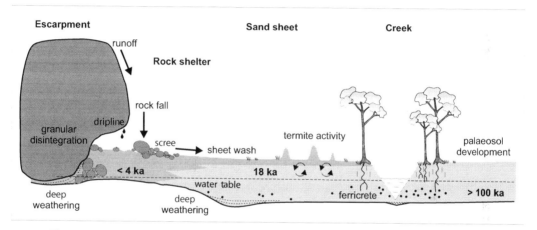

Figure 37.2   Schematic cross-section across Keep River study area (Ward et al. 2005: fig. 12).

This is not the total picture, but rather a means of situating our archaeological knowledge, of specifying some of the things that we do not know. This should not be taken to mean that if we had more dates, or more dots on the map, we could somehow get to a totality. As it happens, these sites are some of the most intensively dated in Australia, since the initial dates were controversial and problematic (Fullagar, Price, and Head 1996; Roberts et al. 1998), but an ever-denser data set does not, in itself, solve the problem identified by Massey of moving from one time slice to the next.

Second, we need to beware of a "zooming in" approach to scale, in which broad scale is implicitly conflated with the long term, and the small scale or local is necessarily conflated with the short term. This would really be the Russian Doll problem. If we go looking for a land surface of any particular age, for example, we won't find it. For one thing, the termites are undermining the possible temporal resolution by processing the sediments. But even if we could find a magical time surface, we would have to acknowledge it as multiply dimensioned. One way to think about how all the different scales are intermixed and held simultaneously is to think again of the sand grains from which the luminescence dates are produced. They hold within them signals of many different times, including the one the archaeologist wants via the luminescence date, the last time they were exposed to heat or light. Their spatial positioning is also crucial to their capacity to pin time in place; if we do not know exactly where they come from in relation to evidence of human activity (for example, below the lowest stone artifact at a particular site), they are archaeologically useless.

Finally, we need to remain conscious of the power of archaeological representations of space and time, such as those evident in the diagrams presented here. These are images that should be understood as having agency in the process of archaeological thinking about space, time, and scale. This is not to suggest that we abandon them, but rather that we maintain a critical reflexivity toward our acts of translating data.

Landscapes of Fruit-Seed Processing—The False Modesty of the Local

A comparative refraction on these issues is provided by the landscape of fruit seed processing, part of the work of Jenny Atchison. We knew that fruits such as *Buchanania obovata* (bush mango, *kilen*) and *Persoonia* (bush pear, *kathan*) provided favored and abundant resources for a short period in the early wet season, and that there was ethnographic evidence of them having been pounded and preserved through drying and cooking for longer term storage (Head, Atchison, and Fullagar 2002). Fortunately, a number of the rockshelter excavations contained abundant seed remains that allowed us to trace the prehistory of these activities.

The necessary labor included examining the chronology and the taphonomy of the cultural plant remains, first then by distinguishing between cultural and noncultural components of the archaeobotanical record. Thus, cultural samples were demonstrated to be dominated by fragmented and burnt seeds, whereas noncultural samples had whole unburnt seeds and a high proportion of grass seeds (Atchison, Head, and Fullagar 2005).

Fruit seed processing spans the 3,500 years before present. It may have existed before that time, but this archaeological window is limited by the preservation factors common to organic remains in tropical Australia. There is considerable spatial and temporal variability after the first evidence of sustained fruit-seed processing at 3,500 B.P. until after European colonization in the late 19th century. In the last 150 to 100 years, there is a decline in fruit seeds at all sites where they had previously been present, possibly as a result of disrupted seasonal movements.

In contrast to the geoarchaeological landscape, the space/time of fruit-seed processing draws attention to its own partiality. It is seasonal; the fruits appear for a few weeks at the beginning of the wet season. The ethnographic evidence suggests that fruit-seed processing is gendered; although collection was undertaken by groups of all ages and both genders, the pounding and processing were likely women's work. The hard (that is, archaeologically well preserved) seeds belong to just a few fruits, which are, in turn, just a subset of the plant foods that are only part of the total diet.

Yet, this is a landscape that is no less embodied for being partial. As we can with the Willandra Lakes footprints, we can imagine the labor, hear the sounds of a group of women pounding, hear the children grizzling or playing, and feel the heat and humidity of the early wet season.

Just as it would be wrong to think of the geoarchaeological landscape as a total landscape, it would be inaccurate to think of the fruit-seed processing landscape as a small and local one. Embedded within it are multiple connections: to later in the year (when the stored fruit seed cakes were used); distant places (they may have facilitated travel for ceremonies); and decadal patterns of seasonality across the Pacific (the role of the El Niño Southern Oscillation and its influence on Australia's northwest monsoon), to name just a few.

Conclusions: The Power of Partiality and Situated Knowledge

Landscape archaeology can benefit from theoretical debates about scale in geography, but it also has dilemmas of its own, particularly those to do with a partially preserved material record. I have used the example of a geoarchaeological landscape that has delivered very significant findings, particularly about what is missing—or potentially missing—in the archaeological record if we look at only part of the landscape. Yet, if we try to pin it to any particular space/time, we face two problems, both to do with scale. One is the false promise of holism, the idea that the broad landscape view somehow provides a holistic picture of human landscapes. The other would be to try and undo the Russian Dolls to a point in space and time where the "human" dimensions of landscape are thought to reside. In contrast, we have no problem seeing the partiality of the space/time of fruit-seed processing. But if we have asked too much of the geoarchaeological overview, we have perhaps expected too little of the more situated engagements of runners on the Willandra hardpan and pounders sitting in the shade in the Keep River. As the luminescence sand grain gets its explanatory power from the constellation of spatial and temporal processes that it pins together, so the embodied ephemeral fruitscape holds within it and is in turn embedded in space/times of many different scales.

All our landscape knowledge is at its most powerful when we recognize its partiality. Thus, situated knowledge is understood as knowledge that, rather than pretending it can see everything, is able to be very specific about where it sees from and thus what it does not see as well as what it does see. In Australian archaeology at least, this will provide better understanding of prehistoric human landscapes as fields of human engagement than the sort of broader models that we have been building and then critiquing. Developing a more dynamic understanding of scale that draws on recent discussions in geography makes an important contribution.

Acknowledgments

For ongoing discussion and insights into issues in the landscape archaeology of the East Kimberley I am indebted to my collaborators and coauthors Richard Fullagar, Jenny Atchison, the late Paddy Carlton, Ken Mulvaney, Biddy Simon, Paul Taçon, Polly Wandanga, and Ingrid Ward. None of them should be held responsible for the arguments presented in this chapter.

References

Akerman, K., Fullagar, R., and van Gijn, A. 2002. Weapons and Wunan: Production, function and exchange of spear points from the Kimberley, northwestern Australia. *Australian Aboriginal Studies* 1: 13–42.

Atchison, J., Head, L., and Fullagar, R. 2005. Archaeobotany of fruit seed processing in a monsoon savanna environment: Evidence from the Keep River region, Northern Territory, Australia. *Journal of Archaeological Science* 32: 167–81.

Bowler, J. M. 1998. Willandra Lakes revisited: Environmental framework for human occupation. *Archaeology in Oceania* 33: 120–55.

Fullagar, R. L. K., Price, D. M., and Head, L. M. 1996. Early human occupation of northern Australia: Archaeology and thermoluminescence dating of Jinmium rockshelter, Northern Territory. *Antiquity* 70: 751–73.

Gille, Z., and O'Riain, S. 2002. Global ethnography. *Annual Review Sociology* 28: 271–95.

Haraway, D. 1991. *Simians, Cyborgs and Women*. London: Free Association Books.

Head, L. M., and Fullagar, R. L. K. 1997. Hunter-gatherer archaeology and pastoral contact: Perspectives from the northwest Northern Territory, Australia. *World Archaeology* 28: 418–28.

Head, L., Atchison, J., and Fullagar, R. 2002. Country and garden: Ethnobotany, archaeobotany and Aboriginal landscapes near the Keep River, northwestern Australia. *Journal of Social Archaeology* 2: 173–96.

Howitt, R. 1993. "A world in a grain of sand": Towards a reconceptualisation of geographical scale. *Australian Geographer* 24: 33–44.

Johnston, R. J., Gregory, D., Pratt, G., and Watts, M. (eds.). 2000. *The Dictionary of Human Geography* (4th ed.). Oxford: Blackwell.

Marston, S. A. 2000. The social construction of scale. *Progress in Human Geography* 24: 219–42.

Massey, D. 1999. Space-time, "science" and the relationship between physical geography and human geography. *Transactions of the Institute of British Geography* NS 24: 261–76.

———. 2001. Globalisation: What does it mean for geography? Keynote lecture, Geographical Association Annual Conference, UMIST.

McGuirk, P. 1997. Multiscaled interpretations of urban change: The federal, the state and the local in the Western Area Strategy of Adelaide. *Environment and Planning D: Society and Space* 15: 481–98.

Raper, J., and Livingstone, D. 1995. Development of a geomorphological spatial model using object-oriented design. *International Journal of Geographical Information Systems* 9: 359–83.

Roberts, R., Bird, M., Olley, J., Gallagher, R., Lawson, E., Laslett, G., Yoshida, H., Jones, R., Fullagar, R., Jacobsen, G., and Hua, Q. 1998. Optical and radiocarbon dating at Jinmium rock shelter, northern Australia. *Nature* 393: 358–62.

Taçon, P., Fullagar, R., Ouzman, S., and Mulvaney, K. 1997. Cupule engravings from Jinmium-Granilpi (northern Australia) and beyond: Exploration of a widespread and enigmatic class of rock markings. *Antiquity* 71: 942–65.

Ulm, S. 2004. Themes in the archaeology of mid-to-late Holocene Australia, in T. Murray (ed.), *Archaeology from Australia*, pp. 187–208. Melbourne: Australian Scholarly Publications.

Ward, I., Fullagar, R., Boer-Mah, T., Head, L., Taçon, P., and Mulvaney, K. 2006. Comparison of sedimentation and occupation histories inside and outside rock shelters, Keep River Region, Northwestern Australia. *Geoarchaeology* 21: 1–27.

Ward, I., Nanson, G., Head, L., Fullagar, R., Price, D., and Fink, D. 2005. Late Quaternary landscape evolution in the Keep River region, northwestern Australia. *Quaternary Science Reviews* 24: 1906–922.

Webb, S., Cupper, M. L., and Robins, R. 2006. Pleistocene human footprints from the Willandra Lakes, southeastern Australia. *Journal of Human Evolution* 50: 405–13.

Wishart, D. 2004. Period and region. *Progress in Human Geography* 28: 305–19.

38

LANDSCAPE AND CLIMATE CHANGE

Michael J. Rowland

Landscape change can range from the patter of tiny raindrops and the shuffling of beetles and rodents on hillslopes, to major slope failures that generate colossal debris flows, to cratering of the earth's surface caused by the impact of meteorites (Allen 2005: 961). The scale of these examples encapsulates processes of change that range from uniformitarianism to catastrophism. Archaeologists have generally tended to discuss change in a gradualistic or uniformitarian way. It is increasingly recognized, however, that various types of discontinuity occur in environmental change, and identifying these discontinuities is critical to understanding their impacts on landscapes and their potential human impacts. Allen (2005) notes that physical landscapes are a result of two interacting systems: an internal system driven by tectonic fluxes of rock and an external system dominated by climate. Landscapes are perturbed by both these internal and external mechanisms at a range of temporal and spatial scales.

In studying past landscapes archaeologists must be conscious of the fact that landscapes can change in response to a range of internal and external factors and at a number of time scales. This is demonstrated in this chapter by reference to climate change. Three periods of change—the Younger Dryas, the 8.2 cal ka event, and the Medieval Warm Period/Little Ice Age—of different magnitude and extent are described, and their implications for landscape modification are considered.

The Nature of Climate Change

Climate is an interlocked system involving the atmosphere, oceans, and cosmic environments with inherent quasi-periodic oscillations. The earth's climate has never been static and never will be. Climate change occurs on a number of time scales, from glacial cycles when temperatures varied by as much 10°C or more to the Little Ice Age when changes of around 1 to 2°C occurred. Climate change is marked by a range of frequency variations and positive and negative feedbacks (e.g., Bell and Walker 1992; Hugget 1997). For example, climate change and related feedbacks in the coastal zone can affect the physical, biological, and biogeochemical characteristics of oceans and coasts, modifying their ecological structure. Large-scale impacts on the oceans might increase or decrease sea level and sea-surface temperatures, increase or decrease ice cover, and change salinity, alkalinity, wave climate, and patterns of ocean circulation. Feedbacks can then occur through changes in ocean mixing, deep water production, and coastal upwelling. Together these changes can have a profound impact on the status, sustainability, productivity, and biodiversity of the coastal zone and marine ecosystem. Climate and ocean regime shifts, for example, may significantly affect fish populations. Marine mammals and birds may also then be affected. The adaptive capacity of marine and coastal ecosystems varies considerably; while some may adapt, others move. Other impacts on the coast include inundation and storm flooding, coastal erosion, seawater intrusion into fresh groundwater, encroachment of tidal waters into estuaries and river systems, and elevated sea-surface and ground temperatures (e.g., Intergovernmental Panel on Climate Change 2001: ch. 6; *National Research Council* [NRC] [U.S.] Committee on Abrupt Climate Change 2002).

A substantial paradigm shift in understanding climate change has occurred in recent years (e.g., Alley 2000a; Alley et al. 1997, 2003; Berger and Labeyrie 1987; Clark, Webb, and Keigwin 1999; Haberle and Chepstow-Lusty 2000; Haberle and David 2004; Jones and Mann 2004; Lockwood 2001; Rial 2004; Wunsch 2006). Prior to the 1990s, the dominant view of past climate change emphasized the slow gradual swings of the ice ages. However, it is now apparent that climate has often changed relatively abruptly, locally varying by as much as 10°C in 10 years. The implications of these changes for past human populations are only now being recognized (e.g., Bawden and Reycraft 2000; Dalfes, Kukla, and Weiss 1997). The discovery of millennial scale variability in climate is considered to be among the "top ten" discoveries of the century (Charles 1998), and such variability is still in the process of being fully understood. It has also been recognized that the interrelationship between climatic factors is marked; that tropical areas also played an important role in climate change; and that impacts occurred over broad areas and that correlations in the timing of many events occurred on a worldwide basis.

Abrupt climate change occurs when the climate system is forced to cross some threshold, triggering a transition to a new state at a rate determined by the climate system itself. Chaotic processes in the climate system may allow even small changes to have significant impacts, and small-scale climate changes can be translated into extreme ecological changes throughout all layers of an ecosystem (e.g., Catterall, Poiner, and O'Brien 2001; Easterling et al. 2000; Hilbert, Ostendorf, and Hopkins 2001; Peteet 2000; Walther et al. 2002) and to genetic change in diverse groups of organisms (e.g., Bradshaw and Holzapfel 2006). There are several causes of abrupt climatic change (Clark, Webb, and Keigwin 1999), and some of these are mentioned below. Climate variability in the Holocene may have been small relative to the formidable changes that characterize transitions from glacials to interglacials, but "the *rate* of change is often similar or greater" (CRC 1995: 6).

At least some of the major cultural changes seen in the archaeological record must reflect adaptive responses to climatic and other environmental changes (e.g., Huntley 1999; *Quaternary International* 2006 Volume 150 and 151), but during the Holocene people also began to affect climate (e.g., Nocete et al. 2005; Zolitschka, Behre, and Schneider 2003). In fact, Ruddiman (2003, 2005) has gone as far as suggesting that human impact on climate began 8,000 years ago with the beginnings of farming.

Human Response to Climate Change

Discussions that include climate change in human history must avoid environmentally deterministic explanations. Research must focus on the relationship between historical, social, economic, and political factors, as well as on the type and the magnitude of the climate change itself (e.g., Kolata 2000; Rowland 1999). The link between climate change and culture change has long been discussed (e.g., Bryson and Murray 1977), and a call for an updated and reformulated explanatory culture history in archaeology strongly influenced by population dynamics and linked to climate change can be supported (e.g., Shennan 2000). Climate change has been considered a cause of human population "bottlenecks" and "releases" (e.g., Ambrose 1998) and a trigger in the movement of humans out of Africa (e.g., Walter et al. 2000). Climate change continues to be seen as a major trigger for societal collapse on a worldwide basis and over many time scales (e.g., Beets et al. 2006; Hunt and Elliott 2005; Weiss et al. 1993) and also for rapid technological and cultural innovation (Gill 2000; Hodell, Curtis, and Brenner 1995; Wells and Noller 1999). However, debates continue over cause and effect and the extent of "environmental determinism" in explanations of change (e.g., see discussion of reasons for the collapse of the Tiwanaku State in Calaway 2005; Kolata 2000; Williams 2002).

Although Holocene climates have been seen as more benign than those of the Pleistocene (Burroughs 2005), researchers have recognized for some time (Denton and Karlén 1973) that the postglacial period was punctuated by a number of sudden shifts back toward glacial conditions, and such fluctuations are now supported by substantial empirical evidence (e.g., Berger and Labeyrie 1987; Broecker and Denton 1990). Changes of several degrees in glacial surface air and sea temperatures apparently occurred multiple times within years to decades (e.g., Alley et al. 1993; Grootes et al. 1993; Mann, Park, and Bradley 1995; Taylor et al. 1993). Such environmental variations raise the important question for archaeologists as to how humans (as individuals or groups) perceived and dealt with landscapes (and hence resource bases) that were visibly changing within the timeframe of a few generations or maybe even a single generation (Straus 1996).

Climate Change Events

The Younger Dryas

The Younger Dryas is the best known of a number of cooling events following the last Ice Age and marks the end of the Pleistocene. It was a millennium-long cooling event at ca. 12,700–11,500 cal B.P. that interrupted the transition from the last glacial to present interglacial (Holocene) period (e.g., Alley 2000b; Berger 1990; Vacco et al. 2005). It featured rapid shifts in North Atlantic atmospheric temperature (e.g., Alley et al. 1993; Taylor et al. 1993) and surface ocean circulation (Lehman and Keigwin 1992), glacier readvance, woodland retreat, and other floral and faunal perturbations and is associated with fluctuations in sea levels (e.g., Fairbanks 1989; Fleming et al. 1998). The abruptness of the transition from the Younger Dryas cold spell suggests a rapid reorganization of the global climate system (Gasse and Van Campo 1994; Wei and Gasse 1999). As originally defined, the Younger Dryas was restricted to the temperate zone of Western Europe (e.g., Troelstra, van Hinte, and Ganssen 1996) but is now recognized as having worldwide effects (e.g., Alley 2000a; Andres et al. 2003; Corrège et al. 2004; Magny and Bégeot, 2004; Sima, Paul, and Schulz 2004; Tarasov and Peltier 2005; Wurth et al. 2004). It is recognized throughout North and South America, though with a weaker signal (e.g., Baker et al. 2001; Rodbell and Seltzer 2000; van't Veer, Islebe, and Hooghiemstra 2000; but see Bennett et al. 2000). Cooling was also evident in the western tropical Pacific, off Japan in the northern Pacific Ocean, the Gulf of Mexico, central North America, the Gulf of California, South Africa, Antarctica, and New Zealand (e.g., Abell and Plug 2000; Hajdas et al. 2006, but see Singer, Shulmeister, and McLea 1998, Turney et al. 2006). In parts of the tropics, the Younger Dryas is recognized as a period of marked climatic instability (e.g., Gasse 2000; Gasse and van Campo 1994). In North Africa, Central America, and on the Tibetan Plateau, an abrupt return to dry conditions occurred at ca. 11,000–9,500 B.P. that is regarded as coeval with the Younger Dryas event (e.g., Street-Perrott and Perrott 1990; Van Campo and Gasse 1993).

Cooling into the Younger Dryas occurred in a few decade(s)-long steps, whereas warming at the end of it occurred primarily in one large step of about 8°C, resulting in the climate of the northern Atlantic region turning to a milder, less stormy regime in less than 40 years (Alley 2000b; Taylor et al. 1997), and perhaps less than five years (Smith et al. 1997). It has been proposed that during the Younger Dryas biotic responses may have been telescoped together into relatively short periods (50–150 years), perhaps disrupting functional interactions among species and thus destabilizing ecosystems (Ammann et al. 2000). The Younger Dryas is linked with a number of cultural changes, in particular the beginning of farming (e.g., Blumler 1996; Harris 1996; Sherratt 1997). Bar-Yosef (1998), for example, claims that cultivation increased as hunting and gathering groups were forced to adapt to decreased annual yields of wild cereals caused by rapid drying and cooling.

The 8.2 cal ka Event

An abrupt climate change at ca. 8,200 years ago saw temperatures drop by 4–8°C in central Greenland and by 1.5–3°C at marine and terrestrial sites around the northeast North Atlantic Ocean (e.g., Alley et al. 1993, 1997; Dansgaard et al. 1993; Johnson et al. 2002). More recent investigations suggest that two separate, abrupt climatic events occurred at this time (e.g., Alley et al. 1997; Ellison, Chapman, and Hall 2006; Hu et al. 1999). Many climatic records from both tropical and subtropical regions also show potentially correlative events at this time (e.g., Alley and Ágústsdóttir 2005; Clarke et al. 2004; Gasse and Van Campo 1994; Johnson et al. 2002; Kurek, Cwynar, and Spear 2004; O'Brien et al. 1995; Seppä, Cwynar, and MacDonald 2003; Street-Perrott and Perrott 1990; Van Campo and Gasse 1993; Xia, Zhao, and Collerson 2001). The 8.2 cal ka cooling event is attributed to a massive outflow of freshwater from Hudson Strait when the glacial lakes Agassiz and Ojibway drained catastrophically into the Labrador Sea (Barber et al. 1999). It is proposed that the 8.2 cal ka event was only half as strong as the Younger Dryas, and whereas the Younger Dryas continued for more than a millennium, the maximum 8.2 cal ka deviations are thought to have lasted less than a century (von Grafenstein et al. 1998). Jonathan Adams and associates (1999: 15) consider the 8.2 cal ka year event to be the most striking cooling event of the Holocene, producing widespread cool, dry conditions lasting about 150 years (Kobashi et al. 2007) before a rapid return to climates warmer and generally moister than present. The effects of the 8.2 cal ka year event on human populations and landscapes are not well studied, but Weiss (2000), for example, argues that in the Middle East some settlements were abandoned, and a new dependence on sheep pastoralism developed as a populations expanded into areas of decreased rainfall.

Medieval Warm Period/Little Ice Age

The Medieval Warm Period/Little Ice Age is one of the most interesting periods of climate change, since it spans the period (ca. A.D. 1850–1300) from the Renaissance through the Enlightenment, including the birth of modern science and the modern European exploration and settlement of the world. It is also the subject of an enormous literature from specialist academic volumes and papers (e.g., Hendy et al. 2002; Qian and Zhu 2002) to major reviews (Matthews and Briffa 2005) and comprehensive popular accounts (Fagan 2000). Although there is no consensus on what caused the Little Ice Age nor whether it constituted a single event or a continuous episode, it remains an important period in defining the effects of climate change on landscapes, since both proxy records and written accounts are available (e.g., D'Arrigo et al. 2003; Nordli 2001). Compared to other climate events of the Holocene, such as the 8.2 cal ka year event, the Little Ice Age was slight. Global temperatures dropped only 0.5–1°C and in the north Atlantic by 3°C. Of course, temperature is just a convenient marker of periods of climate change. It is the associated floods, droughts, high winds, dust storms, unseasonable weather, and other changes that ruin harvests and cause famines.

Major reviews suggest that the Little Ice Age and the Medieval Warm Period had a worldwide impact (Matthews and Briffa 2005; Soon and Baliunas 2003; Soon et al. 2003) but with different regional signatures (e.g., Allen 2006). Specific evidence, for example, comes from Iceland, Greenland, Norway, North America, Canada, South America, Germany, Southern Oman, Russia, China, Africa, New Zealand, and Australia. In this review, I generally make note of the effects of the Little Ice Age. Note, however, that the Medieval Warm Period, which started ca. A.D. 800, was a period of mild weather that enabled many populations to grow and expand (see Fagan 2000: 3–22).

The range of effects resulting from the Little Ice Age was considerable. Increased storminess associated with the Little Ice Age had a major impact on dunes systems over a wide area of the northern hemisphere (e.g., Clarke et al. 2002; Curry 2000; Dawson et al. 2004; Havholm et al. 2004; Knight et al. 2002; Mayer and Mahan 2004; Wilson, McGourty, and Bateman 2004). The village of Santon Downham, East Anglia, for example, was overrun by windblown sand sometime around A.D. 1650. Sand buried walls approximately 2.75 m high and destroyed houses (Bateman and Godby 2004: 581). Erosion was general and widespread and particularly affected high-altitude soils (Dearing et al. 2001). In Norway, historical records document advancing glaciers destroying pastures and even farmhouses (Nordli et al. 2003). Large floods, droughts, and erosion have been identified across the world (e.g., Belloti et al. 2004; Grove 2001; Heine 2004; Hereford 2002; Pišut 2002; Stankoviansky 2003). In some areas isostatic emergence associated with the end of the Little Ice Age melting has resulted in the development of new wetlands and marshes containing new wildlife (e.g., Larsen et al. 2004).

Wide-ranging vegetation changes occurred in both hemispheres, including bushfires with related effects, such as famine (e.g., Andreev et al. 2002; Behling et al. 2004; Payette and Delwaide 2004; Pierce, Meyer, and Jull 2004) and other significant human involvements (e.g., Chepstow-Lusty et al. 1998; Kitagawa et al. 2004). Climate change affected the productivity of oceans (e.g., Hsieh et al. 2005; Wollenburg, Knies, and Mackensen 2004), including both the distribution and the abundance of fish (e.g., Perry et al. 2005; Rose 2004) and shellfish (e.g., Carbotte et al. 2004).

The Little Ice Age affected fauna as diverse as insects (e.g., Brooks and Birks 2004) and birds (e.g., Kinzelbach 2004) and in some cases may have caused genetic variation leading to a variety of changes, including extinctions (e.g., Hadly et al. 2004). Climate change, in general, can affect biodiversity in the broader sense, causing extinctions and leading to the introduction of species and spread of diseases (e.g., Brooks and Ferrao 2005). There is evidence of widespread increase in famine and disease at the time of the Little Ice Age (e.g., Appleby 1980; Reiter 2000). The Little Ice Age also had an impact on cultural landscapes; for example, soapstone buildings in Norway were weathered owing to a higher frequency of damaging frosts (Storemyr 2004).

Aridity-associated demographic stress and related impacts are argued to have affected societies across north and South America (e.g., Neff et al. 2006; Wahl et al. 2006), Africa (Robertshaw and Taylor 2000), and the Pacific (Nunn 2001). The Norse who arrived in Greenland during the Medieval Warm Period succumbed to the climate changes of the early stages of the Little Ice Age (e.g., McGovern 1994).

Conclusions

Many references (both old and new) relating to climate change (e.g., Alley 2000a; Bell and Walker

1992; Berger and Labeyrie 1987; Bradley 1999; Clark et al. 1999; Climate Research Committee 1995; Hugget 1997; IPCC 2001; Jones et al. 1996; Mayewski and White 2002; NRC (U.S.) Committee on Abrupt Climate 2002; Rampino et al. 1987; Troelstra et al. 1996) stress in particular the importance of abrupt climate change. There is a growing literature on the effects of abrupt climate change on humans (e.g., Bawden and Reycraft 2000; Bryson and Murray 1977; Crumley 1994; Dalfes et al. 1997; Gill 2000; Harris 1996; Hassan 2002; Issar and Brown 1998; Lamb 1988). In understanding the impact of climate change on landscapes, archaeologist have yet to fully utilize the results of work on natural disaster research, including El Niño events and Long-Term Ecological Research Sites (e.g., Fagan 1999; Greenland et al. 2003). Archaeologists also have much to learn and to contribute to current debates about climate change.

The significant impact of climate change on prehistoric populations is finding increasing general support among anthropologists and archaeologists (see, in particular, Crumley 1994; Harris 1996; Headland 1997; Potts 1996; Sherratt 1997; Straus 1996), and many proposed scenarios are receiving substantive empirical support from the rapid growth in climatic data (e.g., Arnold and Tissot 1993; Kolata 2000; Weiss et al. 1993). Most recent reviews of climate change (e.g., Rohling and Pälike 2005) suggest that the Younger Dryas, the 8.2 cal ka event, and the Medieval Warm Period/Little Ice Age were not unique events. In the Holocene, for example, climatic events also occurred at 9,000–8,000, 6,000–5,000, 4,200–3,800, 3,500–2,500, 1,200–1,100, and 600–150 cal B.P. Several of these intervals coincide with major disruptions of civilizations, but a number have yet to be fully investigated. Writing in 2001, the National Research Council of the United States (1, 17) noted that although the "new paradigm" of abrupt climate change had been established over the last decade, it was scarcely known in the wider community of natural and social scientists and policymakers. The Council also noted that there is virtually no research on the economic and ecological effects of abrupt climate change; most research has concentrated on gradual change. Culture and nature change together as a result of diverse interactions, and the outcomes are historically contingent. Archaeologists are therefore well placed to investigate these issues, which involve the diachronic, multidisciplinary analysis of landscapes.

References

Abell, P. I., and Plug, I. 2000. The Pleistocene/Holocene transition in South Africa: Evidence for the Younger Dryas event. *Global and Planetary Change* 26(1–3): 173–79.

Adams, J., Maslin, M., and Thomas, E. 1999. Sudden climate transitions during the Quaternary. *Progress in Physical Geography* 23(1): 1–36.

Allen, P. 2005. Striking a chord. *Nature* 434: 961.

Allen, M. S. 2006. New ideas about Holocene climatic variability in the central Pacific. *Current Anthropology* 47: 521–35.

Alley, R. B. 2000a. *The Two-Mile Time Machine. Ice Cores, Abrupt Climate Change, and Our Future.* Princeton, NJ: Princeton University Press.

———. 2000b. The Younger Dryas cold interval as viewed from central Greenland. *Quaternary Science Reviews* 19: 213–26.

Alley, R. B., and Ágústsdóttir, A. M. 2005. The 8k event: Cause and consequences of a major Holocene abrupt climate change. *Quaternary Science Reviews* 24: 1123–49.

Alley, R. B., Marotzke, J., Nordhaus, W. D., Overpeck, J. T., Peteet, D. M., Pielke, R. A., Jr., Pierrehumbert, R. T., Rhines, P. B., Stocker, T. F., Talley, L. D., and Wallace, J. M. 2003. Abrupt climate change. *Science* 299: 2005–10.

Alley, R. B., Mayewski, P. A., Sowers, T., Stuiver, M., Taylor, K. C., and Clark, P. U. 1997. Holocene climatic instability: A prominent, widespread event 8200 yrs ago. *Geology* 25: 483–86.

Alley, R. B., Meese, D. A., Shuman, C. A., Gow, A. J., Taylor, K. C., Grootes, P. M., White, J. W. C., Ram, M., Waddington, E. D., Mayewski, P. A., and Zielinski, G. A. 1993. Abrupt increase in Greenland snow accumulation at the end of the Younger Dryas event. *Nature* 262: 527–29.

Ambrose, S. H. 1998. Late Pleistocene human population bottlenecks, volcanic winter, and differentiation of modern humans. *Journal of Human Evolution* 34: 623–51.

Ammann, B., Birks, H. J. B., Brooks, S. J., Eicher, U., von Grafenstein, U., Hofmann, W., Lemdahl, G., Schwander, J., Tobolski, K., and Wick, L. 2000. Quantification of biotic responses to rapid climatic changes around the Younger Dryas: A synthesis. *Palaeogeography, Palaeoclimatology, Palaeoecology* 159: 313–47.

Andreev, A. A., Siegert, C., Klimanov, V. A., Derevyagin, A. Y., Shilova, G. N., and Melles, M. 2002. Late Pleistocene and Holocene vegetation and climate on the Taymyr Lowland, Northern Siberia. *Quaternary Research* 57: 138–50.

Andres, M. S., Bernasconi, S. M., McKenzie, J. A., and Röhl, U. 2003. Southern Ocean deglacial record supports global Younger Dryas. *Earth and Planetary Science Letters* 216: 515–24.

Arnold, J. E., and Tissot, B. N. 1993. Measurement of significant marine paleotemperature variation using Black Abalone shells from prehistoric middens. *Quaternary Research* 39: 390–94.

Appleby, A. B. 1980. Epidemics and famine in the Little Ice Age. *Journal of Interdisciplinary History* 10: 643–63.

Baker, P. A., Seltzer, G. O., Fritz, S. C., Dunbar, R. B., Grove, M. J., Tapia, P. M., Cross, S. L., Rowe, H. D., and Broda, J. P. 2001. The history of South American tropical precipitation for the past 25,000 years. *Science* 640–43.

Bar-Yosef, O. 1998. On the nature of transitions: The Middle to Upper Paleolithic and the Neolithic revolution. *Cambridge Archaeological Journal* 8: 141–63.

Barber, D. C., Dyke, A., Hillaire-Marcel, C., Jennings, A. E., Andrews, J. T., Kerwin, M. W., Bilodeau, G., McNeely, R., Southon, J., Morehead, M. D., and Gagnon, J.-M. 1999. Forcing of the cold event of 8,200 years ago by catastrophic drainage of Laurentide lakes. *Nature* 400: 344–48.

Bateman, M. D., and Godby, S. P. 2004. Late–Holocene inland dune activity in the UK: A case study from Breckland, East Anglia. *The Holocene* 14: 579–88.

Bawden, G., and Reycraft, R. M. 2000. *Environmental Disaster and the Archaeology of Human Response.* Albuquerque: Maxwell Museum of Anthropology. Anthropological Papers No. 7, University of New Mexico.

Beets, C. J., Troelstra, S. R., Grootes, P. M., Nadeau, M. J., van der Borg, K., de Jong, A. F. M., Hofman, C. L., and Hoogland, M. L. P. 2006. Climate and the Pre-Columbian settlement at Anse á la Gourde, Guadeloupe, northeastern Caribbean. *Geoarchaeology* 21: 271–80.

Behling, H., DePatta Pillar, V., Orlóci, L., and Bauermann, S. G. 2004. Late Quaternary *Araucaria* forest, grassland (Campos), fire and climate dynamics, studied by high-resolution pollen, charcoal and multivariate analysis of the Cambará do Sul core in southern Brazil. *Palaeogeography, Palaeoclimatology, Palaeoecology* 203: 277–97.

Bell, M., and Walker, M. J. C. 1992. *Late Quaternary Environmental Change. Physical and Human Perspectives.* New York: John Wiley and Sons.

Belloti, P., Caputo, C., Davoli, L., Evangelista, S., Garzanti, E., Pugliese, F., and Valeri, P. 2004. Morpho-sedimentary characteristics and Holocene evolution of the emergent part of the Ombrone River delta (southern Tuscany). *Geomorphology* 61: 71–90.

Bennett, K. D., Haberle, S. G., and Lumley, S. H. 2000. The last Glacial-Holocene transition in Southern Chile. *Science* 290: 325–28.

Berger, W. H. 1990. The Younger Dryas cold spell: A quest for causes. *Global and Planetary Change* 3: 219–37.

Berger, W. H., and Labeyrie, L. D. 1987. *Abrupt Climatic Change. Evidence and implications.* Dordrecht: D. Reidel.

Bradley, R. S. 1999. *Paleoclimatology. Reconstructing climates of the Quaternary* (2nd ed.). London: Academic Press.

Bradshaw, W. E., and Holzapfel, C. M. 2006. Evolutionary response to rapid climate change. *Science* 312: 1477–78.

Brooks, D. R., and Ferrao, A. L. 2005. The historical biogeography of co-evolution: Emerging infectious diseases are evolutionary accidents waiting to happen. *Journal of Biogeography* 32: 1291–99.

Brooks, S. J., and Birks, H. J. B. 2004. The dynamics of Chironomidae (Insecta: Diptera) assemblages in response to environmental change during the past 700 years on Svalbard. *Journal of Paleolimnology* 31: 483–98.

Bryson, R. A., and Murray, T. J. 1977. *Climates of Hunger. Mankind and the World's Changing Weather.* Canberra: Australian University Press.

Burroughs, W. J. 2005. *Climate Change in Prehistory. The End of the Reign of Chaos.* Cambridge: Cambridge University Press.

Calaway, M. J. 2005. Ice-cores, sediments and civilization collapse: A cautionary tale from Lake Titicaca. *Antiquity* 79: 778–90.

Carbotte, S. M., Bell, R. E., Ryan, W. B. F., McHugh, C., Slagle, A., Nitsche, F., and Rubenstone, J. 2004. Environmental change and oyster colonization within the Hudson River estuary linked to Holocene climate. *Geo-Marine Letters* 24: 212–24.

Catterall, C. P., Poiner, I. R., and O'Brien, C. J. 2001. Long-term population dynamics of a coral reef gastropod and responses to disturbance. *Austral Ecology* 26: 604–17.

Chambers, F. M., and Brain, S. A. 2002. Paradigm shifts in late-Holocene climatology? *The Holocene* 12: 239–49.

Charles, C. 1998. Palaeoclimatology: The ends of an era. *Nature* 394: 422–23.

Chepstow-Lusty, A. J., Bennett, K. D., Fjeldsa, J., Kendall, A., Galiano, W., and Herrera, A. T. 1998. Tracing 4,000 years of environmental history in the Cuzo area, Peru, from the pollen record. *Mountain Research and Development* 18: 159–72.

Clark, P. U., Webb, R. S., and Keigwin, L. D. 1999. Mechanisms of Global Climate Change at Millennial Time Scales. *Geophysical Monograph Series,* Vol. 112. Washington, DC: American Geophysical Union.

Clarke, G. K. C., Leverington, D. W., Teller, J. T., and Dyke, A. S. 2004. Paleohydraulics of the last outburst flood from glacial Lake Agassiz and the 8200 B.P. cold event. *Quaternary Science Reviews* 23: 389–407.

Clarke, M., Rendell, H., Tastet, J.-P., Clavé, B., and Massé, L. 2002. Late-Holocene sand invasion and North Atlantic storminess along the Aquitaine Coast, southwest France. *The Holocene* 12: 231–38.

Climate Research Committee (CRC). 1995. Introduction. *Natural Climate Variability on Decade-to-Century Time Scales,* pp. 6–8. Washington, DC: National Academy Press.

Corrège, T. M., Gagan, K., Beck, J. W., Burr, G. S., Cabioch, G., and Le Cornec, F. 2004. Interdecadal variation in the extent of South Pacific tropical waters during the Younger Dryas event. *Nature* 428: 927–29.

Crumley, C. L. 1994. *Historical Ecology: Cultural Knowledge and Changing Landscapes*. Santa Fe, NM: School of American Research Press.

Curry, A. M. 2000. Holocene reworking of drift-mantled hillslopes in Glen Docherty, Northwest Highlands, Scotland. *The Holocene* 10: 509–18.

D'Arrigo, R., Buckley, B., Kaplan, S., and Woollett, J. 2003. Interannual to multidecadal modes of Labrador climate variability inferred from tree rings. *Climate Dynamics* 20: 219–28.

Dalfes, H. N., Kukla, G., and Weiss, H. 1997. *Third Millennium B.C. Climate Change and Old Worlds Collapse*. NATO ASI Series 149 Springer-Verlag.

Dansgaard, W. 1987. Ice core evidence of abrupt climate changes, in W. J. Berger and L. Labeyrie (eds.), *Abrupt Climatic Change: Evidence and Implications*, pp. 223–33. Dordrecht: D. Reidel.

Dansgaard, W., Johnsen, S. J., Clausen, H. B., Dahl-Jensen, D., Gundestrup, N. S., Hammer, C. U., Hvidberg, C. S., Steffensen, J. P., Sveinbjörnsdottir, A. E., Jouzel, J., and Bond, G. 1993. Evidence for general instability of past climate from a 250-kyr ice-core record. *Nature* 364: 218–20.

Dansgaard, W., White, J. W. C., and Johnsen, S. J. 1989. The abrupt termination of the Younger Dryas climate event. *Nature* 339: 532–33.

Dawson, S., Smith, D. E., Jordan, J., and Dawson, A. G. 2004. Late Holocene coastal sand movements in the Outer Hebrides, N.W. Scotland. *Marine Geology* 210: 281–306.

Dearing, J. A., Hu, Y., Doody, P., James, P. A., and Brauer, A. 2001. Preliminary reconstruction of sediment-source linkages for the past 6,000 yrs at the Petit Lac d'Annecy, France, based on mineral magnetic data. *Journal of Paleolimnology* 25: 245–58.

Denton, G. H., and Karlén, W. 1973. Holocene climatic variations: Their pattern and possible cause. *Quaternary Research* 3: 155–74.

Easterling, D. R., Meehl, G. A., Parmesan, C., Changon, S. A., Karl, T. R., and Mearns, L. O. 2000. Climate extremes: Observations, modeling, and impacts. *Science* 289: 2068–74.

Ellison, C. R. W., Chapman, M. R., and Hall, R. 2006. Surface and deep ocean interactions during the cold climate event 8200 yrs ago. *Science* 312: 1929–32.

Fagan, B. 1999. *Floods, Famines and Emperors: El Niño and the Fate of Civilizations*. New York: Basic Books.

———. 2000. *The Little Ice Age. How Climate Made History 1300–1850*. New York: Basic Books.

Fairbanks, R. G. 1989. A 17,000-year glacio-eustatic sea level record: Influence of glacial melting rates on the Younger Dryas event and deep-ocean circulation. *Nature* 342: 637–42.

Fleming, K., Johnston, P., Zwartz, D., Yokoyama, Y., Lambeck, K., and Chappell, J. 1998. Refining the eustatic sea-level curve since the Last Glacial Maximum using far- and intermediate-field sites. *Earth and Planetary Science Letters* 163: 327–42.

Gasse, F. 2000. Hydrological changes in the African tropics since the Last Glacial Maximum. *Quaternary Science Reviews* 19: 189–211.

Gill, R. B. 2000. *The Great Maya Droughts: Water, Life and Death*. Albuquerque: University of New Mexico Press.

Greenland, D., Goodin, D. G., and Smith, R. C. 2003. *Climate Variability and Ecosystem Response at Long-Term Ecological Research Sites*. Oxford: Oxford University Press.

Grootes, P. M., Stuiver, M., White, J. W. C., Johnsen, S., and Jouzel, J. 1993. Comparison of oxygen isotope records from the GISP2 and GRIP Greenland ice cores. *Nature* 366: 552–54.

Grove, A. T. 2001. The "Little Ice Age" and its geomorphological consequences in Mediterranean Europe. *Climatic Change* 48: 121–36.

Haberle, S. G., and Chepstow-Lusty, A. 2000. Can climate influence cultural development? A view through time. *Environment and History* 6: 349–69.

Haberle, S. G., and David, B. 2004. Climates of change; human dimensions of Holocene environmental change in low latitudes of the PEP11 transect. *Quaternary International* 118–119: 165–79.

Hadly, E. A., Ramakrishnan, U., Chan, Y. L., van Tuinen, M., O'Keefe, K., Spaeth, P. A., and Conroy, C. J. 2004. Genetic response to climatic change: Insights from ancient DNA and Phylochronology. *PLoS Biology* 2: 1600–09.

Hajdas, I., Lowe, D. J., Newnham, R. M., and Bonani, G. 2006. Timing of the late-glacial reversal in the Southern Hemisphere using high resolution radiocarbon chronology for Kaipo bog, New Zealand. *Quaternary Research* 65: 340–45.

Harris, D. R. 1996. *The Origins and Spread of Agriculture and Pastoralism in Eurasia*. London: UCL Press.

Hassan, F. A. 2002. *Droughts, Food and Culture: Ecological Change and Food Security in Africa's Later Prehistory*. New York: Plenum.

Havholm, K. G., Ames, D. V., Whittecar, G. R., Wenell, B. A., Riggs, S. R., Jol, H. M., Berger, G. W., and Holmes, M. A. 2004. Stratigraphy of back-barrier coastal dunes, northern North Carolina and southern Virginia. *Journal of Coastal Research* 20: 980–99.

Headland, T. N. 1997. Revisionism in ecological anthropology. *Current Anthropology* 38: 605–30.

Heine, K. 2004. Flood reconstructions in the Namib Desert, Namibia and Little Ice Age climatic implications: Evidence from slackwater deposits and desert soil sequences. *Journal Geological Society of India* 64: 535–47.

Hendy, E. J., Gagan, M. K., Alibert, C. A., McCulloch, M. T., Lough, J. M., and Isdale, P. J. 2002. Abrupt decrease in tropical sea surface salinity at end of Little Ice Age. *Science* 295: 1511–14.

Hereford, R. 2002. Valley-fill alluviation during the Little Ice Age (ca. A.D. 1400–1880), Paria River Basin and southern Colorado Plateau, United States. *Geological Society of America Bulletin* 114: 1550–63.

Hilbert, D. W., Ostendorf, B., and Hopkins, M. S. 2001. Sensitivity of tropical forests to climate change in the humid tropics of north Queensland. *Austral Ecology* 26: 590–603.

Hodell, D. A., Curtis, J. H., and Brenner, M. 1995. Possible role of climate in the collapse of Classic Maya civilization. *Nature* 375: 391–94.

Hsieh, C.-h., Glaser, S. M., Lucas, A. J., and Sugihara, G. 2005. Distinguishing random environmental fluctuations from ecological catastrophes for the North Pacific Ocean. *Nature* 435: 336–40.

Hu, F. S., Slawinski, D., Wright, H. E., Jr., Ito, E., Johnson, R. G., Kelts, K. R., McEwan, R. F., and Boedigheimer, A. 1999. Abrupt changes in North American climate during early Holocene times. *Nature* 400: 437–40.

Hugget, R. J. 1997. *Environmental Change: The Evolving Ecosphere*. London: Routledge.

Hunt, B. G., and Elliott, T. I. 2005. Simulation of the climatic conditions associated with the collapse of the Maya civilization. *Climatic Change* 69: 393–407.

Huntley, B. 1999. Climatic change and reconstruction. *Journal of Quaternary Science* 14: 513–20.

Intergovernmental Panel on Climate Change (IPCC). 2001. *Climate Change 2001: Impacts, Adaptation, and Vulnerability*. Cambridge: Cambridge University Press.

Issar, A. S., and Brown, N. 1998. *Water, Environment and Society in Times of Climatic Change*. London: Kluwer Academic Publishers.

Johnson, T. C., Brown, E. T., McManus, J., Barry, S., Barker, P., and Gasse, F. 2002. A high-resolution paleoclimate record spanning the past 25,000 years in southern East Africa. *Science* 296: 113–32.

Jones, P. D., Bradley, R. S., and Jouzel, J. 1996. *Climatic Variations and Forcing Mechanisms of the Last 2000 years*. Berlin: Springer-Verlag.

Jones, P. D., and Mann, M. E. 2004. Climate over past millennia. *Reviews of Geophysics* 42(2): 1–42.

Kinzelbach, R. K. 2004. The distribution of the serin (*Serinus serinus* L., 1766) in the 16th century. *Journal of Ornithology* 145: 177–87.

Kitagawa, J., Nakagawa, T., Fujiki, T., Yamaguchi, K., and Yasuda, Y. 2004. Human activity and climate change during the historical period in central upland Japan with reference to forest dynamics and the cultivation of Japanese horse chestnut (*Aesculus turbinata*). *Vegetation History Archaeobotany* 13: 105–13.

Knight, J., Orford, J. D., Wilson, P., and Braley, S. M. 2002. Assessment of temporal changes in coastal sand dune environments using the log-hyperbolic grain-size method. *Sedimentology* 49: 1229–52.

Kobashi, T., Severinghaus, J. P., Brook, E. J., Barnola, J.-M., and Grachev, A. M. 2007. Precise timing and characterization of abrupt climate change 8,200 years ago from air trapped in polar ice. *Quaternary Science Reviews* 26: 1212–22.

Kolata, A. L. 2000. Environmental thresholds and the "natural history" of an Andean civilization, in G. Bawden and R. M. Reycraft (eds.), *Environmental Disaster and the Archaeology of Human Response*, pp. 163–78. Maxwell Museum of Anthropology. Anthropological Papers No. 7. Albuquerque: University of New Mexico.

Kurek, J., Cwynar, L. C., and Spear, R. W. 2004. The 8,200 cal yr B.P. cooling event in eastern North America and the utility of midge analysis for Holocene temperature reconstructions. *Quaternary Science Reviews* 23: 627–39.

Lamb, H. H. 1988. *Weather, Climate and Human Affairs: A Book of Essays and Other Papers*. London: Routledge.

Larsen, C. F., Motyka, R. J., Freymueller, J. T., Echelmeyer, K. A., and Ivins, E. R. 2004. Rapid uplift of southern Alaska caused by recent ice loss. *Geophysical Journal International* 158: 1118–33.

Lehman, S. J., and Keigwin, L. D. 1992. Sudden changes in North Atlantic circulation during the last deglaciation. *Nature* 356: 757–62.

Lockwood, J. G. 2001. Abrupt and sudden climatic transitions and fluctuations: A review. *International Journal of Climatology* 21: 1153–79.

McGovern, T. H. 1994. Management for extinction in Norse Greenland, in C. Crumley (ed.), *Historical Ecology: Cultural Knowledge and Changing Landscapes*, pp. 127–54. Santa Fe, NM: School of American Research Monograph.

Magny, M., and Bégeot, C. 2004. Hydrological changes in the European midlatitudes associated with freshwater outbursts from Lake Agassiz during the Younger Dryas event and the early Holocene. *Quaternary Research* 61: 181–92.

Mann, M. E., Park, J., and Bradley, R. S. 1995. Global interdecadal and century-scale climate oscillations during the past five centuries. *Nature* 378: 266–70.

Matthews, J. A., and Briffa, K. R. 2005. The "Little Ice Age": Re-evaluation of an evolving concept. *Geografiska Annaler, Series A: Physical Geography* 87(1): 17–36.

Mayer, J. H., and Mahan, S. A. 2004. Late Quaternary stratigraphy and geochronology of the western Killpecker Dunes, Wyoming, USA. *Quaternary Research* 61: 72–84.

Mayewski, P. A., and White, F. 2002. *The Ice Chronicles: The Quest to Understand Global Climate Change*. Biddeford, ME: University of New England Press.

National Research Council (NRC) (U.S.) Committee on Abrupt Climate Change. 2002. *Abrupt Climate Change: Inevitable Surprises*. Washington, DC: National Academy Press.

Neff, H., Pearsall, D. M., Jones, J. G., Arroyo de Pielers, B., and Friedel, D. E. 2006. Climate change and population history in the Pacific lowlands of southern Mesoamerica. *Quaternary Research* 65: 390–400.

Nocete, F., Álex, E., Nieto, J. M., Sáez, R., and Bayona, M. R. 2005. An archaeological approach to regional environmental pollution in the south-western Iberian Peninsula related to third millennium B.C. mining and metallurgy. *Journal of Archaeological Science* 32: 1566–76.

Nordli, P. Ø. 2001. Reconstruction of nineteenth century summer temperatures in Norway by proxy data from farmers' diaries. *Climatic Change* 48: 201–18.

Nordli, P. Ø., Lie, Ø., Nesje, A., and Dahl, S. O. 2003. Spring-summer temperature reconstruction in western Norway 1734–2003: A data-synthesis approach. *International Journal of Climatology* 23: 1821–41.

Nunn, P. D. 2001. Human-environment relationships in the Pacific Islands around A.D. 1300. *Environment and History* 7(1): 3–22.

O'Brien, S. R., Mayewski, P. A., Meeker, L. D., Meese, D. A., Twickler, M. S., and Whitlow, S. I. 1995. Complexity of Holocene climate as reconstructed from a Greenland ice core. *Science* 270: 1962–64.

Payette, S., and Delwaide, A. 2004. Dynamics of subarctic wetland forests over the past 1,500 years. *Ecological Monographs* 74: 373–91.

Perry, A. L., Low, P. J., Ellis, J. R., and Reynolds, J. D. 2005. Climate change and distribution shifts in marine fishes. *Science* 308: 1912–15.

Peteet, D. 2000. Sensitivity and rapidity of vegetational response to abrupt climate change. *Proceedings of the National Academy of Sciences of the United States of America*. 97: 1359–61.

Pierce, J. L., Meyer, G. A., and Jull, A. J. T. 2004. Fire-induced erosion and millennial-scale climate change in northern ponderosa pine forests. *Nature* 432: 87–90.

Pišut, P. 2002. Channel evolution of the pre-channelized Danube River in Bratislava, Slovakia (1712–1886). *Earth Surface Processes and Landforms* 27: 369–90.

Potts, R. 1996. *Humanity's Descent: The Consequences of Ecological Instability*. New York: William Morrow and Company.

Qian, W., and Zhu, Y. 2002. Little Ice Age climate near Beijing, China, inferred from historical and stalagmite records. *Quaternary Research* 57: 109–19.

Rampino, M. R., Sanders, J. E., Newman, W. S., and Königsson, L. K. 1987. *Climate, History, Periodicity, and Predictability*. New York: Van Nostrand Reinhold.

Reiter, P. 2000. From Shakespeare to Defoe: Malaria in England in the Little Ice Age. *Emerging Infectious Diseases* 6(1): 1–11.

Rial, J. A. 2004. Abrupt climate change: Chaos and order at orbital and millennial scales. *Global and Planetary Change* 41: 95–109.

Rodbell, D. T., and Seltzer, G. O. 2000. Rapid ice margin fluctuations during the Younger Dryas in the Tropical Andes. *Quaternary Research* 54: 328–38.

Robertshaw, P., and Taylor, D. 2000. Climate change and the rise of political complexity in Western Uganda. *Journal of African History* 41: 1–28.

Rohling, E. J., and Pälike, H. 2005. Centennial-scale climate cooling with a sudden cold event around 8,200 years ago. *Nature* 434: 975–79.

Rose, G. A. 2004. Reconciling overfishing and climate change with stock dynamics of Atlantic cod (*Gadus morhua*) over 500 years. *Canadian Journal of Fishing and Aquatic Science* 61: 1553–57.

Rowland, M. J. 1999. Holocene environmental variability: Have its impacts been underestimated in Australian pre-History? *The Artefact* 22: 11–48.

Ruddiman, W. F. 2003. The anthropogenic greenhouse era began thousands of years ago. *Climatic Change* 61: 261–93.

———. 2005. *Plows, Plagues and Petroleum: How Humans Took Control of Climate*. Princeton, NJ: Princeton University Press.

Seppä, H., Cwynar, L. C., and MacDonald, G. M. 2003. Post-glacial vegetation reconstruction and a possible 8,200 cal yr B.P. event from the low artic of continental Nunavut, Canada. *Journal of Quaternary Science* 18: 621–29.

Shennan, S. 2000. Population, culture history, and the dynamics of culture change. *Current Anthropology* 41: 811–35.

Sherratt, A. 1997. Climatic cycles and behavioural revolutions: The emergence of modern humans and the beginning of farming. *Antiquity* 71: 271–87.

Sima, A., Paul, A., and Schulz, M. 2004. The Younger Dryas: An intrinsic feature of late Pleistocene climate change at millennial timescales. *Earth and Planetary Science Letters* 222: 741–50.

Singer, C., Shulmeister, J., and McLea, B. 1998. Evidence against a significant Younger Dryas cooling event in New Zealand. *Science* 281: 812–14.

Soon, W., and Baliunas, S. 2003. Proxy climatic and environmental changes of the past 1,000 years. *Climate Research* 23: 89–110.

Soon, W., Baliunas, S., Idso, C., Idso, S., and Legates, D. R. 2003. Reconstructing climatic and environmental changes of the past 1,000 years: A reappraisal. *Energy and Environment* 14: 233–96.

Smith, J. E., Risk, M. J., Schwarcz, H. P., and McConnaughey, T. A. 1997. Rapid climate change in the North Atlantic during the Younger Dryas recorded by deep-sea corals. *Nature* 386: 818–20.

Stankoviansky, M. 2003. Historical evolution of permanent gullies in the Myjava Hill Land, Slovakia. *Catena* 51: 223–39.

Storemyr, P. 2004. Weathering of soapstone in historical perspective. *Materials Characterization* 53: 191–207.

Straus, L. G. 1996. The world at the end of the last Ice Age, in L. G. Straus, B. V. Eriksen, J. M. Erlandson, and D. R. Yesner (eds.), *Humans at the End of the Ice Age: The Archaeology of the Pleistocene-Holocene Transition*, pp. 3–9. New York: Plenum Press.

Street-Perrott, F. A., and Perrott, R. A. 1990. Abrupt climate fluctuations in the tropics: The influence of Atlantic Ocean circulation. *Nature* 343: 607–12.

Tarasov, L., and Peltier, W. R. 2005. Artic freshwater forcing of the Younger Dryas cold reversal. *Nature* 435: 662–65.

Taylor, K. C., Lamorey, G. W., Doyle, G. A., Alley, R. B., Grootes, P. M., Mayewski, P. A., White, W. C., and Barlow, L. K. 1993. The "flickering switch" of late Pleistocene climate change. *Nature* 361: 432–36.

Taylor, K. C., Mayewski, P. A., Alley, R. B., Brook, E. J., Gow, A. J., Grootes, P. M., Meese, D. A., Saltzman, E. S., Severinghaus, J. P., Twickler, M. S., White, J. W. C., Whitlow, S., and Zielinski, G. A. 1997. The Holocene-Younger Dryas transition recorded at Summit, Greenland. *Science* 278: 825–27.

Troelstra, S. R., van Hinte, J. E., and Ganssen, G. M. (eds.). 1996. *The Younger Dryas. Proceedings of a Workshop at the Royal Netherlands Academy of Arts and Sciences on 11–13 April 1994*. Amsterdam: Royal Netherlands Academy of Arts and Sciences.

Turney, C. S. M., Kershaw, A. P., Lowe, J. J., van der Kaars, S., Johnston, R., Rule, S., Moss, P., Radke, L., Tibby, J., McGlone, M. S., Wilmshurst, J. M., Vandergoes, M. J., Fitzsimons, S. J., Bryant, C., James, S., Branch, N. P., Cowley, J., Kalin, R. M., Ogle, N., Jacobsen, G., and Fifield, L. K. 2006. Climatic variability in the southwest Pacific during the Last Termination (20–10 k yr B.P.). *Quaternary Science Reviews* 25: 876–85.

Vacco, D. A., Clark, P. U., Mix, A. C., Cheng, H., and Edwards, R. L. 2005. A speleothem record of Younger Dryas cooling, Klamath Mountains, Oregon, USA. *Quaternary Research* 64: 249–56.

Van Campo, E., and Gasse, F. 1993. Pollen-and diatom-inferred climatic and hydrological changes in Sumxi Co basin (Western Tibet) since 13,000 yr B.P. *Quaternary Research* 39: 30013.

van't Veer, R., Islebe, G. A., and Hooghiemstra, H. 2000. Climatic change during the Younger Dryas chron in northern South America: A test of the evidence. *Quaternary Science Reviews* 19: 1821–35.

von Grafenstein, U., Erlenkeuser, H., Müller, J., Jouzel, J., and Johnsen, S. 1998. The cold event 8,200 years ago documented in oxygen isotope records of precipitation in Europe and Greenland. *Climate Dynamics* 14: 73–81.

Wahl, D., Byrne, R., Schreiner, T., and Hansen, R. 2006. Holocene vegetation change in the northern Pecten and its implications for Maya prehistory. *Quaternary Research* 65: 380–89.

Walter, R. C., Buffler, R. T., Bruggemann, J. H., Guillaume, M. M. M., Berhe, S. M., Negassi, B., Libsekal, Y.,

Cheng, H., Edwards, R. L., Von Cosel, R., Néraudeau D., and Gagnon, M. 2000. Early human occupation of the Red Sea coast of Eritrea during the last interglacial. *Nature* 405: 65–69.

Walther, G.-R., Post, E., Convey, P., Menzel, A., Parmesan, C., Beebee, T. J. C., Fromentin, J.-M., Hoegh-Guldberg, O., and Bairlein, F. 2002. Ecological responses to recent climate change. *Nature* 416: 389–95.

Wei, K., and Gasse, F. 1999. Oxygen isotopes in lacustrine carbonates of West China revisited: Implications for post glacial changes in summer monsoon circulation. *Quaternary Science Reviews* 18: 1315–34.

Weiss, H. 2000. Beyond the Younger Dryas: Collapse as adaptation to abrupt climate change in ancient west Asia and the eastern Mediterranean, in G. Bawden and R. M. Reycraft (eds.), *Environmental Disaster and the Archaeology of Human Response*. Maxwell Museum of Anthropology. Anthropological Papers No. 7, pp. 75–98. Albuquerque: University of New Mexico.

Weiss, H., Courty, M.-A., Wetterstrom, W., Guichard, F., Senior, L., Meadow, R., and Curnow, A. 1993. The genesis and collapse of third millennium north Mesopotamian civilization. *Science* 261: 995–1004.

Wells, L. E., and Noller, J. S. 1999. Holocene coevolution of the physical landscape and human settlement in northern coastal Peru. *Geoarchaeology* 14: 755–89.

Williams, P. R. 2002. Rethinking disaster-induced collapse in the demise of the Andean highland states: Wari and Tiwanaku. *World Archaeology* 33:.361–74.

Wilson, P., McGourty, J., and Bateman, M. D. 2004. Mid-to late-Holocene coastal dune event stratigraphy for the north coast of Northern Ireland. *The Holocene* 14: 406–16.

Wollenburg, J. E., Knies, J., and Mackensen, A. 2004. High-resolution paleoproductivity fluctuations during the past 24 kyr as indicated by benthic foraminifera in the marginal Arctic Ocean. *Palaeogeography, Palaeoclimatology, Palaeoecology* 204: 209–38.

Wurth, G., Niggemann, S., Richter, D. K., and Mangini, A. 2004. The Younger Dryas and Holocene climate record of a stalagmite from Hölloch Cave (Bavarian Alps, Germany) *Journal of Quaternary Science* 19: 291–98.

Wunsch, G. 2006. Abrupt climate change: an alternative view. *Quaternary Research* 65: 191–203.

Xia, Q., Zhao, J.-X., and Collerson, K. D. 2001. Early-mid Holocene climatic variations in Tasmania, Australia: Multi-proxy records in a stalagmite from Lynds Cave. *Earth and Planetary Science Letters* 194: 177–87.

Zolitschka, B., Behre, K.-E., and Schneider, J. 2003. Human and climatic impact on the environment as derived from colluvial, fluvial and lacustrine archives-examples from the Bronze Age to the Migration period, Germany. *Quaternary Science Reviews* 22: 81–100.

39

Human Behavioral Ecology and the Use of Ancient Landscapes

Douglas Bird and Brian Codding

The most basic components of human life revolve around how we utilize landscapes. We create and move across space in order to find and use resources to interact with and avoid conspecifics and to evaluate and evade risks. Landscapes are thus the context for decisions that critically impinge on individuals' survival and reproductive success; they form the ecology of human life—the social history, individual development, and local environmental circumstances that constrain our behavior. Moreover, behavior relative to these constraints is usually temporally and spatially patterned, and it often has material consequences whose traces archaeologists use as a matter of routine to reconstruct the human past. The issue at hand is how to account for the patterns thus reconstructed. Human behavioral ecologists with an archaeological eye attempt to do so by employing hypotheses about behavioral adaptation and its material effects under specific ecological conditions (e.g., O'Connell 1995).

Human behavioral ecology (HBE) is concerned with the *functional* significance of behavior, beginning with the proposition that, as with all animals, human behavioral capacities are designed by natural selection (see Table 39.1). "Functional" here refers to Tinbergen's (1963) designation of four different, nonmutually exclusive levels of behavioral analysis: the *proximate* mechanisms that cause a

particular type of response at a given moment, the *ontogeny* of those responses as an individual develops, the *historical* pattern of behavioral systems, and the survival and reproductive value of a given trait (adaptive *function*). HBE focuses on functional variability; it applies natural selection theory to the study of human behavior in socioecological context. The approach draws on formal optimization and game theoretic models to develop hypotheses about solutions to the fitness-related trade-offs individuals face in variable environmental and social circumstances (Smith and Winterhalder 1992). It has been used now for over three decades as a framework for proposing testable explanations of behavior and their archaeological consequences.

In this chapter, we define and explain key concepts in HBE and illustrate the scope of their application to problems of landscape archaeology. The breadth of studies we use to illustrate HBE applications is partly a product of the way "landscapes" are defined—they can potentially include anything from the smallest of domestic spheres, to the layout of social networks, to the colonization of continents. Here we limit our focus to studies that deal mostly with the daily business of acquiring and using resources—that is, those attempting to understand physical movement, the mundane landscapes of individual lives, and the archaeological patterns that they can produce. Interested readers can find more comprehensive treatment of HBE, its archaeological applications and associated problems, in numerous recent sources (e.g., Bird and O'Connell 2006; Kennett and Winterhalder 2006; Shennan 2002).

HBE Fundamental Principles and Models

The foundations of behavioral ecology were formalized in the 1960s and 1970s by ecologists studying sociality, parental investment, and foraging patterns in animals (e.g., Alexander 1974; Brown 1964; Charnov 1976; Hamilton 1964; Hutchinson 1965; MacArthur and Pianka 1966; Trivers 1971, 1972; Williams 1966; see Parker 2005 for recent overview). Soon thereafter, ethnographers interested in selectionist approaches to cultural behavior adopted and developed the approach for investigating questions about human reproductive and subsistence strategies (e.g., Blurton Jones and Sibly 1978; Hawkes, Hill, and O'Connell 1982; see contributions to Chagnon and Irons 1979 and Winterhalder and Smith 1981). By that time, a few archaeologists were already using models from behavioral ecology, initially focusing on questions about subsistence (e.g., Beaton 1973; Wilmsen 1973). The number and scope of archaeological applications expanded greatly in the 1980s and 1990s, addressing a broad array of issues including patterns in resource transport, subsistence-related changes in technology, the origin and diffusion of agriculture, the material correlates of social status, early human social organization, the development of social hierarchies, and the evolution of human life history (see Bird and O'Connell 2006).

Theoretical and Methodological Individualism

HBE assumes that complex cultural patterns emerge from individual interactions and decisions, that these emergent patterns are what need to be explained and not the explanation *per se*. The approach rests on two basic assumptions: first, that resources are always finite and individuals make trade-offs in the face of constraints; and second, that the capacity to evaluate the costs and benefits associated with those trade-offs is designed by natural selection (Hawkes 1996). This is the basis for the expectation that individuals are likely to behave as if they have the propensity to maximize their inclusive-fitness relative to the trade-offs they face in particular ecological settings (see Table 39.1).

The focus of HBE on *individual* decision making relative to the opportunity costs of alternative strategies is partly theoretical, partly heuristic, and partly empirical. The theoretical justification is founded in part on Williams's (1966) watershed book, which established the standard for evaluating "adaptive" behavior: individuals generally reproduce and die at higher rates than groups, and variability is almost always greater within groups than between them. Thus, selection at the level of the group is constantly swamped by selection for traits among individuals.

Although there has been some renewed debate about the significance of group selection as an evolutionary force (Wilson 1998), HBE begins with the assumption that selection will operate on the differential reproductive success of individuals, favoring traits that increase the inclusive fitness of the individuals that carry them. This process cannot generally spread characteristics that contribute to the survival of the group at the expense of individual reproductive success. Unless differential extinction of groups is frequent enough, or under special conditions that promote the differential propagation of group-advantageous traits (Rogers 1990), selection *within* groups will relentlessly favor individual advantage relative to other conspecifics (Smith and Winterhalder 1992).

A methodological and empirical focus on the individual clearly distinguishes HBE from other ecological approaches in anthropology. Take, for example, neofunctional cultural ecology explanations of the long birth spacing among many ethnographically known hunter-gatherers. This spacing has been typically viewed as a mechanism to insure ecosystem stability by keeping populations low and in balance with resource supply (e.g., Freeman 1971; Lee 1979, 1980). Conversely, in a classic application of HBE, Blurton Jones (1986, 1987; Blurton Jones and Sibly 1978) demonstrated that, relative to the weight of carrying more children and the additional food required to feed them, Ju/'hoansi women maximize the number of *surviving* offspring by spacing births at about four years, thus limiting family size. Thus, mothers face trade-offs in the quality and the quantity of offspring, and individual reproductive success is often maximized by reducing fertility. From this perspective, the group benefits are epiphenomenal to conditional strategies designed relative to the "real time" trade-offs individuals confront in fluctuating local contexts.

While the results of selection (in this case, behavioral design or adaptation) are related to individual fitness (survival and reproductive success), "analyzing design is not the same as measuring

Table 39.1   Common HBE Terms

• Human behavioral ecology:	The branch of evolutionary ecology that studies variability in the adaptive design of human behavior in specific ecological contexts.
• function:	The survival and reproductive value of a trait.
• inclusive fitness:	An individual's propensity to survive and reproduce in a given context, inclusive of weighted effects on closely related individuals.
• trade-offs:	The fitness-related costs and benefits of a particular course of action relative to other proposed options. These are measured with respect to marginal value and opportunity cost.
• marginal value:	The change in initial benefit with an additional unit of investment.
• opportunity cost:	The cost of an activity in terms of the opportunity that is forgone by continuing it. The decision to change behavior is assumed to be set by its marginal value relative to benefits of doing something else.
• conditional strategies:	Abstractions of adaptive "decision rules" that specify covariation in behavior and context—"in order to maximize fitness in context X do a, in context Y do b."
• the phenotypic gambit:	A working assumption that proposes that however individuals acquire their behavior, the evolved mechanisms of phenotypic adaptation allow for flexible responses to variable fitness-related constraints.
• optimization:	A premise that suggests that as a result of natural selection, behavior will tend toward optimal solutions to trade-offs faced in the allocation of life-history effort; used to generate predictions about how individuals may relate to their environment relative to a stipulated fitness-related goal. This is not a premise that behavior is routinely optimal in any absolute sense.
• goal:	The variable in HBE models that specifies a hypothetical purpose for a given behavior—for example, maximizing foraging efficiency, minimizing foraging time, maximizing the number of surviving offspring.
• decision:	The variable in HBE models that specifies the alternative choices associated with a particular goal—for example, maximizing the number of surviving offspring involves decisions about quality vs. quantity.
• currency:	The criterion in HBE models used to compare alternative values of a stipulated decision—for instance, energy or time.
• constraints:	The variables in HBE models used to specify extrinsic and intrinsic factors that limit and define an actors opportunities—for example, the distribution of resources or the size of the forager.
• ESS:	Evolutionary stable strategies; the predictive outcome of game theoretic formulations of design problems, whereby the marginal payoff of a given strategy depends on its own frequency relative to the frequency of other strategies in the population.

fitness" (Stephens and Krebs 1986: ix). The central question in the study of behavioral adaptation is not whether individuals survive and reproduce but rather how design functions relative to expected fitness-related goals and trade-offs (Williams 1966: 159). To operationalize this question, HBE draws on two sets of conceptual tools: the phenotypic gambit and simple optimization or game theoretic models.

Phenotypic Gambit

HBE is expressly concerned with behavioral responses to variable environments (Smith 2000: 30). As a heuristic for evaluating these, HBE often employs a basic assumption referred to as the "phenotypic gambit" (Grafen 1984; see also Smith and Winterhalder 1992: 33). The approach begins with the proposition that natural selection will lead to the evolution of mechanisms that provide the phenotypic *capacity* to respond flexibly to fitness-related problems. Such responses need not have a particularly close link to genetic transmission to have important adaptive consequences. Behavior is *caused* by complex proximate mechanisms under particular ontogenetic and historical circumstances. In many cases, it is neither necessary nor feasible to account for all these causes to propose functional explanations of behavior. This is sometimes a calculated risk, but it allows the investigator to pose functional hypotheses while avoiding questions about whether a trait is "instinctive" or "learned," "biological" or "cultural." Thus, models in HBE do not specify a mode of inheritance; instead, they are tools that allow an investigator to develop tractable hypotheses about how individuals will relate to variable environments *if* they behave in ways that maximize a proposed currency relative to a stated fitness-related goal.

Optimization and Game Theoretic Models

To develop context-specific predictions about behavior, HBE makes use of formal models drawn mainly from optimization analysis (Maynard Smith 1978) and evolutionary game theory (Dugatkin and Reeve 1998; Maynard Smith 1974, 1982). The optimization premise of HBE is that, as a result of natural selection, individuals tend to relate to their social and physical environments in ways that optimize specific fitness-related trade-offs. This premise does not imply that behavior is fully optimal or that individuals are narrowly self-interested. Individuals often have conflicting interests, and "optimal" solutions are often contingent on compromises among conflicting goals in very local circumstances. In fact, the principle of optimization is simply a framework for generating hypotheses; optimality models are tools to characterize solutions to a limited set of hypothetical allocation problems associated with specific contexts (Stephens and Krebs 1986). The models predict optimal outcomes relative to stipulated goals, decisions, currencies, and constraints (Table 39.1). The analyst can then evaluate with naturalistic tests whether actual behavior corresponds with the derived expectations. The extent to which behavior violates the predictions is then of special interest (Bird, Bliege Bird, and Richardson 2004; Bliege Bird and Smith 2005).

Game theory (or in the parlance of behavioral ecology, "evolutionary stable strategy" [ESS] theory) adds a social dimension to the ecology of behavior. It allows the analyst to model interactions among individuals with partially competing interests. These interactions produce "frequency dependant" outcomes, where the behavior of one individual depends on that of others (Smith and Winterhalder 1992). Most applications of BE in archaeology make use of optimality models, although game theory has been deployed to good effect as well (see Bird and O'Connell 2006).

Landscape-Use Intensification

Most explicit HBE studies of prehistoric subsistence have focused on evaluating variability in faunal and floral assemblages relative to the predictions of various models grouped under the rubric of foraging theory (Stephens and Krebs 1986). The best known of these models is encounter-contingent prey choice model (PCM).

Prey Choice

The prey choice model (PCM) focuses on a decision about whether to handle (pursue, capture, harvest, and process) an encountered item of a given prey type or pass it over to continue searching for other prey. Either way, the decision engages an opportunity cost: while handling and encountered resource, a forager trades off the opportunity to search for other resources. If the return per unit time handling a resource is less than the expected return from continuing to search for other prey to handle, the optimizing forager will pass over the resource regardless of how frequently it is encountered.

The optimal solution to the preceding tradeoff is provided by the prey algorithm given in Stephens and Krebs (1986: 22). The model assumes that a forager's goal is to maximize overall foraging efficiency, calculated in terms of a specified currency per unit time spent searching and handling prey in a given patch. Potential prey types are then ranked according to their profitability—that is, their post-encounter return per unit time required to pursue, harvest, and process. The model predicts that the top ranked prey type will always be handled on encounter and that less profitable types will be added to the array selected in descending rank order until the post-encounter return from the next lowest ranked type falls below the expected return from searching for and handling all resources of higher rank. Since taking resources of that type will by definition *reduce* the average overall foraging efficiency, all items of that type and any of lower rank will be passed over to continue searching for more profitable prey.

Counter-intuitively, the PCM predicts that the inclusion of a given prey type will depend *not* on its own abundance but *only* on its post-encounter profitability and the expected encounter rate with higher ranked prey. If the encounter rate with higher ranked prey increases, then overall average foraging efficiency will increase as well. If overall efficiency increases beyond the post-encounter return rate available from relatively low ranked items previously taken, then those items should now be dropped to the diet. If, however, the encounter rate for high ranked items decreases, less profitable prey should added in the diet in rank order as overall efficiency declines.

Patch Utilization

If an encounter with a particular prey type changes the expectation of encountering more items of that type, and resources are temporally or spatially clumped, a forager's decision is *not* "should I search or handle" but "relative to my other options, should I enter a patch, and if so, how long should I stay." Patch models predict that potential foraging locales (or activity types) will be exploited in order of the return rate expected from searching and handling within a given patch vs. the overall return rate expected from traveling to search for and handle resources in more productive patches. Patches will be added to or dropped from the array as a function of fluctuations in the overall return rate from the habitat. However, resources in a given patch are often harvested at a diminishing rate—that is, return rates generally decline with increasing patch residence time. Relative to such diminishing returns, the optimal point at which a forager should leave one patch to travel to another is given by the marginal value theorem (Charnov 1976; Charnov and Orians 1973; Stephens and Krebs 1986: 24–32). The marginal value theorem (see Table 39.1) specifies that in order to maximize the return rate from foraging in a given habitat, the forager will leave the present patch when the return rate has declined to the average efficiency for the habitat as a whole, calculated as return per unit time handling, searching, and traveling. Shorter travel times to patches thus increase the opportunity costs of remaining in a patch, and the optimal residence time is often too short for forgers to completely deplete a patch. A rate-maximizing forager will choose the residence time for each patch type so that the marginal rate of gain at the time of leaving equals the overall return rate for the habitat (Charnov 1976).

Archaeological Applications

The implications and problems of these models for understanding prehistoric subsistence variability, especially relative to broad changes toward more intensified economies (including the transitions to agricultural systems), are far too numerous to discuss here, and many have been reviewed in detail elsewhere (e.g., Bettinger 1991; Bird and O'Connell 2006; Broughton 2002; Kelly 1995; Kennett 2005; Kennett and Winterhalder 2006). Here we highlight some of the advantages and problems of this work with examples related to prehistoric landscape use in the Mediterranean and west coast of North America.

Stiner and colleagues (1999, 2000; Stiner and Munro 2002) have recently argued that an initial shift toward "broad spectrum" diets in the Mediterranean Basin occurred much earlier than previously suggested. As discussed above, the PCM predicts that the range of prey selected should increase with declines in overall foraging efficiency. Stiner and associates argue that as a result of increasing human populations, slow-moving, easily captured items (for example, tortoises, shellfish) common in the Middle Paleolithic components, decline in frequency and mean body size beginning in the Upper Paleolithic (< 45,000 B.P.). They suggest that these prey, because they incorporate relatively little pursuit costs, were likely to have been highly ranked in terms of their post-encounter return rate, and depression in these should intensify efforts in pursuing more costly prey. As expected, the frequency of prey that are more difficult to acquire (for instance, birds, lagomorphs) increases in the later deposits. The analysis suggests that sharp increases in population densities during the Upper Paleolithic took place in several distinct pulses throughout the Mediterranean Basin, each coincident with a shift in small-animal exploitation toward prey that are likely to be associated with higher pursuit costs. Their argument challenges the notion that an increase in the range of occupied landscapes, accompanied by archaeological changes associated with the beginning of the Upper Paleolithic, are best explained by a florescence of modern human behavioral capabilities (e.g., Binford 1984; Klein 1999: 454–63; cf. McBrearty and Brooks 2000).

Some of the earliest explicit use of models in behavioral ecology came in the form of similar attempts to understand variability in coastal landscape use and subsistence intensification in California (e.g., Beaton 1973; Botkin 1980; Jones 1991). Californian archaeologists have debated for decades the causes of Holocene changes and the human impact on prehistoric coastal faunal communities (see Raab 1992). In a recent form of the argument, Kennett (2005; Kennett and Kennett 2000) has argued that people responded to late Holocene climatic instability in southern California by intensifying coastal resource exploitation, which increased sedentism, sociopolitical complexity, and trade. Although the detailed dynamic effects of changing subsistence and palaeoclimates are a long way from being sorted out, both theory and the empirical evidence strongly suggest that intensifying subsistence in coastal California was a response to local resource depression and instability driven by the relationship between climate change and overexploitation (Erlandson and Moss 2001; Rick and Erlandson 2000). These studies suggest that as local resources were depressed, people responded in ways consistent with the predictions of both the PCM and the marginal value theorem: they broadened their diet in rank order and increased the intensity of production within resource patches. The California case demonstrates that these changes are unlikely to reflect variable strategies *designed* to mitigate long-term resource depletion and/or increase general biodiversity (see Broughton 2002, 2004 for a review).

While the archaeological applications discussed above illustrate the value of the approach, one of the primary problems with these and similar studies of resource use is that we know very little about the actual profitability ranking of different prey types for foragers under different landscape constraints (Ugan 2005). Most researchers have assumed that post-encounter return rates scale with prey size. There are at least two reasons to suspect that this might not always be the case. First, large mobile game often requires very long pursuits that often fail, an aspect of handling time that is not usually included in archaeological estimates of post-encounter profitability (Bliege Bird and Bird 2005; Hawkes et al. 1991; O'Connell et al. 1988; Smith 1991: 230–31; Stiner and Munro 2002; Winterhalder 1981: 95–96). Second, larger prey are more likely to be shared to a larger audience, in some cases resulting in very low post-encounter returns for the acquirer (Bliege Bird et al. 2001; Hawkes and Bliege Bird 2002; Hawkes et al. 2001). The pursuit of these may still be especially attractive relative to other currencies (for example, display value), but there is nothing inherent in prey size that will always denote high profitability. Thus, modeling different pursuit strategies, goals, and currencies will enhance the sophistication and realism of common applications of the PCM (e.g., Elston and Zeanah 2002; Hildebrandt and McGuire 2002, 2003; McGuire and Hildebrandt 2005).

Moving Resources across Landscapes

One of the clearest ways in which the assumptions of the prey and patch models discussed above are routinely violated involves the fact that people generally operate from a home base utilizing different hunt types and resource patches that require mutually exclusive technologies. As such, a forager's goal is often something other than maximizing the energetic return rate while in a homogeneous habitat: for example, it may often be the case that the goal of foraging is to maximize the rate at which resources from various patches can be delivered to a central place (e.g., Cannon 2000).

Transport Models

If an acquirer's goal is to maximize the rate at which resources can be delivered to a central locale, when constrained by load size, she or he can either cull parts of lower utility (which increases the value of a load at a cost to foraging-transport time) or transport more loads from the site of acquisition (which means more time to transport more loads of lower quality). Building on seminal work by Orians and Pearson (1979), Metcalfe and Barlow (1992) have explored this decision variable with a formal model that predicts the point at which field processing as opposed to bulk transport will maximize delivery rates. Components of the model, especially the costs and benefits of field processing, have been shown to be critically important for explaining archaeological variability in anatomical part and prey type representation in resources ranging from large game (O'Connell, Hawkes, and Blurton Jones 1988, 1990; O'Connell and Marshall 1989; Zeanah 2000), to seeds (Barlow and Metcalfe 1996), to tool-stone (Beck, Taylor, and Jones 2002; Elston and Raven 1992) to shellfish and acorns (Bettinger, Mahli, and McCarthy 1997; Jones and Richman 1995).

Cannon (2003) has investigated these models and their implications for arguments about land-use in the American Southwest. As predicted by the transport model, foragers should increase investment in field processing low utility elements from an acquired resource if transport distances increase. A diachronic decrease in the proportional representation of low utility parts at a central place should thus indicate that foragers were traveling further to acquire these resources (Metcalfe and Barlow 1992). Based on these predictions, Cannon (2003) suggests that foragers in the Mimbres Valley were experiencing local depression of large game resources and responded by traveling further to more distant foraging locals in order to obtain the desired prey. Counter-intuitively, this response is consistent with the results that indicate a proportional *increase* in high utility parts at residential, possibly in response to increasing investment in culling parts of lower utility prior to transport.

Similar models have also been used to investigate toolstone procurement on the landscape. Beck and associates (2002) examine terminal Pleistocene/early Holocene tool-stone acquisition in the Great Basin of Western North America by assessing the stages of biface production at quarry sites compared to residential base camps that were located at different distances from the quarry. Stone tool production is a reductive process whereby low utility stone is removed at some cost in the process of increasing the utility of a load of lithic material. The model thus predicts that biface tools will be proportionally reduced at the quarry site in relation to the distance that individuals must travel to return to their residential base or central place. The analysis suggests that rather than indicating site function, intersite variability in the proportion of bifaces in different stages of reduction is predicted by the distance between a quarry and a residential locale. Palaeoarchaic residential sites more distant from a toolstone source are characterized by increased biface reduction.

Examining the extent to which variables such as toolstone quality and intended product might affect processing and transport decisions will require repeated experiments. Nevertheless, the implications of these and other HBE studies of transport are clear: variability in resource and land use is often reflected only through a heavy filter of differential transport. The conditions that structure variability in central place foraging and transport strategies (of potentially any resource) should, at least in principle, be predictable with data on the age and gender-specific changes in load utility with investment in field processing (see Barlow and Metcalfe 1996; Elston and Zeanah 2002; Lupo 2001; O'Connell and Marshall 1989; O'Connell, Hawkes, and Blurton Jones 1988, 1990; Rogers and Broughton 2001; Thomas 2002; Zeanah 2000).

Age and Gender Differences in Landscape Use

In a recent application of HBE transport models, Bird and colleagues (Bird 1996, 1997; Bird and Bliege Bird 1997, 2000; Bird et al. 2002; Bird, Bliege Bird, and Richardson, 2004) have investigated prey/patch exploitation, processing, and transport tradeoffs with ethnographic and archaeological data from the Meriam Islands of the Eastern Torres Strait, Australia (see similar applications in de Boer 2000; Thomas 2002). The work has focused on striking differences between the frequency of different shellfish species in Meriam middens (contemporary and prehistoric) and the frequency at which contemporary Meriam adults and children

harvest and transport different shellfish species. Meriam shellfishers are highly selective, and as long as we consider age-linked variability in foraging constraints (for example, walking speed differences between children and adults), shellfishers' prey choice and time allocation are consistent with a hypothesized goal of maximizing energy per unit time searching and handling (Bird 1996, 1997; Bird and Bliege Bird 2002; see PCM predictions above). But this is not reflected archaeologically: the most profitable shellfish, which are always harvested on-encounter, are rare in the middens relative to shellfish that are commonly passed over (especially by adults, see Bird and Bliege Bird 2000, 2002). The analysis demonstrates that this pattern is predicted only when we consider age-linked investment in field processing to increase the utility of a load. Contemporary Meriam gatherers routinely process some types of shellfish at the spot where they harvest, whereas other types are always transported in bulk (Bird and Bliege Bird 1997). As such, we can account for much of the variability in the frequency of different prey types in the middens only when we devalue the observed prey choice (of both adults *and children*) by the costs and benefits of age-linked differences in the costs of field processing (Bird et al. 2002; Bird, Bliege Bird, and Richardson 2004). The effects that other processes (pre- and postdeposition) might have on this pattern are highlighted in the variability that violates these predictions, and are now being investigated with detailed chrono-stratigraphic data.

Ethnographers have long noted that men and women often have different goals with respect to foraging, which sometimes result in clear violations of certain predictions from basic foraging models (Hawkes 1990). While they overlap sometimes, women generally seek reliable resources for provisioning purposes, whereas men pursue riskier strategies, often relative to prestige and political goals (e.g., Bliege Bird, Smith, and Bird 2001; Hawkes 1991; Hawkes and Bliege Bird 2002). These tendencies pose an important problem for central place foragers operating in landscapes with patchy resources: if there are differences in foraging goals, the range of resources and optimal residential locale for women may be different from those for men. Where then should central place foragers locate their residential bases?

Elston and Zeanah (2002) and Zeanah (2004) take up this question relative to Pleistocene-Holocene transitions in two regions of the Great Basin. They begin by identifying those resources that ethnographic data suggest are most likely to be targeted in pursuit of men's and women's foraging goals. They then plot the distribution of these resources across a wide range of different habitat types and use these measures to generate predictions about which locations would maximize the delivery rate of gender-specific resources. Zeanah's (2004) results from the Carson Desert suggest that site locations near reliable resources are most consistent with women's predicted foraging goals. This appears to have reduced residential mobility in general and increased logistical mobility (*sensu* Binford) for men. With more tethered residences, the optimal strategy relative to men's goals would have involved working from logistical camps located at some distance from residential bases in areas with increased possibilities to encounter game, a result borne out in the archaeological landscape.

Elston and Zeanah (2002) suggest that similar differences in foraging goals may account for the differences in the spatial distribution of Great Basin archaeological sites over time. For the early Holocene archaeological patterns in Railroad Valley, female foraging goals largely predict site locations near reliable resources at marshes and lakes (possibly associated with *both* women's and men's goals). During the middle Holocene, however, lakes disappear and in response to the changing environment, female strategies are suggested to have turned to seed storage. As food storage decreased residential mobility, men's foraging goals could be maximized only by increasing logistical mobility, increasing the frequency of task-specific field camps. By most accounts, this seems broadly consistent with rele-vant features of the archaeological record (e.g., Beck and Jones 1997).

Landscape Use, Waste, and Symbolic Resources

Recently a number of human behavioral ecologists have begun to utilize signaling theory to investigate seemingly "irrational" symbolic and social behavior, such as ceremonial feasting, religious commitment, monumental architecture, conspicuous consumption, public goods provisioning, and their material expressions (e.g., Bliege Bird, and Smith 2005; Bliege Bird, Smith, and Bird 2001; Boone 1998; Hawkes and Bliege Bird 2002; Neiman 1997; Smith and Bliege Bird 2000; Smith, Bliege Bird, and Bird 2003; Sosis 2000, 2003; Sosis and Bressler 2003). Signaling theory is based on the premise that expensive phenotypic traits can serve as a means to convey honest information of mutual interest to individuals with competing interests (Grafen 1990; Johnstone 1997;

Zahavi 1975; Zahavi and Zahavi 1997). Behaviors that at first glance may seem absurd, such as when a gazelle stots in front of a lion instead of running away, may be evolutionarily stable when there are mutual benefits to be gained from truthful communication. The information signaled when a gazelle stots ("I can jump high in a very peculiar and standardized fashion"), if it is an honest indication of the gazelle's underlying quality, is of interest to the lion, and as a result both can realize a collective benefit by avoiding a pointless chase. In the language of game theory, such communication can be evolutionarily stable only when it is the best move for both signaler and recipient; that is, a display must be an "honest" indicator of the quality the signaler wishes to convey, thereby reducing the possibility of bluffing (Getty 1998). While the costs of producing the display are one way for the recipient to be certain of the information's validity, honesty can be assured through alternative pathways as well, such as intrinsic links between the signal and the quality it represents (Maynard Smith and Harper 2003).

Hunting Strategies

Where *relative* position is critical to success, very minor differences in capacity or intent will determine large differences in the rewards of competition. This can mean that costly displays of small differences in underlying ability and motivation often result in large power inequalities and runaway evolutionary outcomes. Something akin to this should account for the fact that in many human societies status is maintained and prestige granted only through routine displays of waste, extreme altruism, conspicuous consumption, and particular aspects of gender specific activities (Bliege Bird, Smith, and Bird 2001; Bliege Bird and Hawkes 2002; Bliege Bird and Smith 2005; Boone 1998; Smith and Bliege Bird 2000; Smith, Bliege Bird, and Bird 2003; and Sosis 2000). For example, rather than a strategy of specialization to maximize household food intake, these researchers argue that particular aspects of gender-linked differences in subsistence behavior (such as hunting and widespread sharing) are maintained relative to the way such activities serve as symbolic displays of distinction in prestige competitions.

In a provocative argument, William Hildebrandt and Kelly McGuire (2002, 2003; McGuire and Hildebrandt 2005) have applied similar logic to account for variability in faunal assemblages in many parts of California and the western Great Basin. They suggest that key changes in the archaeological record are consistent with escalating benefits in the signaling value of large game hunting associated with increasing opportunities for hunters to advertise their qualities to larger, more concentrated local audiences. In support, they demonstrate a correlation between marked human population growth and a proportional *increase* in large mammal remains of Middle Archaic (4,000–1,000 B.P.) faunal assemblages. The pattern is opposite to what we might expect from the standard prey choice model described above, where if prey size scales with profitability, increasing population pressure would lead to a *decrease* in large game. If hunting intensity did increase in the Middle Archaic, evaluating whether or not this is the result of changing goals related to social displays (or changes in the encounter rates and/or nutritional benefits of exploiting these or other resources) will require better estimates of the costs involved in large game hunting and palaeoclimatic conditions. Nevertheless, Hildebrandt and McGuire make an important point: social displays (and their material consequences) are often made honest and maintained by the ways in which the costs indicate underlying qualities of the signaler. Such signaling provides an important check on indiscriminate use of "optimal foraging theory" and insight into how different (sometimes conflicting) goals might contribute to social complexity (Bliege Bird 1999).

Material Display and Political Landscapes

In an early application of CST, Neiman (1997) hypothesized that variability in Classic Maya monumental advertisements is a function of conditions that concentrate audiences in space and increase differential access to and control of resource catchment areas. The analysis suggests that monument size and complexity should scale with the benefits of social display made honest by the ability of elites to bear the costs of conspicuous and wasteful advertising. The construction of Mayan calendrical monuments required tremendous expenditure for elites, as well as massive costs for maintaining the intellectual and craft-specialist infrastructure. The ability to bear these costs, because the signals are then impossible to fake, established hegemonic control of the landscape and its surrounding economic support zone. Neiman argues that investment in monuments is likely to have varied directly with both political influence across the landscape and the intensity of interpolity competition, which in turn were related to such factors as agricultural intensification, changing climatic conditions, and demographic pressures. He uses this model as a basis for evaluating how a decreasing signaling effect might be responsible for political disintegration, population decentralization and the collapse of monumental construction. It is not clear whether this particular hypothesis will gain much purchase in the ongoing debate on the Mayan collapse, but similar applications might lead to counterintuitive insights about certain cases of social elaboration and/or disintegration. For example, it may be that enduring hierarchy and despotic power are sometimes a *consequence* of symbolic wasteful display (such as elaborate architecture) rather than a cause.

Conclusions

As Shennan (2002) demonstrates, an archaeology reduced to arguments about how prehistoric people may have "perceived" their ancient landscapes, and who owns representations of the landscapes so perceived, fails to assess the validity of knowledge about the human past and its organization. Assessing the validity of reconstructions and interpretations of past landscape use requires an investigation of fundamental factors that influence the cumulative effects of human decision making. Human behavioral ecology does so with a Darwinian framework for evaluating hypotheses about patterned responses to socioecological variability. Because these patterns often leave material traces, archaeologists have been able to employ models from HBE to good effect for both describing the record and evaluating explanations of the behavior that may account for it (O'Connell 1995). In this chapter we have highlighted some HBE applications that, for over three decades, have provided a systematic framework for identifying and evaluating problems associated with understanding ancient landscape use.

Acknowledgments

Our thanks are due to Peter Dwyer for valuable critique of an earlier draft of this chapter and to Bruno David for his encouragement and patience. We were also greatly helped by discussion, ideas, and comments from James O'Connell, Eric A. Smith, Rebecca Bliege Bird, and Sarah Robinson.

References

Alexander, R. D. 1974. The evolution of social behavior. *Annual Review of Ecology and Systematics* 5: 325–83.

Barlow, K. R., and Metcalfe, D. 1996. Plant utility indices: Two Great Basin examples. *Journal of Archaeological Science* 23: 35–71.

Beaton, J. M. 1973. The nature of aboriginal exploitation of mollusk populations in southern California. Unpublished Master's thesis, University of California, Los Angeles.

Beck, C., and Jones, G. T. 1997. The terminal Pleistocene/early Holocene archaeology of the Great Basin. *Journal of World Prehistory* 11: 161–236.

Beck, C., Taylor, A. K., and Jones, G. T. 2002. Rocks are heavy: Transport costs and Paleoarchaic quarry behavior in the Great Basin. *Journal of Anthropological Archaeology* 21: 481–507.

Bettinger, R. L. 1991. *Hunter-Gatherers: Archaeological and Evolutionary Theory*. New York: Plenum Press.

Bettinger, R. L., Mahli, R., and McCarthy, H. 1997. Central place models of acorn and mussel processing. *Journal of Archaeological Science* 24: 887–99.

Binford, L. R. 1984. *Faunal Remains from Klasies River Mouth*. Orlando, FL: Academic Press.

Bird, D. W. 1996. Intertidal foraging strategies among the Meriam of the Torres Strait islands, Australia: An evolutionary ecological approach to the ethnoarchaeology of tropical marine subsistence. Unpublished Ph.D. thesis, University of California, Davis.

———. 1997. Behavioral ecology and the archaeological consequences of central place foraging among the Meriam, in C. M. Barton and G. A. Clark (eds.), *Rediscovering Darwin: Evolutionary Theory in Archaeological Explanation*, pp. 291–306. Washington, DC: Archaeological Papers of the American Anthropological Association 7.

Bird, D. W., and Bliege Bird, R. 1997. Contemporary shellfish gathering strategies among the Meriam of the Torres Strait Islands, Australia: Testing predictions of a central place foraging model. *Journal of Archaeological Science* 24: 39–63.

———. 2000. The ethnoarchaeology of juvenile foraging: Shellfishing strategies among Meriam children. *Journal of Anthropological Archaeology* 19: 461–76.

———. 2002. Children on the reef: Slow learning or strategic foraging? *Human Nature* 13: 269–98.

Bird, D. W., Bliege Bird, R., and Richardson, J. L. 2004. Meriam ethnoarchaeology: Shellfishing and shellmiddens. *Memoirs of the Queensland Museum*, Cultural Heritage Series 3(1): 183–97.

Bird, D. W., and O'Connell, J. F. 2006. Behavioral ecology and archaeology. *Journal of Archaeological Research* 14: 143–88.

Bird, D. W., Richardson, J. L., Veth, P. M., and Barham, A. J. 2002. Explaining shellfish variability in middens on the Meriam islands, Torres Strait Australia. *Journal of Archaeological Science* 29: 457–69.

Bliege Bird, R. 1999. Cooperation and conflict: The behavioral ecology of the sexual division of labor. *Evolutionary Anthropology* 8: 65–75.

Bliege Bird, R., and Bird, D. W. 2005. Human hunting seasonality in savannas and deserts, in D. Brockman and C. van Schaik (eds.), *Primate Seasonality*, pp. 243–66. Cambridge: Cambridge University Press.

Bliege Bird, R., and Smith, E. A. 2005. Signaling theory, strategic interaction, and symbolic capital. *Current Anthropology* 46: 221–48.

Bliege Bird, R., Smith, E. A., and Bird, D. W. 2001. The hunting handicap: Costly signaling in male foraging strategies. *Behavioral Ecology and Sociobiology* 50: 9–19.

Blurton Jones, N. G. 1986. Bushman birth spacing: A test for optimal interbirth intervals. *Ethology and Sociobiology* 7: 91–105.

———. 1987. Bushman birth spacing: Direct tests of some simple predictions. *Ethology and Sociobiology* 8: 183–203.

Blurton Jones, N. G., and Sibly, R. 1978. Testing adaptiveness of culturally determined behavior: Do Bushmen women maximize their reproductive success by spacing births widely and foraging seldom? in N. G. Blurton Jones and V. Reynolds (eds.), *Comparative Socioecology: The Behavioural Ecology of Humans and Other Mammals*, pp. 367–90. Oxford: Blackwell Scientific Publications.

Boone, J. L. 1998. The evolution of magnanimity: When is it better to give than to receive? *Human Nature* 9: 1–21.

Botkin, S. 1980. Effects of human exploitation on shellfish populations at Malibu Creek, California, in T. Earle and A. L. Christenson (eds.), *Modeling Change in Prehistoric Subsistence Economies*, pp. 121–39. New York: Academic Press.

Broughton, J. M. 2002. Prey spatial structure and behavior affect archaeological tests of optimal foraging models: Examples from the Emeryville Shellmound vertebrate fauna. *World Archaeology* 34: 60–83.

———. 2004. *Prehistoric human impacts on California birds: Evidence from the Emeryville Shellmound avifauna.* Ornithological Monographs 56. Washington, DC: American Ornithologists' Union.

Brown, J. L. 1964. The evolution of diversity in avian territorial systems. *Wilson Bull* 76: 160–69.

Cannon, M. D. 2000. Large mammal relative abundance in Pithouse and Pueblo Period archaeofaunas from southwestern New Mexico: Resource depression in the Mimbres-Mogollon? *Journal of Anthropological Archaeology* 19: 317–47.

Cannon, M. D. 2003. A model of central place forager prey choice and an application to faunal remains from the Mimbres Valley, New Mexico. *Journal of Anthropological Archaeology* 22: 1–25.

Chagnon, N. A., and Irons, W. G. 1979. *Evolutionary Biology and Human Social Behavior: An Anthropological Perspective*. North Scituate, MA: Duxbury Press.

Charnov, E. L. 1976. Optimal foraging, the marginal value theorem. *Theoretical Population Biology* 9: 367–90.

Charnov, E. L., and Orians, G. H. 1973. *Optimal Foraging: Some Theoretical Explorations*. Mimeo, Department of Biology, University of Utah, Salt Lake City.

de Boer, W. F. 2000. *Between the Tides: The Impact of Human Exploitation on an Intertidal Ecosystem, Mozambique*. Veenendaal: Universal Press.

Dugatkin, L. A., and Reeve, H. K. 1998. *Game Theory and Animal Behavior*. Oxford: Oxford University Press.

Elston, R. G., and Raven, C. (eds.). 1992. *Archaeological Investigations at Tosawihi, a Great Basin Quarry*. Silver City, NV: Intermountain Research.

Elston, R. G., and Zeanah, D. W. 2002. Thinking outside the box: A new perspective on diet breadth and sexual division of labor in the Pre-archaic Great Basin. *World Archaeology* 34: 103–30.

Erlandson, J. M., and Moss, M. L. 2001. Shellfish eaters, carrion feeders, and the archaeology of aquatic adaptations. *American Antiquity* 66: 413–32.

Freeman, M. 1971. A social and ecologic analysis of systematic female infanticide among the Netsilik Eskimo. *American Anthropologist* 73: 1011–18.

Getty, T. 1998. Handicap signaling: When fecundity and variability do not add up. *Animal Behavior* 56: 127–30.

Grafen, A. 1984. Natural selection, kin selection and group selection, in J. H. Krebs and E. B. Davies (eds.), *Behavioral Ecology: An Evolutionary Approach* (2nd ed.), pp. 62–84. Oxford: Blackwell Scientific.

———. 1990. Biological signals as handicaps. *Journal of Theoretical Biology* 144: 517–46.

Hamilton, W. D. 1964. The genetical evolution of social behavior. *Journal of Theoretical Biology* 7: 1–52.

Hawkes, K. 1990. Why do men hunt? Some benefits for risky strategies, in E. Cashdan (ed.), *Risk and Uncertainty in Tribal and Peasant Economies*, pp. 145–66. Boulder, CO: Westview Press.

———. 1996. Behavioral ecology, in D. Levinson and M. Ember, *Encyclopedia of Cultural Anthropology*, Vol. I., pp. 121–25. New York: Henry Holt and Company.

Hawkes, K., and Bliege Bird, R. 2002. Showing off, handicap signaling, and the evolution of men's work. *Evolutionary Anthropology* 11: 58–67.

Hawkes, K., O'Connell, J. F., and Blurton Jones, N. G. 1991. Hunting income patterns among the Hadza: Big game, common goods, foraging goals, and the evolution of the human diet. *Philosophical Transactions of the Royal Society*, series B, 334: 243–51.

———. 2001. Hadza hunting and the evolution of nuclear families. *Current Anthropology* 42: 681–709.

Hawkes, K., Hill, K., and O'Connell, J. F. 1982. Why hunters gather: Optimal foraging and the Ache of eastern Paraguay. *American Ethnologist* 9: 379–98.

Hildebrandt, W. R., and McGuire, K. R. 2002. The ascendance of hunting during the California Middle Archaic: An evolutionary perspective. *American Antiquity* 67: 231–56.

———. 2003. Large game hunting, gender-differentiated work organization and the role of evolutionary ecology in California and Great Basin prehistory. *American Antiquity* 68: 790–92.

Hutchinson, G. E. 1965. *The Ecological Theater and the Evolutionary Play*. New Haven, CT: Yale University Press.

Johnstone, R. A. 1997. The evolution of animal signals, in J. R. Krebs and N. B. Davies (eds.), *Behavioural Ecology: An Evolutionary Approach*, pp. 155–78. Oxford: Blackwell Publishing.

Jones, T. L. 1991. Marine resource value and the priority of coastal settlement: A California perspective. *American Antiquity* 56: 419–43.

Jones, T. L., and Richman, J. R. 1995. On mussels: *Mytilus californicus* as a prehistoric resource. *North American Archaeologist* 16: 33–58.

Kelly, R. L. 1995. *The Foraging Spectrum: Diversity in Hunter-Gatherer Lifeways*. Washington, DC: Smithsonian Institution Press.

Kennett, D. J. 2005. *The Island Chumash: Behavioral Ecology of a Maritime Society*. Berkeley and Los Angeles: University of California Press.

Kennett, D. J., and Kennett, J. P. 2000. Competitive and cooperative responses to climatic instability in southern California. *American Antiquity* 65: 379–95.

Kennett, D. J., and Winterhalder, B. (eds.). 2006. *Human Behavioral Ecology and the Origins of Food Production*. Berkeley and Los Angeles: University of California Press.

Klein, R. G. 1999. *The Human Career: Human Biological and Cultural Origins*. Chicago: University of Chicago Press.

Lee, R. B. 1979. The *!Kung San: Men, Women, and Work in a Foraging Society*. Cambridge: Cambridge University Press.

———. 1980. Lactation, ovulation, infanticide, and women's work: A study of hunter-gatherer population, in M. Cohen, R. Malpass, and H. Klien (eds.), *Biosocial Mechanisms of Population Regulation*, pp. 321–48. New Haven, CT: Yale University Press.

Lupo, K. D. 2001. On the archaeological resolution of body part transport patterns: An ethnoarchaeological example from East African hunter-gatherers. *Journal of Anthropological Archaeology* 20: 361–78.

MacArthur, R., and Pianka, E. 1966. On optimal use of a patchy environment. *American Naturalist* 100: 603–09.

Maynard Smith, J. 1974. *Models in Ecology.* Cambridge: Cambridge University Press.

———. 1978. Optimization theory in evolution. *Annual Review of Ecology and Systematics* 9: 31–56.

———. 1982. *Evolution and the Theory of Games.* Cambridge: Cambridge University Press.

Maynard Smith, J., and Harper, D. 2003. *Animal Signals.* Oxford: Oxford University Press.

McBrearty, S., and Brooks, A. S. 2000. The revolution that wasn't: A new interpretation of the origin of modern human behavior. *Journal of Human Evolution* 39: 453–563.

McGuire, K. R., and Hildebrandt, W. R. 2005. Rethinking Great Basin foragers: Prestige hunting and costly signaling during the Middle Archaic Period. *American Antiquity* 70: 695–712.

Metcalfe, D., and Barlow, K. R. 1992. A model for exploring the optimal tradeoff between field processing and transport. *American Anthropologist* 94: 340–56.

Neiman, F. 1997. Conspicuous consumption as wasteful advertising: A Darwinian perspective on spatial patterns in Classic Maya terminal monument dates, in C. M. Barton and G. A. Clark (eds.), *Rediscovering Darwin: Evolutionary Theory in Archaeological Explanation*, pp. 267–90. Washington, DC: Archaeological Papers of the American Anthropological Association 7.

O'Connell, J. F. 1995. Ethnoarchaeology needs a general theory of behavior. *Journal of Archaeological Research* 3: 205–55.

O'Connell, J. F., Hawkes, K., and Blurton Jones, N. G. 1988. Hadza hunting, butchering and bone transport and their archaeological implications. *Journal of Anthropological Research* 44: 113–61.

———. 1990. Reanalysis of large animal body part transport among the Hadza. *Journal of Archaeological Science* 17: 301–16.

O'Connell, J. F., and Marshall, B. 1989. Analysis of kangaroo body part transport among the Alyawara of central Australia. *Journal of Archaeological Science* 16: 393–405.

Orians, G. H., and Pearson, N. E. 1979. On the theory of central place foraging, in D. J. Horn, R. D. Mitchell, and C. R. Stairs (eds.), *Analysis of*

Ecological Systems, pp. 154–77. Columbus: Ohio State University Press.

Parker, G. A. 2005. In press. Behavioral ecology: Natural history as science, in L. W. Simmons and J. Lucas (eds.), *Essays in Animal Behavior.* Oxford: Elsevier.

Raab, L. M. 1992. An optimal foraging analysis of prehistoric shellfish collecting on San Clemente Island, California. *Journal of Ethnobiology* 12: 63–80.

Rick, T. C., and Erlandson, J. M. 2000. Early Holocene fishing strategies on the California coast: Evidence from CA-SBA-2057. *Journal of Archaeological Science* 27: 621–33.

Rogers, A. R. 1990. Group selection by selective emigration: The effects of migration and kin structure. *American Naturalist* 135: 398–413.

Rogers, A. R., and Broughton, J. 2001. Selective transport of animal parts by ancient hunters: A new statistical method and an application to the Emeryville Shellmound fauna. *Journal of Archaeological Science* 28: 763–73.

Shennan, S. 2002. *Genes, Memes and Human History: Darwinian Archaeology and Cultural Evolution.* London: Thames and Hudson.

Smith, E. A. 1991. *Inujjuamiut Foraging Strategies: Evolutionary Ecology of an Arctic Hunting Economy.* Hawthorne: Aldine de Grutyer.

———. 2000. Three styles in the evolutionary study of human behavior, in L. Cronk, W. Irons, and N. Chagnon (eds.), *Human Behavior and Adaptation: An Anthropological Perspective*, pp. 27–46. Hawthorne: Aldine de Grutyer.

Smith, E. A., and Bliege Bird, R. 2000. Turtle hunting and tombstone openings: Generosity and costly signaling. *Evolution and Human Behavior* 21: 245–61.

Smith, E. A., Bliege Bird, R., and Bird, D. W. 2003. The benefits of costly signaling: Meriam turtle hunters and spearfishers. *Behavioral Ecology* 14: 116–26.

Smith, E. A., and Winterhalder, B. (eds.). 1992. *Evolutionary Ecology and Human Behavior.* Hawthorne: Aldine de Grutyer.

Sosis, R. 2000. Costly signaling and torch fishing on Ifaluk Atoll. *Evolution and Human Behavior* 21: 223–44.

———. 2003. Why aren't we all Hutterites? Costly signaling theory and religious behavior. *Human Nature* 14: 91–127.

Sosis, R., and Bressler, E. 2003. Cooperation and commune longevity: A test of the costly signaling theory of religion. *Cross-Cultural Research* 37: 211–39.

Stephens, D. W., and Krebs, J. R. 1986. *Foraging Theory.* Princeton, NJ: Princeton University Press.

Stiner, M. C., and Munro, N. D. 2002. Approaches to prehistoric diet breadth, demography and prey ranking systems in time and space. *Journal of Archaeological Method and Theory* 9: 181–214.

Stiner, M. C., Munro, N. D., and Surovell, T. A. 2000. The tortoise and the hare: Small game use, the broad-spectrum revolution, and paleolithic demography. *Current Anthropology* 41: 39–73.

Stiner, M. C., Munro, N. D., Surovell, T. A., Tchernov, E., and Bar-Yosef, O. 1999. Paleolithic population growth pulses evidenced by small animal exploitation. *Science* 283: 190–94.

Thomas, F. R. 2002. An evaluation of central-place foraging among mollusk gatherers in Western Kiribati, Micronesia: Linking behavioral ecology with ethnoarchaeology. *World Archaeology* 34: 182–208.

Tinbergen, N. 1963. On aims and methods of ethology. *Zeitschrift fur Tierpsychologie* 20: 404–33.

Trivers, R. L. 1971. The evolution of reciprocal altruism. *Quarterly Review of Biology* 46: 35–57.

———. 1972. Parental investment and sexual selection, in B. Campbell (ed.), *Sexual Selection and the Decent of Man*, pp. 139–79. Chicago: Aldine.

Ugan, A. 2005. Does size matter? Body size, mass collecting, and their implications for understanding prehistoric foraging behavior. *American Antiquity* 70: 75–90.

Williams, G. C. 1966. *Adaptation and Natural Selection: A Critique of Some Current Evolutionary Thought*. Princeton, NJ: Princeton University Press.

Wilmsen, E. N. 1973. Interaction, spacing behavior and the organization of hunting bands. *Journal of Anthropological Research* 29: 1–31.

Wilson, D. S. 1998. Hunting, sharing, and multilevel selection: The tolerated-theft model revisited. *Current Anthropology* 39: 73–97.

Winterhalder, B. 1981. Foraging strategies in the boreal forest: An analysis of Cree hunting and gathering, in B. Winterhalder and E. A. Smith (eds.), *Hunter-Gatherer Foraging Strategies: Ethnographic and Archaeological Analyses*, pp. 66–98. Chicago: University of Chicago Press.

Winterhalder, B., and E. A. Smith (eds.). 1981. *Hunter-Gatherer Foraging Strategies: Ethnographic and Archaeological Analyses*. Chicago: University of Chicago Press.

Zahavi, A. 1975. Mate selection: Selection for a handicap. *Journal of Theoretical Biology* 53: 205–14.

Zahavi, A., and Zahavi, A. 1997. *The Handicap Principle: A Missing Piece of Darwin's Puzzle*. Oxford: Oxford University Press.

Zeanah, D. 2000. Transport costs, central place foraging and hunter-gatherer alpine land use strategies, in D. Madsen and M. Metcalf (eds.), *Intermountain Archaeologies*, pp. 1–14. Salt Lake City: University of Utah Anthropological Papers 122.

———. 2004. Sexual division of labor and central place foraging: A model for the Carson Desert of Western Nevada. *Journal of Anthropological Anthropology* 23: 1–32.

40

Desert Landscapes in Archaeology: A Case Study from the Negev

Steven A. Rosen

Harsh and extreme environments by definition, deserts pose special challenges to the humans living in and passing through them. In ancient times, they constituted barriers and borders, often delineating the edges of states and culture areas and offering refuge and isolation both to marginal persons and to societies on the periphery of larger polities. Although deserts are defined environmentally, sedentary societies have reified desert peoples, on the one hand exalting the exotic, autonomous, and romantic in deserts and their societies, and on the other, fearing and reviling them for their perceived barbarity. Deserts have served as the mythological crucible for all three monotheistic religions, and the inhabitants of the deserts have been the noble savages of historical writing since early times. The challenge of a landscape archaeology of deserts is in finding a balance between the reductionism of human ecology in an extreme environment (cf. Layton and Ucko 1999) and the mythological cultural perceptions imposed by Western society (Bender 1993) on desert societies, to understand desert cultural landscapes without reducing them to mechanistic adaptations. Although aridity is the primary defining characteristic, classifications of deserts have emphasized different parameters, usually drawn from physical geography or climatology. Thus, various indices, such as those devised by Köppen, Thornthwaite, Meigs and others (see summaries in Bruins and Berliner 1998; Evenari, Shanan, and Tadmor 1982: 29–31; Spellman 2000) based on relative measures of rainfall, evaporation, and temperature, have been used both to define aridity thresholds and to categorize different types of deserts (for example, hyperarid, arid, semiarid, subhumid). Depending on the specific parameters, the total dry lands constitute up to 33% of the world's land surface (e.g., Bruins and Berliner 1998), making the study of human exploitation of deserts an important endeavor both for understanding the past and planning for the future.

Two primary factors determine desert formation. The desert belts of the subtropics, including the Saharo-Arabian deserts, those of Asia, the American Southwest, and southern Africa, and Australia, are the result of global atmospheric pressure systems, leaving the subtropic landmasses beneath long-term low pressure systems and reducing precipitation. Rainshadow deserts, such as those of the west coast of Chile and the Jordan Rift system, result from atmospheric pressure differentials caused by local or regional topography (e.g., Amiran and Wilson 1973; Hills 1966; Spellman 2000). The basic landforms themselves are, more often than not, relics of earlier environmental systems (e.g., Cook, Warren, and Goudie 1993).

In addition to these physical environmental approaches, desertification (the transformation of previously nondesert zones into deserts), has also focused on ecological and social factors, especially various forms of overexploitation of peripheral lands, affecting basic ecological balances and resulting in dynamic instabilities and environmental deterioration. Since such effects are rarely reflected in the indices described above, measures of ecological diversity, biomass, and economic productivity are usually employed to define and measure desertification (e.g., Blaikie and Brookfield 1987; Glantz 1994; Thomas and Middleton 1994; and references in all).

For archaeologists, a good working definition, based on social and economic attributes directly tied to climate and geography, is the nonpracticability of dry farming, that is, farming without irrigation systems. The standard threshold for wheat dry farming is the 250 mm isohyet (rainfall line) (Grigg 1974: 262), and somewhat lower for barley, defining subsistence levels of agriculture without irrigation.

The goal of this chapter is the presentation of a case study from the Negev, exploring some of the issues specific to landscape archaeology in the desert. Emphasis is placed on the difference between indigenous desert settlement, the adaptation to the desert, and what might be termed intrusive settlement, where the basic settlement system derives from some other cultural and environmental system.

The Protohistoric Negev as a Case Study in Desert Landscape Archaeology

Environmental Background

The Negev can be defined geographically (e.g., Danin 1983; Evenari, Shanan, and Tadmor 1982: 29–76; Orni and Efrat 1980: 15–34) both as the linkage between the Arabian Desert in the east and the Sinai and Saharan Deserts in the west, and as the transition from the Mediterranean and steppe zones in the north to the Saharo-Arabian true desert zone in the south (Figure 40.1). It comprises an elongate isosceles triangle with its apex in the south at Aqaba/Eilat and the Gulf of Aqaba and its base in the north running from the southeast corner of the Mediterranean Sea to roughly the middle of the Dead Sea. Its eastern edge is defined by the deep Syro-African Rift (Wadi Araba, the Arava Valley), extending from the Dead Sea to the Red Sea; the western edge is defined geographically by the transition from the highlands of the Central Negev to the plains of Central Sinai. A clear climatic gradient can be traced from north to south, reflected in yearly average rainfall which declines from ca. 200 mm/year in Beersheva in the northern Negev, to ca. 100 mm/year in the central Negev, and to ca. 25 mm/year in the Eilat region, a linear distance of only about 200 km. Similarly, phytogeographic zones shift from the Mediterranean zone, in the hills just north of Beersheva, to the steppic Irano-Turanian zone characteristic of the central Negev, and to the Saharo-Arabian zone in the southern Negev. Enclaves of tropical Sudano-Deccanian vegetation are found in the Rift valley, relics surviving in microenvironments around natural springs. Local topography and distance from the Mediterranean also affect rainfall and vegetation patterns. Notably, virtually all rainfall originates in Mediterranean climatic systems and rains fall exclusively in the winter months; Indian Ocean summer monsoons rarely penetrate as far north as the Negev.

These physical aspects of the desert are crucial for understanding both long-term trends in human settlement, and for comprehending the specifics of each settlement system. Settlement in the Negev can be seen as a shifting mosaic, adapting to varying trends in social, technological, and environmental factors.

Basic Themes in Landscape Archaeology in the Negev

Two general patterns of landscape-culture relations can be defined in the Negev, and perhaps for deserts in general. The first comprises the set of local or indigenous adaptations, seemingly more integrated and perhaps more consonant with local conditions, and the second, those imported from other areas, in the case of the Negev, especially the Mediterranean zone, and imposed on the desert landscape. In defining these ends of the landscape-culture continuum, one risks crude simplification in application to any specific cultural complex; however, as a starting point such an approach can provide critical insights into the nature and development of cultural landscapes in the desert.

Indigenous cultural landscapes in the Negev are marked by three patterns or trends:

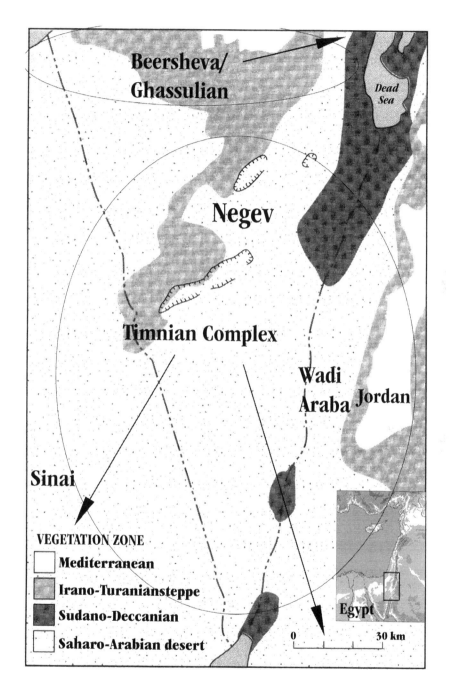

Figure 40.1 Map of Negev and surrounding areas with phytogeographic zonation and culture areas.

1. Habitation sites are relatively small, tending to the extensive rather than intensive and reflecting the small size and the mobility of the human populations;

2. Sites and settlement systems are specifically adapted to local conditions, both in terms of raw materials and general construction and layout. Indigenous landscapes reflect corporate organization of space, with less evidence for private ownership. Mythological landscapes (cf. Bradley 2000: 13) integrate the natural environment with the constructed one;

3. Settlement and culture systems show long-term continuities, reflected in architecture, site placement, and material culture. This continuity is not to be confused with stasis (cf. David 2002: 1–9) but rather to be seen as process (cf. Hirsch 1995).

In contrast, *imported* landscapes tend to show contrasting patterns:

1. Sites are often (but not always) of a larger scale, reflecting greater economic input from outside the desert, capable of supporting larger and denser populations, usually sedentary;

2. Sites and landscapes reflect cultural perceptions deriving from other regions, for example, in exotic architectural types, road grids, and symbol systems. Adaptations to local conditions are secondary in terms of the conception of the features. Corporate ownership is not a standard feature of these landscapes;

3. Settlement systems tend to be episodic, depending on the long-term commitment of the donor society, rather than on the internal viability of the desert population.

Again, it is important to stress that particular societies will reflect these trends to greater or lesser degrees, depending on their specific historical circumstances. Chronologically, the younger the period or culture, the greater the impact of the external world on the desert systems, with the consequent blurring of the distinctiveness of these patterns over time.

Culture and Landscape in the Proto-Historic Periods

The archaeological record of the Negev extends back to Lower Palaeolithic times in the presence of both chopping tools scatters and surface hand axe concentrations in various locales. *In situ* sites from later periods of the Palaeolithic (Middle, Upper, and Epi-) are common, reflecting early hunter-gatherer societies. Although the first hints of desert-sown contrasts can be traced between desert and Mediterranean zone Natufian settlements, distinctive desert adaptations appear most obviously in the Pre-Pottery Neolithic B (PPNB), well evident in the apposition between village agricultural societies in the north and the continued dominance of hunting gathering in the desert.

The Timnian culture complex (e.g., Kozloff 1972/3, 1981; Ronen 1970; Rothenberg 1979; Rothenberg and Glass 1992) dominates the central and southern Negev and Sinai, and southern Jordan in the period following the final collapse of the Pre-Pottery Neolithic B system ca. 6,900 cal B.C. It emerges roughly a millennium later in the evolution of a set of traits deriving especially from the adoption of domestic sheep and goat in the 7th millennium cal B.C. and includes pen-and-attached-room architecture, fields of tumuli and *nawami*s (cylindrical corbel arched burial structures), open air shrines and shrine complexes, and a specific material culture assemblage characterized especially by small arrowheads, transverse arrowheads, tabular scrapers and knives, simple blade technologies, a dominant *ad hoc* tool industry, and globular holemouth jar cooking vessels (Table 40.1: e.g., Henry 1992, 1995: 353–73; Rosen 2002). Although material culture evolves and transforms, and settlement distributions expand and contract over the course of the next three millennia, a general Timnian continuum can be traced through the end of the 3rd millennium cal B.C.

Archaeologically, the Timnian landscape can be divided roughly into three parts: (1) tribal space; (2) domestic space; and (3) ritual space. Tribal space comprises the general division of the desert landscape into tribal territories, presumably a form of corporate ownership or control, marked by the construction of fields of tumuli and *nawami*s (Figure 40.2; cf. Bar-Yosef et al. 1983; Haiman 1993; Rosen and Rosen 2003). To judge roughly by the general distribution of such sites, territories would have covered hundreds of square kilometers. The apparent formalization of tribal territoriality contrasts with the absence of any such markers in the preceding desert Pre-Pottery Neolithic B, notably a hunter-gatherer society. The construction of these mortuary monuments (and the often associated shrine complexes), in some cases achieving megalithic proportions, indeed suggests levels of political

Table 40.1 Chronology and basic features of the Timnian Complex as opposed to northern complexes.

Absolute Chronology B.C. Cal	Culture	Material Culture						Features	Northern culture-chronology
		axes	tabular Scrapers	arrowheads large	small	transverse	bifacial knives		
2000	terminal		(incised)					large cluster architecture	EBIV/MBI
					(lunates)				III
3000	late								II Early Bronze Age
									I
4000	Timnian							Nawamis	
	middle								Ghassulian
								metal ceramics	Besorian Qatifian
5000	early							tumuli	Wadi Raba
	~~ Early Pottery Neolithic ~~~~							pen and room structures shrines, kites	Lodian
6000								domestic sheep/goat	Yarmukian
	Tuwailan ~~~~								PPNC
7000	PPNB								PPNB

and social organization higher than in preceding periods in the deep desert (cf. European megalithic systems [e.g., Kinnes 1982; Renfrew 1984]). Thus, the creation of landscapes of territoriality can be best attributed to the adoption of domestic herd animals and the ramifications of this basic shift in subsistence (cf. Ingold 1980). The porosity of the tribal landscape is reflected in the movement of material culture between different regions and the assumed seasonal movements of herds across tribal areas.

Domestic space is characterized by a new architectural type, the pen-and-attached-room complex or compound (Figure 40.3). Although room construction, consisting of shallow pits, stone foundations or lining, and organic superstructures, is similar to that of the preceding PPNB, the enclosures to which the rooms are attached are a new feature and have usually been interpreted as animal pens (e.g., Haiman 1992; Kozloff 1981; Rosen 2003), although their functions are probably multifaceted. Conceptually, the contrast between the PPNB clusters of pit structures and the centrality of the enclosures

in the succeeding Timnian suggests basic contrasts in the use of space. Speculating, the new spatial pattern perhaps reflects the centrality of the herds in Timnian society, either ideologically, or perhaps simply from the perspective of protection of the animals. Furthermore, the actual construction of functional distinctions and specific architectural areas also suggests increasing organizational complexity. The incorporation of tumuli, not necessarily mortuary in function in these contexts, into Timnian domestic architecture (Haiman 1992, 1993), at least in the later stages of the complex, adds to the contrast with preceding periods, again suggesting changes in conceptions of domestic space perhaps tied to evolving social norms and technologies (cf. Kent 1984).

Ritual space is reflected archaeologically in the construction of mortuary and shrine complexes. Fields of mortuary structures have already been mentioned above in defining tribal landscapes, but they require more elaboration as ritual space. *Nawamis* are apparently family tombs clustered in fields suggesting tribal affinities, located

(A)

Figure 40.2 *Nawami*s field from (A) Gebel Gunna, (B) Sinai, (C) Middle Timnian (photographs by B. Saidel).

in central Sinai and dating to the early-middle 4th millennium cal B.C., and perhaps somewhat earlier. Entrances align with the setting sun, reflecting ideological aspects, while measurement of the specific azimuths of the openings indicate seasonal modalities of use (Bar-Yosef et al. 1983). The tumulus and shrine field at Ramat Saharonim, in the central Negev (Porat et al. 2006; Rosen and Rosen 2003; Rosen et al. 2007) dates to the late 6th and early 5th millennia cal B.C., somewhat earlier than the *nawami*s fields, but shows a similar clustered aspect, as do other tumulus fields in the Negev surveyed by Haiman (1993) and dated to the Early Bronze Age, the late 4th and early 3rd millennium cal B.C. All these mortuary fields project in the landscape, built on plateaus, cliffs, or ridges, and visible from a distance. Within the fields themselves, no (contemporary) domestic sites or even artifacts are found, suggesting activity distinctions, although the *nawami*s contain grave goods which are indeed domestic. Haiman (1993) has noted a geographical association between occupation sites and some tumulus fields, suggesting group identity linkages, but the spatial distinctions

are clear. The presence of tumuli in some Early Bronze Age occupation sites suggests either a different function, or different symbolic meaning in these particular contexts. Notably, analyses of the material contents of small rooms attached to tumuli within habitation sites indicate functional distinctions (Rosen 1997: 124–25).

Like the mortuary systems, and occasionally directly linked to them (e.g., Rosen and Rosen 2003), the shrine systems reflect a cultural landscape in marked contrast to the domestic sphere (Figure 40.4). Shrines appear in the desert in the second half of the 6th millennium cal B.C. (Avner 1998; Avner, Carmi, and Segal, 1994; Porat et al. 2006). They seem to continue in use for hundreds of years, and the construction of later cultic structures in association with them (Rosen et al. 2007; cf. Avner 1984, 1998 for cult continuities) implies long-term retention of cult status for many of the sites. The total absence of domestic assemblages from the cult complexes confirms the conceptual polarity vis-à-vis domestic structures, reflecting major differences in activities carried out in the different spheres. In particular, the episodic nature of cult activities and the use of special

(B)

(C)

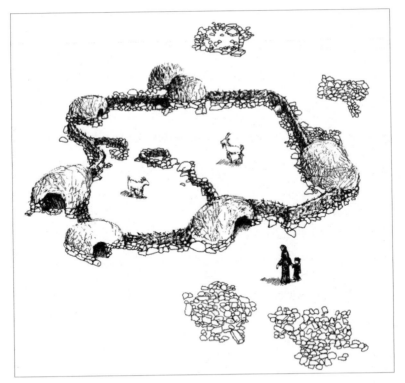

(A)

(B)

Figure 40.3 (A) Pen and room architecture from the Camel site (drawing by Helena Sokolskaya);
(B) photograph after excavation with tumulus locus 49 in foreground, room locus 41 in center, and enclosure
loci 31 and 34 on either side (S. Rosen); (C) site plan (drawing by Patrice Kaminsky).

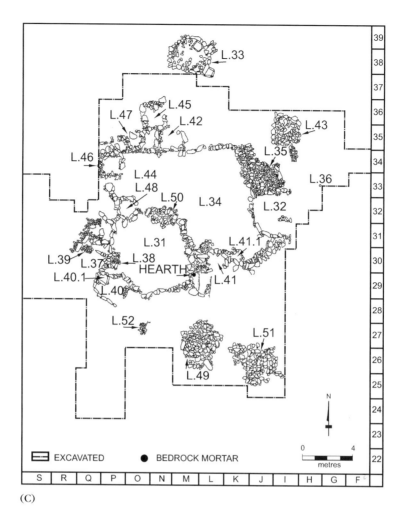

(C)

implements, probably removed from the site after use, contrast with the intensive and longer term occupation of domestic sites with domestic discard patterns and accumulation of waste. Furthermore, the shrines are clearly and deliberately constructed to integrate and coordinate with both features in the landscape, and with astronomical phenomena (Figure 40.5). Many of the shrines from this period are generally aligned with the setting sun of the summer solstice (Rosen and Rosen 2003; Rosen et al. 2007; *per contra* Avner 2002), or in some cases with cardinal directions (Avner 1984). No less important, specific landscape alignments are also integrated into the construction of the shrines. The directionality and placement of Shrine 3 at Ramat Saharonim, directly facing an extinct black volcano, with the view set up between two low hills and the summer solstice setting sun offset just to the south, is a good example of such considerations. Such constructed landscapes obviously reflect complex cosmologies well integrated into local landscapes and features.

Contemporary imported landscapes exhibit important contrasts with those of the Timnian complex. The Ghassulian culture complex overlaps with only part of the Timnian chronologically but provides a good comparative perspective for viewing the different types of desert cultural landscapes. The expansion of this culture complex in the Chalcolithic period (ca. 4400–4000 cal B.C.) into the Beersheva Basin in the northern Negev is not merely the result of a climatic amelioration that permitted the expansion; it also reflects the expansion of a Mediterranean conception of landscape and settlement into the steppe zone. These villages, consisting of above-ground rectilinear stone architecture and subterranean rooms and galleries, were founded on primary stream terraces and were based economically on simple floodwater

Figure 40.4   Shrine 3 at Ramat Saharonim, Early Timnian (S. Rosen). Note contrast between the microenvironment of the shrine area and surrounding areas.

farming and tethered sheep/goat pastoralism (Figure 40.6).

Landscape zonation in the Ghassulian takes a different form from that of the Timnian. Levy (1986, 1995) has suggested that the northern Negev landscape was one of hierarchically organized central villages with satellite sites, which he interprets as reflecting a chiefdom level of social organization. Thus, he envisages a patchwork of territories of a significantly lesser spatial order than that of the Timnian. Even rejecting Levy's reconstructions (e.g., Gilead 1988, 1993), the sedentary landscape of the Ghassulian village system differed from that of the Timnian, especially in the permanence of the occupation sites and all that this implies. Beyond this basic contrast, a pastoral hinterland can also be defined for the Beersheva-Ghassulian horizon, most especially seen in the small encampments 20 km south of Beersheva in the Nahal Sekher area (e.g., Gilead and Goren 1986; Goren and Gilead 1988). Thus, where the indigenous Timnian landscape at its largest scale seems to be divided tribally into large regions, the Ghassulian shows both a functional distinction between pastoral hinterland and agricultural interior, and political or social distinctions between villages.

On a smaller scale, habitation sites are located exclusively along major drainages (Levy 1983), and floodwater farming seems to have been practiced primarily on wadi floodplains and low terraces (Rosen 1987; Rosen and Weiner 1994). Thus, within the agricultural interior, space can be divided into domestic building space, agricultural fields, and the interfluves, on which grazing and other activities probably occurred. In particular, the presence of a Chalcolithic cemetery at Mezad Aluf (Levy and Alon 1985), on the bluff across the wadi from the large village at Shiqmim, attaches mortuary ritual directly to central places, in contrast to at least some of the tumulus and *nawami*s fields, removed significantly from their habitation sites. In fact, the large-scale spatial distinction between cult and domestic as present in the Timnian is present only in the temple at Ein Gedi (Ussishkin 1980), in the Judean Desert, for the Ghassulian, and cult space for the most part seems to be incorporated into the settlements, as at Gilat (Alon and Levy 1989).

Figure 40.5   The solstice setting sun opposite Ramat Saharonim Shrine 4. Note the low hills on either side of the view with cliff walls of the Ramon Crater in the far background. Early Timnian (S. Rosen).

At this smallest scale of domestic landscape, or perhaps better, village landscape, the broad room architecture of Ghassulian villages is a clear import from earlier Late Neolithic architectures of the Mediterranean zone. The underground rooms and galleries at virtually all the primary village sites have been interpreted in various ways. Without entering the debate on their function(s), their integration into the village space indicates either that domestic functions such as storage (e.g., Gilead 1988), habitation (e.g., Perrot 1955), and sanctuary (Levy 1995: 240) were part and parcel of the village landscape, or that ritual and mortuary functions were also integrated into that landscape. Either way, the village landscape seems to reflect a broader range of more disparate functions than the domestic landscape of the Timnian, clearly a result of the fundamental differences between sedentary village and mobile pastoral lifeways.

Finally, the Ghassulian in the northern Negev is ultimately a settlement episode, with virtually no descendents in the vicinity. In the first centuries of the 4th millennium cal B.C., the villages of the Beersheva Basin are abandoned (Gilead 1994), and region is not resettled until the end of the millennium, well into the Early Bronze Age. In this context, it is worth noting that the Timnian complex extends through the Early Bronze Age and that Mediterranean Early Bronze Age urban societies expanded into the deep desert zone (e.g., Beit-Arieh 1986), even more so than the Ghassulian. Without detailing this expansion, this society, too, created its own cultural landscapes in the desert, one consisting of both a desert urban center at Arad (e.g., Amiran 1978; Amiran, Ilan, and Sebbane 1997; Finkelstein 1995), and far-flung trade outposts (Amiran, Ilan, and Sebbane 1973). Like the Ghassulian, it was short-lived.

Conclusions

In the succeeding millennia, with the development of sophisticated technologies of water management and the rise of states and empires, incursions into the desert increased both in number and amplitude. The hill fortresses and farming hamlets that dot the Iron Age landscape in the Central Negev (Cohen 1980) constitute another

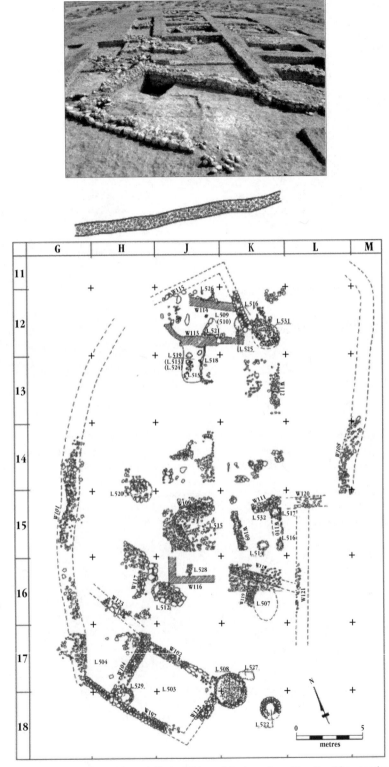

Figure 40.6 Rectilinear Beersheva-Ghassul Chalcolithic architecture from Nevatim (drawing by Isaac Gilead).

short-lived episode superimposed on the desert landscape. Similarly, the rise of desert urbanism, in the middle of the 1st millennium A.D., was based both on the expansion of the Roman Empire into the desert and the development of special irrigation technologies enabling the adoption of Mediterranean agricultural complexes into desert farming systems (e.g., Shershefsky 1991). These incursions into the desert recreated urban and village landscapes derived from the Mediterranean zone, especially evident not only in domestic architecture, such as the Iron Age pillared building or the Greek/Roman villa, but also in the imposition of a conceptually different order on the environment. Thus, the Roman/Byzantine classification of Shivta (Subeita) as a village, but Halutza (Elusa) as a city, derives from a Mediterranean perception of urban political hierarchy, with little relation to the actuality of desert settlement.

However, indigenous landscapes complement these imported landscapes in the form of tribal territories, local settlement systems, indigenous architectural forms, and the likely scheduling of seasonal movement through the settled zone. And with the economic dependence of pastoral nomadic societies on their settled cousins, these indigenous pastoral landscapes, too, collapsed with the decline in imperial fortunes.

Modern times show similar cultural landscape contrasts in the desert, the marked difference between modern government-sponsored Israeli settlements and the camps and villages of the Bedouin being a prime example. These contrasts and disputes reflect fundamental differences in the conception of ownership, struggles over control of resources, and general ethnic tensions, and all are reflected in perceptions of landscape.

Acknowledgments

I am grateful to Hagit Nol for her help organizing and reviewing the research for this paper. Isaac Gilead, Benajmin Saidel, and Yuval Yekutieli made comments on early drafts. Both Isaac Gilead and Benjamin Saidel graciously allowed me use of some of the illustrative materials. David Ilan discussed aspects of Chalcolithic mortuary behavior with me.

References

Alon, D., and Levy, T. E. 1989. The archaeology of cult and the Chalcolithic sanctuary at Gilat. *Journal of Mediterranean Archaeology* 2: 163–221.

Amiran, D. H. K., and Wilson, A. W. (eds.). 1973. *Coastal Deserts: Their Natural and Human Environments.* Tucson: University of Arizona.

Amiran, R. 1978. *Early Arad.* Jerusalem: Israel Exploration Society.

Amiran, R., Beit Arieh, I., and Glass, J. 1973. The Interrelationship between Arad and Sites in the Southern Sinai in the Early Bronze II. *Israel Exploration Journal* 23: 33–38.

Amiran, R., Ilan, O., and Sebbane, M. 1997. *Canaanite Arad: Gateway City to the Wilderness.* Jerusalem: Israel Antiquities Authority.

Avner, U. 1984. Ancient cult sites in the Negev and Sinai Deserts. *Tel Aviv* 11: 115–31 (in Hebrew).

———. 1998. Settlement, agriculture, and paleoclimate in 'Uvda Valley, southern Negev Desert, 6th–3rd millennia B.C., in A. Issar and N. Brown (eds.), *Water, Environment and Society in Times of Climatic Change*, pp. 147–202. Amsterdam: Kluwer.

———. 2002. Studies in the material and spiritual culture of the Negev and Sinai populations, during the 6th–3rd millennia B.C. Unpublished Ph.D. thesis, Hebrew University.

Avner, U., Carmi, I., and Segal, D. 1994. Neolithic to Bronze Age settlement of the Negev and Sinai in light of radiocarbon dating: A view from the southern Negev, in Bar- O. Yosef and R. Kra (eds.), *Late Quaternary Chronology and Paleoclimates of the Eastern Mediterranean*, pp. 26–300. Journal of Radiocarbon in Association with the American School of Prehistoric Research, Peabody Museum, Harvard University.

Bar-Yosef, O., Hershkovitz, I., Arbel, G., and Goren, A. 1983. Orientation of Nawamis entrances in southern Sinai: Expressions of religious belief or seasonality? *Tel Aviv* 10: 52–60.

Beit-Arieh, I. 1986. Two cultures in south Sinai in the third millennium B.C. *Bulletin of the American Schools of Oriental Research* 263: 27–54.

Beit-Arieh, I., and Gophna, R. 1976. Early Bronze II sites in Wadi el-Qudeirat (Kadesh Barnea). *Tel Aviv* 3: 142–50.

Bender, B. 1993. Introduction: Landscape—Meaning and action, in B. Bender (ed.), *Landscape: Politics and perspectives*, pp. 1–27. Oxford: Berg.

Blaikie, P., and Brookfield, H. (eds.). 1987. *Land Degradation and Society.* London: Methuen.

Bradley, R. 2000. *An Archaeology of Natural Places.* Routledge: London.

Bruins, H. J., and Berliner, P. R. 1998. Bioclimatic aridity, climatic variability, drought and desertification: Definitions and management options, in H. J. Bruins and H. Lithwick (eds.), *The Arid Frontier: Interactive management of environment and development*. Dordrecht: Kluwer Academic Publishers.

Cohen, R. 1980. The Iron Age fortresses in the central Negev. *Bulletin of the American Schools of Oriental Research* 236: 61–79.

Cook, R., Warren, A., and Goudie, A. 1993. *Geo-morphology in Deserts*. London: University College London Press.

Danin, A. 1983. *Desert Vegetation of Israel and Sinai*. Jerusalem: Cana.

David, B. 2002. *Landscapes, Rock Art, and the Dreaming*. London: Leicester University Press.

Evenari, M., Shanan, L., and Tadmor, N. 1982. *The Negev: The Challenge of a Desert*. Cambridge, MA: Harvard University Press.

Finkelstein, I. 1995. *Living on the Fringe*. Sheffield: Sheffield Academic Press.

Gilead, I. 1988. The Chalcolithic period in the Levant. *Journal of World Prehistory* 2: 397–443.

———. 1993. Socio-political organization in the northern Negev at the end of the Chalcolithic period, in A. Biran and J. Aviram (eds.), *Biblical Archaeology Today*. 1990, Pre-Congress Symposium Supplement, pp. 82–97. Jerusalem: Israel Exploration Society.

———. 1994. The history of Chalcolithic settlement in the Nahal Beer Sheva area: The radiocarbon aspect. *Bulletin of the American Schools of Oriental Research* 296: 1–13.

Gilead, I., and Goren, Y. 1986. Stations of the Chalcolithic period in Nahal Sekher, northern Negev. *Paleorient* 12: 83–90.

Glantz, M. H. (ed.). 1994. *Drought Follows the Plough*. Cambridge: Cambridge University Press.

Goren, Y., and Gilead, I. 1988. Quaternary environment and man at Nahal Sekher, northern Negev. *Mitkeufat Haeven* 19: 66–79.

Grigg, D. B. 1974. *The Agricultural Systems of the World: An Evolutionary Approach*. Cambridge: Cambridge University Press.

Haiman, M. 1992. Sedentism and pastoralism in the Negev highlands in the early Bronze Age: Results of the Western Negev Highlands Emergency Survey, in O. Bar-Yosef and A. M. Khazanov (eds.), *Pastoralism in the Levant: Archaeological Materials in Anthropological Perspective*, pp. 93–105. Madison, WI: Prehistory Press.

———. 1993. An early Bronze Age cairn field at Nahal Mitnan. *Atiqot* 22: 49–61.

Henry, D. O. 1992. Seasonal movements of fourth-millennium pastoral nomads in Wadi Hisma. *Studies in the History and Archaeology of Jordan* IV: 137–41.

———. 1995. *Prehistoric Cultural Ecology and Evolution: Insights from Southern Jordan*. New York: Plenum Press.

Hirsch, E. 1995. Introduction, landscape: Between place and space, in E. Hirsch and M. O'Hanlon (eds.), *The Anthropology of Landscape*, pp. 1–30. Oxford: Clarendon Press.

Ingold, T. 1980. *Hunters, Pastoralists and Ranchers*. Cambridge: Cambridge University Press.

Kent, S. 1984. *Analyzing Activity Areas: An Ethno-archaeological Study of the Use of Space*. Albuquerque: University of New Mexico Press.

Kinnes, I. 1982. Les fouaillages and megalithic origins. *Antiquity* 56: 24–30.

Kozloff, B. 1972/3. A brief note on the lithic industries of Sinai. *Museum Ha'aretz Yearbook* 15/16: 35–49.

———. 1981. Pastoral nomadism in Sinai: An ethnoarchaeological study. *Production Pastorales et Societe: Bulletin d'Ecologie et d'Anthropologie des Societes Pastorales* 8: 19–24.

Layton, R., and Ucko, P. J. 1999. Introduction: Gazing on the landscape and encountering the environment, in P. J. Ucko and R. Layton (eds.), *The Archaeology and Anthropology of Landscape*, pp. 1–20. London: Routledge.

Levy, T. E. 1983. The Emergence of specialized pastoralism in the southern Levant. *World Archaeology* 15: 15–36.

———. 1986. The Chalcolithic Period. *Biblical Archaeologist* 49: 82–108.

———. 1995. Cult, metallurgy and rank societies: The Chalcolithic period, in T. E. Levy (ed.), *The Archaeology of Society in the Holy Land*, pp. 226–44. London: Leicester University Press.

Levy, T. E., and Alon, D. 1985. Shiqmim: A Chalcolithic village and mortuary center in the northern Negev. *Paleorient* 11: 71–83.

Porat, N., Rosen, S. A., Boaretta, E., and Avni, Y. 2006. Dating the Ramat Saharonim late Neolithic desert cult site. *Journal of Archaeological Science* 33.

Renfrew, C. 1984. Megaliths, territories and populations, in C. Renfrew (ed.), *Approaches to Social Archaeology*, pp. 165–99. Edinburgh: Edinburgh University.

Ronen, A. 1970. Flint implements from south Sinai: Preliminary report. *Palestine Exploration Quarterly* 102: 30–41.

Rosen, A. M., and Weiner, S. 1994. Identifying ancient irrigation: A new method using opaline phytoliths from emmer wheat. *Journal of Archaeological Science* 21: 132–35.

Rosen, S. A. 1987. Phytolith studies at Shiqmim, in T. E. Levy (ed.), *Shiqmim I*, pp. 243–50. Oxford: BAR International Series 356.

Rosen, S. A. 1997. *Lithics after the Stone Age*. Walnut Creek, CA: AltaMira Press.

———. 2002. The evolution of pastoral nomadic systems in the southern Levantine periphery, in E. van den Brink and E. Yannai (eds.), *In Quest of Ancient Settlements and Landscapes*, pp. 23–44. Tel Aviv: Tel Aviv University, Ramot Publishing.

———. 2003. Early multi-resource nomadism: Excavations at the Camel site in the central Negev. *Antiquity* 77: 750–61.

Rosen, S. A., Avni, Y., Bocquentin, F., and Porat, N. 2007. Investigations at Ramat Saharonim: A desert Neolithic sacred precinct in the central Negev. *Bulletin of the American Schools of Oriental Research* 346: 31–57.

Rosen, S. A., and Rosen, Y. J. 2003. The shrines of the setting sun. *Israel Exploration Journal* 53: 1–19.

Rothenberg, B. 1979. Badiat el-Tih, the Desert of the wanderings: Archaeology of central Sinai, in B. Rothenberg et al. (eds.), *Sinai*, pp. 109–27. Bern: Kummerly and Frey.

Rothenberg, B., and Glass, J. 1992. The beginnings and development of early metallurgy and the settlement and chronology of the western Arabah from the Chalcolithic period to the early Bronze IV. *Levant* 24: 141–57.

Shershefski, J. 1991. *Byzantine Urban Settlements in the Negev Desert*. Beersheva, Beersheva V, Studies by the Department of Bible and Ancient Near East, Ben Gurion University Press.

Spellman, G. 2000. The dynamic climatology of drylands, in G. Barker and D. Gilbertson (eds.), *The Archaeology of Drylands*. London: Routledge.

Thomas, D. S. G., and Middleton, N. J. 1994. *Desertification: Exploding the Myth*. New York: Wiley.

Ussishkin, D. 1980. The Ghassulian shrine at En-gedi. *Tel Aviv* 7: 1–44.

41

LANDSCAPES OF FIRE: ORIGINS, POLITICS, AND QUESTIONS

Christian A. Kull

"Tany misy afo misy olona—Where there is fire, there are people" says a proverb in Madagascar. The mastery of fire defines humans more than any other trait. Fire is the most long-term and widespread way in which humans have shaped the surface of the earth. If it were not for fires lit by people, landscapes from the Great Plains of America to the savannas of Africa would look fundamentally different. Australia's eucalypt savannas, Madagascar's grasslands, Venezuela's *gran sabana*, Norway's coastal islands—what would they be without fire either today or in the past?

This chapter reviews the importance of anthropogenic fire as a key human tool for landscape management. It shows how the contrasting land management goals of different interest groups, shaped by long-term processes of ideological, political, and economic change, can lead to conflict over landscape firing. Case studies from Australia, Madagascar, and North America illustrate both the long history of anthropogenic burning and the way fire—and our reactions to it—have rapidly changed in recent times. The chapter then discusses a number of tools that researchers can use to investigate the material expressions of anthropogenic fires, and concludes with several questions for further investigation.

Humans and Fire

Over a million years ago, our ancestors observed that the animals they hunted congregated on the flush of new grass after lightning-strike fires, or that different useful plants grew in burned areas. They picked up burning sticks and carried fire to new places, and learned to maintain and stoke fires. By 350,000 to 400,000 years ago, evidence of home bases and domestic fires suggest that humans had fully mastered fire tending, including ignition. Ever since, throughout the world, throughout history, humans have relied on fire as a simple and effective tool to manage their environment (James 1989; Rolland 2004; Sauer 1975; Schüle 1990; Stewart 1956; Westbroek et al. 1993).

If the land could burn, *Homo sapiens* have burned it. In seasonally dry places prone to lightning, people often beat nature to the task, setting fires as soon as plants were dry enough. Elsewhere they created the right conditions by unleashing livestock or slashing the vegetation. They burned for many reasons, unified by the desire to control and to shape landscapes for a better life. Fires renew and expand grasses crucial to both wild game and domestic grazers. They clear brushy vegetation and thus facilitate cultivation as well as travel, visibility, and security. Frequent, small, early fires are the best way to control wildfires, by avoiding fuel build-up. People also burned to encourage (or discourage) specific plant types, to flush out animals for hunting or bees for hive collection, to better see mineral outcrops or wild tubers, or for a variety of social reasons not directly related to resource management (Bartlett 1956; Jones 1969; Kull, 2004; Laris 2002; Pyne 1995).

The result is fire-shaped landscapes. At its simplest level, increased burning favors fire-adapted species; woody plants give way to grasses. Yet, actual outcomes vary enormously depending on the timing and frequency of fires. For example, early dry season burns every few years can, in the savanna environments of Africa, promote tree cover (Bassett and Koli Bi 2000; Campbell 1996). Seed availability, grazing intensity (whether by domestic livestock, wild mammals, or even insects), soil type, annual variability in timing and amount of precipitation—each of these also affects the outcomes (Pyne et al. 1996). Land managers learn by experience the probable outcomes of certain burning regimes in their particular landscapes.

People not only *set* fires, they can also *control* fires: they created, and create, *fireless* landscapes. For example, people strictly circumscribe free-ranging fire in any intensive farming area. They protect certain places from flames, whether to protect economic resources (such as timber or settlements) or for cultural ideals (for example, sacred groves or nature reserves). As a result, landscapes can be shaped by not just the presence of fire—in varied rhythms and intensities—but also by the absence of fire.

Political Fire

Not everyone benefits from fire. The pastoralist with his wandering herds and free-burning fires inevitably clashed with those whose homes, crops, or fallow fields he burned. Conflict over fire, however, has escalated in the modern age. Scientific advances, growing capitalist economies, and new, powerful state bureaucracies in the 19th century moved the management of many resources out of the hands of rural villagers. In both Europe and its colonies (or ex-colonies), resources were separated into categories (forestry, agriculture, livestock) to which "rational" management strategies were applied. Research sought to increase yields within each sector, leading to plantation forests, intensive single-crop agriculture, and improved pastures. There was no room for fire. Ecological theories of the day supported this view, in particular the theory of succession, which viewed change in vegetation communities as an orderly, staged progression from bare soil to a climax, usually forest (Odum 1969). Fire was seen as an outside disturbance working against succession.

These ideas gained power as technological advances (for example, the internal combustion engine in tractors and bulldozers) and political changes (for instance, growing state bureaucracies with colonial empires) facilitated their wide-scale implementation. As a result, fire landscapes changed. Rich industrialized countries such as France and the United States used impressive fire-fighting technology to suppress burning in most forested landscapes. In tropical countries, rulers sought to replace firestick farming with intensive agriculture and state forests (Pyne 1997). Such changes in resource management were inevitably politicized, because they favored certain landscapes and certain livelihoods over others. This is why villagers in Kumaon, India, repeatedly torched the woodlands in the 1920s (Agrawal 2005) and why farmers in highland Madagascar burned forestry stations in 1947, 1992, and 2005 (Kull 2004).

One way to understand fire conflicts builds on Pyne's (2001) scheme of "first fire," "second fire," and "third fire." "First fire" refers to fires lit by lightning and other nonhuman sources. "Second fire" is anthropogenic biomass burning, which includes slash-and-burn agriculture or rangeland fires. "Third fire' is the combustion of fossil biomass such as coal and oil, characteristic of the industrial, modern age. Following this scheme, when humans domesticate fire or arrive in new lands, "second fire" replaces "first fire."

With industrialization, "second fire" is replaced by "third fire"—fields are fertilized using chemicals, not slash-and-burn rotations, and production moves to the cities. This industrialization frees up some marginal lands, which are then redefined as nature reserves, and sometimes subjected to "first fire" again.

Conflict occurs at the temporal and spatial boundaries between these types of fire—in the turbulent moment when one fire regime replaces another, or at the geographical margin between two types. For example, the arrival of people (carrying "second fire") on unsettled ("first-fire") islands—as frequently occurred in the Malayo-Polynesian diaspora—led to important shifts in vegetation communities. While this shift was historically uncontested, more recently it fuels high profile controversies, such as the tension between farmers and foresters over rainforest loss in Madagascar. Likewise, the juxtaposition today of nature reserves managed for "first fire" located alongside "third fire" landscapes (like industrial pine plantations or suburbia) inevitably sparks anxiety.

Case Studies

The complex relationships of people, fire, and landscape are best illustrated with examples. Three cases are highlighted below—Australia, Madagascar, and North America—including their implications on an archaeology of fire landscapes.

Australia

Australia is perhaps the most famous home of fire. Annual fires burn a broad swath of the northern third of the country, while hot fires periodically blaze through the forests of the southern fringe. Fires burn infrequently in the arid lands in the middle and in the agricultural zones. These fire regimes—and their accompanying vegetation—result from long term shifts in climate, ancient rhythms of Aboriginal burning, and recent changes in land uses and fire policy (Jones 1969; Pyne 1991).

The ancestral Aborigines, who arrived on the continent at least 40,000 years ago, brought anthropogenic fire with them. Their burning contributed to a vegetation shift already long underway owing to climatic drying—from broadly "rainforest" (*Nothofagus*-dominated) vegetation to *Eucalyptus* sclerophyllous forest. Much of the currently dominant vegetation exhibits typical signs of fire adaptation, like *Banksia*'s fire-cured seeds and *Eucalyptus*' prodigious sprouting (Bradstock, Williams, and Gill 2002). Many classic Australian landscapes, like eucalypt woodlands or savanna grasslands, may in one sense be seen as artifacts of Aboriginal culture.

European colonization, beginning in 1788, led to new shifts in fire regimes. Introduced livestock replaced some fires in the role of vegetation removal and altered vegetation communities (with consequences on fire regimes). Agricultural expansion necessitated the clearance of much vegetation cover. Finally, state-led forest management introduced a variety of new forces, from airplane-based prescriptive broadcast burning to highly organized and mechanized fire fighting (Pyne 2006).

Australia is unique in that the government (perhaps out of pragmatic necessity) never completely abandoned "second fire." "Burning off" remains a key management tool in many regions. Key current debates, particularly in the management of state owned lands such as parks and reserves in the southeast, center on whether one should seek to emulate "first fire" (natural) or "second fire" (Aboriginal) burning patterns, what the "second fire" patterns actually were, and whether either approach is pragmatically possible

(Cary, Lindenmayer, and Dovers 2003; Low 2003; Pyne 2006; Whittaker and Mercer 2004).

Madagascar

Madagascar was settled unusually late for a landmass of its size. Before the first human contact, thought to be 2,300 years ago, landscapes ranged from humid rainforests in the east, to dry deciduous forests in the west, to a mosaic of riparian forests, woodlands, and open savannas in the central highlands (Burney 2003; Burney et al. 2004). Vegetation types changed over the longer term in response to climatic variations, over the short term in response to lightning fires and cyclone disturbances, and always in concert with herbivores such as grazing megafauna.

The island's early visitors and eventual settlers followed the worldwide pattern Pyne (1997) has called *landnam*—burning, grazing, and clearing to reshape a landscape to their use. Their fires did not penetrate the humid rainforests, but the highlands were dramatically cleared of almost all woody vegetation. Frequent human-set fires replaced most lightning wildfires, resulting in wide-open stretches dominated by pan-tropical grasses like *Aristida*, *Heteropogon*, and *Hypparhennia*. These landscapes are perhaps the key relic of the original "Vazimba" settlers of the highlands, and continue to be maintained by modern Malagasy herders and farmers despite a century of state repression (Kull 2004).

Other types of fire landscapes abound on the island. Pockets of *Uapaca bojeri* woodland persist due to the tree's fire adaptation and its usefulness to people (hence a tendency to protect it). There are fire exclusion landscapes—the intensive agricultural lands shaped by the hard labor of spades and ploughs and protected from burning. Finally, there are unusual new fire landscapes, where, for example, naturalized imported trees like *Acacia*, *Pinus*, and *Eucalyptus* submit to both "constructive" fires that prepare seedbeds or clear undergrowth, and "destructive" fires like politically-motivated arson of forest plantations (Kull 2004).

North America

People spread rapidly into the North American continent following the last Ice Age, carrying fire with them. The vegetation of a rewarming continent evolved in their presence and in many cases was shaped by their fires. In southern New England, scattered parklike woodlands dominated by sprouting hardwoods such as oak and chestnuts marked the impact of Amerindian fires (Abrams 1992; Cronon 1983; Parshall and Foster 2002; Russell 1983). The tallgrass prairies in the middle of the continent owed much of their expanse to both large-scale grazing by bison and frequent fires set by people and lightning (Howe 1994; Hulbert 1988). In California, up to 13% of nondesert lands burned yearly before white settlement (Martin and Sapsis 1991). Amerindians burned to improve ungulate habitats, to clear underbrush and protect settlements, to manage forests (for example, redwood, Douglas firs, and oaks), to promote plants useful for basketry and weaving, and to improve hunting (Biswell 1989; Huntsinger and McCaffrey 1995; Keeley 2002; Stewart 2002). In northwest British Columbia, native people continue to burn patches of berry-rich land to stimulate new growth and impede conifer invasion (Gottesfeld 1994). Each of these landscapes—or its legacy today in overgrown plots found at the edge of crop fields or sub-urbs—can be seen as an ephemeral artifact of the people that came before.

This influence of Native American fire on landscapes was long ignored by scholars and policymakers who preferred an idea of a pristine, pre-Columbian wilderness (Denevan 1992). Denevan's and others' research on indigenous fire landscapes (e.g., Stewart 2002), however, is debated by scholars (e.g., Vale 2002) who question the extent of Amerindian influence. A landscape archaeology of fire may be required to resolve this debate.

Researching Fire Landscapes

Fire is a tool, but it is ephemeral. Unlike arrow-heads, fish traps, and spades, it does not await the archaeologist's trowel. Even fire sticks or matches decompose rapidly. The attention of a landscape archaeologist must instead focus on oral and archival history, or on the landscapes, sediments, and vegetation that are the outcome of fires. Researchers from a variety of fields have over the past decades assembled an interdisciplinary tool kit of use in documenting fire landscapes.

Ecological analysis of vegetation community change can help to find and to understand anthro-pogenic fire landscapes. Such analysis might focus on the characteristics of individual plant species that condition their responses to fire. Botanists have long recognized certain vegetation traits as fire-adapted or tolerant, including thick bark, seeds that require fire to open or ash beds to sprout, the ability to resprout from epicormic buds, or the placement of significant plant parts underground. Ecological analysis also focuses on disturbance

regimes—such as fire—and how these influence the evolution of particular plant communities. The vegetation that results will depend on the particular management and disturbance regime interacting with different species characteristics, conditioned by the path dependencies of particular places—the inherited soil characteristics, seed availabilities, and current vegetation cover (Cary, Lindenmayer, and Dovers 2003; Goldammer 1990; Goldammer and de Ronde 2004; Veblen et al. 2003; Whelan 1995). Based on such ecological analysis, one can build an understanding of the origins and maintenance of specific vegetation communities. Kepe and Scoones (1999), for example, show specifically how different vegetation communities result from specific management practices—including grazing and fire, in various combinations—in coastal South Africa.

A number of tools are useful for investigating historical fire landscapes—or the vegetation communities they created. *Remote sensing* provides broad-scale evidence in the shorter term (Laris 2002), while repeat photography of *archival photographs* (e.g., Veblen and Lorenz 1991) and investigation of *archival documents* (e.g., Preece 2002)—whether traveler's accounts or ecological monographs—provide evidence going back centuries. A particularly well-developed technique in temperate forests is *dendrochronology*, or the use of tree rings and associated fire scars to create forest fire histories (e.g., Swetnam and Baisan 2003). *Palaeoecologists* analyze pollen and charcoal in lake or ocean sediments to provide a window into much older vegetation communities (see chapters by Porch, Rowe, and Kershaw, Haberle, this volume; Westbroek et al. 1993).

The *stories and experiences of people* are further, rich resources to understanding fire landscapes. No account of Australia is complete without reference to the complex ideas embodied in the "Dreaming" (Langton 1998). In Madagascar, scholars cite the legend of a "Great Fire" that burned the center of the island bare (Battistini and Vérin 1967). Stories and experiences sometimes emerge in surprising places, even in popular song (e.g., Kuhlken 1999: 343; Kull 2004: 179). *Interviews* of people about their fire practices past and present can sometimes be politically touchy (owing to repressive fire politics) but are crucial sources for documenting fire landscapes (Kull 2004; Lewis 1989;).

Finding fire landscapes in stories and song, let alone in buried lake sediments, emphasizes the fact that they can be rather ephemeral artifacts of past human ecologies. To unearth them is not easy; one must look for clues wherever possible.

Conclusions

Seeing fire adaptations and fire legacies in the landscape opens a window into the lives and livelihoods of people through the ages. Some useful questions emerge for further investigation. First, can we read changes in social organization, economy, and technology from the fire record in the landscape in ways analogous to our study of settlement sites? For instance, how clearly recorded in today's landscape are the changes to fire regimes occasioned by the conquering of a pastoral people by agriculturalists 1,000 years ago?

Second, the question of causality and proof of the complex adaptations between fire-wielding humans, plants, herbivores both wild and domestic, and soils remains insufficiently probed. As Pyne (2001: 18) notes, proving adaptation demands circular logic. Did the chicken or the egg come first? What forms of proof are sufficient? If, for example, we demonstrate that *Uapaca bojeri* woodlands in highland Madagascar display common adaptations to fire, that woodland composition is modified from prehuman times, and that humans rely on the woodlands for parts of their livelihoods, then can we argue that this landscape is a material manifestation of human economies layered on certain ecological characteristics (Kull 2004)? Is adaptation the best metaphor, or are there less circular, less problematic ways of understanding the complex path-dependent interactions of soils, plants, animals, and people?

Finally, if some landscapes are the material expressions of the social and economic lives of disappeared or changing cultures—think of African savannas and resident pastoralists—then should we preserve these landscapes as we do other vestiges of past civilizations, whether Roman ruins or Aboriginal burial grounds? What management practices does this imply? Led by whom? Current debates over fire regimes in Australia reflect some of these difficulties: what is "the" Aboriginal fire regime? If Aboriginals are the archetypal "fire stick farmers" (Jones 1969), then is the maintenance of fire stick landscapes the ultimate in cultural recognition?

The anthropogenic use and control of fire on the lands we inhabit is a crucial force in shaping vegetation patterns. People use fire to manipulate the land socially, ecologically, and politically. Current-day landscapes may be artifacts of previous cultures' fire practices. By analyzing regional fire histories, landscape archaeologists—working with both social and ecological research tools—can tell us about the human societies that used the fires, and contribute to debates about modern day fire management.

References

Abrams, M. D. 1992. Fire and the development of oak forests. *BioScience* 42(5): 346–53.

Agrawal, A. 2005. *Environmentality*. Durham, NC: Duke University Press.

Bartlett, H. H. 1956. Fire, primitive agriculture, and grazing in the tropics, in W. L. J. Thomas (ed.), *Man's Role in Changing the Face of the Earth*, pp. 692–720. Chicago: University of Chicago Press.

Bassett, T. J., and Koli Bi, Z. 2000. Environmental discourses and the Ivorian savanna. *Annals of the Association of American Geographers* 90(1): 67–95.

Battistini, R., and Vérin, P. 1967. Ecologic changes in protohistoric Madagascar, in P. S. Martin and H. E. J. Wright (eds.), *Pleistocene Extinctions: The Search for a Cause*, pp. 407–24. New Haven, CT: Yale University Press.

Biswell, H. H. 1989. *Prescribed Burning in California Wildlands Vegetation Management*. Berkeley and Los Angeles: University of California Press.

Bradstock, R. A., Williams, J. E., and Gill, M. A. (eds.). 2002. *Flammable Australia*. Cambridge: Cambridge University Press.

Burney, D. A. 2003. Madagascar's prehistoric ecosystems, in S. M. Goodman and J. P. Benstead (eds.), *The Natural History of Madagascar*, pp. 47–51. Chicago: University of Chicago Press.

Burney, D. A., Burney, L. P., Godfrey, L. R., Jungers, W. L., Goodman, S. M., Wright, H. T., and Jull, A. J. T. 2004. A chronology for late prehistoric Madagascar. *Journal of Human Evolution* 47: 25–63.

Campbell, B. (ed.). 1996. *The Miombo in Transition*. Bogor: Center for International Forestry Research (CIFOR).

Cary, G., Lindenmayer, D., and Dovers, S. (eds.). 2003. *Australia Burning*. Collingwood, Victoria: CSIRO Publishing.

Cronon, W. 1983. *Changes in the Land*. New Haven, CT: Yale University Press.

Denevan, W. M. 1992. The pristine myth: The landscape of the Americas in 1492. *Annals of the Association of American Geographers* 82: 369–85.

Goldammer, J. G. (ed.). 1990. *Fire in the Tropical Biota*. Berlin: Springer-Verlag.

Goldammer, J. G., and de Ronde, C. (eds.). 2004. *Wildland Fire Management Handbook for Sub-Sahara Africa*. Freiburg: Global Fire Monitoring Center (GFMC).

Gottesfeld, L. M. J. 1994. Aboriginal burning for vegetation management in northwest British Columbia. *Human Ecology* 22: 171–88.

Howe, H. F. 1994. Managing species diversity in tallgrass prairie: Assumptions and implications. *Conservation Biology* 8: 691–704.

Hulbert, L. C. 1988. Causes of fire effects in tallgrass prairie. *Ecology* 69: 46–58.

Huntsinger, L., and McCaffrey, S. 1995. A forest for the trees: Forest management and the Yurok environment, 1850 to 1994. *American Indian Culture and Research Journal* 19(4): 155–92.

James, S. R. 1989. Hominid use of fire in the lower and middle Pleistocene: A review of the evidence. *Current Anthropology* 30: 1–26.

Jones, R. 1969. Fire-stick farming. *Australian Natural History* 16: 224–28.

Keeley, J. E. 2002. Native American impacts on fire regimes of the California coastal ranges. *Journal of Biogeography* 29: 303–20.

Kepe, T., and Scoones, I. 1999. Creating grasslands: Social institutions and environmental change in Mkambati Area, South Africa. *Human Ecology* 27: 2–53.

Kuhlken, R. 1999. Settin' the woods on fire: Rural incendiarism as protest. *Geographical Review* 89: 343–63.

Kull, C. A. 2004. *Isle of Fire*. Chicago: University of Chicago Press.

Langton, M. L. 1998. *Burning Questions*. Darwin: Northern Territory University.

Laris, P. 2002. Burning the seasonal mosaic: Preventative burning strategies in the wooded savanna of southern Mali. *Human Ecology* 30: 155–86.

Lewis, H. T. 1989. Ecological and technological knowledge of fire: Aborigines versus park rangers in northern Australia. *American Anthropologist* 91: 940–61.

Low, T. 2003. *The New Nature*. Camberwell: Penguin.

Martin, R. E., and Sapsis, D. B. 1991. *Fire as Agents of Biodiversity: Pyrodiversity promotes biodiversity*. Proceedings of the Symposium on Biodiversity of Northwestern California, Santa Rosa, California.

Odum, E. 1969. The strategy of ecosystem development. *Science* 164: 260–72.

Parshall, T., and Foster, D. R. 2002. Fire on the New England landscape: Regional and temporal variation, cultural and environmental controls. *Journal of Biogeography* 29: 1305–17.

Preece, N. 2002. Aboriginal fires in monsoonal Australia from historical accounts. *Journal of Biogeography* 29: 321–36.

Pyne, S. J. 1991. *Burning Bush*. New York: Henry Holt.

———. 1995. *World Fire*. New York: Henry Holt.

———. 1997. *Vestal Fire*. Seattle: University of Washington Press.

———. 2001. *Fire: A Brief History*. Seattle: University of Washington Press.

———. 2006. *The Still-Burning Bush*. Melbourne: Scribe.

Pyne, S. J., Andrews, P. L., et al. 1996. *Introduction to Wildland Fire*. New York: Wiley.

Rolland, N. 2004. Was the emergence of home bases and domestic fire a punctuated event? A review of the middle Pleistocene record in Eurasia. *Asian Perspectives* 43: 248–80.

Russell, E. W. B. 1983. Indian-set fires in the forests of the northeastern United States. *Ecology* 64(1): 78–88.

Sauer, C. O. 1975. Man's dominance by use of fire. *Geoscience and Man* 10: 1–13.

Schüle, W. 1990. Landscapes and climate in prehistory: Interactions of wildlife, man, and fire, in J. G. Goldammer (ed.), *Fire in the Tropical Biota*, pp. 273–318. Berlin: Springer Verlag.

Stewart, O. C. 1956. Fire as the first great force employed by man, in W. L. J. Thomas (ed.), *Man's Role in Changing the Face of the Earth*, pp. 115–33. Chicago: University of Chicago Press.

———. 2002. *Forgotten Fires*. Norman: University of Oklahoma Press.

Swetnam, T. W., and Baisan, C. H. 2003. Tree-ring reconstructions of fire and climate history in the Sierra Nevada and Southwestern United States, in T. T. Veblen, W. L. Baker, G. Montenegro, and T. W. Swetnam (eds.), *Fire and Climatic Change in Temperate Ecosystems of the Western Americas*, pp. 158–95. New York: Springer-Verlag.

Vale, T. R. (ed.). 2002. *Fire, Native Peoples, and the Natural Landscape*. Washington, DC: Island Press.

Veblen, T. T., Baker, W. L., Montenegro, G., and Swetnam, T. W. (eds.). 2003. *Fire and Climatic Change in Temperate Ecosystems of the Western Americas*. Ecological Studies, Vol. 160. New York: Springer-Verlag.

Veblen, T. T., and Lorenz, D. C. 1991. *The Colorado Front Range: A Century of Ecological Change*. Salt Lake City: University of Utah Press.

Westbroek, P., Collins, M. J., Jansen, J. H. F., and Talbot, L. M. 1993. World archaeology and global change: Did our ancestors ignite the Ice Age? *World Archaeology* 25: 122–31.

Whelan, R. J. 1995. *The Ecology of Fire*. Cambridge: Cambridge University Press.

Whittaker, J., and Mercer, D. 2004. The Victorian bushfires of 2002–2003 and the politics of blame: A discourse analysis. *Australian Geographer* 35: 259–87.

42

MICROBOTANICAL REMAINS IN LANDSCAPE ARCHAEOLOGY

Cassandra Rowe and Peter Kershaw

Botanical microfossil remains are defined as plants, plant fragments, or products invisible to the naked eye and thus requiring magnification for study (Braiser 1980; Dincauze 2000). Although plant microfossil research incorporates a variety of components, including determination of taphonomic processes, field recovery, laboratory preparation, identification, data presentation, and interpretation (Coil et al. 2003), this chapter restricts itself to general research practice in landscape archaeology. Archaeologists should acquire an understanding of the basic principles of plant microfossil data in order to recognize how any particular fossil type can become a useful part of excavation and research programs. At the same time, archaeologists should be aware and appreciative of the expertise required in microbotanical fossil analysis, when addressing problems at various levels of complexity in particular. The purpose of this chapter is to provide an introduction into microfossil analysis and associated literature base so that a knowledgeable step toward expertise and collaboration can be taken.

Landscape archaeology may be described as a discipline seeking out the material manifestation of the relation between humans and the environment (Knapp and Ashmore 1999). Importantly, not all relations can be expected to be represented in material, archaeologically detectable ways, thus highlighting the importance of microbotanical fossil analysis to the discipline. Plants have always been essential resources to people, and they are, at the same time, important indicators of the condition of habitats. Research on plant micro-remains yields important, complementary knowledge to archaeology on diet/nutrition, economy, the presence and activities of people in the wider environment, including agricultural practices, as well as on the environment itself (Knapp and Ashmore 1999; Willerding 1991). The palaeoecological methods that are ultimately chosen should be appropriate to the particular set of archaeological questions (Dincauze 2000).

This chapter encourages readers to question their own research and ask what type of plant microfossil may best strengthen their interpretation. As a guide, a review of seven microbotanical fossil types and their applications is presented (see Table 42.1 for a summary). We focus on physical characteristics, particularly those that may affect microfossil preservation, as well as the type and level of information each microfossil type can be expected to provide. A list of seven microfossil types is by no means exhaustive, but it incorporates the more widely used plant microfossil types associated with archaeological study.

Pollen and Spores

Traditionally, palynology, or pollen analysis, has focused on the study of pollen and also spores, the latter predominantly from ferns and fern allies. Pollen grains are the reproductive male gametes of angiosperms (flowering plants) and gymnosperms (conifers) responsible for the transfer of genetic material, the onset of fertilization, and seed production. Spores are the asexual reproductive cells of fungi, pteridophytes (ferns and fern allies), bryophytes (mosses), and algae that give rise directly to the next generation (Raven, Evert, and Eichhorn 1992; Williams et al. 1998). Few pollen and spores exceed 100 microns (μm) in diameter; most fall within the size range 15 μm to 30 μm.

Fossil pollen grains are represented, in the fossil state, by their outer layer, or exine, which consists of a waxy coat of material known as sporopollenin. The protective, resistant nature of sporopollenin has the effect of preserving pollen grain exines in sediments. This outer wall is also ornamented and perforated in various ways relating largely to the mode of dispersal of the grains as well as to taxonomic affinity. Faegri and associates (1989) present a detailed discussion of the terminology used to describe the structure, sculpturing, and apertures (openings) of the exine that, along with features such as size and shape, assist with identification, typically at the level of genus or family, but within some plant groups, at species level (Bennett and Willis 2001; Lowe and Walker 1984; Pearsall 2000). Spore walls have a similar sporopollenin structure but generally fewer sculptural features. However, they often have a loose additional coating, an exosporium, that allows refined identification when preserved.

The principles of pollen analysis[1] make it clear that there is a long progression between the release of pollen or spores, their incorporation into sediment, and the production of the final pollen assemblage; plants vary greatly in the amount of pollen they produce and in the distance to which pollen grains are dispersed. Preservation of pollen and spores is affected by characteristics related to both taxonomic origin and depositional environments (Bennett and Willis 2001; Coil et al. 2003). Optimal preservation requires anaerobic (oxygen free) or acidic environments that hinder decomposing bacteria (however, dry conditions may also allow preservation). Wetlands, lake sediments, dry cave earths, and even some soils are conducive to the preservation of pollen and spores (Dincauze 2000). Pearsall (2000: 263) provides a list of site types successfully studied using pollen analysis.

Reviews of the history of pollen analysis, especially its application in archaeology, can be found in Bryant and Holloway (1983, 1996) and Pearsall (2000). Walker (1990) provides a brief history of Quaternary palynology. Five major research emphases are represented in contemporary palynological literature: (1) vegetation history; (2) climatic history; (3) biogeography (phytogeography); (4) plant ecology; and (5) human use of and alteration of natural systems (Dincauze 2000). An early emphasis on the use of pollen analysis for regional correlation in the Quaternary has declined with the establishment and progressive refinement of radiometric dating but is still important in pre-Quaternary studies and has recently been revived in correlation of Quaternary straigraphic schemes derived from marine and terrestrial environments (e.g., Tzedakis et al. 1997). Stratigraphic studies focused on vegetation history mainly provide information on the regional environmental setting of human occupation and impact, whereas those from archaeological sites generally focus on specific land uses and activities (Dincauze 2000; Pearsall 2000). In those parts of the world where agriculture has been a major feature of the landscape, initial impacts and subsequent land-use changes on a regional scale can be identified by reductions in the percentages of tree pollen, as a result of deforestation, and specific indicators of disturbance increase either in native opportunists or introduced plants of arable or pastoral systems. For example, Behre (1986) provides a valuable list of anthropogenic indicators in pollen records from Europe, while Maloney (1994) assesses indicators of human disturbance associated with agriculture in Southeast Asia. Changes in composition of forest may also be used as evidence of people, owing to selective use of species that may enhance or reduce their representation, but, because of other agents of change such as climate, volcanic activity, landslides, and so forth, there is often uncertainty in determination of initial or small-scale human impacts on the landscape. One classic debate involves the decline of elm trees in the mid Holocene of northwest Europe that had been attributed to climate change and disease as well as human disturbance. Resolution of the question on the side of human impact was finally determined to the satisfaction of most researchers via a very high (annual) sub-sampling interval of an optimal site (Peglar 1993). Studies such as this have largely revolutionized pollen analysis, and temporally precise studies, particularly where they can be supported by high dating resolution, are now a major component of the subdiscipline.

Table 42.1 Summary of plant microfossil types commonly associated with archaeological studies.

Proxy	Variable Measured	Laboratory Procedures	Preservation	Deposition	Major applications	Advantages	Limitations
Pollen and Spores	Type, relative abundance	Moore et al. (1991) Bennett and Willis (2001)	Acidic and/or anaerobic environments; geological and archaeological (cultural) archives	Local to regional	Past vegetation composition, structure, and dynamic; climatic change; human land use and impact	Widely adopted technique with a strong history of use.	Differential production, dispersal, and preservation may limit data interpretation
Cellular tissue	Presence, abundance	Chemical digestive techniques as used for pollen preparation	Sediments from which fossil pollen grains are typically recovered. Lake and swamp sediments	Local. Not adapted to long-distance transport	Reconstruction of past vegetation: to infer the presence and absence of different plant taxa	Auxiliary technique to pollen analysis	Present in low abundances; occurs too infrequently to infer absolute or relative abundance of plant taxa
Starch	Presence, abundance	Pearsall (2000) Korstanje (2003)	A variety of contexts (possibly) reduced when left in open conditions; geological and archaeological (cultural) archives	Local	Investigation of subsistence strategies and plant use within prehistoric populations	Direct evidence for plant collection and use, the plant food component of early human diets, as well as tool function	Current knowledge of processes affecting morphology and fossil survival of starch grains is limited

Phytoliths	Type, relative abundance	Pearsall (2000) Lentfer and Boyd (1998, 1999)	A variety of contexts, including geological and archaeological (cultural) archives	Predominantly local (decay-in-place model), but can be a component of wind-transported dust	Reconstruction of past environments; identification of agricultural systems (land use) and crop types; information on subsistence and diet	Durability. Preserved where other micro-fossils are commonly absent (including dry, alkaline, anaerobic conditions)	Lack of diagnostic features and range of types within individual plants presently inhibits application to a few taxa, e.g., Poaceae
Diatoms	Type, relative abundance	Battarbee et al. (2001)	Aquatic and damp environments; best preserved in cold, soft water bodies	Predominantly local	Indicators of environmental change, particularly water quality, in lakes and flowing waters, marine and estuarine environments	Cosmopolitan and identifiable to refined taxonomic level; well-defined environmental optima and tolerances	Total or differential dissolution in some squatic environments, especially in carbonate-rich or ephemeral lakes
Chrysophytes	Type and relative abundance of Chrysophyte scale and cyst remains	Zeeb and Smol (2001); same preparation can be used as for diatom analysis	Freshwater environments (exclusively marine taxa are rare); most common in acidic and/or nutrient poor lakes	Local	As with diatoms, indicators of aquatic environmental conditions: lake-level changes, water and habitat availability, salinity, and pollution	Well-defined environmental optima and tolerances	A proxy approach still in early development; limited information on chrysophyte cyst morphology

They are especially important in documenting and explaining rapid changes that have occurred in land use over the last one or two centuries.

In contrast to their value in documenting disturbance and indicating the establishment of pastoral and arable agriculture, pollen analytical studies have been limited in their ability to identify particular crop types. One reason for this is that pollen cannot be identified to an appropriate taxonomic level. For example, cereals form the basis of many agricultural systems and, although cereals can often be separated from native grasses, it is very difficult to make identifications at the species, never mind the cultivar level—another reason is the palynological "invisibility" of root crops, the other major basis of agricultural systems, due mainly to lack of preservation (Maloney 1994).

In those regions and those time periods dominated by hunter-gatherer societies, identification of human impact from pollen evidence can be extremely tentative or even impossible. More certainty can be introduced by examination of changes in charcoal records, as a direct measure of fire, in addition to pollen. At Lynch's Crater in northeastern Australia, Kershaw (1994) interpreted the replacement of fire-sensitive drier rainforest by sclerophyll vegetation around 40,000 years ago as a response to Aboriginal burning with the first arrival of people on the continent. This interpretation was treated with some skepticism until the addition of a charcoal curve provided the required support for the hypothesis (see discussions in Turney et al. 2001).

Greater certainty regarding human activities is derived from pollen analysis of archaeological sites. Because archaeological sites are the locus of human disturbance (Davis 1994: 2), direct pollen analysis of collected sediments can prove to be a reliable method of determining changes in landscape and vegetation as a result of known human presence and activity. Palaeovegetational changes may be linked temporally and spatially to the human changes inferred from archaeologically derived data, and subsequently translated to non-archaeological or "off-site" sediment and fossil sequences. Rowe (2005, 2006) provides examples where pollen analysis of an archaeological rockshelter site in northern Australia highlighted local decline in forest cover, as a result of the onset of permanent occupation of the site, providing a standard from which to interpret swamp sequences in the same region. A number of case studies that address the use and interpretation of archaeological pollen data are presented in Davis (1994).

Non-Pollen Palynomorphs

In recent years, palynological study has expanded to incorporate other botanical entities composed of sporopollenin-like material. Although studies on the gross composition of organic matter in palynological samples, known as "palynodebris" or "palynofacies" analyses, have been undertaken for many years to contribute to a fuller sedimentary picture (Traverse 1988), little attempt was made at botanical sourcing of material. A full or selected identification of non-pollen palynomorphs found on pollen slides may result in substantially greater palaeoenvironmental information (van Geel 2001). Van Geel (1986, 2001) and Cronberg (1986) highlight efforts made to combine the analysis of pollen with the study of all "extra" microfossils in Quaternary deposits in northwest Europe. Among the extra fossils recorded were the remains of algae and cyanobacteria. The recognition of spores, including the spores of fungi, was also expanded.

Van Geel (2001) refers to a number of papers in which the description and illustration of non-pollen palynomorph types and their indicator value are discussed. Of note, spores of the family Zygnemataceae (filamentous green algae) are common in shallow, stagnant freshwater deposits and based on fossil spore records characterize different habitat types (for example, specific sediment types, different trophic conditions). Fossil spores of Zygnemataceae have been described and utilized in an archaeological context by van Geel (1976). Factors resulting in the presence of *Pediastrum* (colonial green algae) spores in sediments include changes in catchment erosion, water turbidity, nutrient status, and pH. Remains of cyanobacteria function as indicators of nitrogen-poor conditions, possibly occurring where there is a general low level of nutrients. Among the fossil fungi are parasitic types that indicate the presence of their host plant (for example, *Amphisphaerella* for the presence of *Populus*), indicators for dung (*Chaetomium*), indicators for fire (*Neurospora*), and soil-inhabiting taxa whose presence in lake deposits points to erosion (*Glomus*) (van Geel 2001).

Cellular Tissue Remains

Undecomposed plant cells, or plant tissues, can form a significant component of the nonmineral or "palynodebris" fraction of sediments (Coil et al. 2003). Microfossils of tricomes (an outgrowth, such as a root hair, from plant epidermal cells; Bailey 1999) or stomata cells, for example, form a beneficial auxiliary technique to palynology. Cellular tissues are similar to plant macrofossils, only to be found in sediments more frequently. They are not prone to long-distance transport and provide a good indication of the local presence, absence, and abundance of different plant taxa (McDonald 2001). For example, Nymphaeaceae have mucilaginous hairs. The basal cells of these hairs, with their central pore and concentric ring pattern, are common in pollen slides from water bodies where *Nymphaea* is present in the local vegetation. The frequency of these basal cells is in sharp contrast to typically rare Nymphaeaceae (entomophilous) pollen (van Geel, 2001).

McDonald (2001) outlines the identification of conifer stomata and suggests taxonomic resolution is as good as for pollen. Stomatal records have been constructed from late Quaternary lake sediments in Europe, Siberia, and North America, and from peatlands in Canada (see McDonald 2001 for a recent review). In alpine and polar regions, fossil stomata have been utilized as indicators of tree fluctuations during the Holocene. The stomata of *Picea abies* (European spruce), *Pinus,* and *Larix decidua* (European larch), for example, have been used to reconstruct changes in the position of the upper tree line, whereas stomata from *Juniperus, Pinus,* and *Abies alba* (fir) signal vegetation change at lower elevations. McDonald (2001) highlights that both natural and anthropogenic vegetation change have been detected using stomatal analysis.

Archaeological samples may also contain undecayed cellular material. Cellulose rings from primary vascular tissues, notably xylem cell walls (Raven, Evert, and Eichhorn 1992), were applied as one of a range of microfossils to address research questions of prehistoric agricultural and pastoral production in Argentina's Valle del Bolsón (Coil et al. 2003). Briuer (1976) found plant residues on stone tools from rockshelter sites in Arizona. Pollen grains, cell walls, cell lumen, raphides, tracheids, and vessal elements were observed as clues to plant collection and use, as well as stone tool function, across the southwestern United States.

Starch

Starch grains are the predominant food storage units of plants, formed within specialized organs (plastids) that occur within individual cells. Of the two kinds of plastids, the chloroplasts occur primarily in leaves and green stems and produce generally transient starches. The more commonly observed starches are produced by amyloplasts, within roots, rhizomes, tubers and seeds. These

structures provide a reservoir of energy for the plant (Bailey 1999; Cortella and Pochettino 1994; Loy 1994). Starch consists of two organic polymers, amylose and amylopectin, that form a series of laminated layers facilitating preservation. The relative amounts of amylose and amylopectin in a given starch grain depend on the species of plant from which the starch grain was obtained. The shape and size (< 100 μm) of starch grains formed by different plant taxa also differ (Bailey 1999; Coil et al. 2003). Czaja (1978) provides an overview of the structure of starch grains in relation to the classification of plant families. Korstanje (2003) and Pearsall (2000) touch on the laboratory preparation of starch samples. Procedures focusing on locating and removing starch residues adhering to stone artifacts are described by Loy (1994).

In an analysis of archaeological sediments from agricultural field sites in northwestern Argentina, Korstanje (2003) outlines conditions considered advantageous for starch preservation, namely semiarid environments, sandy-type soils, and an average pH of seven. Within a subtropical context, an experiment on starch grain preservation by Lu (2003) revealed preservation was reduced significantly when left in open environments and survived better in buried or sheltered conditions. Nonetheless, Piperno and Holst (1998) and Piperno and associates (2000) highlight the presence of starch grains on prehistoric stone tools, as an indicator of early tuber use and root-crop agriculture in the lowland tropics of Panama. Starch was recovered from stone grinding tools from both cave and open-air sites, dating to 8,000 years ago.

Starch granules, recovered from artifact surfaces in particular, are receiving increasing attention from archaeologists, because they can reveal aspects of subsistence and domestic activities not accessible via other proxy methods. As starch-rich plants formed an important component of many prehistoric diets, the study of starch grains provides a direct means of plant use reconstruction. Haslam (2003) reports on the identification of *Zea mays* starch grains on 2,000-year-old artifacts from Maya sites in Honduras. Similarly, Urgent (1987) identifies potato remains from a Pleistocene settlement in south-central Chile. Loy and associates (1992) utilize *Colocasia* and *Alocasia* taro starch, dating to 28,000 years ago, as evidence of plant collection and use in the northern Solomon Islands. Likewise, starch granules on artifacts provide evidence of taro and yam processing at Bitukara in New Guinea (Fullagar et al. 1998 cited in Haslam 2003).

Silica Phytoliths

Opaline phytoliths are formed when silica, dissolved in groundwater as monosilicic acid, is absorbed through the roots of plants and transported through the vascular system to be deposited in epidermal and other plant cells. In many plant taxa, phytoliths take on distinctive shapes (Pearsall 1994; Piperno 2001) and Coil and associates (2003) describe silica phytoliths as microscopic "casts" of cells, aggregates of cells or intercellular spaces. Aerial plant structures, including leaves, fruits, seeds, and inflorescence bracts, accumulate silica deposits more readily than do subterranean roots and tubers, as a possible deterrent to herbivores and pathogens and as a function of plant structure and support. Following death and decay of the plant, phytoliths retain cell shapes and are ultimately deposited into soils and sediments as discrete particles (decay-in-place model, see Pearsall 2000: 392).

Phytolith identification is based primarily on shape and size (5–50 μm). Distinctive phytoliths are increasingly known to be produced in many plant families; genus-level diagnostics are becoming increasingly common. Identifiable phytoliths are produced in quantity in a considerable number of angiosperms, perhaps the best known among the grasses. Gymnosperms and pteridophytes also possess some diagnostic forms (Pearsall, 2000). Regional summaries of plant phytolith production and morphology from locations such as Africa (Runge 1999), North America (Bozarth 1992), the South American tropics (Piperno 1988), New Guinea (Boyd, Lentfer, and Torrence 1998), and Australia (Bowdery 1998; Wallis 2000) are available. Species-specific identification is also possible in a number of domesticated crop species. Investigation of the potential of phytolith analysis for the purpose of crop detection began with the search for a method of identifying maize (Pearsall 1994). Piperno (1984) and Pearsall (1978) utilized the characteristics of cross-shaped phytoliths to separate maize from wild grasses and to demonstrate its presence in archaeological sediments from Ecuador and Panama. Pearsall (2000) summarizes, by geographic region, crops identifiable using phytolith analysis (southeast Asia/east Asia, southwest Asia, sub-Saharan Africa, New World temperate, and New World tropical regions).

Piperno (2001) argues that silica phytoliths are among the most durable plant fossils. Since phytoliths are inorganic, they are resistant to oxidation and are well preserved in many depositional environments, including dry and alkaline conditions (Piperno 2001). In this respect, phytolith analysis is often promoted as a solution to palynological constraints (e.g., Boyd, Lentfer, and Torrence 1998). Methods used to extract phytoliths from sediments and cultural residues are discussed in detail in Pearsall (2000) and Lentfer and Boyd (1998, 1999).

In application, phytolith analysis focused on nonhabitation cultural contexts, including agricultural field or garden site areas, provides a means of obtaining information on subsistence, cropping techniques and other questions of people-plant interrelationships, yielding more direct data than samples from site occupation areas (for example, home floors or storage pits as secondary preparation areas) (Pearsall 2000). For example, Fujiwara (1993) provides evidence of early rice cultivation in Japan, and Fei and Minchang (2003) highlight several distinct phases of rice paddy field development during the Neolithic in southern China. In tropical regions, Pearsall and Trimble (1984) utilize soil phytolith analysis in investigating the intensification of past agricultural activity in Hawai'i, demonstrating vegetation modification associated with both cropping and fallow techniques. Similarly, Pearsall (1994) discusses the introduction and maintenance of agricultural systems to the Ecuadorian coast, incorporating maize, root crops, and palms. In the historical context, Miller and associates (1990, cited in Bowdery 1998) examine garden soils; phytoliths assisting in the identification of 18th–19th century lawns, and ornamental and kitchen gardens in New Jersey. Examples of analyzing phytoliths from non-archaeological sites, for the purpose of reconstructing vegetation, climate, and regional human impacts on environments, include Kealhofer and Penny (1998) in documenting 14,000 years of vegetation change in Thailand, and Zhao and Piperno (1999) and Piperno (2001) in the discrimination of forest type expansion and grassland formations in late Quaternary environments in southern China. Clarkson and Wallis (2003) extend the use of phytoliths from vegetation composition to the reconstruction of climate, suggesting greater mid-Holocene aridity across northern Australia incorporating the possible onset of El Niño-Southern Oscillation conditions. Phytoliths have also been recovered from contexts other than sediments, including mud-wasp nest remains (Roberts et al. 1997), dental calculus (Middleton and Rovner 1994), ceramic cooking and storage vessel residues (Jones 1993), and the working edges of stone tools (Dincauze 2000).

While the focus of phytolith analysis lies with opaline, silica-based phytoliths, some plants form deposits of alternative mineral compositions. Calcium phytoliths may be produced and deposited in living plant cells and tissues, based on the uptake of calcium carbonate or calcium oxylate (Coil et al. 2003; Pearsall 2000). Distinctive calcium phytoliths are produced by plants in the Cactaceae family, for example (Jones and Bryant 1999, cited in Pearsall 2000). Although composed of calcium, as opposed to silica, phytoliths are less likely to be recovered intact in sediments (or may be difficult to extract in the laboratory, [see Pearsall 2000]), they may be preserved in coprolites, storage or cooking vessels, tool residues, or similar protected contexts, thus providing an additional avenue for the identification of plant use (Pearsall 2000). Loy and associates (1992) were able to utilize calcium phytoliths from stone artifact residues as evidence for the Pleistocene human use of plants (root vegetables) in the Solomon Islands.

Diatoms

Diatoms are single-celled algae ranging in size from 5 μm to 200 μm. The diatom cell possesses a siliceous outer wall, or frustule, divided into two overlapping valves serving to enclose the protoplasmic cell mass. The siliceous nature of the frustule facilitates diatom preservation in the fossil record. The intricate detail of the outer silica wall also provides the characteristics most commonly used in diatom taxonomy (Battarbee et al. 2001; Mannion 1987). Frustules are either radially symmetrical (centric diatoms) or bilaterally symmetrical (pennate diatoms) and are perforated by small apertures (punctae). The arrangement of the punctae is particularly relevant to the classification of diatoms (Bailey 1999; Lowe and Walker 1984). Round and associates (1990) discuss the biology, ecology, and taxonomy of diatoms, and an outline of analytical and interpretative methods is provided by Battarbee and colleagues (2001) and Battarbee (1986).

Diatoms occupy almost all aquatic or damp environments. They exist in benthic (bottom-dwelling) and planktonic (free-floating) forms in lakes and marine environments. Epiphytic (attached) diatoms occur in soil, clay deposits, and tree trunks as well as on aquatic plants (Lowe and Walker 1984). Diatoms are sensitive to a range of water chemistry parameters including nutrient concentration, salinity, ionic composition, temperature, and pH; community composition is also influenced by biotic factors such as the relative proportion of littoral to open water environments. Accordingly, diatom records have been used to provide insights into climate change, land-use history, catchment processes, lake successional sequences, and aquatic pollution (Battarbee et al. 2001; Kershaw et al. 2000;

see also Stoermer and Smol 1999, for a review of the application and uses of diatoms).

Examples of diatoms as indicators of environmental change in flowing water and lakes include Gasse and Fontes (1989), providing evidence of Holocene water-level fluctuations and salinity changes in closed lakes throughout tropical Africa. Wolin and Duthie (1999) document alternating cycles of marshland and open water in the north American Great Lakes, in response to fluctuating water levels over the last 1,000 years, and Kershaw and associates (2000) provide a reconstruction of past salinities, as they affect billabongs and other river-related sites in southeastern Australia. The utilization of diatoms as indicators of changes in marine and estuarine environments may include data on whether or not shifts in land/sea level have occurred. As Mannion (1987; see also Denys and Wolf 1999) points out, where archaeological sites are situated in coastal regions, diatom analysis can greatly assist in interpreting the sedimentary sequence and the occupation of the area in question; marine transgressive episodes, in particular, may result in the loss of freshwater habitats and possible abandonment of settlements.

Juggins and Cameron (1999), Mannion (1987), and Battarbee (1986) examine the role of diatoms within archaeology, including direct applications to provenancing individual artifacts, particularly pottery (clay) sourcing and typology. Diatoms also provide the means of assessing human impacts on the environment, notably lake-water quality (for example, Fritz [1989] examined lake development and limnological responses to prehistoric land-use in Norfolk, England, and Osborne and Polunin [1986] examined changes in water depth and nutrient status in Waigani Swamp, New Guinea, as a result of urban runoff in the surrounding catchment).

Chrysophyte Scales and Cysts

Chyrsophytes are a diverse group of algae most commonly referred to as "golden brown algae" that exist as single flagellated cells, or motile, spherical colonies of flagellated cells. Chrysophytes are represented as microfossils by two forms of siliceous, often species-specific remains: the endogenously formed cysts (stomatocysts, statospores, or statocysts) that characterize this group as a whole, or the often sculptured and ornamented scales that characterize such as *Mallomonas* and *Synura*. Bristles and spines from scaled chrysophytes are also sometimes used (Coil et al. 2003; Zeeb and Smol 2001). Numerous texts focus on chrysophytes (e.g., Kristiansen and Andersen 1996; Sandgren,

Smol, and Kristiansen 1995) summarizing aspects of biology, taxonomy, and palaeoenvironmental application (see also Zeeb and Smol 2001). Smol (1986) and Zeeb and Smol (2001) provide numerous drawings and photographs highlighting some of the main chrysophyte features distinguishable with light microscopy.

Most known chrysophytes live in freshwater with well-defined environmental optima and tolerances and can therefore be used as palaeoenvironmental indicators of water pH, trophic status, and salinity/conductivity. Like diatoms, chrysophytes serve as indicators of general environmental conditions, as evidence of hydrological change, flooding, soil transport, and land-use effects (Coil et al. 2003). Both chrysophyte scales and cysts have been included in several palaeolimnological projects, designed to assess long-term trends in lake water pH and metal concentrations in Canada (Dixit et al. 1992, cited in Zeeb and Smol 2001), and on the eutrophication and extent of anthropogenic acidification in Europe (Steinberg, Hartmann, and Krause-de Iiin 1988) and the United States (Cummings et al. 1994, cited in Zeeb and Smol 2001).

Conclusions

As Dincauze (2000: 18, 401) points out for environmental archaeology in general, data integration is possible; it is never easy, but it is essential for any success in the search for knowledge of the past. Botanical microfossils are presented in this chapter as providing a substantial and diverse set of reinforcing data on human activity and associated environments, with the ability to extend discussions and answer questions that often cannot be answered by the archaeological artifact record alone. No single technique, however, can provide all the evidence needed to fully understand the nature of past environments or human landscapes in all their manifestations. Each of the botanical microfossils summarized above offers a slightly different perspective (Coil et al. 2003; Lowe and Walker 1984). Consequently, we stress the potential of multiproxy botanical microfossil studies, and their employment with related proxy studies of macro-plant remains, charcoal, and faunal fossil types as outlined elsewhere in this volume, in order to strengthen archaeological interpretations of landscape.

Notes

1. For a general description of palynological methods, cautions, and errors, several texts may be referenced (e.g., Birks and Birks 1980; Faegri and Iverson 1989; Moore, Webb, and Collinson 1991; Williams et al. 1998).

References

Bailey, J. 1999. *Dictionary of Plant Sciences*. London: Penguin.

Battarbee, R. W. 1986. Diatom Analysis, in B. E. Berglund (ed.), *Handbook of Holocene Palaeoecology and Palaeohydrology*, pp. 527–70. Chichester: John Wiley and Sons.

Battarbee, R. W., Jones, V. J., Flower, R. J., Cameron, N. G., Bennion, H., Carvalho, L., and Juggins, S. 2001. Diatoms, in J. P. Smol, H. J. B. Birks, and W. M. Last (eds.), *Tracking Environmental Change Using Lake Sediments, Vol. 3: Terrestrial, Algal and Siliceous Indicators*, pp. 155–202. Dordrecht: Kluwer Academic Publishers.

Bennett, K. D., and Willis, K. J. 2001. Pollen, in J. P. Smol, H. J. B. Birks, and W. M. Last (eds.), *Tracking Environmental Change Using Lake Sediments, Vol. 3: Terrestrial, Algal and Siliceous Indicators*, pp. 5–32. Dordrecht: Kluwer Academic Publishers.

Birks, H. J. B., and Birks, H. H. 1980. *Quaternary Palaeoecology*. London: Edward Arnold.

Bowdery, D. 1998. *Phytolith Analysis Applied to Pleistocene-Holocene Archaeological Sites in the Australian Arid Zone*. Oxford: BAR International Series 695.

Boyd, W. E., Lentfer, C. J., and Torrence, R. 1998. Phytolith analysis for a wet tropics environment: Methodological issues and implications for the archaeology of Garua Island, west New Britain, Papua New Guinea. *Palynology* 22: 213–28.

Bozarth, S. R. 1992. Classification of opal phytoliths formed in selected dicotyledons native to the great plains, in G. Rapp and S. C. Mulholland (eds.), *Phytolith Systematics: Emerging Issues*, pp. 193–214. New York: Plenum Press.

Braiser, M. D. 1980. *Microfossils*. London: Chapman and Hall.

Briuer, F. L. 1976. New clues to stone tool function: Plant and animal residues. *American Antiquity* 41: 478–84.

Clarkson, C., and Wallis, L. A. 2003. The search for El Nino/Southern Oscillation in archaeological sites: Recent phytolith analysis at Jugali-ya rock shelter, Wardaman Country, Australia, in D. M. Hart and L. A. Wallis (eds.), *Phytolith and Starch Research in the Australian-Pacific-Asian Regions: The State of the Art*, pp. 137–52. Terra Australis 19. Canberra: Pandanus Books.

Coil, J., Korstanje, M. A., Archer, S., and Hastorf, C. A. 2003. Laboratory goals and considerations for multiple microfossil extraction in archaeology. *Journal of Archaeological Science* 30: 991–1008.

Cortella, A. R., and Pochettino, M. L. 1994. Starch grain analysis as a microscopic diagnostic feature in the identification of plant material *Economic Botany* 48: 171–81.

Cronberg, G. 1986. Blue-green algae, green algae and chrysophyceae in sediments, in B. E. Berglund (eds.), *Handbook of Holocene Palaeoecology and Palaeohydrology*, pp. 507–26. Chichester: John Wiley and Sons.

Czaja, A. 1978. Structure of starch grains and the classification of vascular plant families. *Taxon* 27: 463–70.

Davis, O. K. 1994. *Aspects of Archaeological Palynology: Methodology and Applications*. Dallas: AASP Contributions Series Number 29, American Association of Straitgraphic Palynologists Foundation.

Denys, L., and Wolf, H. 1999. Diatoms as indicators of coastal paleoenvironmental and relative sea level change, in E. F. Stoermer and J. P. Smol (eds.), *The Diatoms: Applications for the Environmental and Earth Sciences*, pp. 277–97. Cambridge: Cambridge University Press.

Dincauze, D. F. 2000. *Environmental Archaeology: Principles and Practice*. Cambridge: Cambridge University Press.

Faegri, K., Kalans, P. E., and Krzywinski, K. 1989. *Textbook of Pollen Analysis*. Chichester: Wiley.

Fei, H., and Minchang, L. 2003. Pollen and phytolith records from primitive paddy fields during the Neolithic at Caoxieshan, Taihu Plain, southern China, in D. M. Hart and L. A. Wallis (eds.), *Phytolith and Starch Grain Research in the Australian-Pacific-Asian Regions: The State of the Art*, pp. 163–75. Terra Australis 19. Canberra: Research School of Pacific and Asian Studies, Australian National University.

Fritz, S. C. 1989. Lake development and limnological response to prehistoric land-use in Diss, Norfolk, U.K. *Journal of Ecology* 77: 182–202.

Fujiwara, H. 1993. Research into the history of rice cultivation from plant opal analysis, in D. M. Pearsall and D. R. Piperno (eds.), *Current Research in Phytolith Analysis: Applications in Archaeology and Paleoecology*, pp. 147–58. MASCA Research Papers in Science and Archaeology 10. Philadelphia: University of Pennsylvania.

Gasse, F., and Fontes, J. C. 1989. Palaeoenvironments and palaeohydrology of a tropical closed lake (L. Asal, Djibouti) since 10,000 yr B.P. *Palaeogeography, Palaeoclimatology, Palaeoecology* 69: 67–102.

Hall, J., Higgins, S., and Fullagar, R. 1989. Plant residues on stone tools. *Tempus* 1: 136–55.

Haslam, M. 2003. Evidence for maize processing on 2,000-year-old obsidian artifacts from Copán, Honduras, in D. M. Hart and L. A. Wallis (eds.), *Phytolith and Starch Grain Research in the Australian-Pacific-Asian Regions: The State of the Art*, pp. 153–62. Terra Australis 19. Canberra: Research School of Pacific and Asian Studies, Australian National University.

Ireland, S. 1987. The Holocene sedimentary history of the coastal lagoons of Rio de Janeiro State, Brazil, in I. Shennan and M. J. Tooley (eds.), *Sea-Level Changes*, pp. 25–66. Oxford: Basil Blackwell.

Jones, J. R. 1993. Analysis of pollen and phytoliths in residue from a Colonial Period ceramic vessal, in D. M. Pearsall and D. R. Piperno (eds.), *Current Research in Phytolith Analysis: Applications in Archaeology and Paleoecology*, pp. 31–35. MASCA Research Papers in Science and Archaeology 10. Philadelphia: University of Pennsylvania.

Juggins, S., and Cameron, N. G. 1999. Diatoms and archaeology, in E. F. Stoermer and J. P. Smol (eds.), *The Diatoms. Applications for the Environmental and Earth Sciences*, pp. 389–401. Cambridge: Cambridge University Press.

Kealhofer, L., and Penny, D. 1998. A combined phytolith and pollen sequence for 14,000 years of vegetation change in northeast Thailand. *Review of Palaeobotany and Palynology* 103: 83–93.

Kershaw, A. P. 1994. Pleistocene vegetation of the humid tropics of northeastern Queensland, Australia. *Palaeogeography, Palaeoclimatology, Palaeoecology* 109: 399–412.

Kershaw, A. P., Quilty, P. G., David, B., van Huet, S., and McMinn, A. 2000. Palaeobiogeography of the Quaternary of Australasia. *Memoir of the Association of Australasian Palaeoontologists* 23: 471.

Knapp, A. B., and Ashmore, W. 1999. Archaeological landscapes: Constructed, conceptualised, ideational, in W. Ashmore and A. B. Knapp (eds.), *Archaeologies of Landscape: Contemporary Perspectives*, pp. 1–32. Massachusetts: Blackwell Publishers.

Korstanje, M. A. 2003. Taphonomy in the laboratory: Starch damage and multiple microfossil recovery from sediments, in D. M. Hart and L. A. Wallis (eds.), *Phytolith and Starch Grain Research in the Australian-Pacific-Asian Regions: The State of the Art*, pp. 105–18. Terra Australis 19. Canberra: Research School of Pacific and Asian Studies, Australian National University.

Kristiansen, J., and Andersen, R. 1996. *Chrysophytes: Aspects and problems*. Cambridge: Cambridge University Press.

Lentfer, C. J., and Boyd. W. E. 1998. A comparison of three methods for the extraction of phytoliths from sediments. *Journal of Archaeological Science* 25: 1159–83.

———. 1999. An assessment of techniques for the deflocculation and removal of clays from sediments used in phytolith analysis. *Journal of Archaeological Science* 26: 31–44.

Lowe, J. J., and Walker, M. J. C. 1984. *Reconstructing Quaternary Environments*. London: Longman.

Loy, T. H. 1994. Methods in the analysis of starch residues on prehistoric stone tools, in J. G. Hather (ed.), *Tropical Archaeobotany: Applications and New Developments*, pp. 86–114. London: Routledge.

Loy, T. H., Spriggs, M., and Wickler, S. 1992. Direct evidence for human use of plants 28,000 years ago: Starch residues on stone artifacts from the northern Solomon Islands. *Antiquity* 66: 898–912.

Lu, T. 2003. The survival of starch residue in a subtropical environment, in D. M. Hart and L. A. Wallis (eds.), *Phytolith and Starch Grain Research in the Australian-Pacific-Asian Regions: The State of the Art*, pp. 119–26. Terra Australis 19. Canberra: Research School of Pacific and Asian Studies, Australian National University.

Mannion, A. M. 1987. Fossil diatoms and their significance in archaeological research. *Oxford Journal of Archaeology* 6: 131–47.

McDonald, G. M. 2001. Conifer Stomata, in J. P. Smol, H. J. B. Birks, and W. M. Last (eds.), *Tracking Environmental Change Using Lake Sediments, Vol. 3: Terrestrial, Algal and Siliceous Indicators*, pp. 33–48. Dordrecht: Kluwer Academic Publishers.

Middleton, W., and Rovner, I. 1994. Extraction of opal phytoliths from herbivore dental calculus. *Journal of Archaeological Science* 21: 469–73.

Moore, P. D., Webb, J. A., and Collinson, M. E. 1991. *Pollen Analysis*. Oxford: Blackwell Scientific Publishers.

Pearsall, D. M. 1978. Phytolith analysis of archaeological soils: Evidence for maize cultivation in formative Ecuador. *Science* 199: 177–78.

———. 1994. Investigating New World tropical agriculture: Contributions from phytolith analysis, in J. G. Hather (ed.), *Tropical Archaeobotany: Applications and New Developments*, pp. 115–35. London: Routledge.

———. 2000. *Paleoethnobotany: A Handbook of Procedures*. San Diego: Academic Press.

Pearsall, D. M., and Trimble, M. K. 1984. Identifying past agricultural activity through soil phytolith analysis: A case study from the Hawaiian Islands. *Journal of Archaeological Science* 11: 119–33.

Piperno, D. R. 1984. A comparison and differentiation of phytoliths from maize and wild grasses: Use of morphological criteria. *American Antiquity* 49: 361–83.

———. 1988. *Phytolith Analysis: An Archaeological and Geological Perspective*. San Diego: Academic Press.

———. 1993. Phytolith and charcoal records from deep cores in the American tropics, in D. M. Pearsall and D. R. Piperno (eds.), *Current Research in Phytolith Analysis: Applications in Archaeology and Paleoecology*, pp. 58–71. Philadelphia:

The University Museum of Archaeology and Anthropology.

———. 2001. Phytoliths, in J. P. Smol, H. J. B. Birks, and W. M. Last (eds.), *Tracking Environmental Change Using Lake Sediments, Vol. 3: Terrestrial, Algal and Siliceous Indicators*, pp. 235–52. Dordrecht: Kluwer Academic Publishers.

Piperno, D. R., and Holst, I. 1998. The presence of starch grains on prehistoric stone tools from the humid Neotropics: Indications of early tuber use and agriculture in Panama. *Journal of Archaeological Science* 25: 765–76.

Piperno, D. R., Ranere, A. J., Holst, I., and Hansell, P. 2000. Starch grains reveal early root crop horticulture in a Panamanian tropical forest. *Nature* 407: 894–97.

Raven, P. H., Evert, R. F., and Eichhorn, S. E. 1992. *Biology of Plants*. New York: Worth Publishers.

Roberts, R., Walsh, G., Murray, A., Olley, J., Jones, R., Morwood, M., Tuniz, C., Lawson, E., Macphail, M., Bowdery, D., and Numann, I. 1997. Luminescence dating of rock art and past environments using mud-wasp nests in northern Australia. *Nature* 387: 696–99.

Rowe, C. 2005. A Holocene history of vegetation change in the western Torres Strait region, Queensland, Australia. Unpublished Ph.D. Thesis, School of Geography and Environmental Science, Monash University.

———. 2006. Landscapes in Torres Strait Prehistory, in B. David, B. Barker, and I. J. McNiven (eds.), *The Social Archaeology of Australian Indigenous Societies*, pp. 270–86. Canberra: Aboriginal Studies Press.

Runge, F. 1999. The opal phytolith inventory of soils in central Africa: Quantities, shapes, classification and spectra. *Review of Palaeobotany and Palynology* 107: 23–53.

Sandgren, C. D., Smol, J. P., and Kristiansen, J. 1995. *Chrysophyte Algae: Ecology Phylogeny and Development*. Cambridge: Cambridge University Press.

Smol, J. P. 1986. Chrysophycean microfossil as indicators of lakewater pH, in J. P. Smol (ed.), *Diatoms and Lake Acidity*, pp. 275–87. Dordrecht: Dr. W. Junk Publishers.

Snoeijs, P. 1999. Diatoms and environmental change in brackish waters, in E. F. Stoermer and J. P. Smol (eds.), *The Diatoms: Applications for Environmental and Earth Sciences*, pp. 298–333. Cambridge: Cambridge University Press.

Steinberg, C., Hartmann, H., and Krause-de Iiin, D. 1988. Paleoindicators of acidification in Kleiner Abersee (Federal Republic of Germany, Bavarian Forest) by chydorids, chrysophytes, and diatoms. *Journal of Paleolimnology* 6: 123–40.

Traverse, A. 1988. *Paleopalynology*. Winchester: Allen and Unwin.

Turney, C. S. M., Kershaw, A. P., Moss, P., Bird, M. I., Fifield, L. K., Cresswell, R. G., Santos, G. M., Tada, M. L. D., Hausladen, P. A., and Zhou, Y. 2001. Redating the onset of burning at Lynch's Crater (north Queensland): Implications for human settlement in Australia. *Journal of Quaternary Science* 16: 767–71.

Tzedakis, P. C., Andrieu, V., de Beaulieu, J.-L., Crowhurst, S., Follieri, M., Hooghiemstra, H., Magri, D., Reille, M., Sadori, L., Shackleton, N. J., and Wijmstra, T. A. 1997. Comparison of terrestrial and marine records of changing climate of the last 500,000 years. *Earth and Planetary Science Letters* 150: 171–76.

Ugent, D., Dillehay, T., and Ramirez, C. 1987. Potato remains from a late Pleistocene settlement in south-central Chile. *Economic Botany* 41: 401–15.

van Geel, B. 1976. Fossil spores of Zygnemataceae in ditches of a prehistoric settlement in Hoogkarspel (The Netherlands). *Review of Palaeobotany and Palynology* 55: 337–44.

———. 1986. Diatom Analysis, in B. E. Berglund (ed.), *Handbook of Holocene Palaeoecology and Palaeohydrology*, pp. 527–70. Chichester: John Wiley and Sons.

———. 2001. Non-pollen palynomorphs, in J. P. Smol, H. J. B. Birks, and W. M. Last (eds.), *Tracking Environmental Change Using Lake Sediments, Vol. 3: Terrestrial, Algal and Siliceous Indicators*, pp. 99–120. Dordrecht: Kluwer Academic Publishers.

Walker, D. 1990. Purpose and method in Quaternary palynology. *Review of Palynology and Palaeobotany* 64: 13–27.

Wallis, L. A. 2000. Phytoliths, late Quaternary environments and archaeology in tropical semi-arid northwest Australia. Unpublished Ph.D. thesis, Australian National University, Canberra.

Willerding, U. 1991. Presence, preservation and representation of archaeological plant remains, in W. Van Zeist, K. Wasylikowa, and K. E. Behre (eds.), *Progress in Old World Palaeoethnobotany: A retrospective view on the occasion of 20 years of the international work group for palaeoethnobotany*, pp. 25–52. Rotterdam: A. A. Balkema.

Williams, M., Dunkerley, D., de Deckker, P., and Kershaw, P. 1998. *Quaternary Environments*. London: Hodder Education.

Wolin, J. A., and Duthie, H. C. 1999. Diatoms as indicators of water level change in freshwater lakes, in E. F. Stoermer and J. P. Smol (eds.), *The Diatoms. Applications for the Environmental and Earth Sciences*, pp. 183–204. Cambridge: Cambridge University Press.

Zeeb, B. A., and Smol, J. P. 2001. Chrysophyte scales and cysts, in J. P. Smol and H. H. Birks (eds.), *Tracking Environmental Change Using Lake Sediments, Vol. 3: Terrestrial, Algal and Siliceous Indicators*, pp. 203–24. Dordrecht: Kluwer Academic Publishers.

Zhao, Z., and Piperno, D. R. 1999. Late Pleistocene/Holocene environments in middle Yangtze River Valley, China and rice (Oryza sativa) domestication: The phytolith evidence. *Geoarchaeology* 15: 203–22.

43

BEYOND ECONOMY: SEED ANALYSIS IN LANDSCAPE ARCHAEOLOGY

Andrew Stephen Fairbairn

Forest, field, pasture, moor, hedgerow, garden, plantation, and bog—all are landscape elements partially defined on the basis of the plants growing there; all have a specific set of environmental, ecological, and structural characteristics; all have a color, texture, and shape; all are living and changing elements of the material world in which human experience is situated; all dominate the visible form of land on which they are found; and several (for example, plantation, forest, garden, field, pasture) carry specific economic, social, and political meanings, depending on historical and geographical context. Like it or not, plants are key landscape elements, being directly, or through the land on which they grow, valuable, useful, negotiable, and politically active economic and social resources. Plants also may carry symbolic as well as use values, (for example, the four-leaf clover as a sign of good luck); sometimes plants even stand as symbols of cosmology or identity, appearing as actors in creation beliefs (for instance, the Grass Seed Dreaming of Aboriginal peoples in central Australia; cf. David 2002: 63) and being used in the construction of group identity (for example, the Scottish thistle; the Irish shamrock; the Australian waratah).

Among the most useful archaeological sources for directly investigating the complex web of relationships involving plants and people are seeds (see "Material and Method," below). Seeds are commonly used for numerous purposes (for example, food, oils, drugs), widely preserved in many archaeological contexts, and are also identifiable to a high level of specificity. Seed analysis has been vital for understanding past diet, economy, and the origins of agriculture (e.g., Hillman 2000; Weiss, Kislev, and Hartmann 2006; Wetterstrom 1986) but has played a less important role than it should in landscape archaeology. This, in part, reflects the narrow theoretical base that analysts often bring to their work (see Thomas 1990) and also the perception of "science-based" archaeology in the broader archaeological world. Here I use examples from several geographical regions, but especially southwest Asia, to review how the analysis of seeds from archaeological sites has been used in landscape archaeology and why it could make a more substantial contribution than it usually does. This discussion is followed by a brief introductory guide to the methods of seed analysis, which focuses on the analysis of seeds from archaeological contexts, rather than from natural sediments.

Theory, Practice, and the Construction of Archaeobotanical Knowledge

Seeds are directly studied within the subdiscipline of archaeobotany (here considered synonymous with palaeoethnobotany), which generally continues to interpret the human career in the functionalist adaptive terms of cultural evolution. Key research themes mostly reflect that theoretical stance and include subsistence (that is, the quest for food and other essential items); environmental adaptation; and economic change (especially relating to agricultural and technical innovation). Archaeobotany's approach to landscape archaeology has been largely framed within these terms, describing the environment and the resources it contains, then mapping people onto it through the identification of activities represented in the seeds, fruits, charcoals, and other botanical remains found on archaeological sites. Location of gathering/production activities can be identified through the environmental preferences of the plants involved, with reference to reconstruction of the local environment and modern analogues. Production, processing, and consumption activities are identified with reference to ethnographic observations of plant use, in the tradition of "middle range" research. Archaeobotany also follows its own ethnobotanical and ecological interests that have a bearing on landscape, such as the development of agricultural techniques and human modification of environments. The subject has developed a powerful set of tools and in less than 30 years has gone from a marginal interest to a central, high profile part of archaeological enquiry that has huge time depth (e.g., Goren-Inbar et al. 2004) and is involved in several key debates about the origins and historical trajectories of the human condition (see also Bar-Yosef, Terrell and Hart, both this volume).

Plant-based activities, including agriculture, gathering, and forestry, are major human engagements with the nonhuman material world, requiring social cooperation for their success, and are part of the suite of activities through which people experience and create their world. Archaeobotany has usually assumed that material needs drive human engagement with the nonhuman world and that the logical need for resources determines social strategies for their acquisition. In the standard logic of evolutionary cultural ecology, resources to fulfill material needs are determined by the environment so that society ultimately reflects adaptation to the environment via the economy, which is the means of producing, extracting, distributing, and consuming these resources. It is generally assumed that economic exploitation of plants is practiced optimally and that there is little explicit discussion within the subject of how knowledge about plants and places was achieved or created in the past—often a high level of knowledge concerning the utility and the distribution of plants by past populations is assumed. Many archaeobotanical treatments, though not all, continue to follow this theoretical approach, which, in the positivist tradition, denies the relevance of social, ideological, and cultural phenomena for understanding human affairs and assumes that economic and social phenomena can be analyzed in isolation from each other.

Although cultural ecology often underpins conceptions of human-plant relationships, detailed knowledge of which plants were used and what they were used for relies on the application to seed data of "real-world" knowledge drawn from ethnographic, experimental, and ecological studies. Such applications of seed data can range from use of the ecological tolerances of a species to understand past environments, to complex predictive models that seek to identify human behaviors via compositional analysis of their material traces. Such models, a product of middle range theory (see Charles and Halstead 2001), have been particularly useful in understanding crop processing and husbandry practices as manifest in seed remains (e.g., Bogaard 2004; Hillman 1984).

Archaeobotanical knowledge is thus largely created and articulated in a particular way, which is not only difficult for nonspecialists to understand but also has explored only some of the interpretative potential plant-centered studies offer for understanding past human experience. Furthermore, because archaeology as a whole, and landscape archaeology in particular, have moved beyond cultural ecology and middle range theory, the themes studied in many seed analyses are themselves of marginal concern to current debates. This lack of engagement with contemporary theory and focus on archaeobotanical interests have led to a cursory treatment of plants and vegetation in many archaeologies of landscape, a trend exacerbated by the marginalization of "science-based" archaeology by postprocessual polemic. This is unfortunate, because, as I have attempted to articulate above, plants are key and ever-changing components of landscape, and knowledge about them is essential for understanding both the landscape's changing material form over time, even if used as simply a backdrop for human agents, and its role in the social creation of the human world. Exclusion to a scientific margin denies landscape archaeology the full use of the important information archaeobotany already provides about human actions and the largely untapped social, visual, structural, and ideological properties of plants and vegetation. Full exploitation of this information requires a broadening of archaeobotanical theory and a less exclusive attitude by landscape archaeologists to "scientific" subdisciplines and the information they provide. However, there are encouraging signs that debate on this issue has shifted from a negative to a positive direction, recognizing that science-based studies provide valuable information relevant to various interpretative approaches (Charles and Halstead 2001; Evans 2003; Jones 2002).

Themes in Landscape Archaeology

Mapping and Experiencing Landscapes

Contemporary theory questions the relevance of the objective geographically informed notions of space assumed to be relevant in most science-based archaeology as a basis for understanding how people in the past understood their world (e.g., Tilley 1994). Emphasis instead has moved toward investigating personal engagement with places as the basis for constructing landscapes and worldviews. Seed analysis can provide information to inform these readings at several levels.

First, analysis of archaeological seeds—those preserved in archaeological features by human hand—provides direct evidence of the plant species encountered in the past (see below). Identification of species allows the plant type (physical form, biological group, and life history) to be established and, when preserved together, the composition and form of vegetation (groups of plants) encountered in past landscapes to be understood. This information itself can be of use for identifying the presence and maintenance of particular plant ecological formations, including human-created ones such as the grasslands of southern England (Greig 1984), managed forests in the tropics (Kennedy and Clarke 2004), and "wild orchards" of southwest Asia (Woldring and Cappers 2001). Furthermore, as plants and vegetation often have specific environmental preferences, those plants and vegetation types can be mapped spatially onto a landscape if the details of its soil distribution, landforms, and microenvironments are known. This type of information is routinely used as a basis for understanding the distribution of resources in relation to excavated sites, often using modern vegetation analogues to fill in the gaps in archaeological data, and is directly derived from the tradition of site catchment analysis. Seed analysis, as part of multidisciplinary investigations, provides an important source of information for mapping plants in space.

Seed analysis also allows the direct identification of human engagements with plants and has been of great use in identifying activities related to the production, procurement, processing, and consumption of plant resources. For example, crop-processing stage (that is, a particular step in the removal of edible grain from inedible chaff) is routinely identified in Old World archaeobotany via quantitative comparison to ethnographic models of plant use (e.g., Bogaard 2004). Largely, this has been done to identify the selective effects of crop processing on crop weed seed presence, a key means of identifying the detail of agricultural techniques used, such as sowing, ploughing, and manuring, which are important for understanding how crops were grown and the spread of production systems (see Bogaard 2004). Effective means of identifying processing techniques have also been formulated for gathered plants (e.g., Head, Atchison, and Fullagar 2002). Other studies have used similar data to reconstruct patterns of regional trade and exchange (e.g., Weiss and Kislev 2003).

We can make an interpretative leap from landscape mapping and identifying plant-based activities to understanding where encounters with plants took place in the past. Plants grow in particular spots in the landscape with particular environmental characteristics, identifiable via the tolerances of the plant found in archaeological assemblages. For example, weed seeds found with crop seed provide information about where in the landscape fields were situated, based on the soil preference of those weedy plants. Thus, seed analysis provides a means of sensing human encounters with place and thus the creation of the world through lived experience. The spatial detail of place that seed analysis can bring depends largely on the floral structure of the landscape and the environmental specificity of the preserved plants; it would be impossible to spatially differentiate the site of an encounter in a landscape covered by a single vegetation type, but it would be possible where steep environmental gradients caused changes in the composition and structure of vegetation. For example, the landscape of Neolithic Çatalhöyük in Turkey was one of sharply defined environmental gradients from the immediate landscape of lake and wetland to the more distant calcareous arid steppe to distant well-watered hills (Fairbairn, Atchison, and Fullagar 2005). The presence of hackberry seeds in rubbish dumps at the site, a plant of the hill zone providing edible fruits, immediately indicates movement to, from, and within that area; wetland plant seeds in sheep/goat dung indicate gathering of fodder or grazing on the wetland closer to the site.

We can thus use a contextual seed analysis to construct an experienced landscape of paths and places where particular activities were carried out or, to use Ingold's term, a taskscape (Ingold 1993). Agricultural techniques, such as harvesting and threshing, take place somewhere in the landscape—harvesting takes place in the crop field; threshing may take place there or in a village threshing ground; removing the last seeds before preparing a meal may take place in the home. Also, the identification of a particular activity usually implies others—harvesting in the field and threshing in the village implies that the crops were transported from one place to another. At Çatalhöyük, the evidence suggests a particularly complex construction and use of the landscape for many kilometers around the site, including crop production, which may have included management of fields at great distance from the site itself (Fairbairn 2005a). This study challenged the optimizing assumptions inherent in interpretations of agriculture informed by cultural ecology and instead provided a culturally specific and historically grounded interpretation of praxis. Archaeological analysis in Australia has also suggested that gathering fruits may be as much about seeing country as about gathering food (Head, Atchison, and Fullagar 2002), while a restraint on seed grinding was at times as much about the way the seed grinder felt about seed grinding as it was about food production (David 2002: chapter 8).

Landscapes also have visual and structural qualities that vary with conditions, such as the weather, and also with the life and growth cycles of plants. Plants, especially trees, are important structural elements of the inhabited world and could be considered to define places in a similar way to stone monuments (e.g., Barrett 1996). Plants have greatly differing growth forms that vary with the vegetation of which they are part, because trees growing in an open field will be more compact than the specimens growing in closed woodland, which tend to be taller and more open in structure. They also vary in color and openness, depending partly on the season in temperate zones, where autumn brings a loss of leaves and an opening of vegetation. Traditionally, archaeobotany ignores these properties, but that does not mean they can't be established from seed lists and may perhaps add additional elements to phenomenological interpretations (see Tilley this volume).

Sociality and Social Order

An understanding of plants and plant-based activities is relevant to understanding the sociality of landscape for human agents and landscape as an active element in the negotiation of social order. As

pointed out by Evans (2003), humans rarely have direct social relationships with plants. Rather, it is the places and activities involving plants that provide foci for social discourse. Plants grow on areas of land, the holding of and access to which is usually mediated through social negotiation. Identifying that a plant was gathered from a place implies that a social process allowed that to happen. Finding, growing, producing, processing, and using plants requires also decisions to be made about which activities should be carried out, the location of those tasks, the timing of events, and the organization of people (that is, labor) to ensure they are completed satisfactorily. Many plant-based tasks require the cooperation of more than one person in their practical application or in obtaining permission to carry them out. They are, therefore, social as well as economic activities subject to the disagreement, political negotiation and unintended consequences that accompany just about any decision involving land, resources, and people in any society.

The social power of plant foods has been widely explored (e.g., Hastorf 1991); less attention has been given to understanding this aspect of plants and plant use in the landscape. At Çatalhöyük, tasks identified from seed analyses were not uniformly distributed evenly across wetland and dryland landscapes; parts of the landscape came in and out of social focus depending on the time of year and type of activity—large groups in the distant hill zone focused on fields during harvest and perhaps dispersed in the surrounding woodland when collecting fruits; a greater intensity of social engagement within houses focused on preparing seed stores before winter (Fairbairn, Near, and Martinoli 2005). At the same site, intense social collaboration, including crop production and distribution, and a strong sense of group identity, manifest in part on the basis of landownership, may have supported an "inefficient" production system that produced internal tensions leading to collapse of the system when arable land availability increased owing to environmental change (Fairbairn 2005a).

Time and Memory

If landscape is time materializing (Bender 2002), plants provide some of the key material signs by which time is experienced and understood. In seasonal climates, plants have cyclical growth patterns: spring with its flush of new greenery, autumn with its fruit and leaf-fall and winter with its bare branches. In tropical climates, change may be much more subtle and less predictable, but seasonal change in plant appearance occurs nonetheless and may be an important means of conceptualizing time. Seasonal change has mainly appeared in archaeobotanical landscape readings with regard to understanding resource schedules, which detail the resources available, or actually used as shown in archaeological studies, in parts of the landscape throughout the year. Schedules are important tools for understanding some of the limits of human action within seasonal landscapes, though they are often used prescriptively and assume that people have the knowledge to use an environment optimally. They can, however, be used to understand how different people experienced different parts of the landscape at different times of the year. For example, at Neolithic Çatalhöyük, the distant hills became key places during the harvest season for active adults but may have been contained to memory during the winter and spring, when other activities nearer the houses were more pressing and access was more difficult (Fairbairn et al. 2006). For the very old or disabled, some distant places may have become excluded from direct experience and kept alive only in the memory. Usually, long-term, extra-annual patterns of vegetation change have been constructed to understand patterns of changing resources and have been important for understanding the origins of agriculture in Asia (e.g., Hillman 1996). The patterns of plant growth so created may have been experienced and understood in a number of ways and could be interpreted accordingly. Evidence of famine, crop failure, and environmental catastrophe, as in archaeological and other evidence of economic, social, and landscape change, caused by large-scale changes in weather or human agency could be read in terms of times of plenty, despair, and recovery (see Evans 2003) as manifest in landscapes of plenty, despair, and recovery.

Landscape Creation and Domestication

Seed analysis is a key means of understanding the human modification and creation of landscapes through selecting, removing, planting, and translocating plants growing there, in association with burning and effects of grazing livestock. Human modification of vegetation dates from at least the Late Pleistocene and was well advanced in many regions by the mid-Holocene, being most closely associated with agriculture and the development of Classical civilizations. At one extreme, humans formally domesticated some plant species—that is, crops—by establishing control over reproduction, selection, and dispersal mechanisms, and seed analysis has been fundamentally important

for identifying those processes (see Denham et al. 2003; Fuller et al. 2004; Kislev, Hartmann, and Bar-Yosef 2006; papers in Colledge and Conolly 2007; Weiss, Kislev, and Hartmann 2006; Zohary and Hopf 2000). Human-modified, domesticated landscapes also developed in most regions of the world in the distant past, and their identification through archaeobotany provides perhaps the only way of investigating plant domestication in some tropical areas (Fairbairn 2005b; Terrell et al. 2003).

Material and Method

The following section provides a very brief and general overview of some key methods and materials for archaeological seed analysis. For more details consult the cited texts, especially Pearsall (2000), and other published seed analyses in the journals detailed below.

What Are Seeds?

In archaeobotany, *seed* is a convenient shorthand term that includes the following plant structures (which I here define in their strict botanical sense):

1. Fruits—found only in flowering plants (Angiosperms), developed from the plant ovary and include nutshells (for example, hazelnut, or *Canarium*, nutshells) and legume pods, but may also be fused to the seed, as in the case of grass grains (caryopses). Seeds in the strict sense develop within these fruiting structures and may be ejected from the fruit when dispersed (for instance, in legumes) or dispersed attached to fruiting structures (for example, cereal grains, plum or *Canarium* seeds dispersed within their stones or endocarps).

2. Seeds—developed from the ovule, the product of sexual reproduction in the conifers (Gymnosperms) and flowering plants (Angiosperms), consisting of an embryonic plant, and in many cases storage tissue, held within a protective coat (testa); legumes produce seeds (for example, a pea) within pods (the fruits); Bunya pine (a conifer) nuts, are, in fact, seeds, since they contain no ovarian tissue; *Pandanus* keys, *Canarium* nuts, and hazelnuts all consist of a hard shell (the fruit) containing seeds within (the kernels).

Seed analysis also usually includes the analysis of chaff, which consists of the leaflike structures (glumes, rachis fragments, lemmas, awns) found in the cereals and other grasses that protect the grain and is removed by processing techniques before the grain is eaten.

Seed is thus a useful way of referring to the many plant parts associated with production and dispersal of new plants by sexual reproduction. It cuts across the confusing mix of such scientific and common English definitions of terms as *seeds, nuts,* and *fruits*. Seeds are diverse in shape and size and are produced to disperse and to protect embryonic plants in their journey from parent to growth site; hence, they are usually compact and robust. Seed production rate varies widely between plant species and depends in part on reproductive strategy: low in plants that successfully propagate by other means (for example, rhizomes) and higher in those short-lived plants, such as many arable weeds, that rely solely on this means of propagation.

Where and How Are They Preserved?

Most plant parts decay but are preserved when they (1) are naturally resistant to decay; (2) are deposited in an environment where decay is suppressed (for example, an anaerobic swamp); or (3) are transformed into a state that resists decay (for instance, charring [partial burning] or mineralization [fossilization]). Seeds are naturally robust and have a greater chance than most other plant parts of surviving in archaeological deposits in an identifiable state. As with production and distribution, preservation potential varies between plant species. Some small and fragile seeds, such as those of willow (*Salix* spp.), are rarely, if ever, preserved, whereas the silica rich seeds of the arable weed corn gromwell (*Lithospermum arvense*) are often preserved regardless of preservation conditions. Most seeds are preserved on dry archaeological sites by charring, with anaerobic conditions in wells, ditches, pits, and latrines providing important loci for preservation of waterlogged remains.

Field Techniques

Seeds are usually recovered from soil samples collected from archaeological contexts. Sampling strategy and sample size are subjects of some debate and vary with analytical aims and resources (Pearsall 2000; van der Veen and Fieller 1982). Seeds are only systematically recovered from soil samples by sieving, most being far too small to extract from archaeological sediments by hand. Charred seeds are commonly recovered using a combination of wet sieving—sieving sediment samples in water—and flotation, in which water is used to separate buoyant charred particles from the soil matrix and collected in a sieve. Sieve mesh should be no larger than 250 μm in diameter to collect the smallest commonly found seeds. Various sampling strategies, sample size, and flotation methods are commonly used (Pearsall 2000).

Laboratory Methods

Analysis involves separation of the seeds from other preserved materials using low-powered microscopy, usually after sieving into size fractions (for example, > 4 mm, 2–4 mm). Very large samples can be split into subsamples before analysis. Because small subsamples may be unrepresentative of the original sample population, care should be taken when developing subsampling strategies. Experimentation has shown that subsampling is most representative when using a geological sample splitter (also known as a riffle-box), although other methods can be almost as accurate (van der Veen and Fieller 1982).

Seeds are at the sharp-end of evolutionary selection pressures, their form being directly related to a species' reproductive success. They are thus often highly differentiated in form and are easier to identify to species level than many other plant fossil groups (for example, wood and pollen) and so provide a greater level of detail about environmental conditions and plant community composition. Identification follows uniformitarian principles, relying on comparison of ancient specimens to those from known plant species. Identification is a key analytical step, assigning a plant identity to the fossil type and with it all the characteristics of that plant type. A collection of modern seed reference specimens is vital for accurate identification (see Nesbitt, Colledge, and Murray 2003), though seed manuals are also of great value (e.g., Cappers, Bekker, and Jans 2006; Jacomet 2006; for a good list of older sources see Nesbitt and Greig 1989), especially those with well-designed keys (e.g., Nesbitt 2006). Although some seeds are identifiable with full Linnean species, many can be identified only at a lower level of taxonomic specificity, such as family, subfamily, genus, and type, depending on biological differentiation and preservation. Furthermore, identifications vary in their certainty as reflected in commonly used notation.

Identification uses (1) morphology (that is, seed shape size and form visible under low and high-powered microscopy) and (2) anatomy, (that is, the plant structure, visible using high powered microscopy, including scanning electron microscopy

[SEM]). Shape and features such as surface cell pattern should be described using standard terminology (e.g., Nesbitt 2006), and shape may also be described using measurements of length, breadth, and thickness, which are especially useful in analysis of plant groups with similar seed forms (for example, grasses, including cereals). Description and preferably illustration of specimens are important to help validate identifications, especially in poorly known regions. High quality photographs using SEM or digital imaging on light microscopes are now routine. Drawings are more time consuming and difficult to produce but are perhaps the best form of visual recording, since they can be drawn to emphasize key diagnostic features.

Quantification allows analysis of patterns in seed data and comparison to quantitative models identifying particular plant-base activities. Seeds are counted, though large fragmented items may be quantified by weight. Fragmentation can be accommodated by (1) generating a minimum number of individuals (MNI), counting only parts that appear once in every seed or (2) dividing the fragment weight by a single seed equivalent, based on seeds preserved in the same way from the same site or region ("thousand grain weight"). Abundance data can be used in exploratory statistical investigations—for example, using correspondence analysis—to look for temporal or spatial trends. Sample data can also be compared to external ethnobotanical and ecological data using discriminant analysis or canonical correspondence analysis, as successfully applied to crop weed data to remove the effects of crop processing on sample composition (see example given by Bogaard 2004). Presence and ubiquity (% presence) data are also commonly generated to look at general trends in plant appearance.

Identifying Ecology and Activities

Seed analysis uses the known form and environmental characteristics of modern examples of identified species as a basis for identifying past environmental conditions and vegetation composition/structure. Seed assemblage composition is commonly used to identify aspects of crop processing and husbandry practices, as well as the effect of the former on sample composition and comparability (e.g., Bogaard 2004; Hillman 1984). Assemblage taphonomy—that is, the processes affecting the source of fossils and sample composition— are vitally important for understanding the interpretative limits of analytical comparisons made using these analogues (see Wilkinson and Stevens 2003).

Conclusions

Seeds are widely preserved in archaeological sites, and seed analysis provides a means of empirically investigating the use of plants by ancient people and thus a direct means of sensing ancient landscapes. Methods for recovering, identifying, analyzing, and interpreting seed assemblages are well established using standard scientific principles and with reference to modern ecology and ethnographic analogues. Though most interpretation has been framed historically within the positivist traditions of cultural ecology, seed analysis has the potential to provide information suited to landscape archaeology studies informed by a wide range of theoretical positions. Thus the subject should in the future move beyond the narrow confines of palaeoeconomy and palaeoecology to address a diverse range of contemporary interests in landscape archaeology.

Resources for Further Study

1. Key general texts

   The most comprehensive recent overview of archaeobotany in English is Pearsall (2000). Other overviews include Renfrew (1973), papers in Hastorf and Popper (1988), Greig (1989), and Hastorf (1999), as well as Evans's (2003) groundbreaking interpretative text. Wilkinson and Stevens (2003) provide a good introduction to seed analysis within a broader discussion of environmental archaeology.

2. Journals

   a. *Vegetation History and Archaeobotany*

   b. *Environmental Archaeology; Journal of Human Palaeoecology*

   c. *Journal of Archaeological Science*

   d. *Review of Palaeobotany and Palynology*

3. Online databases of archaeological finds and reference collections from the Old World

   a. Literature on archaeological remains of cultivated plants 1981–2004 (www.archaeobotany.de/)

   b. Archaeobotanical database of Eastern Mediterranean and Near Eastern sites (www.cuminum.de/archaeobotany/)

   c. Naomi Miller's database of Near Eastern archaeobotanical publications (www.sas.upenn.edu/~nmiller0/)

d. George Willcox's website for southwest Asia (http://perso.wanadoo.fr/g.willcox/)

e. The digital seed atlas of the Netherlands (www.seedatlas.nl/)

4. Relevant scientific associations

a. International Workgroup in Palaeoethnobotany (IWGP) (www.palaeoethnobotany.com/)

b. Association for Environmental Archaeology (AEA) (www.envarch.net/)

c. Society for Archaeological Sciences (SAS) (www.socarchsci.org/)

References

Barrett, J. 1996. *Fragments from Antiquity*. London: Blackwell.

Bender, B. 2002. Time and landscape. *Current Anthropology* 43: S103–12.

Bogaard, A. 2004. *Neolithic Farming in Central Europe*. London: Routledge.

Cappers, R. T. J., Bekker, R. M., and Jans, J. E. A. 2006. *Digital Seed Atlas of the Netherlands*. Groningen Archaeological Studies 4.

Charles, M., and Halstead, P. 2001. Biological resource exploitation: Problems of theory and method, in D. R. Brothwell and A. M. Pollard (eds.), *Handbook of Archaeological Sciences*, pp. 365–78. Chichester: J. Wiley.

Colledge, S., and Conolly, J. 2007. *The Origin and Spread of Domestic Plants in Southwest Asia and Europe*. Walnut Creek, CA: Left Coast Press.

David, B. 2002. *Landscapes, Rock Art and the Dreaming: An Archaeology of Preunderstanding*. London: Leicester University Press.

Denham, T. P., Haberle, S. G., Lentfer, C., Fullagar, R., Field, J., Therin, M., Porch, N., and Winsborough, B. 2003. Origins of agriculture at Kuk Swamp in the Highlands of New Guinea. *Science* 301: 189–93.

Evans, J. 2003. *Environmental Archaeology and the Social Order*. London: Routledge.

Fairbairn, A. 2005a. A history of agriculture at Çatalhöyük East, Turkey. *World Archaeology* 37: 197–210.

———. 2005b. An archaeobotanical perspective on Holocene plant use practices in lowland northern New Guinea. *World Archaeology* 37: 487–502.

Fairbairn, A., Asouti, E., Russell, N., and Swogger, J. 2006. Seasonality, in I. Hodder (ed.), *Çatalhöyük Perspectives: Themes from the 1995–1999 seasons*, pp. 93–108. Cambridge/Ankara: McDonald Institute for Archaeological Research/British Institute of Archaeology in Ankara.

Fairbairn, A., Near, J., and Martinoli, D. 2005. Macrobotanical investigations of the North, South and KOPAL areas at Çatalhöyük, in I. Hodder (ed.), *Inhabiting Çatalhöyük: Reports from the 1995–1999 Seasons*, pp. 137–201. Cambridge/Ankara: McDonald Institute for Archaeological Research/British Institute of Archaeology at Ankara.

Fuller, D., Korisettar, R., Venkatasubbaiah, P. C., and Jones, M. K. 2004. Early plant domestications in southern India: Some preliminary archaeobotanical results. *Vegetation History and Archaeobotany* 13: 115–29.

Goren-Inbar, N., Alperson, N., Kislev, M. E., Simchoni, O., Melamed, Y., Ben-Nun, A., and Werker, E. 2004. Evidence of hominin control of fire at Gesher Benot Ya`aqov, Israel. *Science* 304: 725–27.

Greig, J. 1984. The palaeoecology of some British hay meadow types, in W. van Zeist and W. A. Casparie (eds.), *Plants and Ancient Man*, pp. 213–26. Rotterdam: Balkema.

———. 1989. *Archaeobotany*. Handbooks for Archaeologists No. 4. Strasbourg: European Science Foundation.

Hastorf, C. A. 1991. Gender, space and food in prehistory, in J. M. Gero and M. W. Conkey (eds.), *Engendering Archaeology*, pp. 132–59. London: Blackwell.

———. 1999. Recent research in paleoethnobotany. *Journal of Archaeological Research* 7: 55–103.

Hastorf, C. A., and Popper, V. S. (eds.). 1988. *Current Palaeoethnobotany: Analytical Methods and Cultural Interpretations of Archaeological Plant Remains*. Chicago: University of Chicago Press.

Head, L., Atchison, J., and Fullagar, R. 2002. Country and garden: Ethnobotany, archaeobotany and Aboriginal landscapes near the Keep River, northwestern Australia. *Journal of Social Archaeology* 2: 173–96.

Hillman, G. C. 1984. Interpretation of archaeological plant remains: Application of ethnographic models from Turkey, in W. A. Casparie and W. van Zeist (eds.), *Plants and Ancient Man*, pp. 1–41. Rotterdam: Balkema.

———. 1996. Late Pleistocene changes in wild plantfoods available to hunter-gatherers of the northern Fertile Crescent: Possible preludes to cereal cultivation, in D. R. Harris (ed.), *The Origins and Spread of Agriculture and Pastoralism in Eurasia*, p. 159. London: UCL Press.

———. 2000. Abu Hureyra 1: The Epipaleolithic, in A. M. T. Moore, G. C. Hillman, and A. J. Legge (eds.), *Village on the Euphrates*, pp. 327–99. Oxford: Oxford University Press.

Ingold, T. 1993. The temporality of the landscape. *World Archaeology* 25: 152–74.

Jacomet, S. 2006. *Identification of Cereal Remains from Archaeological Sites* (2nd ed.). Basel: IPNA. See: http://pages.unibas.ch/arch/archbot/index.htm for free download.

Jones, A. 2002. *Archaeological Theory and Scientific Practice*. Cambridge: Cambridge University Press.

Kennedy, J., and Clarke, W. 2004. *Cultivated Landscapes of the Southwest Pacific*. Canberra: Australian National University. See http://eprints.anu.edu.au/archive/00002531/

Kislev, M. E., Hartmann, A., and Bar-Yosef, O. 2006. Early Fig domestication in the Jordan Valley *Science* 312: 132–37.

Nesbitt, M. 2006. *Identification Guide for Near Eastern Grass Seeds*. London: Institute of Archaeology.

Nesbitt, M., Colledge, S., and Murray, M. A. 2003. Organisation and management of seed reference collections. *Environmental Archaeology* 8: 77–84.

Nesbitt, M., and Greig, J. 1989. A bibliography for the archaeobotanical identification of seeds from Europe and the Near East. *Circaea* 7: 11–30.

Pearsall, D. M. 2000. *Palaeoethnobotany: A Handbook of Procedures*. New York: Academic Press.

Renfrew, J. 1973. *Palaeoethnobotany*. London: Methuen.

Terrell, J. E., Hart, J. P., Barut, S., Cellinese, N., Curet, A., Denham, T., Kusimba, C. M., Latinis, K., Oka, R., Palka, J., Pohl, M. E. D., Pope, K. O., Williams, P. R., Haines, H., and Staller, J. E. 2003. Domesticated landscapes: The subsistence ecology of plant and animal domestication. *Journal of Archaeological Method and Theory* 10: 323–68.

Thomas, J. 1990. Silent running: The ills of environmental archaeology. *Scottish Archaeological Review* 7: 2–7.

Tilley, C. 1994. *The Phenomenology of Landscape*. Oxford: Berg.

van der Veen, M., and Fieller, N. 1982. Sampling seeds. *Journal of Archaeological Science* 9: 287–98.

Weiss, E., and Kislev, M. 2003. Plant remains as indicators for economic activity: A case study from Iron Age Ashkelon. *Journal of Archaeological Science* 31: 1–13.

Weiss, E., Kislev, M. E., and Hartmann, A. 2006. Autonomous cultivation before domestication. *Science* 312: 1608–10.

Wetterstrom, W. 1986. *Food, Diet and Population at Arroyo Hondo Pueblo, New Mexico*. Santa Fe: School of American Research Press.

Wilkinson, K., and Stevens, C. 2003. *Environmental Archaeology: Approaches, Techniques and Applications*. Stroud: Tempus.

Woldring, H., and Cappers, R. 2001. The origin of the "wild orchards" of central Anatolia. *Turkish Journal of Botany* 25: 1–9.

Zohary, D., and Hopf, M. 2000. *The Domestication of Plants in the Old World*. Oxford: Oxford University Press.

44

THE USE OF WOOD CHARCOAL IN LANDSCAPE ARCHAEOLOGY

Nic Dolby

Charcoal analysis has traditionally concerned the identification, quantification, and analysis of woody plants present as charcoal in archaeological and other sediments. Charcoal is common in many archaeological sites and represents material remains of human activity. Although a source of dating, charcoal also provides identification of the wood of trees and shrubs utilized as firewood and should be used to give insight into larger questions within archaeology beyond construction of chronologies and the past environment. Some scholars have extended the charcoal analysis to include fragmentation of the charcoal as an additional facet (Dolby In preparation; Hesse and Rosen 1988), while the integration of studies on charcoal taphonomy and the ensuing implications for archaeological and paleoenvironmental studies has proceeded fitfully (see Dolby 1995, In preparation; Prior and Alvin 1983, 1986; Prior and Gasson 1993; Rossen and Olson 1985). This chapter outlines ways in which charcoal analyses have been incorporated into landscape archaeology and suggests new potential approaches. I begin with a brief history and outline of methodologies.

History of Charcoal Analysis

Charcoal analysis has been carried out on archaeological charcoal for over a century. Heer and Passerini (Passerini 1864; Heer 1866; Heer and Passerini 1865, all cited in Castelletti 1990; see also Asouti 2006) are widely regarded as pioneers in the field. Charcoal analyses during these early years usually entailed the construction of "shopping lists" of taxa present, with at best a random citation of ethnographic or recent uses of the given taxa. In 1940, Salisbury and Jane (1940) published a paper on the charcoal from Maiden Castle, Dorset (England). Here they argued for Neolithic Oak-Hazel closed woodland that became more open through human exploitation. They further concluded on the basis of charcoal ring widths that past climates during this period of the Neolithic were similar to those of the present day.

This appears to be the earliest charcoal analysis to present details of an assemblage (with relative frequencies of identified taxa) and to seriously extrapolate information on environmental conditions and human activity. Godwin and Tansley published a criticism the following year (1941) that, while praising the "extensive data" presented, argued strongly against the inferences drawn from the Maiden Castle study. They pointed out that the taphonomy of burning and human selection may skew the assemblage away from the living population of trees and argued for more exact quantification of data. These cogent arguments are still being confronted some 65 years later.

Perhaps Godwin and Tansley made their point too forcefully, Anglophone researchers did not return to published charcoal analysis until Western's (1971) *The Ecological Interpretation of Ancient Charcoals from Jericho.* However, Francophone researchers, especially the founder of the Montpellier school, Jean-Louis Vernet, had a growing scholarly momentum, which continues to the present day and developed into now-established and formal schools of Anthracology (*l'Anthracologie*) (see Badal, Bernabeu, and Vernet 1994; Heinz 1991; Ludemann, Michiels, and Nölken 2004; Santa 1958–1959; Vernet 1967, 1973, 1992, In press; Vernet and Thiébault 1987; Vernet et al. 2001). In the United States, South Africa, and Germany, researchers also developed sizable scholarship, though much of the United States research is contract-archaeology-based, with the result often hidden in the "grey" (restricted distribution and usually unpublished and uncited) literature. Australian research started relatively early by general world standards (Bamber 1966), but has progressed slowly (Boyd, Collins, and Bell 2000; Dolby 1995; Donoghue 1989; Hope 1998; Smith, Vellen, and Pask 1995). In the last decade, strong material has come out of Britain, Germany, South Africa, and France (see Asouti 2006; Willcox 2006 for details; see also Cartwright 1999; Neumann 1999).

Methods

The first step in charcoal analysis is the determination of taxa. Wood has a distinctive arrangement of cells. Softwoods (gymnosperms, conifers) have different cells from hardwoods (angiosperms, broad-leaved trees). Cells and internal features can have distinctive qualitative or quantitative arrangements, allowing wood type to be ascertained (to sometimes specific taxonomic level, sometimes less precise). Further details on these can be gained from wood anatomy books (e.g., Carlquist 2001; Wilson and White 1986; for features used in identification, see Richter et al. 2004; Wheeler, Baas, and Gasson 1989). The arrangement of cells is preserved in carbonization (the combustion of wood in a reducing atmosphere). While water and volatiles are driven off by heat, without sufficient oxygen, the carbon structures of the cellulose and lignin-rich wood cell walls remain, leaving charcoal behind, often all but pure carbon. Some aspects of wood do not tend to preserve during carbonization (for example, color, fragrance, cell inclusions), while the wood structure loses mass and shrinks, making some wood identification quantitative measures imprecise (for example, cell dimensions, density). Furthermore,

some charcoal becomes extremely distorted, fragile, or homogenized during the deposition or sedimentation process, and identification becomes all but impossible. The presence of a good reference collection and of anatomical keys and literature is essential. Nonetheless, identification of charcoal to species level is regularly carried out.

There are two major strategies for sediment sampling in charcoal analysis. The Montpellier school favors sampling a "floor," collecting scattered charcoal that represents an accumulation of varying fires and that therefore gives an overview of the utilized vegetation taxa during the period of that surface. However, criticism of this approach includes the view that the targeting of individual archaeological features (representing individual past events), and in particular individual fireplaces, allows us to gain greater details on past human behavior than an analysis of broad and therefore more generalized charcoal assemblages. The selection of wood from individual fires is likely to shed greater light on the cultural and technical requirements of the burning activity. Certainly, a fire or fireplace's use may vary over its life, but the tighter spatial and temporal frame represented by a single hearth may allow greater insight into the past (for further details on such aspects of methodology, see Donoghue 1989; Figueiral 1999; Figueiral and Willcox 1999; Hastorf and Popper 1988; Pearsall 2000; Smart and Hoffman 1988). Thompson (1994) and Neumann (1999) make strong contributions to the methodology of identification, quantification, and interpretation, and Thiébault (2002) has compiled a broad range of charcoal analyses.

Approaches

What follows is a discussion of some of the areas in which charcoal analysis has made or could make a valuable contribution. The reconstruction of past vegetation and human use of the landscape has been widely carried out, but the reconstruction of cultural activities, domestication, and cultural history are areas that remain underdeveloped or unrealized.

Past Vegetation

Charcoal can be used to investigate the past vegetation of a landscape, but there are limitations and caveats. The reconstruction of past vegetation communities via charcoal analysis is limited generally to the woody flora. A caveat on reconstruction attempts is that an archaeological assemblage will tend to be biased toward larger trees, and toward trees that provide good firewood (as opposed to an assemblage representing the range of woody vegetation). The latter point comes about because people tend to select wood from trees that have desirable firewood characteristics, including the production of dead and dry wood (ideally, dead branches still attached to a tree) and good burning properties (including fragrance, heat output, smoky or smoke-free fire, depending on cultural desirability). Certain taxa have these properties more than others.

The skewing of an archaeological charcoal assemblage toward larger trees comes from human selection, from taphonomy, and from analysis. It is more cost effective to gather larger pieces of wood, not twigs but the larger branches that are more available on larger plants; that is, trees are more commonly foraged for firewood than are shrubs, and larger trees rather than smaller trees. Taphonomic bias causing larger sources of wood to dominate the charcoal record is due to the greater biomass of the larger taxa, the increased likelihood of incomplete combustion of larger wood, and the creation of larger sizes of charcoal by larger logs. Laboratory analysis is most easily carried out on larger pieces of charcoal: they are easily picked out and, more important, are easier to section, thus increasing the taxonomic precision achievable. One way around this laboratory bias is to analyze all charcoal present from an excavated assemblage, which is rarely possible with a large assemblage. Another way to reduce this bias is to subsample based on size—that is, by passing the collected charcoal through Endicott-sieves, creating size-classed subsamples. The use of saturation (or accumulation) curves (the graphing of increasing identified taxa against increasing sample size) will help determine the appropriate size of a subsample while further reducing bias.

In the quantification of an assemblage, numeric count (how many pieces) will tend to bias toward smaller taxa, while total taxon weight will reflect contributing-biomass more. Use of both numerical and weight quantification gives a more accurate picture. The resulting fragmentation analysis can also address issues of differential taxonomic or spatial representation at a given period of time, as well as provide indications of temporal changes in charcoal sequences.

Human Use of the Landscape

Charcoal analysis is not only concerned with the environments in which people lived but attempts also to determine patterns of land-use by people in the past. Neumann and associates (1998) carried out an intensive examination of the charcoal from a Medieval Burkina Faso village along with

the current vegetation surrounding the site. From this, they were able to demonstrate the preferred exploitation for firewood of neighboring dune woodlands and river gallery forests over the more abundant park savanna. However, it is worth remembering that although environments of the recent past are able to be recreated from the present habitats of plants, environments and communities of the more distant past may not have present analogues. The discussion between Willis and colleagues (2000, 2001) and Carcaillet and Vernet (2001) on the presence, identification, and nature of forests in Glacial Maximum Europe is indicative of some of the complexities of pre-Holocene vegetation reconstructions.

Cultural Activities

Assemblages of charcoal give information not only on the natural environment but also on the cultural landscape. The use of certain woods for certain purposes allows an understanding of the social differentiation of the place into a culturally conceptualized landscape. For example, the identification of charcoal used in the historic past for high status ceremonies allowed Kolb and Murakami (1994) to identify ritual fireplaces in an Hawai'ian complex. The ceremonial use of pine trees in ancient Lowland Mayan communities and the recognition of places of ceremony and residences of high status is revealed in the work of Lentz, Morehart, and their colleagues (Lentz et al. 2005; Morehart, Lentz, and Prufer 2005).

Domestification

In the northwest of the Mediterranean region, Terral (1996, 1997a, 1997b, 2000; Terral and Arnold-Simard 1996; Terral and Durand 2005) has demonstrated an ability to differentiate between wild and cultivated olive wood, the latter's introduction to the area, and the use of irrigation practices; similar research may be possible with other taxa and in other regions. The importance of arboriculture in the prehistory of western Asia, Southeast Asia, and in the Pacific region has long been a focus of research; the question remains as to whether or not charcoal studies will eventually prove useful to these regional research projects?

Culture History

Of course, charcoal—or rather the firewood culture evident from the charcoal assemblage—may give indications of the histories of specific cultural practices and the changing practices of the occupants of a point or space within a landscape. Examples of this could include change in available wood types, the avoidance or use of certain taxa, and the changing needs of fuel for fires—all of these might become apparent in archaeological charcoal assemblages. To date, charcoal analysis has been poorly employed in the elucidation of such questions anywhere in the world. Intersite comparisons and interpretations, or the analysis of complementary sets of assemblages from past communities or regions, have simply not been undertaken (although some of the researchers cited above have now begun such research programs). The work by Lowell (1999) on Mogollon (or prehistoric western Pueblo) fireplaces alerts us to changes in the conceptualization and use of fire and fireplaces in social practice. That social notions and values of fire and "hearth" (along with their broader sociopolitical contexts) have changed through time can also be seen in the recent past of Britain, where fuels have shifted from wood to coal and, in the present day, to wood again, resulting in changed techniques and notions of food preparation (David 1994; Hartley 1999). It remains for future research to examine the changing nature and context of fuel use in the past, including how such information may inform us as to past social links and movements.

Conclusions

Charcoal analysis has proved to be a strong tool for investigating the past vegetation around archaeological sites and for helping to reconstruct people's use of resources within the landscape. It has, however, greater potential than this and more to offer than the listing of woods that have ended up burnt in archaeological deposits. Aspects of culture history may be revealed by charcoal analysis, as illustrated by some of the approaches used to examine cultural activities surrounding fire, wood use, and domestication, all of which are tied to locations within the landscape. Analysis of archaeological charcoals has the potential to inform us about technology, economy, cultural, and social activities and how they vary across space, thus shedding greater light on the people and landscapes in the past.

References

Asouti, E. 2006. *Charcoal Analysis Web*, School of Archaeology, Classics and Egyptology, University of Liverpool, accessed at http://pcwww.liv.ac.uk/~easouti/, on 20th August 2006.

Badal, E., Bernabeu, J., and Vernet, J.-L. 1994. Vegetation changes and human action from the Neolithic to the Bronze Age (7,000–4,000 B.P.) in Alicante, Spain, based on charcoal analysis. *Vegetation History and Archaeobotany* 3: 155–66.

Bamber, R. K. 1966. Examination of wood charcoals from the Gymea Bay midden material, in J. V. S. Megaw, The excavation of an Aboriginal rock-shelter on Gymea Bay, Port Hacking, N.S.W. *Archaeology and Physical Anthropology in Oceania* 1(1): 23–50, appendix IV, p. 48.

Boyd, W. E., Collins, J. P., and Bell, J. 2000. The accumulation of charcoal within a midden at Cape Byron, northern New South Wales, during the last millennium. *Australian Archaeology* 51: 21–27.

Carcaillet, C., and Vernet, J.-L. 2001. Comments on "The full-glacial forest of Central and Southeastern Europe," Willis et al. *Quaternary Research* 55: 385–87.

Carlquist, S. J. 2001. *Comparative Wood Anatomy: Systematic, Ecological, and Evolutionary Aspects of Dicotyledon Wood*. Berlin: Springer-Verlag.

Castelletti, L. 1990. Legni e carboni in archeologia, in T. Mannoni and A. Molinari (eds.), *Scienze in Archeologia: Il ciclo de lezioni sulla ricerca applicata in archeologia, Certosa di Pontignano (Siena), 7–19 Novembre 1988*, pp. 391–424.

David, E. 1994 [1977]. *English Bread and Yeast Cookery*. Newton: Biscuit Books.

Dolby, N. 1995. A unique perspective: Charcoal, environment and fuel use at the Pleistocene archaeological site of Nunamira, southwest Tasmania. Unpublished B.A. (Hons) thesis, School of Archaeology, LaTrobe University, Bundoora.

———. In preparation. On charcoal (*de carbone*): A thesis on the study of archaeological charcoal based on an assemblage from Ngarrabullgan Cave, Cape York Peninsula, Australia. Unpublished M.A. thesis, School of Geography and Environmental Science, Monash University, Clayton.

Donoghue, D. 1989. Carbonised plant macrofossils, in W. Beck, A. Clarke, and L. Head (eds.), *Plants in Australian Archaeology*, pp. 90–110. Tempus 1, Archaeology and Material Culture Studies in Anthropology. St Lucia: Anthropology Museum, University of Queensland.

Figueiral, I. 1999. Lignified and charcoalified fossil wood, in T. P. Jones and N. P. Rowe (eds.), *Fossil Plants and Spores*, pp. 92–96. London: Geological Society.

Figueiral, I., and Willcox, G. 1999. Archaeobotany: Collection and analytical techniques for sub-fossils, in T. P. Jones and N. P. Rowe (eds.), *Fossil Plants and Spores*, pp. 290–94. London: Geological Society.

Godwin, H., and Tansley, A. G. 1941. Prehistoric charcoals as evidence of former vegetation and climate. *Journal of Ecology* 29: 117–26.

Hartley, D. 1999 [1954]. *Food in England*. London: Little, Brown.

Hastorf, C. A., and Popper, V. S. (eds.). 1988. *Current Paleoethnobotany: Analytical Methods and Cultural Interpretations of Archaeological Plant Remains*. Chicago: University of Chicago Press.

Hather, J. G. 1991. The identification of charred archaeological remains of vegetative parenchymous tissue. *Journal of Archaeological Science* 18: 661–75.

———. 1993. *An Archaeobotanical Guide to Root and Tuber Identification, Vol. 1: Europe and South West Asia*. Oxford: Oxbow Books.

Heer, O. 1866. Die Pflanzen der Pfahlbauten. *Neujahrsblatt, herausgegeben von der Naturforschende Gesellschaft Zürich*, pp. 1–54.

Heer, O., and Passerini, G. 1865. In L. Pigorini, Le abitazioni palustri di Fontanello all'epoca del ferro. *Bulletino di Paletnologia Italiana* 11: 7–11.

Hesse, B., and Rosen, A. 1988. The detection of chronological mixing in samples from stratified archaeological sites, in R. E. Webb (ed.), *Recent Developments in Environmental Analysis in Old and New World Archaeology*, pp. 117–29. Oxford: BAR International Series 416.

Heinz, C. 1991. Upper Pleistocene and Holocene vegetation in the south of France and Andorra. Adaptations and first rupture: New charcoal analysis data. *Review of Palaeobotany and Palynology*. 69: 299–324.

Hope, G. (ed.). 1998. *Identifying Wood Charcoal Remains as Palaeo Evidence for Regions of Central and Northeast Australia*. Canberra: ANH Publications, RSPAS, The Australian National University.

Jones, T. P., and Rowe, N. P. (eds.). 1999. *Fossil Plants and Spores*. London: Geological Society.

Kolb, M. J., and Murakami, G. M. 1994. Cultural dynamics and the ritual role of woods in Pre-Contact Hawai'i. *Asian Perspectives* 33: 57–78.

Lentz, D. L., Yaeger, J., Robin, C., and Ashmore, W. 2005. Pine, prestige and politics of the Late Classic Maya at Xunantunich, Belize. *Antiquity* 79: 573–85.

Lowell, J. C. 1999. The fires of grasshopper: Enlightening transformations in subsistence practices through fire-feature analysis. *Journal of Anthropological Archaeology* 18: 441–70.

Ludemann, T., Michiels, H.-G., and Nölken, W. 2004. Spatial patterns of past wood exploitation, natural wood supply and growth conditions: Indications of natural tree species distribution by anthracological studies of charcoal-burning remains. *European Journal of Forest Research* 123: 283–92.

Morehart, C. T., Lentz, D. L., and Prufer, K. M. 2005. Wood of the gods: The ritual use of pine (*Pinus* spp.) by the ancient lowland Maya. *Latin American Antiquity* 16: 255–74.

Neumann, K. 1999. Charcoal from West African savanna sites: Questions of identification and interpretation,

in M. van der Veen (ed.), *The Exploitation of Plant Resources in Ancient Africa*, pp. 205–19. New York: Kluwer Academic/Plenum Publishers.

Neumann, K., Kahlheber, S., and Uebel, D. 1998. Remains of woody plants from Saouga, a medieval west African village. *Vegetation History and Archaeobotany* 7: 57–77.

Passerini, G. 1864. In P. Strobel and L. Pigorini, Le terramare e le palafitte nel Parmense. *Atti Società Italiana Scienze Naturali* 7: 27–33.

Pearsall, D. M. 2000. *Paleoethnobotany: A Handbook of Procedures* (2nd ed.). San Diego: Academic Press.

Prior, J., and Alvin, K. L. 1983. Structural changes on charring woods of Dichrostachys and Salix from southern Africa. *International Association of Wood Anatomists (IAWA) Bulletin* (New Series) 4: 197–206.

———. 1986. Structural changes on charring woods of *Dichrostachys* and *Salix* from southern Africa: The effect of moisture content. *International Association of Wood Anatomists (IAWA) Bulletin* (New Series) 7: 243–50.

Prior, J., and Gasson, P. 1993. Anatomical changes on charring six African hardwoods. *International Association of Wood Anatomists (IAWA) Journal* 14: 77–86.

Richter, H. G., Grosser, D., Heinz, I., and Gasson, P. 2004. IAWA list of microscopic feature for softwood identification. *IAWA Journal* 25: 1–70.

Rossen, J., and Olson, J. 1985. Controlled carbonization and archaeological analysis of SE U.S. wood charcoals. *Journal of Field Archaeology* 12: 445–56.

Salisbury, E. J., and Jane, F. W. 1940. Charcoals from Maiden Castle and their significance in relation to the vegetation and climate conditions in prehistoric times. *Journal of Ecology* 28: 310–25.

Santa, S. 1958–1959. Essai de reconstitution de paysages végétaux quaternaires d'Afrique de Nord. *Libyca* 6–7: 37–77.

Smart, T. Lee, and Hoffman, E. S. 1988. Environmental interpretation of archaeological charcoal, in C. A. Hastorf and V. S. Popper (eds.), *Current Paleoethnobotany: Analytical Methods and Cultural Interpretations of Archaeological Plant Remains*, pp. 167–205. Chicago: University of Chicago Press.

Smith, M. A., Vellen, L., and Pask, J. 1995. Vegetation history from archaeological charcoals in central Australia: The late Quaternary record from Puritjarra rock shelter. *Vegetation History and Archaeobotany* 4: 171–77.

Terral, J.-F. 1996. Wild and cultivated olive (*Olea europea* L.): A new approach to an old problem using inorganic analyses of modern wood and archaeological charcoal. *Review of Palaeobotany and Palynology*. 91: 383–97.

———. 1997a. Les débuts de la domestication de l'olivier (*Olea europaea* L.) en Méditerannée nord-occidentale, mise en évidence par l'analyse morphométrique appliquée à du matériel anthrac-ologique. *Comptes Rendus de l'Académie des Sciences de Paris* 324(5), Serie IIa: 417–25.

Terral, J.-F. 1997b. Domestication de l'olivier (*Olea europaea* L.) en Mediterranee nord-occidentale: Approche morphométrique et implications paleo-climatiques. Unpublished Ph.D. thesis, Université Montpellier II, Montpellier.

———. 2000. Exploitation and management of the olive tree during prehistoric times in Mediterranean France and Spain. *Journal of Archaeological Science* 27: 127–33.

Terral, J.-F., and Arnold-Simard, G. 1996. Beginnings of olive cultivation in eastern Spain in relation to Holocene bioclimatic changes. *Quaternary Research* 46: 176–85.

Terral, J.-F., and Durand, A. 2005. Bio-archaeological evidence of olive tree (*Olea europaea* L.) irrigation during the Middle Ages in Southern France and North Eastern Spain. *Journal of Archaeological Science* 33: 718–24.

Thiébault, S. (ed.). 2002. *Charcoal Analysis: Methodological Approaches, Palaeoecological Results and Wood Uses.* Proceedings of the Second International Meeting of Anthracology, Paris, September 2000. Oxford: BAR International Series 1063.

Thompson, G. B. 1994. Wood charcoal from tropical sites: A contribution to methodology and interpreta-tion, in J. G. Hather (ed.), *Tropical Archaeobotany: Applications and New Developments*, pp. 9–33. London: Routledge.

Vernet, J.-L. 1967. Premiers résultats de l'étude anatomique to charbons de bois préhistoriques. *Bulletin Assoc. Française Étude du Quaternaire* 12: 211–22.

———. 1973. Étude sur l'histoire de la végétation du sud-est de la France au Quaternaire, d'après les charbons de bois principalement. *Paléobiologie Continentale* 1 (t. IV) [4(1)]: 1–90.

———. (ed.). 1992. Les charbons de bois, les anciens écosystemes et le rôle de l'homme. Colloque organ-ise à Montpellier du 10 au 13 septembre 1991. *Bulletin de la Société Botanique de France* 139, actualites botaniques (2/3/4).

———. In press. History of the *Pinus sylvestris* and *Pinus nigra* ssp. *salzmanni* forest in the Sub-Mediterranean mountains (Grands Causses, Gaint-Guilhem-le-Désert, southern Massif Central, France) based on charcoal from limestone and dolomitic deposits. *Vegetation History and Archaeobotany.*

Vernet, J.-L., Ogereau, P., Figueiral, I., del C. Machado Yanes, M., and Uzquiano, P. 2001. *Guide d'Identifi-cation des Charbons de Bois Prehistoiriques du Sud-Ouest de l'Europe.* Paris: CNRS Editions.

Vernet, J.-L., and Thiébault, S. 1987. An approach to northwestern Mediterranean recent prehistoric vegetation and ecologic implications. *Journal of Biogeography* 14: 117–27.

Western, A. C. 1971. The ecological interpretation of ancient charcoals from Jericho. *Levant* 3: 31–40.

Wheeler, E. A., Baas, P., and Gasson, P. E. 1989. IAWA list of microscopic feature for hardwood identifica-tion. *IAWA Bulletin* 10: 219–332.

Willcox, G. 2006. *Charcoal Remains*, part of *George Willcox (Archaeobotanist)* website, accessed at http://perso.orange.fr/g.willcox/charcoalremains. htm, on 20th August 2006.

Willis, K. J., Rudner, E., and Sümegi, P. 2000. The full-glacial forests of central and southeastern Europe. *Quaternary Research* 53: 203–13.

———. 2001. Reply to Carcaillet and Vernet. *Quaternary Research* 55: 388–89.

Wilson, K., and White, D. J. B. 1986. *The Anatomy of Wood: Its Diversity and Variability.* London: Stobart and Son Ltd.

45

TERRESTRIAL INVERTEBRATES IN LANDSCAPE ARCHAEOLOGY

Nick Porch

This chapter introduces the role of invertebrates in landscape archaeology by providing a synopsis of their analysis and interpretation in terms of climate and local environment. It draws on the rapidly expanding literature of palaeoecology and environmental archaeology dealing with nonmarine mollusks and insects, and cites sources for further reading. A brief overview is provided for several other taxonomic groups less frequently encountered in the archaeological record or in other contexts that may be informative regarding the nature of archaeological landscapes. The role of invertebrates in physically structuring the archaeological and palaeoecological record is not examined (see Carter 1990; Stein 1983, 2001).

The term *invertebrates* refers to a wide range of unrelated animal taxa; "terrestrial invertebrates" commonly refers to "nonmarine" taxa and thus also includes a range of taxa found in saline and freshwater environments on land. Important terrestrial invertebrate groups include: the annelids (for example, earthworms, leeches); the mollusks (for example, snails, slugs); and the arthropods, which include millipedes and centipedes, crustaceans (for instance, crabs, crayfish, prawns, ostracods), arachnids (for example, spiders, ticks, mites, scorpions), and insects (for instance, beetles, bugs, butterflies/moths, earwigs, cockroaches) (for details see Barnes et al. 2001).

In terms of species richness, the invertebrates dominate the biota of both temperate and tropical regions (Wilson 1992); estimates suggest that more than 70% of the planet's diversity is invertebrate and that about 20% comprises species belonging to one insect order alone, the beetles (Hammond 1995). Of the groups listed above, however, few are frequently preserved in the archaeological or palaeoecological record. Two groups that are commonly preserved and often species-rich, especially in alkaline and anaerobic sediments, respectively, are the nonmarine gastropods—regularly referred to as "land snails"—and the insects. Both are frequently utilized to provide data on the nature of the archaeological sites and their settings. In particular, they provide evidence on the environmental setting of archaeological sites including climatic reconstructions, the nature and extent of altered and constructed landscapes, and the specific environs in which people lived.

Nonmarine Mollusks

Background

Nonmarine mollusks include pulmonate ("lunged") gastropods, prosobranch ("gilled") gastropods, and bivalves. Contrary to expectations, not all pulmo-nates are terrestrial, with a range being truly aqua-tic, and several families of prosobranchs contain mainly terrestrial taxa. In common usage, the term *snail* refers to the coiled gastropods; however, a range of pulmonate taxa take other forms with uncoiled limpet-like shells or the complete reduc-tion of the shell to an internal "cyst," (as present in slugs of several families). A few bivalve families are common in freshwater lentic (still water) and/or lotic (flowing water) habitats.

Preservation and Taphonomy

Nonmarine mollusks are generally preserved as aragonite and/or calcite shells and require deposi-tion in carbonate-rich environments for preserva-tion (Goodfriend 1992); strongly acidic sediments will rapidly dissolve even the most robust non-marine mollusk. Preece and Bridgland (1999: figs. 22–24) illustrate a sequence from Hollywell Coombe in southern England that clearly shows the influence of depositional environment on preservation: nonmarine mollusks are consist-ently present in chalkrich sediments and tufa but are absent from a thin organic layer where they have not been preserved. An advantage of nonmarine mollusks is that they are commonly preserved in settings where most, if not all, other types of organic remains have been removed through aerobic microbiological activity. This is especially true of both open sites (in alluvium, colluvium and soils, and so on) and rock-shelter and cave sites (Evans 1972; Thomas 1985). In organic sediments where nonmarine mollusks may be absent (as in the example above), other taxa, (for example, plant and insect macrofossils) are likely to be preserved, providing an alterna-tive means of environmental reconstruction (see Fairbairn, this volume; Rowe and Kershaw, this volume).

Sampling, Recovery, and Identification

Nonmarine mollusk assemblages can be sampled from cores and open sections or during excava-tion. Sediments are disaggregated and washed through sieves, usually 500 um or 1 mm, depend-ing on the nature of the fauna, then dried, after which the mollusk remains picked from the con-centrate: Figure 45.1 shows examples of typical nonmarine mollusk assemblages from prehuman and historic levels of Makauwahi Cave, Kaua'i, Hawai'i. Identification is by comparison with identified reference material, using keys or illus-trations. For further detail on methodology, see

Bell and Walker (1992), Evans (1972), Lozek (1986), Shackley (1981), and Wilkinson and Stevens (2003).

Environmental Interpretation

The modern origin of the use of nonmarine mol-lusks in archaeology and palaeoecology can be traced to the pioneering work of Sparks (1961, 1969), expanded and refined by Evans (1967), culminating in the publication of the seminal *Land Snails in Archaeology* (Evans 1972). Evans's primary interest was in the role that land snails played in reconstructing vegetation struc-ture and, therefore, potentially, the impact of people on local and regional environments. The absence of other forms of biological evidence in many archaeological sites and their regional settings has meant that the use of land snails has become commonplace in many parts of the world, at least when compared with the use of other invertebrate remains.

In mesic temperate and tropical regions, most land snails and slugs are physiologically tied to certain micro-environments by their inability to withstand desiccation. This is less true, however, in arid areas where many species have adapta-tions for coping with long dry periods. Latitudinal, altitudinal, and regional scale patterns of dis-tribution also illustrate the influence of thermal environment, with significant levels of turnover from cold to warmer habitats. Recognition of cli-matic influences on the distribution of nonmarine mollusks has culminated in the development of methods for the quantification of past climates, especially temperature, using nonmarine mol-lusk assemblages (Moine et al. 2002; Rousseau 1991; Rousseau et al. 1994; Rousseau, Preece, and Limondint-Lozouet 1998) and examination of the climatic significance of morphometric variability of individual taxa (e.g., Rowe et al. 2001). At the local scale, micro-environmental conditions are determined by the interaction of effective precipi-tation (reflecting temperature and rainfall), sub-strate type, and vegetation structure, especially its density. The distribution of nonmarine aquatic taxa is influenced by water regime (including flow rate, chemistry, turbidity, substrate, macro-phyte coverage) and climate, although, for obvi-ous reasons, moisture regimes are not directly influential.

Early work on the nature of southern English landscapes based on data from valley bottoms and, in the latter half of the Holocene, archaeological sites—including ditch fills and soils from under barrows—clearly illustrated the contribution land

Figure 45.1   Contrasting nonmarine mollusk assemblages from Makauwahi Cave, Kaua'i. *Left*: Prehuman levels: high diversity of exclusively indigenous terrestrial taxa. *Right*: Early historic levels: low diversity of mainly introduced and indigenous estuarine/marginal aquatic taxa.

snail analysis could make to the archaeology of chalk landscapes (Evans 1971, 1972; Kerney 1966; Kerney, Chandler, and Brown 1964). The replacement of postglacial open country by woodland, which became increasingly dense and diverse by the middle Holocene, was disrupted by the late Mesolithic and especially Neolithic clearing of the forest for conversion to grazing and ultimately, usually later, to arable land. The rich woodland land snail faunas, dominated by shade-loving taxa, were replaced by less diverse assemblages, including taxa characteristic of open habitats like grassland. Thus, the Neolithic landscape, which included monuments such as Stonehenge, was largely open rather than wooded and contrasted with the contemporary view that these monuments were constructed in wooded landscapes (Wilkinson and Stevens 2003). Recent research has modified this story somewhat, adding regional and temporal complexity by indicating, for some sites, that the major episode of forest clearance occurred during the Late Bronze Age, whereas at others open vegetation was created and maintained from the Mesolithic (for examples, see Allen 1997; Bell 1983; French and Lewis 2005; Preece and Brigdland 1999).

A similar but probably less well known story has been revealed on islands across the Pacific, where studies of nonmarine mollusk assemblages were initiated during the 1970s with the publication

by Kirch (1973, 1975) of assemblages from sites from the Marquesas and Moloka'i in Hawai'i, and later from Tikopia (Christensen and Kirch 1981). Across the Pacific, where nonmarine mollusk analyses have been undertaken, several trends recur. Many islands, especially the larger and high islands, have (had) extensive endemic radiations that were heavily affected by the activities of prehistoric people to the extent that many species disappear from the record (and often the planet) following Polynesian arrival (Brook 1999; Burney et al. 2001; Christensen and Kirch 1986; Dye and Tuggle 1998; Neuweger, White, and Ponder 2001; Preece 1998; Solem 1990). Although the exact causes of these extinctions and extirpations are far from certain, a range of factors have been suggested, including: predation by the Pacific Rat (*Rattus exulans*), especially for larger taxa; forest clearance for agriculture and other activities; massively increased fire regimes; and possibly predation by Polynesian-introduced arthropods, especially ants (Burney et al. 2001; Dixon, Soldo, and Christensen 1997; Kirch 1982; Preece 1998). In prehistory, a range of species, common in early archaeological sites, but absent from pre-human-arrival sediments, dispersed across the Pacific in association with people (Cowie and Grant-Mackie 2004; Kirch 1973, 1993; Preece 1998; Rolett 1992). Essentially, this recurring history is one of extensive modification of the prehistoric landscape resulting in a cascade of extinction that is repeated, with increased intensity, following the arrival of Europeans across the Pacific (Cowie 2001; Cowie and Robinson 2003).

Insects

Background

The insects are an incredibly diverse group of organisms that are found in all regions of the world, increasing in diversity towards the tropics. The class Insecta contains around 30 orders, some with fewer than 100 species and others with tens to hundreds of thousands of species (for example, beetles [Coleoptera]; ants, wasps, and bees [Hymenoptera]; butterflies and moths [Lepidoptera]; flies [Diptera]; and true bugs [Heteroptera]). Flies belonging to the family Chironomidae, generally referred to as chironomids, are an abundant and informative group in aquatic settings where they are represented by their identifiable head capsules (Brooks and Birks 2001; Hofmann 1986; Walker et al. 1991). They provide evidence for past water temperature, salinity, and nutrient status, although they are infrequently used in specifically archaeological settings.

Preservation and Taphonomy

Most archaeological sediments do not contain insect assemblages. They have frequently been oxidized under continuous or intermittent aerobic conditions (cf. nonmarine mollusks). There are two primary contexts, however, in which insect fossils are commonly preserved in archaeological settings: (1) anaerobic sediments from waterlogged contexts and (2) perpetually dry sediments such as those occasionally found in caves, rock-shelters, dwellings, and in extremely arid regions (Buckland 1976; Elias 1994). In some circumstances, associated noncultural deposits may exist that provide the opportunity to examine assemblages coeval with the archaeological deposits providing an environmental context that would be otherwise lacking (Elias 1986).

One factor that is readily apparent from the composition of fossil insect assemblages is that some groups preserve far more commonly than others (for example, Coleoptera, Hymenoptera, Heteroptera, and Trichoptera—caddis-flies), whereas others are extremely rare (for instance, Lepidoptera, Orthoptera —crickets and grasshoppers, Thysanoptera—thrips). This difference in preservation is a consequence of the relative robusticity of the groups: beetles are generally strongly sclerotized, whereas butterflies are not. A range of studies have examined potential taphonomic influences on archaeological assemblages by detailed analysis of modern assemblages in relation to their context (e.g., Carrott and Kenward 2001; Kenward 1975, 1985, 1997, 2006; Kenward and Carrott 2006; Osborne 1983; Smith 1996).

Sampling, Recovery, and Identification

Sampling and extraction of insect fossils from archaeological sediments is best undertaken in concert with, or by, the insect analyst. In general, the larger size of individual insect fossils and their lower density (at least compared with microfossils such as pollen, diatoms, and phytoliths, for example) means that relatively large samples are required, usually more than a kilogram of sediment and often much more. The variability in the density of fossils and the possibility that they may not be preserved means that it can be wise to assess small samples to provide a clearer regard for further sampling requirements. For waterlogged sediments, recovery of insect fossils is primarily through the use of kerosene floatation after disaggregation. Further details regarding sampling, processing, sorting, and storage of the recovered material can be found in Ashworth (1979), Coope (1986), Elias (1994), and (for specifically archaeological contexts) Buckland (1976), Buckland and Sadler (2000), and Kenward, Hall, and Jones (1980). Identification of insect fossils is a specialist task principally undertaken by fossil insect researchers through comparison with identified modern reference specimens: examples of insect fossils from Pacific contexts are shown in Figure 45.2.

Environmental Interpretation

The insects are not only species-rich but also diverse in terms of their environmental requirements. Thus, when the ecology and distribution of the taxa in the archaeological record are well known, the local environment can be reconstructed in detail. Insects provide evidence of a wide range of critical environmental parameters including information on the vegetation (including its specific composition), the soil-surface interface, aquatic environments, presence and nature of organic debris (dead wood, refuse, dung, carrion), and climate (see Ashworth 1979; Atkinson et al. 1986; Coope 1977, 1987; Elias 1994; Porch and Elias 2000). A range of

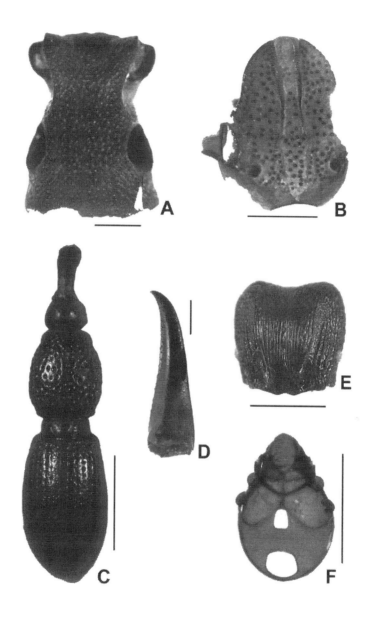

Figure 45.2 Late Holocene arthropod fossils from Pacific contexts. (A) *Rhyncogonus sp.* weevil head—Makauwahi Cave, Kaua'i, prehuman sediments: this species is extirpated following Polynesian arrival. (B) Pentatomid bug head—Santa Cruz highlands, Galapagos. (C) Unidentified weevil—Rimatara, French Polynesia, prehuman sediments. (D) Unidentified dermapteran (earwig) forcep—Makawahi Cave, Kaua'I, Polynesian age sediments. (E) *Pheidole* cf. *fervens* (ant) head, Rimatara, French Polynesia, Polynesian age sediments. (F) Unidentified oribatid mite, Santa Cruz highlands, Galapagos.

synanthropic taxa (including stored product pests) are important and informative components in many archaeological assemblages (Buckland 1981, 1990; Panagiotakopulu 2000). Panagiotakopulu (2001) reviewed records for insect pests, including human ectoparasites, from Egyptian contexts, and noted there is much potential for understanding the origin and the spread of synanthropic taxa. Although current research in this field is focused on northern Europe and the Mediterranean region, there exists much potential for tracing the origin and the spread of pest taxa associated with prehistoric and historic human migrations.

Insect fossils, principally beetles, are routinely utilized in reconstruction of archaeological environments in Western Europe, especially in Britain (see Elias 1994, for a detailed review, and papers in Buckland, Coope, and Sadler 2004). Most research, however, has focused on post-Palaeolithic sites (although see Buckland 1984; Coope 1993, 2006 and references therein; Coope and Elias 2000 for analysis of assemblages from Palaeolithic sites). Assemblages from Palaeolithic contexts provide details of the nature of the local environment of deposition and regional climate, principally in terms of summer warmth and seasonality (Coope 2006). For North America, a range of studies have examined assemblages from open and cave sites, providing a context for several Paleoindian sites (Elias 1986, 1990; Elias and Nelson 1989) although most North American studies have focused on more recent assemblages from habitation sites (for example, Pueblo, Anasazi) or dry-cave sediments (reviewed in Elias 1994). Hoganson and associates (1989) provide an environmental context for Monte Verde, an early Palaeoindian site in Chile; the insect assemblage, dominated by beetles, showed that at the time of deposition the setting was rainforest with areas of open ground beside a creek that included rapid, flowing water and areas of still vegetation-rich water.

The majority of European studies focus on the reconstruction of the local environment of sites from the Iron Age to the recent past. The settings of these sites may be essentially natural or heavily modified in the case of habitation/urban settings (for examples see Buckland 1974; Buckland and Kenward 1972; Buckland, Holdsworth, and Monk 1976; Buckland, Beal, and Heal 1990; Kenward and Large 1997; Osborne 1969, 1971; Ponel et al. 2000). In a review of British insect fossil faunas from the past 10,000 years, Dinnin and Sadler (1999: 545) divide the development of the insect fauna into five phases. These phases, not surprisingly, parallel the pattern of landscape change

inferred from the nonmarine mollusk record noted above: (1) late-glacial-Holocene warming and reinvasion of the thermophilous taxa; (2) early Holocene afforestation and increasing faunal diversity; (3) forest climax with diverse forest obligate community; (4) Neolithic deforestation from the middle Holocene and accelerating from the Bronze Age; and (5) the creation of a "culture steppe." The final two phases represent the culmination of a temporal and spatial increase in the clearance of forest, extirpation of forest elements, especially those associated with primary forest, and expansion of treeless landscapes. Whitehouse (2006) reviewed the Holocene history of human impact in British and Irish forests and noted that about 40 species of beetles were extirpated from Britain (mainly based on data from England) prior to the period of modern collection. Most of these beetles are characteristically associated with old-growth forest habitats, especially dead wood within these forests, and disappear from the fossil record (and the region) mostly in the period 5,000–2,000 years B.P., reflecting the pattern of increasing clearance and modification of primary forest.

Evidence for human impact on regional biodiversity and its composition has been recovered from the analysis of insect faunas associated with Norse settlement of Greenland and Iceland (Barlow et al. 1997; Böcher 1997; Buckland et al. 1986, 1996; Perry, Buckland, and Snæsdóttir 1985; Sadler 1991; Sadler and Skidmore 1995). Norse settlement in these regions resulted in the introduction of a range of synanthropic taxa that probably remained closely tied to human occupation sites. In Greenland, for example, many died out with the collapse and abandonment of Norse settlements although others apparently survived until the modern era (Böcher 1997).

Examination of waterlogged sediments from archaeological sites in the Pacific has revealed abundant insect, spider, and mite remains in samples that predate human arrival, as well as in samples associated with Polynesian, and subsequently, historic presence. Preliminary analysis of such samples (Porch, unpublished data) indicates a scenario of considerable impact on the biota with Polynesian settlement that accelerated after European arrival. This impact includes both the extinction of a range of taxa and the import of human and agricultural commensals. Such results, albeit preliminary, parallel the land snail results for the Pacific, described above. They foster the belief that there is much greater potential in utilizing invertebrate remains in the process of reconstructing prehistoric and historic human impact on the ecosystems of both continents and especially islands than has hitherto been realized.

Other Taxa

The remains of mites (see Figure 45.2f) are commonly encountered in fossil insect samples from archaeological and other contexts (Denford 1978; Erickson 1988; Schlevis 1990, 1997). Like insects, mites are informative in regard to the nature of local environments, although their use tends to provide less informative environmental details than those based on, for example, beetles. Studies have indicated, however, that it may be possible to determine the presence and specific origin of dung in sediments on the basis of the encapsulated predatory mite assemblage (Schlevis 1992). Ostracods and Cladocera, calcareous-bivalved and chitinous crustaceans, respectively, are little utilized in archaeological studies; however, like chironomids, they may be present in abundance in water-lain sediments (Frey 1986). The types of information that cladocerans and ostracods provide (and chironomids)—data regarding the nature of water bodies—is generally not a key interest of those interested in archaeological landscapes. It is not until late in prehistory, more often the industrial age, that humans have sufficiently affected these aquatic ecosystems to cause change in the ecology, and therefore biota, of such systems.

Conclusions

The invertebrates are a diverse group of organisms in terms of their taxonomy, ecology, and preservation in archaeological contexts. The contrasting preservation potential of different invertebrate taxa (such as land-snails and insects) means that in sediments that may be devoid of one group others may be present that can yield information regarding the nature and environmental setting of sedimentary sequences. Their ecological diversity means a wide range of general and specific questions about archaeological landscapes can be addressed, including issues about human landscape modification that leave little, if any, traditional archaeological record. They are almost certainly underutilized in landscape archaeology, partly reflecting the ignorance of many archaeologists regarding their potential utility, and mainly because of the specialist nature of invertebrate analysis.

References

Allen, M. J. 1997. Landscape, land-use and farming, in R. Smith, F. Healy, M. Allen, E. Morris, I. Barnes, and P. Woodward (eds.), *Excavations along the Route of the Dorchester By-Pass, Dorset, 1986–1988*, pp. 166–83, 277–83. Salisbury: Wessex Archaeology.

Ashworth, A. C. 1979. Quaternary Coleoptera studies in North America: Past and present, in T. L. Erwin, G. E. Ball, and D. R. Whitehead (eds.), *Carabid Beetles: Their Evolution, Natural History, and Classification*, pp. 395–406. The Hague: W. Junk.

Atkinson, T. C., Briffa, K. R., Coope, G. R., Joachim, M., and Perry, D. 1986. Climatic calibration of coleopteran data, in B. E. Berglund (ed.), *Handbook of Holocene Palaeoecology and Palaeohydrology*, pp. 851–58. New York: John Wiley and Sons.

Barlow, L. K., Sadler, J. P., Ogilvie, A., Buckland, P. C., Amorosi, T., Ingimundarsson, J. H., Skidmore, P., Dugmore, A. J., and McGovern, T. H. 1997. Ice core and environmental evidence for the end of Norse Greenland. *The Holocene* 7: 489–99.

Barnes, R. S. K., Calow, P. P., Olive, P. J. W., Golding, D. W., and Spicer, J. 2001. *The Invertebrates: A Synthesis* (3rd ed.). Oxford: Blackwell Science.

Bell, M. 1983. Valley sediments as evidence of prehistoric land-use on the South Downs. *Proceedings of the Prehistoric Society* 49: 119–50.

Bell, M. G., and Walker, M. J. C. 1992. *Late Quaternary Environmental Change: Physical and Human Perspectives*. London: Longman.

Böcher, J. 1997. History of the Greenland insect fauna with emphasis on living and fossil beetles, in A. C. Ashworth, P. C. Buckland, and J. P. Sadler (eds.), *Studies in Quaternary Entomology: An Inordinate Fondness for Insects, Quaternary Proceedings* 5, pp. 35–48. Chichester: John Wiley and Sons.

Brook, F. J. 1999. Changes in the landsnail fauna of Lady Alice Island, northeastern New Zealand. *Journal of the Royal Society of New Zealand* 29: 139–57.

Brooks, S. J., and Birks, H. J. B. 2001. Chironomid-inferred air temperatures from Lateglacial and Holocene sites in north-west Europe: Progress and problems. *Quaternary Science Reviews* 20: 1723–41.

Buckland, P. C. 1974. Archaeology and environment in York. *Journal of Archaeological Science* 1: 303–16.

———. 1976. The use of insect remains in the interpretation of archaeological environments, in D. A. Davidson and M. L. Shackley (eds.), *Geoarchaeology: Earth Science and the Past*, pp. 360–96. Boulder, CO: Westview Press.

———. 1981. The early dispersal of insect pests of stored products as indicated by archaeological records. *Journal of Stored Product Research* 17: 1–12.

———. 1984. North-west Lincolnshire 10,000 years ago, in N. Field and A. White (eds.), *A Prospect* of Lincolnshire, pp. 11–17. Lincoln: Field and White.

Buckland, P. C. 1990. Granaries stores and insects: The archaeology of insect synanthropy, in D. Fournier and F. Sigaut (eds.), La Préparation Alimentaire des Cereals, pp. 69–81. Rapports présentés à la Table ronde, Ravello au Centre Universitaire pour les Biens culturels, avril 1988. PACT, Rixensart.

Buckland, P. C., Amorosi, T., Barlow, L. K., Dugmore, A. J., Mayewski, P. A., McGovern, T. H., Ogilvie, A. E. J., Sadler, J. P., and Skidmore, P. 1996. Bioarchaeological and climatological evidence for the fate of Norse farmers in medieval Greenland. Antiquity 70: 88–96.

Buckland, P. C., Beal, C. J., and Heal, S. V. E. 1990. Recent work on the archaeological and environmental context of the Ferriby boats, in S. Ellis and D. R. Crowther (eds.), Humber Perspectives: A Region through the Ages, pp. 131–46. Hull University Press.

Buckland, P. C., Coope, G. R., and Sadler, J. P. 2004. Bibliography of Quaternary Entomology, accessed March 2006, www.bugs2000.org/qbib.html

Buckland, P. C., Holdsworth, P., and Monk, M. 1976. The interpretation of a group of Saxon pits in Southampton. Journal of Archaeological Science 3: 61–69.

Buckland, P. C., and Kenward, H. K. 1972. Thorne Moor: A palaeoecological study of a Bronze Age site. Nature 241: 405–06.

Buckland, P. C., Perry, D. W., Gilason, G. M., and Dugmore, A. J. 1986. The pre-Landnám fauna of Iceland: A palaeontological contribution. Boreas 15: 173–84.

Buckland, P. C., and Sadler, J. P. 2000. Animal remains, identification and analysis: Insects, in L. Ellis (ed.), Archaeological Method and Theory: An Encyclopaedia, pp. 21–26. New York: Garland.

Burney, D. A., James, H. F., Piggott Burney, L., Olson, S. L., Kikuchi, W., Wagner, W. L., Burney, M., McCloskey, D., Kikuchi, D., Grady, F. V., Gage, R., II, and Nishek, R. 2001. Fossil evidence for a diverse biota from Kaua'i and its transformation since human arrival. Ecological Monographs 71: 615–41.

Carrott, J., and Kenward, H. 2001. Species associations among insect remains from urban archaeological deposits and their significance in reconstructing the past human environment. Journal of Archaeological Science 28: 887–905.

Carter, S. P. 1990. The stratification and taphonomy of shells in calcareous soils: Implications for land snail analysis in archaeology. Journal of Archaeological Science 17: 495–508.

Christensen, C. C., and Kirch, P. V. 1981. Nonmarine mollusks from archaeological sites on Tikopia,

southeastern Solomon Islands. Pacific Science 35: 75–88.

———. 1986. Nonmarine mollusks and ecological change at Barbers Point, Oah'u, Hawai'i. Bishop Museum Occasional Papers 26: 52–80.

Coope, G. R. 1977. Quaternary Coleoptera as aids in the interpretation of environmental history, in F. W. Shotton (ed.), British Quaternary Studies: Recent advances, pp. 55–68. Oxford: Clarendon Press.

———. 1986. Coleopteran analysis, in B. E. Berglund (ed.), Handbook of Holocene Palaeoecology and Palaeohydrology, pp. 703–11. Chichester: John Wiley and Sons.

———. 1987. The response of Late Quaternary insect communities to sudden climate changes, in J. H. R. Gee and P. S. Giller (eds.), Organization of Communities, Past and Present, pp. 233–45. Oxford: Blackwell Scientific Publications.

———. 1993. Late-Glacial (Anglian) and Late-Temperate (Hoxnian) Coleoptera, in R. Singer, B. G. Gladfelter, and J. J. Wymer (eds.), The Lower Palaeolithic Site at Hoxne, England, pp. 156–62. Chicago: University of Chicago Press.

———. 2006. Insect faunas associated with Palaeolithic industries from five sites of pre-Anglian age in central England. Quaternary Science Reviews 25: 1738–54.

Coope, G. R., and Elias, S. A. 2000. The environment of Upper Palaeolithic (Magdalenian and Azilian) hunters at Hauterive-Champréveyres, Neuchâtel, Switzerland, interpreted from coleopteran remains. Journal of Quaternary Science 15: 157–75.

Cowie, R. H. 2001. Decline and homogenization of Pacific faunas: The land snails of American Samoa. Biological Conservation 99: 207–22.

Cowie, R. H., and Grant-Mackie, J. A. 2004. Land snail fauna of Mé Auré Cave (WMD007), Moindou, New Caledonia: Human introductions and faunal change. Pacific Science 58: 447–60.

Cowie, R. H., and Robinson, A. C. 2003. The decline of native Pacific island faunas: Changes in the status of the land snails of Samoa through the 20th century. Biological Conservation 110: 55–65.

Denford, S. 1978. Mites and their potential use in archaeology, in Research Problems in Zooarchaeology, pp. 77–83. Occasional Publication 3. London: University of London, Institute of Archaeology.

Dinnin, M. H., and Sadler, J. P. 1999. 10,000 years of change: The Holocene entomofauna of the British Isles, in J. P. Sadler and K. J. Edwards (eds.), Holocene Environments of Prehistoric Britain, Quaternary Proceedings 7, pp. 545–62. Chichester: John Wiley and Sons.

Dixon, B., Soldo, D., and Christensen, C. C. 1997. Radiocarbon dating land snails and Polynesian land use of the islands of Kaua'i, Hawai'i. *Hawaiian Archaeology* 6: 52–62.

Dye, T., and Tuggle, H. D. 1998. Land snail extinctions at Kalaeloa, Oah'u. *Pacific Science* 52: 111–40.

Elias, S. A. 1986. Fossil insect evidence for Late Pleistocene paleoenvironments of the Lamb Spring site, Colorado. *Geoarchaeology* 1: 381–86.

———. 1990. The timing and intensity of environmental changes during Paleoindian period in western North America: Evidence from the fossil insect record, in L. D. Agenbroad, J. I. Mead, and L.W. Nelson (eds.), *Megafauna and Man*, pp. 11–14. Hot Springs and Flagstaff: Mammoth Site of Hot Springs and Northern Arizona University.

———. 1994. *Quaternary Beetles and Their Environments*. Washington, DC: Smithsonian Institution Press.

Elias, S. A., and Nelson, A. R. 1989. Fossil invertebrate evidence for late Wisconsin environments at the Lamb Spring site, Colorado. *Plains Anthropologist* 34: 309–26.

Erickson, J. M. 1988. Fossil oribatid mites as tools for Quaternary paleoecologists: Preservation quality, quantities and taphonomy, in R. S. Laub, N. G. Miller, and D. W. Steadman (eds.), *Late Pleistocene and Early Holocene Paleoecology and Archaeology of the Eastern Great Lakes Region. Bulletin of the Buffalo Society of Natural Sciences* 33: 207–26.

Evans, J. G. 1967. The stratification of Mollusca in chalk soils and their relevance to archaeology. Unpublished Ph.D. thesis, University of London, London.

———. 1971. Habitat change on the calcareous soils of Britain: The impact of Neolithic man, in D. D. A. Simpson (ed.), *Economy and Settlement in Neolithic and Early Bronze Age Britain and Europe*, pp. 27–73. Leicester: Leicester University Press.

———. 1972. *Land Snails in Archaeology*. London: Seminar Press.

French, C., and Lewis, H. 2005. New perspectives on Holocene landscape development in the southern English chalklands: The upper Allen valley, Cranbourne Chase, Dorset. *Geoarchaeology* 20: 109–34.

Frey, D. G. 1986. Cladocera analysis, in B. E. Berglund (ed.), *Handbook of Holocene Palaeoecology and Palaeohydrology*, pp. 667–92. Chichester: John Wiley and Sons.

Goodfriend, G. A. 1992. The use of land snail shells in palaeoenvironmental reconstruction. *Quaternary Science Reviews* 11: 665–85.

Hammond, P. 1995. The current magnitude of biodiversity, in V. H. Heywood and R. T. Watson (eds.), *Global Biodiversity Assessment*, pp. 113–38. Cambridge: Cambridge University Press.

Hoffecker, J. F., and Elias, S. A. 2003. Environment and archaeology in Beringia. *Evolutionary Anthropology* 12: 34–49.

Hofmann, W. 1986. Chironomid analysis, in B. E. Berglund (ed.), *Handbook of Holocene Palaeoecology and Palaeohydrology*, pp. 715–27. Chichester: John Wiley and Sons.

Hoganson, J. W., Gunderson, M., and Ashworth, A. C. 1989. Fossil beetle analysis, in T. D. Dillehay (ed.), *Monte Verde: A Late Pleistocene settlement in Chile, 1. Palaeoenvironment and Site Context*, pp. 211–26. Washington, DC: Smithsonian Institution Press.

Kenward, H. K. 1975. Pitfalls in the environmental interpretation of insect death assemblages. *Journal of Archaeological Science* 2: 85–94.

———. 1985. Outdoor-indoors? The outdoor component of archaeological insect assemblages, in N. R. J. Fieller, D. D. Gilbertson, and N. G. A. Ralph (eds.), *Palaeobiological Investigations: Research Design, Methods and Data Analysis*, pp. 97–104. Oxford: British Archaeological Reports S266.

———. 1997. Synanthropic decomposer insects and the size, remoteness and longevity of archaeological occupation sites: Applying concepts from biogeography to 'islands' of human occupation, in A. C. Ashworth, P. C. Buckland, and J. P. Sadler (eds.), *Studies in Quaternary Entomology: An Inordinate Fondness for Insects, Quaternary Proceedings 5*, pp. 135–51. Chichester: John Wiley and Sons.

———. 2006. The visibility of past trees and woodland: Testing the value of insect remains. *Journal of Archaeological Science* 33: 1368–80.

Kenward, H., and Carrott, J. 2006. Insect species associations characterise past occupation sites. *Journal of Archaeological Science* 33: 1452–73.

Kenward, H., and Large, F. 1997. Insects in urban waste pits in Viking York: Another kind of seasonality. *Environmental Archaeology* 3: 35–53.

Kenward, H. K., Hall, A. R., and Jones, A. K. G. 1980. A tested set of techniques for the extraction of plant and animal macrofossils from waterlogged archaeological deposits. *Science and Archaeology* 22: 315.

Kerney, M. P. 1966. Snails and man in Britain. *Journal of Conchology* 26: 3–14.

Kerney, M. P., Chandler, E. H., and Brown, T. J. 1964. The late-glacial and post glacial history of the Chalk escarpment near Brook, Kent. *Philosophical Transactions of the Royal Society, Series B* 248: 135–204.

Kirch, P. V. 1973. Prehistoric subsistence patterns in the northern Marquesas Islands, French Polynesia. *Archaeology and Physical Anthropology in Oceania* 8: 24–40.

———. 1975. Excavations at Sites A1-3 and A1-4: Early settlement and ecology in Halawa Valley,

Molokai, in P. V. Kirch and M. Kelly (eds.), *Prehistory and Ecology in a Windward Hawaiian Valley: Halawa Valley, Moloka'i*, pp. 17–70. Pacific Anthropological Records 24. Honolulu: Bernice P. Bishop Museum.

Kirch, P. V. 1982. Man's role in modifying tropical and subtropical Polynesian ecosystems. *Archaeology in Oceania* 18: 26–31.

———. 1993. Non-marine molluscs from the To'aga site sediments and their implications for environmental change, in P. V. Kirch and T. L. Hunt (eds.), *The To'Aga Site: Three millennia of Polynesian occupation in the Manu'a Islands, American Samoa*, pp. 115–21. Berkeley: Contributions of the University of California Archaeological Research Facility 51.

Lain Ellis, G., Goodfriend, G. A., Abbott, J. T., Hare, P. E., and Von Endt, D. W. 1996. Assessment of integrity and geochronology of archaeological sites using amino acid racemization in land snail shells: Examples from central Texas. *Geoarchaeology* 11: 189–213.

Lozek, V. 1986. Mollusca analysis, in B. E. Berglund (ed.), *Handbook for Holocene Palaeoecology and Palaeohydrology*, pp. 729–41. Chichester: John Wiley.

Moine, O., Rousseau, D. D., Jolly, D., and Vianey-Liaud, M. 2002. Paleoclimatic reconstruction using mutual climatic range on terrestrial molluscs. *Quaternary Research* 57: 162–72.

Neuweger, D., White, P., and Ponder, W. F. 2001. Land snails from Norfolk Island sites. *Records of the Australian Museum Supplement* 27: 115–22.

Osborne, P. J. 1969. An insect fauna of Late Bronze Age date from Wilsford, Wiltshire. *Journal of Animal Ecology* 38: 555–66.

———. 1971. An insect fauna from the Roman site at Alcester, Warkwickshire. *Brittania* 2: 156–65.

———. 1983. An insect fauna from a modern cesspit and its comparison with probable cesspit assemblages from archaeological sites. *Journal of Archaeological Science* 10: 453–63.

Panagiotakopulu, E. 2000. *Archaeology and Entomology in the Eastern Mediterranean: Research into the History of Insect Synanthropy in Greece and Egypt*. Oxford: British Archaeological Reports International Series 836.

———. 2001. New records for ancient pests: Archaeoentomology in Egypt. *Journal of Archaeological Science* 28: 1235–46.

Perry, D. W., Buckland, P. C., and Snæsdóttir, M. 1985. The application of numerical techniques to insect assemblages from the site of Stóraborg, Iceland. *Journal of Archaeological Science* 12: 335–45.

Ponel, P., Matterne, V., Coulthard, N., and Yvinec, J.-H. 2000. La Tène and Gallo-Roman natural environments and human impact at the Touffréville rural settlement, reconstructed from Coleoptera and plant macroremains (Calvados, France). *Journal of Archaeological Science* 27: 1055–72.

Porch, N., and Elias, S. 2000. Quaternary Beetles: A review and issues for Australian studies. *Australian Journal of Entomology* 39: 1–9.

Preece, R. C. 1998. Impact of early Polynesian occupation on the land snail fauna of Henderson Island, Pitcairn group (South Pacific). *Philosophical Transactions of the Royal Society of London, Series B* 353: 347–68.

Preece, R. C., and Bridgland, D. R. 1999. Hollwell Coombe, Folkestone: A 13,000-year history of an English Chalkland Valley. *Quaternary Science Reviews* 18: 1075–1125.

Rolett, B. V. 1992. Faunal extinctions and depletions linked with prehistory and environmental change in the Marquesas Islands (French Polynesia). *Journal of the Polynesian Society* 101: 86–94.

Rousseau, D.-D. 1991. Climatic transfer functions from Quaternary molluscs in European loess deposits. *Quaternary Research* 36: 195–209.

Rousseau, D.-D., Limondin, N., Magnin, F., and Puissegur, J.-J. 1994. Temperature oscillations over the last 10,000 years in western Europe estimated from terrestrial mollusc assemblages. *Boreas* 23: 66–73.

Rousseau, D.-D., Preece, R., and Limondint-Lozouet, N. 1998. British late glacial and Holocene climatic history reconstructed from land snail assemblages. *Geology* 26: 651–54.

Rowe, C., Stanisic, J., David, B., and Lourandos, H. 2001. The helicinid land snail *Pleuropoma extincta* (Odhner, 1917) as an environmental indicator in archaeology. *Memoirs of the Queensland Museum* 46: 741–70.

Sadler, J. 1991. Beetles, boats and biogeography. *Acta Archaeologica* 61: 199–211.

Sadler, J., and Skidmore, P. 1995. Introductions, extinctions or continuity? Faunal change in the North Atlantic islands, in R. A. Butlin and N. Roberts (eds.), *Ecological Relations in Historical Times: Human Impact and Adaptation*, pp. 206–55. Institute of British Geographers Special Publication 32. Oxford: Blackwell Publishing.

Shackley, M. 1981. *Environmental Archaeology*. London: George Allen and Unwin.

Schlevis, J. 1990. The reconstruction of environments on the basis of remains of oribatid mites (Acari; Oribatidae). *Journal of Archaeological Science* 17: 559–71.

———. 1992. The identification of archaeological dung deposits on the basis of remains of predatory mites (Acari; Gamasida). *Journal of Archaeological Science* 19: 677–82.

———. 1997. Mites in the background: Use and origin of remains of mites (Acari) in Quaternary deposits, in A. C. Ashworth, P. C. Buckland,

and J. P. Sadler (eds.), *Studies in Quaternary Entomology: An Inordinate Fondness for Insects*, pp. 233–36. Quaternary Proceedings 5. Chichester: John Wiley and Sons.

Solem, A. 1990. How many Hawaiian land snails are left? And what can we do for them? *Bishop Museum Occasional Papers* 30: 27–40.

Smith, D. N. 1996. Thatch, turves and floor deposits: A survey of Coleoptera in material from abandoned Hebridean backhouses and the implications for their visibility in the archaeological record. *Journal of Archaeological Science* 23: 161–74.

Sparks, B. W. 1961. The ecological interpretation of Quaternary non-marine Mollusca. *Proceedings of the Linnean Society of London* 172: 71–80.

———. 1969. Non-marine Mollusca in archaeology, in D. Brothwell and E. Higgs (eds.), *Science in Archaeology*, pp. 395–406. London: Thames and Hudson.

Stein, J. K. 1983. Earthworm activity: A source of potential disturbance of archaeological sediments. *American Antiquity* 48: 277–89.

———. 2001. A review of site formation processes and their relevance to geoarchaeology, in P.

Goldberg, V. Holliday, and C. R. Ferring (eds.), *Earth Sciences and Archaeology*, pp. 37–51. New York: Kluwer Academic/Plenum Publishers.

Thomas, K. D. 1985. Land snail analysis in archaeology: Theory and practice, in N. R. J. Fieller, D. D. Gilbertson, and N. G. A. Ralph (eds.), *Palaeobiological Investigations. Research Design, Methods and Data Analysis*, pp. 131–56. AEA Symposia volume 5B. Oxford: BAR International Series 266.

Walker, I. R., Smol, J. P., Engstrom, D. R., and Birks, H. J. B. 1991. An assessment of the Chironomidae as quantitative indicators of past climatic change, *Canadian Journal of Fisheries and Aquatic Science* 48: 975–87.

Whitehouse, N. J. 2006. The Holocene British and Irish ancient forest fossil beetle fauna: Implications for forest history, biodiversity and faunal colonisation. *Quaternary Science Reviews* 25: 1755–89.

Wilkinson, K., and Stevens, C. 2003. *Environmental Archaeology: Approaches, techniques and applications*. Stroud: Tempus.

Wilson, E. O. 1992. *The Diversity of Life*. Cambridge: Belknap Press.

46

ENVIRONMENTAL ARCHAEOLOGY: INTERPRETING PRACTICES-IN-THE-LANDSCAPE THROUGH GEOARCHAEOLOGY

Tim Denham

Environmental archaeology is concerned with "the physical and biological elements and relationships that impinge on" the activities of people in the past (Dincauze 2000: xxiv). Although definitions vary, environmental archaeologists consider human activities in the past within their environmental contexts and apply techniques and interpretations derived from the biological and geophysical sciences to archaeological problems. Thus the term "environmental archaeology" encapsulates geoarchaeology, which is concerned with the "application of concepts and methods of the geosciences to archaeological research" (Waters 1992: 3).

Other chapters in this Handbook address specifically how specialist biological knowledge is used in landscape archaeology (e.g., see chapters by Dolby; Fairbairn; Mainland; Matisoo-Smith; Pate; Porch; Rowe and Kershaw). In this chapter, the geoarchaeological aspect of environmental archaeology is characterized, and its relevance to landscape archaeology explored. Sections address key issues associated with the practice of environmental archaeology, namely multidisciplinary teamwork, geoarchaeological investigations of site stratigraphy, scalar problems, and complementarity of techniques. Each issue is illustrated using examples drawn from my own research in the Upper Wahgi valley in the highlands of Papua New Guinea.

Conceptualizing Human-Environment Interactions

Before characterizing the techniques and methods applied by environmental archaeologists, I consider the concepts used to frame and interpret people-environment relations. Often the technical aspects of environmental archaeology have been prioritized over the conceptual. Indeed, and like other branches of archaeological science, technique and method can become ends-in-themselves rather than means-to-an-end, which, in this case is to shed light on people living in the past. Furthermore, a focus on method can be accompanied by a circumscribed critical stance toward the interpretation of results (Dincauze 2000).

Concept and method are interdependent. Concepts provide a framework within which environmental archaeological issues can be addressed; similarly, the results and scope of investigation afforded by the application of innovative techniques can expand and inform conceptual frameworks. Difficulties encountered when seeking to unravel social-environmental interactions in the past can, in part, arise from the ways in which these interactions are conceptualized.

A tendency in archaeology and other disciplines has been to dissemble the human and the environmental as separate spheres, or entities, of study. Subsequently, either the degrees to which people's activities and way of life are enabled or constrained by the biological and physical environment are assessed (characteristic of culture historic approaches), or human-environment interactions are conceptualized in systemic terms (characteristic of New Archaeology). The former approach echoes early-to-mid-20th century concerns of environmental possibilism, probabilism, and determinism (see review in Lewthwaite 1966), whereby human activities are freely chosen, environmentally influenced, and determined by the environment, respectively. The systemic approach is more recent and characteristic of cultural ecology, human ecology, and processualism, whereby human and environmental components are conceived as part of one interrelated system (Binford 1965; Clarke 1973), although there is still much debate about the degrees to which social systems are influenced or determined by the environment (e.g., Binford 1965, 1967). These types of approaches have predominated within environmental archaeology (see Dincauze 2000), although there has been some movement toward more socially informed perspectives (Evans 2003).

In recent years, alternative perspectives focusing on people living-in-the-world have emerged within archaeology. Here, the world is not considered in terms of its empirical reality but as a meaningful realm of human activity. Advocates of these perspectives acknowledge, as well as sometimes abuse, the engagement and the creativity of the archaeologist in portraying people living-in-a-past (Tilley 1994; cf. Fleming 1999). These phenomenologically informed conceptions have had limited adoption by environmental archaeologists.

Environmental archaeology can, in principle, serve as a bridge between human activities and environmental processes in the past. Although the techniques of the subdiscipline are predominantly derived from the biological and geophysical sciences, interpretations need not become trapped by a naïve empiricism or neofunctional, mechanistic, and systemic thinking. Rather, environmental archaeology has great potential to contribute to our understanding of the complex, mutually transformative, and recursive nature of human-environment interactions in the past.

Practices-in-the-Landscape

The landscape represents a "human-scale" of lived-in places. People live in landscapes, whether they are urban, suburban, rural, coastal, and so on; landscapes are a fundamental realm of lived experience. But how do we begin to understand living-in-a-*past*-landscape? How do we begin to access that past? One starting point particularly relevant to environmental archaeologists seeking to unravel the complexities of people living-in-the-landscape is the concept of "practice" (Barrett 1994).

The archaeological record provides evidence of past practices, whether the practices were the result of enacted structural influences, dispositions, or individual improvisation (after Bourdieu 1990). Practices are effectively what people do. They constitute the nexus of human-environmental interactions; practices are inscribed in the landscape, thereby providing the visible evidence of past interactions (Figure 46.1).

If practices are taken as a focus, four overlapping and pervasive themes of environmental archaeology emerge:

1. reconstructing human-environmental interactions in the past as manifest through practice;

2. placing past practices within broader social and environmental contexts, for example, attempting to understand why people engaged in certain practices;

3. taking into account the social forms and environmental processes that influence our ability to construct knowledge of past practices, such as issues of site formation processes, taphonomy, preservation, and destruction (Binford 1981; Brain 1981; Mountain and Bowdery 1999; Schiffer 1987);

4. reflecting on the ways in which we come to know the past, effectively comparable to Bourdieu's (1990) theory of practical knowledge.

The Multidisciplinary Scope of Environmental Archaeology

A multiplicity of approaches and techniques derived from the biological and geophysical sciences are potentially available in the pursuit of a particular problem (see Tables 46.1 and 46.2; also reviews in Dincauze 2000 and Evans and O'Connor 1999). Despite this potential range of techniques, archaeological projects are usually conducted under fiscal constraints and tight schedules. Therefore, the methods, disciplines, and specialists with the greatest potential to address the project aims need to be identified and informed of the relevant financial, scheduling, and other logistical constraints at an early stage of planning. Choices have to be made. The first questions to

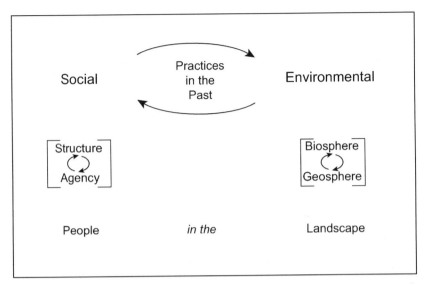

Figure 46.1   Conceptual overview of a practice-centred approach to environmental archaeology and landscape archaeology.

ask are these: "What questions or purpose is the project designed to address?" and "Which methods and techniques are most relevant to the investigation of this particular problem?"

Teamwork

Environmental archaeology is increasingly conducted in a multidisciplinary environment. Most environmental archaeologists are specialized in the application of a particular field of the biological or earth sciences to archaeological concerns. Just as archaeologists in other fields of archaeology, they draw on a range of specialists to augment their work and increasingly have to learn how to manage multidisciplinary teams. There are continual problems in trying to get collaborators to think outside their discipline, to get them to understand the goals of the overall project and their role in it, to refashion the meaning of their work and channel it to different aims (without denigrating the work or the aims), to provide results in a timely fashion, and so on. These are not incidental concerns, since they continually undermine projects and the production of results in a timely fashion, but yet they are rarely discussed constructively or taught.

Geoarchaeology and Site Stratigraphy

Numerous techniques are employed by geoarchaeologists to reveal information about past environments and how people may be implicated in their formation, maintenance, or disturbance. I outline

below a broad range of such techniques, identifying in particular the kinds of information that they are able to address. Geoarchaeological investigations can focus on the stratigraphy either at a single site or compare results from multiple sites across a landscape. Geoarchaeological analyses are used by archaeologists for a multitude of reasons, the most common being these:

1. To disentangle site formation processes, thereby enabling the reconstruction of depositional (sedimentary) and pedogenic (soil formation) processes through time, with particular reference to archaeological evidence of past practices and differentiating postformation and postburial transformations. Distinct stages in the development of each stratigraphic unit and archaeological context are differentiated: (a) original deposition represented by inherited sedimentary stratification; (b) pedogenic alteration of the sediment and, potentially, the formation of archaeologically significant palaeosols and deposits; (c) pedogenic transformations of contexts prior to and after burial (after Barham 1995: 161).

2. To identify direct and indirect evidence of past practices, most often those associated with agriculture, pastoralism, and occupation.

3. To determine postdepositional modifications and context reliability for subsequent dating and palaeoecological analyses.

Table 46.1 Overview of geophysical sciences utilized by environmental archaeologists (that is, geoarchaeology). Note: major overviews generally integrate geology and geomorphology.

Science	Environmental Archaeological Significance	Key References
Climatology	Understanding the influence of past climates (palaeoclimates) and climate change on past environments (palaeo-environments) and human practices, especially with respect to biotic distributions, glacial cycles, and sea level change; there is an awareness of the cumulative human influences on local, regional, and global climates during the Holocene to the present.	Lamb 1995; Burroughs 2005
Geochemistry	Applying geochemical techniques (elemental, crystalline, structural, stable, isotopic, palaeomagnetic, and so on) to understand biological, geophysical, and human interrelationships.	Thompson and Oldfield 1986; Sparks 1995; Pollard and Heron 1996
Geochronology	Applying dating techniques to generate a chronology for archaeological materials, environmental proxies, and associated stratigraphy; the selection of dating technique is dependent on the material being dated and anticipated age range.	Walker 2003
Geology	Understanding the influence of tectonic and volcanic processes and products on landscape formation and human practice.	Brown 1997; Rapp and Hill 1998; Williams et al. 1998; Garrison 2003; Goldberg and MacPhail 2006
Geomorphology	Understanding the influence of earth surface processes and products on human practice, often within specific environments such as alluvial, arid, coastal, estuarine, glacial, lacustrine, and marine; there is an awareness of the increasing human influences on geomorphological processes from the late Pleistocene to the present.	
Pedology	Formation of soils on stable substrates (whether of biogenic, geological, or sedimentary origin); significance for understanding past soils (palaeosols), human practices associated with palaeosols (such as cultivation), and the effects of pedogenesis on the alteration, disturbance, and preservation of buried archaeological and palaeo-ecological remains (namely, site formation processes and taphonomy).	Limbrey 1975; Courty et al. 1989; Barham and MacPhail 1995

Table 46.2 Overview of biological sciences utilized by environmental archaeologists. Note: in several tropical regions archaeobotanists are more reliant on microfossils, especially phytoliths (Piperno 2006) and starch grains (Torrence and Barton 2005) than on traditional macrobotanical identifications of seeds and wood (see Fairbairn, this volume).

Science	Subdivision (with Commonly Used Examples)	Environmental Archaeological Significance	Key References
Biological anthropology	Macrofossil: bones, teeth, collagen, soft tissues	Examining human remains and past diet, disease, and populations (palaeodiet, palaeopathology, palaeodemography) with reference to lifestyle (and associated practices) and environment	Cohen and Armelagos 1984; Hoppa and Vaupel 2002; Roberts and Manchester 2005
	Microfossil: blood, cellular, and subcellular analyses (biochemistry, isotopes, DNA)		
Botany	Macrofossil: (macrobotanical): seeds, wood, kernels, plant parts, and fibres	Analyzing plant remains to reconstruct vegetation histories (palaeoecology) and to understand people's exploitation of plants (archaeobotany)	Dimbleby 1985; Bennett and Willis 2001; Pearsall 2000
	Microfossil: pollen, phytoliths, starch grains, diatoms, parenchyma		
Zoology	Macrofossil: bones, teeth, shells, soft tissues	Analyzing animal remains to reconstruct past environments and to understand people's exploitation of animals (zooarchaeology)	Reitz and Wing 1996; Maltby 2006
	Microfossil: blood, bones, insects, shells, tissue fragments, and residues		
Genetics	DNA	Analyzing the genetic traits of contemporary populations (for phylogenies, significant genetic loci) or ancient biological material (for genetic fingerprinting, phylogenies, comparisons with modern populations)	Jones 2001; Gugerli et al. 2005; Willerslev and Cooper 2005
	aDNA (ancient DNA)		

Geoarchaeological and sedimentological techniques are often differentiated into two groups; those undertaken on disturbed samples and those on undisturbed samples (Canti 1995: 183–6; Courty, Goldberg, and MacPhai 1989: 40–3; Hodgson 1978: 125–32). Disturbed, disaggregated, or bulk samples are used for chemical and physical compositional analyses (for example, elemental, magnetic, mineral, and particle size analyses), whereas undisturbed samples are used for meso- and micromorphological investigations (for example,X-radiograph and thin section description), respectively(Table 46.3).

Disturbed Samples: Multitechnique Assessments

"The exact chemical conditions under which soil minerals form are not known at present" (Bohn et al. 1985: 91). Soils are "open" systems in which transformations are driven by a continuous flux of matter and energy acting on a progressively altered parent material (Chadwick and Chorover 2001: 324). Most soils contain mixed clay assemblages, some of which retain relict features and reflect varying rates and reversibility of chemical processes in the soil (Velde 1985: 30). Soil water

Table 46.3 Uses and limitations of geoarchaeological techniques used in stratigraphic analysis (with permission from Canti 1995: 185; see Courty et al. 1989; Goldberg and MacPhail 2006).

Technique	Uses and Advantages	Limitations
Disturbed Samples		
Age-depth comparisons	Sedimentation rates	Problematic application of uniformitarian principles if subject to periods of soil formation and human disturbance
Calcium carbonate (CaCO$_3$)	Pedogenesis, taphonomy, and site formation	Strongly influenced by postdepositional processes
Major and minor elemental analyses (AAS, ICP-MS, XRF)	Elemental composition, elemental mapping, weathering, and trace element distributions	Extreme detail often not needed and of limited use if minerals composed of common elements
Mineral magnetism	Identify sediment sources (tephra, topsoil, subsoil), burning, and human practices at sites/across the landscape	Problematic to disentangle fermentation effects, limited value in some mineralogical environments
Particle size analysis	Physical composition, general stratigraphic, provenance, and depositional information	Averages microstratigraphy (no information on aggregates), limited use for soils with high clay and organic contents
pH (acidity/alkalinity)	Soil processes, taphonomy, and site formation	Strongly influenced by contemporary processes
Phosphate (PO$_4^{3-}$)	Human and animal additions at sites/across the landscape	Debatable theoretical base except for simplest comparative uses
Scanning electron microscopy (SEM)	Morphology of mineral grains and clay crystals, weathering stages/phases	Extreme detail often not needed in archaeological studies and can be difficult to meaningfully relate to archaeological information
X-ray diffraction (XRD)	Mineral composition and weathering	Unreliable if large amounts of X-ray amorphous material, such as organic material and poorly ordered clay minerals
Undisturbed Samples		
Scanning electron microscopy (SEM)	Microstructure, microcomposition, and weathering	Extreme detail often not needed in archaeological studies or to understand site formation processes
Thin-section description (transmitted/cross-polarized/reflected light)	Microstratigraphy, pedofeatures, taphonomy, and some land uses (e.g., cultivation techniques, animal husbandry, settlement use)	Expensive and time-consuming, qualitative information, evidence of former human activities not always preserved
Thin-section analysis using energy dispersive X-ray analysis (EDXRA)	Elemental compositions, elemental mapping, and weathering	Limited use among minerals composed of common elements, such as highly weathered clays
X-radiography of sediment slices and large thin sections	Comparative analysis of sedimentary and pedogenic processes, potential for land use studies, invaluable in wetlands and waterlogged deposits	Variable value depending on contexts and sample stability

chemistry is particularly significant in determining weathering pathways and rates, with some reactions between secondary clay minerals being reversible given changing pH, cation concentrations, and REDOX potential. Several other factors, including climate, drainage, and vegetation, complicate the sequence of clay mineral formation, as noted for Highland New Guinea (Chartres and Pain 1984: 147–48).

As Bohn and associates (1985: 68) state: "Soil development, in the chemical sense, is roughly synonymous with weathering." The extent of chemical weathering is reflected in the mineralogy of different fine earth fractions (that is, clay, silt, and sand). Through time, primary minerals (mainly sand and silt fractions) are transformed to secondary minerals (mainly clay fraction) with increasing alteration with time. The relative stabilities and susceptibilities of minerals to chemical weathering are generally known (Bohn et al. 1985: 79) and are dependent on soil water geochemistry, principally solute concentrations and pH (Nemecz 1981: 494). A combination of mineralogical, chemical, and magnetic assessments is usually sufficient for the archaeologist to reconstruct major sedimentary events and pedogenic periods, but the inherent complexity of soil chemistry and the use of bulk samples often prevent more sophisticated interpretations.

Undisturbed Samples: Multiscale Investigations

Hodgson has stated that comprehensive field descriptions of soils and sediments "provide an almost indispensable background to laboratory studies whatever their purpose" (Hodgson 1978: 1–2). However, within archaeology, and inscribed within single-context recording systems in particular, is a presumption that the "units of stratification" (Harris 1989: 42) and interfaces (Brown and Harris 1993: 14) identifiable in the field represent significant relict characteristics: "It is often regarded as axiomatic that the properties viewed in a freshly cleaned and exposed archaeological deposit represent the information required for primary archaeological archive and subsequent interpretation" (Barham 1995: 147).

Barham (1995: 147–51, 155–58) has criticized this position by highlighting the "methodological divergence" and practical and theoretical problems of unit and interface identification in different branches of soil science and sedimentology. His main concern is an inability to differentiate original formation (primary) and successive (secondary) attributes in the field. However, meso-scale and microscale investigations can be undertaken to verify field-based macro-scale interpretations, to determine the preservation of primary attributes, and to investigate the palaeoenvironmental and archaeological significance of each unit and interface.

Mixed-Method Geoarchaeological Analysis

Canti (1995: 186) advocates the adoption of mixed-method approaches to geoarchaeological analysis because "overlaid" multitechnique investigations greatly improve interpretative resolution. For example, the archaeologist will want to know what may have caused decreases and increases in sedimentation rates at a site through time. Were increases in sedimentation rate due to people disturbing and clearing forest in the catchment, or were they the result of climate change or a volcanic eruption? Only through the application of additional geoarchaeological analyses, such as mineral magnetics and thin section description, can the types and origins of increased sediment input be determined (for example, whether they are derived from subsoil or topsoil, and whether they consist of particles or aggregates). Furthermore, a combination of geoarchaeological and biological analyses enables more complex interpretations of landscape change; for example, phytolith, pollen, and charcoal (macro- and micro-) studies can be used to reconstruct burning and vegetation histories for the landscape that can be compared to sedimentation and soil formation histories. More detailed examples of mixed-method geoarchaeological investigations are presented below.

Scales of Practice

Depending on the nature of the project, we may seek to understand an individual practice that is fixed within a local situation and limited temporal span (for example, the digging of a ditch or the use of a stone tool [Haslam 2006]), or we may seek to understand the cumulative effects of individual actions through time across a landscape (for instance, the creation of an "agricultural landscape" [Haberle 2003]). Perhaps, though, the ultimate goal of archaeology is to understand the indeterminate relations between individual actions and broader social structures and environmental processes (after Hodder 1999). For example, the interrelatedness of individual practices, such as digging a ditch, and larger-scale processes of landscape and social formation need to be understood and characterized (e.g., Bayliss-Smith and Golson 1999).

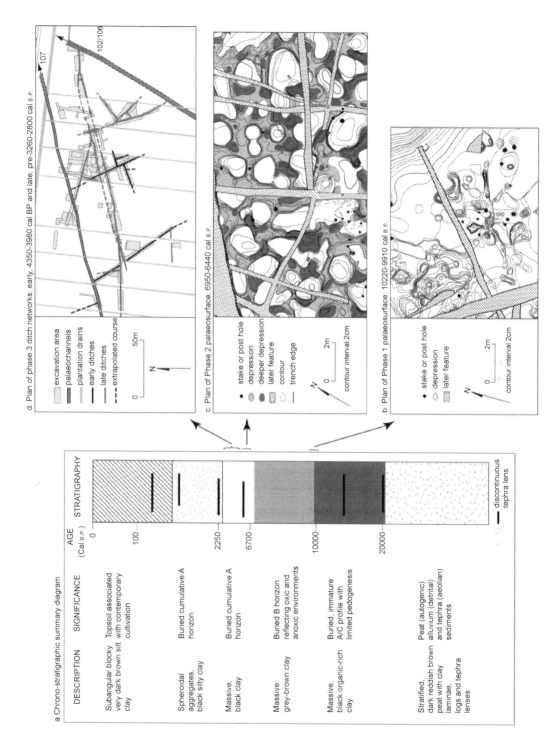

Figure 46.2 Geoarchaeological interpretation of early and middle Holocene plant exploitation at Kuk Swamp, Upper Wahgi valley, Papua New Guinea (reproduced with permission from Denham et al. 2003: fig. 2).

Environmental archaeology, especially geo-archaeology, provides a means to understand and manage scalar issues of time and place. A geoarchaeological approach directs us to situate archaeological remains within their environmental context. The intention is to integrate archaeological evidence representing discrete periods of former human activity to long-term practices and environmental processes in the landscape. Different issues of temporal and spatial resolution are associated with understanding landscape change through time (diachronic) and with understanding practices-in-the landscape at points in time (synchronic) (see also Head, this volume; Heilen et al., this volume). To exemplify this, episodes of wetland manipulation and drainage during the early and middle Holocene at Kuk Swamp in Highland New Guinea can be depicted in such a way that archaeological remains of former (synchronic) practices (see Figure 46.2b–d) articulate with the chrono-stratigraphy, which represents (diachronic) processes of long-term landform development along the wetland margin (see Figure 46.2a). However, each discrete set of practices documented on the wetland margin can be examined in greater or lower historico-geographical resolution.

If we consider practices dating to ca. 10,000 years ago at Kuk, it becomes clear that we can employ a range of techniques to provide both more general and more specific understandings of what was happening at that time. From the features documented in a single excavation trench, we can hone in on practices and environmental processes occurring at higher resolution. The field recording of stratigraphy documents major (macro-scale) and minor (microscale) stratigraphic units as well as the edges and fills of features (Figure 46.3a-b; see also Harris 1989); it is fundamental to and orients more detailed analyses (see Stern, this volume). Higher resolution (meso-scale) techniques such as X-radiography and thin section description can be used to investigate soil formation within feature fills (Figure 46.3b; see also Krinitzsky 1970) and potentially to investigate the mode of aggregate composition, formation, and microfossil deposition within individual feature fills (Figure 46.3c-d; see also Courty, Goldberg, and MacPhail 1989). The analytical precision of higher resolution techniques was intended to identify pedological evidence of past plant exploitation practices. In this case, there is evidence for a relatively shorter period of drier conditions on the wetland margin and the formation of an immature soil profile, which, in turn, is augmented by archaeological and archaeobotanical evidence for the exploitation and processing of

tuberous plants on adjacent surfaces at this time (Denham 2005).

From the same starting point, an excavation trench, we can broaden our spatial focus to consider the relationship of the archaeological finds exhibited in plan (Figure 46.4a) to the landform at that time (Figure 46.4b). Features associated with digging, staking, and microtopographic manipulation occur on areas of higher ground adjacent to a palaeochannel. For some, this 10,000 year-old palaeochannel is artificial and indicative of agricultural practices at this time, whereas I am more circumspect (see debate in Denham, Golson, and Hughes, 2004: 267–78). Irrespective of the exact type of plant exploitation practices in this locale, we can draw on archaeological and palaeoecological findings from across the Upper Wahgi valley to envisage the landscape at this time (Figure 46.4c). These landscape-scale reconstructions creatively draw on empirical information, albeit with some chronological conflation of evidence and inferences drawn from ethnography, to develop interpretations of people living across, in and through this landscape 10,000 years ago.

Complementarity, Integration, and Specificity

Landscapes are the product, in part, of a way of life. The naming of a landscape in terms of a particular lifeway (for example, agricultural, industrial, pastoral) may be as misleading as it can be illuminating. Numerous and diverse types of practice with highly variable spatial and temporal manifestations become inscribed on a landscape through time. How can this diversity and complexity of co-occurring and superimposed practices in a landscape be encapsulated by archaeological research?

The historical development of the Upper Wahgi valley has been likened to the production of an agricultural landscape (Haberle 2003); it has been produced through the interplay and co-occurrence of numerous types of practice that have their own temporality, spatiality, and visibility. For example, some burning, clearing, gathering, tending, and transplanting practices have a greater antiquity than others do and occurred widely across the landscape; these are only generally inferred from palaeoenvironmental proxies (for instance, charcoal, phytolith, and pollen analyses and rates of sedimentation). In contrast, other practices, such as ditch construction, mound construction, processing of plant parts, and tilling the soil are more recent, have more restricted distributions (in terms of both location of practice and preservation), and are often more visible through archaeological,

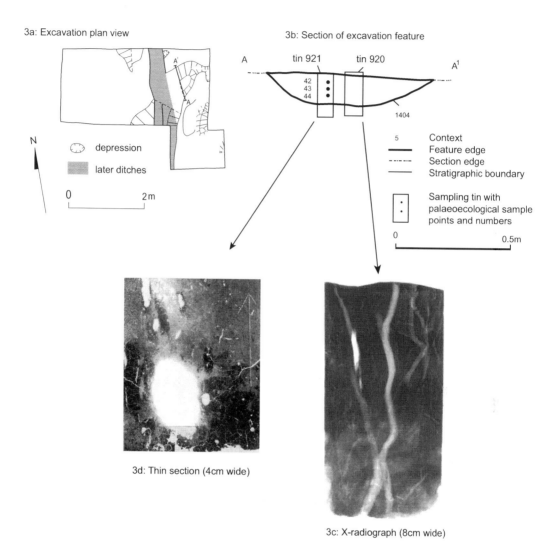

3a: Excavation plan view

3b: Section of excavation feature

Figure 46.3 Issues of geoarchaeological resolution I: from excavation trench to palaeosol: (a) field recording of excavation trench in plan view; (b) field recording of stratigraphy in the walls of an excavated trench; (c) X-radiograph of feature fill showing limited aggregate development at the base (courtesy of Dr. Alain Pierret); and (d) thin section photomicrograph of pedofeatures.

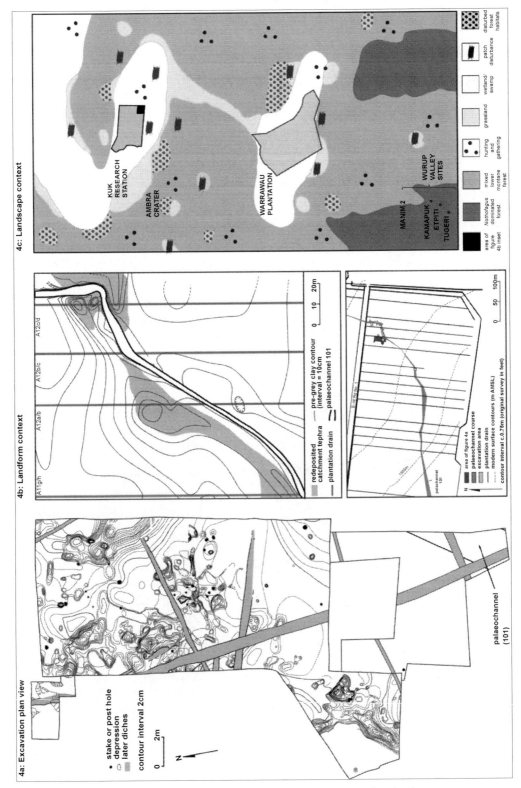

Figure 46.4 Issues of geoarchaeological resolution II: from excavation trench to landscape:
(a) archaeological features recorded in the base of excavated trenches; (b) archaeological finds in landform context; and (c) interpretation of human-environment interactions across the landscape.

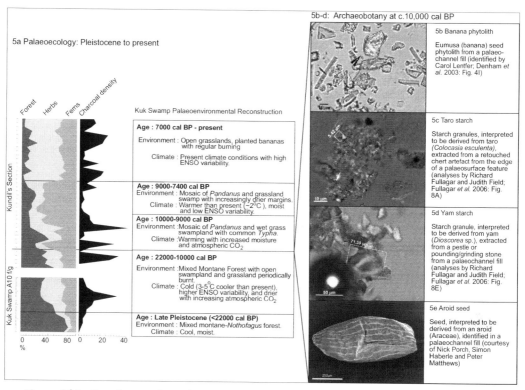

Figure 46.5 Complementarity and integration of archaeobotanical and palaeoecological information: (a) summary pollen and microcharcoal diagram with descriptions (reproduced with permission from Denham et al. 2004b: fig. 6); (b) photomicrographs of phytoliths, starch grains and a seed of starch-rich plants used or present in the landscape 10,000 years ago (reproduced with permission from Denham et al. 2003: fig. 4; Fullagar et al. 2006: fig. 8).

archaeobotanical, and stratigraphic investigation. However, the integration of more general palaeoenvironmental information (relatively low historico-geographic resolution) with archaeological findings (greater specificity and historico-geographic resolution) can contribute to more complex scenarios of practices in the past.

In the Upper Wahgi valley, palaeoecological data provides an impression of how the vegetation has changed through time (Figure 46.5). The causes of those transformations can be only inferred from general patterns of how the forest cover has declined, disturbed habitats expanded and persisted, and burning increased through time. These palaeoecological signals are usually considered representative of an anthropic landscape (Haberle 1994). For example, the lowland montane forests in the Kuk vicinity were already disturbed in the early Holocene. Pollen taxa frequencies suggest persistent, if spatially variable, disturbance using fire and the creation of vegetation mosaics comprising secondary forest, regrowth and grassland within the forest, as well as along the wetland margin.

But what practices opened up the forest at this time? Faunal analyses from sites across the Highlands suggest hunting was ubiquitous (Mountain 1991), but other inter-montane valleys do not witness forest clearance at this time. Archaeobotanical data from Kuk suggest broad-spectrum plant exploitation was important at this time, which included the exploitation of a starch rich aroid and yam (Fullagar et al. 2006), as well as possibly *Eumusa* bananas that were in the landscape from 10,000 years ago (Denham et al. 2003). These findings implicate plant exploitation practices as drivers of landscape change, potentially through the deliberate creation of habitats conducive to the growth, transplanting and (possibly) cultivation of edible plants. Gaps within the forest may have been enhanced or created through ring-barking, clearance, and burning. Here the complementarity of multidisciplinary approaches is significant. Higher resolution archaeological information, primarily representative of on-site practices, grounds general palaeoecological trends and enables more specific interpretations of landscape formation.

Conclusions

Given the vast range of subdisciplines and techniques encapsulated under the labels of environmental archaeology and geoarchaeology, this appraisal has been selective and designed to complement other contributions to this Handbook. I have drawn on conceptual and methodological issues arising during my research, but these are of general relevance and applicability. These issues are not restricted to one field or one technique, rather they cross-cut traditional spheres of concern and are relevant to the interpretation of past practices-in-the landscape.

Acknowledgment

Many thanks to Kara Valle, School of Geography and Environmental Science, Monash University, for drafting new illustrations.

References

Barham, A. J. 1995. Methodological approaches to archaeological context recording: X-radiography as an example of a supportive recording, assessment and interpretative technique, in A. J. Barham and R. I. MacPhail (eds.), *Archaeological Sediments and Soils: Analysis, Interpretation and Management*, pp. 145–82. London: Institute of Archaeology, UCL Press.

Barham, A. J., and MacPhail, R. I. (eds.). 1995. *Archaeological Sediments and Soils: Analysis, Interpretation and Management*. London: Institute of Archaeology, UCL Press.

Barrett, J. 1994. *Fragments from Antiquity: An Archaeology of Social Life 2900–1200* B.C. Oxford: Blackwell Publishing.

Bayliss-Smith, T. P., and Golson, J. 1999. The Meaning of ditches: Deconstructing the social landscapes of drainage in New Guinea, Kuk, Phase 4, in C. Gosden and J. Hather (eds.), *The Prehistory of Food: Appetites for Change*, pp. 199–231. London: Routledge.

Bennett, K. D., and Willis, K. J. 2001. Pollen, in J. P. Smol, H. J. B. Birks, and W. M. Last (eds.), *Tracking Environmental Change Using Lake Sediments, Vol. 3: Terrestrial, Algal, and Siliceous Indicators*, pp. 5–32. Dordrecht: Kluwer Academic Publishers.

Binford, L. R. 1965. Archaeological systematics and the study of cultural process. *American Antiquity* 31: 203–10.

Binford, L. R. 1967. Smudge pits and hide smoking: The use of analogy in archaeological reasoning. *American Antiquity* 32: 1–12.

———. 1981. *Bones: Ancient Men and Modern Myths*. Academic Press: New York.

Bohn, H. L., McNeal, B. L., and O'Connor, G. A. 1985. *Soil Chemistry*. New York: John Wiley.

Bourdieu, P. 1990. *The Logic of Practice*. Cambridge: Polity Press.

Brain, C. K. 1981. *The Hunters or the Hunted? An Introduction to African Cave Taphonomy*. Chicago: University of Chicago Press.

Brown, A. G. 1997. *Alluvial Geoarchaeology: Floodplain Archaeology and Environmental Change*. Cambridge: Cambridge University Press.

Brown, M. R., and Harris, E. C. 1993. Interfaces in archaeological stratigraphy, in E. C. Harris, M. R. Brown, and G. J. Brown (eds.), *Practices of Archaeological Stratigraphy*, pp. 7–20. London: Academic Press.

Burroughs, W. J. 2005. *Climate Change in Prehistory: The End of the Reign of Chaos*. Cambridge: Cambridge University Press.

Canti, M. 1995. A mixed-method approach to geo-archaeological analysis, in A. J. Barham and R. I. MacPhail (eds.), *Archaeological Sediments and Soils: Analysis, Interpretation and Management*, pp. 183–90. London: Institute of Archaeology, UCL Press.

Chadwick, O. A., and Chorover, J. 2001. The chemistry of pedogenic thresholds. *Geoderma* 100: 321–53.

Chartres, C. J., and Pain, C. F. 1984. A climosequence of soils in Late Quaternary volcanic ash in Highland New Guinea. *Geoderma* 32: 131–55.

Clarke, D. L. 1973. Archaeology: The loss of innocence. *Antiquity* 47: 6–18.

Cohen, M. N., and Armelagos, G. J. 1984. *Paleopathology at the Origins of Agriculture*. Orlando: Academic Press.

Courty, M.-A., Goldberg, P., and MacPhail, R. 1989. *Soils and Micromorphology in Archaeology*. Cambridge: Cambridge University Press.

Denham, T. P. 2005. Envisaging early agriculture in the Highlands of New Guinea: Landscapes, plants and practices. *World Archaeology* 37: 290–306.

Denham, T. P., Haberle, S. G., and Lentfer, C. 2004. New evidence and revised interpretations of early agriculture in Highland New Guinea. *Antiquity* 78: 839–57.

Denham, T. P., Haberle, S. G., Lentfer, C., Fullagar, R., Field, J., Therin, M., Porch, N., and Winsborough, B. 2003. Origins of agriculture at Kuk Swamp in the Highlands of New Guinea. *Science* 301: 189–93.

Denham, T. P., Golson, J., and Hughes, P. J. 2004. Reading early agriculture at Kuk (Phases 1–3), Wahgi Valley, Papua New Guinea: The wetland archaeological features. *Proceedings of the Prehistoric Society* 70: 259–98.

Dimbleby, G. W. 1985. *The Palynology of Archaeological Sites*. New York: Academic Press.

Dincauze, D. F. 2000. *Environmental Archaeology: Principles and Practice*. Cambridge: Cambridge University Press.

Evans, J. G. 2003. *Environmental Archaeology and the Social Order*. London: Routledge.

Evans, J. G., and O'Connor, T. 1999. *Environmental Archaeology: Principles and Methods*. Stroud: Sutton Publishing.

Fleming, A. 1999. Phenomenology and the megaliths of Wales: A dreaming too far? *Oxford Journal of Archaeology* 18: 119–25.

Fullagar, R., Field, J., Denham, T. P., and Lentfer, C. 2006. Early and mid Holocene tool-use and processing of taro (*Colocasia esculenta*), yam (*Dioscorea* sp.) and other plants at Kuk Swamp in the highlands of Papua New Guinea. *Journal of Archaeological Science* 33: 595–614.

Garrison, E. G. 2003. *Techniques in Archaeological Geology*. Berlin: Springer-Verlag.

Goldberg, P., and MacPhail, R. I. 2006. *Practical and Theoretical Geoarchaeology*. Oxford: Blackwell Publishing.

Gugerli, F., Parducci, L., and Petit, R. J. 2005. Ancient plant DNA: Review and prospects. *New Phytologist* 166: 409–18.

Haberle, S. G. 1994. Anthropogenic indicators in pollen diagrams: Problems and prospects for late Quaternary palynology in New Guinea, in J. G. Hather (ed.), *Tropical Archaeobotany: Applications and New Developments*, pp. 172–201. London: Routledge.

———. 2003. The emergence of an agricultural landscape in the Highlands of New Guinea. *Archaeology in Oceania* 38: 149–58.

Harris, E. C. 1989. *Principles of Archaeological Stratigraphy* (2nd ed.). London: Academic Press.

Haslam, M. 2006. An archaeology of the instant? Action and narrative in microscopic archaeological residue analysis. *Journal of Social Archaeology* 6: 402–24.

Hodder, I. 1999. *The Archaeological Process: An Introduction*. Oxford: Blackwell Publishing.

Hodgson, J. M. 1978. *Monographs on Soil Survey: Soil Sampling and Soil Description*. Oxford: Clarendon Press.

Jones, M. K. 2001. *The Molecule Hunt: Archaeology and the Search for Ancient DNA*. New York: Arcade Publishing.

Krinitzsky, E. L. 1970. *Radiography in the Earth Sciences and Soil Mechanics*. New York: Plenum Press.

Lamb, H. H. 1995. *Climate, History and the Modern World* (2nd ed.). London: Routledge.

Lewthwaite, G. R. 1966. Environmentalism and determinism: A search for clarification. *Annals of the Association of American Geographers* 56: 1–23.

Maltby, M. 2006. *Integrating Zooarchaeology*. Oxford: Oxbow Books.

Mountain, M.-J. 1991. Landscape use and environmental management of tropical rainforest by preagricultural hunter-gatherers in northern Sahulland. *Bulletin of the Indo-Pacific Prehistory Association* 11: 5–68.

Mountain, M.-J., and Bowdery, D. (eds.). 1999. *Taphonomy: The Analysis of Processes from Phytoliths to Megafauna*. Canberra: Archaeology and Natural History Publications, Australian National University.

Nemecz, E. 1981. *Clay Minerals*. Budapest: Alcadémiai Kiadó.

Pearsall, D. M. 2000. *Paleoethnobotany: A Handbook of Procedures* (2nd ed.). Orlando: Academic Press.

Piperno, D. R. 2006. *Phytoliths: A Comprehensive Guide for Archaeologists and Palaeoecologists*. Walnut Creek, CA: AltaMira.

Pollard, A. M., and Heron, C. 1996. *Archaeological Chemistry*. Cambridge: The Royal Society of Chemistry.

Rapp, G., and Hill, C. L. 1998. *Geoarchaeology*. New Haven, CT: Yale University Press.

Reitz, E. J., and Wing, E. S. 1996. *Zooarchaeology*. Cambridge: Cambridge University Press.

Roberts, C., and Manchester, K. 2005. *The Archaeology of Disease* (3rd ed.). Ithaca: Cornell University Press.

Schiffer, M. B. 1987. *Site Formation Processes of the Archaeological Record*. Albuquerque: University of New Mexico Press.

Sparks, D. L. 1995. *Environmental Soil Chemistry*. San Diego: Academic Press.

Thompson, R., and Oldfield, F. 1986. *Environmental Magnetism*. London: Allen and Unwin.

Tilley, C. 1994. *A Phenomenology of Landscape: Places, Paths and Monuments*. Oxford: Berg.

Torrence, R., and Barton, H. (eds.). 2005. *Ancient Starch Research*. Walnut Creek, CA: Left Coast Press.

Velde, B. 1985. *Clay Minerals: A Physico-Chemical Explanation of Their Occurrence*. Developments in Sedimentology 40. New York: Elsevier.

Walker, M. J. C. 2005. *Quaternary Dating Methods*. Chichester: John Wiley and Sons.

Waters, M. R. 1992. *Principles of Geoarchaeology: A North American Perspective*. Tuscon: The University of Arizona Press.

Willerslev, E., and Cooper, A. 2005. Ancient DNA. *Proceedings of the Royal Society B* 272: 3–16.

Williams, M., Dunkerley, D., De Deckker, P., Kershaw, P., and Chappell, J. 1998. *Quaternary Environments* (2nd ed.). Sydney: Arnold.

47

THE ARCHAEOLOGY OF WETLAND LANDSCAPES: METHOD AND THEORY AT THE BEGINNING OF THE 21ST CENTURY

Robert Van de Noort

The etymology of the word *wetland* shows that it is relatively modern, originating in the middle of the 20th century in the United States as a generic term signifying landscapes that support migratory bird populations, and it was internationalized at the UNESCO-sponsored International Convention on Wetlands in Ramsar, Iran, in 1970. The Ramsar Convention defines wetlands as "areas of marsh, fen, peatland or water, whether natural or artificial, permanent or temporary, with water that is static or flowing, fresh, brackish or salt, including areas of marine water the depth of which at low tide does not exceed 6 meters"; this definition is widely accepted by archaeologists around the world. However, some wetland archaeologists (e.g., Nicholas 2001) distinguish between "wet sites" (as locales where wet-preservation occurs) and "wetland sites" (as archaeological sites in ecological units that are wetlands), but the usefulness of this distinction has been questioned by others—including me—because it promotes the concept that past peoples' chosen interaction with wetlands was restricted to the economically functional character of the landscape, frequently leading to overtly environmental deterministic approaches. It is unquestionable that people in the past would not have recognized a broad definition of wetlands; rather, native ecologies would have defined specific topographical wetland features

or places (e.g., Bradley 2000; Harris 1998; Ingold 1995; Lopez 1986). This is shown, for example, in English place names with suffixes such as "-ings," "-hay," "-moor," "-dyke," "-fen," "-levels," "-fleet," "-pool," "-mere," "-beach," "-ford," "-bridge," or "-on-the-water," and "-on-the-Marsh," recognizing specific values and functions in the landscape including, but not exclusively, economic values and functions.

Nevertheless, there are good reasons for archaeologists to consider wetlands as an entity in their approach to landscapes. The effect of high groundwater tables on burial environments and archaeological preservation is common to all types of wetlands, albeit to varying degrees, and the study of this creates research opportunities that can be realized only through the application of particular techniques and methodologies. These, in turn, have led to the development of the field of wetland archaeology (e.g., Coles 1984, 2001). More recently, it has been argued that wetland archaeology has become too isolated and that future research must be more contextualized and theorized, addressing major questions in landscape research; this new approach includes the deconstruction of the concept of wetlands as a meta-narrative (Van de Noort and O'Sullivan 2006 and below). Past wetland research has also been accused of being too focused on the economic relationship of people with their environment, and a better understanding of the diversity of these relationships is now required.

This chapter considers the research potentials in wetland research, the wetland-specific methods and techniques employed in current research, and the changing theoretical agenda of wetland archaeology at the beginning of the 21st century.

Research Potentials

In landscapes where the water table is permanently high and soils are saturated with water, oxygen is largely absent, and reducing (as opposed to oxidizing) conditions prevail. Under such conditions, the microbial and chemical processes contributing to humification of organic matter is decelerated, and dead organic remains may survive for millennia. A true equilibrium is never achieved, but the pace of change can be so slow that 3,000-year-old timbers look as if these were cut down the previous day. Alongside the wooden structures and artifacts, woven baskets, and leather objects created by people in the (distant) past, environmental macro- and microfossils, including tree and plant remains, pollen, insect, diatoms, testate amoebae, and foraminifera, all have a greater change of long-term survival in wetlands. In exceptional circumstances, such as in an acidic raised bog, human skin, hair, and nails may be preserved as well. This better-than-normal preservation of organic remains presents three, wide-ranging, research opportunities.

The first lies in the prospect of studying *organic* material culture, which is on the whole impossible in archaeological studies of free-draining landscapes. It has been argued, on the basis of cross-cultural comparisons, that the invisibility of women and children in much archaeological research from drylands relates directly to the dearth of organic artifacts, which are used predominantly by them, especially in nomadic societies (e.g., Soffer, Adovasio, and Hyland 2001). Similarly, organic material culture, such as baskets, may tell us more about the everyday life of "ordinary" past people than the study of stone and metal artifacts, or earthen monuments, will ever be able to achieve. The research of the Bronze Age timber circle at Holme in Norfolk, England, popularly known as "Seahenge," illustrates this potential and tells us something about the carpenters' skill and agency in the construction of the site, with 50 different axes used, and split timbers placed on opposite sides of the circle, perhaps to balance the personalities of the builders (Brennand and Taylor 2003).

The second research opportunity lies in the potential to integrate the study of people within their (natural) landscape to a degree that is unimaginable elsewhere. This integration is realized, especially, where a range of environmental proxies is used, which provide perspectives on the human-environment interaction on different geographical scales. Thus, the regional, off-site, environmental reconstruction based on pollen evidence can be enhanced by local, on-site, information from plant and insect macrofossils (Figure 47.1). An example of such research is the study of the peatlands in the Dümmer region of Lower Saxony, Germany (Bauerochse 2003). Here, the integrated analysis of pollen and plant macrofossils, linked to excavations of prehistoric trackways and settlements, and a radiocarbon program, showed that climate changes did not affect the settlement pattern, but did influence the construction and lifespan of trackways.

The third research opportunity, and possibly the most important, lies in the potential to date artifacts, deposits, and archaeological phases through radiocarbon and dendrochronological assay with much higher resolution and greater precision than is possible on non-wetland sites and, thus, can reveal something of the real-time dynamic of prehistoric life. For example, dendrochronology revealed the dynamic sequence of site construction, occupation, and contraction of a Neolithic lake village at Charvines, Lac de Paladru, France (Bacquet and Houot 1994: 21–24; Figure 47.2), and many other wetland sites, such as the crannogs in Scotland (Crone 2000), show patterns of occupation and abandonment each lasting several years or a few decades. We can assume that many non-wetland sites experienced similar dynamic patterns but, without high-resolution dating, this cannot be shown and phases tend to extend over centuries.

Wetland Archaeological Methods and Techniques

The high water table, and the concealment of archaeology beneath waterlain sediments of younger date, prevent identification of most wetland sites through aerial photography, geophysical survey or surface investigations, such as field walking (see Van de Noort 2002). The principal archaeological tool for investigating wetland landscapes is through the multiproxy palaeo-environmental analysis from samples taken from boreholes, which can be used to reconstruct past environments and some aspects of the impact of humans on that landscape, such as woodland clearance as evidenced by pollen (e.g., French 2002). It is, therefore, unsurprising that the overwhelming majority of wetland sites known to us have been

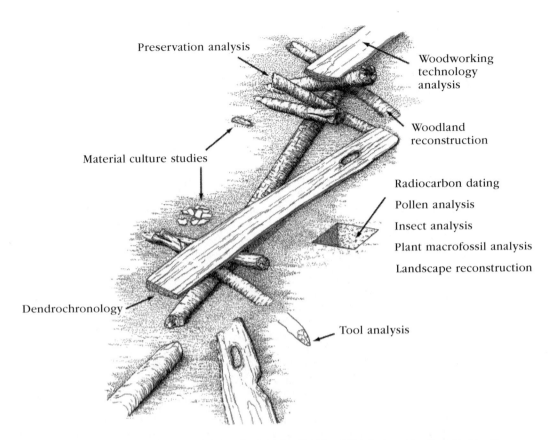

Preservation analysis

Woodworking
technology
analysis

Woodland
reconstruction

Material culture studies

Radiocarbon dating

Pollen analysis

Insect analysis

Plant macrofossil analysis

Landscape reconstruction

Dendrochronology

Tool analysis

Figure 47.1 Wetland archaeology's richness (after B. Coles 1995: 2; drawn by Sean Goddard).

discovered through activities unrelated to archaeology. Prominent examples include peat extraction (for example, the Sweet Track in the United
Kingdom; the many bog bodies from Denmark,
Ireland, Germany, the Netherlands, and the United
Kingdom), drainage works (for instance, Star Carr
and Flag Fen in the United Kingdom; Freisack in
Germany), ploughing or construction work in former wetland landscapes, (for example, Windover
in the United States; many urban wetland sites
across the world), or through natural erosion of
coasts, lakes, and riverbanks, and in estuaries
(for instance, "Seahenge" and the Ferriby Bronze
Age boats in the United Kingdom; the Alpine lake
settlements; the crannogs in Ireland and Scotland;
Ozette in the United States).

Nevertheless, the majority of wetlands across
the world, still widely perceived as unusable wil-
dernesses despite the recent work by nature conservation organizations, are currently threatened
by drainage and are converted into farmland or
urban and industrial landscapes. These developments are usually accompanied by the digging of
drainage ditches, ploughing, and the rapid erosion

of the desiccated organic sediments, which create
opportunities for systematic survey and prospection, through surface collection of material, dike/
drain survey, aerial photography of soil marks of
former river beds, and resistivity and magnetometer
survey of the drying landscape. There are also a
number of promising, but not yet widely used,
survey techniques that may be able to identify
archaeological wetland sites before the remains are
at or near the surface, such as multispectral aerial
reconnaissance (e.g., Donoghue 2001), the use of
ground-penetrating radar (e.g., Clarke et al. 2001),
and micro-topography survey using differential
GPS (e.g., Chapman and Van de Noort 2001). We
should also recognize the early drainage of wetlands as historic relics, and the medieval and early
modern archaeology of wetland exploitation and
inhabitation can be investigated through analysis
of maps and the current landscapes (e.g., Rippon
2005).

There are many good examples of landscape-
scale archaeological studies of wetlands, which
have frequently taken a compare-and-contrast
wetlands with "drylands" approach in order to

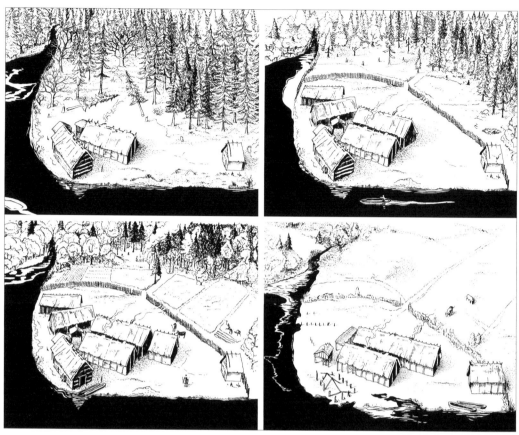

Figure 47.2   Charavines, Lac de Paladru, France. Reconstruction sequence of this Neolithic lake-settlement as revealed by dendrochronology. *Top left*: years 1 and 2; *top right*: years 3 to 8; *bottom left*: years 9 to 17; *bottom right*: years 18 to 22 (after Bocquet and Houet 1994: 21–24; drawn by Sean Goddard).

understand the greater diversity of past peoples lives—for example, the prolonged existence of hunter-gatherer types of subsistence in resource-rich wetlands such as the Ertebølle in Denmark and the Swifterband complexes in the western Netherlands. Within Europe, the most notable projects include those of the Dutch Delta (Louwe Kooijmans 1993), the Assendelver Polder in the Netherlands (Brandt, Groenman-van Waateringe, and Van der Leeuws 1987), the Somerset Levels (Coles and Coles 1986), The Fenlands (Hall and Coles 1994), and the landscape-scale excavations in Fengate and at Flag Fen (Pryor 2001), the North West Wetlands and the Humber Wetlands (Van de Noort 2004) in England, the Severn Estuary (published in the Severn Estuary Levels Research Committee papers), the extended surveys of the Irish Midlands (e.g., Gowen, Ó Néill, and Phillips 2005; Raftery 1990) and the Shannon Estuary (O'Sullivan 2001a), and a range of projects in southern Scandinavia (Larsson 2001). Elsewhere,

notable wetland landscape projects include studies of the Great Basin wetland in North America (Kelly 2001) and in northern Australia (Haynes, Ridpath, and Williams 1991; Williams 1988).

Excavations of wetland sites tend to be small-scale operations. The size of the trenches is limited by several factors, including the friable nature of the organic materials, which must be kept wet throughout the period of exposure, the need to recover all organic remains once exposed to oxygen, and the habitually unstable nature of the surrounding soils (typically peat or alluvial sediments). Digging with metal tools should be avoided when dealing with organic remains, and wooden spatulas or water jets or sprays should be used instead. On-site or nearby wet-tanks should be available for the temporary storage of organic remains once taken out of context. The use of motorized pumps on wetland excavations should be avoided, as this will draw-down the water from the area outside the excavation trench, and extensive damage through desiccation of un-excavated remains will be the consequence. All wetland excavations need to include integrated palaeoenvironmental research and, as a minimum, relevant specialists in geoarchaeology, palynology, plant and insect macrofossils, dendrology, and dendrochronology should be included, ideally from the onset of the design of the project and throughout the excavation and post-excavation stages. It is often argued that wetland excavations are expensive because of high conservation costs and involvement of many specialists. The director of excavation must therefore ensure that the different aspects of the work are closely integrated in order to achieve an overall result that is greater than the individual parts of the research.

Challenges for the 21st Century

Despite the significant results to date and future research potentials, wetland archaeology has essentially failed to influence mainstream archaeology, and thus the great promise of wetland archaeology remains to a large extent unfulfilled (Coles 2001). This failure to influence and inform the broader archaeological debates can be attributed to three aspects of current research in the landscape archaeology of wetlands (see Evans 1990; Gearey 2002; Haselgrove et al. 2001; Scarre 1989; Tilley 1991). First, many research projects remain decontextualized geographically, as if wetlands were islands rather than surrounded by non-wetland landscapes. Second, wetland archaeology frequently appears as being decontextualized in time, as if wetlands were timeless landscapes, disconnected from the changes surrounding them. Third, most wetland landscape projects are disconnected from current theoretical debates in archaeology. The aim of this section is to consider how such a (re-)engagement with mainstream landscape archaeology could be achieved, and it is based on the recent publication *Rethinking Wetland Archaeology* (Van de Noort and O'Sullivan 2006). It suggests seven changes to the present way of doing wetland landscape archaeology.

First, the landscape archaeology of wetlands has to be contextualized. This includes geographical contextualization, since interactions between wetland and non-wetland landscapes are omnipresent, both in the physical (for example, the runoff of nutrients-rich water from hills into a bog) and cultural (for instance, the use of stone axes to build trackways) spheres. Contextualization should extend to include the passing of time and the cultural changes surrounding them, and it should also include the sociopolitical context of

the researchers, who should make their theoretical stance explicit, since we always interpret our data through a "cloud of theory" (Johnson 1999). It should also be recognized that landscapes cannot be simply "read"—landscape archaeology, and thus wetland archaeology, is hermeneutic. Recent studies that present genuinely contextualized wetland studies include Francis Pryor's (2001) work on Flag Fen in the Fenlands of England and Helmut Schlichtherle's (1997) work on the Alpine lake settlements. Martin Bell's research in the Severn Estuary merits also citing (Bell, Caseldine, and Neumann 2000).

Second, the meta-narrative of wetlands must be deconstructed, accepting that this term had no significance for people in the past. The term is useful as a shorthand for the mosaic of ecosystems of wet and damp places, or for defining the area where wet-preserved archaeological and palaeoenvironmental remains may survive, but this should not become the basis for cultural analysis. Instead, the type of wetland should be defined, such as raised mire, valley floodplain, or alder carrland. Studies from continental Europe, where the generic concept of "wetlands" has not been established as well as in other parts of the world, provide some good examples of the focus on specific types of wetland landscapes (e.g., Besteman 1990; Brinkkemper 1991; Koch 1999; Larsson 2001).

Third, the significance of approaching landscapes from the perspective of the people we study should be recognized (e.g., Lopez 1986). Developing a cultural understanding of the significance and meaning of trackways, bog bodies, lake settlements, and so forth cannot be based on a modern Western, functionalist, perspective. Furthermore, it must also be recognized that the perception of wetlands (and other type of landscapes) differs between insiders and outsiders. Two recent studies, both on Ireland, that have put the perception of past people central are Christine Fredengren's (2002) work on crannogs and Aidan O'Sullivan's (2001a) study of fishing in the Shannon estuary.

Fourth, nature-culture interactions should be recognized more clearly. For example, enculturing nature (and the spirits within them) forms a key theme of human behavior, which can be favorably studied in wetland landscapes with its high-resolution dating and close association with palaeoenvironmental source material (e.g., Ingold 1995; Nelson 1983; Tilley 1994). The best example of a study in wetlands that develops this theme is the most recent work on Star Carr in the Vale of Pickering in northeast England (Conneller and Schadla-Hall 2003).

Fifth, special attention should be given to the boundaries and edges of the landscapes. From our observations of the perceived dynamic nature of the natural environment, it follows that the boundaries of edges of these landscapes are often given particular significance, for example, as "natural places" in the sense used by Richard Bradley (2000). A wonderful example of the way in which such boundaries retained their significance for millennia is presented by David Stocker and Paul Everson (2003) for the Witham valley in Lincolnshire, England. Similarly, Naomi Field and Mike Parker Pearson's (2003) work on the trackways in the same valley shows remarkable insights into the events that reinforced the ritual meanings of this wetland as a natural place. Wijnand van der Sanden's (1996) study of bog bodies must also be mentioned in this context.

Sixth, marginality and liminality should not be coalesced. The concept of liminality is frequently invoked where wetlands are traversed. Liminality is a notoriously fluid concept. Originally proposed by Van Gennep (1908), the concept is linked to "rites of passage" to describe the formalized rituals and practices that accompany one's transition from one particular state into another, especially the rites associated with birth, reaching adulthood, marriage, and death. As part of these rituals, symbolic or real "thresholds" needed to be crossed, with the thresholds constituting liminal zones. As economic and ritual activities are not, on a landscape level, mutually exclusive, the recurrent equation of liminality with marginality is often mistaken. Although some liminal zones were to be found in what were considered marginal landscapes, others are located within settlements or within areas in economic use. To date, few wetland studies have made the explicit distinction between the two concepts, but I have sought to develop this issue explicitly in the study of the Humber Wetlands (Van de Noort 2004).

Seventh, the importance of many wetland landscapes as taskscapes, areas where the rhythm of daily life determines the significance of how these wetland landscapes are perceived, should not be underplayed. The phrase "taskscape" was coined by Tim Ingold (1993) to focus on the concept that the manner in which landscapes are experienced and perceived is closely related to the activities or tasks that are undertaken in particular landscapes at particular times. The insiders' view of wetlands is one that offers myriad resources, ranging from eels, fish, and shellfish, to peat for fuel, reed for roofing, to summer pastures and haylands. Raised bogs can also be used intermittently for short-term seasonal grazing by burning the top layer of the bog, for the preservation of butter, the seasoning

of wood, and the curing of leather. It should be recognized that these activities, though seemingly economic practices, are things that people do every day, albeit in specific cultural and social conditions. Few wetland studies have used the concept of taskscape explicitly, but North American studies, especially those looking at the role of fishing for local communities, have addressed this aspect of research well (e.g., Croes 2001), and this theme has also been brought out in some recent studied and reappraisals of lake-settlements in the Alpine region (e.g., Arnold 1986; Bocquet and Huot 1994; Magny 1993; Menotti 2003), Scotland (e.g., Crone 2000), and Ireland (e.g., O'Sullivan 2000, 2001b).

Conclusions

The archaeological study of wetland landscapes offers three major advantages over non-wetland landscapes: the survival of organic material culture, which has the potential to address existing (gender and status) biases in archaeological research; the coexistence of archaeological and palaeoenvironmental data, which provides opportunities for the closely integrated study of nature-culture interrelationships; and the high-resolution and precision dating of organic remains, especially through dendrochronology, which offer opportunities for the study of real-time dynamics as experienced by people in the past. However, these research potentials can be realized only if and when wetland archaeology becomes fully geographically and theoretically contextualized.

References

Arnold, B. 1986. *Cortaillod-Est, un village du Bronze final.* Saint Blaise: Archeologie Neuchateloise 1.
Bauerochse, A. 2003. Environmental change and its influence on trackway construction and settlement in the south-west Dümmer area, in A. Bauerochse and H. Hassmann (eds.), *Peatlands—Archaeological Sites—Archives of Nature—Nature Conservation—Wise Use*, pp. 68–78. Rahden: Verlag Marie Leidorf.
Bell, M., Caseldine, A., and Neumann, H. 2000. *Prehistoric Intertidal Archaeology in the Welsh Severn Estuary.* York: CBA Research Report 120.
Besteman, J. C. 1990. North Holland A.D. 400–1200: Turning tide or tide turned? in J. C. Besteman, J. M. Bos, and H. A. Heidinga (eds.), *Medieval Archaeology in the Netherlands. Studies Presented to H.H. van Regteren Altena*, pp. 91–120. Assen/Maastricht: Van Gorcum.
Bradley, R. 2000. *An Archaeology of Natural Places.* London: Routledge.

Brandt, R. W., Groenman-van Waateringe, W., and Van der Leeuws, S. (eds.). 1987. *Assendelver Polder Papers 1*. Amsterdam: Cingula 10.

Brennand, M., and Taylor, M. 2003. The survey and excavation of a Bronze Age timber circle at Holme-next-the-sea, Norfolk, 1998–1999. *Proceedings of the Prehistoric Society* 69: 1–84.

Brinkkemper, O. 1991. *Wetland Farming in the Area to the South of the Meuse Estuary during the Iron Age and Roman Period: An Environmental and Palaeo-Economic Reconstruction*. Leiden: Analecta 24.

Bocquet, A., and Huot, A. 1994. *Charavines il y a 5000 ans*. Dijon: Editions Faton.

Chapman, H. P., and Van de Noort, R. 2001. Wetland prospection using GPS and GIS: Recent work at Meare (Somerset) and Sutton Common (South Yorkshire). *Journal of Archaeological Science* 28: 365–75.

Clarke, C. M., Utsi, E., and Utsi, V. 2001. Ground penetrating radar investigations at North Ballachulish Moss, Highland, Scotland Archaeological Prospection 6: 107–21.

Coles, B. J., and Coles, J. M. 1986. *Sweet Track to Glastonbury*. London: Thames and Hudson.

Coles, J. M. 1984. *The Archaeology of Wetlands*. Edinburgh: UP.

———. 2001. Energetic activities of commoners. *Proceedings of the Prehistoric Society* 67: 19–48.

Conneller, C., and Schadla-Hall, T. 2003. Beyond Star Carr: The Vale of Pickering in the 10th millennium B.C. *Proceedings of the Prehistoric Society* 69: 85–105.

Crone, A. 2000. *The History of a Scottish Lowland Crannog: Excavations at Buiston, Ayrshire 1989–1990*. Edinburgh: Scottish Trust for Archaeological Research.

Donoghue, D. 2001. Multispectral remote sensing for archaeology. *Consiglia nazionale Delle Ricerche Universita Degli Studi Di Siena* 181–92.

Evans, C. 1990. Review of Purdy, B. A. (ed.). 1988. *Wet Site Archaeology. Proceedings of the Prehistoric Society* 56: 339–40.

Field, N., and Parker Pearson, M. 2003. *Fiskerton: An Iron Age timber Causeway with Iron Age and Roman Votive Offerings: The 1981 Excavations*. Oxford: Oxbow Books.

Fredengren, C. 2002. *Crannogs: A Study of People's Interaction with Lakes, with Particular Reference to Lough Gara in the North-West of Ireland*. Bray: Wordwell.

French, C. 2002. *Geoarchaeology in Action: Studies in Soil Micromorphology and Landscape Evolution*. London: Routledge.

Gearey, B. R. 2002. "Foule and flabby quagmires": The archaeology of wetlands. *Antiquity* 76: 896–900.

Gowen, M., Ó Néill, J., and Phillips, M. 2005. *The Lisheen Mine Archaeological Project 1996–1998*. Dublin: Wordwell.

Hall, D., and Coles, J. 1994. *Fenland Survey: An Essay in Landscape and Persistence*, London: English Heritage.

Harris, M. 1998. The rhythm of life: Seasonality and sociality in a riverine village. *Journal of the Royal Anthropological Institute* 4(1): 65–82.

Haselgrove, C., Armitt, I., Champion, C., Creighton, J., Gwilt, A., Hill, J. D., Hunter, F., and Woodward, A. 2001. *Understanding the British Iron Age: An Agenda for Action*. Salisbury: Wessex Archaeology.

Haynes, C. D., Ridpath, M. G., and Williams, M. A. J. 1991. *Monsoonal Australia: Landscape, Ecology and Man*. London: Taylor and Francis.

Ingold, T. 1993. The temporality of the landscape. *World Archaeology* 25: 152–74.

———. 1995. Building, dwelling, living: How animals and people make themselves at home in the world, in M. Strathern (ed.), *Shifting Contexts: Transformations in Anthropological Knowledge*. London: Routledge.

Johnson, M. 1999. *Archaeological Theory: An Introduction*. London: Blackwell.

Kelly, R. 2001. Prehistory of the Carson Desert and Stillwater Mountains: Environment, mobility, and subsistence in a Great Basin wetland. University of Utah Anthropological Paper 123, Salt Lake City, Utah.

Koch, E. 1999. Neolithic offerings from the wetlands of eastern Denmark, in B. Coles, J. Coles, and M. Shou Jørgenson (eds.), *Bog Bodies, Sacred Sites and Wetland Archaeology*, pp. 125–32. Exeter: WARP.

Larsson, L. 2001. South Scandinavian wetland sites and finds from the Mesolithic and the Neolithic, in B. Purdy (ed.), *Enduring Records: The Environmental and Cultural Heritage of Wetlands*, pp. 158–71. Oxford: Oxbow Books.

Lopez, B. 1986. *Arctic Dreams: Imagination and Desire in a Northern Landscape*. New York: Harvill Press.

Louwe Kooijmans, L. P. 1993. Wetland exploitation and upland relations of prehistoric communities in the Netherlands, in J. Gardiner (ed.), *Flatlands and Wetlands: Current Themes in East Anglian Archaeology*, pp. 71–116. East Anglian Archaeology 50.

Magny, M. 1993. Une nouvelle mise en perspective des sites archéologiques lacustres: Les fluctuations holocènes des lacs jurassiens et subalpins. *Gallia-Préhistoires* 35: 253–82.

Menotti, F. 2003. Cultural response to environmental change in the Alpine lacustrine regions: The displacement model. *Oxford Journal of Archaeology* 22: 375–96.

Nelson, R. K. 1983. *Make Prayers to the Raven: A Koyukon view of the northern forest*. Chicago: University of Chicago Press.

Nicholas, G. P. 2001. Wet sites, wetland sites, and cultural resource management strategies, in B. Purdy (ed.), *Enduring Records: The Environmental and Cultural Heritage of Wetlands*, pp. 262–70. Oxford: Oxbow Books.

O'Sullivan, A. 2000. *Crannogs: Lake-Dwellings of Early Ireland*. Dublin: Town House.

———. 2001a. *Foragers, Farmers and Fishers in a Coastal Landscape: An Intertidal Archaeological Survey of the Shannon Estuary*. Dublin: Royal Irish Academy.

———. 2001b. Crannogs: Places of resistance in the contested landscapes of early modern Ireland, in B. Bender and M. Winer (eds.), *Contested Landscapes: Landscapes of Movement and Exile*, pp. 87–101. Oxford: Berg.

Pryor, F. 2001. *The Flag Fen Basin: Archaeology and Environment of a Fenland Landscape*. Swindon: English Heritage.

Raftery, B. 1990. *Trackways through Time: Archaeological Investigations on Irish Bog Roads, 1985–1989*. Rush: Headline.

Rippon, S. 2005. *Historic Landscape Analysis: Deciphering the Countryside*. York: Council of British Archaeology.

Scarre, C. 1989. Review of J. M. Coles and A. J. Lawson (eds.). 1987. *European Wetlands in Prehistory*. *Proceedings of the Prehistoric Society* 55: 274–75.

Schlichtherle, H. 1997. *Pfahlbauten rund die Alpen*. Stuttgart: Theiss.

Soffer, O., Adovasio, J. M., and Hyland, D. C. 2001. Perishable technologies and invisible people: Nets, baskets and "Venus" wear ca. 26,000 B.P., in B. Purdy (ed.), *Enduring Records: The Environmental and Cultural Heritage of Wetlands*, pp. 233–45. Oxford: Oxbow Books.

Stocker, D., and Everson, P. 2003. The straight and narrow way: Fenland causeways and the conversion of the landscape in the Witham valley, Lincolnshire, in M. Carver (ed.), *The Cross Goes North: Processes of Conversion in Northern Europe* A.D. *300–1300*, pp. 271–88. York: Woodbridge Medieval Press.

Tilley, C. 1991. Review of B. and J. Coles (1989). *People of the Wetlands: Proceedings of the Prehistoric Society* 57: 214–15.

———. 1994. *A Phenomenology of Landscape: Places, Paths and Monuments*. London: Berg.

Van de Noort, R. 2002. Flat, flatter, flattest: The English Heritage wetland surveys in retrospect, in T. Lane and J. Coles (eds.), *Through Wet and Dry: Essays in Honour of David Hall*. Sleaford and Exeter: Lincolnshire Archaeology and Heritage Report Series No. 5 and WARP Occasional Paper 17: 87–95.

———. 2004. *The Humber Wetlands: The Archaeology of a Dynamic Landscape*. Bollington: Windgather Press.

Van de Noort, R., and O'Sullivan, A. 2006. *Rethinking Wetland Archaeology*. London: Duckworth.

Van der Sanden, W. 1996. *Through Nature to Eternity: The Bog Bodies of Northwest Europe*. Amsterdam: Batavia Lion International.

Van Gennep, A. 1908. *Les Rites de Passage*. Paris: Emile Nourry.

Williams, E. 1988. Complex Hunter-Gatherers: A late Holocene example from temperate Australia. BAR International Series 423, Oxford.

48

LITHICS AND LANDSCAPE ARCHAEOLOGY

Chris Clarkson

Stone is a reductive medium. When broken by heat, percussion, or pressure, the parent rock is inexorably altered, and visible and durable reminders of that activity are typically left behind in the form of flakes, cores, discarded implements, and other fragments. By this means, lithic residues come to mark places with histories of human visitation, association, activity, and memory that implore us to read the landscape as humanly inhabited and experienced in the past. Because the stone suited to this intentional shaping of tools and objects of trade or display derives from sources that are sometimes distinctive, unevenly distributed, and of varying quality, it provides us with a record of material transport and selection that is vital to connecting places, choices, and material residues to past human movements and people's technological and social concerns. We also cannot downplay the important functional role that stone artifacts played in enabling people to make a living from the landscape—and hence the constraints on design, efficiency, and organization of production, use, and maintenance that lend lithic assemblages much of their character. These same artifacts also likely held meanings to past artisans beyond their mundane history or potentiality for use as tools, and the ethnography of Aboriginal Australia, for instance, provides many illustrations of the social, mythical, and genealogical meanings

that inhere in particular stone outcrops, distinctive colors, and various artifact forms (Cane 1992; Gould 1968; Gould and Saggers 1985; Harrison 2002; Jones and White 1988; Paton 1994; Taçon 1991; Thomson 1949). Patterns of association between distinctive and nondistinctive lithic objects with significant or unique features of the natural or built environment can also point to the existence of symbolic or ritual connections to places in past societies.

Although stone artifacts theoretically stand to inform us about many dimensions of human experience, purpose, and creativity, it is a truism that extracting such information from lithic assemblages can be difficult. Yet this task is not impossible, and given the vast quantity of lithics in the archaeological record, their survival where other cultural traces have long since vanished, and their potential contribution in answering many questions about past societies' engagement with landscapes, we must seek and assimilate such information into our narratives about the past. However, as with any kind of archaeological evidence, it is easier to glean information and build stronger narratives of past lifeways when multiple lines of evidence are available. Nevertheless, the purpose of this chapter is to identify and illustrate ways that stone artifacts can be used (on their own) to investigate the past cultural geographies of landscapes.

Four kinds of lithic evidence can be considered particularly useful in reconstructing past human engagement with the landscape: transportation, accumulation, association, and alteration. These evidential strands provide a basis on which to connect people to places and people to people at regional scales, and thereby reconstruct networks of social, technological, and economic relations across space and time. First, I review these concepts and particular methods that elucidate them as facets of human engagement with landscapes. These four lines of evidence are then illustrated in a case study from northern Australia in which multiple strands of lithic analyses are used to build an interpretation of past human engagement with the landscape that integrates ecological, economic, and social dimensions of human existence.

Using Lithics to Explore Past Cultural Landscapes

Transportation

Humans, and especially hunter-gatherers and early farmers, are rarely completely sedentary and must move and interact socially and economically over areas of varying size to obtain the materials needed to survive and to attain various goals. For most of human evolution, flakeable stone was one such resource that was often critical to survival, and people often journeyed far, or participated in exchange networks, to obtain it. Lithic assemblages can therefore be highly informative about the range of places visited in the landscape—particularly when lithic sources are distinctive—or the nature and direction of the social relations that enabled transfer of raw materials over distances.

Since one of the primary modes of human engagement with landscape is through individual or group mobility, stone artifact transport is a key line of enquiry in reconstructing the range, frequency, and predictability of residential moves in past societies (Blades 1999; Byrne 1980; Kelly 1992; Parry and Kelly 1987; Shott 1986; but see Brantingham 2003). Determining the exact pathways humans traveled, or even whether artifacts were transported by individuals or traded and exchanged between groups, is, however, a difficult task (Close 2000; Kelly 1988, 1992).

A number of methods are commonly used to build broad understandings of raw material transportation between stone sources and places of discard. These include a battery of chemical (XRF, NAA, INAA, ICPMS, PIXE/PIGME) and visual (color and texture, thin sectioning, microfossil identification) identification techniques for sourcing stone to its original outcrop (Clarkson and Lamb 2005; Shackley 1998; see Summerhayes, this volume, for descriptions of these chemical techniques; see also Weisler, this volume), refitting artifacts at sites across the landscape (Close 2000), and studying transformations in artifact form that result from curation of artifacts and rationing of raw materials as they are moved away from stone sources (see transformation below).

The existence of a limited range of known and spatially separate and discrete raw material sources can greatly aid the identification of patterns of movement and even social differentiation. McNiven (1999), for instance, interpreted the gradual drop out of raw materials from known sources resulting in use of more and more local outcrops to support a model of group fissioning among coastal Aborigines under increasing population pressure in eastern Australia. Group fissioning gave rise over a period of several thousand years to small, well-defined group territories, linguistic differences, and formalized exchange networks that crossed rigid social boundaries. Another example of the value of tracing the transport of distinctive raw materials includes the reconstruction of enormous group territories among early north American societies and the contrast this enables us to draw between the way they used and experienced landscapes compared to later native American societies (Jones et al. 2003; Kelly and Todd 1988). The gradual accumulation of extensive knowledge about the sources of raw materials used to make flaked and polished Neolithic axes in Britain has also been used to reconstruct patterns of trade and exchange over vast areas, including potential centers of redistribution that point to the emergence of new social arrangements (Edmonds 1990).

Higher-level arguments are often formed about the nature of mobility and anticipatory use of landscapes from spatial patterning in lithic assemblages. For instance, many researchers have explored the various ways that patterns of raw material procurement, transportation, stockpiling, and husbanding can provide information about the frequency and nature of group mobility (Bleed 1986; Kuhn 1995; Nelson 1991; Shott 1986). Kuhn (1995) applies the term *provisioning* to the system of ensuring that cores or finished tools are available where and when they are needed.

Synthesizing many ethnographic observations of hunter-gatherer technology, researchers have found that people moving frequently through landscapes with little certainty about the places to be visited or opportunities for foraging and raw material replacement typically provision themselves with portable,

multifunctional, long use-life tools that can be easily maintained using the transported toolkits to
hand. Kuhn (1995) calls this strategy *provisioning of individuals*. Such toolkits often tend to be
made from high-quality stone in order to increase
the performance and use-life of tools that may have
to remain functional for long periods (Goodyear
1989). In contrast, people moving periodically
and predictably between locations tend to provision those places with raw materials rather than
transport small toolkits. This enables the stockpiling of raw materials in anticipation of future needs
as well as greater flexibility in tool production that
allows tools to be specifically designed to suit current tasks and discarded in favor of new tools when
they cease to be as efficient. Kuhn (1995) calls this
strategy *provisioning of places*. Admixtures of both
systems can coexist in the landscape, allowing different place usehistories to be determined. Hence,
predictably occupied places that are used to stage
long-distance logistical foraging trips might see the
anticipatory production of mobile toolkits amid
more expedient technological activities, and places
that see infrequent or unpredictable use might at
times be stockpiled with raw materials and spare
tools cached to guard against future shortfall
(Binford 1979). Although the principles on which
these interpretations of the strategic organization
of technology are derived primarily from hunter-
gatherer societies, they have proved equally relevant in some cases to the way stone was used
by early farmers (Bamforth and Woodman 2004;
Edmonds 1990).

Identifying the archaeological correlates of
these various strategies of artifact transport and
supply can therefore help detect the different
kinds of place use histories and the ways people
likely regarded locations as intimate/domestic or
less familiar/specialized use places. Thus, mobility
might also play a key role in the way landscapes
were experienced and visualized (Edmonds 1990,
1999; Thomas 2001; Whittle 1997).

Accumulation

The frequency with which people fracture stone at
the same place in the landscape will result in varying sized accumulations of lithic debris remaining
at those locations. The more stone is knapped, the
more the archaeological record will be enlarged,
and the more those events are spatially constrained,
the denser those accumulations will be. The record
of past visitations and the kinds of activities represented are routinely used by archaeologists to infer
the intensity with which locations were occupied
and reoccupied, although this can be misleading

in some cases. Typically, this measure of occupation intensity is expressed as numbers, weight,
or volume of stone artifacts per square or cubic
meter. Of course, accurate interpretation of what
caused such accumulations can be complicated,
and the nature of flaking, the duration of occupations, and the number of people involved can all
affect the record in similar ways (Hiscock 1981).
Discriminating between these alternatives is difficult without detailed analyses and multiple lines
of evidence. Nevertheless, we can say that varying
accumulations of lithic debris point to the degree
to which locations served as a locus for human
activity, whether that equates to intensity of occupation or activity or both.

When varying accumulations of artifacts are
viewed at landscape scales, they can inform us
about the choices of individuals and groups about
how much time to spend at a place, what and how
much material to bring, and how much of it to
discard there. In quantifying the amount of lithic
debris found at sites, we must also be mindful that
stone shatters to varying degrees according to raw
material type and different manufacturing (for
instance, bipolar vs. pressure) and taphonomic
processes (for example, heating, freezing, trampling, snapping, plough damage, erosion, and
slopewash), and we must therefore factor in as
much as possible the effects of these processes in
our calculations of occupation intensity. Hiscock
(2002) and Shott (2000) have both devised means
of calculating how much actual flaking was performed to arrive at more accurate descriptions of
the intensity of occupation or of activities conducted at sites.

Lithic accumulations probably also played an
important symbolic role in marking to passing
or returning people that a place has a history. In
some cases, lithic scatters may evoke mythological
or creation stories about a place that might relate
to the tool types or raw materials present, conjure
the memory of close kin who stayed at the site,
or, in cases where access to places is not tightly
controlled, even prompt enquiry about who else
has visited such places. Alternatively, lithic scatters may simply serve as reminders of past events
or activities conducted at sites that could emphasize continuities in site use over time (Edmonds
1999). The scale and content of lithic accumulations might also signify the desirability of a location for habitation or serve a mnemonic purpose in
reminding people of the raw materials that can be
obtained locally or the social relationships that led
to the accumulation of exotic or rare raw mater-
ials/objects visible on the surface. When episodes of
stone artifact manufacture, use, and discard result

in ever-increasing accumulations of lithic material, we might expect these locations to become "increasingly symbolically charged, patterned, and contextualized," to use Paul Taçon's (1999: 41) words. Visible concentrations of lithic material can also serve as source zones for raw material acquisition and often show signs of scavenging and reuse that might suggest continuous revisitation, long periods of residence, or even dedicated visits to procure stone (Camilli and Ebert 1992).

In some cases, it is the subtraction of lithic material rather than accumulation that can lead to significant changes in the appearance and the experience of landscapes (Edmonds 1999; Field 1997). The extensive flint mines of Neolithic Britain are a classic example of the large-scale transformation of the landscape which in some cases appears to have resulted in the acquisition of symbolic significance, leading to frequent construction of high status barrows at mine sites (Field 1997).

Approaches that concentrate on the interpretation of artifact distributions, accumulations, and assemblage composition are ideal for the study of cultural landscapes; as a result, open stone artifact scatters commonly contribute the bulk of the data to landscape scale analyses (Torrence 2002). This can, however, give rise to significant interpretative and contextual problems, the most common of these being varying degrees of surface visibility leading to sampling bias and poor chronological resolution, particularly where temporal markers and datable materials are not present in exposed assemblages and where sites represent palimpsests and time-averaging of assemblages over long periods. These can prove problematic if the palaeoenvironmental and cultural contexts of different occupation episodes are significantly different, leading to quite different assemblage characteristics (Dooley 2004; Zvelebil, Green, and Macklin 1992).

Association

Archaeologists routinely look to patterns of association among lithic assemblages, distinctive artifacts, and physical and cultural features as a means of illuminating past human economic, social, and symbolic engagement with landscapes. Humans often center their activities on visually impressive and unique places (Bradley 2000; Taçon 1999), places of practical utility, as well as features of the built environment. Such features often reveal distinctive patterns of association with lithic assemblages, depending on the nature of the activities and meanings assigned to those places. The nature of assemblage composition and accumulations of

lithic items around such features are therefore likely to be informative of the nature and importance of those places. Such "place use" histories tell us something about the nature of social engagement within defined locales that may differ from the broader socioeconomic use of landscapes, which I choose to call "land-use" in this chapter.

Burials, domestic structures, visually impressive edifices, monuments requiring huge investment of human labor to build or maintain, distinctive landscape features (such as caves, mountains, and unusual rock formations), astrological devices, and places offering sanctuary are all places we would commonly expect to find lithic accumulations resulting from activities, performances, or acts of deliberate destruction that might point to ritual, symbolic, and other important social roles for these places. Examples include the repeated association between polished axes and causewayed enclosures in Neolithic Britain (Edmonds 1990), ceremonial knives and elite burials in Mesoamerica, cylindroconical stones and stone arrangements in Australia (Cundy 1985), and Gerzian ripple flaked knives and burials in pre-Dynastic Egypt (Savage 2001). At a more mundane level, even evidence for the ceremonial knapping of stone in association with the opening and filling of funerary pits (Edmonds 1999), or the presence of simple flakes or distinctive raw materials in grave goods (Haglund 1976), or just lithic accumulations in association with distinctive natural features might point to particular significance for those contexts in the landscape.

Association is therefore a means of linking accumulations or unique objects to unusual contexts in the landscape, offering the potential to explore behavioral disjunctions that are suggestive of aberrations from more typical behavior or the emergence of new belief systems that might not necessarily be apparent from other aspects of material culture.

Alteration

Stone-working is a reductive process that alters parent stone through the successive and irreversible removal of material. How much a piece of stone has been altered can therefore tell us something about the amount of time and energy invested in the production of an artifact, the level of departure from the original form, the amount of material likely to have been created as a byproduct of the alteration process, and the position in the sequence at which changes in manufacturing strategies took place and their likely effects on artifact morphology. At landscape scales, archaeologists are often interested in how much and in what ways stone have been altered (typically in relation to distance from source or proximity to important landscape features) in order to understand systems of time budgeting, mobility, and land-use. This perspective allows investigation of the complex life histories of artifacts from procurement to discard so that the distribution of types and sequence of technological activities and the movement of artifacts in the landscape can be determined. This is because degree and nature of alteration (that is, the differential distribution of sequential steps and stages of manufacture and reduction in space and time) will often reflect aspects of planning, land-use, ecology, ideology, and settlement and subsistence patterns affecting people's daily lives (Kuhn 1995; Nelson 1991).

A large literature has recently emerged developing methods for determining the extent of artifact alteration through reduction. It has mostly focused on artifact alteration through retouching, but studies of other kinds of flaking and grinding also exist, and techniques range from measurement of flake and retouch features (Barton 1988; Blades 2003; Clarkson 2002; Clarkson and Lamb 2005; Close 1991; Eren et al. 2005; Kuhn 1990), estimation of original size or mass (Dibble et al. 2005; Dibble and Pelcin 1995; Shott 1994), and refitting of artifacts to understand changing forms and reduction techniques (Cziesla 1990).

Ecological and Social Dimensions of Land-Use in Wardaman Country, Northern Australia

As an example of the above-mentioned potentialities of lithic analyses to landscape archaeology, the following example presents my own perspective on the changing nature of human engagement with landscape in Wardaman Country, northern Australia, over the last 10,000 years. Patterns of lithic transportation, accumulation, association, and alteration are explored to reconstruct economic, social, and symbolic aspects of place-use over this period (Clarkson 2006). Wardaman Country sits on the edge of the semi-arid zone in the Northern Territory and is a region divided between resource-rich, fertile alluvial plains and gorges (predominantly in the north of the region) that contain diverse high-ranked resources and abundant permanent water, and flat sand plains and black soil plains that are relatively poor in food resources with few and widely dispersed permanent waterholes (Clarkson 2004). These poorer, generally southerly areas typically contain abundant stone outcrops in the form of large habitable rockshelters and quartzite outcrops suitable for stone artifact

manufacture, whereas more northerly, richer areas lack rock overhangs but contain numerous chert sources.

The location of critical resources such as water, shelter, raw materials, and dry locations above flood plains strongly constrains where activities took place in the landscape in Wardaman Country. As Figure 48.1 shows, the majority of lithic accumulations are focused around these features, such that almost no sites occur more than 2 km to any of these features (Figure 48.1).

Stone artifact accumulations also indicate that much more activity was focused in areas with abundant permanent waterholes, and rich plant and animal resources in northern parts of the study area. Furthermore, blanks for implement production (for example, pointed blades suited to the manufacture of points, tulas, side- and end-scrapers, and burins) and cores of nonlocal stone were often procured in the raw material-rich southern land systems and transported and stockpiled for use at spatially extensive sites in these northern richer areas of the study region. As previously noted, Kuhn calls this strategy of stockpiling regularly visited places *place provisioning*. These large sites also contain typically dense and highly diverse assemblages compared to those found in other parts of the study region, with abundant signs of implement manufacture, retooling, and recycling of implements.

Sites situated in these richer northerly land systems also frequently contained site furniture in the form of grindstones, hammerstones, and anvils imported from surrounding regions and kept at large camps where a range of tasks was predicted to occur on a regular basis. Sites located in poorer land systems tended to be either quarries at which cores were typically reduced to obtain pointed blade blanks and other flakes, or sites containing a high proportion of retouched implements (often extensively reduced and made of high-quality raw materials) typical of the kind of "individual provisioning" expected of highly mobile foraging trips targeting specific resources or searching for mobile or dispersed prey. The differences in overall land-use in Wardaman country can therefore be characterized as low-mobility, long-term occupation of numerous, large, predictably and regularly reoccupied base camps in the north, and high-mobility, short-term occupation of special-purpose camps while quickly moving through the landscape in search of resources or while procuring stone in the south. Northern land systems probably therefore represent landscapes that were familiar and "lived in" in comparison to those more transiently visited places in the south of the region.

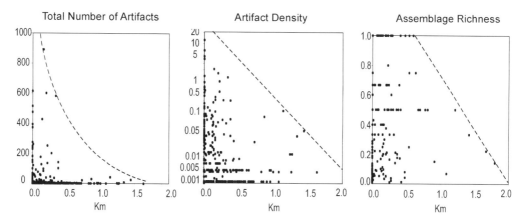

Figure 48.1 Patterns of association between lithic accumulation and the nature of assemblages and important landscape features. X-axes are kilometers from closest permanent water, flakeable stone, shelter, and elevated terrain.

Figure 48.2 Deep rockshelters in the central and southern parts of Wardaman Country.

Interestingly, it is also in these land systems of lower ranking that large rockshelters are abundant, often close to sources of raw material. These rockshelters are usually shady, secluded places with soft sandy floors suited to habitation and flat sandstone walls ideal for the creation of rock art (Figure 48.2). From the impressive accumulations of artifacts and rock art, it is clear that these places have been a significant locus of human occupation, tool-making, and artistic activities on a massive scale, with sometimes enormous and complex painted freezes of Dreaming beings (such as animals and the massive "Lightning Brothers" motifs) and extensive figurative and nonfigurative engravings present in these places. These shelters provide private spaces and are known ethnographically to have been used as places for the performance of restricted rituals and artistic activities.

Rockshelters are also the focus of unusual lithic associations with symbolic significance, such as exotic and mythologically charged Kimberley points, trade "Leilira" blades (Allen 1997; Thomson 1949), and cached items that may have been stored for times of shortfall and that are rarely or never found at the large open sites in northern parts of the study region (Figure 48.3). These items are usually hidden in narrow declivities, often close to human remains, and even include decorated boards of totemic and clan significance. However, the limited economic resources surrounding these rockshelters probably constrained

Figure 48.3 (A) Pressure-flaked Kimberley Points; (B) Leilira blades, cached at rockshelters in Wardaman Country in the last 1,000 years.

their use to specific activities such as raw material procurement, ritual activity, and occasional habitation. In a landscape otherwise traversed quickly, and perhaps without the familiarity of more intensively occupied areas to the north, these outliers must have stood out as highly incongruous, visually distinctive, and perhaps daunting repositories of cultural capital and tradition that were to be either actively avoided or maintained and embraced, depending on one's social status in Wardaman society. They contained the direct linkages to the Dreaming—powerful objects and depictions, the remains of the ancestors and the material significations of past ritual and domestic activities that simultaneously attracted and repelled human engagement (Merlan 1989).

The use of these places was not static in time, and it is clear from numerous excavations

at rockshelters throughout Wardaman Country (for example, Nimji, Garnawala 2, Gordolya, and Jagoliya) that the system of land-use and place-use changed dramatically over the last 15,000 years (Clarkson 2004, 2007). The frequency, predictability, and range of residential moves seems to have dramatically changed around 7,000 years ago, and occupational intensities greatly increased in rockshelters throughout the region (Figure 48.4). Stone was stockpiled at these shelters in the form of cores and large flakes; heavily retouched toolkits were rare, and raw materials were rarely imported from distant sources. This period was followed by a huge drop in occupational intensity between 5,000 and 2,000 years ago, at which time extreme individual provisioning seems to characterize the use of rockshelters. Exotic raw materials indicative of long-distance transport, heavily retouched implements with long complex use-lives including morphological transformations and recycling, and the use of small, standardized, portable, and presumably hafted toolkits dominate assemblages at these sites at this time. Between 2,000 and 1,000 years ago, occupational intensity peaks a second time, with reduced reliance on individual provisioning, greater use of local stone, and some discard of cores rather than just retouched implements. In the last 1,000 years, occupational intensity diminishes once more, and significant items identifying formal, long-distance social networks appear, such as Leilira trade blades and Kimberley points.

Given these changing levels of occupational intensity, and therefore probably also visitation frequency and familiarity as domestic places, we see that ochre deposition peaks at times of least occupational intensity (Figure 48.4). This suggests that places may have gained greater ritual significance and were more intensively decorated and maintained when visited less frequently, as certain activities could be more easily conducted in private and powerful ritual items stored without fear of loss or exposure to unsuitable eyes.

Major changes in rock art styles have also been documented in Wardaman Country that coincide with changes in land-use and occupational intensity as reconstructed from stone artifact assemblages and other lines of evidence (Figure 48.5). These include an early dominance of engraved rock art prior to 3,000 years ago (David, Chant, and Flood 1992; Mulvaney 1969; Watchman et al. 2000), followed by a change to the use of large figurative art panels (perhaps around 3,000 years ago) (Watchman et al. 2000) coincident with infrequent and highly mobile use of rockshelters, and finally the creation of large striped anthropomorphs and

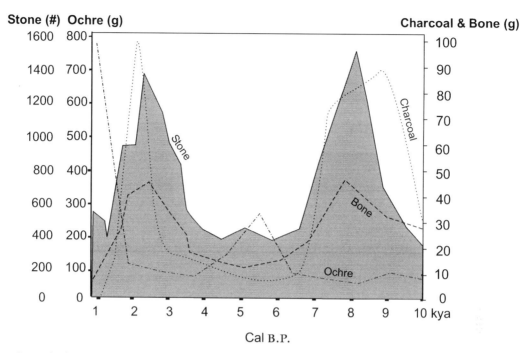

Figure 48.4   Changes in occupation intensity and the deposition of ochre averaged from four rockshelters (Nimji, Garnawala 2, Gordolya, and Jagoliya).

other changes in rock art styles in the last 300 years (Attenbrow, David, and Flood 1995; David 2002), coincident with the appearance of Leilira trade blades and Kimberley points.

It is not surprising that major readjustments in land-use and engagement with landscape should be marked by equivalent changes in other spheres of human life, such as worldview, ritual, and art. David (2002) has recently argued, for instance, that ontology—or the system of meaning and preunderstanding with which people interpret the world and their own place in it—is fundamentally shaped by our relationship to and experience of landscape, material objects, and other people, and by our historical perspective of engaged landscapes, such that a change in any one of these variables will likely also result in a change in systems of belief and preunderstandings about the world. Such changes in worldview are likely signaled by major alterations in land-use and the experience of places based on frequency and nature of use, and are expressed in Wardaman Country through changing rock art styles among other things.

What stimulated these major changes in systems of land-use and place-use in Wardaman Country is not entirely clear, but one possible explanation is that overall population size waxed and waned, or more marginal land systems were more intensively

used depending on changes in rainfall and inter-annual variability (Clarkson and Wallis 2003). The early Holocene optimum dated to between 9,000 and 6,000 years ago was manifested in northern Australia as wetter, more stable, and more productive environments that probably allowed bigger populations and more prolonged habitation and use of marginal land systems, such as those southern land systems containing large rockshelters. The later onset of severe El Niño events between 5,000 and 2,000 years ago may have seen contraction to better watered and more productive areas with less frequent use of poorer land systems under high-mobility regimes (Gagan, Chivas, and Isdale 1994; Gagan et al. 2004; Haberle and David 2004; Kershaw 1995; Koutavas et al. 2002; McGlone, Kershaw, and Markgraf 1992; Nott and Price 1999; Schulmeister and Lees 1995). El Niño events reduced in severity markedly 1,500 years ago, and populations perhaps made more use of marginal land systems again. The last 1,000 years saw diminishing use of the landscape again as El Niño settled into its modern pattern, with very high interannual variation still characterizing the region today (Dewar 2003). People consolidated territorial boundaries to protect critical resources at these times of greater subsistence risk, giving rise to new artistic styles that might have served

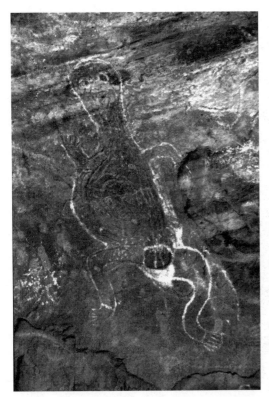

Figure 48.5 Changes in rock art styles are evident at the same time as lithic technology and land-use changes in Wardaman Country (used with permission of Wardaman Elders 2001).

as territorial markers (cf. David and Chant 1995; David and Lourandos 1998; Haberle and David 2004), as well as initiating extensive formal inter-group networks as social storage mechanisms to offset risk, giving rise to long-distance exchange networks that reified social relationships (Cashdan 1985; Myers 1982; Paton 1994).

Conclusions

Landscape has become an important methodological and theoretical concern in contemporary archaeology, and analysis of patterns of lithic transportation, accumulation, association, and alteration provide an important means of exploring past cultural landscapes that should form a central component of any serious landscape study. Stone artifacts stand to shed a great deal of light on many facets of human engagement with landscape, including: mobility; occupational intensity; the nature of planning and anticipatory provisioning for use of different land systems; experience of landscape; worldviews; and broader social relationships. The case study highlights the ways various threads of meaning can be elicited from lithic assemblages

at regional scales and the ways these can be woven with other lines of evidence to create detailed studies of past human land-use, the shifting nature of place use histories, and people's social and economic engagement with places as environments, technologies, and social systems changed.

References

Allen, H. 1997. The distribution of large blades (leilira): Evidence of recent changes in Aboriginal ceremonial exchange networks, in P. McConvell and N. Evans (eds.), *Archaeology and Linguistics: Aboriginal Australia in Global Perspective*. Oxford: Oxford University Press.

Attenbrow, V., David, B., and Flood, J. 1995. Menngeya and the origins of points: New insights into the appearance of points in the semi-arid zone of the Northern Territory. *Archaeology in Oceania* 30: 105–19.

Bamforth, D. B., and Woodman, P. C. 2004. Tool hoards and Neolithic use of the landscape in north-eastern Ireland. *Oxford Journal of Archaeology* 23: 21–44.

Barton, C. M. 1988. *Lithic Variability and Middle Paleolithic Behaviour: New Evidence from the*

Iberian Peninsula. Oxford: British Archaeological Reports.

Binford, L. R. 1979. Organizational and formation processes: Looking at curated technologies. *Journal of Anthropological Research* 35: 255–73.

Blades, B. 1986. The optimal design of hunting weapons: Maintainability or reliability. *American Antiquity* 51: 737–47.

———. 1999. Aurignacian settlement patterns in the Vezere Valley. *Current Anthropology* 40: 712–23.

———. 2003. End scraper reduction and hunter-gatherer mobility. *American Antiquity* 68: 141–56.

Bradley, R. 2000. *An Archaeology of Natural Places*. London: Routledge.

Brantingham, P. J. 2003. A neutral model of stone raw material procurement. *American Antiquity* 68: 487–509.

Byrne, D. 1980. Dynamics of Dispersion: The place of silcrete in archaeological assemblages from the Lower Murchison, Western Australia. *Archaeology and Physical Anthropology in Oceania* 15: 110–19.

Camilli, E. L., and Ebert, J. I. 1992. Artifact reuse and recycling in continuous surface distributions and implications for interpreting land use patterns, in J. Rossignol and L. Wandsnider (eds.), *Space, Time and Archaeological Landscapes*, pp. 113–36. New York: Plenum Press.

Cane, S. 1992. Aboriginal perceptions of their stone tool technology: A case study from the Western Desert, Australia. *Australian Archaeology* 35: 11–31.

Cashdan, E. 1985. Coping with risk reciprocity among the Basara of northern Botswana. *Man* 20: 454–74.

Clarkson, C. 2002. An index of invasiveness for the measurement of unifacial and bifacial retouch: A theoretical, experimental and archaeological verification. *Journal of Archaeological Science* 1: 65–75.

———. 2004. Technological provisioning and assemblage variation in the eastern Victoria River region, Northern Australia: A Darwinian perspective. Unpublished Ph.D. thesis: Australian National University, Canberra.

———. 2006. Interpreting surface assemblage variation in Wardaman country, Northern Territory: An ecological approach, in I. Lilley and S. Ulm (eds.), *An Archaeological Life: Essays in Honour of Jay Hall*. Brisbane: Aboriginal and Torres Strait Islander Studies Unit, The University of Queensland.

———. 2007. *Lithics in the Land of the Lightning Brothers: The archaeology of Wardaman country, Northern Territory*. Terra Australis. Canberra: ANU E Press.

Clarkson, C., and Lamb, L. (eds.). 2005. *Lithics "Down Under": Australian Perspectives on Lithic Reduction, Use and Classification*. BAR International Series S1408. Oxford: Archaeopress.

Clarkson, C., and Wallis, L. A. 2003. The search for El Niño/Southern Oscillation in archaeological sites: Recent phytolith analysis at Jugali-ya rockshelter, Wardaman Country, Australia, in D. M. Hart and L. A. Wallis (eds.), *Phytolith and Starch Research in the Australian-Pacific-Asian Regions: The State of the Art*, pp. 137–52. Terra Australis 19. Canberra: Pandanus Books.

Close, A. 1991. On the validity of Middle Paleolithic tool types: A test case from the eastern Sahara. *Journal of Field Archaeology* 18: 256–69.

———. 2000. Reconstructing movement in prehistory. *Journal of Archaeological Method and Theory* 7: 49–77.

Cundy, B. J. 1985. The secondary use and reduction of cylindro-conical stone artifacts from the Northern Territory. *The Beagle* 2: 115–27.

Cziesla, E. 1990. *The Big Puzzle: International Symposium on Refitting Stone Artefacts*. Bonn: Holos.

David, B. 2002. *Landscapes, Rock Art and the Dreaming: An Archaeology of Preunderstanding*. London: Leicester University Press.

David, B., and Chant, D. 1995. Rock art regionalisation in North Queensland prehistory. *Memoirs of the Queensland Museum* 37: 357–528.

David, B., Chant, D., and Flood, J. 1992. Jalijbang 2 and the distribution of pecked faces in Australia. *Memoirs of the Queensland Museum* 32: 61–77.

David, B., and Lourandos, H. 1998. Rock art and socio-demography in northeastern Australian prehistory. *World Archaeology* 30: 193–219.

David, B., and Wilson, M. 1999. Re-reading the landscape: Place and identity in NE Australia during the Late Holocene. *Cambridge Archaeological Journal* 9: 163–88.

Dewar, R. 2003. Rainfall variability and subsistence systems in Southeast Asia and the Western Pacific. *Current Anthropology* 44: 369–88.

Dibble, H., and Pelcin, A. 1995. The effect of hammer mass and velocity on flake mass. *Journal of Archaeological Science* 22: 429–39.

Dibble, H., Shurmans, U. A., Iovita, R. P., and McLaughlin, M. V. 2005. The measurement and interpretation of cortex in lithic assemblages. *American Antiquity* 70: 545–60.

Dooley, W. A. 2004. Long-term hunter-gatherer land use in central North Dakota: An environmental analysis. *Plains Anthropologist* 49: 105–27.

Edmonds, M. 1990. *Stone Tools and Society*. London: Bastford.

———. 1999. *Ancestral Geographies of the Neolithic*. London: Routledge.

Eren, M. I., Dominguez-Rodrigo, M., Kuhn, S. L., Adler, D. S., Le, I., and Bar-Yosef, O. 2005. Defining and measuring reduction in unifacial stone tools. *Journal of Archaeological Science* 32: 1190–201.

Field, D. 1997. The landscape of extraction: Aspects of the procurement of raw material in the Neolithic, in P. Topping (ed.), *Neolithic Landscapes*, pp. 55–68. Oxford: Oxbow Books.

Gagan, M. K., Chivas, A. R., and Isdale, P. J. 1994. High-resolution isotopic records of the mid-Holocene tropical Western Pacific. *Earth and Planetary Sciences* 121: 549–58.

Gagan, M., Hendy, E. J., Haberle, S. G., and Hantoro, W. S. 2004. Post-glacial evolution of the Indo-Pacific warm pool and El Niño-Southern Oscillation. *Quaternary International* 118–119: 127–43.

Goodyear, A. C. 1989. A hypothesis for the use of crypto-crystalline raw materials among Paleoindian groups of North America, in C. G. Ellis and J. C. Lothrop (eds.), *Eastern Paleoindian Lithic Resource Use*, pp. 1–9. Boulder, CO: Westview Press.

Gould, R. A. 1968. Chipping stones in the outback. *Natural History* 77: 42–48.

Gould, R. A., and Saggers, S. 1985. Lithic procurement in Central Australia: A closer look at Binford's idea of embeddedness in archaeology. *American Antiquity* 50: 117–36.

Haberle, S. G., and David, B. 2004. Climates of change: Human dimensions of Holocene environmental change in low latitudes of the PEPII transect. *Quaternary International* 118–119: 165–79.

Haglund, L. 1976. *An Archaeological Analysis of the Broadbeach Aboriginal Burial Ground*. Brisbane: University of Queensland Press.

Harrison, R. 2002. Archaeology and the colonial encounter, Kimberley spearpoints, cultural identity and masculinity in the north of Australia. *Journal of Social Archaeology* 2: 352–77.

Hiscock, P. 1981. Comments on the use of chipped stone artifacts as a measure of "intensity of site usage." *Australian Archaeology* 13: 20–34.

———. 2002. Quantifying the size of artifact assemblages. *Journal of Archaeological Science* 29: 251–58.

Jones, G. T., Beck, C., Jones, E. E., and Hughes, R. E. 2003. Lithic source use and palaeoarchaic foraging territories in the Great Basin. *American Antiquity* 68: 5–38.

Jones, R., and White, N. 1988. Point blank: Stone tool manufacture at the Ngilipitji Quarry, Arnhem Land, 1981, in B. Meehan and R. Jones (eds.), *Archaeology with Ethnography: An Australian Perspective*, pp. 51–87. Canberra: Department of Prehistory, RSPacS, Australian National University.

Kelly, R. L. 1988. The three sides of a biface. *American Antiquity* 53: 717–34.

———. 1992. Mobility/sedentism: Concepts, archaeological measures and effects. *Annual Review of Anthropology* 21: 43–66.

Kelly, R. L., and Todd, L. C. 1988. Coming into the country: Early palaeoindian hunting and mobility. *American Antiquity* 53: 231–44.

Kershaw, A. P. 1995. Environmental change in Greater Australia. *Antiquity* 69: 656–76.

Koutavas, A., Lynch-Steiglitz, J., Marchitto, T. M., and Sachs, J. P. 2002. El Niño-like pattern in Ice Age tropical Pacific sea surface temperature. *Science* 297: 226–31.

Kuhn, S. 1990. A geometric index of reduction for unifacial stone tools. *Journal of Archaeological Science* 17: 585–93.

———. 1995. *Mousterian Lithic Technology*. Princeton, NJ: Princeton University Press.

Mackay, A. 2005. Informal movements: Changing mobility patterns at Ngarrabullgan, Cape York Australia, in C. Clarkson and L. Lamb (eds.), *Lithics "Down Under": Australian Perspectives on Stone Artefact Reduction, Use and Classification*, pp. 95–108. BAR International Series S1408. Oxford: Archaeopress.

McGlone, M. S., Kershaw, A. P., and Markgraf, V. 1992. El Niño/Southern Oscillation climatic variability in Australasian and South American palaeoenvironmental records, in H. F. Diaz and V. Markgraf (eds.), *El Niño: Historical and Palaeoclimatic Aspects of the Southern Oscillation*, pp. 435–62. Cambridge: Cambridge University Press.

McNiven, I. 1999. Fissioning and regionalisation: The social dimensions of change in Aboriginal use of the Great Sandy region, southeast Queensland, in J. Hall and I. McNiven (eds.), *Australian Coastal Archaeology*, pp. 157–68. Research Papers in Archaeology and Natural History 31. Canberra: Archaeology and Natural History Publications.

Merlan, F. 1989. The interpretive framework of Wardaman rock art: A preliminary report. *Australian Aboriginal Studies* 1989: 14–24.

Mulvaney, D. J. 1969. *The Prehistory of Australia*. London: Thames and Hudson.

Myers, F. 1982. Always ask: Resource use and land ownership among Pintupi Aborigines, in N. Williams and E. S. Hunn (eds.), *Resource Managers*, pp. 173–96. Canberra: Australian Institute of Aboriginal Studies.

Nelson, M. C. 1991. The study of technological organization. *Archaeological Method and Theory* 3: 57–100.

Nott, J., and Price, D. 1999. Waterfalls, floods and climate change: Evidence from tropical Australia. *Earth and Planetary Science Letters* 171: 267–76.

Parry, W. J., and Kelly, R. L. 1987. Expedient core technology and sedentism, in J. K. Johnson and C. A. Morrow (eds.), *The Organization of Core Technology*, pp. 285–304. Boulder, CO: Westview Press.

Paton, R. 1994. Speaking through stones: A study from northern Australia. *World Archaeology* 26: 172–84.

Savage, S. H. 2001. Some trends in the archaeology of predynastic Egypt. *Journal of Archaeological Research* 9: 101–55.

Schulmeister, J., and Lees, B. 1995. Pollen evidence from tropical Australia for the onset of ENSO-dominated climate at c. 4,000 B.P. *The Holocene* 5: 10–18.

Shackley, M. S. 1998. Gamma rays, X rays and stone tools: Some recent advances in archaeological geochemistry. *Journal of Archaeological Science* 25: 259–70.

Shott, M. J. 1986. Technological organization and settlement mobility: An ethnographic examination. *Journal of Anthropological Research* 42: 15–51.

———. 1994. Size and form in the analysis of flake debris: Review of recent approaches. *Journal of Archaeological Method and Theory* 1.

———. 2000. The quantification problem in stone-tool assemblages. *American Antiquity* 65: 725–38.

Taçon, P. S. C. 1991. The power of stone: Symbolic aspects of stone use and tool development in Western Arnhem Land, Australia. *Antiquity* 65: 192–207.

———. 1999. Identifying ancient sacred landscapes in Australia: From physical to social, in W. Ashmore and A. B. Knapp (eds.), *Archaeologies of Landscape: Contemporary Perspectives*, pp. 33–57. Malden: Blackwell Publishing.

Thomas, J. 2001. Archaeologies of place and landscape, in I. Hodder (ed.), *Archaeological Theory Today*, pp. 165–86. Cambridge: Polity Press.

Thomson, D. F. 1949. Arnhem Land: Exploration among an unknown people. *Geographical Journal* 114: 53–67.

Torrence, R. 2002. Cultural landscapes on Garua Island, Papua New Guinea. *Antiquity* 76: 766–76.

Watchman, A. L., David, B., McNiven, I., and Flood, J. 2000. Micro-archaeology of engraved and painted rock surface crusts at Yiwarlarlay (The Lightning Brothers site), Northern Territory, Australia. *Journal of Archaeological Science* 27: 315–25.

Whittle, A. W. R. 1997. Moving on and moving around: Neolithic settlement mobility, in P. Topping (ed.), *Neolithic Landscapes*, pp. 14–22. Oxford: Oxbow Books.

Zvelebil, M., Green, S. W., and Macklin, M. G. 1992. Archaeological landscapes, lithic scatters, and human behaviour, in J. Rossignol and L. Wandsnider (eds.), *Space, Time and Archaeological Landscapes*, pp. 193–226. New York: Plenum Press.

49

THE USE OF HUMAN SKELETAL REMAINS IN LANDSCAPE ARCHAEOLOGY

F. Donald Pate

Studies of skeletal remains provide important information regarding the use of past landscapes by human populations. The physical and chemical characteristics of human skeletons record valuable data about interactions between individuals and their physical and social environments at various life stages. Furthermore, analyses of large skeletal populations derived from burial grounds allow inferences regarding intergroup social relations across various landscapes. Thus, analyses of human skeletal remains offer archaeologists a unique resource that produces spatial-temporal information that is often difficult to obtain from conventional archaeological methods, especially in relation to pre-European contact archaeology.

This chapter provides an overview of the key methods employed by archaeologists to examine past human landscape use via analyses of human skeletal remains. Research involving archaeological human skeletal remains is grounded within the sub-discipline of bioarchaeology. According to Larsen (1997: 3), bioarchaeology "emphasizes the human biological component of the archaeological record" with a focus on analyses of excavated human skeletal remains. In relation to landscape archaeology, a range of analytical methods involving bones and teeth can contribute to an improved understanding of past human relationships within and between

various natural and cultural environments. These methods include chemical analyses (isotopic and elemental), statistical assessment of metric and non-metric traits (biological distance analysis), ancient DNA analysis, and palaeopathology.

Studies of archaeological skeletal remains employing these methods have been used to address a range of past human behaviors including sedentism, territoriality, inclusive vs. exclusive social relations, residential mobility, migration patterns (Beard and Johnson 2000; Budd et al. 2001, 2003, 2004; Hodell et al. 2004; Müller et al. 2003; Pardoe 1994, 1995; Pate 1995, 2000, 2006; Pate, Brodie, and Owen 2002; Price, Burton, and Bentley 2000; Price, Grupe, and Schroter 1994; Price et al. 1994; Sealy, Armstrong, and Shrire 1995; Sealy and van der Merwe 1986; Verano and DeNiro 1993), and also health patterns associated with hunter-gatherer vs. agricultural lifeways (Cohen and Armelagos 1984a, 1984b; Goodman 1993; Goodman et al. 1984a; Larsen 1995).

Cemeteries and Past Social Relations

To examine past social relations and landscape-use employing analyses of human skeletal remains, we generally require large samples of skeletons from sites distributed across a range of environmental zones. Large skeletal samples ensure a good distribution of age, sex, and social groups and improve the reliability of inferences regarding spatial and temporal variability in past human behaviors.

The appearance of cemeteries and the presence of mortuary differentiation within those cemeteries have been used by archaeologists as indicators of increased sedentism and organizational complexity in past societies. Large cemeteries are generally associated with semipermanent or permanent settlements (Bird and Monahan 1995; Chapman, Kinnes, and Randsbor 1981; Chatters 1987; O'Shea 1984; Pardoe 1988; Pate 2006; Price and Brown 1985; Rothschild 1979). Accordingly, research involving archaeological cemetery sites has played a central role in the examination of past social complexity in hunter-gatherer and agricultural societies. Hypotheses regarding social and behavioral differentiation within and between different environmental zones where large cemeteries occur can be further tested employing bioarchaeological methods.

Thus, research designs involving analyses of archaeological skeletal remains should focus on the availability of large skeletal samples with good chronological control. Furthermore, excavated cemetery sites included in the research design should represent the key environmental zones present in particular landscapes. If different segments of a past population were buried in distinct cemeteries, then a number of cemetery sites within each environmental zone need to be included in the sample in order to represent the various social groups that were present in past societies (Beck 1995; Buikstra, Konigsberg, and Bullington 1986; Larsen 1987; Milner, Wood, J. and Boldsen 2000).

Skeletal Chemistry

The chemical composition of bones and teeth represent dietary intake and environment of residence for different stages of the human life cycle (Sealy, Armstrong, and Shrire 1995). The amino acids and elements liberated from ingested foods are incorp-orated into the organic collagen and inorganic hydroxyapatite components of bones and teeth relative to their quantities in the human diet (Katzenberg 2000; Katzenberg and Harrison 1997; Pate 1994; Sandford and Weaver 2000; Schoeninger 1995; Schoeninger and Moore 1992; Schwarcz and Schoeninger 1991). Consequently, foods and environments with distinct chemical signatures are reflected in the chemical composition of human bones and teeth.

In an adult skeleton, teeth provide a record of childhood diet and residence (Hillson 1996, 2000;

Smith 1991), whereas the chemical composition of bone relates to long-term adult dietary composition and habitat-use. In rare cases where well-preserved frozen or mummified human remains are recovered, tissues that record shorter periods of dietary intake—for example, fingernails and hair—can provide evidence regarding short-term dietary variability and landscape-use (Katzenberg and Krouse 1989; Macko et al. 1999; Müller et al. 2003; Richards et al. 2003; Sandford and Kissling 2005; Schoeninger, Iwaniec, and Nash 1998; Sponheimer et al. 2002; West et al. 2004).

Inferences about past human diet and landscape use can be based on variability in both the organic and inorganic chemical constituents of bones and teeth. Bone consists of organic collagen and inorganic hydroxyapatite components (Pate 1994). Whole cortical bone is approximately 69% inorganic, 22% organic, and 9% water (Triffitt 1980). The highly insoluble hydroxyapatite, $Ca_5(PO_4)_3OH$, is the dominant calcium phosphate phase in bone mineral. Tooth crowns are composed of a hard outer layer of enamel that overlies a softer dentine. The inner parts of tooth roots consist primarily of dentine, whereas the outer portions are composed of cementum. Inorganic hydroxyapatite is the primary chemical component of enamel, dentine, and cementum. Enamel consists of approximately 95% hydroxyapatite, 4% water, and 1% organic matter, whereas dentine is 70% hydroxyapatite, 18% organic matter (mainly collagen), and 12% water (Hillson 1996; Mann 2001; Woelfel and Scheid 2002).

In healthy individuals, the development of permanent tooth crowns takes place in three phases (Hillson 2000: 249; Smith 1991).

1. Incisors, canines, and first molars are initiated during the first year after birth (or just before birth) and are completed between 3 and 7 years of age;

2. Premolars and second molars start formation during the second and third years after birth and are completed between 4 and 8 years of age;

3. Third molars are initiated any time between age 7 and 12 years and are completed sometime between 10 and 18 years of age.

The roots of teeth continue to form after the crowns are completed. Following completion of development, the chemical constituents of teeth do not change during the remaining lifetime of the individual.

In contrast, bone continues to change its chemical composition throughout the lifetime of the individual (Pate 1994: 164–65). Bone turnover associated with skeletal development is referred to as *modeling*. Modeling proceeds most rapidly in the infant, declines during subsequent growth, and approaches zero at skeletal maturity. After age 20 years, remodeling accounts for over 95% of human bone tissue turnover. This remodeling continues until the time of death. During remodeling, the annual turnover rate for compact bone averages 2.5% and that for trabecular bone 10%. The mean annual cortical replacement percentage for adult bones is 8.3% for vertebrae, 4.7% for ribs, 2.9% for femora, and 1.8% for the skull (Glimcher 1976; Marshall et al. 1973; Neuman and Neuman 1958; Tanaka, Kawamura, and Nomura 1981). Thus, owing to slow metabolic turnover rates, the chemical composition of normal adult human bone provides a long-term record of dietary intake and landscape-use over a period of decades (Libby et al. 1964; Stenhouse and Baxter 1979).

Consequently, the chemical analysis of teeth and bones provides an opportunity to address changes in diet and residence from childhood to adulthood. Analyses of different teeth from the same individual allow a refinement of childhood dietary reconstruction and habitat use to three stages of maturation. Finally, a combination of bone and tooth chemistry values allows comparisons between the last several decades of life and various subadult stages.

Owing to observed chemical alterations (that is, diagenesis) in tooth dentine and bone mineral in the postmortem burial environment (Budd et al. 2000a; Kohn, Schoeninger, and Barker 1999; Montgomery et al. 1999, 2000; Nelson et al. 1986; Pate and Hutton 1988; Pate, Hutton, and Norrish 1989; Pate et al. 1991), most chemical studies are now focusing on tooth enamel (inorganic hydroxyapatite component) and bone protein (organic collagen) that appear to be less susceptible to diagenesis. Isotopes of strontium ($^{87}Sr/^{86}Sr$), oxygen ($^{18}O/^{16}O$), carbon ($^{18}C/^{12}C$), and lead ($^{206}Pb/^{204}Pb$, $^{207}Pb/^{206}Pb$, $^{208}Pb/^{206}Pb$) in tooth enamel and isotopes of carbon ($^{13}C/^{12}C$), nitrogen ($^{15}N/^{14}N$), and sulphur ($^{34}S/^{32}S$) in bone collagen are the principal chemical components employed in skeletal chemistry studies relating to landscape archaeology (Budd et al. 1999, 2001, 2002; Fricke, Clyde, and O'Neil 1998; Montgomery, Budd, and Evans 2000; Montgomery, Evans, and Neighbour 2003; Montgomery et al. 1999; Pate 1998a, 1998b; Pate, Brodie, and Owen 2002; Richards, Fuller, and Hedges 2001, 2003). Oxygen isotopes extracted from the phosphate position in bone mineral also appear to provide reliable biological signatures (Longinelli 1984; Luz, Kolodny, and Horowitz

1984; Müller et al. 2003). Isotopic studies examining the geographic origins of archaeological fauna (Hall-Martin et al. 1991; Hobson 1999) may provide information regarding the sources of human foods and artifacts constructed from animal tissues. In addition, elemental analyses of tooth enamel barium (Ba) and strontium (Sr) have been employed to address geographic origins of human populations (Burton et al. 2003).

Some recent studies have employed a combination of three or more different isotopes in order to improve the reliability of inferences regarding past uses of various geographic localities (Budd et al. 2001; Fernandez, Panarello, and Schobinger 1999; Müller et al. 2003).

Chemical Variability across Landscapes

The chemical composition of water, sediments, plants, and animals vary across different marine and terrestrial environments. These chemical differences are recorded in the bones and teeth of the human inhabitants of these distinct landscapes. The chemical signatures of local or exotic water and foods consumed by humans are incorporated into their body tissues relative to the biochemical turnover rates of various tissues. Thus, chemical analyses of archaeological skeletal remains may provide information regarding past human access to various foods/water resources and the habitats where they occurred.

The use of chemical variability in archaeological human skeletal remains as a means to reconstruct past cultural landscapes is dependent on the establishment of baseline chemical signatures for various marine and terrestrial habitats located in the region of study. Once distinct chemical signatures have been demonstrated for particular habitats, then hypotheses regarding landscape use and past social relations can be tested employing the chemical composition of archaeological skeletal remains.

Previous chemical analyses of human skeletal remains in archaeology have involved both elemental and isotopic measurements. Because most landscape analyses now focus on isotopic methods, the remainder of this chapter concentrates on isotopic analyses.

Most elements exist as mixtures of two or more isotopes. Isotopes are variants of the same element with differing numbers of neutrons in their nuclei. They possess almost identical chemical properties but have different atomic masses. For example, the element carbon (C) consists of three isotopes, the heavier radioactive ^{14}C and the lighter stable isotopes ^{13}C and ^{12}C. These differences in atomic mass result in variable chemical reaction rates. Kinetic variability allows plants and animals to preferentially incorporate one isotope over others during biochemical reactions associated with photosynthesis and food metabolism. Such changes in isotope ratios are referred to as fractionation. Fractionation enriches or depletes the stable isotopic composition of plant and animal tissues relative to atmospheric, seawater bicarbonate, or dietary ratios (Pate 1994: 171).

Carbon and Nitrogen Isotopes

Carbon and nitrogen isotope values in bone collagen are often used in tandem to examine use of foods from different habitats within a regional landscape (Coltrain, Hayes, and O'Rourke 2004; Katzenberg and Weber 1999; Keegan and DeNiro 1988; Larsen et al. 1992; Pate 1998a, 1998b; Schoeninger and DeNiro 1984; Schoeninger, Iwaniec, and Nash 1998). Distinct carbon isotope ratios are associated with foods derived from (1) marine vs. terrestrial habitats and (2) terrestrial plants employing different photosynthetic mechanisms. Most trees use the C_3 photosynthetic pathway regardless of geographic locality, whereas grasses and smaller understorey plants generally use C_3 photosynthesis in

temperate and cool-season localities and C_4 photosynthesis in tropical and warm-season localities. Owing to the distinct isotopic values and geographic distributions observed for C_3 and C_4 plants (O'Leary 1988; Pate and Noble 2000), the consumption of plants by animals and humans produces bone collagen carbon isotope values indicative of the distinct landscapes.

Because bone collagen stable carbon isotope values for animals feeding on marine foods overlap with those feeding on terrestrial C_4 plants, carbon isotopes can be used to examine marine vs. terrestrial dietary input only in regions dominated by C_3 plants. As nitrogen isotope ratios distinguish between marine foods and terrestrial foods in most coastal regions, nitrogen isotopes can be employed to address marine vs. terrestrial dietary components in many coastal regions with C_4 plants. However, in arid-land habitats with mean annual rainfall less than 350–400 mm (Ambrose 1991; Anson 1997; Pate and Anson 2008; Pate et al. 1998; Schwarcz, Dupras, and Fairgrieve 1999; Sealy et al. 1987), terrestrial herbivores show elevated bone collagen nitrogen isotope values that overlap with those for marine mammals.

Research by Walker and DeNiro (1986) in coastal southern California and Pate and coworkers

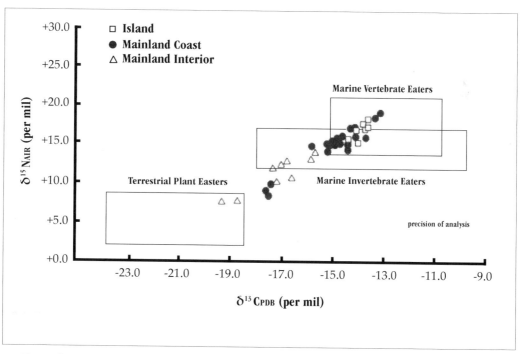

Figure 49.1 Plot of stable nitrogen vs. carbon isotope values for human bone collagen of individuals excavated from archaeological sites on Santa Cruz Island, the mainland coast, and the mainland interior, southern California (adapted from Walker and DeNiro 1986: 55).

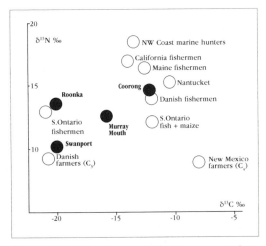

Figure 49.2 Plot of mean stable nitrogen vs. carbon isotope values for human bone collagen from archaeological sites located in four distinct geographic zones in southeastern South Australia: coastal marine (Coorong), coastal river mouth (Lake Alexandrina), inland riverine (Swanport), and inland riverine (Roonka Flat) (adapted from Pate et al. 2002: 3).

(2002) in coastal southeastern South Australia provide examples of the use of bone collagen stable carbon and nitrogen isotope values to demonstrate the use of distinct geographic localities by past human populations. In California, individuals inhabiting the offshore Channel Islands were distinguished from those on the coast and others from inland sites (Figure 49.1). In South Australia, four distinct geographic zones were identified (Figure 49.2): coastal marine, coastal river mouth, inland riverine (20 km from coast), and inland riverine (120 km from coast).

Strontium Isotopes

If strontium (Sr) isotope ratios of the sediments of local geochemical catchments vary significantly, then movement between particular localities can also be inferred from variations in $^{87}Sr/^{86}Sr$ ratios in tooth enamel (Ericson 1985, 1989; Hoppe et al. 1999; Katzenberg and Harrison 1997; Knudson et al. 2004; Montgomery, Evans, and Neighbour 2003; Price, Burton, and Bentley 2002; Sealy et al. 1991).

Strontium isotope ratios in granites, the most common continental rocks, vary from 0.700 to 0.737 but are generally > 0.715. Carbonates typically have a $^{87}Sr/^{86}Sr$ ratio of < 0.710. In contrast, ocean basalts exhibit little variation with a mean isotopic ratio of 0.7037 ± 0.0001, and modern seawater has a $^{87}Sr/^{86}Sr$ ratio of 0.7091. Because

seawater Sr is well mixed, the $^{87}Sr/^{86}Sr$ composition is relatively constant throughout the oceans of the world (Pate 1994: 168). Consequently, the bone Sr isotope ratios of humans consuming terrestrial food can be distinguished from those using marine foods in areas where uplifted marine strata are not prominent components of the terrestrial landscape. In addition, diets derived from discrete terrestrial geochemical environments can be differentiated. If strontium isotope ratios in bone mineral have not been altered significantly by postmortem chemical changes, then comparisons of tooth and bone values can be employed to examine changes in juvenile vs. adult residence (Beard and Johnson 2000; Hodell et al. 2004; Price, Grupe, and Schroter 1994; Sealy, Armstrong, and Shrire 1995).

Comparisons of bone and tooth strontium isotope ratios for past inhabitants of the 14th-century Grasshopper Pueblo, Arizona, were employed to differentiate life-long residents from immigrants (Ezzo, Johnson, and Price 1997; Price et al. 1994). The local strontium isotope signature for the site was determined using analyses of relatively sedentary field mice. Beard and Johnson (2000) summarizes the findings of the study (Figure 49.3).

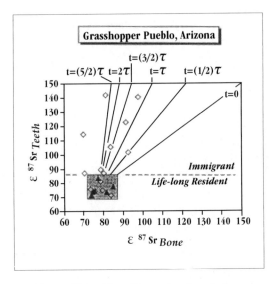

Figure 49.3 Plot of bone strontium isotope values in tooth enamel vs. bone mineral for the 14th century Grasshopper Pueblo, Arizona. Open diamonds represent tooth-bone pairs from the same human skeletons. The gray box indicates isotopic compositions for local Sr where the black triangles show values for field mice (adapted from Beard and Johnson 2000: 1058).

Lead Isotopes

A number of recent studies have employed radiogenic lead isotopes to enhance and to extend strontium-based examinations of residency and mobility (Montgomery, Budd, and Evans 2000: 372). These studies include those of Budd and associates (1998, 1999, 2000b), Gulson and colleagues (1997), Montgomery and associates (1999, 2000), and Müller and others (2003).

In preindustrial societies, most lead incorporated into skeletal tissues was derived from the diet. Dietary lead would reflect the lead composition of local soils as determined by the geology of the region of study. Lead isotope ratios vary in a systematic manner in various geographic regions as a "result of radiogenic isotopic evolution and therefore as a function of age, parent isotope abundance and subsequent remodeling of crustal material" (Montgomery, Budd, and Evans 2000: 372).

Lead is incorporated into bone via the bloodstream, and lead turnover rates in bone vary according to bone tissue remodeling rates, bone density, and bone type. Lead turnover rates in both cortical bone and dentine are estimated at 1% per annum, whereas rates in trabecular bone are somewhat higher (Gulson, Jameson, and Gillings 1997; Rabinowitz et al. 1991). Consequently, lead concentrations in skeletal tissues provide a record of long-term dietary intake and habitat use. Analyses of lead isotopes in deciduous and permanent teeth from individuals provide information regarding dietary intake at different stages of subadult life (Montgomery, Budd, and Evans 2000).

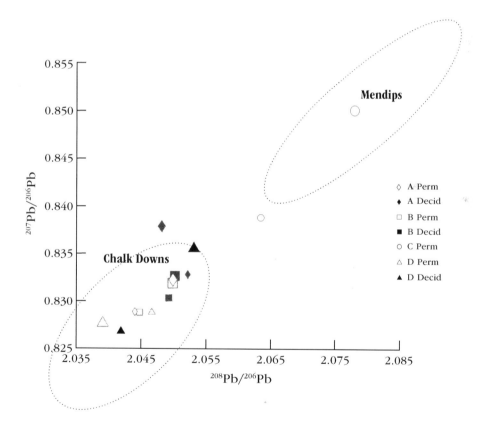

Figure 49.4 Plot of lead isotope values in deciduous and permanent tooth enamel (large symbols) and dentine (small symbols) for three juveniles (A, B, and D) and permanent tooth enamel and dentine for an adult female (C) excavated from a group burial pit at Monkton-up-Wimbourne, Dorset in southern England (adapted from Montgomery et al. 2000: 378).

Montgomery and associates (2000) employed lead and strontium isotope values in tooth enamel to examine mobility and geographic origin for four Neolithic individuals (one adult female and three juveniles) excavated from a group burial pit at Monkton-up-Wimbourne, Dorset, in southern England. The combination of lead and strontium isotope values provided a considerable degree of resolution with which to identify the adult female's place of origin. Isotope values suggest that the adult female originated from the Mendips region at least 80 km northwest of the burial site. Lead isotope values clearly distinguish the adult female from the juveniles (Figure 49.4).

Oxygen Isotopes

Stable oxygen isotope ratios in tooth enamel and bone mineral phosphate have also been employed to examine use of different geographic localities (Budd et al. 2001, 2004; Kohn, Miselis, and Fremd 2003; White et al. 1998). The oxygen isotope values of ingested drinking water are directly related to the tooth and bone values of consumers. Because oxygen isotope ratios vary according to the isotopic composition of local drinking water, human and faunal migrations can be inferred from tooth and bone $^{18}O/^{16}O$ ratios. For example, precipitation in continental interiors is depleted significantly in the heavier ^{18}O when compared to sea water (Longinelli 1984; Luz, Kolodny, and Horowitz 1984).

Differences in oxygen isotope values are also related to variations in relative humidity. Ayliffe and Chivas (1990) report a decrease in macropod bone phosphate oxygen isotope values that correlates with increased relative humidity. Bone phosphate isotope values ranged from 30.1‰ to 16.8‰ for habitats with relative humidities ranging from 39% to 82%. As macropods (kangaroos and wallabies) obtain a large proportion of their water from ingested vegetation, leaf water fractionation processes should be reflected in the stable oxygen isotope composition of macropod bone phosphate. The more positive oxygen isotope values observed in leaf water in arid-land habitats (Gat 1980) is translated into more positive bone phosphate oxygen isotope values in macropod consumers (Pate 1994: 165).

Luz and associates (1990) examined oxygen isotope variability in white-tailed deer in a range of geographic localities across North America and concluded that bone phosphate oxygen isotope values are positively correlated with both relative humidity and the isotopic composition of rainfall. Physiological differences between humans and

other animals must also be considered in relation to models addressing relations between water and food intake and the resulting oxygen isotope composition of bones and teeth (Kohn 1996; Kohn, Schoeninger, and Valley 1996, 1998).

Because climatic change can alter the oxygen isotope composition of precipitation and result in changes in relative humidity, differences in the oxygen isotope ratios of skeletal elements may be influenced by both climatic change and human residence in particular landscapes. Consequently, oxygen isotope studies examining geographic locality should normally be restricted to burial samples from similar temporal periods (Budd et al. 2004).

White and colleagues (1998) employed oxygen isotope analysis of ancient human bone phosphate in Mexico to demonstrate that there was limited movement of individuals between the Tlajinga archaeological site at Teotihuacan (Valley of Mexico) and the Monte Alban site (Valley of Oaxaca). The Tlajinga sites date from ca. A.D. 250 to A.D. 650, whereas the Monte Alban site spans a somewhat longer time period from ca. 500 B.C. to A.D. 1520. The oxygen isotope values for Tlajinga and Monte Alban are completely separated isotopically (Figure 49.5). In addition, there is no temporal variability in oxygen isotope values at any of the sites studied.

Sulphur Isotopes

There have been limited applications of sulphur isotope analysis to archaeological skeletal tissues owing to the time-consuming laboratory preparations and the large sample sizes required.

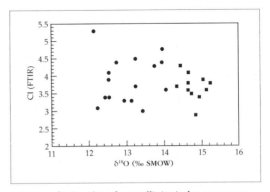

Figure 49.5 Plot of crystallinity index vs. oxygen isotope values of ancient human bone phosphate for individuals excavated at the Tlajinga (solid squares) archaeological site, Teotihuacan (Valley of Mexico) and the Monte Alban (solid circles) site (Valley of Oaxaca), Mexico (adapted from White et al. 1998: 648).

However, recent advances in mass spectrometry allowing simpler preparation methods and relatively small sample sizes have resulted in greater attention to sulphur isotopes in landscape archaeology (Fuller et al. 2003; Richards, Fuller, and Hedges 2001).

Sulphur is introduced to the human body via dietary intake. It is incorporated into bone mineral as calcium sulphate ($CaSO_4$) and into bone collagen as the amino acid methionine. There are four stable isotopes of sulphur (Trust and Fry 1992): ^{32}S (95.02%), ^{34}S (4.21%), ^{33}S (0.75%), and ^{36}S (0.02%). Ratios of the two most abundant isotopes ($^{34}S/^{32}S$) are employed in palaeodietary studies. Like strontium and lead, sulphur isotope values reflect local geology and soil processes, and isotopic ratios in human skeletal tissues record past use of different landscapes.

There are significant differences between the sulphur isotope ratios of plants and animals inhabiting marine vs. freshwater habitats (Hobson 1999; Krouse and Herbert 1988; Peterson and Fry 1987). The soils of coastal regions may have sulphur isotope ratios similar to those of marine water due to the inland movement of sulphur particles by sea spray. In some cases, this sea spray effect extends only a few kilometers inland from the ocean, whereas in other cases large land masses are affected—for example, islands such as New Zealand (Richards et al. 2003: 39). Stable carbon and nitrogen isotope analyses can be employed alongside sulphur isotope analyses to distinguish between proximity to the ocean vs. the consumption of marine foods.

Archaeological applications of sulphur isotopes in bone collagen to address marine vs. terrestrial diet and habitat-use and the identification of immigrants include those of Leach and associates (1996) in the south Pacific and Richards and colleagues (2001) in Europe. The Richards group (2001) reports sulphur isotope values for ancient bone collagen extracted from humans ($n = 23$) and fauna ($n = 4$) excavated from various European archaeological sites. A plot of average sulphur vs. carbon isotope values for five of the regions are presented in Figure 49.6. The coastal Greek sites were excluded from further consideration owing to anomalous C/S ratios that were higher than those of modern collagen. All bone collagen samples from the remaining coastal sites (< 20 km from the sea) showed clear sulphur isotope marine signatures, whereas only one coastal sample had a marine carbon isotope signature. Thus, bone collagen sulphur isotope marine values do not necessarily reflect the consumption of marine protein but do indicate proximity to the sea (Richards et al. 2001: 189). In relation to the inland sites, neither sulphur nor carbon isotope values showed marine signals. In conclusion, inland and coastal sites were clearly distinguished from each other by the bone collagen stable carbon and sulphur isotope values of human and animal inhabitants.

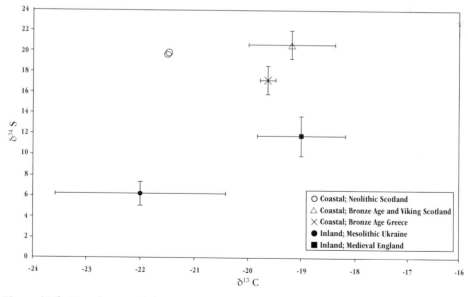

Figure 49.6 Plot of mean sulphur vs. carbon isotope values for ancient bone collagen extracted from humans ($n = 23$) and fauna ($n = 4$) excavated from various European archaeological sites in coastal and inland regions (adapted from Richards et al. 2001: 188).

Metric and Nonmetric Analyses

Biological distance or biodistance involves the measurement and interpretation of relatedness or divergence between populations or subgroups within populations based on analysis of polygenic skeletal and dental traits (Buikstra, Frankenberg, and Konigsberg 1990; Larsen 1997: 302; Pardoe 1994, 1995; Pietrusewsky 2000). Skeletal populations that share attributes are considered to be more closely related than are populations showing many morphological differences. In relation to landscape archaeology, "biodistance analysis has the potential to identify population boundaries, post-marital residence patterns, familial and kin groupings, social groupings, and the presence of individuals from other populations" (Larsen 1997: 304).

Larsen (1997) provides a detailed overview of the application of biodistance research in bioarchaeology. Biodistance analysis is complex, especially in relation to the identification of meaningful patterns of biological variation that distinguish between populations, either in temporal succession or geographic distribution. The skeletal traits employed are influenced by a number of factors including intrinsic genetic, local and general epigenetic, and environmental variables.

There are two classes of biodistance data employed in archaeological skeletal analysis: metric and nonmetric traits. *Metric* traits are continuous variables obtained from linear measurements (for example, length, breadth, radius) that are used to characterize the size and shape of skeletal elements, especially the skull or cranium. In contrast, *nonmetric* traits are discrete or quasi-continuous anatomical features that are recorded along a scale from absence to full expression (for instance, metopic sutures, tooth shoveling). More than 200 cranial nonmetric traits have been identified in humans, and there are approximately 30 standard nonmetric dental traits employed in biodistance analysis.

Multivariate statistical analyses of metric and nonmetric data derived from studies of various skeletal populations distributed across landscapes are employed to examine relationships between past human populations. Digital three-dimensional imaging software has improved the precision of quantitative metric data (Richtsmeier, Cheverud, and Lele 1992)

Because there is a high degree of variation in cranial shape in human populations, craniometric data have played a major role in the identification of cultural-historical and biological relationships (Birdsell 1993; González-José et al. 2003; Howells 1973, 1989, 1995; Macho and Freedman 1987; Macintosh and Larnach 1972; Pietrusewsky 1990, 1994, 2000; Spence 1994; Wright 1992). In addition, variations in tooth size, shape, and occlusal surface complexity have been important in the examination of population structure and history (Brace and Hinton 1981; Brown 1989; Lukacs and Hemphill 1993; Turner 1986, 1987, 1991).

Pietrusewsky (2000) provides an informative case study regarding the application of cranial metric traits to an examination of the population history of the modern and early inhabitants of Japan and the surrounding Australasian and Pacific regions. He employed 29 cranial measurements of 53 comparative cranial series (2,518 male crania) from representative sites. The cranial series include 10 from Japan and 43 from surrounding east Asia, southeast Asia, Australia, and the Pacific. The results of submitting the cranial metric data to stepwise discriminant function analysis and Mahalanobis's generalized distance are reported in Figure 49.7 (Pietrusewsky 2000: 407). Five major constellations of related populations emerged, including (1) Australia and Melanesia, (2) Polynesia and Guam, (3) Southeast Asia, (4) Northern Asia, and (5) the Japanese Archipelago.

Donlon (2000) reviews the use of infracranial (postcranial) nonmetric skeletal variation in the study of population relationships and provides a case study involving population affinities within Australia and between Aboriginal Australians and other major groups. Whereas the utility of cranial nonmetric traits in biodistance studies is well established (Hauser and De Stefano 1989), there has been limited analysis employing postcranial bones. Donlon used a battery of 19 infracranial nonmetric traits on a sample of Holocene male and female adult skeletons whose ancestors originated from five major geographic regions: Australia, Africa, East Asia, Europe, and Polynesia. Population relationships that emerged from the nonmetric analyses are compared with those based on traditional craniometric analyses and genetic markers.

The results of Donlon's nonmetric analyses (Donlon 2000: 360–61) suggest that affiliations of the Aboriginal Australians lie most strongly with the Chinese and the Hawaiians. The close affinity with the Chinese is consistent with dendrograms based on genetic markers and with the model of regional continuity for the origins of the Australians (Kirk 1981; Thorne and Wolpoff 1981). However, the observed heterogeneity in male Aboriginal Australians suggests that there is a significant degree of within-sample variation and little between-sample variation. On the basis of this study, Donlon concluded that infracranial

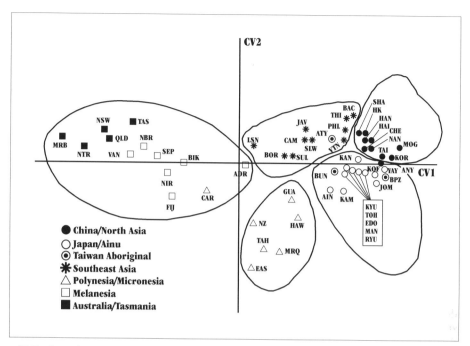

Figure 49.7 Cranial metric data for modern and early inhabitants of Japan and the surrounding Australasian and Pacific regions showing plot of 53 male group means on the first two canonical variates using 29 cranial variates. Five major constellations of related populations emerge including: (1) Australia and Melanesia, (2) Polynesia and Guam, (3) Southeast Asia, (4) Northern Asia, and (5) the Japanese Archipelago (adapted from Pietrusewsky 2000: 407).

nonmetric traits may be most useful in biodistance studies between populations that are closely related, such as groups within a region, rather than unrelated groups, such as those from different continents.

Ancient DNA Analysis

There have been a number of reviews of research involving DNA extracted from ancient human remains (Kaestle and Horsburgh 2002; Kolman and Tuross 2000; O'Rourke, Hayes, and Carlyle 2000; Stone 2000; see Matisoo-Smith, this volume, for a detailed treatment of this subject). Applications involving ancient nuclear and mitochondrial DNA have examined the biological sex of individuals, genetic relationships between individuals within burial populations, origins of migrant individuals and populations, presence of genetic diseases, identification of pathogens in skeletal remains, and phylogenetic relationships between modern and extinct species (for example, Neanderthals and modern humans). Ancient DNA studies addressing past population relationships in various geographic regions of the world (Haak et al. 2005; Hagelberg and Clegg 1993; Izagirre and de la Rúa

1999; Kaestle and Smith 2001; Lalueza-Fox et al. 2004; Wang et al. 2000) have been supplemented by analyses of DNA from modern populations in these regions (Keyeux et al. 2003; Merriwether et al. 1999; Redd and Stoneking 1999; Rubicz et al. 2003; Torroni et al. 1993, 1998; Weiss Bolnick and Smith 2003).

Following initial extractions of human DNA from ancient mummified tissues (Pääbo 1985a, 1985b, 1986), preserved brains recovered from wetland sites (Doran et al. 1986; Pääbo 1986; Pääbo, Gifford, and Wilson 1988), bones (Hagelberg, Sykes, and Hedges 1989), and teeth (Ginther, Issel-Tarver, and King 1992) in the 1980s and early 1990s, there have been an extensive number of applications of ancient DNA data to studies of past human population dynamics. Ancient DNA research was bolstered by the development of PCR (polymerase chain reaction), which allowed the replication of millions of copies of DNA target sequences from extracted DNA fragments.

Major limitations associated with ancient DNA research include preservation of nucleic acids in various post-mortem environments, difficulties associated with amplification of extracted samples, and concerns regarding authenticity of extracted samples (Cooper et al. 2001; O'Rourke, Hayes, and Carlyle 2000: 218). In post-mortem environments, nucleic acids are degraded and modified in various ways (Höss et al. 1996). In general, environments with cooler temperatures, neutral or slightly alkaline pH and dry conditions are the best for DNA preservation, although frozen human remains and remains recovered from wet anoxic conditions have also yielded DNA (Stone 2000: 353). In addition, contamination of samples in the field and laboratory can introduce foreign DNA (Yang and Watt 2005). A number of criteria have been adopted in relation to the authentication of DNA from ancient and forensic human remains (Poinar 2003; Richards, Sykes, and Hedges et al. 1995; Yang, Golenberg, and Shoshani 1997). The authenticity of supposed 60,000-year-old mitochondrial DNA extracted from the Lake Mungo 3 skeleton from southeastern Australia (Adcock et al. 2001) has been challenged by a number of research groups (see Cooper et al. 2001).

Research by Rubicz and associates (2003) provides an example of an application of DNA analysis to further elucidate past human population origins and movements. Mitochondrial DNA was extracted from cheek cells taken from 179 Aleuts living in five small Alaskan communities and in one city (Anchorage). Results indicated that Aleut mtDNA belonged to two of the four haplogroups (A and D) common among Native Americans. Within the Aleut sample, haplogroup D occurred at a very high frequency and unique HVS-I sequences were identified. These findings distinguished Aleuts from Eskimos, Athapaskan Indians, and other northern Amerindian populations. An R-matrix plot of the mtDNA data (Figure 49.8) suggested that Aleuts were most closely related genetically to Chuckchi and Siberian Eskimos rather than to Native Americans and Kamchatkan populations (Koryaks and Itel'men). The mtDNA data support the hypothesis that ancestral Aleuts crossed the Bering Land Bridge and entered the Aleutian Islands from the east, rather than island hopping from Kamchatka into the western Aleutians.

Palaeopathology

The human skeleton records a range of information about stress experienced during the life of an individual (Aufderheide and Rodriguez-Martin 1998; Hillson 2000; Larsen 1997; Ortner 2002; Ortner and Putschar 1985; Steinbock 1976). Physiological disruptions caused by infections or nutritional deficiencies can result in abnormal matrix loss or accumulation in bones and teeth. Osteoarthritis and trauma associated with lifestyle variables, such as workload and violence, also leave characteristic lesions on the skeleton. The general health of individuals as determined by palaeopathological analysis of skeletal remains may be employed as an additional line of evidence regarding human relationships with past landscapes.

For example, the increased physical labor/biomechanical stress, increased population densities, and reduced dietary breadth associated with sedentary hunter-gatherer and agricultural lifeways are reflected in the skeletons of these societies. In general, palaeopathological analyses indicate the presence of chronic stress and reduced health status in sedentary populations when compared with mobile and semisedentary hunter-gatherers (Cohen and Armelagos 1984a, 1984b; Larsen 1995; Milner, Wood, and Boldsen 2000; Ortner and Aufderheide 1991; Verano and Ubelaker 1992; Walker and Lambert 1989; Webb 1995).

To provide a framework for assessing the causes of various skeletal pathologies, Goodman and associates (1984b) assign stress indicators to three major categories: (1) indicators of general, cumulative stress; (2) indicators of general, episodic stress; and (3) indicators of stress associated with specific diseases or conditions. General indicators refer to nonspecific responses to various perturbations, cumulative indicators provide a summation of stress experienced over long periods of time, and episodic indicators are confined to a more restricted time period. Specific indicators can be used to identify particular diseases or conditions.

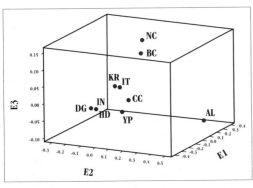

Figure 49.8 R-matrix plot based on mitochondrial DNA halotype frequencies for cheek cells taken from 179 modern Aleuts living in five small Alaskan communities and in one city (Anchorage). Al = Aleut, BC = Bella Coola, CC = Chukchi, DG = Dogrib, HD = Haida, IN = Inuit, IT = Itel'men, KR = Koryak, NC = Nuu-Chah-Nulth, YP = Siberian Yupik Eskimo (adapted from Rubicz et al. 2003: 822).

Webb's (1995) regional overview of palaeopathology in hunter-gatherer Australia and horticultural New Guinea provides a case study in relation to geographic variability in health patterns (Pate 1996). According to the pathological indicators employed in the study, Holocene Aboriginal populations that occupied the arid continental interior were generally the healthiest, whereas those from the Central Murray River region were the least healthy. Arid-land inhabitants showed the lowest incidence of cribra orbitalia (anemia), dental enamel hypoplasia (growth disruption in teeth), and Harris lines (growth disruption in long bones). Thus, chronic malnutrition, parasitic infection, and population aggregation would have been minimal in the arid zone. The occurrence of Harris lines in arid populations suggests the presence of acute stress such as short-term food shortages.

In contrast, the Central Murray River region shows a high incidence of cribra orbitalia and dental enamel hypoplasia in children and adults, but a low incidence of Harris lines. This pathological profile indicates that Central Murray populations were subject to chronic stress. Webb argues that Harris line formation was most likely suppressed by this persistent stress. In addition, high frequencies of dental caries and dental calculus suggest the use of carbohydrate staples (for example, *Typha* root). After considering the archaeological evidence relating to the dense occurrence of large oven mounds in the Central Murray, Webb suggests that the health pattern for this region is indicative of large, sedentary popu-lations with high frequencies of malnutrition, parasitism, nonspecific infection, and endemic (nonvenereal) trepanemotoses.

Papua New Guinea and the eastern and southwestern Australian coasts show similar pathological patterns to the Central Murray. However, the stress levels appear to be less severe in the coastal samples. The major health difference between the horticultural Papua New Guinean and the Australian hunter-gatherer samples is the higher incidence of infectious disease among the former. According to Webb, the large villages of closely spaced permanent dwellings found in Papua New Guinea would have facilitated the transmission of infectious diseases such as endemic trepanemotoses. In contrast, results from the Australian southern coast indicate the presence of regular acute stress, such as seasonal food shortages. The southern coast had the highest percentage of subadults displaying Harris lines and the highest incidence of multiple lines. The regular spacing of the Harris lines suggests periodic stress.

In relation to biomechanical stress, osteoarthritis among Aboriginal groups affected males more than females and occurred in elbows more than knees. The right elbow was the predominant site of upper limb involvement. The Central Murray and desert populations showed the highest frequencies of arthritis. Webb argues that osteoarthritis in these hunter-gatherer populations resulted from physically hard and repetitive activities, with upper limbs bearing the majority of the stress. Long-term and regular use of tools such as spearthrowers, digging sticks, and grinding stones would produce these arthritic patterns. Obtaining adequate food supplies in the densely populated Central Murray and the marginal desert areas would have required a high frequency of these intense, repetitive activities.

In relation to violence, severe cranial trauma was common throughout Holocene Australia, and females generally displayed more fractures than did males. The most common cranial injury was the depressed fracture. The high incidence of these severe head injuries suggests that aggressive behavior was common in most Australian hunter-gatherer societies. In contrast, postcranial fractures were relatively rare. However, 76% of those observed occurred in the upper limbs and on the left side of the body. Most of these were defensive parrying fractures to the ulna. Thus, the types of fractures observed in post-cranial remains also indicate aggressive behavior.

Webb's palaeopathological study provides further evidence that in some cases sedentary hunter-gatherers had health patterns similar to those of settled agriculturalists. The primary limitation of such studies in Australia is the lack of large, well-dated skeletal samples representative of a range of geographic regions.

Conclusions

Archaeological research addressing past landscape use by human populations will benefit significantly from the employment of a range of analytical methods involving human skeletal remains. The reliability of inferences regarding past cultural landscapes is improved by the use of a combination of these bioarchaeological methods (Verano and DeNiro 1993). Bioarchaeology can supplement and extend data obtained from conventional archaeological methods. Furthermore, studies of past population movements and social relationships will benefit from multidisciplinary approaches, including archaeological, linguistic, anthropological, historical, genetic, and ecological data (Hurles et al. 2003; Kirk 1981).

References

Adcock, G. J., Dennis, E. S., Easteal, S., Huttley, G. A., Jermiin, L. S., Peacock, W. J., and Thorne, A. 2001. Mitochondrial DNA sequences in ancient Australians: Implications for modern human origins. *Proceedings of the National Academy of Sciences* 98: 537–42.

Ambrose, S. H. 1991. Effects of diet, climate and physiology on nitrogen isotope abundances in terrestrial foodwebs. *Journal of Archaeological Science* 18: 293–318.

Anson, T. J. 1997. The effect of climate on stable nitrogen isotope enrichment in modern South Australian mammals. Unpublished M.A. thesis, Department of Archaeology, Flinders University, Adelaide.

Aufderheide, A. C., and Rodriguez-Martin, C. 1998. *The Cambridge Encyclopedia of Human Paleopathology*. Cambridge: Cambridge University Press.

Ayliffe, L. K., and Chivas, A. 1990. Oxygen isotope composition of the bone phosphate of Australian kangaroos: Potential as a palaeoenvironmental recorder. *Geochimica et Cosmochimica Acta* 54: 2603–09.

Beard, B. L., and Johnson, C. M. 2000. Strontium isotope composition of skeletal material can determine the birth place and geographic mobility of humans and animals. *Journal of Forensic Sciences* 45: 1049–61.

Beck, L. (ed.). 1995. *Regional Approaches to Mortuary Analysis*. New York: Plenum Press.

Bird, B. F., and Monahan, C. M. 1995. Death, mortuary ritual, and Natufian social structure. *Journal of Anthropological Archaeology* 14: 251–87.

Birdsell, J. B. 1993. *Microevolutionary Patterns in Aboriginal Australia*. Oxford: Oxford University Press.

Brace, C. L., and Hinton, R. J. 1981. Oceanic tooth-size variation as a reflection of biological and cultural mixing. *Current Anthropology* 22: 549–69.

Brown, P. 1989. *Coobool Creek: A Morphological and Metrical Analysis of the Crania, Mandibles and Dentitions of a Prehistoric Australian Human Population*. Terra Australis 13. Canberra: Department of Prehistory, Research School of Pacific Studies, Australian National University.

Budd, P., Chenery, C., Montgomery, J., and Evans, J. 2003. You are where you eat: Isotopic analysis in the reconstruction of prehistoric residency, in M. Parker Pearson (ed.), *Food and Identity*, pp. 69–78. Oxford: Archaeopress. British Archaeological Reports International Series 1117.

Budd, P., Gulson, B. L., Montgomery, J., Rainbird, P., Thomas, R. G., and Young, S. M. M. 1999. The use of Pb- and Sr-isotopes for the study of Pacific islander population dynamics, in J.-C. Galipaud and I. Lilley (eds.), *The Pacific from 5,000 to 2,000 B.P.: Colonisation and Transformations*, pp. 301–11. Paris: Institut de Recherche pour le Developpement.

Budd, P., Millard, A., Chenery, C., Lucy, S., and Roberts, C. 2004. Investigating population movement by stable isotope analysis: A report from Britain. *Antiquity* 78: 127–42.

Budd, P., Montgomery, J., Barreiro, B., and Thomas, R. G. 2000a. Differential diagenesis of strontium in archaeological dental tissues. *Applied Geochemistry* 15: 687–94.

Budd, P., Montgomery, J., Cox, A., Krause, P., Barreiro, B., and Thomas, R. G. 1998. The distribution of lead in ancient and modern human teeth: Implications for long-term and historical exposure monitoring. *The Science of the Total Environment* 220: 121–36.

Budd, P., Montgomery, J., Evans, J., and Barreiro, B. 2000b. Human tooth enamel as a record of the comparative lead exposure of prehistoric and modern people. *The Science of the Total Environment* 263: 1–10.

Budd, P., Montgomery, J., Evans, J., and Chenery, C. 2001. Combined Pb-, Sr- and O-isotope analysis of human dental tissue for the reconstruction of archaeological residential mobility, in J. G. Holland and S. D. Tanner (eds.), *Plasma Source Mass Spectrometry: The New Millennium*, pp. 311–26. Cambridge: Royal Society of Chemistry. *Special Publication* 267.

Budd, P., Montgomery, J., Evans, J., Chenery, C., and Powlesland, D. 2002. Reconstructing Anglo-Saxon immigration and residential mobility from O-, Sr- and Pb-isotope analysis. *Goldschmidt Conference Abstracts 2002*, p. A109.

Buikstra, J. E., Frankenberg, S. R., and Konigsberg, L. W. 1990. Skeletal biological distance studies in American physical anthropology: Recent trends. *American Journal of Physical Anthropology* 82: 1–7.

Buikstra, J. E., Konigsberg, L. W., and Bullington, J. 1986. Fertility and the development of agriculture in the prehistoric Midwest. *American Antiquity* 51: 528–46.

Burton, J. H., Price, T. D., Cahue, L., and Wright, L. E. 2003. The use of barium and strontium abundances in human skeletal tissues to determine their geographic origins. *International Journal of Osteoarchaeology* 13: 88–95.

Chapman, R., Kinnes, I., and Randsbor, K. (eds.). 1981. *The Archaeology of Death*. Cambridge: Cambridge University Press.

Chatters, J. C. 1987. Hunter-gatherer adaptations and assemblage structure. *Journal of Anthropological Archaeology* 6: 336–75.

Cohen, M. N., and Armelagos, G. J. (eds.). 1984a. *Paleopathology at the Origins of Agriculture*. Orlando, FL: Academic Press.

Cohen, M. N., and Armelagos, G. J. (eds.). 1984b. Disease and death at Dr. Dickson's Mounds. *Natural History* 9: 12–18.

Coltrain, J. B., Hayes, M. G., and O'Rourke, D. H. 2004. Sealing, whaling and caribou: The skeletal isotope chemistry of Eastern Arctic foragers. *Journal of Archaeological Science* 31: 39–57.

Cooper, A., Rambaut, A., Macaulay, V., Willerslev, E., Hansen, A. J., and Stringer, C. 2001. Human origins and ancient human DNA. *Science* 292: 1655–56.

Donlon, D. 2000. The value of infracranial nonmetric variation in studies of modern *Homo sapiens*: An Australian focus. *American Journal of Physical Anthropology* 113: 349–68.

Doran, G. H., Dickel, D. N., Ballinger, W. E., Agee, O. F., Laipis, P. J., and Hauswirth, W. W. 1986. Anatomical, cellular and molecular analysis of 8,000-year-old human brain tissue from the Windover archaeological site. *Nature* 323: 803–06.

Ericson, J. E. 1985. Strontium isotope characterization in the study of prehistoric human ecology. *Journal of Human Evolution* 14: 503–14.

———. 1989. Some problems and potentials of strontium isotope analysis for human and animal ecology. In P. W. Rundel, J. R. Ehleringer, and K. A. Nagy (eds.), *Stable Isotopes in Ecological Research*, pp. 252–59. New York: Springer-Verlag.

Ezzo, J. A., Johnson, C. M., and Price, T. D. 1997. Analytical perspective on prehistoric migration: A case study from east-central Arizona. *Journal of Archaeological Science* 24: 447–66.

Fernandez, J., Panarello, H. O., and Schobinger, J. 1999. The Inka mummy from Mount Aconcagua: Decoding the geographic origin of the "Messenger to the Deities" by means of stable carbon, nitrogen and sulfur isotope analysis. *Geoarchaeology* 14: 27–46.

Fricke, H. C., Clyde, W. C., and O'Neil, J. R. 1998. Intra-tooth variations in delta O-18 (PO_4) of mammalian tooth enamel as a record of seasonal variation in continental climate variables. *Geochimica et Cosmochimica Acta* 62: 1839–50.

Gat, J. R. 1980. The isotopes of oxygen and hydrogen in precipitation, in P. Fritz and J. Ch. Fontes (eds.), *Handbook of Environmental Isotope Geochemistry*, pp. 21–47. Amsterdam: Elsevier.

Ginther, C., Issel-Tarver, L., and King, M.-C. 1992. Identifying individuals by sequencing mitochondrial DNA from teeth. *Nature Genetics* 2: 135–38.

Glimcher, M. J. 1976. Composition, structure, and organization of bone and other mineralized tissues and the mechanism of calcification, in *Handbook of Physiology, Endocrinology*. Baltimore: Williams and Wilkins.

González-José, R., González-Martin, A., Hernández, M., Pucciarelli, H. M., Sardi, M., Rosales, A., and van der Molen, S. 2003. Craniometric evidence for Palaeoamerican survival in Baja California. *Nature* 425: 62–65.

Goodman, A. H. 1993. On the interpretation of health from skeletal remains. *Current Anthropology* 34: 281–88.

Goodman, A. H., Lallo, J., Armelagos, G. J., and Rose, J. C. 1984a. Health changes at Dickson Mounds, Illinois (A.D. 950–1300), in M. N. Cohen and G. J. Armelagos (eds.), *Palaeopathology at the Origins of Agriculture*, pp. 271–305. New York: Academic Press.

———. 1984b. Indicators of stress from bone and teeth, in M. N. Cohen and G. J. Armelagos (eds.), *Palaeopathology at the Origins of Agriculture*, pp. 13–49. New York: Academic Press.

Gulson, B. L., Jameson, C. W., and Gillings, B. R. 1997. Stable lead isotopes in teeth as indicators of past domicile: A potential new tool in forensic science? *Journal of Forensic Sciences* 42: 787–91.

Haak, W., Forster, P., Bramanti, B., Matsumura, S., Brandt, G., Tänzer, M., Villems, R., Renfrew, C., Gronenborn, D., Werner Alt, K., and Burger, J. 2005. Ancient DNA from the first European farmers in 7,500-year-old Neolithic sites. *Science* 310: 1016–18.

Hagelberg, E., and Clegg, J. B. 1993. Genetic polymorphisms in prehistoric Pacific Islanders determined by analysis of ancient bone DNA. *Proceedings of the Royal Society of London Biological Sciences* 252: 163–70.

Hagelberg, E., Sykes, B., and Hedges, R. 1989. Ancient bone DNA amplified. *Nature* 342: 485.

Hall-Martin, A. J., van der Merwe, N. J., Lee-Thorp, J. A., Armstrong, R. A., Mehl, C. H., Struben, S., and Tykot, R. 1991. Determination of species and geographic origin of rhinoceros horn by isotopic analysis and its possible application to trade control, in *International Rhino Conference: Rhinoceros Biology and Conservation*, pp. 123–35. San Diego.

Hauser, G., and De Stefano, G. F. 1989. *Epigenetic Variants of the Human Skull*. Stuttgart: Nägele und Obermiller.

Hillson, S. W. 1996. *Dental Anthropology*. Cambridge: Cambridge University Press.

———. 2000. Dental pathology, in M. A. Katzenberg and S. R. Saunders (eds.), *Biological Anthropology of the Human Skeleton*, pp. 249–86. New York: Wiley.

Hobson, K. A. 1999. Tracing origins and migration of wildlife using stable isotopes: A review. *Oecologia* 120: 314–26.

Hodell, D. A., Quinn, R. L., Brenner, M., and Kamenov, G. 2004. Spatial variation of strontium isotopes ($^{87}Sr/^{86}Sr$) in the Maya region: A tool for tracking ancient human migration. *Journal of Archaeological Science* 31: 585–601.

Hoppe, K. A., Koch, P. L., Carlson, R. W., and Webb, S. D. 1999. Tracking mammoths and mastodons: Reconstruction of migratory behavior using strontium isotope ratios. *Geology* 27: 439–42.

Höss, M., Jaruga, P., Zastawny, T. H., Dizdaroglu, M., and Pääbo, S. 1996. DNA damage and DNA sequence retrieval from ancient tissues. *Nucleic Acids Research* 24: 1304–1307.

Howells, W. W. 1973. Cranial variation in man: A study by multivariate analysis of patterns of difference among human populations. Papers of the Peabody Museum of Archaeology and Ethnography 67. Cambridge: Harvard University.

———. 1989. Skull shapes and the map: Craniometric analysis in the dispersion of modern *Homo*. Papers of the Peabody Museum of Archaeology and Ethnography 79. Cambridge: Harvard University.

———. 1995. Who's who in skulls: Ethnic identification of crania from measurements. Papers of the Peabody Museum of Archaeology and Ethnology 82. Cambridge: Harvard University.

Hurles, M. E., Matisoo-Smith, E., Gray, R. D., and Penny, D. 2003. Untangling Oceanic settlement: The edge of the knowable. *Trends in Ecology and Evolution* 18: 531–40.

Izagirre, N., and de la Rúa, C. 1999. An mtDNA analysis in ancient Basque populations: Implications for haplogroup V as a marker for a major Paleolithic expansion from southwestern Europe. *American Journal of Human Genetics* 65: 199–207.

Kaestle, F. A., and Smith, D. G. 2001. Ancient mitochondrial DNA evidence for prehistoric population movement: The Numic expansion. *American Journal of Physical Anthropology* 115: 1–12.

Kaestle, F. A., and Horsburgh, K. A. 2002. Ancient DNA in anthropology: Methods, applications and ethics. *Yearbook of Physical Anthropology* 45: 92–130.

Katzenberg, M. A. 2000. Stable isotope analysis: A tool for studying past diet, demography and life history, in M. A. Katzenberg and S. R. Saunders (eds.), *Biological Anthropology of the Human Skeleton*, pp. 305–27. New York: Wiley.

Katzenberg, M. A., and Harrison, R. G. 1997. What's in a bone? Recent advances in archaeological bone chemistry. *Journal of Archaeological Research* 5: 265–93.

Katzenberg, M. A., and Krouse, H. R. 1989. Application of stable isotope variation in human tissues to problems in identification. *Canadian Society of Forensic Science Journal* 22: 7–19.

Katzenberg, M. A., and Weber, A. 1999. Stable isotope ecology and paleodiet in the Lake Baikal region of Siberia. *Journal of Archaeological Science* 26: 651–59.

Keegan, W. F., and DeNiro, M. J. 1988. Stable carbon and nitrogen isotope ratios of bone collagen used to study coral-reef and terrestrial components of prehistoric Bahamian diet. *American Antiquity* 53: 320–36.

Keyeux, G., Rodas, C., Gelvez, N., and Carter, D. 2003. Possible migration routes into South America deduced from mitochondrial DNA studies in Columbian Amerindian populations. *Human Biology* 74: 211–33.

Kirk, R. L. 1981. *Aboriginal Man Adapting: The Human Biology of Australian Aborigines*. Oxford: Oxford University Press.

Knudson, K. J., Price, T. D., Buikstra, J. E., and Blom, D. E. 2004. The use of strontium isotope analysis to investigate Tiwanaku migration and mortuary ritual in Bolivia and Peru. *Archaeometry* 46: 5–18.

Kohn, M. J. 1996. Predicting animal $\delta^{18}O$: Accounting for diet and physiological adaptation. *Geochimica et Cosmochimica Acta* 60: 4811–29.

Kohn, M. J., Miselis, J. L., and Fremd, T. J. 2003. Oxygen isotope evidence for progressive uplift of the Cascade Range, Oregon. *Earth and Planetary Science Letters* 204: 151–65.

Kohn, M. J., Schoeninger, M. J., and Valley, J. W. 1996. Herbivore tooth oxygen isotope compositions: Effects of diet and physiology. *Geochimica et Cosmochimica Acta* 60: 3889–96.

———. 1998. Variability in herbivore tooth oxygen isotope compositions: Reflections of seasonality or developmental physiology. *Chemical Geology* 152: 97–112.

Kohn, M. J., Schoeninger, M. J., and Barker, W. W. 1999. Altered states: Effects of diagenesis on fossil tooth chemistry. *Geochimica et Cosmochimica Acta* 63: 2737–47.

Kolman, C. J., and Tuross, N. 2000. Ancient DNA analysis of human populations. *American Journal of Physical Anthropology* 111: 5–23.

Krouse, H. R., and Herbert, H. K. 1988. Sulphur and carbon isotope studies of food webs, in B. V. Kennedy and G. M. LeMoine (eds.), *Diet and Subsistence: Current Archaeological Perspectives*, pp. 315–22. Calgary: University of Calgary Archaeology Association.

Lalueza-Fox, C., Sampietro, M. L., Gilbert, M. T. P., Castri, L., Facchini, F., Pettener, D., and Bertranpetit, J. 2004. Unravelling migrations in the steppe: Mitochondrial DNA sequences from ancient Central Asians. *Proceedings of the Royal Society of London B* 271: 941–47.

Larsen, C. S. 1987. Bioarchaeological interpretations of subsistence economy and behavior from human skeletal remains, in M. B. Schiffer (ed.), *Advances in Archaeological Method and Theory*, pp. 339–445. San Diego: Academic Press.

———. 1995. Biological changes in human populations with agriculture. *Annual Review of Anthropology* 24: 185–213.

———. 1997. *Bioarchaeology: Interpreting Behavior from the Human Skeleton*. Cambridge: Cambridge University Press.

Larsen, C. S., Schoeninger, M. J., van der Merwe, N. J., Moore, K. M., and Lee-Thorp, J. A. 1992. Carbon and nitrogen stable isotopic signatures of human dietary change in the Georgia Bight. *American Journal of Physical Anthropology* 89: 197–214.

Leach, B. F., Quinn, C. J., and Lyon, G. L. 1996. A stochastic approach to the reconstruction of prehistoric human diet in the Pacific region from bone isotope signatures. *Tahingu: Records of the Museum of New Zealand Te Papa Tongarewa* 8: 1–54.

Libby, W. F., Berger, R., Mead, J. F., Alexander, G. V., and Ross, J. F. 1964. Replacement rates for human tissue from atmospheric radiocarbon. *Science* 146: 1170–72.

Longinelli, A. 1984. Oxygen isotopes in mammal bone phosphate: A new tool for palaeohydrological and palaeoclimatological research? *Geochimica et Cosmochimica Acta* 48: 385–90.

Lukacs, J. R., and Hemphill, B. E. 1993. Odontometry and biologic affinities among south Asians: An analysis of three ethnic groups from northwest India. *Human Biology* 65: 279–325.

Luz, B., Kolodny, Y., and Horowitz, M. 1984. Fractionation of oxygen isotopes between mammalian bone phosphate and environmental drinking water. *Geochimica et Cosmochimica Acta* 48: 1689–93.

Luz, B., Cormie, A. B., and Schwarcz, H. P. 1990. Oxygen isotope variations in phosphate of deer bones. *Geochimica et Cosmochimica Acta* 54: 1723–28.

Macho, G., and Freedman, L. 1987. *A Re-Analysis of the Andrew A. Abbie Morphometric Data on Australian Aborigines.* Canberra: Australian Institute of Aboriginal Studies. *Occasional Papers in Human Biology* 4: 1–80.

Macintosh, N. W. G., and Larnach, S. L. 1972. The persistence of *Homo erectus* traits in Australian Aboriginal crania. *Archaeology and Physical Anthropology in Oceania* 7: 1–7.

Macko, S. A., Engel, M. H., Andrusevich, V., Lubec, G., O'Connell, T. C., and Hedges, R. 1999. Documenting the diet in ancient human populations through stable isotope analysis of hair. *Philosophical Transactions of the Royal Society of London, Series B* 354: 65–76.

Mann, S. 2001. *Biomineralisation: Principles and Concepts in Bioinorganic Materials Chemistry.* Oxford: Oxford University Press.

Marshall, J. H., Liniecki, J., Lloyd, E. L., Marotti, G., Mays, C. W., Rundo, J., Sissons, H. A., and Snyder, W. S. 1973. Alkaline earth metabolism in adult man. *Health Physics* 24: 125–221.

Merriwether, D. A., Friedlaender, F., Mediavilla, J., Mgone, C., Gentz, F., and Ferrell, R. E. 1999. Mitochondrial DNA variation as an indicator of Austronesian influence in island Melanesia. *American Journal of Physical Anthropology* 110: 243–70.

Milner, G. R., Wood, J. W., and Boldsen, J. L. 2000. Palaeodemography, in M. A. Katzenberg and S. R. Saunders (eds.), *Biological Anthropology of the Human Skeleton*, pp. 467–97.

Montgomery, J., Budd, P., Cox, A., Krause, P., and Thomas, R. G. 1999. LA-ICP-MS evidence for the distribution of lead and strontium in Romano-British, medieval and modern human teeth: Implications for life history and exposure reconstruction, in S. M. M. Young, A. M. Pollard, P. Budd, and R. A. Ixer (eds.), *Metals in Antiquity*, pp. 258–61. British Archaeological Reports International Series 792. Oxford: Archaeopress.

Montgomery, J., Budd, P., and Evans, J. 2000. Reconstructing the lifetime movements of ancient people: A Neolithic case study from southern England. *European Journal of Archaeology* 3: 370–85.

Montgomery, J., Evans, J., and Neighbour, T. 2003. Sr isotope evidence for population movement within the Hebrides Norse community of NW Scotland. *Journal of the Geological Society, London* 160: 1.

Müller, W., Fricke, H., Halliday, A. N., McCulloch, M. T., and Wartho, J.-A. 2003. Origin and migration of the alpine Iceman. *Science* 302: 862–65.

Nelson, B. K., DeNiro, M. J., Schoeninger, M. J., DePaolo, D. J., and Hare, P. E. 1986. Effects of diagenesis on strontium, carbon, nitrogen, and oxygen concentration and isotopic composition of bone. *Geochimica et Cosmochimica Acta* 50: 1941–49.

Neuman, W. F., and Neuman, M. W. 1958. *The Chemical Dynamics of Bone Mineral.* Chicago: University of Chicago Press.

O'Leary, M. H. 1988. Carbon isotopes in photosynthesis. *BioScience* 38: 328–36.

O'Rourke, D. H., Hayes, M. G., and Carlyle, S. W. 2000. Ancient DNA studies in physical anthropology. *Annual Review of Anthropology* 29: 217–42.

Ortner, D. J. 2002. *Identification of Pathological Conditions in Human Skeletal Remains* (2nd ed.). New York: Academic Press.

Ortner, D. J., and Aufderheide, A. C. (eds.). 1991. *Human Paleopathology: Current Syntheses and Future Options.* Washington, DC: Smithsonian Institution Press.

Ortner, D. J., and Putschar, W. G. J. 1985. *Identification of Pathological Conditions in Human Skeletal Remains.* Washington, DC: Smithsonian Institution Press.

O'Shea, J. M. 1984. *Mortuary Variability: An Archaeological Investigation.* New York: Academic Press.

Pääbo, S. 1985a. Molecular cloning of ancient Egyptian mummy DNA. *Nature* 314: 644–45.

———. 1985b. Preservation of DNA in ancient Egyptian mummies. *Journal of Archaeological Science* 12: 411–17.

Pääbo, S. 1986. Molecular genetic investigations of ancient human remains. *Cold Springs Harbor Symposium in Quantitative Biology* LI: 441–46.

Pääbo, S., Gifford, J. A., and Wilson, A. C. 1988. Mitochondrial DNA sequences from a 7,000-year-old brain. *Nucleic Acids Research* 16: 9775–87.

Pardoe, C. 1988. The cemetery as symbol: The distribution of Aboriginal burial grounds in southeastern Australia. *Archaeology in Oceania* 23: 1–16.

———. 1994. Bioscapes: The evolutionary landscape of Australia. *Archaeology in Oceania* 29: 182–90.

———. 1995. Riverine, biological and cultural evolution in southeastern Australia. *Antiquity* 69: 696–713.

Pate, F. D. 1994. Bone chemistry and paleodiet. *Journal of Archaeological Method and Theory* 1: 161–209.

———. 1995. Stable carbon isotope assessment of hunter-gatherer mobility in prehistoric South Australia. *Journal of Archaeological Science* 22: 81–87.

———. 1996. Review of S. Webb, *Palaeopathology of Aboriginal Australians: Health and Disease across a Hunter-Gatherer Continent*. Cambridge University Press, 1995. *Australian Archaeology* 43: 54–56.

———. 1998a. Bone collagen stable nitrogen and carbon isotopes as indicators of prehistoric diet and landscape use in southeastern South Australia. *Australian Archaeology* 46: 23–29.

———. 1998b. Stable carbon and nitrogen isotope evidence for prehistoric hunter-gatherer diet in the lower Murray River Basin, South Australia. *Archaeology in Oceania* 33: 92–99.

———. 2000. Bone chemistry and palaeodiet: Bioarchaeological research at Roonka Flat, lower Murray River, South Australia 1983–1999. *Australian Archaeology* 50: 67–74.

———. 2006. Hunter-gatherer social complexity at Roonka, South Australia, in B. David, B. Barker, and I. J. McNiven (eds.), *The Social Archaeology of Australian Indigenous Societies*, pp. 226–41. Canberra: Aboriginal Studies Press.

Pate, F. D., and Anson, T. J. 2008. Stable nitrogen isotope values in arid-land kangaroos correlated with mean annual rainfall: Potential as a palaeoclimatic indicator. *International Journal of Osteoarchaeology* 18: 317–26.

Pate, F. D., Anson, T. J., Schoeninger, M. J., and Noble, A. H. 1998. Bone collagen stable carbon and nitrogen isotope variability in modern South Australian mammals: A baseline for palaeoecological inferences. *Quaternary Australasia* 16: 43–51.

Pate, F. D., Brodie, R., and Owen, T. D. 2002. Determination of geographic origin of unprovenanced Aboriginal skeletal remains in South Australia employing stable isotope analysis. *Australian Archaeology* 55: 1–7.

Pate, F. D., and Hutton, J. T. 1988. The use of soil chemistry data to address postmortem diagenesis in bone mineral. *Journal of Archaeological Science* 15: 729–39.

Pate, F. D., Hutton, J. T., Gould, R. A., and Pretty, G. L. 1991. Alterations of *in vivo* elemental dietary signatures in archaeological bone: Evidence from the Roonka Flat Dune, South Australia. *Archaeology in Oceania* 26: 58–69.

Pate, F. D., Hutton, J. T., and Norrish, K. 1989. Ionic exchange between soil solution and bone: Toward a predictive model. *Applied Geochemistry* 4: 303–16.

Pate, F. D., and Noble, A. H. 2000. Geographic distribution of C^3 and C^4 grasses recorded in bone collagen stable carbon isotope values of South Australian herbivores. *Australian Journal of Botany* 48: 203–07.

Peterson, B. J., and Fry, B. 1987. Stable isotopes in ecosystem studies. *Annual Review of Ecology and Systematics* 18: 293–320.

Pietrusewsky, M. 1990. Craniofacial variation in Australasian and Pacific populations. *American Journal of Physical Anthropology* 82: 319–40.

———. 1994. Pacific-Asian relationships: A physical anthropological perspective. *Oceanic Linguistics* 33: 407–29.

———. 2000. Metric analysis of skeletal remains: Methods and applications, in M. A. Katzenberg and S. R. Saunders (eds.), *Biological Anthropology of the Human Skeleton*, pp. 375–415. New York: Wiley.

Poinar, H. N. 2003. The top 10 list: Criteria of authenticity for DNA from ancient and forensic samples. *International Congress Series* 1239: 575–79.

Price, T. D., and Brown, J. A. (eds.). 1985. *Prehistoric Hunter-Gatherers: The Emergence of Cultural Complexity*. New York: Academic Press.

Price, T. D., Burton, J. H., and Bentley, R. A. 2002. The characterization of biologically available strontium isotope ratios for the study of prehistoric migration. *Archaeometry* 44: 117–35.

Price, T. D., Grupe, G., and Schroter, P. 1994. Reconstruction of migration patterns in the Bell Beaker period by stable strontium isotope analysis. *Applied Geochemistry* 9: 413–17.

Price, T. D., Johnson, C. M., Ezzo, J. A., Ericson, J., and Burton, J. 1994. Residential mobility in the prehistoric southwest United States: A preliminary study using strontium isotope analysis. *Journal of Archaeological Science* 21: 315–30.

Price, T. D., Manzanilla, L., and Middleton, W. D. 2000. Immigration and the ancient city of Teotihuacan in Mexico: A study using strontium isotope ratios in human bone. *Journal of Archaeological Science* 27: 903–13.

Rabinowitz, M. B., Bellinger, D., Leviton, A., and Wang, J. D. 1991. Lead levels among various deciduous tooth types. *Bulletin of Environmental Contamination and Toxicology* 47: 602–08.

Redd, A. J., and Stoneking, M. 1999. Peopling of Sahul: Mitochondrial DNA variation in Australian aboriginal and Papua New Guinean populations. *American Journal of Human Genetics* 65: 808–28.

Richards, M. P., Fuller, B. T., and Hedges, R. 2001. Sulphur isotopic variation in ancient bone collagen from Europe: Implications for human palaeodiet, residence mobility, and modern pollutant studies. *Earth and Planetary Science Letters* 191: 185–90.

Richards, M. P., Fuller, B. T., Sponheimer, M., Robinson, T., and Ayliffe, L. 2003. Sulphur isotopes in palaeodietary studies: A review and results from a controlled feeding experiment. *International Journal of Osteoarchaeology* 13: 37–45.

Richards, M. P., Sykes, B., and Hedges, R. 1995. Authenticating DNA extracted from ancient skeletal remains. *Journal of Archaeological Science* 22: 291–99.

Richtsmeier, J. T., Cheverud, J. M., and Lele, S. 1992. Advances in anthropological morphometrics. *Annual Review of Anthropology* 21: 283–305.

Rothschild, N. A. 1979. Mortuary behavior and social organization. *American Antiquity* 44: 658.

Rubicz, R., Schurr, T. G., Babb, P. L., and Crawford, M. H. 2003. Mitochondrial DNA variation and the origins of the Aleuts. *Human Biology* 75: 809–35.

Sandford, M. K., and Weaver, D. S. 2000. Trace element research in anthropology: New perspectives and challenges, in M. A. Katzenberg and S. R. Saunders (eds.), *Biological Anthropology of the Human Skeleton*, pp. 329–50. New York: Wiley.

Sandford, M. K., and Kissling, G. E. 2005. Multivariate analyses of elemental hair concentrations from a medieval Nubian population. *American Journal of Physical Anthropology* 95: 41–52.

Schoeninger, M. J. 1995. Stable isotope studies in human evolution. *Evolutionary Anthropology* 4: 83–98.

Schoeninger, M. J., and DeNiro, M. J. 1984. Nitrogen and carbon isotopic composition of bone collagen from marine and terrestrial animals. *Geochimica et Cosmochimica Acta* 48: 625–39.

Schoeninger, M. J., DeNiro, M. J., and Tauber, H. 1983. Stable nitrogen isotope ratios of bone collagen reflect marine and terrestrial components of prehistoric human diet. *Science* 220: 1381–83.

Schoeninger, M. J., Iwaniec, U. T., and Nash, L. T. 1998. Ecological attributes recorded in stable isotope ratios of arboreal prosimian hair. *Oecologia* 113: 222–30.

Schoeninger, M. J., and Moore, K. M. 1992. Bone stable isotope studies in archaeology. *Journal of World Prehistory* 6: 247–96.

Schwarcz, H. P., Dupras, T. L., and Fairgrieve, S. I. 1999. 15N enrichment in the Sahara: In search of a global relationship. *Journal of Archaeological Science* 26: 629–36.

Schwarcz, H. P., and Schoeninger, M. J. 1991. Stable isotope analyses in human nutritional ecology. *Yearbook of Physical Anthropology* 34: 283–321.

Sealy, J. C., Armstrong, R., and Shrire, C. 1995. Beyond life-time averages: Tracing life histories through isotopic analysis of different calcified tissues from archaeological human skeletons. *Antiquity* 69: 290–300.

Sealy, J. C., and van der Merwe, N. J. 1986. Isotope assessment and the seasonal mobility hypothesis in the southwestern Cape, South Africa. *Current Anthropology* 27: 135–50.

Sealy, J. C., van der Merwe, N. J., Lee-Thorp, J. A., and Lanham, J. L. 1987. Nitrogen isotopic ecology in southern Africa: Implications for environmental and dietary tracing. *Geochimica et Cosmochimica Acta* 51: 2707–17.

Sealy, J. C., van der Merwe, N. J., Sillen, A., Kruger, F. J., and Krueger, H. W. 1991. $^{87}Sr/^{86}Sr$ as a dietary indicator in modern and archaeological bone. *Journal of Archaeological Science* 18: 399–416.

Smith, H. B. 1991. Standards of human tooth formation and dental age assessment, in M. A. Kelley and C. S. Larsen (eds.), *Advances in Dental Anthropology*, pp. 143–68. New York: Wiley-Liss.

Spence, M. W. 1994. Human skeletal material from Teotihuacan, in M. L. Sempowski and M. W. Spence (eds.), *Mortuary Practices and Skeletal Remains from Teotihuacan*, pp. 315–427. Salt Lake City: University of Utah Press.

Sponheimer, M., Robinson, T., Ayliffe, L., Roeder, B., Hammer, L., Passey, B., West, A., Cerling, T., Dearing, D., and Ehleringer, J. 2002. Nitrogen isotopes in mammalian herbivores: Hair $\delta^{15}N$ values from a controlled-feeding study. *International Journal of Osteoarchaeology* 13: 80–87.

Steinbock, R. T. 1976. *Palaeopathological Diagnosis and Interpretation*. Springfield, IL: Charles C Thomas.

Stenhouse, M. J., and Baxter, M. S. 1979. The uptake of bomb ^{14}C in humans, in R. Berger and H. E. Suess (eds.), *Radiocarbon Dating*, pp. 324–41. Berkeley and Los Angeles: University of California Press.

Stone, A. C. 2000. Ancient DNA from skeletal remains, in M. A. Katzenberg and S. R. Saunders (eds.), *Biological Anthropology of the Human Skeleton*, pp. 351–71.

Tanaka, G. I., Kawamura, H., and Nomura, E. 1981. Reference Japanese man-II: Distribution of strontium in the skeleton and in the mass of mineralized bone. *Health Physics* 40: 601–14.

Thorne, A. G., and Wolpoff, M. H. 1981. Regional continuity in Australasian Pleistocene hominid evolution. *American Journal of Physical Anthropology* 55: 337–49.

Torroni, A., Schurr, T. G., Cabell, M. F., Brown, M. D., Neel, J. V., Larsen, M., Smith, D. G., Vullo, C. M., and Wallace, D. C. 1993. Asian affinities and continental radiation of the four founding Native American mtDNAs. *American Journal of Human Genetics* 53: 563–90.

Torroni, A., Bandelt, H.-J., D'Urbano, L., Lahermo, P., Moral, P., Sellitto, D., Rengo, C., et al. 1998. mtDNA analysis reveals a major late Paleolithic population expansion from southwestern to northeastern Europe. *American Journal of Human Genetics* 62: 1137–52.

Triffitt, J. T. 1980. The organic matrix of bone tissue, in M. R. Urist (ed.), *Fundamental and Clinical Bone Physiology*, pp. 45–82. Philadelphia: J. B. Lippincott.

Trust, B. A., and Fry, B. 1992. Stable sulphur isotopes in plants: A review. *Plant, Cell and Environment* 15: 1105–10.

Turner, C. G. 1986. The first Americans: The dental evidence. *National Geographic Research* 2: 37–46.

———. 1987. Late Pleistocene and Holocene population history of east Asia based on dental variation. *American Journal of Physical Anthropology* 73: 305–21.

———. 1991. *The Dentition of Arctic Peoples*. New York: Garland.

Verano, J. W., and Ubelaker, D. H. (eds.). 1992. *Disease and Demography in the Americas*. Washington, DC: Smithsonian Institution Press.

Verano, J. W., and DeNiro, M. J. 1993. Locals or foreigners? Morphological, biometric and isotopic approaches to the question of group affinity in human skeletal remains recovered from unusual archaeological contexts, in M. K. Sandford (ed.), *Investigations of Ancient Human Tissue: Chemical Analyses in Archaeology*, pp. 361–86. Langhorne: Gordon and Breach.

Walker, P., and DeNiro, M. J. 1986. Stable nitrogen and carbon isotope ratios in bone collagen as indices of prehistoric dietary dependence on marine and terrestrial resources in southern California. *American Journal of Physical Anthropology* 71: 51–61.

Walker, P., and Lambert, P. 1989. Skeletal evidence for stress during a period of cultural change in prehistoric California, in L. Capasso (ed.), *Advances in Paleopathology*, pp. 207–12. *Journal of Paleopathology, Monographic Publications* 1.

Wang, L., Oota, H., Saitou, N., Jin, F., Matsushita, T., and Ueda, S. 2000. Genetic structure of a 2,500-year-old human population in China and its spatiotemporal changes. *Molecular Biology and Evolution* 17: 1396–1400.

Webb, S. G. 1995. *Palaeopathology of Aboriginal Australians: Health and Disease across a Hunter-Gatherer Continent*. Cambridge: Cambridge University Press.

Weiss Bolnick, D. A., and Smith, D. G. 2003. Unexpected patterns of mitochondrial DNA variation among Native Americans from the southeastern United States. *American Journal of Physical Anthropology* 122: 336–54.

West, A. G., Ayliffe, L. K., Cerling, T. E., Robinson, T. F., Karren, B., Dearing, M. D., and Ehlringer, J. R. 2004. Short-term diet changes revealed using stable carbon isotopes in horse tail-hair. *Functional Ecology* 18: 616–24.

White, C. D., Spence, M. W., Stuart-Williams, H. L. Q., and Schwarcz, H. P. 1998. Oxygen isotopes and the identification of geographical origins: The Valley of Oaxaca versus the Valley of Mexico. *Journal of Archaeological Science* 25: 643–55.

Woelfel, J. B., and Scheid, R. C. 2002. *Dental Anatomy: Its Relevance to Dentistry* (6th ed.). Philadelphia: Lippincott Williams and Wilkins.

Wright, R. V. S. 1992. Correlation between cranial form and geography in *Homo sapiens*: CRANID—a computer program for forensic and other applications. *Archaeology in Oceania* 27: 128–35.

Yang, H., and Watt, K. 2005. Contamination controls when preparing archaeological remains for ancient DNA analysis. *Journal of Archaeological Science* 32: 331–36.

Yang, H., Golenberg, E. M., and Shoshani, J. 1997. A blind testing design for authenticating ancient DNA sequences. *Molecular Pylogenetics and Evolution* 7: 261–65.

50

USING DNA IN LANDSCAPE ARCHAEOLOGY

Elizabeth Matisoo-Smith

From the mid-1960s and the first application of a "molecular clock" to human evolution (Sarich and Wilson 1967, 1969), the worlds of molecular biology and prehistory were brought together. Methodological developments in the 1980s, such as Polymerase Chain Reaction (PCR) and its application to mitochondrial DNA (mtDNA) studies, made obtaining large genetic data sets on human genetic variation possible (Cann, Stoneking, and Wilson 1987; Vigilant et al. 1991). As a result, biologists and archaeologists were increasingly asking similar or related questions regarding population origins and dispersals. In November 1989, when Erika Hagelberg and colleagues announced in a paper in *Nature* that they were able to successfully extract DNA from bone (Hagelberg, Sykes, and Hedges 1989), molecular biology and archaeology were linked in ways never before possible and thus began a new sub discipline within archaeology—molecular archaeology.

Today, DNA and ancient DNA labs are relatively common facilities found in anthropology and archaeology departments in universities and museums around the world, as it is becoming increasingly clear that the tools of molecular biology are very useful for addressing archaeological questions. However, the application of a tool is valuable only if applied appropriately, and it has been argued that genetic methods have often

been applied inappropriately to archaeological problems (Terrell, Kelly, and Rainbird 2001). However, as archaeologists have and continue to become more involved in both directing the application of genetic methods to archaeological questions and the interpretation of genetic data within an archaeological framework, the value of molecular archaeology increases. Landscape archaeology is one perspective that particularly benefits, as the rapidly accumulating corpus of genetic information allows us to consider a wide range of questions regarding biological change in the landscape through both space and time.

This chapter discusses and presents examples of both the direct and indirect applications of DNA to landscape archaeology. Direct applications include understanding how and when various populations of people and animals moved across the landscape and ways in which they actually interacted with those landscapes. DNA can also be a valuable tool for more traditional landscape studies, for example in providing key species identifications for faunal analysts. Despite its many applications, however, it is most important that archaeologists recognize the limitations of DNA analyses for archaeological studies and engage with the data and the methods in such a way that they can recognize both the powers and the pitfalls of the use of DNA in archaeological analyses.

521

Population Origins and Migrations

Probably the most notable contribution of DNA to landscape archaeology has been in using DNA variation to track population origins and dispersals. Since the first "out of Africa" mitochondrial DNA (mtDNA) studies (Cann, Stoneking, and Wilson 1987), researchers have been collecting DNA samples from human populations around the world. The most recent and perhaps most ambitious attempt of a global population genetic study is the Genographic Project, run by National Geographic and IBM (Behar et al. 2007). Mitochondrial DNA studies, which provide evidence of maternal genealogies, have in the last few years been joined by analyses of Y chromosome variation, which allows us to track the movements of men (Jobling and Tyler-Smith 1995; Underhill et al. 2000). This combination of mitochondrial and Y-chromosome data allows a much more balanced perspective for assessing population movement and allows for the possibility of addressing questions regarding differences between male and female dispersal (e.g., Hage and Marck 2003). Comparisons of the variation within and between populations are generally used to construct phylogenetic trees and other depictions of evolutionary relationships between populations. As a result of these comparisons, researchers have identified several "population specific" genetic markers and have tracked their evolutionary history, thus theoretically tracing population migrations. This approach has been particularly valuable for understanding the migrations of humans into the New World (for reviews, see Eshleman, Malhi, and Smith 2003 and Schurr 2004 for the Americas, and Matisoo-Smith 2006 for the Pacific) but has also been applied to Old World populations (see Salas et al. 2002, 2004 and Watson et al. 1997 for Africa; Renfrew and Boyle 2000 for Europe; Kivisild et al. 2002 and Palanichamy et al. 2004, and Tanaka et al. 2004 for Asia). Analyses of genetic variation in modern populations have also been used to address classic archaeological questions such as whether the spread of agriculture and various associated major language families were the result of diffusion or migration (Richards et al. 1996; Richards, Oppenheimer, and Sykes 1998; Semino et al. 1996).

Domestication and Other Impacts of Humans on the Environment

While human geneticists add to the debates regarding the impact of agriculture on human population histories, others are using DNA techniques to address the impact of humans on the landscape, for example, by addressing the domestication of plants and animals (see Diamond 2002). Genetic methods have been applied to identify the location(s) of origin for the domestication of plants, including wheat (Huen et al. 1997), maize (see Doebly 2004 for a review), rice (Li, Zhou, and Sang 2006), and taro and yams (Matthews and Terauchi 1994). Similarly, analyses of genetic variation in domestic animals and comparisons with their wild relatives have lead to discussions on the timing and process of animal domestication by humans.

Over the last ten years, studies have been conducted on a range of animals, including dogs (Savolainen et al. 2002, 2004; Vilà, Maldonado, and Wayne 1999; Vilà et al. 1997), horses (Vilà et al. 2001), donkeys (Beja-Pereira 2004), cattle (Troy et al. 2001), goats (Luikart et al. 2001), sheep (Hiendleder et al. 2002), chickens (Liu et al. 2006) and pigs (Guiffra et al. 2000; Kijas and Andersson 2001). The majority of these animal domestication studies present evidence for multiple domestication events. The question of the timing of the split between the domesticated populations and their wild progenitors, however, has been somewhat controversial, as estimated dates of divergence between wild and domesticated populations tend to fall in the range of 100,000 to 200,000 years ago or more (e.g., Guiffra et al. 2000; Luikart et al. 2001; Vilà et al. 1997) and are therefore at odds with archaeological evidence for animal domestication. Advances in phylogenetic methods and calculation of mutation rates (for example, see Ho et al. 2005 and Drummond et al. 2006 for recent developments) used in combination with increasing data from ancient DNA studies (e.g., Lambert et al. 2002) may eventually allow us to resolve these timing issues.

Ancient DNA

By studying DNA in living populations, we can see the end result of how humans and their associated plants and animals have moved across the landscape. However, as the issue of molecular dates for domestication show, interpreting the past based on the evidence of the present is fraught with difficulties. This is, of course, where ancient DNA (aDNA) becomes a particularly useful tool. Although the term "ancient DNA" conjures up images of Jurassic Park, technically it refers to the DNA obtained from anything other than fresh tissue. In an archaeological context, it most commonly refers to DNA obtained from archaeological bone, teeth, or other preserved tissues; however, aDNA is also now commonly recovered from preserved plant remains, soils, coprolites, and a range of other organic remains.

Unfortunately, one of the drawbacks of most aDNA studies is that, because DNA begins to decay and break down as soon as an organism dies, over time and depending on the environmental conditions, the DNA may be so degraded that it is no longer recoverable. This is particularly the case for nuclear DNA (DNA found in the nucleus of the cell), because each cell has only one nucleus and that contains only two copies of DNA (one inherited from the mother and one from the father). However, each cell has hundreds of mitochondria (or in the case of plants, chloroplasts), and those organelles contain numerous copies of their own DNA—or mtDNA. The higher copy number of mtDNA (1,000s of copies per cell) therefore means that it is more likely to survive and be recoverable over time. So, for analysis of material obtained from anything but the most perfect preservation conditions (for example, from samples preserved in permafrost or in cold, dry conditions), most aDNA studies focus on mtDNA. In the right conditions, researchers have been able to obtain mtDNA from bones that are tens of thousands of years old, (for instance, Neanderthal remains [Krings et al. 1997]) and, in some cases, from fauna over a hundred thousand years old (Shapiro et al. 2004). Y chromosome and other nuclear DNA studies have so far been more limited in the age of material, from which DNA can be obtained, with most focusing on historic samples. However, technological developments are increasing the likelihood of the recovery of ancient nuclear DNA and are consistently pushing back the boundaries of time.

Increasingly, DNA obtained from archaeological remains is being incorporated with data from modern populations to establish some chronological control and test theories developed based on modern population samples (e.g., see Brown 1999; Freitas et al. 2003 and Erickson et al. 2005 for aDNA in plant remains; Edwards et al. 2004 for cattle; Leonard et al. 2002 for dogs; Larson et al. 2005 for pigs; and Burger et al. 2005 for humans and the Neolithic farming dispersal). In some cases, ancient DNA is the only source of data to understand the impact of humans on particular environments, for example, when studying extinct animals. Ancient DNA has been studied in bison remains from Northern Eurasia and North America to assess the impact of human predation on the species. Results showing decreasing genetic diversity through time indicated that the bison population declined significantly before humans arrived in the region and that climate change rather than hunting was the most likely cause of that decline (Shapiro et al. 2004).

Ancient DNA analyses can also aid traditional faunal and osteological studies to produce important data for landscape studies. It can, for example, provide key information regarding species identification or sex determination when these are not possible based on morphological variation (see Bunce et al. 2003; Butler and Bowers 1998; Newman et al. 2002; Nicholls et al. 2003).

Pacific Case Studies in the Use of DNA and Landscape Approaches

The Pacific is an ideal region in which to apply a landscape perspective, particularly in regard to the use of genetic data. In terms of human population dispersals, the settlement of the Pacific involved both some of the earliest migrations of humans, with the settlement of Sahul which occurred some 40,000 to 60,000 years ago, and one of the last major human migrations—the settlement of Polynesia—taking place over the last few thousand years. While island populations are not isolates *per se*, the relative isolation (compared to most continental populations), makes tracking population movement and interaction across the Pacific landscapes/seascapes more identifiable (Hurles et al. 2003).

Researchers have now identified a number of mitochondrial and Y chromosome markers for Pacific populations that allow them to track migrations and origins (see Friedlaender et al. 2005; Hurles et al. 2002; Kayser et al. 2000; Su et al. 2000; Trejaut et al. 2005). Mitochondrial lineages P and Q and Y chromosome haplotypes with the M9 mutation are ancient and appear to be markers carried into the region by the earliest inhabitants of Near Oceania (the New Guinea region through the main Solomon Island chain). Later, at some point(s) in the Holocene, peoples moved into the Pacific carrying markers from Island Southeast Asia, specifically mitochondrial lineages belonging to haplogroup B and Y chromosomes belonging to haplotypes O1 and O3. The appearance of these "Asian" markers in the Pacific has been linked with the spread of the Oceanic subgroup of languages, part of the widespread Austronesian expansion from Taiwan. Interestingly, in Polynesian populations, there seems to be a much stronger signal for the Asian-derived maternal lineages, compared to the paternal markers which appear to be more influenced by Near Oceanic lineages. This discrepancy between mtDNA and Y chromosome markers has been interpreted by Hage and Marck (2003) as being the "effect of matrilocal residence and matrilineal descent in Proto-Oceanic society" (2003: S121).

The initial human settlement of Remote Oceania (the islands east of the main Solomon Island chain, including those of the Polynesian Triangle) is clearly associated with the spread of the Lapita Cultural Complex, which first appeared in the Bismarck Archipelago some 3,200 years ago. It is from this region that the expansion of the Lapita Culture occurred as Lapita colonists moved not only through the already inhabited islands of the Bismarcks and the Solomon Islands but out into the previously uninhabited islands of Remote Oceania. Lapita settlements appear in Vanuatu, New Caledonia, Fiji, Samoa, and Tonga within 200–300 years of their first appearance in Near Oceania. The rest of the Polynesian triangle was settled by the descendants of these Lapita peoples from western Polynesia some 1,200 to 2,000 years later (Kirch 1997, 2000).

As Kirch (1997, 2000), among others, has described, "transported landscapes" were a key adaptive strategy for the Lapita settlement of the islands of Remote Oceania and similarly for the people who went on to settle the Polynesian triangle. Lapita peoples carried with them and introduced a range of plant and animal species to the islands they settled. These Pacific "transported landscapes" have been the subject of several genetic studies over the last ten years, particularly in regard to the development of the commensal model for studying human migrations (Matisoo-Smith 1994; Matisoo-Smith et al. 1998, 2004).

Commensal formally means, "to eat at the same table" but refers to the close and dependent relationship between humans and the animals that accompanied them, in the Pacific case, as they crossed the Pacific Ocean. Because the animals transported by the Lapita peoples and their descendants, the dog (*Canis familiaris*), pig (*Sus scrofa*), chicken (*Gallus gallus*), and rat (*Rattus exulans*) could not self-disperse across major water gaps, it was recognized that identification of the origins of the particular island populations of the animals would indicate the immediate origin of the canoes in which they traveled. Thus, we can track the movement of people across the Pacific. The origins of those animal populations could be identified through analyses of the genetic variation within and between the various island populations (Matisoo-Smith 1994).

The first of the commensal studies in the Pacific focused on genetic analyses of the Pacific rat (*Rattus exulans*) (Matisoo-Smith 1994; Matisoo-Smith et al. 1998). These initial studies focused on extant populations of *R. exulans*, which was chosen as a proxy for tracing human settlement across the Pacific for several reasons: (1) it is the one animal that made it to all islands settled by Lapita peoples and their descendants; (2) it is a species distinct from the later European-introduced rats (*R. rattus* and *R. norvegicus*) and thus does not interbreed with them; (3) its remains are found in the earliest archaeological layers throughout Remote Oceania, suggesting it was on colonizing canoes; and (4) it is a particularly adaptable species and therefore maintained its presence throughout prehistory into the historic period. These features mean that extant populations can be studied and the identified relationships used to infer population origins and provide evidence of human mobility. Indeed, the results of the analyses of genetic variation within and between Polynesian populations of *R. exulans* identified several aspects of prehistoric human mobility that were remarkably congruous with archaeological and linguistic evidence. New Zealand populations were particularly diverse, suggesting multiple introductions from locations in a general homeland region encompassing the Southern Cook and Society Islands. In fact, two major interaction spheres within Polynesia were identified: one, encompassing this central homeland region with islands to the south (New Zealand and the Kermadec Islands); the other, linking it with the archipelagos in the north (Hawaii and the Marquesas).

Once the value of the commensal approach was demonstrated, it was then expanded both through time and space with the inclusion of analyses of aDNA obtained from archaeological and museum samples of *R. exulans*. Modern and archaeological rat bones from Chatham Island were studied and the DNA obtained compared to extant populations in New Zealand and the Kermadec Islands to assess the question of rates of mutation and to compare the degree of variation in two archipelagos likely to have experienced two very different levels of prehistoric interaction (Matisoo-Smith et al. 1999). Archaeological evidence and Maori migration traditions suggest that the Kermadecs were a likely stepping stone for voyages between New Zealand and central East Polynesia (Leach et al. 1986; Te Rangi Hiroa 1949), where the Chatham Islands were very difficult to access, given winds and sailing conditions (Irwin 1992), and were thus more isolated. The DNA evidence was thoroughly consistent with this scenario, with the rats from Raoul Island, in the Kermadecs, showing significantly more genetic variation than those from Chatham Island. More recently, analyses of ancient DNA from Easter Island populations of *R. exulans* have also provided genetic evidence consistent with archaeological evidence for the relative isolation of the island (Barnes et al. 2006).

The use of *R. exulans* and the commensal approach was then extended beyond Polynesia to address the issue of Lapita origins (Matisoo-Smith and Robins 2004). In this study, aDNA was obtained from archaeological and museum samples representing *R. exulans* populations across the Pacific and through Island Southeast Asia. Three very distinct mitochondrial DNA haplogroups were identified, which had very clear geographical limits suggesting three likely spheres of human interaction through which rats were being dispersed: Group I encompassed the Philippines, Borneo, and Sulawesi; Group II included rats from the Philippines, the Moluccas, New Guinea, and the Solomon Islands; Group III contained all the rats from Remote Oceania. These results were compared to those expected given several models of Lapita origins (Green 2003). Given the acceptance of *R. exulans* dispersal as being a signature of the Lapita cultural complex, two often-cited models for Lapita origins could be rejected. The DNA evidence clearly showed that the Remote Oceanic *R. exulans* populations did not originate in Taiwan or in Near Oceania but that there was clear phylogenetic signal in the mtDNA variation. This result therefore allowed for the rejection of the so-called Express Train from Taiwan and the Bismarck Archipelago Indigenous Inhabitants models. Instead, it appears that Lapita peoples picked up the rats somewhere in Wallacea. Thus, models that stress a reasonable degree of interaction and integration of indigenous and intrusive components are much more consistent with both the rat and the human genetic data regarding Lapita origins (Hurles et al. 2003).

The commensal approach has also been applied to some of the other animals transported by Pacific peoples, such as pigs (Allen et al. 2001; Larson et al. 2005, 2007), dogs (Savolainen et al. 2004), and chickens (Storey 2004; Storey et al. 2007) and to stowaways such as lizards (Austin 1998). In addition to animals, genetic analyses of the plants transported by Pacific peoples are also providing evidence of human movement within and transformation of the Pacific environment. Hinkle (2004) used genetic variation in *Cordyline* to track the initial west to east movement of people into the Pacific; Clarke (2006) has studied genetic variation in bottlegourd (*Lagenaria siceraria*), an important Pacific plant species believed to have a South American origin, to test hypothesis of Polynesian contact and interaction with the Americas.

Ethics and DNA studies

One of the key issues that must be considered in any DNA study is, of course, ethics. The ethical issues associated with DNA studies are not too different from ethical issues most archaeologists deal with in regard to obtaining appropriate permissions, and issues relating to ownership of samples and ownership of resulting data. The ethical issues of ownership of genetic data however can be particularly touchy in many communities (not surprising given issues such as the patenting of DNA), and therefore full discussion of possible plans for genetic analyses must be undertaken and consent granted before any DNA studies proceed. One added ethical issue in dealing with ancient DNA studies is the fact that most ancient DNA extraction protocols are destructive, and DNA analyses can be very expensive; thus the value of the DNA data obtained must be weighed against the cost of the destruction of or damage to the sample.

Future Development for DNA Studies and Landscape Archaeology

As the development of new methods continues to push the current boundaries for ancient DNA recovery, new possibilities for its application to landscape issues emerge. Researchers are beginning to explore the preservation of aDNA in soil and other sediments, which will contribute to environmental reconstructions and the ability to track changes in the environment through time (Willerslev et al. 2003, 2004). In the last few years there have been numerous accounts of the extraction of the pathogen DNA from human remains to test morphological identifications and historical accounts of the presence of diseases such as tuberculosis, leprosy and the plague (Drancourt et al. 1998; Drancourt and Raoult 2004; Taylor et al. 2005). Like the earlier reports of DNA extracted from residues on artifacts (e.g., Loy and Dixon 1998), many of these pathogen results have been controversial (Gilbert et al. 2004a, 2004b); yet, like the potential for residue analyses (Burger et al. 2000), they remain exciting topics of research for the future.

Conclusions

It is clear that the last 25 years have seen a dramatic increase in the use of DNA in archaeological and landscape studies in general, but the ability to extract DNA from ancient remains really brought the fields of molecular biology and archaeology together. The diachronic perspective of molecular archaeology (ancient DNA) is particularly valuable for landscape archaeology. But, perhaps more importantly, the perspective of landscape archaeology is necessary for the interpretation of any molecular data being used to address questions of the past. It is only through looking at genetic variation through time and space, and interpreting that variation in the appropriate cultural and historical context, that such data are useful for, and will truly contribute to, our understanding of past human behavior.

References

Allen, M. S., Matisoo-Smith, E., and Horsburgh, K. A. 2001. Pacific "babes": Pig origins, dispersals and the Potentials of mitochondrial DNA analyses. *International Journal of Osteoarchaeology* 11: 4–13.

Austin, C. C. 1999. Lizards took the express train to Polynesia. *Nature* 397: 113–14.

Barnes, S. S., Matisoo-Smith, E., and Hunt, T. 2006. Ancient DNA of the Pacific Rat (*Rattus exulans*) from Rapa Nui (Easter Island). *Journal of Archaeological Science* 33: 1536–40.

Behar, D. M., Rosset, S., Blue-Smith, J., Balanovsky, O., and Tzur, S., et al. 2007. The Genographic Project public participation mitochondrial DNA database. *Public Library of Science, Genetics* 3(6): e104.doi:10.1371/journal.pgen.0030104.

Beja-Pereira, A., England, P. R., Ferrand, N., Jordan, S., Bakhiet, A. O., Abdalla, M. A., Mashkour, M., Jordana, J., Taberlet, P., and Luikart, G. 2004. African origins of the domestic donkey. *Science* 304: 1781.

Brown, T. 1999. How ancient DNA may help in understanding the origin and spread of agriculture. *Philosophical Transactions of the Royal Society London, Series B* 354: 89–98.

Bunce, M., Worthy, T. H., Ford, T., Hoppitt, W., Willerslev, E., Drummond, A., and Cooper, A. 2003. Extreme reversed sexual dimorphism in the extinct New Zealand moa Dinornis. *Nature* 425: 172–75.

Burger, J., Forster, P., Bramanti, B., Matsumura, S., Brandt, G., Tanzer, M., Villems, R., Renfrew, C., Gronenborn, D., Alt, K. W., and Burger, J. 2005. Ancient DNA from the first European farmers in 7,500-year-old Neolithic sites. *Science* 310: 1016–18.

Burger, J., Hummel, S., and Herrmann, B. 2000. Palaeogenetics and cultural heritage. Species determination and STR-genotyping from ancient DNA in art and artifacts. *Thermochim Acta* 365: 141–46.

Butler, V. L., and Bowers, N. J. 1998. Ancient DNA from salmon bones: A preliminary study. *Ancient Biomolecules* 2: 17–26.

Cann, R. L., Stoneking, M., and Wilson, A. C. 1987. Mitochondrial DNA and human evolution. *Nature* 344: 288–89.

Clarke, A. C., Burtenshaw, M. K., McLenachan, P. A., Erickson, D. L., and Penny, D. 2006. Reconstructing the origins and dispersal of the Polynesian bottle-gourd (*Lagenaria siceraria*). *Molecular Biology and Evolution* 23: 893–900.

Diamond, J. 2002. Evolution, consequences and future of plant and animal domestication. *Nature* 418: 700–07.

Drancourt, M., Aboudharam, G., Signoli, M., Dutour, O., and Raoult, D. 1998. Detection of 400-year-old Yersinia pestis DNA in human dental pulp: An approach to the diagnosis of ancient septicaemia. *Proceedings of the National Academy of Sciences, USA* 95(21): 12637–40.

Drancourt, M., and Raoult, D. 2004. Molecular detection of *Yersinia pestis* in dental pulp. *Microbiology* 150: 263–64.

Drummond, A. J., Ho, S. Y. W., Phillips, M. J., and Rambaut, A. 2006. Relaxed phylogenetics and dating with confidence. *PLoS Biology* 4(5): e88.

Doebly, J., 2004. The genetics of maize evolution. *Annual Reviews in Genetics* 38: 37–59.

Edwards, C. J., MacHuch, D. E., Dobney, K. M., Matin, L., Russell, N., Horwitz, L. K., McIntosh, S. K., MacDonald, K. C., Helmer, D., Tresset, J.-D., Vigne, A., and Bradley, D. G. 2004. Ancient DNA analysis of 101 cattle remains: Limits and prospects. *Journal of Archaeological Science* 31: 695–710.

Erickson, D. L., Smith, B. D., Clark, A. C., Sandweiss, D. H., and Tuross, N. 2005. An Asian origin for a 10,000-year-old domesticated plant in the Americas. *Proceedings of the National Academy of Sciences, USA* 102: 18315–20.

Eshleman, J. A., Malhi, R. S., and Smith, D. G. 2003. Mitochondrial DNA studies of Native Americans: Conceptions and misconceptions of the population prehistory of the Americas. *Evolutionary Anthropology* 12: 7–18.

Forester, P., and Renfrew, C. 2003. The DNA chronology of prehistoric human dispersals, in P. Bellwood and C. Renfrew (eds.), *Examining the Farming/Language Dispersal Hypothesis*, pp. 89–97. Cambridge: McDonald Institute for Archaeological Research.

Freitas, F. O., Bendel, G., Allaby, R. G., and Brown, T. A. 2003. DNA from primitive maize landraces and archaeological remains: Implications for the domestication of maize and its expansion into South America. *Journal of Archaeological Science* 30: 901–08.

Friedlaender, J., Schurr, T., Gentz, F., Koki, G., Friedlaender, F., Horvat, G., Babb, P., Cerchio, S., Kaestle, F., Schanfield, M., Deka, R., Yanagihara, R., and Merriwether, D. A. 2005. Expanding southwest Pacific mitochondrial haplogroups P and Q. *Molecular Biology and Evolution* 22: 1506–17.

Gilbert, M. T. P., Cuccui, J., White, W., Lynnerup, N., Titball, R. W., Cooper, A., and Prentice, M. B. 2004a. Response to Drancourt and Raoult. *Microbiology* 150: 264–65.

Gilbert, M. T. P., Cuccui, J., White, W., Lynnerup, N., Titball, R. W., Cooper, A., and Prentice, M. B. 2004b. Absence of Yersinia pestis-specific DNA in human teeth from five European excavations of putative plague victims. *Microbiology* 150: 341–54.

Green, R. C. 2003. The Lapita horizon and traditions: Signature for one set of Oceanic migrations, in C. Sand (ed.), *Pacific Archaeology: Assessments and Anniversary of the First Lapita Excavation (July 1952)*, pp. 95–120. Noumea, New Caledonia: Le cahiers de l'Archeologie en Nouvelle-Caledonie, Vol 15.

Guiffra, E., Kijas, J. M. H., Amarger, V., Carlborg, O., Jeon, J.-T., and Andersson, L. 2000. The origin of the domestic pig: Independent domestication and subsequent introgression. *Genetics* 154: 1785–91.

Hage, P., and Marck, J. 2003. Matrilineality and the Melanesian origin of Polynesian Y chromosomes. *Current Anthropology* 44: 121–27.

Hagelberg, E., Sykes, B., and Hedges, R. 1989. Ancient bone DNA amplified. *Nature* 342: 485.

Hiendleder, S., Kaupe, B., Wassmuth, R., and Janke, A. 2002. Molecular analysis of wild and domestic sheep questions current nomenclature and provides evidence for domestication from two different subspecies. *Proceedings of the Royal Society B* 269: 893–904.

Hinkle, A. E. 2004. Distribution of a male sterile form of "Ti" (*Cordyline fruticosa*) in Polynesia and its ethnobotanical significance. *Journal of the Polynesian Society* 113: 263–90.

Ho, S. Y., Phillips, M. J., Cooper, A., and Drummond, A. J. 2005. Time dependency of molecular rate estimates and systematic overestimation of recent divergence times. *Molecular Biology and Evolution* 22: 1561–68.

Huen, M., Schafer-Pregl, R., Klawan, D., Castagna, R., Accerbi, M., Borghi, B., and Salamini, F. 1997. Site of einkorn wheat domestication identified by DNA fingerprinting. *Science* 278: 1312–14.

Hurles, M. E., Matisoo-Smith, E., Gray, R. D., and Penny, D. 2003. Untangling Pacific settlement: On the edge of the knowable. *Trends in Ecology and Evolution* 18: 531–40.

Hurles, M. E., Nicholson, J., Bosch, E., Renfrew, C., Sykes, B. C., and Jobling, M. A. 2002. Y chromosomal evidence for the origins of Oceanic-speaking peoples. *Genetics* 160: 289–303.

Irwin, G. 1992. *The Prehistoric Exploration and Colonisation of the Pacific*. Cambridge: Cambridge University Press.

Jobling, M. A., and Tyler-Smith, C. 1995. Fathers and sons: The Y chromosome and human evolution. *Trends in Genetics* 11: 449–55.

Kayser, M., Brauer, S., Weiss, G., Underhill, P. A., Roewer, L., Schiefenhövel, W., and Stoneking, M. 2000. Melanesian origin of Polynesian Y chromosomes. *Current Biology* 10: 1237–46.

Kijas, J. M. H., and Andersson, L. 2001. A phylogenetic study of the origin of the domestic pig estimated from the near-complete mtDNA genome. *Journal of Molecular Evolution* 52: 302–08.

Kirch, P. V. 1997. *The Lapita Peoples: Ancestors of the Oceanic World*. Oxford: Blackwell Publishing.

———. 2000. *On the Road of the Winds: An Archaeological History of the Pacific Islands before European Contact*. Berkeley and Los Angeles: University of California Press.

Kivisild, T., Tolk, H.-V., Parik, J. Y., Wang, Y., Papiha, S. S., Bandelt, H.-J., and Villems, R. 2002. The emerging limbs and twigs of the East Asian mtDNA tree. *Molecular Biology and Evolution* 19: 1737–51.

Krings, M., Stone, A., Schmitz, R. W., Krainitzki, H., Stoneking, M., and Paabo, S. 1997. Neanderthal DNA sequences and the origin of modern humans. *Cell* 90: 19–30.

Lambert, D. M., Ritchie, P. A., Millar, C. D., Holland, B., Drummond, A. J., and Baroni, C. 2002. Rates of evolution in ancient DNA from Adelie penguins. *Science* 295: 2270–73.

Larson, G., Dobney, K., Albarella, U., Fang, M., Matisoo-Smith, E., Robins, J., Lowden, S., Finlayson, H., Brand, T., Willerslev, E., Rowley-Conwy, P., Andersson, L., and Cooper, A. 2005. Worldwide phylogeography of wild boar reveals multiple centres of pig domestication. *Science* 307: 1618–21.

Larson, G., Cucchi, T., Fujita, M., Matisoo-Smith, E., Robins, J., Anderson, A., Rolett, B., Spriggs, M., Dolman, G., Kim, T.-H., Thuy, N. T. D., Randi, E., Doherty, M., Due, R. A., Bollt, R., Djubianto, T., Griffin, B., Intoh, M., Keane, E., Kirch, P., Li, K.-T., Morwood, M., Pedriña, L. M., Piper, P. J., Rabett, R. J., Shooter, P., Van den Bergh, G., West, E., Wickler, S., Yuan, J., Cooper, A., and Dobney, K. 2007. Phylogeny and ancient DNA of *Sus* provides new evidence for human dispersal routes in Island Southeast Asia and the Pacific. *Proceedings of the National Academy of Sciences, USA* 104(12): 4834–39.

Leach, B. F., Anderson, A. J., Sutton, D. G., Bird, R., Duerden, P., and Clayton, E. 1986. The origin of prehistoric obsidian artefacts from the Chatham and Kermadec Islands. *New Zealand Journal of Archaeology* 8: 143–70.

Leonard, J. A., Wayne, R. K., Wheeler, J., Valadez, R., Guillen, S., and Vilà, C. 2002. Ancient DNA evidence for Old World origin of New World dogs. *Science* 298: 1613–16.

Li, C., Zhou, A., and Sang, T. 2006. Rice domestication by reducing shattering. *Science* 311: 1936–39.

Liu, Y.-P., Wu, G.-S., Yao, Y.-G., Miao, Y.-W., Luikart, G., Baig, M., Beja-Pereira, A., Ding, Z.-L., Palanichamy, M. G., and Zhang, Y.-P. 2006. Multiple maternal origins of chickens: Out of the Asian jungles. *Molecular Phylogenetics and Evolution* 38: 12–19.

Loy, T. H., and Dixon, E. J. 1998. Blood residues on fluted points from eastern Beringia. *American Antiquity* 63: 21–46.

Luikart, G., Gielly, L., Excoffier, L., Vigne, J.-D., Bouvet, J., and Taberlet, P. 2001. Multiple maternal origins and weak phylogeographic structure in domestic goats. *Proceedings of the National Academy of Sciences, USA* 98: 5927–32.

Matisoo-Smith, E. 1994. The human colonisation of Polynesia: A novel approach—Genetic analyses of the Polynesian Rat (*Rattus exulans*). *Journal of the Polynesian Society* 103: 75–87.

———. 2006. The peopling of Oceania, in M. H. Crawford (ed.), *Anthropological Genetics: Theory, Methods and Applications*. Cambridge: Cambridge University Press.

Matisoo-Smith, E., Roberts, R. M., Allen, J. S., Irwin, G. J., Penny, D., and Lambert, D. M. 1998. Patterns of human colonisation in Polynesia revealed by mitochondrial DNA from the Polynesian Rat. *Proceedings of the National Academy of Sciences, USA* 95: 15145–50.

Matisoo-Smith, E., and Robins, J. 2004. Origins and dispersals of Pacific peoples: Evidence from mtDNA phylogenies of the Pacific rat. *Proceedings of the National Academy of Sciences, USA* 101: 9167–72.

Matisoo-Smith, E., Sutton, D. G., Ladefoged, T. N., Lambert, D. M., and Allen, J. S. 1999. Prehistoric mobility in Polynesia: MtDNA variation in *Rattus exulans* from the Chatham and Kermadec Islands. *Asian Perspectives* 38: 186–99.

Matthews, P. J., and Terauchi, R. 1994. The genetics of agriculture: DNA variation in taro and yam, in J. G. Hather (ed.), *Tropical Archaeobotany: Applications and New Developments*, pp. 251–70. London: Routledge.

Newman, M. E., Parboosing, J. S., and Bridge, P. J. 2002. Identification of archaeological animal bone by PCR/DNA analysis. *Journal of Archaeological Science* 29: 77–84.

Nicholls, A., Matisoo-Smith, E., and Allen, M. S. 2003. Novel application of molecular techniques to Pacific archaeofish remains. *Archaeometry* 45: 121–35.

Palanichamy, M. G., Sun, C., Agrawal, S., Bandelt, H.-J., Kong, Q.-P., Khan, F., Wang, C.-Y., Chaudhuri, T. K., Palla, V., and Zhang, Y. P. 2004. Phylogeny of mitochondrial DNA macrohaplogroup N in India, based on complete sequencing: Implications for the Peopling of South Asia. *American Journal of Human Genetics* 75: 966–78.

Renfrew, C., and Boyle, K. (eds.). 2000. *Archaeogenetics: DNA and the Population Prehistory of Europe*. Cambridge: McDonald Institute for Archaeological Research.

Richards, M., Corte-Real, H., Forster, P., Macaulay, V., Wilkinson-Herbots, H., Demaine, A., Papiha, S., Hedges, R., Bandelt, H.-J., and Sykes, B. C. 1996. Palaeolithic and Neolithic lineages in the European mitochondrial gene pool. *American Journal of Human Genetics* 59: 185–203.

Richards, M., Oppenheimer, S., and Sykes, B. 1998. MtDNA suggests Polynesian origins in eastern Indonesia. *American Journal of Human Genetics* 63: 1234–36.

Salas, A., Richards, M., De la Fe, T., Lareu, M. V., Sobrino, B., Sanchez-Diz, P., Macaulay, V., and Carracedo, A. 2002. The making of the African mtDNA landscape. *American Journal of Human Genetics* 71: 1082–111.

Salas, A., Richards, M., Lareu, M. V., Scozzari, R., Coppa, A., Torroni, A., Macaulay, V., and Carracedo, A. 2004. The African diaspora: Mitochondrial DNA and the Atlantic slave trade. *American Journal of Human Genetics* 74: 454–65.

Sarich, V. M., and Wilson, A. C. 1967. Immunological time scale for hominid evolution. *Science* 158: 1200–03.

———. 1969. A molecular time scale for human evolution. *Proceedings of the National Academy of Sciences, USA* 63: 1088–93.

Savolainen, P., Zhang, Y.-P., Luo, J., Lundeberg, J., and Leitner, T. 2002. Genetic evidence for an East Asian Origin of domestic dogs. *Science* 298: 1610–13.

Savolainen, P., Leitner, T., Wilton, A. N., Matisoo-Smith, E., and Lundeberg, J. 2004. A detailed picture of the Origin of the Australian Dingo, obtained from the study of Mitochondrial DNA. *Proceedings of the National Academy of Sciences, USA* 101: 12387–90.

Schurr, T. G. 2004. The peopling of the New World: Perspectives from molecular anthropology. *Annual reviews in Anthropology* 33: 551–83.

Semino, O., Passarino, G., Brega, A., Fellous, M., and Sanachiara-Benerecetti, A. S. 1996. A view of the Neolithic demic diffusion in Europe through two Y-chromosome-specific markers. *American Journal of Human Genetics* 59: 964–68.

Shapiro, B., Drummond, A. J., Rambaut, A., Wilson, M. C., Matheus, P. E., Sher, A. V., Pybus, O. G., Gilbert, M. T. P., Barnes, I., Binladen, J., Willerslev, E., Hansen, A. J., Baryshnikov, G. F., Burns, J. A., Davydov, S., Driver, J. C., Froese, D. G., Harington, C. R., Keddie, G., Kosintsev, P., Kinz, M. L., Martin, L. D., Stephenson, R. O., Storer, J., Tedford, R., Zimov, S., and Cooper, A. 2004. Rise and fall of the Beringian Steppe bison. *Science* 306: 1561–65.

Storey, A. A. 2004. Save Me a Drumstick: Molecular taphonomy, differential preservation and ancient DNA from the Kingdom of Tonga. Unpublished M.A. thesis, Simon Fraser University, Canada.

Storey, A. A., Ramírez, J. M., Quiroz, D., Burley, D. V., Addison, D. J., Walter, R., Anderson, A. J., Hunt, T. L., Athens, J. S., Huynen, L., and Matisoo-Smith, E. 2007. Radiocarbon and DNA evidence for a pre-Columbian introduction of Polynesian chickens to Chile. *Proceedings of the National Academy of Sciences, USA* 104(25): 10335–39.

Su, B., Jin, L., Underhill, P., Martinson, J., Saha, N., McGarvey, S. T., Shriver, M. D., Chu, J., Oefner, P., Chakraborty, R., and Deka, R. 2000. Polynesian origins: Insights from the Y-chromosome. *Proceedings of the National Academy of Sciences, USA* 97: 8225–28.

Tanaka, M., Cabrera, V. M., Ganzalez, A. M., Larruga, J. M., Takeyasu, T., Fuku, N., Guo, L.-J., Hirose, R., Fujita, Y., Kurata, M., Shinoda, K., Umetsu, K., Yamada, Y., Oshida, Y., Sato, Y., Hattori, N., Mizuno, Y., Arai, Y., Hirose, N., Ohta, S., Ogawa, O., Tanaka, Y., Kawamori, R., Shamoto-Nagai, M., Maruyama, W., Shimokata, H., Suzuki, R., and Shimodaira, H. 2004. Mitochondrial genome variation in Eastern Asian and the peopling of Japan. *Genome Research* 14: 1832–50.

Taylor, G. M., Young, D. B., and Mays, S. A. 2005. Genotypic analysis of the earliest known prehistoric case of tuberculosis in Britain. *Journal of Clinical Microbiology* 43: 2236–40.

Te Rangi Hiroa (P. H. Buck). 1949. *The Coming of the Maori.* Wellington: Whitcome and Tombs.

Terrell, J. E., Kelly, K. M., and Rainbird, P. 2001. Foregone conclusions? In search of "Papuans" and "Austronesians." *Current Anthropology* 42: 97–124.

Trejaut, J. A., Kivisild, T., Loo, J. H., Lee, C. L., He, C. L., Hsu, C. J., Li, Z. Y., and Lin, M. 2005. Traces of archaic mitochondrial lineages persist in Austronesian-speaking Formosan populations. *PLoS Biology* 3(8): e247.

Troy, C. S., MacHugh, D. E., Bailey, J. F., Magee, D. A., Loftus, R. T., Cunningham, P., Chamberlain, A. T., Sykes, B. C., and Bradley, D. G. 2001. Genetic evidence for Near-Eastern origins of Eurpean cattle. *Nature* 410: 1088–91.

Underhill, P. A., Shen, P., Lin, A. A., Jin, L., Passarino, G., Yang, W. H., Kauffman, E., Bonnet-Tamir, B., Bertranpetit, J., Francalacci, P., Ibrahim, M., Jenkins, T., Kidd, J. R., Mehdi, S. Q., Seielstad, M. T., Wells, R. S., Piazza, A., Davis, R. W., Feldman, M. A., Cavalli-Sforza, L. L., and Oefner, P. J. 2000. Y-chromosome sequence variation and the history of human populations. *Nature Genetics* 26: 358–61.

Vigilant, L., Stoneking, M., Harpending, H., Hawkes, K., and Wilson, A. C. 1991. African populations and the evolution of mitochondrial DNA. *Science* 232: 1140–42.

Vilà, C., Leonard, A. L., Gotherstrom, A., Marklunnd, S., Sandberg, K., Liden, K., Wayne, R. K., and Ellegren, H. 2001. Widespread origins of domestic horse lineages. *Science* 291: 474–77.

Vilà, C., Maldonado, J. E., and Wayne, R. K. 1999. Phylogenetic relationships, evolution and genetic diversity of the domestic dog. *Journal of Heredity* 90: 71–77.

Vilà, C., Savolainen, P., Maldonado, J. E., Amorim, I. R., Rice, J. E., Honeycutt, R. L., Crandall, K. A., Lundeberg, J., and Wayne, R. K. 1997. Mulitple and ancient origins of the domestic dog. *Science* 276: 1687–89.

Watson, E., Forster, P., Richards, M., and Bandelt, H. J. 1997. Mitochondrial footprints of human expansions in Africa. *American Journal of Human Genetics* 61: 691–704.

Willerslev, E., Hansen, A. J., Binladen, J., Brand, T. B., Gilbert, M. T. P., Shapiro, B., Bunce, M., Wiuf, C., Gilichinsky, D. A., and Cooper, A. 2003. Diverse plant an animal genetic records from Holocene and Pleistocene sediments. *Science* 300: 791–95.

Willerslev, E., Hansen, A. J., and Poinar, H. N. 2004. Isolation of nucleic acids and cultures from fossil ice and permafrost. *Trends in Ecology and Evolution* 19: 141–47.

51

SOURCING TECHNIQUES IN LANDSCAPE ARCHAEOLOGY

Glenn R. Summerhayes

Identifying the movement of archaeological materials across the landscape at given points in time is a powerful tool for understanding the socioeconomic processes of the past. The aim of this chapter is, first, to provide a background for identifying the physical distribution of archaeological materials and, second, to outline some of the more powerful analytical techniques used by archaeologists to identify such movements of materials.

The movement of materials *must be demonstrated* rather than assumed. Such a demonstration has been made easier for the archaeologist with recent advances in science and technology. The sourcing of archaeological material is not new. Perhaps the most famous early study was by Anna Shepard who, in the 1930s and 1940s, postulated the exchange of pottery over wide areas of the Southwest United States by analyzing mineral inclusions in thin sections to pinpoint the origins of such minerals (Shepard 1965). The modern chemical analyses were much later. Chemical techniques have been used since the mid-19th century on archaeological metal artifacts from Central Europe (Schwab et al. 2006). Neutron activation analysis was applied to coins from the Louvre in 1952, while in 1957 Oriental ceramics were analyzed using nondestructive methods of X-ray fluorescence spectrometry (XRF) and X-ray diffraction (XRD) analysis (Young and Whitmore 1957).

Access to these techniques was restricted and costly.

A major advance in the chemical analysis of archaeological material came from the initiative of Dr. Robert Oppenheimer. On the 8th of March, 1956, he assembled a group of archaeologists and chemists at the Institute of Advanced Studies, Princeton, to discuss the possibility of applying methods of nuclear research to the study of archaeology (Sayre and Dodson 1957). As a result of this meeting, work was undertaken at two laboratories—the Brookhaven National Laboratory in the United States and the Research Laboratory for Archaeology and the History of Art at Oxford in England. Techniques deployed included Neutron Activation Analysis (NAA) and Spectrographic methods. These studies were reasonably successful, being able to separate pottery wares from Asia Minor, Greece, and Italy, as well as different factories of Samian ware. For the next three decades, these studies laid the foundations for chemical analyses in which thousands of varying techniques were carried out on many types of objects, including pottery, stone (obsidian, marble, chert, volcanic rocks), amber, and metals, including coins. Apart from NAA, XRF, XRD and spectrographic methods, techniques currently in use include Proton Induced X-ray Emission and Proton Induced Gamma Ray Emission (PIXE-PIGME) (Bird, Duerden, and Wilson

1983), Inductively Coupled Plasma Emission spectrometry (ICP) (Kennett et al. 2004), lead isotope analysis (Webb et al. 2006; Weisler and Woodhead 1995), and electron microscopy (Summerhayes 1997, 2000), to mention a few. Major changes in the instrumentation of these techniques over the last 30 years have meant that more elements can be analyzed with a higher precision. The choice of technique depends on availability to the archaeologist and cost. It is on these techniques that the rest of this chapter focuses.

Why Undertake These Analyses?

The aim of many physico-chemical analyses is to discern the geological origin of the object under study. Ideally, the sources of material with restricted distribution could be characterized and "fingerprinted" chemically or petrographically. Obsidian, for instance, can be chemically homogeneous within its source flow yet chemically discernable between source areas; its sources can thus be chemically fingerprinted. Obsidian found in archaeological contexts can be analyzed and related back to the geologic source. Earlier work by Colin Renfrew had demonstrated its importance in modeling exchange patterns (Renfrew 1969, 1977; Renfrew, Dixon, and Cann 1966). Identifying the potential source areas of a material requires an intensive search within a known region. For instance, in a recent study on obsidian exchange within the western Pacific over a 20,000-year time span, archaeologists have concentrated on the source areas within West New Britain, Papua New Guinea. Over a number of years, obsidian flows were mapped and samples taken for characterization with the result that a finer discrimination of sources was possible with five source localities chemically defined (Summerhayes et al. 1998). By analyzing obsidian found in archaeological sites (some artifacts more than 3,700 km from their source), it was possible not only to trace obsidian to its source localities but also to define changes in the selection of different sources over time. (For an overview of archaeological approaches to obsidian studies, see Shackley 1998.)

Pottery, however, is more difficult to characterize or fingerprint. Often archaeologists use what is called a "criterion of relative abundance," which infers that most pottery from a production center is to be found locally. Yet, if pottery was produced for exchange then only little would be found locally, thus limiting the application of this concept. The "criterion of relative abundance" is useful, however, in a physico-chemical analysis for provenance studies; that is, relate the pottery under study back to its raw materials: nonplastic inclusions (for example, minerals, sands) and clay. Mineral inclusions found within pottery can be related to local geology and both river and beach sands. Clays, in contrast, vary greatly in respect of the underlying geology, and a study of the elemental composition of pottery may give a clue to their origin. The chemical analysis of clay is complex. Bishop and associates (1982) note that clay minerals not only depend on their parent material but are also affected by weathering or hydrothermal activity, climate, and geomorphology. They state that "even if a region is considered to be homogeneous in its gross geologic characteristics, significant mineralogical and chemical differences may be discerned between clay deposits" (Bishop, Rands, and Holley 1982).

The relationship of other material to its origin depends on the materials under study—chert, volcanic stone, amber, glass, metals, and so on; each has different parameters to take into account when an attempt is made to identify origins. In any provenance study of archaeological objects, a cautionary statement from Garman Harbottle, a nuclear physicist, best be heeded:

> archaeologists love the term *sourcing*, with its upbeat, positive thrust—that you analyze or examine an artifact and, by comparison with the material of known origin, "source" it. In point of fact, with very few exceptions, you cannot unequivocally source anything. What you can do is characterize the object, or better, groups of similar objects found in a site or archaeological zone by mineralogical, thermoluminescent, density, hardness, chemical and other tests, and also characterize the equivalent source materials, if they are available, and look for the similarities to generate attributions. A careful job of chemical characterization, plus a little numerical taxonomy and some auxiliary archaeological and/or stylistic information, will often do something as useful: It will produce groupings of artifacts that make archaeological sense. This, rather than absolute proof of origin, will often necessarily be the goal. (Harbottle 1982)

Popular Techniques Available

The following review looks at five techniques. The first three (PIXE-PIGME, the electron microscope, XRF) work on the basis that each atom emits a characteristic X-ray wavelength and energy. All one has to do is (1) generate the X ray and (2) measure the emitted secondary X rays. This allows the identification of elements present and their quantity in weight. The fourth technique uses nuclear technology (NAA), and the last technique is ICP-MS.

PIXE-PIGME

Proton Induced X-ray Emission and Proton Induced Gamma Ray Emission have been used extensively for obsidian analysis and, to a lesser extent, on pottery and other substances, such as bone, metal, and glass. It is the preferred technique in the Pacific obsidian sourcing program, in which obsidian from archaeological contexts in the Pacific region are sourced (see Summerhayes et al. 1998). The sample is hit with a high energy beam and measured first for X rays emitted by protons and then for gamma rays generated. (For a detailed description of the technique and its uses, see Bird, Duerden, and Wilson 1983.)

Accessibility. This technique has been used only on archaeological samples from less than 25 centers worldwide. PIXE-PIGME is dependent on the generation of a high energy ion beam. Accelerators to produce such beams are not found at every university. At the Australian Nuclear Science and Technology Organisation (ANSTO), a higher energy ion beam of 2.5 MeV for PIXE-PIGME was generated by a 3 MeV de Graff accelerator. This accelerator has now been replaced with a Tandetron STAR accelerator. These accelerators are not common; accessibility to PIXE-PIGME as such is thus a problem. Furthermore, a license is needed to generate the proton beam. An archaeologist would not be expected to train to operate such a machine.

Sample Preparation. Samples such as pottery are crushed into a powder and pressed into a pellet for analysis. At ANSTO, obsidian samples are attached onto a specially prepared aluminum sample holder, holding up to 60 samples. The area of obsidian to be analyzed is exposed by a hole in the holder. The obsidian selected for analysis should have some of its surface flat for attachment to the sample holder. One of the major advantages of analyzing obsidian using PIXE-PIGME is that it is nondestructive in the real sense. Many techniques (see below) are said by their practitioners to be nondestructive. Yet, *nondestructive* means something different to the archaeologist than it does to the chemist. For the chemist, it means that a sample residue is left after the analysis. Thus, samples that are crushed or melted, such as pottery, are regarded by the chemist to be nondestructive. Fortunately, the only sample preparation for obsidian undergoing PIXE-PIGME is cleaning using ethanol.

Data Provided. A major advantage of PIXE-PIGME is that it can provide data on over 60 elements (major, minor, and trace), with many samples being capable of analysis per day. At ANSTO, up to 20 samples per day can be analyzed within a normal working day (see Summerhayes et al. 1998 for machine conditions). If the machine is run around the clock, this figure is closer to 60 samples per day.

Emitted X rays are detected and counted by a Si (Li) detector; a higher energy Ge (Li) detector is used for counting gamma rays. Elements to be analyzed for obsidian analyses include F, Na, and Al using PIGME, and Si, K, Ca, Ti, Mn, Fe, Cu, Zn, Ca, Rb, Sr, Y, Zr, Nb, and Pb using PIXE. PIXE-PIGME can, however, analyze for most of the range of major, minor, and trace elements within detection limits.

Thus, in a week, it is possible to have 300 samples analyzed using PIXE-PIGME and, in the case of obsidian, the flakes returned without modification. The same number can be achieved for pottery as well. Ochre has also been similarly analyzed.

Electron Microscope and Microprobe

This technique basically measures X rays that are generated from the surface of a sample by a low energy (10–20 keV) electron beam directed onto a sample—both the Electron Microprobe and the Scanning Electron Microscope (SEM) proving popular in characterizing archaeological material. They are easily accessible, relatively inexpensive to use, and easy for archaeologists to learn to operate. They can analyze micro-sections of metals, glazes, and slips on pottery and can allow the imagery of macro- and micro-materials. In addition, the probe is important in analyzing coarse-ware pottery. A major problem in the use of pottery chemical characterization concerns the effect of manually added mineral inclusions on elemental concentrations. Mineral inclusions either occur naturally in the clays or are artificially added to counteract shrinkage during the pot's drying process. Their manual addition to the clay will affect the chemical profile of a pot and, if not compensated for, will result in erroneous data used for modeling production, exchange, and consumption. For instance, pots made from the same clay source could have different chemical elemental concentrations owing to the addition of either varying amounts of similar minerals or different mineral inclusions. This, in turn, could result in the pottery being attributed to different loci of production with misleading exchange and consumption patterning resulting. To get around this problem, the electron microprobe is of importance. By positioning the specimen under the electron beam, the microprobe can discriminate between nonplastic inclusions and the clay matrix.

Accessibility. This technique is very accessible. Nearly every university or major research organization has a scanning electron microscope or a specialist electron microprobe. Most research centers and departments housing this equipment will allow archaeologists to use it after adequate training.

Sample Preparation. For a full qualitative analysis, all samples must be perfectly flat. This means a lengthy preparation for rocks, pottery, and metals. For pottery, this involves impregnation with an epoxy resin, and the sample's placement into a resin pellet. For rocks, a thick section on a glass slide would suffice. For obsidian, chips are cut from the obsidian piece and placed into a specialist sample holder. No matter what the material is, the samples are highly polished as surface irregularities will affect the results. Extreme caution must also be taken when preparing pottery to remove oxygen within the sample. Oxygen will cause problems when the sample goes into the vacuum for analysis. Unlike with rocks and minerals, it can take 20 minutes for the probe to pump down the vacuum prior to analysis.

Lastly, samples are cleaned with ethanol and freon in an ultrasonic cleaner and coated with carbon under vacuum. A conductive coating is needed to "provide a path for the probe current to flow to earth" (Reed 1977: 178). Carbon is selected because of its low atomic number, "which ensures that the coating does not significantly absorb the energy X-rays" (Reed 1977: 179).

Data Provided. Both a wavelength dispersive spectrometer (WDS—whereby crystals separate out the X rays by wavelength) and energy dispersive spectrometer (EDS—a black box detector measuring wavelength energy) are used with the electron microprobe; an EDS is mostly used only with a SEM (EDXA). Both WDS and EDS separate and measure the intensity of the X ray, and both are used with geological standards. WDS, however, provides a better separation and resolution of many of the elements. The electron microprobe and SEM will provide most major and minor elements, and the former can even analyze trace elements. However, the SEM's resolution below 1% is not good. For pottery analysis, the following elements (as oxides) are often used: Mg, Al, Si, K, Ca, Ti, and Fe. In previous research, these elements were found more useful than the trace elements in discriminating not only prehistoric pottery groupings but also pots made from a single production center (see Summerhayes 1997 for further discussion).

In summary, these microscopes are perfect for pottery, obsidian rock, and metal analyses. They are relatively easy to use (even an archaeologist could be trained), and easy to access. The probe is, however, much more costly by a factor of four over the SEM.

X-Ray Fluorescence Spectroscopy (XRF)

XRF has been one of the most popular techniques for the past 40 years (Best et al. 1992; Weisler 1998; Weisler and Kirch 1996; see Weisler 1997 for a detailed background to its use in identifying trade in the Pacific). Every geology department has one, plus it is easy to learn and easy to maintain. It basically involves the sample being hit with a primary X ray—the resulting characteristic elements are measured either by a WDS or an EDS (see above). Like PIXE-PIGME and the Electron microscope, it is fast, efficient, and effective. For archaeological materials, it has been used mostly for rocks, pottery, obsidian and glasses, and metals.

Accessibility. This is probably the most accessible of all techniques. As noted above, it is found in most geology departments and is the favorite machine of instruction to undergraduates.

Sample Preparation. The sample preparation depends on the machine detectors. When one is using a WDS, two archaeological samples are required for a complete coverage of major, minor, and select trace elements. The first, for major and minor elements, is crushed into a powder. The second, for trace elements, is melted and fused into a pellet. When one is using an EDS, the sample does not have to be either powdered or melted—but a word of caution: the sample must be polished for good results. It is possible to analyze a complete artifact if the sample chamber will allow it, but the results would be semiquantitative in nature.

Data Provided. The distribution of elemental data is commensurate with the electron microprobe. It is obvious that XRF is a fast and easy to access technique. It works well with rocks (stone tools), glass, and metals. It is adequate for fine pottery, but for coarse ware it suffers from the same problem with temper outlined above.

Neutron Activation Analysis (NAA)

Neutron activation analysis is a technique that involves the production of radioactivity that decays on an observable time scale. Basically, a sample is hit with neutrons. The neutrons interact with the neutrons of the sample resulting in elements to form radioactive isotopes. The decay for each element is already known and thus, when measured, informs us of the element and its concentration. This is the preferred technique by most archaeologists for sourcing.

Accessibility. NAA is not an easily accessible technique as it needs a nuclear reactor. There are a number of laboratories that specialize in archaeological work, the best known being at the University of Missouri–Columbia. Samples can be sent there for analysis on a commercial basis.

Sample Preparation. Sample preparation is minimal. A crushed sample or, if small enough, an entire object can be analyzed directly. NAA can also analyze liquid and gaseous samples.

Data Provided. NAA provides a variety of elements, with little material needed (10 mg). It has the best detection limits of sensitive elements (down to parts per billion), making it the premier technique available. It is perfect for rock (stone tools), obsidian, and glass. As well as for pottery, since it can measure the rare earths found within the clay matrix, which are undetectable using other techniques.

This is the premier technique for sourcing. (See Neff 1992 for detailed examples of the application of NAA to pottery.)

Inductively Coupled Plasma-Mass Spectrometer (ICP-MS)

ICP-MS is a technique on the increase. Although it was developed in the late 1980s, the number of applications on archaeological data has risen steeply over the last few years. There are many reasons for this; one, of course, is its purchase in science-based university departments and research centers for geological research, and its subsequent "big sell" for archaeological studies. ICP-MS has been promoted as faster, more sensitive, and having superior detection limits on many elements (see Kennett et al. 2004). The technique basically involves an argon plasma to atomize and ionize the elements in a sample. A mass spectrometer is used to measure the elements in a sample.

Accessibility. ICP-MS is becoming very accessible, with most universities investing in one. Unlike XRF or the probe, which are common in geology departments, this technique would be more at home in a chemistry department. This is not a technique that an archaeologist would be able to use by him- or herself without extensive training.

Sample Preparation. Sample preparation is lengthy. There are two types of ICP-MS: one using laser ablation, the other using microwave digestion. The preparation for both differs, with microwave digestion being more labor intensive with the samples crushed into a powder, mixed with

acids, evaporated and mixed again, decanted and purified with water, and decanted again. This is a lengthy and complicated procedure. The other preparation involving laser ablation of the sample is faster and less messy. (See Kennet et al. 2004 for an example of ICP-MS with microwave digestion.)

Data Provided. ICP-MS is argued to measure the whole range of elements—major, minor, and trace down to parts per billion. For example, in a recent analysis on Lapita pottery, 37 elements were measured: Be, Mg, Al, K, Sc, V, Cr, Mn, Fe, Co, Ni, Cu, Zn, Ga, Rb, Sr, Y, In, Cs, Ba, La, Ce, Pr, Nd, Sm, Eu, Gd, Tb, Dy, Ho, Er, Yb, Lu, Pb, Bi, Th, and U.

This is a new technique that needs more street credibility before it becomes popular. The crushing of the sample, or even using laser ablation, would be ill-advised for coarse-ware pottery, but for materials with a homogeneous matrix, such as obsidian, it would be ideal.

Conclusions

The choice of technique used by archaeologists has unfortunately often rested on what is close at hand. Instead the choice should be dependent on the questions asked by the archaeologist, the material being analyzed, the elements needed to discriminate between sources, and the availability of the technique. There are many techniques available, too numerous to list them here. Some are old and well known, such as X-ray diffraction analysis (XRD), which is commonly used for mineral analysis (in particular clay minerals); others are rarely used, such as mineral magnetics, which has been used for ochre characterization (Mooney, Geiss, and Smith, 2003) still others are becoming popular such as Raman Spectroscopy (both dispersive and interferometric Fourier transform) on mineral analyses, which has a wide variety of uses. (See Edwards and Chalmers's [2005] edited volume; and Smith and Clark's [2004] review article.) With a bit of common sense, and wisdom from geologists and archaeologists experienced with these techniques, the task of choosing between techniques will be a much easier one.

References

Bird, J. R., Duerden, P., and Wilson, D. J. 1983. Ion beam techniques in archaeology and the arts. *Nuclear Science Applications* 1: 357–516.

Bishop, R., Rands, R., and Holley, G. 1982. Ceramic compositional analyses in archaeological perspective. *Advances in Archaeological Method and Theory* 5: 275–320.

Best, S., Sheppard, P., Green, R., and Parker, R. 1992. Necromancing the stone: Archaeologists and adzes in Samoa. *Journal of the Polynesian Society* 101: 45–85.

Edwards, H. G. M., and Chalmers, J. M. 2005. *Raman Spectroscopy in Archaeology and Art History*. Cambridge: Royal Society for Chemistry.

Harbottle, G. 1982. Chemical characterization in archaeology, in Ericson J. Earle T. (eds.), *Contexts for Prehistoric Exchange*. New York: Academic Press.

Kennett, D. J., Anderson, A. J., Cruz, M. J., Clark, G. R., and Summerhayes, G. R. 2004. Geochemical characterisation of Lapita pottery via Inductively Coupled Plasma Mass Spectroscopy (ICP-MS). *Archaeometry* 46: 35–46.

Mooney, S. D., Geiss, C., and Smith, M. 2003. The use of mineral magnetic parameters to characterize archaeological ochres. *Journal of Archaeological Science* 30: 511–23.

Neff, H. 1992. *Chemical Characterization of Ceramic Pastes in Archaeology*. Monographs in World Archaeology 7. Madison: Prehistory Press.

Reed, S. 1977. *The Electron Microprobe*. Cambridge: Cambridge University Press.

Renfrew, C. 1969. Trade and cultural process in European prehistory. *Current Anthropology* 10: 151–69.

———. 1977. Alternative models for exchange and spatial distribution, in T. K. Earle and J. T. Ericson (eds.), *Exchange Systems in Prehistory*, pp. 71–90. New York: Academic Press.

Renfrew, C., Dixon, J. E., and Cann, J. R. 1966. Obsidian and early cultural contact in the Near East. *Proceedings of the Prehistoric Society* 32: 30.

Sayre, E., and Dodson, R. 1957. Neutron activation study of Mediterranean potsherds. *American Journal of Archaeology* 61: 35–41.

Schwab, R., Heger, D., Hoppner, B., and Pernicka, E. 2006. The provenance of iron artefacts from Maching: A multi-technique approach. *Archaeometry* 48: 433–52.

Shackley, S. 1998. *Advances in Archaeological Volcanic Glass Studies*. New York: Plenum Press.

Shepard, A. 1965. Rio Grande glaze-paint pottery: A test of petrographic analysis, in F. Matson (ed.), *Ceramics and Man*. New York: Viking Fund Publications in Anthropology No. 41.

Smith, G. D., and Clark, R. J. H. 2004. Raman microscopy in archaeological science. *Journal of Archaeological Science* 31: 1137.

Summerhayes, G. R. 1997. Losing your temper: The effect of mineral inclusion on pottery analysis, *Archaeology in Oceania* 32: 108–18.

———. 2000. *Lapita Interaction*. Terra Australia 15. Canberra: Research School of Pacific and Asian Studies, Australian National University.

Summerhayes, G. R., Bird, R., Fullagar, R., Gosden, C., Specht, J., and Torrence, R. 1998. Application of PIXE-PIGME to archaeological analysis of changing patterns of obsidian use in west New Britain, Papua New Guinea, in S. Shackley (ed.), *Advances in Archaeological Volcanic Glass Studies*. New York: Plenum Press.

Webb, J. M., Frankel, D., Stos, Z. A., and Gale, N. 2006. Early Bronze Age metal trade in the eastern Mediterranean: New compositional and lead isotope evidence from Cyprus. *Oxford Journal of Archaeology* 25: 261–88.

Weisler, M. 1997. *Prehistoric Long-Distance Interaction in Oceania: An Interdisciplinary Approach*. New Zealand Archaeological Association Monograph 21. Auckland: New Zealand Archaeological Association.

———. 1998. Hard evidence for prehistoric interaction in Polynesia. *Current Anthropology* 39: 521–32.

Weisler, M., and Kirch, P. V. 1996. Inter-island and inter-archipelago transfer of stone tools in prehistoric Polynesian. *Proceedings of the National Academy of the USA* 93: 1381–85.

Weisler, M., and Woodhead, J. 1995. Basalt Pb isotope analysis and the prehistoric settlement of Polynesia. *Proceedings of the National Academy of Sciences of the USA* 92: 1881–85.

Young, W., and Whitmore, F. 1957. Analysis of oriental ceramics by non-destructive methods. *Far Eastern Ceramic Bulletin* 9: 1–27.

52

TRACKING ANCIENT ROUTES ACROSS POLYNESIAN SEASCAPES WITH BASALT ARTIFACT GEOCHEMISTRY

Marshall Weisler

Much has been written about landscape archaeology from a multitude of perspectives—all extolling a holistic approach to understanding humans within their environmental, social, and ideological contexts at various geographic scales and throughout many time periods. In considering the human colonization of Oceania, nowhere do more "scapes" interweave to form the inseparable whole of islandscapes, seascapes, and starscapes. In the Pacific, the open ocean was anchored by nearby and distant islands to form spheres of interaction between parent and daughter communities for trade and exchange of goods, services, and marriage partners (Kaeppler 1978; Kirch 1988; Weisler 1995). In Remote Oceania (Green 1991), where landfalls are generally smaller and spaced wider than islands in the western Pacific, the starscape was of unparalleled importance, because it was the sequence of stars rising on the horizon that guided voyagers to their destination and return (Finney 1979: 332–35; Lewis 1976). During the day, skilled navigators could read other important "seascape" signposts such as permanent patches of seaweed, submerged reefs, currents that carry colder or discolored water, and turbulent patches and swell and wind patterns. By observing sea birds returning to island roosts after traveling more than 100 kilometers to feed, navigators expanded the location of islands and could follow the birds

to land (Di Piazza and Pearthree 2004: 19; Finney 1996: 100). In essence, it is thus not possible to talk about island colonization detached from the three "scapes" of land, sea, and sky.

This chapter reviews the development of interaction studies utilizing the petrographic and geochemical characteristics of Polynesian basalt adze material and source rocks for documenting prehistoric long-distance interaction (so-called trade and exchange) and colonization routes; it then briefly outlines some significant results. Mapping this "seascape" and determining colonization routes by plotting the spatial and temporal distribution of adze material is of no small significance, since scholars have not always accepted that the colonization of the Pacific was purposeful and intentional; during the past nearly 200 years, debates have polarized between accidental drift voyages and those that were planned (for recent summaries, see Finney 1979; Irwin 1992: 13–16). Sharp (1956) was a proponent of accidental, one-way voyaging; he believed that ancient mariners could not account for drifting off course (set) and that they did not have the ability to navigate through different wind belts to small island targets. However, the innovative computer simulations of Levison and associates (1973) demonstrated that island colonization must have been purposeful. They considered current, wind direction and speed, length of voyage, and time of year from different islands, demonstrating that it was impossible to settle all the Pacific islands by chance or accidental drift voyages.

When Captain Cook entered Polynesian waters two centuries ago, he marveled at the similarities of language, custom, and physical appearance across the region and pondered the origin of the "Indians of the South Seas." Thor Heyerdahl popularized the American origin hypothesis of the Polynesians with his *Kon Tiki* drift voyage from South America to Polynesia (1950), although there is little support today for an eastern origin for Pacific islanders. Conversely, there is overwhelming evidence for west-to-east settlement from the radiocarbon corpus, where archaeological sites are progressively younger from west to east across Oceania (Irwin 1990; Kirch 1997). Physical characteristics of the Polynesians (Howells 1970), routes of the commensal Pacific rat (Matisoo-Smith and Robins 2004), connections based on human mtDNA (Hill and Serjeanston 1989; Lum and Cann 1998), language origins and dispersals (Kirch and Green 2001; Pawley and Ross 1993), and the spread of agriculture (Bellwood 2005) all chart the west to east dispersal of Polynesian peoples.

Voyaging in an easterly direction against the dominant tradewinds may seem a deterrent, but sailing close to the wind was an effective exploration and colonization strategy (Finney 1988; Irwin 1989, 1992). Experimental voyages using replica canoes and traditional voyaging techniques such as celestial navigation have recreated ancient sailing routes (Finney 1977, 1979; Finney et al. 1986).

One way to empirically determine settlement routes, to understand colonization strategies, and to document the extent of postcolonization interaction is by charting the scale, frequency, and duration of exotic artifacts found in distant habitation sites (Weisler 1998). In Polynesia, the woodworking adze was a fundamental component of the prehistoric tool kit (Figure 52.1). Although whole finished forms are somewhat rare in the archaeological record, the byproducts of its manufacture—stone flakes, unfinished and reworked adzes—are relatively common artifacts found in habitation sites. There is another reason why fine-grained basalt adze material occupies an important niche in Polynesian interaction studies: obsidian and pottery, used the world over for charting prehistoric interaction, have a limited occurrence in East Polynesia (Weisler and Woodhead 1995: 1881). Indeed, only New Zealand and Easter Island have obsidian deposits, and their isolation at the far corners of the region makes it unlikely that this resource was traded to tropical Polynesia. Although pottery is common in West Polynesia,

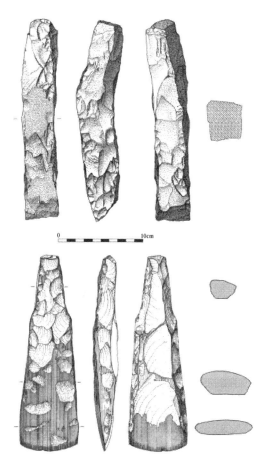

Figure 52.1 Fine-grained basalt adzes: upper, a preform from Moloka'i, Hawaiian Islands and lower, a finished late prehistoric adze from Pitcairn Island.

only a few sherds have been recovered from East Polynesia. Consequently, fine-grained adze material occupies a prominent role in East Polynesian interaction studies.

Some Working Definitions and the Notion of Scale

Trade and exchange are commonly referred to in the Polynesian archaeological literature, yet the necessary and sufficient conditions for these ethnographically charged terms is the identification of exotic artifacts in archaeological contexts that represent the two-way movement of material—a condition that is usually assumed and less often documented empirically (Renfrew 1969: 152; Weisler, Kirch, and Endicott. 1994: 214). The two-way movement of artifacts has been conclusively documented only in the Pitcairn Group-Mangareva region of southeast Polynesia

(Weisler 1997a; Weisler and Woodhead 1995; Woodhead and Weisler 1997). For this reason, it is more accurate to talk about an *interaction sphere* (Caldwell 1964), where the movement of material within a geographically defined region can be archaeologically documented.

The identification of exotic materials in habitation sites distant from their geological origin is generally referred to as *sourcing studies*; *interaction studies* and *provenance studies* are synonyms (see Summerhayes, this volume). Within every interaction sphere, at least two kinds of sites are identified. In the Polynesian context, *quarries* are geological sources of fine-grained basalt that is surface collected, extracted from excavated pits, or removed from outcrops. The Tataga-matau, Samoa, quarry is one of the few places where extraction pits have been identified (Leach and Witter 1987). Tabular material was removed from outcrops atop Mauna Kea, Hawai'i Island (McCoy 1990), but most often source rock is simply surface collected as erosional products near the geological source (Weisler and Sinton 1997: 180). *Sources* are secondary geological deposits such as an accumulation of cobbles downslope in a gulch bottom, river mouth, or seashore (Weisler 2004a: 114). Sources can be archaeologically unknown but identified by artifacts from a geologically well-known area. In this instance, artifacts can be traced to a particular volcano, flow, dike, or outcrop by the geochemistry provided in the geological literature (Weisler and Sinton 1997: 180).

Because most sourcing studies in East Polynesia use fine-grained basalt adze material to track prehistoric interaction, it is necessary to understand the term *fine-grained basalt* from archaeological and geological perspectives. Basalt is a dark-colored igneous rock, commonly extrusive, and fine-grained. There are many kinds of basalt that are distinguished further by their chemistry (for example, alkali vs. tholeiitic) and common location (oceanic island, mid-ocean ridge, and continental). Nunn (1994) provides a good discussion on the origins and development of ocean islands and their rocks. Perhaps the most confusion when describing fine-grained basalt is the term *fine-grained*. Geologists define fine-grained as particles that have an average diameter of < 1 mm (Bates and Jackson 1984: 183). However, in a review of the petrographic characteristics of eight Hawaiian basalt adze quarries, Cleghorn and associates (1985) defined very fine-grained as > 130 grains/mm, fine-grained as ~ 100–130 grains/mm, and coarse-grained as < 90 grains/mm. In geologic-al perspective, all these values refer to fine-grained basalt.

Scale is a useful term in Polynesian sourcing studies, because it defines the geographic limits of the study (for a general discussion of "scale" in landscape archaeology, see also Head, this volume). In geological terms, scale can refer to the petrologic province (oceanic island basalt), archipelago (Mangareva), island (Pitcairn), volcano (Mauna Kea), geologic feature (cone or dike swarm), geological event (flow, dike), or geologic sample (rock). This is a hierarchical relationship, whereby each unit is contained within the larger scale above it (Weisler 1993a: 62–63; Weisler and Sinton 1997: 180–82). Scale is helpful in archaeological sourcing studies as certain unique hand specimen; petrographic or geochemical characteristics are useful at assigning artifacts to source at different spatial scales. For example, a high abundance of yttrium (Y) is found only at one quarry when one is considering the eight fine-grained basalt sources of west Moloka'i, a region composed of a single volcano. However, if the geographic scale is enlarged to the whole of Polynesia, then high Y is not necessarily unique.

Analytical Techniques

Whenever the scale of interest is expanded, more powerful analytical methods might be required. Petrographic thin-sections were the first exploratory method used to determine the characteristics of Hawaiian basalt quarries and sources and to assign artifacts to source (Cleghorn et al. 1985; Lass 1994; Powers 1939). However, not all characteristics are described consistently, rendering comparisons between studies problematic. X-ray fluorescence (XRF) analysis was quickly adopted as the preferred method, because results are quantitative, analytical accuracy is routinely evaluated to worldwide standards (e.g., Govindaraju 1989), and precision can be controlled by careful sample preparation and close attention to analytical procedures (Weisler 1993a: 64). The XRF method is described in Summerhayes (this volume; Weisler and Sinton 1997: 175–77), so it will not be discussed here in detail. Briefly, samples are bombarded with X rays, and fluorescent X rays are emitted from the sample that is characteristic of the kind and amount of the sample's chemical composition. Specimen sample sizes can be as low as ~ 3–5 grams for destructive wavelength dispersive XRF, where the results are fully quantitative. Energy-dispersive XRF can accommodate whole specimens (an obvious advantage when using museum artifacts or borrowing specimens from traditional owners), but results are often best used as ratio-level data (Weisler 1993a: 73–74).

Each analytical method comes with its limitations and XRF, despite its obvious advantages when assigning artifacts to source at the scale of archipelago (Allen and Johnson 1997; Rolett et al. 1997; Weisler, Conte, and Kirch 2004), encounters problems when considering the entire petrologic province of Polynesia. As summarized by Weisler and Woodhead (1995: 1882; Woodhead and Weisler 1997), variation in oxides and trace elements is largely controlled by near-surface magma-chamber processes such as fractional crystallization and crustal assimilation. Because crystal fractionation history is broadly similar between most oceanic islands, individual parameters are rarely unique to a single volcano. Ratios of highly incompatible trace elements are relatively insensitive to these processes and can be used to discriminate; however, there can be considerable overlap between volcanic centers. Importantly, radiogenic isotope variations of lead (Pb), strontium (Sr), and neodymium (Nd) are a function of both parent/daughter elemental ratios in the mantle source of oceanic island basalts (OIB) and the age of this source. The 1–2 billion year age of OIB has resulted in a diversity of isotopic compositions for volcanoes (quarries and sources), facilitating source assignments of artifacts at the Polynesia-wide scale. Additionally, only 100 mg of sample is needed for routine analysis. The first application of lead isotope analysis of source rocks and artifacts was in southeast Polynesia (Mangareva and the Pitcairn Group), where adze material from the Tautama, Pitcairn Island source was identified in Mangareva, 400 km to the west (Weisler and Woodhead 1995).

Polynesian Interaction Studies

There is ample evidence accumulating from basalt sourcing for interaction at the intra-island, intra-archipelago, and inter-archipelago scales for Tonga (Weisler 2004b), Samoa (Best et al. 1992; Clark, Wright, and Herdrich 1997; Weisler 1993b; Winterhoff et al. 2007), the Phoenix and Line archipelagos (Di Piazza and Pearthree 2001, 2004), the Cook Islands (Allen 1996; Allen and Johnson 1997; Sheppard, Walter, and Parker 1997; Walter 1998; Walter and Sheppard 1996, 2001; Weisler and Kirch 1996; Weisler, Kirch, and Endicott 1994), the Marquesas (Rolett 1998; Rolett et al. 1997), Hawai'i (Mills and Lundblad 2006; Weisler 1990), the Australs (Bollt, in press), the Society Islands (Weisler 1998), Mangareva (Weisler 1996, 2004b; Weisler and Green 2001; Weisler, Conte, and Kirch 2004), the Pitcairn Group (Weisler 1995, 1997a, 1998, 2002; Weisler and Woodhead 1995; Woodhead and Weisler 1997), and Norfolk Island (Anderson et al. 1997). At one fundamental level, the results demonstrate contact

between source locations and habitation sites at different spatial scales, but more sophisticated interpretations are emerging.

Interaction as a Colonization Strategy

In the most detailed study to date, Weisler (1995, 1997a, 2002; Diamond 2005: 120–35) identified not only fine-grained basalt but also imported volcanic glass, oven stones, and artifact styles, as well as transferred plants and animals over ca. 800 years in the Pitcairn Group-Mangareva interaction sphere (Weisler 1998: fig. 1). The resource-poor, raised limestone (makatea) island of Henderson is 100 km north of the volcanic Pitcairn Island. Both landfalls are isolated and small, with limited pot-able water and arable land, and narrow encircling reefs. Under the best conditions, islands such as these tested the limits of permanent Polynesian occupation. Settled from the more ecologically diverse Mangareva, some 400 km west, the culture-historical record of Henderson shows six centuries of importation of goods (and undoubtedly services and marriage partners) from parent communities on Mangareva. When imports cease in Henderson's sequence at about A.D. 1500, local Tridacna clamshell is fashioned into inferior adzes to replace those of stone. For making fishhooks, small local pearlshell replaced the large black-lipped pearlshell from the Mangareva lagoon. Most devastating was probably the inability to communicate with Mangareva for planting stock and marriage partners to add genetic diversity to the small population. Consequently, the long-term provisioning from Mangareva to the isolated outposts in the Pitcairn Group was a colonization strategy for resource-poor islands, and the cessation of interaction was clearly tied to changes in the material culture record and eventual abandonment of the Pitcairn Group.

Intensity of Interaction

The changing abundance of fine-grained basalt during a culture-historical sequence can also be tied to the relative contribution of various stone sources over time, signaling changing relationships with different social groups. Allen and Johnson (1997) have shown that the changing frequency of fine-grained basalt from the Ureia coastal habitation site on Aitutaki, southern Cook Islands, was related to the intensity of use of different extra-archipelago sources over time. In the Marquesas, Rolett and associates (1997) identified the decline of intra-archipelago voyaging in the Marquesas as signaled by an absence of imports in habitation sites around A.D. 1500.

Assessing the Frequency and Scale of Postcolonization Interaction

It is important to point out that basalt adzes were only a small component of ethnographically known Polynesian interaction networks (Weisler 1997b: fig. 1.2). In the Fiji-Tonga-Samoa interaction sphere (Kaeppler 1978), fewer than 10% of the items circulated have preserved in archaeological contexts. The interaction sphere between the high volcanic Society Islands and the northwest Tuamotu atolls, as recorded historically (Oliver 1974: 1148 n.2), contained, most notably, shells and white dog hair but no basalt adzes. Changing from presence-absence to frequency-ranked data brings increased uncertainty for interpretations. For example, the cessation of inter-island voyaging is assumed when imports are no longer present in the excavated sequence, and this seems reasonable. Changing frequencies, however, are more problematic. Issues of quantification must also be addressed since artifact counts (and not weights) may reflect reworking of imported whole adzes.

Other factors should also be considered, such as the distance that an imported object is moved. One would expect that less exotic adze material might be found as distance increases from the source (Renfrew 1969). The number of archipelagos where exotic adze material is found is a good ubiquity measure that defines the geographic scale of imports that might relate to the frequency of voyaging. Large-scale interaction systems might develop over longer periods of time than smaller ones.

The largest interaction sphere in East Polynesia is centered at the Eiao quarry in the northern Marquesas (Figure 52.2). Basalt from Eiao has been identified from four archipelagos: the Line Islands, some 2,400 km to the northwest; the Tuamotus (1,100 km southwest), the Society Islands (1,425 km southwest), and Mangareva (1,750 km to the south-southeast) (Di Piazza and Pearthree 2001; Weisler 1998 and unpublished). This is clear evidence of postcolonization interaction, especially since the finished Eiao adzes sourced to the Tuamotus, Societies, and Mangareva are all late

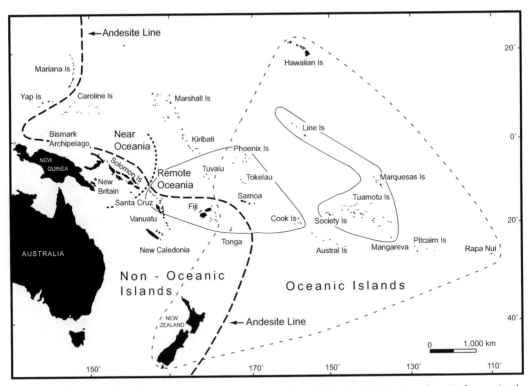

Figure 52.2 The largest Polynesian interaction spheres identified by the distribution of exotic fine-grained basalt adze material. The West Polynesian sphere is centered at the adze quarries on Tuapila Island, Samoa. The East Polynesian sphere is centered on the Eiao quarry, northwest Marquesas Islands. Both interaction spheres encompass ca. 4,000 km from end to end. Several smaller interaction networks are not plotted here (Weisler 1998: fig. 1).

prehistoric forms (Weisler 1998: fig. 2). What the *intensity* of this interaction represents is uncertain, but voyagers were traveling great distances after initial colonization. Anderson suggests that "interaction between archipelagos was limited in extent" (2003: 173), but the *geographic* extent of the Eiao source covers ca. 4,000 km from end to end—a remarkable achievement for prehistoric voyagers.

Conclusions

There is much to be learned from the geochemical sourcing of fine-grained basalt adze material in Polynesia, which has provided insights into how Pacific navigators used the seascape as a bridge to reach even the most distant and isolated landfalls. Determining the scale and the frequency of exotic adze materials by geochemical characterization of artifacts and source rocks is a powerful means for understanding colonization strategies and the changing dynamics of trade and exchange. Acquiring additional source rocks from the leeward Society Islands and the Australs remains a priority, and determining the intrasource variability of each quarry or source is integral to unequivocally assigning artifacts to sources. Highly accurate isotopic studies (Weisler and Woodhead 1995) have shown great promise in determining artifact sources using small (~ 100 mg) samples. This method has also been particularly useful when suspected prehistoric interaction regions are large, resulting in the greater likelihood of geochemical overlap between volcanoes. In these circumstances, artifact assignments using XRF are most problematic.

The methods and protocols described here should be effective in volcanic regions worldwide (Japan is an excellent example), thus providing archaeologists another tool for tracking ancient routes across land and seascapes. Future studies will become only more sophisticated and address a wider range of issues that are only briefly canvassed here.

Acknowledgments

I thank Les O'Neill and Judith Ogden for the adze illustrations. Michael Haslam amended Figure 52.2, originally drafted by Les.

References

Allen, M. 1996. Patterns of interaction in southern Cook Island prehistory. *Bulletin of the Indo-Pacific Prehistory Association* 15: 13–21.
Allen, M., and Johnson, K. 1997. Tracking ancient patterns of interaction: recent geochemical studies in the Southern Cook Islands, in M. Weisler (ed.), *Prehistoric Long-Distance Interaction in Oceania: An Interdisciplinary Approach*, pp. 111–33. New Zealand Archaeological Association Monograph 21.
Anderson, A. 2003. Entering uncharted waters: Models of initial colonization in Polynesia, in M. Rockman and J. Steele (eds.), *Colonization of Unfamiliar Landscapes: The Archaeology of Adaptation*, pp. 169–89. London: Routledge.
Anderson, A., Ambrose, W., Leach, F., and Weisler, M. 1997. Material sources of basalt and obsidian artefacts from a prehistoric settlement site on Norfolk Island, South Pacific. *Archaeology in Oceania* 32: 39–46.
Bates, R., and Jackson, J. (eds.). 1984. *Dictionary of Geological Terms*. New York: Anchor Press/Doubleday.
Bellwood, P. 2005. *First Farmers: The Origins of Agricultural Societies*. Oxford: Blackwell Publishing.
Best, S., Sheppard, P., Green, R. C., and Parker, R. 1992. Necromancing the stone: Archaeologists and adzes in Samoa. *Journal of the Polynesian Society* 101: 45–85.
Bollt, R. In press. *Peva: The Archaeology of a Valley on Rurutu, Austral Islands, East Polynesia*. Bishop Museum Bulletins in Anthropology.
Caldwell, J. 1964. Interaction spheres in prehistory, in J. Caldwell and R. Hall (eds.), *Hopewellian Studies*, pp. 134–43. Springfield: Illinois State Museum Scientific Papers 12.
Clark, J., Wright, E., and Herdrich, D. 1997. Interactions within and beyond the Samoan archipelago: Evidence from basaltic rock geochemistry, in M. Weisler (ed.), *Prehistoric Long-Distance Interaction in Oceania: An Interdisciplinary Approach*, pp. 68–84. New Zealand Archaeological Association Monograph 21.
Cleghorn, P., Dye, T., Weisler, M., and Sinton, J. 1985. A preliminary petrographic study of Hawaiian stone adze quarries. *Journal of the Polynesian Society* 94: 235–51.
Diamond, J. 2005. *Collapse: How Societies Choose to Fail or Survive*. London: Allen Lane.
Di Piazza, A., and Pearthree, E. 2001. Voyaging and basalt exchange in the Phoenix and Line archipelagos: The viewpoint from three Mystery islands. *Archaeology in Oceania* 36: 146–52.
———. 2004. *Sailing Routes of Old Polynesia, the Prehistoric Discovery, Settlement and Abandonment of the Phoenix Islands*. Bishop Museum Bulletin in Anthropology 11.
Finney, B. 1977. Voyaging canoes and the settlement of Polynesia. *Science* 196: 1277–85.
———. 1979. Voyaging, in J. Jennings (ed.), *The Prehistory of Polynesia*, pp. 323–51. Cambridge: Harvard University Press.
———. 1988. Voyaging against the direction of the trades: A report of an experimental canoe voyage

from Samoa to Tahiti. *American Anthropologist* 90: 401–05.

Di Piazza, A., and Pearthree, E. 1996. Colonizing an island world, in W. Goodenough (ed.), *Prehistoric Settlement of the Pacific*, pp. 71–116. Philadelphia: American Philosophical Society.

Finney, B., Kilonsky, B., Somsen, S., and Stroup, E. 1986. Re-learning a vanishing art. *Journal of the Polynesian Society* 95: 41–90.

Govindaraju, K. 1989. *1989 Compilation of Working Values and Sample Description for 272 Geostandards*. Geostandards Newsletter Special Issue 13: 1–113.

Green, R. 1991. Near and Remote Oceania: Disestablishing "Melanesia" in culture history, in A. Pawley (ed.), *Man and a Half: Essays in Pacific Anthropology and Ethnobiology in Honour of Ralph Bulmer*, pp. 491–502. Auckland: The Polynesian Society.

Heyerdahl, T. 1950. *The Kon-Tiki Expedition by Raft Across the South Seas*. London: George Allen and Unwin.

Hill, A., and Serjeanston, S. (eds.). 1989. *The Colonization of the Pacific: A Genetic Trail*. Oxford: Clarendon Press.

Howells, W. 1970. Anthropometric Grouping analysis of Pacific peoples. *Archaeology and Physical Anthropology in Oceania* 5: 192–217.

Irwin, G. 1989. Against, across and down the wind: A case for the systematic exploration of the remote Pacific Islands. *Journal of the Polynesian Society* 98: 167–206.

———. 1990. Human colonization and change in the remote Pacific. *Current Anthropology* 31: 90–94.

———. 1992. *The Prehistoric Exploration and Colonization of the Pacific*. Cambridge: Cambridge University Press.

Kaeppler, A. 1978. Exchange patterns in goods and spouses: Fiji, Tonga and Samoa. *Mankind* 11: 246–52.

Kirch, P. 1988. Long-distance exchange and island colonization: The Lapita case. *Norwegian Archaeological Review* 21: 103–17.

———. 1997. *The Lapita Peoples, Ancestors of the Oceanic World*. Oxford: Blackwell Publishing.

Kirch, P., and Green, R. 2001. *Hawaiki, Ancestral Polynesia: An Essay in Historical Anthropology*. Cambridge: Cambridge University Press.

Lass, B. 1994. *Hawaiian Adze Production and Distribution: Implications for the Development of Chiefdoms*. Monograph 37. Los Angeles: Institute of Archaeology, University of California.

Leach, H., and Witter, D. 1987. Tataga-matau "rediscovered." *New Zealand Journal of Archaeology* 9: 33–54.

Levison, M., Ward, R., and Webb, J. 1973. *The Settlement of Polynesia: A Computer Simulation*. Minneapolis: University of Minnesota Press.

Lewis, D. 1976. A return voyage between Puluwat and Saipan using Micronesian navigational techniques, in B. Finney (ed.), *Pacific Navigation and Voyaging*, pp. 15–28. Wellington: The Polynesian Society.

Lum, J., and Cann, R. 1998. mtDNA and language support a common origin of Micronesians and Polynesians in Island Southeast Asia. *American Journal of Physical Anthropology* 105: 109–19.

Matisoo-Smith, E., and Robins, J. 2004. Origins and dispersals of Pacific peoples: Evidence from mtDNA phylogenies of the Pacific rat. *Proceedings of the National Academy of Sciences, USA* 101: 9167–72.

McCoy, P. 1990. Subsistence in a "non-subsistence" environment: Factors of production in a Hawaiian alpine desert adze quarry, in D. Yen and J. Mummery (eds.), *Pacific Production Systems: Approaches to Economic Prehistory*, pp. 85–119. Occasional Papers in Prehistory 18. Canberra: Department of Prehistory, Research School of Pacific Studies, The Australian National University.

Mills, P., and Lundblad, S. 2006. *Preliminary Field Report: The Geochemistry of the Ko`oko`olau Complex, Mauna Kea Adze Quarry (50-10-23-4136) TMK: 4-4-15:10*. Prepared for the Hawaii Natural Area Reserves System (NARS) Commission. Hilo: University of Hawai'i.

Nunn, P. 1994. *Oceanic Islands*. Oxford: Blackwell Publishing.

Oliver, D. 1974. *Ancient Tahitian Society*. 3 volumes. Honolulu: University of Hawai'i Press.

Pawley, A., and Ross, M. 1993. Austronesian historical linguistics and culture history. *Annual Review of Anthropology* 22: 425–59.

Powers, H. 1939. Hawaiian adz materials in the Haleakala section of Hawai'i National Park. *Proceedings of the Hawai'ian Academy of Science*, 14th Annual Meeting 1938–1939: 24. B. P. Bishop Museum Special Publication 34. Honolulu.

Renfrew, C. 1969. Trade and culture process in European prehistory. *Current Anthropology* 10: 151–69.

Rolett, B. 1998. *Hanamiai, Prehistoric Colonization and Cultural Change in the Marquesas Islands (East Polynesia)*. New Haven, CT: Yale University Publications in Anthropology 81.

Rolett, B., Conte, E., Pearthree, E., and Sinton, J. 1997. Marquesan voyaging: Archaeometric evidence for inter-island contact, in M. Weisler (ed.), *Prehistoric Long-Distance Interaction in Oceania: An Interdisciplinary Approach*, pp. 134–48. New Zealand Archaeological Association Monograph 21.

Sharp, A. 1956. *Ancient Voyagers in the Pacific*. Wellington: The Polynesian Society.

Sheppard, P., Walter, R., and Parker, R. 1997. Basalt sourcing and the development of Cook Island exchange systems, in M. Weisler (ed.), *Prehistoric Long-Distance Interaction in Oceania: An Interdisciplinary*

Approach, pp. 85–110. New Zealand Archaeological Association Monograph 21.

Walter, R. 1998. *Anai'o: The Archaeology of a Fourteenth-Century Polynesian Community in the Cook Islands*. New Zealand Archaeological Association Monograph 22.

Walter, R., and Sheppard, P. 1996. The Ngati Tiare adze cache: Further evidence of prehistoric contact between West Polynesia and the southern Cook Islands. *Archaeology in Oceania* 31: 33–39.

———. 2001. Cook Islands basalt sourcing: current issues and directions, in M. Jones and P. Sheppard (eds.), *Australasian Connections and New Directions: Proceedings of the 7th Australasian Archaeometry Conference*, pp. 383–99. Research in Anthropology and Linguistics Number 5. Department of Anthropology, University of Auckland.

Weisler, M. I. 1990. A technological, petrographic, and geochemical analysis of the Kapohaku adze quarry, Hawai'ian Islands. *New Zealand Journal of Archaeology* 12: 29–50.

———. 1993a. The provenance of Polynesian adze material: A review and suggestions for improving regional data bases. *Asian Perspectives* 32(1): 61–83.

———. 1993b. Chemical characterization and provenance of Manu'a adz material using a non-destructive x-ray fluorescence technique, in P. Kirch and T. Hunt (eds.), *The To'aga Site: Three millennia of Polynesian Occupation in the Manu'a Islands, American Samoa*, pp. 167–88. Berkeley and Los Angeles: University of California, Archaeological Research Facility Contributions 51.

———. 1995. Henderson island prehistory: Colonization and extinction on a remote Polynesian island. *Biological Journal of the Linnean Society, London* 56: 377–404.

———. 1996. An archaeological survey of Mangareva: Implications for regional settlement models and interaction studies. *Man and Culture in Oceania*. 12: 61–85.

———. 1997a. Prehistoric long-distance interaction at the margins of Oceania, in M. Weisler (ed.), *Prehistoric Long-Distance Interaction in Oceania: An Interdisciplinary Approach*, pp. 149–72. New Zealand Archaeological Association Monograph 21.

———. 1997b. Introduction, in M. Weisler (ed.), *Prehistoric Long-Distance Interaction in Oceania: An Interdisciplinary Approach*, pp. 7–18. New Zealand Archaeological Association Monograph 21.

———. 1998. Hard evidence for prehistoric interaction in Polynesia. *Current Anthropology* 39: 521–32.

———. 2002. Centrality and the collapse of long-distance voyaging in East Polynesia, in M. D. Glascock (ed.), *Geochemical Evidence for Trade and Exchange*, pp. 257–73. Westport, CT: Bergin and Garvey.

———. 2004a. A stone tool basalt source on 'Ata, southern Tonga. *New Zealand Journal of Archaeology* 25: 113–20.

———. 2004b. Contraction of the southeast Polynesian interaction sphere and resource depression on Temoe Atoll. *New Zealand Journal of Archaeology* 25: 57–88.

Weisler, M. I., Conte, E., and Kirch, P. V. 2004. Material culture and geochemical sourcing of basalt artifacts, in E. Conte and P. V. Kirch (eds.), *Archaeological Investigations in the Mangareva Islands, French Polynesia*, pp. 128–48. Archaeological Research Facility Contribution No. 62. Berkeley: University of California.

Weisler, M. I., and Green, R. C. 2001. Holistic approaches to interaction studies: A Polynesian example, in M. Jones and P. Sheppard (eds.), *Australasian Archaeometry 2001*, pp. 417–57. Research Papers in Anthropology and Linguistics. Auckland: University of Auckland.

Weisler, M. I., and Kirch, P. V. 1996. Interisland and interarchipelago transfer of stone tools in prehistoric Polynesia. *Proceedings of the National Academy of Sciences, U. S. A.* 93: 1381–85.

Weisler, M. I., Kirch, P. V., and Endicott, J. M. 1994. The Mata'are basalt source: Implications for prehistoric interaction studies in the Cook Islands. *Journal of the Polynesian Society* 103: 203–16.

Weisler, M. I., and Sinton, J. 1997. Towards identifying prehistoric interaction in Polynesia, in M. Weisler (ed.), *Prehistoric Long-Distance Interaction in Oceania: An Interdisciplinary Approach*, pp. 173–93. New Zealand Archaeological Association Monograph 21.

Weisler, M. I., and Woodhead, J. D. 1995. Basalt Pb isotope analysis and the prehistoric settlement of Polynesia. *Proceedings of the National Academy of Sciences, U.S.A.* 92: 188–95.

Winterhoff, E., Wozniak, J., Ayres, W., and Lash, E. 2007. Intra-island source variability on Tutuila, Samoa and prehistoric basalt adze exchange in Western Polynesia-Island Melanesia. *Archaeology in Oceania* 42: 65–71.

Woodhead, J. D., and Weisler, M. I. 1997. Accurate sourcing of basaltic artifacts by radiogenic isotope analysis, in M. Weisler (ed.), *Prehistoric Long-Distance Interaction in Oceania: An Interdisciplinary Approach*, pp. 212–23. New Zealand Archaeological Association Monograph 21.

53

THE USES OF ARCHAEOLOGICAL FAUNAL REMAINS IN LANDSCAPE ARCHAEOLOGY

Ingrid L. Mainland

Animals inhabit the landscape, are affected by it, and are often catalysts for its modification or destruction. The presence of particular animals in a landscape will be limited by a species' biological and ecological adaptation to specific habitats and environments, while the terrain encountered by an animal may bring about skeletal adaptation and modification. For domesticates, human intervention determines the lifestyle of an animal within the landscape, by restricting its mobility and foraging through, at the one extreme, stalling, penning or hobbling, and at the other, supervised herding, transhumant systems, and range-managed animals. Moreover, the procurement of fodder and grazing for domesticates is one of the primary uses of the landscape by stock-holders, and the mismanagement of such resources has, and continues to have, a devastating impact on the environment in many areas of the world.

Archaeological animal remains are typically found within "on-site" contexts such as middens, ditches, and pits. More occasionally, non-anthropogenic deposits of animal bone are found within the landscape, in, for example, lake or river sediments, cave deposit, peat bogs, and tar pits. Although the recovery of faunal remains is primarily a "site-based" activity, analysis of animal bones can provide useful insight into both the nature of the landscape in the past and the use of the landscape by ancient herders

and their animals. This chapter outlines the potential of archaeological animal remains to allow insight into the ancient landscape both in terms of its biotic and abiotic nature (that is, characterizing past environments) and the articulation and impact of pastoral activities within it.

Animals as Indicators of Past Physical and Biotic Landscapes

The distribution of many species is limited by *inter alia* physiology, foraging behavior, climate, and habitat with the result that some species are associated with very specific environments. Furthermore, the terrain traversed habitually by an animal can affect its skeletal morphology. This section explores how ecological, behavioral, and climatic factors, as well as the landscape itself, affect the animals living within a specific environment and explains how such information can be used to characterize past landscapes.

Ecological, Behavioral, and Climatic Variables Affecting the Distribution of Vertebrates in the Landscape

Any consideration of the distribution of vertebrate species—whether mammals, birds, fish, or

544

herpetofauna—serves to indicate one basic fact: the geographic range of a species is limited; no species is everywhere (e.g., Stuart 1982; Yalden 1999). Moreover, some species have broad ranges, whereas others have very narrow ranges. This variability in vertebrate distribution can be attributed to various limiting factors, both abiotic (that is, physical and chemical factors) and biotic (living, generally the influence of other living organisms) (Krebs 1994). Of the abiotic factors, temperature and humidity are perhaps the most important, affecting distribution either directly through the physiological tolerances of particular species or indirectly by limiting the growth of vegetation on which a species, or its prey, is dependent. The winter distribution of many North American passerines, for example, typically correlates with average minimum January temperatures as, being endotherms, their metabolism cannot meet energetic requirements at lower temperatures (Root 1988). Similar gradients are found in other species, both endothermic and exothermic. For example, the roe deer (*Capreolus capreolus)* requires regular water intake, and hence the distribution of this species is limited by aridity; the presence of breeding populations of the European pond tortoise (*Emys orbialaris*) is indicative of summer temperatures of ca. > 18–20°C, the temperature required for its eggs to hatch (Stuart 1979).

The distribution of vertebrate species also reflects habitat preferences. Here, one of the main factors governing distribution is foraging behavior, though other variables such as nesting and denning behavior and the need for camouflage are also important. Small mammals (for example, rodents, voles) are often highly habitat-specific (stenotopic) and thus are considered good indicators of past environments. Examples here include (in Europe) indicators of grassland such as the field vole (*Microtus agrestis*) and common vole (*Microtus arvalis*), and of woodland, the woodmouse (*Apodemus sylvaticus*) and the bank vole (*Clethrionomys glareous*). In general, larger mammalian fauna tolerate a broader range of environmental conditions (eurytopic) and consequently tend to be less useful as palaeoenvironmental indicators. Nevertheless, correlations do exist between habitat type and the modern distribution of many of the larger mammalian fauna, and these can be used to give some insight into local environmental conditions in the past; fauna such as the wild cat (*Felis sylvestris*) and wild boar (*Sus scofa*) all preferentially inhabit temperate woodland, whereas nondomesticated horses (*Equus* sp.) are typically found in open grassland. Like small mammals, many species of bird have specific ecological

requirements for both feeding and nesting and thus permit high-resolution palaeoenvironmental reconstruction (Baird 1989; Morales Muñíz 1993). Reptiles and amphibians tend also to be stenotopic and as such can be very informative sources of evidence for past ecosystems (Stuart 1982).

The observation that many vertebrate species are limited by climatic conditions and/or habitat has played a key role in the analysis of faunal assemblages from Quaternary deposits (e.g., Benecke 1999; Gardeisen 1999; Lowe and Walker 1997; Montuire 1999; Roberts and Parfitt 1999; Schmitt and Lupo 2004; Schmitt, Madsen, and Lupo 2002; Stuart 1982). Palaeoenvironmental interpretation relies on analogy with the modern distribution and/or climatic and ecological tolerances of animals. Analysis typically considers presence/absence or the relative frequencies of indicator species. In their study of faunal assemblages in the Bonneville Basin, Utah, Schmitt and associates (2002; Schmitt and Lupo 2004; 2002) used several indices of species frequency to explore local environments and climate change during the Late Pleistocene and early-mid-Holocene. The lower ratio of cottontails (*Sylvilagus* sp.), which require dense perennial shrubs for cover, to jackrabbits (*Lepus* sp.), an indicator of open desert habitats, along with reduced proportions of seven species favoring cool montane or moist conditions and/or grass/sagebrush cover indicated a rapid onset of desertification during the middle Holocene (ca. 8,300 B.P.). More recently, studies have begun to focus on the structure and the evolution of mammalian communities. Montuire (1999), for example, has demonstrated an increasingly arid environment in Spain at the Pliocene-Quaternary boundary using species richness in Arvicolinae and Murinae.

Faunal Palaeodietary Analysis and Ancient Landscapes

As outlined above, foraging behavior is one of the key limiting factors determining an animal's presence/location within a given landscape. Moreover, the diet of an animal will reflect the nature of the biomass available in its home range. Faunal palaeodiet thus provides a powerful tool for environmental reconstruction and has the advantage of being applicable to extinct fauna for which no modern analogues exist. The analysis of dental morphology is an established technique for assessing dietary adaptation (Janis 1990; Kay and Covert 1984; Van Valkenburgh 1988). This approach, which relies on the fact that the form and structure of teeth will be adapted to the main foodstuffs consumed, has been widely used to assign palaeodietary behavior and thus habitat preferences in many extant and extinct species, including ruminants, primates, and carnivores. It is, however, limited in that it assumes that the diet to which teeth are adapted is "current" and is a rather broad-brush approach that does not allow the detection of minor dietary adaptations or seasonal shifts in diet, both of which can be very useful for understanding the nature of past landscapes.

A more refined approach, which is becoming increasingly common within palaeoenvironmental studies, is dental microwear analysis. The food consumed by an individual leaves recognizable traces, microwear features, on tooth surfaces. Analyses of microwear patterning in modern animals have demonstrated that both broad dietary adaptations (for example, browsing vs. grazing, folivory vs. frugivory) and seasonal shifts in diet can be detected in diverse species. This technique has principally been used to explore the dietary adaptation of hominines and other extinct primates (Rose and Ungar 1998). Microwear analyses by Grine (1986) suggested that *Australopithecus* (*A. africanus*) and *Paranthropus* (*A. bosei*) had very different diets. *Paranthropus* consumed a much more abrasive, or, grit-laden diet than did *Australopithecus,* which subsisted on a softer diet of folivorous or frugivorous material; this indicates that *Paranthropus* is likely to have lived in a much drier environment. Ungulate microwear has been used to identify landscapes characterized by C4 grasslands, high-altitude grasslands, and woodland and shrub environments (e.g., Merceron et al. 2005; Solounias and Semprebon 2002; Solounias et al. 2000), although with a few exceptions (e.g., Mainland 2006; Rivals and Deniaux 2003, 2005) such studies have been restricted to Tertiary contexts.

Palaeodietary evidence obtained from the isotopic composition of ungulate bone and teeth is a further useful palaeoenvironmental indicator (Bocherens et al. 1999; Price 1989; Sealy et al. 1991). Bone and apatite biochemistry will reflect the isotopic composition of food and water ingested, which, in turn, will be dependent on factors such as plant metabolism, climate (for example, temperature, aridity, degree of continentality), and the geology of the landscape (Bentley et al. 2003; Price 1989). In continental tropical zones, stable carbon isotopes ($^{13}C/^{12}C$) in herbivores can be used to distinguish between the consumption of terrestrial plants following C3 (flowering shrubs, trees, grasses growing in cool temperatures/high altitudes) and C4 photosynthetic pathways (mainly xeric environment grasses), effectively the ratio of C4 grasslands to C3 woodlands or of grasslands in winter rainfall vs. summer rainfall zones (DeNiro and Epstein 1978; Lee-Thorp and Beaumont 1995). In such regions as temperate Europe, where C4 plants are not present, Drucker and associates (2003) have suggested that ^{13}C ratios are depleted in animals consuming vegetation from the understorey of woodlands. They have used this effect in red deer bone to identify the development of wooded landscapes in continental western Europe during the Late Glacial/Early Holocene (but cf. Stevens, Lister, and Hedges et al. 2006). Isotopic evidence from higher trophic levels can also provide useful palaeoenvironmental evidence. Bocherens and colleagues (1999) have suggested that $\delta^{13}C$ and $\delta^{15}N$ values of carnivores from Scladina Cave, an interglacial Upper Pleistocene cave site in Belgium, indicate consumption of herbivores from open environments.

Postcranial Functional Morphology and Environmental Characterization

An animal's locomotor anatomy will demonstrate adaptation to the substrate that it habitually traverses (DeGusta and Vrba 2005). By identifying characteristics of the skeleton that relate to locomotion in specific substrates/habitats, palaeoenvironmental reconstruction can thus be achieved. Kappelman (1988; Kappelman et al. 1991), for example, identified morphological criteria in the femur of African bovids, which enabled distinction between open (plains) and wooded environments with various degrees of cover. Similarly, DeGusta and Vrba (2005) have recently proposed that the morphology of phalanges in African bovids are indicative of habitat; species that inhabit forests and areas of light cover (light bush, tall grass and hilly areas) tend to have longer phalanges, whereas those in habitats with heavy cover (bush, woodland, swamp) or in open ecotones/arid zones exhibit shorter phalanges.

Faunal Remains, Herds, and Herders in the Landscape

With the advent of animal domestication and the subsequent dominance of domesticates in many archaeological assemblages, the potential of faunal data to provide insight into past landscapes is much reduced; humans have considerably extended the natural range of domesticated species through artificial feeding, landscape modification, and selective breeding—sheep, for example, originally an inhabitant of central Asia, today live in Arctic Greenland. Nevertheless, the distribution of domesticates is still limited to a certain degree by their basic physiology and dietary adaptations and can provide an indication, albeit rather generalized, of past environments. Dairy cattle need to drink water regularly and hence indicate access to water (for example, springs, wells, rivers) in the vicinity of a settlement where cattle were kept for milk. Sheep do not thrive in damp, marshy conditions or woodland (e.g., Halstead 1981), and the absence of suitable woodland habitats is often cited as a barrier to successful pig rearing (e.g., Church et al. 2005).

A more interesting application of faunal remains in postdomestication contexts is in allowing insight into the use of and impact on the landscape by herders and their stock. How a landscape is used by domestic herbivores depends on the mobility of the society concerned. Three broad strategies can be identified: (1) animal herding/husbandry by sedentary agropastoralists; (2) seminomadic pastoralism (often termed transhumance); and (3) specialized pastoralism (Arnold and Greenfield 2004; Chang 1993; Halstead 1996). Where sedentary farmers rear domestic herbivores as part of a mixed agropastoral economy, stock movements within the landscape are typically localized and of short distance. In seminomadic transhumance, a varied degree of arable agriculture is practiced, and herders will move between seasonally differentiated pastures (for example, summer pastures in the uplands and winter in the lowlands or vice versa). Differences traversed may be large or only a few kilometers. Specialized pastoralists do not undertake arable agriculture and will rely heavily on one or two species of herd animals. Movements in the landscape are typically extensive and usually involve whole communities.

The relative frequencies of the different herd animals reared and mortality/harvest profiles are two aspects of faunal assemblages that have been used to explore the mobility of herders. Halstead (1996) has argued that in the Mediterranean region, specialized pastoralists may potentially be identified by an emphasis on single species because of the difficulties of herding mixed groups of species together across vast areas. Mortality profiles in which neonates and young are emphasized are considered indicative of specialized milk production (reflecting the culling or sale of lambs and calves at about 6 weeks to maximize milk production) and hence specialized pastoralists (Halstead 1998). Likewise, Arnold and Greenfield (2004) have suggested that under transhumant pastoralism in the Balkans, where herders move between lowland winter and upland summer pastures, a complementary set of mortality profiles from upland and lowland sites would be anticipated; for example, assuming a spring birth and migration into the uplands in early spring, mortality profiles for sheep would demonstrate a cull of 0–2 months old but lack 2–6 months old at lowland sites, whereas at upland summer pasture sites, 0–2 months would be absent and 2–6 months present.

These approaches demonstrate degree of mobility within a given landscape but not where within that landscape the herders were located. Palaeodietary studies (microwear, fecal studies, isotope analysis) have begun to address this question and are allowing more specific insights into the use of past landscapes. Charles and Bogaard's (2005) study of sheep/goat dung-derived archaeobotanical deposits from the Neolithic settlement of Jeitun in Turkmenistan has suggested from the presence of summer fruiting taxa that transhumance was not practiced but that sheep were grazed in the vicinity of the site during summer. Bentley and associates (2003) have demonstrated mobility of cattle at LBK period sites in Germany using strontium isotopic ratios in teeth. Strontium isotopes vary according to the geology of the region. Levels recovered in cattle teeth from Vaihingen were lower than the local range expected, indicating the possibility that this species had been herded in a region of basaltic volcanic geology 50–60 km to the south. Oxygen isotopes in dental tissue are also potentially useful in detecting movement across the landscape (d'Angela et al. 1991; Fricke and O'Neil 1996; Mashkour et al. 2005). The ratio of $^{18}O/^{16}O$ in mammalian tissue reflects that within ingested water, which in turn varies with temperature, degree of continentatilty, evaporation, and altitude (Mashkour, Bocherens, and Moussa 2005). Movement of grazing animals across regions in which clear variability is evident within oxygen isotope ratios can thus potentially allow detection of where an animal had grazed and, if dental tissue are sampled, the season of grazing within that habitat.

The impact of domestic ungulates on the landscape can be profound (Evans 1998). Pastoralists and farmers remove woodland to provide grazing for their animals, and overstocking and overgrazing can bring about further modification. Dental microwear analysis has recently been used to explore the impact of ovicaprines on the marginal environments of the North Atlantic islands (Mainland 2006). In Medieval Norse Greenland, microwear patterns indicative of overgrazing were identified in the inland region of the Western and Eastern Settlements from A.D. 1150 onward and were also detected during the later phases of occupation in the Western settlement (14th and 15th centuries A.D.). These results provide evidence to support suggestions that maladaptive grazing practices led to a decline in the viability of pastoral farming in Greenland and may have contributed to its eventual demise.

Conclusions

Vertebrate remains within archaeological contexts are typically used for the reconstruction of economic strategies, diet, and subsistence (Charles and Halstead 2000). As the preceding indicates, faunal assemblages can, however, also provide useful insight into past physical and biotic landscapes and the use of the landscape by herders and their animals. The reliability of such evidence will, however, be limited by a number of factors. For many of the approaches outlined above, palaeoenvironmental characterization is based on the modern distribution, habitat preferences, and physiological tolerances of vertebrates and assumes that these will have remained unchanged through time. This assumption will not always be valid, and care must be taken when one is using modern data in this way (Lowe and Walker 1997; Stuart 1982). Distribution and habitat preference will, for example, be affected by predation and competition: the lion, which today has a very restricted African distribution owing to competition with humans, in the past was found in Eurasia (Stuart 1982). Taphonomy is a further complicating factor. Diverse processes bring about the formation of a faunal assemblage, including human activity, death, and accidental inclusion and the activities of predators (Lowe and Walker 1997). These will affect the species represented in an assemblage and how useful it is for landscape studies. Postdepositional factors, soil acidity, redeposition of deposits, weathering, and the like will further modify both species presence and anatomical representation. Any analysis of the nature and use of past landscapes using faunal remains must, therefore, make allowance for the various taphonomic processes that may have affected the assemblage.

References

Arnold, E. R., and Greenfield, H. J. 2004. A zooarchaeological perspective on the origins of vertical transhumant pastoralism and the colonisation of marginal habitats in temperate southeastern Europe, in M. Mondini, S. Muñoz, and S. Winkler (ed.), *Colonisation, Migration and Marginal Areas: A Zooarchaeological Approach*, pp. 296–317. Oxford: Oxbow Books.

Baird, R. F. 1989. Fossil bird assemblages from Australian caves: Precise indicators of late Quaternary environments? *Palaeogeography Palaeoclimatology Palaeoecology* 69: 241–44.

Benecke, N. 1999. *The Holocene History of the European Vertebrate Fauna*. Rahden: Westfalen: Verlag Marie Leidorf.

Bentley, R. A., Krause, R., Price, T. D., and Kaufmann, B. 2003. Human mobility at the early Neolithic settlement of Vaihingen, Germany: Evidence from strontium isotope analysis. *Archaeometry* 45: 471–86.

Bocherens, H., Billiou, D., Patou-Mathis, M., Otte, M., Bonjean, D., Toussaint, M., and Mariott, A. 1999. Palaeoenvironmental and palaeodietary implications of isotopic biogeochemistry of late interglacial Neanderthal and mammal bones in Scladina Cave (Belgium). *Journal of Archaeological Science* 26: 599–607.

Chang, C. 1993. Pastoral transhumance in the Southern Balkans as social ideology: Ethnoarchaeological research in Northern Greece. *American Anthropologist* 95: 687–703.

Charles, M. and Bogaard, A. 2005. Identifying livestock diet from charred plant remains: A neolithic case study from southern Turkmenistan in diet and health in past animal populations, in J. Davis, M. Fabis, I. Mainland, M. Richards, and R. Thomas (ed.), *Diet and Health in Past Animal Populations: Current Research and Future Directions*, pp. 93–103. Oxford: Oxbow Books.

Charles, M. and Halstead, P. 2000. Biological resource exploitation: Problems of theory and method, in D. R. Brothwell and A. M. Pollard (eds.), *Handbook of Archaeological Sciences*, pp. 365–78. Chichester: Wiley.

Church, M. J., Arge, S. V., Brewington, S., McGovern, T. H., Woollett, J. H., Perdikaris, S., Lawson, I. T., Cook, G. T., Amundsen, C., Harrison, R., Krivogorskaya, K., and Dunbar, E. 2005. Puffins, pigs, cod, and barley: Palaeoeconomy at Undir Junkarinsfløtti, Sandoy, Faroe. *Environmental Archaeology* 10: 17–97.

d'Angela, D., Maggi, R., Nisbet, R., and Barker, G. 1991. Palaeoclimatic conditions in the Po Plain during the Neolithic, Chalcolithic and Bronze Ages: First results. *Rivista di studi liguri* 57: 119–26.

DeGusta, D., and Vrba, E. 2005. Methods for inferring palaeohabitats from the functional morphology of bovid phalanges. *Journal of Archaeological Science* 32: 1099–113.

DeNiro, M. J., and Epstein, S. 1978. Influence of diet on distribution of carbon isotopes in animals. *Geochimica et Cosmochimic a Acta* 42: 495–506.

Drucker, D., Bocherens, H., Bridault, A., and Billiou, D. 2003. Carbon and nitrogen isotopic composition of Red Deer (*Cervus elaphus*) collagen as a tool for tracking palaeoenvironmental change during Lateglacial and Early Holocene in northern Jura (France). *Palaeogeography Palaeoclimatology Palaeoecology*: 375–88.

Evans, R. 1998. The erosional impacts of grazing animals. *Progress in Physical Geography* 22: 251–68.

Fricke, H. C., and O'Neil, J. R. 1996. Inter- and intra-tooth variation in the oxygen isotope composition of mammalian tooth enamel phosphate: Implications for palaeoclimatological and palaeobiological research. *Palaeogeography Palaeoclimatology Palaeoecology* 126: 91–100.

Gardeisen, A. 1999. Middle Palaeolithic subsistence in the West Cave of "Le Portel"(Pyrénées, France). *Journal of Archaeological Science* 26: 1145–58.

Grine, F. E. 1986. Dental evidence for dietary differences in Australopithecus and Paranthropus: A quantitative analysis of permanent molar microwear. *Journal of Human Evolution* 15: 783–822.

Halstead, P. 1981. Counting sheep in Neolithic and Bronze Age Greece, in I. Hodder, G. Isaac, and N. Hammond (eds.), *Pattern of the Past: Studies in Honour of David Clarke*, pp. 307–39. Cambridge: Cambridge University Press.

———. 1996. Pastoralism or household herding? Problems of scale and specialization in early Greek husbandry. *World Archaeology* 28: 20–42.

———. 1998. Mortality models and milking: Problems of uniformitarianism, optimality and equi-finality reconsidered. *Anthropozoologica* 27: 3–20.

Janis, C. M. 1990. Correlation of cranial and dental variables with dietary preferences in mammals: A comparison of macropodoids and ungulates. *Memoirs of the Queensland Museum* 28: 349–66.

Kappelman, J. 1988. Morphology and locomotor adaptations of the bovid femur in relation to habitat. *Journal of Morphology* 198: 119–30.

Kappelman, J., Plummer, T., Bishop, L., Duncan, A., and Appleton, S. 1991. Bovids and indicators of Plio-Pleistocene palaeoenvironments in East Africa. *Journal of Human Evolution* 32: 229–56.

Kay, R. F., and Covert, H. H. 1984. Anatomy and behavior of extinct primates, in D. J. Chivers, B. A. Wood, and A. Bilsborough (eds.), *Food Acquisition and Processing in Primates*, pp. 467–508. New York: Plenum Press.

Krebs, C. J. 1994. *Ecology: The Experimental Analysis of Distribution and Abundance* (4th ed.). New York: Harper Collins.

Lee-Thorp, J. A., and Beaumont, P. B. 1995. Vegetation and seasonality shifts during the Late Quaternary deduced from 13C/12C ratios of grazers at Equus Cave, South Africa. *Quaternary Research* 43: 426.

Lowe, J. J., and Walker, M. J. C. 1997. *Reconstructing Quaternary environments* (2nd ed.). London: Longman.

Mainland, I. 2006. Pastures lost? A dental microwear study of ovicaprine diet and management in Norse Greenland. *Journal of Archaeological Science* 33: 238–52.

Mashkour, M., Bocherens, H., and Moussa, I. 2005. Long-distance movement of sheep and goats of Bakhtiari nomads tracked with intra-tooth variations of stable isotopes, in J. Davis, M. Fabis, I. Mainland, M. Richards, and R. Thomas (eds.), *Diet and Health in Past Animal Populations: Current Research and Future Directions*, pp. 113–24. Oxford: Oxbow Books.

Merceron, G., d. Bonis, L., Viriot, L., and Blonde, C. 2005. Dental microwear of the late Miocene bovids of northern Greece: Vallesian/Turolian environmental changes and disappearance of Ouranopithecus macedoniensis? *Bulletin de la Société Géologique de France* 176: 475–84.

Montuire, S. 1999. Mammalian faunas as indicators of environmental and climatic changes in Spain during the Pliocene-Quaternary Transition. *Quaternary Research* 52: 129–37.

Morales Muñíz, A. 1993. Ornithoarchaeology: The various aspects of the classification of bird remains from archaeological sites. *Archaeofauna* 2: 1–13.

Price, T. D. 1989. *The Chemistry of Prehistoric Human Bone*. Cambridge: Cambridge University Press.

Rivals, F., and Deniaux, B. 2003. Dental microwear analysis for investigating the diet of an argali population (*Ovis ammon antiqua*) of mid-Pleistocene age, Caune de l'Arago cave, eastern Pyrenees, France. *Palaeogeography, Palaeoclimatology, Palaeoecology* 193: 443–55.

———. 2005. Investigation of human hunting seasonality through dental microwear analysis of two Caprinae in late Pleistocene localities in Southern France. *Journal of Archaeological Science* 32: 1603–12.

Roberts, M. B., and Parfitt, S. A. 1999. *Boxgrove: A Middle Pleistocene Hominid Site at Eartham Quarry, Boxgrove, West Sussex*. London: English Heritage.

Root, T. 1988. *Atlas of Wintering North American birds: An Analysis of Christmas Bird Count Sata*. Chicago: University of Chicago Press.

Rose, J. C., and Ungar, P. S. 1998. Gross dental wear and dental microwear in historical perspective, in K. W. Alt, F. W. Rösing, and M. Teschler-Nicola (eds.), *Dental Anthropology*. Vienna: Springer-Verlag.

Schmitt, D. N., and Lupo, K. D. 2004. Worst of times, the best of times: Jackrabbit hunting by middle Holocene human foragers in the

Bonneville Basin of western North America, in M. Mondini, S. Muñoz, and S. Winkler (eds.), *Colonization, Migration, and Marginal Areas: A Zooarchaeological Approach*, pp. 86–95. Oxford: Oxbow Books.

Schmitt, D. N., Madsen, D. B., and Lupo, K. D. 2002. Small-mammal data on early and middle Holocene climates and biotic communities in the Bonneville Basin, USA. *Quaternary Research* 58: 255–60.

Sealy, J. C., van der Merwe, N. J., Sillen, A., Kruger, F. J., and Krueger, H. W. 1991. 87Sr/86Sr as a dietary indicator in modern and archaeological bone. *Journal of Archaeological Science* 18: 399–416.

Solounias, N., McGraw, W. S., Hayek, L.-A., and Werdelin, L. 2000. The paleodiet of the giraffidae, in E. S. Vrba and G. B. Schaller (eds.), *Antelopes, Deer, and Relatives*, pp. 84–95. New Haven, CT: Yale University Press.

Solounias, N., and Semprebon, G. 2002. Advances in the reconstruction of ungulate ecomorphology with application to early fossil equids. *American Museum Novitates* 1–49.

Stevens, R. H., Lister, A. M., and Hedges, R. E. M. 2006. Predicting diet: Tropic level and palaeoecology from bone stable isotope analysis. A comparative study of five red deer populations. *Oecologia* 149: 12–21.

Stuart, A. J. 1979. Pleistocene occurrences of the European pond tortoise (*Emy orbicularis* L.) in Britain. *Boreas* 8: 359–71.

———. 1982. *Pleistocene Vertebrates in the British Isles*. London: Longman.

Van Valkenburgh, B. 1988. Trophic diversity in past and present guilds of large predatory mammals. *Palaeobiology* 7: 162–82.

Yalden, D. W. 1999. *The History of British Mammals (Poyser Natural History S.)*. London: T. and A. D. Poyser Ltd.

54

Survey Strategies in Landscape Archaeology

Thomas Richards

This chapter considers *survey strategies*, a critically important component of research design that largely determines whether or not a useful contribution will be made to the problem under investigation. A survey strategy is a master plan that articulates a regional research question with a specific combination of sensing, survey, recording, collection, and analysis methods and techniques designed to capture the information necessary to address the problem (see also Cheetham, this volume; Conolly, this volume). Careful thought put in at this critical planning stage of a survey allows greater flexibility during its execution where, on multiyear projects in particular, the focus of the research may shift, owing to the discovery of un-expected data, changing local conditions, or even through the influence of new team members bringing in different ideas (Barker 1995: 40).

In many cultural heritage management (CHM) contexts, there is often only one chance to get it right. Paradoxically, although planning is even more important than usual when much of the archaeological record on a landscape is scheduled to be destroyed, it is precisely in such contexts that adequate time for consideration of survey strategy is sometimes lacking, resulting in hastily planned surveys. The topics covered in this chapter should thus be considered by anyone faced with developing a strategy to survey a landscape, regardless of the overarching reasons for the study.

History of Regional Archaeological Landscape Studies

The modern archaeological regional study of landscapes is generally agreed to have started with Willey's (1953) South American settlement pattern surveys. Ten years later, Binford's (1964) seminal paper on research design championed both the regional approach, plus probability sampling, in the context of providing a research program for processual archaeologists to follow. Over the next 15 years, archaeologists working mostly in the arid parts of the United States and Mexico undertook such regional studies, frequently with probabilistic survey sampling (e.g., Flannery 1976; Mueller 1975; Schiffer and Gumerman 1977; Thoms 1988). It is this regional approach, along with similar British studies around the Mediterranean from the mid-1970s onward (e.g., Barker et al. 1995; Cherry, Gamble, and Shennan 1978; Keller and Rupp 1983) that firmly established the validity and usefulness of the study of human behavior across the landscape so that it has, in the last two decades, permeated archaeological research of virtually any theoretical persuasion in many parts of the world (e.g., Alcock and Cherry 2004a; Allen, Green, and Zubrow 1990; Ashmore and Knapp 1999; Bender 1993; Birks et al. 1988; Gillings, Mattingly, and van Dalen 1999; Nash 1997; Rossignol and Wandsnider 1992; Tilley 1994; Wilkinson 2004).

Distributional ("non-site," "siteless," "off-site") approaches, in which past human behavior is conceived as having occurred across entire landscapes, view the surface archaeological record as more or less continuous but variable in organization and density. These studies, which focus on the discovery and the recording of individual artifacts, started to be undertaken in the late 1960s (e.g., Bintliff and Snodgrass 1988; Cherry 1983; Dunnell and Dancey 1983; Ebert 1992; Foley 1981a, 1981b; Gallant 1986; Lewarch and O'Brien 1981; Schadla-Hall and Shennan 1978; Stoddart and Whitehead 1991; Thomas 1975; Wilkinson 1982), but site-based surveys are still common. Many studies either combine off-site with *site-based* approaches (Alcock and Cherry 2004b: 3) or define "sites" in the laboratory after conducting an off-site survey, for modeling purposes or to satisfy the documentation requirements of government cultural heritage authorities (e.g., Keay and Millett 1991; Peterson and Drennan 2005; Richards 1998; Thomas 1988).

Off-site landscape archaeological studies rapidly expanded from their initial focus on stone artifact distributions to a broader range of elements, including pottery (e.g., Bintliff and Snodgrass 1988; Wilkinson 1982), faunal remains (e.g., McNiven 1992), and even modified trees (e.g., Rhoads 1992; Stryd and Eldridge 1993; Webber and Burns 2004). Discussion in this chapter is compatible with siteless and/or site-based approaches to the study of archaeological landscapes.

Selection of a Landscape

Once a research question is developed, a suitable landscape must be selected, except in CHM contexts, where it is often the case that meaningful research questions have to be devised to fit a landscape threatened by development.

Research problems often require a landscape to be representative of a larger area, so the distributional patterns of the surface record and associated human behavioral interpretations can be widely extrapolated. This issue was assessed by David Hurst Thomas for the first regional probabilistic sampling study (also the initial non-site survey), the Reese River Valley, Nevada survey, undertaken in the late 1960s:

We were very careful to collect a 10% random sample from *within* the study area, and I felt sanguine about generalizing the results for the entire 25 x 30 km area. But the ultimate objective of this fieldwork . . . is to generate statements of anthropological rather than statistical relevance. One must wonder: How far can the Reese River pattern be generalized beyond the 1,400 square tracts in the initial population? After all, three years is no small investment in such labor-intensive fieldwork, and mere description of 800 km² hardly seems worth the effort. Could one legitimately extend the observed settlement pattern to the 160-km length of the Reese River Valley? . . . to the entire central Nevada region? . . . across the entire Great Basin? . . . What is the population against which to project these sample results? (Thomas 1988: 160)

Thomas (1988: 162) took the issue of regional representativeness seriously enough to undertake a second large scale regional landscape study, in the 30-km distant Monitor Valley, to test whether the archaeological patterning observed within the Reese River Valley study area was present.

A research problem may sometimes only be able to be addressed in a study area containing unique characteristics, so that the research results can inform on a process of great general interest and import-ance. An example of this approach to regional landscape selection is MacNeish's decades-long research into the development of agriculture in Mesoamerica, which saw him undertake extensive landscape reconnaissance surveys in several locations, until he found a landscape in the Tehuacan Valley that contained an archaeological record suitable for answering his research question (MacNeish 1978).

There is a current trend toward the study of smaller and smaller landscapes in more and more detail (Terrenato 2004: 47); this is both a logical and desirable outgrowth from the much more spatially extensive surveys of past decades in many parts of the world, providing complementary datasets suitable for differing scales of analysis.

Landscape Formation Considerations

Following selection of a regional landscape study area, the formation of its present surface characteristics needs to be considered in developing the survey strategy.

Geomorphology

If a specific archaeological phenomenon or period is the subject of interest, then the survey strategy needs to have a strong geoarchaeological emphasis, so that survey is directed to landforms of the appropriate age (Butzer 1982: 262; Terrenato 2004: 39; see Denham, this volume; Stern, this volume). It is still important to know the age of land surfaces across the region for more broad-based, comprehensive surveys, to promote understanding of observed patterning of the surface record. Of course, it is not only the age of landforms that is of interest, but formation processes of the archaeological record on a landscape scale (Foley 1981a; Zvelebil, Green, and Macklin 1992: 204; see Head, this volume; Heilen, Schiffer, and Reid, this volume). Understanding landscape-wide minor processes that could obscure or expose artifacts on specific survey tracts should result in provision for measuring and recording their effects during survey, thus providing data for post-survey analysis of site formation and "ground surface visibility" effects (e.g., Barton et al. 2002; Fanning and Holdaway 2004).

Another factor affecting visibility of the ground surface, and one considered by most archaeologists, is vegetation, discussed in more detail below.

Surface Vegetation

Logically, there should be an inverse linear relationship between the proportion of a land surface obscured by vegetation and the number of artifacts or sites observed. Recent research is demonstrating that this is not the case (Bevan and Conolly 2004; Fanning and Holdaway 2004; Thompson 2004). These studies support long-held intuitive beliefs by some archaeologists that the relationship between vegetation and surface visibility is complex and nonlinear.

A landscape survey project on the island of Kythera, Greece, examined the relationship between both artifact and site discovery and surface visibility and found that "although mean tract densities of artifacts do increase with ground visibility, there is no linear correlation between individual tract artifact densities and tract visibility (r^2 = 0.06)" and, further, that ground visibility has little predictive effect on site discovery (Bevan and Conolly 2004: 127–28).

Similarly, Thompson (2004: 74) found for the Metaponto survey in southern Italy that "no rela-

tionship exists between vegetation density and site discovery." He also noted that there are "higher than expected artifact densities at lower visibility ranges" in the Metaponto and several other Mediterranean surveys (Thompson 2004: 76). A possible explanation forwarded for this patterning is that the recording of vegetation cover on a ploughed surface is actually a proxy measure for surface weathering:

> Following plowing, both sown crops and/or weeds gradually establish themselves, while at the same time the plowed surface weathers, settles, and deflates. One of the consequences of this weathering process is that, in addition to being cleaned by rain and irrigation water, artifacts previously lying at least partially *within* the surface gradually come to rest *on* the surface . . . Although weeds and crops may grow up through this surface, partially obscuring it in the process, weathering of the surface soils actually acts to make the archaeological component of these soils more visible. (Thompson 2004: 77)

Because of its potential broad applicability, Thompson's hypothesis is worth testing empirically by others (cf. Clark and Schofield 1991: 102).

There may also be "archaeologist effects" that partially counteract the masking of surfaces by vegetation, including these: (1) archaeologists try harder to find artifacts or sites when vegetation reaches a certain density threshold—this may take the form of walking just a little more slowly across an overgrown tract than across a freshly ploughed one, resulting in more items being found than expected; (2) when vegetation cover is dense, but patchy, the archaeologist's gaze is directed away from the vegetation and onto any expanses of bare ground, so these are scrutinized with greater intensity than they would be under more open conditions, again resulting in unexpectedly high artifact, or more particularly, site discovery rates; (3) archaeologists will discover large, dense scatters almost regardless of surface vegetation cover—skilled eyes will not miss such scatters if virtually any ground surface is visible.

Human Land Use

The other side of landscape formation is, of course, human land-use history, particularly large-scale ground disturbance activities. The effects of these activities are often as profound as general geomorphological processes, or even more damaging. In some cases, landscape-wide cultural formation processes are directly relevant to the subject of interest, rather than simply something to be measured as a negative effect on the record. For example, when research involves study of the development of agriculture, then evidence of land-clearing and cultivation, terracing, construction of water control systems, and the like are phenomena that the survey will probably be designed to record. To another archaeologist interested only in nonagricultural systems, these activities may be of interest mainly in regard to their effects on transforming the targeted record.

After decades of study and experimentation, archaeologists have some understanding of the effects of ploughing (including tilling, disking, and harrowing) on the surface record (e.g., Ammerman 1985; Lewarch and O'Brien 1981; O'Brien and Lewarch 1981; Odell and Cowan 1987; see Brooks, this volume). Essentially, what has been learned is that: (1) some horizontal dispersion of clusters of items does occur after initial ploughing, so that they are found archaeologically as somewhat larger, less dense clusters or scatters on a ploughed surface (Ammerman 1985: 40; Clark and Schofield 1991: 96–99; Lewarch and O'Brien 1981: 309; Odell and Cowan 1987: 481); (2) around 5% of the artifacts in the ploughzone will be exposed on the surface after a ploughing event (Ammerman 1985: 39; Clark and Schofield 1991: 99–100; Odell and Cowan 1987: 480); (3) larger items come to the surface disproportionately frequently (Clark and Schofield 1991: Tab. 8.2; Lewarch and O'Brien 1981: 310; Odell and Cowan 1987: 480). It has been suggested that cultivated surfaces should be surveyed more than once, each time following a ploughing episode, to obtain a representative sample of the contents of the ploughzone (Lewarch and O'Brien 1981: 311; Shott 1995: 488), although this may not always be practical, particularly in CHM situations.

Correction Formulae

Various factors obscure, expose, or remove aspects of the surface archaeological record. The effects of surface vegetation (e.g., David 1996: 44–45), small-scale sedimentation (e.g., Fanning and Holdaway 2004), surface sediment type (for example, gravel versus sand), ploughing (e.g., Schott 1995), or amateur artifact collecting (e.g., Richards and Jordan 1999: 83–112) on a survey sample should be considered by archaeologists, but one approach, particularly in CHM applications, involves the

calculation of "effective survey" and related correction formulae. This is a highly fraught road to take in that the effects of vegetation or sedimentation on the visibility of the surface archaeological record are highly landscape-specific, locally variable (Fanning and Holdaway 2004), and changeable over short durations (Burger et al. 2004: 418). The application of simple cookbook formulas risks introducing more distortion and uncertainty than they are likely to "correct" for, especially if formation issues have not been adequately considered analytically for the landscape under study (Bevan and Conolly 2004: 127; Thompson 2004: 83).

Survey Considerations

Once the characteristics of the selected landscape are understood, choosing the other elements of the survey strategy largely involves decisions based on sampling considerations (Tables 54.1 and 54.2).

Intensity and Coverage

Survey intensity is "the degree of detail with which the ground surface of a given survey unit is inspected" (Plog, Plog, and Wait 1978: 389) and is commonly measured by the spacing between individual surveyors walking parallel paths (commonly referred to as "transects") (Schiffer, Sullivan, and Klinger 1978: 13). Survey intensity has generally increased over the past several decades, to the point where field walkers are regularly spaced only 5 m or 10 m apart. However, survey intensity needs to be precisely geared to the nature of the archaeological record of interest (that is, its size, abundance, and obtrusiveness), the specific research goals, and ground visibility within the landscape. Situations to avoid involve survey that is not intense enough to recover data necessary to adequately address the research question, or unnecessarily intensive coverage, which is wasteful of time and resources and could have profoundly negative effects on the overall sample size. For example, if large earth mounds are the target of the survey, the spacing between survey transects can be much wider in comparison to a focus on finding and recording all potsherds on the landscape. Also, walking parallel transects may not be the most appropriate approach in some situations—zigzag or other patterns may be more effective for certain purposes (for example, when targeting discrete, obtrusive features).

Traditionally, and simplistically, all regional landscape studies are dualistically divided into sample surveys vs. complete (or full-coverage) surveys, or probability sampling surveys vs. judgment

Table 54.1 Sampling design for survey unit or artifact selection (Haggett 1965; Nance 1983; Orton 2000; Read 1975; Thomas 1975).

Sampling Design	Description
Simple Random	Random selection of sample elements, each of which must be known in advance.
Cluster	Random selection of sample units (e.g., quadrats or transects), each of which contains clusters of elements (e.g., artifacts or sites).
Stratified Random	Landscape divided into two or more sub-areas on basis of assumed greater homogeneity of variables of interest within each stratum than across whole landscape.
Systematic Random	Origin point of grid of equally spaced sample units is randomly chosen.
Stratified Systematic Unaligned	Landscape is gridded into units (actually strata) and a smaller sample unit is randomly chosen from within each unit.
Adaptive Cluster	Initial sample units are randomly selected, but additional neighboring units are surveyed when a condition within a sample unit is met (i.e., contains artifacts) and further neighboring units will be surveyed adjacent to any of the first set of neighboring units meeting the condition, and so on until no units meet the condition.
Systematic	Nonrandom, equally spaced sample units, such as parallel transects or a grid pattern of quadrats.
Judgment	Archaeologist uses expertise and experience to survey locations where sites are likely to be located or to select "diagnostic" artifacts for recording attributes in survey units.
Targeted	Specific components of the archaeological record are of interest only.
Opportunistic	Survey is undertaken on basis that is neither random nor based on judgment, e.g., all fields with good surface visibility, tracts for which permission to survey is granted, etc.
Grab	Selection of first items which come to hand from a sample unit.

surveys. In particular, the classic New Archaeology probabilistic sampling surveys of the American Southwest and Great Basin from the late 1960s onward have long been regarded in many quarters as exemplary sample surveys. There is, however, a vast, worldwide literature of regional landscape studies that are based on nonprobabilist-ic sampling or that incorporate some randomizing or systematic sampling procedures in combination with judgment and/or opportunistic sampling. This has to do with the highly variable adoption of processual archaeology and its methods outside the United States, including English-speaking countries where the classic texts of the New Archaeology have been most widely available. Probabilistic and other forms of survey sampling have limitations, especially in regard to informing on the spatial structure of the archaeological record, which many have suggested can be overcome by complete surveys (e.g., Wandsnider and Camilli 1992: 171).

What is a full-coverage survey anyway? Certainly, a limited objective survey may provide a complete inventory of discrete and obtrusive aspects of the archaeological record such as architectural or monumental ruins, but full-coverage implies that the entire landscape has been walked and all visible components of the surface archaeological record recorded. As noted earlier, intensive survey generally is acknowledged to involve the use of 5–10 m spaced parallel pedestrian transects. Regardless of spacing, if all other factors influencing artifact discoverability are optimal, then the area thoroughly observed by a surveyor walking a straight line would at most be a 2-m wide window. Thus, with 10-m transects, only 20% of the available ground surface will be systematically searched. If an entire study was looked at in this way, a 20% systematic survey would be effectively undertaken, not a full-coverage survey. If 10-m transects within a sampling unit are walked, a 20% systematic subsample

Table 54.2 Sampling decisions and scale (Burger et al. 2004; Cherry and Shennan 1978; Doelle 1977; Flannery 1976; Foley 1978; Orton 2000; Schiffer, Sullivan, and Klinger 1978; Thomas 1975, 1988; Whalen 1990).

Scale	Sampling Considerations	Type of Sampling
Study Landscape	Where and how big; how representative in a general sense	Judgment
Survey Sample Unit	Coverage: whole landscape or some of it: how much; how to choose sample; representative sample of artifacts or sites; spatial relationships of sites	100%; probabilistic; systematic; opportunistic; judgment
Ground Surface	Intensity—how closely is the landscape going to be examined by walking, crawling, or subsurface testing; how closely spaced will the surveyors be; how many times will the sample unit be traversed; how many postploughing examinations will there be	Multiscale, nested intensity; systematic transects
Artifact	Which material classes, "types," "diagnostic items"; how many; how to choose; collect or record in field	100%; judgment; grab; systematic; probabilistic
Attributes	Which artifact or site attributes to record	Judgment

survey of the landscape surface is achieved. This point has been made once every ten years or so since the 1970s (e.g., Cowgill 1990: 254; Orton 2000: 90; Plog, Plog, and Wait 1978: 389–94), although its implications have not been fully digested by many archaeologists.

Although intensity may be the most critical element in a survey strategy, a major problem is this: although it is known that increasing the intensity for a given survey area results in more sites being found, it is still not understood what proportion of the total population of artifacts or sites that samples found by even the most intensive surveys represent (Terrenato 2004: 44). One way of addressing this problem is by undertaking a *crawling* survey, where every square centimeter of land surface is eyeballed. Odd as this may seem, such a survey has recently been completed, with highly provocative results. A team of archaeologists in northwestern Nebraska employed a "Modified-Whittaker multiscale sampling plot, which gathers observations . . . on the same sample units at multiple resolutions" (Burger et al. 2004: 409).

Walking survey of the 1,000-square-m plots involved a 70-cm-wide spacing between surveyors (literally shoulder to shoulder), which is much more intensive than most other surveys. Nevertheless, the subsequent *crawling* survey of subplots within the 1,000-square-m plots found much more lithic material (of similar size) than the

walking survey covering the same area (Burger et al. 2004: 416–19). The authors concluded:

> The results of the crawling survey are considered a close estimate of the total number of artifacts on the surface at the time of the survey and can be used to evaluate the accuracy of coarser-grained coverage of the same area . . . Comparison of the techniques indicates that the crawling survey increased the total number of chipped stone artifacts discovered per square meter by 362% . . . The magnitude of this increase was greater than many archaeologists (ourselves included) might have assumed and the implications need to be considered. (Burger et al. 2004: 418–19)

The crawling survey results provide a closer picture of the *actual* surface record (or artifact population) within the sample units, allowing the effectiveness and representativeness of more traditional pedestrian survey methods to be reliably measured (Burger et al. 2004: 418). It would therefore be useful to include similar nested intensity surveying during the first field season of many landscape studies, so that a sampling strategy for the subsequent seasons can be based on detailed knowledge of the characteristics of the surface artifact population.

Subsurface Testing

Commonly, the surface of a landscape is at least partially obscured by dense pasture grass, or covered in thick leaf litter, loose, drifting sand or flood-deposited silts. Shovel probing, augering, excavation of test pits or mechanical trenching are commonly employed to survey obscured surfaces, the shallow subsurface or even relatively deeply buried surfaces, particularly in CHM contexts (e.g., Alexander 1983; Lovis 1976; Lynch 1980; McManamon 1984; Plog, Weide, and Stewart 1977; Richards 1998; Richards and Johnston 2004; Stone 1981; Thomas and Kelly 2007). Given the premise that the surface record is usually a reflection of the shallow subsurface record, and subsurface testing is a means whereby the shallow subsurface record is looked at as a substitute for examining the surface, it must be concluded that this whole approach is actually getting away from surface surveying, and the deeper the testing, the more this is so.

Subsurface testing is time-consuming and expensive, and its proper use is not as a primary regional survey tool but rather as a supplementary approach to primarily pedestrian regional surveys to overcome surface visibility problems (e.g., Lovis 1976; Shott 1989) or for evaluating the concordance between the surface and near-surface records (Burger et al. 2004: 419; Lightfoot 1986). Also, many of the considerations in evaluating surface survey intensity apply to finding subsurface artifacts, either singly or in variable density scatters and clusters (that is, sites) (Cowgill 1990; Kintigh 1988; Lightfoot 1986: 492–95; Nance and Ball 1986: 479; Wandsnider and Camilli 1992). For example, with a staggered grid pattern of test pits to find buried sites, there is a 0.94 probability of a test pit being within the boundary of a site with a diameter equal to the test pit interval (Krakker, Shott, and Welch 1983: 472). That site, however, will be found only if the test pit encounters artifacts, so it is critical to match test pit size to artifact density (Lightfoot 1986: 493). Chances of encountering an artifact, and thus discovering a site with an artifact density of one artifact per square meter, are slim if 20-cm shovel probes are employed (Krakker et al. 1983; Nance and Ball 1986; Orton 2000). Knowing site or artifact scatter/cluster size and artifact density parameters prior to undertaking the testing may not be possible in some CHM applications, so estimates must be made on the basis of the results of surveys on comparable landscapes and applied to the subsurface testing design.

Finally, shovel probes should be undertaken *only away* from known sites; any subsurface sampling *within* an archaeological site should follow only detailed excavation methods, to avoid unacceptable damage to archaeological sites without properly recording the relationship of archaeological materials to one another and to chronostratigraphic details.

Recording and Collection

A big decision to be made in developing a survey strategy is whether to collect surface artifacts or record attribute states in the field, although it is much more common these days not to collect. Field recording has been facilitated by global positioning systems and small portable computers, but the mere availability of useful technology doesn't help to decide which attributes on which items to record so as to obtain data critical to the research problem. Anything less than total recording of all items detected on the ground surface adds another level of sampling, sometimes of dubious representativeness—it was once common to collect a "grab sample" of items from each transect within a survey tract while field walking and to examine them back in the laboratory. Such sampling has, however, been shown to seriously bias the record toward selected artifact types, without the integrity of those types being adequately understood or assessed. More recent approaches have included: (1) the targeting of all temporally sensitive items, particularly diagnostic potsherds or projectile points, for collection or field recording; (2) using a "clicker" to record counts of a ubiquitous class of artifact; (3) recording or collecting all items encountered during survey; (4) collecting all artifacts within a small and well-defined sample area (for example, 1 m²) within a broader surveying area or site; (5) employing a separate team of specialists to return to all or selected artifact concentrations/sites to record some or all items in detail.

The long and the short of it is, in the development of a survey strategy careful decisions need to be made regarding which attributes should be recorded to address the targeted research question(s) and whether to measure these on all or a sample of surface artifacts while being aware of the potential impact of sampling decisions on the ultimate research results (Foley 1978: 60).

Predictive Modeling

Predictive modeling arose in the late 1970s and 1980s in North America as a cultural heritage management planning tool involving modeling of the distribution and density of archaeological sites or artifacts on a regional landscape (Kohler and Parker

1986: 400). The models frequently were based on correlations between environmental variables and archaeological distributions, but there was always a tension between the goals of modeling relationships between site location and environmental variables and modeling the human behavior behind site location (Kohler and Parker 1986: 442).

Despite this tension, many highly worthwhile studies were undertaken (e.g., Judge and Sebastian 1988), but ultimately, predictive models have largely come to be perceived as costly tools of limited utility by cultural heritage managers on the one hand, and of environmentally deterministic character and limited explanatory power to research archaeologists on the other. Predictive modeling has, however, received something of a minor revival with the recent broadly available desktop PC geographic information system computer programs (e.g., Westcott and Brandon 2000; Wheatley and Gillings 2002: 165–81), although the fundamental problems of relating correlations between sites or artifacts and environmental variables to past human behavior remain (Ebert 2000: 130).

Conclusions

The first decision to be made in relation to survey strategy, and probably the most important, is which landscape to study. Today, choice of a landscape is frequently dictated by CHM exigencies, but for archaeologists with the luxury of being able to select their study area, the location and the extent are determined by the nature of the research question, as well as the resources available. It is apparent, however, that when archaeologists are able to choose, they usually pick a landscape with good surface visibility—at least partially to avoid having to employ subsurface testing methods on a large scale and also because of the larger surface sample readily available for study.

Once research goals are defined, a study landscape selected, and initial considerations such as geomorphology and landscape formation have been considered, development of a survey strategy is largely a sampling exercise. A critically important sampling issue to be considered is survey intensity, or how closely the ground surface is to be scrutinized. While Burger and associates (2004) caution against using their crawling survey results in other regions, this only highlights the desirability of similar surveys to be undertaken by other archaeologists so the surface record of additional landscapes may be equally well understood. The resulting knowledge of actual surface artifact population characteristics should stimulate

the posing of more informed research questions and result in better-designed survey strategies.

References

Alcock, S. E., and Cherry, J. F. (ed.). 2004a. *Side-by-Side Survey: Comparative Regional Studies in the Mediterranean World*. Oxford: Oxbow Books.

———. 2004b. Introduction, in S. E. Alcock and J. F. Cherry (ed.), *Side-by-Side Survey: Comparative Regional Studies in the Mediterranean World*, pp. 1–9. Oxford: Oxbow Books.

Alexander, D. 1983. The limitations of traditional surveying techniques in a forested environment. *Journal of Field Archaeology* 10: 177–86.

Allen, K. M. S., Green, S. W., and Zubrow, E. B. W. (eds.). 1990. *Interpreting Space: GIS and Archaeology*. London: Taylor and Francis.

Ammerman, A. J. 1985. Plow-zone experiments in Calabria, Italy. *Journal of Field Archaeology* 12: 33–40.

Ashmore, W., and Knapp, A. B. 1999. *Archaeologies of Landscape: Contemporary Perspectives*. Oxford: Blackwell Publishers.

Barker, G. 1995. The Biferno Valley Survey: Methodologies, in G. Barker (ed.), *A Mediterranean Valley: Landscape Archaeology and Annales History in the Biferno Valley*, pp. 40–61. London: Leicester University Press.

Barker, G., Hodges, R., Hunt, C., Lloyd, J., Suano, M., Taylor, P., and Wickham, C. 1995. *A Mediterranean Valley: Landscape Archaeology and Annales History in the Biferno Valley*. London: Leicester University Press.

Barton, C. M., Bernabeu, J., Emili Aura, J., Garcia, O., and La Roca, N. 2002. Dynamic landscapes, artifact taphonomy, and landuse modeling in the Western Mediterranean. *Geoarchaeology* 17: 155–90.

Bender, B. (ed.). 1993. *Landscape, Politics and Perspectives*. Oxford: Berg.

Bevan, A., and Conolly, J. 2004. GIS, archaeological survey, and landscape archaeology on the Island of Kythera, Greece. *Journal of Field Archaeology* 29: 123–38.

Binford, L. R. 1964. A consideration of archaeological research design. *American Antiquity* 29: 425–41.

Bintliff, J., and Snodgrass, A. 1988. Off-site pottery distributions: A regional and interregional perspective. *Current Anthropology* 29: 506–13.

Birks, H. H., Birks, H. J. B., Kaland, P. E., and Moe, D. (eds.). 1988. *The Cultural Landscape: Past, Present and Future*. Cambridge: Cambridge University Press.

Burger, O., Todd, L. C., Burnett, P., Stohlgren, T. J., and Stephens, D. 2004. Multi-scale and nested-intensity sampling techniques for archaeologic-al survey. *Journal of Field Archaeology* 29: 409–23.

Butzer, K. W. 1982. *Archaeology as Human Ecology: Method and Theory for a Contextual Approach.* Cambridge: Cambridge University Press.

Cherry, J. F. 1983. Frogs around the pond: Perspectives on current archaeological survey projects in the Mediterranean region, in D. R. Keller and D. W. Rupp (eds.), *Archaeological Survey in the Mediterranean Area,* pp. 375–416. British Archaeological Reports, International Series 155. Oxford: BAR.

Cherry, J. F., Davis, J. L., and Mantzourani, E. 1991. *Landscape Archaeology as Long-Term History: The Keos Survey.* Los Angeles: UCLA Institute of Archaeology.

Cherry, J. F., Gamble, C., and Shennan, S. (eds.). 1978. *Sampling in Contemporary British Archaeology.* British Archaeological Reports, British Series 50. Oxford: BAR.

Cherry, J. F., and Shennan, S. 1978. Sampling cultural systems: Some perspectives on the application of probabilistic regional survey in Britain, in J. F. Cherry, C. Gamble, and S. Shennan (eds.), *Sampling in Contemporary British Archaeology,* pp. 17–48. British Archaeological Reports, British Series 50. Oxford: BAR.

Clark, R. H., and Schofield, A. J. 1991. By experiment and calibration: An integrated approach to archaeology of the ploughsoil, in A. J. Schofield (ed.), *Interpreting Artefact Scatters: Contributions to Ploughzone Archaeology,* pp. 93–105. Oxford: Oxbow Books.

Cowgill, G. L. 1990. Toward refining concepts of full-coverage survey, in S. K. Fish and S. A. Kowalewski (eds.), *The Archaeology of Regions: A Case for Full-Coverage Survey,* pp. 249–59. Washington, DC: Smithsonian Institution Press.

David, B. 1996. The 1984 Chillagoe surveys. *Queensland Archaeological Research* 10: 36–53.

Doelle, W. H. 1977. A multiple survey strategy for cultural resource management studies, in M. B. Schiffer and Gummerman (eds.), *Conservation Archaeology: A Guide for Cultural Resource Management Studies,* pp. 201–09. New York: Academic Press.

Dunnell, R. C., and Dancey, W. S. 1983. The siteless survey: A regional scale data collection strategy, in M. B. Schiffer (ed.), *Advances in Archaeological Method and Theory* 6, pp. 267–87. New York: Academic Press.

Ebert, J. I. 1992. *Distributional Archaeology.* Albuquerque: University of New Mexico Press.

———. 2000. State of the art in "inductive" predictive modeling: Seven big mistakes (and lots of smaller ones), in K. L. Wescott and R. J. Brandon (eds.), *Practical Applications of GIS for Archaeologists: A Predictive Modeling Toolkit,* pp. 129–34. London: Taylor and Francis.

Fanning, P. C., and Holdaway, S. J. 2004. Artifact visibility at open sites in western New South Wales, Australia. *Journal of Field Archaeology* 29: 255–71.

Flannery, K. V. (ed.). 1976. *The Early Mesoamerican Village.* New York: Academic Press.

Foley, R. 1978. Incorporating sampling into initial research design: Some aspects of spatial archaeology, in J. F. Cherry, C. Gamble, and S. Shennan (eds.), *Sampling in Contemporary British Archaeology,* pp. 49–65. British Archaeological Reports, British Series 50. Oxford: BAR.

———. 1981a. Off-site archaeology: An alternative approach for the short-sited, in I. Hodder, G. Isaac, and N. Hammond (eds.), *Pattern of the Past: Studies in Honour of David Clarke,* pp. 157–83. Cambridge: Cambridge University Press.

———. 1981b. *Off-Site Archaeology and Human Adaptations in Eastern Africa.* British Archaeological Reports, International Series 97. Oxford: BAR.

Gallant, T. W. 1986. "Background noise" and site definition: A contribution to survey methodology. *Journal of Field Archaeology* 13: 403–18.

Gillings, M., Mattingly, D., and van Dalen, J. (eds.). 1999. *Geographic Information Systems and Landscape Archaeology.* Oxford: Oxbow Books.

Haggett, P. 1965. *Locational Analysis in Human Geography.* London: Edward Arnold.

Judge, J. W., and Sebastian, L. (eds.). 1988. *Quantifying the Present and Predicting the Past: Theory, Method, and Application of Archaeological Predictive Modelling.* Denver: U.S. Department of the Interior, Bureau of Land Management.

Keay, S. J., and Millett, M. 1991. Surface survey and site recognition in Spain: The Ager Tarraconensis survey and its background, in A. J. Schofield (ed.), *Interpreting Artefact Scatters: Contributions to Ploughzone Archaeology,* pp. 129–39. Oxford: Oxbow Books.

Keller, D. R., and Rupp, D. W. (eds.). 1983. *Archaeological Survey in the Mediterranean Area.* British Archaeological Reports, International Series 155. Oxford: BAR.

Kintigh, K. W. 1988. The effectiveness of subsurface testing: A simulation approach. *American Antiquity* 53: 686–707.

Kohler, T. A., and Parker, S. A. 1986 Predictive models for archaeological resource location, in M. B. Schiffer (ed.), *Advances in Archaeological Method and Theory,* Vol. 9, pp. 397–452. New York: Academic Press.

Krakker, J. J., Shott, M. J., and Welch, P. D. 1983. Design and evaluation of shovel-test sampling in regional archaeological survey. *Journal of Field Archaeology* 10: 469–80.

Lewarch, D. E., and O'Brien, M. J. 1981. The expanding role of surface assemblages in archaeological research, in M. B. Schiffer (ed.), *Advances in Archaeological Method and Theory*, Vol. 4, pp. 297–342. New York: Academic Press.

Lightfoot, K. G. 1986. Regional surveys in the eastern United States: The strengths and weaknesses of implementing subsurface testing programs. *American Antiquity* 51: 484–504.

Lovis, W. A. 1976. Quarter sections and forests: An example of probability sampling in the northeastern woodlands. *American Antiquity* 41: 364–72.

Lynch, M. B. 1980. Site artifact density and the effectiveness of shovel probes. *Current Anthropology* 21: 516–17.

MacNeish, R. S. 1978. *The Science of Archaeology?* North Scituate, MA: Duxbury Press.

McManamon, F. P. 1984. Discovering sites unseen, in M. B. Schiffer (ed.), *Advances in Archaeological Method and Theory*, Vol. 7, pp. 223–92. New York: Academic Press.

McNiven, I. J. 1992. Shell middens and mobility: The use of off-site faunal remains, Queensland, Australia. *Journal of Field Archaeology* 19: 495–508.

Mueller, J. W. (ed.). 1975. *Sampling in Archaeology.* Tucson: University of Arizona Press.

Nance, J. 1983. Regional sampling in archaeological survey: The statistical perspective, in M. B. Schiffer (ed.), *Advances in Archaeological Method and Theory*, Vol. 6, pp. 289–356. New York: Academic Press.

Nance, J., and Ball, B. 1986. No surprises? The reliability and validity of test pit sampling. *American Antiquity* 51: 457–83.

Nash, G. (ed.). 1997. *Semiotics of Landscape: Archaeology of Mind.* British Archaeological Reports, International Series 661. Oxford: BAR.

O'Brien, M. J., and Lewarch, D. E. (eds.). 1981. *Plowzone Archaeology: Contributions to Theory and Technique.* Publications in Archaeology 27. Nashville: Vanderbilt University.

Odell, G. H., and Cowan, F. 1987. Estimating tillage effects on artifact distributions. *American Antiquity* 52: 456–84.

Orton, C. 2000. *Sampling in Archaeology.* Cambridge Manuals in Archaeology. Cambridge: Cambridge University Press.

Peterson, C. E., and Drennan, R. D. 2005. Communities, settlements, sites, and surveys: Regional-scale analysis of prehistoric human interaction. *American Antiquity* 70: 5–30.

Plog, F., Weide, M., and Stewart, M. 1977. Research design in the SUNY-Binghamton contract program, in M. B. Schiffer and G. J. Gumerman (eds.), *Conservation Archaeology: A Guide for Cultural Resource Management Studies,* pp. 107–27. New York: Academic Press.

Plog, S., Plog, F., and Wait, W. 1978. Decision making in modern surveys, in M. B. Schiffer (ed.), *Advances in Archaeological Method and Theory*, Vol. 1, pp. 383–421. New York: Academic Press.

Read, D. W. 1975. Regional sampling, in J. W. Mueller (ed.), *Sampling in Archaeology,* pp. 45–60. Tucson: University of Arizona Press.

Rhoads, J. W. 1992. Significant sites and non-site archaeology: A case-study from south-east Australia. *World Archaeology* 24: 198–217.

Richards, T. 1998. *A Predictive Model of Aboriginal Archaeological Site Distribution in the Otway Range.* Occasional Report No. 49. Melbourne: Aboriginal Affairs Victoria.

Richards, T., and Johnston, R. 2004. Chronology and evolution of an Aboriginal landscape at Cape Bridgewater, south west Victoria, in T. Richards and Y. Kerridge (eds.), *Archaeology in the Indigenous Community*, pp. 97–112. *The Artefact* 27.

Richards, T., and Jordan, J. 1999. *Aboriginal Archaeological Investigations in the Barwon Drainage Basin.* Occasional Report No. 50. Melbourne: Aboriginal Affairs Victoria.

Rossignol, J., and Wandsnider, L. (eds.). 1992. *Space, Time, and Archaeological Landscapes.* New York: Plenum Press.

Schadla-Hall, R. T., and Shennan, S. J. 1978. Some suggestions for a sampling approach to archaeological survey in Wessex, in J. F. Cherry, C. Gamble, and S. Shennan (eds.), *Sampling in Contemporary British Archaeology,* pp. 87–104. British Archaeological Reports, British Series 50. Oxford: BAR.

Schiffer, M. B., and Gumerman, G. J. 1977. *Conservation Archaeology: A Guide for Cultural Resource Management Studies.* New York: Academic Press.

Schiffer, M. B., Sullivan, A. P., and Klinger, T. C. 1978. The design of archaeological surveys. *World Archaeology* 10: 1–28.

Shott, M. J. 1989. Shovel-test sampling in archaeological survey: Comments on Nance and Ball, and Lightfoot. *American Antiquity* 54: 396–404.

———. 1995. Reliability of archaeological records on cultivated surfaces: A Michigan case study. *Journal of Field Archaeology* 22: 475–89.

Stoddart, S. K. F., and Whitehead, N. 1991. Cleaning the Iguvine stables: Site and off-site analysis from a central Mediterranean perspective, in A. J. Schofield (ed.), *Interpreting Artefact Scatters: Contributions to Ploughzone Archaeology,* pp. 141–48. Oxford: Oxbow Books.

Stone, G. D. 1981. On artifact density and shovel probes. *Current Anthropology* 22: 182–83.

Stryd, A. H., and Eldridge, M. 1993. CMT archaeology in British Columbia: The Meares Island studies. *BC Studies* 99: 184–234.

Terrenato, N. 2004. Sample size matters! The paradox of global trends and local surveys, in S. E. Alcock and J. F. Cherry (eds.), *Side-by-Side*

Survey: Comparative Regional Studies in the Mediterranean World, pp. 36–48. Oxford: Oxbow Books.

Thomas, D. H. 1975. Nonsite sampling in archaeology: Up the creek without a site? in J. W. Mueller (ed.), *Sampling in Archaeology*, pp. 61–81. Tucson: University of Arizona Press.

———. 1988. The Monitor Valley Survey: Rationale, strategy, and tactics, in D. H. Thomas and contributors, *The Archeaology of Monitor Valley, Vol. 3. Survey and Additional Excavations*, pp. 155–70. *Anthropological Papers of the American Museum of Natural History* Vol. 66, Pt 2.

Thomas, D. H., and Kelly, R. L. 2007. *Archaeology Down to Earth* (3rd ed.). Belmont, CA: Thomson Wadsworth.

Thompson, S. 2004. Side-by-side and back-to-front, in S. E. Alcock and J. F. Cherry (eds.), *Side-by-Side Survey: Comparative Regional Studies in the Mediterranean World*, pp. 36–48. Oxford: Oxbow Books.

Thoms, A. V. 1988. A survey of predictive locational models: Examples from the late 1970s and early 1980s, in J. W. Judge and L. Sebastian (eds.), *Quantifying the Present and Predicting the Past: Theory, Method, and Application of Archaeological Predictive Modelling*, pp. 581–645. U.S. Department of the Interior, Bureau of Land Management, Denver.

Tilley, C. 1994. *A Phenomenology of Landscape.* Oxford: Berg.

Wandsnider, L., and Camilli, E. L. 1992. The character of surface archaeological deposits and its influence on survey accuracy. *Journal of Field Archaeology* 19: 169–88.

Webber, H., and Burns, A. 2004. Seeing the forest through the trees: Aboriginal scarred trees in Barrabool Flora and Fauna Reserve, in T. Richards and Y. Kerridge (eds.), *Archaeology in the Indigenous Community*, pp. 36–45. *The Artefact* 27.

Westcott, K. L., and Brandon, R. J. (eds.). 2000. *Practical Applications of GIS for Archaeologists.* London: Taylor and Francis.

Whalen, M. E. 1990. Sampling versus full-coverage survey: An example from western Texas, in S. K. Fish and S. A. Kolwalewski (eds.), *The Archaeology of Regions: A Case for Full-Coverage Survey*, pp. 219–36. Washington, DC: Smithsonian Instituition Press.

Wheatley, D., and Gillings, M. 2002. *Spatial Technology and Archaeology: The Archaeological Applications of GIS.* London: Taylor and Francis.

Wilkinson, T. J. 1982. The definition of ancient manured zones by means of extensive sherd-sampling techniques. *Journal of Field Archaeology* 9: 323–33.

———. 2004. The archaeology of landscape, in J. Bintliff (ed.), *A Companion to Archaeology*, pp. 334–56. Oxford: Blackwell Publishing.

Willey, G. R. 1953. *Prehistoric Settlement Patterns in the Viru Valley, Peru.* Bulletin of the Bureau of American Ethnology, No. 155. Washington, DC: Government Printing Office.

Zvelebil, M., Green, S. W., and Macklin, M. G. 1992. Archaeological landscapes, lithic scatters, and human behaviour, in J. Rossignol and L. Wandsnider (eds.), *Space, Time, and Archaeological Landscapes*, New York: Plenum Press.

55

Noninvasive Subsurface Mapping Techniques, Satellite and Aerial Imagery in Landscape Archaeology

Paul N. Cheetham

The future appears bright for the landscape archaeologist as we are entering into a period that will see enormous advances in our understanding of archaeological landscapes. New technologies, some still in their toddler state, if not infancy, will increasingly provide landscape archaeologists with currently unimagined levels of detail regarding human exploitation of landscapes over time. Some survey and mapping techniques such as earthwork surveys and oblique aerial photography, mainstays of archaeology in the 20th century, while still contributing, will take their rightful role as complementary elements in what is becoming GIS-based research spaces dominated and underpinned by geophysical, geochemical, and laser altimetry and hyperspectral imagery data, the last two collected from varying altitudes and platforms. This chapter tries to assess what these technologies are currently contributing to landscape archaeology, as well as their future potential.

What is clear at the outset is that what has been done to date will increasingly be seen to be paltry in comparison with what will come. For example, all the archaeological geophysical surveys ever undertaken in the United States up to the year 2000 amounted to less than the number undertaken in just one year in the United Kingdom (Kvamme 2003), where it still continues at similar rates. And it is not just that more geophysical surveys will

be undertaken in the future as prices of specialist equipment drop in real terms and become more widely available; it is, rather, that year after year increasingly larger contiguous areas are being covered, and so for the first time total coverage of individual landscapes, at least by magnetometry, has become a viable proposition (cf. Gaffney, Gaffney, and Corney 1998; Gater 2005).

This great increase in detailed coverage is due to improved survey technology employing multiple sensors for increased resolution and/or coverage and multiple sensor-type platforms allowing, for example, ground-penetrating radar (GPR) and earth resistivity to be undertaken simultaneously (e.g., Leckebusch 2003). When such extensive geophysical surveys are combined with the high coverage rates and ease of repeatability of multispectral airborne/satellite techniques, we will be in a position to have the consistent and broad spatial and temporal coverage required to enhance major programs of landscape analysis that decades of conventional approaches have struggled to provide (Donohue 2005:558). Despite this potential, the failure to incorporate multispectral imagery survey in the techniques section of the recently published research agenda of a World Heritage site (Downes, Foster, and Wickham-Jones 2005) seems rather out of step with current knowledge, with only conventional film aerial photography being included (Brophy 2005), so awareness may be an issue. Therefore, one major aim of this chapter is to raise awareness of the complementary nature of these technologies and so promote a more integrated approach to their use.

The landscape archaeologist has always been able to obtain information from surface scatters, shovel and test pits,[1] standing earthworks and structures. However, much information is now obscured from immediate view either by virtue of being buried or simply because the scale is such that remnants of features and structures surviving within the modern landscape can be identified and mapped effectively only from an elevated viewpoint. Therefore, this chapter also reviews the potential of these techniques, which allow landscape archaeologists efficient access to information that cannot otherwise be obtained without recourse to extensive excavation combined with extensive and highly intensive surface surveys. It cannot, and does not, intend to provide an introduction or manual for the practical use of the individual techniques, and the reader should follow up the sources cited and where possible collaborate with experienced practitioners when employing any in a landscape project.

Development

For present purposes only, a condensed history is required to contextualize the different strands. The reader is directed to Gaffney and Gater (2003), Heron (2001), and Giardino and Hayley (2006) for recent concise histories of geophysical survey, geochemical prospection, and airborne/satellite remote sensing, respectively, which are the sources for the following material in this section.

The archaeological survey utility of geophysical phenomena was recognized in the late 19th century when "bosing" (an acoustic prospecting technique involving hammering the ground) was successfully used to detect unseen ditches on chalk downland in southern Britain. However, it was half a century later, in the 1940s, before technology was harnessed to create the first successful electrical surveys of archaeological sites, followed a decade later by the first magnetic surveys. Electromagnetic (EM) techniques were also added to the armory of archaeological geophysics at the same time, although the specific EM variant that is ground penetrating radar was not employed for archaeological work until the 1970s. Despite this early work, the frequent and widespread use of geophysical survey had to wait until the 1990s, when more efficient systems became available, equipment costs dropped and microcomputing

equipment to process the data became widely available. Current advances that are affecting data collection methodologies are the use of global position systems (GPS), mechanized rapid coverage systems, and multisensor platforms.

Aerial photography, despite being early 20th century and so later in origin, developed rather more quickly but only began to make a significant contribution after the Second World War, with the increased availability of suitable aircraft, camera equipment, pilots and resources dedicated to the technique in the United Kingdom. Initially utilizing visible spectrum black-and-white panchromatic film, both color and false-color near-infrared film were increasingly used, followed by aircraft becoming platforms for a range of multispectral digital sensor systems together with radar and laser altimetry. Satellite imagery had to wait until the Space Race of the 1960s and 1970s, and although data from early systems such as LANDSAT were applied to archaeological applications, it was really the turn of the new century that saw the release of higher resolution declassified intelligence imagery and the introduction of a new generation of high-resolution multi- and hyperspectral systems that have much greater potential for archaeological applications.

The Techniques as a Group

As stated above, this short chapter cannot even begin to cover each technique from a technical stance in the depth required to become competent to employ these in practice. That said, it is not an approach the author would wish to take anyway, because a rather more overarching discussion reviewing these techniques together in terms of what they can and, in practice, do contribute to landscape archaeology in combination is what is missing from the current literature. This chapter thus provides an opportunity to address this deficiency, filling a gap in the landscape archaeologist's essential knowledge base.

Forming such an overarching view can be difficult to achieve given that non-invasive subsurface mapping, aerial imagery, and satellite imagery are so distinct in terms of the skill and knowledge base required to use each effectively, and more importantly, the differing types of evidence—and so the contribution to landscape archaeology that each makes (or potentially could make). The result of this situation is perhaps inevitable, in that it is hard to find general texts that integrate these specialist techniques effectively. For example, Bowden (1999) primarily focuses on analytical earthwork surveys. This book does include a short chapter on aerial photography together with an even shorter chapter that covers geophysical survey and surface collection, grouped together as "other" survey techniques, but it does not consider satellite and airborne multispectral imagery. Similarly, one of the latest texts to be published—entitled *Remote Sensing in Archaeology: An explicitly North American Perspective* (Johnson 2006)—is almost exclusively a study of ground-based geophysical survey methods, with only one chapter to cover all airborne and satellite remote sensing (see also David 2006 for a recent broad survey that also covers underwater techniques). Although this source should be praised for integrating all these techniques together in one volume, as with Bowden's (1999) volume focusing on earthwork survey, the book's purpose, and so outcome, cannot be considered a balanced worldview of the techniques covered taking the amount of literature on airborne and satellite techniques now available. In fact, considering that any grouping of techniques in relative isolation of other techniques is counterproductive; full integration of all the relevant data must be the aim. Remote sensing results that are not combined with data from alluvial, colluvial, and aeolian overburden surveys may well be flawed to an unknown extent (e.g., Powlesland et al. 2006).

Unfortunately, it is not simply that the literature fragments all of these specialist areas, and in the process fails to assist in bringing them together in the service of landscape archaeology. Archaeological organizations need to ensure that the full range of disparate landscape survey techniques and approaches are consistently and effectively integrated into major landscape investigation projects.[2]

Choice of Technique and Survey Methodology

The selection of survey techniques and survey methodologies to meet the aims of a particular landscape project can be problematic. The approach of providing tables or simplistic guidelines for the application of these techniques may lead to inappropriate techniques being employed; at the same time, adhering to rigid standard operating procedures (SOPs) inevitably stifles innovative approaches. While such an approach has been taken in the past for geophysical survey in a commercial archaeological assessment context (David 1995), Schmidt (2002) stresses that although there are scholarly sources and published guidelines, specialist advice and specialist practitioners should

be employed at all stages, particularly since any of the techniques covered in this chapter will be but one element in a much wider program of work. Schmidt then lists the most important variables that determine the choice of geophysical survey technique (Schmidt 2002: 9). To encompass all the techniques covered in this chapter, an adaptation of this list of variables that should be considered is suggested to be

- the survey objectives;
- archaeological questions;
- previous remotely sensed evidence and results;
- current land-use;
- former land-use;
- underlying solid and drift geology;
- other local geomorphological and topographic factors;
- degree of access to the land;
- time, money, personnel, and equipment available for the survey.

In many situations, such desktop evaluations need to be followed by pilot studies or other methods of assessment based on field results. Powlesland and associates (2006) describe the use of "reference fields" where crop marks form on a regular basis and so provide a check on crop-mark responses obtained using conventional oblique aerial photography. The approach highlighted gaps in otherwise extensive and contiguous crop-mark complexes that required investigation using multispectral imagery and geophysical techniques.

For the discussion of the individual techniques, they have been grouped into the ground-based noninvasive subsurface mapping techniques followed by the remotely sensed imagery techniques that use airborne and satellite platforms.

Ground-Based Noninvasive Subsurface Mapping Techniques

Ground-based noninvasive subsurface mapping techniques can be divided into geophysical and geochemical techniques.

Geophysical Survey

The most important techniques for landscape survey are, in order of importance, geomagnetic (magnetometry and topsoil magnetic susceptibility), geoelectrical (earth resistivity and electrical imaging), and electromagnetic, which includes georadar (ground penetrating radar). The techniques are described in a broad survey of the specific methods and their archaeological applications in Gaffney and Gater (2003), with further detail in Clark (1996), Scollar (1990), and Becker and Fassbinder (2001) for magnetometry specifically; and, for ground penetrating radar, Conyers (2004). The quarterly journal *Archaeological Prospection* is now the main source for more detailed technical research into archaeological geophysical survey, although this journal also includes papers on airborne/satellite imagery and geochemical survey. Geochemical survey, usually considered separately as it is here, has a very close association with geophysical survey, especially regarding magnetic geophysical effects resulting from the variations in iron oxides that arise from inorganic and organic chemical processes. There is also a close affinity between those geophysical methods that exploit differences in moisture content (for example, earth resistivity) and crop-mark/parch-mark aerial photography, which also exploit such moisture differentials.

Unfortunately, the take up of geophysical methods and their application is quite variable across the profession when considered globally (see Kvamme 2003: 436, and cf. Cheetham 2005). This inevitably results from both a combination of the differing starting points and pace of development, and so expertise and directions in different countries, but also the actual relationship between geophysicists and archaeologists. This can vary from a very high level of integration with highly experienced professional archaeogeophysicists—in the United Kingdom both have been working as one since the 1950s (Clark 1996: 16–20)—or with science-trained archaeologists leading projects, to more of a service situation of geological, engineering, and environmental geophysicists with little archaeological experience working for the less geophysically literate arts educated archaeologists. Although the latter can cause difficulties, such collaborations can lead to new and imaginative approaches. In particular, the geomorphological history of an area is unlikely to be fully appreciated without collaboration between archaeologist and geologists who together are aiming to understand landscape change and geological events that influence postdepositional events and factors, and past settlement patterns (e.g., Similox-Tohon et al. 2004).

So while the use of geophysical methods for investigating individual archaeological sites is well established at least in some regions, it is less well known how geophysical survey contributes to studies of the wider landscape. From a landscape archaeological perspective, little has been written specifically about the peculiar nature of geophysical evidence; and, as with many archaeological prospection and detailed investigation techniques, although there has been a recognition that off-site survey is desirable (Gaffney and Tingle 1984), a site-focused approach may still predominate. This is often simply because of the effort required to conduct large-scale geophysical surveys, and the starting point is inevitably an identifiable site of some nature. However, there have been some quite amazing feats of manual survey that have broken the tyranny of site focused survey: 1,000 hectares have now been surveyed during work in the Vale of Pickering in the United Kingdom (Lyall 2006), and the "landscapes" of entire ancient towns and cities are now almost routinely being revealed to us by geophysics (e.g., Gaffney et al. 2000). These large area geophysical surveys allow the challenging of basic ideas about the density and the character of activity that could not be addressed previously (e.g., Powlesland et al. 2006).

What is peculiar about geophysical survey (and also with respect to some phenomena detectable with aerial and satellite imagery), and the one reason archaeologists so enthusiastically use it, is that it can discover archaeological traces where there are no visible surface indications because the archaeology has essentially been erased from the landscape. This may seem obvious, but it can be considered from a slightly different perspective. Although geophysical survey can be used in situations where standing remains exist, it is in situations where all traces of former landscape features have been erased that geophysical survey results raise interesting questions about past landscapes. For example, during large-area magnetic surveys to investigate the landscape around the multiperiod site at Billown (Isle of Man; Cheetham and Stocks 2004; Cheetham et al. 2000), a large and currently largely featureless pasture field produced evidence for a number of periods of landscape use (Figure 55.1). What is striking is that this is not seen to be the adaptation over time of one system but appears to be the total erasure of one system replaced by another. Although modern destruction and organization can be put down to the practicalities of modern, mechanized intensive cultivation, this would seem not to be the case in the past. What would possess a past occupier of this area to fill in hundreds of meters of ditches and replace them with an equal length of new ditches on a slightly different alignment? Something required a change on a vast scale that must be related to how people in the past perceived and related to the landscape

Figure 55.1   Ballahot, Isle of Man. This magnetometry (fluxgate gradiometry) survey of the largest field on the Island reveals an unexpectedly dense palimpsest of archaeological landscapes from the prehistoric to recent past. Of note is that the extensive ditched systems running NW to SE in the center NW of the plot seem to have been replaced by a similarly extensive system on a slightly different alignment. As it was in this case, sometimes only geophysical survey can reveal and document effectively such long-erased and large-scale events in the exploitation of past landscapes. The field has not produced any evidence from other forms of survey that would suggest such a complex landscape history (courtesy of the Billown Neolithic Landscape Project, Bournemouth University).

they inhabited. That this happened a number of times is the most compelling evidence of the effort that people will go to acculturate the landscape to their satisfaction. No other approach demonstrates this as effectively and consistently in its results as geophysical survey does, time after time. In contrast, surveys of just the surviving earthwork may provide a very distorted and partial picture of past landscapes and be particularly poor at evidencing landscape change where that change has obliterated any upstanding evidence, if it ever existed.

Magnetometry

The technique of magnetometry (Becker and Fassbinder 2001) is the technique of choice for large-scale landscape survey. In the context of the use of lightweight fluxgate gradiometers, it was regarded as

"the workhorse—and racehorse—of British archaeological prospecting" (Clark 1996: 69). Today large multiple sensor platforms, the latest of which is the vehicle-towed superconducting quantum interference device (SQUID) sensor systems (Schultze et al. 2005 and Figures 55.2 and 55.3), are providing rates of coverage at spatial resolutions and sensor sensitivities unimagined a decade ago (Merali 2006). The great strength of magnetometry is not, however, just the rate of coverage attainable (multi-antenna GPR can also provide high rates of coverage at high traverse resolutions—e.g., Finzi, Francese, and Morelli 2005) but the wide range of types of archaeological features and deposits that the technique responds to in the context of landscape archaeology. So effective is it as an archaeological prospection technique it has been described by one practitioner as "Nature's gift to archaeology" (Kvamme 2003).

Figure 55.2 Magnetometry, which has always been the most rapid ground-based archaeological geophysical survey technique, can now be undertaken at speeds of up to 30 km/h. The sensors employed here are super conducting quantum interference devices (SQUIDs) configured as a pair of gradiometers on a cart system that can be either manually or mechanically propelled (note the space for a third gradiometer not yet fitted owing to the high cost of these sensors). The tiny sensors are mounted in flasks that hold the liquid helium required to achieve the low temperature needed for the SQUID sensors to operate (courtesy of the Department of Quantum Electronics, IPHT Jena, Germany).

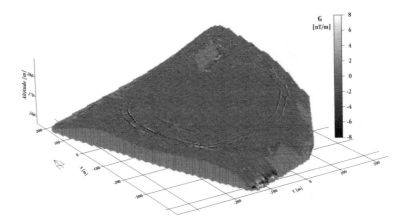

Figure 55.3 Survey of the hilltop enclosure of Niederzimmern (Thuringia, Germany) undertaken with the SQUID sensor system illustrated in Figure 55.2. This 27 hectare survey was undertaken in only three short winter days. It provides not only magnetic data but also a high-resolution topographic survey as a consequence of the sophisticated GPS navigation system needed to guide the survey. The constraints that the sensor design has on the sensor orientation results in the response characteristics of SQUID sensors being different from that of more conventional vertical gradiometers (courtesy of the Department of Quantum Electronics, IPHT Jena, Germany).

Magnetometry is a passive technique that measures anomalies in the Earth's magnetic field caused by archaeological features having differing magnetic properties to the natural background. The causes of these magnetic differences are varied but often result from burning, whereas the instruments that can be used to detect and map them are also varied in cost, sensitivity, and ease of use. Vertical gradiometers (two sensor instruments) with sensor separations of 1 m or less filter out deeper geological and larger geomorphological features (but have problems mapping thin shallow deposits—see below) and so focus on shallow substantive remains that result from the processes of magnetic enhancement (both natural and anthropogenic) and the redistribution or *in situ* formation of magnetically enhanced deposits that are or become the fills of pits, ditches, gullies, post holes. Burning, such an aspect of many domestic, ritual, and technological practices and processes in preindustrial societies, contributes greatly to this magnetic enhancement and extends the range of detectable substantive features to hearths, ovens, kilns, furnaces, and the locations of bonfires, pyres, and structures destroyed by fire. In terms of landscape archaeology, magnetic survey can show where agriculture was undertaken by mapping field systems and can detect the grain pits and silos that food was stored in and the hearths and ovens where food was cooked, within the buildings in which it was consumed. It can also help to identify organized higher-intensity production activities

(for example, salt production, ceramic production, and metallurgy, the latter an agent for more widespread landscape change owing to deforestation resulting from charcoal production).

Topsoil Magnetic Susceptibility

Where activity in the landscape does not involve the construction of structures, it may still result in the anthropogenic magnetic enhancement of the immediate topsoil over areas of occupation. Topsoil magnetic susceptibility survey, although subject to many limitations, can provide a rapid coverage of large areas in order to target more intensive surveys either by using specialized instruments or the in-phase mode of electromagnetic conductivity meters. Figure 55.4 demonstrates how a 10-m sampling interval field coil surface topsoil magnetic susceptibility survey has delimited an area of strong enhancement that magnetometry and earth resistivity have been able to resolve into detailed archaeological features of a Roman villa and its surrounding enclosures.

Despite its successes, magnetic survey does have substantial limitations. Magnetic susceptibility enhancement requires longer periods of intense occupation to develop, and so short-lived or non-settlement sites (for example, ritual structures such as mortuary enclosures) may not show up unless there is a strong natural topsoil subsoil magnetic contrast. An excellent example of the limitations of magnetic survey is shown in Gaffney and Gater

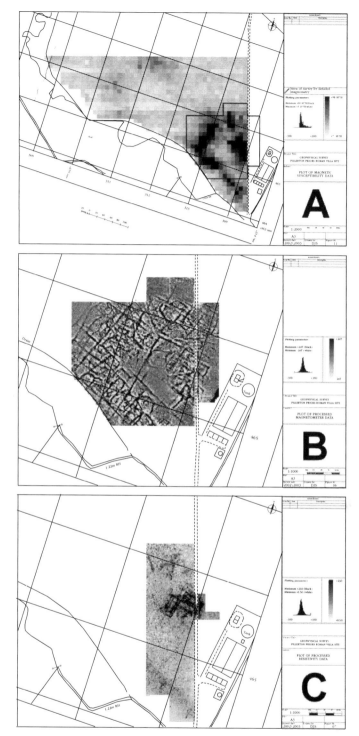

Figure 55.4 Pillerton Priors, U.K. Geophysical techniques, like many other landscape prospection techniques, are often found to be complementary, but some have higher rates of coverage and so are more appropriate for initial assessment with less rapid techniques being employed selectively to target specific classes of evidence. In this example, topsoil magnetic susceptibility surveys at 10m reading intervals (A) have located an area of magnetic enhancement that has been further investigated with magnetic gradiometry (B) and finally earth resistivity (C) to locate and map enclosure systems and stone buildings respectively (reproduced with the permission of David Sabin and Kerry Donaldson).

(2003: 127, fig. 60), where the magnetic response of ditches falls off rapidly to invisibility from the focus of "habitation," the ditches presumably forming a contemporary field system of the settlement, the full extent of which cannot be mapped by magnetic methods.

Earth Resistivity

As evident in Figure 55.4, earth resistivity survey is most effective over stone structures. It works by passing an electric current through the ground and mapping areas of high electrical resistance or low electrical resistance compared with background levels. Dense dry features such as stone will resist the electric current, whereas porous wet features such as ditches will conduct the electric current more readily. Until now, earth resistivity has not been used to provide areal coverage approaching anything like that of magnetic susceptibility

or magnetometry. However, it has gained a new lease of life with the introduction of hand- or vehicle-powered rapid survey versions exploiting the square array (Aspinall and Saunders 2005), making large-area survey feasible, if at a somewhat lower resolution (Figures 55.5 and 55.6). Additionally, it can become the essential technique when either the archaeology or the environment does not produce the magnetic contrast that the two previously discussed techniques require to function. It can also be used to provide three-dimensional survey in situations that do not favor ground penetrating radar and, importantly, depth profiles that allow the geomorphological history of areas of the landscape to be understood (e.g., Challis and Howard 2006). However, such three-dimensional imaging comes at the expense of areal coverage because of the time taken to take multiple readings at each reading station to provide data at different depths.

Figure 55.5 Employment of the square array has allowed mechanized systems to greatly increase the rates of coverage for earth resistivity surveys. The automatic resistivity profiler (ARP) system shown utilizes three differently sized square arrays and so also provides three different depths of survey simultaneously. At a traverse interval of 1 m and taking readings every 20 cm along the traverse, it can survey up to 10 hectares per day under ideal conditions (reproduced with the permission of Michel Dabas, Geocarta).

Figure 55.6 A 50-hectare area surveyed by the ARP system shown in Figure 55.5 undertaken in 10 days. In this survey (high resistivity shown in black, low in white) the large scale sinuous dark anomalies represent evidence of palaeochannels. At this scale, many smaller archaeological features that are present are not evident in the plot, but traces of earlier field systems are visible in the south east. The ability to cover such large areas economically at high resolution and at multiple depths allows both small and large scale natural, as well as anthropogenic, features to be identified, which can then be incorporated into the archaeological analysis of the landscape (survey plot courtesy of Geocarta and Prologis-GSE).

Electromagnetic Systems

Electromagnetic (EM) systems that use an electromagnetic field generated by one coil and detected by another can provide information similar to earth resistivity, topsoil magnetic susceptibility, and magnetometry, all from one instrument. This is possible by varying the instrument coil orientation and recording the in-phase (or real) and the 90° out-of-phase (quadrature) parts of the instrument's response simultaneously. The coil orientation determines the depth response characteristics, while these two phases can detect magnetic and conductive anomalies respectively. The interpretation of EM data can be difficult due to the complexity of the instrument's response, particularly regarding changes with depth, while magnetic and conductive effects can "stray" into the other's response phase and in some situations predominate. Again, mechanized systems with EM instruments mounted on multisensor platforms can provide coverage at rates allowing the mapping of larger-scale landscape features such as field systems, scoring over earth resistivity in situations where obtaining readings by inserting electrodes is not possible (for example, under very dry conditions), while simultaneously collecting magnetic data. Used for extensive surveys more frequently by North American than by European practitioners, the volume edited by Johnson (2006) is a good starting point for information about EM techniques together with example surveys. Despite the great potential of EM survey for landscape survey it remains a rather under exploited technique with Jordan (2000) commenting on the lack of use of EM in Mediterranean landscape archaeology.

Ground Penetrating Radar (GPR)

Ground penetrating radar employs electromagnetic pulses directed down into the ground that reflect back from interfaces that represent changes in the electromagnetic properties of the subsurface. Most effective for mapping in three dimensions stone structures and voids, it can also detect ditches and pits if the conditions are favorable. Although GPR is currently not used frequently for larger-scale landscape surveys, it has proved its worth in some difficult landscape survey situations. Utsi (2004) demonstrates the effectiveness of GPR in wetland peat bogs that are generally not regarded as environments offering high potential using conventional archaeological electrical and magnetic survey methodologies owing to the waterlogged nature and depth of the deposits. Mechanized GPR survey systems are now making an impact (Finzi, Francese, and Morelli

2005; Leckebusch 2003), and so GPR's contribution to landscape archaeology is likely to increase greatly as these systems become more widely used.

Impact on Landscape Archaeology

The technological innovation in archaeological geophysical survey, and in particular magnetometry, has been such that very large area surveys are now frequently undertaken and total surveys of entire landscapes are not only proposed but also represent a viable aim. Gater (2005) puts a case for total coverage of the landscape that constitutes the Orkney World Heritage site, where initial geophysical surveys of the area around the Stones of Stenness produced important findings regarding this landscape: "we did not imagine how dramatic our results would be. Almost from the first day it became clear that the known sites represented only a tiny percentage of what was actually there" (Card 2005: 343). With over 100 hectares of magnetometry supplemented by targeted resistivity undertaken already, the aim of total geophysical survey coverage of such landscapes is clearly realistic. However, more importantly in this case, on the strength of the results obtained so far, such extensive coverage has transformed our view of areas previously considered and described to be "sterile" into a vision of landscape development representing almost every period from the Neolithic onward. The result is a staggering increase in our knowledge of, and the importance of, this World Heritage Site (Card 2005). Multisensor platforms (Figure 55.7) incorporating both magnetometry and resistivity can increase coverage rates making larger surveys more viable.

It is important that geophysical data is correlated and so interpreted with the topography of the area of the survey. Although geoelectrical imaging and GPR are routinely corrected for topography (e.g., Goodman et al. 2006; Similox-Tohon et al. 2004), this is less often done with large area surveys of resistivity and magnetometry, or done only at a coarse level when combined with topographical data surveyed for wide coverage purposes (that is, reading interval of 1 m or more). Both resistivity and magnetometry will record false anomalies as a result of relatively small but abrupt changes in topography encountered when one is surveying areas with upstanding earthworks or features such as revetted terraces in garden landscapes or arising from agricultural terracing. In such circumstances, a high-resolution topographical survey created by ground-based laser scanners integrated with the geophysical surveys within a GIS environment (e.g., Neubauer 2004) identifies, documents, and

Figure 55.7 Multisensor platforms that employ a number of geophysical techniques simultaneously to provide comprehensive coverage more efficiently are becoming more prevalent. In this illustration earth resistivity and fluxgate gradiometry are being undertaken using a manually pulled Geoscan Research MSP40, but mechanized systems that employ various combinations of ground penetrating radar, electromagnetic, magnetic and earth resistance instruments are available for larger area surveys (photograph by Paul Linford, English Heritage).

assists in the correct interpretation of geophysical anomalies that result from topographical effects or simply correlate with them.

Geochemical Survey

Geochemical survey in a landscape archaeological context has generally not been undertaken anywhere near as frequently or spatially as extensively as geophysical and airborne and satellite methods. Reviewed by Heron in 2001 and concluded to be the "Cinderella of archaeological prospecting," this partially invasive technique has the potential to turn a landscape not to one that simply reveals its physical features but, directly, the use of that landscape in ways that the archaeologist is interested in (Entwistle, Dodgshon, and Abrahams 2000). As such, geochemical survey offers the prospect of moving beyond features and artifacts to the wider landscape and so the "environmental" setting of sites.

Phosphate surveys, the most established category of archaeological geochemical survey, are still regularly being undertaken, but as Heron

(2001) points out, establishing standardization of extraction and detection protocols while also facing up to the identified uncertainties in the method are both required to push this technique forward. The current brief review suggests this analysis is still valid in 2006. Multi-element survey has become more prevalent, but the "life cycles" of many of the elements are not understood, nor is how to interpret identified enhancements in terms of the anthropogenic activity that produced them. Phosphate is still regarded as the most easily identified and interpretable evidence of large inputs of organic matter, even if the exact form of the original organic matter is unknown, and although there is potential for lipid and amino acid survey to input into interpretation, this refinement of the technique is even less established in basic research (for example, establishing the difference between manured and simply grazed land) and in extensive application in landscape archaeological contexts. Although the method offers great potential, understanding what are anthropogenic geochemical enhancements between site and "off-site" samples is extremely problematic when we move to the land-scape situation where the whole landscape becomes the "site". It has already been alluded to that some aspect of geophysical properties, particularly mag-netic susceptibility, result from land-use and "indus-trial" processes, and so it is not surprising that geochemical surveys are often combined with mag-netic susceptibility surveys, providing evidence for the distinct zonation of activities (e.g., Clark 1996: 113, fig. 88).

For useful results to be obtained, there is a need to sample sites carefully at appropriate points in the soil profile and establish natural variations by adopting appropriate sampling strategies. Physically collecting, recording, and bagging the individual soil samples and transporting them to a laboratory for pretreatment prior to actual analy-sis make geochemical surveys slow, complex, and expensive in comparison with other prospection techniques. Few extensive area surveys have been undertaken. Aston and associates (1998) is one such more extensive multi-element survey.

The recent availability of portable but expensive X-ray fluorescence (XRF) instruments that allow direct multi-element analysis of soils in the field may make a significant impact on geochemical sur-vey. The utility of portable XRF instruments has been demonstrated in environmental soil analysis applications (Clark et al. 1999), and although these can be used to take direct readings on soil surfaces *in situ*, better results can be obtained if the sam-ples are sieved before analysis. This is a balance between rapid survey against quality of individual determinations; however, for many archaeological applications less precise *in situ* measurement may well be adequate. A few protostudies have been undertaken, such as Wager and associates (1998), in which Bronze Age copper ore processing sites were identified and the instrument used deemed to be "highly effective" for this application, par-ticularly if used in conjunction with topographical and geophysical surveys.

Remotely Sensed Imagery from Airborne and Satellite Platforms

This section includes traditional film aerial photog-raphy, laser altimetry, and airborne and satellite multispectral imaging.

Aerial Photography

Aerial photography is recognized as the most productive archaeological prospection technique, which in favorable circumstances can identify a wide range of site types, landscape management, and environmental information and often provides the principal record and hence framework for the types of archaeology and spatial relationships within particular landscapes. However, its value to individual landscape archaeology projects needs to be questioned strongly. In comparison with geo-physical survey, aerial photography is much more biased in terms of the sample of the past it pro-vides owing to its coverage being dependent on many more variables, some of which are not pos-sible for even the most diligent archaeologist and well-funded project to overcome. For example, woodland or desert sand will not produce crop marks, whereas they will often produce excellent geophysical survey results. Crop regimes, cultiva-tion patterns, and restrictions on flying near sen-sitive military and civilian sites all hamper aerial photography's providing a consistent sample of a landscape's archaeology. The magnetic effects that make magnetometry the most successful geo-physical survey technique is a phenomenon not accessible via aerial imaging, so whole swathes of archaeological landscapes can go undetected. This pattern contrasts with geoelectrical methods that will often produce results similar to those provided by crop and parch marks. What is often championed as the major advantage of aerial photography is probably its most problematic in terms of being able to understand how the land-scape was perceived and exploited by those who peopled it in the past: this is that an aerial view makes it easier to make "sense" of things that when viewed at ground level would be extremely difficult for the archaeologist to identify and to categorize. If it is to be used simply as a tool for archaeological recording, then this aspect of aerial photography is not a problem. However, if this is extended to making the assumption that people in the past also had a comparable aerial overview concept of the landscape that in some way deter-mined its organization and ritual activities, then we may run into serious problems when forming our interpretations.

It is not the role of this section to go into the detail of individual aerial photographic methods but to consider the contribution of the technique as a whole to landscape archaeology to date and to identify any emerging strengths and weak-nesses. There is a large body of excellent gen-eral and more detailed sources, including Riley (1997) and Brophy and Cowley (2005), and with a Mediterranean perspective, Jones (2000). For keep-ing abreast of ongoing technical developments, the newsletter of the Aerial Archaeology Research Group is the most useful source.

It is very hard to balance the often-stunning clarity and sheer quantity of the archaeology that is frequently revealed by aerial photographic methods against the fundamental weaknesses of the method, in that the archaeologist has little or no control over the uniformity of the aerial archaeological data set over wide areas. Such uniformity of basic data is a fundamental requirement of landscape archaeological approaches. It was quite obvious to the author back in his undergraduate days, when looking at the aerial photography records of the North Yorkshire (United Kingdom) County Sites and Monuments Record, that although the Vale of Pickering had been flown to the point of near-exhaustion in terms of the ratio of new site to repeat sites photographed, less than 40 km to the west in the Vale of York there was a much higher proportion of new to repeat sites, because of the lack of an equivalent level of sustained and systematic survey (Cheetham 1985). The consequences for any useful comparisons of the settlement of these two areas based on the aerial photographic evidence are obvious. In the United Kingdom, such deficiencies in coverage are being addressed by English Heritage's National Mapping Programme but will remain a factor in the historic coverage.

Featherstone and associates (1999) reported 1996 as the best year in the United Kingdom for recording new sites since the extreme drought year of 1976. That the 1996 campaign resulted in half of the sites photographed being new to the record may seem to be a triumph. In fact, if anything, this highlights the very poor recovery in almost a century of flying and pulls into sharp focus the lack of any control over the sample of archaeology that is recovered as crop-mark evidence. The consequences for landscape archaeology of relying on such randomly produced evidence are clearly demonstrated by the fact that during the 1996 campaign, a very major site, that of a Roman legionary fortress, was discovered in Norfolk. Again, a great triumph for aerial photography, but how valuable would be any study of the Roman period landscape of the area around this site without knowledge of the presence of such a major socioeconomic center, and one that may have remained unknown for another century if we relied solely on aerial photography? Any landscape analysis that relies on sites located by aerial photography is in danger of being rewritten ad nauseum as the unpredictability of oblique crop-mark survey introduces new sites into the equation at random. The debate between the effectiveness of selective oblique photography against the rather more objective vertical coverage has been discussed by Doneus (2000) and

considered more recently by Mills (2005), both of whom suggest that a more systematic and so better overall landscape coverage is afforded by vertical surveys as long as they are undertaken at appropriate times of the year.

Streamlining the management of oblique aerial photographic surveys has been focused on by Leckebusch (2005), who proposes a totally automated system that links high-resolution digital cameras with orientation sensor, GPS, and PC, so enabling the geographical coordinates of the frame to be automatically established, thereby allowing more time to be spent searching for, observing, and photographing rather than recording. This system would also allow the seamless porting of the resulting primary record into a GIS system and on to a digital archive for wide access and integration with other sources of data, the latter being a principal requirement of the landscape archaeologist (e.g., Campana and Francovich 2003) as the basis for well-supported interpretation (see also Conolly, this volume).

Laser Altimetry

One problem in aerial photography employing natural sunlight as the illuminating source is that, as is the case with many of aerial photography's other parameters, it is difficult, if not impossible, to control shadow in any meaningful way other than being there at the right time when the sun's angle reveals the targeted site, which may occur only for a few days or weeks in any year. One way out of this problem is the use of airborne laser altimetry, also known as LiDAR (Light Detection And Ranging), to provide a high-resolution digital terrain model that can be lit from any angle including directly from the north, a sun direction never available to aerial archaeologists (or the south, if they are working in the southern hemisphere). Such an approach led to the discovery that ramparts were standing as low earthworks at Newton Kyme (North Yorkshire, United Kingdom), despite the site being previously heavily photographed owing to its excellent crop-mark response (Bewley 2003).

Within landscape archaeology, airborne LiDAR sits somewhere between ground-based survey techniques of earthwork and high-resolution topographical surveys (both important field techniques for recording and understanding archaeological landscapes, particularly settlement layout and associated field systems) and true *remote* sensing techniques, the distinction being somewhat blurred by the increasing use of surface or tower-mounted LiDAR scanners (Neubauer 2004).

Challis (2006) looks at landscape archaeological applications of LiDAR data that were originally collected by the United Kingdom Environment Agency for flood prediction and management, and concludes that even this relatively coarse data (2 m ground resolution; ground-based platform mounted LiDAR can be subdecimeter resolution over restricted areas) is particularly effective for mapping mature middle-reach flood plains in terms of providing the geomorphological context for understanding past cultural landscapes, defining areas of low or high potential for further study and cultural resource management, and so parameters for predictive modeling. Challis and Howard (2006) also note that the intensity of the reflected LiDAR pulse operating in the near-infrared can also be used and that this is currently being explored. Crutchley (2006) tried to assess low-ground resolution (2 m) height data for the definition of archaeological features and found limitations, whereas higher ground resolution (1 m) data used at Stonehenge exceeded expectations (Bewley, Crutchely, and Shell 2005).

Airborne Multispectral Imagery

Despite the use of increasingly sophisticated satellite-based imagery systems and the use of declassified military imagery for evaluating landscape change in both environmental and Cultural Resource Management (CRM) temporal studies, multispectral imagery (MSS) sensing from lower altitude airborne platforms offers the greatest potential for archaeological applications. The increased ground resolution offered by airborne MSS compares favorably with conventional film aerial photography, and the detection of crop marks at wavelengths beyond the visible allows greater latitude in observing such effects than does the more restricted window of identification offered by just the visible and near-infrared wavelengths. This approach also has the advantage of providing a more uniform coverage requiring fewer repeat visits during optimal conditions for crop-mark formation. As discussed above, Powlesland and associates (1997) and Donoghue (2001) highlight these advantages of airborne multispectral imagery over conventional aerial photographic approaches when used in landscape projects. Barnes (2003) also demonstrates its combination with LiDAR data to provide information for managing entire archaeological landscapes, which underpins future archaeological analysis of such landscapes. Airborne thermography, although having potential for archaeological survey (e.g., Ben-Dor et al. 2001), has failed to make any significant impact, and few surveys have been reported in the last five

years. Until national and international agencies that are responsible for the archaeological resource undertake airborne multispectral scanning routinely to provide the appropriate data to underpin landscape archaeology, commissioning flights to obtain such data may, in practice, be beyond the resources of individual projects.

Satellite Imagery

Ground-resolution, bandwidth, spatial-coverage, and image-processing techniques are constantly improving, and for information on these, the landscape archaeologist is directed to the latest sources, such as the *International Journal of Remote Sensing,* to keep abreast of developments in the more technical aspects of satellite imagery. Currently, ground resolutions of 0.61 m panchromatic and 2.44 m multispectral are reported for Digital Globe's QuickBird system (Giardino and Haley 2006), making such systems now capable of comparing favorably with low-resolution, high-altitude airborne surveys. Although such higher-resolution imagery is always more useful, lower-resolution imagery can be used to provide landscape environmental information that gives wider coverage and so build on more limited coverage ground and airborne surveys. As is the case with aerial photography and airborne imagery, although satellite remote sensing can be used as a very effective prospection and landscape-assessment tool, it must be used very carefully for making interpretations about past landscape perception from such an artificial and large-scale generalizing viewpoint. That said, a number of projects employing remote sensing illustrate the effectiveness of this technique in parts of the earth that are not accessible to either extensive ground-based or lower-altitude aerial survey. Satellite imagery can also play an important part in providing information for areas of the world that lack detailed conventional topographical mapping, and as such provide a supporting role in archaeological surveys. It can also be the route into landscape work in countries that have restrictions on airborne and ground-based surveys owing to political and legal issues. For example, Altaweel (2005) demonstrates the utility of analyzing ASTER multispectral data from 2000 in conjunction with 1960s and 1970s CORONA imagery for studies of hollow route ways, canal systems, and tell sites in Iraq.

Satellite imagery is also useful for providing information in areas where ground survey is physically difficult to undertake, such as in wetlands and tropical forests, but has proved particularly successful in low-vegetation desert environments (e.g., Campana and Francovich 2003). However, Fowler and Fowler (2005) suggest that equal success can be obtained in temperate regions, demonstrating that CORONA KH-4B medium-resolution imagery can be used to detect crop marks in southern England. It follows from this that even where other remote-sensing techniques are available, satellite imagery may still have a productive role in landscape archaeology projects in all environments and on all continents.

Perhaps one of the most useful examples of the effective use of satellite imagery that is tied into an established landscape archaeology program is given in Clark and associates (1998). Satellite imagery of southern Madagascar was found to be particularly useful in helping derive maps of primary, secondary, or regenerative forest types. Through this approach it has been possible to identify former settlement zones and small patches of primary forest thought to represent sacred sites. Also revealed in this study was evidence of palaeodune systems that are indicative of climatic and environmental change. The authors suggest that it may be possible to link these environmental changes to long-term social change in southern Madagascar.

To give an overall judgment on the long-term effect that satellite imagery will have on landscape archaeology is challenging. Despite decades of published studies, these are on examination mostly evaluations, often not attached to wider landscape projects and so hard to assess in terms of their impact on the subject. In many cases the results have not been followed up by ground truthing and so remain unevaluated. Even some of the latest work is still at the stage of evaluating systems (e.g., Lasaponara and Masini 2005). From the literature it is evident that multispectral remote sensing data have not been fully exploited for archaeological purposes, with expense and availability remaining issues. The processing of all satellite data, and in particular processing that exploits multispectral and multiple-source data, remains a highly specialist task and so a barrier to its routine use. It is likely that only projects with appropriate budgets and supporting infrastructure will be in a position to fully exploit the available satellite imagery resource (see below). So, although satellite imagery has been demonstrated to be useful in a narrow range of applications, particularly identifying large-scale lineations in a range of landscapes, its role so far can be seen more as providing the synoptic underpinning for landscape archaeology projects employing more conventional approaches. In the future, there should be an increasing role for satellite imagery to provide detailed baseline data for large-scale landscape projects as availability increases and costs drop, while expertise in using such satellite imagery effectively to address more relevant archaeological questions will, we hope, also improve.

Integration

One aspect that links all the techniques is their integration within geographical information systems (GIS) as the basis for moving on to higher levels and, arguably, more relevant interpretation at the landscape level than any individual geophysical survey, aerial photograph, or satellite image can hope to do. GIS, covered in this volume by Conolly, incorporates the relationship of GIS to remotely sensed data, the spatial technologies offered by GIS being useful in order to visualize and so to inspect the spatial relationships between landforms and archaeological data (see Conolly, this volume, Figure 55.2). It is, however, naive to simply consider the integration of remotely sensed data without incorporating surface survey, excavation, and environmental, historical, and other relevant information, which is beyond the scope of this chapter. Powlesland and colleagues (2006) explore the relative effectiveness and complementary nature of aerial photography, airborne multispectral imagery, and magnetometry (Figure 55.8), but these are considered along with an almost 2,500 auger core survey of the aeolian deposits. Lock (2003) also contextualizes many of the survey and prospection techniques covered here together with other non-invasive survey techniques within the specialism of archaeological computing and its wide range of applications. This source reviews specific computer-based techniques, such as the rectification of oblique aerial photography and the presentation of geophysical data together with data integration, to create "digital landscapes" (Lock 2003: 164–82) that GIS can exploit. As with rectification, the conversion of conventional film airborne images into digital form is required to integrate and exploit effectively the information they hold. Forghani and Gaughwin (n.d.) used conventional color and black- and-white film aerial photography, which was scanned, rectified in a GIS, and followed by supervised image classification to identify and map an historic road network in a forested area of Tasmania (Australia). Schmidt (2004) provides a concise overview of informatics requirements of geophysical prospection and airborne and satellite remote sensing specifically for archaeological prospection. To integrate such diverse data sets requires the landscape archaeologist to be conversant with these informatics aspects as gigabytes of diverse data turn into terabytes and

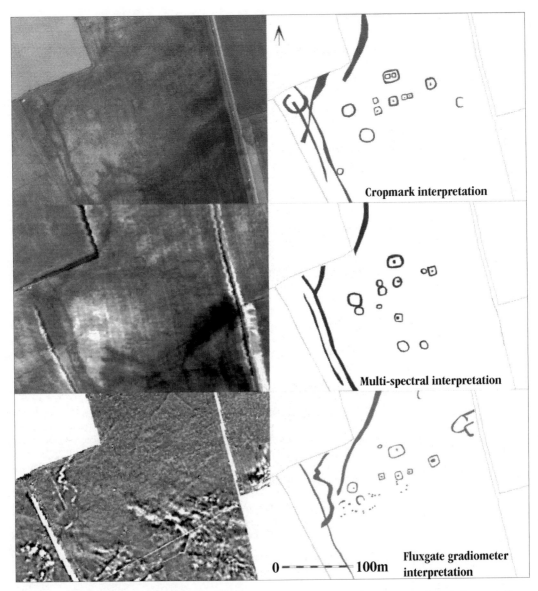

Figure 55.8 Color-rectified oblique aerial, airborne multispectral (Band 11- thermal) and fluxgate gradio-metry imagery of an area of the River Derwent Floodplain (North Yorkshire) with interpretations alongside. Although the three are broadly comparable, each has revealed features of archaeological significance that the others have not. In this example, the only airborne multispectral survey has detected ring ditches in the southern part of the survey area, which appears less responsive to both conventional aerial photographic and magnetic techniques (reproduced with the permission of Dominic Powlesland, The Landscape Research Centre).

on to petabytes. The abstraction of archaeologically meaningful results from such large data sets without assistance from automated classification systems may well be beyond human interpretative abilities or time scales, and so it is in this area that major advances need to be made if we are to reap the benefits of the comprehensive subsurface, airborne, and satellite remote-sensing data that are increasingly becoming available to the landscape archaeologist.

Conclusions

Particularly in the areas of geophysics and airborne LiDAR and multispectral scanning, technical advances and the wider availability of these techniques are making their use not only viable but also arguably essential to getting the comprehensive and uniform levels of data recovery over large areas to underpin any serious landscape archaeology research project. Unfortunately, although individual techniques can be demonstrated to have significant impacts and potential yet to be fully exploited, the employment in any one project of anything like the full range of techniques that are available is still a rarity. As suggested in this chapter, this is in part due to the organization of these highly specialized techniques into individual and somewhat isolated centers resulting in a lack of coordination in their development and application, and so integration in the service of landscape archaeology. Without this coordination and integration, these highly effective tools of landscape archaeology will continue to be applied piecemeal with only the most well-resourced projects being able to benefit in full from the excellent development, knowledge base, and specialized expertise that is now available to the landscape archaeology community. Landscape archaeology must not accept second best in this area of the discipline, because all higher levels of analysis and interpretation should be based on the best data available; otherwise, the discovery of one missed important site could require the whole to be revised, or worse if important sites still unknown through lack of appropriate prospection efforts being lost forever to landscape archaeology by development, agriculture, or erosion.

Notes

1. The inefficiency and ineffectiveness of test pitting in comparison to geophysical prospection is discussed in Kvamme (2003).

2. A more useful example and source of integrated programs or survey and research can be found at the Laboratory of Geophysical-Satellite Remote Sensing and Archaeo-environment (www.ims. forth.gr/lab_index.html) of the Institute for Mediterranean Studies (I.M.S.)/Foundation of Research and Technology (F.O.R.T.H.). Since 1996, this laboratory has combined and so integrated geophysical prospection, satellite remote sensing, and archaeo-environmental and geographical information systems (GIS) within one facility, which is clearly the way forward both from a research and application perspective.

Acknowledgments

The author would like to thank all those who provided illustrative material for this chapter (cited individually with the figures) and the editors, referees, and Kayt Armstrong (Bournemouth University) for their comments on the draft text.

References

Altaweel, M. 2005. The use of ASTER satellite imagery in archaeological contexts. *Archaeological Prospection* 12: 151–66.

Argote-Espino, D., and Chávez, R. 2005. Detection of possible archaeological pathways in central America through digital processing of remote sensing images. *Archaeological Prospection* 12: 105–14.

Aspinall, A., and Saunders, M. 2005. Experiments with the square array. *Archaeological Prospection* 12: 115–29.

Aston, M., Martin, M., and Jackson, A. 1998. The potential for heavy metal soil analysis on low status archaeological sites at Shapwick, Somerset. *Antiquity* 72: 838–47.

Bailey, J. (ed.). 1998. *Science in Archaeology*. London: English Heritage.

Barnes, I. 2003. Aerial remote-sensing techniques used in the management of archaeological monuments on the British army's Salisbury Plain training area, Wiltshire. *Archaeological Prospection* 10: 83–90.

Becker, H., and Fassbinder, J. 2001. *Magnetic Prospecting in Archaeological Sites*. Monuments and Sites VI, München: International Council on Monuments and Sites (ICMOS) and Bavarian State Conservation Office.

Ben-Dor, E., Kochavi, M., Vinizki, L., Shionin, M., and Portugali, J. 2001. Detection of buried ancient walls using airborne thermal video radiometry. *International Journal of Remote Sensing* 22: 3689–702.

Bewley, R. H. 2003. Aerial survey for archaeology. *The Photogrammetric Record* 18: 273–92.

Bewley, R. H., Crutchely, S. P., and Shell, C. A. 2005. New light on ancient landscapes: LiDAR survey in the Stonehenge World Heritage Site. *Antiquity* 79: 636–47.

Bewley, R., Donoghue, D., Gaffney, V., van Leuden, M., and Wise, A. 1999. *Archiving Aerial Photography and Remote Sensing Data: A Guide to Good Practice*. Oxford: Archaeological Data Service and Oxbow (also available online at http://ads.ahds. ac.uk/project/goodguides/apandrs).

Bowden, M. (ed.). 1999. *Unravelling the Landscape: An Inquisitive Approach to Archaeology*. Stroud: Tempus.

Brophy, J. 2005. Aerial survey, in J. Downes, S. M. Foster, and C. R. Wickham-Jones. 2005. *The Heart of Neolithic Orkney World Heritage Site Research Agenda*, pp. 104–05. Edinburgh: Historic Scotland (also available from www.historic-scotland.gov. uk/index/publications/worldhsitespublications/ orkneyresearch.htm).

Brophy, K., and Cowley, D. (eds.). 2005. *From the Air: Understanding Aerial Archaeology*. Stroud: Tempus.

Brothwell, D. R., and Pollard, A. M. (eds.). 2001. *Handbook of Archaeological Sciences*. Chichester: Wiley.

Campana, S., and Francovich, R. 2003. Landscape archaeology in Tuscany: Cultural resource management, remotely sensed, GIS based data integration and interpretation, in M. Forte, P. R. Williams, F. El Baz, and J. Wiseman (eds.), *The Reconstruction of Archaeological Landscapes through Digital Technology*, pp. 15–28. Oxford: BAR S1151.

Card, N. 2005. The heart of Neolithic Orkney. *Current Archaeology* 199: 342–47.

Cavalli, R., Marino, C., and Pignatti, S. 2000. Environmental studies through active and passive airborne remote sensing systems, in M. Pasquinucci and F. Trément (eds.), *Nondestructive Techniques Applied to Landscape Archaeology*, pp. 31–37. Oxford: Oxbow Books.

Challis, K. 2005. Airborne laser altimetry in alluviated landscapes. *Archaeological Prospection* 13: 103–27.

Challis, K., and Howard, A. J. 2006. A review of trends within archaeological remote sensing in alluvial environments. *Archaeological Prospection* 13: 231–40.

Cheetham, P. N. 1985. The archaeological database: Applied? in *Proceedings of Computer Applications and Statistical Methods in Archaeology 1985*. Southampton: University of Southampton.

———. 2005. Forensic geophysics, in J. R. Hunter and M. Cox, *Forensic Archaeology: Advances in Theory and Practice*, pp. 62–95. London: Routledge.

Cheetham, P. N., Darvill, T. D., Doonan, R., and Russell, B. 2000. Mann's landscapes revealed. *Archaeolgia Polona* 41: 137–40.

Cheetham, P. N., and Stocks, A. 2004. Geophysical survey of Ballahot, in T. D. Darvill, *Billown Neolithic Landscape Project. Eighth Report: 2003*, pp. 23–28. Research Report 12, Bournemouth and Douglas: Bournemouth University and Manx National Heritage.

Clark, A. J. 1996. *Seeing Beneath the Soil*. London: Batsford.

Clark, C. D., Garrod, S. M., and Parker Pearson, M. 1998. Landscape archaeology and remote sensing in southern Madagascar. *International Journal of Remote Sensing* 19: 1461–77.

Clark, S., Menrath, W., Chen, M., Roda, S., and Succop, P. 1999. Use of a field portable X-ray fluorescence analyzer to determine the concentration of lead and other metals in soil samples. *Annals of Agricultural and Environmental Medicine* 6: 27–32.

Conyers, L. B. 2004. *Ground-Penetrating Radar for Archaeology*. Walnut Creek, CA: AltaMira Press.

Crutchley, S. 2006. Light detection and ranging (lidar) in the Witham Valley, Lincolnshire: An assessment of new remote sensing techniques. *Archaeological Prospection* 13: 251–57.

Darvill, T. D. 2004. *Billown Neolithic Landscape Project. Eighth Report: 2003*. Research Report 12, Bournemouth and Douglas, Bournemouth University and Manx National Heritage.

David, A. 1995. *Geophysical Survey in Archaeological Field Evaluation*. Research and Professional Services Guideline No.1, London: English Heritage.

———. 2006. Finding sites, in J. Balme and A. Patterson (eds.), *Archaeology in Practice: A student Guide to Archaeological Analyses*, pp. 1–38. Malden: Blackwell Publishing.

Doneus, M. 2000. Vertical and oblique photographs. *AARGnews* 20 (March 2000): 33–39 (also available online from http://aarg.univie.ac.at/aargnews/pdf/ AARGnews20.PDF).

Donoghue, D. N. M. 2001. Remote Sensing, in D. R. Brothwell and A. M. Pollard (eds.), *Handbook of Archaeological Sciences*, pp. 555–63. Chichester: Wiley.

Downes, J., Foster, S. M., and Wickham-Jones, C. R. 2005. *The Heart of Neolithic Orkney World Heritage Site Research Agenda*. Edinburgh: Historic Scotland [also available from www.historicscotland.gov. uk/index/publications/worldhsitespublications/ orkneyresearch).

Entwistle, J. A., Dodgshon, R. A., and Abrahams, P. W. 2000. An investigation of former land-use activity through the physical and chemical analysis of soils from the Isle of Lewis, Outer Hebrides. *Archaeological Prospection* 7: 171–88.

Featherstone, R., Horne, P., Macleod, D., and Bewley, R. 1999. Aerial reconnaissance over England in summer 1996. *Archaeological Prospection* 6: 47–62.

Finzi, E., Francese, R. G., and Morelli, G. 2005. High-resolution geophysical investigation of the archaeological site of "Le Pozze" in the surroundings of the town of Lonato (Brescia, Northern Italy), in S. Piro (ed.), *Sixth International Conference on Archaeological Prospection: Proceedings Extended Abstracts*, pp. 215–19. Rome: National Research Council.

Forghani, A., and Gaughwin, D. (n.d.). Spatial information technologies to aid archaeological site mapping, www.gisdevelopment.net/application/archaeology/site/archs0005pf.htm, accessed 13 November 2006.

Fowler, M. J. F. 2002. Satellite remote sensing and archaeology: A comparative study of satellite imagery of Figsbury Ring, Wiltshire. *Archaeological Prospection* 9: 55–69.

Fowler, M. J. F, and Fowler, Y. M. 2005. Detection of archaeological crop marks on declassified CORONA KH-4B intelligence satellite photography of southern Britain. *Archaeological Prospection* 12: 257–64.

Gaffney, V., Gaffney, C. F., and Corney, M. 1998. Changing the Roman landscape, in J. Bailey (ed.), *Science in Archaeology*, pp. 145–56. London: English Heritage.

Gaffney, C., and Gater, J. 2003. *Revealing the Buried Past: Geophysics for Archaeologists*. Stroud: Tempus.

Gaffney, C. F., Gater, J. A., Linford, P., Gaffney, V. L., and White, R. 2000. Large-scale systematic fluxgate gradiometry at the Roman city of Wroxeter. *Archaeological Prospection* 7: 81–99.

Gaffney, V., and Tingle, M. 1984. The tyranny of the site: Method and theory in field survey. *Scottish Archaeological Review* 3: 134–40.

Gater, J. 2005. Geophysics, in J. Downes, S. M. Foster, and C. R. Wickham-Jones, *The Heart of Neolithic Orkney World Heritage Site Research Agenda*, pp. 98–100. Edinburgh: Historic Scotland (also available from www.historicscotland.gov.uk/index/publications/worldhsitespublications/orkneyresearch.htm).

Goodman, D., Nishimura, Y., Hongo, H., and Higashi, N. 2006. Correcting for topography and the tilt of ground-penetrating radar antennae. *Archaeological Prospection* 13: 157–61.

Heron, C. 2001. Geochemical prospecting, in D. R. Brothwell and A. M. Pollard (eds.), *Handbook of Archaeological Sciences*, pp. 565–73. Chichester: Wiley.

Holden, N. 2001. The use of laser scanning and multi spectra imager in environmental mapping, in M. Donus, A. Elder-Hinterleitner, and W. Neubauer (eds.), *Archaeological Prospection: Fourth International Conference on Archaeological Prospection*, pp. 114–16. Vienna: Prehistoric Commission of the Austrian Academy of Sciences.

Hunter, J. R., and Cox, M. 2005. *Forensic Archaeology: Advances in Theory and Practice*. London: Routledge.

Jones, B. 2000. Aerial archaeology around the Mediterranean, in M. Pasquinucci and F. Trément (eds.), *Non-Destructive Techniques Applied to Landscape Archaeology*, pp. 49–60. Oxford: Oxbow Books.

Johnson, J. K. (ed.). 2006. *Remote Sensing in Archaeology: An Explicitly North American Perspective*. Tuscaloosa: University of Alabama Press.

Jordan, D. 2000. Magnetic techniques applied to archaeological survey, in M. Pasquinucci and F. Trément (eds.), *Non-Destructive Techniques Applied to Landscape Archaeology*, pp. 114–24. Oxford: Oxbow Books.

Kvamme, K. L. 2003. Geophysical surveys as landscape archaeology. *American Antiquity* 68: 435–57.

Lasaponara, R., and Masini, M. 2005. Evaluation of potentialities of satellite Quickbird imagery for archaeological prospection: Preliminary results, in S. Piro (ed.), *Sixth International Conference on Archaeological Prospection: Proceedings Extended Abstracts*, pp. 392–95. Rome: National Research Council.

Leckebusch, J. 2003. Ground-penetrating radar: A modern three-dimensional prospection method. *Archaeological Prospection* 10: 213–40.

———. 2005. Aerial archaeology: A full digital workflow for aerial archaeology. *Archaeological Prospection* 12: 235–44.

Lock, G. 2003. *Using Computers in Archaeology*. London: Routledge.

Lyall, J. 2006. It's been a long walk: 1,000 hectares of magnetic surveying in the Vale of Pickering. Abstracts of "Recent Work in Archaeological Geophysics," Environmental and Industrial Geophysics Group, London, 19 December.

Merali, Z. 2006. Dig here for treasures of the ancient world. *New Scientist* 2561: 30–31.

Mills, J. 2005. Bias and the world of the vertical aerial photograph, in K. Brophy and D. Cowley (eds.), *From the Air: Understanding Aerial Archaeology*, pp. 117–26. Stroud: Tempus.

Neubauer, W. 2004. GIS in Archaeology: The interface between prospection and excavation. *Archaeological Prospection* 11: 159–66.

Pasquinucci, M., and Trément, F. (eds.). 2000. *Non-Destructive Techniques Applied to Landscape Archaeology*. Oxford: Oxbow Books.

Powlesland, D., Lyall, J., and Donoghue, D. 1997. Enhancing the record through remote sensing: The application and integration of multi-sensor, non-invasive remote sensing techniques for the enhancement of the sites and monuments record.

Heslerton Parish Project, North Yorkshire, England. *Internet Archaeology* 2, http://intarch.ac.uk/journal/issue2/pld_index.html

Powlesland, D., Lyall, J., Hopkinson, G., Donoghue, D., Beck, M., Harte, A., and Stott, D. 2006. Beneath the sand: Remote sensing, archaeology, aggregates, and sustainability: A case study from Heslerton, the Vale of Pickering, North Yorkshire, UK. Archaeological Prospection 13: 291–99.

Sarris, A., Topouzi, S., Chatziiordanou, E., Liu, J., and Xu, L. 2002. Space technologies in archaeological research and CRM of semi-arid and desertification affected regions: Examples from China and Greece, in B. Warmbein (ed.), *Proceedings of Space Applications for Heritage Conservation Conference* (ESA SP-515), Strasbourg, France, 5–8 November 2002, published on CD-ROM (also available online from www.ims.forth.gr/Journals/publications/Euricy_2002/Sarris_ESA_2.pdf).

Schmidt, A. 2002. *Geophysical Data in Archaeology: A Guide to Good Practice.* Oxford: Archaeological Data Service and Oxbow (also available online from http://ads.ahds.ac.uk/project/goodguides/geophys).

———. 2004. Remote sensing and geophysical prospection. *Internet Archaeology*, http://intarch.ac.uk/journal/issue15/schmidt_index.html

Schultze, V., Chwala, A., Stolz, R., Schulz, M., Linzen, S., Meyer, H. G., and Schüler, T. 2005. A SQUID System for Geomagnetic Archaeometry, in S. Piro (ed.), *Sixth International Conference on Archaeological Prospection: Proceedings Extended Abstracts*, pp. 245–48. Rome: National Research Council.

Scollar, I. 1990. *Archaeological Prospecting and Remote Sensing.* Cambridge: Cambridge University Press.

Showalter, P. S. 1993. Thematic mapper analysis of the prehistoric Hokokam canal system, Phoenix Arizona. *Journal of Field Archaeology* 20: 77–90.

Similox-Tohon, D., Vanneste, K., Sintubin, M., Muchez, P., and Waelkens, M. 2004. Two-dimension resistivity imaging: A tool in archaeoseismology. An example from ancient Sagalassos (southwest Turkey). *Archaeological Prospection* 11: 1–18.

Stanjek, H., and Fabinder, J. W. E. 1995. Soil aspects affecting archaeological detail in aerial photographs. *Archaeological Prospection* 2: 91–101.

Utis, E. 2004. Ground penetrating radar time-slices from north Ballachulish Moss. *Archaeological Prospection* 11: 65–75.

Wager, E. C. W., Jenkins, D. A., and Ottaway, B. S. 1998. X-ray fluorescence as a tool for the identification of copper ore processing sites on the Great Orme, North Wales, U.K. Abstracts of the *31st International Symposium on Archaeometry*, Budapest, Hungary, 27 April–1 May.

Wilson, D. R. 2000. *Air Photo Interpretation—for Archaeologists.* Stroud: Tempus.

Winterbottom, S. J., and Dawson, T. 2005. Airborne multi-spectral prospection for buried archaeology in mobile sand dominated systems. *Archaeological Prospection* 12: 205–19.

56

GEOGRAPHICAL INFORMATION SYSTEMS AND LANDSCAPE ARCHAEOLOGY

James Conolly

A Geographical Information System (GIS) is a computer-based tool for collecting, managing, integrating, visualizing, and analyzing geographically referenced information. The use of GIS is scale-independent and may be applied to the study of archaeological data at the continental (or smaller) scale (e.g., Gkiasta et al. 2003; Holmes et al. 2006), at the regional or landscape scale (e.g., Bevan 2003; Howley 2007; Winterbottom and Long 2006), or for intrasite or larger-scale analyses (e.g., Bird et al. 2007; Craig et al. 2006; Marean et al. 2001; Moyes 2002). A number of related technologies overlap with GIS, including remote sensing, geodesy, and digital cartography, but are sufficiently distinct in their methods to warrant separate treatment. However, all these computer technologies can be grouped under the umbrella term of "spatial technologies." These are linked to the emerging discipline of Geographical Information Science (GISc), which is more broadly concerned with developing integrated method and theory of the use of computer-based tools for building understanding of natural and social spatial processes. A parallel development in archaeology, better described as "computational archaeology" or "archaeoinformatics," is similarly concerned with developing an integrated method and theory for archaeological computing.

The growth of GIS and its early expansion from geography into other social sciences, including archaeology, has been described as nothing short of a revolution. There are now a significant number of archaeologists—both commercial and academic—whose primary activity is using and developing GIS and related spatial technologies. This is partly expected and arises from our shared interest with geographers in understanding and explaining the spatial organization of human behavior and its complex, long-term, and multiscalar interrelationship with the natural world.

The purpose of this contribution is to outline the value and application of GIS to landscape archaeology and to outline a conceptual rather than practical introduction to the forms of GIS analyses that archaeologists have found useful. It is, in fact, impossible in a review paper to offer practical guidance on the implementation of the various techniques or indeed on the use of GIS itself: for this, readers are recommended to consult textbook-length works, such as Conolly and Lake (2006) or Wheatley and Gillings (2000). GIS software vendors also provide introductory learning tools for novice users: among the more helpful ones are those included in ESRI's ArcGIS suite of programs. An alternative excellent free and OpenSource GIS is GRASS (Geographic Resources Analysis Support System: http://grass.itc.it/). This has a slightly steeper learning curve than some other desktop packages, but is a very powerful tool used by many academic and government agencies worldwide.

History and Development

The history of archaeology's engagement with GIS is well-documented and covered elsewhere (e.g., Lock 2003), so here it is necessary to highlight only a few pivotal moments. The archaeological use of GIS (as distinct from spatial analysis, vis-à-vis Hodder and Orton 1976) can be traced initially to a series of articles published in the 1980s on predictive modeling (Kohler and Parker 1986; Kvamme 1983, 1986, 1989). GIS was also being highlighted as a new method by Judge and Sebastian (1988) while computers were being used to visualize spatial data, such as artifact distributions and settlement patterns (e.g., Aspinall and Haign 1988; Boismier and Reilly 1988; Harris 1988).

Through the 1990s and into the new millennium, GIS entered its postpioneer phase and expanded dramatically. It became the new plaything of the technologically literate archaeological community, spawning mainstream publications in settlement studies, predictive modeling, pattern analysis, movement, and visibility (see, among others, Aldenderfer and Maschner 1996; Allen et al. 1990; Brandt et al. 1992; Fisher et al. 1997; Gaffney and Stančič 1991; Gillings et al. 1999; Hunt 1992; Lake et al. 1998; Lee and Llobera 1996; Lock and Stančič 1995; Stucky 1998; Westcott and Brandon 2000; Wheatley 1993, 1995).

In more recent years, archaeological GIS entered a more self-reflective and critical phase that has addressed many concerns raised about its contribution to knowledge (as distinct from "information") (cf. Taylor 1990). For example, the relevance and meaning of viewsheds (Llobera 2001; Tschan et al. 2000), the validity of energetic least cost-surfaces for predicting movement (Bell and Lock 2000; Llobera 2000), issues of environmental determinism (Gaffney and van Leusen 1995), and the potential statistical errors in predictive modeling (Wheatley 2004; Woodman and Woodward 2002) have all received significant critical evaluation.

The end result is that, toward the end of the first decade of the new millennium, GIS is both sufficiently mainstream that it is no longer a unique selling point—the days of being told somewhat meaninglessly that a project will "undertake a GIS analysis" are thankfully gone—and its contributions to the study of past human behavior are also more substantive and theoretically aware.

Recent applications that have weathered the critical storm of the 1990s reflect these developments: for example, studies of visibility and movement are methodologically and theoretically more sophisticated (Bell et al. 2002; Lake and Woodman 2003; Llobera 2003; Ogburn 2006), and issues of scale are being properly grappled with (Molyneaux and Lock 2006). As a parallel development, sources of bias and error in region-scale datasets, which are the bread and butter of a majority of GIS-based studies, are being addressed (Banning et al. 2006; Hawkins et al. 2003).

Applications

It is difficult to pigeon-hole GIS applications in archaeology neatly, but one useful division is to divide current usage into four areas: (1) collection and management of spatial data; (2) data visualization; (3) spatial analysis; and (4) quantitative modeling. In reality, GIS users may dart between all four areas seamlessly and have difficulty distinguishing when, for example, management becomes data visualization or when analysis becomes modeling. This grouping is thus used here only as a convenience to organize the range of applications that GIS offers to landscape archaeology.

Spatial Data Collection and Management. The first category concerns the use of GIS as a form of spatial database to manage archaeological information that possesses a strong spatial component (for example, arising from landscape survey). Much primary data in archaeology are now collected in digital form—(for instance, from a total station or global positioning system [GPS] survey)—or are converted into digital format by scanning or digitizing paper records. A large proportion of GIS work, especially within the framework of Cultural Resource Management (CRM), concerns the management and integration of these datasets (Conolly and Lake 2006: 33–50; Garcia-Sanjuan and Wheatley 1999). Although this may seem a pedestrian area of study, only through properly structured and managed databases can we identify relationships and patterns in a digital dataset. The near ubiquity of GIS usage for "Sites and Monument Records"—the term used in the UK for the archaeological records maintained by local government bodies—exemplifies the value of GIS for the maintenance of the primary record (Bevan and Bell 2004). Research projects are also increasingly reliant on GIS for spatial data management, to the extent that the research aims of many complex projects depend on GIS to scaffold the building of understanding of the complex interrelationships between multiscalar spatial data (e.g., Barton et al. 2002; Bevan and Conolly 2004).

Much of the ability of GIS to manipulate spatial data so effectively is founded on its core geodatabase that stores and relates data within a common spatial framework (that is, within a specific map projection, such as Universal Transverse Mercator [UTM], or a national, regional, or locally defined Euclidean grid system). Spatial data representation, which constitutes the backbone of GIS, can be implemented in a variety of different ways, depending on the software in question. Spatial data may be represented by using a *vector data model*, in which spatial objects are defined discretely (that is, they are defined by their precise geometric location in space) using a combination of points, lines, or polygons constructed by one or more *x*, *y* coordinate pairs and the topological connections between them; or by using a *raster data model*, in which spatial objects are defined by a matrix of pixels (cells) of a defined size and shape. Virtually all modern GIS systems are able to manipulate both types of data models, and the choice of vector data or raster data is now based on the appropriateness of the model for representing the phenomenon in question, and the forms of analysis to which each is suited.

Accessing and retrieving data from a GIS typically involves a spatial query, which retrieves data on the basis of a spatial location, or relationship, which may be combined with additional nonspatial attribute criteria (for example, "find all survey units that contain bronze age ceramics that are located between 100 and 200 m elevation and within 1 km of the coast"). The interface for such queries is usually called from within the GIS program itself, and it has also increasingly common for archaeological organizations to mount their data on bespoke web interfaces to permit public access and interrogation of primary data. These may be built on web-GIS mapping software such as ESRI's ArcIMS (www. esri.com/software/arcgis/arcims/index.html), the open source equivalent, MapServer (http://mapserver.gis.umn.edu), or through a combination of other technologies. An early but good example of the last type is the spatial search tool of the Archaeobotanical Computer Database developed by Tomlinson and Hall (1996). A more recent example is York Archaeological Trust's The Archaeology of York Web Series, which uses an innovative mix of interactive maps and databases to publish excavation reports (www.yorkarchaeology.co.uk).

Finally, GIS users must be aware of the potential spatial errors that can arise during the collection of spatial data and/or the integration of data from different map projections and/or different scales of recording. A necessary part of spatial data management is the keeping of properly structured metadata that records such details as acquisition methods and spatial errors. Further details on these methodological issues can be found in Conolly and Lake (2006) and Wheatley and Gillings (2002).

Spatial Data Visualization. This area of GIS deserves its own category, given the tremendous importance of the visual display of spatial data in archaeology. The many varieties of visualization techniques that arise from spatial technologies have provided archaeologists with new ways of examining and thinking about landscape data. It is useful to consider GIS as a tool for *scientific visualization*, because these basic yet critical techniques are instrumental for gaining understanding and insight into data patterns—archaeologists have always plotted data on maps to illuminate patterns, structure, and process. GIS and other spatial technologies offer a range of techniques in this regard, such as by facilitating the creation of distribution maps and then allowing these to be draped on aerial photographs; by manipulating remote sensing imagery to inspect the relationship between landforms and archaeological data; by viewing the structure in a network analysis of connections between settlements (Figure 56.1); by visualizing the spatial variability in soil type and its relationship to archaeological settlements (Figure 56.2); or simply by inspecting the patterning of artifacts across a survey area (Figure 56.3). The popularity of Google Earth for exploring the regional context of archaeological data (as well as for distributing data) is one of the more recent examples of the uses of this form of visual interrogation.

A further application afforded by GIS is the ability to visualize temporal change, either through "time slice" techniques or through more dynamic means, such as those developed by the University of Sydney's TimeMap (www.timemap.net; Johnson and Wilson 2003). The latter provides a toolkit for visualizing temporal datasets through a variety of means, such as by filtering objects so that they are visible only within specific time ranges.

Remote sensing has created additional sources for the scientific visualization of archaeological

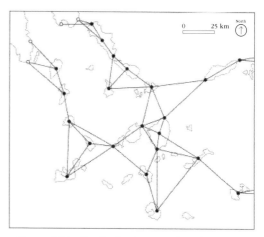

Figure 56.1 Proximal point network analysis of the Cycladic islands (redrawn after Broodbank 2000: fig. 75).

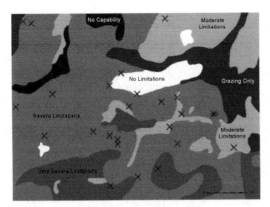

Figure 56.2 Location of archaeological sites in relation to soil agricultural potential.

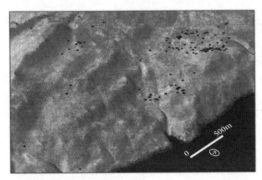

Figure 56.3 QuickBird imagery draped over a digital elevation model (DEM). Black dots represent the location of prehistoric pottery (source: Antikythera Survey Project, www.tuarc.trentu.ca/asp).

data, leading to a number of publications that have exploited its ability to reveal regional scale patterning of archaeological data. For example, De Laet and associates (2007) have explored the potential of DigitalGlobe's high-resolution QuickBird Imagery to identify archaeological features in the southwest Turkey; Ur (2003) has used Corona imagery to identify road networks in Mesopotamia; and tell locations and the structure of ancient cities have been mapped from imagery in the Near East by Philip and associates (2002), Challis and colleagues (2004), and Menze and associates (2006). Remotely sensed data may also be an important source of information about the environmental characteristics of a region, such as its vegetational, hydrological, and mineralogical composition, but this is a more specialized area of study that falls beyond the scope of this chapter.

Finally, there has long been interest in integrating 3D imaging with GIS to create a more subjective and experiential encounter with digital landscapes (e.g., Exon et al. 2000; Gillings 2002, 2005; Gillings and Goodrick 1996; Winterbottom and Long 2006).

Although considerable insight can be gained through these methods, they have yet to be developed sufficiently to have generated more widespread application than the few published examples cited here.

Spatial Analysis. GIS is often an entry point for landscape archaeologists into more sophisticated forms of spatial analysis than are typically encountered in desktop statistical packages. Although it is possible to work within GIS and not confront spatial statistics—that is, the engagement with a spatial dataset may begin and end with its compilation and visualization—this ignores a powerful set of tools for identifying and understanding spatial patterns and structure.

The form of archaeological engagement with the tools of spatial analysis can be loosely divided into three categories: those that are concerned with (1) spatial pattern analysis, (2) spatial structure within a spatial dataset, and (3) multivariate locational analysis. All the techniques described below are accessible from within most GIS packages or are available as plug-ins or extensions using third-party statistical tools.

Spatial Pattern Analysis. The first category pertains to questions about the type and degree of spatial arrangement (that is, deviation from random arrangement toward clustering or regularity), typically applied to point distributions (of artifacts, features, sites, and so on). Questions of this variety have a long history in landscape archaeology: early applications of Clark and Evan's nearest-neighbor R statistic (Clark and Evans 1954) include Hodder and Hassell (1971) and Whallon (1974), and it has been used more recently by Ladefoged and Pearson (2000) and Perlès (2001: 134–38). More sophisticated techniques of point-pattern analysis include Ripley's K (Ripley 1977), which deals more effectively with irregularly shaped sampling regions and multiscalar distributions that are often encountered in landscape archaeology (Bevan and Conolly 2006). Membership of clusters can be defined by using techniques such as k-means, hierarchical cluster, or kernel density estimates: these and related clustering methods are defined in more detail in textbooks such as Conolly and Lake (2006) and Wheatley and Gillings (2002). Getis's G* statistic (Ord and Getis 1995) can be used to identify "hot spots" in data compiled by enumeration units, such as sherds of a specific date within survey or excavation units. This has relevance for archaeological applications in which basic exploratory questions of the form "is my data clustered?" need to be answered. Contributions to the statistic (that is, the statistic's Z score) can also be plotted to view the clustering and to aid in the identification of the patterning of the attribute in question (see Figure 56.4).

Spatial Structure. Examination of the spatial structure within the attributes of spatially located objects is also important. An important test in this category is the degree of spatial autocorrelation in a spatial dataset. This term refers to the influence that proximity has on the similarity between pairs of observed values in spatial phenomenon (that is, the extent to which there is "patterned variation"). It has been applied in a variety of regional analyses from the size of artifacts (Hodder and Orton 1976) to radiocarbon dates (Kvamme 1990; Premo 2004; Williams 1993), although its wide applicability for understanding spatial structure in the archaeological record has yet to be fully exploited. A further tool for building understanding of spatial structure is offered by Geographically Weighted Regression (GWR), which provides a method for identifying and exploring the manner in which local regions deviate from global trends (Fotheringham et al. 2002). This relatively advanced technique is considered in further detail below, under the heading of statistical modeling.

Locational Analysis. Locational analysis is related to "site-catchment" analysis developed in the 1970s (Higgs and Vita-Finzi 1972; Vita-Finzi and Higgs 1970), insofar as the approach seeks to understand relationships between sites and their locational characteristics and how these change through time and space. GIS-based methods differ from early formations in their more rigorous and quantitative approach to isolating significant variables from those that are potentially significant. Locational analysis may be undertaken to build understanding of the factors that influence choice of site location (for example, to compare the logic of site choice in two periods of settlement in a defined region), or it may be the preliminary stage of a predictive model (described in the following section) to determine the probability of site locations in unsampled areas.

The form of quantitative analysis used in locational studies is distinct from spatial-pattern analysis and analysis of spatial structure insofar as it uses standard parametric and nonparametric statistics to assess the probability of an association between the phenomena in question. A simplistic example is an analysis that seeks to clarify the relationship between site location and soil type: numbers of sites found on different soils may be obtained from the GIS (as a basic query function) and this observation can then be compared either to several sets of random samples of points (that is, as in a Monte-Carlo simulation) or by deriving an expected distribution based on the amount of each soil type found in the study area. Parametric or nonparametric tests (such as Student's t or Kolmogorov-Smirnov, see Shennan 1997) are then used to establish the probability that the sites are not randomly located with regard to the variable in question.

Although analysis of site patterning against a single variable like this is rarely wholly informative of past behavior, the examination of many potentially significant variables against site location can provide insight into the combined factors that have measurable influence on the choice of location for settlements. For example, Woodman (2000) used

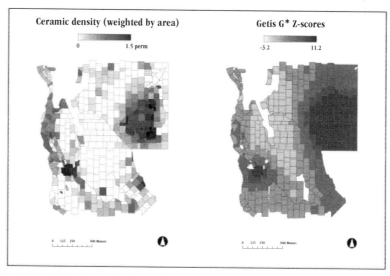

Figure 56.4 *Left:* weighted ceramic density by survey unit; *right:* "hot spot" spatial analysis via Getis G*. Darker colors indicate greater probability unit lies in an area of locally dense values (source: Antikythera Survey Project, www.tuarc.trentu.ca/asp).

14 topographic and environmental variables, such as relief, aspect, angle of view, and exposure, as part of her study in Mesolithic site location. These (and other) variables that have been shown to be useful in locational analysis, such as terrain curvature, hydrology, and watersheds (Bevan 2003), are derived in GIS from digital elevation models. The creation of these datasets from digital elevation base maps, although straightforward, does require some understanding of sources of error in elevation models, and so further advice should be obtained from the sources cited previously.

Locational analysis may also encompass less formulaic variables, such as visual characteristics of a landscape, the presence of contemporaneous settlements (that is, "communities"), landscape potential for different types of subsistence activities, and spatial relationships to other cultural features, such as roads, known (or modeled) pathways, burial mounds, and other potentially significant features. Although archaeologists accept that experiential variables are also significant in influencing the choice and characteristics of settlement, they have yet to be widely used in GIS because of the difficulty of implementation (but see Wheatley and Gillings 2002: 166–68).

Modeling

One of the most exciting and least-developed areas of research in GIS-based studies of landscape is (mathematical) modeling. There are an enormous variety of contemporary approaches that can be characterized in this way, and only a sample that relate specifically to GIS and landscape studies can be highlighted. Current archaeological uses can summarily be reviewed in three categories: (1) statistical modeling, including predictive models; (2) cellular models, often based on topography, including visibility, movement, erosion, hydrology; and (3) agent-based models. There are additional approaches to modeling, such as social network analysis (e.g., Allen 1990: 175–210; Bentley and Shennan 2003; Broodbank 2000; Conolly and Lake 2006: chapter 11; Mackie 2001) and phylogenetic/geographic analysis (e.g., Coward et al. 2007; O'Brien et al. 2001) that have significance to regional studies in archaeology, but as these endeavors often take place independently of GIS, they are not considered here.

Statistical Models. Two related techniques that mathematically model the relationship between a dependent and independent (spatial) variable are spatial regression and geographically weighted regression (GWR). Like many of the approaches described here, the former has a long history, achieving its most widely cited application in the definition of the strong correlation between the quantity of obsidian found at archaeological sites in southwest Asia and their linear distance from the geological source of material (Renfrew and Dixon 1976). Spatial regression has also been used to examine such phenomena as the spread of agriculture (Ammerman and Cavalli-Sforza 1971; Gkiasta et al. 2003; Pinhasi et al. 2005); the effect of changing surface visibility on artifact density in survey data (Fanning and Holdaway 2004); and the relationship between terrain curvature and prehistoric site location (Bevan 2003). GWR is a relatively new statistical technique that has yet to find widespread applicability in archaeology but is of growing importance in geography (Fotheringham et al. 2002). It corrects a source of error in standard regression when one is working with a spatially autocorrelated variable, by identifying local variability in global patterns, which is of potential importance in a wide range of landscape-based studies in archaeology. It also may deal effectively with "Simpson's Paradox," when two or more distinct local patterns are combined and collectively produce a potentially spurious global relationship. For example, the local variability within the global regression of early Neolithic radiocarbon dates against distance from Jericho (Gkiasta et al. 2003) is likely to hold considerable value for interpretations that seek to understand regional variations in the uptake of farming, and how these relate to different topographic and social contexts (e.g., Davison et al. 2006).

Trend surface analysis is another common GIS-based statistical model that has been used in archaeology, which takes as its input the values of a set of observations that are located in space and then attempts to generalize the rate of change across that space. Mathematically it is identical to a regression analysis in three dimensions, and the output can thus be viewed as a map that depicts the rate of change. Archaeological examples include Allen and Fulford (1996), Kvamme (1990), and Neiman (1997).

Finally, predictive modeling has been a traditional concern of GIS, partly because GIS is well-suited to the logic of this type of empirical investigation (Kvamme 1992). The principal of this approach is that there is explanatory value in the (typically environmental) variables related to site location and that their relative importance can be empirically derived (following the methods explained in "locational analysis," above). Variables that are determined to have been important can then be combined quantitatively, usually through a modeling technique called *linear logistic regression*, which results in a polynomial equation that expresses the contribution of each variable to site location. Then, given any location on the landscape in which those variables can be measured, one can calculate a probability for site presence. Provided a training sample has been withheld, the accuracy of the model can also be determined.

Predictive modeling was one of the first applications of GIS-based statistical investigations of landscape data, but it has also come under the most scrutiny for its apparent environmental determinism (e.g., Gaffney and van Leusen 1995; Wheatley 2004, among others). However, it is still one of the most widely used approaches to the formal modeling of site location and has spawned a specialized literature within the CRM community (e.g., Westcott and Brandon 2000).

Cellular Models. Within this category are a range of diverse approaches that include environmental, movement, and visibility models. The manner in which these are constructed in GIS may be as simple as clicking on a tool-bar button (for example, as is the case if one wishes to derive a map of slope values from a digital elevation model/DEM, itself a cellular model of landscape) or by defining a few options such as the minimum size catchment for a watershed model before clicking on a menu item. Other models may require more extensive input variables, such as erosion models that may use a dozen or more parameters to calculate the rate of sediment flow across a landsurface.

Cellular models take one or more raster map layers as inputs and use a combination of neighborhood functions and/or matrix algebra (that is, "map algebra" in GIS terminology) to produce a new map. Each cell value in the output map may be derived from several input values. For example, the end product of a predictive model may be a raster map in which each cell expresses a value reflecting the local probability of site presence based on a large number of input maps (such as location of know sites, local soils, geology, elevation, slope, hydrology, and so on). Neighborhood calculations on raster maps work by deriving new values for each cell by looking to values of the surrounding cells. Neighborhood calculations can also be used to construct models of spatial phenomenon: for example, the slope of a landsurface may be derived from a DEM by establishing the maximum difference in elevation between one cell and its neighbors, and thus the vertical angle between them, and repeating the calculation. Hydrological models work in a similar fashion, but look for neighboring cells with the lowest value to establish flow direction. Other neighborhood operations that are not models, but are still

important in GIS, include filtering, smoothing, and enhancing visual patterns in image data by combining surrounding values in various ways (for example, such as running a moving window filter to replace locally high or low values in an image with the mean value of the cell's surrounding eight neighbors). For example, a "high-pass filter" for emphasizing areas of change (that is, "edge detection") in raster datasets can be created by subtracting a measure of local variability (typically the mean value of cells in a 3 x 3 cell window) from the original map (Conolly and Lake 2006: 200–01).

Other forms of cellular models use distance from one or more points to construct a new map layer. This is the basis for Theissen polygon models, commonly, if uncritically, used to estimate territorial boundaries around sites (e.g., Perlès 2001: 139–40). Cells are simply allocated to the nearest site, thereby creating a model of equidistant boundaries (e.g., Wilkinson 1998: fig. 6; Perlès 2001: fig. 7.9). (Note, however, that Theissen polygon modeling is now more usually performed as a vector than a raster operation, but the principle is the same.) Such models of "territoriality" are obviously a great abstraction of a more complicated past social reality, but they nevertheless provide landscape archaeologists with a starting set of possibilities about the social organization of settlements (Conolly and Lake 2006: 208–33).

A further application of cellular models to landscape archaeology concerns "viewsheds," which are calculations of the potentially visible area from a defined spot on an elevation model, given a specified height of an observer and target. Viewsheds are derived from line-of-sight calculations that determine whether a target point is visible from an observer's location. This calculation is repeated for every potential target within a defined radius, providing a map that identifies the visible areas (Figure 56.5). These are straightforward to calculate, with most GIS providing viewshed and/or line-of-sight tools, but there are some sampling biases that need to be taken into account with viewshed analysis, particularly the issue of edge effects (Conolly and Lake 2006: 229). In addition, background sampling of the visual characteristics of the landscape, usually by quantitatively comparing the viewsheds of several sets of random locations against the locations under investigation, is crucial to justify any claims that sites are preferentially located with respect to their visual affordances (e.g., see Jones 2006 for an example of where the lack of any background sampling seriously undermines claims of patterning). These issues are discussed in Lake and Woodman (2000,

2003), as well as in the general textbooks previously cited. Some of the most successful applications of viewshed analysis are drawn from the open moorlands of northern Britain (e.g., Fisher et al. 1997; Lake and Woodman 2000; Winterbottom and Long 2006).

A related approach to visibility is cumulative viewshed analysis (CVA), which calculates the number of "times seen" of a given location (Wheatley 1995), which has been applied to intervisibility studies of monuments (Lake et al. 1998; Wheatley 1995). Other approaches have developed methods for examining the morphology of the horizon from stone circles (Lake and Woodman 2003); for creating "fuzzy viewsheds," which record an uncertainty value that expresses the potentiality of a target being seen (Fisher 1992; Ogburn 2006); and for extending the concept of CVA to a "total viewshed," which calculates the "times seen" value for every cell in a study region (Llobera 2003). Total viewsheds are extremely computationally intensive, and have yet to see any practical archaeological application, but they do have potential significance for understanding the interrelationship between "viewscapes" and landscape features such as ancient monuments and pathways.

Finally, the derivation of movement has also been the subject of considerable interest, with several studies using GIS to construct "least-cost" path models. These are technically complex, because energy expenditure in movement is determined at a minimum by terrain and slope, with a host of other variables, such as headwind, contributing to difficulty. In GIS-based studies of movement, slope is usually used as the basis for generating a "cost-surface," and these can be either *isotropic* (cost is the same in all directions) or *anisotropic* (cost varies depending on direction of movement). Anisotropic cost surfaces are more complicated to construct, because they require factoring the direction of movement in order to determine costs. Both models, however, have been used in archaeological investigations (e.g., Bell et al. 2002; Bell and Lock 2000; Harris 2000; Madry and Rakos 1996; Silva and Pizziolo 2001). Theoretical issues of deriving and integrating movement into archaeological interpretation have also been considered by Llobera (2000).

Agent-Based Modeling. An increasingly important area of study is the integration of GIS with agent-based modeling (ABM), which has roots in the pioneering studies of modeling and simulation in the 1970s and 1980s (Clarke 1972; Renfrew and Cooke 1979; Sabloff 1981). Early applications of ABM used artificial landscapes within which agents with defined characteristics (for example, goals, movement, reactions) would interact and their collective behavior examined, often in order to understand the emergent properties of complex social systems (Kohler and Gumerman 2000). More recent applications, however, allow GIS maps to be used as the setting for agent interaction, permitting the agents to react and respond to real variability in the landscape and environment. Good examples of this approach include Lake (2000), who constructed a simulation model of hunter-gatherer land-use and its anticipated archaeological patterning to compare against the results of fieldwork, as well as papers in the volumes edited by Gimblett (2002) and Kohler and van der Leeuw (2007).

A highly specialized area of research that requires some understanding of computer programming, the linkage of ABM to GIS is a potential breakthrough in modeling of past human behavior. It provides a means of examining how the combined effect of many individuals' behavior (albeit in a simplified and abstract form) may lead to the sorts of emergent patterns we see in the archaeological record, such as changes in settlement patterns, the emergence of settlement hierarchies, or even changes in subsistence strategies. These "bottom-up" approaches to complexity suffer less from the criticisms that were leveled at the "top-down" systemic models of earlier decades (Bentley and Maschner 2003). Their application is a specialized area of archaeological investigation, but they may eventually provide significant insights into some long-standing archaeological problems that seek to understand the dynamics of, for example, settlement morphogenesis and emergent social complexity.

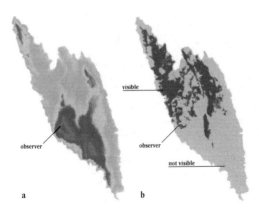

Figure 56.5 Viewshed analysis. (a) Location of observer on digital elevation model (darker colors indicate increasing altitude); (b) output map displaying areas of landscape visible and not visible to observer (source: Antikythera Survey Project, www.tuarc.trentu.ca/asp).

Conclusions

This brief review of the applicability of GIS to landscape studies should have provided sufficient detail on the range of archaeological questions and problems to which this technology is well-suited. It remains to be noted, however, that GIS has been subject to a fair amount of theoretical criticism from proponents of a landscape archaeology that are at times at odds with GIS. Specific oft-heard criticisms include GIS's tendency to focus on environmental variables, leading to a revisitation of 1970s-style functionalism (Wheatley 2004); its conceptualization of space in a dehumanized, 2D Euclidean framework such that relationships between phenomena, especially cultural phenomena, are conceived of as mathematical (rather than experiential, contextual, or social) (see Llobera 1996; Thomas 1993: 25); a tendency of archaeologists interested in GIS to focus on explanations that are processual and behavioralist, rather than experiential and symbolic (Gaffney et al. 1996); the inability to represent in mathematical (that is, digital) format the subjective and experiential landscape with which past peoples are said to have participated (e.g., Tilley 1994: 9–11); and a tendency to use positivist approaches for developing understanding of archaeological data (Gaffney et al. 1996; Wheatley 1993).

Although it is true that environmental variables provide the starting point of much GIS work, many of these criticisms can be countered by demonstrating that understanding of the past does flow from the GIS-based approaches described above, as the many examples cited in this paper attest. There are, in addition, a number of researchers interested in integrating GIS to aid understanding of socially textured and politicized space: the work of Boaz and Uleberg (2000), Chapman (2000), Hamilton and associates (2007), Howley (2007), Maschner (1996), Zubrow (1994), and, outside archaeology, Robbins (2003), are particularly interesting in this regard. In addition, experiential space, particularly as related to vision and movement, is also developing from the pioneering work of Gillings (2005), Lake and Woodman (2003), Llobera (2000, 2003), Wheatley (1995), and Wheatley and Gillings (2000). Yet, it remains true that most of these examples and almost all other GIS work are undertaken within a strongly quantitative and empirical framework. Whether this is because GIS constrains the types of analyses that can be undertaken or because GIS attracts researchers who prefer to work within a more rigorously empirical paradigm is debatable, but it is likely to be a mix of the two. In any event, the importance of GIS to landscape archaeology cannot be overemphasized. As this brief review shows, GIS provides both a powerful tool to think with and an intellectual environment for developing innovative forms of analysis that collectively help build understanding of the human past.

References

Aldenderfer, M., and Maschner, H. D. G. (eds.). 1996. *Anthropology, Space and Geographic Information Systems*. Oxford: Oxford University Press.

Allen, J. R. L., and Fulford, M. G. 1996. The distribution of southeast Dorset Black Burnished Category 1 pottery in southwest Britain. *Britannia* 27: 223–81.

Allen, K. M. S. 1990. Modelling early historic trade in the eastern Great Lakes using geographic information systems, in K. M. S. Allen, S. W. Green, and E. B. W. Zubrow (eds.), *Interpreting Space: GIS and Archaeology*, pp. 319–29. London: Taylor and Francis.

Allen, K. M. S., Green, S. W., and Zubrow, E. B. W. (eds.). 1990. *Interpreting Space: GIS and Archaeology*. London: Taylor and Francis.

Ammerman, A. J., and Cavalli-Sforza, L. L. 1971. Measuring the rate of spread of early farming in Europe. *Man* N.S. 6: 674–88.

Aspinall, A., and Haign, J. G. B. 1988. A review of techniques for the graphical display of geophysical data, in S. P. Q. Rahtz (ed.), *Computer and Quantitative Methods in Archaeology*, pp. 295–307. BAR International Series 446. Oxford: British Archaeological Reports.

Banning, T., Hawkins, A. L., and Stewart, S. T. 2006. Detection functions for archaeological survey. *American Antiquity* 71: 723–42.

Barton, C. M., Bernabeu, J., Aura, E., Garcia, O., and Roca, N. 2002. Dynamic landscapes, artifact taphonomy, and landuse modeling in the western Mediterranean. *Geoarchaeology* 17: 155–90.

Bell, T., and Lock, G. 2000. Topographic and cultural influences on walking the Ridgeway in later prehistoric times, in G. Lock (ed.), *Beyond the Map: Archaeology and Spatial Technologies*, pp. 85–100. Amsterdam: IOS Press.

Bell, T., Wilson, A., and Wickham, A. 2002. Tracking the Samnites: Landscape and communication routes in the Sangro Valley, Italy. *American Journal of Archaeology* 106: 169–86.

Bentley, R. A., and Maschner, H. D. G. (eds.). 2003. *Complex Systems and Archaeology: Foundations of Archaeological Inquiry*. Salt Lake City: University of Utah Press.

Bentley, R. A., and Shennan, S. J. 2003. Cultural transmission and stochastic network growth. *American Antiquity* 68: 459–85.

Bevan, A. 2003. The rural landscape of Neopalatial Kythera: A GIS perspective. *Journal of Mediterranean Archaeology* 15: 217–56.

Bevan, A., and Bell, T. 2004. A survey of standards for the English archaeological record community: A report on behalf of English heritage. London: English Heritage.

Bevan, A., and Conolly, J. 2004. GIS, archaeological survey, and landscape archaeology on the Island of Kythera, Greece. *Journal of Field Archaeology* 29: 123–38.

———. 2006. Multiscalar approaches to settlement distributions, in G. Lock and B. Molyneaux (eds.), *Confronting Scale in Archaeology: Issues of Theory and Practice*. London: Springer.

Bird, C., Minichillo, T., and Marean, C. W. 2007. Edge damage distribution at the assemblage level on Middle Stone Age lithics: An image-based GIS approach. *Journal of Archaeological Science* 34: 771–80.

Boaz, J., and Uleberg, E. 2000. Quantifying the non-quantifiable: Studying hunter-gatherer landscapes, in G. Lock (ed.), *Beyond the Map: Archaeology and Spatial Technologies*, pp. 101–15. Amsterdam: IOS Press.

Boismier, W. A., and Reilly, P. 1988. Expanding the role of computer graphics in the analysis of survey data, in C. L. N. Ruggles and S. P. Q. Rahtz (eds.), *Computer and Quantitative Methods in Archaeology*, pp. 221–25. BAR International Series 393. Oxford: British Archaeological Reports.

Brandt, R., Groenewoudt, B. J., and Kvamme, K. L. 1992. An experiment in archaeological site location: Modelling in the Netherlands using GIS techniques. *World Archaeology* 24: 268–82.

Broodbank, C. 2000. *An Island Archaeology of the Early Cyclades*. Cambridge: Cambridge University Press.

Challis, K., Priestnall, G., Gardner, A., Henderson, J., and O'Hara, S. 2004. CORONA remotelysensed imagery in dryland archaeology: The Islamic city of alRaqqa, Syria. *Journal of Field Archaeology* 29: 139–53.

Chapman, H. 2000. Understanding wetland archaeological landscapes: GIS, environmental analysis and landscape reconstruction: Pathways and narratives, in G. Lock (ed.), *Beyond the Map: Archaeology and Spatial Technologies*, pp. 49–59. Amsterdam: IOS Press.

Clark, P., and Evans, F. 1954. Distance to nearest neighbour as a measure of spatial relationships in populations. *Ecology* 35: 445–53.

Clarke, D. L. (ed.). 1972. *Models in Archaeology*. London: Methuen.

Conolly, J., and Lake, M. 2006. *Geographical Information Systems*. Cambridge: Cambridge University Press.

Coward, F., Shenna, S., Colledge, S., Conolly, J., and Collard, M. 2007. The spread of Neolithic plant economies from the Near East to northwest Europe: A phylogenetic analysis. *Journal of Archaeological Science 2007*. http://dx.doi.org/10.1016/j.jas.2007.02.022

Craig, N., Aldenderfer, M., and Moyes, H. 2006. Multivariate visualization and analysis of photo-mapped artifact scatters. *Journal of Archaeological Science* 33: 1617–27.

Davison, K., Dolukhanov, P., Sarson, G. R., and Shukurov, A. 2006. The role of waterways in the spread of the neolithic. *Journal of Archaeological Science* 33: 641–52.

De Laet, V., Paulissen, E., and Waelkens, M. 2007. Methods for the extraction of archaeological features from very high-resolution Ikonos-2 remote sensing imagery, Hisar (southwest Turkey). *Journal of Archaeological Science* 34: 830–41.

Exon, S., Gaffney, V., Woodward, A., and Yorston, R. 2000. *Stonehenge Landscapes: Journeys through real and imagined worlds*. Oxford: Archaeopress.

Fanning, P. C., and Holdaway, S. J. 2004. Artifact visibility at open sites in western New South Wales, Australia. *Journal of Field Archaeology* 29: 255–71.

Fisher, P. 1992. First experiments in viewshed uncertainty: Simulating fuzzy viewsheds. *Photogrammetric Engineering and Remote Sensing* 58: 345–52.

Fisher, P. F., Farrelly, C., Maddocks, A., and Ruggles, C. 1997. Spatial analysis of visible areas from the Bronze Age cairns of Mull. *Journal of Archaeological Science* 24: 581–92.

Fotheringham, A. S., Brunsdon, C., and Charlton, M. 2002. *Geographically Weighted Regression: The Analysis of Spatially Varying Relationships*. London: John Wiley and Sons.

Gaffney, V., and Stančič, Z. 1991. *GIS Approaches to Regional Analysis: A Case Study of the Island of Hvar*. Ljubljana: Znanstveni institut Filozofske fakultete.

Gaffney, V., Stančič, Z., and Watson, H. 1996. Moving from catchments to cognition: Tentative steps toward a larger archaeological context for GIS, in M. Aldenderfer and H. D. G. Maschner (eds.), *Anthropology, Space and Geographic Information Systems*, pp. 132–54. Oxford: Oxford University Press.

Gaffney, V., and van Leusen, P. M. 1995. Postscript: GIS, environmental determinism and archaeology: A parallel text, in G. R. Lock and Z. Stančič (eds.), *Archaeology and Geographical Information Systems: A European Perspective*, pp. 367–82. London: Taylor and Francis.

Garcia-Sanjuan, L., and Wheatley, D. 1999. The state of the Arc: Differential rates of adoption of GIS for European heritage management. *European Journal of Archaeology* 2: 201–28.

Gillings, M. 2002. Virtual archaeologies and the hyperreal: Or, what does it mean to describe something as virtually real? in P. Fisher and D. Unwin (eds.), *Virtual Reality in Geography*, pp. 17–34. London: Taylor and Francis.

———. 2005. The real, the virtually real, and the hyperreal: The role of VR in archaeology, in

S. Smiles and S. Moser (eds.), *Envisioning the Past: Archaeology and the Image*, pp. 223–39. Oxford: Blackwell Publishing.

Gillings, M., and Goodrick, G. T. 1996. Sensuous and reflexive GIS: Exploring visualisation and VRML. *Internet Archaeology* 1. http://intarch.ac.uk/journal/issue1/gillings_index.html, accessed March 8 2004.

Gillings, M., Mattingly, D., and van Dalen, J. (eds.). 1999. *Geographical Information Systems and Landscape Archaeology 3. The Archaeology of Mediterranean Landscapes*. Oxford: Oxbow Books.

Gimblett, H. R. (ed.). 2002. *Integrating Geographic Information Systems and Agent Based Modeling Techniques for Simulating Social and Ecological Processes*. Oxford: Oxford University Press.

Gkiasta, M., Russell, T., Shennan, S., and Steele, J. 2003. Neolithic transition in Europe: The radiocarbon record revisited. *Antiquity* 77: 45–63.

Hamilton, S., and Whitehouse, R. 2007. Phenomenology in practice: Towards a methodology for a "subjective" approach. *European Journal of Archaeology* 9: 31–71.

Harris, T. 1988. Digital terrain modelling and three dimensional surface graphics for landscape and site analysis in archaeology and regional planning, in C. L. N. Ruggles and S. P. Q. Rahtz (eds.), *Computer Applications and Quantitative Methods in Archaeology*, pp. 12–16. BAR International Series 393. Oxford: British Archaeological Reports.

———. 2000. Moving GIS: Exploring movement within prehistoric cultural landscapes using GIS. In G. Lock (ed.), *Beyond the Map: Archaeology and spatial technologies*, pp. 116–23. Amsterdam: IOS Press.

Hawkins, A., Stewart, S. T., and Banning, E. B. 2003. Inter-observer bias in enumerated data from archaeological survey. *Journal of Archaeological Science* 30: 1503–12.

Higgs, E. S., and Vita-Finzi, C. 1972. Prehistoric economies: A territorial approach, in E. S. Higgs (ed.), *Papers in Economic Prehistory: Studies by Members and Associates of the British Academy Major Research Project in the Early History of Agriculture*, pp. 27–36. Cambridge: Cambridge University Press.

Hodder, I., and Hassell, M. 1971. The non-random spacing of Romano-British walled towns. *Man* N.S. 6: 391–407.

Hodder, I., and Orton, C. 1976. *Spatial Analysis in Archaeology*. Cambridge: Cambridge University Press.

Holmes, K. M., Robson Brown, K. A., Oates, W. P., and Collins, M. J. 2006. Assessing the distribution of Asian Palaeolithic sites: a predictive model of collagen degradation. *Journal of Archaeological Science* 33: 971–86.

Howley, M. C. L. 2007. Using multi-criteria cost surface analysis to explore past regional landscapes: A case study of ritual activity and social interaction in Michigan, A.D. 1200–1600. *Journal of Archaeological Science 2007.* http://dx.doi.org/10.1016/j.jas.2007.01.002

Hunt, E. D. 1992. Upgrading site catchment analysis with the use of GIS: Investigating the settlement patterns of horticulturalists. *World Archaeology* 24: 283–309.

Johnson, I., and Wilson, A. 2003. The Time-Map project: Developing timebased GIS display for cultural data. *Journal of GIS in Archaeology* 1: 125–35.

Jones, E. E. 2006. Using viewshed analysis to explore settlement choice: A case study of the Onondaga Iroquois. *American Antiquity* 71: 523–38.

Judge, W. J., and Sebastian, L. (eds.). 1988. *Quantifying the Present and Predicting the Past: Theory, Method and Application of Archaeological Predictive Modeling*. Washington, DC: U.S. Bureau of Land Management, Department of Interior, U.S. Government Printing Office.

Kohler, T. A., and van der Leeuw, S. E. (eds.). 2007. *Model-Based Archaeology of Socionatural Systems*. Sante Fe, NM: SAR Press.

Kohler, T. A., and Gumerman, G. J. (eds.). 2000. *Dynamics in Human and Primate Societies: Agent-Based Modelling of Social and Spatial Processes*. Oxford: Oxford University Press.

Kohler, T. A., and Parker, S. C. 1986. Predictive modelling for archaeological resource location, in M. B. Schiffer (ed.), *Advances in Archaeological Method and Theory* 9, pp. 397–452. New York: Academic Press.

Kvamme, K. L. 1983. Computer processing techniques for regional modeling of archaeological site locations. *Advances in Computer Archaeology* 1: 26–52.

———. 1986. The use of Geographic Information Systems for modelling archaeological site distributions, in B. K. Opitz (ed.), *Geographic Information Systems in Government 1*, pp. 345–62. Hampton: Deepak Publishing.

———. 1989. Geographic Information Systems in regional archaeological research and data management, in M. B. Schiffer (ed.), *Archaeological Method and Theory* 1, pp. 139–203. Tucson: University of Arizona Press.

———. 1990. Spatial autocorrelation and the Classic Maya collapse revisited: Refined techniques and new conclusions. *Journal of Archaeological Science* 17: 197–207.

———. 1992. A predictive site location model on the High Plains: An example with an independent test. *Plains Anthropologist* 37: 19–40.

Ladefoged, T. N., and Pearson, R. 2000. Fortified castles on Okinawa Island during the Gusuku Period, A.D. 1200–1600. *Antiquity* 74: 404–12.

Lake, M. W. 2000. MAGICAL computer simulation of Mesolithic foraging, in T. A. Kohler and G. J. Gumerman (eds.), *Dynamics in Human and Primate Societies: Agent-Based Modelling of Social and Spatial Processes*, pp. 107–43. New York: Oxford University Press.

Lake, M. W., and Woodman, P. E. 2000. Viewshed analysis of site location on Islay, in S. J. Mithen (ed.), *Hunter-Gatherer Landscape Archaeology: The Southern Hebrides Mesolithic Project, 1988–1998, Volume 2: Archaeological Fieldwork on Colonsay, Computer Modelling, Experimental Archaeology, and Final Interpretations*, pp. 497–503. Cambridge: The McDonald Institute for Archaeological Research.

———. 2003. Visibility studies in archaeology: A review and case study. *Environment and Planning B: Planning and Design* 30: 689–707.

Lake, M. W., Woodman, P. E., and Mithen, S. J. 1998. Tailoring GIS software for archaeological applications: An example concerning viewshed analysis. *Journal of Archaeological Science* 25: 27–38.

Lee, J., and Stucky, D. 1998. On applying viewshed analysis for determining leastcost paths on digital elevation models. *International Journal of Geographical Information Science* 12: 891–905.

Llobera, M. 1996. Exploring the topography of mind: GIS, social space and archaeology. *Antiquity* 70: 612–22.

———. 2000. Understanding movement: A pilot model towards the sociology of movement, in G. Lock (ed.), *Beyond the Map: Archaeology and Spatial Technologies*, pp. 65–84. Amsterdam: IOS Press.

———. 2001. Building past landscape perception with GIS: Understanding topographic prominence. *Journal of Archaeological Science* 28: 1005–14.

———. 2003. Extending GIS-based visual analysis: The concept of "visualscapes." *International Journal of Geographical Information Science* 17: 25–48.

Lock, G. 2003. *Using Computers in Archaeology*. London: Routledge.

Lock, G. R., and Stančič, Z. (eds.). 1995. *Archaeology and Geographic Information Systems: A European Perspective*. London: Taylor and Francis.

Mackie, Q. 2001. *Settlement Archaeology in a Fjordland Archipelago: Network Analysis, Social Practice and the Built Environment of Western Vancouver Island, British Columbia, Canada since 2000 B.P.* British Archaeological Reports International Series 926. Oxford: Archaeopress.

Madry, S., and Rakos, L. 1996. Line of sight and cost surface techniques for regional archaeological research in the Arroux river valley, in H. D. G. Maschner (ed.), *New Methods, Old Problems: Geographic Information Systems in Modern Archaeological Research*, pp. 104–26.

Carbondale: Southern Illinois University Center for Archaeological Investigations.

Marean, C. W., Abe, Y., Nilssen, P. J., and Stone, E. C. 2001. Estimating the minimum number of skeletal elements (MNE) in zooarchaeology: A review and a new image analysis GIS approach. *American Antiquity* 66: 333–48.

Maschner, H. D. G. 1996. The politics of settlement choice on the northwest coast: Cognition, GIS, and coastal landscapes, in M. Aldenderfer and H. D. G. Maschner (eds.), *Anthropology, Space and Geographic Information Systems*, pp. 175–90. Oxford: Oxford University Press.

Menze, B. H., Ur, J. A., and Sherratt, A. G. 2006. Detection of ancient settlement mounds: Archaeological survey based on the SRTM terrain model. *Photogrammetric Engineering and Remote Sensing* 72: 321–27.

Molyneaux, B., and Lock, G. (eds.). 2006. *Confronting Scale in Archaeology: Issues of Theory and Practice*. New York: Springer-Verlag.

Moyes, H. 2002. The use of GIS in the spatial analysis of an archaeological cave site. *Journal of Cave and Karst Studies* 64: 9–16.

Neiman, F. D. 1997. Conspicuous consumption as wasteful advertising: A Darwinian perspective on spatial patterns in Classic Maya terminal monument dates, in M. C. Barton and G. A. Clark (eds.), *Rediscovering Darwin: Evolutionary Theory and Archaeological Explanation*, pp. 267–90. Archaeological Papers of the American Anthropological Association 7. Washington, DC: American Anthropological Association.

O'Brien, M. J., Darwent, J., and Lyman, R. L. 2001. Cladistics is useful for reconstructing archaeological phylogenies: Palaeoindian points from the southeastern United States. *Journal of Archaeological Science* 28: 1115–36.

Ogburn, D. E. 2006. Assessing the level of visibility of cultural objects in past landscapes. *Journal of Archaeological Science* 33: 405–13.

Ord, J. K., and Getis, A. 1995. Local spatial autocorrelation statistics: Distributional issues and an application. *Geographical Analysis* 27: 286–306.

Perlès, C. 2001. *The Early Neolithic in Greece*. Cambridge: Cambridge University Press.

Philip, G., Donoghue, D., Beck, A., and Galiatsatos, N. 2002. CORONA satellite photography: An archaeological application from the Middle East. *Antiquity* 76: 109–18.

Pinhasi, R., Fort, J., and Ammerman, A. J. 2005. Tracing the origin and spread of agriculture in Europe. *PLoS Biology* 3/12: 2220–28.

Premo, L. 2004. Local spatial autocorrelation statistics quantify multiscale patterns in distributional data: An example from the Maya Lowlands. *Journal of Archaeological Science* 31: 855–66.

Renfrew, C., and Cooke, K. L. (eds.). 1979. *Transformations: Mathematical Approaches to Culture Change*. New York: Academic Press.

Renfrew, C., and Dixon, J. 1976. Obsidian in Western Asia: A review, in G. d. G. Sieveking, I. H. Longworth, and K. E. Wilson (eds.), *Problems in Economic and Social Archaeology*, pp. 137–50. London: Duckworth and Co.

Ripley, B. D. 1977. Modelling spatial patterns. *Journal of the Royal Statistical Society*, B41: 368–74.

Robbins, P. 2003. Beyond ground truth: GIS and the environmental knowledge of herders, professional foresters, and other traditional communities. *Human Ecology* 31: 233–53.

Sabloff, J. A. (ed.). 1981. *Simulations in Archaeology*. Albuquerque: University of New Mexico Press.

Shennan, S. 1997. *Quantifying Archaeology*. Edinburgh: Edinburgh University Press. 2nd edition.

Silva, M. D., and Pizziolo, G. 2001. Setting up a "human calibrated" anisotropic cost surface for archaeological landscape investigation, in Z. Stančič and T. Veljanovski (eds.), *Computing Archaeology for Understanding the Past: Proceedings of the CAA2000 Conference*, pp. 279–86. British Archaeological Reports International Series 931. Oxford: Archaeopress.

Taylor, P. J. 1990. GKS. *Political Geography Quarterly* 9: 211–12.

Thomas, J. 1993. The politics of vision and the archaeologies of landscape, in B. Bender (ed.), *Landscape: Politics and Perspectives*, pp. 19–48. Oxford: Berg.

Tilley, C. 1994. *A Phenomenology of Landscape*. Oxford: Berg.

Tomlinson, P., and Hall, A. R. 1996. A review of the archaeological evidence for food plants from the British Isles: An example of the use of the archaeobotanical computer database (ABCD). *Internet Archaeology* 1. http://intarch.ac.uk/journal/issue1/tomlinson toc.html, accessed 20060313.

Tschan, A. P., Raczkowski, W., and Latałowa, M. 2000. Perception and viewsheds: Are they mutually inclusive? in G. Lock (ed.), *Beyond the Map: Archaeology and Spatial Technologies*, pp. 28–48. Amsterdam: IOS Press.

Ur, J. 2003. CORONA satellite photography and ancient road networks: A northern Mesopotamian case study. *Antiquity* 77: 102–16.

Vita-Finzi, C., and Higgs, E. S. 1970. Prehistoric economy in the Mount Carmel area of Palestine: Site catchment analysis. *Proceedings of the Prehistoric Society* 36: 1–37.

Westcott, K. L., and Brandon, R. J. (eds.). 2000. Practical Applications of GIS for Archaeologists: A predictive modeling kit. London: Taylor and Francis.

Whallon, R. 1974. Spatial analysis of occupation floors, II: The application of nearest neighbour analysis. *American Antiquity* 39: 16–34.

Wheatley, D. 1993. Going over old ground: GIS, archaeological theory and the act of perception, in J. Andresen, T. Madsen, and I. Scollar (eds.), *Computing the Past: Computer Applications and Quantitative Methods in Archaeology*, pp. 133–38. Aarhus: Aarhus University Press.

———. 1995. Cumulative viewshed analysis: A GIS-based method for investigating intervisibility, and its archaeological application, in G. Lock and Z. Stančič (eds.), *Archaeology and Geographical Information Systems*, pp. 171–86. London: Taylor and Francis.

———. 2004. Making space for an archaeology of place. *Internet Archaeology* 15. http://intarch.ac.uk/journal/issue15/10/toc.html

Wheatley, D., and Gillings, M. 2000. Vision, perception and GIS: Developing enriched approaches to the study of archaeological visibility, in G. Lock (ed.), *Beyond the Map: Archaeology and Spatial Technologies*, pp. 1–27. Amsterdam: IOS Press.

———. 2002. *Spatial Technology and Archaeology: The Archaeological Applications of GIS*. London: Taylor and Francis.

Wilkinson, T. J. 1998. Water and human settlement in the Balikh Valley, Syria: Investigations from 1992–1995. *Journal of Field Archaeology* 25: 63–87.

Williams, J. T. 1993. Spatial autocorrelation and the Classic Maya collapse: One technique, one conclusion. *Journal of Archaeological Science* 20: 705–09.

Winterbottom, S. J., and Long, D. 2006. From abstract digital models to rich virtual environments: Landscape contexts in Kilmartin Glen, Scotland. *Journal of Archaeological Science* 33: 1356–67.

Woodman, P. E. 2000. A predictive model for Mesolithic site location on Islay using logistic regression and GIS, in S. Mithen (ed.), *Hunter-Gatherer Landscape Archaeology: The Southern Hebrides Mesolithic Project 1988–1998, Vol. 2: Archaeological Fieldwork on Colonsay, Computer Modelling, Experimental Archaeology, and Final Interpretations*, pp. 445–64. Cambridge: McDonald Institute for Archaeological Research.

Woodman, P. E., and Woodward, M. 2002. The use and abuse of statistical methods in archaeological site location modelling, in D. Wheatley, G. Earl, and S. Poppy (eds.), *Contemporary Themes in Archaeological Computing*. University of Southampton Department of Archaeology Monographs 39–43. Oxford: Oxbow Books.

Zubrow, E. B. W. 1994. Knowledge representation and archaeology: A cognitive example using GIS, in C. Renfrew and E. B. W. Zubrow (eds.), *The Ancient Mind: Elements of a Cognitive Archaeology*, pp. 107–18. Cambridge: Cambridge University Press.

57

PLOUGHZONE ARCHAEOLOGY IN HISTORICAL ARCHAEOLOGY

Alasdair Brooks

Ploughzone (also *plowzone/plow zone* in North American usage) refers to the portion of a site disturbed by post-occupation ploughing or other agricultural activity (such as cattle grazing). In the past, and in some rare cases also in the present, archaeologists often treated the agriculture-related surface sediments of a site as "disturbed," thereby either discarding or not analyzing in any meaningful way those surface contexts. Over the last 30 to 40 years, however, a broad literature on the importance of ploughzone archaeology has emerged, conclusively demonstrating the research value of these upper, disturbed stratigraphic levels.

This chapter examines the use of archaeological studies of the ploughzone in historical archaeology, focusing on examples from the United States and Australia. However, this should not obscure the fact that studies of the ploughzone in landscape archaeology have a rich tradition and background in studies of other periods and regions (particularly the United Kingdom where specialized studies of ploughzone archaeology have a relatively long and distinguished history, especially with respect to the Neolithic and subsequent periods [e.g., Schofield 1991; Steinberg 1996]). Although the specific examples used in this chapter are most directly relevant to archaeology of the more recent past, many of the issues explored here are applicable to ploughzone archaeology more broadly.

Ploughzone Methodology

The precise nature and depth of a ploughzone will vary according to the type of agricultural activity in question, but the central challenge of plough-zone archaeology remains broadly consistent: that agricultural activity significantly disturbs the soil in which it takes place, destroying subsurface arch-aeological features and changing the location of artifacts. This is equally true whether the agricultural activity in question is current or the plough-zone is itself an archaeological stratigraphic layer related to past activity. Accordingly, most research papers in ploughzone archaeology have focused on the manner in and the extent to which artifacts within the ploughzone can be said to inform on original site definition, both temporally and spatially (e.g., see papers in Schofield 1991).

Perhaps the most important question in ploughzone archaeology is the extent to which ploughing and other agricultural activity disrupts artifact distributions within the ploughed soil (see, for example, Steinberg 1996). It has been recognized for some time that although ploughing results in a destruction of *vertical* stratigraphy and the mixing of materials from originally separate stratigraphic contexts, *horizontal* relationships usually remain broadly intact (O'Brien and Lewarch 1981). Thus, if a

17th-century layer was located underneath an 18th-century layer, the stratigraphic relationship between those layers will be, for the most part, destroyed by subsequent ploughing; but if a 17th-century refuse pit was located *next to* a separate 18th-century refuse pit, then ploughing will not usually destroy the spatial relationship of adjacent deposits. Similarly, although ploughing can and does move individual artifacts a considerable distance, the majority of artifacts will stay reasonably close to their original horizontal point of deposition, allowing identification and mapping of artifact concentrations. Riordan (1988: 3) stresses that such identification and mapping are best undertaken for larger-scale artifact scatters such as middens and architectural scatters rather than small-scale features such as postholes. These basic principles remain broadly true across archaeological periods.

The question of how to sample and map the ploughzone has been the subject of considerable discussion (e.g., Custer 1992; Riordan 1988). In addition to surveys across ploughed areas and their surrounding areas, a variety of subsurface sampling strategies have been used, with common practice consisting of the excavation of a series of test units at regular intervals across the site complementing whatever excavation takes place at the site's core (though it is certainly not unknown for ploughzone excavations to consist solely of test units where there is no core structure). The soils are then sieved through a screen in order to recover the artifacts. The size of and interval between test units vary, but most American historical archaeologists use units measuring between 1 x 1 m and 1.5 x 1.5 m (though this being the United States, these are usually expressed as between 3 x 3 ft and 5 x 5 ft). The spacing of units across the site is also crucial, with most archaeologists favoring the systematic spacing of units across a site, because this is more likely to recover information about different archaeological distributions at a site. King (2004) recommends spacing the units no more than 4.5 m (15 ft) apart, from center point to center point. After the artifacts have been catalogued, common practice is to enter the data into a computer program that can use the sampling data to map distributions across the entire site.

The Development of Ploughzone Studies in Historical Archaeology

The history of ploughzone studies in historical archaeology dates back at least to the early 1960s. In 1962, MacCord excavated the ploughzone

from a 17th-century Native American postcontact site, demonstrating the research potential of ploughzone artifact distributions (MacCord 1969), but this approach was neither immediately nor universally adopted. King (2004) notes, for example, that Ivor Noël Hume (one of the founding fathers of American historical archaeology) publicly questioned the archaeological value of the ploughzone and continued to strip his sites' topsoil by machine into the 1980s while nonetheless acknowledging that ploughzone artifacts were "in approximately their original locations" (Noël Hume 1982: 9–11).

By the end of the 1980s, however, the potential research value of ploughzone archaeology was broadly acknowledged. Extensive statistical studies of the nature of ploughzone were undertaken by American archaeologists in the 1970s (Ammerman and Feldman 1978; Baker 1978; O'Brien and Lewarch 1981; Redman and Watson 1970; Roper 1976), and historical archaeologists built on this work, culminating in a sequence of papers in the journal *Historical Archaeology* conclusively demonstrating the value of ploughzone artifact distributions to archaeological studies of the more recent past (King 1988; King and Miller 1987; Pogue 1988; Riordan 1988). More recently, the *Journal of Middle Atlantic Archaeology* featured a debate on "Issues in plow zone archaeology" (based on a forum held at the 2006 Atlantic Archaeology Conference), with several papers discussing not only methodological dimensions but also the value and the political and scientific constraints and opportunities of ploughzone archaeology.

Ploughzone studies in Australian historical archaeology are not as advanced, and as of this writing very few ploughzone sites have been the subject of archaeological study specifically acknowledging the ploughed component. This, however, has more to do with the nature of Australian historical sites than does any failure on the part of Australian archaeologists; the comparative brevity of European settlement in Australia means that abandoned historical sites that have since been subjected to agricultural activity are at present relatively uncommon. Furthermore, arable land forms a much smaller component of the Australian landscape than is the case in the United States, where a comparatively larger area of the landscape was ploughed during the 19th and 20th centuries.

Some Historical Archaeology Ploughzone Case Studies

King and Miller's (1987) classic study of artifact distributions at the 17th- and 18th- century Van

Sweringen site in St. Mary's City, Maryland, offers an excellent introduction into many of the issues relevant to ploughzone archaeology. St. Mary's City was the capital of the Maryland colony from settlement in 1634 through to 1695, after which the former capital soon became a small agricultural village. Dutch immigrant Garret Van Sweringen had operated a lodging house on his St. Mary's City property throughout the late 17th century. After Garret's death in 1698, his son Joseph occupied the site until his own death in 1723. After passing through the hands of the subsequent husband of Joseph's widow, occupation ended ca. 1745.

The Van Sweringen site had been extensively ploughed in the more than 200 years since the end of occupation. While major subsurface structural features were located beneath the ploughzone and recorded, it was the artifact distributions within the ploughzone that offered additional information about site structure and use through time. By plotting horizontal distributions of dateable artifacts across the site's ploughzone, King and Miller were able to observe changes in depositional behavior and to identify two separate phases of deposition. These two phases coincided in date with occupation of the site by the two different Van Sweringen family members. Although the artifacts dating to Garret's occupation were largely located on the original street in front of the site, the artifacts dating to Joseph's occupation were largely located behind the main house, in the yard space enclosed by the site's original fence. Plotting the ploughzone artifacts demonstrated that a significant shift in depositional location had occurred at the site in a single generation. In addition, whereas the earlier deposit associated with Garret Van Sweringen appeared to have consisted of a single undifferentiated midden, the later deposits associated with Joseph were characterized by unmistakable differentiation in deposition. In other words, different types of artifacts appear to have been discarded in different areas. King and Miller interpreted these changes in discard behavior as indicating both an increase in local complexity as Maryland's colonial society matured, and the rise of a Georgian emphasis on greater formality.

In the case of the Van Sweringen site, subsurface architectural remains—including brickwork that survived the post-occupation ploughing. But ploughzone archaeology can also help identify the location of structural features subsequently destroyed by agricultural activity. The Quarter Site at Thomas Jefferson's Poplar Forest in

Virginia, a late 18th- to early 19th-century slave settlement at the retreat home of the third president of the United States, featured two separate identified ploughing episodes—one predating occupation of the site, and the more relevant mid-19th-century episode that destroyed much of the structural evidence at the slave quarter (Heath 1999: 32–33). Although the lower parts of cellars and some structural posts survived, the locations of doors, fencelines, and paths across the site were all partially identified through the mapping of ploughzone distributions. Ceramic distributions were also used to suggest different periods of occupation and differences in status among the three identified structures at the site (Heath 1999: 38, 41).

As important as the ploughzone was for the interpretation of the Quarter Site, the site was also methodologically important for being only partially ploughed. A third of the site had not been subject to post-occupation ploughing, providing a rare opportunity to compare artifact distributions in ploughed and unploughed sections of the same site. A study of ceramics and glass crossmends (conjoins) across both components of the site demonstrated that it is possible to compare data across differentially disturbed components of the site and that horizontal distributions in the ploughzone maintain a certain level of coherence. The line between the ploughed and unploughed components of the site also divided two of the structures at the site. The refined earthenwares and bottle glass were almost entirely contained north of the "zone of transition" between the two structures. That this was a meaningful distribution, and not the result of ploughing, was demonstrated by the fact that the coarse earthenware and stoneware mends *crossed* the zone of transition (Brooks 1996). This enabled Poplar Forest archaeologists to interpret artifact distributions across the site as the result of human behavior rather than primarily caused by post-occupation activity.

Not all studies of ploughzone sites involve actual excavation. The combination of surface collection and modern geophysical remote sensing techniques (see Cheetham, this volume) can make it possible to gain a picture of a site prior to, or even entirely without, excavation work. The Willoughby Bean site, a rare example of an Australian historic site with a ploughzone component—and that is the subject of ongoing research—is located in Gippsland, the southeast corner of the state of Victoria. It was the residence of the first permanent Anglican minister in Gippsland between 1848 and 1858, and the structure was destroyed by fire in 1861. The site is believed to have been used as farmland ever since. While the general location of the site was known from historic maps, the specific location in what is now an open grazing paddock was identified through a surface artifact scatter. Surface scatters are not uncommon on ploughzone sites, and the visibility of surface artifacts is often improved by recent ploughing activity or rain. The surface artifacts at the Bean site were mapped *in situ* in early 2006. A 100 x 100 m section of the grazing paddock was then subjected to geophysical survey using a fluxgate gradiometer. By superimposing the artifact map with the geophysical survey, the research team was able to confirm that the surface artifact scatter was still archaeologically meaningful, to confirm that archaeological features survived beneath the ploughzone, and to engage in preliminary site interpretation prior to excavation.

As with the Van Sweringen and the Quarter Sites, distinct areas of artifact deposition were identifiable at the Bean site. The bricks were concentrated behind a structure and over a subsurface anomaly, probably indicating a chimney. The ceramics were largely concentrated in an area of subsurface disturbance believed to be a garden of some sort. The bottle glass seems to follow a fenceline dividing the two identified structures. Although excavation of the site (forthcoming) will no doubt help to refine this data, the combination of remote sensing technology and surface collection has proven to be a powerful research tool at Bean's residence.

Conclusions

Although this discussion has focused on the research potential and use of ploughzone archaeology at three historic sites in the United States and Australia, the basic principles are applicable across periods and geographic regions. Ploughing and other agricultural activities are often an integral part of site formations processes, whether that site is large scale or small scale, single period or multiperiod. Archaeologists, whether historical archaeologists or not, should not dismiss the potential research value of the ploughzone by stripping and discarding surface layers. The artifact distributions within ploughzone layers often turn out to considerably add to an understanding of sites located within that landscape, as demonstrated here by how archaeologists were able to gain new insight into both site occupation and the impact of post-occupation agricultural activity the Van Sweringen, Quarter, and

Bean sites through their in-depth analysis of the ploughzone component. While remote-sensing techniques, as used at Bean, may transform historical archaeologists' approach to ploughzone sampling in coming years, the spatial analysis of ploughzone deposits will remain integral to an understanding of those sites where postdepositional agricultural activity has occurred.

Acknowledgments

Thanks are due Bruno David for inviting me (at very short notice!) to contribute to this volume. Special thanks to Julie King and Barbara Heath for their assistance with their own research—particularly Julie for pointing me in the direction of her excellent short summary of Chesapeake ploughzone archaeology. The Willoughby Bean site discussion in this chapter is based on ongoing, still unpublished, research by a team consisting of this author, Susan Lawrence, and Jane Lennon, and funded by an Australian Research Council Discovery Grant. The geophysical survey at the Bean site was undertaken by Hans Dieter Bader of New Zealand-based consultancy firm Geometria.

References

Ammerman, A. T., and Feldman, M. W. 1978. Replicated Collection of Site Surfaces. *American Antiquity* 43: 734–40.

Baker, C. M. 1978. The size effect: An explanation of variability in surface artifact assemblage content. *American Antiquity* 43: 288–93.

Brooks, A. M. 1996. Across the great divide: Crossmend distributions on a partially plowed site. Unpublished paper presented at the 1996 meeting of the Council for Northeastern Historical Archaeology, Albany, New York.

Custer, J. F. 1992. A Simulation of plow zone excavation sampling designs: How much is enough? *North American Archaeologist* 13(3): 263–80.

Heath, B. J. 1999. *Hidden Lives: The Archaeology of Slave Life at Thomas Jefferson's Poplar Forest.* Charlottesville: University Press of Virginia.

King, J. A. 1988. A comparative midden analysis of a seventeenth-century household and inn in St. Mary's City, Maryland. *Historical Archaeology* 22(2): 17–39.

———. 2004. A review and assessment of archaeological investigations at 44RD183, Warsaw, Virginia. Unpublished report prepared for the Council of Virginia Archaeologists. www.chesapeakearchaeology.org/HTM_Main/PZ%20Archaeology.htm

King, J. A., and Miller, H. M. 1987. The view from the midden: An analysis of midden distribution and composition at the van Sweringen site, St. Mary's City, Maryland. *Historical Archaeology* 21(2): 37–59.

MacCord, H. 1969. Camden: A postcontact Indian site in Caroline County. *Quarterly Bulletin of the Archaeological Society of Virginia* 24(1): 1–55.

Noël Hume, I. 1982. *Martin's Hundred*. New York: Alfred A. Knopf.

O'Brien, M. J., and Lewarch, D. E. 1981. *Plowzone Archaeology: Contributions to Theory and Technique*. Nashville: Vanderbilt University Publications in Archaeology 27.

Pogue, D. J. 1988. Spatial analysis of the King's Reach Homelot. *Historical Archaeology* 22(2): 40–56.

Redman, C. L., and Watson, P. J. 1970. Systematic, intensive surface collection. *American Antiquity* 35: 279–91.

Riordan, T. B. 1988. The interpretation of 17th-century sites through plow zone surface collections: Examples from St. Mary's City, Maryland. *Historical Archaeology* 22(2): 2–16.

Roper, D. 1976. Lateral displacement of artifacts due to plowing. *American Antiquity* 41: 372–74.

Schofield, A. J. (ed.). 1991. *Interpreting Artefact Scatters: Contributions to Ploughzone Archaeology*. Oxford: Oxbow Books.

Steinberg, J. M. 1996. Ploughzone sampling in Denmark: Isolating and interpreting site signatures from disturbed contexts. *Antiquity* 70: 368–90.

58

LANDSCAPE FORMATION PROCESSES

Michael P. Heilen, Michael Brian Schiffer,
and J. Jefferson Reid

In the early 1970s, two University of Arizona graduate students and a University of Arizona junior faculty member founded "behavioral archaeology." Behavioral archaeology was formulated as a new approach to investigating people-material interactions in all times and places (Reid 1995; Reid et al. 1974; Reid et al. 1975; Schiffer 1975, 1999). Since that time, behavioral archaeologists have made important advances in the study of prehistory, history, social change, technology, and communication (LaMotta 2001; LaMotta and Schiffer 2001; Reid and Whittlesey 1987 1997, 1999; Schiffer 1988, 1996, 2002; Schiffer and Miller 1999; Schiffer and Skibo 1987, 1997; Walker 1999; Walker et al. 2000; Zedeño and Stoffle 2003; Zedeño 2000). One major focus of behavioral archaeology has been the study of *formation processes* (Binford 1979; Reid 1985; Schiffer 1983, 1987; Shott 1998, 2006). As an essential component of behavioral archaeology—not its totality—the conceptual place of formation processes in landscape archaeology is the focus of this chapter.

As a result of an emphasis on formation processes and the success of behavioral approaches, behavioral archaeology is sometimes inaccurately equated with the study of formation processes (see Shaw and Jameson 1999: 113). A recent compilation of important contributions to formation theory can be found in Shott (2006). As formulated by behavioral archaeologists, the study of formation processes includes investigation of both environmental and cultural formation processes. Cultural formation processes include reuse, discard, reclamation, and cultural disturbance. Examples of specific cultural formation processes are lateral cycling, secondary refuse disposal, scavenging, and trampling. Environmental formation processes include deterioration, decay, environmental disturbance processes, and earth surface processes, such as patination, fungal decay, cryoturbation, and volcanism (Schiffer 1987).

A misconception in archaeology is the use of "taphonomy" as a synonym for "formation processes." Taphonomy is "the study of the processes . . . that affect animal and plant remains as they become fossilized" (*Merriam Webster's Collegiate Dictionary*, 10th edition). When used appropriately, "taphonomy" is not a synonym for "formation processes" but a small subset of formation processes having to do specifically with the transformation of organisms from the biosphere to the lithosphere.

Formation Processes of the Archaeological Record

In *Formation Processes of the Archaeological Record*, Schiffer (1987) presented a behavioral

framework for understanding the formation of artifacts, deposits, and sites and brought together in
one volume much of what was known about how
the archaeological record is formed. *Formation
Processes* was a contribution relevant at every level
of archaeological investigation, from the design of
surveys to the collection, analysis, interpretation,
and modeling of data. Schiffer is careful to point
out, however, that *Formation Processes* was only
an introduction to the study and urged the reader
to consult other sources. Most important, Schiffer
stressed that archaeologists need to incorporate
understandings of formation processes at every
level of archaeological inference. The study of
formation processes involves not only the identification of potential transformations of the archaeological record but also continuous evaluation
of potential relationships to systemic behaviors
and the effects of formation processes on specific
artifacts, deposits, and sites.

Since *Formation Processes* was published,
investigations of formation processes have contributed to many different kinds of archaeological
inquiry. Formation processes are fundamental to
the building of middle-range theory and the construction of behaviorally and archaeologically
meaningful inference. By informing on how the
archaeological record is formed, formation process studies are crucial to learning about people-
material interactions. As Schiffer (1987: xviii) puts
it, "the cultural past is knowable, but only when
the nature of the evidence is understood."

Formation Processes in Landscape Archaeology

Formation processes have traditionally been investigated at the level of artifacts, deposits, sites, and
regions. Although investigation of formation processes at the level of landscapes is rare (Bintliff
and Snodgrass 1988; Stafford 1995; Stafford and
Hajic 1992), development of frameworks for modeling landscape formation processes is essential
to the implementation of a scientific landscape
archaeology. Recently, Heilen (2005) formulated
a behavioral approach to landscape archaeology.
Heilen's (2005) framework is founded on one of
the most fundamental distinctions in archaeology,
the distinction between systemic context and
archaeological context. Elements (people, tools,
facilities, places) participating in a behavioral system are in systemic context. Artifactual materials
no longer participating in systemic context, such
as those recovered during the course of archaeological fieldwork, are in archaeological context.
Reuse and reclamation processes allow elements

in archaeological context to be reincorporated into
systemic context (Schiffer 1972).

From a behavioral perspective, there are two
kinds of landscapes: (1) archaeological landscapes and (2) systemic landscapes (Figure 58.1).
Archaeological landscapes are arrays of archaeological materials—artifacts, features, deposits, and
sites. *Systemic landscapes* are networks of people,
places, materials, and activities connected through
the exchange of matter, energy, and information (cf. Basso 1996; Bentley 2003; Bentley and
Maschner 2003; Binford 1982; Buchanan 2002;
Heilen 2005; Newman 2003; Schlanger 1992;
Strogatz 2001; Thomas 2001; Tilley 1994; Watts
1999; Watts and Strogatz 1998; Whittlesey 1997;
Wilkinson 2003; Zedeño 2000).

From a transformation perspective, archaeological landscapes and systemic landscapes are related but are not isomorphic structures. In the course
of fieldwork and analysis, archaeologists routinely
measure and record attributes of archaeological
landscapes. The properties of systemic landscapes
can only be inferred from the properties of archaeological landscapes through careful development
and application of middle-range theory. Middle-
range theory is a set of deductively or inductively
established principles, corollaries, and material
correlates that can be used to specify relationships
between systemic processes and archaeological
patterns. Middle-range theory includes principles
of formation processes, such as c-transforms and
n-transforms. All too often, archaeologists attempt
to "read" systemic landscapes directly from archaeological landscapes (Heilen 2005).

Landscape Formation Processes and Scale

The issue of scale should be considered a central focus of landscape archaeology. In landscape

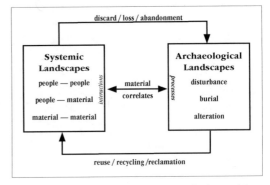

Figure 58.1 A schematic diagram of relationships
between systemic and archaeological landscapes.

archaeology, scale can be understood in terms of spatial, temporal, and behavioral dimensions. In the analysis of archaeological landscapes, the most behaviorally meaningful scales are those that match the inferred scales of systemic landscapes under investigation.

Spatial Scale

Spatial scale can be defined by *extent* and *grain*. *Extent* refers to the absolute size of a study area or landscape. *Grain* refers to the absolute size of the smallest analytical spatial unit. Changes in grain or extent can cause changes in landscape metrics, some of which are predictable according to scaling relations (Figure 58.2). Incompatibilities and disagreements between landscape studies can result from incongruent spatial scales. Results of landscape studies conducted at incongruent spatial scales can be rendered comparable through the application of empirically or theoretically derived scaling relations (Ebert 1992; O'Neill et al. 1989; Turner and Gardner 1991; Turner et al. 2001; Wiens 1989; Wu 2004; Wu and Qi 2000).

Spatial scaling is an important component of landscape archaeology that has bearing on how landscapes form and how they are analyzed. Drawing upon ecological and ethnographic frames of reference, Heilen (2005) argues that the extent and grain of systemic landscapes depends on how organisms exchange matter, energy, and information with their environments (Chust et al. 2004;

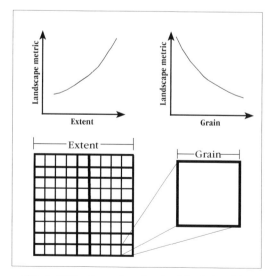

Figure 58.2 Examples of potential relationships between landscape metrics, such as diversity or fragmentation, and changes in grain or extent.

Ritchie 1998). The spatial scaling of human systemic landscapes is related to the specific ways in which human groups perceive and interact with their environments and can vary along economic, technological, and ecological dimensions (Heilen 2005; Heilen and Reid In press).

Temporal Scale

Temporal scale can be understood in terms of *span* and *interval*. Here, *span* refers to the absolute time frame in which relevant events or processes are understood to have occurred. *Interval* refers to the smallest resolvable temporal unit. The time depth of individual temporal units, such as a geological layer or archaeological feature deposit, can be highly variable. An archaeological interval often depends on processes that are independent of behavioral processes of archaeological interest (Dean 1978; McPherron et al. 2005; Ramsey 2003). An interval can be a geological layer or a deposit, depth range within a deposit, or a combination of temporal markers such as diagnostic artifacts, sedimentary facies, or chronometric age determinations that constrain the time dimension of archaeological manifestations. Understanding the relationships between time, archaeological pattern formation, and behavior is a central problem in archaeology and will continue to be so in landscape archaeology. For instance, despite compelling and informative ethnoarchaeological studies of land use and site formation, few observations on living landscape behaviors come anywhere close to approximating the interval of many archaeological deposits. Since many archaeological landscapes at least partly consist of surface archaeology, and many intervals are incongruous or ambiguous, controlling for time in landscape archaeology is a vexing problem. Establishing the sequencing or contemporaneity of events and processes and especially, the temporal scale at which events or processes can be ascertained, is essential to assessing relationships among components of archaeological landscapes.

Pioneering efforts in establishing temporal dimensions for surface archaeological landscapes were made by southwestern archaeologists Malcolm Rogers (1939, 1945, 1958, 1966) and Julian D. Hayden (1961, 1965, 1967, 1976a, 1976b, 1994, 1998). Rogers developed the concept of horizontal stratigraphy, a method of establishing chronology by finding relationships between complimentary temporal markers—artifact weathering patterns, overprinting and erasure of landscape elements and features, and relative and absolute ages of surface deposits. Hayden applied and refined Rogers's methods in the Sierra Pinacate of northern Mexico, where he developed a continuous, but controversial, culture history based on his intricate interpretation of surface archaeological landscape patterns (Heilen 2001, 2004). Other more recent work has established the temporality of landscape units by combining surface and subsurface archaeology over large areas. Ultimately, a combination of surface and subsurface evidence, geological investigation, and rule-based systems for establishing temporal relationships between weathering processes, artifact and feature types, surface ages, and rates of deposition and erosion will be required to develop realistic models of archaeological landscape formation (Fanning and Holdaway 2001; Holdaway et al. 2002; Stafford 1994, 1995; Stafford and Hajic 1992; Waters 1988).

Behavioral Scale

Behavioral scale can be understood in terms of *interactions*, *activities*, and *behavioral systems*. Low-level behavioral scales are hierarchically nested within higher-level behavioral scales. *Interaction scale* refers to discrete mechanical, thermal, chemical, acoustic, or visual interactions between people and materials. *Activity scale* refers to discrete activities performed by individuals, households, and task groups. Activities consist of a finite number of interactions. *Systemic scale* refers to one or more behavioral systems. Behavioral systems are networks of activities (LaMotta and Schiffer 2001; Schiffer 1987).

A major focus of behavioral landscape archaeology is to investigate the topological structure of complex landscape networks. To understand landscape networks, we need to develop archaeologically relevant quantities that can be used to model networked exchanges, perhaps by simulating the interplay between simple interactions and activities. In this way, complex network structures and their resulting archaeological consequences can be modeled as the outcome of specific performances, strategies, and recipes for action (Heilen 2005).

Discussion

The archaeological record is the residual consequence of an enormous array of interacting natural and cultural processes occurring across a broad range of spatial, temporal, and behavioral scales. The archaeological record is created by behavior, refracted, or distorted by postdepositional (or postbehavioral) processes, and

sampled according to both theoretical and methodological biases. In order to draw reliable inferences about systemic behavior, archaeologists need to establish frameworks that conceptualize the nature and organization of formation processes, assess their relationship to past behaviors, and empirically identify relationships between formation processes and specific deposits (paraphrasing Schiffer 1983: 676).

Schiffer (1983) outlines several models of the nature and organization of formation processes in archaeology: (1) Ascher's (1968) entropy model; (2) Cowgill's (1970) sampling bias model; and (3) the transformation model of behavioral archaeology. In the entropy model, the effects of time degrade the quantity and quality of artifactual evidence, leading to loss of information or increase in entropy over time. In the sampling bias model, natural and cultural processes progressively sample behavioral systems and archaeological deposits. Archaeological manifestations (the "physical finds population") represent a progressively biased sample of the material consequences of systemic behavior (the "physical consequences population").

Behavioral archaeology uniquely offers the perspective that formation processes act differentially and selectively on the archaeological record. The transformationist view, the most empirically and theoretically grounded of the three, stresses that "diverse processes transform or distort materials, and the many ways they do so: formally, spatially, quantitatively, and relationally" (Schiffer 1983: 677). Schiffer (1987) has argued that formation processes should be assessed case-by-case at the level of artifacts and deposits. In *Formation Processes*, Schiffer (1987) argued that at the artifact and deposit level, fine-scale local variation accounts for the differential survival or preservation of archaeological elements such as artifacts, not broad-scale variation. If this is true, how are we to understand the relationships among landscape patterns, formation processes, and behavior? Can we use broadly scaled environmental characteristics to understand broadly scaled archaeological patterns?

Landscape archaeology is typically conducted at scales that transcend the scales at which most formation process studies have been conducted. Studies in geomorphic landscape evolution notwithstanding (Holliday 2004; Waters 1992), archaeologists have only limited working knowledge of the relationships between archaeological landscape patterns and landscape-level formation processes. Major questions in the study of landscape formation processes include the following.

Do archaeological landscape patterns result from top-down or bottom-up processes? What are the relevant quantities for investigating archaeological landscape patterns and systemic landscape processes? How do these quantities relate to each other? What results from interactions between bottom-up and top-down landscape formation processes? (Brown and Witschey 2003; Brown et al. 2005; Corning 1998; Lansing 2003; Layton 2003; Turcotte and Rundle 2002; West et al. 1999).

Heilen (2005) advances the concept of landscape hierarchy in the study of landscape formation processes. Hierarchy theory specifies that higher-level landscape properties or processes guide or constrain lower-level properties or processes (Kotliar and Wiens 1990; O'Neill et al. 1989). Some broadly scaled archaeological landscape patterns may result from broadly scaled environmental or cultural processes (Bintliff and Snodgrass 1988). For instance, the availability of raw materials or the size and shape of major landscape features may exert high-level hierarchical controls over lower-level phenomena, such as the behavior of individuals or the distribution of species (Milewski and Diamond 2000). There are many earth surface processes that transform and distort patterning in archaeological deposits or even create distinct patterns independent of artifact depositing behaviors (Schiffer 1987). Consequently, fundamental properties of archaeological landscapes—such as the characteristics of artifact-bearing deposits—may be partly organized or biased according to broadly scaled environmental regimes or attributes.

Conversely, bottom-up processes involving the interplay of many simple interactions and activities could result in complex, broad-scale patterns. Simulation modeling is one potential approach to identifying kinds of archaeological patterns that could be generated through the repeated interplay of simple interactions and activities. An important task in landscape archaeology will be discovering how systemic processes and archaeological patterns are related across a wide range of scales.

Conclusions

Behavioral archaeologists have established that archaeological patterns are the product of numerous interacting formation processes. However, it is impossible to control for (or even identify) them all. Instead, archaeologists can only hope to focus their attention on formation processes that are most relevant to particular research problems.

Ultimately, investigation of landscape formation processes is crucial to understanding relationships between archaeological and systemic landscapes. After all, identifying important sources of variation in the archaeological record is key to inferring past behavior. Landscape formation processes must be investigated according to a framework that actively incorporates the multiple spatial, temporal, and behavioral scales involved in the formation of the archaeological record. We suggest that a behavioral framework is the most appropriate to understanding human landscape behavior in all times and places. In doing so, archaeologists are able to draw robust conclusions about landscapes of the past, present, and future.

References

Ascher, R. 1968. Time's arrow and the archaeology of a contemporary community, in K. C. Chang (ed.), *Settlement Archaeology*, pp. 43–52. Palo Alto: National Press.

Basso, K. 1996. *Wisdom Sits in Places: Landscape and Language among the Western Apache.* Albuquerque: University of New Mexico Press.

Bentley, R. A. 2003. Scale-free network growth and social inequality, in R. A. Bentley and H. D. G. Maschner (eds.), *Complex Systems and Archaeology*, pp. 27–45. Salt Lake City: The University of Utah Press.

Bentley, R. A., and Maschner, H. D. G. (eds.). 2003. *Complex Systems and Archaeology.* Salt Lake City: The University of Utah Press.

Binford, L. R. 1979. Organization and formation processes: Looking at curated technologies. *Journal of Anthropological Research* 35: 172–97.

———. 1982. The archaeology of place. *Journal of Anthropological Archaeology* 1: 5–31.

Bintliff, J., and Snodgrass, A. 1988. Off-site pottery distributions: A regional and interregional perspective. *Current Anthropology* 29: 506–13.

Brown, C. T., and Witschey, W. R. T. 2003. The fractal geometry of ancient Maya settlement. *Journal of Archaeological Science* 30: 1619–32.

Brown, C. T., Witschey, W. R. T., and Liebovitch, L. S. 2005. The broken past: Fractals in archaeology. *Journal of Archaeological Method and Theory* 12: 37–78.

Buchanan, M. 2002. *Nexus: Small Worlds and the Groundbreaking Theory of Networks.* New York: W. W. Norton and Company.

Chust, G., Pretus, J. L., Ducrot, D., and Ventura, D. 2004. Scale dependency of insect assemblages in response to landscape pattern. *Landscape Ecology* 19: 41–47.

Corning, P. A. 1998. "The synergism hypothesis": On the concept of synergy and its role in the evolution of complex systems. *Journal of Social and Evolutionary Systems* 21: 133–72.

Cowgill, G. I. 1970. Some sampling and reliability problems in archaeology, in *Archéologie et Calculateurs: Problèmes semiologiques et mathematiques*, pp. 161–75. Paris: Editions du CNRS. Colloque International du CNRS.

Dean, J. S. 1978. Independent dating in archaeological analysis, in M. B. Schiffer (ed.), *Advances in Archaeological Method and Theory* 1, pp. 223–55. New York: Academic Press.

Ebert, J. I. 1992. *Distributional Archaeology*. Salt Lake City: The University of Utah Press.

Fanning, P. C., and Holdaway, S. J. 2001. Stone artifact scatters in western NSW, Australia: Geomorphic controls on artifact size and distribution. *Geoarchaeology* 16: 667–86.

Hayden, J. D. 1961. Malcolm Jennings Rogers, 1890–1960. *American Anthropologist* 63: 1323–24.

———. 1965. Fragile-pattern areas. *American Antiquity* 31: 272–76.

———. 1967. A summary prehistory and history of the Sierra Pinacate, Sonora, Mexico. *American Antiquity* 32: 335–44.

———. 1976a. Changing climate in the Sierra Pinacate of Sonora, Mexico, in P. Paylore and R. B. J. Haney (eds.), *Desertification: Process, Problems, Perspectives*, pp. 70–86. Tucson: University of Arizona Arid/Semi-Arid Natural Resources Program.

———. 1976b. Pre-altithermal archaeology in the Sierra Pinacate, Sonora, Mexico. *American Antiquity* 41: 274–89.

———. 1994. The Sierra Pinacate, the legacy of Malcolm Rogers, and the archaeology of the Lower Colorado River, in J. A. Ezzo (ed.), *Recent Research along the Lower Colorado River*. Tucson: Statistical Research Technical Series 51.

———. 1998. *The Sierra Pinacate*. Tucson: Southwest Center Series, University of Arizona Press.

Heilen, M. P. 2001. A re-examination of Julian Hayden's Malpais model: Field notes, formation processes and the Clovis vs Pre-Clovis debate. Unpublished Master's thesis, Department of Anthropology, University of Arizona, Tucson.

———. 2004. Julian Hayden's Malpais model: A Pre-Clovis claim from the American Southwest. *Kiva* 69: 305–31.

———. 2005. An archaeological theory of landscapes. Unpublished Ph.D. thesis, Department of Anthropology, University of Arizona.

Heilen, M. P., and Reid, J. J. In press. A landscape of gamble and guts: Commodification of land on the Arizona frontier, in B. Bowser and M. N. Zedeño (eds.), *The Archaeology of Meaningful Places*. Salt Lake City: University of Utah Press.

Holdaway, S. J., Fanning, P. C., Witter, D. C., Jones, M., Nicholls, G., and Shiner, J. 2002. Variability in the chronology of late Holocene Aboriginal occupation on the arid margin of southeastern Australia. *Journal of Archaeological Science* 29: 351–63.

Holliday, V. T. 2004. *Soils in Archaeological Research*. New York: Oxford University Press.

Kotliar, N. B., and Wiens, J. A. 1990. Multiple scales of patchiness and patch structure: A hierarchical framework for the study of heterogeneity. *Oikos* 59: 253–60.

LaMotta, V. 2001. Behavioral variability in mortuary deposits: A modern material culture study. *Arizona Anthropologist* 14: 53–80.

LaMotta, V. M., and Schiffer, M. B. 2001. Behavioral archaeology: Toward a new synthesis, in I. Hodder (ed.), *Archaeological Theory Today*, pp. 14–64. Malden: Blackwell Publishing.

Lansing, J. S. 2003. Complex adaptive systems. *Annual Review of Anthropology* 32: 183–204.

Layton, R. 2003. Agency, structuration, and complexity, in R. A. Bentley and H. D. G. Maschner (eds.), *Complex Systems and Archaeology*, pp. 103–09. Salt Lake City: The University of Utah Press.

McPherron, S. J. P., Dibble, H. L., and Goldberg, P. 2005. *Geoarchaeology* 20: 243–62.

Milewski, A. V., and Diamond, R. E. 2000. Why are very large herbivores absent from Australia? A new theory of micronutrients. *Journal of Biogeography* 27: 957–78.

Newman, M. E. J. 2003. The structure and function of complex networks. *Society for Industrial and Applied Mathematics* 45: 167–256.

O'Neill, R. V., Johnson, A. R., and King, A. W. 1989. A hierarchical framework for the analysis of scale. *Landscape Ecology* 3(3/4): 193–205.

Ramsey, C. B. 2003. Punctuated dynamic equilibria: A model for chronological analysis, in R. A. Bentley and H. D. G. Maschner (eds.), *Complex Systems and Archaeology*, pp. 85–92. Salt Lake City: The University of Utah Press.

Reid, J. J. 1985. Formation processes for the practical prehistorian: An example from the southeast, in R. S. Dickens, Jr., and H. T. Ward (eds.), *Structure and Process in Southeastern Archaeology*. Tuscaloosa: University of Alabama Press.

———. 1995. Four strategies after twenty years: A return to basics, in J. M. Skibo, W. H. Walker, and A. E. Nielsen (eds.), *Expanding Archaeology*, pp. 15–21. Salt Lake City: University of Utah Press.

Reid, J. J., Rathje, W. L., and Schiffer, M. B. 1974. Expanding archaeology. *American Antiquity* 39: 125–26.

Reid, J. J., Schiffer, M. B., and Rathje, W. L. 1975. Behavioral archaeology: Four strategies. *American Antiquity* 77: 864–69.

Reid, J. J., and Whittlesey, S. M. 1997. *The Archaeology of Ancient Arizona*. Tucson: University of Arizona Press.

———. 1999. *Grasshopper Pueblo: A Story of Archaeology and Ancient Life*. Tucson: University of Arizona Press.

Ritchie, M. E. 1998. Scale-dependent foraging and patch choice in fractal environments. *Evolutionary Ecology* 12: 309–30.

Rogers, M. J. 1939. *Early Lithic Industries of the Lower Basin of the Colorado River and Adjacent Desert Areas*. San Diego Museum Papers No. 3.

———. 1945. An outline of Yuman prehistory. *Southwestern Journal of Anthropology* 1: 167–98.

———. 1958. San Dieguito implements from the terraces of the Rincon-Pantano and Rillito drainage system. *The Kiva* 24: 1–23.

———. 1966. *Ancient Hunters of the Far West*. San Diego: Copley Press.

Schiffer, M. B. 1972. Archaeological context and systemic context. *American Antiquity* 37: 156–65.

———. 1975. Archaeology as behavioral science. *American Anthropologist* 77: 836–48.

———. 1983. Toward the identification of formation processes. *American Antiquity* 48: 675–706.

———. 1987. *Formation Processes of the Archaeological Record*. Albuquerque: University of New Mexico Press.

———. 1988. The structure of archaeological theory. *American Antiquity* 53: 461–85.

———. 1996. Some relationships between behavioral and evolutionary archaeologies. *American Antiquity* 61: 643–62.

———. 1999. Behavioral archaeology: Some clarifications. *American Antiquity* 64: 166–68.

———. 2002. Studying technological differentiation: The case of 18th-century electrical technology. *American Anthropologist* 104: 1148–61.

Schiffer, M. B., and Miller, A. R. 1999. *The Material Life of Human Beings: Artifacts, Behavior, and Communication*. London: Routledge.

Schiffer, M. B., and Skibo, J. 1987. Theory and experiment in the study of technological change. *Current Anthropology* 28: 595–622.

———.1997. The explanation of artifact variability. *American Antiquity* 62: 27–50.

Schlanger, S. H. 1992. Recognizing persistent places in Anasazi settlement systems, in J. Rossignol and L. Wandsnider (eds.), *Space, Time, and Archaeological Landscapes*, pp. 91–112. New York: Plenum Press.

Shaw, I., and Jameson, R. (eds.). 1999. *A Dictionary of Archaeology*. Oxford: Blackwell Publishers.

Shott, M. J. 1998. Status and role of formation theory in contemporary archaeological practice. *Journal of Archaeological Research* 64: 299–329.

———. 2006. *Formation Theory in Archaeology: Readings from American antiquity and Latin American Antiquity*. Washington, DC: SAA Press.

Stafford, C. R. 1994. Structural changes in Archaic landscape use in the dissected uplands of Southwestern Indiana. *American Antiquity* 59: 219–37.

———. 1995. Geoarchaeological perspectives on paleolandscapes and regional subsurface archaeology. *Journal of Archaeological Method and Theory* 2: 69–104.

Stafford, C. R., and Hajic, E. R. 1992. Landscape scale: Geoenvironmental approaches to prehistoric settlement strategies, in J. Rossignol and L. Wandsnider (eds.), *Space, Time, and Archaeological Landscapes*, pp. 137–61. New York: Plenum Press.

Strogatz, S. H. 2001. Exploring complex networks. *Nature* 410: 268–76.

Thomas, J. 2001. Archaeologies of Place and Landscape, in I. Hodder (ed.), *Archaeological Theory Today*, pp. 165–86. Malden: Polity Press.

Tilley, C. 1994. *A Phenomenology of Landscape: Places, Paths, and Monuments*. Oxford: Berg.

Turcotte, D. L., and Rundle, J. B. 2002. Self-organized complexity in the physical, biological, and social sciences. *Proceedings of the National Academy of the Sciences* 99: 2463–65.

Turner, M. G., and Gardner, R. H. 1991. Quantitative methods in landscape ecology: An introduction, in M. G. Turner and R. H. Gardner (eds.), *Quantitative Methods in Landscape Ecology*, pp. 3–14. New York: Springer-Verlag.

Turner, M. G., Gardner, R. H., and O'Neill, R. V. 2001. *Landscape Ecology in Theory and Practice: Pattern and Process*. New York: Springer-Verlag.

Walker, W. 1999. Ritual, life histories, and the afterlives of people and things. *Journal of the Southwest* 41: 383–405.

Walker, W., Lucero, L., Dobres, M. A., and Robb, J. 2000. The depositional history of ritual technology and power. *Agency in Archaeology*, pp. 130–47. London: Routledge.

Waters, M. R. 1988. Implications of the alluvial record of the Santa Cruz River to the discovery of Paleoindian and early Archaic sites in the Tucson basin, Arizona. *Current Research in the Pleistocene* 5: 98–99.

———. 1992. *Principles of Geoarchaeology: A North American perspective*. Tucson: The University of Arizona Press.

Watts, D. J. 1999. Networks, dynamics and the small-world phenomenon. *American Journal of Sociology* 105: 493–527.

Watts, D. J., and Strogatz, S. H. 1998. Collective dynamics of small world networks. *Nature* 393: 440–42.

West, G. B., Brown, J. H., and Enquist, B. J. 1999. The fourth dimension of life: Fractal geometry and allometric scaling of organisms. *Science* 284: 1677–79.

Whittlesey, S. M. 1997. Archaeological landscapes: A methodological and theoretical discussion, in S. M. Whittlesey, R. Ciolek-Torello, and J. H. Altschul (eds.), *Vanishing River: Landscapes and Lives of the Lower Verde Valley*, pp. 17–28. Tucson, AZ: SRI Press.

Wiens, J. A. 1989. Spatial scaling in ecology. *Functional Ecology* 3: 385–97.

Wilkinson, T. J. 2003. *Archaeological Landscapes of the Near East*. Tucson: University of Arizona Press.

Wood, W. R., and Johnson, D. L. 1978. A survey of disturbance processes in archaeological site formation, in M. B. Schiffer, *Advances in Archaeological Method and Theory*, pp. 315–81. New York: Academic Press.

Wu, J. 2004. Effects of changing scale on landscape pattern analysis: Scaling relations. *Landscape Ecology* 19: 125–38.

Wu, J., and Qi, Y. 2000. Dealing with scale in landscape analysis: An overview. *Geographic Information Sciences* 6(1): 1–5.

Zedeño, M. N. 2000. On what people make of places: A behavioral cartography, in M. B. Schiffer (ed.), *Social Theory in Archaeology*, pp. 97–111. Foundations of Archaeological Inquiry. Salt Lake City: University of Utah Press.

Zedeño, M. N., and Stoffle, R. W. 2003. Tracking the role of pathways in the evolution of a human landscape: The St. Croix riverway in ethnohistorical perspective, in M. Rockman and J. Steele (eds.), *Colonization of Unfamiliar Landscapes: The Archaeology of Adaptation*, pp. 59–80. London: Routledge.

59

COUNTER-MAPPING IN THE ARCHAEOLOGICAL LANDSCAPE

Denis Byrne

Maps ostensibly depict geographical reality, but we know that they also have a hand in *creating* that reality. Maps and mapping were instrumental, for instance, in configuring the exotic landscapes of the West's colonial dominions according to a Western frame of knowledge that enabled them to be understood, administered, and exploited (Carter 1987; Pratt 1992). The mapping of archaeological landscapes is not exempt from the influence of this history. Although it is true that great strides have been taken in the decolonization of archaeological practice over the last few decades (e.g., Hall 2005; Lilley & Williams 2005; McNiven and Russell 2005; Meskell 2005; Smith 2005), it would still be rash to assume that our mapping practices are value-free or ideologically neutral.

My primary concern in this chapter is the contention that we still pay remarkably little attention to the associations that contemporary people have with the archaeological sites that are an integral component of the landscapes in which they live. There are many reasons for this neglect, but my focus is on ways in which our mapping practices both collude in creating the illusion of "unsocialized" archaeological landscapes (that is, landscapes where archaeological traces are not enmeshed in contemporary social practice) and

help to operationalize heritage management strategies that ignore social context.

The line of thought in this chapter was triggered by the reading of some of the recent literature by anthropologists working with Indigenous "tribal minorities" in Southeast Asia. In the course of their work, these scholars had become engaged in their subjects' struggle to preserve the integrity of their forest habitats against encroachment by logging, mining, and even natural conservation interests. Through their engagement, the anthropologists came to appreciate the importance of being able to produce maps of the culturally inscribed landscapes of the forest dwelling groups as a counter, or antidote, to those existing maps that ignored or misrepresented these landscapes. The term *counter-mapping* has been adopted for what might be described as this tactical deployment of cultural mapping (Brosius et al. 2005; Cooke 2003; Peluso 1995). Anyone who has operated in the fields of cultural heritage or natural resource management knows that interest groups do need maps that describe their interests. To arrive mapless at the decision-making tables at which all the other key players—the loggers and miners, for instance—have unfurled their own maps of your homeland is virtually to invite your own marginalization.

Distancing Discourses

"How do people become aware that they are strangers in their own lands? Sometimes they are forcibly removed. Sometimes they are just reclassified." These words by anthropologist Anna Tsing (1993: 154) are a starting point for thinking about the way that archaeological mapping can make strangers out of the people who live in and around what we refer to as "archaeological landscapes." Tsing's comment was made with reference to the groups of shifting agriculturalists who occupy many of Southeast Asia's tropical forests. On colonial era maps, these forests were typically classified as "wasteland" or "barren" land and, as such, were appropriated by the state as unoccupied natural resource zones (Roseman 2003; Sowerwine 2004; Tsing 2003). The field systems and fixed villages of the lowland agricultural areas, by contrast, typically did find their way onto colonial maps, and the people of these areas were recognized as holding traditional title to their fields and plantations. On some early maps, the "blank" areas of forest surrounding zones of settled agriculture are reminiscent of those vacant spaces beyond the edge of the sea that are shown on medieval European maps (Roseman 2003: 114).

The condition of being "off the map" was one in which the forest people, from an official-cartographic point of view, were left floating in a noncartographic space with little or no recognition of their belonging to or in their own country. To be made the subject of colonial mapping may have been onerous for minority groups (Anderson 1991); to be left unmapped could be positively dangerous. The postcolonial governments of Malaysia, Indonesia, and Vietnam, rather than rejecting these maps, further elaborated them. The classification of tribal habitats as primary forest simplified the process of treating them as state resources that could be allocated as logging concessions and mining leases to national and international companies without reference to the people who inhabited them. Meanwhile, the "cultural villages" set up by the region's postcolonial states as tourist attractions purport to offer visitors an authentic experience of ethnic minority cultures in theme-park miniaturized landscapes (Yea 2002). The culture of minority groups is thus celebrated off-site at the same time that it is being threatened in its home terrain. Rather than mapping the cultural landscapes of these peoples *in situ* in their forest habitats the state chooses to map them elsewhere in microcosm.

If it is understandable that postcolonial governments would adopt colonialism's mapping practices as they consolidated the "geo-body" (Thongchai 1994) of the new national states, it is perhaps more surprising to find these practices so often adopted uncritically by nature conservationists. It has been observed that nature conservation, originating in the 19th-century Western world as a reaction to the devastating impact of industrialism and extractive capitalism on the environment, tended to "ghetto-ize nature in enclaves of bio-authenticity" (Campbell 2005: 283). This is to say that, turning from the blighted landscape of industrialization, nascent conservationism sought its opposite in an idealized pure nature, a nature perceived to be uninscribed by culture and situated in enclaves remote from the urban setting. The "distant-nature conservationist mindset" (Campbell 2005: 285) remains deeply embedded. In the 20th century, the discourse of international conservation, when it directed its attention to tropical zone forests, readily fell in with existing land classification systems in places such as Indonesia that classified forests as existing in a "natural/primary" state or as having been in this state before being "vandalized" by shifting agriculturalists. The colonial mapping regimes thus found new subscribers among those who longed for unacculturated natural landscapes that were available to be saved. In this regard, as Ben Campbell (2005: 301) asserts in a recent provocative paper, uncritical conservationism can be seen as "belong[ing] to a colonial genealogy of perceiving foreign lands as *terra nullius*."

It is only comparatively recently that anthropologists living with and studying forest people such as the Meratus Dayaks (Tsing 1993) and the Bagak Salako (Peluso 1992, 2003) of Kalimantan and the Temiars of Peninsula Malaysia (Roseman 2003) have begun to produce maps that show the real subtlety and complexity of the cultural landscapes these people construct and inhabit in areas that previously were mapped as "barren." These counter-maps show forests where former swiddens are regarded as storied historical sites, where individual trees are often known by their own names, and where ridges and gullies are intricately inscribed with the territories of spirits and deities. The Western construction of nature does not exist for people such as these; their forest habitat—their "nature," in our terms—is permeated with tangible and intangible traces of their own presence and their own history. Nature is never distant here; it is close, unreified, and inside.

It need hardly be said that archaeological traces present in any landscape are also part of this total experience. Although nationalist heritage discourse would have it that the archaeological record is handed down to us as part of our patrimony, in reality it is always-already there as a dimension of the habitat we learn to live in. Archaeological traces are continuous with a whole universe of marks in the landscape that we learn to read in the course of our daily lives and in the context of our particular cultural worldview (e.g., Bradley 2002). Archaeologists, however, have been inclined to regard the archaeological record for any one period of the past as part of a landscape that belongs *in* that period and *to* that period. We have been inclined to think of it as belonging to the society that produced it. This ignores plentiful evidence that people in the present narrate these traces into their lives through myth or song, that they weave them into their own accounts of who they are, or that they apprehend them as being animated with the presence of spirits or deities. Although the archaeologist is often interested simply in origins, local peoples tend to absorb "archaeological" traces into the lived reality of contemporary lives that are lived in a past-present-future continuum. Distant-nature might thus be said to have a counterpart in a discourse of "distant-traces." If the former denotes a pure nature that is always "out there" (Campbell 2005: 289) rather than meshed into the contemporary lived environment, then the latter denotes an archaeological record that is always "back there," rather than integrated by contemporary culture.

Campbell juxtaposes the conservationist's "distant-nature" mindset with Tim Ingold's (1992) "dwelling perspective," which posits a physical environment that is never conceptually separable from our daily lives. Ingold argues that rather than being born into a physical environment that they gradually adapt to using inherited knowledge and skills, people become culturally human by developing knowledge and skills in the course of everyday activities that occur in an environment that is conceptually and practically inseparable from culture. He sees human skills "not as transmitted from generation to generation but [as] . . . regrown in each" (Ingold 2000: 139). Under the conventional "genealogical" model, "the land itself can be no more than a kind of stage" on which culture is enacted (Ingold 2000: 139). In the dwelling perspective, by contrast, the land/environment is part of the same integrated continuum as culture. Both are part of the total experience of dwelling in the world.

Similarly, I suggest we should think of archaeological sites and objects as being an integrated part of the world people dwell in; that we should consider them inseparable from the dwelling experience. These archaeological traces unquestionably derive from past behavior and events but they are continually recycled back into the contemporary social environment. Places and objects may thus derive from the past, but they are "regrown" in the lived environment of each new generation.

Counter-Mapping and Disenchantment

Postcolonial governments, like their colonial predecessors, make decisions based on land classifications that privilege that which can be *seen* (Peluso 1995). The spiritual landscape of the Indigenous peoples encountered by the colonial West was in the category of the *unseen*. The reason it was unseen by Westerners lay partly in the enduring effect in Western culture of the 16th-century Protestant Reformation. Since the Reformation, the West has taken a relatively narrow view of the ways in which Christianity's God is manifest (Byrne et al. 1996). Whereas in Medieval Catholicism God was a living presence in the landscape, manifest in the miraculous efficacy flowing from saintly people, sacred relics, and sacred places (Bender 1993: 253–55; Geary 1986; Meskell 2004: 43–44), in the Protestant view, particularly the Calvinist view, religion was to be a matter between one's soul and a God who dwelt in heaven (Eire 1986: 312). Europeans, particularly northern Europeans, no longer looked for or believed in a divine presence in the landscape. In repudiating belief in the presence of magical and sacramental forces in the landscape, the Reformation effectively removed God from nature as an active, causal force, opening the natural world up to understanding through learned inquiry via discourses such as natural history and archaeology. It is in this sense that one speaks of the "disenchantment" (Weber 1946: 155) of the post-Reformation world. And it is in this state of disenchantment that the West set out to make sense of and colonize the Other World (Byrne Forthcoming).

A key element of counter-mapping is the necessity to understand what our existing maps exclude, and why. In elaborating on the effacement of the divine from our mapping of archaeological landscapes I turn now to New South Wales, Australia, where a clear distinction has tended to be drawn by archaeologists and heritage practitioners between Aboriginal archaeological sites and Aboriginal sacred sites. The former have been characterized as purely secular entities that are of scientific value, whereas the latter are acknowledged as having a spiritual context within an Aboriginal religious system termed the Dreaming (Creamer 1988). Both classifications are problematic. The secular category precludes contemporary spiritual associations with Aboriginal people, whereas the sacred category is framed by an essentialist understanding of traditional Aboriginal culture in terms of which "real" Aboriginal culture ceased being practiced in New South Wales during the 19th century. It is conceded that "sacred sites," many of which are natural landscape features believed to have been created in the course of the travels of world-creating ancestral beings, can still be of spiritual significance to contemporary Aboriginal people, but it is not acknowledged that contemporary Aboriginal culture has the capacity to produce authentic sacred sites in the present. Both the "archaeological" and the "sacred" classifications thus contribute to the idea that contemporary Aboriginal environments are disenchanted.

Yet, although no specific study has been made of contemporary Aboriginal place-based spirituality in NSW, it is not difficult to find references that are suggestive of recent or contemporary Aboriginal associations with archaeological sites that are spiritual in nature. Some of these relate to beliefs that archaeological sites and objects are imbued with supernatural power. On the New England tablelands in recent times, for instance, there have been Aboriginal people who believe that old ceremonial raised-earth circles (*bora* rings) can cause your feet to swell up if you inadvertently walk on them (Cohen and Somerville 1990: 58). Among contemporary Muruwari Aboriginal people in northwestern NSW, Harrison (2004: 199) notes that "ancestors'" spirits are associated with the objects that they used during their lifetimes" (see also McNiven and Russell 2005: 193–94). When visiting archaeological sites, Muruwari people sometimes rub flaked stone artifacts on their skin. As a Muruwari woman, Vera Dixon, explains: "when you're rubbing the stones over your skin you can get the feeling of . . . you sort of get the feeling of the spirits coming into your skin somehow or another" (Harrison 2004: 199).

Perhaps the most visible expression of such beliefs is seen in the insistence by Aboriginal people that the skeletal remains of their ancestors—remains that are often thousands of years old—be treated not merely as *people* rather than specimens but as spiritually animated subjects (Byrne 2004). The anxiety expressed by the contemporary Aboriginal people of NSW about ancient burials exposed in the course of erosion or earthworks is likely to relate to historic-period beliefs that inappropriate contact with human skeletal remains is a cause of illness. This is illustrated by a case in which Aboriginal people, moved from their traditional country by the government to a camp at Lake Menindee in western NSW in 1933, attributed a number of subsequent

deaths there to exposure to dust blowing off an adjacent eroding ridge where burials were exposed (Kennedy and Donaldson 1982: 17).

Probably the most detailed and thoughtful study of contemporary Aboriginal relationships with an historic period archaeological site in NSW is Rodney Harrison's (2004, 2006) account of the association that Muruwari Aboriginal people have with a former campsite (1870s–1940s) at Dennawan on the Culgoa River in the northwest of the state. During their periodic visits to the abandoned settlement, Muruwari people sometimes have visions of spirits in human form, and they believe that removal of artifacts from the site results in illness (Harrison 2004: 201). Artifacts found at sites like Dennawan are described by Harrison (2004: 205) as "becom[ing] extensions of long-dead relatives: powerful, troubling and at times problematic." Seen in this light, traditional country is an extension of known/named deceased kinfolk, as well as more distant ancestors. It is not, then, that people attribute an other-worldly agency or efficacy to the sort of objects and places described here; the force of these objects and places comes from their being very much of *this* world, the world of living people. This seems very much the sort of "dialogic relationship" that Lynn Meskell (2004: 77) describes for people-object relationships in ancient Egypt. She focuses on the precise social context in which specific objects are mobilized (2004: 54) rather than simply characterizing them as powerful or efficacious. For the most part, such contexts have gone unexamined by archaeologists working in NSW.

In the literature and practice of heritage conservation, the almost complete lack of acknowledgment of Aboriginal spiritual associations with archaeological sites and objects stems, as I have suggested, from a rigidly bipolar secular/sacred classification of Aboriginal heritage. As an instance of exclusion-through-classification, this omission resonates with the land classification and mapping practices in Southeast Asia mentioned earlier. In both cases, classification and mapping have rendered contemporary culture invisible in the landscape. In the Southeast Asian case, the classification of rainforest habitats as primary forest wilderness becomes a charter for the resettlement of shifting agriculturalists out of their forest homes (McElwee 2001; McWilliam 2003; Sowerwine 2004). In the case of NSW, Aboriginal people were "resettled" from almost the entire landscape within a century or so of British settlement in 1788. There is a real sense in which, as living Aboriginal people were displaced from the landscape, they were replaced by the archaeological traces of their own past

(Byrne 1998, 2003: 77–78). This occurred by way of the following transposition. As settler interest in Aboriginal culture in NSW declined through the 19th and 20th centuries, in response to a perception that Aboriginal people had lost their traditional culture, interest in Aboriginal archaeological sites correspondingly increased until, toward the end of this period, they came to be incorporated as part of the national heritage. But the sites were perceived by white society as representing past Aboriginal society rather than the society of the Aboriginal people now living on small government reserves and in "fringe camps" on the outskirts of settler towns. By the 1960s, the conceptual disconnection was so embedded that any claim for continuity between these sites and living Aboriginal people was likely to be met "with shock and disbelief" (Sullivan 1985: 144).

Mapping from the Inside

For Indigenous minorities in settler colonies, including the Aboriginal people of NSW, the experience of becoming "strangers in their own land" (Tsing 1993: 154) has been accompanied by the experience of seeing themselves replaced in the land(scape) by their own archaeology (Byrne 2003a). Aboriginal archaeological traces became familiar to settlers as one of the attributes of the land they now owned but living Aboriginal people became strangers in that land.

By 2006, over 50,000 Aboriginal archaeological sites had been recorded and inventoried in NSW. Some of these recordings date to the early 20th century, but most were triggered by the legislation enacted in 1969 to protect Aboriginal heritage. Further legislation in 1980 required that such sites be identified in the course of environmental impact assessments (EIA) for certain categories of land development. But we should note that this mapping of Aboriginal archaeological sites has occurred partly because it has been *possible* to map them. They have a physical presence in the landscape, and archaeologists have possessed the methodologies for mapping that presence.

Archaeologists have highly tuned visual skills but, as Nick Shepherd (2006: 1) observes in relation to colonial archaeology, these skills constitute a "particular optic" that has "involved *not* seeing . . . [a]s much as it involved seeing." What we have been inclined not to see includes the sort of associations, illustrated in the previous section, that living Aboriginal people have with archaeological remains. The physical remains have a transcultural "legibility" (Scott 1998) that associations lack. It is the task of making these associations legible that

a counter-mapping strategy in Aboriginal heritage might be expected to apply itself to.

An Aboriginal counter-mapping strategy might also attempt to depict the landscape according to an Aboriginal worldview. Conventionally, maps of Aboriginal heritage places in NSW have consisted simply of recorded archaeological sites plotted from heritage inventories onto topographic survey maps to form constellations of dots scattered across the terrain (this applies irrespective of whether the recorder has used site-based or "off-site" archaeological methodologies; see Thomas Richards, this volume). But it is not, of course, how contemporary Aboriginal people experience their heritage landscape. The archaeological maps provide the kind of flattened-out, bird's-eye view that the Western cartographic tradition has developed, a view that has little or nothing to do with the way that people (who, after all, do not dwell in the sky) experience their environment (Ingold 2000: 241). In everyday life, people apprehend their environment from within; they experience it in the course of moving *through* a three-dimensional reality rather than *across* a two-dimensional surface (Ingold 2000: 241). To pick up on an earlier point: an aspect of this three-dimensionality that is absent from conventional maps is the "depth" the landscape comes to possess by way of it being an extension of deceased kin and more remote ancestors. This extension is actualized by people in the present who are able to "see" or to apprehend past lives as they move through the landscape.

Such an understanding of human landscape perception highlights the key role of pathways as constituting the trajectories that we take through the world and along which our bodies and minds experience the world (Tilley 1994; see also Tilley, this volume). The experiential reality of our life in the environment, as Ingold (2000: 242) observes, is "laid down along paths of movement, of action and perception. Every living being, accordingly, grows and reaches out into the environment along the sum of its paths." Recent attempts in NSW to bring the mapping of Aboriginal cultural heritage closer to lived Aboriginal experience have included projects where archaeologists and historians have worked with Aboriginal communities on the north coast of the State to record their patterns of movement through local landscapes over the last century or so (Byrne and Nugent 2004; English 2002). In the course of these projects, local people drew the routes of historical pathways onto enlarged aerial photographs or walked the pathways with heritage professionals. The resulting pathway maps were contextualized within local culture and animated by the pathway narratives that were recorded during oral history interviews, but it remains true that ultimately the pathways were inscribed on conventional bird's-eye-view topographic maps. The challenge that lies ahead is for us to venture right off these conventional maps and attempt to depict local Aboriginal environments in ways that approximate the way they are perceived by those who dwell inside them (this is important also for researchers aiming to historicize Indigenous *pasts*—that is, the ways that Indigenous peoples perceived their worlds in the past). If people perceive and experience the environment by moving along pathways—Ingold and Tilley, among others, encourage us to think of these as "paths of view" (Ingold 2000: 239)—then it is these views that we should be striving to map.

One component of such a viewing experience comprises the archaeological sites people may see, think about, and talk about as they move through their landscape (see also Rainbird, this volume for a discussion of various sensual experiences of cultural places). For Indigenous minorities who have been dispossessed of most or all of their land, the heritage inventory map that shows the spread of archaeological sites over the terrain of their local region is, to an extent, quite unreal. It is unreal in that they will never be able to view or to visit more than a small fraction of those sites for the simple reason that they don't have access to the privately owned land on which they lie. For them, the archaeological landscape consists of those portions of public land (for example, road corridors, riverbank reserves, town commons, recreation reserves, national parks) that they do have access to. This is their viewable heritage, their visitable past. In addition to precolonial archaeological heritage sites, these public lands are also likely to be rich in archaeological traces of postcontact occupation. Living as they have for the last two centuries inside an alien cadastral grid, the pockets and strips of public land have provided a means by which Aboriginal people in NSW have been able to move around inside the colonized landscape. They have thus long been expert readers of the map of public lands, lands that can usefully be thought of as constituting gaps and openings in the cadastral grid of settler properties (Byrne 2003b; Goodall 2006).

It is not just the larger proportion of the archaeological landscape that is locked up inside the cadastral grid of private property, it is also the greater proportion of fishing places and other wild resource locales. The mapping of such places has a "counter" aspect to it in that it potentially unsettles the colonial mapping of resources that, like colonial mapping in Southeast Asia, classified the landscape in terms of its usefulness in the framework of the colonial, not the Indigenous economy (Byrne and Nugent 2004: 15–16). Given the modern nation state's constant appetite for more national heritage, it would seem that archaeological mapping in places like Australia, unless it documents the association of contemporary Aboriginal people with archaeological landscapes, is simply a continuation of the colonial project of taking possession of new lands. This is a dimension of mapping that all archaeologists should reflect on before developing survey strategies.

Conclusions: Mapping Value

As maps proliferate and become accessories to more of the activities of everyday life there is, in some quarters, a counter-current of nervousness about the power that maps wield (e.g., Monmonier 1991; Wood 1992). In the field of landscape archaeology, we have cause to feel uneasy about the potential for the increased representability of archaeological landscapes via maps to come at the expense of a diminution of attention to the meaning these landscapes have for the contemporary people who live in them. Those of us who have engaged in large-scale environmental assessments are sometimes, for instance, disturbed by the insistence of biological scientists on referring to archaeological heritage sites as "values." Our task, as they appear to see it, is to go into the field to map these "values" in a conceptually equivalent way to that by which biodiversity "values" are mapped. We find ourselves having to insist that the value archaeological materials have is not intrinsic to them but rather is ascribed to them by past and present people. Values in this sense, however, are perceived as being much harder to document and interpret than archaeological sites in their pure physicality, and there is often reason to worry that once the coordinators of environmental assessments have the archaeological field data in hand they will incorporate these data without the contextualizing social value data. As Meskell (Forthcoming) writes, "*things* trump people at every turn."

Fortunately, archaeologists are not alone in the struggle against such tendencies. The counter-mapping initiative in environmental anthropology, reviewed at the beginning of this chapter, is a source of support as also are the numerous projects in which Indigenous people have themselves taken charge of mapping their own landscape interests. Not that counter-mapping doesn't also have its downside. As Indigenous peoples are only too well aware, this includes the fact that an increasing volume of Indigenous knowledge is becoming public domain. Another side effect is the embedding of maps as privileged forms of spatial knowledge (Harris and Harrower 2006: 7) as distinct, for example, from storytelling. A mud-map or sand-map is erased by nature soon after being inscribed; it "belongs" to the map-maker in the sense that its materiality often lasts only for the duration of a performance. It belongs, in a sense, to the story, which in turn belongs to the teller. A digital or printed map, in contrast, can be reproduced at will and consumed without reference to the original knowledge-holder.

All these concerns, rather than being an argument against counter-mapping, take me back to my earlier point that mapping is a technology of power and that, for marginalized peoples, being mapped is as potentially harmful as being left unmapped. Their best option is to take the mapping process into their own hands. A similar point might be made about landscape archaeology: our mapping, indeed our whole practice, is implicated in relations of power. This is not a reason to cease; it is a reason to be watchful and careful.

References

Anderson, B. 1991 [1983]. *Imagined Communities*. London: Verso.

Bender, B. 2003. Stonehenge: Contested landscapes (medieval to present day), in B. Bender (ed.), *Landscape: Politics and Perspectives*, pp. 245–78. Oxford: Berg.

Bradley, R. 2002. *The Past in Prehistoric Societies*. London: Routledge.

Brosius, P. J., Tsing, A. L., and Zerner, C. (eds.). 2005. *Communities and Conservation*. Walnut Creek, CA: AltaMira Press.

Byrne, D. 1996. Deep nation: Australia's acquisition of an indigenous past. *Aboriginal History* 20: 82–107.

———. 2003a. The ethos of return: Erasure and reinstatement of Aboriginal visibility in the Australian historical landscape. *Historical Archaeology* 37(1): 73–86.

———. 2003b. Nervous landscapes: race and space in Australia. *Journal of Social Archaeology* 3: 169–93.

———. 2004. Archaeology in reverse, in Nick Merriman (ed.), *Public Archaeology*, pp. 240–54. London: Routledge.

———. Forthcoming. The fortress of rationality: Archaeology and Thai popular religion, in L. Meskell (ed.), *Cosmopolitan Archaeologies*. Durham, NC: Duke University Press.

Byrne, D., Goodall, H., Wearing, S., and Cadzow, A. 2006. Enchanted parklands. *Australian Geographer* 37(1): 103–15.

Byrne, D., and Nugent, M. 2004. *Mapping Attachment: A Spatial Approach to Aboriginal Post-Contact Heritage*. Sydney: Department of Environment and Conservation (NSW).

Campbell, B. 2005. Changing protection policies and ethnographies of environmental engagement. *Conservation and Society* 3(2): 280–322.

Carter, P. 1987. *The Road to Botany Bay*. London: Faber and Faber.

Cohen, P., and Somerville, M. 1990. *Ingelba and the Five Black Matriarchs*. Sydney: Allen and Unwin.

Cooke, F. M. 2003. Maps and counter-maps: Globalised imaginings and local realities of Sarawak's plantation agriculture. *Journal of Southeast Asian Studies* 34(2): 265–84.

Creamer, H. 1988. Aboriginality in New South Wales: Beyond the image of cultureless outcasts, in J. Beckett (ed.), *Past and Present*, pp. 45–62. Canberra: Australian Institute of Aboriginal Studies.

Eire, C. M. 1986. *War Against the Idols*. Cambridge: Cambridge University Press.

English, A. 2002. *The Sea and the Rock Give Us a Feed*. Sydney: NSW National Parks and Wildlife Service.

Geary, P. 1986. Sacred commodities: The circulation of medieval relics, in A. Appadurai (ed.), *The Social Life of Things*, pp. 169–91. Cambridge: Cambridge University Press.

Goodall, H. 2006. Indigenous peoples, colonialism, and memories of environmental injustice, in S. H. Hood, P. C. Rosier, and H. Goodall (eds.), *Echoes from the Poisoned Well: Global Memories of Environmental Injustice*, pp. 73–95. Oxford: Lexington.

Hall, M. 2005. Situational ethics and engaged practice: The case of archaeology in Africa, in L. M. Meskell and P. Pels (eds.), *Embedding Ethics: Shifting the Boundaries of the Anthropological Profession*, pp. 169–94. Oxford: Berg.

Harrison, R. 2004. *Shared Landscapes: Archaeologies of Attachment and the Pastoral Industry in New South Wales*. Department of Environment and Conservation (NSW) and University of NSW Press.

———. 2006. "It will always be set in your heart": Archaeology and community values at the former Dennawan Reserve, northwestern NSW, Australia, in N. Agnew and J. Bridgeland (eds.), *Of the Past, For the Future: Integrating Archaeology and Conservation*, pp. 94–101. Los Angeles: Getty Conservation Institute.

Ingold, I. 1992. Culture and the perception of the environment, in D. Parkin and E. Croll (eds.), *Bush Base: Forest Farm—Culture, Environment and Development*, pp. 39–50. London: Routledge.

———. 2000. *The Perception of the Environment: Essays in Livelihood, Dwelling and Skill*. London: Routledge.

Kennedy, E., and Donaldson, T. 1982. Coming up out of the Nhaalya: Reminiscences of the life of Eliza Kennedy. *Aboriginal History* 6(1): 5–27.

Lilley, I., and Williams, M. 2005. Archaeological and indigenous significance: A view from Australia, in C. Mathers, T. Darvill, and B. Little (eds.), *Heritage of Value, Archaeology of Renown: Reshaping Archaeological Assessment and Significance*, pp. 227–47. Gainesville: University of Florida Press.

McElwee, P. 2001. Parks or people: Exploring alternative explanations for protected areas development in VietNam. *Workshop 2001: Conservation and Sustainable Development—Comparative Perspectives*, pp. 1–24, http://research.yale.edu/CCR/environment/EnvironmentWorkshop2001.htm

McNiven, I. J., and Russell, L. 2005. *Appropriated Pasts: Indigenous Peoples and the Colonial Culture of Archaeology*. Walnut Creek, CA: AltaMira Press.

McWilliam, A. 2003. New beginnings in East Timor forest management. *Journal of Southeast Asian Studies* 34(2): 307–27.

Meskell, L. 2004. *Object Worlds in Ancient Egypt*. Oxford: Berg.

———. 2005. Archaeological ethnography: Conversations around Kruger National Park. *Archaeologies: Journal of the World Archaeology Congress* 1: 83–102.

———. Forthcoming. Cosmopolitan heritage ethics, in L. Meskell (ed.), *Cosmopolitan Archaeologies*. Durham, NC: Duke University Press.

Monmonier, M. 1991. *How to Lie with Maps*. Chicago: University of Chicago Press.

Peluso, N. L. 1992. *Rich Forests, Poor People: Resource Control and Resistance in Java*. Berkeley and Los Angeles: University of California Press.

———. 1995. Whose woods are these? Counter-mapping forest territories in Kalimantan, Indonesia. *Antipode* 27(4): 383–406.

———. 2003. Fruit trees and family trees in an anthropogenic forest: Forest zones, resource access, and environmental change in Indonesia, in Charles Zerner (ed.), *Culture and the Question of Rights*, pp. 184–218. Durham, NC: Duke University Press.

Pfeiffer, E. W. 1984. The conservation of nature in Vietnam. *Environmental Conservation* 11(3): 217–21.

Pratt, M. L. 1992. *Imperial Eyes: Travel Writing and Transculturation*. London: Routledge.

Roseman, M. 2003. Singers of the landscape: Song, history, and property rights in the Malaysian rainforest, in Charles Zerner (ed.), *Culture and the Question of Rights*, pp. 111–41. Durham, NC: Duke University Press.

Rugendyke, B., and Son, N. T. 2005. Conservation costs: Nature-based tourism as development at Cuc Phuong National Park, Vietnam. *Asia Pacific Viewpoint* 46(2): 185–200.

Scott, J. C. 1998. *Seeing Like a State: How Certain Schemes to Improve the Human Condition Have Failed*. New Haven, CT: Yale University Press.

Shepherd, N. 2006. Local practise, global discipline. *Archaeologies* 2(1): 1–2.

Sowerwine, J. 2004. Territorialisation and the politics of highland landscapes in Vietnam: Negotiating property relations in policy, meaning and practice. *Conservation and Society* 2(1): 97–136.

Smith, C. 2005. *Indigenous Archaeologies: Decolonising Theory and Practice*. London: Routledge.

Sullivan, S. 1985. The Custodianship of Aboriginal sites in Southeastern Australia, in I. McBryde (ed.), *Who Owns the Past?*, pp. 139–56. Melbourne: Oxford University Press.

Tilley, C. 1994. *A Phenomenology of Landscape*. Oxford: Berg.

Thongchai, W. 1994. *Siam Mapped: A History of the Geo-Body of a Nation*. Honolulu: University of Hawai'i Press.

Tsing, A. L. 1993. *In the Realm of the Diamond Queen*. Princeton, NJ: Princeton University Press.

Weber, M. 1946. Science as a Vocation, in H. H. Gerth and C. W. Mills (ed.), *From Max Weber: Essays in Sociology*, pp. 129–57. New York: Oxford University Press.

Wood, D. 1992. *The Power of Maps*. New York: The Guildford Press.

Yea, S. 2002. Sarawak on stage: The Sarawak Cultural Village and the colonization of cultural space in the making of state identity, in B. David and M. Wilson (eds.), *Inscribed Landscapes: Marking and Making Place*, pp. 240–52. Honolulu: University of Hawai'i Press.

NONLEVEL PLAYING FIELDS: DIVERSITIES, INEQUALITIES, AND POWER RELATIONS IN LANDSCAPE ARCHAEOLOGY

Doing archaeology implies the right, the power, and the ability to investigate the past, and in many parts of the world, this involves *other peoples'* pasts. As an academic discipline, archaeology has its own performative rules and regulations. The outcome is a disciplined, and more or less elitist (academic) approach to how our own and other peoples' histories come to be understood. Simply put, knowledge of the past tends to be sanitized by what scientific procedures permit, at the expense of other ways of understanding history (for example, religious faiths, oral traditions). History makes us who we are, and archaeologists play a privileged role in creating a sense of history. People thus come to know themselves at least in part by how others (in this context, archaeologists) construct *their* histories. The challenge is, as Ian McNiven (1998: 47) aptly puts it, that writing the past "is more than a clash of belief systems—it is a clash of powers to control constructions of identity." This clash of powers expresses itself in varied aspects of social life, including education (for example, the cultural, social, political, and economic circumstances that limit or encourage people to take up archaeology at university), cultural choice (for instance, the way we engage with our own cultural preunderstandings), employment (for example,

the social forces that lead one person to obtain a historicising job over another), and representation (for instance, in Western society, the social prioritizing of professional archaeological versions of the past and the marginalization of Indigenous notions of history).

The chapters in this section explore varied dimensions of this clash, negotiation, and accommodation of powers as they concern the ways we come to construct meaningful notions of past lived landscapes. They discuss key aspects of landscape archaeology, from ethical questions of how "others" are represented in discourses of past and ancestral archaeological landscapes, to the positioning of the self in historical space, to imaginative renderings of the historical and always—at least in part—mysterious world in which we all live.

Reference

McNiven, I. J. 1998. Shipwreck saga as archaeological text: Reconstructing Fraser Island's Aboriginal past, in I. J. McNiven, L. Russell, and K. Schaffer (eds.), *Constructions of Colonialism: Perspectives on Eliza Fraser's Shipwreck*, pp. 37–50. London: Leicester University Press.

60

LANDSCAPES OF POWER, INSTITUTION, AND INCARCERATION

Eleanor Conlin Casella

How do people experience confinement? With scholars, reformists, philanthropists, social engineers, clinicians, and politicians writing about incarceration since the late 18th century, a vast interdisciplinary literature exists on the institutional landscape. While historians and architects have examined how early communal forms of social welfare and punishment transformed into the stark penitentiaries and fortified compounds of the 20th century (Evans 1982; Ignatieff 1978), criminologists, legal theorists, and philosophers have debated the relative civic effects of imprisonment as a mode of punishment, deterrence, and retribution (Garland 1990; Howe 1994).

Others from sociology, anthropology, and culture studies have considered the lived experience of institutionalization by exploring the psychological impact of the custodial environment on inmates (Clemmer 1940; Goffman 1961), staff (Liebling and Price 2001), dependent children and families (Owen 1998), and even the researchers themselves (Fleisher 1989). Finally, archaeological perspectives have illuminated the material and spatial conditions of the modern institution. This work has revealed a profound dissonance between ideal designed landscapes of disciplinary intention and embodied landscapes of insubordination and compromise. Ultimately, places of confinement are fabricated through the interplay of three distinct modes of social power: domination, resistance, and negotiation.

Disciplinary Space

The years between 1770 and 1850 witnessed a rapid emergence of institutional confinement as a uniquely modern form of social management (Casella 2007; Foucault 1977; Markus 1993). The movement began with John Howard, an English county sheriff who conducted inspection tours of existing jails and debtor's houses across England, Wales, and Ireland. His influential 1792 report *The State of the Prisons* offered a meticulous account of the scandalous conditions behind the perimeter walls of Britain's prisons: subterranean dungeons contaminated with human filth; male and female prisoners freely associating in a state of perpetual drunkenness; desperate paupers starving in chains, unable to earn the bribes required by corrupt jailers. Governed primarily by local customs and medieval laws, the vast majority of traditional civic punishments assumed a corporeal form—involving periods of public humiliation administered through the stocks or pillory, or sanguinary retribution such as flogging, branding, and, increasingly over the 18th century, public hanging.

Howard's relentless exposure of these penal horrors to Parliamentary Committees eventually

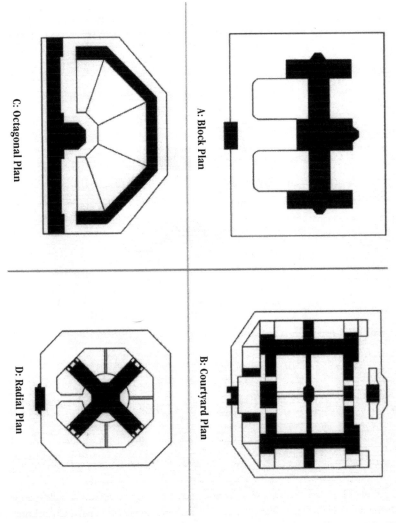

Figure 60.1   English "reformed" penal designs. (a) Exeter County Gaol, Devon (1790–1794); (b) Shrewsbury County Gaol, Shropshire (1787–1793); (c) Chester Castle County Gaol, Cheshire (1785);(d) Ipswich County Gaol, Suffolk (1786–1790) (redrawn from Brodie et al. 2002).

generated a new "reformed" penitentiary architecture. Working in close collaboration with Howard, the English architect William Blackburn perfected four "reformed" designs intended not only to improve the ventilation and sanitation of prisons but also to introduce a strict regime of spatial order, classification, and segregation on all inmates (Figure 60.1). A decade later, the early industrialist and utilitarian philosopher Jeremy Bentham published his radical designs for the Panopticon—a cylindrical model devised to emphasize a disciplinary self-reform of the prisoner's soul over corporal punishment of his flesh. Based on new technologies of surveillance fabricated through the spatial medium of architecture, the Panopticon subjected the male inmate to constant (yet unverifiable) judgmental observation. Encased within a ring of cells around a central observational hub, prisoners were exposed to "a state of conscious and permanent visibility that assures the automatic functioning of power" (Foucault 1977: 201). Further, Bentham's penitentiary introduced the solitary cell as a primary mechanism for both isolating inmates from contaminating associations and encouraging rehabilitative moral self-reflection. By the 1790s, Bentham's fearful design offered a rational, humane, and yet entirely brutal machine for "grinding rogues honest" (Evans 1982: 198).

When Bentham's principles of surveillance and isolation were merged with Blackburn's radial plan, a dreadful carceral landscape was born. Consisting of a series of cellblock wings arranged around a centralized custodial hub, penitentiaries of the early 19th century were open internally from ground floor to skylight roof, thereby providing unhindered visual and auditory surveillance over all inhabitants. As guards perambulated the cast-iron balconies of these silent wards, their footfalls muted by the soft leather soles of their specially designed boots, all stray noises were amplified along the long empty corridors. Spy holes were installed into each cell door. Covered by a hinged metal flap on the external side, the mechanism exposed the cell interior to routine inspection while limiting views of the adjoining corridor. Walls and grated windows circumscribed all sensory experiences of the external world. A perpetual disciplinary regime choreographed all movement throughout the institution, with segmented stalls and enclosed exercise yards maintaining inmate solitude even during daily periods of recreation and chapel attendance.

Textures remained similarly prescribed. To both humiliate and discipline the male inmate, expressions of self-identity were restricted through the provision of an identity number and institutional uniform of coarse wool and cotton. Sparsely furnished with an identical set of artifacts, prisoner cells each contained a tin cup, bowl and spoon, an iron or wooden cot, a wool blanket, a white earthenware chamber pot, a broom, a Bible, and a framed list of institutional rules and regulations.

Over the 1820s, as a "carceral enthusiasm" swept the young American Republic (Hirsch 1992), two distinct and competing models of penal organization achieved international acclaim. The "Separate System" of the Eastern State Penitentiary at Cherry Hill, Pennsylvania (1829), assigned inmates to solitary labor at leather boot manufacture within their isolated cells. Conversely, the "Congregate System" of New York's Auburn State Penitentiary (1823) collected inmates into communal workshops for silent assembly-line work. Two decades later, Imperial Britain established its own infamous "Separate System" penitentiaries for men at Pentonville, England (1842), Port Arthur, Tasmania (1847), and Mountjoy, Ireland (1850). Thus, by the 1850s, the institution had emerged as a rehabituative landscape, one designed to forge a progressive and internalized transformation of the male criminal.

Britain retained its Victorian era prisons throughout the 20th century (Brodie et al. 2002). Reflecting the gradual modernization of living standards and social rationale behind "imprisonment," penal facilities have been periodically updated with new security features (reinforced skylight and window glazing, CCTV cameras, high-tension wire mesh between floors) and social amenities (expanded visitation rooms, gymnasiums, multifaith chapels). Nonetheless, as the prison population reached crisis levels over the 1990s, incarceration all too frequently transformed into a daily routine of 23 hours of lock-down within a dangerously overcrowded cell.

In the United States, as state authority became increasingly centralized over the early 20th century, modern technologies of imprisonment continued to perfect the construction of disciplinary space (Friedman 1993). Established under the Department of Justice in 1891, the federal prison system developed a particularly severe form of penitentiary architecture. These forbidding monuments consisted of two separate structures: a three-to-five-storey block of adjoining rows of individual cells, all encased within a massive stone, steel, and concrete façade (Figure 60.2). A landscape of complete surveillance, iron bars (and later, clear reinforced plastic) replaced solid cell doors, and free-standing watch towers guarded the fortified perimeter boundaries. In a stark departure from the optimistic rehabilitative philosophies of the 19th century, these "total institutions" (Goffman 1961) were designed to enforce imprisonment as a painful form of civic retribution. Currently in operation, Leavenworth Penitentiary, Kansas (1895), continues to serve as the largest American maximum-security facility, with approximately 2,000 inmates incarcerated.

Resistance and Insubordination

Despite the disciplinary weight of this carceral landscape, not all inhabitants yield to institutional conditions. Since power exists as both forces of compliance and forces of action, resistance is born at the same moment as domination (Foucault 1977). Further, the shared experience of incarceration frequently cultivates a unique social cohesion among inmates, with various studies revealing a distinct "society of captives" within the penal environment (Clemmer 1940; Fleisher 1989; Giallombardo 1966; Owen 1998; Sykes 1958). Through these alternative social worlds, inmates actively challenge the penal order by materially deploying acts of both individual and collective resistance.

Although recalcitrance does take the extreme form of riots and open rebellions, typical expressions are carefully designed to thwart, rather than to conquer, systems of domination (Scott 1990).

Figure 60.2   Cell Block "A," Alcatraz Island Federal Penitentiary, California (Library of Congress, HABS CAL, 38-ALCA, 1-A-17).

Providing means for a gradual erosion of authority, resistance operates as a loose constellation of daily activities undertaken by inmates for "working the system to their minimum disadvantage" (Hobsbawm 1973: 7). As a result, insubordination tends to address the worst "pains" of imprisonment: deprivation of liberty and freedom of movement, deprivation of goods and services, deprivation of personal identity, deprivation of autonomy, and deprivation of personal security (Sykes 1958).

Archaeological studies have observed that institutional zones related to "unfree labor" (Nicholas 1988) frequently provide a focal site for inmate subversion (Casella 2007). Originally established in 1838, the first Rhode Island State Prison adopted the "Congregate System," with the 1845 addition of a communal industrial workshop to its fortified compound. Through archival research, James Garman linked the failure of an ambitious scheme for the prison manufacture of decorative ladies' fans to intentional inefficiencies, or "foot-dragging strategies," adopted by inmate workers along the assembly line (Garman 2005: 146). Additionally, his work mapped collective patterns of resistance across excavated architectural features by

locating "intra-institutional" offenses from 1872 through 1877 according to specific activity zone. Results demonstrated a clear focus of recalcitrance. Ranging from challenges to the code of silence and refusing to work, to outright destruction of prison property, approximately 60% of the infractions occurred within the penitentiary workshops—that exact institutional space, in other words, specifically designated for inmate congregation and communal labor.

Of course, the most admired form of spatial resistance entails a total rejection of the penal landscape. Material evidence of escape attempts can be found throughout places of confinement. At Johnson's Island (1862–1865), an American Civil War prison camp for captured Confederate Army officers, archaeological excavation of the latrine features revealed numerous escape tunnels dug into the rear of privy vaults, particularly those nearest the stockade's western perimeter wall. Probable escape tools were additionally recovered in association with some latrine tunnels, these objects including a large iron bar, a table knife, and the worn distal end of a bovine long bone (Bush 2000: 71).

A similar escape attempt was recovered from Kilmainham Gaol, Dublin. With the incarceration of anti-Treaty and Irish Republican Army (IRA) activists during Ireland's Civil War (1922–1923), female political prisoners were confined within the recently decommissioned prison on the western edge of Dublin (McCoole 2004). By March 1923, "B" Wing inmates developed plans for an escape tunnel. After establishing a roster, and disguising their digging activities with noisy handball games in the adjoining exercise yard, the women commenced excavation with spoons stolen from the prison kitchen (McCoole 1997: 50). When a matron discovered their plot one month later, the inmates had created a hole 1.2 m—an "archaeological" feature still preserved within the Kilmainham Gaol museum. To pass on the benefits of their stymied efforts to future prisoners, inmate and dedicated nationalist Sighle Humphreys inscribed the plaster at the base of her cell wall with a penciled message (Figure 60.3):

> Tunnel begun
> in basement laundry
> inside door on left
> may be of use to successors
> good luck, S.

Requiring a substantial degree of organization and subterfuge on the part of inmates, these escape attempts materially represented a fermentation of collective resistance and inmate solidarity, as communicated through the dramatic rejection of the institutional landscape.

Negotiated Space

The experience of incarceration cannot be reduced to a simple oppositional struggle between "staff" and "inmates." Recognizing the limits of traditional binary models, an increasing number of scholars have interpreted power as a social relationship characterized by plural, varying, and circumstantial moments of opportunity. Offering the term *heterarchy* (Ehrenreich et al. 1995) to emphasize the lateral, nested, and transient structures of power, this theoretical approach supports an exploration of how the austere penal landscape itself becomes negotiated, modified, and compromised (Casella 2007).

Within the carceral setting, a primary arena of negotiation involves the architecture and basic layout of the institution. As extensively demonstrated by Michel Foucault (1977), disciplinary technologies function by standardizing institutional inhabitants—separating them into isolated, yet fully identical, units. Thus, elements of the built environment that deviate from the standard institutional template represent a form of compromise, an acknowledgment of diversity, and a limit to disciplinary power. The presence of "Secure Wards" within modern penitentiaries demonstrates one such architectural negotiation. Established for the "protective custody" of disenfranchised inmates

Figure 60.3 Sighle Humphreys, cell graffiti, August 2004. Kilmainham Gaol, Ireland (photograph courtesy Niamh O'Sullivan and the Office of Public Works, Dublin).

(such as former police or prison employees; disabled, elderly, young, or gay prisoners; recovering addicts; informers; and pedophiles), these "prisons within prisons" reveal the hierarchies, violence, tensions, and vulnerabilities that internally fracture inmate society (Toch 1977: 206–23; Wikberg and Rideau 1995).

Gender has also necessitated a compromise of the ideal disciplinary landscape. Since its origins in the 18th century, the carceral landscape has functioned as a distinctly masculine environment—with the presence of women as both inmates and custodial staff posing an enduring set of difficulties (Carlen 1983; Liebling and Price 2001; Owen 1998). In particular, penal administrators have struggled to provide secure and hygienic accommodation for the dependent children of female inmates, with various solutions proposed and rejected over the last two centuries. From the 1830s, when the first dedicated female prisons were established in the British penal colonies of Australia, designs for women's institutions were modified to include separate Nursery Wards (Casella 2002). By the 1870s, this architectural practice was globally extended as independent female prisons were constructed in Britain and the United States (Rafter 1990; Zedner 1991).

A second arena of spatial negotiation has involved the presence of illicit black market networks across the institutional landscape. This "subrosa" exchange of contraband mobilizes four generalized types of desire (Williams and Fish 1974): the embodied longing for food, personal safety, or sexual activity inadequately provided through official channels; the addictive craving for cigarettes, alcohol, and drugs; the social desire for solidarity, reciprocity, and obligation among inmates and complicit staff members; and the strategic quest for influence and social status within the penal environment.

Requiring a degree of staff collusion, black market networks circulate valuable commodities through both recreational and functional modes of transaction (Casella 2000). Representing the first centralized state apparatus in the New World, the Walnut Street Prison of Philadelphia was established during the early 1790s to cultivate inmate rehabilitation through constant industry, religious instruction, and moral supervision. During the spring of 1973, excavations sampled from two of the prison workshops (Cotter et al. 1988). Evidence from the bone assemblage indicated a frequent co-option of institutional resources for clandestine forms of production, with 14 small fragments carefully worked into cubic and rectangular shapes. Since two artifacts had been inked with dots, the

items demonstrated that a covert manufacture of bone dice occurred within the prison workshops. Providing a mechanism for both personal amusement and prohibited gaming activities among inmates, these illicit objects suggested that alternative social networks cut across the disciplinary landscape.

Similar contraband was recovered from Hyde Park Barracks (1819), an early Australian accommodation and administrative facility for male felons in the British penal colony of New South Wales. This assemblage consisted of handcrafted bone and ceramic gaming tokens (Figure 60.4) excavated from underfloor deposits located below the stairway landings. While gaming served as a recreational diversion, it also provided a structured functional mechanism for the illicit circulation of desired goods and services throughout the penal environment.

A distinct spatial focus of these clandestine activities was archaeologically revealed during excavations at the Ross Female Factory (1848), an Australian female prison established in the penal colony of Van Diemen's Land (Tasmania). Although contraband appeared throughout the inmate dormitories of the main penal compound, the greatest concentrations of illicit artifacts (coins, olive glass alcohol bottles, and tobacco pipes) were recovered from the earthen floors of the Solitary Cells (Casella 2000, 2002). As places of ultimate punishment, these isolation cells were architecturally fabricated to discipline repeat offenders—those women located at the apex of the "subrosa" economy who were best able to exploit its operation to their own benefit. Thus, the high frequencies of contraband indicated the shadowy dynamics of an alternative inmate landscape within this institution, with covert pathways of internal trade negotiating the disciplinary force of incarceration.

Figure 60.4   Convict game tokens, Hyde Park Barracks, Sydney, Australia (photograph by E. C. Casella).

Conclusions

A uniquely modern human experience, incarceration reveals the simultaneous operation of three spatial forms of social power. From the 18th century, penal architecture has sought to elaborate, if not perfect, the imposition of self-discipline and social control. Institutional inmates have responded in kind, undertaking spatially located acts of insubordination designed to reject the penal landscape. But binary models of domination and resistance limit our understandings of incarceration. With the ideal disciplinary template architecturally modified to accommodate a myriad of diverse inhabitants, inmates further negotiate penitential structures by forging their own alternative landscapes of collusion, exploitation, obligation, and material exchange. Analysis of both architectural and archaeological elements of these penal sites has exposed how dynamics of spatial order, social practice, and insubordinate agency shape these landscapes of incarceration. Thus, the carceral landscape ultimately represents a complex world of built intention perpetually negotiated by lived compromise.

References

Brodie, A., Croom, J., and Davies, J. O. 2002. *English Prisons*. Swindon: English Heritage.

Bush, D. 2000. Interpreting the latrines of the Johnson's island Civil War military prison. *Historical Archaeology* 34(1): 62–78.

Carlen, P. 1983. *Women's Imprisonment*. London: Routledge.

Casella, E. C. 2000. "Doing Trade": A sexual economy of nineteenth-century Australian female convict prisons. *World Archaeology* 32: 209–21.

———. 2002. *Archaeology of the Ross Female Factory*. Records of the Queen Victoria Museum, No. 108. Launceston (Australia): QVMAG Publications.

———. 2007. *The Archaeology of Institutional Confinement*. Gainesville: University Press of Florida.

Clemmer, D. 1940. *The Prison Community*. Boston: Christopher Publishing House.

Cotter, J. L., Moss, R. W., Gill, B. C., and Kim, J. 1988. *The Walnut Street Prison Workshop*. Philadelphia: The Athenæum of Philadelphia.

Ehrenreich, R. M., Crumley, C. L., and Levy, J. E. 1995. *Heterarchy and the Analysis of Complex Societies*. Archaeological Paper No. 6. Arlington: American Anthropological Association.

Evans, R. 1982. *The Fabrication of Virtue*. Cambridge: Cambridge University Press.

Fleisher, M. S. 1989. *Warehousing Violence*. Newbury Park, CA: Sage Publications.

Foucault, M. 1977. *Discipline and Punish*. New York: Vintage Books.

Friedman, L. M. 1993. *Crime and Punishment in American History*. New York: Basic Books.

Garland, D. 1990. *Punishment and Modern Society*. Oxford: Clarendon Press.

Garman, J. C. 2005. *Detention Castles of Stone and Steel*. Knoxville: University of Tennessee Press.

Giallombardo, R. 1966. *Society of Women*. New York: Wiley.

Goffman, E. 1961. *Asylums*. New York: Anchor Books.

Hirsch, A. J. 1992. *The Rise of the Penitentiary*. New Haven, CT: Yale University Press.

Hobsbawm, E. 1973. Peasants and Politics. *Journal of Peasant Studies* 1(1): 3–22.

Howe, A. 1994. *Punish and Critique*. London: Routledge.

Ignatieff, M. 1978. *A Just Measure of Pain*. New York: Pantheon Books.

Liebling, A., and Price, D. 2001. *The Prison Officer*. Leyhill: Prison Service Journal.

Markus, T. A. 1993. *Buildings and Power*. London: Routledge.

McCoole, S. 1997. *Guns and Chiffon*. Dublin: Government of Ireland.

———. 2004. *No Ordinary Women*. Dublin: The O'Brien Press.

Nicholas, S. 1988. *Convict Workers*. Cambridge: Cambridge University Press.

Owen, B. 1998. *In the Mix*. New York: State University of New York Press.

Rafter, N. H. 1990. *Partial Justice*. New Brunswick: Transaction Publishers.

Scott, J. 1990. *Domination and the Arts of Resistance*. New Haven, CT: Yale University Press.

Sykes, G. M. 1958. *The Society of Captives*. New York: Rinehart.

Toch, H. 1977. *Living in Prison*. New York: Free Press.

Walmsley, R. 2001. World Prison Populations: an attempt at a complete list, in D. van Zyl Smit and F. Dünkel (eds.), *Imprisonment Today and Tomorrow*, pp. 775–95. The Hague: Kluwer Law International.

Wikberg, R., and Rideau, W. 1995. Protective custody, in B. Foster, W. Rideau, and D. Dennis (eds.), *The Wall Is Strong* (3rd ed.), pp. 174–92. Lafayette: Center for Louisiana Studies, University of Southwestern Louisiana.

Williams, V. L., and Fish, M. 1974. *Convicts, Codes, and Contraband*. Cambridge: Ballinger.

Zedner, L. 1991. *Women, Crime and Custody in Victorian England*. Oxford: Clarendon Press.

61

Cultural Resource Management and the Protection of Valued Tribal Spaces: A View from the Western United States

Diane L. Teeman

I have been working within the field of cultural resource management (CRM) since 1988. As a Northern Paiute woman, working in this profession has been both challenging and rewarding. My experiences include working for various Federal and state agencies, state and university museums, and my own tribal community. From these experiences, I have gained a relatively broad understanding of the complex relationships guiding the interactions between CRM personnel, academic archaeologists, and Indigenous descendant communities. Over the course of the last two decades, I have watched and participated in the various interactions my tribal community has had with cultural resource managers working for Federal and state agencies. These cultural resource managers now control the majority of my Tribe's traditional lands. I have monitored the manner in which my Tribe has been able to negotiate protections of our traditional landscapes with the State Historic Preservation Office and other state and Federal agencies. Although many Federal and state land and "resource" managers have worked diligently to find ways to incorporate tribal needs, often the understanding of both what tribal needs entail and the means to incorporate those needs effectively are absent. Because of this, the success of adequate protections for my community's important places has been limited. For the most part, CRM law has

not developed inclusive of Indigenous perspectives (Tsosie 1997), and discussion addressing the concerns of descendant communities is often never initiated beyond the cursory consultation that Federal law requires.

In this chapter, I discuss some of the primary points of misunderstanding and miscommunication among cultural resource managers, archaeologists, and Native American descendant groups within the Western Great Basin of the United States. From this discussion, I hope to further illuminate some of the landscape issues cultural resource managers and archaeologists working in the Great Basin and elsewhere should consider as they continue to work in the homeland of Indigenous descendant communities. My continued goal within my community is to find ways of incorporating our Tribe's traditional values and concerns into the Federal and state holders' cultural resource management practices. From this discussion, it is my hope that further efforts conducive to improving collaborative working relationships between cultural resource managers, archaeologists, and Indigenous descendant communities will be fostered. As a professional archaeologist, my ethical responsibilities include working to improve efforts toward greater collaboration between Native American Tribes and archaeologists, both in CRM and academic research. Although a majority of my work as an archaeologist has included working with tribes of the Pacific Northwest, the issues I discuss are general enough in content and scope to be applicable for archaeologists and cultural resource managers working anywhere there are incongruent and/or competing values guiding land-use practices.

Land-Use Theory in the United States

A primary issue of contention between mainstream American culture and at least some North American Indigenous communities is the manner in which relationships between people and the landscapes they inhabit are perceived and evaluated.

Policies guiding land use in the United States have developed from early notions of America as an untamed wilderness along with ideological assertions about land and land ownership founded in Western philosophy. Prevailing ideas such as the "right of discovery," "eminent domain," and, later, "manifest destiny" allowed the majority culture to justify the appropriation of Indigenous people's lands. Under American land-use policy, lands were for the most part parceled and deeded to individual land owners. Individual land ownership was a primary factor in the status given to any person, and group ownership was not usually socially acceptable or legally possible. Race and gender were primary factors guiding access to land ownership in early America; because of this, the rights of Native American peoples were seldom considered. As a result of prevailing early land-ownership and -use policies guiding efforts to take control of land in North America, all tribes Indigenous to the United States lost some of their land base, and some tribes lost control and access to their homelands entirely. Forced removal and relocation of Native Americans in the United States began in earnest with the Indian Removal Act of 1830 and continued until all "hostiles" were effectively silenced in the late 19th century (Higham 2003; O'Neil 2003). In the aftermath of such turmoil, surviving Indigenous communities have rallied to work toward repatriation of their traditional homelands, as well as working toward greater control and consideration in land-use policy and practice. Within these struggles, misconceptions about the unique relationships Indigenous North American cultures have with their homelands prevail. In time, however, increased communication will allow for achievement of increased understanding of the importance of place to this country's first peoples.

A residual problem facing some Indigenous North American communities is that they have been virtually silenced by the legal process they must now use to gain greater control over their traditional lands. Unfortunately, the "experts" who now have greatest authority in speaking to the nature of the relationship between any given Indigenous community and their land are not the people themselves but the anthropologists who study their culture (Fuller 1997; Rosen 1977). Although every relationship between anthropologists and the Indigenous communities they study is unique, and some relationships are much closer than others, the question remains whether an anthropologist's etic perspective can ever fully account for the cultural beliefs of a community that differs from their own. I offer an example from the Great Basin to further explore the differing ways anthropologists and archaeologists have interpreted relationships of Indigenous communities with their landscape, from that of how a member of one of those communities views those relationships.

Through Another's Eyes

Palaeoecology has been an important aspect of archaeological interpretations in the Great Basin since the 1930s (Grayson and Cannon 1999; Steward 1938). However, since the 1960s, there has been an almost unyielding emphasis toward ecological and materialist models (Baumhoff and Heizer 1965; Bettinger 1977; Bettinger and Baumhoff 1983; Broughton and Grayon 1993; Christianson 1980; Fowler 1972; Zeanah 2002). Because of this, many of the archaeological and anthropological explanations of Great Basin cultures' lifeways have a mechanistic quality about them. Predator/prey models, optimal foraging theory, and other such cost-benefit analyses of the subsistence strategies of Native American Great Basin cultures have diminished the primacy of the role cultural values play in guiding the actions of Indigenous Great Basin peoples.

Beyond ethnographic accounts portraying Indigenous peoples as informants, alternative discussions or explanations by Great Basin descendant communities are all but absent from archaeological or anthropological literature. Because of this, the voices of Indigenous Great Basin peoples have been silenced by the archaeological and anthropological interpretations of those peoples, and the intimate relationships between people and places have been for the most part overlooked (Brewster 2006).

Many problems have resulted from these oversights, including the present need for some descendant communities to fight against the archaeological and anthropological stories that have been created about them in order to have their lands and their ancestors repatriated to their communities. Within these grass roots battles, a movement toward making Indigenous perspectives not only known but also accepted as valid alternative interpretations has developed. Although I offer only one standpoint from within Northern Paiute culture, I suggest what such an alternative interpretation of the relationship between the people of my community and their landscape should include.

Looking beyond Subsistence Strategies: When Land Is Family

The landscapes of the Great Basin of the western United States are sometimes as harsh and foreboding as they are beautiful. The Northern Paiute peoples who have inhabited these spaces since time immemorial (Burns Paiute Tribe 1997) have honed their skills and practices to meet the challenges of this highly variable climate. A close examination of the Northern Paiute people's interactions with the landscapes they inhabit would illustrate the complexity of that relationship, although ethnographic and archaeological literature often refers to our cultural system as "simple."

The primary connection between Northern Paiutes and their homelands should not simply be seen as that of a culture existing on a geographic location as a backdrop. Nor should it merely be seen as a relationship of humans exploiting the resources available to them. Rather, the nature of the relationship between landscape and Northern Paiute peoples is more accurately described from my own emic perspective as familial. The people are the land and the land is the people. The reasons for this are numerous, but I highlight some of the most apparent.

Within the oral history of most Great Basin cultures, an era is recounted. There was a time when the relationships between Indigenous Great Basin peoples and the other beings of the land shared such a close relationship that all species could communicate with one another. Since then, it has been carried down from generation to generation so that we are today as, Great Basin peoples, related to the other animals who share our landscape. Likewise, other living beings, including flora and geographic places, are also given due consideration and respect. The reciprocal relationships observed by anthropologists between Northern Paiute peoples are also practiced between humans, animals, and other things and places in our Great Basin homelands.

A perusal of the ethnographic data from the Great Basin could support this claim, although the standard anthropological interpretations rarely if ever conclude an assertion of the existence of a familial relationship between humans and other species. The term *animism* is used to describe beliefs akin to the Northern Paiute practice I describe above, but defining a practice in anthropological terms or determining cultural beliefs as a "worldview" does not diminish the need to take such cross-cultural realities into account when considering appropriate land-use policy and practice.

Additionally, it is common knowledge among Indigenous Great Basin descendant communities that the landscape is the keeper of our history. The acts and events occurring on the landscape become part of it. A component of the people who were part of those acts and events is forever intermingled with those places; in effect, an action on a landscape is also an action on not only prior acts and events but also the people who were involved in those activities. Moreover, the actual physical remains of the ancestors of Indigenous Great Basin peoples are intermingled with the soil of the landscapes of our homelands. Because of this fact, we are the landscape from which we came and will each some day return. Our success at protection of important places is vital to this cycle of individual life and death, and the health and well being of our communities depends on our ability to actively participate in a healthy relationship with our lands.

The complex nature of the relationship between Northern Paiute communities and their landscapes is little understood by people practicing traditional anthropology. The near-exclusive use of standardized anthropological foraging models diminishes the value of the critical relationships between Northern Paiute people and their homelands to that of exploitation, or acts of instinctual animal response to the landscape, thereby dehumanizing the Indigenous people of the Great Basin. To correct this problem, many changes must occur within the way archaeologists, anthropologists, and cultural resource managers go about their business. Most pressing, however, is the need for incorporation of Indigenous cosmologies and epistemologies to that of current CRM land-use policies and practice, so that the concerns and needs of all affected communities are addressed. Until such issues are taken seriously, the cultural resource management of lands in the Great Basin will not meet the needs of Indigenous Great Basin descendant communities.

The Business of CRM

In the United States, Federal legal mandate requires cultural resource management inventories of all lands involved in Federal undertakings. Section

106 process of the National Historic Preservation Act (NHPA) is a primary tool for CRM evaluation, and this Act focuses principally on the tangible remnants of past human activity still available on the landscape. Federal and sometimes state protections for Native American human remains, funerary objects, objects of cultural patrimony, and sacred items, as well as places of significant religious importance to Native American tribes, also exist,[1] although the application of these protections through the cultural resource management process is often difficult and highly controversial. Within Section 106 process, inventories, locations with tangible remains of past human activity, are documented. The cultural properties identified as a result of cultural resource inventories are then protected, usually by measures toward avoidance, until such time when they can be further evaluated. The process for providing permanent preservation protections to cultural properties requires an evaluation of a property's eligibility for listing on the National Register of Historic Places (NRHP). But most locations identified during the cultural resource management inventory process are never evaluated for their eligibility to the NRHP, because of cost and time constraints. This often leaves identified cultural properties in a legal limbo. Additionally, the NRHP evaluation process evaluates the importance of providing permanent protections to a property based on whether the property possesses one or more of the following criteria outlined in the NHPA: It (1) is associated with an important event; (2) is associated with an important person; (3) embodies distinctive characteristics; (4) has the potential of yielding important data for major research questions. But differences often exist between the import Northern Paiute communities place on particular locations on the landscape and the significance criteria of the Section 106 process.

CRM law has developed within a predominantly Western ideological system, and because of this, the land and the "resources" available on it are given a value based from within that particular cultural system. For instance, within the language and accepted meaning of CRM law, archaeological sites are viewed as "properties" (16 U.S.C. §§ 431–433) with an attached intrinsic value and rights of ownership (Tsosie 1997). This approach differs from the views held by Great Basin Indigenous communities in that, although a particular place may indeed have an intrinsic value for one or more descendant communities, the value is seldom related to Western "research" interests, unless a community has chosen to combat ideologically Western-based research with the same in support of the community's aims. Further discussion is needed to effectively meet the needs of Indigenous descendant communities. Discussion of efforts to differentiate cultural resource management from cultural heritage management further exemplify differences between how the tangible remnants of past activities are defined cross-culturally.

Cultural Resource Management vs. Cultural Heritage Management

Within mainstream American culture, it is popularly assumed that the "archaeological record" is *America*'s heritage and is, therefore, owned by all Americans (White Deer 1997). While this instilled attitude of "national pride" has been actively fostered both by initiatives to educate the public, and to gain public support for the protection of archaeological sites, the overall assertion of public domain and public dominion (White Deer 1997) of these sites often creates tensions between the understandings of the public, the needs of archaeologists, and the rights of Indigenous descendant communities.

The term *culture heritage management* (CHM) is currently causing controversy among some Great Basin communities because of who is using the term and what cultural heritage management means in relationship to cultural resource management. The descendant communities tied most closely to the physical remains comprising "archaeological" sites often feel their heritage is being appropriated by archaeologists and others who profess legal and other claims of ownership to the material remains from such places. The notion that tangible materials from the precontact era are claimed as belonging to anyone's "heritage" beyond that of the descendant community or communities Indigenous to that particular region is offensive. Because of this, some indigenous descendant communities propose to define cultural resource management and to clearly separate it from cultural heritage management. Further awareness of the implications of the adoption of this term is sure to cause additional tension between Indigenous communities and the agencies now claiming to be "heritage" managers. The use of the term *heritage* as part of program management also implies that the agency using the term is aware of all aspects of the heritage of the landscape they are managing, which is arguably not the case in terms of nontribal agencies managing the homelands of Indigenous communities.

CHM is a term that has been adopted by some tribal agencies combining deference for CRM law, while practicing the culturally appropriate treatment of landscapes for the communities the office serves. In such a setting, CRM is incorporated into the CHM program but represents only a small subset of considerations for these communities. However, Howard (2002) cautions that use of the term *heritage*, even for tribal agencies, may lead to misconceptions of exactly what is valued and deemed worthy of protection, because *heritage* narrowly defined most commonly refers only to tangible items.

There are several primary differences between traditional CRM practice and the CHM methodologies being implemented by some Indigenous descendant communities. Although also bound by Federal law in evaluating the significance of places the law considers cultural properties, tribal communities often consider the same landscapes either more or less important or significant for very different reasons. For example, Indigenous descendant communities may not necessarily value the tangible attributes of a cultural property for its potential to yield Western scientific data, especially if the data and the knowledge yielded fall outside the bounds of that culture's cosmological and/or epistemological system. Additionally, the value of a particular landscape for Indigenous communities may or may not be identifiable through the inventory of tangible evidence associated with that particular landscape: lack of visible physical remains on or within a particular landscape does not necessarily equate to lack of value for Indigenous descendant communities. In fact, it is often just as likely that the value of a landscape could not be assessed by anyone outside the particular Indigenous community that values it. But often because of the current lack of meaningful communication between CRM personnel and Native American tribes, descendant communities have little or no voice in defining the "value" of most places within their own traditional lands. Because cultural resource management law has primarily concerned protecting places and things of importance to majority culture without regard for or understanding of the Indigenous and Indigenous descendant worldviews, frustrations frequently arise when Indigenous Native American communities try to use the existing legal system, most usually CRM legislation, to protect those landscapes important within their own communities.

Recent attempts to incorporate Indigenous perspectives within state and Federal CRM laws have had mixed success, especially when conflicts arise between the desires of Indigenous communities and the public and/or archaeologists. Conversely, these communities have sometimes had great success at working together to protect important landscapes. Occasionally, because of other factors, such as the Federal legal primacy of historic mining claims, no amount of lobbying, negotiating, protest, or collaboration can protect an area (Cannon

1996; *In the Light of Reverence* 2001). When an issue arises pitting some within the archaeological community against descendant communities, problems increase exponentially. Nowhere is this more apparent than in cases involving human remains and/or sacred sites (Van Pelt et al. 1997).

Consultation vs. Collaboration

An additional issue that continues to plague the development of mutually respectful and mutually informed relationships between Native Americans and archaeologists is the limited amount of meaningful collaboration occurring between and among members of these respective groups. Current CRM law requires that formal consultation occur between Native American Tribes and the CRM offices overseeing protection of those tribe's traditional lands. The legal protocol, however, requires only limited interaction between these groups, and at least some archaeologists and the land managers they report to don't see a need for further interaction with descendant communities beyond what they interpret legislation to require (Coahran 2006). Because of this situation, and because of the fact that many CRM departments are operating with limited budgets and numerous time constraints, the level of open discussion that leads to relationship-building and cross-cultural understanding is often lacking. More often than not, the immediate tensions among archaeologists, CRM personnel, and Indigenous communities are due to the dysfunctional and disjointed relationships among these groups. At least some archaeologists and CRM personnel go about their careers oblivious to the cross-cultural mayhem they may be creating. Building collaborative relationships can also be very time-consuming, but those land managers and archaeologists who have taken the time and effort to seek collaboration with descendant communities have often found their efforts toward collaboration effective and most rewarding (Cannon 1996; Fuller 1997; U.S. Fish and Wildlife Service and the Burns Paiute Tribe 1992). The issues I now introduce are common within the Great Basin but are also points of consideration for anyone approaching the question of collaborating with Indigenous communities as conversations relate to "archaeological sites."

The Sacred, the Secular, and the Profane

Many descendant communities hold the greatest of respect for the traditional landscapes that birthed their people. Somehow, however, this relationship between Indigenous peoples and their traditional lands has been translated in the minds of the public and some archaeologists as a "sacred" relationship. While some Native Americans might speak of their relationship with the landscape of their people as "sacred," others would not. To be certain, most if not all tribal groups have places within their homelands that hold great spiritual value, but the idea that every location within a given groups' traditional landscape is sacred seems highly unlikely. The notion of "sacredness" as a constant companion to the identity of any Native American developed from age-old stereotypes of the supraspiritual Indian. Such stereotypes portray Native Americans as having a heightened mystical spiritual relationship with everything around them. Why Native Americans have been portrayed this way and why this stereotype has been perpetuated is somewhat unclear, but this preconception sometimes forces descendant communities to speak in terms of the sacred in order to be heard. Without support based in the public's perception of a sacred relationship between Indians and nature, and sometimes even in spite of it, majority society does not feel much need to recognize Indigenous descendant communities' rights to protect areas of importance to them. Support from the public for protection of places important to Indigenous communities, usually gains momentum only when the place also holds value for mainstream society. Of course, there have been exceptions, and it is within these exceptions that Indigenous communities and their supporters find hope for future change.

Conclusions: Bending without Breaking

Each of the issues I have introduced in this chapter can be effectively addressed by increasing meaningful communication among archaeologists, cultural resource managers, cultural heritage managers, and descendant communities. It sometimes seems impossible to hope for fulfilling collaborative relationships considering the present difficulties of even achieving mutually acceptable consultation from various agencies. The reality is, although effective collaboration is happening in some places, discussion leading to a paradigmatic shift toward collaborative work is occurring in academic and other settings (Ferguson et al 1997; Harrison 1997; White Deer 1997). Through continued efforts to increase awareness, the thought processes and practices that have emerged from early successful collaborations can and will eventually spread. Further awareness by land managers

of cross-cultural differences will also, we hope, increase the breadth of considerations they are willing to give to the views of descendant communities, as well as allow land managers to see the value of advocating for collaboration with those communities. Although mutually agreeable compromises may not always be possible, it is important to at the very least make an earnest attempt at understanding the viewpoints of others. Nowhere is this truer than in cases wherein the traditional landscapes of descendant communities are being managed by others. It would be far easier to continue to buy into the environment of misunderstanding, thereby maintaining the status quo. But if cultural management is to survive as a method of protecting the places of importance for all communities within the United States, it must evolve in a manner that protects the interests of all citizens while giving proper due to this land's Indigenous cultures.

Note

1. The Native American Graves Protection and Repatriation Act (1990); National Historic Preservation Act (amended 1992); The Religious Freedom Restoration Act (1993); The American Indian Religious Freedom Act (amended 1994); Executive Order 13007 (1996).

References

Baumhoff, M. A., and Heizer, R. F. 1965. Postglacial climate and archaeology in the desert west, in H. Wright and D. Frey (eds.), *Quaternary of the United States*, pp. 697–707, Princeton, NJ: Princeton University Press.

Bettinger, R. L. 1977. Aboriginal human ecology in Owens Valley: Prehistoric changes in the Great Basin. *American Antiquity* 42: 3–17.

Bettinger, R. L., and Baumhoff, M. A. 1983. Return rates and intensity of resource use in Numic and Pre-Numic adaptive strategies. *American Antiquity* 48: 830–34.

Broughton, J. M., and Grayson, D. K. 1993. Diet breadth, adaptive change, and the White Mountain faunas. *Journal of Archaeological Science* 20: 331–36.

Burns Paiute Tribe. 1997. *Wadatika Ma-Ni-Pu-Neen*. Burns, Oregon: Burns Paiute Tribe.

Brewster, M. 2006. Knowledge, power, and authority in Numic studies: From Yosemite to Eagle Lake, the erasure of Northern Paiutes from the Sierra Nevada. Paper presented at the 30th Biennial Great Basin Anthropological Conference, October, 2006, Las Vegas, Nevada.

Cannon, W. 1996. Tucker Hill Base survey atlas mining environmental assessment report. Unpublished Cultural Resources Survey Report. Lakeview District, Lakeview Resource Area. Lakeview: Bureau of Land Management, Department of the Interior.

Coahran, E. 2006. Professional Indigenous eyes: How have we faired in the Last quarter century? Paper presented at the Chacmool Conference, University of Calgary, Department of Archaeology, Calgary Canada.

Christianson, A. L. 1980. Change in the human food niche in response to population growth, in T. K. Earle and A. L. Christenson (eds.), *Modeling Change in Prehistoric Subsistence Economies*, pp. 31–72. New York: Academic Press.

Ferguson, T. J., Watkins, J., and Pullar, G. 1997. Native Americans and archaeologists: Commentary and personal perspectives, in N. Swidler, N. K. E. Dongoske, R. Anyon, and A. S. Downer (eds.), *Native Americans and Archaeologists: Stepping Stones to Common Ground*, pp. 237–52. Walnut Creek, CA: AltaMira Press.

Fowler, C. 1972. Some ecological clues to Proto-Numic homelands, in D. D. Fowler (ed.), *Great Basin Cultural Ecology: A Symposium*, pp. 105–22. Reno and Las Vegas: Desert Research Institute Publications in the Social Sciences 8.

Fuller, R. 1997. Aspects of Consultation for the Central Sierran MeWuk, in N. Swidler, N. K. E. Dongoske, R. Anyon, and A. S. Downer (eds.), *Native Americans and Archaeologists: Stepping Stones to Common Ground*, pp. 143–48. Walnut Creek, CA: AltaMira Press.

Grayson, D., and Cannon, M. 1999. Human paleo-ecology and foraging theory in the Great Basin, in C. Beck (ed.), *Models for the Millennium: Great Basin Anthropology Today*, pp. 141–51. Salt Lake City: University of Utah Press.

Harrison, F. 1997. Ethnography as politics, in F. Harrison (ed.), *Decolonizing Anthropology: Moving toward an Anthropology for Liberation* (2nd ed.). Arlington, VA: American Anthropological Association.

Higham, D. 2003. Indian-white Relations: U.S. 1831–1870, in C. Barnett (ed.), *American Indian History*, Vol. 1, pp. 278–84. Pasadena, CA: Salem Press Inc.

Howard, T. 2002. Untitled paper presented at the 28th Biennial Great Basin Anthropological Conference, Elko, Nevada.

In the Light of Reverence. 2001. A film directed by Christopher McCleod, produced by Christopher McLeod and Malinda Maynor, narrated by Peter Coyote and Tantoo Cardinal. A Production of the Sacred Land Film Project of Earth Island Institute, a presentation of the Independent Television Service in association with Native American Public Telecommunications.

O'Neil, P. 2003. Indian-white Relations: U.S. 1871–1933, in C. Barnett (ed.), *American Indian History*, Vol. 1, pp. 284–92. Pasadena, CA: Salem Press Inc.

Rosen, L. 1977. The anthropologist as expert witness. *American Anthropologist* (New Series) 79: 555–78.

Steward, J. H. 1938. Basin-Plateau aboriginal socio-political groups. Washington, DC: *Bureau of American Ethnology Bulletin* 120.

Tsosie, R. 1997. Indigenous rights and archaeology, in N. Swidler, N. K. E. Dongoske, R. Anyon, and A. S. Downer (eds.), *Native Americans and Archaeologists: Stepping Stones to Common Ground*, pp. 64–76. Walnut Creek, CA: AltaMira Press.

U.S. Fish and Wildlife Service and the Burns Paiute Tribe Memorandum of Understanding Concerning Human Remains, November 1992. On file at the Malheur National Wildlife Refuge, Burns, Oregon.

Van Pelt, J., Burney, M., and Bailor, T. 1997. Protecting Cultural Resources on the Umatilla Indian Reservation, in N. Swidler, N. K. E. Dongoske, R. Anyon, and A. S. Downer (eds.), *Native Americans and Archaeologists: Stepping Stones to Common Ground*, pp. 167–71. Walnut Creek, CA: AltaMira Press.

White Deer, G. 1997. Return of the sacred: Spirituality and the scientific imperative, in N. Swidler, N. K. E. Dongoske, R. Anyon, and A. S. Downer (eds.), *Native Americans and Archaeologists: Stepping Stones to Common Ground*, pp. 37–43. Walnut Creek, CA: AltaMira Press.

Zeanah, D. 2002. Central place foraging and prehistoric pinyon utilization in the Great Basin, in B. Fitzhugh and J. Habu (eds.), *Beyond Foraging and Collecting: Evolutionary Change in Hunter-Gatherer Settlement Systems*, pp. 231–56. New York: Kluwer Academic/Plenum Publishers.

62

WHEN A STONE TOOL IS A DINGO: COUNTRY AND RELATEDNESS IN AUSTRALIAN ABORIGINAL NOTIONS OF LANDSCAPE

John J. Bradley

Archaeology has long sought to understand the past by studying the "things" left behind by people. There are times, however, when these "things," although no longer used, still have a significance that is beyond their utilitarian use. This chapter explores how stone tools, although no longer used by the Yanyuwa people of the southwest Gulf of Carpentaria in northern Australia, are still an important part of the landscape and demonstrates the connectedness of "things" to a place and space in the landscape. What is evident is that "things" such as stone tools have an important cosmological place in the landscape and important connections also to people and other living things.

Journey through Country

A line on a map marks the 20-km journey we have made along the edge of the dry creek known as Fletcher Creek, northern Australia. As we traveled, the senior Yanyuwa elders for this country named the places that we knew not, one by one: Nyindiyanantha, Marrawi, Warraba, Warrkala, Yirrinjini, and, finally, our destination, Kalkaji. There are no English names for these places; they are all subsumed on the map by the reference to Fletcher Creek. In the minds of the Yanyuwa men and women traveling in the four-wheel drive vehicle, each named place is distinguishable by

topography, vegetation, and their knowledge of the movements of their old people and, perhaps more important on this day, the movements of the Dingo ancestral being.

Kalkaji is a dry creek bed. It has some quartzite outcrops on its eastern bank, with scatterings of completed and partly completed stone blades, as well as numerous pieces of chert (a stone that has been brought in from elsewhere), both worked and unworked. It is the quartzite outcrops, however, and the presence of the worked pieces of this stone that give this locality its other name, *Ma-wudawudawiji* ["the place of the stone tools"].

Jerry Ngarnawakajarra, the most senior elder, stands alone and cries. He has not seen this place for a long time, having last visited it as a boy, with his father, and he is now probably in his mid-70s. Jerry calls out to the place, apologizing for not having returned for so long. He reminds the place that he is a child of this particular country and asks the spirits of the deceased, who he knows are watching, not to be ignorant toward him and his family that are with him.

Jerry leans down and picks up three well-made quartzite blades, and some smaller chert flakes, and says to those who are now assembled around him (Jerry Ngarnawakajarra, personal communication 1992):

633

Ma-ja ma-jamurimuri ma-ngatha,
Wurrundurla ma-ja, Wurrundurla na-
maynyngul barra, yiwa ambirrijungu
kilu-yabimanthaninya kulu nganu
li-ngulakaringu, ma-ja ma-wudawuda
wurrbingu yiku ki-Wurrundurlawu kujika
barra miku kujika nganinya:
Warrakiwarraki
Warrakiwarraki
Kakami *kakamayi*
Warndama
Warndamayi
Warrakiwarraki
Warrakiwarraki
Kakami kakamayi

["This stone tool is my most senior paternal
grandfather, this stone tool is the Dingo, it is
his fat, he was first to make these things, and
we people came behind, these stone tools are
truly the Dingo, the song verses for him we
sing like this:
Well-made stone blades
Discarded flakes lay scattered
Flaking the well-made stone blade
The well-made blade
Discarded flakes lay scattered."]

Stone tool technology has not been used among
the Yanyuwa since the early 1900s, owing mainly
to early contact with European explorers and,
more important, to contact with Macassan traders
from Southeast Asia (Baker 1999; Macknight 1976).
Leichhardt also commented that in his initial meeting with the Yanyuwa in 1844 they had a strong
sense of understanding the value of steel knives;
he associated this with their contact with "Malays"
(Roderick 1988: 355). It is still possible to find over
Yanyuwa country large caches of glass and steel
items put aside to be later turned into tools; stone
tools when found are often obliquely referred to as
ki-wankalawu ["for the ancestors"].

The short narrative described above took place
in 1992, and it is quite clear that the old man still
had a knowledgeable relationship to the stone
tools. But this was not a relationship based on
technological prowess or the making and using of
such items. It was, rather, a relationship of power,
kinship, and emotion. At the heart of the old man's
statements are two main considerations: one is of
"country" and the other, which is related to the
first, is the supervital nature of the stone source
and of the tools derived from it.

When Jerry Ngarnawakajarra arrived at Kalkaji
he had cried out, spoken, and then sung verses
from the ritual songline that flows through this
area. To sing the song of the country is one of the
most powerful demonstrations of knowledge that
can be shown publicly—such songs have come to
be known in popular imagination and literature as
"song lines." Known as *kujika* in Yanyuwa, they
are multiverse, sung narratives that track country. These are songs that the Yanyuwa describe as
"bringing everything into line." All living and nonliving things, material objects, peoples' names, the
names of the land, the winds, and other seasonal
events are all given a place in these songs.

Landscape Becomes "Country"

At one level, one can call these songs *environmental narratives*, but that would underestimate their
purpose and content. The terms *environment* and
landscape are words that the teachers of Western
knowledge use to describe the places in which we
find ourselves living and working and spending
our lived existences. In Yanyuwa, the same word
would be *awara*, a term that can mean earth, dirt,
land, place, soil, possessions, sea, reef, and home.
Awara in Aboriginal English becomes the word
"country," a term used by many other Indigenous
groups throughout Australia (Povinelli 1993; Rose
1992, 1996). However, the meaning given to this
word by Indigenous people is much different from
that usually attributed to its English version.

For Indigenous people, "country" is spoken
about in the same way that people talk about their
living human relatives. People cry about country,
they worry about country, they listen to country, and
they visit country and long to visit country. Objects
such as the stone blades are not just a part of country; they are also themselves called "country," they
are part of the kinship and emotional wealth of the
country. In return, country can feel, hear, and think;
country can also accept and reject, and be hard or
easy, just as living people can be to one another.
So, it is no surprise that sometimes people also
address each other as "country"—that is, as close
relatives who bring to that relationship all of their
past experience, their now, and their future. When
Aboriginal people talk about their country and sing
their country, all of these relationships are presented
(see also Myers 1986).

An important feature of Jerry Ngarnawakajarra's
journey to the stone quarry site is his ability to
name the country where it is located, to name the
ancestral being responsible for bringing it into
being, and to be able to name the objects themselves; they are all understood to be part of one
single numinous event or presence whereby everything is related to all others. A Western reading of
that same presence could break the landscape into
atomized ritual, song, language, kinship, material

culture, and archaeological evidence. A Yanyuwa understanding of this landscape does not concern such categories; it is rather about the potential of human and nonhuman organisms and objects to move between them, to be related to everything else, and to be recontextualized and reclassified according to context. In other words, people are concerned about and value the possibility (and authority) of potentially endless explorations of webs of interconnectedness (Rose 1997).

Having or not having a name according to context is a way by which people transpose and recontextualize animals and things from the vital (everyday pragmatic sense) to the supervital domains (everyday sense in which the thing/object/being becomes a phenomenon supercharged with meaning). The following example well illustrates this shift and brings to the fore the underlying cultural and moral aspects that are part of any form of classification. As will become apparent, stressing that a living or nonliving thing has a name challenges and blurs the boundaries between these two categories, vital and supervital (Bradley and Tamisari 2004).

The Dingo central to the quarry site of Kalkaji may at once be considered with Law by virtue of its name. The Law in this sense then can be taken to mean a body of moral, jural, and social rules and correct practices that are believed to derive from the cosmogonic actions by which ancestral beings—with the ability of changing from animal and phenomenal forms into humans—shaped and named the land, transforming parts of their bodies into landscape features, natural phenomena, and plants. Along their journeys they also gave life to people at particular places, bestowed these places upon them, and taught each group the correct manner of doing things: from hunting and foraging, processing of food, and the making of tools to the performance of paintings, songs, and dances. Each cosmogonic action of bodily transformation thus establishes a consubstantial relationship between an ancestral being, a place and a group of people who identify with the land and own it (Bradley 2006). The image of the journey is held to be the mechanism that orders, distributes, and differentiates groups' rights to and ownership of particular tracts of land or "countries." The names given to the Dingo and his fat, the latter being the source of the stone tools that are still his fat in another form, connect both the ancestral Dingo, the dingoes still to be found in this country today, and the stone tools transformed in locally emplaced cosmogonic actions during those events popularly known in English as "the Dreaming." Together, Dingo, stone tools, and place are associated with other aspects of the land, and together they remain property of their owner. A dingo can be seen in the landscape and has ancestral Law; likewise, a stone tool can be seen—and in the past it was used in daily life—and it, too, can be kin, because it has Law by having a name. What is important here is that the shift from the vital to the supervital is carried out according to context. In some circumstances, dingoes are simply dingoes, stone tools are just stone tools; but at other times they are relatives, and the action of singing the country and the stone tools' immediacy and presence in place relates them to the manifestations of ancestral beings associated with country (see also Williams 1986).

Thus, the names used by the Yanyuwa elder derive from a specific place and have associated with them a number of meanings, any one of which may be called on, depending on need and context. The name links the individual to species and to place. At one and the same time, these names become an expression of the supervital qualities of both the Dingo and the stone tools. In the example above, the notion of stone also being the fat of the Dingo is also seen as an expression of health, vitality, and power in Yanyuwa cosmology (see also Jones and White 1988).

What becomes apparent, then, for the Yanyuwa people is that the reality of objects such as a stone quarry and associated evidence of its use, even in a contemporary context, is based on a logic that allows for the oscillation between human and nonhuman, intention and nonintention, social and nonsocial, moral and amoral, poetic and nonpoetic—divisions that are not binaries but the points *between which* the vital and supervital may be observed in action. It is the vitality and the supervitality of living worlds as known by the Yanyuwa that point to all classifications, encompassing not only different logics or ways of reasoning but also, most importantly, the fact that such classifications are grounded in social action.

Angles of Perception

The observations above illustrate several critical points for disciplines such as archaeology and social anthropology. The first concerns the personal, cultural, and theoretical expectations that every field researcher brings to his or her tasks. Whether the focus is a search for objects or language and culture, these expectations will influence what questions are asked, how they are asked, and whom they are asked of (Peacock 1987; Watson 1987). Such an argument applies to any academic discipline, and, of course, the answers given are subject to similar consideration. Consider the results if an archaeologist had come across the quarry at Kalkaji without the presence of any Yanyuwa people, or whether an anthropologist would have even noticed the feature. Importantly, our findings always sit within wider contexts of debate, so although our personal preferences may be attributed to a particular history and life experience, these preferences are also subject to a prevailing social climate in which we see what we have been trained to see (Peacock 1987).

However, as the case study above demonstrates, a geographical, or historicized, archaeological version of the apparently "real landscape" is not the only version, since for some people, such as Indigenous Australians, an archaeologically derived landscape may not even be credible. Nonetheless, archaeological views, if left embedded in the dominant, and seemingly powerful, Western discourses and "tradition," may blind us to other cultural alternatives, something very much at issue when as archaeologists or social anthropologists we are often taught to perform within the pervasive logic of a so-called objective enquiry. This all suggests that what we might call multiple subjectivities, and consider "obstacles to be overcome" (Rose 1997: 73) in the quest for an objective, universal, and abiding truth, may in reality be understood as invaluable and integral elements in Indigenous Australian systems of knowledge. Thus Aboriginal "subjectivities" represent other "angles of perception," just as much as conventional archaeological landscapes also remain "subjectivities" and other "angles of perception," all contributing to the sum of what is known.

Conclusions

So often academic discussion concerning Indigenous attachment to country is defined solely in terms of functional or material outcomes, as the object of practice, and only sparse (if any) reference is made to the particular social perceptions that inform both the historical and the contemporary understandings of these activities. It is, therefore, not surprising that Indigenous people sometimes consider the "findings" of such observers with degrees of anger, acceptance, and sometimes humor, and their translations of what has been observed have been rendered to childlike or quaint interpretations (Povinelli 1993: 695). As can be seen by the landscape perceptions that the Yanyuwa applied to the quarry site at Kalkaji, understanding is multilayered and not easily reduced to the language of objectivism whereby object and subject, language and speech, place and people exist as separate and autonomous

entities. Therefore, to gain entry into those worlds, attempts must be made to examine or even at times to deconstruct some of the basic and taken-for-granted assumptions that underpin Western knowledge systems in which landscapes can be measured as external contexts of human presence and human action. Indigenous approaches to landscape present new archaeological opportunities to understanding country beyond specifying just processes, toward examining the subjective, emotional, and "imagined" phenomena from which landscapes are created (see Tamisari and Wallace 2006; see also Russell, this volume).

The example given at the beginning of this chapter illustrates how for people like the Yanyuwa their "country" or landscape is full of meaning—meaning that is derived from intersecting trajectories of history, kinship, and spirituality (Kearny and Bradley 2006). Time and place collapse at such locations as the quarry and remind us that when an archaeologist or anthropologist enters such a place he or she is historicizing both the engaged landscape and the objects and environments that may be found there. Such places as the quarry at Kalkaji are thick with meaning and force us to rethink how we may view history. People such as the Yanyuwa, prior to modern times, never had a word for history; they have since borrowed the English term. History as we know it from Western ontology and epistemology seeks to retain an impersonal, exact record of past events. Instead, the Yanyuwa have always had a word for "remembering" (*linginmantharra*), which requires past events to be experienced from a personal perspective. Once people live through certain events, their significance becomes etched into their memories. It is these memories, full of meaning that the Yanyuwa have constantly sought to maintain. The power of knowing one's "master story," the dominant details that give meaning to large and small moments alike, is that the story can never be relegated to the back of a historical bookshelf or an academic text.

References

Baker, R. 1999. *Land Is Life: From Bush to Town: The Story of the Yanyuwa People*. St. Leonards: Allen and Unwin.

Bradley, J. 2006. The social, economic and historical construction of cycad palms among the Yanyuwa, in B. David, I. J. McNiven, and B. Barker (eds.), *The Social Archaeology of Australian Indigenous Societies*. Canberra: Aboriginal Studies Press.

Bradley, J., and Tamisari, F. 2004. Place and event, in A. Minelli, G. Ortalli, and G. Sanga (eds.), *Animal*

Names. Venice: Instituto Veneto Di Scienze Lettere Ed Arti, Plazzo Loredan,Campo Santo Stefano.

Jones, R., and White, N. 1988. Point blank: Stone tool manufacture at Ngilipitji quarry, Arnhem Land, 1981, in B. Meehan and R. Jones (eds.), *Archaeology with Ethnography: An Australian Perspective*, pp. 51–87. Canberra: Research School of Pacific Studies, Australian National University.

Kearny, A., and Bradley, J. 2006. Landscape with shadows of once living people: Kundawira and the challenge for archaeology to understand, in B. David, I. J. McNiven, and B. Barker (eds.), *The Social Archaeology of Australian Indigenous Societies*. Canberra: Aboriginal Studies Press.

Macknight, C. 1976. *The Voyage to Marege*. Melbourne: Melbourne University Press.

Myers, F. 1986. *Pintupi Country, Pintupi Self: Sentiment, Place, and Politics among Western Desert Aborigines*. Canberra: Australian Institute of Aboriginal Studies.

Peacock, J. 1987. *The Anthropological Lens*. Cambridge: Cambridge University Press.

Povinelli, E. 1993. Might be something: The language of indeterminacy in Australian Aboriginal land use. *Man* 28: 697–704.

Roderick, C. 1988. *Leichhardt: The Dauntless Explorer*. Sydney: Angus and Robertson.

Rose, D. 1992. *Dingo Makes Us Human: Life and Land in an Australian Aboriginal Culture*. Cambridge: Cambridge University Press.

———. 1996. *Nourishing Terrains: Australian Aboriginal views of Landscape and Wilderness*. Canberra: Australian Heritage Commission.

———. 1997. Indigenous ecological knowledge and the scientific community, in *Bushfire '97 Proceedings: Australian Bushfire Conference 8–10 July*, pp. 69–74. Darwin: CSIRO Tropical Ecosystems Research Centre.

Tamisari, F., and Wallace, J. 2006. Towards an experiential archaeology of place: From location to situation through the body, in B. David, I. J. McNiven, and B. Barker (eds.), *The Social Archaeology of Australian Indigenous Societies*. Canberra: Aboriginal Studies Press.

Watson, H. 1987. Make me reflexive—but not yet: Strategies for managing essential reflexivity in ethnographic discourse. *Journal of Anthropological Research* 43: 29–41.

Williams, N. 1986. *The Yolngu and Their Land: A System of and Tenure and the Fight for Its Recognition*. Canberra: Australian Institute for Aboriginal Studies.

63

IMAGINED LANDSCAPES: EDGES OF THE (UN)KNOWN

Lynette Russell

Connections to landscapes often rely on entirely imagined relationships, particularly when these represent locations that are spatially or chronologically remote or detached from those with which we are most familiar. Ancient Greeks and Romans imagined that beyond the edges of their known world, there was a land frequented by grotesque, monster-like Barbarians who behaved in crude and aberrant ways (cf. McNiven and Russell 2005). During the Renaissance of Europe, the yet-to-be-discovered "New World" was figured to be populated with hideous and deformed versions of humanity in an invented and imagined landscape. Such visions of fantastic landscapes are not simply a thing of times past; they continue into the present. The latest trend in Aotearoa/New Zealand,[1] based on Peter Jackson's trilogy of films, is a series of *Lord of the Rings* tours in which thousands of tourists visit the film locations. All of these imaginary places also exist as geographic features and as part of Maori and Pakeha historical landscapes. Similar tourist ventures in Papua New Guinea include Kokoda Track tours that offer school groups the opportunity to relive and imagine key moments in World War II history and to create in young Australians a sense of national pride and palpable national spirit. In this latter venture, the cultural landscapes of the Papua New Guineans are overlain with an imagined war-scape. For archaeologists, a number of issues arise: how to interpret, to present, and to conserve a landscape or sites that have cultural values that are "of the mind," and how to accommodate the concerns of people who believe they have a relationship with that landscape that is not empirically demonstrable (cf. Everson and Williamson 1998).

This chapter explores some of these relationships between people and place as they relate to both (historically) real and imaginary landscapes. The case studies include Aboriginal Australian's traditions of passing down connections to (and stories associated with) land and place, even when colonialism has ensured that there is no longer an opportunity to maintain an actual physical relationship. Contrasting this are Anglo-Australian concerns for the integrity of the World War I site of ANZAC Cove at Gallipoli, in Turkey, as recently documented in controversy over whether or not Australia should have a say in the development of the site for tourism (see below). Australians generally imagined that they had a connection to and relationship with a landscape that most will never visit and that this should allow an official level of intervention and engagement with development issues.

The exploration of imagined relationships to real and invented landscapes enables the development of theoretical models that can assist in appreciating both contemporary and historic relationships to land and place and provide a heuristic device for understanding change over time (e.g., David 2002; Everson and Williamson 1998). Interrogating contemporary concerns for landscapes of the mind or imagination forces us to recognize that place-making includes imaginative processes of connection to place. Such connectivities are intuitive rather than empirically demonstrable, and imaginative processes of this kind relate to the past as much as they relate to the present (cf. Aston 1985; Nash 2000).

Knowing Your Place

It is a commonly held (and somewhat romantic) assertion that Australian Aboriginal people do not own land but that they belong to it. Aboriginal associations with country, their particular country, are usually passed down through the generations. Even when actual visitation was impossible and missions and other reserves had removed the opportunity to live "on country" (see Bradley, this volume), Aboriginal families passed on their land's stories and memories. For many Aboriginal people, these landscapes of the mind—what others might call imagined places—represent real, viable, and tangible links to their heritage. Contemporary Federal legal provisions such as those of the Native Title Act (Australia) continuously struggle to appreciate, accommodate, and recognize these connections. Overly pessimistic analyses of Aboriginal people's connections to land based on simplistic readings of physical presence and access, led the geographers Stephen Davis and Victor Prescott (1992: 134) to argue that:

> rights . . . accorded to Aboriginal groups by the ancestral beings are contingent upon the Aboriginal custodians continuing to care for the territory by singing the songs and performing the ceremonies associated with the territory as well as by caring for the sacred objects and places.

In the absence of physical proximity Aboriginal people were therefore doomed to lose such connections, because "traditional knowledge that allows territories to be precisely defined will continue to be lost. Fewer traditional ceremonies will be performed, songs will no longer . . . sung [and so on]" (Davis and Prescott 1992: 142).

Fortunately, the legal courts, particularly those associated with Native Title cases, have not applied these standards, and significant flexibility has been applied in judging connection to land. This situation has not, however, resulted in legal victory

for Indigenous Australians. As Bruno David (2006: 123) has remarked, the difference between the court's view and that of Davis and Prescott is in "practice and theory . . . more apparent than real."

As part of a large project involving the Aboriginal communities of Victoria, in southern Australia, I have been a member of a team that has undertaken some 100 interviews structured around the question of "storytelling" and storymaking—that is, the construction of narratives of history, personal, familial, and communal.[2] In this process, many Aboriginal people have revealed that they have maintained the stories and narratives associated with their country, in some cases despite being restricted from visiting the actual locations. Reasons for not visiting country included being raised and living in a different state or city; difficulties with transport; land being in private ownership and the current owners not allowing access; and a desire to preserve the significance of memories as the locations had changed so dramatically that there was a preference for remembering a place rather than visiting it. In each of these cases, the relationship to country and the sense of belonging were not perceived to be diminished by the absence of visitation. The challenge for the archaeologist is to consider how these people who have never visited their country might be enabled to exercise their rights as traditional owners, stakeholders who are entitled to express their concerns for development, management, or research.

Recognizing and accepting connections to land never visited have also occupied Melbourne-based academic and performance artist Mark Minchinton. In 2003, he walked from Busselton (Western Australia), birthplace of his Aboriginal great-grandparents, to Kellerberrin, where his mother was raised as a white person. He kept a web-diary, which he updated daily. As he saw it, his grandmother had moved from being an Aboriginal to white in the process of moving across the landscape from Busselton to Kellerberrin. Minchinton hoped that he could attempt a kind of reversal involving reinstating his family's Aboriginal identity as he moved back across the same land. The process of walking his country enabled him to articulate a particularly poignant engagement with imagined landscapes. He wrote:

> I want to be claimed. I want to feel the land with my feet, my body. I want the land to be written on my body . . . I want to know, in some way, this place I might have known already had my life been different, my family been different, the history of this country been different. (Minchinton 2004)

The desire to belong to land, to know it intimately, is a powerful force. However, as Minchinton (2004: 5) reflects, he does not "pretend that by walking [the landscape] I will become 'Aboriginal.'"[3] Nonetheless, he demonstrates both the power and the desire to know "your place" and, however imaginatively, to understand where you belong.

Anthropologist Paul Basu (2005) has explored similar issues with reference to Scottish diasporic landscapes and the sense of belonging that emerges from the popular trend of "roots-tourism." Basu found that many diasporic Scotts had settled in lands such as the United States, Australia, and South Africa, where dispossessed Indigenous peoples impinged on the newcomers' capacity to feel that they belong. Traveling to the Scottish Highlands, memorizing "myths" and stories, and identifying genealogical clan connections enabled an "appeal of indigenousness." This was a "sense of unproblematic territorial belonging that has become impossible in their diasporic home countries" (Basu 2005: 147). That this belonging is often entirely mythical and imagined, based on 19th-century popular accounts of clan histories that bear little resemblance to historical fact or process, seems irrelevant. Instead, for those roots-tourists the (re)discovery of identity is a deeply meaningful experience, and the connection between soil and blood, however illusory, is significant.

War Sites and Remembrance

Perhaps nowhere is the connection between blood (spilled rather than inherited) and soil more keenly felt than in discussions of war sites and memorials. In April 2005, on the eve of the 90th anniversary of the landing of Imperial forces at Gallipoli, a controversy arose that in many ways exemplified the connection that people can feel toward places distant from or remote to them. Gallipoli in Turkey was, during World War I, where the ANZAC legend formed. This narrative of loss and sacrifice, of betrayal and heroism, has become a key feature of public discussions of Australian national identity. The narrative itself has enjoyed fluctuating fortunes. After flagging interest was shown in the ANZAC story from the 1960s onward,[4] there has been increasing popularity since the 75th anniversary in 1990. ANZAC Cove at Gallipoli has become a fashionable visitation site for backpackers and other tourists, many of whom aimed to be present at the dawn ceremony on the 25th of April, 2005. These tourists had put increasing pressure on the site, and the Turkish government heritage agency sought to upgrade the facilities at the site by widening the access road. These road works were requested by the Australian Federal government, led by conservative Prime Minister John Howard. As a result, a ridge, which was the location of army headquarters, mobile hospitals, and first-aid stations, was cut into and fundamentally changed (Grattan 2005).[5]

One of the key points to emerge from the issues surrounding the controversial road works was that the Australian public believed that they had a fundamental right, indeed were stakeholders in, the Turkish landscape at Gallipoli. Even though the overwhelming majority of people will never visit the site, there was a tacit assumption that Australia should be entitled to decide what happens to it and how any development is managed.

Another war site of both memory and imagination is the Kokoda Track in Papua New Guinea. Walking the track has become a popular pastime with both Australian school students and tourists.[6] The walk has also been used in a number of television documentaries to help "straighten out" troubled teenagers.[7] In each of these cases, there was an expectation that proximity to the track, and the heroic deeds that took place there during World War II, would have a positive impact on the young people. It is as if the organizers of these tours hope that landscape itself will imprint its history onto the contemporary trekkers. Yet, surely, such connections are illusory. Young teens from inner-city suburbs, many from multicultural backgrounds, have little connection with the World War II sites of Papua New Guinea. It is as a feature of a national discourse that celebrates masculinity, "heroic war deeds," and mateship that the Kokoda Track imparts its power. The Track itself, devoid of these signifiers, has no power.

Perhaps most interestingly of all is that Kokoda and Gallipoli, both situated on other nations' sovereign soil, form part of an imagined national landscape that defies contemporary geopolitical borders. One wonders: if the Japanese people and government sought to visit Darwin, the site of significant World War II bombings, and celebrate it as a site of Japanese war-time achievement, would the Australian public and government officials welcome them?

Layers of meaning, entangled and competing, add to our understanding of people's engagement with landscapes. Real or imagined, the relationships that visitors perceive that they have with Kokoda and Gallipoli should play a significant role in how that landscape is managed, presented, and interpreted—and, most important of all, how theoretical discussions of belonging are developed.

Hobbits' Houses and Maori Sites

After the Pakeha[8] film director Peter Jackson filmed his *Lord of the Rings* trilogy in Aotearoa/New Zealand, many of the film locations became sought-after tourist sites. Over 40 tourism companies advertise *Lord of the Rings* tours.[9] As the tour buses travel to the mythical Middle-Earth locations of Rivendell, Lothlorien, and Helms Deep, the tourists travel through a palimpsestic landscape comprising overlays of geologic, geographic, Maori, and Pakeha narratives. The south island of Aotearoa/New Zealand comprises an extraordinarily diverse landscape of snow-capped mountains, glaciers, fjord lands, grassy plains, high-energy coastlines, and roaring rivers. Tourist brochures emphasize its isolation, remoteness, history, and beauty. Recently, Aotearoa/New Zealand generally and the South Island in particular are promoted as (Tolkien's) "Middle-Earth," where "the story is fiction, but the place real."

Traveling the imaginary landscape is not merely the domain of organized tours. Maps and popular books are available, so that the self-guided *Lord of the Rings* enthusiast can also locate the key sites of Middle-Earth. Interestingly, in neither the advertising brochures nor the maps, or even the book on *Lord of the Rings* locations, is there mention of the Maori landscape over which these imaginary places were built. Maori values and even the historical values that Pakeha New Zealanders ascribe to the land are absent. It seems that the mythical and imaginary landscape of Middle-Earth has superseded the actual, real landscape comprising history and geography. In July 2006, I observed that a new series of tours had emerged—Narnia Tours. These tours are based on the film of C. S Lewis's *The Chronicles of Narnia*. In the Narnia tours, there is an opportunity to visit the "Chariot Run Gully," the site of the "Death of the Witch," and "Aslan's Stand." Many of these sites are the same sites that can also be visited as part of the Lord of the Rings tours. Ascribing a cultural heritage or archaeological value on these sites and landscape features means weighing up the competing claims for connection and meaning. In a landscape where Indigenous values now compete with geographic, historical, and even imaginary interpretations, the archaeologist faces significant challenges that will force a move beyond positivist and measurable approaches and delve into something much more ephemeral and difficult to fully appreciate.

These competing interpretive claims can also extend to discussion about intellectual property rights, raising questions such as: can a landscape be owned, patented, or copyrighted? For decades, the Aboriginal community of Mutitjulu has requested that tourists and other visitors refrain from climbing the monolith Uluru (previously known as Ayers Rock). Uluru is currently managed by the Australian Government's National Parks, in close consultation with the traditional Aboriginal owners; however, banning the climbing of Uluru has not been undertaken, since these activities generate significant money from tourists. The government controlled website states:

Nganana Tatintja Wiya—"We Never Climb"

The Uluru climb is the traditional route taken by ancestral Mala men upon their arrival to Uluru. Anangu do not climb Uluru because of its great spiritual significance.
Anangu have not closed the climb. They prefer that you—out of education and understanding—choose to respect their law and culture by not climbing. Remember that you are a guest on Anangu land.[10]

Recently, in response to government intervention with regard to self-determination, health, and a number of other issues, the Mutitjulu community have indicated that traditional owners of Uluru Mutitjulu residents were considering a ban from climbing Uluru as part of a civil disobedience action plan. An elder was recorded as stating: "The tourist industry brings a lot of dollars into the territory, and tourists all come to Uluru. . . . Obviously civil disobedience can come in protest form."[11] It is clear from these responses that the intellectual and cultural property rights associated with the site of Uluru are regarded as inalienable by the traditional owners even if such status is less clear within a legal framework. Reflecting on similar issues, Aboriginal legal scholar Terri Janke (1998: 3) observed: "Indigenous cultural and intellectual property rights . . . comprise all objects, sites, and knowledge." The importance of archaeology to these discussions should not be understated; as Nicholas and Bannister (2004: 329) observe, archaeology needs to "understand the underlying issues of ownership and control of material and intellectual property as related to cultural knowledge and heritage." In the case of Uluru, and perhaps by extension the hobbit's houses and Maori sites, the tension among different ways of knowing, imagining, and 'owning' sites is complex, and multiple perspectives yield multiple approaches.

A Long Time Ago, in a Galaxy Far Away

Although the imaginary geography of Middle-Earth and Narnia overlies the actual terrain of Aotearoa/New Zealand, there are even more extreme examples of imaginary landscapes. This is demonstrated by a place known as "Tatooine."

There have been few film series as well received as George Lucas's *Star Wars* (episodes 1–6). Opening nights have seen fans decked out in carefully constructed costumes, so as to imitate their favorite characters. The costumes and artifacts they produce represent many hours of painstaking work. When the posters were released for the first in the second series of films (the prequels to the films of the late 1970s and early 1980s), *The Phantom Menace*, many fans also demonstrated a deep understanding and familiarity with the geography of Lucas's imaginary worlds. The poster depicted the landscape of the planet Tatooine, where young Anakin Skywalker casts a shadow that is shaped like his future metamorphosis Darth Vader. Fans maintained their deep understanding of the landscape and geography of this (imaginary) planet when they objected to the single shadow, because the planet is *known* to have two suns. These discussions took place primarily in Internet chat rooms, bulletin boards, and electronic magazines. The core concern of each of the contributors was whether or not two suns (a binary system) would throw a double shadow or if, depending on which sun was in its zenith, a single shadow would form. The one thing that none of the discussants raised was that Tatooine did not exist; that it was a fantasy.[12]

Discussion

People want to belong, they want to know a geography and unproblematically fit into a landscape—even if that relationship or landscape itself is imagined. Memory and imagination play important roles in our connections to landscapes and places. Anyone who, as an adult, has visited a place of his or her childhood is usually surprised by how small everything is. Windows are closer to the ground, shelves are lower than remembered, houses, paddocks, even trees are recalled as having been larger, rather than the self remembered as having been smaller. Our remembered landscapes belong to our imagination, but this does not diminish their importance or significance, however personal or idiosyncratic that might be. Understanding engagements with a childhood landscape, or places that are seen

to signify national narratives of loss and heroism, or even imaginary locations from far-away galaxies, offers a means for comprehending the complexities of human interactions with their environments. Although ascribing heritage or archaeological values to such places would be difficult, these connections (and belongings) should not be trivialized. As Stuart Hall (1990: 224) remarked (see also Rutherford 1990): "We should not, for a moment, underestimate or neglect the importance of the act of imaginative rediscovery which this conception of a rediscovered, essential identity entails."

Similar acts of imaginative rediscovery can be seen in the actions of modern Druids, who have claimed Stonehenge as a site of their heritage, even though archaeological understandings affirm that the Megalithic monument vastly predates Druid culture. Modern Druid celebrations of the *summer* solstice at Stonehenge today proceed despite overwhelming evidence that Stonehenge actually marked for its builders the *winter* solstice (see Chippindale 2004: 236). It is important that such contemporary imaginary and imagined relations to the landscape of Stonehenge not be ignored or trivialized, because to do so would both deny the contemporary relevance of historical sites to people today (whether or not they are reinterpreted through the imagination) and possibly pose a threat to the site (if Druid activity was not realistically acknowledged as meaningful to some, and carefully controlled). Social interactions with sites are real, contemporary, and, for the people involved, utterly meaningful, thus adding an important social layer to the historically and archaeologically complex and incomplete understanding of the Stonehenge landscape.

Imagining landscapes and imagining relationships to landscapes is part of the performativity of belonging (Bell 1999). The specialization of this process brings us closer to understanding the link between imagination, land, and identity and the resonances and connections among various ways of knowing.

Conclusions

This chapter considered the role that creative imagining plays in people's relationships to landscapes and places—past, present, remembered, and even fictional. These connections, even when imaginary, are real for the participants and must be considered; when one is attempting to undertake an archaeological reading of landscapes, these issues need to be given due attention.

Notes

1. Maori people prefer to call New Zealand *Aotearoa*; however, because most readers outside the Oceania region are unfamiliar with this term, I use Aotearoa/New Zealand as an alternative.

2. This project, entitled *Trust and Technology*, has sought to uncover the relationships that Aboriginal people have with the records held in the Public Trust by government and other agencies. This is a large collaborative and multidisciplinary project, involving various university researchers and government and nongovernment agencies, that is funded through an Australian Research Council Linkage Grant and industry contributions.

3. The reader is also directed to his performance and journey as these appear in his web diary entitled *Void: Journey to Kellerberrin*, www.soca.ecu.edu.au/mailman/listinfo/void.

4. This was particularly the case during and immediately after the Vietnam War.

5. See also the Media Release issued by the Prime Ministers Office, entitled ANZAC cove, 23rd April 2005, available at www.pm.gov.au/news/media_releases/media_Release1346.html

6. I am especially grateful to Tim Russell-Cook, who walked the track in 2003 as part of a three-week school trip to Papua New Guinea. His observations of the impact that this trek had on him and his school friends were invaluable in crystallizing my own thoughts on this issue.

7. For example, Channel Seven network in Melbourne in March 2005 showed a group of Muslim youths trekking the Kokoda Trail and argued that this experience had inspired them to become leaders in their community and work against terrorism. This story aired as part of the program *Today Tonight*.

8. *Pakeha* is the Maori term for non-Maori New Zealanders.

9. An Internet search was conducted using the search engine Google and the search terms "*Lord of the Rings* tours New Zealand." More than 200 hits were recorded; however, careful examination revealed that these represented more than 40 companies.

10. www.environment.gov.au/parks/uluru/no-climb.html, accessed 9 November 2007.

11. www.news.com.au/dailytelegraph/story/0,22049,21970291-5001021,00.html, accessed 9 November 2007.

References

Aston, M. 1985. *Interpreting the Landscape: Local Archaeology in Landscape Studies*. London: Batsford.

Basu, P. 2005. Macpherson Country: Genealogical identities, spatial histories and the Scottish diasporic clanscape. *Cultural Geographies* 12: 123–50.

Bell, V. 1999. Performativity and belonging: An introduction, in V. Bell (ed.), *Performativity and Belonging*, pp. 1–10. London: Sage Publications.

Chippindale, C. 2004. *Stonehenge Complete*. London: Thames and Hudson.

David, B. 2002. *Landscapes, Rock Art and the Dreaming: An Archaeology of Preunderstanding*. London: Leicester University Press.

———. 2006. Indigenous rights and the mutability of cultures: Tradition, change and the politics of recognition, in L. Russell (ed.), *Boundary Writing: An Exploration of Race, Culture, Gender and Sexuality Binaries in Contemporary Australia*, pp. 122–48. Honolulu: University of Hawai'i Press.

Davis, S. L., and Prescott, J. R. V. 1992. *Aboriginal Frontiers and Boundaries in Australia*. Melbourne: Melbourne University Press.

Everson, P., and Williamson, T. (ed.). 1998. *The Archaeology of Landscape: Studies Presented to Christopher Taylor*. Manchester: Manchester University Press.

Grattan, M. 2005. Howard must defend the indefensible. *The Age* April 24: 5.

Hall, S. 1990. A place called home: Identity and the cultural politics of difference, in J. Rutherford (ed.), *Identity: Community, Culture, Difference*, pp. 222–37. London: Lawrence and Wishart.

Janke, T. 1998. *Our Culture. Our Future. Report on Australian Indigenous Cultural and Intellectual Property Rights*. Surrey Hills: Australian Institute of Aboriginal and Torres Strait Islander Commission/Michael Frankel.

McNiven, I. J., and Russell, L. 2005. *Appropriated Pasts: Indigenous People and the Colonial Culture of Archaeology*. Walnut Creek, CA: AltaMira Press.

Minchinton, M. 2004. I was born white. *New Internationalist* 364, www.newint.org/features/2004/02/01/born-white.

Nash, G. (ed.). 2000. *Signifying Place and Space: World Perspectives of Rock Art and Landscape*. Oxford: Archaeopress.

Nicholas, G., and Bannister, K. P. 2004. Copyrighting the past? Emerging intellectual property rights issues in archaeology. *Current Anthropology* 45: 327–50.

Rutherford, J. 1990. Introduction, in J. Rutherford (ed.), *Identity: Community, Culture, Difference*, pp. 9–27. London: Lawrence and Wishart.

64

Topographies of Values: Ethical Issues in Landscape Archaeology

Marisa Lazzari

The Spatial and Temporal Dimensions of Ethics

The decisions that people make based on arguments of moral or ethical evaluation vary significantly across cultural and geographical orders (D. Smith 2000: 5). Social practices of judging and deciding thus have a spatial dimension that is relevant for an understanding of not only how people create and configure dwelling spaces but also how the work of archaeologists or heritage managers can severely modify both the conceptualization and the material configuration of particular places and landscapes.

The spatial dimensions of ethical judgment create substantial material effects with enduring affective power. Landscapes, in particular, are complex spatial formations produced by people and nonhuman agents in the process of inhabiting, that is, practicing, valuing, and imagining particular topographies over the long term (Ingold 2000: 190–93; Roepstorff and Bubandt 2004: 15). Material and immaterial aspects of human practices flesh out particular landscapes, which in this way become perfect examples of the "trialectic" dynamic that underwrites all human spatial formations: simultaneously physical, conceptualized, and lived (Lefebvre 1991: 33–39).

Landscapes are indeed value-laden and thus shape the consciousness of people who inhabit them. However, they are also spaces perpetually subject to change. This change results not only from shifting social meanings and concomitant shifts in political and interpretive strategies but also from the physical and affective presence of a multitude of nonhuman agents, from atmospheric conditions to ancestors and deities. It is this collective character of landscapes (in the sense of Latour 1993), this way in which they call attention to the ontological equality of all aspects of life, that makes them a unique empirical and theoretical space to think about ethical judgment and action in the field of archaeological inquiry.

In the following sections, the chapter discusses some of the main issues in ethical thought and their implications for landscape. It will be proposed that although codes of practice—connected with analytical philosophical traditions—are certainly necessary, archaeological ethics could benefit from a relational perspective inspired by continental philosophy and the actual experience of people involved in particular landscapes as part of their professional practice and/or their daily lives.

Landscapes provide unique possibilities for developing a truly relational approach to ethics. The practice of landscape archaeology, which forces scholars to transcend sites and locales as well as narrow temporal scales, often takes place in a climate of environmental and identity politics, Indigenous land claims, heritage conservation, and community development. This context for professional practice presents many challenges to some of the more pervasive and problematic disciplinary assumptions, which presume that archaeology's subjects are mainly dead or that archaeology can be a depoliticized, objective field of inquiry (see discussions in McGuire 2003; Meskell 2002, 2005a, 2005b; Pluciennik 2001: 1).

The question of ethics in archaeology has been generally focused on the analysis and discussion of the rights and duties of archaeologists in their specific professional involvement in the world (e.g., Lynott and Wylie 2000, see below). These deliberations and explorations are never carried out detached from political concerns. Although some authors have argued that ethical deliberations are distinct from actual political issues (e.g., Groarke and Warrick 2006: 169), in practice, ethical judgments routinely shape decisions and lead to action (Singer 1994: 8). Moreover, the discussion over principles for action and decisions involves—or excludes—interpretative communities with different degrees of social power (Smith and Burke 2003). Conversely, peoples' political actions have increasingly been caught in multiple and competitive normative frameworks, from human rights to religious law, social justice, and citizen security, to name just a few (Goodale 2006: 34; see also Cowan 2006: m15). This complex intertwining becomes more apparent when addressing past landscapes in the context of contemporary land claims or cultural and environmental struggles, particularly when those landscapes were partly shaped by a history of violence from conquerors or settlers towards a previous population whose descendants seek to redress in the present.

The practice of landscape archaeology has lead scholars to realize that far from being mere records of past lives, landscapes are contemporary entities, embedded in living traditions with deep temporal dimensions (e.g., Balée and Erickson 2006; Bender 1993, 1998; Daehnke 2007; Knapp and Ashmore 1999; Layton and Ucko 1999; Nicholas 2006; Thomas 2001). Additionally, although initially inspired by critical geography's textual approach that examined landscape as a way of representing the world (e.g., Cosgrove 1998; Daniels and Cosgrove 1989; see Thomas 1993), current social archaeology seems to be moving toward emphasizing how landscapes are the results of habitual practice and interaction, in turn actively shaping competing social projects and experiences (Bradley 2000; Edmonds 1999; Van Dyke 2003).

Because landscapes render palpable the intricacies of human existence built by both durable and fleeting moments—the ebbs and flows in the making and unmaking of memory through the sensuous engagement with landforms, places, resources, atmospheric conditions, and so on—landscape archaeology is not only a subfield for which guidelines of professional conduct may be produced, but more importantly, it constitutes an adequate framework for interrogating archaeological and ethical practice in general.

Ethical Theory and Archaeology: Implications for Landscape Research

Originally concerned with regulating archaeological practice in the face of the new demands placed by the development of Cultural Resource Management (known as CRM in the United States; subsequently Cultural *Heritage* Management in some parts of the world) as well as with establishing professional boundaries with nonscholarly practitioners and collectors, the search for ethical guidelines grew slowly, but steadily, into ways of reflecting on the legitimacy of archaeologists´ authority as sole producer and guardian of the past (Wylie 2005). This transition toward a reflexive mood has been partly the result of the political intervention of Indigenous peoples in various areas of the world who started to claim control over the production of knowledge about their past, and of the objects associated to that past, as part of their rights to self-determination (Dogonske et al. 2002; Moser 1995; Swindler et al. 1997; Watkins 2002, 2003; Watkins et al. 2000). This trend was also stimulated by early postprocessual archaeologists, who argued for archaeological practice as a political practice situated in the present (e.g., Shanks and McGuire 1996; Shanks and Tilley 1987). With differences, the critical inquiry over academic cultural production and representation (of both past and present peoples) has been largely seen as a necessary requirement for a politically and ethically informed social archaeology (compare Hamilakis 2001; Hodder 1999; Meskell 2002; Meskell and Preucel 2004; Tarlow 2001; Thomas 2004).

The ongoing dialogue within the archaeological community and with several interlocutors outside its professional boundaries has established important discussion forums and yielded numerous publications concerned with the multiple levels of this debate (e.g., Lynott and Wylie 2000; Mathers et al. 2005; McNiven and Russell 2005; Meskell and Pels 2005; Pluciennik 2001; Scarre and Scarre 2006; Smith and Wobst 2005; Vitelli 1996; Zimmerman et al. 2003). Accepting that archaeological practice is situated at the intersection of multiple discourses, domains of action, and value regimes, several professional societies and other disciplinary associations around the world have established guidelines that tend to coincide on most of the main issues. However, their language and structuring often reveal deep contradictions and conflicts over what should define archaeology's accountability. Such a major disciplinary fault line traverses—rhizome-like, to borrow Deleuze and Guattari's (1987) metaphor—subdisciplines, institutions, and countries (e.g., Endere 2004; Funari 2006; Rosenswig 1997; Smith and Burke 2003).

It has been argued that these codes convey little consensus about which responsibilities should be first addressed (Pluciennik 2001; Smith and Burke 2003; Tarlow 2001). For instance, the Society of American Archaeology (SAA) has long espoused an ethics of "stewardship of the archaeological record" as its first principle, while the World Archaeology Congress (WAC) lists eight principles to abide and several rules to adhere, all concerned with the responsibilities of archaeologists toward Indigenous communities. Likewise, the issues raised by the Vermillion and Tamaki Makau-rau Accords on human remains and display of human remains and sacred objects, both sponsored by WAC, are rarely acknowledged by other codes and guidelines. These differences reveal the extent of the ongoing discussion and renegotiation of disciplinary values and question the likelihood of ever reaching international ethical standards beyond some general agreements such as respect for others and for archaeological materials (Smith and Burke 2003: 192). It should be noted that this concern for the difficulties in reaching international standards, although praiseworthy, may also facilitate a dangerous managerial attitude— akin to what characterizes many international aid agencies—toward highly diverse local situations. Codes of practice have been shown to favor hegemonic values and domains of signification deemed "universal," a concept that often hides very particular cultural and political agendas (Korstanje and García Azcárate 2007; Labadi 2007; Meskell 2002, 2005a; Omland 2006).

In a similar vein, procedural codes of ethics and general principles of conduct have been criticized for imposing abstract principles in what otherwise are complex situations needing to be resolved on a case-by-case basis (Tarlow 2006). On some occasions, these codes and guidelines merely provide practitioners with a series of formalities to be followed in order to comply with regulations and avoid potential legal problems and research constraints. As it has been rightly noted, local participation often means something very different from "local agreement" (Meskell 2005b: 90).

Lawlike codifications may also work to legitimize and perpetuate professional authority and exclusivity (Meskell and Pels 2005: 3). The existence of archaeological codes of ethics should not dispose practitioners to forego the constant reflection and revision of their principles and assumptions. Ethical practice means that principles and assumptions cannot be unilaterally defined but have to be the product of an open process of constant negotiation with those affected by studies of particular past and present cultural configurations.

It has also been argued that discursive formations concerned with self-reflexivity and the improvement of practice are deeply implicated with a wider culture of auditing and accountability, shaped by the migration of ideas and practices across transnational managerial domains (Strathern 2000: 58). Similarly, ethical discourse has been criticized for attempting to remove social analysis from the consideration of assymetries in knowledge production (Bourdieu and Wacquant 1999: 43; Pels 2000: 136). Although such claims are themselves loaded with moral values (Kauppi 2000), they raise the valid point that ethics, if isolated from the political field, can be instrumental only to the perpetuation of inequalities (see McNiven and Russell 2005: 236). As Povinelli (2002: 14–17) has cogently argued, scholars cannot be naïve about the contradictory logic of moral projects that while aiming at "recognizing" the rights of Indigenous peoples or other disadvantaged groups, do so by means of discourses and practices that demand their self-fashioning into "acceptable" citizens, thereby erasing the very possibility of difference at the heart of contemporary multicultural democracies. A further twist on this logic is that modern states appropriate pre-Colonial history to build seemingly inclusive narratives of nationhood, while the same nation disavows the complex histories and ignores the present experiences of Indigenous groups (Byrne 1996, 2003; see Byrne, this volume).

In a more philosophical sense, the examination of the limits and responsibilities of individuals in professional groups tends to reify the modern category of the autonomous self; the all-encompassing academic "I" that sees everything, including the darkest motives and desires behind her/his moral decisions. The desire to bring to light, to incorporate into language, to make public, partakes of the revelatory project of the Enlightenment (Latour 1993: 142; Walker 1997: 73). In this way, the conception of ethics as the rational dilemmas of an idealized subject constitutes more an example of Foucault's (1988: 18) "technologies of the self" than a true engagement with the conflictive nature of social reality (Pels 2000: 138). Reflexivity—a dialogical process of mutually constituent subjects and objects—has been considered the alternative to the philosophies of reflection where the subject speculates about the nature of an opposing, detached object (Sandywell 1999). But what qualifies an adequate "constitutive reflexivity" as distinct from a mere self-satisfying introspection? And are standardized reflexive devices helpful at all (Woolgar 1988: 31)? Rather than denying the need for reflexivity, this questioning draws attention to the reality that judgments and decisions are made not by isolated individuals, but by people embedded in multiple and overlapping networks of signification and social values (Lynch 2000).

At the heart of these debates lies the philosophical problem (and its anthropological modulation) of the universal and the particular. This problem underlies the arguments of traditional ethical theory, whether espousing reasoning based on rules and duties (deontology), stressing the consideration of potential outcomes of actions (consequentialism), or aiming at the fulfillment of virtues such as trust (see Colwell-Chanthaphonh and Ferguson 2006a; Singer 1994; Wylie 2000). Beyond differences, traditional ethical theory has sought to develop the rational justification for a number of principles that may guide human action beyond the particular constraints of history and culture.

In Continental philosophy, Levinas, while rejecting founding values and principles, found the source for an ethical stance in the encounter with the irreducible "Other" (Bernasconi 1990: 75; Busch 1992). Ethical existence is prompted by the face of the Other, a phenomenological force that places a calling on us and by this calling reveals the existence of countless others (Levinas 1979: 234). This calling is by no means an opening for reciprocity: ethical action is independent of retribution (Bernasconi 1988: 234). For Levinas, the ethical signification of the other contests the primacy of ontology and makes of ethics a "first philosophy" (Levinas 1990: 75–87, 1988: 128). Yet Merleau-Ponty's (1995: 130–35) phenomenology challenges the self/other binary while offering elements for a profoundly ethical practice. By disclosing the ultimate intersubjective nature of the world and its products, Merleau-Ponty unveils the situated, constituted contingency of subjects whose "lived intelligibility" does not require eternal essences or principles (Flay 1990: 157) as it is constructed and experienced through the interaction with other beings and phenomena, whether visible or invisible.

Perhaps we should neither fear the absence of consensus in the field of ethics nor harbor the fantasy that conflictive situations may ever achieve a final moral equilibrium. The universal and the particular need each other for existence, and political life is about defining what the exact content of the "universal" may be under changing circumstances (Laclau 1993: 30–35, 49–51). Ethical existence means embracing conflict and contradiction as the fertile field for new and ever-changing social creations (de Beauvoir 1994: 120, 154). One may ask what landscape archaeology can do to reshape this terrain.

Landscape archaeology is a unique medium to develop practices toward a truly relational ethics, because it requires the understanding of historically specific ecological ways of knowing the world that unfold from the transformation of material forms by human practice (Balée and Erickson 2006: 9; Tilley 1994: 23). Furthermore, the archaeology involved with landscapes looks not only at landforms and human constructions but also at the material links between places and beings (imaginary or concrete) that build the specific fabric of a landscape as a deep-time spatial formation (Andrews et al. 2000; see Lazzari 2005).

It should be noted that the critical review of the principled and professionalized approach to ethics does not imply the rejection of guiding principles as reflexive aids (Hall 2005: 185; Meskell 2005b: 83; Meskell and Pels 2005: 8–9). Professional guidelines for engaging ethically with landscape research are particularly necessary in contexts where institutions that guarantee social justice are weak or underdeveloped for political and/or economic reasons, with the subsequent undermining of community, environmental and scientific values. As it can be seen in northwestern Argentina these days, it is often because of the damage caused by modern mining exploitation by transnational companies that consciousness about landscape as a cultural-natural spatial unity of deep temporality tends to become explicit, both among local communities and archaeologists. Such guidelines are also necessary in countries with long histories of institutional stability, since it is often in these contexts where alternative decision-making processes and social practices are usually at odds with national legal apparatuses (see papers Harrison et al. 2005; Lilley 2000; Watkins 2002). In this global context, the collective production and ongoing revision of professional guidelines can contribute to the development of legal and academic practices that are more sensitive than national and international legal apparatuses are to the multiple ways of managing and valuing landscapes that coexist in modern states (e.g., Goodall 2006; Goodall et al. 2004).

The Interfaces of Landscape

Landscapes mediate the dialectical relationship between various domains of reality (Palang and Fry 2003: 2) and may therefore help convert such notions as decisions, rights, and duties into fully intersubjective social phenomena. Useful for the naturalization of social orders based on the yearning for older ways of life (Dorrian and Rose 2003: 15; A. Smith 1986), the "aura" of landscapes certainly helps both hegemonic and counter-hegemonic projects (Scham 2003). However, landscapes are neither homogeneous nor static. Not only may multiple spatialities coexist in any given landscape (Erickson 2006: 348; Lefebvre 1991: 86–87; Munn 1996: 462; see Strang's chapter on "uncommon ground," this volume, and Russell, this volume), but landscapes are also always infused by movement. Moving bodies (for example, people, objects, animals, and so on) perpetually shape the boundaries of landscapes and their conceptualizations (Bender 2001: 7). Like "culture" (Cowan et al. 2001), landscapes may be better viewed as a hybrid spaces of belonging with very porous boundaries.

Tracing the lived experience of people in particular landscapes provides evidence of how places and identities elude taxonomies (Byrne 2003; McBryde 1997; Meskell 2005b). This conclusion is often at odds, though, with heritage laws designed under positivist legal paradigms. In the political economy of heritage (Carman 2005), such rules promote the performance of authenticity, expressed in the compulsive requirement of demonstrating "proof" of cultural inheritance, purity, and attachment (Lilley and Williams 2005). Additionally, the removal of locations and features deemed "irrelevant" from heritage listings (Byrne 1996, 1997; Harrison and Williamson 2004; Korstanje and García Azcárate 2007; Little 2005; see Byrne, this volume) helps perpetuate dominant historical narratives and exemplifies the selective nature of hegemonic memory practices that fashion particular landscapes. Non-academic concerns about landscapes (e.g., see Bradley, this volume) may not easily be translated into standard research frameworks, in turn complicating the building of archaeological methodologies that are sensitive to those concerns (Budhwa 2005). Difficulties notwithstanding, the addressing of non-academic concerns about landscape may be one of the most significant endeavors of contemporary archaeology (Colwell-Chanthaphonh and Ferguson 2006b; Endere and Curtoni 2006: 200–365; Smith and Wobst 2005; Watkins 2003). This may be particularly so in countries where violent technologies of governance alienated native peoples from their lands. In countries such as Australia, addressing this challenge has resulted in fruitful alternatives to the "stakeholders of the past" model, through different instances of partnership that look for points of commensurability between Indigenous and Western epistemologies (McNiven and Russell 2005: 235–36). Although commensurability might be an elusive quest, intelligibility is possible *across* frameworks (see Lucas 2005: 66), as people participate simultaneously in multiple orders of signification and value. Archaeological practice understood as a "partnership" between "host and guests" refuses to assign equal value to all stakeholders by establishing the primacy of Indigenous interests and claims. Simultaneously, this approach challenges the assumption that the perspectives of archaeologists and Indigenous peoples are essentially different; an assumption that may lead to the unproductive expectation that archaeological research will necessarily be different if conducted by Indigenous people (McNiven and Russell 2005: 10; see also Korstanje 2005).

Probably nothing exemplifies better the coexistence of multiple understandings than the intertwining of landscape and collective memory. Küchler (1993) distinguishes between landscapes *of* memory and landscapes *as* memory. The first consists of inscriptions of the land by means, for instance, of the construction of monuments. The second involves the shaping of a land through incorporation of habitual practices of memory-work. This distinction does not necessarily equate with the Western/non-Western dichotomy, because these kinds of memory-work may coexist in any topography, past or present. We should also be alert to the interplay of these two aspects under specific circumstances, and to the kind of social projects they have sustained.

A case study from Argentina (Curtoni et al. 2003) may serve to illustrate this interplay. The various phases of the ongoing struggle over the reinterpretation of the 19th-century war with the native population of the Pampa and Patagonia areas (wastelands by European and *Criollo* standards) have shaped the landscape where one of the decisive battles of that war has been assumed to have occurred. Artifacts of memory have been placed there by different social groups with competing claims: a monolith, a monument, and a reburial mausoleum in the shape of a pyramid. The last is an ambiguous artifact built by the reemergent Rankülche, an Indigenous group that had until recently been considered almost "extinct" (A. Lazzari 2003). The artifact/mausoleum inscribes the land with the skull of Chief Mariano Rosas, the recovered ancestor, thereby reestablishing the Rankülche's severed link with the land by means of fixing the meaning of the surroundings. It may be argued that the Ranülche are participating in the logic of the landscape *of* memory, trapped in the genealogical and "positive proof" discourse that the state imposes onto those who claim recognition. However, landscapes are fields of relations, textured by connections, ebbs and flows anchored in strong points—themselves the effect of a bundle of relations—such as monuments and artifacts (Lefebvre 1991: 85–86, 222). The participants in this commemoration also performed other spatial practices. They undertook a long-distance pilgrimage to bring back to the burial ground the skull of the chief that had so far been kept in a national museum, and they built temporary tentlike structures to present various forms of memory-work during annual festivals performed around the mausoleum. These spatial practices signal that this is a landscape *as* memory, a work of the art of remembrance in progress. However, the first aspect, the performance of the landscape *of* memory, is not less "authentic:" spatial inscriptions are concrete representations (Soja 1996: 46), primary components of Indigenous peoples' lived existence both within and in spite of contemporary nation states.

It should be said that this case is not mentioned here as example of an ethically engaged landscape archaeology. As the authors acknowledged (Curtoni et al. 2003 200–365), the article only aimed at demonstrating the richness and potential of interdisciplinary perspectives on landscapes, hitherto unexplored in their country. Yet it serves to demonstrate that although modernity relegated the landscape to the role of background or scenery, it is only by fully acknowledging its powerful constitutive agency that we may be able to create an ethical project open to difference. Indigenous peoples and archaeologists undertaking collabora-tive projects (e.g., Erickson 2006; Ferguson and Colwell-Chanthaphonh 2006; Fordred Green et al. 2003; Lilley and Williams 2005) find themselves subverting the objectified notions of landscape to which they have been exposed in radically different ways (for example, as subjecting ideology, as toolkit for valid knowledge-building). This joint unlearning and relearning through lived landscape archaeology perhaps best illustrates what an ethical practice may be: the unfolding of an embodied intelligibility that discloses itself after the meeting of various forms of knowledge, wisdom, practices, and imaginations. While agreeing with the dialogical model of ethical practice espoused by recent anthropological works, landscape archaeology moves beyond discursive paradigms by forcefully showing that dialogues are basically embodied transformative acts. These transformative acts are prompted by people's experiences of particular topographies within multiple value regimes, understandings and expectations: academic, communal, religious, political, legal, to name just a few. More than any other spatial formation, landscapes are good examples of what Lefebvre (1991: 33) named "lived spaced" (or "thirdspace" for Soja 1996): the space of possibilities, the best testimony of the indeterminacy of social existence.

As human/nonhuman artifacts, landscapes stress the production of knowledge as part of engagement and relationality, as the outcome, in other words, of practicing bodies immersed in material and meaningful geographies. As metaphor for the theoretical/ethical topography, landscape calls us to move beyond entrenched notions of positions and understandings as comparable to self-contained and essentially different "points on a map" (Palang and Fry 2003). Landscapes call us to see these different understandings as a continuum of affordances that make possible different social projects in the contemporary world. These projects may seek different and even opposite goals, yet the various undercurrents (for example, material, informative, experiential, political) that connect people with one another and with places give landscapes—metaphoric and material—their particular character.

References

Andrews, G., Barrett, J., and Lewis, J. 2000. Interpretation not record: The practice of archaeology. *Antiquity* 74(285): 525–30.

Balée, W., and Erickson, C. 2006. Time, complexity and historical ecology, in W. Balée and C. Erickson (eds.), *Time and Complexity in Historical Ecology: Studies in the Neotropical Lowlands*, pp. 1–17. New York: Columbia University Press.

Bender, B. 1993. Introduction, in B. Bender (ed.), *Landscape, Politics and Perspectives*, pp. 1–18. Oxford: Berg.

———. 1998. *Stonehenge: Making Space (Materializing Culture)*. Oxford: Berg.

———. 2001. Introduction, in B. Bender and M. Winer (eds.), *Contested Landscapes. Movement, Exile, and Place*, pp. 1–18. Oxford: Berg.

Bernasconi, R. 1988. Levinas: Philosophy and beyond, in H. Silverman (ed.), *Philosophy and Non-Philosophy since Merlau-Ponty*, pp. 232–58. London: Routledge.

———. 1990. One-way traffic: The ontology of decolonization and its ethics, in G. A. Johnson and M. B. Smith (eds.), *Ontology and Alterity in Merleau-Ponty*, pp. 67–80. Evanston, IL: Northwestern University Press.

Bourdieu, P., and Wacquant, L. 1999. On the cunning of imperialist reason. *Theory, Culture, and Society* 16: 41–58.

Bradley, R. 2000. *An Archaeology of Natural Places*. London: Routledge.

Budhwa, R. 2005. An alternate model for First Nations involvement in resource management archaeology. *Canadian Journal of Archaeology/Journal Canadien d'Archéologie* 29: 20–45.

Busch, T. W. 1992. Ethics and ontology: Levinas and Merleau-Ponty. *Man and World, Review of Continental Philosophy* 25(2): 195–202.

Byrne, D. 1996. Deep nation: Australia's acquisition of an Indigenous past. *Aboriginal History* 20: 82–107.

———. 1997. The archaeology of disaster. *Public History Review* 5/6: 17–29.

———. 2003. Nervous landscapes: Race and space in Australia. *Journal of Social Archaeology* 3(2): 169–93.

Butler, J. 1990. *Gender Trouble: Feminism and the Subversion of Identity*. London: Routledge.

Carman, J. 2005. Good citizens and sound economics: The trajectory of archaeology in Britain from "heritage" to "resource," in C. Mathers, T. Darvill, and B. J. Little (eds.), *Heritage of Value, Archaeology of Renown: Reshaping Archaeological Assessment and Significance*, pp. 43–57. Gainesville: University Press of Florida.

Colwell-Chanthaphonh, C., and Ferguson, T. J. 2006a. Trust and archaeological practice: Towards a framework of Virtue Ethics, in C. Scarre and G. Scarre (eds.), *The Ethics of Archaeology: Philosophical Perspectives on Archaeological Practice*, pp. 115–30. Cambridge: Cambridge University Press.

———. 2006b. Memory pieces and footprints: Multivocality and the meanings of ancient times and ancestral places among the Zuni and Hopi. *American Anthropologist* 108: 148–62.

Cosgrove, D. 1998 [1984]. *Social Formation and Symbolic Landscape* (with a new introduction). Madison: The University of Wisconsin Press.

Cowan, J. K. 2006. Culture and rights after *Culture and Rights. American Anthropologist* 108(1): 9–24.

Cowan, J. K., Dembour, M.-B., and Wilson, R. 2001. Introduction, in J. K. Cowan, M.-B. Dembour, and R. Wilson (eds.), *Culture and Rights: Anthropological Perspectives*, pp. 1–26. Cambridge: Cambridge University Press.

Curtoni, R. P., Lazzari, A. C., and Lazzari, M. 2003. Middle of nowhere: A place of war memories and aboriginal re-emergence. *World Archaeology* 35: 65–78.

Daehnke, J. D. 2007. A "strange multiplicity" of voices: Heritage stewardship, contested sites and colonial legacies on the Columbia River. *Journal of Social Archaeology* 7(2): 250–75.

Daniels, S., and Cosgrove, D. 1989. Introduction: Iconography and landscape, in D. Cosgrove and S. Daniels (eds.), *The Iconography of Landscape. Essays on the Symbolic Representation, Design and Use of Past Environments*, pp. 1–10. Cambridge: Cambridge University Press.

de Beauvoir, S. 1994 [1948]. *The Ethics of Ambiguity*. New York: Citadel Press.

Deleuze, G., and Guattari, F. 1987. *A Thousand Plateaus: Capitalism and Schizophrenia*. Minneapolis: University of Minnesota Press.

Dogonske, K. E., Aldenderfer, M., and Doehner, K. (eds.). 2002. *Working Together: Native Americans and Archaeologists*. Washington, DC: Society for American Archaeology.

Dorrian, M., and Rose, G. 2003. Introduction. *Deterritorialisation: Revisioning Landscapes and Politics*. London: Black Dog Publishing.

Edmonds, M. 1999. *Ancestral Geographies of the Neolithic: Landscapes, Monuments and Memory*. London: Routledge.

Endere, M. L. 2004. Patrimonios en disputa: Acervos nacionales, investigación arqueológica y reclamos técnicos sobre restos humanos. *Trabajos de Prehistoria* 57(1): 5–17.

Endere, M. L., and Curtoni, R. 2006. Entre lonkos y "ólogos": La participación de la comunidad Rankülche de Argentina en la investigación arqueológica. *Arqueología Suramericana/Arqueología Sul-Americana* 2(1): 72–91.

Erickson, C. E. 2006. Intensification, political economy and the farming community: In defense of a bottom-up perspective of the past, in J. Marcus and C. Stanish (eds.), *Agricultural Strategies*, pp. 233–65. Los Angeles: Cotsen Institute.

Ferguson, T., and Colwell-Chanthaphonh, C. 2006. *History Is in the Land: Multivocal Tribal Traditions in Arizona's San Pedro valley*. Tucson: University of Arizona Press.

Flay, J. C. 1990. Merleau-Ponty and Hegel: Radical essentialism, in G. A. Johnson and M. B. Smith (eds.), *Ontology and Alterity in Merleau-Ponty*, pp. 142–57. Evanston, IL: Northwestern University Press.

Fordred Green, L., Green, D. R., and Góes Neves, E. 2003. Indigenous knowledge and archaeological science: The challenges of public archaeology in the Reserva Uaçá. *Journal of Social Archaeology* 3: 366–39.

Foucault, M. 1988. Technologies of the self, in L. H. Martin, H. Gutman, and P. H. Hutton (eds.), *Technologies of the Self: A Seminar with Michel Foucault*, pp. 16–49. London: Tavistock.

Funari, P. P. A. 2006. Reassessing archaeological significance: Heritage of value and archaeology of renown in Brazil, in C. Mathers, T. Darvill, and B.

J. Little (eds.), *Heritage of Value, Archaeology of Renown: Reshaping Archaeological Assessment and Significance*, pp. 125–36. Gainesville: University Press of Florida.

Goodale, M. 2006. Ethical theory as social practice. *American Anthropologist* 108(1): 25–37.

Goodall, H. 2006. Indigenous people, colonialism and environmental justice, in S. Washington, P. Rosier, and H. Goodall (eds.), *Echoes from the Poisoned Well*, pp. 73–96. Lanham, MD: Lexington.

Goodall, H., Byrne, D., Wearing, S., and Kijas, J. 2004. Recognising cultural diversity: The Georges River project in south-western Sydney, in E. Katz and H. Cheney (eds.), *Sustainability and Social Science*, pp. 159–85. Sydney and Melbourne: The Institute for Sustainable Futures and CSIRO.

Groarke, L., and Warrick, G. 2006. Stewardship gone astray? Ethics and the SAA, in C. Scarre and G. Scarre (eds.), *The Ethics of Archaeology: Philosophical Perspectives on Archaeological Practice*, pp. 163–77. Cambridge: Cambridge University Press.

Hall, M. 2005. Situational ethics and engaged practice: The case of archaeology in Africa, in L. Meskell and P. Pels (eds.), *Embedding Ethics: Shifting Boundaries of the Anthropological Profession*, pp. 169–96. London: Berg.

Hamilakis, I. 2001. Archaeology and the burden of responsibility, in M. Pluciennik (eds.), *Responsibilities of Archaeologists: Archaeology and Ethics*, pp. 91–96. Lampeter Workshop in Archaeology 4. Oxford: BAR International Series 981.

Harrison, R., McDonald, J., and Veth, P. (eds.). 2005. Native Title and archaeology. *Australian Aboriginal Studies* 2005/1.

Harrison, R., and Williamson, C. (eds.). 2004. *After Captain Cook: The Archaeology of the Recent Indigenous Past in Australia*. Indigenous Archaeologies 2. Walnut Creek, CA: AltaMira Press.

Hodder, I. 1999. *The Archaeological Process: An Introduction*. Oxford: Blackwell Publishing.

Ingold, T. 2000. *The Perception of the Environment: Essays on Livelihood, Dwelling and Skill*. London: Routledge.

Kauppi, N. 2000. The sociologist as *moraliste*: Pierre Bourdieu's practice of theory and the French intellectual tradition. *Substance* 93: 7–21.

Knapp, A. B., and Ashmore, W. 1999. Archaeological landscapes: Constructed, conceptualized, ideational, in W. Ashmore and B. Knapp (eds.), *Archaeologies of Landscapes: Contemporary Perspectives*, pp. 1–30. Oxford: Blackwell Publising.

Korstanje, M. A. 2005. Comment on "Dwelling at the margins, action at the intersection? Feminist and Indigenous archaeologies, 2005." *Archaeologies* 1(1): 71–73.

Korstanje, M. A., and García Azcárate, J. 2007. The Qhapac Ñan project: A critical view. *Archaeologies*: 116–31

Labadi, S. 2007. Representations of the nation and cultural diversity in discourses on World Heritage. *Journal of Social Archaeology* 7: 147–70.

Laclau, E. 1993. *Emancipation(s)*. London: Verso.

Latour, B. 1993. *We Have Never Been Modern*. New York: Harvester Wheatsheaf.

Layton, R., and Ucko, P. 1999. Introduction: Gazing on the landscape and encountering the environment, in P. Ucko and R. Layton (eds.), *The Archaeology and Anthropology of Landscape: Shaping Your Landscape*, pp. xxvi–xx. London: Routledge.

Lazzari, A. 2003. Aboriginal recognition, freedom, and phantoms: The vanishing of the Ranquel and the return of the Rankülche in La Pampa. *Journal of Latin American Anthropology* 8(3): 59–83.

———. 2005. The texture of things: Objects, people and landscape in northwestern Argentina (first millennium A.D.), in L. Meskell (ed.), *Archaeologies of Materiality*, pp. 126–61. Oxford: Blackwell Publishing.

Lefebvre, H. 1991 [1974]. *The Production of Space*. Oxford: Blackwell Publishing.

Levinas, E. 1979. *Totality and Infinity: An Essay on Exteriority*. The Hague: M. Nijhoff Publishers.

———. 1989. Is ontology fundamental? *Philosophy Today* 33(2): 121–29.

———. 1990. Ethics as first philosophy, in S. Hand (ed.), *The Levinas Reader*, pp. 75–87. Oxford: Blackwell Publishing.

Lilley, I. (ed.). 2000. *Native Title and the Transformation of Archaeology in the Postcolonial World*. Sydney: University of Sydney Press.

Lilley, I., and Williams, M. 2005. Archaeological and Indigenous significance: A view from Australia, in C. Mathers, T. Darvill, and B. J. Little (eds.), *Heritage of Value, Archaeology of Renown: Reshaping Archaeological Assessment and Significance*, pp. 227–57. Gainesville: University Press of Florida.

Little, B. 2005. The U.S. National Register of Historic Places and the shaping of archaeological significance, in C. Mathers, T. Darvill, and B. J. Little (eds.), *Heritage of Value, Archaeology of Renown: Reshaping Archaeological Assessment and Significance*, pp. 114–24. Gainesville: University Press of Florida.

Lynch, M. 2000. Against reflexivity. *Theory, Culture and Society* 17(3): 26–54.

Lynott, M. J., and Wylie, A. (eds.). 2000. *Ethics in American Archaeology: Challenges for the 1990s* (2nd ed.). Washington, DC: Society for American Archaeology.

Lucas, G. 2005. *The Archaeology of Time*. London: Routledge.

Mathers, C., Darvill, T., and Little, B. (eds.). 2005. *Heritage of Value, Archaeology of Renown: Reshaping Archaeological Assessment and Significance.* Gainesville: University of Florida Press.

McBryde, I. 1997. The cultural landscape of aboriginal long-distance exchange systems: Can they be confined within our heritage registers? *Historical Environment* 13(3&4): 6–14.

McGuire, R. 2003. Foreword, in L. Zimmerman, K. D. Vitelli, and J. Hollowell-Zimmer (eds.), *Ethical Issues in Archaeology,* pp. vii–ix. Walnut Creek, CA: AltaMira Press.

———. 2004. Contested pasts: Archaeology and Native Americans, in L. Meskell and R. Preucel (eds.), *A Companion to Social Archaeology,* pp. 374–95. Oxford: Blackwell Publishing.

McNiven, I. J., and Russell, L. 2005. *Appropriated Pasts: Indigenous Peoples and the Colonial Culture of Archaeology.* Walnut Creek, CA: AltaMira Press.

Merleau-Ponty, M. 1995 [1968]. *The Visible and the Invisible.* Evanston, IL: Northwestern University Press.

Meskell, L. 2002. Negative heritage and past mastering in archaeology. *Anthropological Quarterly* 75: 557–74.

———. 2005a. Sites of violence: Terrorism, tourism and heritage in the archaeological present, in L. Meskell and P. Pels (eds.), *Embedding Ethics. Shifting Boundaries of the Anthropological Profession,* pp. 123–46. London: Berg.

———. 2005b. Archaeological ethnography: Conversations around Kruger National Park. *Archaeologies* 1(1): 81–100.

Meskell, L., and Pels, P. (eds.). 2005. *Embedding Ethics: Shifting Boundaries of the Anthropological Profession.* London: Berg.

Meskell, L., and Preucel, R. W. 2004. Knowledges, in L. Meskell and R. W. Preucel (eds.), *A Companion to Social Archaeology,* pp. 3–22. Oxford: Blackwell Publishing.

Moser, S. 1995. The "Aboriginalization" of Australian archaeology: The contribution of the Australian Institute of Aboriginal Studies to the indigenous transformation of the discipline, in P. Ucko (ed.), *Theory in Archaeology: A World Perspective,* pp. 150–77. London: Routledge.

Munn, N. 1996. Excluded spaces: The figure in the Australian Aboriginal landscape. *Critical Inquiry* 22: 446–65.

Nicholas, G. P. 2006. Decolonizing the archaeological landscape: The practice and politics of archaeology in British Columbia. *The American Indian Quarterly* 30(3&4): 350–80.

Olwig, K. 2002. *Landscape, Nature and the Body Politic: From Britain's Renaissance to America's New World.* Madison: University of Wisconsin Press.

Omland, A. 2006. The ethics of the world heritage concept, in C. Scarre and G. Scarre (eds.), *The Ethics of Archaeology: Philosophical Perspectives on Archaeological Practice,* pp. 242–59. Cambridge: Cambridge University Press.

Palang, H., and Fry, G. 2003. Landscape interfaces: Introduction, in H. Palang and G. Fry (eds.), *Landscape Interfaces: Cultural Heritage in Changing Landscapes,* pp. 1–13. Dordretch: Kluwer Academic Publishers.

Pels, P. 2000. The trickster's dilemma: Ethics and the technologies of the anthropological self, in M. Strathern (ed.), *Audit Cultures: Anthropological Studies in Accountability, Ethics, and the Academy,* pp. 135–72. London: Routledge.

Pluciennik, M. 2001. Introduction: The responsibilities of archaeologists, in M. Pluciennik (ed.), *Responsibilities of Archaeologists: Archaeology and Ethics,* pp. 1–8. Lampeter workshop in Archaeology 4. Oxford: BAR International Series 981.

Povinelli, E. 2002. *The Cunning of Recognition: Indigenous Alterities and the Making of Australian Multiculturalism.* Durham, NC: Duke University Press.

Roepstorff, A., and Bubandt, N. 2004. General introduction: The critique of culture and the plurality of nature, in A. Roepstorff, N. Bubandt, and K. Kull (eds.), *Imagining Nature: Practices of Cosmology and Identity,* pp. 9–30. Aarhus: Aarhus University Press.

Sandywell, B. 1999. Specular grammar: The visual rhetoric of modernity, in I. Heywood and B. Sandywell (eds.), *Interpreting Visual Culture: Explorations in the Hermeneutics of the Visual,* pp. 30–56. London: Routledge.

Scarre, C., and Scarre, G. (eds.). 2006. *The Ethics of Archaeology: Philosophical Perspectives on Archaeological Practice.* Cambridge: Cambridge University Press.

Scham, S. 2003. From the river unto the land of the Philistines: The "memory" of Iron Age landscapes in modern visions of Palestine, in M. Dorrian and G. Rose (eds.), *Deterritorialisations: Revisioning Landscapes and Politics,* pp. 73–79. London: Black Dog Publishing.

Shanks, M., and McGuire, R. 1996. The craft of archaeology. *American Antiquity* 61: 75–88.

Shanks, M., and Tilley, C. 1987. *Social Theory and Archaeology.* Cambridge: Cambridge University Press.

Singer, P. (ed.). 1994. *Ethics.* Oxford: Oxford University Press.

Smith, A. 1986. *The Ethnic Origins of Nations.* Oxford: Blackwell Publishing.

Smith, C., and Burke, H. 2003. In the spirit of the code, in L. J. Zimmerman, K. D. Vitelli, and J. Hollowell-Zimmer (eds.), *Ethical Issues in Archae-*

ology, pp. 177–97. Walnut Creek, CA: AltaMira Press.

Smith, C., and Wobst, M. (eds.). 2005. *Indigenous Archaeologies: Decolonising theory and practice*. London: Routledge.

Smith, D. M. 2000. *Moral Geographies: Ethics in a World of Difference*. Edinburgh: Edinburgh University Press.

Soja, E. 1996. *Thirdspace: Journeys to Los Angeles and Other Real and Imagined Places*. London: Blackwell Publishing.

Strathern, M. 2000. Introduction: New accountabilities, in M. Strathern (ed.), *Audit Cultures: Anthropological Studies in Accountability, Ethics, and the Academy*, pp. 1–18. London: Routledge.

Swindler, N., Dongoske, K. E., Anyon, R., and Downer, A. S. (eds.). 1997. *Native Americans and Archaeologists: Stepping Stones to Common Ground*. Walnut Creek, CA: AltaMira Press.

Tarlow, S. 2001. The responsibilities of representation, in M. Pluciennik (ed.), *Responsibilities of Archaeologists: Archaeology and Ethics*, pp. 57–64. Lampeter workshop in Archaeology 4. Oxford: BAR International Series 981.

———. 2006. Archaeological ethics and the people of the past, in C. Scarre and G. Scarre (eds.), *The Ethics of Archaeology: Philosophical Perspectives on Archaeological Practice*, pp. 199–216. Cambridge: Cambridge University Press.

Thomas, J. 1993. The politics of vision and archaeologies of landscape, in B. Bender (ed.), *Landscape: Politics and Perspectives*, pp. 19–48. Oxford: Berg.

———. 2001. Archaeologies of place and landscape, in I. Hodder (ed.), *Archaeological Theory Today*, pp. 165–86. Cambridge: Polity Press.

———. 2004. *Archaeology and Modernity*. London: Routledge.

Tilley, C. 1994. *A Phenomenology of Landscape: Places, Paths and Monuments*. Oxford: Berg.

Van Dyke, R. 2003. Memory and the construction of Chacoan society, in R. Van Dyke and S. Alcock (eds.), *Archaeologies of Memory*, pp. 151–79. London: Blackwell Publishing.

Vitelli, K. (eds.). 1996. *Archaeological Ethics*. Walnut Creek, CA: AltaMira Press.

Walker, M. 1997. *Moral Understandings: A Feminist Study in Ethics*. London: Routledge.

Watkins, J. 2002. *Indigenous Archaeology: American Indian Values and Scientific Practice*. Walnut Creek, CA: AltaMira Press.

———. 2003. Archaeological ethics and American Indians, in L. J. Zimmerman, K. D. Vitelli, and J. Hollowell-Zimmer (eds.), *Ethical Issues in Archaeology*, pp. 129–41. Walnut Creek, CA: AltaMira Press.

Watkins, J., Goldstein, L., Vitelli, K., and Jenkins, L. 2000. Accountability: Responsibilities of archaeologists to other interest groups, in M. J. Lynott and A. Wylie (eds.), *Ethics in American Archaeology: Challenges for the 1990s* pp. 33–37. Washington, DC: Society for American Archaeology.

Woolgar, S. 1988. Reflexivity is the ethnographer of the text, in S. Woolgar (ed.), *Knowledge and Reflexivity: New Frontiers in the Sociology of Knowledge*, pp. 14–34. London: Sage Publications.

Wylie, A. 2003. On ethics, in L. J. Zimmerman, K. D. Vitelli, and J. Hollowell-Zimmer (eds.), *Ethical Issues in Archaeology*, pp. 3–16. Walnut Creek, CA: AltaMira Press.

———. 2005. The promises and perils of an ethics of stewardship, in L. Meskell and P. Pels (eds.), *Embedding Ethics: Shifting Boundaries of the Anthropological Profession*, pp. 47–68. London: Berg.

Zimmerman, L., Vitelli, K. D., and Hollowell-Zimmer, J. (eds.). 2003. *Ethical Issues in Archaeology*. Walnut Creek, CA: AltaMira Press.

65

CONTESTED LANDSCAPES—RIGHTS TO HISTORY, RIGHTS TO PLACE: WHO CONTROLS ARCHAEOLOGICAL PLACES?

Jane Lydon

Most human conflict is fundamentally about control over land. As the basis for subsistence and society, occupation of landscape determines survival, prosperity, and power over others; it is not merely a symbolic representation but a material fact of dispossession central to colonization and other conflict between peoples. In addition, culturally specific understandings of the past and of landscape are the basis for assertions of collective identity and rights in the present. Archaeological and heritage discourses that give meaning to landscapes are therefore contested by those with differing views of the past, as the basis for opposed interests in the present. Postcolonial critique has shown that the effects of Western imperialism, a historical process predicated on land seizure, remain fundamental within current global geopolitical formations and continue to be expressed and contested within landscape.

In this chapter, I review the historically contingent nature of the Western concept of landscape and, in particular, its complicity with colonialism. I identify the crucial roles of land and archaeological representations of cultural landscape in current identity politics, played out through heritage discourse. Despite the centrality of the liberal principles of equal rights and participatory democracy to heritage discourse, colonial practices and ideas continue to structure heritage regimes that in turn

are deeply implicated in current struggles over the ownership and representation of landscape. At the same time, current conceptions of landscape as constructed and fluid are expressed by current attempts within the heritage system to account for intangible heritage and community values.

Western Constructions of Landscape

It is now acknowledged that the relationship between physical landscape and culture is interactive: culture and nature shape each other, and all landscapes are inherently cultural. Since the early 1990s, critique of the Western concept of landscape has shown how it emerged within an intellectual tradition that privileged visual and spatial forms of knowledge, coming to "equate the knowable with that which can be visualized and logic, the rules of knowledge, with orderly arrangements of pieces of knowledge in space" (Fabian 1983: 116). Sometimes termed "Cartesian perspectivalism" (Jay 1994: 69–70), this culturally specific method of perception became modernity's dominant way of seeing, representing space according to the rules of Euclidean geometry, as homogeneous, neutral, and universal. In this way, Western art since the Renaissance has represented landscape as visible, passive, and "natural," concealing the emergence of this tradition in conjunction with capitalism as a historically specific and politicized imagined geography (Bender 1993; Cosgrove 1984; Hirsch 1995: 2). In this realist scopic regime an increasing abstraction of vision separated observer from observed (Crary 1990), positioning the viewer in a transcendental, dominant relation to this field of view.

Grounds for Dispute

However, this way of seeing denies difference in perspective and accords primacy to a dominant white male observing subject—as feminist and postcolonial scholars, for example, have pointed out (e.g., Blunt and Rose 1994; Rose 1993). Rather than being neutral and universal, this imagined topography has been shown to express and reproduce the inequalities of Western society. As well as naturalizing the Western social order at home it has also been complicit with imperialism, assisting colonizers to map and control foreign lands and peoples (e.g., Ryan 1997). Many colonial administrators deliberately created landscapes designed to reproduce European hierarchy, as archaeologists have shown: for example, Annapolis, the capital city of Maryland, USA, was designed by royal governor Sir Francis Nicholson to aggrandize central authority at a time when the British crown had weak hold over the colony. Drawing on Renaissance principles of perspective, a baroque street plan was created in three-dimensional space to enhance objects of authority, such as the state house and church (Leone 1984, 1987). But within this landscape, invisible to those who controlled it, operated a secret ritual world created by African-American slaves: impelled by diasporic African beliefs and rebellion against domination, through ritual action slaves encoded space such as underfloor cavities, yards, and gardens with hidden meanings, establishing a crossroads between the everyday and spirit worlds in the struggle to preserve spiritual and family values and to fashion New World identities (Ruppel et al. 2003). Such contestations demonstrate that humans have socialized landscape in diverse ways.

As Barbara Bender has argued (2001), this diversity of perspectives means that landscapes are always tensioned, in process, being made and remade by those who inhabit them. Such differences may be cultural (e.g., see Langton, this volume, on Aboriginal Australia; Küchler 1993 on New Ireland), or aligned along class, gender, or other axes of social difference, and intimately linked to relations of power. A range of archaeological studies of "contested landscapes" have demonstrated the historical and cultural specificity

of the Western landscape tradition and its political uses (e.g., Ashmore and Knapp 1999; Bender 1993, 1998; Bender and Winer 2001; Burke 1999; David and Wilson 2002; Fontein 2005; Jarman 2003; Loeffler 2005; Tilley 2006). Archaeologists have also sought to unsettle the primacy of modernist visualism by developing alternative analyses of place that reintegrate subject and lived context in reconstructing past meanings of landscape—for example, drawing on hermeneutic phenomenology (e.g., Ingold 1993; Thomas 1993; Tilley 1994).

Landscape, Identity, and Colonialism

One of the key processes that shapes today's world has been the dismantling of the great world empires of Western Europe that expanded across most of the world from 1492 to 1945. Among colonialism's consequences have been the dominance of the nation state as a political entity since the 19th century (including the more recent emergence of non-Western anticolonial nationalisms) and the movement of people around the globe on an unprecedented scale. These phenomena have promulgated a modern emphasis on conceptions of identity as linked to bounded territory, home, and belonging. Such claims are founded on representations of the past and its material traces in the present, including archaeological representations of cultural landscape expressed through the practice of heritage. Heritage is a form of historical consciousness that produces meanings from objects and locales by constituting them as a focus of social memory and shared narratives; it has become the primary way in which the past is invoked by cultural institutions, enfolding conflicts within a depoliticized, jointly owned national past. As the concrete processes and procedures implemented by public policy makers to regulate the expression of identity, heritage management constitutes a cultural practice through which social groups contest and expand their power (for accounts of this process see Lydon and Ireland 2005; Smith 2004).

Recent analyses have implicated the discipline of archaeology in sustaining colonial interests, especially in settler nations such as the United States and Australia. As a form of "expert knowledge" (Foucault 1991) within government, critics have shown how archaeology has furthered a nationalist agenda, drawn on to promote a democratic, inclusive view of the state past, while subordinating Indigenous and other marginal peoples (e.g., Ireland 2003; McNiven and Russell 2005). Along with this intellectual critique, since the 1960s Indigenous demands for recognition of traditional rights to land and culture have placed archaeological constructions of landscape and the past at the center of identity politics. Native title, for example, has become an important issue in settler nations such as Canada, Australia, the United States, New Zealand, and South Africa (Lilley 2000). In seeking to regain land through the courts, Indigenous peoples have been forced to demonstrate the continuity of their connections to place and to meet expectations of "authenticity" in asserting cultural identity. Representations within heritage discourse that were the basis for conquest persist, such as the rhetoric of "wilderness" that posits the colonized landscape as empty, untouched, and primordial (Adams and Mulligan 2003). Notions of a pristine land and a timeless people live on despite the legal rejection of the concept of *terra nullius* (an empty land) and scientific evidence for long-term interactions between people and their environment, such as fire regimes and the subsequent introduction of new flora and fauna (Head 2000).

"Cultural Landscapes"

The concept of "cultural landscape" has also become increasingly important over recent decades, partly in recognition of heritage management's emphasis on material preservation, and its resulting tendency to reify culture, to divorce it from the natural world, and to privilege fabric over the intangible values attached to landscape by living peoples. To transcend such dichotomies between people and place, nature and culture, and tangible and intangible values, cultural landscapes were included on the World Heritage List in 1992 (Fowler 2003). This reconceptualization reflected developing intellectual approaches toward landscape as fluid and interactive, as well as the need to address the marginalization of non-Western, non-elite landscape traditions in an attempt to acknowledge difference and incorporate diversity. An emphasis on the tangible literally overlooks the experience of marginal or displaced groups such as women, children, immigrants, and the colonized, who are less likely to have shaped the visible environment and often leave behind relatively ephemeral traces of their experience (e.g., Hayden 1996). More recently initiatives such as the UNESCO (2003) *Convention for the Safeguarding of the Intangible Cultural Heritage* explicitly address the need to embed place within culture. The less substantial material culture of nonindustrial peoples was often overwritten by modern European nations. Regimes that emphasize the preservation of tangible heritage therefore

mask historical conflict and valorize the material culture of the conquerors above the memories and experiences of the defeated, with the effect of perpetuating historical subjugation.

Cultural landscapes are now listed against three categories (UNESCO 1996: clauses 23–24, 35–42, 2005, item 47): (1) *designed landscapes,* such as parks and gardens, evoke the visible and tangible principles of Cartesian perspectivalism, being "designed and created intentionally by humans"—such as Lednice-Valtice in the Czech Republic; (2) *organically evolved landscapes* represent "traditional human settlement or land-use that is representative of a culture (or cultures)"—such as the Rice Terraces of the Philippine Cordilleras; (3) *associative cultural landscape*s are distinguished by the "religious, artistic or cultural associations of the natural element rather than the material culture evidences, which may be insignificant or even absent" and may be physical or "mental images embedded in a people's spirituality, cultural tradition and practice"—such as New Zealand's Tongariro National Park. Where conquest in the past displaced the defeated from the landscape, often erasing the traces of their experience, as well, associative landscapes potentially represent a more inclusive vision of the past. They also dissolve the Western distinction between natural and cultural dimensions of place by recording the interdependence of spiritual, ceremonial, harvesting, and kinship relations with landscape.

Community Values

In a similar shift toward inclusiveness, it is now increasingly recognized that the meaning attached by communities to places and things is crucial to understanding their significance. Within international heritage management, the "values-based" approach is becoming prevalent, advocating that all aspects of a place's value, or cultural significance, are assessed as a primary stage in identifying and defining heritage and as a basis for decision making (first codified by the ICOMOS Burra Charter 1999; also see Avrami et al. 2000). The significance-assessment process disaggregates cultural significance into categories of aesthetic, historic, scientific, social, and spiritual value. Like intangible heritage (in fact, these technically distinct categories overlap considerably), community ideas and beliefs about places, or their "social significance," have not always been readily amenable to measurement by western rationalism. In practice, Western heritage discourse has privileged the scientific worldview of archaeology and subordinated the category of "social value." Denis

Byrne (2004), for example, points out that in many Asian countries, the popular investment of monuments and places with religious significance calls for renewal of fabric and so comes into conflict with Western heritage's conservation ethos (but see Sullivan 2005 for a more optimistic view). Innovative approaches toward "mapping attachment" to landscape are now seeking to recover the life experiences of colonized people that can be mapped spatially, including their segregation, avoidance, and transgression of the colonial cadastral grid (Byrne 2003; Byrne and Nugent 2004; Johnston 1992). In the wake of these radical shifts in heritage discourse, landscapes are increasingly understood in terms of the cultural meanings attached to places by diverse social actors, displacing the visibility, tangibility, and elite values characteristic of Cartesian perspectivalism.

However, although UNESCO policy (e.g., World Heritage Convention 1972) promotes the liberal values of participatory democracy and equal rights, seeking to involve local communities as well as to encourage the preservation of heritage considered to be of "outstanding universal value" to humanity, the benefits of world heritage listing are sometimes ambiguous.[1] It has been argued that the inclusion of cultural landscapes would correct a bias toward the monumental European tradition by making the recognition and nomination of heritage "more accessible to regions currently underrepresented on the World Heritage List" (Fowler 2003: 19, 23; Rössler 2000: 33). However, more recently Labadi (Forthcoming, 2006) shows that despite explicit UNESCO policy, national myths centering on the heroic elite male continue to be promoted and reproduced at the level of implementation by the state, excluding marginal groups and, notably, women. Less than a quarter of nomination dossiers mention the active participation of the local population despite the great emphasis placed on this by the Operational Guidelines, and European Christian sites remain most typical. Some argue that increased management of landscapes and more extensive formal protection within national and regional laws have deprived local communities of interaction with, and control over, their own environments (Kirshenblatt-Gimblett 2006).

Such tension between current interests within heritage management is also linked to a contradiction between the relativist notion of culture entailed in UNESCO's cultural heritage policy (UNESCO 1995) and the universal liberal values it espouses (Eriksen 2001). Sometimes, in the discourse termed the "culture versus rights" debate, liberal critics identify a tension between the cultural relativism that seeks to define and preserve cultural traditions and a notion of universal human rights (e.g., Cowan et al. 2001). In their reification of culture such formulations of "cultural diversity" risk clashing with individual human rights and imprisoning indigenous peoples within a state-imposed framework rather than allowing them to transform. But identity politics remain unacknowledged in much heritage discourse, which emphasizes universal and shared values in representing a consensual, collective vision of "authenticity" that continues to serve dominant interests. By contrast, the approach represented by the Australia ICOMOS Burra Charter (1999) is often praised as an example of a code that champions local and indigenous custodians and especially recognizes incommensurable views of the past in managing heritage places (e.g., Meskell 2002; Sullivan 2005). Article 13 states: "Co-existence of cultural values should be recognized, respected and encouraged, especially in cases where they conflict."[2] Acknowledging differences of perspective in analysis and interpretation is imperative.

Conclusions

Increasingly, cultural landscapes are coming to be understood as the site of social action in the present and a basis for community aspirations for the future. Some have suggested that the recent past will therefore come to acquire preeminent heritage significance (e.g., Bradley et al. 2004; Clarke and Johnston 2003). But such immediacy and importance raise some troubling questions: where heritage intersects with anguish, for example, how do we remember? Perhaps "traumascapes" (Tumarkin 2005)—places marked by suffering and loss such as Ground Zero and Sarajevo—become perversely therapeutic heritage, and revisiting them the source of resilience and reconciliation. Perhaps memory is best served by decay, and some landscapes are best left to disappear. In the meantime, aspects of the heritage system that perpetuate past inequalities, whether, for example, those of colonialism, forced migration, class or gender relations, remain the focus of critique and tension. The integral role of landscape in shaping contemporary assertions of identity and culture ensures that archaeological narratives will continue to be contested terrain.

Notes

1. Embodied in international treaty *Convention Concerning the Protection of the World Cultural and Natural Heritage*, or World Heritage Convention, adopted by UNESCO in 1972. To

qualify for inscription on the World Heritage List, nominated properties must have values that are outstanding and universal (World Heritage Criteria www.deh.gov.au/heritage/worldheritage/criteria.html). The purposes of the United Nations are to maintain international peace and security; to develop friendly relations among nations; to cooperate in solving international economic, social, cultural, and humanitarian problems and in promoting respect for human rights and fundamental freedoms; and to be a center for harmonizing the actions of nations in attaining these ends (www.un.org/aboutun/basicfacts/unorg.htm).

2. Article 13: Coexistence of cultural values: coexistence of cultural values should be recognized, respected, and encouraged, especially in cases where they conflict. Note: for some places, conflicting cultural values may affect policy development and management decisions. In this article, the term *cultural values* refers to those beliefs that are important to a cultural group, including but not limited to political, religious, spiritual, and moral beliefs. This is broader than values associated with cultural significance.

References

Adams, W. M., and Mulligan, M. 2003. *Decolonizing Nature: Strategies for Conservation in a Post-Colonial era*. London: Earthscan Publications.

Ashmore, W., and Knapp, A. B. (eds.). 1999. *Archaeologies of Landscape: Contemporary Perspectives*. Oxford: Blackwell Publishing.

Australia ICOMOS Charter for Places of Cultural Significance (Burra Charter). 1999. Australia ICOMOS, Canberra.

Avrami, E. C., Mason, R., and De la Torre, M. 2000. *Values and Heritage Conservation: Research Report*. Los Angeles: Getty Conservation Institute.

Bender, B. (ed.). 1993. *Landscape: Politics and Perspectives*. Oxford: Berg.

———. 1998. *Stonehenge: Making Space*. Oxford: Berg.

———. 2001. Introduction, in B. Bender and M. Winer (eds.). 2001. *Contested Landscapes: Movement, Exile and Place*. Oxford: Berg.

Bender, B., and Winer, M. (eds.). 2001. *Contested Landscapes : Movement, Exile and Place*. Oxford: Berg.

Blunt, A., and Rose, G. 1994. Introduction: Women's colonial and postcolonial geographies, in A. Blunt and G. Rose (eds.), *Writing Women and Space: Colonial and Postcolonial Geographies*, pp. 1–25. New York: Guilford Press.

Bradley, A., Buchli, V., Fairclough, G., Hicks, D., Miller, J., and Schofield, J. 2004. *Change and Creation: Historic Landscape Character, 1950–2000*. London: English Heritage.

Burke, H. 1999. *Meaning and Ideology in Historical Archaeology*. New York: Plenum Press, Global Contributions to Historical Archaeology series.

Byrne, D. 2003. Nervous Landscapes: Race and space in Australia. *Journal of Social Archaeology* 3(2): 169–93.

———. 2004. Chartering heritage in Asia's postmodern world. *Getty Conservation Institute Newsletter* 19(2): 1–6.

Byrne, D., and Nugent, M. 2004. *Mapping Attachment: A Spatial Approach to Aboriginal Post-Contact Heritage*. Hurstville, NSW: Department of Environment and Conservation (NSW).

Clarke, A., and Johnston, C. 2003. Time, memory, place and land: Social meaning and heritage conservation in Australia. *Place-Memory-Meaning: Preserving Intangible Values in Monuments and Sites*, ICOMOS 14th General Assembly and Scientific Symposium, Victoria Falls, Zimbabwe. www.international.icomos.org/victoriafalls2003/papers.htm, accessed 18/4/2006.

Cosgrove, D. 1984. *Social Formation and Symbolic Landscape*. Madison: University of Wisconsin Press.

Cowan, J., Dembour, M., and Wilson, R. (eds.). 2001. *Culture and Rights: Anthropological Perspectives*. Cambridge: Cambridge University Press.

Crary, J. 1990. *Techniques of the Observer: On Vision and Modernity in the Nineteenth Century*. Cambridge, MA: October Books, MIT Press.

David, B., and Wilson, M. 2002. Spaces of resistance: Graffiti and Indigenous place markings in the early European contact period of Northern Australia, in B. David and M. Wilson (eds.), *Inscribed Landscapes: Marking and Making Place*, pp. 44–45. Honolulu: University of Hawai'i Press.

Eriksen, T. 2001. Between universalism and relativism: A critique of the UNESCO concepts of culture, in J. Cowan, M. Dembour, and R. Wilson (eds.), *Culture and Rights: Anthropological Perspectives*, pp. 127–48. Cambridge: Cambridge University Press.

Fabian, J. 1983. *Time and the Other: How Anthropology Makes Its Object*. New York: Columbia University Press.

Fontein, J. 2005. *The Silence of Great Zimbabwe: Contested Landscapes and the Power of Heritage*. London: UCL Press.

Foucault, M. 1991. Governmentality, in G. Burchell, C. Gordon, and P. Miller (eds.), *The Foucault Effect*, pp. 87–104. London: Wheatsheaf Harvester.

Fowler, P. J. 2003. *World Heritage Cultural Landscapes 1992–2002*. World Heritage Papers 6, UNESCO World Heritage Centre.

Hayden, D. 1996. *The Power of Place: Urban Landscapes as Public History*. Cambridge, MA: The MIT Press.

Head, L. 2000. *Second Nature: The History and Implications of Australia as Aboriginal Landscape.* Syracuse, NY: Syracuse University Press.

Hirsch, E. 1995. Landscape: Between place and space, in E. Hirsch and M. O'Hanlon (eds.), *The Anthropology of Landscape*, pp. 1–30. Oxford: Clarendon Press.

Ingold, T. 1993. The temporality of the landscape. *World Archaeology* 25(2): 152–74.

Ireland, T. 2003. "The Absence of Ghosts": Landscape and identity in the archaeology of Australia's settler culture. *Historical Archaeology* 37(1): 56–72.

Jay, M. 1994. *Downcast Eyes: The Denigration of Vision in Twentieth-Century French Thought.* Berkeley and Los Angeles: University of California Press.

Johnston, C. 1992. *What Is Social Value?.* Australian Government Publishing Service, Canberra.

Kirshenblatt-Gimblett, B. 2006. World heritage and cultural economics, in I. Karp and C. Kratz (eds.), *Museum Frictions: Public Cultures/Global Transformations.* Durham, NC: Duke University Press.

Küchler, S. 1993. Landscape as memory: The mapping of process and its representation in a Melanesian society, in B. Bender (ed.), *Landscape: Politics and Perspectives.* Oxford: Berg.

Labadi, S. 2006. From exclusive to inclusive representations of the nation and cultural identity and diversity within discourses on world heritage. *Cultures of Contact* conference. Palo Alto, CA: Stanford University.

———. (ed.). Forthcoming. State of World Heritage Report, Paris: UNESCO World Heritage Centre.

Leone, M. P. 1984. Interpreting ideology in historical archaeology: Using the rules of perspective in the William Paca Garden in Annapolis, Maryland, in C. Tilley and D. Miller (eds.), *Ideology, Representation and Power in Prehistory*, pp. 25–35. Cambridge: Cambridge University Press.

———. 1987. Rule by ostentation: The relationship between space and sight in eighteenth-century landscape architecture in the Chesapeake region of Maryland, in S. Kent (ed.), *Method and Theory for Activity Area Research: An Ethnoarchaeological Approach*, pp. 604–33. New York: Columbia University Press.

Lilley, I. (ed.). 2000. *Native Title and the Transformation of Archaeology in the Postcolonial World.* Sydney: University of Sydney, Oceania Monograph 50.

Loeffler, D. 2005. *Contested Landscapes/Contested Heritage: History and Heritage in Sweden and Their Archaeological Implications Concerning the Interpretation of the Norrlandian Past.* Doktorsavhandling, Umeå Universitet 2005. Institutionen för arkeologi och samiska studier, 901 87 Umeå.

Lydon, J., and Ireland, T. (eds.). 2005. *Object Lessons: Archaeology and Heritage in Australian society.* Melbourne: Australian Scholarly Publishing.

McNiven, I., and Russell, L. 2005. *Appropriated Pasts: Indigenous Peoples and the Colonial Culture of Archaeology.* Walnut Creek, CA: AltaMira Press.

Meskell, L. 2002. Negative heritage and past mastering in archaeology. *Anthropological Quarterly* 75(3): 557–74.

Rose, G. 1993. *Feminism and Geography: The Limits of Geographical Knowledge.* Cambridge: Polity Press.

Rössler, M. 2000. Landscape stewardship: New directions in conservation of nature and culture. *The George Wright Forum* 17(1): 27–34.

Ruppel, T., Neuwirth, J., Leone, M. P., and Fry, G. 2003. Hidden in view: African spiritual spaces in North American landscapes. *Antiquity* 77(296): 321–36.

Ryan, J. 1997. *Picturing Empire: Photography and the Visualization of the British Empire.* Chicago: University of Chicago Press.

Smith, L. 2004. *Archaeological Theory and the Politics of Cultural Heritage.* London: Routledge.

Sullivan, S. 2005. Loving the ancient in Australia and China, in J. Lydon and T. Ireland (eds.), *Object Lessons: Archaeology and Heritage in Australia.* Melbourne: Australian Scholarly Publishing.

Thomas, J. 1993. The politics of vision and the archaeologies of landscape, in B. Bender (ed.), *Landscape: Politics and Perspectives*, pp. 19–48. Oxford: Berg.

Tilley, C. 1994. *A Phenomenology of Landscape: Places, Paths and Monuments.* Oxford: Berg.

———. 2006. Introduction: Identity, place, landscape and heritage. *Journal of Material Culture*, Special issue: Landscape, Heritage and Identity. 11(1–2): 7–32.

Tumarkin, M. 2005. *Traumascapes: The power and Fate of Places Transformed by Tragedy.* Carlton: Melbourne University Press.

United Nations Educational, Scientific and Cultural Organisation (UNESCO). 1972. *Convention Concerning the Protection of the World Cultural and Natural Heritage.* Paris: World Heritage Centre.

———. 1995. *Our Creative Diversity: Report of the World Commission on Culture and Development.* Paris: Culture and Development Co-ordination Office.

———. 1996. *The World Heritage Convention. Operational Guidelines for the Implementation of the World Heritage Convention.* Paris: World Heritage Centre.

———. 2003. *Convention for the Safeguarding of the Intangible Cultural Heritage.* Paris: World Heritage Centre.

———. 2005. *Intergovernmental Committee for the Protection of the World Cultural and Natural Heritage.* Paris: World Heritage Centre.

ABOUT THE AUTHORS

LISA ARAHO is a student in community history at the University of Papua New Guinea. She has participated in international archaeological research programs along the Kikori River of Papua New Guinea.

SUSANA NAHOE ARELLANO is a *licenciada* in anthropology (Universidad de Chile) with experience in Rapa Nui archaeology and was formerly Director of SERNATUR, National Tourism Office (Rapa Nui).

WENDY ASHMORE is Professor of Anthropology at the University of California, Riverside. She is co-editor of *Archaeologies of Landscape: Contemporary Perspectives* (1999) and author of *Settlement Archaeology at Quirigua, Guatemala* (2006).

BRYCE BARKER is Associate Professor in Anthropology/Archaeology and Head of the School of Humanities and Communication at the University of Southern Queensland. He is the author of *The Sea People: Late Holocene Maritime Specialisation on the Central Queensland Coast* (2004) and co-editor (with Bruno David and Ian McNiven) of *The Social Archaeology of Australian Indigenous Societies* (2006).

OFER BAR-YOSEF is the MacCurdy Professor of Prehistoric Archaeology in the Department of Anthropology at Harvard University. He has conducted Palaeolithic and Neolithic excavations in the Levant with numerous colleagues, as well as in the Republic of Georgia, and has recently been involved in research on late period prehistoric archaeology of China.

DOUGLAS BIRD is an Assistant Professor in the Anthropology Department and the Archaeology Center at Stanford University. He has published widely on hunter-gatherer ecology and the ethnoarchaeology of subsistence. He currently directs a long-term project in collaboration with Martu Aborigines on the socioecology of land-use, foraging strategies, prestige, and cooperation in Australia's Western Desert.

JOHN BRADLEY has worked for over three decades with the Yanyuwa people of the southwest Gulf of Carpentaria. As an anthropologist, he has concentrated on issues of language, land, sea, and kin relationships, ethnoecology, and ethnobiology, as well as issues of memory as the point of emotional engagement with country. He has produced an Indigenous Atlas of Yanyuwa country and is presently working on new understandings of song lines with a view to animation.

ALASDAIR BROOKS is the Finds and Environmental Officer for CAM ARC, Cambridgeshire County Council's archaeology unit. He is the author of *An Archaeological Guide to British Ceramics in Australia, 1788–1901* (2005) and has worked as an historical archaeologist in the United Kingdom, the United States, and Australia.

DENIS BYRNE leads the research program in cultural heritage at the Department of Environment and Climate Change N.S.W. in Sydney. He is author of *Surface Collection: Archaeological Travels in Southeast Asia* (2007).

ELEANOR CONLIN CASELLA is a Senior Lecturer at the University of Manchester. She is the author of *The Archaeology of Institutional Confinement* (2007) and co-editor of *The Archaeology of Plural and*

Changing Identities: Beyond Identification (with Chris Fowler, 2005).

EDWARD CASEY is Distinguished Professsor of Philosophy at SUNY, Stony Brook. He is the author of a series of books on the subject of place: *Getting Back into Place* (1993), *The Fate of Place* (1997), *Representing Place* (2002), and *Earth-Mapping* (2005). His most recent book is *The World in a Glance* (2007).

JEAN-CHRISTOPHE CASTEL is a specialist in archaeozoology at the Natural History Museum of Geneva (Switzerland). He is currently directing several excavations of Palaeolithic and Mesolithic sites in southwestern France and Switzerland and completing experimental reference bases on the taphonomy of faunal remains.

JEAN-PIERRE CHADELLE, Co-Director with Jean-Michel Geneste of research at Combe-Saunière, specializes in Upper Palaeolithic lithic technology. He is responsible for preventive archaeological excavations of historic and prehistoric sites for the General Council of the Department of Dordogne, France.

ANDREW CHAMBERLAIN is Professor of Biological Anthropology in the Department of Archaeology at the University of Sheffield. His research interests focus on the study of the structure and dynamics of past human populations.

JOHN CHAPMAN is a Reader in Archaeology at Durham University and Vice President of the Prehistoric Society. His principal specialties are fragmentation theory, landscape archaeology, and later Balkan prehistory.

PAUL CHEETHAM is Senior Lecturer in Archaeological Science at Bournemouth University specializing in archaeological and forensic geophysical survey, including work at World Heritage Sites. He has co-authored "Excavation and Recovery in Forensic Archaeological Investigations" in the sister World Archaeology Congress volume, *The Handbook of Forensic Anthropology and Archaeology* (2008).

CHRIS CLARKSON received his Ph.D. from the Australian National University in 2004 on long-term technological and cultural change in the Northern Territory, Australia. He is currently a Lecturer in the School of Social Science at the University of Queensland, where he continues his research into lithic technology in Australia, India, France, and Africa.

BRIAN CODDING is a doctoral student in the Department of Anthropology at Stanford University. His current research is focused on understanding how spatial variation in zooarchaeological assemblages relates to human-environment interactions.

JAMES CONOLLY is the Canada Research Chair in Archaeology at Trent University. His interests are human palaeoecology, the origins and spread of agriculture, neolithization, and settlement and landscape archaeology.

JOE CROUCH is a Ph.D. candidate with the Centre for Australian Indigenous Studies, Monash University. He has practiced partnership archaeology with Indigenous Torres Strait Islander communities for the past seven years, with a focus on marine specialization and nonresidential islands.

VICKI CUMMINGS is a Lecturer in Archaeology at the University of Central Lancashire. She is the co-author of *Places of Special Virtue: Megaliths in the Neolithic Landscapes of Wales* (with Alasdair Whittle, 2004) and co-editor of *The Neolithic of the Irish Sea: Materiality and Traditions of Practice* (with Chris Fowler, 2004).

TIMOTHY DARVILL is Professor of Archaeology and Director of the Centre for Archaeology, Anthropology, and Heritage in the School of Conservation Sciences, Bournemouth University. His interests lie with archaeological resource management and the Neolithic of northwest Europe. Recent publications include *Concise Oxford Dictionary of Archaeology* (2nd ed., 2008) and *Stonehenge: The Biography of a Landscape* (2006).

BRUNO DAVID is Co-Director of the Programme for Australian Indigenous Archaeology at Monash University. His latest books are *Landscapes, Rock Art, and the Dreaming* (2002); *The Social Archaeology of Australian Indigenous Societies* (2006); and *Gelam's Homeland* (2008).

TIM DENHAM is ARC/Monash Research Fellow at Monash University, Melbourne. His research focuses on the history of plant exploitation and agriculture in the highlands of Papua New Guinea. He is co-editor of *Rethinking Agriculture: Archaeological and Ethnoarchaeological Perspectives* (with José Iriarte and Luc Vrydaghs, 2008).

NIC DOLBY is a sessional academic within the School of Geography and Environmental Science and the Faculty of Medicine at Monash University and also works at the Aboriginal Community Elders Service, Brunswick East, Victoria, Australia.

JOHN DOP works at the Papua New Guinea National Museum and Art Gallery, where he is actively involved in cultural heritage management. He has participated in international archaeological research programs along the Kikori River of Papua New Guinea.

ANDREW FAIRBAIRN is Lecturer in Archaeology at the University of Queensland, Brisbane. He is an archaeobotanist investigating the origins of food production and state-level economies, with ongoing research projects in Turkey, Jordan, China, Italy, and Papua New Guinea.

CHRIS FOWLER is Lecturer in Prehistoric Archaeology at Newcastle University, where he specializes in British Neolithic and Early Bronze Age archaeology. He is the author of *The Archaeology of Personhood: An Anthropological Approach* (2004) and co-editor of a volume with Vicki Cummings and another with Eleanor Casella (see above).

CLIVE GAMBLE is Professor of Geography at Royal Holloway, University of London. His most recent publication is *Origins and Revolutions: Human Identity in Earliest Prehistory* (2007).

JEAN-MICHEL GENESTE is a specialist in lithic industries of the European Middle and Upper Palaeolithic, of which he has excavated and studied numerous sites. He directs archaeological studies at Chauvet Cave and Lascaux, and the Centre National de Préhistoire in France.

MICHAEL GREEN is Head of the Indigenous Cultures at Museum Victoria, Melbourne. His Ph.D. research investigated patterns of prehistoric cranial variation in Papua New Guinea and their congruence with the archaeological and linguistic record.

FRANCISCO TORRES H. is Director of the Museo Antropologico P. Sebastian Englert and Co-Director of the "Landscapes of Construction Project," both in Rapa Nui.

SUE HAMILTON is Reader in Prehistory at the Institute of Archaeology, University College London. Her research covers British and European Bronze and Iron Age societies and ceramics, issues of field practice, and landscape archaeology—particularly from a sensory perspective.

JOHN HART earned his Ph.D. in anthropology from Northwestern University. He is Director of the Research and Collections Division at the new York Museum in Albany. His research focuses on the evolution of agriculture, particularly on maize-bean-squash polycropping in northeastern North America.

LESLEY HEAD is Professor of Geography and Head of the School of Earth & Environmental Sciences at the University of Wollongong. Her research interests focus on long-term changes in the Australian landscape and the interactions of both prehistoric and contemporary cultures with these environments.

MICHAEL P. HEILEN is Research Director at Statistical Research, Inc. in Tucson, Arizona. His major research interests are landscape archaeology, behavioral archaeology, and spatial modeling. His dissertation, *An Archaeological Theory of Landscapes* (2005), presents a general integrative framework for landscape archaeology.

RUSSELL HILL is a Reader in Evolutionary Anthropology in the Department of Anthropology at Durham University. His research interests span behavioral ecology, conservation biology, evolutionary psychology, and theoretical modeling with field studies on primates and other large mammals.

ZENOBIA JACOBS is an archaeologist and Research Fellow in the School of Earth and Environmental Sciences at the University of Wollongong. Her specialty is single-grain optically stimulated luminescence dating of archaeological sediments, especially at sites concerning modern human evolution in Africa.

AMANDA KEARNEY is a Lecturer in the Centre for Australian Indigenous Studies, Monash University. Her work bridges archaeology and anthropology and has been developed in working with Yanyuwa people, the Indigenous owners of land and sea in the southwest Gulf of Carpentaria, Australia.

PETER KERSHAW is Professor of Geography and Environmental Science and Director of the Centre for Palynology and Palaeoecology at Monash University. His research is focused on the reconstruction of past vegetation, climate, fire, and human impact within the Australasian-Southeast Asian region.

THOMAS KOKENTS is a student in cultural heritage management at the University of Papua New Guinea. He has participated in international archaeological research programs along the Kikori River of Papua New Guinea.

CHRISTIAN KULL is Senior Lecturer at the School of Geography and Environmental Science, Monash University, Melbourne, and author of the book *Isle of Fire: The Political Ecology of Landscape Burning in Madagascar* (2004).

PAUL LANE is a Senior Lecturer in the Department of Archaeology, University of York. He specializes in the later Holocene archaeology and material culture of sub-Saharan Africa. His research interests include ethnoarchaeology, the archaeology of slavery, landscape historical ecology, and Indigenous concepts of "the past."

MARISA LAZZARI is a Lecturer in the Department of Archaeology at the University of Exeter. She specializes in the archaeology of social landscapes and exchange in northwestern Argentina, where she also works on contemporary cultural heritage issues.

JANE LYDON is a Postdoctoral Fellow at the Centre for Australian Indigenous Studies at Monash University. Her recent publications include *Eye Contact: Photographing Indigenous Australians* (2005) and a co-edited collection, *Object Lessons: Archaeology and Heritage in Australia* (2005).

INGRID MAINLAND is a Senior Lecturer in Environmental Archaeology at the University of Bradford, U.K. Her research focuses animal-human interactions in the North Atlantic islands and palaeodietary analysis, and she has pioneered the use of dental microwear analysis in the study of domestic animal diet.

BEN MARLER is a graduate student in Anthropology at Idaho State University and co-author of four major surveys of Darwinian theory in archaeological practice.

HERBERT D. G. MASCHNER is Research Professor of Anthropology at Idaho State University, a Senior Scientist at the Idaho Accelerator Center, and Associate Editor of the *Journal of World Prehistory*.

ELIZABETH (LISA) MATISOO-SMITH is an Associate Professor in the Department of Anthropology at the University of Auckland and a Principal Investigator in the Allan Wilson Centre for Molecular Ecology and Evolution. Her main research focuses on the use of genetic data to address issues of Pacific prehistory.

MIKE MCCARTHY is Senior Lecturer in Archaeology at the University of Bradford and has also worked extensively in contracting archaeology and in the field of medieval ceramics.

LESLEY MCFADYEN is a prehistorian currently undertaking research on a Leverhulme Early Career Fellowship entitled "Archaeological Architectures" that seeks to explore areas of interdisciplinarity and dialogue between archaeology and architecture. With architect Matthew Barac, she recently guest edited a special issue of the interdisciplinary journal *Home Cultures* (2007).

ROD MCINTOSH has excavated in West Africa for over thirty years in Mali, Senegal, and Ghana, and is currently a Visiting Professor at the University of Pretoria in South Africa. He is at Yale University, where he teaches archaeology and palaeoclimate.

IAN J. MCNIVEN is Reader in Archaeology and Co-Director of the Programme for Australian Indigenous Archaeology, School of Geography & Environmental

Science, Monash University. His recent books include *Appropriated Pasts* (2004, with Lynette Russell) and the co-edited *Social Archaeology of Australian Indigenous Societies* (2006, with Bruno David and Bryce Barkerr).

DONALD PATE is a bioanthropologist in the Department of Archaeology at Flinders University in Adelaide. He was a member of the Australian Archaeological Association Executive and editor of its journal *Australian Archaeology* from 1999–2006.

THOMAS C. PATTERSON is Distinguished Professor and Chair of Anthropology, University of California, Riverside. He is the author of *Marx's Ghost: Conversations with Archaeologists* (2003).

MAX PIVORU is a Rumu man and leader of the Himaiyu clan. He lives in Kopi village along the Kikori River of Papua New Guinea.

WILLIAM PIVORU is a Rumu man of the Himaiyu clan. He lives in Kopi village along the Kikori River of Papua New Guinea and is active in local community issues.

NICK PORCH is a Research Associate in the Department of Archaeology and Natural History at the Australian National University. He specializes in the use of insect remains in palaeoecology and is pioneering archaeological applications in the Indo-Pacific region.

PAUL RAINBIRD is Senior Lecturer and Head of the Department of Archaeology and Anthropology at the University of Wales, Lampeter. He is author of *The Archaeology of Micronesia* (2004) and *The Archaeology of Islands* (2007), both published by Cambridge University Press.

J. JEFFERSON REID is Professor of Anthropology and University Distinguished Professor at the University of Arizona, Tucson. He directed the university's archaeological field school at Grasshopper, Arizona (1979–1992) and was editor of *American Antiquity* (1990–1993).

COLIN RICHARDS is a Reader in Archaeology at the University of Manchester. He has worked extensively on the Orcadian Neolithic (*Dwelling Among the Monuments*, 2005) and has just begun a five-year project on "construction" in Rapa Nui.

THOMAS RICHARDS is undertaking graduate studies at Monash University while employed as Manager, Metropolitan Heritage Programs at Aboriginal Affairs Victoria, Melbourne. His work has focused on research-oriented cultural heritage management and community archaeology in Canada and Australia.

RICHARD "BERT" ROBERTS is a geomorphologist and Professor in the School of Earth and Environmental

Sciences at the University of Wollongong. His research interests include thermoluminescence and optically stimulated luminescence dating of archaeological and megafauna sites in Africa, Asia, and Australia.

STEVE ROSEN is Professor of Archaeology at Ben-Gurion University, Israel. His primary foci of research over the past 25 years have been the archaeology of pastoral nomadism (with a focus on the Negev in Israel) and the analysis of lithic industries from the Bronze and Iron Ages.

CASSANDRA ROWE currently holds a palaeoecological postdoctoral position with the School of Geographical Sciences, Bristol University. Having worked across Northern Australia, she maintains an interest in monsoonal, tropical vegetation communities, and past to present human-environment relationships.

MIKE ROWLAND is Principal Archaeologist with the Queensland Department of Natural Resources and Water. He has undertaken archaeological fieldwork in New Zealand, Fiji, and Australia. His focus has been on coasts and islands with an emphasis on the relationship between environmental change and variation in the archaeological record.

LYNETTE RUSSELL holds the Chair in the Centre for Australian Indigenous Studies at Monash University. Her research is in the area of colonial frontiers and early contact in Australia and cross-cultural interactions. Most of her research is interdisciplinary, using anthropological, archaeological, and historical research methods and theories.

MICHAEL BRIAN SCHIFFER is Fred A. Riecker Distinguished Professor of Anthropology at the University of Arizona. His major research interests are behavioral archaeology, ceramic technology, history of electrical and electronic technologies, and technology and society. His latest book is *Power Struggles: Scientific Authority and the Creation of Practical Electricity Before Edison* (2008).

DOUGLAS SIMALA from Samberigi village in the Erave district of the Southern Highlands Province, Papua New Guinea, works for Esso Highlands as a Community Affairs Officer and deals with the local communities in the project area and is also an active member in local community issues along the Kikori River of Gulf Province.

NICOLA STERN is a Senior Lecturer in the Archaeology Program at La Trobe University, Australia. In collaboration with the Traditional Owners, she is currently engaged in a palaeolandscape study of the Mungo Basin in the Willandra Lakes World Heritage Area. She is co-author of *A Record in Stone: The Study of Australia's Flaked Stone Artefacts* (2004).

VERONICA STRANG is Professor of Social Anthropology at the University of Auckland. An environmental anthropologist, she has written extensively on water, land, and resource issues in Australia and the U.K., and is the author of *Uncommon Ground: Cultural Landscapes and Environmental Values* (1997) and *The Meaning of Water* (2004).

GLENN R. SUMMERHAYES is Head of Anthropology at the University of Otago. He specializes in the archaeology of the western Pacific, in particular of Papua New Guinea, and is the author of *Lapita Interaction* (2000).

PAUL S. C. TAÇON is Professor of Archaeology and Anthropology, School of Arts, Griffith University, Queensland. He is co-editor of *The Archaeology of Rock Art* (1998) and author of over 130 academic papers on art, archaeology, and Aboriginal studies.

DIANE TEEMAN is a Ph.D. candidate in the Department of Anthropology, University of Oregon, U.S.A. She is currently also Director of the Burns Paiute Tribe's Culture and Heritage Department.

JOHN EDWARD TERRELL is the Regenstein Curator of Pacific Anthropology at Chicago's Field Museum of Natural History and Director of the museum's New Guinea Research Program.

JULIAN THOMAS is Chair of Archaeology at Manchester University and a Vice President of the Royal Anthropological Institute. His primary research interests are with the Neolithic period in Britain and northwest Europe and with theory and philosophy of archaeology. His recent publications include *Understanding the Neolithic* (1999), *Archaeology and Modernity* (2004), and *Place and Memory: Excavations at the Pict's Knowe, Holywood, and Holm Farm* (2007).

CHRISTOPHER TILLEY is Professor of Material Culture in the Department of Anthropology, University College London. He is a series editor of the *Journal of Material Culture*. Recent books include *Handbook of Material Culture* (ed. 2006), *The Materiality of Stone* (2004), *Metaphor and Material Culture* (1999), and *An Ethnography of the Neolithic* (1996).

ROBIN TORRENCE is a Principal Research Scientist at the Australian Museum, Sydney. Based primarily on archaeological fieldwork in West New Britain, Papua New Guinea, she is currently researching obsidian exchange, the impact of natural disasters on long-term history, the shaping of cultural landscapes, and the use of ethnographic collections to monitor Indigenous agency.

ROBERT VAN DE NOORT is Professor of Wetland Archaeology at the University of Exeter, U.K. He conducts his research in the terrestrial and intertidal wetlands around the North Sea Basin, with particular emphasis on late prehistoric perceptions of coastal and wetland landscapes, prehistoric maritime archaeology, theoretical and methodological developments and the politics, management, and conservation of wetlands.

RUTH M. VAN DYKE is Associate Professor of Anthropology at Colorado College. She is author of *The Chaco Experience: Landscape and Ideology at the Center Place* (2007) and co-editor (with Sue Alcock) of *Archaeologies of Memory* (2003).

JAMES F. WEINER is a consulting anthropologist in native title and Indigenous landowner issues in Australia and Papua New Guinea, and Leverhulme Professor-elect at the University of St. Andrews. He has carried out ethnographic research among the Foi of Papua New Guinea since 1979 and is the author of *The Empty Place* (1991) and *Tree Leaf Talk* (2001).

MARSHALL WEISLER is Head of the archaeology program at the University of Queensland, Australia. He has worked throughout the Pacific Islands for nearly 30 years and is best known for his work on tracing patterns of prehistoric interaction.

MARÍA NIEVES ZEDEÑO is an Associate Research Anthropologist at the Bureau of Applied Research in Anthropology and Associate Professor at the Department of Anthropology, University of Arizona, Tucson. She maintains an active research program on Native American cultural preservation and revitalization.

AUTHOR CONTACTS

LISA ARAHO
Social Research Institute
P.O. Box 172
University Post Office
National Capital District
Papua New Guinea

SUSANA NAHOE ARELLANO
Licenciado en Antropologa y
Aróveoloeria
Universidao de Chile, Santiago
Chile

WENDY ASHMORE
Department of Anthropology
1334 Watkins Hall
University of California, Riverside
Riverside, CA 92521-0418
U.S.A.
wendy.ashmore@ucr.edu

BRYCE BARKER
Anthropology/Archaeology
Department of Humanities and
International Studies
University of Southern Queensland
Toowoomba, Queensland 4350
Australia
barker@usq.com.au

OFER BAR-YOSEF
Department of Anthropology
Harvard University
Cambridge, MA 02138
U.S.A.
obaryos@fas.harvard.edu

DOUGLAS BIRD
Anthropological Sciences
450 Serra Mall, Building 360
Stanford University
Palo Alto, CA 94305-2117
U.S.A.

JOHN BRADLEY
School of Political and Social Inquiry
Monash University
Clayton, Victoria 3800
Australia
john.bradley@arts.monash.edu.au

ALASDAIR BROOKS
CAM ARC
15 Trafalgar Way

Bar Hill
Cambridgeshire
CB23 8SQ
United Kingdom
alasdair@provocateur.co.uk

DENIS BYRNE
Research Section
Culture & Heritage Division
Department of Environment and
Conservation
P.O. Box 1967, Hurstville
N.S.W. 2220
Australia
Denis.Byrne@environment.nsw.gov.au

ELEANOR CONLIN CASELLA
School of Arts, Histories, and Cultures
Humanities, Mansfield-Cooper Building
University of Manchester
Manchester M13 9PL
United Kingdom
e.casella@manchester.ac.uk

EDWARD CASEY
The Philosophy Department
SUNY at Stony Brook
Stony Brook, NY 11794
U.S.A.
Escasey3@aol.com

JEAN-CHRISTOPHE CASTEL
Muséum d'Histoire Naturelle
Département d'archéozoologie
Genève
Suisse
jean-christophe.castel@ville-ge.ch
Currently affiliated with:
PACEA UMR 5199 du CNRS
Université de Bordeaux 1

JEAN-PIERRE CHADELLE
Conseil Général de la Dordogne
Service d'Archéologie
Périgueux
France
chadelle@archaeologist.com
Currently affiliated with:
PACEA UMR 5199 du CNRS
Université de Bordeaux 1

ANDREW CHAMBERLAIN
Department of Archaeology
University of Sheffield

Northgate House
West Street, Sheffield, S1 4ET
United Kingdom
A.Chamberlain@sheffield.ac.uk

JOHN CHAPMAN
Department of Archaeology
Durham University
South Road
Durham, DH1 3LE
United Kingdom
j.c.chapman@durham.ac.uk

PAUL CHEETHAM
School of Conservation Sciences
Bournemouth University
Poole, Dorset
United Kingdom
PCheetham@bournemouth.ac.uk

CHRIS CLARKSON
School of Social Science
The University of Queensland
Brisbane, Queensland 4072
Australia
c.clarkson@uq.edu.au

BRIAN CODDING
Anthropological Sciences
450 Serra Mall, Building 360
Stanford University
Palo Alto, CA 94305-2117
U.S.A.

JAMES CONOLLY
Trent University Archaeological Research
Centre
Trent University
Peterborough, Ontario K9J 7B8
Canada
jamesconolly@trentu.ca

JOE CROUCH
Centre for Australian Indigenous Studies
Monash University
Clayton, Victoria 3800
Australia
joe.crouch@arts.monash.edu.au

VICKI CUMMINGS
School of Forensic and Investigative
Science
University of Central Lancashire
Preston, PR1 2HE
United Kingdom
VCummings1@uclan.ac.uk

TIMOTHY DARVILL
Archaeology and Historic Environment
Group

School of Conservation Sciences
Bournemouth University
Fern Barrow, Poole
Dorset BH12 5BB
United Kingdom
tdarvill@bournemouth.ac.uk

BRUNO DAVID
School of Geography and Environmental
Science
Monash University
Clayton, Victoria 3800
Australia
Bruno.David@arts.monash.edu.au

TIM DENHAM
Programme for Australian Indigenous
Archaeology
School of Geography and Environmental
Science
Monash University
Clayton, Victoria 3800
Australia
Tim.Denham@arts.monash.edu.au

NIC DOLBY
School of Geography and Environmental
Science
Monash University
Clayton, Victoria 3800
Australia
nic.dolby@arts.monash.edu.au

JOHN DOP
Anthropology Department
PNG National Museum and Art Gallery
P.O. Box 5560
Boroko
National Capital District
Papua New Guinea
jokundop@hotmail.com

ANDREW FAIRBAIRN
Archaeology Program, School of Social
Science
The University of Queensland
St. Lucia, Queensland 4072
Australia
a.fairbairn@uq.edu.au

CHRIS FOWLER
School of Historical Studies
University of Newcastle Upon Tyne
Armstrong Building
Newcastle Upon Tyne
NE1 7RU
United Kingdom
C.J.Fowler@newcastle.ac.uk

CLIVE GAMBLE
Department of Geography
Royal Holloway, University of London
Egham TW20 0EX
United Kingdom
Clive.Gamble@rhul.ac.uk

JEAN-MICHEL GENESTE
Centre National de Préhistoire
Ministère de la Culture et de la
Communication
Périgueux
France
jean-michel.geneste@culture.gouv.fr
Currently affiliated with:
PACEA UMR 5199 du CNRS
Université de Bordeaux 1

MICHAEL GREEN
Indigenous Cultures Department
Museum Victoria
G.P.O. Box 666
Melbourne, Victoria 3001
Australia
mgreen@museum.vic.gov.au

FRANCISCO TORRES H.
Museo Antropologico P. Sebastian Englert
Rapa Nui

SUE HAMILTON
Institute of Archaeology
University College London
31-34 Gordon Square
London
WC1H 0PY
United Kingdom
s.hamilton@ucl.ac.uk

JOHN HART
Director, Research & Collections, Chief
Scientist (Archaeology)
New York State Museum
Albany, NY 12230
U.S.A.
jhart@mail.nysed.gov

LESLEY HEAD
GeoQuEST Research Centre
School of Earth and Environmental
Sciences
University of Wollongong
N.S.W. 2522
Australia
lhead@uow.edu.au

MICHAEL P. HEILEN
Statistical Research Inc.
P.O. Box 31865
Tucson, AZ 85751-1865

U.S.A.
mheilen@sricrm.com

RUSSELL HILL
Evolutionary Anthropology Research
Group
Department of Anthropology
Durham University
43 Old Elvet
Durham DH1 3HN
United Kingdom
r.a.hill@durham.ac.uk

ZENOBIA JACOBS
GeoQuEST Research Centre
School of Earth and Environmental
Sciences
University of Wollongong
Wollongong, New South Wales 2522
Australia
zenobia@uow.edu.au

AMANDA KEARNEY
Centre for Australian Indigenous Studies
Monash University
Clayton, Victoria 3800
Australia
Amanda.Kearney@arts.monash.edu.au

PETER KERSHAW
Centre for Palynology and Palaeoecology
School of Geography and Environmental
Science
Monash University
Clayton, Victoria 3800
Australia
Peter.Kershaw@arts.monash.edu.au

THOMAS KOKENTS
Social Research Institute
P.O. Box 172
University Post Office
National Capital District
Papua New Guinea

CHRISTIAN KULL
School of Geography and Environmental
Science
Monash University
Melbourne, Victoria 3800
Australia
christian.kull@arts.monash.edu.au

PAUL LANE
Department of Archaeology
University of York
King's Manor
York YO1 7EP
United Kingdom
pjl503@york.ac.uk

MARISA LAZZARI
    Archaeology Department
    University of Exeter,
    Laver Building, North Park Road,
    Exeter EX4 4QE
    Devon
    United Kingdom and
    Museo Etnográfico "J. B. Ambrosetti"
    Moreno 350, (1091) Buenos Aires
    Argentina
    M.Lazzari@exeter.ac.uk

JANE LYDON
    Centre for Australian Indigenous Studies
    Monash University
    Clayton, Victoria 3800
    Australia
    jane.lydon@arts.monash.edu.au

INGRID MAINLAND
    Department of Archaeological Sciences
    University of Bradford
    Bradford BD7 1DP
    United Kingdom
    I.L.Mainland@Bradford.ac.uk

BEN MARLER
    Department of Anthropology
    Campus Box 8005
    Idaho State University
    Pocatello, ID 83209
    U.S.A.
    marlben3@isu.edu

HERBERT D. G. MASCHNER
    Department of Anthropology
    Campus Box 8005
    Idaho State University
    Pocatello, ID 83209
    U.S.A.
    maschner@isu.edu

ELIZABETH (LISA) MATISOO-SMITH
    Allan Wilson Centre for Molecular
    Ecology and Evolution and Department of
    Anthropology
    University of Auckland
    New Zealand
    e.matisoo-smith@auckland.ac.nz

MIKE MCCARTHY
    Department of Archaeological Sciences
    University of Bradford
    Bradford, BD20 6EU
    United Kingdom
    M.Mccarthy@Bradford.ac.uk

LESLEY MCFADYEN
    School of Archaeology and Ancient History
    University of Leicester
    University Road
    Leicester, LE1 7RH
    United Kingdom
    lm127@leicester.ac.uk

ROD MCINTOSH
    Department of Anthropology
    Yale University
    51 Hillhouse Avenue
    New Haven, CT 06520
    U.S.A.
    Roderick.McIntosh@yale.edu

IAN J. MCNIVEN
    Programme for Australian Indigenous
    Archaeology
    School of Geography and Environmental
    Science
    Monash University
    Clayton, Victoria 3800
    Australia
    ian.mcniven@arts.monash.edu.au

DONALD PATE
    Department of Archaeology
    Flinders University
    G.P.O. Box 2100, Adelaide, South
    Australia 5001
    Australia
    donald.pate@flinders.edu.au

THOMAS C. PATTERSON
    Department of Anthropology
    University of California, Riverside
    Riverside, CA 92521
    U.S.A.
    thomas.patterson@ucr.edu

MAX PIVORU
    Kopi Village
    Kikori District
    Gulf Province
    Papua New Guinea

WILLIAM PIVORU
    Kopi Village
    Kikori District
    Gulf Province
    Papua New Guinea

NICK PORCH
    Archaeology and Natural History
    RSPAS
    Australian National University
    Canberra, ACT 0200
    Australia
    nicholas.porch@anu.edu.au

PAUL RAINBIRD
    Department of Archaeology &
    Anthropology

University of Wales, Lampeter
Wales
United Kingdom
p.rainbird@lamp.ac.uk

J. Jefferson Reid
Department of Anthropology
The University of Arizona
1009 E. South Campus Drive
P.O. Box 210030
Tucson, AZ 85721-0030
U.S.A.
jreid@email.arizona.edu

Colin Richards
School of Arts, Histories, and Cultures
Humanities Bridgeford Street Building
University of Manchester
Oxford Road
Manchester M13 9PL
United Kingdom
colin.richards@manchester.ac.uk

Thomas Richards
Programme for Australian Indigenous
Archaeology
School of Geography and Environmental
Science
Monash University
Clayton, Victoria 3800
Australia and
Aboriginal Affairs Victoria
Department for Victorian Communities
Melbourne, Victoria 3001
Australia
thomas.richards@dvc.vic.gov.au

Richard (Bert) Roberts
GeoQuEST Research Centre
School of Earth and Environmental
Sciences
University of Wollongong
Wollongong, New South Wales 2522
Australia
rgrob@uow.edu.au

Steve Rosen
Department of Archaeology
Ben-Gurion University of the Negev
P.O. Box 653
Beer-Sheva 84105
Israel
rosen@bgu.ac.il

Cassandra Rowe
Centre for Palynology and Palaeoecology
School of Geography and Environmental
Science
Monash University

Clayton, Victoria 3800
Australia
Cassandra.Rowe@arts.monash.edu.au

Mike Rowland
Cultural Heritage Coordination Unit
Locked Bag 40 Coorparoo Delivery Centre
Brisbane, Queensland 4151
Australia
mike.rowland@nrw.qld.gov.au

Lynette Russell
Centre for Australian Indigenous Studies
Monash University
Clayton, Victoria 3800
Australia
Lynette.Russell@arts.monash.edu.au

Michael Brian Schiffer
Department of Anthropology
The University of Arizona
1009 E. South Campus Drive
P.O. Box 210030
Tucson, Arizona 85721-0030
U.S.A.
schiffer@email.arizona.edu

Douglas Simala
Community Affairs-PNG Gas
Esso Highlands Ltd
P.O. Box 842
Port Moresby
National Capital District
Papua New Guinea
PNGESS09@osl.com

Nicola Stern
Archaeology Program
La Trobe University
Bundoora, Victoria 3086
Australia
n.stern@latrobe.edu.au

Veronica Strang
Department of Anthropology
University of Auckland
Private Bag 92019
Auckland 1001
New Zealand
v.strang@auckland.ac.nz

Glenn R. Summerhayes
Department of Anthropology
University of Otago
P.O. Box 56, Dunedin
New Zealand
glenn.summerhayes@stonebow.otago.ac.nz

Paul S. C. Taçon
School of Arts
Gold Coast Campus

Griffith University
Queensland 4222
Australia
p.tacon@griffith.edu.au

DIANE TEEMAN
Department of Anthropology
University of Oregon
U.S.A.
dteeman@uoregon.edu

JOHN EDWARD
Regenstein Curator of Pacific
Anthropology
Field Museum of Natural History
1400 South Lake Shore Drive
Chicago, IL 60605
U.S.A.
terrell@fieldmuseum.org

JULIAN THOMAS
School of Arts, Histories, and Cultures
Humanities Bridgeford Street Building
University of Manchester
Oxford Road
Manchester M13 9PL
United Kingdom
julian.thomas@manchester.ac.uk

CHRISTOPHER TILLEY
Department of Anthropology
University College London
Gower Street, London WC1E 6BT
United Kingdom
c.tilley@ucl.ac.uk

ROBIN TORRENCE
Australian Museum
6 College Street
Sydney, N.S.W. 2010
Australia
robin.torrence@austmus.gov.au

ROBERT VAN DE NOORT
University of Exeter
Exeter
Devon
United Kingdom
r.van-de-noort@ex.ac.uk

RUTH M. VAN DYKE
Department of Anthropology
Colorado College
Colorado Springs, CO 80903
U.S.A.
rvandyke@coloradocollege.edu

JAMES F. WEINER
Visiting Fellow, Resource Management in
Asia Pacific
Research School of Pacific and Asian
Studies
Australian National University
Canberra, ACT 0200
Australia
james.weiner@anu.edu.au

MARSHALL WEISLER
Archaeology Program
School of Social Science
Michie Building
The University of Queensland
St. Lucia, Queensland 4072
Australia
m.weisler@uq.edu.au

MARÍA NIEVES ZEDEÑO
Bureau of Applied Research in
Anthropology
Haury Building 30
The University of Arizona
Tucson, AZ 85721
U.S.A.
mzedeno@email.arizona.edu

CREDITS

Figure 6.1, p. 89 by Rod McIntosh (drafted by Kara Rasmanis); Figure 11.1, pp. 134–135 by Joe Crouch (drafted by Gary Swinton and Phil Scamp); Figure 13.1, p. 153 courtesy John Bradley; Figure 14.1, p. 164 by Bruno David, with kind permission of Max Pivoru; Figure 16.1, p. 178 modified from Lipo and Hunt 2005; Figure 16.2, p. 179 modified from Van Tilburg (1994) and Love (1993); Figure 16.3, p. 180 by Colin Richards; Figure 16.4, p. 181 by Colin Richards; Figure 16.5, p. 181 courtesy Claudio Christino and Patricia Vargas; Figure 16.6, p. 182 by Colin Richards; Figure 16.7, p. 182 by Colin Richards; Figure 16.8, p. 184 by Colin Richards; Figure 17.1, p. 190 after Bradley and Ford (2004) (drafted by Kara Rasmanis); Figure 17.2, p. 191 after Scheer (1990) (drafted by Kara Rasmanis); Figure 17.3, p. 192 after Schaller-Åhrberg (1990) (drafted by Kara Rasmanis); Figure 17.4, p. 193 after Bradley and Ford (2004) (drafted by Kara Rasmanis); Figure 17.5, p. 195 from Gaydarska et al. (in press); Figure 17.6, p. 196 after Todorova et al. (2002) (drafted by Kara Rasmanis); Figure 17.7, p. 197 by John Chapman (drafted by Kara Rasmanis); Figure 20.1, p. 219 by Paul Taçon; Figure 20.2, p. 220 by Paul Taçon; Figure 20.3, p. 222 by Paul Taçon; Figure 20.4, p. 223 by Paul Taçon; Figure 21.1, p. 233 by Jean-Michel Geneste, Jean-Christophe Castel and Jean-Pierre Chadelle; Figure 24.1, p. 257 by Clive Gamble (from Gamble 2007); Figure 24.2, p. 260 by Clive Gamble (from Gamble 2007); Figure 25.1, p. 267 by Paul Rainbird; Figure 28.1, p. 286 by Vicki Cummings; Figure 28.2, p. 288 by Vicki Cummings from data produced by kind permission of Ordnance Survey. Crown copyright NC/03/20516; Figure 28.3, p. 289 by Vicki Cummings; Figure 29.1, p. 294 after Cummings, Jones, and Watson (2002), reproduced with kind permission of Vicki Cummings and the *Cambridge Archaeological Journal* (drafted by Kara Rasmanis); Figure 29.2, p. 295 by Chris Fowler (drafted by Kara Rasmanis); Figure 31.1, p. 308 by Marisa Clements (drafted by Kara Rasmanis); Figure 31.2, p. 309 after Ashbee, Smith, and Evans (1979), courtesy of The Prehistoric Society (drafted by Kara Rasmanis); Figure 31.3, p. 311 after Ashbee, Smith, and Evans (1979), courtesy of The Prehistoric Society (drafted by Kara Rasmanis); Figure 31.4, p. 312 after Pollard (2005), courtesy of John Pollard (drafted by Kara Rasmanis); Figure 32.1, p. 317 by Ofer Bar-Yosef (drafted by Kara Rasmanis); Figure 32.2, p. 322 by Ofer Bar-Yosef (drafted by Kara Rasmanis); Figure 32.3, p. 323 by Ofer Bar-Yosef (drafted by Kara Rasmanis); Figure 34.1, p. 334 by Robin Torrence (drawn by Trudy Doelman, drafted by Kara Rasmanis); Figure 34.2, p. 336 by Robin Torrence; Figure 34.3, p. 336 by Robin Torrence; Figure 34.4, p. 337 by Robin Torrence (drawn by Trudy Doelman, drafted by Kara Rasmanis); Figure 34.5, p. 339 by Robin Torrence (drafted by Kara Rasmanis); Figure 35.1, p. 348 by Richard G. Roberts and Zenobia Jacobs; Figure 36.1, p. 366 after Rand McNally (1979); Figure 36.2, p. 367 by Nicola Stern (after Isaac 1997); Figure 36.3, p. 369 after Feibel et al. (1989); Figure 36.4, p. 370 after Rogers et al. (1994); Figure 36.5, p. 371 after Brown and Feibel (1985); Figure 36.6, p. 372 by Nicola Stern; Figure 36.7, p. 373 by Nicola Stern; Figure 36.8, p. 374 by Nicola Stern; Figure 36.9, p. 374 by Nicola Stern; Figure 36.10, p. 375 by Nicola Stern, photo by Rudy Frank; Figure 36.11, p. 376 by Nicola Stern; Figure 37.1, p. 383 from Ward et al. (2005), reproduced with kind permission of *Quaternary Science Reviews*; Figure 37.2, p. 383 from Ward et al. (2005), reproduced with kind permission of *Quaternary Science Reviews*; Figure 40.1, p. 411 by Steven A. Rosen; Figure 40.2, p. 414, A, B, C, by Benjamin Saidel; Figure 40.3, p. 416, A by H. Sokolskaya, B by Steven A. Rosen, and C by P. Kaminsky; Figure 40.4, p. 418 by Steven A. Rosen; Figure 40.5, p. 419 by Steven A. Rosen; Figure 40.6, p. 420 drawing courtesy of Isaac Gilead (redrafted by Kara Rasmanis); Figure 45.1, p. 459 by Nick Porch; Figure 45.2, p. 461 by Nick Porch; Figure 46.1, p. 470 by Tim Denham (drafted by Kara Rasmanis); Figure 46.2, p. 475 from Denham et al. (2003), reproduced with kind permission of *Science*; Figure 46.3, p. 476 by Tim Denham, with X-radiograph courtesy of Alain Pierret (drafted by Kara Rasmanis); Figure 46.4, p. 477 by Tim Denham (drafted by Kara Rasmanis); Figure 46.5, p. 478 by Tim Denham, with individual images reproduced with kind permission of *Journal of Archaeological Science, Science,* and The Prehistoric Society (drafted by Kara Rasmanis); Figure 47.1, p. 484 after Coles (1995), drawn by Sean Goddard; Figure 47.2, p. 485 from Bocquet and Houot (1994), drawn by Sean Goddard; Figure 48.1, p. 495 by Chris Clarkson; Figure 48.2, p. 495 by Chris

Clarkson; Figure 48.3, p. 496 by Chris Clarkson; Figure 48.4, p. 497 by Chris Clarkson; Figure 48.5, p. 498 by Chris Clarkson, courtesy of Wardaman Elders; Figure 49.1, p. 505 after Walker and DeNiro (1986); Figure 49.2, p. 506 after Pate et al. (2002); Figure 49.3, p. 506 after Beard and Johnson (2000); Figure 49.4, p. 507 after Montgomery et al. (2000); Figure 49.5, p. 508 after White et al. (1998); Figure 49.6, p. 509 after Richards et al. (2001); Figure 49.7, p. 511 after Pietrusewski (2000); Figure 49.8, p. 512 after Rubicz et al. (2003); Figure 52.1, p. 537 by Les O'Neill and Judith Ogden; Figure 52.2, p. 540 by Marshall Weisler (drafted by Kara Rasmanis); Figure 55.1, p. 566 courtesy of the Billown Neolithic Landscape Project, Bournemouth University; Figure 55.2, p. 567 courtesy of the Department of Quantum Electronics, IPHT Jena, Germany; Figure 55.3, p. 568 courtesy of the Department of Quantum Electronics, IPHT Jena, Germany; Figure 55.4, p. 569 by David Sabin and Kerry Donaldson; Figure 55.5, p. 570 courtesy of Michel Dabas, Geocarta; Figure 55.6, p. 571 courtesy of Geocarta and Pologis-GSE; Figure 55.7, p. 573 by Paul Linford, English Heritage; Figure 55.8, p. 578 courtesy of Dominic Powlesland, The Landscape Research Centre; Figure 56.1, p. 585 after Broodbank (2000); Figure 56.2, p. 586 by James Conolly; Figure 56.3, p. 586 courtesy of Antikythera Survey Project; Figure 56.4, p. 587 courtesy of Antikythera Survey Project; Figure 56.5, p. 590 courtesy of Antikythera Survey Project; Figure 58.1, p. 602 by Michael Heilen, Michael B. Schiffer, and J. Jefferson Reid; Figure 58.2, p. 603 by Michael Heilen, Michael B. Schiffer, and J. Jefferson Reid; Figure 60.1, p. 620 after Brodie et al. (2002); Figure 60.2, p. 622 courtesy of Library of Congress, HABS CAL, 38-ALCA, 1-A-17; Figure 60.3, p. 623 by Niamh O'Sullivan, Office of Public Works, Dublin; Figure 60.4, p. 624 by Eleanor Casella (drafted by Kara Rasmanis).

INDEX

44326021R00400

Made in the USA
Middletown, DE
03 June 2017